Lecture Notes in Computer Science 4131

Commenced Publication in 1973
Founding and Former Series Editors:
Gerhard Goos, Juris Hartmanis, and Jan van Leeuwen

Stefanos Kollias Andreas Stafylopatis
Włodzisław Duch Erkki Oja (Eds.)

Artificial Neural Networks – ICANN 2006

16th International Conference
Athens, Greece, September 10 – 14, 2006
Proceedings, Part I

Springer

Volume Editors

Stefanos Kollias
Andreas Stafylopatis
National Technical University of Athens
School of Electrical and Computer Engineering
157 80 Zographou, Athens, Greece
E-mail: {stefanos,andreas}@cs.ntua.gr

Włodzisław Duch
Nicolaus Copernicus University
Department of Informatics
ul. Grudziadzka 5, 87-100 Torun, Poland
E-mail: wduch@phys.uni.torun.pl

Erkki Oja
Helsinki University of Technology
Laboratory of Computer and Information Science
P.O. Box 5400, 02015 Hut, Finland
E-mail: erkki.oja@hut.fi

Library of Congress Control Number: 2006931797

CR Subject Classification (1998): F.1, I.2, I.5, I.4, G.3, J.3, C.2.1, C.1.3

LNCS Sublibrary: SL 1 – Theoretical Computer Science and General Issues

ISSN 0302-9743
ISBN-10 3-540-38625-4 Springer Berlin Heidelberg New York
ISBN-13 978-3-540-38625-4 Springer Berlin Heidelberg New York

Springer is a part of Springer Science+Business Media

springer.com

Typesetting: Camera-ready by author, data conversion by Scientific Publishing Services, Chennai, India
Printed on acid-free paper SPIN: 11840817 06/3142 5 4 3 2 1 0

Preface

This book includes the proceedings of the International Conference on Artificial Neural Networks (ICANN 2006) held on September 10-14, 2006 in Athens, Greece, with tutorials being presented on September 10, the main conference taking place during September 11-13 and accompanying workshops on perception, cognition and interaction held on September 14, 2006.

The ICANN conference is organized annually by the European Neural Network Society in cooperation with the International Neural Network Society, the Japanese Neural Network Society and the IEEE Computational Intelligence Society. It is the premier European event covering all topics concerned with neural networks and related areas. The ICANN series of conferences was initiated in 1991 and soon became the major European gathering for experts in these fields.

In 2006 the ICANN Conference was organized by the Intelligent Systems Laboratory and the Image, Video and Multimedia Systems Laboratory of the National Technical University of Athens in Athens, Greece.

From 475 papers submitted to the conference, the International Program Committee selected, following a thorough peer-review process, 208 papers for publication and presentation to 21 regular and 10 special sessions. The quality of the papers received was in general very high; as a consequence, it was not possible to accept and include in the conference program many papers of good quality.

A variety of topics constituted the focus of paper submissions. In regular sessions, papers addressed topics such as learning algorithms, hybrid architectures, neural dynamics and complex systems, self-organization, computational neuroscience, connectionist cognitive science, neural control, robotics and planning, data analysis, signal and time series processing, image and vision analysis, pattern recognition and applications to bioinformatics, market analysis and other real-world problems.

Special sessions, organized by distinguished researchers, focused on significant aspects of current neural network research, including cognitive machines, Semantic Web technologies and multimedia analysis, bridging the semantic gap in multimedia machine learning approaches, feature selection and dimension reduction for regression, learning random neural networks and stochastic agents, visual attention algorithms and architectures for perceptional understanding and video coding, neural computing in energy engineering, bio-inspired neural network on-chip implementation and applications, computational finance and economics.

Prominent lecturers provided key-note speeches for the conference. Moreover, tutorials were given by well-known researchers. John Taylor was the honorary Chair of the conference.

Three post-conference workshops, on intelligent multimedia, semantics, interoperability and e-culture, on affective computing and interaction and on cognitive machines, concluded the focus of ICANN 2006 on the state-of-the-art research on neural networks and intelligent technologies in relation to the domains of cognition, perception and interaction. In-depth discussion was made on the prospects and future developments of the theoretical developments and applications of neural network models, algorithms and systems in the fields of cognition, neurobiology, semantics, perception and human computer interaction.

We would like to thank all members of the organizing laboratories for their contribution to the organization of the conference. In particular we wish to thank Lori Malatesta and Eleni Iskou, who greatly helped in handling a variety of technical and administrative problems related to the conference organization. Finally, we wish to thank Alfred Hofmann and Christine Guenther from Springer for their help and collaboration in the publication of the ICANN proceedings.

July 2006 Stefanos Kollias, Andreas Stafylopatis

Organization

General Chair

Stefanos Kollias,
National Technical University of Athens

Co-Chair

Andreas Stafylopatis, NTUA, Greece

Program Chair

Wlodzislaw Duch, Torun, Poland and Singapore
ENNS President
Erkki Oja, Helsinki, Finland

Honorary Chair

John G. Taylor, Kings College, London, UK; ENNS Past President

International Program Committee

- **Hojat Adeli,** Ohio State University, USA
- **Peter Andras,** University of Newcastle, UK
- **Marios Angelides,** Brunel University, UK
- **Panos Antsaklis,** University of N. Dame, USA
- **Bruno Apolloni,** University of Milan, Italy
- **Nikolaos Bourbakis,** Wright State University, USA
- **Peter Erdi,** University of Budapest, Hungary and Kalamazoo
- **Georg Dorffner,** University of Vienna, Austria
- **Patrick Gallinari,** Université Paris 6, France
- **Christophe Garcia,** France Telecom
- **Erol Gelenbe,** Imperial College, UK
- **Stan Gielen,** University of Nijmegen, The Netherlands
- **Pascal Hitzler,** University of Karlsruhe, Germany
- **Nikola Kasabov,** Kedri, Australia, New Zealand
- **Janusz Kacprzyk,** Warsaw, Poland
- **Okyay Kaynak,** Bogazici University, Turkey
- **Chris Koutsougeras,** Tulane University, USA
- **Thomas Martinetz,** University of Luebeck, Germany
- **Evangelia Micheli-Tzanakou,** Rutgers University, USA

- **Lars Niklasson,** Skøvde University, Sweden
- **Andreas Nuernberger,** University of Magdeburg, Germany
- **Marios Polycarpou,** University of Cyprus
- **Demetris Psaltis,** Caltech, USA
- **Branimir Reljin,** University of Belgrade, Serbia
- **Olli Simula,** Technical University of Helsinki, Finland
- **Alessandro Sperduti,** University of Padova, Italy
- **Lefteris Tsoukalas,** Purdue University, USA
- **Michel Verleysen,** Louv.-la-Neuve, Belgium
- **Alessandro Villa,** University of Grenoble, France

Local Organizing Committee

- **Yannis Avrithis,** ICCS-NTUA
- **Christos Douligeris,** Piraeus University
- **George Dounias,** Aegean University
- **Kostas Karpouzis,** ICCS-NTUA
- **Aris Likas,** University of Ioannina
- **Konstantinos Margaritis,** University of Macedonia
- **Vassilis Mertzios,** DUTH
- **Stavros Perantonis,** NCSR Demokritos
- **Yannis Pitas,** AUTH, Salonica
- **Costas Pattichis,** University of Cyprus
- **Apostolos Paul Refenes,** AUEB
- **Christos Schizas,** University of Cyprus
- **Giorgos Stamou,** ICCS-NTUA
- **Sergios Theodoridis,** UoA
- **Spyros Tzafestas,** NTUA
- **Nicolas Tsapatsoulis,** University of Cyprus
- **Mihalis Zervakis,** TUC, Crete

Reviewers

Abe	Shigeo	Kobe University
Adamczak	Rafal	Nicholas Copernicus University
Aiolli	Fabio	University of Pisa
Akrivas	George	National Technical University of Athens
Albrecht	Andreas	University of Hertfordshire
Alhoniemi	Esa	University of Turku
Andonie	Razvan	Central Washington University
Anguita	Davide	University of Genoa
Angulo-Bahon	Cecilio	Univ. Politecnica de Catalunya, Spain
Archambeau	Cedric	Université Catholique de Louvain
Atencia	Miguel	Universidad de Malaga

Aupetit	Michael	Commissariat à l`Energie Atomique
Avrithis	Yannis	National Technical University of Athens
Bedoya	Guillermo	Technical University of Catalonia, Spain
Bianchini	Monica	Università di Siena
Boni	Andrea	University of Trento
Caputo	Barbara	Royal Institute of Technology
Caridakis	George	National Technical University of Athens
Cawley	Gavin	University of East Anglia
Chetouani	Mohamed	Université Paris
Chortaras	Alexandros	National Technical University of Athens
Cichocki	Andrzej	RIKEN
Clady	Xavier	Université Pierre et Marie Curie
Corchado	Emilio	Applied Computational Intelligence Unit
Cottrell	Marie	Université Paris I
Crook	Nigel	Oxford Brookes University
Dablemont	Simon	Université Catholique de Louvain
Delannay	Nicolas	Université Catholique de Louvain
Derpanis	Kostas	York University
Dimitrakakis	Christos	IDIAP
Dominguez Merino	Enrique	E.T.S.I. Informatica, Spain
Dorronsoro	Jose	Universidad Autónoma de Madrid
Douligeris	Christos	Piraeus University
Dounias	George	Aegean University
Drosopoulos	Nasos	National Technical University of Athens
Duch	Wlodzislaw	Nicolaus Copernicus University
Elizondo	David	De Montfort University
Ferles	Christos	National Technical University of Athens
Flanagan	Adrian	Nokia Research Center
Francois	Damien	Université Catholique de Louvain
Fyfe	Colin	University of Paisley
Garcia-Pedrajas	Nicolas	University of Cordoba
Gas	Bruno	LISIF-UPMC
Gonzales Abril	Luis	Faculdad Ciencias Economicas y Empresari
Goser	Karl	Universitaet Dortmund
Gosselin	Bernard	Faculté Polytechnique de Mons
Grana	Manuel	Univ. Pais Vasco
Grothmann	Ralph	University of Bremen
Hammer	Barbara	University of Osnabrueck
Haschke	Robert	Bielefeld University
Hatziargyriou	Nikos	National Technical Univesity of Athens
Heidemann	Gunther	Bielefeld University
Hollmen	Jaakko	Technical University of Helsinki

Honkela	Antti	Helsinki University of Technology
Hryniewicz	Olgierd	Systems Research Institute PAS
Huang	Di	City University of Hong Kong
Huang	Te-Ming	The University of Auckland
Huelse	Martin	Fraunhofer Institut
Igel	Christian	Ruhr-Universitaet Bochum
Indiveri	Giacomo	UNI-ETH Zurich
Isasi	Pedro	Universidad Carlos III de Madrid
Ishii	Shin	Nara Institute of Science and Technology
Ito	Yoshifusa	Aichi-Gakuin University
Jirina	Marcel	Acad. of Sciences of the Czech Republic
Kaban	Ata	University of Birmingham
Kalveram	Karl Theodor	Institute of Experimental Psychology
Karpouzis	Kostas	ICCS-NTUA
Kasderidis	Stathis	Institute of Computer Science - FORTH
Kim	DaeEun	Max Planck Institute for Psychological Research
Kollias	Stefanos	National Technical University of Athens
Korbicz	Jozef	UZG
Koronacki	Jacek	IPI PAN
Koskela	Markus	Technical University of Helsinki
Kosmopoulos	Dimitris	National Centre for Scientific Research
Kounoudes	Anastasios	SignalGenerix Ltd
Kouropteva	Olga	University of Oulu
Kurfess	Franz	California Polytechnic State University
Kurzynski	Marek	Wroclaw University of Technology
Laaksonen	Jorma	Technical University of Helsinki
Lang	Elmar	University of Regensburg
Leclercq	Edouard	Université du Havre
Lee	John	Université Catholique de Louvain
Lehtimaki	Pasi	Helsinki University of Technology
Leiviska	Kauko	University of Oulu
Lendasse	Amaury	Helsinki University of Technology
Likas	Aris	University of Ioannina
Loizou	Christos	Intercollege, Limassol Campus
Madrenas	Jordi	Technical University of Catalunya
Malatesta	Lori	National Technical Univesity of Athens
Mandziuk	Jacek	Warsaw University of Technology
Marchiori	Elena	Vrije Universiteit Amsterdam
Marcu	Teodor	University of Duisburg-Essen
Markellos	Raphael	Athens University of Economics and Business
Markowska-Kaczmar	Urszula	Wroclaw University of Technology

Martin-Merino	Manuel	University Pontificia of Salamanca
Masulli	Francesco	Polo Universitario di La Spezia G.Marco
Micheli	Alessio	University of Pisa
Morra	Lia	Politecnico di Torino
Moutarde	Fabien	Ecole des Mines de Paris
Mueller	Klaus-Robert	University of Potsdam
Muresan	Raul	SC. NIVIS SRL
Nakayama	Minoru	CRADLE
Nikolopoulos	Konstantinos	Lancaster University Management School
Ntalianis	Klimis	National Technical University of Athens
Oja	Erkki	Helsinki University of Technology
Olteanu	Madalina	Université Paris 1
Ortiz Boyer	Domingo	University of Cordoba
Osowski	Stanislaw	Warsaw University of Technology
Parra	Xavier	Technical University of Catalonia
Pateritsas	Christos	National Technical University of Athens
Pattichis	Marios	University of New Mexico
Pattichis	Costas	University of Cyprus
Paugam-Moisy	Helene	Institut des Sciences Cognitives
Pedreira	Carlos	Catholic University of Rio de Janeiro
Pelckmans	Kristiaan	K.U.Leuven
Perantonis	Stavros	NCSR Demokritos
Pertselakis	Minas	National Technical University of Athens
Peters	Gabriele	Universitaet Dortmund
Piegat	Andrzej	Uniwersytet Szczecinski
Pitas	Yannis	Aristotle University of Thessaloniki
Polani	Daniel	University of Hertfordshire
Porrmann	Mario	Heinz Nixdorf Institute
Prevost	Lionel	Lab. Instr. et Systèmes d'Ile de France
Prevotet	Jean-Christophe	Université Pierre et Marie Curie, Paris
Raivio	Kimmo	Helsinki University of Technology
Raouzeou	Amaryllis	National Technical University of Athens
Rapantzikos	Konstantinos	National Technical University of Athens
Refenes	Apostolos Paul	Athens University Economics & Business
Risto	Risto	Tampere University of Technology
Rocha	Miguel	Universidade do Minho
Romariz	Alexandre	Universidade de Brasilia
Rossi	Fabrice	INRIA Rocquencourt
Rovetta	Stefano	University of Genova
Rutkowska	Danuta	Technical University of Czestochowa
Rynkiewicz	Joseph	Université Paris 1
Salojarvi	Jarkko	Technical University of Helsinki

Schrauwen	Benjamin	Universiteit Gent
Schwenker	Friedhelm	University of Ulm
Seiffert	Udo	Leibniz Institute of Plant Genetics
Sfakiotakis	Michael	Institute of Computer Science FORTH
Sierra	Alejandro	Universidad Autónoma de Madrid
Siivola	Vesa	Technical University of Helsinki
Skodras	Thanos	University of Patras
Stafylopatis	Andreas	National Technical University of Athens
Stamou	Giorgos	ICCS-NTUA
Steil	Jochen J.	University of Bielefeld
Steuer	Michal	University of the West of England
Stoilos	Giorgos	National Technical Univesity of Athens
Strickert	Marc	University of Osnabrueck
Suárez	Alberto	Universidad Autónoma de Madrid
Sugiyama	Masashi	Fraunhofer FIRST
Suykens	Johan	Katholieke Universiteit Leuven
Szczepaniak	Piotr	TUL
Tadeusiewicz	Ryszard	AGH
Tagliaferri	Roberto	Univ. Salerno
Taylor	John	King's College London
Terra	Marco	University of Sao Paulo
Theodoridis	Sergios	UoA
Tomas	Ana Maria	Universidade Aveiro
Trentin	Edmondo	Università di Siena
Tsakiris	Dimitris	University of Crete
Tsapatsoulis	Nicolas	University of Cyprus
Tsotsos	John	York University
Tzouvaras	Vassilis	National Technical Univesity of Athens
Usui	Shiro	RIKEN
Van Looy	Stijn	Universiteit Gent
Vannucci	Marco	Scuola Superiore Sant`Anna
Venetis	Anastassios	National Technical Univesity of Athens
Venna	Jarkko	Helsinki University of Technology
Verbeek	Jakob	University of Amsterdam
Viet	Nguyen Hoang	Polish Acacdemy of Sciences
Villmann	Thomas	Clinic for Psychotherapy
Vitay	Julien	INRIA
Wallace	Manolis	National Technical Univesity of Athens
Watanabe	Norifumi	Keio University
Wennekers	Thomas	University of Plymouth
Wiegerinck	Wim	Radboud University Nijmegen
Wira	Patrice	Universitede Haute-Alsace

Wyns	Bart	Ghent University
Yang	Zhijun	University of Edinburgh
Yearwood	John	University of Ballarat
Zervakis	Mihalis	TUC
Zimmermann	Hans-Georg	Siemens AG

Table of Contents – Part I

Feature Selection and Dimension Reduction for Regression (Special Session)

Learning Algorithms (I)

Learning Algorithms (II)

Advances in Neural Network Learning Methods (Special Session)

Ensemble Learning

Learning Random Neural Networks and Stochastic Agents (Special Session)

Hybrid Architectures

Self Organization

Connectionist Cognitive Science

Cognitive Machines (Special Session)

Neural Dynamics and Complex Systems

Computational Neuroscience

Neural Control, Reinforcement Learning and Robotics Applications

Robotics, Control, Planning

Bio-inspired Neural Network On-Chip Implementation and Applications (Special session)

Table of Contents – Part II

Neural Networks, Semantic Web Technologies and Multimedia Analysis (Special Session)

Bridging the Semantic Gap in Multimedia Machine Learning Approaches (Special Session)

Signal and Time Series Processing (I)

Signal and Time Series Processing (II)

Data Analysis (I)

Data Analysis (II)

Pattern Recognition

Visual Attention Algorithms and Architectures for Perceptional Understanding and Video Coding (Special Session)

Vision and Image Processing (I)

Vision and Image Processing (II)

Computational Finance and Economics (Special Session)

Neural Computing in Energy Engineering (Special Session)

Applications to Biomedicine and Bioinformatics

Applications to Security and Market Analysis

Real World Applications (I)

Real World Applications (II)

Dimensionality Reduction Based on ICA for Regression Problems

Nojun Kwak[1] and Chunghoon Kim[2]

[1] Samsung Electronics, Suwon P.O. Box 105, Gyeonggi-Do, 442-742 Korea,
nojunk@ieee.org,
http://csl.snu.ac.kr/~nojunk

[2] School of Electrical Engineering and Computer Science, Seoul National University,
#047, San 56-1, Sillim-dong, Gwanak-gu, Seoul 151-744, Korea
spinoz@csl.snu.ac.kr

Abstract. In manipulating data such as in supervised learning, we often extract new features from the original features for the purpose of reducing the dimensions of feature space and achieving better performance. In this paper, we show how standard algorithms for independent component analysis (ICA) can be applied to extract features for regression problems. The advantage is that general ICA algorithms become available to a task of feature extraction for regression problems by maximizing the joint mutual information between target variable and new features. Using the new features, we can greatly reduce the dimension of feature space without degrading the regression performance.

1 Introduction

In regression problems, one is given an array of attributes to predict the target value. These attributes are called *features*, and there may exist irrelevant or redundant features to complicate the learning process, thus leading to incorrect prediction. Even when the features presented contain enough information about the target variable, they may not predict the target correctly because the dimension of feature space may be so large that it may require numerous instances to determine the relationship between input features and target. This problem can be avoided by selecting only the relevant features or extracting new features containing the maximal information about the target variable from the original ones. The former methodology is called feature selection or subset selection, while the latter is named feature extraction which includes all the methods that compute any functions, logical or numerical, from the original.

This paper considers the feature extraction problem since it often results in improved performance by extracting new features from the original, especially when small number of dimensions is required. Among the various approach, we focus on finding an appropriate subspace spanned by a set of new features that are arbitrary linear combinations of original. PCA (principle component analysis) [1], ICA (independent component analysis) [2] [3], and LDA (linear discriminant analysis) [4] are the most popular subspace methods. However,

S. Kollias et al. (Eds.): ICANN 2006, Part I, LNCS 4131, pp. 1–10, 2006.

most of these methods cannot be used for regression problems since some of them such as PCA and ICA focus on finding features by unsupervised manner and others such as LDA have been mainly developed for classification problems.

In our previous work, we have developed ICA-FX (feature extraction based on independent component analysis) [5], a supervised feature extraction method for classification problems, by modifying the learning rule of original ICA. Like ICA, it utilizes higher order statistics, while unlike ICA, it was developed as a supervised method in that it includes the output class information to find an appropriate feature subspace. This method is well-suited for classification problems in the aspect of constructing new features that are strongly related to output class.

In this paper, the ICA-FX for classification problems is extended to regression problems. Because the output class label was coded as a numerical value in the ICA-FX algorithm, the method can be easily applied to regression problems without changing much from original ICA-FX for classification problems.

This paper is organized as follows. In Section 2, we briefly review the ICA algorithm. In Section 3, we develop ICA-FX for regression problems. This follows almost the same steps as we did for classification problems. Experimental results showing the advantages of the proposed algorithm are presented in Section 4 and conclusions are provided in Section 5.

2 Review of ICA

The problem setting of ICA is as follows. Assume that there is an L-dimensional zero-mean non-Gaussian source vector $\boldsymbol{s}(n) = [s_1(n), \cdots, s_L(n)]^T$, such that the components $s_i(n)$'s are mutually independent, and an observed data vector $\boldsymbol{x}(n) = [x_1(n), \cdots, x_N(n)]^T$ is composed of linear combinations of sources $s_i(n)$ at each time point n, such that

$$\boldsymbol{x}(n) = A\boldsymbol{s}(n) \tag{1}$$

where A is a full rank $N \times L$ matrix with $L \leq N$. The goal of ICA is to find a linear mapping W such that each component of an estimate $\boldsymbol{u}$ of the source vector

$$\boldsymbol{u}(n) = W\boldsymbol{x}(n) = WA\boldsymbol{s}(n) \tag{2}$$

is as independent as possible. The original sources $\boldsymbol{s}(n)$ are exactly recovered when W is the pseudo-inverse of A up to some scale changes and permutations. For a derivation of an ICA algorithm, one usually assumes that $L = N$, because we have no idea about the number of sources. In addition, sources are assumed to be independent of time n and are drawn from independent identical distribution $p_i(s_i)$.

To find W in (2), Bell and Sejnowski [2] have developed the Infomax algorithm, one of the popular algorithms for ICA, in which they used a feed-forward neural processor. This neural processor takes $\boldsymbol{x}$ as an input vector. The weight W is multiplied to the input $\boldsymbol{x}$ to give $\boldsymbol{u}$ and each component u_i goes through a bounded invertible monotonic nonlinear function $g_i(\cdot)$ to match the cumulative distribution of the sources.

From the view of information theory, maximizing the statistical independence among variables u_i's is equivalent to minimizing mutual information among u_i's. This can be achieved by minimizing mutual information between $y_i = g_i(u_i), i = 1 \cdots L$, since the nonlinear transfer function $g_i(\cdot)$ does not introduce any dependencies.

In [2], it has been shown that by maximizing the joint entropy $H(\boldsymbol{y})$ of the output $\boldsymbol{y} = [y_1, \cdots, y_N]^T$ of a processor, we can approximately minimize the mutual information among the output components y_i's

$$I(\boldsymbol{y}) = \int p(\boldsymbol{y}) \log \frac{p(\boldsymbol{y})}{\prod_{i=1}^{N} p_i(y_i)} d\boldsymbol{y}. \tag{3}$$

Here, $p(\boldsymbol{y})$ is the joint probability density function (pdf) of a vector $\boldsymbol{y}$, and $p_i(y_i)$ is the marginal pdf of the variable y_i.

The joint entropy of the outputs of this processor is

$$H(\boldsymbol{y}) = -\int p(\boldsymbol{y}) \log p(\boldsymbol{y}) d\boldsymbol{y} = -\int p(\boldsymbol{x}) \frac{p(\boldsymbol{x})}{\log |\det J(\boldsymbol{x})|} d\boldsymbol{x} \tag{4}$$

where $J(\boldsymbol{x})$ is the Jacobian matrix whose (i, j)th element is partial derivative $\partial y_j / \partial x_i$. Note that $J(\boldsymbol{x}) = W$. Differentiating $H(\boldsymbol{y})$ with respect to W and applying natural gradient by multiplying $W^T W$ on the right, we get the learning rule for ICA:

$$\Delta W \propto [I - \boldsymbol{\varphi}(\boldsymbol{u}) u^T] W \tag{5}$$

where

$$\boldsymbol{\varphi}(\boldsymbol{u}) = \left[-\frac{\frac{\partial p_1(u_1)}{\partial u_1}}{p_1(u_1)}, \cdots, -\frac{\frac{\partial p_N(u_N)}{\partial u_N}}{p_N(u_N)} \right]^T. \tag{6}$$

In this paper, we adopt the extended Infomax algorithm [3] because it is easy to implement with less strict assumptions on source distribution.

3 ICA-FX for Regression

ICA outputs a set of maximally independent vectors that are linear combinations of the observed data. Although these vectors might have some applications in such areas as blind source separation and data visualization, it is not suitable for feature extraction for supervised learning, because it does not make use of the output target information. The effort to incorporate the standard ICA with supervised learning has been made in our previous work [5], where a new feature extraction algorithm, ICA-FX for classification problems was proposed. In this paper, the original ICA-FX algorithm for classification problems is extended for regression problems.

Before we present our algorithm, we formalize the purpose of feature extraction for regression problems.

The success of a feature extraction algorithm depends critically on how much information about the output class is contained in the newly generated features. This can be formalized as follows.

(Problem statement) *Assume that there are a normalized input feature vector, $\boldsymbol{x} = [x_1, \cdots, x_N]^T$, and target variables, $\boldsymbol{t} = [t_1, \cdots, t_{N_t}]^T$. The purpose of feature extraction for regression problems is to extract $M (\leq N)$ new features $\boldsymbol{f_a} = [f_1, \cdots, f_M]^T$ from $\boldsymbol{x}$, containing the maximum information on the output target variables $\boldsymbol{t}$. Here N_t is the number of target variables.*

In information theory, the information between random variables is measured by *mutual information* and this problem statement can be formalized using this information theoretical term as follows:

(Problem statement - information theoretic view) *The purpose of a feature extraction for regression problem is to extract $M (\leq N)$ features $\boldsymbol{f}_a$ from $\boldsymbol{x}$, such that $I(\boldsymbol{f}_a;\boldsymbol{t})$, the mutual information between newly extracted features $\boldsymbol{f}_a$ and target output variables $\boldsymbol{t}$ becomes maximum.*

The main idea of the proposed method is simple. It tries to apply the standard ICA algorithms to feature extraction for regression problems by making use of the target variables to produce two sets of new features; features that carry as much information on the target variables (these features will be useful for regression) as possible and the others that do not (these will be discarded). The advantage is that the general ICA algorithms can be used for feature extraction by maximizing the joint mutual information between the target variables and new features.

Now consider the structure shown in Fig. 1. Here, the original feature vector $\boldsymbol{x}$ is fully connected to $\boldsymbol{u} = [u_1, \cdots, u_N]$, the target vector $\boldsymbol{t}$ is connected only to $\boldsymbol{u}_a = [u_1, \cdots, u_M]$, and $u_{N+l} = t_l$, $l = 1, \cdots, N_t$. In the figure, the weight matrix $\boldsymbol{W} \in \Re^{(N+N_t)\times(N+N_t)}$ becomes

$$\boldsymbol{W} = \left(\begin{array}{c|c} W & V \\ \hline \mathbf{0}_{N_t,N} & I_{N_t} \end{array}\right) = \left(\begin{array}{c|ccc} & w_{1,N+1} & \cdots & w_{1,N+N_t} \\ & \vdots & & \vdots \\ W & w_{M,N+1} & \cdots & w_{M,N+N_t} \\ & & \mathbf{0}_{N-M,N_t} & \\ \hline \mathbf{0}_{N_t,N} & & I_{N_t} & \end{array}\right). \quad (7)$$

where $W \in \Re^{N\times N}$ and $V = [V_a^T, \mathbf{0}_{N-M,N_t}^T]^T \in \Re^{N\times N_t}$. Here the first nonzero M rows of V is denoted as $V_a \in \Re^{M\times N_t}$.

As stated before, in information theoretic view, the aim of feature extraction is to extract M new features $\boldsymbol{f}_a$ from the original N features, $\boldsymbol{x}$, such that

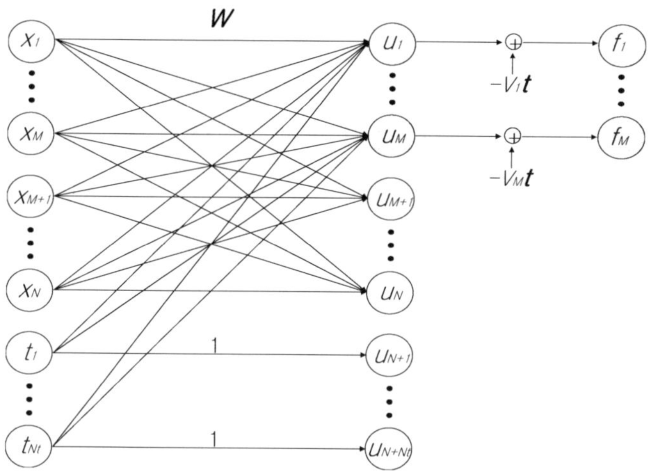

Fig. 1. Feature extraction algorithm based on ICA (ICA-FX)

$I(\boldsymbol{f}_a; t)$, the mutual information between newly extracted features $\boldsymbol{f}_a$ and the target variables $\boldsymbol{t}$ is maximized.

Because of the data processing inequality it is maximized when $I(\boldsymbol{f}_a; \boldsymbol{t})$ becomes equal to $I(\boldsymbol{x}; \boldsymbol{t})$, the mutual information between the original features $\boldsymbol{x}$ and the target variables $\boldsymbol{t}$.

This can be satisfied if we can separate the input feature space $\boldsymbol{x}$ into two linear subspaces: one that is spanned by $\boldsymbol{f}_a = [f_1, \cdots, f_M]^T$, which contains the maximum information on the target variables $\boldsymbol{t}$, and the other spanned by $\boldsymbol{f}_b = [f_{M+1}, \cdots, f_N]^T$, which is independent of $\boldsymbol{t}$ as much as possible.

The condition for this separation can be derived as follows. If it is assumed that $\boldsymbol{W}$ is nonsingular, then $\boldsymbol{x}$ and $\boldsymbol{f} = [f_1, \cdots, f_N]^T$ span the same linear space, which can be represented with the direct sum of $\boldsymbol{f}_a$ and $\boldsymbol{f}_b$, and then by the data processing inequality,

$$I(\boldsymbol{x}; \boldsymbol{t}) = I(W\boldsymbol{x}; \boldsymbol{t}) = I(\boldsymbol{f}; \boldsymbol{t}) = I(\boldsymbol{f}_a, \boldsymbol{f}_b; \boldsymbol{t}) \geq I(\boldsymbol{f}_a; \boldsymbol{t}). \quad (8)$$

The first equality holds because W is nonsingular. The second and the third equalities are from the definitions of $\boldsymbol{f}$, $\boldsymbol{f}_a$ and $\boldsymbol{f}_b$. In the inequality on the last line, the equality holds if $I(\boldsymbol{f}_b; \boldsymbol{t}) = I(u_{M+1}, \cdots, u_N; t) = 0$.

If this is possible, the dimension of the input feature space can be reduced from N to $M(< N)$ by using only $\boldsymbol{f}_a$ instead of $\boldsymbol{x}$, without losing any information on the target variables.

To solve this problem, the feature extraction problem is interpreted in the structure of the blind source separation (BSS) problem as shown in Fig. 2. The detailed description of each step is as follows:

(Mixing) *Assume that there are N independent sources $\boldsymbol{s} = [s_1, \cdots, s_N]^T$ which are also independent of the target variables t. Assume also that the ob-*

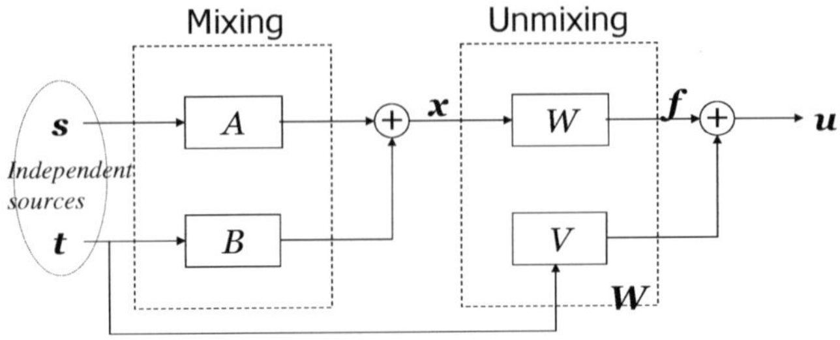

Fig. 2. Interpretation of Feature Extraction in the BSS structure

served feature vector $\boldsymbol{x}$ *is a linear combination of the sources* $\boldsymbol{s}$ *and* $\boldsymbol{t}$ *with the mixing matrix* $A \in \Re^{N\times N}$ *and* $B \in \Re^{N\times N_t}$*; i.e.,*

$$\boldsymbol{x} = A\boldsymbol{s} + B\boldsymbol{t}. \tag{9}$$

(Unmixing) *The unmixing stage is slightly different from the BSS problem as shown in Fig. 1. In the figure, the unmixing equation becomes*

$$\boldsymbol{u} = W\boldsymbol{x} + V\boldsymbol{t}. \tag{10}$$

Suppose $\boldsymbol{u}$ *is somehow made equal to* $\boldsymbol{e}$*, the scaled and permuted version of the source* $\boldsymbol{s}$*; i.e.,*

$$\boldsymbol{e} \triangleq \Lambda\Pi\boldsymbol{s} \tag{11}$$

where Λ *is a diagonal matrix corresponding to an appropriate scale and* Π *is a permutation matrix. The* u_i*'s* $(i = 1, \cdots, N)$ *are then independent of the target variables* $\boldsymbol{t}$ *by the assumption. Among the elements of* $\boldsymbol{f} = W\boldsymbol{x}(= \boldsymbol{u} - V\boldsymbol{t})$*,* $\boldsymbol{f}_b = [f_{M+1}, \cdots, f_N]^T$ *will be independent of* t *because the* i*th row of* V*,* $V_i = [w_{i,N+1}, \cdots, w_{i,N+N_t}] = \mathbf{0}$ *and* $f_i = u_i$ *for* $i = M+1, \cdots, N$*. Therefore, the* $M(< N)$ *dimensional new feature vector* $\boldsymbol{f}_a$ *can be extracted by a linear transformation of* $\boldsymbol{x}$ *containing the most information on the class if the relation* $\boldsymbol{u} = \boldsymbol{e}$ *holds.*

The learning rule for the ICA-FX for regression is obtained by the same way as that of ICA-FX for classification problem using the MLE (maximum likelihood estimation) approach as follows.

If it is assumed that $\boldsymbol{u} = [u_1, \cdots, u_N]^T$ is a linear combination of the source $\boldsymbol{s}$; i.e., it is made equal to $\boldsymbol{e}$, a scaled and permutated version of the source, $\boldsymbol{s}$, as in (11), and that each element of $\boldsymbol{u}$ is independent of the other elements of $\boldsymbol{u}$, which is also independent of the target vector $\boldsymbol{t}$, the log likelihood of the data for a given $\boldsymbol{W}$ becomes the following:

$$L(\boldsymbol{u}, \boldsymbol{t}|\boldsymbol{W}) = \log|\det \boldsymbol{W}| + \sum_{i=1}^{N} \log p_i(u_i) + \log p(\boldsymbol{t}) \tag{12}$$

because

$$p(\boldsymbol{x}, \boldsymbol{t}|\boldsymbol{W}) = |\det \boldsymbol{W}|\ p(\boldsymbol{u}, \boldsymbol{t}) = |\det \boldsymbol{W}| \prod_{i=1}^{N} p_i(u_i)\ p(\boldsymbol{t}). \tag{13}$$

Now, L can be maximized, and this can be achieved by the steepest ascent method. Because the last term in (12) is a constant, differentiating (12) with respect to $\boldsymbol{W}$ leads to

$$\begin{aligned}\frac{\partial L}{\partial w_{i,j}} &= \frac{adj(w_{j,i})}{|\det \boldsymbol{W}|} - \varphi_i(u_i)x_j \qquad 1 \le i,j \le N \\ \frac{\partial L}{\partial w_{i,N+j}} &= -\varphi_i(u_i)t_j \qquad 1 \le i \le M, 1 \le j \le N_t\end{aligned} \tag{14}$$

where $adj(\cdot)$ is adjoint and $\varphi_i(u_i) = -\frac{dp_i(u_i)}{du_i}/p_i(u_i)$.

It can be seen that $|\det \boldsymbol{W}| = |\det W|$ and $\frac{adj(w_{j,i})}{|\det \boldsymbol{W}|} = W_{i,j}^{-T}$. Thus the learning rule becomes

$$\begin{aligned}\Delta W &\propto W^{-T} - \boldsymbol{\varphi}(\boldsymbol{u})\boldsymbol{x}^T \\ \Delta V_a &\propto -\boldsymbol{\varphi}(\boldsymbol{u}_a)\boldsymbol{t}^T.\end{aligned} \tag{15}$$

Here $\boldsymbol{\varphi}(\boldsymbol{u}) \triangleq [\varphi_1(u_1), \cdots, \varphi_N(u_N)]^T$ and $\boldsymbol{\varphi}(\boldsymbol{u}_a) \triangleq [\varphi_1(u_1), \cdots, \varphi_M(u_M)]^T$.

Applying a natural gradient on updating W, by multiplying $W^T W$ on the right side of the first equation of (15), the following is obtained.

$$\begin{aligned}W^{(n+1)} &= W^{(n)} + \mu_1[I_N - \boldsymbol{\varphi}(\boldsymbol{u})\boldsymbol{f}^T]W^{(n)} \\ V_a^{(n+1)} &= V_a^{(n)} - \mu_2\boldsymbol{\varphi}(\boldsymbol{u}_a)\boldsymbol{t}^T.\end{aligned} \tag{16}$$

Here μ_1 and μ_2 are the learning rates that can be set differently. By this weight update rule, the resulting u_i's will have a good chance of fulfilling the assumption that u_i's are not only independent of one another but also independent of the target variables $\boldsymbol{t}$.

Note that the learning rule for W is the same as the original ICA learning rule [2], and also note that $\boldsymbol{f}_a$ corresponds to the first M elements of $W\boldsymbol{x}$. Therefore, the optimal features $\boldsymbol{f}_a$ can be extracted by the proposed algorithm when it finds the optimal solution for W by (16).

4 Experiment Results

In this section, we have applied ICA-FX to several regression problems and show the performance of the method. For all the problems below, we have compared the performance of ICA-FX with those of PCA and original features.

As a regression tool, standard multi-layer perceptron (MLP) with one hidden layer was used. The numbers of nodes in input, hidden, and output layers were set to the number of extracted features, six, and one respectively. As a transfer functions of hidden and output layers, *tansig (tangent sigmoid)* and *purelin (pure linear)* were used respectively. As a training rule of the MLP, *trainlm (Levenberg-Marquardt)* was used. The weight update rule of the method is

$$W_{mlp}(k+1) = W_{mlp}(k) - (J^T J + \mu I)^{-1} J^T \boldsymbol{e}_{mlp}.$$

Here, J is the Jacobian matrix that contains first derivatives of the network errors with respect to the weights, and $\boldsymbol{e}_{mlp}$ is a vector of network errors. For adaptive value μ, default settings of the *Matlab* were used.

4.1 Artificial Problems

Linear Case. Consider the simple problem of the following:

Suppose we have five independent input features $x_1 \sim x_5$ which have normal distribution with zero mean and variance of 1. Also suppose that the target output variable t has the following relationship with the input $\boldsymbol{x}$: $t = 2x_1 + 3x_3$.

For this problem, 1000 samples were generated which were divided into 500 training data and 500 test data. On the training data, ICA-FX was applied where the number of extracted feature M was set to 1. The first row of weight matrix $\boldsymbol{W}$ in (7) after ICA-FX was [7.0904, 0.0004, 10.9933, 0.0002, 0.0003, -13.1630]. Thus the newly extracted feature is $f = 7.0904x_1 + 0.0004x_2 + 10.9933x_3 + 0.0002x_4 + 0.0003x_5$. Note that the coefficients for x_2, x_4 and x_5 are very small compared to those of x_1 and x_3. In addition, note that the ratio of coefficient of x_1 and that of x_3 is approximately 2:3. The extracted weights show that the ICA-FX performs quite well for linear regression problems.

For this problem, we have also performed PCA and compared the performance of the ICA-FX, PCA, and the original 5 features in Table 1. In this experiment, the number of extracted features for ICA-FX was varied from 1 to 5. The performance is the root mean square (rms) error of the test data. Averages of 10 experiments with standard deviations are reported here.

Table 1. Performance for the simple linear dataset (rms error). Averages of 10 experiments. Numbers in the parentheses are the standard deviations.

	no. of features	rms error
Original	5	0.00 (0.00)
PCA	1	2.85 (0.64)
	3	2.17 (0.68)
	5	0.00 (0.00)
ICA-FX	1	0.01 (0.02)
	3	0.00 (0.00)
	5	0.00 (0.00)

The table also shows that the ICA-FX performs well with a small number of features.

Nonlinear Case. Consider the following problems:

As the problem above, suppose we have five independent input features $x_1 \sim x_5$ which have normal distribution with zero mean and variance of 1. Now, *suppose that the target output variable t has the following nonlinear relationship with the input* $\boldsymbol{x}$: $t = sin(x_2 + 2x_4)$.

For this problem, 1000 samples were generated which were divided into 500 training data and 500 test data. On the training data, ICA-FX was applied where the number of extracted feature M was set to 1. The first row of weight matrix $\boldsymbol{W}$ in (7) after ICA-FX was [-0.0176, -0.5146, -0.0982, -0.9607, 0.0046, 0.4886]. Thus the newly extracted feature is $f = -0.0176x_1 - 0.5146x_2 - 0.0982x_3 - 0.9607x_4 + 0.0046x_5$. Note that the coefficients for x_1, x_3 and x_5 are very small compared to those of x_2 and x_4. This indicates that the resultant weights after ICA-FX can be used to select appropriate features. In addition, note that the ratio of coefficient of x_2 and that of x_4 is approximately 1:2 which is the ratio of the corresponding coefficients inside *sin* term. The extracted weights show that the ICA-FX performs well for this nonlinear regression problem too.

As in the linear case, we have performed PCA for this dataset and compared the rms error of the ICA-FX with those of PCA and the original 5 features in Table 2 The number of extracted features for ICA-FX was varied from 1 to 5.

Table 2. Performance for the simple nonlinear dataset (rms error). Averages of 10 experiments. Numbers in the parentheses are the standard deviations.

	no. of features	rms error
Original	5	0.14 (0.17)
PCA	1	0.73 (0.02)
	3	0.57 (0.20)
	5	0.08 (0.05)
ICA-FX	1	0.15 (0.03)
	3	0.19 (0.27)
	5	0.08 (0.06)

The table shows that the ICA-FX performs better than PCA and the original data for this dataset.

4.2 UCI Dataset - Housing (Boston)

In this section, we have applied ICA-FX to *Housing (Boston)* dataset in UCI Machine Learning Repository [6].

The dataset contains 13 input features, 12 continuous and 1 binary, and one continuous target variable. There are total 506 instances. We have randomly divided this dataset into 90% training and 10% test sets 10 times and reported the average rms error on the test data in Table 3. In applying ICA-FX, we have normalised each input feature to have zero mean and unit variance and varied the number of extracted features M from 1 to 13.

Note that ICA-FX performs better than PCA generally and the performance was robust irrespective of the number of extracted features. Note also that the ICA-FX performs better than using 13 original features.

Table 3. Performance for the Housing dataset (rms error). Numbers in the parentheses are the standard deviations.

no. of features	original	PCA	ICA-FX
1	–	7.36 (0.68)	3.59 (0.60)
3	–	4.70 (1.17)	3.60 (0.54)
5	–	4.16 (1.72)	3.60 (0.65)
7	–	3.49 (0.51)	3.50 (0.45)
9	–	4.00 (0.91)	3.80 (1.01)
11	–	3.67 (0.71)	3.42 (0.63)
13	4.20 (0.99)	4.41 (1.73)	3.37 (0.63)

5 Conclusions

In this paper, we have extended the feature extraction algorithm ICA-FX to regression problems. The proposed algorithm is based on the standard ICA and can generate very useful features for regression problems.

Although ICA can be directly used for feature extraction, it does not generate useful information because of its unsupervised learning nature. In the proposed algorithm, we added output target information in training ICA. With the additional target information we can extract new features containing maximal information about the target. The number of extracted features can be arbitrarily chosen.

Since it uses the standard feed-forward structure and learning algorithm of ICA, it is easy to implement and train. Experimental results for several data sets show that the proposed algorithm generates good features that outperform the original features and other features extracted from other methods. Because the original ICA is ideally suited for processing large datasets such as biomedical ones, the proposed algorithm is also expected to perform well for large-scale regression problems.

References

1. I.T. Joliffe, *Principal Component Analysis*, Springer-Verlag, 1986.
2. A.J. Bell and T.J. Sejnowski, "An information-maximization approach to blind separation and blind deconvolution," *Neural Computation*, vol. 7, no. 6, June 1995.
3. T-W. Lee, M. Girolami, and T.J. Sejnowski, "Independent component analysis using an extended infomax algorithm for mixed sub-gaussian and super-gaussian sources," *Neural Computation*, vol. 11, no. 2, Feb. 1999.
4. K. Fukunaga, *Introduction to Statistical Pattern Recognition*, Academic Press, second edition, 1990.
5. N. Kwak and C.-H. Choi, "Feature extraction based on ICA for binary classification problems," *IEEE Trans. on Knowledge and Data Engineering*, vol. 15, no. 6, pp. 1374–1388, Nov. 2003.
6. http://www.ics.uci.edu/~mlearn/MLSummary.html

A Functional Approach to Variable Selection in Spectrometric Problems*

Fabrice Rossi[1], Damien François[2], Vincent Wertz[2], and Michel Verleysen[3]

[1] Projet AxIS, INRIA, Domaine de Voluceau, Rocquencourt, B.P. 105, 78153 Le Chesnay Cedex, France
[2] Université catholique de Louvain - Machine Learning Group, CESAME, 4 av. G. Lemaître, 1348 Louvain-la-Neuve, Belgium
[3] Université catholique de Louvain - Machine Learning Group, DICE, 3 place du Levant, 1348 Louvain-la-Neuve, Belgium

Abstract. In spectrometric problems, objects are characterized by high-resolution spectra that correspond to hundreds to thousands of variables. In this context, even fast variable selection methods lead to high computational load. However, spectra are generally smooth and can therefore be accurately approximated by splines. In this paper, we propose to use a B-spline expansion as a pre-processing step before variable selection, in which original variables are replaced by coefficients of the B-spline expansions. Using a simple leave-one-out procedure, the optimal number of B-spline coefficients can be found efficiently. As there is generally an order of magnitude less coefficients than original spectral variables, selecting optimal coefficients is faster than selecting variables. Moreover, a B-spline coefficient depends only on a limited range of original variables: this preserves interpretability of the selected variables. We demonstrate the interest of the proposed method on real-world data.

1 Introduction

In many real-world problems, objects are described by sampled functions rather than by vectors. In the simplest case, an object is given by a function f, from $\mathbb{R}$ to $\mathbb{R}$, specified by a list of m input/output pairs, $((x_j, f(x_j)))_{1\leq j\leq m}$ (m and $(x_j)_{1\leq j\leq m}$ depend on the object). Examples of such situation include applications in which temporal evolution of objects is monitored (and therefore where each object is described by one or several time series) and others in which objects are characterized by spectra (near infrared transmittance for instance).

* M. Verleysen is Research Director of the Belgian F.N.R.S. (National Fund for Scientific Research). D. François is funded by a grant from the Belgian F.R.I.A. Parts of this research result from the Belgian Program on Interuniversity Attraction Poles, initiated by the Belgian Federal Science Policy Office. The scientific responsibility rests with its authors.

S. Kollias et al. (Eds.): ICANN 2006, Part I, LNCS 4131, pp. 11–20, 2006.

One of the main problems in spectrometry is regression: one wants to predict physical or chemical properties of a sample via its spectrum. Because chemical and physical analyses are long, difficult and expensive, we have generally a low number of examples, typically a few hundreds. On the contrary, spectra are generally sampled at a high resolution, up to thousands of wavelengths per spectrum. It is therefore quite common to have more sampling points (spectral variables) than spectra.

As the sampling is generally fixed (i.e. m and $(x_j)_{1\leq j\leq m}$ are fixed for the whole data set), each spectrum can be considered as a high-dimensional vector. However, because of the low number of spectra, even simple linear methods are difficult to apply directly to many spectrometric problems. In practice, the standard solution is to rely on dimension reduction methods coupled with linear regression, mainly Principal Component Regression (PCR) and Partial Least Squares Regression (PLSR). PCR for instance consists in a simple linear model constructed on a few Principal Components of the original data. PLSR consists in finding linear projections that have maximum correlation with the target variable; a linear regression is then built on the projected coordinates. While those methods give generally satisfactory results, they are unfortunately difficult to interpret: the linear model is constructed on projected features whose dependency to the original spectral variables, albeit linear, can be quite complex. In general, one cannot determine from the model which spectral range is useful for the regression problem under focus. Moreover, PCR and PLSR are intrinsically limited by their linear nature.

Another solution consists in using variable selection methods to keep only a small number of the original spectral variables and then to build a nonlinear model on those data [1,2,3]. When a small number of spectral variables are selected, this approach both avoids overfitting and eases interpretation. Even if the processing of the selected variables is nonlinear, the model emphasizes generally the dependency of the target variable to a small number of original spectral variables. However, those methods suffer from two problems. They are generally quite slow, even when filter selection methods are used (i.e., when the relevance of a group of variables is estimated via a simpler model than the nonlinear model). Moreover, while filter methods tend to be less sensitive to overfitting than the nonlinear models used for the second part of the analysis, they nevertheless face the difficulty of dealing with high-dimensional data and can select redundant or useless variables.

This paper proposes to use a functional representation approach as a preprocessing step before variable selection for regression problems. The main idea is to leverage the functional nature of spectra to replace high-resolution representations by a low number of variables that keep almost all the original information and still allow one to assess the importance of some wavelength ranges in the regression task. This prior reduction of the data dimensionality eases the variable selection both in terms of computational requirement and in terms of statistical significance.

2 Functional Data Analysis

The idea is this paper is based on the concepts of Functional Data Analysis (FDA [4]). The main idea of FDA is to adapt standard data analysis methods such that they make explicit use of the functional aspect of their inputs. There are two standard approaches in FDA: *regularization* and *filtering*.

The regularization approach addresses the overfitting problem via complexity control. Let us consider for example the problem of linear regression on functional data. In a theoretical and perfect setting, we have a random variable X with values in L^2 (the Hilbert space of square integrable functions from $\mathbb{R}$ to $\mathbb{R}$) and a target random variable Y with values in $\mathbb{R}$. The functional linear model is given by $Y = \langle h, X \rangle$, where $\langle \cdot, \cdot \rangle$ denotes the inner product in L^2, i.e.

$$Y = \int hX \mathrm{d}\lambda.$$

Finding h via observations, i.e. via realizations of the pair (X, Y), is difficult: because of the infinite dimension of L^2, the problem is ill-posed. It has generally an infinite number of solutions and they are difficult to estimate. The problem is solved by looking for smooth candidates for the function h, for instance twice differentiable functions with minimal curvature, see e.g. [5,6]. This can be considered as a functional version of the standard ridge regression [7]. The regularization approach has been applied to other data analysis problems, such as Principal Component Analysis [8]. As shown in [9] in the case of discriminant analysis, when the data are sampled functions, ridge regularization leads to worse solutions than a functional regularization.

In the filtering approach, each list of input/output pairs is considered as a function approximation problem for which a simple truncated basis solution is chosen: the list $((x_j, f(x_j)))_{1 \leq j \leq m}$ is replaced by the vector $(u_1, \ldots, u_p)$ obtained as the minimizer of

$$\sum_{j=1}^{m} \left(f(x_j) - \sum_{k=1}^{p} u_k \phi_k(x_j) \right)^2,$$

i.e., the square reconstruction error of f by the basis functions $(\phi_k)_{1 \leq k \leq p}$. Filtering can therefore be considered as a preprocessing step in which functional data are consistently transformed into vector data. It has been used as a simple way to adapt many data analysis methods to functional data, see for instance [10] for linear regression, [11,12] for Multi-Layer Perceptrons and Radial Basis Function Networks, [13] for k-nearest neighbors and [14] for Support Vector Machines.

3 B-spline Representation

3.1 B-splines

Obviously, the filtering approach of FDA can be used to reduce the dimensionality of spectra: one can freely choose p (the number of basis functions) and therefore greatly reduce the number of spectral variables. In fact, this idea of using function representation for spectra has been used in the field of chemometrics since [15]. The idea of this early work was to compress the spectra in order to speed up linear methods (PCR and PLSR for instance). We use the same idea to speed up variable selection. As in [16,17] we use B-splines, however, our additional motivation is the locality of B-splines, that allows us to maintain interpretability.

Let us consider an interval $[a, b]$ and p sub-intervals, defined by the $p+1$ values, $t_0, \ldots, t_p$, called *knots*, such that $t_j < t_{j+1}$, $t_0 = a$ and $t_p = b$. We recall that splines of order d are C^{d-2} piecewise polynomial functions given by a polynomial of degree $d-1$ on each interval $[t_j, t_{j+1}[$ (the last interval is $[t_{p-1}, t_p]$). The vector space of such functions has a basis of $p-1+d$ *B-splines*, $B_1^d, \ldots, B_{p-1+d}^d$ (see [18] for details).

We consider n spectra, $(s_i)_{1 \le i \le n}$ which are functions from $\mathbb{R}$ to $\mathbb{R}$, observed at m wavelengths, $(w_j)_{1 \le j \le m}$. We denote a the smallest wavelength and b the largest. Given $p+1$ knots as above, we associate to a spectrum s_i the coordinates $\mathbf{c}(s_i)$ of its best approximation by a spline of order d on the associated B-spline basis. The proposed method consists in replacing the m original variables by the $p-d+1$ B-spline coordinates.

Those coordinates are the solution of the standard least square optimization problem:

$$\mathbf{c}(s_i) = \arg \min_{\mathbf{c} \in \mathbb{R}^{p-1+d}} \sum_{j=1}^{m} \left(s_i(w_j) - \sum_{k=1}^{p-1+d} c_k B_k^d(w_j) \right)^2 . \tag{1}$$

This quadratic problem leads to a linear solution, i.e. there is a $(p-1+d) \times m$ matrix R, that depends only on d, p and $(w_j)_{1 \le j \le m}$, such that

$$\mathbf{c}(s_i) = R\mathbf{s}_i, \tag{2}$$

where $\mathbf{s}_i$ denotes the vector representation of s_i, i.e. $\mathbf{s}_i = (s_i(w_1), \ldots, s_i(w_m))$.

3.2 Interpretability

Of course, this type of linear relationship applies to any filtering approach that consists in projecting the considered functions on the sub-vector space spanned by some well-chosen basis functions. An interesting and quite specific aspect of B-splines however it that R is approximately localized. Let us consider more precisely the case of a single spectrum s. The coordinates of its projection are given by:

$$c(s_i)_l = \sum_{j=1}^{m} R_{lj} s_i(w_j).$$

A remarkable property of B-splines is that most coefficients in R have a very small magnitude. In practice, this means that the value of a new variable depends only on some of the original variables. Moreover, dependencies are localized: the

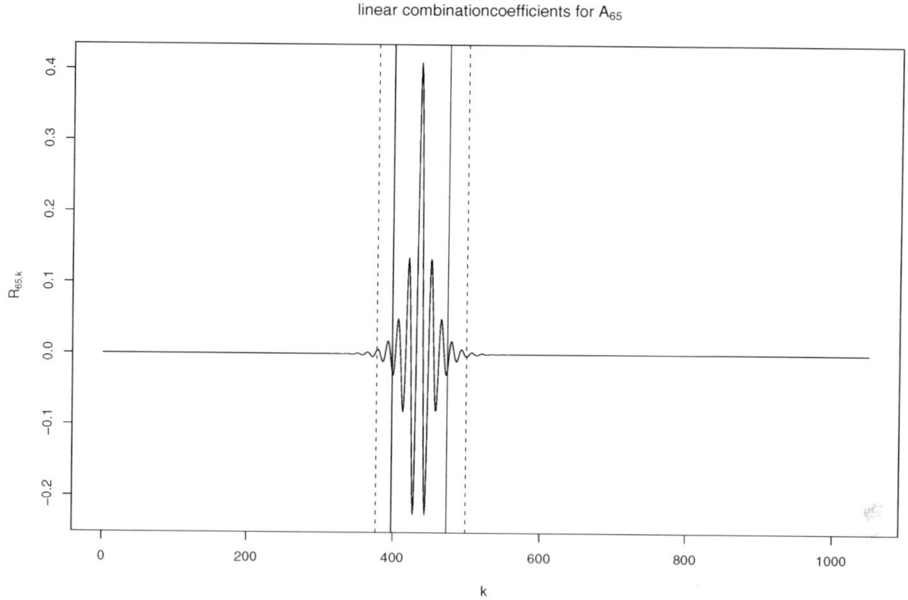

Fig. 1. Graphical representation of $R_{65,k}$ for splines of order 5 with 155 B-splines calculated for 1050 original spectral variables

coordinate of s_i on the B-spline B_k^d depends only on a sub-interval of the original wavelength interval $[a, b]$, as shown on an example in Figure 1.

In theory, values in R are nonzero because when we construct an orthogonal basis of the space of splines of order d, the functions of this basis do not have a smaller support than interval $[a, b]$. Compactly supported wavelet bases [19] provide alternative solutions in which some lines of R have actual zero entries. However, in this case, the low-resolution wavelet spans the full original interval and therefore some lines of R do not have any negligible coefficient. Up to a small approximation, B-splines offer on the contrary a localized basis.

In practice, a wavelength range can be associated to each new variable $c(s)_l$. If we assume the list of wavelengths $(w_j)_{1\leq j\leq m}$ to be in increasing order, and given a precision ratio $\epsilon > 0$, the indexes of the bounds of the interval are

$$l_i = \max\left\{1 \leq j \leq m \,\middle|\, \max_{1\leq k<j} |R_{ik}| < \epsilon \max_{1\leq k\leq m} |R_{ik}|\right\}, \tag{3}$$

$$u_i = \min\left\{1 \leq j \leq m \,\middle|\, \max_{j<k\leq m} |R_{ik}| < \epsilon \max_{1\leq k\leq m} |R_{ik}|\right\}, \tag{4}$$

with the convention that $\max_{1\leq k<1} |R_{ik}| = \max_{m<k\leq m} |R_{ik}| = 0$. The lower bound w_{l_i} corresponds to the largest index j such that all coefficients R_{ik} for $k < j$ are smaller than ϵ times the maximal coefficient. The upper bound w_{u_i}

is defined in a symmetric way. Figure 1 displays two wavelength intervals: the vertical solid lines give the bounds of the interval calculated for $\epsilon = 0.05$ and the dashed lines correspond to $\epsilon = 0.01$.

3.3 Optimal B-splines Basis

Obviously, the quality of the new variables depends both on d and on p. For instance $d = 1$ corresponds to a piecewise constant approximation that has generally a low quality. In order to compare possible choices for d and p, we use the leave-one-out error estimate described in [11]. This estimate is based only on the spectra themselves and does not take into account the regression task. It can be implemented very efficiently : for a single spectrum, the cost is $O(p^2 + pm)$. Moreover, because most of the calculation does not depend on the spectrum, the cost for n spectra is $O(p^2 + pmn)$. It should be noted that in spectrometric applications, spectra do not exhibit very strong differences and it is therefore possible to select the optimal basis by using only a small subset of the original data set.

4 Experimental Results

4.1 Methodology

In this section we apply the general idea of using a B-spline representation to a spectrometric regression problem. The actual method consists in the following steps:

1. Extraction of the B-spline coefficients for each spectrum. The number and the order of the B-splines is chosen by the leave-one-out error estimate.
2. Selection of the B-spline coefficients through mutual information (MI) maximization with a forward-backward search (as in [3]). Any other variable selection method could be used.
3. Calculation of the wavelength ranges associated to the selected variables, as explained in Section 3.2, with $\epsilon = 0.01$.
4. Construction of a nonlinear model (Radial Basis Function Network, RBFN) on the coefficient selected by the previous step. The meta-parameters of the RBFN are chosen by a 3-fold cross validation technique.

In order to assess the performances of the proposed method, its results are compared to the performances of linear models namely a principal component regression (PCR) and a partial least square regression (PLSR). The numbers of components in the PLSR and in the PCR model are chosen with the same 3-fold cross-validation method used to choose the meta-parameters of the nonlinear model. To motivate the use of a nonlinear model, we also include the results of a standard linear regression (LR) built on the selected variables.

The comparison of the models is done according to the Normalized Mean Square Error (NMSE) they reach on an independent test set.

Finally, we use for comparison a simple method to extract the wavelengths that play a significant role in the prediction of the target variable by the best linear model obtained with PCR or PLSR. The output of such a model can be written

$$y = \alpha_0 + \sum_{j=1}^{m} \alpha_i X(w_j), \qquad (5)$$

where $X(w_j)$ is a scaled version of the original input variable $s(w_j)$ (i.e., $X(w_j)$ has zero mean and unit variance). As in section 3.2, we consider that wavelength w_j is important if $|\alpha_j| > \epsilon \max_{1\leq l\leq m} |\alpha_l|$.

4.2 Results

We use the data set from the software contest organized at the International Display Research Conference held in 1998. It consists of scans and chemistry gathered from fescue grass (*Festuca elatior*). The grass was bred on soil medium with several nitrogen fertilization levels. The aim of the experiments was to try to find the optimum fertilization level to maximize production and to minimize the consequences on the environment. In this context, the problem to address is the following: can NIR spectrometry measure the nitrogen content of the plants?

Although the scans were performed on both wet and dry grass samples, we only consider wet samples here (i.e., the scans were performed directly after harvesting). The dataset contains 141 spectra discretized to 1050 different wavelengths, from 400nm to 2498nm. The nitrogen level goes from 0.8 to 1.7 approximately. The data can be obtained from the Analytical Spectroscopy Research Group of the University of Kentucky[1].

We have split randomly the dataset into a test set containing 36 spectra and a training set with the remaining 105 spectra. The random split has been done in a way that preserve roughly the distribution of the target variable (the nitrogen level).

The leave-one-out error calculation leads to the selection of an optimal basis of 155 B-splines of order 5 (the optimal number of B-splines is chosen in $[50, 500]$) and achieves therefore a good compression ratio. The forward-backward mutual information procedure selects ten coordinates. Both phases take a few minutes on a personal computer, whereas the same variable selection procedure would have taken several hours on the original variables.

The results on the test set (NMSE) for the studied methods are given in Table 1. The 10 variables selected by maximizing the mutual information cannot be used to construct a linear model with performances comparable to the ones of the optimal linear models. The nonlinear model constructed on those variables has clearly the best performances.

[1] `http://kerouac.pharm.uky.edu/asrg/cnirs/shoot_out_1998/`

Table 1. Normalized mean square error on the test set for the nitrogen content prediction problem

Method	Variables	NMSE (test)
PCR	10	$1.57\,10^{-1}$
PLSR	9	$1.51\,10^{-1}$
MI + RBFN	10	$1.21\,10^{-1}$
MI + LR	10	$2.59\,10^{-1}$

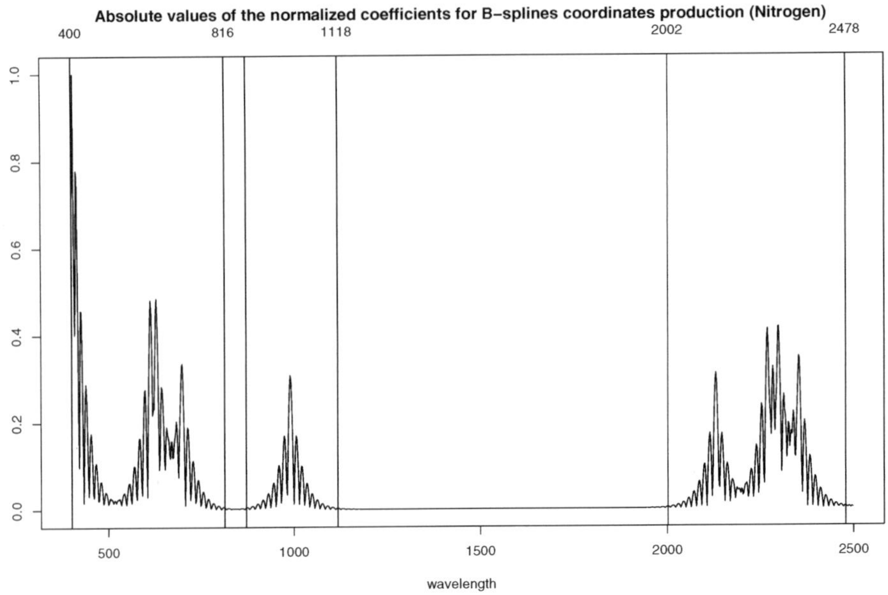

Fig. 2. Normalized absolute value of the coefficients used to compute the selected variables from the original spectral variables

While the mutual information maximization leads to the selection of 10 variables, they are calculated using only three intervals of the original wavelength range: [400, 816], [874, 1118] and [2002, 2478]. Figure 2 represents the normalized coefficients used to compute the new variables. It appears clearly that only some original wavelengths are used.

It is not possible to select a few wavelength ranges from the linear model induced by the PLSR: only 17 weights out of 1050 are smaller than $\epsilon = 0.01$ times the higher one in this linear model. As illustrated by Figure 3, the PLSR uses almost the full wavelength range.

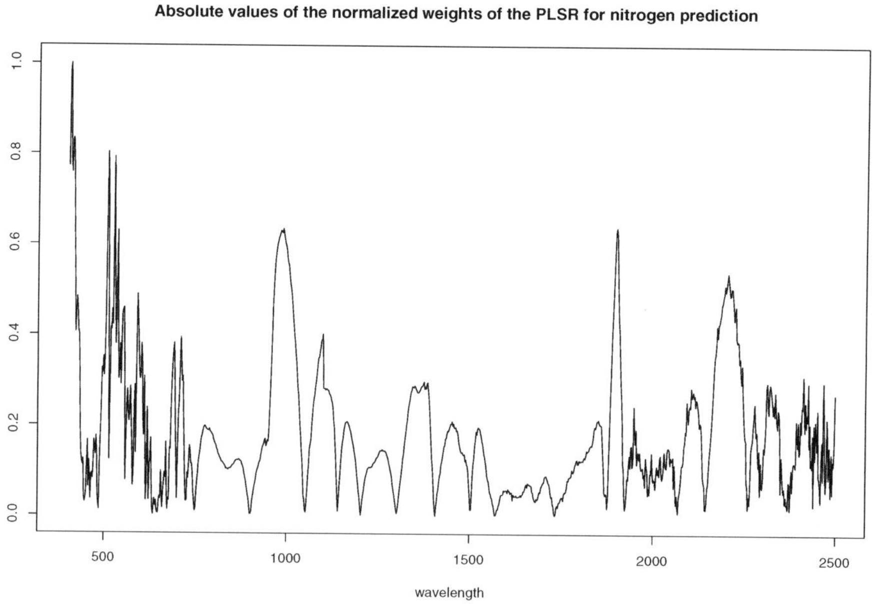

Fig. 3. Normalized absolute values of the coefficients of the linear model induced by PLSR

5 Conclusion

This paper proposes a simple and generic approach for variable selection in spectrometric regression problems. The method is based on the standard filtering approach of Functional Data Analysis: the original spectral variables are replaced by coordinates of the corresponding functions on a B-spline basis. The new variables are still interpretable as each of them depends only on a limited sub-range of the original wavelength interval. The optimal basis is selected by a fast leave-one-out procedure. The prior reduction of the number of variables allows one to use time consuming variable selection methods such as the forward-backward mutual information maximization used in the present paper. The performances obtained on a real-world benchmark are very good. Constructing the full regression model takes a few minutes, compared to several hours that would be needed to run the chosen variable selection procedure on the original spectral variables. On the chosen benchmark, the obtained model outperforms linear models. While these linear techniques are faster, they induce complex dependencies between the target variable and almost all considered wavelengths, and provide therefore no insight on the data. On the contrary, the selected variables are based on only 3 interpretable sub-intervals of the initial wavelength range.

References

1. Benoudjit, N., Cools, E., Meurens, M., Verleysen, M.: Chemometric calibration of infrared spectrometers: Selection and validation of variables by non-linear models. Chemometrics and Intelligent Laboratory Systems **70**(1) (2004) 47–53
2. Benoudjit, N., François, D., Meurens, M., Verleysen, M.: Spectrophotometric variable selection by mutual information. Chemometrics and Intelligent Laboratory Systems **74**(2) (2004) 243–251
3. Rossi, F., Lendasse, A., François, D., Wertz, V., Verleysen, M.: Mutual information for the selection of relevant variables in spectrometric nonlinear modelling. Chemometrics and Intelligent Laboratory Systems **80**(2) (2006) 215–226
4. Ramsay, J., Silverman, B.: Functional Data Analysis. Springer Series in Statistics. Springer Verlag (1997)
5. Hastie, T., Mallows, C.: A discussion of "A statistical view of some chemometrics regression tools" by I.E. Frank and J.H. Friedman. Technometrics **35** (1993) 140–143
6. Marx, B.D., Eilers, P.H.: Generalized linear regression on sampled signals with penalized likelihood. In A. Forcina, G. M. Marchetti, R.H., Galmacci, G., eds.: Statistical Modelling. Proceedings of the 11th International workshop on Statistical Modelling, Orvietto (1996)
7. Hoerl, A.E., Kennard, R.W.: Ridge regression: Biased estimation for non-orthogonal problems. Technometrics **12**(1) (1970) 55–67
8. Pezzulli, S., Silverman, B.: On smoothed principal components analysis. Computational Statistics **8** (1993) 1–16
9. Hastie, T., Buja, A., Tibshirani, R.: Penalized discriminant analysis. Annals of Statistics **23** (1995) 73–102
10. Cardot, H., Ferraty, F., Sarda, P.: Functional linear model. Statist. & Prob. Letters **45** (1999) 11–22
11. Rossi, F., Delannay, N., Conan-Guez, B., Verleysen, M.: Representation of functional data in neural networks. Neurocomputing **64** (2005) 183–210
12. Rossi, F., Conan-Guez, B.: Theoretical properties of projection based multilayer perceptrons with functional inputs. Neural Processing Letters **23**(1) (2006) 55–70
13. Biau, G., Bunea, F., Wegkamp, M.: Functional classification in Hilbert spaces. IEEE Transactions on Information Theory **51** (2005) 2163–2172
14. Rossi, F., Villa, N.: Support vector machine for functional data classification. Neurocomputing **69**(7–9) (2006) 730–742
15. Alsberg, B.K.: Representation of spectra by continuous functions. Journal of Chemometrics **7** (1993) 177–193
16. Alsberg, B.K., Kvalheim, O.M.: Compression of nth-order data arrays by b-splines. part 1: Theory. Journal of Chemometrics **7**(1) (1993) 61–73
17. Olsson, R.J.O., Karlsson, M., Moberg, L.: Compression of first-order spectral data using the b-spline zero compression method. Journal of Chemometrics **10**(5–6) (1996) 399–410
18. de Boor, C.: A Practical Guide to Splines. Volume 27 of Applied Mathematical Sciences. Springer (1978)
19. Daubechies, I.: Orthonormal bases of compactly supported wavelets. Communications in Pure & Applied Mathematics **41** (1988) 909–996

The Bayes-Optimal Feature Extraction Procedure for Pattern Recognition Using Genetic Algorithm

Marek Kurzynski, Edward Puchala, and Aleksander Rewak

Wroclaw University of Technology, Faculty of Electronics, Chair of Systems and Computer Networks, Wyb. Wyspianskiego 27, 50-370 Wroclaw, Poland
{marek.kurzynski, edward.puchala, aleksander.rewak}@pwr.wroc.pl

Abstract. The paper deals with the extraction of features for statistical pattern recognition. Bayes probability of correct classification is adopted as the extraction criterion. The problem with complete probabilistic information is discussed and Bayes-optimal feature extraction procedure is presented in detail. The case of recognition with learning is also considered. As method of solution of optimal feature extraction a genetic algorithm is proposed. A numerical example demonstrating capability of proposed approach to solve feature extraction problem is presented.

1 Introduction

Feature dimension reduction has been an important and long-stading research problem in statistical pattern recognition. In general, dimension reduction can be defined as a transformation from original high-dimensional space to low-dimensional space where an accurate classifier can be constructed.

There are two main methods of dimensionality reduction ([2], [6]): *feature selection* in which we select the best possible subset of input features and *feature extraction* consisting in finding a transformation (usually linear) to a lower dimensional space. Although feature selection preserves the original physical meaning of selected features, it costs a great degree of time complexity for an exhaustive comparison if a large number of features is to be selected. In contrast, feature extraction is considered to create a new and smaller feature set by combining the original features. We shall concentrate here on feature extraction for the sake of flexibility and effectiveness [7].

There are many effective methods of feature extraction. One can consider here linear and nonlinear feature extraction procedures, particularly ones which ([4], [5]):

1. assume underying Gaussian distribution in the data ([6], [7], [8]),
2. utilize nonparametric sample-based methods when data cannot be described with the Gaussian model ([9]),
3. minimize the empirical probability of Bayes error ([6], [10]),
4. maximize the criteria for the information values of the individual features (or sets of features) describing the objects ([4], [5], [11]).

S. Kollias et al. (Eds.): ICANN 2006, Part I, LNCS 4131, pp. 21–30, 2006.

For the purpose of classifiation, it is sensible to use linear feature extraction techniques which is considered as a linear mapping of data from a high to a low-dimensional space, where class separability is approximately preserved. Construction of linear transformation is based on minimization (maximization) of proper criterion in the transformed space. In other words, in order to define a linear transformation one should determine the values of the transformation matrix components as a solution of an appropriate optimization problem.

As it seems, the Bayes probability of error (or equivalently, the Bayes probability of correct classification) i.e. the lowest attainable classification error is the most appropriate criterion for feature extraction procedure. Unfortunately, this criterion is very complex for mathematical treatment, therefore researches have restored to other criteria like various functions of scatter matrices (e.g. Fisher criterion) or measures related to the Bayes error (e.g. Bhattacharyya distance).

In this paper we formulate the optimal feature extraction problem adopting the Bayes probability of correct classification as an optimality criterion. Since this problem cannot be directly solved using analytical ways (except simple cases including for example multivariate normal distribution), we propose to apply genetic algorithm (GA), which is very-well known heuristic optimization procedure and has been successfully applied to a broad spectrum of optimization problems, including many pattern recognition and classification tasks [12], [13].

The contents of the paper are as follows. In section 2 we introduce necessary background and formulate the Bayes-optimal feature extraction problem. In section 3 and 4 optimization procedures for the cases of complete probabilistic information and recognition with learning are presented and discussed in detail. Section 5 describes numerical example for which both analytical way and genetic algorithm were applied to find optimal solution. Finally, conclusions are presented in section 6.

2 Preliminaries and the Problem Statement

Let us consider the pattern recognition problem with probabilistic model. This means that n-dimensional vector of features describing recognized pattern $x = (x_1, x_2, ..., x_n)^T \in \mathcal{X} \subseteq \mathcal{R}^n$ and its class number $j \in \mathcal{M} = \{1, 2, ..., M\}$ are observed values of a pair of random variables $(\mathbf{X}, \mathbf{J})$, respectively. Its probability distribution is given by *a priori* probabilities of classes

$$p_j = P(\mathbf{J} = j),\ j \in \mathcal{M} \tag{1}$$

and class-conditional probability density function (CPDFs) of $\mathbf{X}$

$$f_j(x) = f(x/j),\ x \in \mathcal{X},\ j \in \mathcal{M}. \tag{2}$$

In order to reduce dimensionality of feature space let consider linear transformation

$$y = Ax, \tag{3}$$

which maps n-dimensional input feature space $\mathcal{X}$ into m-dimensional derivative feature space $\mathcal{Y} \subseteq \mathcal{R}^m$, or - under assumption that $m < n$ - reduces dimensionality of space of object descriptors. It is obvious, that y is a vector of observed values of m dimensional random variable $\mathbf{Y}$, which probability distribution given by CPDFs depends on mapping matrix A, viz.

$$g(y/j; A) = g_j(y; A), \quad y \in \mathcal{Y},\ j \in \mathcal{M}. \tag{4}$$

Let introduce now a criterion function $Q(A)$ which evaluates discriminative ability of features y, i.e. Q states a measure of feature extraction mapping (3). As a criterion Q any measure can be involved which evaluates both the relevance of features based on a feature capacity to discriminate between classes or quality of a recognition algorithm used later to built the final classifier. In the further numerical example the Bayes probability of correct classification will be used, namely

$$Q(A) = Pc(A) = \int_{\mathcal{Y}} \max_{j \in \mathcal{M}} \{p_j\, g_j(y; A)\}\, dy. \tag{5}$$

Without any loss of generality, let us consider a higher value of Q to indicate a better feature vector y. Then the feature extraction problem can be formulated as follows: for given *priors* (1), CPDFs (2) and reduced dimension m find the matrix A^* for which

$$Q(A^*) = \max_A Q(A). \tag{6}$$

3 Optimization Procedure

In order to solve (6) first we must explicitly determine CPDFs (4). Let introduce the vector $\bar{y} = (y, x_1, x_2, ..., x_{n-m})^T$ and linear transformation

$$\bar{y} = \bar{A}\, x, \tag{7}$$

where

$$\bar{A} = \begin{bmatrix} A \\ - - - \\ I \mid 0 \end{bmatrix} \tag{8}$$

is a square matrix $n \times n$. For given y equation (7) has an unique solution given by Cramer formulas

$$x_k(y) = \mid \bar{A}_k(y) \mid \cdot \mid \bar{A} \mid^{-1}, \tag{9}$$

where $\bar{A}_k(y)$ denotes matrix with k-th column replaced with vector $\bar{y}$. Hence putting (9) into (2) and (4) we get CPDFs of $\bar{y}$ ([3]):

$$\bar{g}_j(\bar{y}; A) \;=\; J^{-1} \cdot f_j(x_1(\bar{y}), x_2(\bar{y}), \cdots, x_n(\bar{y})), \tag{10}$$

where J is a Jacobian of mapping (7). Integrating (10) over variables $x_1, ..., x_{n-m}$ we simply get

$$g_j(y;A) = \int_{\mathcal{X}_1}\int_{\mathcal{X}_2} \cdots \int_{\mathcal{X}_{n-m}} \bar{g}_j(\bar{y};A)\, dx_1\, dx_2 \ldots dx_{n-m}. \tag{11}$$

Formula (11) allows one to determine class-conditional density functions for the vector of features y, describing the object in a new m-dimensional space. Substituting (11) into (5) one gets a criterion defining the probability of correct classification for the objects in space $\mathcal{Y}$:

$$Q(A) = Pc(A) = \int_{\mathcal{Y}} \max_{j\in\mathcal{M}} \left\{ p_j \cdot \int_{\mathcal{X}_1}\int_{\mathcal{X}_2} \cdots \int_{\mathcal{X}_{n-m}} J^{-1} \times \right.$$

$$\left. \times f_j(x_1(\bar{y}), x_2(\bar{y}), \cdots, x_n(\bar{y}))\, dx_1\, dx_2 \ldots dx_{n-m} \right\} dy =$$

$$= \int_{\mathcal{Y}} \max_{j\in\mathcal{M}} \left\{ p_j \cdot \int_{\mathcal{X}_1}\int_{\mathcal{X}_2} \cdots \int_{\mathcal{X}_{n-m}} J^{-1} \times \right.$$

$$\left. \times f_j(|\bar{A}_1(y)| \cdot |\bar{A}|^{-1}, \cdots, |\bar{A}_n(y)| \cdot |\bar{A}|^{-1})\, dx_1\, dx_2 \ldots dx_{n-m} \right\} dy. \tag{12}$$

Thus, the solution of the feature extraction problem (6) requires that such matrix A^* should be determined for which the Bayes probability of correct classification (12) is the maximum one.

Consequently, complex multiple integration and inversion operations must be performed on the multidimensional matrices in order to obtain optimal values of A. Although an analytical solution is possible (for low n and m values), it is complicated and time-consuming. Therefore it is proposed to use numerical procedures. For linear problem optimization (which is the case here) classic numerical algorithms are very ineffective. In a search for a global extremum they have to be started (from different starting points) many times whereby the time needed to obtain an optimal solution is very long. Thus it is only natural to use the parallel processing methodology offered by genetic algorithms ([14]).

Fig. 1 shows the structure of a GA-based feature extractor using Bayes probability of correct classification as an evaluation criterion. The GA maintains a

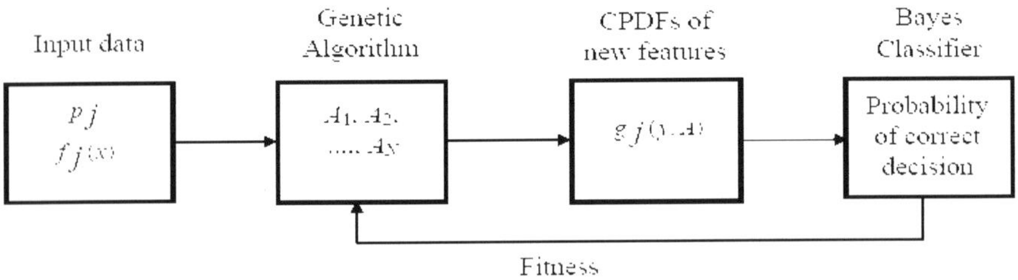

Fig. 1. GA-based Bayes-optimal feature extractor

population of transformation matrices A. To evaluate each matrix in this population, first the CPDFs (11) of features y in transformed space must be determined and next probability of Bayes correct classification (12) is calculated. This accuracy, i.e. fitness of individual is a base of selection procedure in GA. In other words, the GA presented here utilizes feedback from the Bayes classifier to the feature extraction procedure.

4 The Case of Recognition with Learning

It follows from the above considerations that an analytical and numerical solution of the optimization problem is possible. But for this one must know the class-conditional density functions and the *a priori* probabilities of the classes. In practice, such information is rarely available. All we know about the classification problem is usually contained in the so-called learning sequence:

$$S_L(x) = \{(x^{(1)}, j^{(1)}), (x^{(2)}, j^{(2)}), ..., (x^{(L)}, j^{(L)})\}. \tag{13}$$

Formula (13) describes objects in space $\mathcal{X}$. For the transformation to space $\mathcal{Y}$ one should use the relation:

$$y^{(k)} = A \cdot x^{(k)}; \quad k = 1, 2, ..., L \tag{14}$$

and then the learning sequence assumes the form:

$$S_L(y) = \{(y^{(1)}, j^{(1)}), (y^{(2)}, j^{(2)}), ..., (y^{(L)}, j^{(L)})\}. \tag{15}$$

The elements of sequence $S_L(y)$ allow one to determine (in a standard way) the estimators of the *a priori* probabilities of classes p_{jL} and class-conditional density functions $f_{jL}(x)$. Then the optimization criterion assumes this form:

$$Q_L(A) = Pc_L(A) = \int_{\mathcal{Y}} \max_{j \in \mathcal{M}} \left\{ p_{jL} \cdot \int_{\mathcal{X}_1} \int_{\mathcal{X}_2} \cdots \int_{\mathcal{X}_{n-m}} J^{-1} \times \right.$$

$$\left. \times f_{jL}(x_1(\bar{y}), x_2(\bar{y}), \cdots, x_n(\bar{y}))\, dx_1\, dx_2 \ldots dx_{n-m} \right\} dy. \tag{16}$$

Alternatively, in case of recognition with learning, the criterion (16) can be estimated nonparametrically by first estimating CPDFs of features y on the base of samples (15) (e.g. using either k-NN or Parzen procedures [1], [2]) and then classifying the available samples according to the empirical Bayes rule. The number of samples misclassified by the algorithm is counted and the error estimate is obtained by dividing this number by the total number of training samples.

The next section presents a numerical example illustrating proposed approach to Bayes-optimal feature extraction problem.

5 Numerical Example

Let consider two-class pattern recognition task with equal *priors* and reduction problem of feature space dimension from $n = 2$ to $m = 1$. Input feature vector is uniformly distributed and its CPDFs are as follows:

$$f_1(x) = \begin{cases} 0.5 & \text{for } 0 \leq x_1 \leq 2 \text{ and } 0 \leq x_2 \leq x_1, \\ 0 & \text{otherwise,} \end{cases} \quad (17)$$

$$f_2(x) = \begin{cases} 0.5 & \text{for } 0 \leq x_1 \leq 2 \text{ and } x_1 \leq x_2 \leq 2, \\ 0 & \text{otherwise.} \end{cases} \quad (18)$$

Feature extraction mapping (3) has now the form

$$y = [a, 1] \cdot [x_1, x_2]^T = a \cdot x_1 + x_2 \quad (19)$$

and problem is to find such a value a^* which maximize criterion (12).

To illustrate the behavior of the GA as solution method of optimal feature extraction problem, we solve this example in threefold manner: (1) directly, according to the analytical procedure presented in section 3, (2) using GA and assuming that complete probabilistic information is given and (3) using GA procedure for the case of classification with learning.

1. Complete probabilistic information - analytical solution

Since Jacobian of (7) is equal to 1 hence from (9) and (10) for $j = 1, 2$ we get

$$\bar{g}_j(\bar{y}, a) = f_j(x_1, y - a \cdot x_1). \quad (20)$$

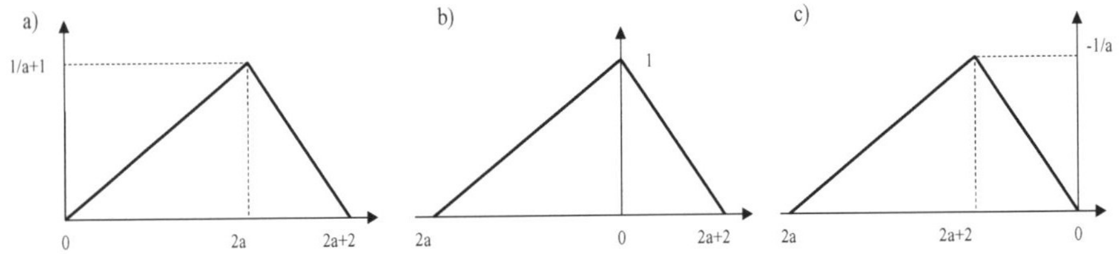

Fig. 2. Illustration of example

The results of integrating (20) over x_1, i.e. CPDFs (11) for $a \geq 1, -1 \leq a \leq 1$ and $a \leq -1$ are presented in Fig.2. a), b) and c), respectively.

Finally, from (5) we easy get:

$$Pc(a) = \begin{cases} \frac{a+1}{4a} & \text{for } a \geq \mid 1 \mid, \\ \frac{a+1}{4} & \text{for } a \leq \mid 1 \mid . \end{cases} \quad (21)$$

The chart demonstrating the Bayes probability of misclassification $P_e(a) = 1 - P_c(a)$ depending on parameter a of feature extraction mapping is depicted in Fig.3. The best result $P_e(a^*) = 0$ (or equivalently $P_c(a^*) = 1$) is obtained for $a^* = -1$.

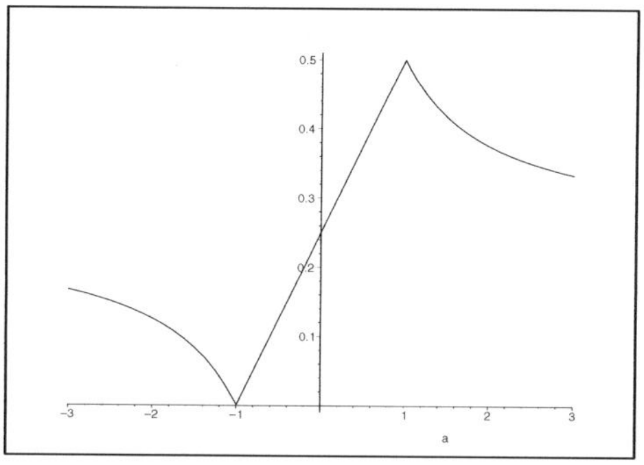

Fig. 3. Probability of misclassification

2. Complete probabilistic information - solution via GA

In order to find parametr a^*, the GA was applied, which was proceeded as follows:

- *Coding method* - Binary representation has been widely used for GA analysis. In our task, the value of parametr a was directly coded to the chromosome. It means, that a is represented by a binary string length:

$$Length = log_2[(a_{max} - a_{min})/\triangle a], \tag{22}$$

  where a_{max}, a_{min} and $\triangle a$ denote the maximum value, the minimum value and the resolution of a, respectively. To avoid irregularities we decided to put $a_{max} = 32.536$, $a_{min} = -33.0$ and $\triangle a = 0.001$ which gave the length of chromosome $Length = 16$ bits (genes).
- *The fitness function* - The Bayes probability of correct classification (12).
- *Initialization* - GA needs an initial individual population to carry out parallel multidirectional search of optimal solution. The initial population of chromosomes with which the search begins was generated randomly. The size of population after trials was set to 40.
- *Selection* - The probability of selecting a specific individual can be calculated by using the individuals fitness and the sum of population fitness. In this research a roulette wheel approach was applied. Additionally, an elitizm policy, wherein the best individuals from the current generation is copied directly to the next generation, was also used for fast convergence.
- *Crossover* - The crossover process defines how genes from the parents have been passed to the offspring. In each generation a standard two-point crossover was used and probability of crossover was equal to 1.

- *Mutation* - The mutation process simulates the natural disturbance during crossover. It was a bit-by-bit operation made with probability 0.01.
- *Stop procedure* - evolution process was terminated after 300 generations. In fact, the fitness value usually converged within this value. Fig. 4. shows the fitness change against generation number in one run of GA.

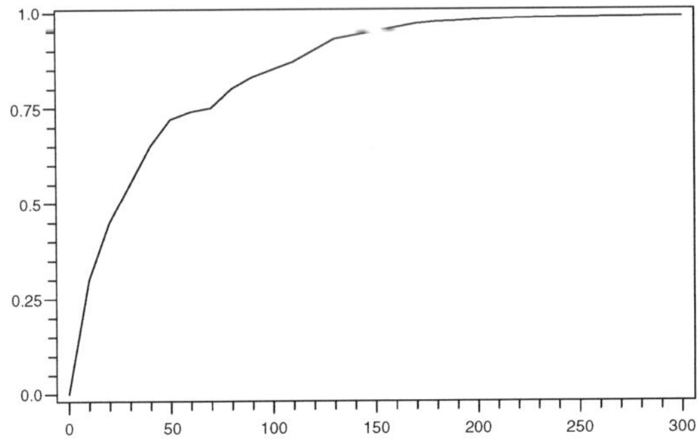

Fig. 4. The example of the course of the fitness value vs. number of generation

Table 1. Results of genetic algorithm applied to the example

Trial	Complete Information	Recognition with Learning		
		L=100	L=200	L=300
1	-0.987	-0.946	-1.075	-0.983
2	-0.983	-0.951	-1.048	-0.971
3	-0.974	-1.102	-0.957	-1.036
4	-1.078	-0.938	-0.939	-0.947
5	-1.023	-1.093	-0.962	-1.056
6	-0.991	-1.084	-1.053	-1.043
7	-0.975	-0.956	-0.961	-1.032
8	-1.052	-1.066	-0.972	-0.953
9	-0.979	-0.941	-1.042	-1.023
10	-1.044	-1.076	-0.951	-1.041
Best	-0.991	-0.956	-0.962	-1.023
Mean	-1.008	-1.015	-0.996	-1.009
SD	0.036	0.07	0.049	0.039

To compare the optimal solution and the performance of GA, ten independent runs of GA were carried out for different random initial populations. The results are shown in Table 1. The values depicted in the Table are those of the best solution obtained at the end of a GA trial.

3. Recognition with learning - solution via GA

For evaluation of GA performance in the case of recognition with learning, three experiments were made on computer generated data with different number of learning patterns ($L = 100, 200, 300$, respectively). Patterns were generated according to the CPDFs (17) and (18) using Maple 10 environment. In each experiment *priors* were calculated in standard way and CPDFs were estimated using Parzen estimator with uniform kernel function [1]. Next, GA was applied as a method of solution of optimization problem presented in section 4. GA was used with the same control parameters as in the prevoius case and the number of trials was qual to 10. The results are depicted in Table 1.

Table 1 contains also the best result, the mean value and standard deviation for each case where GA was applied. Results demonstrate that the proposed GA method can reach value of parametr of extraction mapping (19) very close to optimal solution a^*.

6 Conclusions

Feture extraction is an important task in any practical example that involves pattern classification. In this paper we formulate the optimal feature extraction problem with the Bayes probability of correct classification as an optimality criterion. Since this problem, in general case, cannot be directly solved using analytical methods, we propose to apply genetic algorithm, which is effective heuristic optimization procedure and has been successfully applied to a wide range of practical problems. This proposition leads to the distribution-free Bayes-optimal feature extraction method, which can be applied both in the case of complete probabilistic information and in the case of recognition with learning. A numerical example demonstrates that the GA is capable to solve this optimization problem for both cases.

Many questions of GA application in proposed procedure of feature extraction are still open, e.g. the proper choice of the appropriate GA model, especially the choice of GA control parameters and investigation of their influence on result of optimization process. Our related works are underway and the results will be reported in the near future.

References

1. Devroye L., Gyorfi P., Lugossi G.: A Probabilistic Theory of Pattern Recognition, Springer Verlag, New York, 1996
2. Duda R., Hart P., Stork D.: Pattern Classification, Wiley-Interscience, New York, 2001
3. Golub G., Van Loan C.: Matrix Computations, Johns Hopkins University Press, 1996
4. Guyon I., Gunn S., Nikravesh M, Zadeh L.: Feature Extraction, Foundations and Applications, Springer Verlag, 2004
5. Park H., Park C., Pardalos P.: Comparitive Study of Linear and Nonlinear Feature Extraction Methods - Technical Report, Minneapolis, 2004

6. Fukunaga K.: Introduciton to Statistical Pattern Recognition, Academic Press, 1990.
7. Hsieh P., Wang D., Hsu C.: A Linear Feature Extraction for Multiclass Classification Problems Based on Class Mean and Covariance Discriminant Information, IEEE Trans. on PAMI, Vol. 28 (2006) 223-235
8. Loog M., Duin R., Haeb-Umbach R.: Multiclass Linear Dimension Reduction by Meighted Pairwise Fisher Criteria, IEEE Trans. on PAMI, Vol. 23 (2001) 762-766
9. Kuo B., Landgrebe D.: A Robust Classification Procedure Based on Mixture Classifiers and Nonparametric Weighted Feature Extraction, IEEE Trans. on GRS, Vol. 40 (2002) 2486-2494
10. Buturovic L.: Toward Bayes-Optimal Linear Dimension Reduction, IEEE Trans. on PAMI, Vol. 16 (1994) 420-424
11. Choi E., Lee C.: Feature Extraction Based on the Bhattacharyya Distance, Pattern Recognition, Vol. 36 (2002) 1703-1709
12. Raymer M., Punch W. at al.: Dimensionality Reduction Using Genetic Algorithms, IEEE Trans. on EC, Vol. 4 (2002) 164-168
13. Rovithakis G., Maniadakis M., Zervakis M.: A Hybrid Neural Network and Genetic Algorithm Approach to Optimizing Feature Extraction for Signal Classification, IEEE Trans. on SMC, Vol. 34 (2004) 695-702
14. Goldberg D.: Genetic Algorithms in Search, Optimization and Machine Learning. Adison-Wesley, New York, 1989

Speeding Up the Wrapper Feature Subset Selection in Regression by Mutual Information Relevance and Redundancy Analysis

Gert Van Dijck and Marc M. Van Hulle

Computational Neuroscience Research Group, Laboratorium voor Neuro-en Psychofysiologie, K.U. Leuven, B-3000 Leuven, Belgium
{gert, marc}@neuro.kuleuven.ac.be

Abstract. A hybrid filter/wrapper feature subset selection algorithm for regression is proposed. First, features are filtered by means of a relevance and redundancy filter using mutual information between regression and target variables. We introduce permutation tests to find statistically significant relevant and redundant features. Second, a wrapper searches for good candidate feature subsets by taking the regression model into account. The advantage of a hybrid approach is threefold. First, the filter provides interesting features independently from the regression model and, hence, allows for an easier interpretation. Secondly, because the filter part is computationally less expensive, the global algorithm will faster provide good candidate subsets compared to a stand-alone wrapper approach. Finally, the wrapper takes the bias of the regression model into account, because the regression model guides the search for optimal features. Results are shown for the 'Boston housing' and 'orange juice' benchmarks based on the multilayer perceptron regression model.

1 Introduction

Feature selection and feature construction have been addressed by many researchers in statistics and machine learning, see [1] for a recent overview. Feature construction constructs new features from the original inputs in a linear or non-linear way. Most feature construction techniques are developed for classification problems. However, they are easily adapted for regression problems by first discretizing the continuous target values using class-blind discretization algorithms [2], hence, artificially creating class labels. Feature selection on the other hand considers a selection from the original inputs, without constructing new ones. Both feature construction and feature selection help tackling the curse of dimensionality. In reducing the number of inputs one searches for the optimal bias-variance trade-off: a large number of inputs imply that more parameters need to be estimated and this causes a larger variance, however a too small number of inputs increases the bias. Feature construction has the disadvantage that it does not preserve the semantics of the inputs: combining inputs in a linear or non-linear way, makes the new features hard to interpret and hence makes an understanding of the nature of the problem difficult. Another huge disadvantage is

S. Kollias et al. (Eds.): ICANN 2006, Part I, LNCS 4131, pp. 31 – 40, 2006.

that feature construction does not decrease the measuring cost: all inputs still need to be measured, by possibly very expensive sensors, even when they are non-informative.

Therefore we adopt a feature subset selection approach in this article. Feature selection can be separated in two approaches: the *filter* approach and the *wrapper* approach [3]. In the filter approach the feature subset selection is performed independently of the training of the regression model. In this case feature subset selection is considered as a preprocessing step to induction. This is computationally more efficient, but ignores the fact that an optimal selection of features is dependent on the regression model. As stated before the performance of the regression model is strongly dependent on the size of the feature subset. On the other hand the wrapper approach is computationally more involved, but takes the interactions of the feature subset and the regression model into account. The term 'wrapper' stems from the fact that the feature selection is wrapped around the regression model which is considered as a black-box. In this article we propose a hybrid solution: first irrelevant and largely redundant features are removed, subsequently a search with a wrapper is performed among the features that passed the filter.

2 Filter Preprocessing

In this section we investigate an information-theoretic measure in order to determine irrelevant and redundant features. We use the 'Boston housing' and the 'orange juice' datasets for illustrative purposes. The proposed methods are inherited from [4] where a hybrid approach is proposed for pattern recognition (classification), instead of regression.

2.1 Irrelevance Determination by Permutation

As explained before, a wrapper approach takes the limitations of the particular regression model into account. In the search for optimal feature subsets we need to estimate the performance of the feature sets found so far. This requires the training of the regression model based on the selected features for a chosen training set. The accuracy of the model is then estimated by simulating the trained regression model on a test set. It is common that a lot of features are included in the feature subset search that do not contain any information about the target variable. This information can be described by the concept of mutual information between the regression variable F_i and the target variable T:

$$I(F_i,T) = \int_{f_i}\int_{t} P(f_i,t)\log_2\left(\frac{P(f_i,t)}{P(f_i)P(t)}\right)df_i dt\,. \tag{1}$$

The use of the mutual information in regression is largely motivated by the data inequality theorem, which states that [5]:

$$I(F_i,T) \geq I(g(F_i),T)\,. \tag{2}$$

Hence, a function of the variable F_i cannot increase the information about the target T. If we can show that the original variable T is not dependent on F_i (F_i is not informative about the target T), which implies the mutual information in (1) is equal to 0, we can discard F_i, because any further processing can not increase the information about the target.

In practice we face the problem that we do not know the joint distribution between target variable and input variables, hence, the mutual information needs to be estimated from the data. This finite sample estimate is likely to be different from 0 and in general will depend on the sample size, parameter settings of the estimator and the distributions in (1). Thus, looking whether the estimated mutual information is exactly equal to 0 is not satisfactory. However, we can easily circumvent this problem in the following way. We define a hypothesis test where the null hypothesis H_0 tests the assumption that the feature variable and the target variable are independent. We can easily obtain the distribution of the mutual information conditioned under the particular sample distributions. Therefore, we randomly permute the ordering of the samples of the target variable, hence, removing the dependencies between the target variable and the input variable, relative to the feature samples. Performing this permutation N times provides us with N samples of a sample distribution of the mutual information under the H_0 hypothesis. Note that this strategy contains some resemblance with the creation of surrogate time series in time series analysis [6]: a 'ground-truth' or reference is created by e.g. randomly permuting the phase of the signals under the given sample distribution of the frequency spectrum.

Further on, we will estimate the mutual information with the $I^{(1)}$ estimator of Kraskov et al. [7], which estimates mutual information directly from a K-Nearest Neighbour method. Figure 1 shows the sample distribution of the mutual information between input variable F_5 (nitric oxides concentration, NOX) and the target variable (median value of owner-occupied homes, MEDV) of the 'Boston housing' data set for 1000 permutations. We note that under the H_0 hypothesis the mean (0.1106) of the mutual information is considerably different from 0. This divergence from 0 can be partly explained by the fact that the $I^{(1)}$ is designed for continuous distributed features and target variables. However, the NOX variable appears to have a discrete nature (although in the accompanying housing.names file it is considered as a continuous feature). Based on the sample distribution we can define a threshold for which a feature (when larger than the threshold) can be considered statistically relevant. For the example in figure 1 we have $P_{0.01}$ = 0.1369, the actual MI (without performing the permutation) is equal to 0.1920. When we perform this analysis for all 13 features in the 'Boston housing' dataset we find that all features are statistically relevant, except for input variable F_4 (Charles River dummy variable CHAS, $P_{0.01}$ = 0.02 and actual MI equal to 0.0163). We remark that the CHAS variable is discrete and therefore we did not use the $I^{(1)}$ estimator, but estimated the mutual information by means of the marginal entropy estimator (marginal and conditional entropies estimated from formula (20) in [7]):

$$\hat{I}(F,T) = \hat{H}(T) - \sum_{i=0,\ldots,C} \hat{H}(T \mid F = i).\hat{p}(F = i)\,. \tag{3}$$

In formula (3) the discrete feature F (CHAS for the 'Boston housing' data set) takes different category values: 0,…, C (0 and 1 in this case).

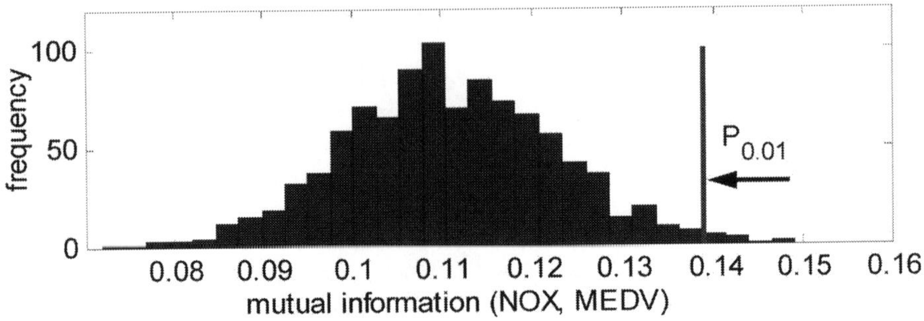

Fig. 1. Sample distribution of the mutual information between input variable NOX and target variable MEDV. The distribution was obtained under null hypothesis (independent input and target variable) by randomly permuting (1000 permutations) the samples. Note that the mean differs considerably from 0. The actual MI is equal to 0.1920, this is larger than $P_{0.01}$ ($P_{0.01}$ = 0.1389) and hence NOX can be considered as a relevant feature for target variable MEDV.

In figure 2 we show the mutual information of all features of the 'orange juice' database and the $P_{0.01}$ thresholds from 100 permutations. Where the MI exceeds the threshold the features are statistically relevant.

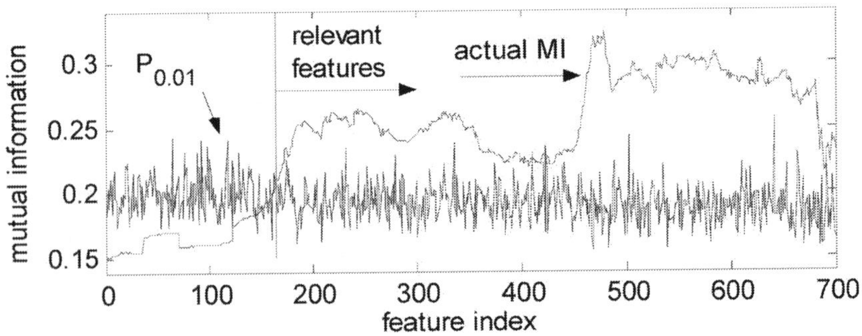

Fig. 2. Feature relevance as determined by the $P_{0.01}$ value of the permutation test (using 100 permutations). The lower noisy curve shows the $P_{0.01}$ value determined from the permutations. The upper curve shows the actual MI (without permutations). Note that starting from approximately feature 165 all features are considered as statistically significant.

Finally, we remark that permutation testing for feature relevance analysis has been described independently in [8].

2.2 Redundancy Detection

Features that are individually relevant might, however contain overlapping information considering the target variable. Therefore in literature [9] the distinction is made between strongly relevant and weakly relevant features. A strongly relevant feature F_i is defined as:

$$P(T \mid F_i, G_i) \neq P(T \mid G_i) \tag{4}$$

$$G_i = F - \{F_i\},$$

with F the complete feature set.

A weakly relevant feature F_i is a feature for which (4) holds for at least one strict subset G_i' of G_i. So weakly relevant features need to be interpreted as relevant features, but for which redundant, i.e. strongly correlated, features or feature sets exist.

A redundancy filter tries to detect and remove the redundant features of the weakly relevant features. Thus, the redundancy filter needs to filter out redundant feature subsets, but needs to retain a representative feature for the redundant subset. From formula (4) it is clear that we need to rely on heuristics for the identification of weakly relevant features:

- We do not dispose of the real underlying distributions in (4),
- It requires that we find at least one subset of G_i for which the inequality in (4) holds, in a worst case scenario this requires considering all possible subsets of G_i. This is of almost the same complexity as solving the FSS problem itself, because this would require finding the smallest possible subset of the complete feature set for which the equality in (4) holds.

The heuristic approach is taken where redundant features are assembled in a cluster and a representative feature is taken out of the cluster. The feature closest to the cluster centroid can act as a representative feature for all features in the cluster. We have following requirements for the clustering procedure:

- A first requirement for the clustering procedure is: strongly relevant features must form a cluster on themselves. Therefore in the clustering procedure it is sound to consider every feature initially as a separate cluster.
- A second requirement is that the maximum distance between any features in a cluster should be limited in order for the feature closest to the centroid to be representative.

In order to achieve these goals clusters are iteratively merged starting from the initial features as seeds. In order to obtain compact clusters, when merging, the distance between 2 clusters D_i and D_j is defined as the maximum distance between any features:

$$d_{\max}(D_i, D_j) = \max_{\substack{\mathbf{F}_i \in D_i \\ \mathbf{F}_j \in D_j}} \left\| \mathbf{F}_i - \mathbf{F}_j \right\|. \tag{5}$$

As a distance measure between features we propose 1 minus the normalized MI between features, this leads to 0 distance for the distance between the same features and a distance of 1between independent features.

Cluster merging is stopped when distance between any clusters exceeds a predefined threshold τ (maximal number of clusters to be formed or a maximal distance that may not be exceeded when merging clusters). The described clustering procedure is known as hierarchical agglomerative clustering with a 'complete' merging strategy of the clusters [10]. The threshold τ heuristically defines the non-redundant features which are represented by the cluster centroids.

A wrapper search is then performed on the features that pass both the relevance and redundancy filter. If we apply this redundancy filtering strategy to the 'orange juice data' and set τ equal to 5 clusters we get following result in figure 3.

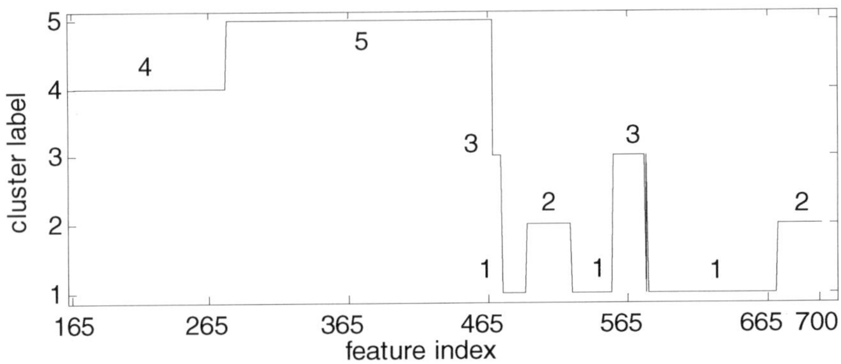

Fig. 3. Redundancy analysis on spectral 'orange juice' data. The figure shows which features are assigned to which cluster if we set τ equal to 5 clusters in the redundancy analysis. It is interesting to observe that contiguous features tend to end up in the same cluster. This could be expected, while small differences in spectral components tend to give rise to redundant features. As a distance measure 1-nMI (normalized mutual information) was used.

From figure 3 we observe the interesting (but expected) result that contiguous features tend to be assigned to the same cluster, hence, features that are obtained from small differences in spectra tend to be redundant. We obtained a similar result for features computed from the continuous wavelet transform in [4]: features obtained for small changes in scale coefficients tend to be strongly dependent and therefore can be approximated by the cluster centroid [4]. Note that this redundancy analysis can be considered as a strategy of sampling from the initial feature space. The sampling strategy has the advantage that where features are strongly redundant we need only a few representative features, while where features are not redundant we need more feature samples.

2.3 Wrapper Search

A supervised search is performed on the features that pass both the relevance and redundancy filter. Given the strong dependency of the regression model on the curse of dimensionality and the assumptions made in the regression model to map input variables to a target variable, these interactions need to be taken into account to achieve optimal performance. By applying filter techniques the wrapper is focused on strongly relevant features. By applying the filtering techniques the wrapper can be applied with decreased computational cost. In the wrapper approach 2 choices need to be made: the regression model and the search among the possible subsets. We opted for the following choices:

- Regression model: we used a widely accepted model for regression: a Multi-layer Perceptron (MLP) neural network [11]. Such MLP models are capable of approximating any function on a finite interval, provided the number of hidden neurons and the training data set are large. The input layer is defined by the number of inputs (D), for the hidden layer we choose 5 sigmoid neurons, the output layer is determined by the number of targets and consists of 1 linear neuron. The Levenberg-Marquardt algorithm was used in batch-mode to train the parameters of the network. To compute the performance of the feature subset, the data set was divided in 3 parts: a training set, a validation set and a test set. The validation set was used to avoid overtraining of the network, hence, when the error on the validation set increased the training was stopped. The intermediate performance of the feature set was then estimated on the test-set. This was repeated 10 times by using a 10-fold cross-validation procedure, the final performance of the test set was obtained from the averages of the intermediate performances.

- Search procedure: several search procedures have been proposed to the feature subset selection problem, although most often research has been focusing on pattern classification. Among the best well-known search procedures in feature selection for pattern classification are: exhaustive search, branch and bound [12], sequential search algorithms (SSA's) [13] and more recently Genetic Algorithms (GA's) [4], [14], [15]. We focus on GA's, because in a comparative study [15] it was shown that GA's can compete with the best search algorithms (SSA's) for feature subset selection and even outperform SSA's for larger feature sets (typically when the number of features is larger than 50). The 'roulette wheel' selection strategy was chosen, where the fitness function was determined by:

$$\text{fitness}(\{F_i\}) = \begin{cases} (\text{Var}(T)-\text{MSE}(\{F_i\}))^n & , \text{if MSE}(\{F_i\}) < \text{Var}(T) \\ 0 & , \text{if MSE}(\{F_i\}) \geq \text{Var}(T). \end{cases} \tag{6}$$

From (6) we observe that any feature subset (F_i) with a mean square error performance (MSE) smaller than the variance of the target variable, gets rewarded. Parameter n controls 'selective pressure' [14]: a higher n will reward good solutions disproportionally. We set n equal to 2. Finally, we used following settings in the GA: the probability of cross-over between individuals (an individual is a particular feature subset) p_c is equal to 0.3, the probability of mutation p_m of a feature within every individual equal to 0.01, the number of individuals per population equal to 30 and the number of populations equal to 100.

Finally, in figure 4 a schematic overview of the overall feature subset selection strategy is presented.

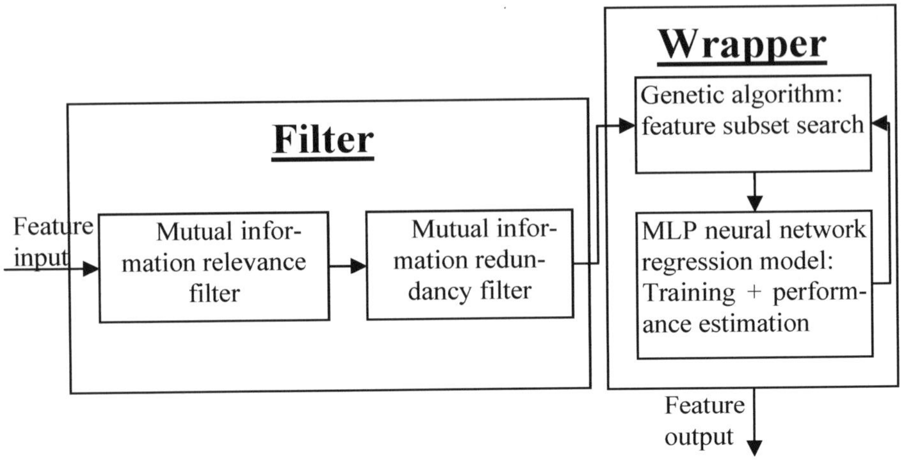

Fig. 4. Schematic overview of the overall feature subset selection strategy for regression. First, irrelevant and redundant features are removed in the filter. Second, the wrapper approach focuses on the smaller set of interesting features.

3 Results

3.1 Boston Housing

We summarize the application of the FSS strategy of figure 4 in table 1 for the 'Boston housing' data for feature subset sizes 1 to 5.

We remark that the relevance analysis showed that only feature 1 is irrelevant (CHAS feature) and that the smallest distance between any features is equal to 0.461 (features RAD: index of accessibility to radial highways and TAX: full-value property-tax rate). We performed further simulations with feature subsets up to all features (13). The best feature subset obtained contained 12 features (feature 4 not included) and has an MSE of 14.83, however, none of the results obtained with more than 3 features could be proven to be statistically significant compared to the result with 3 features. Feature 4 was never included in the smaller subset sizes and thus could have been successfully removed by the relevance filter.

Table 1. Performance of the MLP feature subset strategy on the 'Boston housing' data

Feature subset size	*Feature list (1-13) Best solution*	*Performance (MSE)*
1	[13]	28.08 ± 0.75
2	[6 13]	21.66 ± 1.25
3	[3 6 13]	18.81 ± 1.80
4	[2 5 6 13]	18.96 ± 1.63
5	[6 8 9 11 13]	16.03 ± 1.13

3.2 Orange Juice

Table 2 presents the results of the MLP FSS strategy on the 'orange juice' data set. This data set has been made available by the BNUT unit of the UCL (Université Catholique de Louvain). In performance1 the results of the algorithm of figure 4 without the filter and in performance2 the results with filter (with τ equal to 25 clusters) are tabulated.

Table 2. Performance of the MLP feature subset strategy on the 'orange juice' data

Feature subset size	*Performance1*		*performance2*	
	MSE	*time*	*MSE*	*Time*
5	49.64	301	55.64	254
6	50.91	299	58.36	280
7	69.63	352	53.78	279
8	57.25	360	59.45	325
9	54.76	410	60.57	386
10	57.94	485	46.48	461
11	51.73	641	56.98	520
12	54.71	534	47.28	545
13	38.04	680	49.81	416
14	49.14	750	46.98	463
15	52.34	647	48.66	489

MSE is the performance in 'mean square error' of the best feature subset found, *time* is the total number of times a 10-fold cross-validation procedure (this means: training, validation and testing) was needed over 100 populations to estimate the performance of a feature subset (one has a maximum of 30*100 evaluations). Once this performance for a subset is computed, it can be stored and thus it does not need to be recomputed if the feature subset reappears in future populations. Reappearance of performing subsets is very likely (and expected), due to the fitness selection strategy. The increased performance in speed (lower time) in table 2 can be explained by the reduction of the 700 features to 25 features used in the wrapper: crossover and mutation are more likely to generate previous occurring individuals. Hence, while an increase in speed for a hybrid approach would be evident for search strategies such as: exhaustive search, SSA's, greedy search and so on, it is less evident for GA's, when keeping the number of individuals per population and the number of populations fixed. Furthermore, a paired t-test on the MSE's (mean square errors) shows that the performance of the 2 approaches is equivalent; on the other hand a paired t test shows that the difference in number of evaluations needed is statistically significant. The cost of the filter preprocessing can be ignored if a limited number of permutations are performed.

4 Conclusions

We have shown that relevance and redundancy analysis helps interpreting the data under study. Permutation tests are used to find statistically motivated thresholds that determine statistically relevant features. Finally, it was shown that the filter preprocessing increases the speed of the wrapper approach in the feature subset search.

Acknowledgements

This first author was supported by the Institute for the Promotion of Innovation through Science and Technology in Flanders (IWT Vlaanderen). The second author is funded by the Belgian Fund for Research -- Flanders (G.0248.03, G.0234.04).

References

1. Guyon, I. Elisseeff, A.: An Introduction to Variable and Feature Selection. Journal of Machine Learning Research 3 (2003) 1157-1182.
2. Kurgan, L.A., Cios, K.J.: CAIM Discretization Algorithm. IEEE Transactions on Knowledge and Data Engineering 16 (2004) 145-153.
3. Kohavi, R., John G. H.: Wrappers for Feature Subset Selection. Artificial Intelligence 97 (1997) 273-324.
4. Van Dijck G., Van Hulle M. M., Wevers, M.: Hierarchical Feature Subset Selection for Features Computed from the Continuous Wavelet Transform. 2005 IEEE Workshop on Machine Learning for Signal Processing (2005) 81-86.
5. Cover, T. M., Thomas, J. A.: Elements of information theory. John Wiley & Sons, New York (1991).
6. Schreiber, T., Schmitz, A.: Surrogate Time Series, Physica D 142 (2000) 346 – 382.
7. Kraskov, A., Stögbauer, H., Grassberger, P.: Estimating Mutual Information. Phys. Rev. E. 69 (2004) 066138.
8. Francois, D., Wertz, V., Verleysen, M.: The Permutation Test for Feature Selection by Mutual Information. European Symposium on Artificial Neural Networks (2006) 239-244.
9. John, G., Kohavi, R. Pfleger, K.: Irrelevant Features and the Subset Selection Problem. In Proc. of the Eleventh Int. Conf. on Machine Learning, (1994) 121-129.
10. Duda, R.O., Hart, P.E., Stork, D. G. *Pattern Classification*, John Wiley & Sons Inc., New York (2001).
11. Bishop, C.M. Neural Networks for Pattern Recognition. Oxford University Press Inc., New York (1997).
12. Narendra, P. M., Fukunaga, K.: A Branch and Bound Algorithm for Feature Subset Selection. IEEE Trans. Computers 26 (1977) 917-922.
13. Pudil, P., Novovicova, J., Kittler, J., Floating Search Methods in Feature Selection. Pattern Recognition Letters 15 (1994) 1119-1125.
14. Michalewicz, Z.: Genetic Algorithms + Data Structures = Evolution Programs. 3rd edn. Springer-Verlag, Berlin Heidelberg New York (1996).
15. Kudo, M., Sklansky, J., Comparison of Algorithms that Select Features for Pattern Recognition. Pattern Recognition 33 (2000) 25-41.

Effective Input Variable Selection for Function Approximation

L.J. Herrera[1], H. Pomares[1], I. Rojas[1], M. Verleysen[2], and A. Guilén[1]

[1] Computer Architecture and Computer Technology Department
University of Granada, 18071, Granada, Spain
[2] Machine Learning Group
3 Place du Levant, 1348 Louvain la Neuve, Belgium

Abstract. Input variable selection is a key preprocess step in any I/O modelling problem. Normally, better generalization performance is obtained when unneeded parameters coming from irrelevant or redundant variables are eliminated. Information theory provides a robust theoretical framework for performing input variable selection thanks to the concept of mutual information. Nevertheless, for continuous variables, it is usually a more difficult task to determine the mutual information between the input variables and the output variable than for classification problems. This paper presents a modified approach for variable selection for continuous variables adapted from a previous approach for classification problems, making use of a mutual information estimator based on the k-nearest neighbors.

1 Introduction

Input variable selection is a very important preprocessing step in any supervised or unsupervised learning problem. Having a number of irrelevant or redundant input variables can lead to overfitting and to a poor generalization of the model [3]. Furthermore in models that suffer from the curse of dimensionality in the number of input variables like grid-based fuzzy models [5], input variable selection becomes essential.

Two main trends can be followed to perform this process. Filter methods try to select the variables in a preprocess step with the only information that the I/O values bring. Wrapper methods employ the learning methodology that is going to be used, in order to select the subset of variables that brings the best performance. In both cases, there are two options to perform the "selection" of the variables subset. On the one hand it is possible to select a subset of the original variables (feature selection or input variable selection). On the other hand the initial set of input variables can be replaced by a new subset of variables that are usually obtained by linear or nonlinear transformations of the original ones (feature extraction or input variable extraction).

This paper deals with filter methods for feature selection. Filter methods have the great advantage that the model has no influence on the selected variables.

S. Kollias et al. (Eds.): ICANN 2006, Part I, LNCS 4131, pp. 41–50, 2006.

Thus they can be used as a completely separated preliminary step to the Input/Output (I/O) modelling problem. Several methodologies for input variable selection exist in the literature for both feature selection and feature extraction approaches. Principal component analysis (PCA) and kernel-PCA algorithms are examples of feature extraction methods [2,3,4]. Feature selection methods have the advantage that the meaning or understandability of the input variables of the problem is kept in the model.

For input variable selection, information theory offers a good theoretical environment for variable filtering thanks to the concepts of entropy and mutual information (MI) between variables [11]. Nevertheless, for regression problems it is a harder task to use these concepts. In regression the input and output variables take continuous values, and additional techniques have to be used to estimate the probability distribution [1]. This problem becomes even more pronounced specially when the number of data points is low comparing to the number of input variables (DNA Micro-arrays, etc.). Among the techniques to estimate the probability density functions (PDF) we can find histogram and kernel-based PDF estimators. But those estimators suffer from the curse of dimensionality and can be used for problems with a low number of variables.

A number of estimators for the entropy based on the k-nearest neighbor statistics also exist. Only recently they have been extended to the mutual information estimation by Kraskov et al [9,10]. A nice property of this estimator is that it can be used easily for sets of variables.

Using the concept of mutual information between two or more variables, a number of algorithms could be designed [1,7,8]. This paper presents a modification of the work presented in [6], adapted for continuous variables thanks to the use of the MI estimator based on the k-nearest neighbors [9]. The simulations section presents the application of the new methodology to Least Squares Support Vector Machines (LS-SVMs). It is also compared with other recent input variable selection methods presented in the literature.

The rest of the paper is organized as follows. Section 2 briefly explains the mutual information concept for continuous variables (subsection 2.1); it overviews the k-nearest neighbors estimator that is used (subsection 2.2); and finally presents the proposed algorithm for variable selection based on MI (subsection 2.3). Section 3 shortly reviews the basics of the LS-SVMs. Section 4 presents examples of application of the variable selection methodology. Section V presents the main conclusions drawn from the study.

2 Effective Feature Selection Based on the Mutual Information

In this section the basics of the proposed variable selection algorithm are presented. First the mutual information concept for continuous variables is briefly explained, followed by a k-nearest neighbors procedure to estimate it. Finally the algorithm, which is an adaptation of a previous work for discrete cases in [6], is presented.

2.1 Mutual Information

Given a single-output multiple input (MISO) function approximation or classification problem, with input variables $X = [x_1, x_2, \ldots, x_n]$ and output variable $Y = y$, the main goal of a modelling problem is to reduce the uncertainty on the dependent variable Y. According to the formulation of Shannon, and in the continuous case, the uncertainty on Y is given by its entropy defined as

$$H(Y) = -\int \mu_Y(y) \log \mu_Y(y) dy, \tag{1}$$

considering that the marginal density function $\mu_Y(y)$ can be defined using the joint PDF $\mu_{X,Y}$ of X and Y as

$$\mu_Y(y) = \int \mu_{X,Y}(x, y) dx. \tag{2}$$

Given that we know X, the resulting uncertainty of Y conditioned to known X is given by the conditional entropy, defined by

$$H(Y|X) = -\int \mu_X(x) \int \mu_Y(y|X = x) \log \mu_Y(y|X = x) dy dx. \tag{3}$$

The joint uncertainty on the $[X, Y]$ pair is given by the joint entropy, defined by

$$H(X, Y) = -\int \mu_{X,Y}(x, y) \log \mu_{X,Y}(x, y) dx dy. \tag{4}$$

The mutual information (also called cross-entropy) between X and Y can be defined as the amount of information that the group of variables X provide about Y, and can be expressed as

$$I(X, Y) = H(Y) - H(Y|X). \tag{5}$$

In other words, the mutual information $I(X, Y)$ is the decrease of the uncertainty on Y once we know X. Due to the mutual information and entropy properties, the mutual information can also be defined as

$$I(X, Y) = H(X) + H(Y) - H(X|Y), \tag{6}$$

leading to

$$I(X, Y) = \int \mu_{X,Y}(x, y) \log \frac{\mu_{X,Y}(x, y)}{\mu_X(x)\mu_Y(y)} dx dy. \tag{7}$$

Thus, only the estimate of the joint PDF between X and Y is needed to estimate the mutual information between two groups of variables.

Estimating the joint probability distribution can be performed using a number of techniques. As mentioned already, histograms and kernel density estimators have been used for this purpose [1]. The next subsection will shortly review how to use a k-nearest neighbors methodology to estimate the MI.

2.2 Estimating the Mutual Information Using the k-Nearest Neighbors

There is extensive literature about estimators based on the k-nearest neighbors for the entropy, but it has been only recently extended to the MI [9].

Thanks to that estimator, it is possible to use sets of variables in the estimation of the MI, and thus it will allow to adapt the method presented in [6].

We define the space $Z = X, Y$ and we will use the maximum norm for any pair of points $z = (x, y)$ and $z' = (x', y')$,

$$\|z - z'\| = \max\{\|x - x'\|, \|y - y'\|\}, \tag{8}$$

although any other norm could be used. Denote by $\varepsilon(i)$ the distance from a point z_i to it is k-th nearest neighbor and by $\varepsilon_x(i)$ and $\varepsilon_y(i)$ the distances between the same points projected into the X and Y subspaces. Obviously $\varepsilon(i) = \max\{\varepsilon_x(i), \varepsilon_y(i)\}$.

We will count the number $n_x(i)$ of points x_j whose distance from x_i is strictly less than $\varepsilon(i)$, and similarly for y instead of x. The estimate for MI is then (see [9] for a proof of the convergence of this estimator)

$$\hat{I}_1(X, Y) = \psi(k) - \frac{1}{N} \sum_{i=1}^{N} [\psi(n_x(i) + 1) + \psi(n_y(i) + 1)] + \psi(N), \tag{9}$$

where ψ is the digamma function given by

$$\psi(t) = \frac{\Gamma'(t)}{\Gamma(t)} = \frac{d}{dt} \ln \Gamma(t). \tag{10}$$

Function ψ satisfies the recursion $\psi(x + 1) = \psi(t) + 1/x$ and $\psi(1) = C$ where $C = -0.5772156\ldots$ is the Euler-Mascheroni constant.

Another alternative is to replace $n_x(i)$ and $n_y(i)$ by the number of points with $\|x_i - x_j\| \leq \varepsilon_x(i)/2$ and $\|y_i - y_j\| \leq \varepsilon_y(i)/2$. The estimate for MI is then

$$\hat{I}_2(X, Y) = \psi(k) - \frac{1}{k} - \frac{1}{N} \sum_{i=1}^{N} [\psi(n_x(i)) + \psi(n_y(i))] + \psi(N). \tag{11}$$

In this paper this second estimator is used, which is the one implemented in [10]. Please check [9] for an extended explanation.

As can be noted, this MI estimator has a dependency on the value chosen for k (k-th nearest neighbor). As it is recommended in [12] for a tradeoff between variance and bias, in the examples, a mid-range value for k ($k = 6$) will be used.

2.3 Effective Variable Selection for Function Approximation Problems Using MI

The MI estimator detailed above will allow us to carry out the proposed variable selection method. It also gives the possibility of estimating the MI for groups of variables even when the number of data points we have at disposal is relatively small.

In the following it is reviewed how the MI can be used for variable selection, and it is presented the proposed method for variable selection in function approximation problems.

According to the definition of MI, $I(X,Y)$ gives the information that the group of variables X bring about Y. Any modelling problem would try to use this information and try to predict new values of Y given new values of X. As mentioned before, having unneeded variables can unnecessarily complicate the model. Furthermore the generalization capability can be decreased. Thus it is essential to select a right subset $X_G \subset X$ that comprises the same information that X has of Y. That is, we want to find a subset $X_G \subset X$ such that

$$I(X,Y) \cong I(X_G,Y). \tag{12}$$

We could directly try to evaluate $I(X_G,Y)$ for all the possible subsets X_G of X, and then select the smallest subset X'_G, whose $I(X_G,Y)$ is the highest. In this way irrelevant and redundant variables would not be selected in the optimal X'_G. This approach suggested in [13] and [1], and partially in [8] has two main drawbacks. First the number of possibilities for X_G is exponential in the number of input variables n (2^n possible subsets). Second, as the number of available data is limited, the robustness of the k-nearest neighbor MI estimator is also limited when taking into account too many input variables in X_G.

Other approaches could consider the MI of single input variables over the output variable $I(x_i,Y)$ to perform a ranking and use it as a filter to eliminate variables [7]. This approach is very good for avoiding irrelevant variables but does not consider redundant ones. For example two variables x_i and x_j can have a very high MI with respect the output variable Y, but using both of them can bring no more MI w.r.t the output variable. In this case $I(\{x_i,x_j\},Y)$ is similar to $I(x_i,Y)$ and $I(x_j,Y)$, and thus one could be discarded.

A more robust approach would be to try selecting input variables as the MI with respect to the output variable of the selected subset increases. An iterative process would add a new variable to the current subset such that

$$I(\{X_G \cup x_i\},Y) - I(X_G,Y), \tag{13}$$

is maximum over j. Nevertheless, as mentioned before, if the number of variables to be selected is high, the precision of the MI estimator can be lost. The results offered by the MI estimator, as exposed in [9] are optimal when the number of data points is very high, but in practice this is rarely the case.

Here it is proposed to adapt the method for discrete variables presented in [6] to function approximation problems (i.e. to continuous variables). An iterative backwards variable selection will be performed, starting from the complete set of variables X. The idea is to eliminate a variable x_i in the current selected subset $X_G \cup \{x_i\} \subset X$ if we estimate that $I(\{X_G \cup \{x_i\}\},Y) = I(X_G,Y)$. To help in this iterative procedure, the concept of Markov blanket, adapted for this problem, will be used. We will suppose that this concept can be applied for the specific variable selection problem we deal with. The use of Markov blankets [15] implies strong conditioning between the variables. Nevertheless it will be relaxed to help us performing the variable selection.

Definition: Let M be a set of variables that do not contain x_i. We say that M is a Markov blanket for x_i if $I(\{M \cup x_i\}, X - \{M \cup x_i\}) \cong I(M, X - \{M \cup x_i\})$

Corollary: Let X_G be a subset of variables and x_i a variable in X_G. Assume that a subset M of X_G is a Markov blanket of x_i. Then $I(X_G, Y) \cong I(X_G - \{x_i\}, Y)$.

As we can see, the Markov blanket condition is stronger than the one we desire. It can be even a harder problem to find a Markov blanket of a variable in a set of variables that the variable selection problem itself. However it brings the idea on how to perform a more robust variable selection procedure. The difficult evaluation of $I(\{X_G \cup \{x_i\}\}, Y) = I(X_G, Y)$ to eliminate variables, will be transformed into estimating if x_i has a Markov blanket in X_G. Those x_i will be removed from the current X_G.

As already mentioned, calculating the Markov blanket of a variable in a set, or even trying to know if it exists is a very difficult task. Therefore it will be assumed that the Markov blanket exists, and we will derive a heuristic to guess the variables M that compose the Markov blanket of any variable x_i. The proposed algorithm is the following:

1. Calculate the MI between every two input variables $I(x_i, x_j)$
2. Starting from the complete set of input variables $X_G = X$, iterate:
   (a) For each variable x_i, let the candidate Markov blanket M_i be the set of p variables in X_G for which $I(x_i, x_j)$ is highest.
   (b) Compute for each x_i
   $$Loss_i = I(\{M_i \cup x_i\}, Y) - I(M, Y). \tag{14}$$
   (c) Choose the x_i for which $Loss'_i$ is lowest and eliminate x'_i from X_G.

The procedure may be stopped after a fixed number of input variables are eliminated; alternatively it may be stopped when $Loss'_i$ reaches a certain threshold. This methodology is suboptimal in some aspects, but still offers a robust variable selection methodology. The Markov blanket selected for every variable is just an approximation and the number p of variables is fixed a priori. With respect to parameter p, high values can bring better chances that the pseudo-Markov blankets taken subsume real Markov blankets of the variables. However, the reliability of the MI estimator can be decreased. Considering this trade off, in general, a medium value of p should be considered. For problems with low number of data points, a lower value for p should be taken.

As we will see in the simulation section, the method proposed can outperform the other methods commented in this paper: it can therefore be a good solution for the key problem of variable selection in regression or function approximation problems.

3 Least-Squares Support Vector Machines

This section presents a brief introduction to the learning methodology used in the simulations. LS-SVMs are reformulations to standard SVMs, closely related to regularization networks and Gaussian processes but additionally emphasize and exploit primal-dual interpretations from optimization theory. LS-SVMs are a paradigm specially well suited for function approximation problems [4].

The LS-SVM model [14] is defined in its primal weight space by

$$\hat{y} = W^T \phi(X) + b, \tag{15}$$

where W^T and b are the parameters of the model, $\phi(X)$ is a function that maps the input space into a higher-dimensional feature space and X is the n-dimensional vector of inputs x_i. In Least Squares Support Vector Machines for function approximation, the following optimization problem is formulated,

$$\min_{W,b,e} J(W,e) = \frac{1}{2} W^T W + \gamma \frac{1}{2} \sum_{i=1}^{N} e_i^2, \tag{16}$$

subject to the equality constraints (inequality constraints in the case of SVMs)

$$e_i = y_i - \hat{y}_i(X_i), i = 1 \ldots N. \tag{17}$$

Solving this optimization problem in dual space leads to finding the λ_i and b coefficients in the following solution

$$\hat{y}_i = \sum_{i=1}^{N} \lambda_i K(X, X_i) + b, \tag{18}$$

where the function $K(X, X_i)$ is the kernel function defined as the dot product between the $\phi(X)$ and $\phi(X_i)$ mappings. If we consider Gaussian kernels, the width of the kernel σ_i together with the regularization parameter γ, are the hyper-parameters of the problem. Note that in the case of Gaussian kernels, the obtained model resembles Radial Basis Function Networks (RBFN), with the particularities that there is an RBF node per data point, and that overfitting is controlled by a regularization parameter instead of by reducing the number of kernels [7]. In LS-SVM, the hyper-parameters of the model are usually optimized by cross-validation.

4 Simulations

This section presents the application of the variable selection method proposed in this paper to a significant example. The MI estimator in [10] will be used in all the simulations. A LS-SVM Matlab toolbox can be found in [14]. The error measure used here is the Normalized Mean Square Error, NMSE [7].

The example considered has been taken from [7] and is a spectrometric data set coming from the food industry. This type of data form vectors with a large number of exploitable variables. Usually however, only a small subset of them is required to build a good model. The "tecator meat" data set consist of 100 spectral input variables and one output variable (the original data set has three, but we consider only the fat content). It relates to the determination of the fat content of meat samples analyzed by near infrared transmittance spectroscopy. This data set contains 172 training spectra and 43 test spectra. As in [7], the spectra are reduced to zero mean and unit variance. Also to avoid loosing information, the original mean and standard deviation are kept as two additional variables. A selection of training spectra is shown in Figure 1.

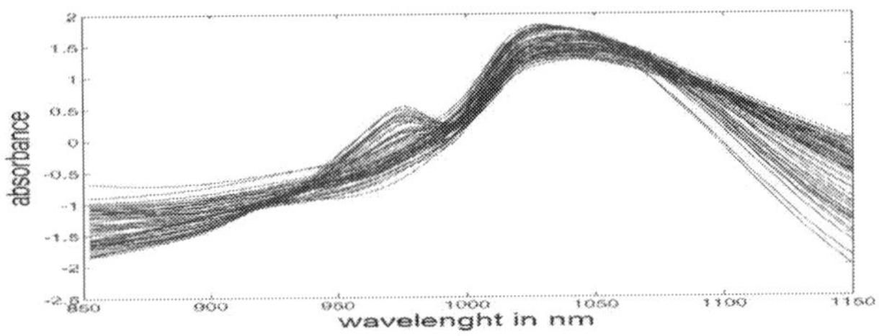

Fig. 1. A selection of the spectra from the "tecator meat" data set

In [7], first an initial subset of 16 variables is selected. Using them, a LS-SVM is optimized using cross-validation. The test NMSE obtained for this case is 0.0040. Next, all 2^{16} possible subsets of variables are tested, checking which subset of those 16 variables brings the highest MI with respect to the output variable. The optimal subset found had 8 variables. The test NMSE on the LS-SVM model is 0.0049. Note that in the comparisons presented in this section, there are some differences with the results shown in [7], since the second estimator $\hat{I}_2(X, Y)$ was used here.

Next we proceed by selecting 16 most significant variables using the approach proposed in this paper (for example using $p = 6$). The test NMSE obtained after the optimization of the LS-SVM is 0.0022. As can be seen, the initial subset selected by our method has a higher performance than the one selected by the approach in [7]. Forcing the number of variables to 8 (with parameter $p = 6$) to compare with the sub-set selected in [7], the test NMSE was 0.0024.

With respect to the value of the parameter p, similar results of NMSE with 16 variables were given by other values of p both in training and in test, showing the efficiency of the method eliminating irrelevant and redundant variables. For very low values of p, the training and test errors were even lower (test NMSE for $p = 1$ is 0.0016). It is noticeable that the variables selected for different values of p are remarkably different. This is due to the high level of redundancies that exist among the input variables and also to the low number of data points that we handle in this problem. In problems like this one, there are usually several possibilities of suboptimal subsets of variables, instead of a single optimal set.

For this problem the results obtained show that lower values of p provide better results both for training and test data sets. But during the filtering process, there are also some details that suggest discarding higher values for p. The loss function (see Eq. 14) for high values of p does not follow an increasing trend. We have a low number of data points and very high redundancies among the input variables. Thus the MI estimator can provide confusing results.

Taking $p = 1$, we will now look for a final pseudo-optimal subset of variables. As mentioned before, we could have in principle two possible stopping criteria in our algorithm. One is to specify a number of input variables to be selected. This number could be chosen according to the results of the model on the subsets of variables. In this case, the method would become a mixture of filter and wrapper. A good stopping criterion would be to select the subset that brought the best training error after the cross-validation optimization of the LS-SVM model.

Nevertheless, in order not to loose the filtering advantage of the method, a possible heuristics is to select a limit in the loss function in Eq. 14. In Figure 2 the evolution of the loss function in Eq. 14 in the iterative process is shown, for $p = 1$. From this graph we can get an idea on how much information is lost as variables are discarded in the iterative process. We saw that selecting a subset of 16 variables leads to good performance. For this value, the $Loss'$ is around 0.23. Consequently we establish a limit of 0.26, that corresponds to the next peak in the graph. From this threshold no more variables will be eliminated. Finally, the optimal subset contains 11 variables and the performances are test NMSE = 0.0016 and training NMSE = 0.0010. Other tests showed that further elimination of variables leads to a small worsening of the performances (both in training and test), increasing as more variables are discarded.

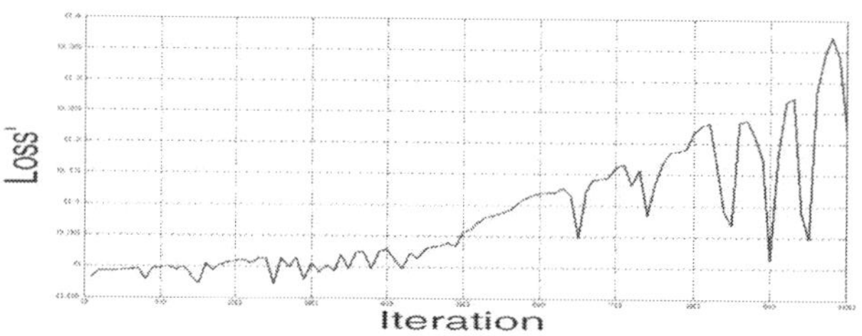

Fig. 2. Evolution of the $Loss'$ function in the run of the algorithm with $p = 1$

5 Conclusions and Further Work

In this paper it was presented an effective backward variable selection method for function approximation problems, based on the concept of MI, adapted from a previous method for classification problems. It is a robust approach compared to other ones from the literature, thanks to the use of the Markov blanket concept. As further work, we intend to design a general methodology to select the number of input variables to be discarded. Furthermore the application of the method in other domains, in particular in time series prediction, will be investigated to help solving the difficult problem of deciding which variables should intervene in the prediction model.

Acknowledgements

This work is partially supported by the Spanish CICYT Project TIN 2004-01419. This work was partially done during the stay of L. J. Herrera in the UCL, under the supervision of M. Verleysen, Research Director of the Belgian FNRS.

References

1. B.V. Bonnlander, A.S. Weigend, "Selecting input variables using mutual information and nonparametric density estimation, in Proc. of the ISANN 2004, Taiwan, 1994, pp. 42-50
2. B. Schoelkopf, A. Smola: Learning with Kernels. Cambridge, MA: MIT Press, 2002
3. S. Haykin, Neural Networks, Prentice Hall, New Jersey, 1998
4. J.A.K . Suykens, T. Van Gestel, J. De Brabanter, J. De Moor, B., Vandewalle: Least Squares Support Vector Machines, World Scientific, Singapore, 2002
5. L.J. Herrera, H. Pomares, I. Rojas, O. Valenzuela, A. Prieto: "TaSe, a Taylor Series Based Fuzzy System Model that Combines Interpretability and Accuracy". Fuzzy Sets and Systems, vol. 153, No.3, 2005, 403-427
6. D. Koller, M. Sahami, "Toward Optimal Feature Selection", in Proc. Int. Conf. on Machine Learning, 1996, pp. 284-292
7. F.Rossi, A. Lendasse, D. Franois, V. Wertz, M. Verleysen: "Mutual Information for the selection of relevant variables in spectrometric nonlinear modeling", Chem. and Int. Lab. Syst., 2005, In Press
8. N. Benoudjit, D. Franois, M. Meurens, M. Verleysen: "Spectrophotometric variable selection by mutual information", Chem. and Int. Lab. Syst., vol. 74, 2004, pp. 243-251
9. A. Kraskov, H. Stgbauer, P. Grassberger, "Estimating mutual information", Phys.Rev.,E 69, 2004, 066138
10. http://www.klab.caltech.edu/~kraskov/MILCA/
11. T.M. Cover, J.A. Thomas, Elements of Information Theory, Wiley, New York, 1991
12. S. Harald, K. Alexander, A.A. Sergey, G. Peter, "Least dependent component analysis based on mutual information", Phys. Rev., E 70, 2004, 066123
13. A. Sorjamaa, J. Hao, A. Lendasse, "Mutual Information and k-Nearest Neighbors Approxi-mator for Time Series Prediction", ICANN 2005, LNCS 3697, pp. 553 - 558
14. http://www.esat.kuleuven.ac.be/sista/lssvmlab/
15. J. Pearl: Probabilistic Reasoning in Intelligent Systems. Morgan Kaufmann, CA, 1988

Comparative Investigation on Dimension Reduction and Regression in Three Layer Feed-Forward Neural Network

Lei Shi and Lei Xu

Chinese University of Hong Kong, Shatin, NT, Hong Kong
{shil, lxu@}cse.cuhk.edu.hk

Abstract. Three layer feed-forward neural network (3-LFFNN) has been widely used for nonlinear regression. It is well known that its hidden layer can be regarded as taking the role of feature extraction and dimension reduction, and that the regression performance relies on how the feature dimension or equivalently the number of hidden units is determined appropriately. There are many publications on determining the hidden unit number for a desired generalization error. However, few comparative studies have been made on different approaches proposed, especially on those typical model selection criteria for this purpose. This paper targets such an aim. Using both simulated data and several real world data sets, a comparative study has been made on the regression performances with the number of hidden units determined by several typical model selection criteria, including Akaike's Information Criterion (AIC), the consistent Akaike's information criterion (CAIC), Schwarz's Bayesian Inference Criterion (BIC) which coincides with Rissanen's Minimum Description Length (MDL) criterion, and the well-known technique cross-validation (CV), as well as the Bayesian Ying-Yang harmony criterion on a small sample size (BYY-S). As shown in experiments on a small size of samples, BIC and CV are better than AIC and CAIC obviously. Moreover, BIC may be better than CV on certain data sets, while CV may be better than BIC on other data sets. Interestingly, BYY-S generally outperforms both BIC and CV.

1 Introduction

As a popular supervised learning model, three layer feed-forward neural network (3-LFFNN) has been widely investigated and applied in many areas [13,15], in which the three layers are input layer, hidden layer and output layer, respectively. The activation function in each hidden unit is usually a sigmoid function with respect to a linear combination of input signals, while the output layer performs a linear combination. In the literature, many methods have been proposed to train a size-predetermined 3-LFFNN and estimate the connection weights so as to decrease the mean squared error (MSE) at the output layer, such as the well-known back-propagation technique [3,4,13]. However, this kind of training approaches can only provide a minimum training MSE but not guarantee a network with desired generalization errors, leaving the problem of determining the

S. Kollias et al. (Eds.): ICANN 2006, Part I, LNCS 4131, pp. 51–60, 2006.

network structure unsolved. Particularly, when widely used in nonlinear regression, 3-LFFNN can approximate an unknown input-output relation underlying a target system based only on a set of training data [1,15], in which the hidden layer can be regarded as playing the role of feature extraction and dimension reduction. Both the quality of regression and the generalization ability depend on the network's architecture. Since both the input and output layers are naturally determined by the input and output patterns, respectively, the abilities of a 3-LFFNN thus depend on the number of the hidden units.

To select the hidden unit number in 3-LFFNN, we are faced with both the under-fitting problem caused by insufficient hidden units, and a high risk of being over-fitting caused by using too many hidden units. In literature, many efforts have been proposed to determine 3-LFFNN's structure [1,12,13], which can be roughly divided into two types. The first type uses some heuristic methods to change and choose the network's structure, such as adding hidden units from a simple network or doing pruning from a complex network until reaching a stable state [13]. The other kind of methods describes this problem from a model selection viewpoint, i.e., the optimal model is determined by a given model selection criterion in a so-called two-phase procedure, during which a set of candidate models are firstly obtained by the minimum MSE training, and then the "optimal" model is selected according to the given criterion. In this paper, we focus on comparing different popular criteria used in this second type.

A variety of criteria for model selection have been proposed in literature, such as Akaike's Information Criterion (AIC) [5], the consistent Akaike's information criterion (CAIC) [6], Schwarz's Bayesian Inference Criterion (BIC) [7] which coincides with Rissanen's Minimum Description Length (MDL) criterion [8], and another well-known technique cross-validation (CV) [10]. Bayesian Ying-Yang (BYY) learning was proposed as a unified statistical learning framework firstly in 1994 and systematically developed in the past decade. Providing a general learning framework, BYY harmony learning consists of a general BYY system and a fundamental harmony learning principle as a unified guide for developing new regularization techniques, a new class of criteria for model selection, and a new family of algorithms that perform parameter learning with automatic model selection. When applied to a 3-LFFNN structure selection, BYY criteria have been proposed and developed [14,15,16]. Especially, when the small sample size cases are considered and a two-phase procedure is implemented too, a better BYY criterion (BYY-S) has been proposed in [15].

There have been some studies to investigate separately AIC, BIC/MDL or CV's performance on 3-LFFNN model selection [1,2,12], however, few systematic ones have been proposed on those different typical criteria for this purpose. This paper intends to provide a comparative investigation on AIC, CAIC, BIC, CV and BYY-S in selecting the optimal hidden unit number for 3-LFFNN on different data sets. In our experiments, two different kinds of results are reported due to the data situations. For data sets with known numbers of hidden units, the number selection results by each criterion are reported for comparison, while for those cases with hidden unit numbers unknown, we compare training and generalization *root mean squared errors* (RMSE).

The rest of this paper is organized as follows. In Section 2, we review the regression problem and the 3-LFFNN model. Section 3 introduces typical model selection criteria for determining 3-LFFNN hidden unit number, including AIC, CAIC, BIC, CV and BYY-S. In Section 4, a series of comparative experiments on both simulated and real data sets are reported. Finally, we present some concluding remarks in Section 5 and further discussion is given in Section 6.

2 Regression and 3-LFFNN

3-LFFNN has been widely investigated and applied in literature [1,2,11,14,15], with the three layers being input layer, hidden layer and output layer, respectively. One of the most important application of 3-LFFNN is to form a non-linear model that can approximate or regress any reasonable continuous input-output relation. Here, we use 3-LFFNN with *binary hidden units* to regress an underlying function f with the dimensionality reduced at the hidden layer [1,2,15]. Given a set of N pairs input-output samples $\{\mathbf{x}_i\}_{i=1}^{N}$, with each $\mathbf{x}_i = [\xi_i, \eta_i]$, where $\eta_i \in \mathcal{R}$ are targets and $\xi_i \in \mathcal{R}^d$ are inputs [1,14], we consider a 3-LFFNN with 1 output unit, k hidden units, and d input units, as shown in Fig.(1), where $y_{i,j}$ is the output of the jth hidden unit for ith input vector ξ_i. That is, we describe a nonlinear regression by

$$\eta_i = \sum_{j=1}^{k} w_j^o y_{i,j} + b^o + \epsilon_i, \quad i = 1, \ldots, N, \tag{1}$$

where $w_j^o \in \mathcal{R}$ is the weight connecting the jth hidden unit and the output unit, $b^o \in \mathcal{R}$ is the output unit's bias, ϵ_i is the independent random noise with a normal distribution $N(0, \sigma^2)$, and $y_{i,j}$ takes binary values according to the Bernoulli probability distribution,

$$\hat{y}_{i,j} = s(\mathbf{W}_j \xi_i + b_j), \quad p(y_{i,j}) = \hat{y}_{i,j}\delta(y_{i,j}) + (1 - \hat{y}_{i,j})\delta(1 - y_{i,j}), \tag{2}$$

where $\mathbf{W}_j \in \mathcal{R}^d$ and $b_j \in \mathcal{R}$ are the weight vector connected to and the bias on the jth hidden unit, respectively. Moreover, $s(\bullet)$ is the sigmoid function, usually in the form of $s(r) = 1/(1 + e^{-r})$.

Given a 3-LFFNN with fixed architecture, training the network for estimating parameters can be usually solved by the well-known back-propagation algorithm [3,4,15] for a minimum training mean squared error (MSE) destination.

3 Estimating Hidden Unit Number

To determine the optimal hidden unit number of a 3-LFFNN, we face a dilemma: although the regression errors for the training examples will reduce to zero with

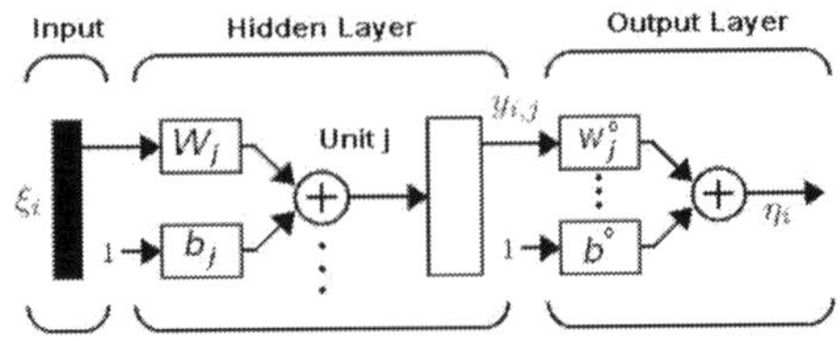

Fig. 1. Structure of a typical three layer feed-forward neural network

an increasing number of the parameters, the errors on testing examples also increase. That is, minimizing training MSE and minimizing testing MSE competes with each other. To determine a desired structure, under the name of model selection, one typical approach is the so-called two-phase procedure by using a model selection criterion $J(\hat{\theta}_k, k)$ [1,2,11,15] as follows:

1: Enumerate k within a range from k_{inf} to k_{sup}, which is assumed to contain the optimal k^*. For each k, the network is trained by the well-developed back-propagation technique to estimate $\hat{\theta}_k$ [3,4,15].
2: Select $k^* = arg\min_{k \in [k_{inf}, k_{sup}]} J(\hat{\theta}_k, k)$.

3.1 Typical Model Selection Criteria

The typical model selection criteria under our consideration include Akaike's Information Criterion (AIC) [5], the consistent Akaike's information criterion (CAIC) [6], Schwarz's Bayesian Inference Criterion (BIC) [7] which coincides with Rissanen's Minimum Description Length (MDL) criterion [8]. These three criteria can be summarized into the following general form:

$$J(\hat{\theta}_k, k) = N \ln(MSE_t) + d(k)c(N), \tag{3}$$

$$MSE_t = \frac{\sum_{i=1}^{N}(\eta_i - \hat{\eta}_i)^2}{N}, \tag{4}$$

where $\hat{\theta}_k$ is an estimation of parameters, k is the hidden unit number, N is the number of observations, MSE_t is the *training mean squared error*. $d(k)$ is the number of free parameters in a specified structure. For a 3-LFFNN with d input units, k hidden units, and p output units, the number of the connection weights from the input layer to the hidden layer is $k(d+1)$, while the number of the weights from hidden layer to the output layer is $p(k+1)$. Thus the total number of free parameters should be $d(k) = k(1+d) + p(k+1)$. Moreover, $c(N)$ is a constant that depends on the number of observations:

* $c(N) = 2$ for AIC,
* $c(N) = \ln(N) + 1$ for CAIC,
* $c(N) = \ln(N)$ for BIC and MDL.

Another well-known model selection technique is cross-validation (CV) [10]. After firstly dividing the original data set into m subsets, for the ith partition, a single subset D_i is retained as the testing data, while the remaining $m-1$ subsets D_{-i} are used for training. This process is repeated m times (folds), with each of the m subsets used exactly once as the validation data. Finally, the model with the smallest mean *generalization mean squared error* MSE_g is selected according to the criterion as follows, $J(\hat{\theta}_k, k) = \sum_{i=1}^{m}(MSE_{gi}|\hat{\theta}_{-i})$, where MSE_{gi} is the MSE_g on the testing data subset D_i, and $\hat{\theta}_{-i}$ are parameters estimated on the training data D_{-i}. Featured by m, this procedure is referred as an m-fold CV. In our following experiments, we use 10-fold CV.

3.2 Improved BYY Criterion on Small Size Samples

Given an enough number of training samples, the hidden unit number k^* can be chosen by using any one of the previously discussed criteria in help of the two-phase procedure. However, as discussed in [16], when N is relatively too small, the performance of model selection criteria might deteriorate. That is, when the training sample size N is small or not large enough, each of those criteria actually provides a rough estimate that can not guarantee to give an appropriate k^*, perhaps even resulting in a wrong number.

Bayesian Ying-Yang (BYY) learning provides a promising tool for learning and model selection [14,15]. Mathematically, the parameter learning can be implemented by maximizing the so-named harmony function. Particularly, for a 3-LFFNN with binary hidden units, BYY learning can select the hidden unit number automatically during parameter learning [14,15]. Furthermore, a better BYY criterion is given in [15,16] for a two-phase implementation on a small sample size. Applied to the 3-LFFNN model in this paper, the BYY model selection criterion $J(k)$ is given as follows (shortly BYY-S):

$$J(k) = 0.5\ln(MSE_t) - H_q + \rho\frac{d(k)}{N}, \quad \text{where } \rho = 0.5 \text{ or } 1.0, \tag{5}$$

$$H_q = \sum_{j=1}^{k}[q_j \ln q_j + (1-q_j)\ln(1-q_j)], \quad q_j = \frac{1}{N}\sum_{i=1}^{N} y_{i,j}, \tag{6}$$

where $d(k)$ is still the effective number of free unknown parameters as discussed previously in Sec 3.1, N is the number of observations, and MSE_t is the training mean squared error. Moreover, ρ takes either 0.5 or 1.0, corresponding to two different approximation ways as discussed in Section 3.5 of [16], which shortly we denote BYY-S$_{0.5}$ and BYY-S$_{1.0}$, respectively.

Although the network training can be implemented by an adaptive Ying-Yang machine algorithm referred in [14], in this paper we keep using back-propagation to train all the 3-LFFNN, in order to compare with the performances of these typical criteria discussed in Sec 3.1 under a same background.

4 Empirical Comparative Experiments

In this part, we conduct the comparative experiments on both simulated data sets and real world data sets by using the criteria based on back-propagation training, including AIC, CAIC, BIC, CV, BYY-$S_{0.5}$ and BYY-$S_{1.0}$. On each training data set, for each k in a range $[k_{inf}, k_{sup}]$ that contains the true number k^*, the network is trained 10 times started with different random parameter initialization, in order to avoid potential local optima. Then, every criterion is calculated so that the estimated $\hat{k}^*$ is selected.

4.1 Simulated Data

In the first experiment, we select the underlying function similar to [2] as $\eta = f(\xi) = \frac{3}{1+exp(-\xi-3)} - \frac{3}{1+exp(-\xi+3)}$, where each sample consists of two variables $\mathbf{x}_i = [\xi_i, \eta_i]$. The samples are generated by adding normal distributed noises from $N(0, \sigma^2)$ to the function. We set the candidate range by $k_{inf} = 1$ and $k_{sup} = 5$, where the underlying hidden unit number is 2. For one experimental situation with predetermined sample size and noise, 100 different sample sets with a size N are generated randomly by adding normal distributed noises to $f(\xi_i)$ to get η_i. To examine each of the generated sets, the 3-LFFNN is trained and criteria are calculated at each $k \in [1, 5]$ based on the trained $\hat{\theta}_k$, then each criterion obtains totally 100 selection results for hidden unit number k^*. We report the selection times with respect to k by different criteria after the 100 simulations.

Focusing on the effects of sample size and noise, we design three series of experimental situations. Firstly, in order to show the effect of sample size, four situations are designed with different sample sizes N to be 32, 64, 128 and 256, respectively, while noise variance is fixed at $\sigma = 1.0$. The results are reported in Table. (1.a). From the experimental results, we can find that, AIC and CV have a risk of overestimating the hidden unit number, especially when the sample size is small, while CAIC has a risk of underestimation. BIC shows better perform, but suffering inaccurate selection when N is small. Interestingly, BYY-$S_{0.5}$ and BYY-$S_{1.0}$ obtain better or comparative results than others, with the advantage being more and more obvious as the sample size goes down.

Further, to investigate the effect of noise, we fix the sample size at $N = 128$ and change the noise to be $\sigma = 0.6, 0.8, 1.0, 1.2$, as reported in Table. (1.b). As indicated, when the noise is small, all criteria prefers to choose the right selection. However, when the noise becomes larger, AIC tends to over select the result, while CAIC has a risk of underestimation. BIC performs better, while CV shows advantage over AIC, CAIC, and BIC when the noise is large. BYY-$S_{1.0}$ and BYY-$S_{0.5}$ still hold their preference to others generally.

Then, to see a rough combined effect on both sample size and noise variance, we artificially choose four combined cases, with information and results shown in Table. (1.c). Roughly, we can receive a impression that BYY-$S_{1.0}$ and BYY-$S_{0.5}$ generally outperforms other criteria as the "estimation situation" becomes difficult, i.e. the noise grows and sample size decreases.

4.2 *Robot Arm* Data

One standard data set mostly commonly used in neural network regression and model selection is the *Robot Arm* data from MacKay [11,4]. In this data set, we

Table 1. Experimental results on the simulated data after 100 repetitions. For each generated data set, the selection rates of corresponding hidden unit number k^* are reported. (a) shows the effect of sample size, with noise variance fixed at $\sigma^2 = 0.09$. (b) indicates the effect of noise variance, with N fixed at 128. (c) shows results about some specified cases with the combined changing on both sample size and noise.

(a) The effect of sample size: $N = 32, 64, 128, 256$, with noise fixed at $\sigma = 1.0$

criteria	AIC				CAIC				BIC/MDL				10-fold CV				BYY-$S_{0.5}$				BYY-$S_{1.0}$			
k^* \ N	32	64	128	256	32	64	128	256	32	64	128	256	32	64	128	256	32	64	128	256	32	64	128	256
1	0	0	0	0	7	0	0	0	2	0	0	0	4	0	0	0	3	0	0	0	3	0	0	0
2*	16	21	42	61	51	62	77	81	51	62	73	75	41	46	69	78	69	68	79	80	71	72	82	78
3	10	16	11	9	31	29	17	16	35	26	20	21	33	31	22	16	18	26	18	17	21	23	14	18
4	29	23	18	11	8	7	4	3	7	9	3	4	19	16	9	6	6	5	3	3	4	5	1	4
5	45	40	29	19	3	2	2	0	5	3	4	0	4	7	0	0	4	1	0	0	1	0	3	0

(b) The effect of noise: $\sigma = 0.6, 0.8, 1.0, 1.2$, with sample size fixed at $N = 128$

criteria	AIC				CAIC				BIC/MDL				10-fold CV				BYY-$S_{0.5}$				BYY-$S_{1.0}$			
k^* \ σ	0.6	0.8	1.0	1.2	0.6	0.8	1.0	1.2	0.6	0.8	1.0	1.2	0.6	0.8	1.0	1.2	0.6	0.8	1.0	1.2	0.6	0.8	1.0	1.2
1	0	0	0	4	0	0	0	5	0	0	0	3	0	0	0	6	0	0	0	3	0	0	0	4
2*	77	47	42	34	92	81	77	56	90	79	73	55	88	82	69	61	90	85	79	64	92	86	82	68
3	6	14	11	7	5	13	17	26	8	13	20	33	7	11	22	21	6	4	18	18	8	5	14	21
4	13	22	18	28	2	6	4	5	2	8	3	2	1	7	9	11	4	9	3	9	0	9	1	4
5	4	17	29	27	0	0	2	8	0	0	4	7	4	0	0	1	0	2	0	6	0	0	3	3

(c) The combined effect of both sample size and noise. Four different cases are chosen and reported, with Case I: $N = 256$, $\sigma = 0.6$; Case II: $N = 128$, $\sigma = 0.8$; Case III: $N = 64$, $\sigma = 1.0$; Case IV: $N = 32$, $\sigma = 1.2$

criteria	AIC				CAIC				BIC/MDL				10-fold CV				BYY-$S_{0.5}$				BYY-$S_{1.0}$			
k^* \ cases	I	II	III	IV	I	II	III	IV	I	II	III	IV	I	II	III	IV	I	II	III	IV	I	II	III	IV
1	0	0	0	11	0	0	0	26	0	0	0	22	0	0	0	18	0	0	0	20	0	0	0	25
2*	82	47	21	46	95	81	62	37	92	79	62	41	89	82	46	57	96	85	68	48	93	86	72	53
3	6	14	16	28	2	13	29	18	5	13	26	23	6	11	31	14	4	4	26	16	5	5	23	14
4	11	22	23	12	3	6	7	6	2	8	9	4	2	7	16	2	0	9	5	12	2	9	5	6
5	1	17	40	3	0	0	2	13	2	0	3	10	3	0	7	9	0	2	1	4	0	0	0	2

predict the robot arm position from two joint angles (θ_1, θ_2) as shown in Fig. (2.a) by $(y_a,\ y_b) = (r_1 \cos\theta_1 + r_2 \cos(\theta_1 + \theta_2) + \epsilon_1,\ r_1 \sin\theta_1 + r_2 \sin(\theta_1 + \theta_2) + \epsilon_2)$, where $\epsilon_i \sim N(0, 0.05^2), i.i.d.$ The values for θ_1 are generated uniformly from the intervals [-1.932, -0.453] and [0.453, 1.932], while those for θ_2 are generated independently and uniformly from [0.534, 3.142]. As in [11], we take $r_1 = 2$ and $r_2 = 1.3$, with data generated and divided into two groups: one training set of 200 observations and a testing set of 200 observations.

The candidate size range is selected as [2,15]. After training the network for each hidden unit number k, the criteria values are calculated. Since there are two output units regressed by the hidden units and the parameter number $d(k)$ here becomes k(1+d+2)+2. For observation convenience, the criteria values have been normalized and shown together in Fig. (2.b). As shown, AIC feels confused to decide the optimal number within this specified range and tends to overestimate, while CAIC and BIC/MDL prefer to choose $k^* = 7$, which confirms with [12]. CV, BYY-$S_{0.5}$ and BYY-$S_{1.0}$ choose the result as 8.

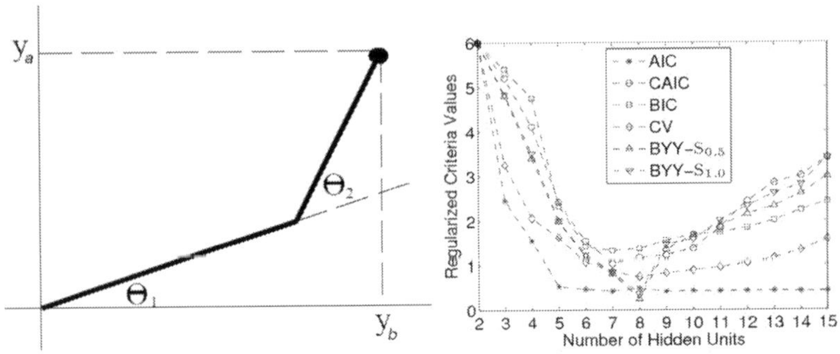

(a) *Robot Arm* (b) Criteria values

Fig. 2. *Robot Arm* and the experimental results. (a) indicates the *Robot Arm* function. (b) shows the regularized criteria values indexed by the hidden unit number.

4.3 Real World Data Sets

We further focus on four commonly used real world data sets collected from UCI Machine Learning Repository[1], including *CPU-performance*, *Housing*, *Auto-mpg*, and *Servo*. *CPU-performance* has 209 instances, 6 continuous attributes to regress 1 attribute; *Housing* has 506 instances, 12 continuous attributes and 1 binary attribute to regress 1 attribute; *Auto-mpg* has 398 cases with 5 continuous attributes and 3 multi-valued discrete attributes to regress 1 attribute; *Servo* has 167 samples with 2 continuous and 2 discrete variables to regress 1 attribute.

Each database is firstly divided evenly into ten blocks. Then p blocks are chosen to form a training data set, while the remaining $10 - p$ blocks form testing data. After training on the p blocks and testing, each criterion is calculated to make selection. In order to investigate the effect of training sample size, especially of small-sample-size cases, a series of experiments are implemented on different training/testing data blocks selection, with $p = 9, 7, 5, 3$, respectively. For example, when $p = 9$, we use 9 blocks for training and $10 - p = 1$ block for testing. The experimental results are reported in Table. 2 in two magnitudes: the *training root mean squared error* ($RMSE_t$) on the training data and the *generalization root mean squared error* ($RMSE_g$) on the testing data.

As shown in Table. 2, when using relatively large training database to model a regression in 3-LFFNN, all the criteria perform well without obvious differences. However, as the sample size decreases, AIC, CAIC and BIC turn to deteriorate quickly and greatly, while CV, BYY-$S_{0.5}$ and BYY-$S_{1.0}$ do not deteriorate that much. This may be not difficult to imagine, because the smaller the sample size is, the less guarantee there would be to make sure the rough estimation can give k^*. Compared to AIC, CAIC and BIC, CV deals well with the case when the sample size is small. Interestingly, BYY-$S_{0.5}$ and BYY-$S_{1.0}$ generally select structures with better generalization performance.

[1] http://www.ics.uci.edu/ mlearn/MLSummary.html.

Table 2. The *training root mean squared errors* ($RMSE_t$) on the training data and the *generalization root mean squared errors* ($RMSE_g$) on testing data for real world data sets. The values are expressed in the form of $RMSE_g(RMSE_t)$. Each ratio in the second column is for the corresponding training data size to testing data size, i.e. $p:(10-p)$.

		$RMSE_g(RMSE_t)$ for each criterion					
data set	ratio	AIC	CAIC	BIC/MDL	10-fold CV	BYY-$S_{0.5}$	BYY-$S_{1.0}$
CPU	9:1	52.8(41.2)	52.9(38.3)	52.4(23.4)	52.3(25.5)	52.4(25.1)	52.4(25.1)
	7:3	53.1(42.4)	52.2(41.2)	52.9(26.5)	52.7(26.9)	51.8(26.0)	52.3(26.8)
	5:5	55.5(43.7)	57.0(40.3)	55.2(27.3)	54.7(26.2)	53.4(25.5)	53.1(25.9)
	3:7	57.9(48.1)	57.7(42.1)	56.3(28.2)	55.9(26.4)	54.2(27.3)	54.2(27.6)
Housing	9:1	1.15(1.07)	1.08(1.08)	1.08(1.08)	1.09(1.11)	1.10(1.12)	1.10(1.12)
	7:3	1.18(1.16)	1.14(1.15)	1.13(1.12)	1.08(1.13)	1.11(1.09)	1.11(1.09)
	5:5	1.27(1.26)	1.34(1.44)	1.30(1.32)	1.25(1.21)	1.28(1.23)	1.26(1.24)
	3:7	1.41(1.43)	1.38(1.29)	1.39(1.28)	1.27(1.35)	1.25(1.30)	1.26(1.27)
Auto	9:1	2.95(2.46)	2.92(2.61)	2.88(2.64)	2.79(2.82)	2.94(2.53)	2.94(2.53)
	7:3	2.97(2.68)	2.89(2.82)	2.93(2.76)	2.79(2.88)	2.95(2.62)	2.99(2.64)
	5:5	3.04(2.53)	3.21(2.99)	3.15(2.82)	3.07(2.79)	2.96(2.78)	2.96(2.79)
	3:7	3.46(2.35)	3.30(2.73)	3.26(2.78)	3.18(2.84)	3.05(2.82)	3.04(2.85)
Servo	9:1	0.26(0.15)	0.24(0.15)	0.23(0.16)	0.21(0.20)	0.21(0.17)	0.21(0.17)
	7:3	0.28(0.31)	0.29(0.19)	0.23(0.21)	0.25(0.23)	0.22(0.24)	0.23(0.18)
	5:5	0.49(0.37)	0.44(0.32)	0.37(0.30)	0.34(0.26)	0.27(0.28)	0.25(0.29)
	3:7	0.47(0.33)	0.52(0.29)	0.45(0.28)	0.33(0.31)	0.31(0.24)	0.30(0.31)

5 Conclusion

This paper has described a connection between regression and 3-LFFNN, based on which we have discussed several model selection criteria to determine the 3-LFFNN's structure, i.e. the hidden unit number. After introducing the conventional model selection methods including AIC, CAIC, BIC, and CV, we briefly illustrated the BYY improved model selection criteria (BYY-S). Comparative experiments were conducted on several data sets, during which we paid more attention to the selection performances on data sets with different sample sizes and different noises. The experimental results show that BIC and CV mostly outperform AIC and CAIC. BYY-S can generally perform as well as, if not better than BIC and CV. Moreover, as the sample size decreases or the noise increases, BYY-S outperforms CV, AIC, CAIC, and BIC interestingly, providing both a desired selection of hidden unit number and a relatively more accurate regression performance.

6 Further Discussion

When sample size is large enough, all these typical criteria including AIC, CAI, BIC, CV and BYY-S do not differentiate from each other much, while leaving high computational performance in choosing among a large set of candidate models. For these cases, we suggest to use another so-called BYY harmony automatic learning approach [15], which can determine the hidden unit number automatically in parallel with the parameter learning. Compared to the two-phase implementation, it owns the time-saving advantage, especially for large-sample-size cases. To save space, this part is not discussed in detail here. Another expected further research is to compare BYY-$S_{0.5}$ with BYY-$S_{1.0}$ and to find a more accurately approximated BYY criterion.

Acknowledgement

The work described in this paper was fully supported by a grant from the Research Grant Council of the Hong Kong SAR (Project No: CUHK4173/06E).

References

1. Arai, M.: Mapping abilities of three-layer neural networks. Proc. IJCNN'89, vol.1, pp.419-423, 1989.
2. Hayasaka, T., Hagiwara, K., Toda, N. and Usui, S.: Determination of the number of hidden units from a statisticalviewpoint. Proc. ICONIP'99, vol.1, pp.240-245, 1999.
3. Rumelhart, D.E. et al.: Learning internal representations by error propagation. Parallel Distributed Processing, vol.1, MIT Press, Cambridge, MA, 1986.
4. Neal, R.M.: Bayesian learning for neural networks. Springer-Verlag, New York, 1996.
5. Akaike, H.: A new look at the statistical model identification. IEEE Trans. Automatic Control, 19, 714-723, 1974.
6. Bozdogan, H.: Model selection and Akaike's information criterion (AIC): the general theory and its analytical extensions. Psychometrika, vol.52(3), pp.345-370, 1987.
7. Schwarz, G.: Estimating the dimension of a model. Annl. of Stat., 6, 461-464, 1978.
8. Rissanen, J.: Modeling by shortest data description. Automation, 14, 456-471, 1978.
9. Moody, J.E.: The effective number of parameters: an analysis of generalization and regularization in nonlinear learning systems. Advances in NIPS, Moody, J.E. et.al eds., vol.4, pp.847-854, MIT Press, Cambridge, MA, 1992.
10. Stone, M.: Cross validation choice and assessment of statistical predictions. Journal of the Royal Statistical Society, B36, 111-147, 1974.
11. MacKay, D.J.C.: A practical bayesian framework for backpropagation networks. Neural Computation, vol.4(3), pp.448-472, 1992.
12. Brake, G.te., Kok, J.N., and Vit'anyi, P.M.B.: Model selection for neural networks: Comparing MDL and NIC. Proc. European Symposium on Artificial Neural Networks, Brussels, Belgium, 31-36, 1994.
13. Xu, L., Klasa, S., and Yuille, A.L.: Recent Advances on Techniques Static Feedforward Networks with Supervised Learning. International Journal of Neural Systems, vol.3, No.3, pp.253-290, 1992.
14. Xu, L.: BYY learning, regularized implementation, and model selection on modular networks with one hidden layer of binary units. Neurocomputing, vol.51, pp.277-301, 2003.
15. Xu, L.: Advances on BYY harmony learning: information theoretic perspective, generalized projection geometry, and independent factor auto-determination. IEEE Trans. Neural Networks, vol.15(5), pp.885-902, 2004.
16. Xu, L.: Trends on Regularization and Model Selection in Statistical Learning: A Perspective from Bayesian Ying Yang Learning. Challenges to Computational Intelligence (in press), Duch, W., Mandziuk, J. and Zurada, J.M. eds, the Springers series - Studies in Computational Intelligence, Springer-Verlag, 2006.

On-Line Learning with Structural Adaptation in a Network of Spiking Neurons for Visual Pattern Recognition

Simei Gomes Wysoski, Lubica Benuskova, and Nikola Kasabov

Knowledge Engineering and Discovery Research Institute,
Auckland University of Technology, 581-585 Great South Rd,
Auckland, New Zealand
{swysoski, lbenusko, nkasabov}@aut.ac.nz
http:www.kedri.info

Abstract. This paper presents an on-line training procedure for a hierarchical neural network of integrate-and-fire neurons. The training is done through synaptic plasticity and changes in the network structure. Event driven computation optimizes processing speed in order to simulate networks with large number of neurons. The training procedure is applied to the face recognition task. Preliminary experiments on a public available face image dataset show the same performance as the optimized off-line method. A comparison with other classical methods of face recognition demonstrates the properties of the system.

1 Introduction

The human brain has been modelled in numerous ways, but these models are far from reaching comparable performance. These models are still not as general and accurate as the human brain despite that outstanding performances have been reported [1] [2] [3]. Of particular interest to this research are the models for visual pattern recognition. Visual pattern recognition models can be divided in two groups according to the connectionist technique applied. Most of the works deal with the visual pattern recognition using neural networks comprised of linear/non-linear processing elements based on the neural rate-based code [4] [5]. Here we refer to these methods as traditional methods. In another direction, a visual pattern recognition system can be constructed through the use of brain-like neural networks.

Brain-like neural networks are networks that have a closer association with what is known about the way brains process information. The definition of brain-like networks is intrinsically associated with the computation of neuronal units that use pulses. The use of pulses brings together the definitions of time varying postsynaptic potential *(PSP)*, firing threshold (ϑ), and spike latencies (Δ), as depicted in Figure 1 [6]. Brain-like neural networks, despite being more biologically accurate, have been considered too complex and cumbersome for modeling the proposed task. However recent discoveries on the information processing capabilities of the brain and technical advances related to massive parallel processing, are bringing back the idea of using biologically realistic networks for pattern recognition. A recent pioneering work has shown that the primate (including human) visual system can analyze complex

S. Kollias et al. (Eds.): ICANN 2006, Part I, LNCS 4131, pp. 61 – 70, 2006.

natural scenes in only about 100-150 ms [7]. This time period for information processing is very impressive considering that billions of neurons are involved. This theory suggests that probably neurons, exchanging only one or few spikes, are able to form assemblies, and process information. As an output of this work, the authors proposed a multi-layer feed-forward network (SpikeNet) of integrate-and-fire neurons that can successfully track and recognize faces in real time [7].

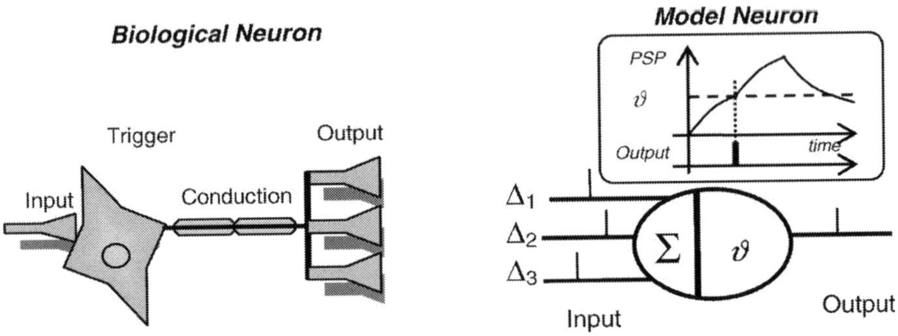

Fig. 1. On the left: Representation of biological neuron. On the right: Basic artificial unit (spiking neuron).

This paper intends to review the network model SpikeNet proposed in [8] and extend its applicability to perform on-line learning. In the next sections the spiking neural network model will be presented and the new learning procedure will be described. The new learning method is applied to the face recognition task. The results are compared with previous work and other models. Discussion and additional required analysis concludes the paper.

2 Spiking Network Model

In this section we describe the steps of the biologically realistic model used in this work to perform on-line visual pattern recognition. The system has been implemented based on the SpikeNet introduced in [7] [8] [9] [10]. The neural network is composed of 3 layers of integrate-and-fire neurons. The neurons have a latency of firing that depends upon the order of spikes received. Each neuron acts as a coincidence detection unit, where the postsynaptic potential for neuron i at a time t is calculated as:

$$PSP(i,t) = \sum \mathrm{mod}^{order(j)} w_{j,i} \quad (1)$$

where $\mathrm{mod} \in (0,1)$ is the modulation factor, j is the index for the incoming connection and $w_{j,i}$ is the corresponding synaptic weight. See [7] [9] for more details.

Each layer is composed of neurons that are grouped in two-dimensional grids forming neuronal maps. Connections between layers are purely feed-forward and each neuron can spike at most once on spikes arrival in the input synapses. The first layer cells represent the ON and OFF cells of retina, basically enhancing the high contrast parts of a given image (high pass filter). The output values of the first layer are encoded to pulses in the time domain. High output values of the first layer are encoded as pulses with short time delays while long delays are given to low output values. This technique is called Rank Order Coding [10] and basically prioritizes the pixels with high contrast that consequently are processed first and have a higher impact on neurons' PSP.

Second layer is composed of eight orientation maps, each one selective to a different direction (0°, 45°, 90°, 135°, 180°, 225°, 270°, and 315°). It is important to notice that in the first two layers there is no learning, in such a way that the structure can be considered simply passive filters and time domain encoders (layers 1 and 2). The theory of contrast cells and direction selective cells was first reported by Hubel and Wiesel [11]. In their experiments they were able to distinguish some types of cells that have different neurobiological responses according to the pattern of light stimulus.

The third layer is where the learning takes place and where the main contribution of this work is presented. Maps in the third layer are to be trained to represent classes of inputs. See Figure 2 for the complete network architecture. In [7], the network has a fixed structure and the learning is done off-line using the rule:

$$\Delta w_{j,i} = \frac{\mathrm{mod}^{order(a_j)}}{N} \tag{2}$$

where $w_{j,i}$ is the weight between neuron j of the 2nd layer and neuron i of the 3rd layer, mod $\in$ (0,1) is the modulation factor, $order(a_j)$ is the order of arrival of spike from neuron j to neuron i, and N is the number of samples used for training a given class.

In this rule, there are two points to be highlighted: a) the number of samples to be trained needs to be known *a priori*; and b) after training, a map of a class will be selective to the average pattern.

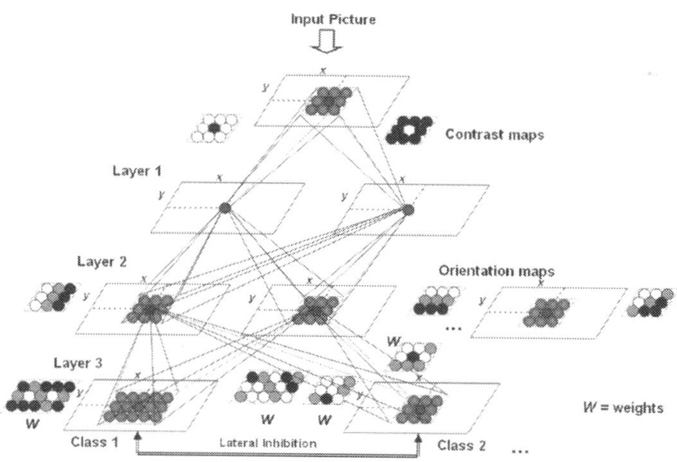

Fig. 2. Adaptive spiking neural network (aSNN) architecture for visual pattern recognition

There are also inhibitory connections among neuronal maps in the third layer, so that when a neuron fires in a certain map, other maps receive inhibitory pulses in an area centred in the same spatial position. An input pattern belongs to a certain class if a neuron in the corresponding neuronal map spikes first.

One of the properties of this system is the low activity of the neurons. It means that the system has a large number of neurons, but only few take active part during the retrieval process. In this sense, through the event driven approach the computational performance can be optimized [8] [12]. Additionally, in most cases the processing can be interrupted before the entire simulation is completed. Once a single neuron of the output layer reaches the threshold to emit a spike the simulation can be finished. The event driven approach and the early simulation interruption make this method suitable for implementations in real time.

3 On-Line Learning and Structural Adaptation

3.1 General Description

Our new approach for learning with structural adaptation aims to give more flexibility to the system in a scenario where the number of classes and/or class instances is not known at the time the training starts. Thus, the output neuronal maps need to be created, updated or even deleted on-line, as the learning occurs. In [13] a framework to deal with adaptive problems is proposed and several methods and procedures describing adaptive systems are presented.

To implement such a system the learning rule needs to be independent of the total number of samples since the number of samples is not known when the learning starts. Thus, in the next section we propose to use a modified equation to update the weights based on the average of the incoming patterns. It is important to notice that, similarly to the batch learning implementation of Equation 2, the outcome is the average pattern. However, the new equation calculates the average dynamically as the input patterns arrive.

There is a classical drawback to learning methods when, after training, the system responds optimally to the average pattern of the training samples. The average does not provide a good representation of a class in cases where patterns have high variance (see Figure 3). A traditional way to attenuate the problem is the *divide-and-conquer* procedure. We implement this procedure through the structural modification

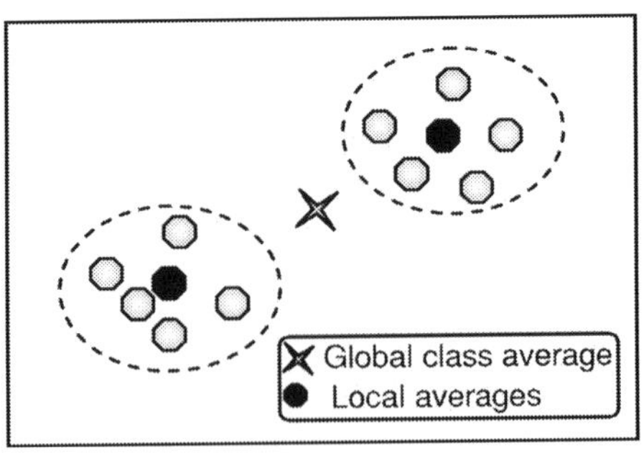

Fig. 3. *Divide and conquer* procedure to deal with high intra class variability of patterns in the hypothetical space of class *K*. The use of multiple maps that respond optimally to the average of a subset of patterns provides a better representation of the classes.

of the network during the training stage. More specifically, we integrate into the training algorithm a simple clustering procedure: patterns within a class that comply with a similarity criterion are merged into the same neuronal map. If the similarity criterion is not fulfilled, a new map is generated. The entire training procedure follows 4 steps described in the next section and is summarized in the flowchart of Figure 4.

3.2 Learning Procedure

The new learning procedure can be described in 4 sequential steps:

1. Propagate a sample *k* of class *K* for training into the layer 1 (retina) and layer 2 (direction selective cells – DSC);
2. Create a new map $Map_{C(k)}$ in layer 3 for sample *k* and train the weights using the equation:

$$\Delta w_{j,i} = \text{mod}^{order(a_j)} \tag{3}$$

   where $w_{j,i}$ is the weight between neuron *j* of the layer 2 and neuron *i* of the layer 3, $\text{mod} \in (0,1)$ is the modulation factor, $order(a_j)$ is the order of arrival of spike from neuron *j* to neuron *i*.

   The postsynaptic threshold ($PSP_{threshold}$) of the neurons in the map is calculated as a proportion $c \in [0,1]$ of the maximum postsynaptic potential (*PSP*) created in a neuron of map $Map_{C(k)}$ with the propagation of the training sample into the updated weights, such that:

$$PSP_{threshold} = c\max(PSP) \tag{4}$$

   The constant of proportionality *c* express how similar a pattern needs to be to trigger an output spike. Thus, *c* is a parameter to be optimized in order to satisfy the requirements in terms of false acceptance rate (FAR) and false rejection rate (FRR).
3. Calculate the similarity between the newly created map $Map_{C(k)}$ and other maps belonging to the same class $Map_{C(K)}$. The similarity is computed as the inverse of the Euclidean distance between weight matrices.
4. If one of the existing maps for class *K* has similarity greater than a chosen threshold $Th_{simC(K)}>0$, merge the maps $Map_{C(k)}$ and $Map_{C(Ksimilar)}$ using arithmetic average as expressed in equation:

$$W = \frac{W_{Map_{C(k)}} + N_{samples} W_{Map_{C(Ksimilar)}}}{1 + N_{samples}} \tag{5}$$

   where matrix W represents the weights of the merged map and $N_{samples}$ denotes the number of samples that have already being used to train the respective map. In similar fashion the $PSP_{threshold}$ is updated:

$$PSP_{threshold} = \frac{PSP_{Map_{C(k)}} + N_{samples} PSP_{Map_{C(Ksimilar)}}}{1 + N_{samples}} \tag{6}$$

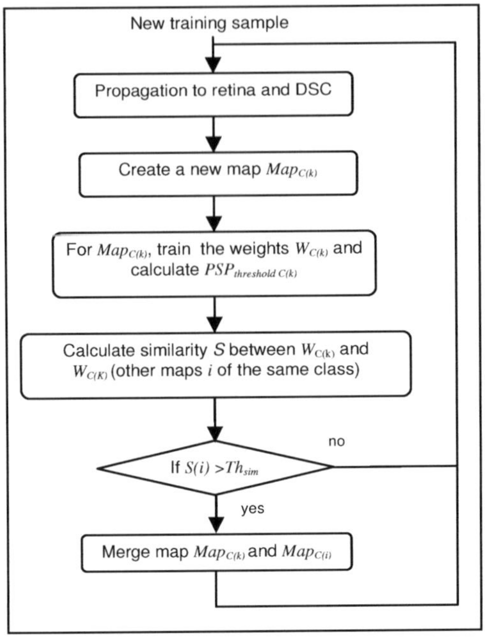

Fig. 4. On-line learning procedure flowchart

4 Experiments and Results

We have implemented the learning procedure proposed in the previous section in a network of spiking neurons as described in section 2. To evaluate the performance and compare with previous work, we used the same dataset as in [7], which is available from [14]. The dataset is composed of 400 faces taken from 40 different people. The frontal views of faces are taken in rotation angles varying in the range of [-30°, 30°].

4.1 Image Preparation

We manually annotated the position of eyes and mouth and used it to centralize the face images. The faces were rotated to align the right and left eyes horizontally. The boundaries of our region of interest (ROI) were then defined as a function of the inter-ocular distance and the distance between the eyes and mouth. The ROI is then normalized to the size 20 x 30 pixels in greyscale. The 2 dimensional array obtained has been used as input to the SNN. No contrast or illumination manipulation has been performed as previous work demonstrated the good response of the network under the presence of noise and illumination changes [7].

4.2 Spiking Network Parameters

The neuronal maps of retina, DSC and output maps have size of 20 x 30. The number of time steps used to encode the output of retina cells to the time domain is set to 100.

The threshold for the direction selective cells is set to 600, chosen in such a way that on average only 20% of neurons emits output spikes. The modulation factor mod ∈ (0, 1) is set to 0.98. In this way the efficiency of the input of a given neuron is reduced to 50% when 50% of the inputs get a spike. The retina filters are implemented using a 5 x 5 Gaussian grid and direction selective filters are implemented using Gabor functions in a 7x7 grid. All these parameters were not optimized. Rather, we tried to reproduce as close as possible the scenario described in [7] for comparison purposes.

4.3 Results

Previous work demonstrated the high accuracy of the network to cope with noise, contrast and luminance changes, reaching 100% in the training set (10 samples for each class) and 97.5% when testing the generalization properties [7]. For the generalization experiment, the dataset was divided in 8 samples for training and the remaining 2 for test. With the adaptive learning method proposed here, we have obtained similar results for the training set.

In another experiment, to test the system ability to add on-line output maps for better generalization, we used only 3 sample images from each person for training. The remaining 7 views of each person were used for test. Among the dataset faces, we chose manually those samples taken from different angles that appeared to be most dissimilar. Thus, the training set was composed mostly of one face view taken from the left side (30°), one frontal view and one face view taken from the right side (-30°), as depicted in Figure 5. The results are shown in Table 1. In column 2 of Table 1, Th_{sim} is set in such a way that only one output map for each class is created. In such condition, the on-line learning procedure becomes equivalent to the original off-line learning procedure described by Equation 2. Tuning of Th_{sim} for performance, it can be clearly seen the advantage of using more maps to represent classes that contain highly variant samples, as the accuracy of face recognition increases by 6% with a reduction on the FAR.

In Table 2 and Table 3 is presented the network performance for different values of PSP threshold that are calculated as a function of the proportionality constant c. In all the experiments the constant c is the same for all maps and chosen prior to the training start. In a batch mode operation the value of c can be optimized independently for each map after the training is completed using, e.g., Genetic Algorithms (GA).

Fig. 5. Example of image samples used for training (30°, frontal and -30°)

Table 1. Results for the test set according to different similarity thresholds Th_{sim}.Three pictures of each class are used for training and the remaining seven for test.

Similarity threshold Th_{sim} (x10^{-3})	0.5	0.833	1.0	1.25	2
Number of output maps	40	47	80	109	120
Accuracy (%)	74.28	77.49	78.57	80.00	80.00
False Acceptance Rate (FAR) (%)	2.32	2.20	2.18	2.26	1.77
False Rejection Rate (FRR) (%)	0.00	0.00	0.00	0.00	0.00

Table 2. Accuracy for different values of *c* keeping Th_{sim} = 0.5x10-3. Output maps = 40

PSP threshold	*c* = 0.30	*c* = 0.35	*c* = 0.40	*c* = 0.45
Accuracy (%)	72.14	74.28	73.21	71.43
False Acceptance Rate (FAR) (%)	3.10	2.32	1.58	1.25
False Rejection Rate (FRR) (%)	0.00	0.00	0.00	2.50

Table 3. Accuracy for different values of *c* keeping Th_{sim} = 2.0x10-3. Output maps = 120

PSP threshold	*c* = 0.30	*c* = 0.35	*c* = 0.40	*c* = 0.45
Accuracy (%)	75.00	78.57	80.00	80.00
False Acceptance Rate (FAR) (%)	2.95	2.49	1.77	1.06
False Rejection Rate (FRR) (%)	0.00	0.00	0.00	3.57

In another comparison, to check how difficult the dataset is and to have a better idea of the performance of our learning algorithm, we compare the face recognition system using adaptive SNN (aSNN) with other three traditional methods of face recognition (Table 4). In these methods, PCA (principal component analysis) is used to extract facial features. The classification is done using SVM (support vector machine), MLP (multi layer perceptron) neural network and ECF (evolving classifier function). MLP and SVM are batch mode methods while ECF present similar adaptive learning characteristics as proposed in this work. ECF can be trained in both one-pass and recursive mode (several epochs)[13]. As expected, the batch mode algorithms over performed the one-pass on-line methods. The reason is that in the batch mode, the training samples are recursively presented to the classification method to minimize the output errors. In the one-pass on-line learning the adjustment of weights occurs only once at the time the training samples are presented to the network. Therefore, the performance of the batch methods can be considered roughly the target or the maximum accuracy that be reached. When comparing both one-pass online methods, the adaptive SNN presented better performance than ECF. Notice that, in this comparison we can not detect if the better performance is due to the learning method or to the different representation of the features.

Table 4. Comparison among different methods of face recognition (experiments using NeuCom [15])

Method	Accuracy (%)	Properties
PCA + SVM	90.7	Batch mode
PCA + MLP	89.6	Batch mode
PCA + ECF	74.0 (120 nodes)	One-epoch on-line method
Adaptive SNN	80.0 (109 maps)	One-pass on-line method

5 Discussion and Conclusion

A simple procedure to perform on-line learning in a network of spiking neurons has been presented. During learning, new output maps are created and merged based on the clustering of intra-class samples. Preliminary experiments have shown that the learning procedure reaches similar levels of performance of the previously presented work, and better performance can be reached in classes where samples have high variability. As a price, one more parameter needs to be tuned, e.g. Th_{sim}. In addition, more output maps require more storage memory.

In terms of normalization, the rank order codes are intrinsically invariant to changes in contrast and input intensities, basically because the neuronal units compute the order of the incoming spikes and not the latencies itself [7]. This can be a reason why adaptive SNN present better result than PCA+ECF as the feature extraction using PCA can degrade performance with illumination changes.

The adaptive SNN doesn't cope well with patterns rotation. In all the experiments presented in this work we aligned the samples in the image preparation stage. Alternatively, a certain degree of rotation invariance can be reached with the use of additional neuronal maps, in which each map need to be trained to cover different angles. In this case, the learning procedure described here, can automatically generate the new maps when it's required.

With respect to the overall system, the computation with pulses, contrast filters and orientation selective cells finds a close correspondence with traditional ways of image processing such as wavelets and Gabor filters [16] that already have proven to be very robust for feature extraction in visual pattern recognition problems. From the biological perspective, despite still being a very simplified representation of what effectively happens in the brain, the use of pulses is a starting point.

In our future work, aiming to improve the use of biologically realistic neural networks for pattern recognition, we intend to add adaptation to layer 1 and layer 2. It has been experimentally proven [17] that neural filters adaptively change to increase the information carried by the neural response. As a result, the contrast and direction selective cells are optimized filters to describe natural scenes. We intend to explore how to adaptively obtain optimal filters in different types of data.

Acknowledgments

The work has been supported by the NERF grant X0201 funded by FRST (L.B., N.K.) and by the Tertiary Education Commission of New Zealand (S.G.W.).

References

1. Fukushima, K.: Active Vision: Neural Network Models. In Amari, S., Kasabov, N. (eds.): Brain-like Computing and Intelligent Information Systems. Springer-Verlag (1997)
2. Mel, B. W.: SEEMORE: Combining colour, shape, and texture histrogramming in a neurally-inspired approach to visual object recognition. Neural Computation 9 (1998) 777-804

3. Wiskott, L., Fellous, J. M., Krueuger, N., von der Malsburg, C.: Face Recognition by Elastic Bunch Graph Matching: In Jain, L.C. et al. (eds.): Intelligent Biometric Techniques in Fingerprint and Face Recognition. CRC Press (1999) 355-396
4. Haykin, S.: Neural Networks - A Comprehensive Foundation. Prentice Hall (1999)
5. Bishop, C.: Neural Networks for Pattern Recognition. University Press, Oxford New York (2000)
6. Gerstner, W., Kistler, W. M.: Spiking Neuron Models. Cambridge Univ. Press, Cambridge MA (2002)
7. Delorme, A., Thorpe, S.: Face identification using one spike per neuron: resistance to image degradation. Neural Networks, Vol. 14. (2001) 795-803
8. Delorme, A., Gautrais, J., van Rullen, R., Thorpe, S.: SpikeNet: a simulator for modeling large networks of integrate and fire neurons. Neurocomputing, Vol. 26-27. (1999) 989-996
9. Delorme, A., Perrinet, L., Thorpe, S.: Networks of integrate-and-fire neurons using Rank Order Coding. Neurocomputing. (2001) 38-48
10. Thorpe, S., Gaustrais, J.: Rank Order Coding. In: Bower, J. (ed.): Computational Neuroscience: Trends in Research. Plenum Press, New York (1998)
11. Hubel, D.H., Wiesel, T.N.: Receptive fields, binocular interaction and functional architecture in the cat's visual cortex. J. Physiol, 160 (1962) 106-154
12. Mattia, M., del Giudice, P.: Efficient Event-Driven Simulation of Large Networks of Spiking Neurons and Dynamical Synapses. Neural Computation, Vol. 12 (10). (2000) 2305-2329
13. Kasabov, N.: Evolving Connectionist Systems: Methods and Applications in Bioinformatics, Brain Study and Intelligent Machines. Springer-Verlag (2002)
14. http://www.cl.cam.ac.uk/Research/DTG/attarchive/facedatabase.html
15. http://www.aut.ac.nz/research/research_institutes/kedri/research_centres/centre_for_novel_methods_of_computational_intelligence/neucom.htm
16. Sonka, M., Hlavac, V., Boyle, R.: Image Processing, Analysis, and Machine Vision, 2nd edn. (1998)
17. Sharpee, T. *et al.*: Adaptive filtering enhances information transmission in visual cortex. Nature, Vol. 439 (2006) 936-942

Learning Long Term Dependencies with Recurrent Neural Networks

Anton Maximilian Schäfer[1,2], Steffen Udluft[1], and Hans Georg Zimmermann[1]

[1] Information & Communications, Learning Systems
Siemens AG, Corporate Technology, 81739 Munich, Germany
{Schaefer.Anton.ext, Steffen.Udluft, Hans_Georg.Zimmermann}@siemens.com
[2] Department Optimisation and Operations Research, University of Ulm, 89069 Ulm, Germany

Abstract. Recurrent neural networks (RNNs) unfolded in time are in theory able to map any open dynamical system. Still they are often blamed to be unable to identify long-term dependencies in the data. Especially when they are trained with backpropagation through time (BPTT) it is claimed that RNNs unfolded in time fail to learn inter-temporal influences more than ten time steps apart.

This paper provides a disproof of this often cited statement. We show that RNNs and especially normalised recurrent neural networks (NRNNs) unfolded in time are indeed very capable of learning time lags of at least a hundred time steps. We further demonstrate that the problem of a vanishing gradient does not apply to these networks.

1 Introduction

Recurrent neural networks (RNNs) allow the identification of dynamical systems in form of high dimensional, nonlinear state space models [1,2]. They offer an explicit modeling of time and memory and allow in principle to model any type of open dynamical system [3]. The basic concept is more than 20 years old, so e.g., unfolding in time of neural networks and related modifications of the backpropagation algorithm can already be found in [4].

Nevertheless, there is often a negative attitude towards RNNs because it has been claimed by several authors that RNNs unfolded in time are unable to identify and learn long-term dependencies of more than ten time steps [5,6,7]. To overcome the stated dilemma new forms of recurrent neural networks, e.g., long short-term memory (LSTM) networks [8], were developed, but these networks do not offer the desirable correspondence between equations and architectures as RNNs unfolded in time do.

Still, the analyses in the mentioned papers [5,6,7] were all based on a very basic architecture of RNNs and, even more important, made from a static perspective. In this paper we therefore disprove the statement that RNNs unfolded in time and trained with backpropagation through time (BPTT) are in general unable to learn long-term dependencies. We outline that RNNs and especially normalised recurrent neural networks (NRNNs) unfolded in time have no difficulty with an identification and learning of past-time information within the data which is more than ten time steps apart. Furthermore we show that by using shared weights training of these networks is not a major problem.

S. Kollias et al. (Eds.): ICANN 2006, Part I, LNCS 4131, pp. 71–80, 2006.

It even helps to overcome the problem of a vanishing gradient as the networks possess a self-regularisation characteristic which adapts the error information flow.

We start with a recapitulation of the basic RNN unfolded in time (sec. 2). Here we especially emphasise the advantage of overshooting and point out that this simple extension regularises the learning with BPTT. We further enhance the basic RNN architecture so that it only possesses one single (high-dimensional) transition matrix. This so called normalised recurrent neural network (NRNN) increases the stability of the learning process (sec. 3). In section 4 we then demonstrate that both NRNN and RNN successfully learn long-term dependencies. In doing an analysis of the backpropagated error flow we finally show that the problem of a vanishing gradient is not a relevant question for both networks. In section 5 we give a conclusion and an outlook on further research.

2 Recurrent Neural Networks Unfolded in Time

The basic time-delay recurrent neural network (RNN) consists of a state transition and an output equation [1,9]:

$$\begin{aligned} s_{t+1} &= \tanh(As_t + c + Bu_t) && \text{state transition} \\ y_t &= Cs_t && \text{output equation} \end{aligned} \tag{1}$$

Here, the state transition equation s_{t+1} ($t = 1, \ldots, T$ where T is the number of available patterns) is a nonlinear combination of the previous state s_t and external influences u_t using weight matrices A and B of appropriate dimension and a bias c, which handles offsets in the input variables u_t. The network output y_t is computed from the present state s_t employing matrix C. It is therefore a nonlinear composition applying the transformations A, B, and C.

Training the RNN of equation 1 is equivalent to solving a parameter optimisation problem, i.e., minimising the error between the network output y_t and the real data y_t^d with respect to an arbitrary error measure, e.g.:

$$\sum_{t=1}^{T} \left(y_t - y_t^d\right)^2 \to \min_{A,B,C,c} \tag{2}$$

It can be solved by finite unfolding in time using shared weight matrices A, B, and C [1,4]. Shared weights share the same memory for storing their weights, i.e., the weight values are the same at each time step of the unfolding and for every pattern $t \in \{1, \ldots, T\}$ [1,4]. This guarantees that we have the same dynamics in every time step. By using unfolding in time the RNN can be trained with error backpropagation through time (BPTT) [1,4], which is a shared weights extension of the standard backpropagation algorithm [10]. Figure 1 depicts the resulting spatial neural network architecture [9].

We extend the autonomous part of the RNN into the future by so-called overshooting [9], i.e., we iterate matrices A and C in future direction (see fig. 1). In doing so we get a sequence of forecasts as an output. More important, overshooting forces the learning to focus on modeling the autonomous dynamics of the network, i.e., it supports the extraction of useful information from input vectors which are more distant to the output.

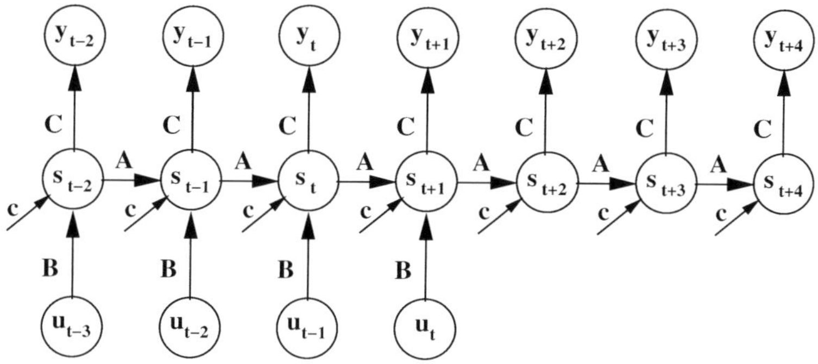

Fig. 1. RNN unfolded in time using overshooting

Consequently overshooting is a very simple remedy to the problem that the backpropagation algorithm usually tries to model the relationship between an output and its most recent inputs because the fastest adaptation takes place in the shortest path [5]. Therefore also the learning of false causalities is decreased. Hence, overshooting regularises the learning and thus improves the model's performance [9]. Note, that due to shared weights no additional parameters are used.

3 Normalised Recurrent Neural Networks

As a preparation for the development of normalised recurrent neural networks (NRNNs) [11] we first separate the state equation of the basic time-delay RNN (eq. 1) into a past and a future part. In this framework s_t is always regarded as the present time state. That means that for this pattern t all states s_τ with $\tau \leq t$ belong to the past part and those with $\tau > t$ to the future part. The parameter τ is thereby always bounded by the length of the unfolding in time m and the length of the overshooting n [9], such that we have $\tau \in \{t-m, \ldots, t+n\}$ for all $t \in \{m, \ldots, T-n\}$. The present time ($\tau = t$) is included in the past part, as these state transitions share the same characteristics. We get the following representation of the optimisation problem:

$$
\begin{aligned}
\tau \leq t: \quad & s_{\tau+1} = \tanh(As_\tau + c + Bu_\tau) \\
\tau > t: \quad & s_{\tau+1} = \tanh(As_\tau + c) \\
& y_\tau \quad = Cs_\tau \\
& \sum_{t=m}^{T-n} \sum_{\tau=t-m}^{t+n} (y_\tau - y_\tau^d)^2 \rightarrow \min_{A,B,C,c}
\end{aligned} \tag{3}
$$

In this model, past and future iterations are consistent under the assumption of a constant future environment. Still, the difficulty with this kind of RNN is the training with BPTT, because a sequence of different connectors has to be balanced. The gradient computation is not regular, i.e., we do not have the same learning behavior for the weight matrices in the different time steps. In our experiments we found, that this problem becomes more important for the training of large RNN. Even the training itself is unstable due to the concatenated matrices A, B, and C. As the training changes weights in all of these matrices, different effects or tendencies, even opposing ones, can influence them and may superpose. This implies, that there results no clear learning direction or change of weights from a certain backpropagated error [11].

NRNNs (eq. 4) avoid the stability and learning problems resulting from the concatenation of the three matrices A, B, and C because they incorporate besides the bias c only one connector type, a single transition matrix A:

$$\begin{aligned} \tau \leq t: \quad & s_\tau = \tanh(As_{\tau-1} + c + \begin{bmatrix} 0 \\ 0 \\ \text{Id} \end{bmatrix} u_\tau) \\ \tau > t: \quad & s_\tau = \tanh(As_{\tau-1} + c) \\ & y_\tau = [\text{Id } 0 \ 0] s_\tau \\ & \sum_{t=m}^{T-n} \sum_{\tau=t-m}^{t+n} (y_\tau - y_\tau^d)^2 \to \min_{A,c} \end{aligned} \tag{4}$$

The corresponding architecture is depicted in figure 2.

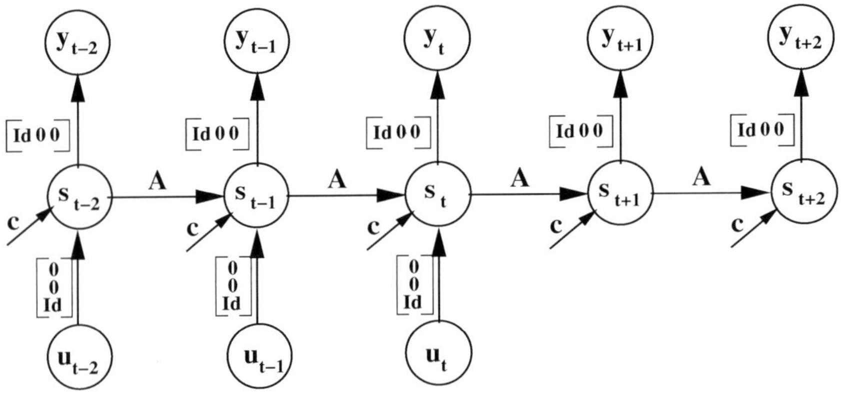

Fig. 2. Normalised recurrent neural network

Using NRNN modeling is solely focused on the transition matrix A. The matrices between input and hidden as well as hidden and output layer are fixed and therefore not changed during the training process. Consequently matrix A does not only code the autonomous and the externally driven parts of the dynamics, but also the (pre-)processing of the external inputs u_τ and the computation of the network outputs y_τ. This implies that all free parameters, as they are combined in one matrix, are now treated the same way by BPTT.

At first view it seems, that in the network architecture (fig. 2) the external input u_τ is directly connected to the corresponding output y_τ. This is not the case, because we enlarge the dimension of the internal state s_τ, such that the input u_τ has no direct influence on the output y_τ. Assuming that we have a number of p outputs, q computational hidden neurons and r external inputs, the dimension of the internal state would be $\dim(s) = p + q + r$. With the matrix [Id 0 0] we connect only the first p neurons of the internal state s_τ to the output layer y_τ. As this connector is not trained, it can be seen as a fixed identity matrix of appropriate size. Consequently, the NRNN is forced to generate its p outputs at the first p components of the state vector s_τ. The last state neurons are used for the processing of the external inputs u_τ. The connector $[0\ 0\ \mathrm{Id}]^T$ between the externals u_τ and the internal state s_τ is an appropriately sized fixed identity matrix. More precisely, the connector is designed such that the input u_τ is connected to the last r state neurons. To additionally support the internal processing and to increase the network's computational power, we add a number of q hidden neurons between the first p and the last r state neurons. This composition ensures, that the input and output processing of the network is separated but implies that NRNNs can only be designed as large neural networks [11].

Our experiments indicate that NRNNs show, in comparison to RNNs, a more stable training process, even if the dimension of the internal state is very large.

4 Learning Long-Term Dependencies

We use a very simple but well-known problem to demonstrate the ability of learning long-term dependencies of RNNs and NRNNs. Similar problems have already been studied in [5] and [8]. In both papers the performance of RNNs trained with BPTT has been tested to be unsatisfactory and the authors concluded that RNNs are not suited for the learning of long-term dependencies.

We created time series of 10000 values which are uniformly distributed on an interval $[-r, r]$ with $r \in \mathbb{R}$ and $0 < r < 1$. Every d-th value, with $d \in \mathbb{N}$ is 1. These are the only predictable values for the network. Consequently, for a successful solution to the problem the network has to remember the occurrence of the last 1, d-time steps afore in the time series data. In other words, it has to be able to learn long-term dependencies. The higher d the longer memory is necessary. We used the first 5000 data points for training and left the other half for generalisation.

4.1 Model Description

We applied an RNN (sec. 2) and an NRNN (sec. 3) with one input neuron per time step in the past and one output neuron per time step in the future. In contrast to the descriptions in sections 2 and 3 we did not implement any outputs in the past part of the networks, as those would not help to solve the problem. This implies that the gradient information of the error function has to be propagated back from the future outputs to all past time steps. It also avoids a superposition of the long-term gradient information with a local error flow in the past. Therefore the omission of outputs in the past also eases the analysis of the error backflow.

The networks were both unfolded a hundred time steps into the past. Whereas the NRNN was unfolded twenty time steps into future direction, we did not implement any overshooting for the RNN. In doing so we kept the RNN as simple as possible to show that even such a basic RNN is able to learn long-term dependencies. The total unfolding therefore amounts to 101 time steps for the RNN and to 120 steps for the NRNN. The dimension of the internal state matrix A is always set to 100, which is equivalent to the amount of past unfolding. We initialised the weights randomly with a uniform distribution on $[-0.2, 0.2]$. In all hidden units we implemented the hyperbolic tangent as activation function. We further used the quadratic error function

$$E := \sum_{t=m}^{T-n} \sum_{\tau=t-m}^{t+n} (y_\tau - y_\tau^d)^2 \tag{5}$$

to minimise the difference between network output and target (eqs. 4 and 3). The networks were trained with BPTT in combination with pattern-by-pattern learning [12]. The learning rate η was set to 10^{-4}.

4.2 Results

Table 1 summarises our results for different time gaps d and several noise ranges r. The error limit shows the optimal achievable error for the given problem plus a 10% tolerance. It is calculated by the variance of the uniform distribution given a certain noise range r and assuming no error for the time indicators in every d-th time step. We give the average number of epochs RNN and NRNN needed to pass this error limit, i.e., the number of learning epochs necessary to solve the problem with a maximum of a 10% error tolerance.

Table 1. Results for different time gaps d and noise ranges r

time gap d	range r	Error limit	# Epochs RNN	# Epochs NRNN
40	0.1	0.003575	19	13
40	0.2	0.0143	19	9
40	0.4	0.0572	50	28
60	0.1	0.00361	39	33
60	0.2	0.01442	437	23
60	0.4	0.05769	389	248
100	0.1	0.00363	65	106
100	0.2	0.01452	353	59
100	0.4	0.05808	96	84

The results demonstrate the ability of NRNNs as well as of basic RNNs to learn long-term dependencies of $d = 40$, 60 and even 100 which is obviously more than the often cited limit of ten time steps [7]. After only a small number of learning epochs both networks were able to solve the problem. Still, in comparison to the RNN, the NRNN in general showed a more stable learning behaviour and needed in most cases slightly shorter to identify the data structure.

As expected, a longer gap d resulted in more learning epochs, the networks needed to succeed. Also a higher noise range r, i.e., a larger uniform distribution of the data, made it more challenging for the networks to identify the time indicators. Still, even in more difficult settings, RNN and NRNN captured the structure of the problem very quickly.

Using smaller dimensions for the single transition matrix A increased the number of epochs necessary to learn the problem (fig. 3). This is probably due to the fact that the network needs a certain dimension to store long-term information. So e.g., with a hundred dimensional matrix the network can easily store a time gap of $d = 100$ in form of a simple shift register. Downsizing the dimension forces the network to build up more complicated internal matrix structures which take more learning epochs to develop.

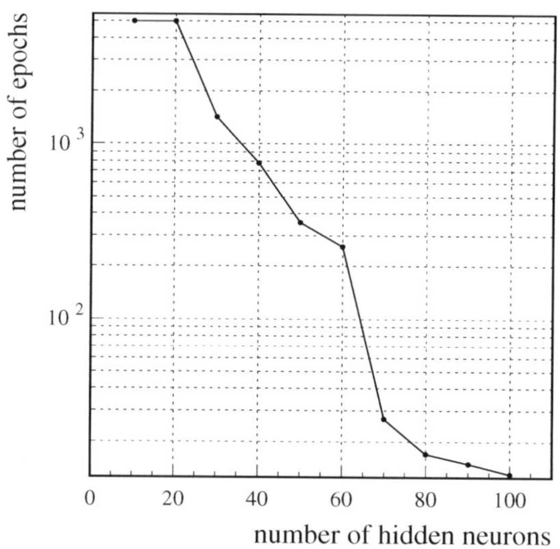

Fig. 3. Number of epochs needed by an NRNN to solve the problem with $d = 40$ and $r = 0.1$ using different numbers of hidden, i.e. internal state, neurons. We stopped training after 5000 epochs which implies that the network was not able to solve the problem for $dim(s) \leq 20$.

4.3 Analysis of the Backpropagated Error

To put the claim of a vanishing gradient in RNNs unfolded in time and trained with BPTT [7] into perspective we analysed the backpropagated error within our networks. We noticed that under certain conditions vanishing gradients do indeed occur, but are only a problem if we put a static view on the networks like it has been done in [5,7]. Studying the development of the error flow during the learning process we observed that the networks themselves have a regularising effect, i.e., they are able to prolong their information flow and consequently solve the problem of a vanishing gradient. We see two main reasons for this self-regularisation behaviour: shared-weights and overshooting (sec. 2). Whereas shared weights constrain the networks to change weights (concurrently) in every unfolded time step, overshooting forces the networks to focus on the autonomous sub-dynamics. Especially the former allows the networks to adapt the gradient information flow.

Similar to the analysis in [5] and [7] we further confirmed that the occurrence of a vanishing gradient is dependent on the values of the weight matrix A. By initialising matrix A with different weight values it turned out, that an initialisation with a uniform distribution in $[-0.2, 0.2]$ is a good choice for our networks (sec. 4.1). We never experienced any vanishing gradient in these cases. In contrary, when initialising the networks only within $[-0.1, 0.1]$, the gradient vanished in the beginning of the learning procedure. Nevertheless, during the learning process the networks themselves solved this problem by changing the weight values. Figure 4 shows an exemplary change of the gradient information flow during the learning process.

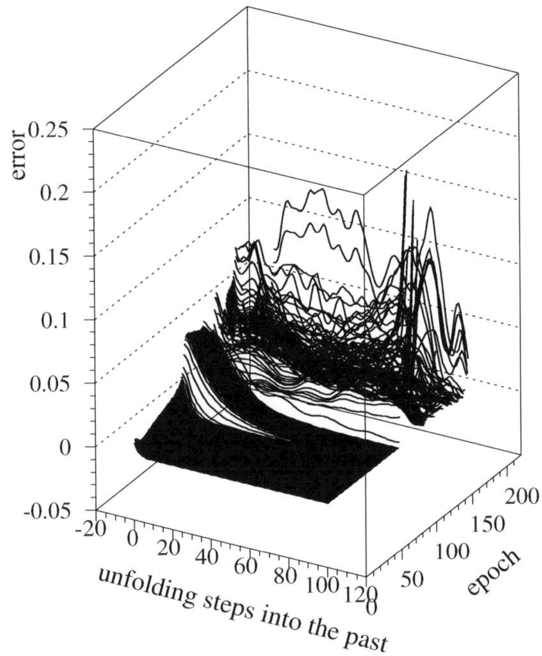

Fig. 4. Exemplary adaptation of the gradient error flow during the learning process of an NRNN which has been initialised with small weights: The graph shows that for a number of learning epochs smaller than approximately 100, the gradient vanishes very quickly. After that the error information distributes more and more over the different unfolding steps, i.e., the networks prolongs its memory span. Finally after about a 150 epochs the error information is almost uniformly backpropagated to the last unfolded time step 100.

5 Conclusion and Outlook

In this paper we demonstrated that NRNNs as well as basic RNNs unfolded in time and trained with BPTT are, in opposition to an often stated opinion, well able to learn long-term dependencies. Using shared weights and overshooting in combination with a reasonable learning algorithm like pattern-by-pattern learning the problem of a vanishing gradient becomes irrelevant. Our results even show that due to shared weights the networks possess an internal regularisation mechanism which keeps the error flow up and allows for an information transport over at least a hundred time steps. Consequently RNNs and especially NRNNs are valuable in time series analysis and forecasting.

Looking at the results and our general experience with recurrent neural networks we further assume that there is a conjunction between the internal state dimension and the weight values in form of an optimal expected row sum of the transition matrix A. The confirmation of this assumption will be part of our future research. Besides that we want to investigate in how far a theoretical analysis of the examined self-regularisation ability of recurrent neural networks is possible.

Acknowledgment

Our computations were performed on our Neural Network modeling software SENN (*Simulation Environment for Neural Networks*), which is a product of Siemens AG.

References

1. Haykin, S.: Neural Networks: A Comprehensive Foundation. Macmillan, New York (1994)
2. Kolen, J.F., Kremer, S.: A Field Guide to Dynamical Recurrent Networks. IEEE Press (2001)
3. Schaefer, A.M., Zimmermann, H.G.: Recurrent neural networks are universal approximators. In: Proceedings of the International Conference on Artificial Neural Networks (ICANN-06), Athens (2006)
4. Rumelhart, D.E., Hinton, G.E., Williams, R.J.: Learning internal representations by error propagation. In Rumelhart, D.E., et al., J.L.M., eds.: Parallel Distributed Processing: Explorations in The Microstructure of Cognition. Volume 1. MIT Press, Cambridge (1986) 318–362
5. Hochreiter, S.: The vanishing gradient problem during learning recurrent neural nets and problem solutions. International Journal of Uncertainty, Fuzziness and Knowledge-Based Systems **6**(2) (1998) 107–116
6. Bengio, Y., Simard, P., Frasconi, P.: Learning long-term dependencies with gradient descent is difficult. IEEE Transactions on Neural Networks **5**(2) (1994) 157–166
7. Hochreiter, S., Bengio, Y., Frasconi, P., Schmidhuber, J.: Gradient flow in recurrent nets: The difficulty of learning long-term dependencies. In Kolen, J.F., Kremer, S., eds.: A Field Guide to Dynamical Recurrent Networks. IEEE Press (2001) 237–243
8. Hochreiter, S., Schmidhuber, J.: Long short-term memory. Neural Computation **9**(8) (1997) 1735–1780
9. Zimmermann, H.G., Neuneier, R.: Neural network architectures for the modeling of dynamical systems. In Kolen, J.F., Kremer, S., eds.: A Field Guide to Dynamical Recurrent Networks. IEEE Press (2001) 311–350

10. Werbos, P.J.: Beyond Regression: New Tools for Prediction and Analysis in the Behavioral Sciences. PhD thesis, Harvard University (1974)
11. Zimmermann, H.G., Grothmann, R., Schaefer, A.M., Tietz, C.: Dynamical consistent recurrent neural networks. In Prokhorov, D., ed.: Proceedings of the International Joint Conference on Neural Networks (IJCNN), Montreal, MIT Press (2005)
12. Neuneier, R., Zimmermann, H.G.: How to train neural networks. In Orr, G.B., Mueller, K.R., eds.: Neural Networks: Tricks of the Trade. Springer Verlag, Berlin (1998) 373–423

Adaptive On-Line Neural Network Retraining for Real Life Multimodal Emotion Recognition

Spiros Ioannou[1], Loic Kessous[2], George Caridakis[1], Kostas Karpouzis[1], Vered Aharonson[2], and Stefanos Kollias[1]

[1] School of Electrical and Computer Engineering, National Technical University of Athens, Politechnioupoli, Zographou, Greece
{sivann, gcari, kkarpou,stefanos}@image.ece.ntua.gr
[2] Tel Aviv Academic College of Engineering
218 Bnei Efraim St. 69107, Tel Aviv, Israel
kessous@post.tau.ac.il, vered@nexsig.com

Abstract. Emotions play a major role in human-to-human communication enabling people to express themselves beyond the verbal domain. In recent years, important advances have been made in unimodal speech and video emotion analysis where facial expression information and prosodic audio features are treated independently. The need however to combine the two modalities in a naturalistic context, where adaptation to specific human characteristics and expressivity is required, and where single modalities alone cannot provide satisfactory evidence, is clear. Appropriate neural network classifiers are proposed for multimodal emotion analysis in this paper, in an adaptive framework, which is able to activate retraining of each modality, whenever deterioration of the respective performance is detected. Results are presented based on the IST HUMAINE NoE naturalistic database; both facial expression information and prosodic audio features are extracted from the same data and feature-based emotion analysis is performed through the proposed adaptive neural network methodology.

1 Introduction

Humans interact with each other in a multimodal manner to convey general messages; emphasis on certain parts of a message is given via speech and display of emotions by visual, vocal, and other physiological means, even instinctively. In the last decade much effort has been directed towards multimodal user interfaces that emulate human to human communication with the goal of enabling computer interfaces with means of natural, expressive and thus more intuitive ways of interaction.

Typical examples of human communication vehicles include auditory channels that carry speech or paralinguistic intonation and visual channels that convey facial expressions or body movements. The related senses of sight and hearing are examples of modalities. Everyday face-to-face communication utilizes many and diverse channels and modalities, increasing the flexibility of a communication scheme. In these situations, failure of one channel is usually recovered by another channel; this kind of behaviour should actually be considered as a model requirement for robust, natural and efficient multimodal HCI [12]. Therefore, the introduction of an emotion analysis

S. Kollias et al. (Eds.): ICANN 2006, Part I, LNCS 4131, pp. 81 – 92, 2006.

system that can analyse intonation and visual cues, to help infer the likely emotional state of a specific user in real life environments, can enhance the affective nature [13] of MMI applications. Adaptive artificial neural network classifiers are proposed in this paper, which can treat both sound and vision cues for emotion analysis, can evaluate their single or multi-modal performance and can adapt their knowledge, through on-line retraining, to real life changing environments.

Probably the most important issue when designing and training artificial neural networks in real life applications is network generalization. Many significant results have been derived during the last few years regarding generalization of neural networks when tested outside their training environment. Examples include algorithms for adaptive creation of the network architecture during training, such as pruning or constructive techniques, modular and hierarchical networks, or theoretical aspects of network generalization, such as the VC dimension. Specific results and mathematical formulations regarding error bounds and overtraining issues have been obtained when considering cases with known probability distributions of the data. Despite, however, the achievements obtained, most real life applications do not obey some specific probability distribution and may significantly differ from one case to another mainly due to changes of their environment. That is why straightforward application of trained networks, to data outside the training set, is not always adequate for solving image recognition, classification or detection problems, as is the case with (multimodal) emotion analysis. Instead, it would be desirable to have a mechanism, which would provide the network with the capability to automatically test its performance and be automatically retrained when its performance is not acceptable. The retraining algorithm should update the network weights taking into account both the former network knowledge and the knowledge extracted from the current input data.

This paper presents an approach for improving the performance of neural networks when handling real life multimodal emotion analysis, based on an automatic decision mechanism, which determines when network retraining should take place, and a retraining - nonlinear programming - algorithm.

Section 2 formulates the retraining problem under investigation. Section 3 presents the retraining technique, while section 4 presents the decision mechanism for activating retraining. Section 5 presents the multimodal emotion recognition problem and the application of the afore-mentioned technologies to the problem, while section 6 summarizes and provides conclusions on the capabilities of the proposed approach.

2 Formulation of the Problem

Let us assume that we seek to classify, to one of, say, p available emotion classes ω, each input vector $\underline{x}_i$ containing the features extracted by one or more input modalities. A neural network produces a p-dimensional output vector $\underline{y}(\underline{x}_i)$

$$\underline{y}(\underline{x}_i) = \left[p^i_{\omega_1} \, p^i_{\omega_2} \cdots p^i_{\omega_p} \right]^T \tag{1}$$

where $p^i_{\omega_j}$ denotes the probability that the ith input belongs to the jth class.

Let us first consider that a neural network has been initially trained to perform the previously described classification task using a specific training set, say, $S_b = \{(\underline{x}'_1, \underline{d}'_1), \cdots, (\underline{x}'_{m_b}, \underline{d}'_{m_b})\}$, where vectors $\underline{x}'_i$ and $\underline{d}'_i$ with $i = 1, 2, \cdots, m_b$ denote the ith input training vector and the corresponding desired output vector consisting of p elements. Let $\underline{y}(\underline{x}_i)$ denote the network output when applied to the ith input outside the training set, corresponding to a new user, or to a change of the environmental conditions; new network weights should be estimated in such cases.

Let $\underline{w}_b$ include all weights of the network before retraining, and $\underline{w}_a$ the new weight vector which is obtained after retraining. A training set S_c is assumed to be extracted from the current operational situation composed of, (one or more), say, m_c inputs; $S_c = \{(\underline{x}_1, \underline{d}_1), \cdots, (\underline{x}_{m_c}, \underline{d}_{m_c})\}$ where $\underline{x}_i$ and $\underline{d}_i$ with $i = 1, 2, \cdots, m_c$ similarly correspond to the ith input and desired output retraining data. The retraining algorithm that is activated, whenever such a need is detected, computes the new network weights $\underline{w}_a$, minimizing the following error criterion with respect to weights,

$$E_a = E_{c,a} + \eta E_{f,a}$$

$$E_{c,a} = \frac{1}{2} \sum_{i=1}^{m_c} \left\| \underline{z}_a(\underline{x}_i) - \underline{d}_i \right\|_2, \qquad E_{f,a} = \frac{1}{2} \sum_{i=1}^{m_b} \left\| \underline{z}_a(\underline{x}'_i) - \underline{d}'_i \right\|_2 \tag{2}$$

where $E_{c,a}$ is the error performed over training set S_c ("current" knowledge), $E_{f,a}$ the corresponding error over training set S_b ("former" knowledge); $\underline{z}_a(\underline{x}_i)$ and $\underline{z}_a(\underline{x}'_i)$ are the outputs of the retrained network, corresponding to input vectors $\underline{x}_i$ and $\underline{x}'_i$ respectively, of the network consisting of weights $\underline{w}_a$. Similarly $\underline{z}_b(\underline{x}_i)$ would represent the output of the network, consisting of weights $\underline{w}_b$, when accepting vector $\underline{x}_i$ at its input; when retraining the network for the first time $\underline{z}_b(\underline{x}_i)$ is identical to $\underline{y}(\underline{x}_i)$. Parameter η is a weighting factor accounting for the significance of the current training set compared to the former one and $\|\cdot\|_2$ denotes the L_2 -norm.

3 The Retraining Approach

The goal of the training procedure is to minimize (2) and estimate the new network weights $\underline{w}_a$, i.e., $\mathbf{W}_a^0$ and $\underline{w}_a^1$ respectively. The adopted algorithm has been proposed by the authors in [2]. Let us first assume that a small perturbation of the network weights (before retraining) $\underline{w}_b$ is enough to achieve good classification performance. Then,

$$\mathbf{W}_a^0 = \mathbf{W}_b^0 + \Delta\mathbf{W}^0, \quad \underline{w}_a^1 = \underline{w}_b^1 + \Delta\underline{w}^1 \tag{3}$$

where $\Delta\mathbf{W}^0$ and $\Delta\underline{w}^1$ are small increments. This assumption leads to an analytical and tractable solution for estimating $\underline{w}_a$, since it permits linearization of the non-linear activation function of the neuron, using a first order Taylor series expansion.

Equation (2) indicates that the new network weights are estimated taking into account both the current and the previous network knowledge. To stress, however, the importance of current training data in (2), one can replace the first term by the constraint that the actual network outputs are equal to the desired ones, that is

$$z_a(\underline{x}_i) = d_i \quad i = 1,\ldots,m_c, \quad \text{for all data in } S_c \tag{4}$$

Equation (4) indicates that the first term of (2), corresponding to error $E_{c,a}$, takes values close to zero, after estimating the new network weights.

Through linearization, solution of (4) with respect to the weight increments is equivalent to a set of linear equations

$$\underline{c} = \mathbf{A} \cdot \Delta\underline{w} \tag{5}$$

where $\Delta\underline{w} = \left[(\Delta\underline{w}^0)^T (\Delta\underline{w}^1)^T \right]^T$, $\Delta\underline{w}^0 = \text{vec}\left\{ \Delta\mathbf{W}^0 \right\}$, with $\text{vec}\left\{ \Delta\mathbf{W}^0 \right\}$ denoting a vector formed by stacking up all columns of $\Delta\mathbf{W}^0$; vector $\underline{c}$ and matrix $\mathbf{A}$ are appropriately expressed in terms of the previous network weights. In particular,

$$\underline{c} = \left[z_a(\underline{x}_1) \cdots z_a(\underline{x}_{m_c}) \right]^T - \left[z_b(\underline{x}_1) \cdots z_b(\underline{x}_{m_c}) \right]^T,$$

expressing the difference between network outputs after and before retraining for all input vectors in S_c. $\underline{c}$ can be written as

$$\underline{c} = \left[d_1 \cdots d_{m_c} \right]^T - \left[z_b(\underline{x}_1) \cdots z_b(\underline{x}_{m_c}) \right]^T \tag{6}$$

Equation (6) is valid only when weight increments $\Delta\underline{w}$ are small quantities. It can be shown [2] that, given a tolerated error value, proper bounds ϑ and ϕ can be computed for the weight increments and input vector $\underline{x}_i$ in S_c

Let us assume that the network weights before retraining, i.e., $\underline{w}_b$, have been estimated as an optimal solution over data of set S_b. Furthermore, the weights after retraining are considered to provide a minimal error over all data of the current set S_c. Thus, minimization of the second term of (2), which expresses the effect of the new network weights over data set S_b, can be considered as minimization of the absolute difference of the error over data in S_b with respect to the previous and the current network weights. This means that the weight increments are minimally modified, resulting in the following error criterion

$$E_S = \left\| E_{f,a} - E_{f,b} \right\|_2 \tag{7}$$

with $E_{f,b}$ defined similarly to $E_{f,a}$, with z_a replaced by z_b in (2).

It can be shown [2] that (7) takes the form of

$$E_S = \frac{1}{2}(\Delta \underline{w})^T \cdot \mathbf{K}^T \cdot \mathbf{K} \cdot \Delta \underline{w} \tag{8}$$

where the elements of matrix **K** are expressed in terms of the previous network weights $\underline{w}_b$ and the training data in S_b. The error function defined by (8) is convex since it is of squared form. The constraints include linear equalities and inequalities. Thus, the solution should satisfy the constraints and minimize the error function in (8). The gradient projection method is adopted to estimate the weight increments.

Each time the decision mechanism ascertains that retraining is required, a new training set S_c is created, which represents the current condition. Then, new network weights are estimated taking into account both the current information (data in S_c) and the former knowledge (data in S_b). Since the set S_c has been optimized over the current condition, it cannot be considered suitable for following or future states of the environment. This is due to the fact that data obtained from future states of the environment may be in conflict with data obtained from the current one. On the contrary, it is assumed that the training set S_b, which is in general provided by a vendor, is able to roughly approximate the desired network performance at any state of the environment. Consequently, in every network retraining phase, a new training set S_c is created and the previous one is discarded, while new weights are estimated based on the current set S_c and the old one S_b, which remains constant throughout network operation.

4 Decision Mechanism for Network Retraining

The purpose of this mechanism is to detect when the output of the neural network classifier is not appropriate and consequently to activate the retraining algorithm at those time instances when a change of the environment occurs.

Let us index images or video frames (similar definitions are used for speech signals) in time, denoting by $\underline{x}(k, N)$ the feature vector of the kth image or image frame, following the image at which the Nth network retraining occurred. Index k is therefore reset each time retraining takes place, with $\underline{x}(0, N)$ corresponding to the feature vector of the image where the Nth retraining of the network was accomplished. Retraining of the network classifier is accomplished at time instances where its performance deteriorates, i.e., the current network output deviates from the desired one. Let us recall that vector $\underline{c}$ expresses the difference between the desired and the actual network outputs based on weights $\underline{w}_b$ and applied to the current data set S_c. As a result, if the norm of vector $\underline{c}$ increases, network performance deviates from the desired one and retraining should be applied. On the contrary, if vector $\underline{c}$ takes small values, then no retraining is required. In the following we denote this vector as $\underline{c}(k,N)$ depending upon feature vector $\underline{x}(k,N)$.

Let us assume that the Nth retraining phase of the network classifier has been completed. If the classifier is then applied to all instances $\underline{x}(0,N)$, including the ones used for retraining, it is expected to provide classification results of good quality. The difference between the output of the retrained network and of that produced by the initially trained classifier at feature vector $\underline{x}(0,N)$ constitutes an estimate of the level of improvement that can be achieved by the retraining procedure. Let us denote by $e(0,N)$ this difference and let $e(k,N)$ denote the difference between the corresponding classification outputs, when the two networks are applied to the feature set of the kth image or image frame (or speech segment) following the Nth network retraining phase. It is anticipated that the level of improvement expressed by $e(k,N)$ will be close to that of $e(0,N)$ as long as the classification results are good. This will occur when input images are similar to the ones used during the retraining phase. An error $e(k,N)$, which is quite different from $e(0,N)$, is generally due to a change of the environment. Thus, the quantity $a(k,N)=|e(k,N)-e(0,N)|$ can be used for detecting the change of the environment or equivalently the time instances where retraining should occur. Thus, no retraining is needed if:

$$a(k,N)<T \tag{9}$$

where T is a threshold which expresses the max tolerance, beyond which retraining is required for improving the network performance. In case of retraining, index k is reset to zero while index N is incremented by one.

Such an approach detects with high accuracy the retraining time instances both in cases of abrupt and gradual changes of the operational environment since the comparison is performed between the current error difference $e(k,N)$ and the one obtained right after retraining, i.e., $e(0,N)$. In an abrupt operational change, error $e(k,N)$ will not be close to $e(0,N)$; consequently, $a(k,N)$ exceeds threshold T and retraining is activated. In case of a gradual change, error $e(k,N)$ will gradually deviate from $e(0,N)$ so that the quantity $a(k,N)$ gradually increases and retraining is activated at the frame where $a(k,N)>T$.

Network retraining can be instantaneously executed each time the system is put in operation by the user. Thus, the quantity $a(0,0)$ initially exceeds threshold T and retraining is forced to take place.

5 Application to Multimodal Emotion Analysis

5.1 How to Combine Modalities

While evaluating the user's emotional state, information on one modality can be used to disambiguate information on the other ones. Two obvious approaches exist of fusing information from different cues: the first is to integrate information at the signal or feature level, whereas the second is to process information and make a decision independently on each modality and finally fuse those decisions at semantic level.

For the first strategy, namely fusion at the signal level, to be meaningful, two conditions must be satisfied: first, modalities must have features that can be handled in a similar way and second the modalities must be synchronized. Such is the case in the combined speech and lip movement analysis. The obvious disadvantages of treating inputs on the signal level include the requirement of large amounts of training data, and the inability to combine the fusion process with possible knowledge about the internal mechanisms present in physical multimodal understanding.

On the other hand, fusion on the decision level, can be applied to modalities which have different time scale characteristics; in this case timing in each modality can be different not only on the frequency of feature extraction but also on the time interval where each decision is valid. For example, an audio prosodic feature concerning some milliseconds of speech could reveal a specific emotional speaker disposition, while the presence of a facial expression could have to be detected for several seconds before it reveals a specific underlying emotion. Decision-level fusion offers several advantages over feature-level fusion. Firstly, each modality is treated independently therefore, they can be both separately trained and their integration does not require excessive computation. A disadvantage of this method is the fact that it does not support mutual disambiguation: using information from one modality to enhance or reject information coming from the other.

In the current approach, a novel technique is proposed, based on the above described adaptive neural network retraining detection. In particular, the proposed approach is applied separately, but synchronised, to the two modalities. The performance of each unimodal classifier is monitored through the decision mechanism of section 4. Whenever a deterioration of performance in one modality is detected, the other one, if still successful, is used to provide the desired outputs for retraining the modality where the problem occurred. The experimental study is presented next.

5.2 The Experimental Study

In this work, we analyzed naturalistic data from the EU IST HUMAINE Network of Excellence [5] naturalistic database. The database includes persons driven to real emotional discourse, being annotated in valence and activity terms by several experts. Both facial expression information in the form of MPEG-4 features [9], and prosodic audio features were extracted from the same data and feature-level classification was employed. Our main synchronization unit has been chosen to be audio tunes, i.e. for the video analysis MPEG-4 FAPs have been extracted on each video frames both at the location of tunes and at the location of silence between tunes (a tune being the portion of the pitch contour that lies between two audio pause boundaries) [11,14]. We observed that in the majority of the cases from a subjective point of view, a tune defined with audio pauses of at least 150 ms seems to be a good segmentation at the sentence level.

Regarding training, testing and performance evaluation of automatic recognizers of multimodal data, a frequent problem is the absence of labelling on separate modalities. The work here is really at its infancy: there are only one or two annotated naturalistic databases, and those have not been annotated separately on each modality, i.e. having human experts produce an emotional annotation by watching only one modality at a time. Moreover, there is the question of the labelling synchronization: when dealing with tune segments, is it proper to reduce continuous labeling to tune labeling instead of first defining tunes and then labeling them?

5.3 Extraction of Visual Features

At first face detection is performed using nonparametric discriminant analysis with a Support Vector Machine (SVM) [6], which classifies face and non-face areas by reducing the training problem dimension to a fraction of the original with negligible loss of classification performance. The face detection step provides us with a rectangle head boundary which includes the whole face area. The latter is segmented roughly using static anthropometric rules [1] into three overlapping rectangle regions of interest which include both facial features and facial background; these three feature-candidate areas include the left eye/eyebrow, the right eye/eyebrow and the mouth. Continuing, we utilize these areas to initialize the feature extraction process. Facial feature extraction performance depends on head pose, thus head pose needs to be detected and the head restored in the upright position; in this work we are mainly concerned with roll rotation, since it is the most frequent rotation encountered in real life video sequences.

Head pose is estimated through the detection of the left and right eyes in the corresponding eye candidate areas. After locating the eyes, we can estimate head roll rotation by calculating the angle between the horizontal plane and the line defined by the eye centers. For eye localization we propose an efficient technique using a feed-forward back propagation neural network with a sigmoidal activation function. The multi-layer perceptron (MLP) we adopted employs Marquardt-Levenberg learning [8] while the optimal architecture obtained through pruning has two 20 node hidden layers and 13 inputs.

We apply the network separately on the left and right eye-candidate face regions. For each pixel in these regions the 13 inputs to the neural network are the luminance Y, the Cr & Cb chrominance values and the 10 most important DCT coefficients (with zigzag selection) of the neighboring 8x8 pixel area. The MLP has two outputs, one for each class, namely eye and non-eye, and it has been trained with more than 100 hand-made eye masks that depict eye and non-eye area in random frames from the ERMIS and HUMAINE [5] databases, in images of diverse quality, resolution and lighting conditions.

Eyes are located with the aid of the aforementioned network while this information is also combined with other feature detectors in a fusion process, to create facial feature masks, i.e. binary maps indicating the position and extent of each facial feature. The left, right, top and bottom–most coordinates of the eye and mouth masks, the left, right and top coordinates of the eyebrow masks as well as the nose coordinates, are used to define the considered feature points (FPs).

For the nose and each of the eyebrows, a single mask is created. On the other hand, since the detection of eyes and mouth can be problematic in low-quality images, a variety of methods is used each resulting in a different mask. In total, we have four masks for each eye and three for the mouth. These masks have to be calculated in near-real time, thus avoiding utilizing complex or time-consuming feature extractors. The use of the afore-mentioned neural network greatly serves this scope. The feature extractors developed for this work are described in [4].

Fig. 1. (a) original frame, (b) final mask for the eyes, (c) detected feature points from the mask

Eyebrows are detected with a procedure involving morphological edge detection and feature selection using data from [1]. Nose detection is based on nostril localization. Nostrils are easy to detect due to their low intensity. Connected objects (i.e. nostril candidates) are labeled based on their vertical proximity to the left or right eye, and the best pair is selected according to its position, luminance and geometrical constraints from [1].

Since, as was already mentioned, the detection of a mask using the applied methods can be problematic, all detected masks have to be validated against a set of criteria. Each one of the criteria examines the masks in order to decide whether they have acceptable size and position for the feature they represent. This set of criteria consists of relative anthropometric measurements, such as the relation of the eye and eyebrow vertical positions, which when applied to the corresponding masks produce a value in the range [0,1] with zero denoting a totally invalid mask. More information about the used expression profiles can be found in [9].

5.4 Extraction of Audio Features

The features used in this work are exclusively based on prosodic features. We consider here features related to pitch and rhythm. All information related to emotion that one can extract from pitch is probably not only in these features, but the motivation of this approach is in the desire to develop and use a higher level of speech prosody analysis than the usual pitch features used in previous studies.

We analyzed each tune with a method employing prosodic representation based on perception called 'Prosogram'. Prosogram is based on a stylization of the fundamental frequency data (contour) for vocalic (or syllabic) nuclei. It gives globally for each voiced nucleus a pitch and a length. According to a 'glissando threshold' in some cases we don't get a fixed pitch but one or more lines to define the evolution of pitch for this nucleus. This representation is in a way similar to the 'piano roll' representation used in music sequencers. This method, based on the Praat environment, offers the possibility of automatic segmentation based both on voiced part and energy maxima. From this model/representation stylization we extracted several types of features: pitch interval based features, nucleus length features and distances between nuclei.

In musical theory, ordered pitch interval is the distance in semitones between two pitches upwards or downwards. For instance, the interval from C to G upward is 7, but the interval from G to C downwards is −7. Using integer notation (and eventually modulo 12) ordered pitch interval, *ip*, may be defined, for any two pitches x and y, as:

$$\begin{aligned} ip\left\langle y, x\right\rangle &= x - y \\ ip\left\langle x, y\right\rangle &= y - x \end{aligned} \tag{1}$$

In this study we considered pitch intervals between successive voiced nuclei. For any two pitches x and y, where x precedes y, we calculate the interval $ip\left\langle x, y\right\rangle = y - x$, then deduce the following features.

For each tune, feature (f1) is the minimum of all the successive intervals in the tune. In a similar way, we extract the maximum (f2), the range (absolute difference between minimum and maximum) (f3), of all the successive intervals in each tune. Using the same measure, we also deduce the number of positive intervals (f4) and the number of negative intervals (f5). Using the absolute value, a measure equivalent to the unordered pitch interval in music theory, we deduce a series of similar features: minimum (f6), maximum (f7), mean (f8) and range (f9) of the pitch interval. Another series of features is also deduced from the ratio between successive intervals, here again maximum (f10), minimum (f11), mean (f12) and range (f13) of these ratios give the related features. In addition to the aforementioned features, the usual pitch features have also been used such as fundamental frequency minimum (f14), maximum (f15), mean (f16) and range (f17). The global slope of the pitch curve (f18), using linear regression, has also been added.

As was previously said, each segment (voiced "nucleus" if it is voiced) of this representation has a length, and this has also been used in each tune to extract features related to rhythm. These features are, as previously, maximum (f19), minimum (f20), mean (f21) and range (f22). Distances between segments have also been used as features and the four last features we used are maximum (f23), minimum (f24), mean (f25) and range (f26) of these distances.

5.5 Adaptive Multimodal Emotion Analysis

In our study, we tested the proposed neural-network-based adaptive classification, evaluation and retraining procedure on the multimodal data sets that were described above. More than 100 tunes of speech and 1000 video frames showing four personalities reacting to an emotion provoking environment named SAL (Sensitive Artificial Listener) developed in the framework of the IST NoE Humaine. The goal was to classify each instant of visual and speech input to one of the quadrants of the emotional wheel, which measures emotion based on a 2-D representation, where dimensions correspond to activation and evaluation of interaction.

While the basic classification rates for each input modality (speech, face) were close to 67%, by implementing the retraining procedure, whenever a change of personality or a lower performance measure was detected, and relying on the cues provided by both modalities, the classification rate was raised to 79%, which illustrates the ability of the proposed method to take advantage of multimodal analysis for improving the obtained results in emotion analysis and classification problems.

6 Conclusions

A novel neural network on line retraining procedure has been proposed in this paper, which is appropriate for real life analysis of multimedia applications. Illustration of the method's ability to achieve multimodal emotion recognition is given in this paper using naturalistic audio and visual data, created in the HUMAINE IST Network of Excellence (2004-2008). The proposed approach is based on neural network architectures which examine each input modality, monitoring the performance of the classification operation and provide a measure of confidence on the achieved accuracy. Whenever this measure gets unacceptable, an efficient on-line retraining of the network knowledge takes place, using the gradient projection method and combining input from all modalities under investigation. Extensive studies are currently under implementation, for further evaluation of the method capabilities.

References

1. J.W. Young, Head and face anthropometry of adult U.S. civilians, FAA Civil Aeromedical Institute, 1993.
2. A. Doulamis, N.Doulamis and S. Kollias, On-line Retrainable Neural Networks: Improving the Performance of Neural Networks in Image Analysis Problems, IEEE Transactions on Neural Networks, vol. 11, no 1, pp. 137-157, 2000.
3. A. Krog, J. Vedelsby, Neural network ensembles, cross validation and active learning, in Tesauro G., Touretzky D., Leen T. (Eds) Advances in neural information processing systems 7, pp. 231-238, Cambridge, MA. MIT Press, 1995.
4. S. Ioannou, A. Raouzaiou, V. Tzouvaras, T. Mailis, K. Karpouzis and S. Kollias, Emotion recognition through facial expression analysis based on a neurofuzzy network, Special Issue on Emotion: Understanding & Recognition, Neural Networks, Elsevier, Volume 18, Issue 4, Pages 423-435, 2005.
5. HUMAINE, Human-Machine Interaction Network on Emotion IST-2002-2.3.1.6 (http://emotion-research.net/)
6. R. Fransens, Jan De Prins, SVM-based Nonparametric Discriminant Analysis, An Application to Face Detection, Ninth IEEE International Conference on Computer Vision Volume 2, October 13 - 16, 2003
7. S. Kollias and D. Anastassiou. "An adaptive least squares algorithm for the efficient training of artificial neural networks". IEEE Transactions on Circuits and Systems, Volume: 36 , Issue: 8 , Aug. 1989 pp:1092 – 1101
8. M.T. Hagan and M. Menhaj, "Training feedforward networks with the Marquardt algorithm". IEEE Transactions on Neural Networks, vol. 5, no. 6, pp. 989-993, 1994.
9. A. Raouzaiou, N. Tsapatsoulis, K. Karpouzis and S. Kollias, "Parameterized facial expression synthesis based on MPEG-4", EURASIP Journal on Applied Signal Processing, Vol. 2002, No. 10, pp. 1021-1038, Hindawi Publishing Corporation, October 2002.
10. Baldonado, M., Chang, C.-C.K., Gravano, L., Paepcke, A.: The Stanford Digital Library Metadata Architecture. Int. J. Digit. Libr. 1 (1997) 108–121

11. Mertens, Piet: The Prosogram: Semi-Automatic Transcription of Prosody based on a Tonal Perception Model. in B. Bel & I. Marlien (eds.) Proceedings of Speech Prosody 2004, Nara (Japan), 23-26 March. (ISBN 2-9518233-1-2)
12. Cowie R., Douglas-Cowie E., Tsapatsoulis N., Votsis G., Kollias S., Fellenz W., Taylor J., Emotion Recognition in Human-Computer Interaction, IEEE Signal Processing Magazine, 2001.
13. R. W. Picard, Affective Computing, MIT Press, Cam-bridge, MA, 2000.
14. R. Cowie, E. Douglas-Cowie, Automatic statistical analysis of the signal and prosodic signs of emotion in speech. Proceedings of the 4th International Conference of Spoken Language Processing (pp. 1989–1992). 1996, Philadelphia, USA.

Time Window Width Influence on Dynamic BPTT(h) Learning Algorithm Performances: Experimental Study

V. Scesa[1], P. Henaff[1], F.B. Ouezdou[1], and F. Namoun[2]

[1] LISV, Université de Versailles St Quentin,
10, 12 avenue de l'Europe, 78140 Vélizy, France.
[2] BIA company,
8 rue de l'Hautil, 78730 Conflans Ste Honorine, France.

Abstract. The purpose of the research addressed in this paper is to study the influence of the time window width in dynamic truncated BackPropagation Through Time BPTT(h) learning algorithms. Statistical experiments based on the identification of a real biped robot balancing mechanism are carried out to raise the link between the window width and the stability, the speed and the accuracy of the learning. The time window width choice is shown to be crucial for the convergence speed of the learning process and the generalization ability of the network. Although, a particular attention is brought to a divergence problem (gradient blow up) observed with the assumption where the net parameters are constant along the window. The limit of this assumption is demonstrated and parameters evolution storage, used as a solution for this problem, is detailed.

1 Introduction

Born from the collaboration between a robotic research laboratory (LISV) and an industrial company (BIA), the supporting project of this study aims to shape a smart architecture able to learn to control non linear multi actuators system under real time constraints. Through the design and the implementation of this controller, we want to evaluate the ability of neural architectures to achieve the non linear control needs in real time. The project experiments are carried out on two structures: the ROBIAN biped robot from the LISV [1], and the BIA road simulator (www.bia.fr). The architecture of the developed algorithm is based on continuous time recurrent neural networks (CTRNN). To shape this architecture, a truncated BPTT algorithm was chosen for its ability of integrating the learning error and for its simplicity to be implemented for real time online applications. This well known gradient algorithm is widely used for system modeling and control [2], optimization applications, speech recognition [3], or meta learning [4]. A detailed description of this algorithm is given in [5], [6] and [7]. Nevertheless, as far as we know, no study aimed to extract the important role of the truncation width on the success of the learning. Obviously, the truncation width influences deeply the learning speed and quality. Our real time experimentations on ROBIAN biped also show us that, when the net parameters are considered as constant along a large window, the learning stability could be hardly spoiled.

S. Kollias et al. (Eds.): ICANN 2006, Part I, LNCS 4131, pp. 93 – 102, 2006.

Thus, through a statistical analysis of learning results, the links between the width of the time window, the stability of the learning, the "constant parameters" assumption, the convergence speed and the generalization abilities of the learned networks, have to be raised. To focus on this influence, a first experiment dealing with direct model identification rather than a controller shaping will be considered.

The next section details the neural model and the BPTT(h) learning algorithm used. The experimental plant is described in section 3. In section 4, the influence of the window width will be studied. The "constant parameters" assumption with its influence on the learning stability will be discussed. A stabilizing modification of the learning equation, to take into account the parameters evolution history, will be proposed. It will allow raising the dependencies between the window width and the learning results. Finally, discussions about the parameters evolution storage and the window choice to find a compromise between learning speed and nets accuracy will be given.

2 Dynamic Truncated Backpropagation Through Time Algorithm

2.1 Neural Model

Following the classical CTRNN equation (1), the input data are propagated in the network to generate the neurons outputs. The net activities belong to the intrinsic neurons and network parameters : weights (w_{ij}), biases (b_j) and time parameters (T_j).

$$T_j \cdot \frac{\partial y_j}{\partial t} = -y_j + f\left(\sum_i \left[w_{ij} \cdot y_i\right] + b_j\right) \tag{1}$$

Where y_j corresponds to the j neuron activity, f is the activation function (tanh). Using a time scale parameter S_j (classically $S_j = \Delta t / T_j$, Δt is the time step), the discrete corresponding equation can be written as follows:

$$y_j(t) = S_j \cdot f\left(\sum_i \left[w_{ij} \cdot y_i(t - \Delta t)\right] + b_j\right) + (1 - S_j) \cdot y_j(t - \Delta t) \tag{2}$$

As the network is fully recurrent, each neuron receives the outputs of all the other ones. The weights, biases and time constants shape the temporal response of the net.

2.2 Learning Algorithm

The objective of the parameters modification consists in minimizing a desired criterion. For an identification process, the criterion would be the gap between the neural model and the taught system. To carry out the adaptation of network parameters, BackPropagation Through Time algorithm, detailed in [5], [6] and [7], can be used.

This algorithm is computing an error function that corresponds to the criterion to be minimized. The error function (E) is defined as the integral of each net output errors. The parameters modification ($\Delta param$) is leaded by the error gradient inverse value. The following equations give the error function and the parameters modification law:

$$E = \int_{t0}^{t} [e_j(\tau) \cdot d\tau] \qquad (3) \qquad \Delta param = -\eta \cdot \frac{\partial E}{\partial param} \qquad (4)$$

Where $e_j(\tau)$ is the output error of the neuron j stored at the current time step in the time window (τ). During the learning, to minimize E, the algorithm modifies the net parameters following the gradient descent (4), where *param* is w_{ij}, b_j or S_j and η is the corresponding learning rate. To compute these "delta" values for continuous time neurons, a dynamic BPTT learning algorithm [5] is needed. The gradient values giving the influence of each parameter are computed with equations (5), (6) and (7):

$$\frac{\partial E}{\partial w_{jk}} = \int_{t0}^{t} [S_k \cdot z_k(\tau) \cdot f'(x_k(\tau - \Delta t)) \cdot y_j(\tau - \Delta t)] \cdot d\tau \qquad (5)$$

$$\frac{\partial E}{\partial b_j} = \int_{t0}^{t} [S_j \cdot z_j(\tau) \cdot f'(x_k(\tau - \Delta t))] \cdot d\tau \qquad (6)$$

$$\frac{\partial E}{\partial S_j} = \int_{t0}^{t} [z_j(\tau) \cdot (f(x_j(\tau - \Delta t)) - y_j(\tau - \Delta t))] \cdot d\tau \qquad (7)$$

The backpropagated errors ($z_j(\tau)$) for a jth neuron can be written as following:

$$z_j(\tau) = \frac{\partial E}{\partial y_j(\tau)} = \sum_k [z_k(\tau + \Delta t) \cdot S_k \cdot f'(x_k(\tau)) \cdot w_{jk}] + z_j(\tau + \Delta t) \cdot (1 - S_j) + e_j(\tau) \cdot \Delta t \qquad (8)$$

Where *f'* is the derivative of the activation function (tanh). For each neuron, the z_j is computed by considering the network states at each time step as successive layers. Thus, the output errors are backpropagated in the network and in time. In BPTT learning, z_j is computed for each neuron and each time step since the starting. To prevent a memory explosion induced by the storage of every network states, a truncated BPTT algorithm (BPTT(h)) is proposed in [8]. This algorithm is keeping in memory only the past states included in a window following the current instant ($t_0 = t - h$ with h the window width). The z_j values are computed along this sliding window ($\tau \in [t-h;t]$). Hence, only a gradient approximation is computed. The previous (5) to (8) equations are constructed on the assumption that the parameters are constant along the time window. This assumption presented in [9] and [10] minimizes the process memory needs as the parameters history is not stored. In our study, these algorithms are experimented on the ROBIAN biped robot described in the next section.

3 ROBIAN Identification Description

3.1 Plant Description and Perturbation Signals

To focus on the time window width influence, a model identification rather than a controller shaping will be carried out. Fig.1 shows how the identification of ROBIAN's torso influence on the ZMP (Zero Moment Point) will be performed. For more details concerning the ROBIAN biped and its torso mechanism see [1].

The ZMP is a key notion in the balance control of walking robot. It corresponds to the position of the center of pressure on the ground. The ZMP algorithm, introduced thirty five years ago [11], consists of controlling the equilibrium of the biped robot by keeping the ZMP inside the polygon defined by the contact points with the ground.

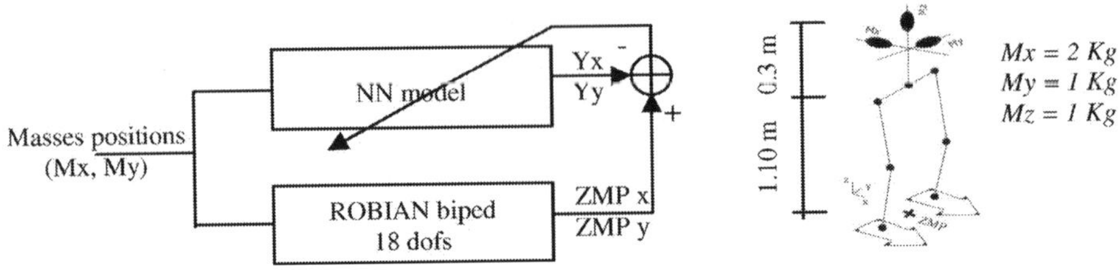

Fig. 1. Learning architecture for the identification (*left*) of ROBIAN robot (*right*)

The motion of the X and Y masses of the torso (Mx and My) are perturbing the robot balancing, leading to variations of the X and Y ZMP positions. During the learning process, the neural network (NN) computes the gap between the measured positions and its own outputs and modifies its parameters to vanish this error, by minimizing a cost function. This function is defined as the normalized squared errors sum on the two axes and the output errors ($e_j(t)$) can be expressed as following:

$$Cost = \left(\frac{ZMP_x - y_x}{maxZMP_x}\right)^2 + \left(\frac{ZMP_y - y_y}{maxZMP_y}\right)^2 \quad (9) \qquad e_{axis}(t) = \frac{\partial Cost}{\partial y_{axis}(t)} = -\frac{2\cdot(ZMP_{axis} - y_{axis}(t))}{maxZMP_{axis}} \quad (10)$$

Where y_x, y_y are the output activities and $maxZMP_{axis}$ is the maximum amplitude on each axis (for normalization).

Representative situations should be given during the learning. They must fully characterize the behavior of the studied system. This implies to give the network patterns that express the dynamic of the ZMP for the X and Y axis. Fig 2 represents the learning pattern adopted.

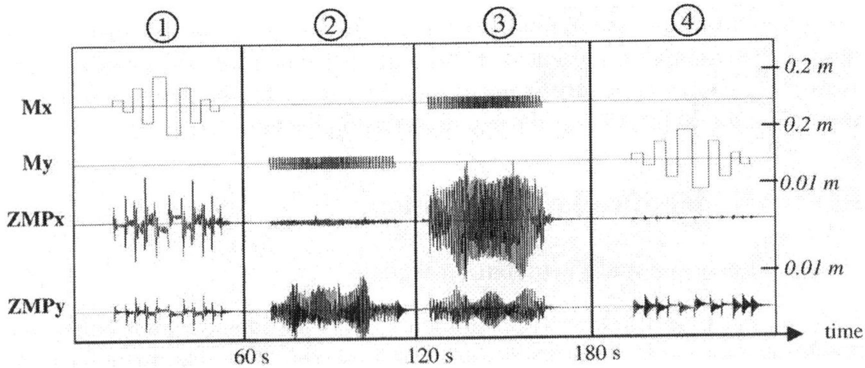

Fig. 2. Masses positions and ZMP positions on the two axis for the learning pattern (sampling rate = 45Hz). In periods 1 and 4, the X and then the Y mass, are submitted to successive steps with various amplitudes. In the 2 and 3 periods, steps with varying frequency are applied.

The learning pattern contains a succession of four different excitations applied to the robot: squared commands with varying amplitudes (0.01m to 0.20m) and with varying frequencies (1Hz-3Hz, containing the resonance of the robot), for both X and Y axes. In the test pattern, a sine command with a varying frequency is applied on both axes.

3.2 Example of Learning

The cost function evolution is depicted on Fig. 3. In this experiment, the network (composed of 2 inputs, 15 hidden neurons and 2 outputs) was taught with one hundred loops on the learning pattern (learning rates: η_1 = 0.025 for weights and biases, η_2 = 1.25 for time constants) and a time window width h=20.

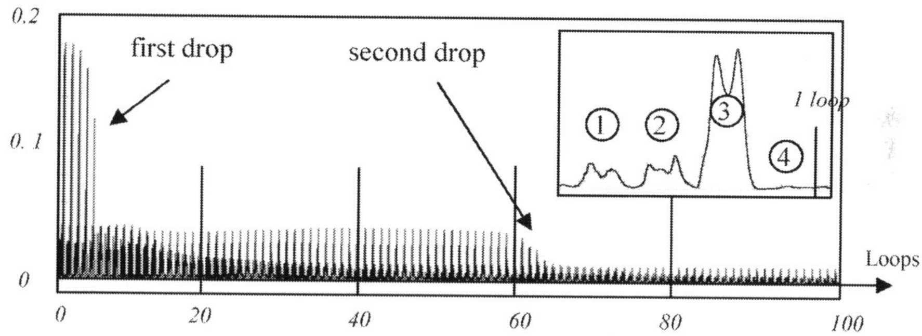

Fig. 3. Cost evolution, and a zoom on the first loop (*top right corner*), 1 loop =10200 iterations

At the beginning of the learning, the network parameters are randomly chosen to generate stable outputs. The time scale parameters (S_j) are taken in the interval [0.5;1], to obtain neurons with speeds close to the dynamic of the system.

Fig.3 gives the evolution of the cost during the learning. In the zoom area, the four peaks are due to the different kinds of command (amplitude changes: peaks 1 and 4, or frequency variation: peaks 2 and 3) of the learning pattern.

During the experiment, two main drops happened. They are respectively linked to a quick decrease of peak 3 (learning of the X direction behavior) and peak 2 (learning on Y direction). The instant when the second drop happened (**SDI** : Second Drop Instant) is a good factor to quantify the convergence speed. It happens when the network identifies correctly the behavior of ROBIAN torso in the both two directions. Here it takes about 64 loops to arise. The SDI value can be interpreted as a physical expression of the convergence for our experiments. It can be related to the choice of a threshold below which the cost is considered as satisfactory.

After the learning, a relevant value for estimating the generalization ability of the learned network is the costs sum along the whole test pattern (**TES** : Test Errors Sum). It expresses the difference between an ideal model of the robot and the learned one. With a perfect model, the TES value is equal to zero. For the considered example, before the learning, TES = 363, and at the end of the learning process (after the 100th loop), the remaining error decreases to TES = 40.

4 Time Window Width Influence

4.1 Stability Limit with the "Constant Parameters" Assumption

In the usual gradient calculus, the parameters are considered as constant along the time window [9], [10]. This approximation can be seen in the gradient equations (5) to (8) where the parameters (w_{ij}, b_j or S_j) are not indexed by time. The time window width reduces the use of this assumption by destabilizing the learning.

The identification of the ROBIAN's torso influence on the ZMP positions is carried out for different time window width values (TW) from 5 to 60 states stored. For each TW, ten learning courses are carried out for a statistical analysis of the results. Each network is evaluated on the test pattern after learning. The learning and test results are depicted on the following figure:

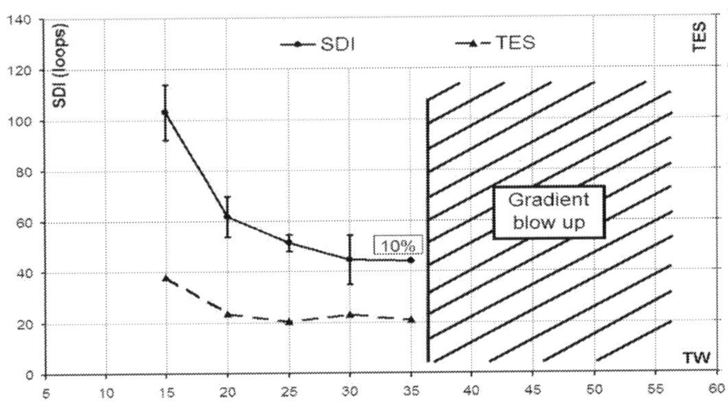

Fig. 4. Learning and test results following TW with η_1=0.025 for weights and biases, η_2=1.25 for time constants. The average value of SDI and TES are plotted. The error bars correspond to the standard deviation measured. They express the iteration range of convergence. If not all the 10 trials lead to a convergence, a mark is added on the graph giving the percentage of success. Here, it happens for TW=35 where there was only 10% of success.

This graph can be decomposed in three parts. When TW is less than 15, the learning is so slow that it doesn't manage to converge (no results on the graph). For TW$\in$[15;35], the SDI, and TES seem to decrease proportionally to TW, thus the convergence speed and the generalization ability are better for larger TW. Finally, if TW is greater than 35 states stored, the algorithm is diverging with a gradient blow up. For a stronger learning rate (η=0.1) the gradient blow up occurs since TW=15.

4.2 Gradient Blow Up Divergence and Parameters Evolution Storage

We called this divergence problem gradient blow up since it's caused by a sudden explosion of the neurons gradient values. In Fig. 5, the evolution in time of the gradient values for each neuron are depicted. In this experiment, the time window width is TW=100 states stored (with a sampling rate=45Hz, it corresponds to an interval of 2.22s). The three graphs represent the initiation of the gradient blow up. Just after the last graph, all the gradient values are exploded.

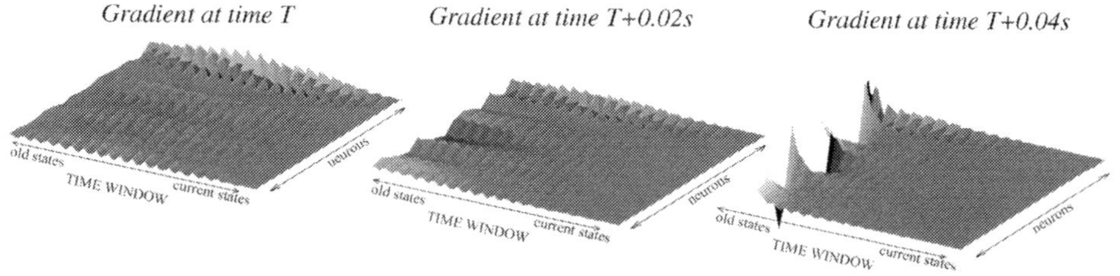

Fig. 5. Gradient blow up initiation. The little waves correspond to the "normal" backpropagation of the error in the time window on the neurons. The strong drops and jumps in the third graph correspond to the "abnormal" divergence called gradient blow up.

The blow up begins with a divergence of the oldest past values. Through the recurrent links raised in the network, the explosion is spread in the entire network. Finally, the divergence is so important that the parameters take infinite values and the algorithm is stopped. Our guess is that the divergence of the farer gradients is due to the "constant parameter" assumption. This assumption means that the parameters are the same along the entire time window. However, if the window width is large compared to the parameters variation, it's obviously false. Actually, if the width is important, the parameter values that contribute to create the farer past states stored could be strongly different from the current ones. This happens when either the first or the second drop is initiated, i.e. when the parameters are strongly modified. That's why the divergence is initiated in the oldest part of the time window. Hence, the assumption could only be used with short window or small learning rates.

A way to avoid this blow up problem is to store the parameters history along the time window. So, the gradient equations will take into account their evolutions as in (5'), (6') and (7'). The weights, biases and time scale parameters become a function of time τ. The storage of the parameter history along the sliding time window implies an increase of memory needs. The number of stored values is multiplied by TW.

$$\frac{\partial E}{\partial w_{jk}} = \int_{t0}^{t} \left[\boxed{S_k(\tau-1)} \cdot z_k(\tau) \cdot f'(x_k(\tau-\Delta t)) \cdot y_j(\tau-\Delta t) \right] \cdot d\tau \qquad (5')$$

$$\frac{\partial E}{\partial b_j} = \int_{t0}^{t} \left[\boxed{S_j(\tau-1)} \cdot z_j(\tau) \cdot f'(x_k(\tau-\Delta t)) \right] \cdot d\tau \qquad (6')$$

$$z_j(\tau) = \sum_k \left[z_k(\tau+\Delta t) \cdot \boxed{S_k(\tau)} \cdot f'(x_k(\tau)) \cdot \boxed{w_{jk}(\tau)} \right] + z_j(\tau+\Delta t) \cdot \left(1 - \boxed{S_j(\tau)}\right) + e_j(\tau) \cdot \Delta t \qquad (8')$$

Equation (7) is not changed since no parameter is involved. This modification of the classical BPTT learning equations will be used for the analysis of TW influence on the speed and accuracy of the learning in next section.

4.3 Influence on the Convergence Speed and on the Accuracy of the Learned Nets

The same identification process is carried out with the modified learning algorithm, for time window width values from 5 to 60 states stored, and with 10 initial random nets for each TW value. The learning and test results are depicted on the fig 6:

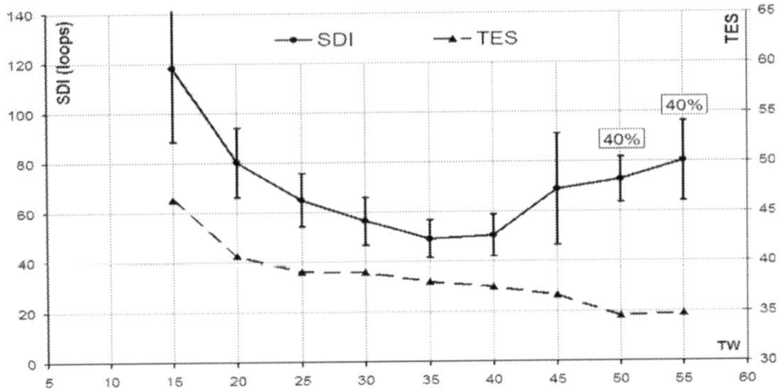

Fig. 6. Learning and test results following the TW values with η_1=0.025 and η_2=1.25. All the learning succeed for TW≤45, for TW= 50 and 55, only 40% of the nets reach a convergence.

The previous graph can be decomposed in three parts depending on the TW value. If TW<15 states stored, the algorithm is not able to converge to a correct solution. Next, for TW varying between from 15 to 35, the convergence speed average is increased while the average TES is decreased. Then, with TW>35, the convergence process is more oscillating. The standard deviations are thus larger, and not all the 10 initial random networks lead to a correct solution. But, for these TW values, in case of convergence, the algorithm finds better solutions (i.e. the TES values are smaller). This is due to the amount of data taken into account. The modified algorithm allows also increasing the learning rates. Figure 7 represents the results obtained with η_1=0.1:

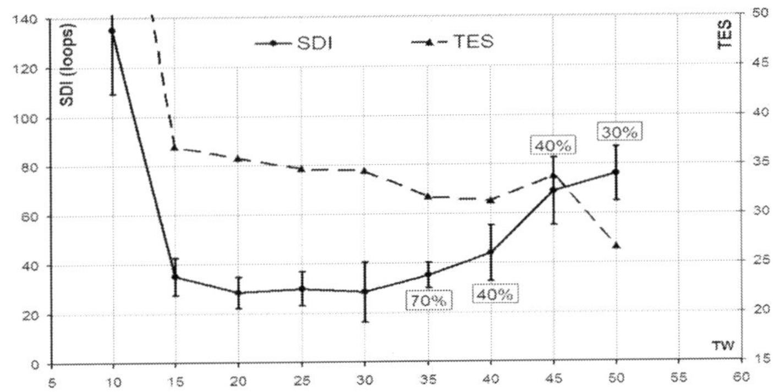

Fig. 7. Learning and test results with η_1=0.1 for weights and biases and η_2=1.25 for time constants. All the learning succeed for TW≤35, for TW= 40, 45, 50 and 55, the percentages of success are respectively 70%, 40%, 40%, 30% and 0% for 55.

Compared to the experiments done with η_1=0.025, the convergence speed gets faster since a TW = 15 states stored. Nevertheless, the learning is more unstable with this higher learning rate, and the algorithm meets convergence difficulties earlier. For larger TW values, the number of success is less important as the learning is oscillating, due to a too important learning rate. But, when they converge, the networks learned are more accurate. Here again, the TES obtained are the best for the biggest TW. As far as our experiment is concerned, the best configuration ensuring maximum speed, best quality and high percentage of success, is TW=30 and η_l=0.1.

5 Discussion and Conclusion

5.1 Storing the Parameters Evolution?

The comparison between the results of the "constant parameters" assumption and those presented for the modified algorithm, shows that the second one is better for our identification experiment. First, the classical one is only valid for small TW values whereas the second one allows larger ones. Next, for the same TW range, the classical method finds networks with worst generalization abilities. The only advantage of the classical one seems to be a faster convergence speed. Nevertheless, as a stronger learning rate is not prohibited with the modified method, this speed advantage can be overcome. In that case, the modified algorithm results are comparable to the ones achieved with a smaller learning rate and a larger TW value. With the modified algorithm, for the largest TW values, the convergence is not always met. The algorithm fails to find a correct solution. But no gradient blow up occurs.

The TW limit that leads to a gradient blow up for the classical method is probably linked to the dynamic of the studied system and the convergence speed defined by the learning rates values. If the learning rates are strong, the parameters modification will be fast and they couldn't be approximated as constant along the time window. In the same way, if the system is fast, the learning will tend to bring the time constants and the network to a faster dynamic. Thus, the errors will be quickly backpropagated in the time window, and the algorithm will behave as if the time window was bigger and will diverge for smaller TW. So, the choice between using or not the "constant parameters" assumption will depend on the dynamic of the system.

5.2 Convergence Speed vs Generalization Ability

The comparison of the convergence speeds and the quality of the learned networks demonstrates that the TW value choice must be a trade-off between these two results. Choosing the largest one, i.e. the best quality, the speed could be strongly decreased or the convergence not guaranteed. Choosing the fastest convergence, with a shorter TW, could lead to non suitable networks. There is no general rule for optimizing the learning. The choice must be done following the needs and criteria fixed for the learning process. Here again, the dynamic of the studied system will act upon the TW that leads to the fastest convergence. For the same criteria, the chosen TW could be different for a slow or fast system. But, allowing the use of larger TW is obviously useful.

6 Conclusions

In this paper, we studied the influence of the time window width parameter upon the stability, the learning speed and the quality of the networks obtained. Based on statistical experiments, aiming to identify the links between the balancing of a biped robot and its torso motion, we discussed the net parameters evolution storage and the choice of an optimal TW value. We found that it could be useful to enlarge the window to reach faster and more accurate learning. As the "constant parameter" assumption classically adopted for BPPTT(h) is not adapted for larger window, a modified algorithm taking into account the parameters history can be advantageous.

In the future, we will first carry out similar experiments on faster or slower systems to find out the influence of the system dynamic on the learning. Next, we will perform a mathematical analysis of the modified algorithm. The learning algorithm will be also applied to carry out inverse model identification and control.

References

1. Mohamed, B., Gravez, F., Ouezdou, F.B.: Emulation of the dynamic effects of human torso during walking gait. In Journal of Mechanical Design, vol. 126, p830-841, Sept 2004.
2. Tsung, F-S.: Modeling Dynamical Systems with Recurrent Neural Networks. PhD thesis, Depart-ment of Computer Science. University of California, San Diego, 1994.
3. Nguyen, M.H., Cottrell, G.W.: Tau Net: A neural network for modeling temporal variability. In Neurocomputing 15 pp. 249-271, 1997.
4. Hochreiter, S., Younger, A.S., Conwell, P.R.: Learning to learn using gradient descent. In lecture notes on Comp. Sci. 2130, proc. Intl. Conf. on Artificial Neural Networks (ICANN-2001), pages 87-94. Springer: Berlin, Heidelberg, 2001.
5. Pearlmutter, B.A.:Gradient calculation for dynamic recurrent neural networks: a survey. In Transactions on Neural Networks, 6(5):1212-1228, 1995.
6. Werbos, P.J.: Backpropagation through time: what it does and how to do it. Proceedings of the IEEE, vol. 78, no. 10, pp. 1550 1560, 1990.
7. Rumelhart, D.E., Hinton, G.E., Williams, R.J.: Learning internal representations by error propagation Parallel distributed processing: explorations in the microstructure of cognition. eds D E Rumelhart, J L Mc- Clelland and the PDP Research Group (MIT Press, Cambridge MA) pp 318–362, 1986.
8. Williams, R. J., Zipser, D.: Gradient-based learning algorithms for recurrent connectionist networks. In Y. Chauvin and D. E. Rumelhart, editors. Backpropagation: Theory, Architectures, and Applications, Erlbaum, Hillsdale, NJ, 1990.
9. Williams, R. J., Peng, J.: An efficient gradient--based algorithm for on--line training of recurrent network trajectories. In Neural Computation, vol 2 p 490-501 (MIT Press). 1990
10. Campolucci, P., Uncini, A., Piazza, F., Rao, B. D.: On-Line Learning Algorithms for Locally Recurrent Neural Networks. In IEEE-NN vol 10 p253. March 1999.
11. Vukobratovic, M. and Borovac, B.:Zero-moment point – thirty five years of its life. In International Journal of Humanoid Robotics. 1(1) p 157-173. 2004.

Framework for the Interactive Learning of Artificial Neural Networks

Matúš Užák and Rudolf Jakša

Department of Cybernetics and Artificial Intelligence
Technical University of Košice, Slovakia
uzak@neuron.tuke.sk, jaksa@neuron.tuke.sk

Abstract. We propose framework for interactive learning of artificial neural networks. In this paper we study interaction during training of visualizable supervised tasks. If activity of hidden node in network is visualized similar way as are network outputs, human observer might deduce the effect of this particular node on the resulting output. We allow human to interfere with the learning process of network, thus he or she can improve the learning performance by incorporating his or her lifelong experience. This interaction is similar to the process of teaching children, where teacher observes their responses to questions and guides the process of learning. Several methods of interaction with neural network training are described and demonstrated in the paper.

1 Introduction

The process of learning of artificial neural network can be visualized, observed, and interactively guided by a human observer. Traditionally, learning of artificial neural networks is treated as adaptation of a black box, although works focused on visualization what happens inside the box can be found dating back to beginnings of the field. On the other side, establishment of the Interactive Evolutionary Computation (IEC) domain brings into forefront the idea of interactive intervention into algorithm by a human observer. This idea of interactive guidance of algorithm has to be explored in the neural networks domain too.

Combination of the learning algorithm, which searches a weight space of the network, and a human observer, which gains an overview over the behavior of algorithm and is able to guide this algorithm, may bring some new possibilities into the field of neural networks. Experience of a human might be usable for the algorithm to escape from the local minima trap. Ability to guide the learning might bring new tasks for neural networks, not strictly defined by a training data set. Basic method for incorporation of a human observer into learning process is the visualization. This is used in IEC domain and also studied in the past in neural networks area. Easiest tasks to visualize are these which are defined in two-dimensional space and naturally have a visual character, although three-dimensional, motion video or these with audio character may by visualized or another way presented to a human observer too. Survey of tasks and also methods studied in IEC area is provided in [1].

S. Kollias et al. (Eds.): ICANN 2006, Part I, LNCS 4131, pp. 103–112, 2006.

The aim of most neural network visualization techniques is to help to understand what neural networks really do. The essence of each technique lies in visualization of some object common to all neural networks: topology, response to processed data, internal mappings. Although in general, the target of all techniques is to visualize internal mappings, different paradigms are used. This is mostly due to fact, that in different applications, different aspect of network performance are studied.

First visualization techniques used were Hinton and Bond diagrams described in [2]. These techniques help to analyze the input units importance through the magnitude of weights connecting them with hidden units. However, the analysis of internal mapping is reduced to analysis of weights. These methods display the topology of network. Craven in [3] also accompanied them with trajectory diagrams. Craven's visualization tool called *Lascaux*, did visualize many objects relevant only for single training vectors: the activation signal, and error signal propagated through the network, but its essence was in modification of Bond diagrams implementation. This modification is later referred to as the network interpretation diagram (NID) used by Olden [4], similar modification implemented in three-dimensional space was done by Edlund [5]. Olden used Garson's diagrams to visualize the input nodes importance to the outputs and the sensitivity analysis as a method to comprehend the inner mappings.

Important is also the work of Streeter [6] where is described a visualization tool that provides opportunities for interaction in the learning process. Learning is realized through the error backpropagation algorithm or using Evolutionary Strategy (ES). Human observer is able to manually adjust weights and their learning rate in the backpropagation case, or mutation rate in the ES case. They used a modification of NID, and a modification of Hinton diagrams to visualize as many networks in the evolution process as possible.

Recently were introduced methods which analyze the performance of network by observing its reaction to processed data. Interesting projection techniques have been described by Duch, [7][8], that should help to analyze the network performance through visual interpretation of decision borders. Tzeng [9] introduced a modification of NID with built in representation of modified Carson's method that analyzed hidden units importance. They analyze the network in data-driven approach.

Our own work is focused to studying reactions of single units to the testing set which represents the task. As weights are a part of the unit they are connected to, the visualized information is reduced compare to methods that visualize weights. This method is described in [3] as a Response Function Visualization. Similar to hyperplane plots [3], this method is constrained to two-dimensional tasks, but it is obvious that combined with methods proposed by Hinton, Tzeng and Duch, this limitation can by surpassed.

Generally, the challenge is to combine the best from mentioned methods to provide the human observer with only the most valuable information about the learning process, and allow him or her to maintain full concentration to interactive interventions.

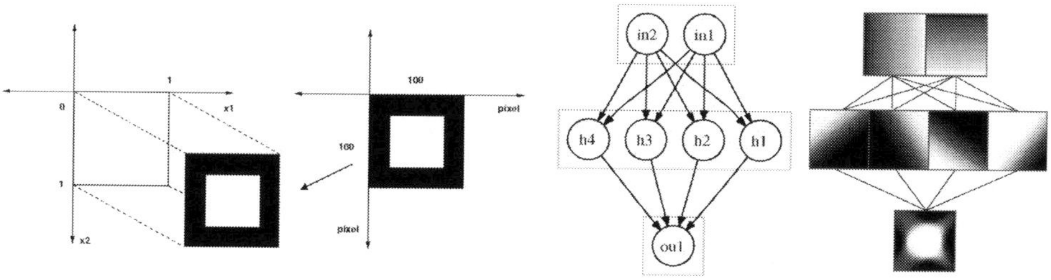

Fig. 1. (*Left*) The image with dimensions 100x100 pixels mapped to space of classification task. (*Right*) Example of the response of network units to the function signal propagating through the network. Images were created by presenting full set of testing vectors from the space of classification task. We classify the vectors inside and outside of given square.

2 Visualization of Learning of Neural Network

We will describe the design of visualization method of learning process using the multilayer perceptron concept. The multilayer perceptron consists of a layer of input units, of one or more hidden layers and an output layer. Computational units of the network are in output layer and in hidden layers. Proper visualization of behavior of these units during the learning should allow to reason about the learning process itself.

We will visualize the learning of a classification task. The space, where the classification takes place is projected into two-dimensional (2D) space for visualization. Training data are randomly chosen from samples generated from classification task space. If the classification task itself is defined in the 2D space, the 2D image representing classification results can be easily generated as a direct mapping of this space into image space. Figure 1 clarifies the procedure.

The goal of the visualization is to capture responses of individual neurons to the function signal during the whole testing phase. The function signal is an input signal – stimulus, that comes in the input end of the network, propagates forward (neuron by neuron) through the network, and emerges at the output end of the network as an output signal [10]. The signal itself is a set of vectors from defined space. Testing phase is a presentation of defined set of input vectors to the network and propagation of function signals through the network. In addition to visualization of responses of active neurons – computational units of network, also input units – entrances of input signal can be visualized to extend the observer's view. Thus the function signal can be completely observed from its entering point to the output point in the network. This method is also referred as the response function visualization [3]. With visualization the observer can track the reactions of individual neurons to the propagating signals. Visualization of input units provides the information about the intensity of the signals which enter the network.

The visualized area of classification space might be extended to cover also places which are far from the training data. By observing behavior of network in these distant areas an observer may reason about extrapolation/generalization qualities of network. Such acquired knowledge should help to uncover possible problems with using the network in situations where it can be confronted with the samples from outside the area of maximum classification confidence [11]. See Fig.2 for an example of extrapolation test.

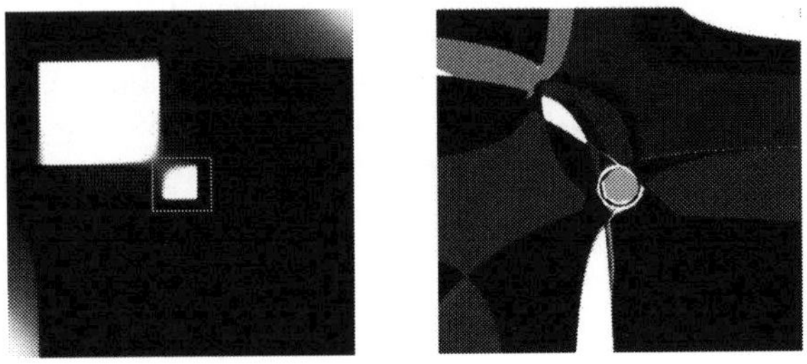

Fig. 2. (*Left*) The network was trained to classify vectors inside the framed square in the middle of image. The another big square is an extrapolation artifact and is related to poor classification performance on the upper left corner of original square. The training data are only from the framed area. (*Right*) Extrapolation artifacts on spiral classification.

3 Interactive Intervention into Learning

Human observer of the learning process of neural network may be allowed to interfere with this learning. Such intervention should allow for incorporation of his or her lifelong experience into the learning process. We will study several alternatives of interactive intervention into learning:

- amplification of outputs of neurons,
- amplification of inputs of network,
- manual adjustment of bias of neuron,
- adjustment of individual learning parameters of neurons,
- reinitialization of individual neurons.

Some of these interventions require small modifications of learning algorithm or modification of structure of neural network. We will use error backpropagation algorithm for the learning of network. Consider multilayer perceptron neural network with neuron activations x_i, link weights w_{ij} which links the j-th neuron into i-th neuron, biases (thresholds) θ_i, and neuron activation functions $f_i(in_i)$, described by (1). The in_i is input into i-th neuron and M is the number of links connecting into i-th neuron. Gradient based error minimizing adaptation of weights follows (2), with the γ learning rate constant, and δ_i error signal. The $f'(in_i)$ is the derivative of activation function $f(in_i)$. Error signal δ_i for output neurons can be computed using (3a), but for neurons in hidden layer the (3b) should be used instead. The N_h is number of links coming from i-th neuron and h is index of these links and corresponding neurons.

$$x_i = f_i(in_i), \quad in_i = \sum_{j=1}^{M} w_{ij}x_j + \theta_i \tag{1}$$

$$\Delta w_{ij} = \gamma \delta_i x_j \tag{2}$$

$$\delta_i = (ev_i - x_i)f'(in_i), \quad \delta_i = f'(in_i)\sum_{h=1}^{N_h} \delta_h w_{hi} \tag{3}$$

Rule (3b) is the error backpropagation rule, it defines the backward propagation of error through network. Rule (2) defines weight changes for minimization of this error, and (3a) defines initial error signal on the network output. The error backpropagation algorithm is defined by rules (1), (2), and (3).

3.1 Amplification of Outputs of Neurons and Inputs of Network

The influence of individual neuron on the function of whole network is weighted by weights on links connected to this neuron. We add another weight onto output of every neuron to further control influence of individual neurons. This weight is manually adjusted by a human observer during the learning. Let's call this weight a master weight (mw). Equation (1) of network description has to be changed to (4), so the activations of neurons are multiplied with this mw. Further, backpropagation algorithm rules (3) have to be modified into Neural network usually converges during the learning to a certain stable point, where the weights change only slightly. During this convergence, sudden amplification of outputs of neurons by a human observer may have an interesting effect, mainly in stability disruption. (5).

$$x_i = f_i(in_i)mw_i \tag{4}$$

$$\delta_i = (ev_i - x_i)f'(in_i)mw_i, \quad \delta_i = f'(in_i)mw_i \sum_{h=1}^{N_h} \delta_h w_{hi} \tag{5}$$

$$in_i = in_i^{orig} ss_i \tag{6}$$

Similarly to amplification of outputs of neurons, inputs of the network might be amplified too. Human observer then gains ability to control the amplitude of signals which enter the network. Optimal amplification for the learning of given task then can be searched for. Let's call this amplification parameter a sensoric strength (ss). We must introduce a new equation (6) to describe this modified amplified input signal, where in_i^{orig} is original input signal without any amplification. The input signal after its amplification or attenuation is further propagated through the network. Learning algorithm will not be affected by the amplification but it is important to consider the amplified signal when computing the weight changes for the neurons of the first hidden layer.

3.2 Manual Adjustment of Bias of Neuron

Manual adjustment of bias value of neuron allows for balancing the output of given neuron – shifting the output of neuron to bigger or lower values. Bias or threshold is one special weight of every neuron in network. It is used to balance the activation of a neuron, responding to its inputs. Let's call additional bias weighting parameter a master threshold (mt). Equation (1) will be changed into (7), and the (2) when used for a bias adaptation must be modified too, into (8).

$$in_i = \sum_{j=1}^{M} w_{ij} x_j + \theta_i mt_i \tag{7}$$

$$\Delta\theta_i = \gamma \delta_i mt_i \tag{8}$$

3.3 Adjustment of Individual Learning Parameters of Neurons

Manual adjustment of individual learning rate parameters of backpropagation algorithm for individual neurons allows for control of adaptation rate of neurons. This allows a human observer to freeze well responding neurons, and set up more aggressive learning rate for these with not so good response. In algorithm we must introduce individual learning rate parameters γ_i instead of fixed learning rate γ in standard error backpropagation algorithm. Equation (2) is then changed into (9).

$$\Delta w_{ij} = \gamma_i \delta_i x_j \tag{9}$$

3.4 Reinitialization of Individual Neurons

More radical form of dealing with not well responding neurons is their reinitialization. Reinitialization of particular neuron is done by setting of weights on its links into random values from interval used for initialization of neural network. Reinitialization of important neuron can damage also the behavior of neurons connected to it, however reinitialization of not well responding neuron should be not so damaging as there is a chance, that other neurons are not tightly connected to this particular poorly behaving neuron. Useful practice is to lower learning rate to all well behaving neurons prior such reinitializations, to mitigate possible damage. Reinitialization of not well responding neurons is quick and easy to do for a human observer. It can be used to deal with neurons in the saturation stage, or to explore the parameter space of neural network during learning by observing convergence of reinitialized neurons.

The spectrum of intervention methods should allow a human observer of learning of neural network to exploit a maximum possibilities of such interaction. However the bigger this spectrum is, the more time for learning how to use them is necessary.

4 Experiments

We will study easily visualizable classification tasks. We will run them through visualized interactive learning with multilayer perceptron networks and error backpropagation algorithm and describe our experience with this interactive learning. Visualization method described in Sect.2 will be used for visualization and methods described in Sect.3 will be used for interaction.

Tasks of classification of two-dimensional geometrical forms of circle, spiral, square, and square frame will be studied. See Fig.3 for exact view of these forms. The classification task itself is to classify whether a given point falls inside the form or outside it. The point is defined by its coordinates.

Fig. 3. Graphical interpretation of studied classification tasks

The aim of experiments is to try to extract the know-how to enable effective solution of classification tasks and also effective usage of interactive interventions. The effectiveness is evaluated in terms of quality of acquired knowledge. The network is evaluated by observing how was the resulting knowledge built, i.e. which building blocks were formed during the learning, connected together and modified for error minimization. These building blocks are represented by visualized responses of hidden layers neurons to function signal.

4.1 Network Inputs Amplification and Reinitialization of Neurons

Low levels of the input signals which enters the network may cause learning problems. The confirmed premise for experiments was, that after increasing of levels of input signals, backpropagation learning should be able to keep the correct direction for approaching the closest minimum in error space from the start of learning, thus avoiding possible saturations. See Fig.5 for an example.

Reinitialization of neurons is the fastest correctional intervention into the learning process out of the interactive interventions described previously in the paper. The experiments showed that it is also probably the most useful interactive intervention. However, when most of the neurons were saturated, reinitializations did not have a required effect. After reinitialization, the weights did return to their original values, and the rest of saturated network did not move its weights in any direction. Actual weight vector can not escape from a local extreme when certain amount of neurons are saturated.

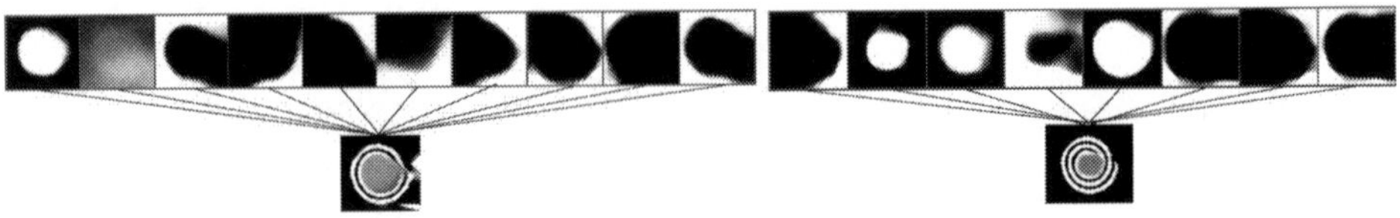

Fig. 4. A larger number of hidden layer neurons does not guarantee for qualitatively better resulting knowledge. More important is, which parts does the resulting knowledge comprise of. Two networks learning the spiral classification were trained in the same number of iterations. Only the second hidden layer is displayed on the figure.

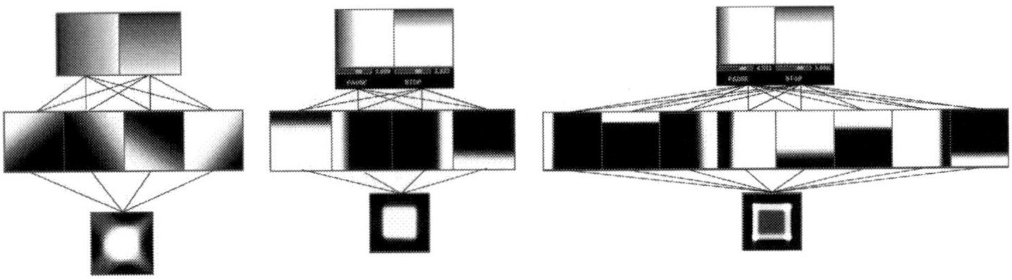

Fig. 5. (*Left*) Average best result acquired by the backpropagation learning for the square classification with a minimal topology without the input signals amplification. (*Right*) Results obtained with optimized input signal amplifications.

4.2 Adjustment of Individual Learning Parameters of Neurons

Adjustment of individual learning rate parameters of neurons has proved itself as a suitable complement for reinitialization of neurons that responded unfavorably to a function signal. It is useful to set the learning rate parameter for well responding neurons to the value close to zero. This fixes the favorable building blocks and only unfavorable blocks are further changing. It enables human observer to reduce the search-space of learning algorithm according to his visual impression. Setting the learning rate parameter to zero causes that the search will not move along the given axis anymore. Left part of Fig.6 represents an example of described procedure.

4.3 Manual Adjustment of Bias of Neuron

Our experiments were focused on question how bias adjustments affects the resulting knowledge. The interactive adjustments were incorporated during the learning, and after the learning too. The goal is to improve acquired knowledge.

Problem with this approach are quick reactions of learning algorithm to user interventions. Learning algorithm quickly balances weights of neurons to compensate interventions, and this limits usability of bias adjustments. Also reactions of other neurons are sometimes dramatic and hard to predict with this approach. However, bias adjustments did not disrupt the stability of learning behavior of the network as much as the signal amplifications. Bias adjustments

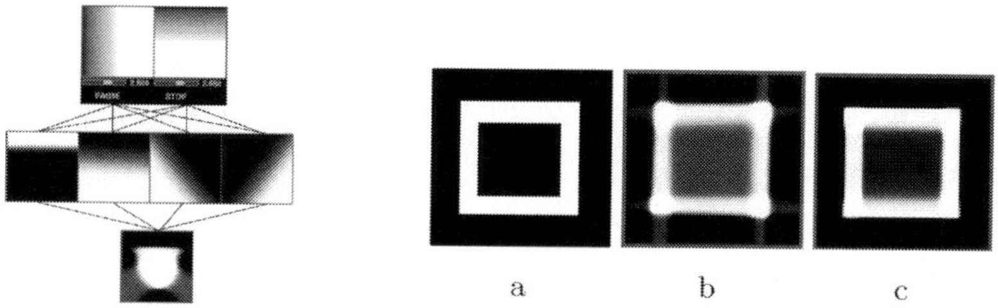

Fig. 6. (*Left*) First two hidden layer neurons respond well to the second input – sensor. To enforce influence of their behavior, their learning rate parameters have to be set to small values, while rest two neurons have to be reinitialized. (*Right*) Example of a result achievable with a proper bias adjustment and signal amplifications, (a) is the optimal result we want to obtain, (b) is result of learning prior interactive intervention, and (c) is almost flawless classification after interactive refinement.

proved useful after the learning was finished, to refine final behavior of network, see right part of Fig.6.

4.4 Amplification of Outputs of Neurons and Virtual Input Units

With various task types and topologies we found interactive amplifications of neurons outputs prone to damaging the knowledge of the network acquired before their application. These amplifications can easily destabilize learning and they can lead into oscillations in learning behavior. Simple amplification of outputs of neurons as described by (5) seems not well suited for interactive interventions. However, as with bias adjustments, amplification of outputs of neurons is useful after the learning is finished, it is demonstrated on the left part of Fig.7.

When solving more complicated tasks, it is reasonable to create functional links [12] feeding the network with virtual signals formed by certain functions applied on real input signals of network. Addition of these virtual inputs makes an aid for neurons on the first hidden layer. These neurons can use this added degree of freedom in the weight space for adjustment of their behavior. See right part of Fig.7 for an example.

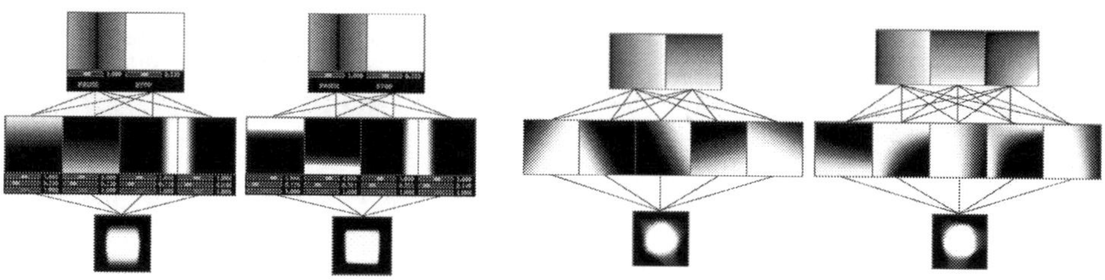

Fig. 7. (*Left*) Decision borders represented by the first two hidden neurons are blurred, which slightly deforms resulting knowledge. After the learning, signal intensities from these neurons were amplified, which qualitatively improved resulting classification. (*Right*) Several neurons from hidden layer respond to added virtual input unit, a result is a smoother decision border for circle classification.

5 Conclusion

Interactive interventions into learning of neural networks shift the focus of learning from simple error correction to a process guided by a human observer – teacher. The goal can be just acceleration of learning pace, but it can move into ability to embed human knowledge into learning process. As it is an interactive process it also allows a human observer to better understand learned behavior of a neural network by observing consequences of his or her interventions.

The framework for interactive learning presented in this paper differs from this used in more established field of interactive evolutionary computation. The main difference is requirement of expert knowledge from neural network area, which makes it a research tool. We hope in future interactive learning of neural networks can be refined enough to be used in real world applications.

Our presented experiments demonstrate usefulness of approach and show the characteristics of particular methods of intervention. Our future research is focused on neural networks with a big number of neurons, where we apply clustering methods to limit the amount of information for a visualization.

References

1. Takagi, H.: Interactive evolutionary computation: Fusion of the capabilities of EC optimization and human evaluation. Proceedings of the IEEE **89**(9) (2001) 1275–1296
2. Weichert, J., Tesauro, G.: Visualizing processes in neural networks. IBM J. Res. Develop. **35** (1991) 244+
3. Craven, M.W., Shavlik, J.W.: Visualizing learning and computation in artificial neural networks. International Journal on Artificial Intelligence Tools (1) (1991) 399–425
4. Olden, J., Jackson, D.: Illuminating the "black box": Understanding variable contributions in artificial neural networks. Ecological Modelling **154** (2002) 135–150
5. Edlund, M., Caudel, T.: Realtime visualization of the learning processes in the lapart neural architecture as it controls a simulated autonomous vehicle. Proceedings of the International Joint Conference on Neural Networks **3** (2000) 41+
6. Streeter, M., Ward, M., Alvarez, S.A.: Nvis: An interactive visualization tool for neural networks. Proceedings of SPIE Symposium on Visual Data Exploration and Analysis **4302**(8) (2001) 234–241
7. Duch, W.: Coloring black boxes: visualization of neural network decision. Proc. of International Joint Conference on Neural Networks (IJCNN) **1** (2003) 1735–1740
8. Duch, W.: Visualization of hidden node activity in neural networks: I. visualization methods. Lecture Notes in Computer Science **3070** (2004) 38–43
9. Tzeng, F.Y., Ma, K.L.: Opening the black box - data driven visualization of neural networks. Proceedings of IEEE Visualization '05 Conference (2005) 383–390
10. Haykin, S.: Neural Networks: A Comprehensive Foundation. Macmillan College Publishing Company, Inc., New York, USA (1994)
11. Užák, M.: Visualization and interaction in the process of neural network learning. Master's thesis, Technical university of Košice (2005) in Slovak.
12. Pao, Y.H.: Adaptive pattern recognition and neural networks. Addison-Wesley Longman Publishing Co., Inc., Boston, MA, USA (1989)

Analytic Equivalence of Bayes a Posteriori Distributions

Takeshi Matsuda and Sumio Watanabe

Department of Computational Intelligence and Systems Science
Tokyo Institute of Technology
matsuken@cs.pi.titech.ac.jp

Abstract. A lot of learning machines which have hidden variables or hierarchical structures are singular statistical models. They have singular Fisher information matrices and different learning performance from regular statistical models. In this paper, we prove mathematically that the learning coefficient is determined by the analytic equivalence class of Kullback information, and show experimentally that the stochastic complexity by the MCMC method is also given by the equivalence class.

1 Introduction

Learning machines such as layered neural networks, normal mixtures, hidden Markov models, Boltzmann machines, Bayes networks and stochastic context-free grammars are not regular statistical models, because their Fisher information matrices are not positive definite. These learning machines are called *singular statistical models* because they are not subject to the conventional statistical theory of regular statistical models. In fact, neither the distribution of the maximum likelihood estimator nor the Bayes a posteriori distribution converges to the normal distribution, even when the number of training samples goes to infinity.

Recently, it was proved that the generalization performance of a singular learning machine in Bayes estimation is determined by the algebraic geometrical structure of the learning machine [5]. The generalization error G, which is defined as the expectation value of the Kullback information from the true distribution to the Bayes predictive distribution, is equal to

$$G = \frac{\lambda}{n} + o(\frac{1}{n}),$$

where n is the number of training samples and λ is the learning coefficient. The constant $(-\lambda)$ is equal to the largest pole of the zeta function of a learning machine,

$$\zeta(z) = \int H(w)^z \varphi(w) dw \quad (z \in \mathbf{C}),$$

where $H(w)$ is the Kullback information from the true distribution to the learning machine with the parameter w and $\varphi(w)$ is the Bayes a priori distribution.

S. Kollias et al. (Eds.): ICANN 2006, Part I, LNCS 4131, pp. 113–121, 2006.

The learning coefficients of some learning machines, for example, a three-layer perceptron and a reduced rank regression, have been obtained by using resolution of singularities [2,3], which clarified that the generalization errors of singular learning machines are smaller than those of regular statistical models, if Bayes estimation is employed in learning.

In this paper, we introduce the concept of analytic equivalence between the Kullback informations, and show the following three facts.

(1) We prove that, if two learning machines are analytically equivalent, then they have the same learning coefficient.
(2) For the case when the Kullback information is defined on the two-dimensional Euclidean space, we derive the concrete learning coefficient of a given equivalence class.
(3) We show experimentally that the the stochastic complexity obtained by the Markov chain Monte Carlo method is also determined by the analytic equivalence class.

In regular statistical models, the asymptotic behavior of a learning machine is completely determined by the Fisher information matrix, whereas in singular learning machines, it is determined by the analytic equivalence class of the Kullback information.

2 Statistical Framework of Machine Learning

In this section, we summarize the well known statistical framework of Bayes estimation.

2.1 Bayes Learning

Let $q(x)$ be a probability density function called as the true distribution which is defined on the N-dimensional Euclidean space, $\mathbf{R}^N$. A set of random variables

$$X^n = (X_1, X_2, ..., X_n)$$

consists of training samples which are independently taken from the probability distribution $q(x)dx$. The integer n is referred to as the number of training samples. A learning machine is represented by a conditional probability density function $p(x|w)$ where w is a d-dimensional parameter. When an a priori probability density function $\varphi(w)$ is given on $\mathbf{R}^d$, the Bayes a posteriori distribution is defined by

$$p(w|X^n) = \frac{1}{Z(X^n)}\varphi(w)\prod_{i=1}^{n} p(X_i|w),$$

where and $Z(X^n)$ is the normalizing constant. The Bayes predictive distribution is also defined by

$$p(x|X^n) = \int p(x|w)\, p(w|X^n)\, dw,$$

which is the estimated probability density function on $\mathbf{R}^N$ by Bayes learning. The Generalization error $G(n)$ is measured by the average Kullback information from the true distribution $q(x)dx$ to the predictive distribution $p(x|X^n)dx$,

$$G(n) = E\Big[\int q(x)\log\frac{q(x)}{p(x|X^n)}dx\Big],$$

where $E[\cdot]$ denotes the expectation value overall sets of X^n. Also we define the stochastic complexity by

$$F(n) = E\Big[-\log Z(X^n)\Big] + n\int q(x)\log q(x)dx.$$

It is easy to show that

$$G(n) = F(n+1) - F(n)$$

holds for an arbitrary natural number n. The stochastic complexity indicates how appropriate the set $p(x|w)$ and $\varphi(w)$ is for a given training sample set X^n.

2.2 Asymptotic Theory

In learning theory, it is important to clarify the asymptotic behaviors of $G(n)$ and $F(n)$. The relation between algebraic geometry of the Kullback information and singular learning machines was clarified, and the following theorem was proved.

Theorem 1. *When n tends to infinity, the generalization error and the stochastic complexity are respectively given by*

$$\begin{aligned} G(n) &= \frac{\lambda}{n} + o(\frac{1}{n}), \\ F(n) &= \lambda\log n - (m-1)\log\log n + O(1), \end{aligned}$$

where $(-\lambda)$ *and* m *are respectively equal to the largest pole and its order of the zeta function,*

$$\zeta(z) = \int H(w)^z\,\varphi(w)\,dw.$$

Here $H(w)$ *is the Kullback information*

$$H(w) = \int q(x)\log\frac{q(x)}{p(x|w)}dx.$$

Proof. The proof is given in [5].

This theorem shows that the learning coefficient is determined by the Kullback information $H(w)$ and the a priori distribution $\varphi(w)$.

3 Analytic Equivalence of Kullback Information

In this section we introduce the concept of analytic equivalence and prove that, if the Kullback informations are analytically equivalent, then they have the same learning coefficient.

Definition 1. *Let U and V be open sets in $\mathbf{R}^d$ whose closures are compact. Two real analytic functions $K(w)$ on U and $H(w)$ on V are said to be analytically equivalent if there exists a bijective analytic map $g : V \to U$*

$$H(w) = K(g(w)) \quad (w \in V)$$

and the Jacobian $|g'(w)|$ satisfies the condition that $\epsilon < |g'(w)| < C$ in V for some $\epsilon, C > 0$.

Then by the definition of the analytic equivalence, the following theorem holds.

Theorem 2. *Assume that two real analytic functions $K(w)$ on U and $H(w)$ on V are analytically equivalent. Then two zeta functions*

$$\zeta_1(z) = \int_U K(w)^z dw$$

$$\zeta_2(z) = \int_V H(w)^z dw$$

have the same largest pole.

Proof. It is well known that $\zeta_1(z)$ and $\zeta_2(z)$ are meromorphic functions and all poles of them are negative and real numbers [4]. From the definition, it follows that

$$\zeta_2(z) = \int_U H(g(w))^z |g'(w)| dw$$

$$= \int_U K(w)^z |g'(w)| dw$$

Let $(-\lambda)$ be the largest pole of $\zeta_1(z)$. When z is real and $z > -\lambda$

$$\epsilon|\zeta_1(z)| < |\zeta_2(z)| < C|\zeta_1(z)|$$

This inequality shows that the largest poles should coincide.

Note that two zeta functions do not have the same second largest pole in general.

Definition 2. *Let $v = (v_1, \ldots, v_n)$ be a set of nonnegative integers. For a monomial $x^u = x_1^{u_1} \cdots x_n^{u_n}$, we define the weighted degree $ord_w(x^u)$ with the weight v by*

$$ord_w(x^u) = < v, u > = v_1 u_1 + \ldots + v_n u_n.$$

A polynomial is said to be guasi-homogeneous if it is a linear combination of the monomials which have the same weighted degree with some weight.

Definition 3. *An analytic function f is said to have an algebraic isolated singularity at O, if the dimension of a real vector space*

$$M(f) = R[[x_1, \cdots, x_n]] / < \frac{\partial f}{\partial x_1}, \cdots, \frac{\partial f}{\partial x_n} >$$

is finite.

The following theorem shows a sufficient condition of the analytic equivalence.

Theorem 3. *Let f be an analytic function*

$$f = f_d + f_{d+1} + f_{d+2} + \cdots, f_d \neq 0$$

where f_d is a quasi-homegeneous polynomial of degree q with weight $\mathtt{v}$. If the weighted degree of $\mathtt{x}^{u_i}$ exceeds d, $c_1, \ldots, c_s$ are constants, then f and $f_d + c_1\mathtt{x}^{u_1} + \cdots + c_s\mathtt{x}^{u_s}$ are analytically equivalent.

Proof. For the proof of this theorem, see [1].

4 Two Dimensional Parameter Space

In the previous section, we have shown that the learning coefficient is determined by the analytic equivalence class. In this section we give the concrete learning coefficients for a given analytic function on two-dimensional space.

Theorem 4. *Let f be an analytic function given by*

$$f(x, y) = \Sigma_{k+l=4} a_{kl} x^k y^l + \Sigma_{2 \leq i+j \leq 3} b_{ij} x^i y^j,$$

where a_{kl} and b_{ij} are the real number coefficients of $x^k y^l$, $x^i y^j$, respectively. We consider the zeta function such that

$$\zeta(z) = \int f^{sz} dx dy.$$

Then largest pole of the zeta function is as follows.

$$\lambda = \begin{cases} \frac{2}{as} & (k' > i', l' = j') or (\Sigma b_{ij} x^i y^j \text{is a symmetric expression.}) \\ \frac{2n+1}{(4n+l')s} & (4 - a < j', k' > i', l' < j') \\ \frac{2}{(4-a)s} & (4 - a \geq j', k' > i', l' > j') \end{cases}$$

where k' is the value of the minimum k which satisfies $a_{kl} \neq 0$, l' is the value of the minimum l which satisfies $a_{kl} \neq 0$, i' is the value of the minimum i which satisfies $b_{ij} \neq 0$, j' is the value of the minimum j which satisfies $b_{ij} \neq 0$.

Proof. Let X be the curve $f(x, y) = 0$. We put $x = x$, $y = xy$ on X_{11}. Then we have $f_{11}^{sz} =$

$$x^{asz+1} y^{sl'z} (\Sigma_{k+l=4} a_{kl} x^{4-a} y^{l-l'} + \Sigma_{2 \leq i+j \leq 3} b_{ij} x^{i+j-a} y^{j-l'})^{sz}.$$

Similarly, we put $x = xy$, $y = y$ on X_{12}. Then we obtain $f_{12}^{sz} =$

$$x^{ai'z}y^{asz+1}(\Sigma_{k+l=4}a_{kl}x^{k-i'}y^{k+l-a} + 1 + \Sigma_{2\leq i+j\leq 3, i\neq i'}b_{ij}x^{i-i'}y^{i+j-a})^{sz}.$$

where $k' > i', l' < j'$.
Hence,

$$\zeta(z) = \int_{X_{11}} f_{11}^{sz}dxdy + \int_{X_{12}} f_{12}^{sz}dxdy.$$

Here, $\Sigma_{k+l=4}a_{kl}x^{4-a}y^{l-l'} + \Sigma_{2\leq i+j\leq 3}b_{ij}x^{i+j-a}y^{j-l'}$ and $x^{4-a} + y^{j'-l'}$ are analytic equvalence. Therefore, we have to consider only about

$$\int_{X_{11}} x^{asz+1}y^{sl'z}(x^{4-a} + y^{j'-l'})^{sz}dxdy.$$

When continuing resolution of singularity, the zeta function $\zeta(z)$ is as follows.

$$\zeta(z) = \int_{X_{n+1,1}} x^{n(4sz+2)+sl'z}y^{4sz+2+sl'z}(1+\cdots)^{sz}dxdy$$
$$+ \int_{X_{n+1,2}} x^{asz+1}y^{n(4sz+2)+sl'z}(x^{k+l-a} + y^{j'-l'-n(k+l-a)})^{sz}dxdy,$$

where $(j' - l') = n(k + l - a)$. Hence, we obtain

$$\lambda = \frac{2n+1}{(4n+l')s}.$$

In the other case, too, it is possible to prove in the same way. Also, if replacing x and y, we can get the value of λ in all cases.

5 Stochastic Complexity by MCMC Method

In the previous section, we have shown that the learning coefficients are determined by the analytic equivalence classes. In this section, by comparing the theoretical results with the numerical results by the Markov chain Monte Carlo method, we show that the stochastic complexities in real applications are also determined by the equivalence classes. Let us study the function,

$$F(n) = -\log \int \exp(-nf(x,y))\varphi(x,y)dxdy.$$

From the theoretical point of view, it has the asymptotic expansion,

$$F(n) = \lambda \log n - (m-1)\log\log n.$$

In the real applications of Bayes estimation, $F(n)$ is numerically calculated by

$$F(n) = \int_0^1 E_t[nf(x,y))]dt$$

where $E_t[\cdot]$ shows the expectation value over the probability distribution,

$$E_t[nf(x,y)] = \int nf(x,y)p_t(x,y)dxdy,$$

where

$$p_t(x,y) \propto \exp(-ntf(x,y))\varphi(x,y).$$

The random samples subject to $p_t(x,y)$ can be generated by the MCMC method. The following tables show the experimental results of $F(n)$ for $n = 10000$.

analytic function	$F(n)$
$y^2 + x^3$	1.900087
$y^2 + x^3 + y^3$	1.902980
$y^2 + x^3 + xy^3$	1.965838
$y^2 + x^3 + x^4$	2.012928
$y^2 + x^3 + y^3 + xy^3 + x^4$	1.862806
$y^2 + x^3 + y^5$	1.930222
$y^2 + x^3 + x^{10}$	1.910901
$y^2 + x^3 + y^5 x^{10}$	1.909953
$y^2 + x^3 + x^{10} + y^5$	1.914395
$y^2 + x^3 + x^{10} + y^5 + y^5 x^{10}$	1.909564
$y^2 + x^3 + x^{100} y^{100}$	1.890016
$y^2 + x^3 + x^{100} y^{100} + x^{100} + y^{100}$	1.895358

analytic function	$F(n)$
$y^3 + x^5$	1.115474
$y^3 + x^5 + x^2 y^2$	1.162772
$y^3 + x^5 + x^{10} y^{10}$	1.115357
$y^3 + x^5 + x^{15} y^{10}$	1.115979
$y^3 + x^5 + x^{10} y^{15}$	1.114232
$y^3 + x^5 + x^{20} y^{20}$	1.115570
$y^3 + x^5 + x^{100} y^{100}$	1.103289
$y^3 + x^5 + x^{100} y^{100} + x^{100} + y^{100}$	1.101518

analytic function	$F(n)$
$y^5 + x^7$	0.563414
$y^5 + x^7 + x^{10} y^2$	0.555947
$y^5 + x^7 + x^2 y^{10}$	0.561365
$y^5 + x^7 + x^{10} y^{10}$	0.562976
$y^5 + x^7 + x^{10} y^{15}$	0.559782
$y^5 + x^7 + x^{15} y^{10}$	0.552454
$y^5 + x^7 + x^{100} y^{100}$	0.566214
$y^5 + x^7 + x^{100} y^{100} + x^{100} + y^{100}$	0.564594

These results show that the numerically calculated stochastic complexities are determined by the analytic equivalence classes.

6 Discussion

In this paper, we have studied the relation between the learning coefficients and analytic equivalence classes. Let us discuss the results from three viewpoints.

Firstly, from the mathematical point of view, the result of this paper is devoted to the case of isolated singularities. In almost all learning machines, their singularities are not isolated, however, there is no simple criterion that can judge the analytic equivalence for non-isolated singularities. To construct the mathematical criterion of the analytic equivalence class in singular learning machines is the problem for the future study.

Secondly, from the statistical point of view, our result is a generalization of Fisher's asymptotic statistics. If two analytic functions have nondenerate Hesse matrices, then they are analytically equivalent. This is the reason why the learning coefficients of regular statistical models are determined by the dimensions of the parameter spaces. In singular learning machines, even if the learning machines have the same-dimensional parameter spaces, they have different learning coefficients in general. Consequently, the concept of the analytic equivalence class is a generalization of the Fisher information matrix.

And lastly, from the learning theoretical point of view, our result shows how the stochastic complexities are determined in the real world applications. The stochastic complexity is important in Bayes learning, which is applied to model selection and hyperpramater optimization. However, it is well known that it requires huge computational costs to calculate the stochastic compelxity. We expect that a new efficient algorithm based on the concept of analytic equivalence class.

7 Conclusion

We proved mathematically that the learning coefficients are determined by the analytic equivalence class of the Kullback information, and showed experimentally that the practical stochastic complexities are also determined by the analytic equivalence class. To construct the mathematical method which enables us to calculate the learning coefficients for the higher dimensional Kullback information is the problem for the future study.

Acknowledgment. This work was supported by the Ministry of Education, Science, Sports, and Culture in Japan, Grant-in-aid for scientific research 15500310.

References

1. V.I.Arnol'd, "Normal forms of functions in neighbourhoods of degenerate critical points," Russian Mathematical Surveys, Vol.29,No.2,pp.10-50,1974
2. M.Aoyagi, S.Watanabe, "Stochastic complexities of reduced rank regression in Bayesian estimation," Neural Networks, Vol.18,No.7,pp.924-933,2005.

3. M.Aoyagi,S.Watanabe, “Resolution of singularities and generalization error with Bayesian estimation for layered neural network,” Vol.J88-D-II,No.10,pp.2112-2124,2005.
4. M.F. Atiyah, “Resolution of singularities and division of distributions,” Communications of Pure and Applied Mathematics, 13, 145-150, 1970.
5. S.Watanabe,”Algebraic Analysis for Non-identifiable Learning Machines,” Neural Computation, Vol.13, No.4, pp.899-933, 2001

Neural Network Architecture Selection: Size Depends on Function Complexity

Iván Gómez, Leonardo Franco, José L. Subirats, and José M. Jerez

Departamento de Lenguajes y Ciencias de la Computación
Universidad de Málaga, 29071 Málaga, Spain
{ivan, lfranco, jlsubirats, jja}@lcc.uma.es

Abstract. The relationship between generalization ability, neural network size and function complexity have been analyzed in this work. The dependence of the generalization process on the complexity of the function implemented by neural architecture is studied using a recently introduced measure for the complexity of the Boolean functions. Furthermore an association rule discovery (ARD) technique was used to find associations among subsets of items in the whole set of simulations results. The main result of the paper is that for a set of quasi-random generated Boolean functions it is found that large neural networks generalize better on high complexity functions in comparison to smaller ones, which performs better in low and medium complexity functions.

1 Introduction

The learning and generalization properties of artificial neural networks confer to these models a wide applicability in pattern recognition and classification tasks. The cornerstone of the neural network modeling is the selection of the network architecture for a determined application, since there is not a theoretical formula giving clear insight to this problem. Nevertheless, some general theoretical results have been published about the size of the network needed to implement a desired function [1, 2, 3], but at the time of the implementation the theory is not always accurate.

In particular, some authors [3, 4] have demonstrated that very large networks perform sometimes better than smaller ones, whereas others [8] state that the generalization process depends mainly on the weight values distribution and not on the network size. Moreover, most of the practical results based on simulations [6, 7] concentrate on the use of very few functions (up to 30 different functions), and also in general the complexity of the analyzed functions is ignored [3, 9].

In fact, the complexity of Boolean functions has been studied for a long time, the aim of that being to have a criterion for deciding if a problem is easier to solve or implement than another [12, 3].

This work studies the relationship among learning, generalization, network architectures, and complexity over a set of one thousand different boolean functions. With this aim, we use a recently introduced measure for the complexity of Boolean functions [10, 11] to analyze how the network architecture affects the

S. Kollias et al. (Eds.): ICANN 2006, Part I, LNCS 4131, pp. 122–129, 2006.

generalization ability in different classes of functions grouped by their levels of complexity. Also, the huge amount of simulation data led the authors to use a data mining technique (association rules) to find an statistical correlation among generalization, network architecture and functions complexity. Association rule discovery (ARD) is a particular technique widely used in data mining problems [19, 20, 17, 18] to find interesting associations and/or correlation relationships among large sets of data items. This algorithm extracts relevant information from the data and provides rules in the form of "if-then" statements, and, unlike the if-then rules of Logic, association rules are probabilistic in nature.

2 Methods

2.1 Measure for the Complexity of Boolean Functions

The measure proposed in [9] is based on the results obtained in [15, 16], where a close relationship between the number of examples needed to obtain valid generalization and the number of neighboring examples with different outputs was found. The complexity measure $\mathcal{C}[f]$ is obtained from the number of pairs of neighboring examples having different outputs. The complexity measure used in this paper is defined as:

$$\mathcal{C}[f] = \mathcal{C}_1[f] + \mathcal{C}_2[f], \tag{1}$$

where $\mathcal{C}_i[f]$, $i = 1, 2$ are the terms taking into account pairs of examples at a Hamming distance one and two. The first term can be written as:

$$\mathcal{C}_1[f] = \frac{1}{N_{ex} * N_{neigh}} \sum_{j=1}^{N_{ex}} \left(\sum_{\{l|Hamming(e_j,e_l)=1\}} (|f(e_j) - f(e_l)|) \right), \tag{2}$$

where the first factor, $\frac{1}{N_{ex} * N_{neigh}}$, is a normalization one, counting for the total number of pairs considered, N_{ex} is the total number of examples equals to 2^N, and N_{neigh} stands for the number of neighbor examples at a Hamming distance of 1. The second term $\mathcal{C}_2[f]$ is constructed in an analogous way. The complexity of the Boolean functions using the measure of Eq. 1 ranges from 0 to 1.5 [9].

2.2 Association Rules Discovery

Data mining, also known as Knowledge Discovery in Databases(KDD), has been defined as the extraction of implicit previously unknown useful information from data. Motivated by the huge amount of information provided by the simulation results, data mining techniques were applied to the process of finding interesting relationships among generalization, complexity and network architecture.

ARD technique is a data mining method used in many applications to discover associations patterns between subsets of items in transactions databases. It detects set of elements that co-occur frequently, creating relations of the form

$X->Y$, which means that when the antecedent of the rule, X, occurs, it is probably for the consequent, Y, also to be occur. This technique has been extensively used in market analysis [19, 20] and analysis of gene expression data [17, 18].

In addition to the antecedent (*if* condition) and the consequent (*then* condition), an association rule provides two numbers that express the degree of uncertainty about the rule. In association analysis, the antecedent and consequent are disjointed items sets. The first number, named support for the rule, is simply the number of rows that include all items in both antecedent and consequent parts of the rule. The second number represents the lift, another parameter of interest in the association analysis, that gives the ratio of confidence to expected confidence. Expected confidence is the number of transactions, that include the consequent divided by the total number of transactions.

In the context of this work, we define transactions that contains function complexity, network architecture (number of hidden neurons) and generalization capability. Each item included in the transaction was discretized into four categories as follows: $[0, 0.25)$, $[0.25, 0.50)$, $[0.50, 0.75)$ and $[0.75, 1]$ for functions complexity; $[0.5, 0.70)$, $[0.70, 0.85)$, $[0.85, 1.0)$ for generalization ability; and $[0, 10)$, $[10, 20)$, $[20, 30)$, $[30, 100]$ for the number of hidden neurons.

One of the limitations of ARD is the large amount of rules that can be generated, which becomes a major problem in many applications. Some post-processing pruning methods have been proposed to reduce the number of rules [18]. One of the limitations of ARD is the large amount of rules that can be generated. Some pruning methods have been proposed to reduce the number of rules. In this work, we have used a filter designed to eliminate those rules that present low confidence or does not have the generalization item in, what reduces drastically the number of association rules.

2.3 Set of Quasi-random Functions

In a previous work [21], the authors analyzed the generalization ability of different network architectures studying how the generalization ability is affected by the complexity of the functions being implemented and by the size of the architecture. More exactly, generalization ability was tested in six different architectures in size, and a set of 512 symmetric boolean functions was considered. The results showed that was necessary the introduction of a second layer of neurons to improve the generalization ability of very complex functions.

In this paper we extend the study to a set of quasi-random functions. There exists 2^{2^N} boolean functions of N inputs, making their general study very complicated except for very simple and small cases. Totally random functions are very complex with an average complexity around 1.0 [10]. To analyze a large set of different functions we generate functions with different complexity by modifying with a certain probability K outputs of the constant function. The value of K is related to the complexity of the generated function, and the most complex generated function is the parity one. This procedure let us to obtain functions that are asymmetric in the number of outputs and with a complexity in the range

1.0 to 0. One hundred functions were analyzed for each of 10 levels of complexities in which the functions where grouped, with average complexity from 0.1 to 1.0 in steps of 0.1.

2.4 Neural Network Architectures and Simulations

To analyze the generalization ability of different architectures and to study how this property change with network size and functions complexities, we carried intensive numerical simulations for quasi-random Boolean functions generated according to the procedure described before. All the networks were trained with back-propagation with momentum. An early stopping procedure was implemented with the aim of avoid overtraining and improve generalization. The validation error was monitored during training and at the end of the maximum number of epochs permitted, the generalization error was taken at the point where the validation error was minimum.

The number of examples available for training was 256 as in all cases we consider input functions of size N=8. We divided the examples in a training set containing 128 examples, and into validation and generalization test sets containing 64 examples each one. The learning rate constant was fixed during training and equal to 0.01 and the momentum term was set to 0.05. The maximum number of epochs allowed for training was set to 10000.

3 Results

We performed numerical simulations on one hidden layer neural networks using the whole set of quasi-random functions (up to 1000 different functions) with $N = 8$ input variables. The number of neurons for the analyzed cases ranged from 5 to 100 in steps of 5 hidden units. The best results for each complexity level, in terms of the generalization ability obtained, were computed simulating for every network architecture and averaging over the 100 quasi-random functions. In the case of a coincidence in the generalization ability from two architectures with a different number of hidden neurons, a simple and general idea about what network size come from Occam's razor: the simpler the solution the better. In Fig. 1 mean values for architecture sizes are represented vs function complexity, and standard deviations were also computed for every complexity group.

It is possible to observe from Fig. 1 that the most appropriate network size for obtaining the best generalization ability remains small for low level complexity groups (from 0.0 to 0.55), noting also that the standard deviation is also relatively low. As the complexity of the functions increase, the number of hidden neurons grows up fast for groups with a complexity between 0.64 and 0.97. In table 1, the average generalization ability obtained for the different architectures used in every complexity group is shown together with the training error at the point in which the minimum validation error was found.

In Fig. 2 the number of iterations as a function of the number of hidden neurons, for each complexities group, is shown. In this figure the lowest level

Table 1. Average generalization ability and final training error (with standard deviations between parenthesis) obtained for the network architectures constructed to compute all Boolean quasi-random functions with N=8 inputs

Complexity group (mean value)	Generalization ability	Training error
0.07	0.976 (0.009)	0.015 (0.015)
0.14	0.968 (0.010)	0.031 (0.017)
0.27	0.929 (0.019)	0.078 (0.033)
0.34	0.898 (0.023)	0.078 (0.029)
0.45	0.875 (0.022)	0.125 (0.038)
0.55	0.835 (0.026)	0.171 (0.039)
0.64	0.804 (0.029)	0.203 (0.050)
0.75	0.757 (0.032)	0.250 (0.052)
0.86	0.695 (0.041)	0.296 (0.056)
0.97	0.585 (0.067)	0.421 (0.072)

complexities groups are located at the top of the figure, being the highest level complexity group labeled as 10. Figure 2 demonstrates that the time of training (number of epochs) in very complex functions (labeled as $8-10$) is proportional to the network size. Moreover, in the case of low complexity functions, the time

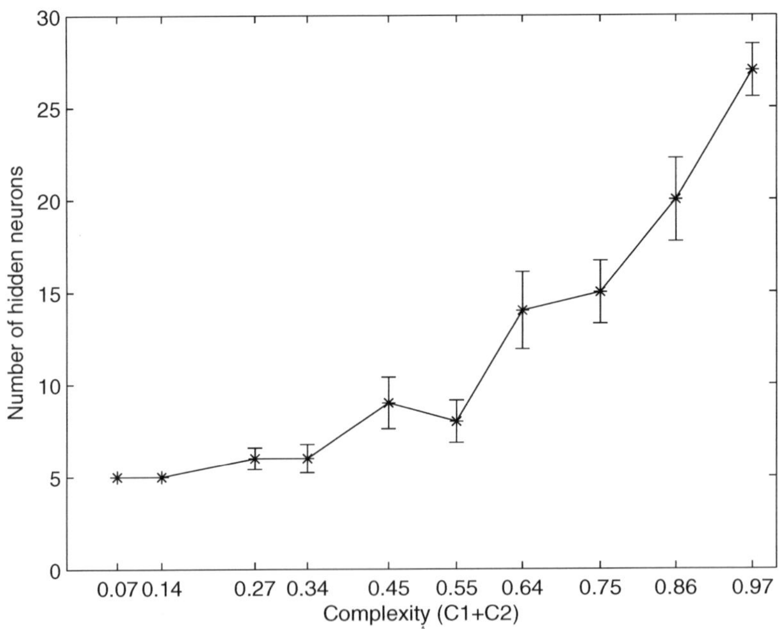

Fig. 1. Number of hidden neurons vs function complexity for quasi-random functions with N=8 inputs. The x-axis labels identify the mean values for each level of complexity.

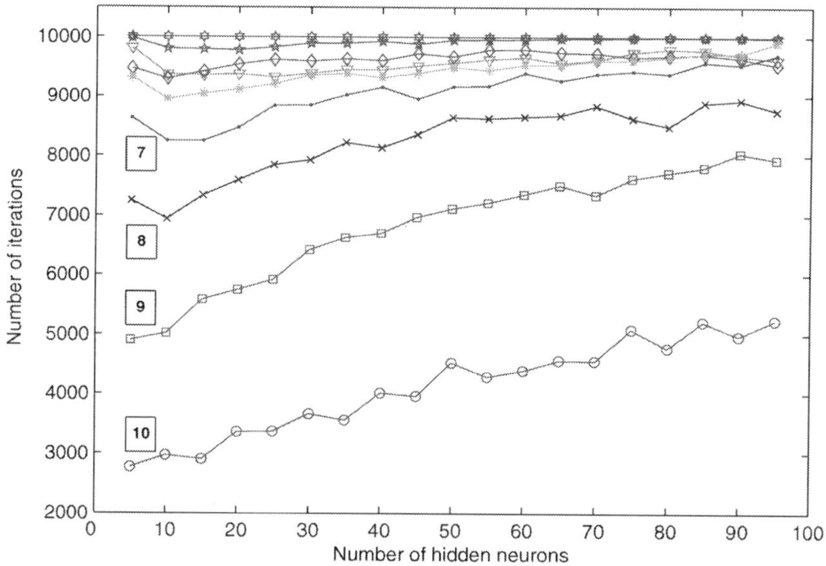

Fig. 2. Training time, in epochs, as a function of the network size, in number of hidden neurons, for every complexity group of Boolean functions

necessary to train the network is approximately constant and independent of the network size selected to implement that function.

The ARD technique was applied to the whole set of 20.000 simulations results. Association rules were extracted with absolute minimum support value of 10, minimum confidence of 40% and minimum improvement of one, obtaining a total of 35 association rules. After filtering and analyzing the set of rules extracted, it is possible to observe that a correlation exists among the architecture size, the complexity level and the generalization ability. That is confirmed through association rules with the form: { low complexity } and { low network size } → { high generalization ability }, with confidence 0.98 and lift 1.66; and others as { high complexity } and { medium network size } → { low generalization ability }, with confidence 0.93 and lift 1.60. In this case, the ARD technique confers statistical precision to graphical results, since the confidence value can be interpreted as the a posteriori probability for the consequent given the antecedent. Moreover, a lift value greater than 1.0 is interpreted as a positive correlation between the antecedent and the consequent.

4 Discussion

We have analyzed through numerical simulations the relationship between neural network size and function complexity. It was found that for groups of low complexity functions, smaller architectures have a better generalization ability than larger ones. Also, it was found that the optimal value for the number of hidden neurons (between the analyzed values) grows with function complexity. This result is what is expected from theoretical reasons but the interestingly, up to our knowledge, is one of the first times that this problem is systematically studied and reported. We have also made use of a extraction rule technique (ARD) to analyze the result of the simulations and this technique confirmed our analysis giving also statistical confidence. Regarding the architecture selection process, much work remains to be done and we are in the process of studying it in multi-layers networks and modular architectures.

Acknowledgement. This work was supported by the "Comisión Interministerial de Ciencia y Tecnología" (Spain) through grant TIN2005-02984, and by FEDER funds. Leonardo Franco acknowledges support from the Spanish Ministry of Education and Science through a "Ramón y Cajal" fellowship.

References

1. Haykin, S. (1994). *Neural Networks: A Comprehensive Foundation.* Macmillan/IEEE Press.
2. Baum, E.B. & Haussler, D. (1989) What size net gives valid generalization ? *Neural Computation, 1*, pp. 151-160..
3. Lawrence, S., Giles, C. L., & Tsoi, A. C. (1996). What Size Neural Network Gives Optimal Generalization ? Convergence Properties of Backpropagation. In Technical Report UMIACS-TR-96-22 and CS-TR-3617, Institute for Advanced Computer Studies, Univ. of Maryland.
4. Caruana, R., Lawrence, S., & Giles, C.L. (2001). Overfitting in Neural Networks: Backpropagation, Conjugate Gradient, and Early Stopping. In Leen, T. K., Dietterich, T. G. & Tresp, V. editors, Advances in Neural Information Processing Systems, MIT Press, *13*, pp. 402-408.
5. Krogh, A. & Hertz,J.A. (1992) A simple weight decay can improve generalization. In J.E. Moody, S. J. Hanson, & R. P. Lippmann editors, Advances in Neural Information Processing Systems Morgan Kaufmann, San Mateo, CA, *4*, pp. 950 957.
6. Prechelt, L. (1998). Automatic Early Stopping Using Cross Validation: Quantifying the Criteria. *Neural Networks, 11*, pp.761-767.
7. Setiono,R. (2001) Feedforward neural network construction using cross-validation, *Neural Computation, 13*, pp. 2865-2877.
8. Bartlett,P.L. (1997). For valid generalization the size of the weights is more important than the size of the network. In M.C. Mozer, M. I. Jordan, & T. Petsche, editors, Advances in Neural Information Processing Systems, MIT Press, *9*, pp. 134-140 .
9. Franco, L. & Anthony, M. (2004). On a generalization complexity measure for Boolean functions. In *Proceedings of the 2004 IEEE International Joint Conference on Neural Networks*, IEEE Press, pp. 973-978.
10. Franco, L. Generalization ability of Boolean functions implemented in feedforward neural networks. Neurocomputing. (2006). In Press.
11. Franco, L. and Anthony, M. The influence of oppositely classified examples on the generalization complexity of Boolean functions. IEEE Transactions on Neural Networks. (2006). In Press.

12. Wegener, I. (1987) *The complexity of Boolean functions.* Wiley and Sons Inc.
13. Siu, K.Y., Roychowdhury, V.P., & Kailath, T. (1991) Depth-Size Tradeoffs for Neural Computation *IEEE Transactions on Computers, 40*, pp. 1402-1412.
14. Franco, L. & Cannas, S.A. (2004). Non glassy ground-state in a long-range antiferromagnetic frustrated model in the hypercubic cell *Physica A, 332*, pp. 337-348.
15. Franco, L. & Cannas, S.A. (2000). Generalization and Selection of Examples in Feedforward Neural Networks. *Neural Computation, 12*, 10, pp. 2405-2426.
16. Franco, L. & Cannas, S.A. (2001). Generalization Properties of Modular Networks: Implementing the Parity Function. *IEEE Transactions on Neural Networks, 12*, pp. 1306-1313.
17. Becquet, C. & Blachon, S. & Jeudy, B. & Boulicaut, J.F. & Gandrillon, O. (2002). Strong association rules mining for large-scale gene-expression data analysis: A case study on human SAGE data. *Genome Biology, 3*, pp. 1-16.
18. Creighton, C. & Hanash, S. (2003). Mining gene expressions databases for association rules. *Bioinformatics, 19*, pp. 79-86.
19. Agrawal, R. & Imielinski, T. & Swami, A. (1993). Mining associations rules between sets of items in large databases. In *Proceedings of the ACM SIGMOD international conference on Management of data*, Washignton D.C., pp. 207-216.
20. Brian, S. & Motwani, R. & Silverstein, C. (1997). Beyond Market baskets: Generalizing associations rules to correlations. In *Proceedings of the ACM SIGMOD conference*, Tucson, pp. 265-276.
21. Franco, L. & Jerez, J.M. & Bravo, J.M (2005). Role of function complexity and network size in the generalization ability of feedforward networks. *LNCS, 3512*, pp. 1-8.

Competitive Repetition-suppression (CoRe) Learning

Davide Bacciu[1,2] and Antonina Starita[2]

[1] IMT Lucca School for Advanced Studies,
Via San Micheletto 3, 55100 Lucca, Italy
davide.bacciu@imtlucca.it,
http://www.imtlucca.it

[2] Dipartimento d'Informatica, Università degli Studi di Pisa,
Largo B. Pontecorvo 3, 56127 Pisa, Italy
starita@di.unipi.it,
http://ciml.di.unipi.it/

Abstract. The paper introduces Competitive Repetition-suppression (CoRe) learning, a novel paradigm inspired by a cortical mechanism of perceptual learning called repetition suppression. CoRe learning is an unsupervised, soft-competitive [1] model with conscience [2] that can be used for self-generating compact neural representations of the input stimuli. The key idea underlying the development of CoRe learning is to exploit the temporal distribution of neurons activations as a source of training information and to drive memory formation. As a case study, the paper reports the CoRe learning rules that have been derived for the unsupervised training of a Radial Basis Function network.

1 Introduction

The present work introduces a novel learning algorithm inspired by a cortical mechanism of perceptual learning called repetition suppression. The fundamental aspects of this cortical mechanism have been modeled in a competitive learning schema, the *Competitive Repetition-suppression (CoRe) Learning*, for the unsupervised generation of compact neural representations of the input stimuli.

We propose the idea that mere stimuli repetition acts as a fundamental resource for efficient memory formation. In particular we explore the neurophysiological hypothesis [3] that the repetition suppression mechanism serves as an unsupervised mean for reducing the size of stimuli representation, that is the number of neurons coding a given stimulus, while strengthening the responses of the most selective neurons, i.e. those showing sharp responses to particular classes of input stimuli.

For the sake of the present paper, we focus on the application of the proposed model to the automated construction of a Radial Basis Function Network (RBFN) [4]. In particular, we show how CoRe learning can be used to optimize the number, position and shape of RBFN gaussian kernels by means of a completely unsupervised training procedure.

S. Kollias et al. (Eds.): ICANN 2006, Part I, LNCS 4131, pp. 130–139, 2006.

In the following sections we introduce the neurophysiological findings that have inspired this work together with the computational model of CoRe learning. The paper ends with the results of the tests conducted on the Iris classification dataset.

2 Neurophysiological Foundations of Repetition Suppression

The ability of humans and animals to learn from experience is recognized to be supported by multiple memory systems with different functional characteristics and neural basis. Our work has been inspired by the characteristics of a particularly interesting learning scheme, called *perceptual learning* [5]. Perceptual learning supports the formation of non-declarative memory at the level of the visual cortex, providing a mean for improving the performance on several sensory tasks following practice.

In particular, we focused on a phenomenon known as *repetition suppression* (RS), which appears to be fundamental for mediating perceptual learning [3]. Repetition suppression induces long-lasting changes to the visual cortex, decreasing the neural activity as a consequence of the repeated presentation of similar stimuli. This cortical mechanism is strictly item-specific and appears at an abstract representational level. For instance, neurophysiological evidences have shown that its effects can be observed also when the repeated stimulus is presented at different retinal locations and in case of variations to the visual stimulus geometry. Repetition suppression does not depend on the behavioral significance of the stimuli, i.e. it is not specifically linked to the active maintenance of a sample in memory, nor it requires any form of response/reward signal. Furthermore, RS is an unconscious process, since its effects can be recorded also in anesthetized subjects [3].

In brief, repetition suppression involves sharpening the neural representation of items by means of an overall reduction of the number of active neurons which is counterbalanced by the steepening of the response of the most item-selective neurons. This process seems to be aimed at the selection of neurons that act as detectors of the most informative features. Moreover it appears that such a process may facilitate novelty detection, since more familiar stimuli experience more suppression than unfrequent items.

We suggest that repetition suppression, and in general perceptual learning, may provide interesting hints for the development of innovative learning schemes and mechanisms. The literature in this field offers very few works that try to exploit the neurophysiological issues described so far. Mozer [6] proposes a computational model reproducing the repetition suppression phenomenon by means of a network of binary-hypothesis neurons which uses blind equalization to suppress the irrelevant inputs and to enhance the most active neurons, by steepening the curve of the sigmoid driving their responses. Another interesting model is that proposed by French and called *activation sharpening* [7]. This algorithm extends the standard backpropagation with an extra step in which activations patterns at the hidden layer are sharpened, i.e., the activation level of the most active hidden nodes is increased slightly for each pattern, while the other nodes activations are decreased. The activation sharpening process, although not mediated by repetition, offered an interesting starting point for the development of our repetition suppression-based model. However, the learning scheme we propose is intended to have a broader scope than those described so far and to be applicable to a wide range of neural networks and learning systems on real world tasks (e.g. machine vision).

3 Competitive Repetition-suppression (CoRe) Learning

3.1 A Soft-Competitive Approach to Learning

The general idea underlying the proposed model is to make the neural population evolve in the direction of maximum selectivity by means of a procedure that penalizes or enhances the responses of the neurons on the basis of the stimuli frequency. This approach falls into the family of the competitive learning [8] schemes, in which units compete to be active during training. Hard-competitive approaches such as Winner Takes All (WTA), allow only the winning unit, i.e. the one with the highest activation, to learn on each case. The soft-competitive approach [1], on the other hand, allows each unit, or the units from a selected subset, to adapt its weights in proportion to its activation strength.

CoRe learning falls into the family of soft-competitive approaches. At each step, it selects the most active neurons to form the *winners pool*, while the remainder of the units forms the *losers pool*. We define the winners pool for the input x_k as the set of units u_i that fires more than θ, that is

$$win_k = \{i \mid y_i(x_k) \geq \theta, \; u_i \in U\} \tag{1}$$

where $y_i(x_k)$ is the activation of the i-th unit on the k-th input pattern. The definition of the losers pool can be obtained by mirroring the inequality in (1). The competition is engaged between the units of two pools: the winners gets rewarded and their activity is strengthened, while the losers are penalized depending on the amount of the repetition suppression generated. In this aspect, CoRe extends the *rival penalized competitive learning* (RPCL) algorithm [9]. The key idea of RPCL is that for each stimulus not only the winner is learned to approach the input pattern, but also the second winner (the rival) is de-learned away from it for a bit. We extended this approach by defining a *soft-rival* algorithm, in which winner and rival refer to pools of units.

3.2 CoRe Learning

The primary issue for implementing a repetition suppression learning scheme is modeling the stimuli repetition. We build our scheme on a parameter, the

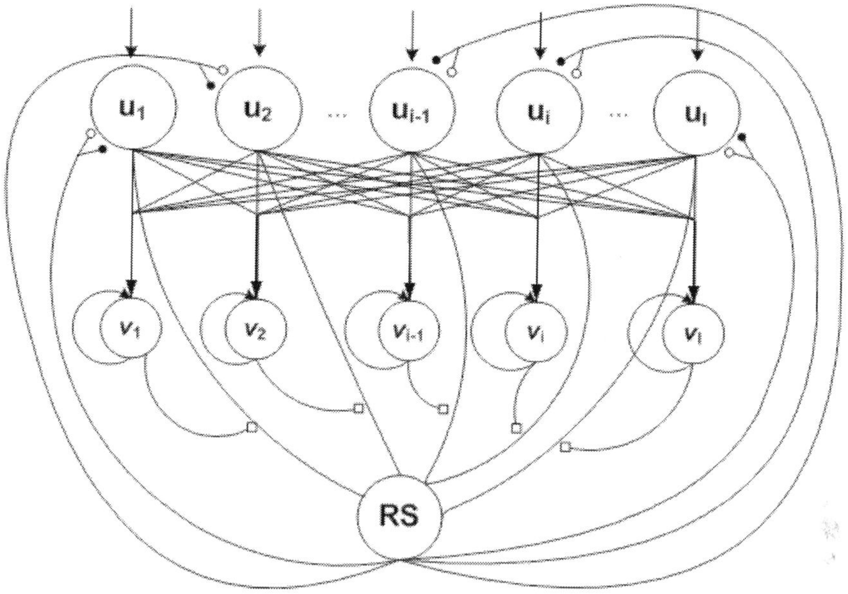

Fig. 1. General form of a CoRe learning layer of neurons: units u_i are the feature detectors; ν_i accumulates the conscience [2] related to unit u_i, while the RS unit generates the repetition suppression effect which inhibits (*empty-dot connections*) or enhance (*filled-dot connections*) the feature detectors.

stimulus predominance, that gives a soft measure of the stimuli frequency. The stimulus predominance at the time t is defined as

$$\nu_i^t = \frac{1}{|\chi_t|} \sum_{x_k \in \chi_t} \frac{y_i(x_k)}{z_U^k}, \tag{2}$$

where χ_t is the set of the input stimuli presented to the network up to time t, while $y_i(x_k)$ is the output of the i-th unit on the k-th input pattern and z_U^k is (an approximation of) the output of the maximally active unit, from the set U, on the pattern x_k. For instance, it can be approximated, by means of the softmax, as

$$z_U^k = \sum_{u_j \in U} y_j(x_k) \frac{e^{q y_j(x_k)}}{\sum_l e^{q y_l(x_k)}}, \tag{3}$$

where q is a parameter which regulates the sharpness of the approximation.

The idea underlying the definition in (2) is to associate each unit $u_i \in U$ with a prototype p_i that identifies the reference stimuli for the neuron. Then, each unit uses eq. 2 to measure the frequency with which stimuli similar to the prototype p_i have been shown to the network. Here we assume that the unit output y_i is an increasing function of the similarity between the input vector x_k and the unit's prototype p_i. This approach resembles competitive learning with *conscience* [2]. The conscience mechanism was proposed for making frequently winning representatives less likely to win in the future because of their heavier conscience [10]. In our model we use a different conscience mechanism for suppressing the responses of the less selective neurons. Bienenstock, Cooper and Munro tackled with the selectivity issue in their notable BCM model [11]. Their Hebb-like learning rule is aimed at training neurons with a maximal response on one particular pattern, while retaining a very low response on the other patterns.

CoRe learning pursues neuron selectivity by defining a penalty factor, the repetition suppression, which is mediated by the stimulus predominance of the winning neurons. The amount of repetition suppression generated at time t in response to the pattern x_k is calculated as

$$RS_k^t = \frac{1}{|win_k|} \sum_{i \in win_k} \nu_i^t y_i(x_k), \tag{4}$$

where the winners pool win_k is calculated as in (1).

Figure 1 gives a visual interpretation of the model. The ν_i units accumulate the history of the feature detector (u_i) activations in order to generate the stimulus predominance for the prototype p_i. The activation of the winning neurons, scaled by their stimulus predominance (the empty-square connections in figure 1), is then used to generate the repetition suppression factor in the RS unit.

The neurons in the losers pool have a different learning rule with respect to those in the winners pool, although in both cases we define a pseudo-target to be used as a reference signal for the training procedure. The target activation for neurons in the losers pool ($u_i \in lose_k$) is defined as $\hat{y}_i^t(x_k) = y_i(x_k)(1 - RS_k^t)$, where x_k is the current input pattern. This expression forces the loser neurons to shrink their activation proportionally to the amount of repetition suppression they receive. The representation error of the i-th loser can be written as

$$\underline{E}_{i,k}^t = \frac{1}{2}(\hat{y}_i^t(x_k) - y_i(x_k))^2 = \frac{1}{2}(-y_i(x_k)RS_k^t)^2. \tag{5}$$

Conversely, the target activation for the neurons $u_i \in win_k$ is set to 1 in order to strengthen their activation (assuming 1 to be the the maximal output of a neuron). The representation error, in this case, is simply

$$\overline{E}_{i,k}^t = \frac{1}{2}(1 - y_i(x_k))^2. \tag{6}$$

The network parameters can be adapted by means of a supervised learning algorithm that minimize the error functions defined in (5) and (6). In section 3.3 we show an example of a Radial Basis Function Network whose gaussian units have been trained by CoRe learning and gradient descent. Notice that although CoRe learning resort to supervised learning for training the network parameters, it is a completely unsupervised algorithm since all the reference signals it uses are self-generated on the basis of the input stimuli distribution over space and time.

The repetition suppression phenomenon produces a compact neural representation by evolving a set of highly selective neurons from a large pool of units. Hence, it is important to define a metric for identifying the most significant neurons which have been produced by the learning process. We define the *relevance factor* for the unit u_i as

$$\hat{\nu}_i^t = \frac{1}{|win_{u_i}^t|} \sum_{x_k \in win_{u_i}^t} \frac{y_i(x_k)}{z_{win_k}^k}, \quad (7)$$

where $z_{win_k}^k$ follows the definition in (3) and $win_{u_i}^t$ is the set of patterns $x_k \in \chi_t$ for which unit u_i was in the winners pool, i.e.

$$win_{u_i}^t = \{x_k \mid y_i(x_k) \geq \theta, \ x_k \in \chi_t\}. \quad (8)$$

In other words, the relevance factor defines a soft-measure of the frequency with which u_i was the most active unit in the winners pool. This measure can be used to prune those neurons which are less significant for the stimuli representation (e.g. see RBF pruning in section 3.3).

3.3 RBFN Structure Optimization

A radial basis function network can be interpreted as a composition of localized receptive fields that measure the similarity of incoming patterns x_k to the prototype p_i they represent. In case of gaussian basis functions, the activation of the i-th RBF unit is defined as

$$y_i(x_k) = \frac{1}{2} e^{\frac{\|x - c_i\|^2}{\sigma_i^2}} \quad (9)$$

where the gaussian center c_i corresponds to the prototype vector p_i, while the gaussian variance σ_i modulates the steepness of the units' response.

We applied CoRe learning to solve the structure optimization problem of a gaussian RBFN, i.e, defining number, position and shape of the radial basis functions. In order to do this, we used gradient descent to derive the CoRe learning rules for the parameters of the gaussian kernels. The parameter increments for the units $u_i \in lose_k$ can be derived by differentiating the error function in (5) with respect to the parameters c_i and σ_i, that is

$$\triangle \underline{c}_{i,k}^t = \frac{\partial \underline{E}_{i,k}^t}{\partial c_i} = -y_i RS_k^t \frac{\partial(-y_i RS_k^t)}{\partial c_i} = \left(\frac{y_i RS_k^t}{\sigma_i}\right)^2 (x_k - c_i) \quad (10)$$

$$\triangle \underline{\sigma}_{i,k}^t = \frac{\partial \underline{E}_{i,k}^t}{\partial \sigma_i} = -y_i RS_k^t \frac{\partial(-y_i RS_k^t)}{\partial \sigma_i} = (y_i RS_k^t)^2 \frac{\|x_k - c_i\|^2}{\sigma_i^3}. \quad (11)$$

Similarly, the parameter increments for the units $u_i \in win_k$ can be calculated as

$$\triangle \overline{c}_{i,k}^t = \frac{\partial \overline{E}_{i,k}^t}{\partial c_i} = -(1 - y_i) y_i \frac{(x - c_i)}{\sigma_i^2} \quad (12)$$

$$\triangle\overline{\sigma}_{i,k}^{t} = \frac{\partial \overline{E}_{i,k}^{t}}{\partial \sigma_i} = -(1-y_i)y_i \frac{\|x_k - c_i\|^2}{\sigma_i^3} \tag{13}$$

where $\overline{E}_{i,k}^{t}$ follows the definition in (6).

The update rules for the RBF parameters are

$$c_i^t = c_i^{t-1} - \alpha_c \triangle c_{i,k}^t \tag{14}$$

$$\sigma_i^t = \sigma_i^{t-1} - \alpha_\sigma \triangle \sigma_{i,k}^t \tag{15}$$

where $\triangle c_{i,k}^t = \triangle \overline{c}_{i,k}^t$ if $u_i \in win_k$ and $\triangle c_{i,k}^t = \triangle \underline{c}_{i,k}^t$ if $u_i \in lose_k$ (similarly for $\triangle \sigma_{i,k}^t$).

The sign of the increments $\triangle c_{i,k}^t$ and $\triangle \sigma_{i,k}^t$ is coherent with the expected repetition suppression effect. Units in the losers pool, for instance, experience the displacement of their centers away from the current input as well as the enlargement of their widths. Conversely, winner neurons have their centers moved closer to the current stimuli and their responses steepened by the reduction of their receptive field's width.

CoRe learning starts with a large RBF network and incrementally trains the RBF units at each pattern presentation. If an input parameter x_k does not produce a sufficient activation in any of the units in the network, then CoRe learning triggers a search procedure to find the neuron with the lowest relevance factor $\hat{\nu}_i^t$ and trains this unit to memorize the stimulus x_k. Moreover, at the end of each learning epoch, the least significant neurons, i.e those with a relevance factor under a certain threshold θ_{pr}, are pruned from the network. This allows to generate networks with a compact structure, while retaining an high representational power.

The outputs of the N RBF units are linearly combined in the output units o_j by the weighted summation

$$o_j(x_k) = \sum_{i=1}^{N} w_{ij} y_i(x_k). \tag{16}$$

The linear parameters w_{ij} are trained by a supervised algorithm independently and in parallel to CoRe learning [12].

4 Results

We evaluated the performance of CoRe learning on a classification task based on the IRIS dataset [13]. This dataset contains 150 samples of dimension 4 equally partitioned into three different IRIS classes: setosa, versicolor, and virginica. Among these, one is linearly separable from the others, while two of them are not. The dataset was split into a training and a test sets, with 75 samples each; 25 samples from each class were randomly selected to be presented to the network and thirty different partitions of the dataset were generated randomly.

Table 1. Performance of various RBF models applied to the IRIS problem: mean classification score on the test set, variance, number of initial RBF (when applicable) and average number of RBF generated during training

Model	Mean Score	Variance	Initial RBF Number	Final RBF Number
CGA-RBF [14]	97.04%	±1.97%	N.A.	6.4
CoRe-RBF	95.91%	±1.19%	50	6.6
RBF-DDA [15]	94.50%	±1.50%	N.A.	11.8

To evaluate the capabilities of the CoRe approach we implemented a radial basis function network with CoRe learning rules for training the gaussian kernels and a standard gradient descent for adapting the output weights w_{ij}. The network was trained and tested separately on each of the thirty random partitions of the dataset.

Table 1 shows a comparison of the results of the test phase for the CoRe-RBF network and for two constructive RBF models. The CoRe-RBF network had an initial population of 50 neurons and, as a result of CoRe learning, reduced its size to an average number of 6 RBF. As it can be seen from Table 1, our model achieved an higher score than the RBF-DDA model [15] and generated a more compact network structure. The evolutive CGA-RBF [14], on the other hand, achieved better results than CoRe-RBF with respect to the test-set error. However, the CGA-RBF learning scheme needs a-priori knowledge about the number of classes, while our model is completely unsupervised and does not require any a-priori information about the distribution of the input patterns. In addition, the CoRe model achieved a very stable learning, demonstrated by the low error variance (see Table 1) over the 30 task repetitions.

Figure 2 shows the performance of CoRe learning on a clustering task: the dataset consists of 125 datapoints generated by four gaussians with different mean, variance and density. The small dots represent the 2-dimensional input stimuli, while the pluses (+) identify the neurons prototypes. The network evolves from the initial random prototype allocation (Fig. 2.a) to a sparser input coverage (Fig. 2.b) as a result of the repetition suppression mechanism. The least active units are pushed away from the stimuli, becoming less significant as learning proceeds. When their relevance factor falls behind a predefined threshold, they are pruned from the network. Conversely, the most significant units are retained and positioned as to cover the four clusters (Fig. 2.c).

The CoRe algorithm was run starting from a pool of 50 neurons and converged to a stable state after only 30 training epochs, reducing the number of prototypes to 5. It is worth noting how 4 units positioned approximately on the four clusters mean, while 1 neuron was allocated to cover the outliers of the 2 bottommost clusters. The same dataset has been used to run the RPCL algorithm (Fig. 2.d): starting from an initial population of 50 neurons, RPCL evolved to a stable state with 36 active neurons, i.e. neurons winning at least one competition. The results of the test show the robustness of CoRe with respect to the choice of the initial units allocation: CoRe is capable of converging to a compact prototype allocation even though it is run starting from an oversized neural population.

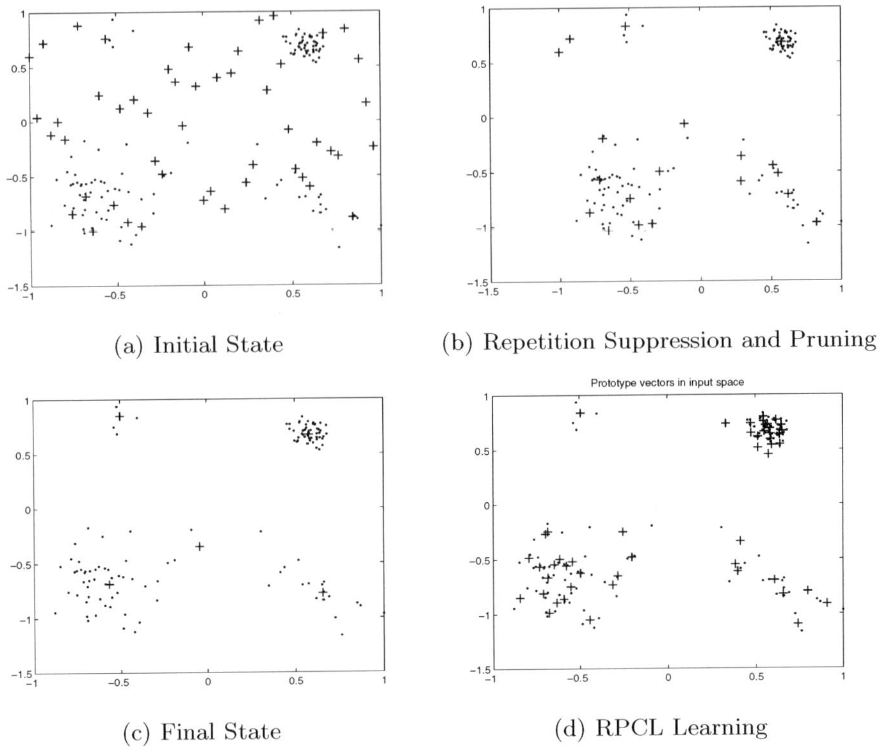

Fig. 2. Clustering test on four gaussian clusters: (a) CoRe starts with a random positioning of the gaussian centers; (b) the least active units are pushed away from the stimuli and network pruning eliminates non-relevant units; (c) the CoRe algorithm converges generating 5 units positioned on the four clusters and on their outliers; (d) prototype positioning by the RPCL algorithm after 200 learning epochs

5 Conclusion

This paper introduces CoRe learning, a soft-competitive learning scheme inspired by a cortical mechanism of implicit visual memory, i.e. repetition suppression. CoRe learning allows unsupervised training of feature detector units (e.g. RBF) without resorting to any explicit information concerning the input pattern distribution (e.g. number of classes), but only on the basis of the stimuli repetition.

We derived the CoRe learning rules for training the gaussian units of a radial basis function network and we tested the effectiveness of the proposed approach on classification and clustering tasks.

The CoRe learning model is part of a work aiming at the development of an articulated model of perceptual learning for machine vision applications. In particular, we believe that the repetition suppression mechanism, and therefore the CoRe learning scheme, may constitute an important tool for generating compact and sparse neural representations of the visual stimuli.

References

1. Nowlan, S.: Soft Competitive Adaptation: Neural Network Learning Algorithms based on Fitting Statistical Mixtures. Phd thesis, Carnegie-Mellon University, Pittsburg (1991)
2. DeSieno, D.: Adding conscience to competitive learning. In: IEEE Annu. Int. Conf. Neural Networks, IEEE Computer Society (1988) 1117–1124
3. Desimone, R.: Neural mechanisms for visual memory and their role in attention. In: Proceedings of Natl. Acad. Sci. USA. Volume 93. (1996) 13494–13499
4. Poggio, T., Girosi, F.: Networks for approximation and learning. In: Proceedings of the IEEE. Volume 78. (1990) 1481–1497
5. Tdodyks, M., Gilbert, C.: Neural networks and perceptual priming. Nature **7010**(431) (2004) 775–781
6. Mozer, M.C., Mytkowicz, T., Zemel, R.S.: Achieving robust neural representations:an account of repetition suppression. Technical report, Computer Science Deparment, University of Colorado, Boulder (2004)
7. French, R.M.: Semi-distributed representations and catastrophic forgetting in connectionist networks. Connection Science **4** (1992) 365 – 377
8. Rumelhart, D., Zipser, D.: Competitive learning. Cognitive Science **9** (1985) 75 – 112
9. Xu, L., Krzyzak, A., Oja, E.: Rival penalized competitive learning for clustering analysis, rbf net, and curve detection. IEEE Transactions on Neural Networks **4**(4) (1993) 636 – 649
10. Banerjee, A., Ghosh, J.: Frequency-sensitive competitive learning for scalable balanced clustering on high-dimensional hyperspheres. IEEE Transactions on Neural Networks **15** (2004) 702 – 719
11. Bienenstock, E.L., Cooper, L.N., Munro, P.W.: Theory for the development of neuron selectivity: orientation specificity and binocular interaction in visual cortex. Neurocomputing: foundations of research (1988) 437–455
12. Karayiannis, N.: Growing radial basis neural networks: merging supervised and unsupervised learning with network growth techniques. IEEE Transactions on Neural Networks **8**(6) (1997) 1492–1506
13. Fisher, R.A.: The use of multiple measurements in taxonomic problems. Ann. Eugenics **7**(2) (1936) 179 – 188
14. de Castro, L.N., Hruschka, E.R., Campello, R.J.G.B.: An evolutionary clustering technique with local search to design RBF neural network classifiers. In: Proceedings of the 2004 IEEE International Joint Conference on Neural Networks. Volume 3. (2004) 2083–2088
15. Paetz, J.: Feature selection for RBF networks. In: Proceedings of the 9th International Conference on Neural Information Processing (ICONIP'O2). Volume 2. (2002) 986–990

Real-Time Construction of Neural Networks

Kang Li[1], Jian Xun Peng[1], and Minrui Fei[2]

[1] School of Electronics, Electronic Engineering & Computer Science
Queen's University Belfast, Ashby Building, Stranmillis Road, Belfast BT9 5AH, UK
{K.Li, J.Peng}@qub.ac.uk

[2] Shanghai Key Laboratory of Power Station Automation Technology,
School of Mechatronics and Automation, Shanghai University, Shanghai 200072, China

Abstract. A stepwise two-stage algorithm is proposed for real-time construction of generalized single-layer networks (GSLNs). The first stage of this algorithm generates a network using a forward selection procedure, which is then reviewed at the second stage to replace insignificant neural nodes. The main contribution of this paper is that these two stages are performed within one regression context using Cholesky decomposition, leading to significantly neural network performance and concise real-time network construction procedures.

1 Introduction

The generalized single-layer networks (GSLN) represent a large class of flexible and efficient structures due to their excellent approximating capabilities [1][4]. A GSLN is a linear combination of some basis functions that are arbitrary (usually nonlinear) functions of the inputs. Examples are RBF networks and Volterra networks.

A critical issue in the application for GSLNs is that a large number of candidate basis functions may have to be considered initially, from which only small subset is selected to represent the data by minimizing a cost function. This however becomes extremely difficult if one has to enumerate all combinations to find the best subset, part of which is often referred to as the curse of dimensionality problem in the literature. To overcome this difficulty, forward subset selection methods seem to be one of very few feasible approaches [2][3][5][6]. Forward subset selection algorithms select the best basis function each time by minimizing the cost function, and this procedure is repeated until the desired number of, say n, basis functions have been selected. If n is unknown a 'prior', some selection criteria could be applied to stop the network construction, such as the Akaike's information criteion (AIC) [7].

However, the major problem with forward approaches is that the resultant network is not necessarily efficient [6][8]. In this paper, a stepwise algorithm is proposed for real time construction of GLSNs based on the Cholesky decomposition. In the proposed method, the network is generated using a forward subset selection procedure, which is then reviewed, and insignificant neurons (basis functions) are replaced, resulting in a network of significantly improved performance.

S. Kollias et al. (Eds.): ICANN 2006, Part I, LNCS 4131, pp. 140–149, 2006.

2 Problem Formulation

Suppose M candidate basis functions $\{\phi_i(t), i = 1,2,\cdots,M\}$, and N samples are used for network construction and training, producing to the following matrices

$$\mathbf{\Phi} = [\phi_1, \phi_2, \cdots, \phi_M],\ \phi_i = [\phi_i(1), \phi_i(2), \cdots, \phi_i(N)]^T, i = 1,2,\cdots,M \tag{1}$$

If one needs to select n significant basis functions, denoted as $\mathbf{p}_1, \mathbf{p}_2, \cdots, \mathbf{p}_n$, which form a selected *regression matrix*

$$\mathbf{P} = [\mathbf{p}_1, \mathbf{p}_2, \cdots, \mathbf{p}_n] \tag{2}$$

and produce the network output of

$$\mathbf{y} = \mathbf{P}\boldsymbol{\theta} + \mathbf{e} \tag{3}$$

which best fits the data in the sense that

$$J = \|\Lambda \mathbf{e}\| = \|\Lambda(\mathbf{y} - \mathbf{P}\boldsymbol{\theta})\| \to \min \tag{4}$$

where $\mathbf{y} = [y(1), y(2), \cdots, y(N)]^T$ is the vector of the target outputs; $\|\bullet\|$ denotes the 2-norm of a vector; $\boldsymbol{\theta}$ is the output weights; Λ is the diagonal *weighting matrix*, $\Lambda = diag(\lambda(1), \cdots, \lambda(N))$; J is the *cost function* which is the sum of the squared weighted errors (SSWE). If $\mathbf{P}$ is fully column ranked, optimum estimation of the output weights is given by

$$\boldsymbol{\theta} = [(\Lambda\mathbf{P})^T(\Lambda\mathbf{P})]^{-1}(\Lambda\mathbf{P})^T(\Lambda\mathbf{y}) \tag{5}$$

There are $M!/(n!/(M-n)!)$ possible combinations. Practically forward subset selection is the most popular approach.

3 Forward Selection

The forward approach selects a basis function each time, and this procedure repeats for n times to produce the network of n basis functions. Obviously a series of intermediate neural networks are generated along the process.

Denote the regression matrices of the intermediate network of k basis functions as

$$\mathbf{P}_k = [\mathbf{p}_1, \mathbf{p}_2, \cdots, \mathbf{p}_k], k = 1,2,\cdots,n \tag{6}$$

The corresponding cost function becomes

$$J(\mathbf{P}_k) = \mathbf{y}^T\Lambda^2\mathbf{y} - \mathbf{y}^T\Lambda^2\mathbf{P}_k(\mathbf{P}_k^T\Lambda^2\mathbf{P}_k)^{-1}\mathbf{P}_k^T\Lambda^2\mathbf{y} \tag{7}$$

where $\Lambda^2 = \Lambda^T\Lambda$. If $\Lambda\mathbf{P}_k$ is of full column rank, then $\mathbf{W} = \mathbf{P}_k^T\Lambda^2\mathbf{P}_k = [w_{i,j}]_{k\times k}$ is symmetric and positive definite, and can be decomposed as

$$\mathbf{W} = \mathbf{P}_k^T \Lambda^2 \mathbf{P}_k = \tilde{\mathbf{A}}^T \mathbf{D} \tilde{\mathbf{A}} \tag{8}$$

where $\mathbf{D} = diag(d_1, \cdots, d_k)$ is a diagonal matrix and $\tilde{\mathbf{A}} = [\tilde{a}_{i,j}]_{k \times k}$ is a unity upper triangular matrix. Define

$$\mathbf{A} \triangleq \mathbf{D}\tilde{\mathbf{A}} = [a_{i,j}]_{k \times k}, a_{i,j} = \begin{cases} 0, & j < i \\ d_i \tilde{a}_{i,j} & j \geq i \end{cases} \tag{9}$$

where $\tilde{a}_{i,i} = 1$ and $d_i = a_{i,i}$ for $i = 1, \cdots, k$.

From (8), it gives

$$a_{i,j} = w_{i,j} - \textstyle\sum_{s=1}^{i-1} a_{s,i} a_{s,j} / a_{s,s}, i = 1, \cdots, k, j = i, \cdots, k \tag{10}$$

Define

$$\mathbf{a}_y \triangleq \mathbf{A}\theta = \mathbf{D}\tilde{\mathbf{A}}\theta = [a_{1,y}, \cdots, a_{k,y}]^{\mathrm{T}} \tag{11}$$

$$\mathbf{w}_y \triangleq \mathbf{P}_k^{\mathrm{T}} \Lambda^2 \mathbf{y} = [w_{1,y}, \cdots, w_{k,y}]^{\mathrm{T}} \tag{12}$$

Then left-multiply $\mathbf{W} = \mathbf{P}_k^{\mathrm{T}} \Lambda^2 \mathbf{P}_k$ on both sides of (5) for $\mathbf{P} = \mathbf{P}_k$, substituting (8) gives

$$\tilde{\mathbf{A}}^{\mathrm{T}} \mathbf{a}_y = \mathbf{w}_y \tag{13}$$

As $\tilde{\mathbf{A}}$ is a unity upper triangular matrix, using (9), $\mathbf{a}_y$ could be computed as

$$a_{i,y} = w_{i,y} - \textstyle\sum_{i=1}^{k-1} a_{s,i} a_{s,y} / a_{s,s}, i = 1, \cdots, k \tag{14}$$

Substituting (8) into (7) and noting (13) gives

$$J(\mathbf{P}_k) = \mathbf{y}^{\mathrm{T}} \Lambda^2 \mathbf{y} - \mathbf{a}_y^{\mathrm{T}} \mathbf{D}^{-1} \mathbf{a}_y = \mathbf{y}^{\mathrm{T}} \Lambda^2 \mathbf{y} - \textstyle\sum_{i=1}^{k} a_{i,y}^2 / a_{i,i} \tag{15}$$

Suppose that one more basis function, denoted as $\mathbf{p}_{k+1}$, is selected, then

$$J(\mathbf{P}_{k+1}) = \mathbf{y}^{\mathrm{T}} \Lambda^2 \mathbf{y} - \textstyle\sum_{i=1}^{k+1} a_{i,y}^2 / a_{i,i} \tag{16}$$

The reduction in the cost function due to the new basis function $\mathbf{p}_{k+1}$ is

$$\Delta J_{k+1}(\mathbf{p}_{k+1}) = J(\mathbf{P}_k) - J(\mathbf{P}_{k+1}) = a_{k+1,y}^2 / a_{k+1,k+1} \tag{17}$$

where $\mathbf{P}_{k+1} = [\mathbf{P}_k, \mathbf{p}_{k+1}]$.

According to (15), the elements $a_{i,j}$ will not change as the new basis function $\mathbf{p}_{k+1}$ is introduced. To minimize the cost function given previously selected basis functions $\mathbf{p}_1, \cdots, \mathbf{p}_k$ are fixed, one needs to select a basis function from the rest candidates which maximizes ΔJ_{k+1}

$$\min\{J([\mathbf{P}_k, \phi])\} = J(\mathbf{P}_k) - \max\{\Delta J_{k+1}(\phi)\}, \quad s.t. \quad \phi \in \{\phi_{k+1}, \cdots, \phi_M\} \tag{18}$$

where $\{\phi_{k+1},\cdots,\phi_M\}$ denotes the candidate basis function pool from the full regression matrix $\mathbf{\Phi}$, i.e. $\mathbf{\Phi}=[\mathbf{p}_1,\mathbf{p}_2,\cdots,\mathbf{p}_k,\phi_{k+1},\cdots,\phi_M]$.

To minimize (18), $\tilde{\mathbf{A}}$, $\mathbf{A}$, and $\mathbf{a}_y$ defined above are augmented as follows to store corresponding information of the selected basis functions and all the candidates.

For a model of k basis functions, augment $\tilde{\mathbf{A}}$ and $\mathbf{A}$ from k-by-k to k-by-M, and $\mathbf{a}_y$ from k-by-1 to M-by-1, respectively. For $\mathbf{A}$ defined in (9), it becomes

$$\mathbf{A}=[a_{i,j}]_{k\times M},\ a_{i,j}=\begin{cases}0, & j<i\\ w_{i,j}-\sum_{s=1}^{i-1}a_{s,i}a_{s,j}/a_{s,s}, & i\le j\le M\end{cases} \tag{19}$$

$$w_{i,j}=\begin{cases}\mathbf{p}_i^T\Lambda^2\mathbf{p}_j, & j\le k\\ \mathbf{p}_i^T\Lambda^2\phi_j, & j>k\end{cases}, \tag{20}$$

Matrix $\tilde{\mathbf{A}}$ defined in (9) becomes

$$\tilde{\mathbf{A}}=[\tilde{a}_{i,j}]_{k\times M},\tilde{a}_{i,j}=a_{i,j}/a_{i,i}. \tag{21}$$

which is unity upper triangular, but not a square matrix. Vector $\mathbf{a}_y$ in (14) becomes

$$\mathbf{a}_y=[a_{i,y}]_{M\times 1},\ a_{i,y}=w_{i,y}-\sum_{s=1}^{i-1}a_{s,i}a_{s,y}/a_{s,s} \tag{22}$$

where $w_{i,y}=\begin{cases}\mathbf{y}^T\Lambda^2\mathbf{p}_i, & i\le k\\ \mathbf{y}^T\Lambda^2\phi_i, & i>k\end{cases}$.

In addition, another M-by-1 vector $\mathbf{b}$ can be defined as

$$\mathbf{b}=[b_i]_{M\times 1},\ b_i=w_{i,i}-\sum_{s=1}^{i-1}a_{s,i}a_{s,i}/a_{s,s} \tag{23}$$

where $w_{i,i}=\begin{cases}\mathbf{p}_i^T\Lambda^2\mathbf{p}_i, & i\le k\\ \phi_i^T\Lambda^2\phi_i, & i>k\end{cases}$. Obviously for $w_{i,i}$ $i=1,\cdots,k$, it holds that

$$b_i=a_{i,i},\ i=1,\cdots,k \tag{24}$$

Based on (17), the contribution of each of the candidate basis functions is given by

$$\Delta J_{k+1}(\phi_i)=a_{i,y}^2/b_i,\quad i=k+1,\cdots,M \tag{25}$$

The one from $\{\phi_{k+1},\cdots,\phi_M\}$ which gives the maximum contribution is then selected as the $(k+1)$'th basis function.

Assume $\Delta J_{k+1}(\phi_j)=\arg\max\{\Delta J_{k+1}(\phi_i), i=k+1,\cdots,M\}$, i.e. candidate ϕ_j is selected and denote $\mathbf{p}_{k+1}=\phi_j$. Other candidates are reordered and $\{\phi_{k+2},\cdots,\phi_M\}$ becomes the new candidate pool. According to (17), the $(k+1)$'th basis function leads to a further reduction of the cost function by $\Delta J_{k+1}(\mathbf{p}_{k+1})$.

Since $\mathbf{p}_{k+1} = \phi_j$, both ϕ_{k+1} and ϕ_j in the original full regression matrix $\mathbf{\Phi}$ have to be interchanged, leading to the change of various intermediate matrices and vectors. For $\mathbf{A}$ and $\tilde{\mathbf{A}}$, columns k+1 and j should also be interchanged as

$$\hat{a}_{i,k+1} = a_{i,j}, \hat{a}_{i,j} = a_{i,k+1}, \hat{\tilde{a}}_{i,k+1} = \tilde{a}_{i,j}, \hat{\tilde{a}}_{i,j} = \tilde{a}_{i,k+1}, \; i = 1, \cdots, k \tag{26}$$

where $\hat{a}_{i,k+1}$ denotes the updated $a_{i,k+1}$. To denote an updated element, a cap is applied to the element here after.

Similarly, elements k+1 and j for $\mathbf{a}_y$ and $\mathbf{b}$ are interchanged as follows

$$\hat{a}_{k+1,y} = a_{j,y}, \hat{a}_{j,y} = a_{k+1,y}, \hat{b}_{k+1} = b_j, \hat{b}_j = b_{k+1}. \tag{27}$$

In addition, as the (k+1)'th basis function is selected, a new row should be appended to $\mathbf{A}$ and $\tilde{\mathbf{A}}$

$$\left.\begin{aligned} a_{k+1,i} &= w_{k+1,i} - \textstyle\sum_{s=1}^{k} \tilde{a}_{s,k+1} a_{s,i} \\ w_{k+1,i} &= \mathbf{p}_{k+1}^T \Lambda^2 \phi_i, \tilde{a}_{k+1,i} = a_{k+1,i} / a_{k+1,k+1} \end{aligned}\right\}, \; i = k+1, \cdots, M \tag{28}$$

Furthermore, for $\mathbf{a}_y$ and $\mathbf{b}$, elements from the (k+2)'th to the last one should be updated as follows according to the definitions (22) and (23)

$$\hat{a}_{i,y} = a_{i,y} - \tilde{a}_{k+1,i} a_{k+1,y}, \hat{b}_i = b_i - \tilde{a}_{k+1,i} a_{k+1,i}, \; i = k+1, \cdots, M \tag{29}$$

Given above details, a forward selection procedure similar to [5] can be proposed.

4 A Two-Stage Algorithm

To overcome the drawbacks of the forward selection algorithm, one solution is to review all the selected basis functions once the forward selection procedure terminates. For each selected basis function (of a of size n), say $\mathbf{p}_i$, $1 \le i \le n$, its contribution to the cost function reduction $\Delta J_n(\mathbf{p}_i)$ is compared with that of the candidate which has the maximum contribution among all the candidates. Denote the maximum candidate contribution as $\Delta J_n(\phi_j)$, and if $\Delta J_n(\mathbf{p}_i) < \Delta J_n(\phi_j)$ then $\mathbf{p}_i$ is said to be insignificant, and will be replaced by ϕ_j, in the meantime $\mathbf{p}_i$ is put back into the candidate pool. Thus, the cost function can be further reduced by $\Delta J_n(\phi_j) - \Delta J_n(\mathbf{p}_i)$. The remaining problem is, to compute the contribution of a previously selected basis function, say $\mathbf{p}_i$ to the cost function and compare its contribution with that of the candidates $\phi_{n+1}, \cdots, \phi_M$, an appropriate regression context should be re-constructed. That is, $\mathbf{A}$, $\tilde{\mathbf{A}}$, $\mathbf{a}_y$ and $\mathbf{b}$ have to updated.

In order to assess the significance of the basis function $\mathbf{p}_i$, the first step is to move $\mathbf{p}_i$ to the n'th position of the full regression matrix $\mathbf{\Phi}$ as if it were the last selected basis function in the forward selection procedure. This is done by a series of interchanges between two adjacent basis functions $\mathbf{p}_x$ and $\mathbf{p}_{x+1}$ for $x = i,\cdots,n-1$ and the regression context is updated correspondingly.

Suppose $\mathbf{p}_1,\cdots,\mathbf{p}_n$ are the selected basis functions in the order of selection. If two adjacent basis functions $\mathbf{p}_x$ and $\mathbf{p}_{x+1}$ is to be changed, $\mathbf{A}$, $\widetilde{\mathbf{A}}$, $\mathbf{a}_y$ and $\mathbf{b}$ have to be updated. Denote the n basis functions in the new selected order as $\mathbf{p}_1,\cdots,\mathbf{p}_{x-1},\widehat{\mathbf{p}}_x,\widehat{\mathbf{p}}_{x+1},\mathbf{p}_{x+2}\cdots,\mathbf{p}_n$, where $\widehat{\mathbf{p}}_x = \mathbf{p}_{x+1}$ and $\widehat{\mathbf{p}}_{x+1} = \mathbf{p}_x$. Then based on (10), for columns 1 to x-1 in $\mathbf{A}$, since

$$\left.\begin{aligned} \widehat{w}_{i,x} &= \mathbf{p}_i^{\mathrm{T}}\Lambda^2\widehat{\mathbf{p}}_x = \mathbf{p}_i^{\mathrm{T}}\Lambda^2\mathbf{p}_{x+1} = w_{i,s+1} \\ \widehat{w}_{i,x+1} &= \mathbf{p}_i^{\mathrm{T}}\Lambda^2\widehat{\mathbf{p}}_{x+1} = \mathbf{p}_i^{\mathrm{T}}\Lambda^2\mathbf{p}_x = w_{i,x} \\ \widehat{w}_{i,k} &= w_{i,k}, k = i,\cdots,x-1,x+2,\cdots,M \end{aligned}\right\}, i = 1,\cdots,x-1 \tag{30}$$

therefore

$$\left.\begin{aligned} \widehat{a}_{i,x} &= a_{i,x+1},\ \widehat{a}_{i,x+1} = a_{i,x} \\ \widehat{a}_{i,k} &= a_{i,k}, k = i,\cdots,x-1,x+2,\cdots,M \end{aligned}\right\}, i = 1,\cdots,x-1 \tag{30}$$

Noting $\widehat{w}_{x,x+1} = \widehat{\mathbf{p}}_x^{\mathrm{T}}\Lambda^2\widehat{\mathbf{p}}_{x+1} = \mathbf{p}_{x+1}^{\mathrm{T}}\Lambda^2\mathbf{p}_x = w_{x+1,x}$, for the x'th row of **A**, gives

$$\widehat{a}_{x,x+1} = a_{x,x+1} \tag{32}$$

Noting $\widehat{w}_{x,x} = \widehat{\mathbf{p}}_x^{\mathrm{T}}\Lambda^2\widehat{\mathbf{p}}_x = \mathbf{p}_{x+1}^{\mathrm{T}}\Lambda^2\mathbf{p}_{x+1} = w_{x+1,x+1}$, it holds that

$$\widehat{a}_{x,x} = a_{x+1,x+1} + a_{x,x+1}a_{x,x+1} / a_{x,x} \tag{31}$$

while for the rest elements in the x'th row, it gives

$$\widehat{a}_{x,j} = a_{x,j} + a_{x,x+1}a_{x,j} / a_{x,x},\ j = x+2,\cdots,n \tag{32}$$

Noting $\widehat{w}_{x+1,x+1} = \widehat{\mathbf{p}}_{x+1}^{\mathrm{T}}\Lambda^2\widehat{\mathbf{p}}_{x+1} = \mathbf{p}_x^{\mathrm{T}}\Lambda^2\mathbf{p}_x = w_{x,x}$, for the ($x$+1)'th row of **A**, yields

$$\widehat{a}_{x+1,x+1} = a_{x,x} - (a_{x,x+1})^2 / \widehat{a}_{x,x} \tag{35}$$

$$\widehat{a}_{x+1,j} = a_{x,j} - a_{x,x+1}\widehat{a}_{x,j} / \widehat{a}_{x,x},\ j = x+2,\cdots,M \tag{36}$$

Noting $\widehat{w}_{i,j} = \widehat{\mathbf{p}}_i^{\mathrm{T}}\Lambda^2\widehat{\mathbf{p}}_j = \mathbf{p}_i^{\mathrm{T}}\Lambda^2\mathbf{p}_j = w_{i,j},\ i,j > x+1$, for elements of **A** in row x+2,

$$\begin{aligned} \widehat{a}_{x+2,j} &= \widehat{w}_{x+2,j} - \textstyle\sum_{s=1}^{x+1}\widehat{a}_{s,x+2}\widehat{a}_{s,j} / \widehat{a}_{s,s} = w_{x+2,j} - \\ &- \textstyle\sum_{s=1}^{x-1}\dfrac{a_{s,x+2}a_{s,j}}{a_{s,s}} - \dfrac{\widehat{a}_{x,x+2}\widehat{a}_{x,j}}{\widehat{a}_{x,x}} - \dfrac{\widehat{a}_{x+1,x+2}\widehat{a}_{x+1,j}}{\widehat{a}_{x+1,x+1}} \end{aligned} \tag{37}$$

Furthermore, it could be derived that

$$\frac{\widehat{a}_{x,x+2}\widehat{a}_{x,j}}{\widehat{a}_{x,x}} - \frac{\widehat{a}_{x+1,x+2}\widehat{a}_{x+1,j}}{\widehat{a}_{x+1,x+1}} = \frac{a_{x,x+2}a_{x,j}}{a_{x,x}} - \frac{a_{x+1,x+2}a_{x+1,j}}{a_{x+1,x+1}} \tag{38}$$

which means that the (*x*+2)'th row of **A** has no change, and row (*x*+2) to row *n* of **A** have no change as well, i.e.

$$\widehat{a}_{i,j} = a_{i,j} \,,\ i = x+2,\cdots,n \,,\ j = x+2,\cdots,n \tag{39}$$

Similarly,

$$\left.\begin{aligned} \widehat{a}_{x,y} &= a_{x+1,y} + a_{x,x+1}a_{x,y} / a_{x,x} \\ \widehat{a}_{x+1,y} &= a_{x,y} - a_{x,x+1}\widehat{a}_{x,j} / \widehat{a}_{x,x} \end{aligned}\right\} \tag{40}$$

For vector **b**, both the *x*'th and the (*x*+1) 'th element are changed also as follows

$$\widehat{b}_x = \widehat{a}_{x,x} ,\ \widehat{b}_{x+1} = \widehat{a}_{x+1,x+1} \tag{41}$$

Now eventually $\mathbf{p}_i$ is moved to the *n* position in the selected basis functions, its contribution can then be computed as follows based on (17)

$$\Delta J_n(\mathbf{p}_i) = \Delta J_n(\widehat{\mathbf{p}}_n) = a_{n,y}^2 / a_{n,n} \tag{42}$$

where $\mathbf{p}_i = \widehat{\mathbf{p}}_n$ indicates that $\mathbf{p}_n$ has been moved to the *n*'th position. To compute the contribution of the candidate basis functions, define:

$$\left.\begin{aligned} a_{s,y}^{(-i)} &= a_{s,y} + a_{n,s}a_{n,y} / a_{n,n} \\ b_s^{(-i)} &= b_s + (a_{n,n+1})^2 / a_{n,n} \end{aligned}\right\},\ s = n+1,\cdots,M \tag{43}$$

which are the corresponding elements in vector $\mathbf{a}_y$ and **b** if the basis function $\mathbf{p}_i$ is pruned from the network. The contribution of all the candidates is given by

$$\Delta J_n(\phi_s) = (a_{s,y}^{(-i)})^2 / b_s^{(-i)} ,\ s = n+1,\cdots,M \tag{44}$$

Now the significance of $\mathbf{p}_i$ can then be checked given the above derivations. First, identify the candidate basis function that gives the maximum contribution

$$\Delta J_n(\phi_j) = \max\{\Delta J_n(\phi_s), s = n+1,\cdots,M\} . \tag{45}$$

If $\Delta J_n(\phi_j) > \Delta J_n(\mathbf{p}_i)$, $\mathbf{p}_i$ becomes insignificant and should be replaced by ϕ_j as the new basis function in the network, while $\mathbf{p}_i$ should be put back into the candidate pool, taking the position of ϕ_j. In this case, the regression context needs to be updated again due to the interchange of $\mathbf{p}_i$ and ϕ_j. For matrix **A**, interchange columns *n* and *j* from row 1 to *n*-1 due to the interchange of $\mathbf{p}_i$ and ϕ_j, and re-calculate the *n*'th row of **A** as

$$\left.\begin{aligned} &\widehat{a}_{n,n} = b_j^{(-i)},\quad \widehat{a}_{n,j} = a_{n,j}, w_{n,k} = \widehat{\mathbf{p}}_n^{\mathrm{T}} \Lambda^2 \phi_k \\ &\widehat{a}_{n,k} = w_{n,k} - \sum_{s=1}^{n-1} \frac{a_{s,n}a_{s,k}}{a_{s,s}}, k = n+1,\cdots,M, k \neq j \end{aligned}\right\} \tag{46}$$

In addition, elements n to M of vector $\mathbf{a}_y$ and $\mathbf{b}$ is updated respectively as

$$\left.\begin{aligned} \widehat{a}_{n,y} &= a_{n,y}^{(-i)}, a_{j,y} = a_{n,y} - \widehat{a}_{n,j}\widehat{a}_{n,y} / \widehat{a}_{n,n}, \\ a_{s,y} &= a_{s,y}^{(-i)} - \frac{\widehat{a}_{n,s}\widehat{a}_{n,y}}{\widehat{a}_{n,n}}, \ s = n+1,\cdots,M, s \neq j \end{aligned}\right\} \tag{47}$$

$$\left.\begin{aligned} \widehat{b}_n &= b_j^{(-i)}, \widehat{b}_j = a_{n,n} - (\widehat{a}_{n,n+1})^2 / \widehat{a}_{n,n} \\ \widehat{b}_s &= b_s^{(-i)} - (\widehat{a}_{n,n+1})^2 / \widehat{a}_{n,n}, \ s = n+1,\cdots,M, s \neq j \end{aligned}\right\} \tag{48}$$

Similarly the corresponding element of $\widetilde{\mathbf{A}}$ can be recalculated based on (21).

The computational complexity analysis for the above proposed algorithm can be performed by referring to [6].

5 Real Time Implementation

To facilitate real time neural network construction using the above algorithm, a weighting scheme for the data samples is applied to reduce the data storage. In this paper, the weights for samples at current time instance t are always set at 1 and $\lambda^{\tau/2}$ for the $t-\tau$, where $0<\lambda<1$ and normally close to 1. In this way, the weights for the past data samples decrease exponentially as time goes on, leading to the gradual removal of past data samples in the real-time neural network construction.

In detail, the weight matrix $\mathbf{W}$ in (8) and $\mathbf{w}_y$ in (12) is applied to all basis functions, i.e.

$$\mathbf{W} = [w_{i,j}]_{M\times M}, w_{i,j} = \boldsymbol{\phi}_i^{\mathrm{T}} \boldsymbol{\Lambda}^2 \boldsymbol{\phi}_j \tag{49}$$

$$\mathbf{w}_y = [w_{i,y}]_{M\times 1}, w_{i,y} = \mathbf{y}^{\mathrm{T}} \boldsymbol{\Lambda}^2 \boldsymbol{\phi}_i \tag{50}$$

Note that $\mathbf{W}$ is symmetric and only the upper triangle, including the diagonal elements, needs to be stored. Applying the exponential weighting scheme, gives

$$\left.\begin{aligned} w_{i,j}(t) &= \textstyle\sum_{\tau=1}^{t} \lambda^{t-\tau} \phi_i(t)\phi_j(t) \\ w_{i,y}(t) &= \textstyle\sum_{\tau=1}^{t} \lambda^{t-\tau} y(t)\phi_j(t) \end{aligned}\right\} \tag{51}$$

In real time implementation, $\mathbf{W}$ and $\mathbf{w}_y$ are dynamically updated as

$$w_{i,j}(t) = \lambda w_{i,j}(t-1) + \phi_i(t)\phi_j(t), \ w_{i,j}(0) = 0 \tag{52}$$

$$w_{i,y}(t) = \lambda w_{i,y}(t-1) + y(t)\phi_j(t), \ w_{i,y}(0) = 0 \tag{53}$$

In this weighting scheme, for $w_{i,j}$, if $|\phi_i(t)\phi_j(t)| \to C$ as $t\to\infty$, then $|w_{i,j}(t)| \to C/(1-\lambda)$. The algorithm proposed in this paper is a two-stage method, therefore the real time implementation is also divided into two stage.

To initialize the algorithm, let $k=0$ and $w_{i,y}=0, w_{i,j}=0,\ j>i$ for $i=1,\cdots,M$, as each sample of data is recorded, following procedure is implemented.

A) With the recorded new sample of data, update **W** and $\mathbf{w}_y$ respectively.

B) Update **A**, $\tilde{\mathbf{A}}$, $\mathbf{a}_y$ and **b** in (19), (21), (22) and (23), respectively

C) If $k<n$, implement step b), c) and d) of the forward selection procedure to select the *k*'th basis function. Otherwise implement the reviewing procedure to check all the *n* selected basis functions.

D) Identify the output weights based on (5). The neural network (of *k* basis functions) can then be used for real time application.

Note that the neural network size *n* is given previously. This procedure is implemented recursively.

6 Simulations

Consider the following non-linear dynamic system

$$y(t)=0.5+0.5y(t-1)-0.05y^2(t-2)+0.8u(t-2)+u^2(t-1)+\xi(t) \tag{54}$$

where *t*, y and *u* represent the time series, the system output and input, respectively; $\xi(t)$ is the system noise given as $\xi \sim N(0,0.05)$. By simulating (54) with $u(t)$ uniformly distributed within the range [-1.0, 1.0], two data sets of 500 samples for each were generated, the first one for neural network construction and training and the other for validation. Full polynomial of orders 0 to 3 of $y(t\text{-}l)$ and $u(t\text{-}l)$ for l = 1, 2, and 3 are used. The network size is set to be 5. Both the proposed two-staged algorithm and the forward-only algorithm were then used to select 5 basis functions. The indices of the selected terms and the corresponding SSWE are listed in Table 1 for both the algorithms. Table 2 compares the normalized one-step-ahead and long-term predictions errors over both the training and validation data sets, respectively, of the two produced neural networks. From the simulation example, it is obvious that the two stage algorithm achieves a neural network of far better performance than forward subset selection methods.

Table 1. Selected nodes and performance

Algorithm	Index of the selected terms	SSWE
Forward	1, 2, 6, 23, 44	5.4425
Two-staged	1, 2, 6, 14, 23	1.3728

Table 2. Normalized prediction error (%)

Algorithm	Training data		Validation data	
	One step	Long term	One step	Long term
Forward	6.78	7.51	6.53	7.28
Two-stage	3.41	3.40	3.23	3.21

7 Conclusion

In this paper, a stepwise two-stage algorithm has been proposed for real-time construction of generalized single-layer networks (GSLNs), which enables both network growing and network modification. The main contribution of this paper is that these two directions of network construction are performed within one regression context using Cholesky decomposition, leading to both significantly neural network performance and concise network construction procedures. Simulation results have demonstrated the effectiveness of this method.

Acknowledgement

Dr K. Li wishes to acknowledge the financial support of the UK Engineering and Physical Sciences Research Council (EPSRC Grant GR/S85191/01). Dr M. Fei wishes to acknowledge the financial support of Shanghai Leading Academic Disciplines (T0103).

References

1. K. M. Adeney, M. J. Korenberg: Iterative fast orthogonal search algorithm for MDL-based training of generalized single-layer networks. Neural Networks, Vol 13 (2000) 787-799.
2. S. Chen, S. A. Billings, W. Luo: Orthogonal Least Squares Methods and Their Application to Non-linear System Identification. International Journal of Control, Vol. 50, No.5 (1989) 1873-1896.
3. S. Chen, J. Wigger: Fast Orthogonal Least Squares Algorithm for Efficient Subset Model Selection. IEEE Transactions on Signal Processing, Vol. 43, No.7, (1995) 1713-1715.
4. B. Igelnik and Y. H. Pao, "Additional Perspectives of feedforward neural-nets and the functional-link", IJCNN '93, Nagoya, Japan, 1993.
5. K. Li, J. Peng, G. Irwin: A fast nonlinear model identification method. IEEE Transactions on Automatic Control. Vol. 50, No. 8 (2005) 1211-1216.
6. K. Li, J. Peng, Er-Wei Bai: A two-stage algorithm for identification of nonlinear dynamic systems. Automatica, Vol. 42, No 7 (2006) (in press).
7. H. Akaike: A New Look at the Statistical Model Identification. J. R. Statist. Soc. Ser. B., Vol.36 (1974) 117-147.
8. Sherstinsky, R. W. Picard: On the Efficiency of the Orthogonal Least Squares Training Method for Radial Basis Function Networks. IEEE Transactions on Neural Networks, Vol. 7, No. 1 (1996) 195-200.

MaxMinOver Regression: A Simple Incremental Approach for Support Vector Function Approximation

Daniel Schneegaß[1,2], Kai Labusch[1], and Thomas Martinetz[1]

[1] Institute for Neuro- and Bioinformatics
University at Lübeck, D-23538 Lübeck, Germany
martinetz@informatik.uni-luebeck.de
[2] Information & Communications, Learning Systems
Siemens AG, Corporate Technology, D-81739 Munich, Germany
daniel.schneegass.ext@siemens.com

Abstract. The well-known MinOver algorithm is a simple modification of the perceptron algorithm and provides the maximum margin classifier without a bias in linearly separable two class classification problems. In [1] and [2] we presented DoubleMinOver and MaxMinOver as extensions of MinOver which provide the maximal margin solution in the primal and the Support Vector solution in the dual formulation by dememorising non Support Vectors. These two approaches were augmented to soft margins based on the ν-SVM and the C2-SVM. We extended the last approach to SoftDoubleMaxMinOver [3] and finally this method leads to a Support Vector regression algorithm which is as efficient and its implementation as simple as the C2-SoftDoubleMaxMinOver classification algorithm.

1 Introduction

The Support-Vector-Machine (SVM) [4], [5] is a very efficient, universal and powerful tool for classification and regression tasks (e.g. [6], [7], [8]). A major drawback, particularly for industrial applications where easy and robust implementation is an issue, is its complicated training procedure. A large Quadratic-Programming problem has to be solved, which requires sophisticated numerical optimisation routines which many users do not want or cannot implement by themselves. They have to rely on existing software packages, which are hardly comprehensive and, in some cases at least, error-free. This is in contrast to most Neural Network approaches where learning has to be simple and incremental almost by definition. The iterative and incremental nature of learning in Neural Networks usually leads to simple training procedures which can easily be implemented. It is desirable to have similar training procedures also for the SVM.

Several approaches for obtaining more or less simple incremental training procedures for Support Vector classification and regression have been introduced so far [9], [10], [11], [12]. We want to mention in particular the Kernel-Adatron by Friess, Cristianini, and Campbell [9] and the Sequential-Minimal-Optimisation algorithm (SMO) by Platt [10]. This is the most widespread iterative training procedure for SVMs. It is fast and robust, but it is still not yet of a pattern-by-pattern nature and well suited as a

S. Kollias et al. (Eds.): ICANN 2006, Part I, LNCS 4131, pp. 150–158, 2006.

starting point for an online learning scheme. It is not yet as easy to implement as one is used from Neural Network approaches.

According to this goal in [2] and [1] we had revisited and extended the MinOver algorithm. The so-called DoubleMaxMinOver [3] method which we briefly revisit in this paper as the combination of DoubleMinOver and MaxMinOver converges provable to the Support Vector solution in the dual representation for classification tasks. Moreover, the angle γ_t between the correct solution $\mathbf{w}^*$ and the interim solution $\mathbf{w}_t$ after t steps is bounded by $O(t^{-1})$.

Furthermore we introduced a soft margin approach based on the C2-SVM error model. In a straightforward way this model can be adapted for regression [13]. We consequently show that we similarly obtain our regression algorithm as an extension of DoubleMaxMinOver for classification as well. We give a proof for its convergence properties and consider benchmark results.

2 The Support Vector Machine

The SVM represents a linear classifier or function approximator, respectively. For given sets of input vectors $X = \{\mathbf{x}_1, \ldots, \mathbf{x}_l\}$ and output vectors $Y = \{y_1, \ldots, y_l\}$ the SV solution for the inference of $f(\mathbf{x}) = y$ is always the best one in the sense that in classification the margin between the separating hyperplane and the two classes is largest and in regression the obtained curve is the flattest one. In both cases one only needs a few input vectors, the so-called Support Vectors, to describe the solution. This property is very important. The lower the number of Support Vectors, the better the expected generalisation capability [14].

The goal is to solve the optimisation problem

$$\frac{1}{2}\mathbf{w}^T\mathbf{w} = \min$$
$$\forall i \in \{1, \ldots, l\} : y_i(\mathbf{w}^T\mathbf{x}_i - b) \geq 1$$

in classification, which is apparently equivalent to the maximisation of the margin holding $||\mathbf{w}||$, and

$$\frac{1}{2}\mathbf{w}^T\mathbf{w} = \min$$
$$\forall i \in \{1, \ldots, l\} : |\mathbf{w}^T\mathbf{x}_i - b - y_i| \leq \epsilon.$$

in regression tasks. Note that classification and regression are geometrically analogous in the following sense. While in classification the Support Vectors are the ones lying on the margin, that is having smallest distance to the separating hyperplane, in regression the Support Vectors lie on the edges of the ϵ-tube around the function values. In classification all other vectors are located more or less far away from the margin. In regression they lie within the ϵ-tube. Hence SV classification can be seen as the perspective from inside to outside the hyperplane area while SV regression is the other way around.

An important advantage of SVMs is furthermore that the classifier or function approximator $f(\mathbf{x}) = \mathbf{w}^T\mathbf{x} - b$ can be formulated in terms of the observations, i.e. as $f(\mathbf{x}) = \left(\sum_{i=1}^{l} \alpha_i \mathbf{x}_i^T\right) \mathbf{x} - b$. Therefore, it is possible to replace the inner product $\mathbf{x}^T\mathbf{z}$ by a positive definite Kernel $K(\mathbf{x}, \mathbf{z}) = \langle \Phi(\mathbf{x}), \Phi(\mathbf{z}) \rangle$ which leads to the implicit use of arbitrary feature spaces, whose definition has not necessarily to be known, making profoundly non-linear problems linear.

3 The C2-SoftDoubleMaxMinOver Algorithm for Classification

The DoubleMinOver algorithm [1] as the first extension of MinOver is a simple maximal margin learning machine. Starting from any interim solution $\mathbf{w}_t$ the method selects two vectors each of both classes having the smallest margin to the hyperplane defined by $V = \{\mathbf{v} | \mathbf{w}^T\mathbf{v} - b = 0\}$, that is $i_{a,\min} = \arg\min_{i,y_i=a} af(\mathbf{x}_i)$, respectively. Afterwards the Lagrange-coefficients α_i will be increased by 1 to change the direction of the weight vector given by $\forall \mathbf{x} : f(\mathbf{x}) = \mathbf{w}^T\mathbf{x} = \sum_{i=1}^{l} y_i \alpha_i K(\mathbf{x}_i, \mathbf{x})$ towards these two input vectors and increase their margins. By induction it can be seen that $\mathbf{w}_t$ indeed tends to $\mathbf{w}^*$ while $t \to \infty$. It has been shown [15] that the angle γ_t between the correct solution $\mathbf{w}^*$ and the interim solution $\mathbf{w}_t$ after t steps is bounded by $O(t^{-1})$. Furthermore the time complexity per iteration is $O(l)$, where l is the number of input vectors.

In order to achieve the dual SV solution, not to decrease the convergence speed and to hold the optimisation constraints, in DoubleMaxMinOver the vectors with the largest margin will be decreased, if it is known that they cannot be Support Vectors while the ones with smallest margin will be increased twice. In addition we introduced a soft margin approach working with an extended kernel $K'(\mathbf{x}_i, \mathbf{x}_j) = K(\mathbf{x}_i, \mathbf{x}_j) + \frac{\delta_{i,j}}{C}$. It can be seen that indeed the data remains linearly separable by construction [13] within this extended feature space. Later we will see that all these statements can straightforwardly be adopted for our regression approach.

The combination of this soft margin, the DoubleMinOver, and the MaxMinOver approaches finally leads to the C2-SoftDoubleMaxMinOver [1,2,3] method, which is as fast as the LibSVM toolbox on standard benchmarks and performs comparable results.

4 The MaxMinOver Regression Approach

More general than the classification task is the one of regression. Now the goal is to approximate a function $g(\mathbf{x})$ by using

$$f(\mathbf{x}) = \sum_{i=1}^{l} \alpha_i K(\mathbf{x}_i, \mathbf{x}) - b.$$

This can be interpreted as linear regression within an appropriate feature space, again defined by the Kernel K. To get a scope for the minimisation of $\|\mathbf{w}\|$ the user defines an ϵ-tube around the real function values $y_i = g(\mathbf{x}_i)$. As in our soft margin classification approach we consequently use the C2-SVM model for regression. If $C \to \infty$ the Support Vector solution is in a way the simplest or the flattest one while making no more error than ϵ. The Support Vectors are the vectors with Lagrange-coefficients $\alpha_i \neq 0$ lying on the edges of the ϵ-tube. The introduction of a finite C is not only necessary as regularisation parameter, but also if it cannot be guaranteed that a regression with an error of at most ϵ is achievable at all. Fortunately the geometrical interpretation is similar to the one in the C2-SVM classification, so it is not difficult to see that also in the regression case an unequivocal Support Vector solution exists, which can be interpreted

Algorithm 1. The MaxMinOver Regression Algorithm

Require: given set of normed input vectors $X = \{(\mathbf{x}_i, y_i), i \in \{1, \ldots, l\}\}$ with normed function values and parameters C and ϵ
Ensure: calculates the minimal hyperplane regressing the input data given in dual representation
  set $\forall i : \alpha_i \leftarrow 0, t \leftarrow 0, f \leftarrow \left(\mathbf{x}_i \to \sum_{j=1}^{l} \alpha_j K(\mathbf{x}_j, \mathbf{x}_i) + \frac{\alpha_i}{C}\right)$
  $R \leftarrow \sqrt{\max_{i,j,y_i=1,y_j=-1} \left(K(\mathbf{x}_i,\mathbf{x}_i) - 2K(\mathbf{x}_i,\mathbf{x}_j) + K(\mathbf{x}_j,\mathbf{x}_j)\right)}$
  while the desired precision is not reached **do**
    set $t \leftarrow t + 1$
    find $i_{\min} = \arg\min_i (f(\mathbf{x}_i) - y_i)$
    find $i_{\max} = \arg\max_i (f(\mathbf{x}_i) - y_i)$
    find $i_{minNSV} = \arg\max_{i,\alpha_i>0} (f(\mathbf{x}_i) - y_i)$
    find $i_{maxNSV} = \arg\min_{i,\alpha_i<0} (f(\mathbf{x}_i) - y_i)$
    if $(f(\mathbf{x}_{i_{\max}}) - y_{i_{\max}}) - (f(\mathbf{x}_{i_{\min}}) - y_{i_{\min}}) > 2\epsilon$ **then**
      if $(f(\mathbf{x}_{i_{minNSV}}) - y_{i_{minNSV}}) - (f(\mathbf{x}_{i_{\min}}) - y_{i_{\min}}) > \frac{4R^2+16(2+3\epsilon^2+4\epsilon)}{t}$ **then**
        set $\alpha_{i_{\min}} \leftarrow \alpha_{i_{\min}} + \frac{1}{t} + \min\left(\frac{1}{t}, \alpha_{i_{minNSV}}\right)$
        set $\alpha_{i_{minNSV}} \leftarrow \alpha_{i_{minNSV}} - \min\left(\frac{1}{t}, \alpha_{i_{minNSV}}\right)$
      else
        set $\alpha_{i_{\min}} \leftarrow \alpha_{i_{\min}} + \frac{1}{t}$
      end if
      if $(f(\mathbf{x}_{i_{\max}}) - y_{i_{\max}}) - (f(\mathbf{x}_{i_{maxNSV}}) - y_{i_{maxNSV}}) > \frac{4R^2+16(2+3\epsilon^2+4\epsilon)}{t}$ **then**
        set $\alpha_{i_{\max}} \leftarrow \alpha_{i_{\max}} - \frac{1}{t} - \min\left(\frac{1}{t}, -\alpha_{i_{maxNSV}}\right)$
        set $\alpha_{i_{maxNSV}} \leftarrow \alpha_{i_{maxNSV}} + \min\left(\frac{1}{t}, -\alpha_{i_{maxNSV}}\right)$
      else
        set $\alpha_{i_{\max}} \leftarrow \alpha_{i_{\max}} - \frac{1}{t}$
      end if
    else
      if $(f(\mathbf{x}_{i_{\min}}) - y_{i_{\min}}) - (f(\mathbf{x}_{i_{minNSV}}) - y_{i_{minNSV}}) > \frac{4R^2+16(2+3\epsilon^2+4\epsilon)}{t}$ **then**
        set $\alpha_{i_{minNSV}} \leftarrow \alpha_{i_{minNSV}} - \min\left(\frac{1}{t}, \alpha_{i_{minNSV}}\right)$
      end if
      if $(f(\mathbf{x}_{i_{\max}}) - y_{i_{\max}}) - (f(\mathbf{x}_{i_{maxNSV}}) - y_{i_{maxNSV}}) > \frac{4R^2+16(2+3\epsilon^2+4\epsilon)}{t}$ **then**
        set $\alpha_{i_{maxNSV}} \leftarrow \alpha_{i_{maxNSV}} + \min\left(\frac{1}{t}, -\alpha_{i_{maxNSV}}\right)$
      end if
    end if
  end while
  set $b \leftarrow \frac{1}{2}((f(\mathbf{x}_{i_{\min}}) - y_{i_{\min}}) + (f(\mathbf{x}_{i_{\max}}) - y_{i_{\max}}))$

as the solution with at most ϵ error within the extended feature space defined by $K'(\mathbf{x},\mathbf{z}) = K(\mathbf{x},\mathbf{z}) + \frac{\delta(\mathbf{x},\mathbf{z})}{C}$. This is as reasonable as the statement that the preconditioned linear equation system $\left(K + \frac{1}{C}E_l\right)\alpha = y$, which is used in kernalized Ridge Regression [13], has a solution for all but finitely many $C < \infty$, even if the bias $b = 0$ and $\epsilon = 0$.

As a first approach consider the direct adaptation of the DoubleMinOver algorithm. As in the classification problem, where we choose one input vector each of both classes with the smallest margin, we now choose the two input vectors making the largest positive and negative regression error, respectively. Apparently, using an appropriate step width, the algorithm converges to any feasible solution, that is all input vectors lie within the ϵ-tube and the above constraint holds true by construction in each iteration.

Still if the Lagrange-coefficient of any of the input vectors will be included faulty as potential Support Vectors, we need an instrumentation to turn back this decision, if they are indeed non Support Vectors. Once again as in the classification task, where we choose the input vectors each of both classes with the largest margin, we now choose the two vectors lying furthest away from its edge of the ϵ-tube and dememorise them by controlling the Lagrange-coefficients back without loosing convergence speed (see algorithm 1). Consequently this is the MaxMinOver Regression algorithm. The difficult looking dememorisation criterion will be explained in the next section.

4.1 On the Convergence of MaxMinOver Regression

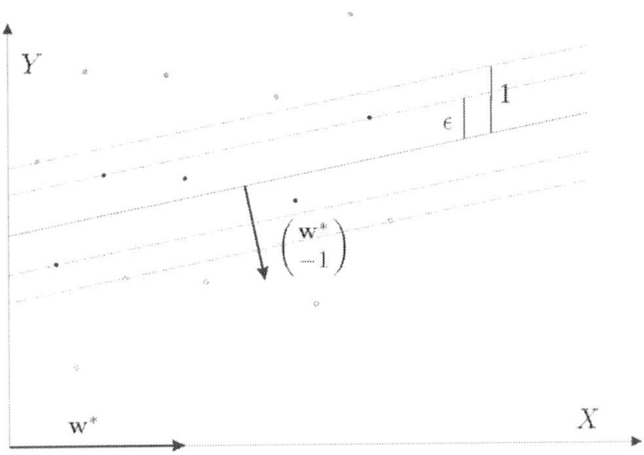

Fig. 1. Construction of the classification problem. The black dots are the true samples which we want to approximate. By shifting them orthogonal to the SVM fit curve in both senses of direction and assigning different classes (bright ($\mathbf{x}_i^1$) and dark gray ($\mathbf{x}_i^{-1}$) dots) we construct a classification problem whose maximal margin separating hyperplane is exactly the regression fit.

In the following without loss of generality we assume $C \to \infty$ and suppose that a solution with no more error than ϵ exists. Otherwise we choose an appropriate $C > 0$ and reset $K'(\mathbf{x},\mathbf{z}) = K(\mathbf{x},\mathbf{z}) + \frac{\delta(\mathbf{x},\mathbf{z})}{C}$. First of all, the Support Vector solution and only the

SV solution is a fixed point of the algorithm. This can be seen as follows. Once we have reached the SV solution, the outer condition $(f(\mathbf{x}_{i_{\max}}) - y_{i_{\max}}) - (f(\mathbf{x}_{i_{\min}}) - y_{i_{\min}}) > \epsilon + \epsilon$ of the algorithm is evaluated to false, all points have to be within the ϵ-tube. The left hand side of the inner conditions is always 0, because only Support Vectors have non-zero Lagrange coefficients and all Support Vectors lie exactly on the edge of the ϵ-tube, while the right hand side is always positive. Hence nothing would change and the algorithm has finished.

On the other hand, it can further be seen that no other configuration of α can be a fixed point. If any pair of vectors lie outside of the ϵ-tube, then the outer condition would be fulfilled. Otherwise there has to be at least one vector within the ϵ-tube and not on its edge having non-zero Lagrange coefficient and therefore leading to a positive fulfillment of the inner condition after a finite number of iterations.

Now we show that the MaxMinOver Regression algorithm indeed approaches the Support Vector solution, by ascribing the regression problem to a classification problem. We construct the classification problem within the extended $(\dim(Z)+1)$-dimensional space, where Z is the feature space and the additional dimension represents the function values y, and demonstrate that the classification and the regression algorithms both work uniformly in the limit. Let $X = \{\mathbf{x}_1, \ldots, \mathbf{x}_l\}$ be the set of input vectors with its function values $Y = \{y_1, \ldots, y_n\}$ and $\mathbf{w}^*$ the Support Vector solution of the regression problem. Of course, if one uses a non-linear feature space, then $\mathbf{x}_i$ has to be replaced by $\Phi(\mathbf{x}_i)$ or $\Phi'(\mathbf{x}_i)$, respectively. We construct the set $\hat{X} = \{\mathbf{x}_1^1, \mathbf{x}_1^{-1}, \ldots, \mathbf{x}_l^1, \mathbf{x}_l^{-1}\}$ from X by substituting

$$\mathbf{x}_i^1 = \begin{pmatrix} \mathbf{x}_i \\ y_i \end{pmatrix} + \frac{1+\epsilon}{\|\mathbf{w}^*\|^2 + 1} \begin{pmatrix} \mathbf{w}^* \\ -1 \end{pmatrix}$$

$$\mathbf{x}_i^{-1} = \begin{pmatrix} \mathbf{x}_i \\ y_i \end{pmatrix} - \frac{1+\epsilon}{\|\mathbf{w}^*\|^2 + 1} \begin{pmatrix} \mathbf{w}^* \\ -1 \end{pmatrix}$$

and assigning the classes $y_i^a = a$. The Support Vector solution of this classification problem is $\hat{\mathbf{w}}^* = \begin{pmatrix} \mathbf{w}^* \\ -1 \end{pmatrix}$ apart from its norm with functional margin 1, because

$$\min_{i,y} \max_b y((\hat{\mathbf{w}}^*)^T \mathbf{x}_i^y - b) = \min_{i,y} \max_b y((\mathbf{w}^*)^T \mathbf{x}_i - y_i - b) \tag{1}$$

$$+1+\epsilon \tag{2}$$

$$= -\epsilon + 1 + \epsilon$$

$$= 1$$

and for each vector $\hat{\mathbf{v}} \neq \hat{\mathbf{w}}^*$ (with $\|\hat{\mathbf{v}}\| = \|\hat{\mathbf{w}}^*\|$) it holds both $\hat{\mathbf{v}}^T \hat{\mathbf{w}}^* \frac{1+\epsilon}{\|\mathbf{w}^*\|^2+1} < (\hat{\mathbf{w}}^*)^T \hat{\mathbf{w}}^* \frac{1+\epsilon}{\|\mathbf{w}^*\|^2+1} = 1+\epsilon$ (compare eqn. part 2) and at least for the Support Vectors either $\mathbf{v}^T \mathbf{x}_i - y_i - b \geq \epsilon$ or $y_i - \mathbf{v}^T \mathbf{x}_i + b \geq \epsilon$ using any bias b (compare eqn. part 1), because $\mathbf{w}^*$ is the Support Vector solution of the actual regression problem.

Now suppose an interim solution $\mathbf{w}$. Then MaxMinOver for classification and for regression choose the same vectors in the next iteration. It holds for the first two find-statements of algorithm 1

$$\begin{aligned}\arg\min_i (\mathbf{w}^T\mathbf{x}_i - y_i) &= \arg\min_i \left(\mathbf{w}^T\mathbf{x}_i - y_i + S\right) \\ &= \arg\min_i \hat{\mathbf{w}}^T\mathbf{x}_i^1\end{aligned} \tag{3}$$

$$\begin{aligned}\arg\max_i (\mathbf{w}^T\mathbf{x}_i - y_i) &= \arg\max_i \left(\mathbf{w}^T\mathbf{x}_i - y_i - S\right) \\ &= \arg\max_i \hat{\mathbf{w}}^T\mathbf{x}_i^{-1} \\ &= \arg\min_i \left(-\hat{\mathbf{w}}^T\mathbf{x}_i^{-1}\right)\end{aligned} \tag{4}$$

$$\text{with } S = \frac{(1+\epsilon)(\mathbf{w}^T\mathbf{w}^* + 1)}{\|\mathbf{w}^*\|^2 + 1}$$

and further for the the second two statements concerning the dememorisation of non Support Vectors

$$\begin{aligned}\arg\max_{i,\alpha_i>0} (\mathbf{w}^T\mathbf{x}_i - y_i) &= \arg\max_{i,\alpha_i>0} \left(\mathbf{w}^T\mathbf{x}_i - y_i + S\right) \\ &= \arg\max_{i,\alpha_i>0} \hat{\mathbf{w}}^T\mathbf{x}_i^1\end{aligned} \tag{5}$$

$$\begin{aligned}\arg\min_{i,\alpha_i<0} (\mathbf{w}^T\mathbf{x}_i - y_i) &= \arg\min_{i,\alpha_i<0} \left(\mathbf{w}^T\mathbf{x}_i - y_i - S\right) \\ &= \arg\min_{i,\alpha_i<0} \hat{\mathbf{w}}^T\mathbf{x}_i^{-1} \\ &= \arg\max_{i,\alpha_i<0} \left(-\hat{\mathbf{w}}^T\mathbf{x}_i^{-1}\right).\end{aligned} \tag{6}$$

Note that the constraints $\alpha_i \geq 0$, respectively $\alpha_i \leq 0$ implicitly hold for eq. 3, respectively eq. 4 while for eqn. 5 and 6 it is necessary to choose only potential Support Vectors of the correct classes for dememorisation.

The presented interrelationship implies that, if the classification algorithm will increment or decrement some $y_i\alpha_i$, then the regression algorithm does so as well. But while $\|\hat{\mathbf{w}}^{classification}\|$ increases linearly in time and converges to $\hat{\mathbf{w}}^*$ (apart from its norm), $\|\hat{\mathbf{w}}^{regression}\|$ must not increase arbitrarily. There exists a time $t_{start} < t$, from which on $\exists C_1, C_2 \in \mathbb{R} : C_1 t < -\hat{\mathbf{w}}_{l+1}^{classification} < C_2 t$ with $C_1 < C_2$. But for $\hat{\mathbf{w}}^{regression}$ is $\hat{\mathbf{w}}_{l+1}^{regression} = -1$ implicitly given. Hence $\|\hat{\mathbf{w}}^{regression}\|$ has to be constrained and therefore we choose an appropriate step width of order $O\left(\frac{1}{t}\right)$ (instead of normalising after each iteration). Although the harmonic series diverges and hence any possible solution vector within the feature space is reachable, we want to emphasize that the convergence speed can be tuned significantly by normalising Y or choosing an appropriate constant factor for the step width.

We derive the criterion for dememorisation of non Support Vectors from the classification method, where $R^2_{classification} \leq \max_{i,j,y_i=1,y_j=-1} (\mathbf{x}_i - \mathbf{x}_j)^2$ has already been proven [1,2]. As far this criterion must only be an upper bound, we estimate

$$\begin{aligned} R^2_{regression} &= \max_{i,j} \left(\left(\begin{pmatrix} \mathbf{x}_i \\ y_i \end{pmatrix} + \frac{1+\epsilon}{\|\mathbf{w}^*\|^2+1} \begin{pmatrix} \mathbf{w}^* \\ -1 \end{pmatrix} \right) \right. \\ &\quad \left. - \left(\begin{pmatrix} \mathbf{x}_j \\ y_j \end{pmatrix} - \frac{1+\epsilon}{\|\mathbf{w}^*\|^2+1} \begin{pmatrix} \mathbf{w}^* \\ -1 \end{pmatrix} \right) \right)^2 \\ &\leq R^2_{classification} + 4 \left(1 + \frac{3\epsilon^2 + 4\epsilon + 1}{\|\mathbf{w}^*\|^2+1} \right) \\ &\leq R^2_{classification} + 4 \left(2 + 3\epsilon^2 + 4\epsilon \right) . \end{aligned}$$

Note that we do not need to know the solution $\mathbf{w}^*$. Hence it is possible to apply this estimation in practice.

5 Experimental Results on Artificial Data

To evaluate the MaxMinOver Regression algorithm we constructed randomly generated artificial datasets and modified the variance, the regression error, the parameters C and ϵ, the number of iteration steps, the distribution of the data points, the number of training examples, and the dimension of the feature space. The table shows the averaged results of these evaluations. It can be seen that the regression error is comparable to the one achieved by the LibSVM Toolbox. The higher the number of iterations the better is the performance of our method and the closer to the results of the LibSVM, whose step width was not changed. Only the number of Support Vectors is sometimes different. More iteration steps in both methods should lead to a convergence of the number of Support Vectors to each other.

Table 1. Averaged regression results obtained with MaxMinOver Regression (B) on artificial data. For comparison averaged results obtained with the ϵ-SVR of the LibSVM Toolbox (A) are listed. The simple MaxMinOver Regression algorithm achieves comparable results with a few training steps. (Caption: $\mathbf{w}$ norm of weight vector, L_2 squared error, $L_{2,\epsilon}$ squared error tolerating error ϵ, $|SV|$ number of Support Vectors)

Parameters			Results								
steps	distr	number	$\|\mathbf{w}_A\|$	$\|\mathbf{w}_B\|$	$L_{2,A}$	$L_{2,B}$	$L_{2,\epsilon,A}$	$L_{2,\epsilon,B}$	$\|SV_A\|$	$\|SV_B\|$	$\frac{\|\mathbf{w}_A - \mathbf{w}_B\|}{\|\mathbf{w}_A\|}$
1200			2.0218	2.0431	0.0052	0.0043	0.0007	0.0008	106	111	0.0368
2400			2.0231	2.0358	0.0052	0.0041	0.0007	0.0008	107	107	0.0289
6000			2.0372	2.0426	0.0051	0.0045	0.0007	0.0007	107	89	0.0134
	uniform		2.3792	2.3922	0.0054	0.0044	0.0003	0.0004	71	102	0.0137
	normal		0.9718	0.9853	0.0045	0.0041	0.0020	0.0020	215	105	0.0639
		50	1.9600	1.9827	0.0080	0.0063	0.0006	0.0006	42	43	0.0342
		500	2.0947	2.0983	0.0024	0.0023	0.0009	0.0010	171	162	0.0183

6 Conclusions

Regression is an important mathematical problem which occurs in a wide variety of practical applications. Support Vector regression achieves results with a high generalisation capability. Different complicated Support Vector regression approaches have been introduced in the past.

The main goal of this paper is hence to show that even Support Vector regression can be dealt with in a simple way with the MaxMinOver Regression approach. In benchmarks the method achieves performances as good as the LibSVM Toolbox. Both, its simplicity and its good performance makes this approach interesting for implementations in industrial environments.

References

1. Martinetz, T., Labusch, K., Schneegass, D.: Softdoubleminover: A simple procedure for maximum margin classification. Proc. of the International Conference on Artificial Neural Networks (2005) 301–306
2. Martinetz, T.: Maxminover: A simple incremental learning procedure for support vector classification. Proc. of the International Joint Conference on Neural Networks (IEEE Press) (2004) 2065–2070
3. Schneegass, D., Martinetz, T., Clausohm, M.: Onlinedoublemaxminover: A simple approximate time and information efficient online support vector classification method. Proc. of the European Symposium on Artificial Neural Networks (2006 (in preparation))
4. Cortes, C., Vapnik, V.: Support-vector networks. Machine Learning **20**(3) (1995) 273–297
5. Vapnik, V.: The Nature of Statistical Learning Theory. Springer-Verlag, New York (1995)
6. LeCun, Y., Jackel, L., Bottou, L., Brunot, A., Cortes, C., Denker, J., Drucker, H., Guyon, I., Muller, U., Sackinger, E., Simard, P., Vapnik, V.: Comparison of learning algorithms for handwritten digit recognition. Int.Conf.on Artificial Neural Networks (1995) 53–60
7. Osuna, E., Freund, R., Girosi, F.: Training support vector machines:an application to face detection. CVPR'97 (1997) 130–136
8. Schölkopf, B.: Support vector learning (1997)
9. Friess, T., Cristianini, N., Campbell, C.: The kernel adatron algorithm: a fast and simple learning procedure for support vector machine. Proc. 15th International Conference on Machine Learning (1998)
10. Platt, J.: Fast Training of Support Vector Machines using Sequential Minimal Optimization. In: Advances in Kernel Methods - Support Vector Learning. MIT Press (1999) 185–208
11. Keerthi, S.S., Shevade, S.K., Bhattacharyya, C., Murthy, K.R.K.: A fast iterative nearest point algorithm for support vector machine classifier design. IEEE-NN **11**(1) (2000) 124–136
12. Li, Y., Long, P.: The relaxed online maximum margin algorithm. Machine Learning **46(1-3)** (2002) 361–387
13. Cristianini, N., Shawe-Taylor, J.: Support Vector Machines And Other Kernel-based Learning Methods. Cambridge University Press, Cambridge (2000)
14. Vapnik, V.N.: Statistical Learning Theory. John Wiley & Sons, Inc., New York (1998)
15. Martinetz, T.: Minover revisited for incremental support-vector-classification. Lecture Notes in Computer Science **3175** (2004) 187–194

A Variational Formulation for the Multilayer Perceptron

Roberto Lopez and Eugenio Oñate

International Center for Numerical Methods in Engineering (CIMNE),
Technical University of Catalonia (UPC),
Barcelona, Spain
rlopez@cimne.upc.edu, onate@cimne.upc.edu
www.cimne.com

Abstract. In this work we present a theory of the multilayer perceptron from the perspective of functional analysis and variational calculus. Within this formulation, the learning problem for the multilayer perceptron lies in terms of finding a function which is an extremal for some functional. As we will see, a variational formulation for the multilayer perceptron provides a direct method for the solution of general variational problems, in any dimension and up to any degree of accuracy. In order to validate this technique we use a multilayer perceptron to solve some classical problems in the calculus of variations.

1 Introduction

Queen Dido of Carthage was apparently the first person to attack a problem that can readily be solved by using the calculus of variations. Dido, having been promised all of the land she could enclose with a bull's hide, cleverly cut the hide into many lengths and tied the ends together. Having done this, her problem was to find the closed curve with a fixed perimeter that encloses the maximum area [1]. The problem is based on a passage from Virgil's Aeneid:

The Kingdom you see is Carthage, the Tyrians, the town of Agenor;
But the country around is Libya, no folk to meet in war.
Dido, who left the city of Tyre to escape her brother,
Rules here-a long a labyrinthine tale of wrong
Is hers, but I will touch on its salient points in order...
Dido, in great disquiet, organised her friends for escape.
They met together, all those who harshly hated the tyrant
Or keenly feared him: they seized some ships which chanced to be ready...
They came to this spot, where to-day you can behold the mighty
Battlements and the rising citadel of New Carthage,
And purchased a site, which was named 'Bull's Hide' after the bargain
By which they should get as much land as they could enclose with a bull's hide.

Despite the circle appears to be an obvious solution to Dido's problem, proving this fact is rather difficult. Zenodorus proved that the area of the circle is larger

S. Kollias et al. (Eds.): ICANN 2006, Part I, LNCS 4131, pp. 159–168, 2006.

than that of any polygon having the same perimeter, but the problem was not rigorously solved until 1838 by Jakob Steiner [1].

Although the history of variational calculus dates back to the ancient Greeks, it was not until the seventeenth century in western Europe that substantial progress was made. A problem of historical interest is the brachistochrone problem, posed by Johann Bernoulli in 1696 [1]. The term brachistochrone derives from the Greek 'brachistos' (the shortest) and 'chronos' (time):

Given two points A and B in a vertical plane, what is the curve traced out by a particle acted on only by gravity, which starts at A and reaches B in the shortest time?

Sir Isaac Newton was challenged to solve the problem, and did so the very next day. In fact, the solution to the brachistochrone problem, which is a segment of a cycloid, is credited to Johann and Jacob Bernoulli, Sir Isaac Newton and Guillaume de L'Hospital [1]. In that context, Dido's problem was called the isoperimetric problem, and it was stated as:

Of all simple closed curves in the plane of a given length l, which encloses the maximum area?

The aim of both the brachistochrone and the isoperimetric problems is to find a function which is the minimal or the maximal value of a specified functional. By a functional, we mean a correspondence which assigns a number to each function belonging to some class [2]. The calculus of variations gives methods for finding extremals of functionals, and problems that consist in finding minimal and maximal values of functionals are called variational problems [2].

While some simple variational problems can be solved analytically, the only practical technique for general problems is to approximate the solution using direct methods [2]. The fundamental idea underlying the so called direct methods is to consider the variational problem as a limit problem for some function optimization problem in many dimensions [2]. Unfortunately, variational problems are difficult to solve, and new numerical methods need to be developed in order to overcome that difficulties.

Neural networks is one of the main fields of artificial intelligence. The multilayer perceptron is an important model of neural network, and much of the literature in the area is referred to that model. Traditionally, the learning problem for the multilayer perceptron has been formulated in terms of the minimization of an error function of the free parameters in the network, in order to fit the neural network to an input-target data set [3]. In that way, the only learning tasks allowed for the multilayer perceptron are data modelling type problems, such as function regression or pattern recognition.

In this work we present a variational formulation for the multilayer perceptron. Within this formulation, the learning problem for the multilayer perceptron lies in terms of solving a variational problem by minimizing a performance functional of the function space spanned by the network. The choice of a suitable performance functional depends on the particular application. On the other hand, the performance functional might need the integration of functions, ordinary differential equations or partial differential equations in order to be evaluated.

As we will see, neural networks are not only able to solve data modelling problems, but also a wide range of mathematical and physical problems. More specifically, a variational formulation for neural networks provides a direct method for solving general variational problems.

In order to validate this numerical method for the solution of variational problems, we use a multilayer perceptron to solve the brachistochrone problem and the isoperimetric problem, and compare the results provided by the neural network to the analytical results.

2 The Multilayer Perceptron Function Space

A neuron model is the basic information processing unit in a neural network, and the perceptron is the characteristic neuron model in the multilayer perceptron [4]. On the other hand, artificial neurons can be combined in a network architecture to form a neural network. The characteristic network architecture in the multilayer perceptron is the so called feed-forward architecture [4]. In this way, a multilayer perceptron can be defined as a feed-forward network architecture composed of perceptron neuron models.

Mathematically, a multilayer perceptron spans a parameterized function space V from an input $X \subseteq \mathbf{R}^n$ to an output $Y \subseteq \mathbf{R}^m$ [5]. Elements of V are parameterized by the free parameters in the network, which can be grouped together in a s-dimensional free parameter vector $\underline{\alpha} = (\alpha_1, ..., \alpha_s)$. The dimension of the function space V is therefore s. The elements of the function space spanned by a multilayer perceptron are of the form

$$\begin{aligned} \mathbf{y} : \mathbf{R}^n &\to \mathbf{R}^m \\ \mathbf{x} &\mapsto \mathbf{y}(\mathbf{x}; \underline{\alpha}). \end{aligned}$$

A multilayer perceptron with as few as one hidden layer of sigmoid neurons and an output layer of linear neurons provides a general framework for approximating any function from one finite dimensional space to another up to any desired degree of accuracy, provided sufficiently many hidden neurons are available. In this sense, multilayer perceptron networks are a class of universal approximators [6].

3 The Variational Problem

Traditionally, the learning problem for the multilayer perceptron has been formulated in terms of the minimization of an error function of the free parameters in the network, in order to fit the neural network to an input-target data set [3]. In that way, the only learning tasks allowed for the multilayer perceptron are data modelling type problems.

In a variational formulation for the multilayer perceptron, the concept of error function, $e(\underline{\alpha})$, is changed by the concept or performance functional, $F[\mathbf{y}(\mathbf{x};\underline{\alpha})]$ [5]. A performance functional for the multilayer perceptron is of the form

$$\begin{aligned} F: \quad & V \to \mathbf{R} \\ & \mathbf{y}(\mathbf{x};\underline{\alpha}) \mapsto F[\mathbf{y}(\mathbf{x};\underline{\alpha})]. \end{aligned}$$

The performance functional defines the task that the network is required to accomplish and provides a measure of the quality of the representation that the network is required to learn. In this way, the choice of a suitable performance functional depends on the particular application. As we will see, changing the concept of error function by the concept of performance functional allows us to extend the number of learning tasks for the multilayer perceptron to any variational problem. Some examples are optimal control problems [5], inverse problems [7] or optimal shape design problems.

The learning problem for the multilayer perceptron can then be formulated in terms of the minimization of a performance functional of the function space spanned by the neural network [5]:

Problem 1 (Variational problem for the multilayer perceptron). Let V be the space of all functions $\mathbf{y}(\mathbf{x};\underline{\alpha})$ spanned by a multilayer perceptron, and let s be the dimension of V. Find a function $\mathbf{y}^*(\mathbf{x};\underline{\alpha}^*) \in V$ for which the functional $F[\mathbf{y}(\mathbf{x};\underline{\alpha})]$, defined on V, takes on a minimum or a maximum value.

A variational problem for the multilayer perceptron can be specified by a set of constraints, which are equalities or inequalities that the solution must satisfy. Such constraints are expressed as functionals. An easy approach is to reduce the constrained problem into an unconstrained problem by adding a penalty term to the original performance functional for each constraint in the problem [5].

4 The Reduced Function Optimization Problem

The performance functional, $F[\mathbf{y}(\mathbf{x};\underline{\alpha})]$, has a performance function associated, $f(\underline{\alpha})$, which is defined as a function of the free parameters in the network [5],

$$\begin{aligned} f: \mathbf{R}^s & \to \mathbf{R} \\ \underline{\alpha} & \mapsto f(\underline{\alpha}). \end{aligned}$$

The minimum or maximum value of the performance functional is achieved for a vector of free parameters at which the performance function takes on a minimum or maximum value, respectively. Therefore, the learning problem for the multilayer perceptron, formulated as a variational problem, can be reduced to a function optimization problem [5]:

Problem 2 (Reduced function optimization problem for the multilayer perceptron). Let $\mathbf{R}^s$ be the space of all vectors $\underline{\alpha}$ spanned by the free parameters of a multilayer perceptron. Find a vector $\underline{\alpha}^* \in \mathbf{R}^s$ for which the function $f(\underline{\alpha})$, defined on $\mathbf{R}^s$, takes on a minimum or a maximum value.

In this sense, a variational formulation for the multilayer perceptron provides a direct method to approximate the solution of general variational problems, in any dimension and up to any desired degree of accuracy [5].

The training algorithm is entrusted to solve the reduced function optimization problem. There are many different training algorithms for the multilayer perceptron, which have different requirements and characteristics. One of the most used is the conjugate gradient [3].

5 The Brachistochrone Problem

In this section we solve the brachistochrone problem by means of a multilayer perceptron, and compare the neural network results to the analytical results. For this example we take the points $A = (a, f_a)$ and $B = (b, f_b)$ to be $A = (0, 1)$ and $B = (1, 0)$. The problem is solved with the Flood library [8].

The first step is to choose a network architecture to represent the curve. Here a multilayer perceptron with a sigmoid hidden layer and a linear output layer is used. This neural network is a class of universal approximator [6]. The curve is to be represented in cartesian coordinates $y = y(x)$, so the network must have one input and one output neuron. As an initial guess, we use six neurons in the hidden layer. Such a multilayer perceptron defines a family V of parameterized functions $y(x; \underline{\alpha})$ of dimension $s = 19$, which is the number of free parameters in the network. Figure 1 is a graphical representation of this network architecture.

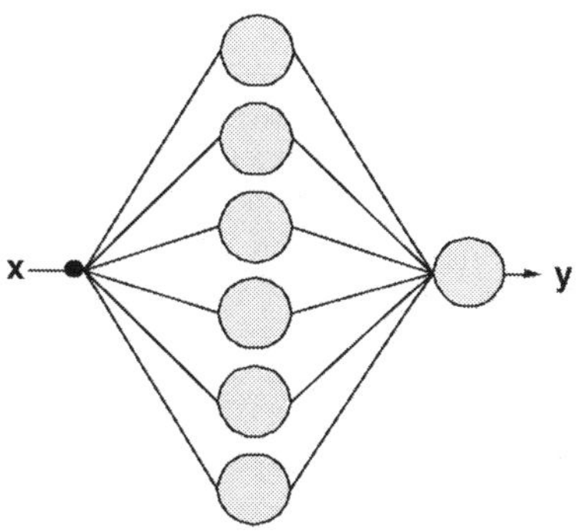

Fig. 1. Network architecture for the brachistochrone problem

The second step is to derive a performance functional for the brachistochrone problem. The time for a particle to travel from point A to point B along a curve $y(x; \underline{\alpha})$ is given by the integral [1]

$$T[y(x; \underline{\alpha})] = \frac{1}{\sqrt{2g}} \int_a^b \sqrt{\frac{1 + [y'(x; \underline{\alpha})]^2}{y_a - y(x; \underline{\alpha})}} dx, \tag{1}$$

where $g = 9.81$ is the gravitational acceleration. The constraints of this problem can be expressed as error functionals,

$$
\begin{aligned}
E_A[y(x;\underline{\alpha})] &= y(a) - y_a \\
&= 0, \qquad (2) \\
E_B[y(x;\underline{\alpha})] &= y(b) - y_b \\
&= 0. \qquad (3)
\end{aligned}
$$

Making use of Equation (1) and considering the constraints (2) and (3), we get

$$
\begin{aligned}
F[y(x;\underline{\alpha})] = {} & \rho_T \frac{1}{\sqrt{2g}} \int_a^b \sqrt{\frac{1+[y'(x;\underline{\alpha})]^2}{y_a - y(x;\underline{\alpha})}}dx \\
& + \rho_A \left(y(a;\underline{\alpha}) - y_a\right)^2 + \rho_B \left(y(b;\underline{\alpha}) - y_b\right)^2, \qquad (4)
\end{aligned}
$$

where ρ_T, ρ_A and ρ_B are penalty term ratios, which are set to 10^{-3}, 1 and 1, respectively. Please note that evaluation of the performance functional in Equation (4) requires a numerical method for the integration of functions. Here we choose the Simpson's composite method [9] with 100 integration intervals. The brachistochrone problem for the multilayer perceptron can then be stated as:

Problem 3 (Brachistochrone problem for the multilayer perceptron). Let V be the space of all functions $y(x;\underline{\alpha})$ spanned by a multilayer perceptron with 1 input, 6 sigmoid neurons in the hidden layer and 1 linear output neuron. The dimension of V is 19. Find a vector of free parameters $\underline{\alpha}^* \in \mathbf{R}^{19}$ that addresses a function $y^*(x;\underline{\alpha}^*) \in V$ for which the functional (4), defined on V, takes on a minimum value.

The third step is to choose a suitable training algorithm for solving the reduced function optimization problem. Here we use a conjugate gradient with Polak-Ribiere search direction and Brent optimal step size methods for training [3]. The tolerance in the Brent's method is set to 10^{-6}. On the other hand, training of the neural network with the conjugate gradient requires the evaluation of the performance function gradient vector $\nabla f(\underline{\alpha})$ [3]. This is carried out by means of numerical differentiation. In particular, we use the symmetrical central differences method [3] with an ϵ value of 10^{-6}. In this example, we set the training algorithm to stop after 1000 epochs of the training process. Table 1 shows the training results for this problem.

Table 1. Training results for the brachistochrone problem

$F[y^*(x;\underline{\alpha}^*)]$	$T[y^*(x;\underline{\alpha}^*)]$	$E_A[y^*(x;\underline{\alpha}^*)]$	$E_B[y^*(x;\underline{\alpha}^*)]$
$3.404 \cdot 10^{-4}$	0.576	$2.889 \cdot 10^{-3}$	$-5.716 \cdot 10^{-5}$

The errors made in the constraints by the multilayer perceptron are of order 10^{-3} and 10^{-5} for points A and B, respectively. The descent time provided by that neural network is 0.576, while that provided by the analytical result is 0.577.

This yields a percentage error of around -0.173%. Results here are good, since the constraint errors made by the network are very small and the descent time provided by the network is very similar to the descent time for the brachistochrone. Figure 2 illustrates the neural network solution to the brachistochrone problem.

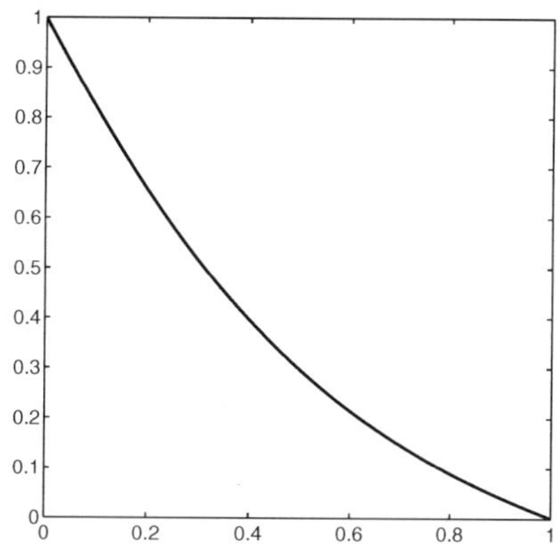

Fig. 2. Neural network solution to the brachistochrone problem

6 The Isoperimetric Problem

In this section we solve the isoperimetric problem by means of a multilayer perceptron, and compare the neural network results to the analytical results. The perimeter goal is set to be $l = 1$. This problem is solved with the Flood library [8].

Here a multilayer perceptron with a sigmoid hidden layer and a linear output layer is used. This neural network is a class of universal approximator [6]. The closed curve is to be represented in polar coordinates $r = r(\theta)$, for $\theta \in [0, 2\pi]$, so the network must have one input and one output neuron. We use 6 neurons in the hidden layer. This multilayer perceptron spans a family V of functions $r(\theta; \underline{\alpha})$ with dimension $s = 19$. Figure 3 is a graphical representation of this network architecture.

In order to derive a performance functional for the isoperimetric problem we first consider the expression for the area enclosed by a polar curve [1],

$$A[r(\theta; \underline{\alpha})] = \frac{1}{2} \int_0^{2\pi} [r(\theta; \underline{\alpha})]^2 d\theta. \tag{5}$$

The perimeter constraint is expressed as an error functional, by considering the arc length of a curve in polar coordinates [1]

$$\begin{aligned} E_P[r(\theta; \underline{\alpha})] &= \int_0^{2\pi} \sqrt{[r(\theta; \underline{\alpha})]^2 + [r'(\theta; \underline{\alpha})]^2 d\theta} - l \\ &= 0. \end{aligned} \tag{6}$$

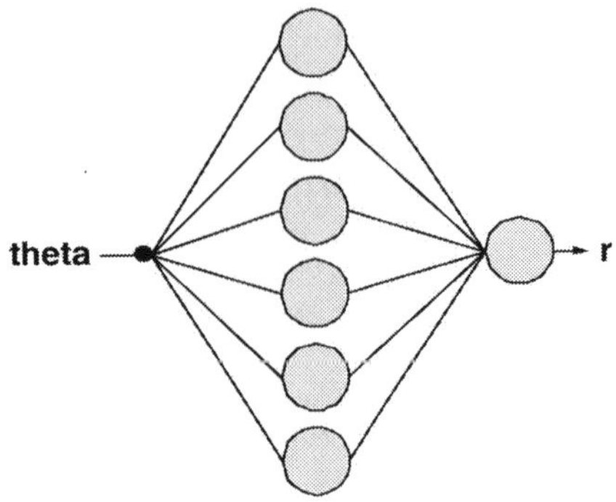

Fig. 3. Network architecture for the isoperimetric problem

Similarly, the join constraint can be expressed as an error functional,

$$\begin{aligned} E_J[r(\theta;\underline{\alpha})] &= r(2\pi;\underline{\alpha}) - r(0;\underline{\alpha}) \\ &= 0. \end{aligned} \tag{7}$$

Making use of Equations (5), (6) and (7), we obtain the performance functional to be minimized,

$$\begin{aligned} F[r(\theta;\underline{\alpha})] = &- \rho_A \frac{1}{2} \int_0^{2\pi} [r(\theta;\underline{\alpha})]^2 d\theta \\ &+ \rho_P \left(\int_0^{2\pi} \sqrt{[r(\theta;\underline{\alpha})]^2 + [r'(\theta;\underline{\alpha})]^2} d\theta - l \right)^2 \\ &+ \rho_J \left(r(2\pi;\underline{\alpha}) - r(0;\underline{\alpha}) \right)^2, \end{aligned} \tag{8}$$

where $\rho_A = 10^{-3}$, $\rho_P = 1$ and $\rho_J = 1$ are penalty term ratios. Evaluation of (8) requires a numerical method for the integration of functions. Here we choose the Simpson's composite method [9] with 100 integration intervals. The isoperimetric problem for the multilayer perceptron is then stated as:

Problem 4 (Isoperimetric problem for the multilayer perceptron). Let V be the space of all functions $r(\theta;\underline{\alpha})$ spanned by a multilayer perceptron with 1 input, 6 sigmoid neurons in the hidden layer and 1 linear output neuron. The dimension of V is 19. Find a vector of free parameters $\underline{\alpha}^* \in \mathbf{R}^{19}$ that addresses a function $r^*(\theta;\underline{\alpha}^*) \in V$ for which the functional (8), defined on V, takes on a minimum value.

Here we use a conjugate gradient with Polak-Ribiere search direction and Brent optimal step size methods for training [3]. The tolerance in the Brent's method is set to 10^{-6}. Evaluation of the performance function gradient vector $\nabla f(\underline{\alpha})$ [3] is carried out by means of the symmetrical central differences method [3], with $\epsilon = 10^{-6}$. The training algorithm is set to stop after 100 epochs.

Table 2. Training results for the isoperimetric problem

$F[r^*(\theta;\underline{\alpha}^*)]$	$A[r^*(\theta;\underline{\alpha}^*)]$	$E_P[r^*(\theta;\underline{\alpha}^*)]$	$E_J[r^*(\theta;\underline{\alpha}^*)]$
$-6.332 \cdot 10^{-6}$	0.079579	$-1.266 \cdot 10^{-5}$	$-4.228 \cdot 10^{-9}$

The perimeter error is of order 10^{-5}, while the join error is of order 10^{-9}. The area of the closed curve provided by the neural network is 0.079579, while that of the circle is 0.079577. This yields a percentage error of around $2.513 \cdot 10^{-3}\%$. Results here are also good, since the error made in all the constraints by the network is very small and the area provided by the network is very similar to that of the circle. Figure 4 illustrates the network solution to the isoperimetric problem.

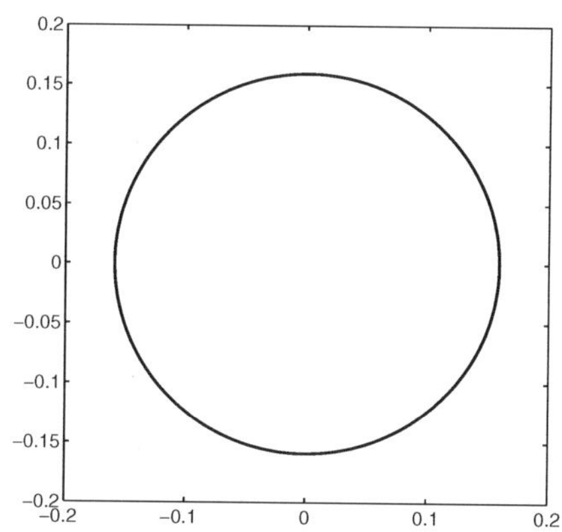

Fig. 4. Neural network solution to the isoperimetric problem

7 Conclusions

A variational formulation for neural networks provides a direct method for the solution of general variational problems, in any dimension and up to any degree of accuracy. This numerical method has been validated for two classical problems in the calculus of variations with analytical solution. Ongoing work focuses on the solution of variational problems in engineering. Some examples are optimal control, inverse or optimal shape design problems.

References

1. Weisstein, E. W.: MathWorld - A Wolfram Web Resource. http://mathworld. wolfram.com (2006).
2. Elsgolc, L. E.: Calculus of Variations. Pergamon Press (1961).

3. Bishop, C.: Neural Networks for Pattern Recognition. Oxford University Press (1995).
4. Šíma, J. and Orponen, P.: General-Purpose Computation with Neural Networks: A Survey of Complexity Theoretic Results. Neural Computation 15 (2003) 2727-2778.
5. Lopez, R., Balsa-Canto, E. and Oñate, E.: Artificial Neural Networks for the Solution of Optimal Control Problems. Proceedings of the Sixth Conference on Evolutionary and Deterministic Methods for Design, Optimisation and Control with Applications to Industrial and Societal Problems (2005).
6. Hornik, K., Stinchcombe, M. and White, H.: Multilayer feedforward networks are universal approximators. Neural Networks 2-5 (1989) 359-366.
7. Dadvand, P., Lopez, R. and Oñate, E.: Artificial Neural Networks for the Solution of Optimal Control Problems. Proceedings of the International Conference ERCOFTAC 2006 (2006).
8. Lopez, R.: Flood: An Open Source Neural Networks C++ Library. www.cimne.com/flood (2005).
9. Stoer, J. and Bulirsch, R.: Introduction to Numerical Analysis. Springer-Verlag (1980).

Natural Conjugate Gradient Training of Multilayer Perceptrons

Ana González and José R. Dorronsoro*

Dpto. de Ingeniería Informática and Instituto de Ingeniería del Conocimiento
Universidad Autónoma de Madrid, 28049 Madrid, Spain

Abstract. For maximum log–likelihood estimation, the Fisher matrix defines a Riemannian metric in weight space and, as shown by Amari and his coworkers, the resulting natural gradient greatly accelerates on–line multilayer perceptron (MLP) training. While its batch gradient descent counterpart also improves on standard gradient descent (as it gives a Gauss–Newton approximation to mean square error minimization), it may no longer be competitive with more advanced gradient–based function minimization procedures. In this work we shall show how to introduce natural gradients in a conjugate gradient (CG) setting, showing numerically that when applied to batch MLP learning, they lead to faster convergence to better minima than that achieved by standard euclidean CG descent. Since a drawback of full natural gradient is its larger computational cost, we also consider some cost simplifying variants and show that one of them, diagonal natural CG, also gives better minima than standard CG, with a comparable complexity.

1 Introduction

The standard approach in Multilayer Perceptron (MLP) training is to minimize the square error function

$$e(W) = \frac{1}{2}\int ||F(X,W) - Y||^2 dP(X,Y),$$

where Y denotes the target associated to a pattern X, $F(X,W)$ is the MLP transfer function and $P(X,Y)$ is the joint (X,Y) probability distribution. In practice, rather than minimizing the global error $e(W)$, one tries to do so for its sample version

$$\hat{e}(W) = \frac{1}{2N}\sum_i ||F(X_i,W) - Y_i||^2.$$

In this light MLP training can be seen as a nonlinear regression problem, but if we assume an error model $Y = F(X,W) + Z$, with Z a multivariate gaussian $g(Z)$ with density

$$g(Z) = \frac{1}{\sqrt{2\pi}\sigma}\exp^{\frac{-||Z||^2}{2\sigma^2}} = \frac{1}{\sqrt{2\pi}\sigma}\exp^{\frac{-||f(X,W)-Y||^2}{2\sigma^2}}.$$

* With partial support of Spain's TIN 2004–07676.

S. Kollias et al. (Eds.): ICANN 2006, Part I, LNCS 4131, pp. 169–177, 2006.

we can alternatively formulate MLP training as a semiparametric maximum log likelihood estimation problem. In fact, the likelihood associated to the sample (X_i, Y_i) is

$$\prod_i g(Z_i) = \prod_i \frac{1}{\sqrt{2\pi}\sigma} \exp^{\frac{-||f(X_i,W)-Y_i||^2}{2\sigma^2}},$$

and therefore

$$-\log\left(\prod_i g(Z_i)\right) = \frac{1}{2}\sum_i ||F(X_i, W) - Y_i||^2 + C,$$

with C a suitable constant.

In this general context of likelihood estimates for parametric probability models, it has been shown by S.I. Amari [2,8] that a a Riemannian structure can be defined in weight space, for which the metric tensor is given by the matrix

$$\begin{aligned} G(W) &= E_{X,\,y}[||f - Y||^2(\nabla_W f)(\nabla_W f)^t] \\ &= \sigma^2 E_X[(\nabla_W f)(\nabla_W f)^t]. \end{aligned} \tag{1}$$

and the inner product to be used in the tangent space at a point W is $\langle u, v\rangle_W = u^t G(W) v$. It turns out [1] that the maximum descent direction of the global error $e(W)$ with respect to the $G(W)$ metric is then given by the "natural" gradient

$$\nabla_G\, e(W) = G(W)^{-1}\nabla e(W).$$

As shown by Amari and his coworkers [1,13], this can be put to advantage when on–line MLP training is considered. In fact, denoting the local error $||f(X, W) - Y||^2$ as $e(X, Y; W)$ and defining natural gradient descent as

$$W_{t+1} = W_t - \eta_t G(W_t)^{-1}\nabla e(X_t, y_t; W_t), \tag{2}$$

one obtains what probably is the fastest converging MLP on–line training method.

The main drawback of on–line natural gradient training is its complexity. For a single hidden layer MLP with input dimension D, H hidden units and C dimensional outputs, and an N pattern sample, the weight–bias dimension is then $\mathcal{D} = H(D+1)+C(H+1)$, which would imply a cost $O(\mathcal{D}^3)$ for G's inversion and an overall cost of $O(\mathcal{D}^3 N)$ for each on–line full sample pass. There are ways in the on–line setting to alleviate this [3,14] and its impact is much smaller if batch natural gradient is considered. In fact, the inversion of G is done only once per batch epoch and if $N \gg \mathcal{D}$, as it happens in most settings, the main cost is then the computation of the matrix G, which is then $O(N\mathcal{D}^2)$.

However, G coincides with the Gauss–Newton approximation to the Hessian of a square error function, and batch natural gradient descent can be seen [5,6] to be closely related to the Levenberg–Marquardt approach to mean square minimization. In turn, this can be used to give another explanation of the speed–up in batch MLP training with respect to standard gradient descent. But for batch MLP training there are other simpler methods such as the conjugate gradient or the variable metric methods, which also have a fast convergence without needing costly Hessian computations.

In any case, the introduction of a Riemannian structure in weight space through the Fisher metric can be done independently [12,8] of the minimization setting described above, and the resulting fast on–line convergence could also be considered as a consequence of the "naturalness" of the Fisher metric. This should also be reflected, for instance, in ways to improve on established batch minimization methods. Some of these methods rely on Hessian computations or approximations, something that may not be easy to do in a Riemannian setting. The situation should be simpler for gradient based methods. Among these, the best known is the conjugate gradient method, a good choice for instance for batch MLP training [7]. In the next section we shall briefly review conjugate gradient and show how to define a natural conjugate gradient and in section 3 we shall numerically illustrate its advantages over its standard counterpart. The paper will finish with a brief review of the paper's results and some concluding remarks.

2 Natural Conjugate Gradient

The standard conjugate gradient (CG) [10] method seeks a fast way to attain the minimum of a general function $f(W)$ by succesively performing for $i = 0, \ldots$, the following steps from an initial W_0 and $g_0 = h_0 = -\nabla\ f(W_0)$:

1. Define $g_{i+1} = -\nabla f(W_{i+1})$, where W_{i+1} is the minimum of f over the line $\{W_i + th_i : t > 0\}$;
2. Set $h_{i+1} = g_{i+1} + \gamma_{i+1} h_i$, with

$$\gamma_{i+1} = \frac{g_{i+1} \cdot g_{i+1}}{g_i \cdot g_i}.$$

The rationale for this approach comes from the fact that, for a quadratic $e(W) = c - b \cdot W + \frac{1}{2} W^t H W$, the above defined g_i, h_i verify for $j < i$

$$g_i \cdot g_j = g_i \cdot h_j = h_i^t H h_j = 0.$$

It thus follows that for such a quadratic e, a minimum W^* is achieved in at most D iterations, with D the dimension of W.

The above formulation can be easily extended when the standard gradient of the mse function $e(W)$ is replaced by its natural counterpart. More precisely, if we denote the natural gradient at W_{i+1} as $\tilde{g}_{i+1} = -G_{i+1}^{-1} \nabla e(W_{i+1})$, with G_{i+1} the natural metric at W_{i+1}, and define

$$\tilde{\gamma}_{i+1} = \frac{\langle \tilde{g}_{i+1}, \tilde{g}_{i+1} \rangle_{G_{i+1}}}{\langle \tilde{g}_i, \tilde{g}_i \rangle_{G_i}},$$

the new conjugate direction is then $\tilde{h}_{i+1} = \tilde{g}_{i+1} + \tilde{\gamma}_{i+1} \tilde{h}_i$. Under some extra assumptions, it can be shown that for the above $\tilde{g}_i$, $\tilde{h}_i$, and a quadratic $e(W)$,

$$\langle \tilde{g}_{i+1}, \tilde{g}_i \rangle_{G_{i+1}} = \langle \tilde{g}_{i+1}, \tilde{h}_i \rangle_{G_{i+1}} = \tilde{h}_{i+1}^t H \tilde{h}_i = 0. \quad (3)$$

Table 1. Training architectures used in the numerical experiments

Problem set	input dim.	hid. units	targ. dim
br. cancer	9	5	2
glass	9	6	6
heart dis.	13	7	5
ionosphere	33	3	2
iris	4	3	3
pima	7	4	2
thyroid	8	5	2
XOR4	3	10	4
abalone	7	4	1
housing	13	5	1

It easily follows from the above discussion that if we do not take into account the line minimization required to obtain the W_i, the cost of standard and natural CG is essentially that of computing the corresponding gradients. We recall that for an N pattern sample and a single hidden layer MLP with input dimension D, H hidden units and C dimensional outputs, the cost of the mse standard gradient is $O(NDHC)$ per batch iteration. When natural gradient is considered and we denote the number of MLP weights as $\mathcal{D} = H(D+1) + C(H+1)$ as done before, this cost is dominated by the rather larger cost $O(N\mathcal{D}^2) = O(N(DH + HC)^2)$ of computing the Fisher matrix. Recall that the $\mathcal{D}^2$ term is due to the neeed to compute about $\mathcal{D}^2/2$ expectations

$$E\left[\frac{\partial e}{\partial w_{lk}}\frac{\partial e}{\partial w_{nm}}\right]. \tag{4}$$

There are several ways to lower this. We may begin by using a block–diagonal version of G, where if denote by w^O_{oh} the hidden–to–output weights and by w^H_{hi} the input–to–hidden weights, we simply assume that

$$E\left[\frac{\partial e}{\partial w^O_{oh}}\frac{\partial e}{\partial w^H_{hi}}\right] \approx 0. \tag{5}$$

The resulting cost would then be $O(N(H^2D^2 + C^2H^2))$. We can further reduce the complexity assuming [5] independence between the output o_k of unit k at a given layer and the generalized error δ_l of unit l of the next layer. Since we have $\partial e(X,Y;W)/\partial w_{lk} = \delta_l o_k$ for the local gradient [4], we can therefore write

$$E\left[\frac{\partial e}{\partial w_{lk}}\frac{\partial e}{\partial w_{nm}}\right] = E\left[\delta_l o_k \delta_n o_m\right] \approx E\left[\delta_l \delta_n\right] E\left[o_k o_m\right].$$

Precomputing the matrices $E\left[\delta_l \delta_n\right]$ and $E\left[o_k o_m\right]$ for the input–to–hidden and hidden–to–output weights has a cost of $O(N(C^2 + H^2))$ for the δ matrices and $O(N(H^2 + D^2))$ for the o matrices. The overall cost of this "independent" natural gradient is then $O(N(D^2+H^2+C^2))$, which now is dominated by the $O(NDHC)$ cost of the standard

Table 2. Final mean mse values and their standard deviation for standard CG (second column), natural CG (third column), diagonal natural CG (fourth column) and line minimization natural gradient. Best final values overall when equality of means is rejected at the 5 % level are given in bold face, second place values in italics and third place values in typewriter type.

Problem set	standard CG	natural CG	diagonal NCG	line min. NCG
breastc	`0.0382 ± 0.0005`	**0.0306 ± 0.0010**	*0.0315 ± 0.0011*	0.0405 ± 0.0009
glass	`0.3499 ± 0.0107`	**0.3357 ± 0.0090**	**0.3383 ± 0.0103**	0.3997 ± 0.0112
heartdis	`0.3444 ± 0.0070`	**0.3373 ± 0.0066**	**0.3350 ± 0.0089**	0.4126 ± 0.0060
ionosphere	`0.0358 ± 0.0025`	**0.0298 ± 0.0037**	**0.0307 ± 0.0035**	0.1026 ± 0.0092
iris	`0.0453 ± 0.0012`	**0.0384 ± 0.0002**	**0.0384 ± 0.0002**	0.0504 ± 0.0028
pima	`0.2263 ± 0.0050`	**0.2189 ± 0.0058**	**0.2205 ± 0.0066**	0.2409 ± 0.0064
thyroid	`0.0488 ± 0.0007`	**0.0400 ± 0.0020**	*0.0416 ± 0.0019*	0.0492 ± 0.0020
xor405	`0.1615 ± 0.0022`	**0.1470 ± 0.0020**	**0.1477 ± 0.0018**	0.1721 ± 0.0057
abalone	0.4172 ± 0.0004	**0.4112 ± 0.0019**	*0.4140 ± 0.0017*	`0.4122 ± 0.0022`
housing	0.0789 ± 0.0029	**0.0706 ± 0.0030**	*0.0771 ± 0.0031*	`0.0735 ± 0.0025`

gradient. Finally, the simplest approach would be to consider what we may call diagonal natural gradient, where we replace the full Fisher matrix $G(W)$ by its diagonal, which results in a cost of $O(N(DH + HC))$, dominated again by the cost of standard CG.

In the following section we shall compare the perfomance against standard CG of natural CG and its pure diagonal variant. Similar results are obtained in the other cases and will be published elsewhere.

3 Numerical Examples

We shall compare natural conjugate gradient MLP training against standard conjugate gradient on 10 datasets. Two of these datasets correspond to regression problems and 8 to classification problems. Nine of the datasets are taken from the UCI database [9]: we shall work with the abalone age and Boston housing regression problems, and the classification problems given by the Wisconsin breast cancer, glass, heart disease, ionosphere, iris, diabetes in Pima indians and thyroid disease datasets. In some instances the UCI repository gives separate training and test sets. Since we are interested only on square error minimization, in these cases we join both sets in a single training set.

The tenth dataset, which we denote XOR4, is a 4 class synthetic problem, an extension of bidimensional XOR to 3 dimensions, where eight 0.5 standard deviation gaussian distributions centered at the opposite corners of the unit cube are considered and four classes are defined pairing diagonally opposite distributions. That is, the gaussian centers of the first class are at $(-1, -1, -1)$ and $(1, 1, 1)$, those of the second are at $(-1, -1, 1)$ and $(1, 1, -1)$ and so on.

In all cases we have normalized input components to zero mean and one variance, and we also have done so for target values in the regression problems. Table 1 shows the training parameters used; the number of hidden units has been set heuristically, but it essentially agrees with values used in other studies.

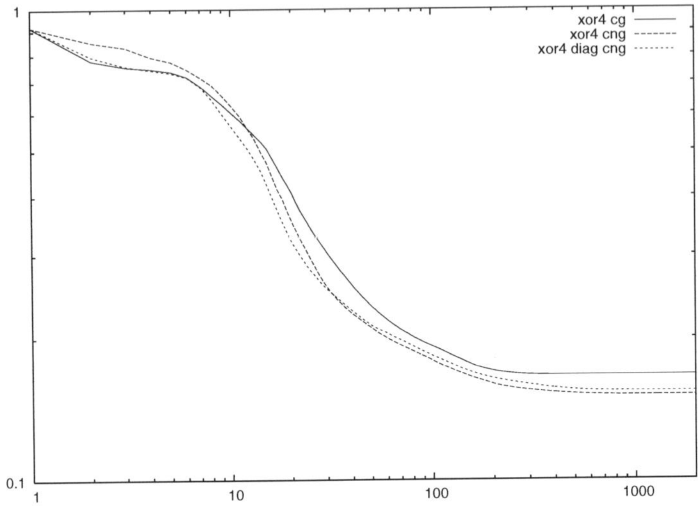

Fig. 1. Mse evolution for the XOR4 problem of standard (solid line), natural (large dash line) conjugate and diagonal natural (small dash line) conjugate gradients

We have used the Numerical Recipes implementation of standard conjugate gradient ([11], section 10.6) and adapted it for natural conjugate gradient. Instead of the Fletcher–Reeves formula for $\tilde{\gamma}_{i+1}$ given in section 2, we have used the Polak–Ribiere variant, as it seems better suited for general function minimization [10,11], namely

$$\tilde{\gamma}_{i+1} = \frac{\langle \tilde{g}_{i+1}, \tilde{g}_{i+1} - \tilde{g}_i \rangle_{G_{i+1}}}{\langle \tilde{g}_i, \tilde{g}_i \rangle_{G_i}},$$

Also, to avoid singularity problems, we invert the matrix $G + \mu I$ instead of G, with I the identity matrix and the scalar μ having an initial value of 0.05 that is decreased by a factor of 0.9 per iteration. In all cases we have run 30 independent trainings starting at different initial weights, with a maximum of 2000 gradient iterations (in many cases the Numerical Recipes implementation makes natural and standard CG descent to stop well before that limit is reached). To avoid instabilities due to training divergence, of all these, only the 20 runs with the best final mean square errors (mse) values are selected and their mean and standard deviations computed. Notice that there are more significant ways to measure MLP performance, such as computing for instance test set accuracies. However we are essentially comparing function minimization procedures, which in the MLP case means to compare final mse values.

Table 2 gives for each data set these final values for standard (second column), and full natural (third) and diagonal (fourth) CG. It also shows results for line mimization based on natural gradient (fifth column). To better compare them we have performed pairwise mean equality tests between all procedures. The table shows in bold face the smaller overall value when equality of means is rejected at the 5% confidence level. It is given in 4 cases by the natural conjugate gradient alone, and in the other 6 cases the performance of natural CG and its diagonal counterpart is similar in the sense that equality

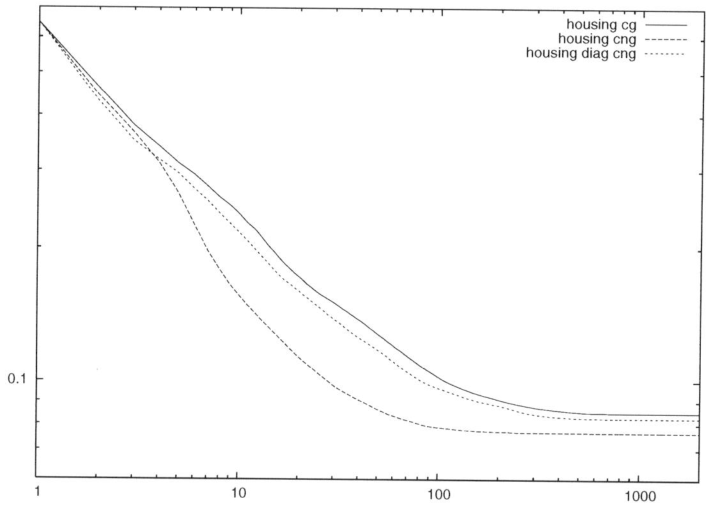

Fig. 2. Mse evolution for the housing problems of standard (solid line), natural (large dash line) conjugate and diagonal natural (small dash line) conjugate gradients

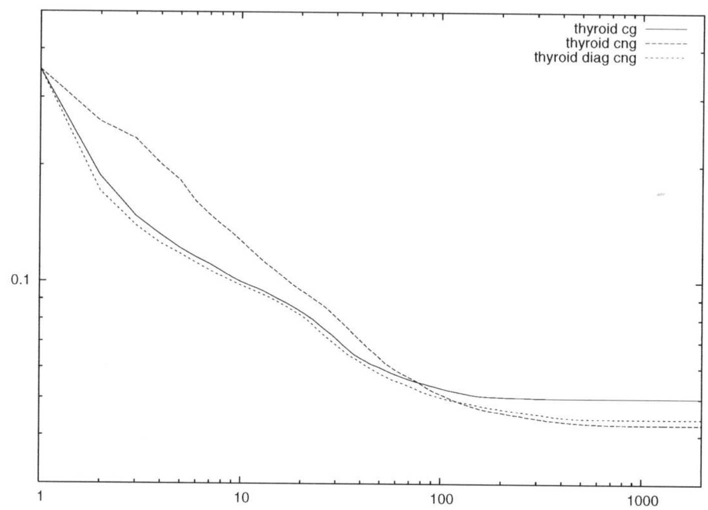

Fig. 3. Mse evolution for the thyroid problem of standard (solid line), natural (large dash line) conjugate and diagonal natural (small dash line) conjugate gradients

of means cannot be rejected. Second overall values are shown in italics and third values in typewriter type. As it can be seen from the table standard and diagonal natural CG beat standard CG in all cases. On the other hand, standard CG beats line minization based natural gradient in all problems but for the abalone and housing datasets. It can

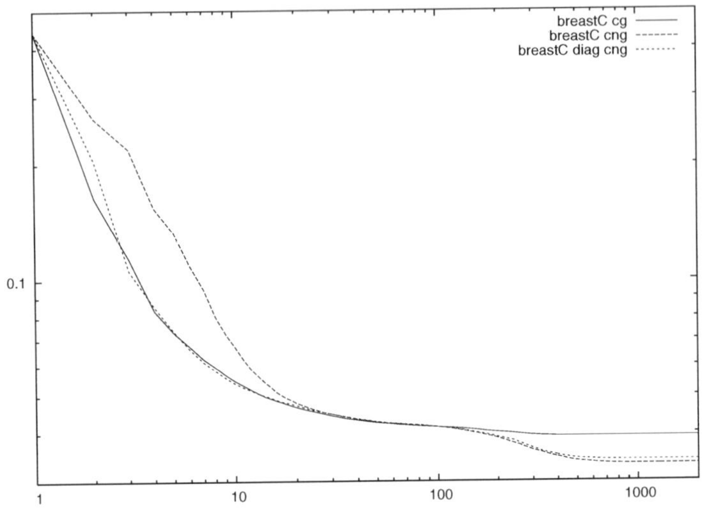

Fig. 4. Mse evolution for the breast cancer problem of standard (solid line), natural (large dash line) conjugate and diagonal natural (small dash line) conjugate gradients

be safely concluded that natural CG, either full or diagonal, yields better minima than ordinary CG for MLP training.

Besides providing better minima, natural conjugate gradient convergence can also be faster than that of ordinary conjugate gradients. This is illustrated in figures 1 and 2 for the XOR4 and housing problems, where natural CG overtakes the standard one at about the tenth iteration and does so for its diagonal variant shortly thereafter (all figures in logarithmic scale on both axes). In other cases this overtaking may happen later, but in all the datasets considered, it takes place before the 100–th iteration. This is shown, for instance, in figures 3 and 4 for the thyroid and breast cancer problems. When comparing convergence speed, one should also take into account the distinct complexity of, say, full natural CG against that of standard CG, something which we are currently studying. In any case, in all datasets diagonal natural CG does overtake standard CG at about the 10–th iteration, while both methods have essentially the same complexity.

4 Conclusions

It was shown by Rao [12] that, in a maximum log–likelihood setting, the Fisher matrix defines a Riemannian metric in weight space alternative to the standard euclidean one. Besides its theoretical advantages, Amari and his coworkers have demonstrated that for on–line MLP training, the resulting natural gradient provides minimization directions that result in a faster convergence.

If batch MLP training is considered, natural gradient descent can be seen as a variant of the Gauss–Newton method, closely related to Levenberg–Marquardt's minimization. A such it may not be competitive with other advanced batch methods, such as for instance, conjugate gradient (CG). In this paper we have shown how natural gradient can be introduced in the conjugate gradient setting and have numerically demonstrated that the performance of the resulting natural CG is consistently better than that of standard CG.

As it is the case for on–line MLP training, a drawback of natural CG is the larger complexity resulting from the required Fisher matrix computations. This can be alleviated by approximating the Fisher matrix under some simplifying assumptions, of which we have considered here the diagonal natural CG. It has essentially the same complexity of standard CG but gives better minima (although not always as good as those achieved by the full natural CG procedure) and a faster convergence.

We finally point out that there might be some interest in further research on the application of natural gradients in general function minimization. A more complete study should be made of full natural CG taking its complexity into account in a precise way. On the other hand, the definition (1) of the natural metric makes sense not only for square error problems but also for other global error functions defined as local error expectation. We are currently considering these and other similar issues.

References

1. Amari, S. (1998). Natural Gradient Works Efficiently in Learning. *Neural Computation*, 10, 251–276.
2. Amari, S., Nagaoka, H. **Methods of information geometry**. American Mathematical Society, 2000.
3. Amari, S., Park, H., Fukumizu, K. (2000). Adaptive Method of Realizing Natural Gradient Learning for Multilayer Perceptrons. *Neural Computation*, 12, 1399–1409.
4. Duda, R., Hart, P., Stork, D. **Pattern classification**. Wiley, 2000.
5. Heskes, T. (2000). On natural Learning and pruning in multilayered perceptrons. *Neural Computation*, 12, 1037–1057.
6. Igel, Ch., Toussaint, M., Weishui, W. (2005). Rprop Using the Natural Gradient, in *Trends and Applications in Constructive Approximation*, International Series of Numerical Mathematics, Vol. 151, Birkhäuser.
7. LeCun, J., Bottou, L., Orr, G., Müller, K.R. Efficient BackProp, in *Neural Networks: tricks of the trade*, 9–50. Springer, 1998.
8. Murray, M., Rice, J.W. **Differential Geometry and Statistics**. Chapman & Hall, 1993.
9. Murphy, P., Aha, D. *UCI Repository of Machine Learning Databases*, Tech. Report, University of Califonia, Irvine, 1994.
10. Polak, F. **Computational Methods in Optimization**. Academic Press, 1971.
11. Press, W., Teukolsky, S., Vetterling, W., Flannery, B. **Numerical Recipes in C**. Cambridge U. Press, 1988.
12. Rao, C.R. (1945). Information and accuracy attainable in estimation of statistical parameters. *Bull. Cal. Math. Soc.*, 37, 81–91.
13. Rattray, M., Saad, D., Amari, S. (1998). Natural gradient descent for on–line learning. *Physical Review Letters*, 81, 5461–5464.
14. Yang, H., Amari, S. (1998). Complexity Issues in Natural Gradient Descent Method for Training Multi-Layer Perceptrons. *Neural Computation*, 10, 2137–2157.

Building Ensembles of Neural Networks with Class-Switching

Gonzalo Martínez-Muñoz, Aitor Sánchez-Martínez, Daniel Hernández-Lobato, and Alberto Suárez

Universidad Autónoma de Madrid,
Avenida Francisco Tomás y Valiente, 11,
Madrid 28049, Spain
gonzalo.martinez@uam.es, aitor.sanchezm@estudiante.uam.es, daniel.hernandez@uam.es, alberto.suarez@uam.es

Abstract. This article investigates the properties of ensembles of neural networks, in which each network in the ensemble is constructed using a perturbed version of the training data. The perturbation consists in switching the class labels of a subset of training examples selected at random. Experiments on several UCI and synthetic datasets show that these class-switching ensembles can obtain improvements in classification performance over both individual networks and bagging ensembles.

1 Introduction

Ensemble methods for automatic inductive learning aim at generating a collection of diverse classifiers whose decisions are combined to predict the class of new unlabeled examples. The goal is to generate from the same training data a collection of diverse predictors whose errors are uncorrelated. Ensembles built in this manner often exhibit significant performance improvements over a single predictor in many regression and classification problems. Ensemble can be built using different base classifiers: decision stumps [1] decision trees [2,3,4,5,1,6,7,8,9,10], neural networks [11,12,13,14,9,15], support vector machines [16], etc.

Generally, ensemble methods introduce a random element somewhere in the process of generating of an individual predictor. This randomization can be introduced either in the algorithm that builds the base models or in the training datasets that these algorithms receive as input.

The rationale behind injecting randomness into the base learning algorithm is that different executions of the randomized training algorithm on the same data should generate diverse classifiers. For example, in *randomization* [17] the base learners are decision trees generated with a modified tree construction algorithm. This algorithm computes the best 20 splits for every internal node and then chooses one at random. Another simple algorithm of this type consists in generating diverse neural networks using different random initializations of the synaptic weights. This simple technique is sufficient to generate fairly accurate ensembles [6].

S. Kollias et al. (Eds.): ICANN 2006, Part I, LNCS 4131, pp. 178–187, 2006.

The randomization of the training dataset can be introduced in different ways: using bootstrap samples from the training data, modifying the empirical distribution of the data (either by resampling or reweighting examples), manipulating the input features or manipulating the output targets. Bagging [4], one of the most widespread methods for ensemble learning, belongs to this group of techniques. In bagging, each individual classifier is generated using a training set of the same size of the original training set, obtained by random resampling with replacement from it. In Boosting [2], the individual classifiers are sequentially built assigning at each iteration different weights to the training instances. Initially the weights of the training examples are all equal. At each iteration of the boosting process these weights are updated according to the classification given by the last generated classifier. The weights of correctly classified examples are decreased and the weights of incorrectly classified ones are increased. In this way the subsequent base learner focuses on examples that are harder to classify. Another strategy consists in manipulating the input features. For instance, one can randomly eliminate features of the input data before constructing each individual classifier. In *random subspaces* [18] each base learner is generated using a different random subset of the input features. Another data randomization strategy consists in modifying the class labels. In particular, in classification problems with multiple classes, one can build each classifier in the ensemble using a different coding of the class labels [19,20]. Other algorithms that manipulate the output targets and that are not limited to multiclass problems are based on randomly switching the class label of a fraction of the training set to generate each classifier (e.g. flipping [21] and class-switching [22]).

Class-switching ensembles composed of a sufficiently large number of unpruned decision trees exhibit a good generalization performance in many classification problems of interest [22]. In this article, the performance of the class-switching algorithm using neural networks as the base learners is analyzed. Because of the different properties of neural networks and decision trees, several modifications of the procedure described in [22] need to be made to generate effective class-switching ensembles composed of neural networks.

Section 2 introduces the class-switching algorithm based on modifying the class labels of the training examples and adapt it to build neural network ensembles. Section 3 presents experiments that compare the classification performance of a single neural network, class-switching and bagging ensembles in twelve datasets. Finally, the conclusions of this research are summarized in Section 4.

2 Class-Switching Ensembles

Switching the class labels to generate ensemble of classifiers was first proposed by Breiman [21]. In this work the class switching procedure described in [22] is adapted to generate ensembles that use neural networks as base learners. Class-switching ensembles are built by generating each classifier in the ensemble using different perturbed versions of the original training set. To generate a perturbed version of the training set, a fixed fraction p of the examples of the original training set are randomly selected and the class label of each of these selected examples is randomly switched to a different one. The class label randomization can be characterized by a transition probability matrix

$$\begin{aligned} P_{j \leftarrow i} &= p/(K-1) \; for \; i \neq j \\ P_{i \leftarrow i} &= 1 - p \, , \end{aligned} \tag{1}$$

where $P_{j \leftarrow i}$ is the probability that an example whose label is i becomes labeled as belonging to class j. K is the number of classes in the problem.

The class-flipping procedure proposed by Breiman [21] is designed to ensure that, on average, the class proportions of the original training set are maintained in the modified training sets. However, for class unbalanced datasets this procedure has proved not to perform efficiently [22]. On the contrary, class-switching ensembles [22] applied to decision trees has proved to be competitive with bagging and boosting ensembles for a large range of balanced and unbalanced classification tasks.

In order for this method to work, the fraction of switched examples p, should be small enough to ensure that there are, for any given class, a majority of correctly labeled examples (i.e. not switched). This condition is fulfilled on the training set (on average) if $P_{j \leftarrow i} < P_{i \leftarrow i}$. Using (1)

$$p < (K-1)/K \, . \tag{2}$$

From this equation, the ratio of the class-switching probability to its maximum value is defined as

$$\hat{p} = p/p_{max} = pK/(K-1) \, . \tag{3}$$

Using values of p over this limit would generate, for some regions in feature space, a majority of examples incorrectly labeled and consequently those regions would be incorrectly classified by the ensemble.

In Ref. [22] class-switching ensembles composed of unpruned decision trees were experimentally tested. Using unpruned decision trees instead of pruned trees was motivated by their better performance when combined in the ensemble. Note that, provided that there are no training examples with identical attributes values belonging to different classes, an unpruned decision tree achieves perfect classification (0 error rate) on the perturbed training set. Under these conditions and in order for class-switching to obtain good generalization errors it is necessary to combine a large number of trees in the ensemble (≈ 1000) and to use relatively high values of $\hat{p}$. Empirically a value of $\hat{p} \approx 3/5$ produced excellent results in all the classification tasks investigated [22].

Preliminary experiments were performed to check whether the prescription used for decision trees (i.e. 0 training error of the trees on the perturbed sets, large number of units in the ensemble and high values of $\hat{p}$) can be directly applied to neural networks. Note that the architecture and training parameters of the neural network have to be tuned for each problem in order to obtain neural models with ≈ 0 error rates in the modified training sets. This is a drawback

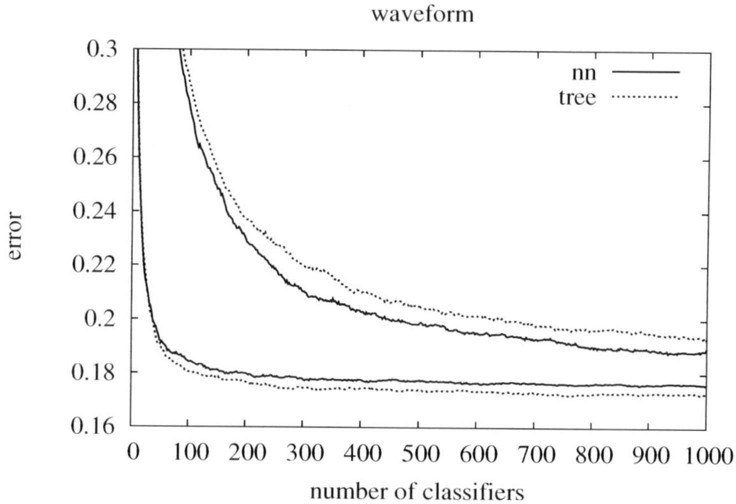

Fig. 1. Average test errors for class-switching ensembles composed of neural networks (*solid lines* in the plot) and decision trees (*trait lines* in the plot) using $\hat{p} = 2/5$ (bottom curves) and $\hat{p} = 4/5$ (*top curves*) for the *Waveform* dataset

with respect to decision trees, where achieving 0-error models is straightforward and problem independent.

Figure 1 displays the evolution with the number of base classifiers of the average generalization error (over 10 executions) for class-switching ensembles composed of neural networks (shown with *solid lines* in the plot) and decision trees (with *trait lines*) for the *Waveform* dataset. In these experiments, the architecture of the network and the training parameters are chosen to achieve 0-error in the perturbed versions training data. In particular, networks with 28 hidden units, trained over 1000 epochs are used. The bottom curves correspond to a $\hat{p}$ value of 2/5 and the top curves correspond to $\hat{p} = 4/5$.

The leaning curves displayed in this figure show that the generalization errors of the class-switching neural ensembles generated in this way are similar to those produced by decision trees class-switching ensembles. However, since the baseline performance given by a single decision tree is different from the performance of a neural net, the conclusions are different for ensembles composed of decision trees and for ensembles of neural networks. The improvement obtained by decision tree class-switching ensembles over a single tree is substantial (the generalization error of a single decision trees is $\approx 30\%$). Hence, for decision trees, the strategy of generating 0-error base learners seems to perform well. The piecewise-constant boundaries produced by single trees evolve to more complex and convoluted decisions boundaries when the decisions of the individual trees are combined in the ensemble. In contrast, the results obtained by the neural ensembles show that this strategy does not lead to ensembles that significantly improve the classification accuracy of a single network. In particular, a single neural network with 7 hidden units and trained with 500 epochs achieves an error rate of 19.3%, a result that is equivalent to the final error of the neural class-switching ensemble of networks with zero error on the perturbed training set with $\hat{p} = 4/5$ (19.0%). In contrast, class-switching or bagging ensembles composed of 100 neural nets with only about 6 hidden units trained in the same conditions obtain a generalization error of $\approx$ 16.4%, which is a significant improvement over the configuration that uses more complex nets and larger ensembles. Note that a neural net with this smaller number hidden units does not necessarily obtain a 0-error model on the modified training data. Nonetheless, this ensemble of simple networks is trained much faster and exhibits better generalization performance than an ensemble of complex networks trained to exhibit zero error on the perturbed versions of the training set.

3 Experiments

To assess the performance of the proposed method experiments are carried out in ten datasets from the UCI repository [23] and in two synthetic datasets (proposed by Breiman *et al.* [24,5]). The datasets are selected to sample a variety of problems from different fields of application. The characteristics of the selected datasets, of the testing method and the networks generated are shown in Table 1.

Table 1. Characteristics of the datasets, testing method, number of input units, average number (± standard deviation) of hidden units and average number of training epochs for the neural networks used in the experiments

Dataset	Instances	Test	Attrib.	Classes	Input units	Hidden units	Training epochs
Breast W.	699	10-fold-cv	9	2	9	4.12±1.49	328
Diabetes	768	10-fold-cv	8	2	8	5.36±1.62	364
German	1000	10-fold-cv	20	2	61	4.98±1.65	173
Heart	270	10-fold-cv	13	2	23	4.84±1.70	201
Labor	57	10-fold-cv	16	2	37	4.42±1.54	405
New-thyroid	215	10-fold-cv	5	3	5	16.2±3.55	618
Sonar	208	10-fold-cv	60	2	60	5.14±1.46	331
Tic-tac-toe	958	10-fold-cv	9	2	27	4.38±1.50	200
Twonorm	300	5000 cases	20	2	20	4.36±1.61	330
Vehicle	846	10-fold-cv	18	4	18	11.7±3.19	810
Waveform	300	5000 cases	21	3	21	5.56±1.45	511
Wine	178	10-fold-cv	13	3	13	5.88±1.43	435

The base classifiers are feedforward neural networks with one hidden layer. Sigmoidal transfer functions for both the hidden and output layers are used. The number of units in the output layer is equal to the number of classes of the classification task and the networks are trained to approximate the posterior probability of each class. The neurons are trained using an improved RPROP batch algorithm [25]. The optimal architecture and number of training epochs for the neural networks is estimated for every partition of the training data using cross validation. The same architecture and number of epochs is used in bagging and class-switching ensembles. For the neural networks, the FANN library [26] implementation is used.

The results given are averages of a 100 experiments for each dataset. In the real-world datasets these experiments consist in the execution of 10×10-fold-cv. For the synthetic datasets (*Twonorm* and *Waveform*) each experiment involves a random sampling to generate the training and testing sets (see Table 1 for the sizes of the sets). In general, each experiment involves the following steps:

1. Obtain the random training/testing datasets from the corresponding fold in the real-world datasets and by random sampling in the synthetic ones.
2. Build a single neural network using the whole training dataset. The configuration of the network is estimated using cross-validation of 10-fold in the training data. Different architectures (3, 5, and 7 hidden units) and different values for the number of epochs (100, 300, 500 and 1000) are explored. The configuration that obtains on average the best accuracy on the separate folds of the training data, is used. For some datasets the range of possible hidden units was incremented. For the *Vehicle* data set it was necessary to test 5, 7, 11, and 15 hidden units and for *New-thyroid* the tested architectures are 7, 11, 15 and 20.
3. Build the neural networks ensembles using class-switching (with $\hat{p}$ values of: 0/5, 1/5, 2/5, 3/5 and 4/5) and bagging and using the configuration obtained for the single net. Note that class-switching with $\hat{p} = 0/5$ can not be considered a class-switching algorithm: the variability in the ensemble is achieved solely by the training process converging to different weight values because of the different random initial values used.
4. Estimate the generalization error of the classifiers (single NN, bagging and class-switching) on the test set.

Figure 2 displays the average generalization error curves for bagging and class-switching ensembles for four datasets (*German, Heart, Labor* and *New-thyroid*). These plots show that the convergence of the error class-switching ensembles is related to the fraction of switched examples (i.e. $\hat{p}$): Higher $\hat{p}$ values result in a slower convergence rate. For most of the ensemble configurations combining 200 networks seems to be sufficient for the error curves to level off. However, in some datasets (see *German* and *Labor* datasets in Fig. 2) with a high class-switching probability (class-switching with $\hat{p} = 4/5$), 200 is not sufficient to reach the asymptotic ensemble error rate. In contrast, random initialized neural networks ensembles ($\hat{p} = 0/5$) reach their asymptotic error level after combining a fairly small number of neural networks (≈ 20 NN).

Table 2 presents the average test errors over the 100 executions for single networks, bagging and class-switching ensembles for the different values of $\hat{p}$. The lowest generalization error for every dataset is highlighted in bold-face. The standard deviations are given after the $\pm$ sign. These results show that

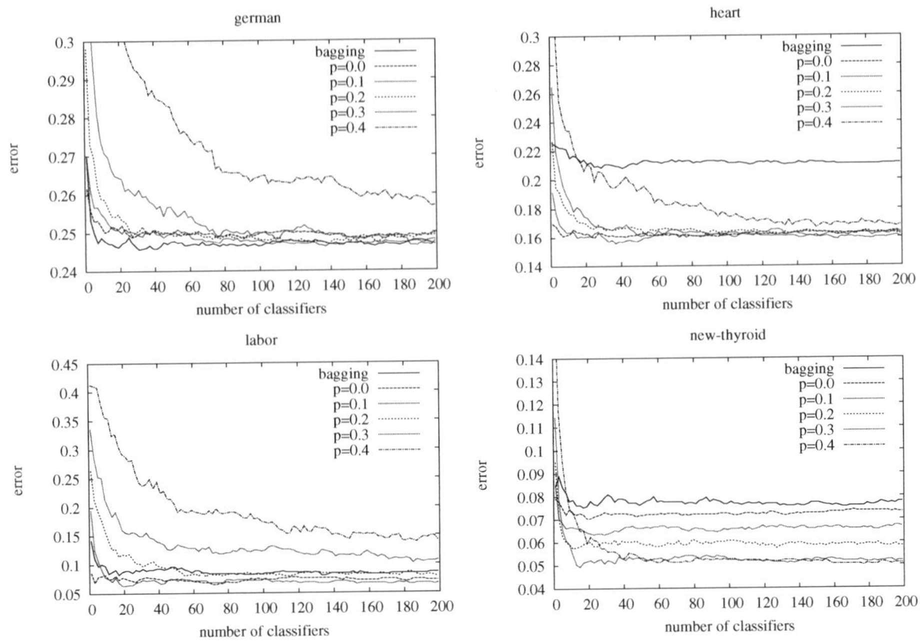

Fig. 2. Average test errors for the *German Credit* (*top left plot*), *Heart* (*top right plot*), *Labor Negotiations* (*bottom left plot*) and *New-thyroid* (*bottom right plot*) datasets

class-switching ensembles exhibit the best results in nine of the twelve problems analyzed ($2 \times \hat{p} = 1/5$, $2 \times \hat{p} = 2/5$, $5 \times \hat{p} = 3/5$ and $3 \times \hat{p} = 4/5$). Bagging has the best performance in two datasets and ensembles with $\hat{p} = 0/5$ also in two dataset. The performance of a single network is suboptimal in all cases investigated and is poorer than most of the different ensembles.

Table 2 shows that most configurations of class-switching ensembles reach similar generalization errors for most datasets. In particular, the same error rate is achieved in *Waveform* by class-switching with $\hat{p} = 1/5, 2/5$ and $3/5$ and nearly the same results (within 0.2 points) are obtained in *Diabetes, German, Tic-tac-toe, Twonorm* and *Wine*. The $\hat{p} = 4/5$ configuration exhibits significantly worse results in *German, Labor, Sonar* and *Tic-tac-toe*. In some cases this is due to the fact that larger ensembles ought to have been used.

Table 3 shows the p-values of the paired t-test for the differences between bagging and class-switching ensembles using the different values of $\hat{p}$. Significant differences against class-switching have been underlined and statistically significant differences in favor of class-switching are high-lighted in bold-face. The last row of Table 3 displays the *win/draw/loss* records, where the first (second / third) numbers displayed in each cell correspond to the number of sets in which the algorithm displayed in the topmost row of the table wins (draws / losses) with respect to bagging. These records show that class-switching ensembles with $\hat{p} = 1/5$ and $\hat{p} = 2/5$ never perform worse than bagging and that they outperform bagging in some of the studied datasets. Class-switching with $\hat{p} = 3/5$ and

Table 2. Average generalization errors

Dataset	NN	Bagging	Class-switching ($\hat{p} =$)				
			0/5	1/5	2/5	3/5	4/5
Breast	3.9±2.5	3.9±2.4	3.8±2.4	3.8±2.4	3.7±2.2	**3.3**±2.3	**3.3**±2.3
Diabetes	25.9±4.7	24.9±4.6	24.8±4.7	24.9±4.5	24.8±4.2	24.8±4.4	**24.6**±4.7
German	26.2±5.4	24.7±5.9	25.0±6.2	24.8±5.9	24.9±5.9	**24.7**±6.0	25.7±6.4
Heart	17.0±7.9	21.1±13	16.4±7.8	**16.0**±7.6	16.2±7.4	16.3±7.3	16.6±7.4
Labor	8.6±12	8.4±12	7.2±11	**6.6**±11	7.5±12	10.8±15	14.5±16
New-thyroid	5.3±4.7	5.8±5.1	5.0±4.6	4.6±4.2	**4.2**±4.0	**4.2**±3.9	4.4±3.8
Sonar	23.5±8.8	**20.2**±8.7	21.3±8.4	21.0±8.5	21.1±9.2	21.6±9.3	23.2±9.6
Tic-tac-toe	2.2±1.8	1.8±1.3	1.9±1.4	1.8±1.3	1.8±1.2	**1.7**±1.2	7.7±5.5
Twonorm	3.8±0.7	3.1±0.4	3.5±0.6	3.1±0.4	**2.9**±0.4	**2.9**±0.5	3.3±1.1
Vehicle	19.4±4.0	17.0±4.1	**15.9**±3.6	16.1±3.5	16.4±3.7	17.1±3.7	17.8±3.3
Waveform	20.6±8.0	**16.4**±1.0	**16.4**±1.0	16.5±1.0	16.5±0.9	16.5±1.0	16.6±0.9
Wine	5.1±5.0	2.2±3.7	2.0±3.4	1.6±2.8	1.4±2.7	1.5±3.0	**1.2**±2.6

Table 3. Results of a paired *t-test* for the differences between the test errors of bagging ensembles and class-switching ensembles

Dataset	**class-switching ($\hat{p} =$)**				
	0/5	**1/5**	**2/5**	**3/5**	**4/5**
Breast	$8.6{\cdot}10^{-1}$	$6.1{\cdot}10^{-1}$	$7.7{\cdot}10^{-2}$	$\mathbf{8.3{\cdot}10^{-6}}$	$\mathbf{8{\cdot}10^{-7}}$
Diabetes	$7.0{\cdot}10^{-1}$	$8.2{\cdot}10^{-1}$	$5.5{\cdot}10^{-1}$	$7.3{\cdot}10^{-1}$	$3.7{\cdot}10^{-1}$
German	$1.0{\cdot}10^{-1}$	$2.7{\cdot}10^{-1}$	$2.5{\cdot}10^{-1}$	$1.0{\cdot}10^{0}$	$3.2{\cdot}10^{-4}$
Heart	$\mathbf{8.6{\cdot}10^{-5}}$	$\mathbf{3.0{\cdot}10^{-5}}$	$\mathbf{6.9{\cdot}10^{-5}}$	$\mathbf{1.0{\cdot}10^{-4}}$	$\mathbf{2.6{\cdot}10^{-4}}$
Labor	$2.6{\cdot}10^{-1}$	$1.1{\cdot}10^{-1}$	$4.4{\cdot}10^{-1}$	$9.2{\cdot}10^{-2}$	$9.5{\cdot}10^{-5}$
New-thyroid	$\mathbf{2.5{\cdot}10^{-2}}$	$\mathbf{4.5{\cdot}10^{-3}}$	$\mathbf{4.5{\cdot}10^{-4}}$	$\mathbf{2.1{\cdot}10^{-4}}$	$\mathbf{1.9{\cdot}10^{-3}}$
Sonar	$3.6{\cdot}10^{-2}$	$1.2{\cdot}10^{-1}$	$1.3{\cdot}10^{-1}$	$4.6{\cdot}10^{-2}$	$2.2{\cdot}10^{-4}$
Tic-tac-toe	$6.0{\cdot}10^{-3}$	$1.8{\cdot}10^{-1}$	$1.0{\cdot}10^{-1}$	$9.6{\cdot}10^{-2}$	$5.1{\cdot}10^{-19}$
Twonorm	$5.1{\cdot}10^{-21}$	$2.3{\cdot}10^{-1}$	$\mathbf{7.6{\cdot}10^{-11}}$	$\mathbf{4.5{\cdot}10^{-7}}$	$2.5{\cdot}10^{-1}$
Vehicle	$\mathbf{1.5{\cdot}10^{-5}}$	$\mathbf{2.4{\cdot}10^{-4}}$	$\mathbf{3.5{\cdot}10^{-3}}$	$7.1{\cdot}10^{-1}$	$4.5{\cdot}10^{-3}$
Waveform	$4.6{\cdot}10^{-1}$	$8.6{\cdot}10^{-2}$	$1.8{\cdot}10^{-1}$	$2.3{\cdot}10^{-1}$	$2.0{\cdot}10^{-2}$
Wine	$4.8{\cdot}10^{-1}$	$\mathbf{1.7{\cdot}10^{-2}}$	$\mathbf{3.9{\cdot}10^{-3}}$	$\mathbf{4.7{\cdot}10^{-2}}$	$\mathbf{3.0{\cdot}10^{-3}}$
	3/6/3	4/8/0	5/7/0	5/6/1	4/2/6

$\hat{p} = 4/5$ and random initialized neural networks ensembles ($\hat{p} = 0/5$) improve the results of bagging in several datasets but also perform significantly worse than bagging in other datasets.

4 Conclusions

In the present article the performance of class-switching ensembles [22] using neural networks as base classifiers is analyzed. The class-switching ensembles generate a diversity of classifiers using different perturbed versions of the training set. To generate each perturbed set, a fraction of examples is selected at random and their class labels are switched also at random to a different label.

The prescription used for decision trees (generate individual classifiers that achieve 0-error in the perturbed training datasets) is found not to be the appropriate configuration for neural networks ensembles constructed with class-switching. Combining neural networks whose architecture is designed by standard architecture selection techniques (and that therefore do not necessarily achieve 0 error in the perturbed training datasets) produces significantly better results than ensembles composed of more complex nets that do achieve 0 error in the perturbed datasets. Since the networks in the ensemble are not forced to have zero error on the perturbed training sets, they seem to avoid overfitting to the noise injected, at least to a certain extent, and perform reasonably well in the original unperturbed problem. As a consequence, the number of base learners needed for the convergence of the ensembles to the asymptotic error level is smaller than in class-switching ensembles composed of decision trees.

Class-switching ensembles of neural networks built according to this prescription exhibit a classification accuracy that is better or equivalent to bagging on the tested datasets. Ensembles generated with a class-switching rate of $\hat{p} = 1/5$ and $2/5$ obtain the best overall results for the datasets investigated. These configurations (i.e. $p = 1/5, 2/5$) never obtain results statistically worse than bagging. This is not the case for other values of $\hat{p}$ (that is, $\hat{p} = 0/5, 3/5$ and $4/5$), where results both better and worse than bagging have been obtained. Further analysis is needed to determine the optimal choice of the class-switching probability, $\hat{p}$.

Acknowledgments

The authors acknowledge financial support from the Spanish *Dirección General de Investigación*, project TIN2004-07676-C02-02. Daniel Hernández-Lobato acknowledges support from the Consejería de Educación de la Comunidad Autónoma de Madrid and from the European Social Fund under an F.P.I. grant.

References

1. Schapire, R.E., Freund, Y., Bartlett, P.L., Lee, W.S.: Boosting the margin: A new explanation for the effectiveness of voting methods. The Annals of Statistics **12**(5) (1998) 1651–1686
2. Freund, Y., Schapire, R.E.: A decision-theoretic generalization of on-line learning and an application to boosting. In: Proc. 2nd European Conference on Computational Learning Theory. (1995) 23–37
3. Quinlan, J.R.: Bagging, boosting, and C4.5. In: Proc. 13th National Conference on Artificial Intelligence, Cambridge, MA (1996) 725–730
4. Breiman, L.: Bagging predictors. Machine Learning **24**(2) (1996) 123–140
5. Breiman, L.: Arcing classifiers. The Annals of Statistics **26**(3) (1998) 801–849
6. Opitz, D., Maclin, R.: Popular ensemble methods: An empirical study. Journal of Artificial Intelligence Research **11** (1999) 169–198
7. Bauer, E., Kohavi, R.: An empirical comparison of voting classification algorithms: Bagging, boosting, and variants. Machine Learning **36**(1-2) (1999) 105–139

8. Dietterich, T.G.: An experimental comparison of three methods for constructing ensembles of decision trees: Bagging, boosting, and randomization. Machine Learning **40**(2) (2000) 139–157
9. Rätsch, G., Onoda, T., Müller, K.R.: Soft margins for AdaBoost. Machine Learning **42**(3) (2001) 287–320
10. Breiman, L.: Random forests. Machine Learning **45**(1) (2001) 5–32
11. Hansen, L.K., Salamon, P.: Neural network ensembles. IEEE Trans. Pattern Anal. Mach. Intell. **12**(10) (1990) 993–1001
12. Wolpert, D.H.: Stacked generalization. Neural Networks **5**(2) (1992) 241–259
13. Perrone, M.P., Cooper, L.N.: When networks disagree: Ensemble methods for hybrid neural networks. In Mammone, R.J., ed.: Neural Networks for Speech and Image Processing. Chapman-Hall (1993) 126–142
14. Sharkey, A.J.C.: Combining Artificial Neural Nets: Ensemble and Modular Multi-Net Systems. Springer-Verlag, London (1999)
15. Cantador, I., Dorronsoro, J.R.: Balanced boosting with parallel perceptrons. In: IWANN. (2005) 208–216
16. Valentini, G., Dietterich, T.G.: Bias-variance analysis of support vector machines for the development of svm-based ensemble methods. Journal of Machine Learning Research **5** (2004) 725–775
17. Kong, E.B., Dietterich, T.G.: Error-correcting output coding corrects bias and variance. In: Proceedings of the Twelfth International Conference on Machine Learning. (1995) 313–321
18. Ho, T.K.: C4.5 decision forests. In: Proceedings of Fourteenth International Conference on Pattern Recognition. Volume 1. (1998) 545–549
19. Dietterich, T.G., Bakiri, G.: Solving multiclass learning problems via error-correcting output codes. Journal of Artificial Intelligence Research **2** (1995) 263–286
20. Fürnkranz, J.: Round robin classification. Journal of Machine Learning Research **2** (2002) 721–747
21. Breiman, L.: Randomizing outputs to increase prediction accuracy. Machine Learning **40**(3) (2000) 229–242
22. Martínez-Muñoz, G., Suárez, A.: Switching class labels to generate classification ensembles. Pattern Recognition **38**(10) (2005) 1483–1494
23. Blake, C.L., Merz, C.J.: UCI repository of machine learning databases (1998)
24. Breiman, L., Friedman, J.H., Olshen, R.A., Stone, C.J.: Classification and Regression Trees. Chapman & Hall, New York (1984)
25. Igel, C., Hüsken, M.: Improving the rprop learning algorithm. In: Proceedings of the Second International Symposium on Neural Computation, ICSC Academic Press (2000) 115–121
26. Nissen, S.: Implementation of a fast artificial neural network library (fann). Technical report, Department of Computer Science, University of Copenhagen (2003)

K-Separability

Włodzisław Duch

Department of Informatics, Nicolaus Copernicus University, Grudziądzka 5, Toruń, Poland,
and School of Computer Engineering, Nanyang Technological University, Singapore
Google: Duch

Abstract. Neural networks use their hidden layers to transform input data into linearly separable data clusters, with a linear or a perceptron type output layer making the final projection on the line perpendicular to the discriminating hyperplane. For complex data with multimodal distributions this transformation is difficult to learn. Projection on $k \geq 2$ line segments is the simplest extension of linear separability, defining much easier goal for the learning process. The difficulty of learning non-linear data distributions is shifted to separation of line intervals, making the main part of the transformation much simpler. For classification of difficult Boolean problems, such as the parity problem, linear projection combined with k-separability is sufficient.

1 Introduction

Many popular classifiers, including MLPs, RBFs, SVMs, decision trees [1], nearest neighbor and other similarity based methods [2,3], require special approaches (architectures, kernels) or cannot handle at all complex problems, such as those exemplified by the parity problem: given a training set of binary strings $\{b_1, b_2 ... b_n\}$ determine if the number of bits equal to 1 is odd or even. In principle universal approximators, such as neural networks, are capable of handling such problems, and there is a whole literature on architectures and neural activation functions that enable the solution of parity problem. However, solutions proposed so far are manually designed to solve this particular problem, and thus will not work well for slightly different problems of similar kind.

Dealing with difficult learning problems like parity off-the-shelf algorithms (for example those collected in Weka [4] or Ghostminer [5] packages) in the leave-one-out or crossvalidation tests for more than 3-bit problems give results at the baserate (50%) level. Knowing beforehand that the data represents parity problem allows for setting an appropriate MLP architecture to solve it [6,7,8,9,10,11,12,13,14], but for large n in real situation it will be very difficult to guess how to choose an appropriate model. Learning Boolean functions similar to parity may indeed be a great test for methods that try to evolve neural architecture to solve a given problem, but so far no such systems are in sight. The reason for this failure is rather simple: neural and other classifiers try to achieve linear separability, and non-linear separable data may require a non-trivial transformation that is very difficult to learn. Looking at the image of the training data in the space defined by the activity of the hidden layer neurons [15,16] one may notice that a perfect solution is frequently found in the hidden space – all data falls into separate clusters – but the clusters are non-separable, therefore the perceptron output layer is

S. Kollias et al. (Eds.): ICANN 2006, Part I, LNCS 4131, pp. 188–197, 2006.

unable to provide useful results. Changing the goal of learning from linear separability to other forms of separability should make the learning process much easier.

It would be very useful to break the notion of non-linearly separable problems into well defined classes of problems with increasing difficulty. This is done in the next section, where the notion of k-separability is introduced. In the third section this notion is combined with linear projections and applied to the analysis of Boolean functions. Algorithms based on k-separability for general classification problems are outlined in section four, with the last section containing a final discussion.

2 k-Separability

Adaptive systems, such as feedforward neural networks, SVMs, similarity-based methods and other classifiers, use composition of vector mappings

$$Y(\mathbf{X}) = M^{(m)}(M^{(m-1)}...(M^{(2)}(M^{(1)}(\mathbf{X}))...)) \tag{1}$$

to assign a label Y to the vector $\mathbf{X}$. To be completely general direct dependence of mappings on inputs and previous transformations should be considered, for example $M^{(2)}(M^{(1)}(\mathbf{X}), \mathbf{X})$, but for simplicity this will be omitted, considering only strictly layered mappings. These mappings may include standardization, principal component analysis, kernel projections, general basis function expansions or perceptron transformations. $\mathbf{X}^{(i)} = M^{(i)}(\mathbf{X}^{(i-1)})$ is the result of mapping after i transformations steps. For dichotomic problems considered below $Y = \mathbf{X}^{(m)} = \pm 1$.

If the last transformation $Y = M^{(m)}(\mathbf{X}^{(m-1)})$ is based on a squashed linear transformation, for example a perceptron mapping $Y = \tanh(\sum_i W_i \mathbf{X}_i^{(m-1)})$, then the values of Y are projections of $\mathbf{X}$ on the $[-1, +1]$ interval, and a perfect separation of classes means that for some threshold Y_0 all vectors from the Y_+ class are mapped to one side and from the Y_- class to the other side of the interval. This means that the hyperplane $\mathbf{W}$ defined in the $\mathbf{X}^{(m-1)}$ space divides samples from the two classes, and $\mathbf{W} \cdot \mathbf{X}^{(m-1)}$ is simply a projection on the line $\mathbf{W}$ perpendicular to this hyperplane, squashed to the $[-1, +1]$ interval by the hyperbolic tangent or similar function.

General parity problems can be solved in many ways. The simplest solution [13] is to look at the sign of the $\prod_{i=1}^{n}(x_i - t_i)$, with $t_i \in (0, 1)$, that is to use a product neuron without any hidden neurons. This solution is very specific to the parity problem and it cannot be generalized to other Boolean functions. Many such solutions that work only for parity problem have been devised [6]–[14], but the challenge is to provide more general solutions that work also for problems of similar or higher difficulty. Many MLP training algorithms have already some difficulties to solve the XOR problem. RBF network with Gaussian hidden units cannot solve it unless special tricks are used. Solutions based on local functions require here a large number of nodes and examples to learn, while non-local solutions may be expressed in a compact way and need only a few examples (this has been already noted in [17]). Consider the noisy version of the XOR problem (Fig. 1). RBF network with two Gaussian nodes with the same standard deviation σ and linear output provides the following two transformations:

$$\mathbf{X} \to \mathbf{X}^{(1)} = (\exp(-|\mathbf{X} - \mu_1|/2\sigma, \exp(-|\mathbf{X} - \mu_2|/2\sigma) \tag{2}$$

$$\mathbf{X}^{(1)} \to Y = \mathbf{X}^{(2)} = \mathbf{W} \cdot \mathbf{X}^{(1)} \tag{3}$$

The solution obtained using maximum likelihood approach [1] placed one basis function in the middle of left-corner cluster, and the other close to the center, as shown in Fig. 1. Although the network fails to achieve linear separability of the data, it is clear from Fig. 1 that all new data will be properly assigned to one of the 3 clusters formed in the hidden space; in crossvalidation test 100% correct answers are obtained on this basis (searching for the nearest neighbor in the hidden space), while the linear output from the network achieves only 50% accuracy (base rate). If the target $Y = \mathbf{X}^{(2)} = \pm 1$ is desired the linear output provides a hyperplane (in this case a line) that tries to stay at a distance one from all data points. If a separate linear output for each class is used lines representing both outputs are parallel, with identical weights but shifted on two units, as shown in the right Fig. 1. Projecting $\mathbf{X}^{(1)}$ data on these lines gives one interval with the data from first class surrounded by two intervals with the data from the second class, separating the data into 3 intervals.

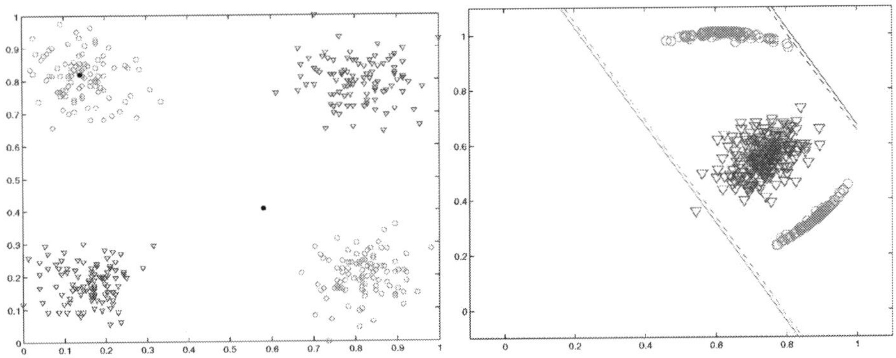

Fig. 1. Noisy XOR problem solved with two Gaussian functions. Left: data distribution and position of two Gaussian functions after training using maximum likelihood principle; right: mapping of the input data to the hidden space shows perfect clusterization, showing lines representing linear weights of the two output units.

In n dimensions a single linear unit $\mathbf{W} \cdot \mathbf{X}$ with all weights $W_i = 1$ easily achieves separation into $n + 1$ groups, with 0, 1, 2 .. n bits equal to 1. This weight vector is the diagonal connecting vertices $[0, 0, \ldots 0]$ and $[1, 1, \ldots 1]$, and $\mathbf{W} \cdot \mathbf{X}$ is the projection on this line. Obviously using a single node with $Y = \cos(\omega \mathbf{W} \cdot \mathbf{X})$ gives for $\omega = \pi$ correct answer to all parity problems, $+1$ for even and -1 for odd number of bits. This is the simplest general solution of the parity problem, using a single node network (this solution has not bee found previously [6]–[14]). The importance of selection of appropriate transfer functions in neural networks is quite evident here (for a taxonomy of transfer functions that may be used in neural networks see [18] and [17]). In the context of Boolean functions periodic projection is useful only for parity and its negation obtained by symmetric transformations of the hypercube with vertices labeled according to their parity. Projections of other Boolean functions may not be periodic but certainly will show several groups of vectors from alternating classes.

There is no particular reason why the target of learning should be linear separability. The last transformation $Y = M^{(m)}(\mathbf{X}^{(m-1)})$ may be designed in any way that will make learning easier. If a projection separating two clusters on a line $Y = \mathbf{W} \cdot \mathbf{X}$ exist the data is linearly separable. If it does not exist a projection forming 3 or more intervals containing clusters from a single class should be sought.

Definition: dataset $\{\mathbf{X}_i\}$ of vectors that belong to two classes is called k-separable if a direction $\mathbf{W}$ exist such that all points $Y_i = \mathbf{W} \cdot \mathbf{X}_i$ are clustered in k intervals, each containing vectors from a single class.

Linearly separable data is called 2-separable, while XOR belongs to the 3-separable category of data distributions, with projection on the $\mathbf{W} = (1, 1)$ line from even, odd, and again even class. This is the simplest extension of separability, replacing the final mapping $M^{(m)}(\cdot)$ by logical rule IF ($Y \in [Y_0, Y_1]$ THEN even ELSE odd, and thus making the non-linearity rather harmless. More sophisticated mappings from one, two or higher number of dimensions may be devised as long as transformation $M^{(m)}(\cdot)$ is easy to set up, providing easier goals for the learning process. The Error Correcting Output Codes (ECOC) [19] tries also to define easier learning targets, but it is still based on linear separability, setting a number of binary targets that define a prototype "class signature" vectors, and comparing the distance from the actual output to these class prototypes. The change of the learning target advocated here is much more powerful.

A dataset that is k-separability may also be $(k + m)$-separable. Strictly speaking the separability index for the data should be taken as the lowest k, but some learning methods may generate solutions with larger number of clusters. For example, if the data is k-separable into clusters with very small number of elements, or if the margin separating the intervals between two such clusters is very small, $(k + 1)$-separability that leads to larger minimum size of small clusters and their margins may be preferred.

Solving a k-separability problem requires finding the direction $\mathbf{W}$ and then setting appropriate $k - 1$ thresholds defining intervals on the projection line.

Conjecture: the complexity of k-separable learning should be much easier then 2-separable learning.

This is rather obvious; transforming the data into k-separability form should be much easier because additional transformations are needed to achieve linear separability, and the number of adaptive parameters may grow significantly. For example, the number of hyperplanes that an MLP network needs for the n-parity problem is of the order of n (see comparison of solutions in [13]), giving altogether $O(n^2)$ parameters, while treating it as a k-separable problem requires only $n + k$ parameters. In general, cases when transformation of decision borders in the original input space $\mathbf{X}$ based on continuous deformations may flatten them linear separability will be sufficient, but if discontinuous transformations are needed, as in the case of learning most Boolean functions, transformations that map data into k-separable form should be easier.

An interesting question is how many Boolean functions belong to the k-separable category. For n-variables there are 2^{2^n} possible functions; only the bounds for the number of separable Boolean functions are known: the number is between $2^{n^2-O(n)}$ and 2^{n^2} [20], a vanishing fraction of all functions. Unfortunately such estimations are not yet known for the k-separable case.

For linearly separable data projections on $\mathbf{W}$ and $-\mathbf{W}$ generate symmetrical solutions (Y_+, Y_-) and (Y_-, Y_+); in case of k-separability additional symmetries and permutations are possible.

3 Boolean Functions

It is instructive to analyze in detail the case of learning Boolean functions with $n = 2$ to $n = 4$ bits with the simplest model based on linear projections. Several interesting questions should be investigated: how many k-separability cases for a given direction $\mathbf{W}$ are obtained; which direction gives the largest separation between projected clusters; how many k-separability cases for each direction $\mathbf{W}$ exist; how many different directions are needed to find all these cases.

The Boolean functions $f(x_1, x_2, \dots x_n) \in \{-1, +1\}$ are defined on the 2^n vertices of n-dimensional hypercube. Numbering these vertices from 0 to $2^n - 1$, they are easily identified converting decimal numbers to bits, for example vertex 3 corresponds to b-bit string 00..011. There are 2^{2^n} possible Boolean functions, each corresponding to a different distribution of ± 1 values on hypercube vertices. There are always two trivial cases corresponding to functions that are always true and always false, that is 1-separable functions. Each Boolean function may be identified by a number from 0 to $2^n - 1$, or a bit string from 00...0 to 11...1, where the value 0 stands for false or -1, and 1 for true or $+1$. For example, function number 9 has 2^n bits 00...1001, and is true only on vertex number 0 and 3.

Values of Boolean functions may be represented as black (-1) or white ($+1$) vertices of the hypercube. Learning a Boolean function is equivalent to separation of projections of the black and white vertices of the hypercube. Separation into small number of well separated clusters should lead to a good generalization when some function values are not known. For two binary variables almost all non-canonical directions (not connecting vertices of the square) avoid mapping vertices of different color to exactly the same point (degeneracy) and give 6, 6 and 2 projections 2, 3 and 4-separated, respectively. It is easy to find two directions that together learn all 12 linearly separable functions (for example $\mathbf{W}_{(1/3,-1/2)}$ and orthogonal direction $\mathbf{W}_{(1/3,2/9)}$). These directions and $\mathbf{W}_{(1,1)}$ that learns two 3-separable functions (XOR and its negation) are sufficient to learn all Boolean functions.

3.1 3-D Case

For 3 bits there are 8 vertices in the cube and $2^8 = 256$ possible Boolean functions. Functions $f(x_1, x_2, x_3)$, and their negations $\neg f(x_1, x_2, x_3)$, are related by the sign reversal symmetry or changing color of all vertices, therefore it is sufficient to consider only 128 functions corresponding to all black vertices (1 case), one black vertex (8 cases), two blacks ($\binom{8}{2} = 28$ cases), three blacks ($\binom{8}{3} = 56$ cases), or four blacks ($\binom{8}{4} = 70$ cases, but only half are unique due to the sign reversal symmetry), so together there are 1+8+28+56+35=128 such unique functions.

Projections on coordinate directions $\mathbf{W}_{(001)}, \mathbf{W}_{(010)}, \mathbf{W}_{(100)}$ separates only three functions $f(x_1, x_2, x_3) = x_k, k = 1, 2, 3$ that belong to the 35 cases with 4 black and 4 white vertices. There are 6 projection directions along the diagonals of the cube's faces: $\mathbf{W}_{(110)}$, $\mathbf{W}_{(1-10)}$, $\mathbf{W}_{(101)}$, $\mathbf{W}_{(10-1)}$, $\mathbf{W}_{(011)}$, $\mathbf{W}_{(01-1)}$, and 4 projection directions along the longest diagonals of the cube: $\mathbf{W}_{(111)}$, $\mathbf{W}_{(11-1)}$, $\mathbf{W}_{(1-11)}$, $\mathbf{W}_{(-111)}$. Together 13 canonical directions should be considered, and a "zero direction" to check if there is only one class.

Consider now direction $\mathbf{W}_{(110)}$. Two points, $(0,0,0)$, $(0,0,1)$ are projected to $Y = 0$, 4 points $(1,0,0)$, $(0,1,0)$, $(1,0,1)$, $(0,1,1)$ are at $Y = \sqrt{2}/2$ and two points $(1,1,0)$, $(1,1,1)$ are at $Y = \sqrt{2}$. Any Boolean function that has the same value for the first 6 points and an opposite value for the last 2 point, or vice versa, will be linearly separable. There are 4 such functions. Any function that has the same value at the middle 4 vertices, and an opposite on the remaining 4 will be 3-separable. There are 2 such functions. However, separation and size of the clusters should also be noted. For example, function 27 (00011011) is separated by $\mathbf{W} = [0.75, 1, -0.25]$ into 000 11 0 11 segments with minimum gap of 1/4 and by $\mathbf{W} = [1, 0.25, -0.75]$ into 0 1 000 111 segments with the same minimum gap. The first projection contains only one group with single 0, while the second contains two such groups, one with 0 and one with 1. For Boolean functions with small number of bits generalization is meaningless (there is no evidence to choose a particular function), but for larger number of bits avoiding small clusters should give a better chance to find most probable functions even if some values are missing. For example, for the n-bit parity if some of the values on vertices with $m \in [3, n-2]$ bits 1 are missing projection on $\mathbf{W} = [11\ldots11]$ will still provide the best explanation of the data separating it into $n+1$ intervals.

If degeneracy is removed by slightly shifting $0, \pm 1$ weight values of canonical directions (adding 0.01 to the first, 0.02 to the second and 0.03 to the weights $\mathbf{W}$ is sufficient) for an arbitrary projection direction always the same number of 1 to 8-separable functions is found: 2, 14, 42, 70, 70, 42, 14, 2. Thus for a projection on an arbitrary direction most functions are 4 or 5-separable. Searching for the best projection for each function using slightly perturbed canonical directions there are 2 cases of 1-separable functions, and 102, 126 and 26 of 2, 3 and 4-separable functions. For more than half of the 3-bit Boolean functions there is no linear projection that will separate the data. Almost half (126) of all functions may be learned using 3-separability. Because there are 102 linearly separable functions and each projection can recognize only 14 of them at least 8 directions are needed to check whether the function is separable.

3.2 4-D Case

For the 4-bit problem there are 16 hypercube vertices, with Boolean functions corresponding to 16-bit numbers, from 0 to 65535, or 64K functions. Projection on each fixed direction gives symmetric distribution of the number of k-separability functions, with the same number of functions for k and $17-k$ separability. Two functions are 1-separable and two are 16-separable, changing periodically all 16 values from 0 to 1. Linear separation (and 15-separability) is found only for 30 functions, 3-separability for 210, 4 to 8 separability for 910, 2730, 6006, 10010 and 12870 functions respectively. Thus a random initialization of a single perceptron has the highest chance of creating 8 or 9 clusters in the 4-bit data.

Checking how many functions are k-separable requires learning the best direction for a given data. For the 4-bit case searching for the best projection along canonical directions ($W_i = 0, \pm 1$) that give lowest k-separability index gives 1228, 6836, 19110, 25198, 12014, 1132 and 16 projections with 2-8 clusters. These are not yet the lowest separability indices for this data, as more detailed search allowing fractional values (multiples of 1/3, 1/4, 1/5 and 1/6 in the $[-1, +1]$ range) of the $\mathbf{W}$ direction coefficients shows that the highest k is 5, confirming the suspicion that $k = n + 1$ is the highest separability index. The number of linearly separable functions is 1880, or less than 3% of all functions, with about 22%, 45% and 29% being 3 to 5-separable. About 188 functions were found that seem to be either 4 or 5-separable, but in fact contain projection of at least two hypercube vertices with different labels on the same point. Although the percentage of linearly separated functions rapidly decreases relatively low k-separability indices resolve most of the Boolean functions.

An algorithm that searches for lowest k but also maximizes minimum distance between projections of points with different labels finds projection directions (with minimum separation of 1/6 or more) that require $k = 6$ for these functions and gives significantly larger separations between intervals containing vectors from a single class. With only 30 linarly separable functions per one direction and 1880 separable functions at least 63 different directions should be considered to find out if the function is really linearly separable. Learning all these functions is already a difficult problem.

For 5-bits the number of all Boolean functions grows to 2^{32}, or over 4 billions (4G). Direct search in 5-dimensional space for each of these functions is already prohibitively expensive. It seems quite likely that for n-bit Boolean functions each projection direction will separate the maximum number of functions for $k \approx 2^n/2$, and that learning the best projection for a given function will give the largest number of functions separated into n clusters, with percentage of linearly separable functions going quickly to zero. The number of elements in most cluster quickly grows, therefore with such as simple model it should be possible to learn them correctly even if only a subset of all values is given.

4 Algorithms Based on k-Separability

Linear projection combined with k-separability already gives quite powerful learning system, but almost all computational intelligence algorithms may implement in some form k-separability as a target for learning. It is recommended to search first for linearly separable solutions, and then to increase k searching for the simplest solution, selecting the best model using crossvalidation or measures taking into account the size and separation between projected clusters. Distribution of $y(bX; \mathbf{W})$ values allows for calculation of $P(y|Y_{\pm})$ class distributions and posterior probabilities using Bayesian rules. Estimation of probability distributions in one dimension is easy and may be done using Parzen-windows kernel methods.

The main difficulty in formulating a learning procedure is the fact that targets are not fully specified; instead of a single target for Y_+ class two or more labels Y_{+1}, Y_{+2} may be needed. This may actually be of some advantage, allowing for a better interpretation of the results. It is clearly visible in the case of parity problems: each group differs not only by the parity but also by different number of 1's, providing an additional label. Learning should therefore combine unsupervised and supervised components. In the first step random initialization is performed several times, selecting the lowest k cluster projection. Centers of these clusters $t_i, i = 1 \ldots k$ are the target variables for learning, and each center has a class label $Y(t_i)$. Slightly modified quadratic error function may be used for learning:

$$E(\mathbf{W}, \mathbf{t}) = \frac{1}{2} \sum_{\mathbf{X}} \left(y(\mathbf{X}; \mathbf{W}) - t_j(\mathbf{X})\right)^2; \tag{4}$$
$$j = \arg\min_i \{||t_i - y(\mathbf{X}; \mathbf{W})||, Y(t_i) = Y(\mathbf{X})\}$$

For each input $\mathbf{X}$ that belongs to the class $Y(\mathbf{X})$ the nearest (on the projected line) cluster center from the same class is taken as the learning target. A more complex cost function may be devised that penalize for the number of clusters, for overlapping of clusters, and for impurity of clusters, but this is beyond the scope of this article.

In the two-class case there are always two possibilities: either the first class vectors are projected to the lowest Y values, leading to clusters $Y_+, Y_-, Y_+, \ldots$, or vice versa, $Y_-, Y_+, Y_-, \ldots$. The 3-separable case is particularly simple and often encountered in practice. If vectors from one of the classes represent unusual objects or states (for example hypo and hiper-activity in some medical problems) projections with clusters Y_-, Y_+, Y_- are fairly common. This may be checked quite easily visualizing distribution of activations for a single perceptron (linear neuron is sufficient). Additional transformations (network layers) are needed to reach linear separability, but 3-separability may often be reached using just one node.

For $k = 3$ these projections are in 3 intervals: $[-\infty, a], [a, b], [b, +\infty]$. Taking $t = (a+b)/2$, and denoting $Y_X = Y(\mathbf{X})$ a linear error function suitable for learning is:

$$\begin{aligned} E(a, b, \mathbf{W}) &= \sum_{\mathbf{X}} [\mathrm{T}(y \le t)\delta(Y_X, Y_+) \max(0, y - a) + \delta(Y_X, Y_-) \max(0, a - y) \\ &+ \mathrm{T}(y > t)\delta(Y_X, Y_+) \max(0, b - y) + \delta(Y_X, Y_-) \max(0, y - b)] \end{aligned} \tag{5}$$

where $\mathrm{T}(y > t)$ is 0 if false and 1 if true. This function admits a trivial $\mathbf{W} = 0$ solution, therefore either a condition $||\mathbf{W}||$ should be introduced, or a distance scale should be fixed by requiring one of the components to be constant. It assumes that Y_+ vectors contribute to errors only outside of the $[a, b]$ interval, with the error growing in a linear way, and that Y_- vectors contribute to error in the linear way only inside this interval. It requires good initialization to map all Y_+ vectors to correct side of t. Using this function for 3-separable Boolean functions with multiple starts to find approximate 3-separability projection quickly learns such functions using a simple gradient method. To avoid threshold functions $\mathrm{T}(y > t)$ may be replaced by a logistic function $\sigma(y - t)$.

3-separable backpropagation learning in purely neural architecture requires a single perceptron for projection plus a combination of two neurons creating a "soft trapezoidal window" type of function $F(Y; a, b) = \sigma(Y + a) - \sigma(Y + b)$ that implements interval $[a, b]$ [21]. These additional neurons (Fig. 2) have fixed weights ($+1$ and -1) and biases a, b, adding only 2 adaptive parameters. An additional parameter determining the slope of the window shoulders may be introduced to scale the Y values. The input layer may of course be replaced by a hidden layer that implements additional mapping.

This network architecture has $n + 2$ parameters and is able to separate a single class bordered by vectors from other classes. For n-dimensional problem with 3-separable

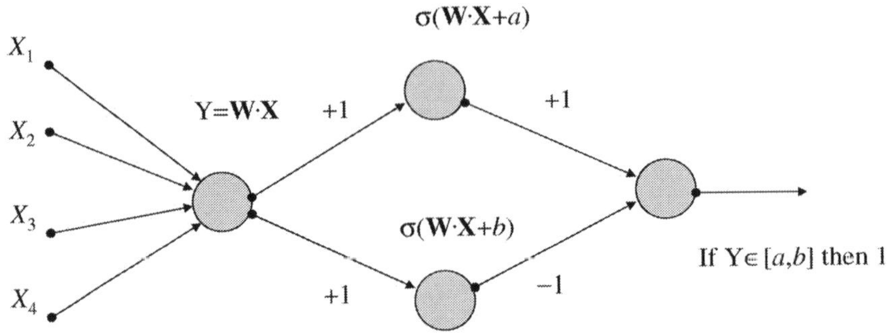

Fig. 2. MLP solution to the 3-separable case of 4-bit Boolean functions

structure standard architecture requires at least two hidden neurons connected to an output neuron with $2(n+1)+3$ parameters. For k-separability case this architecture will simply add one additional neuron for each new interval, with one bias parameter. n-bit parity problems require only n neurons (one linear perceptron and $n-1$ neurons with adaptive biases for intervals), while in the standard approach $O(n^2)$ parameters are needed [13].

5 Discussion and Open Problems

A radically new approach to learning has been proposed, simplifying the process by changing the goal of learning to easier target and handling the remaining nonlinearities with well defined structure. k-separability is a powerful concept that will be very useful for computational learning theory, breaking the space of non-separable functions into subclasses that may be separated into more than two parts. Even the simplest linear realization of k-separability with interval nonlinearities is quite powerful, allowing for efficient learning of difficult Boolean functions. Multiple-threshold perceptrons [22] may implement such intervals, although k-separability learning algorithms require more than multiple-threshold step functions. So far there are no systems that can routinely handle difficult Boolean functions, despite a lot of effort devoted to special Boolean problems, such as the parity problem. Using neural algorithms or special error functions described in the previous chapter almost all $n=4$ Boolean functions have been learned in less than 40 trials (Duch and Adamczak, in preparation).

Redefining the goal of learning may have some biological justification. Neurons in association cortex form strongly connected microcuircuits found in minicolumns, resonating with different frequencies when an incoming signal $X(t)$ appears. A perceptron neuron observing the activity of a minicolumn containing many microcircuits learns to react to signals in an interval around particular frequency. Combination of outputs from selected perceptron neurons is used to discover a category. These outputs may come from resonators of different frequencies, implementing an analogue to the combination of disjoint projections on the $\mathbf{W}\cdot\mathbf{X}$ line.

An interesting concept creates many open problems. How many boolean function each direction k-separates in general case? What minimal k is sufficient for n-bit problems? How will different cost functions perform in practice? What other simple ways to "disarm" linearites, besides projection on a k-segment line, may be used? These and many other questions will be addressed soon.

Acknowledgement. Support by the Polish Committee for Scientific Research, research grant 2005-2007, is gratefully acknowledged.

References

1. Duda, R.O., Hart, P.E., Stork, D.: Patter Classification. J. Wiley & Sons, New York (2001)
2. Duch, W.: Similarity based methods: a general framework for classification, approximation and association. Control and Cybernetics **29** (2000) 937–968
3. Duch, W., Adamczak, R., Diercksen, G.: Classification, association and pattern completion using neural similarity based methods. Applied Math. & Comp. Science **10** (2000) 101–120
4. Witten, I., Frank, E.: Data Mining: Practical machine learning tools and techniques. Morgan Kaufmann, San Francisco (2nd Ed, 2005)
5. Jankowski, N., Grąbczewski, K., Duch, W., Naud, Adamczak, R.: Ghostminer data mining software. Technical report (2000-2005) http://www.fqspl.com.pl/ghostminer/.
6. Stork, D., Allen, J.: How to solve the n-bit parity problem with two hidden units. Neural Networks **5** (1992) 923–926
7. Minor, J.: Parity with two layer feedforward nets. Neural Networks **6** (1993) 705–707
8. Setiono, R.: On the solution of the parity problem by a single hidden layer feedforward neural network. Neurocomputing **16** (1997) 225–235
9. Lavretsky, E.: On the exact solution of the parity-n problem using ordered neural networks. Neural Networks **13** (2000) 643–649
10. Arslanov, M., Ashigaliev, D., Ismail, E.: N-bit parity ordered neural networks. Neurocomputing **48** (2002) 1053–1056
11. Liu, D., Hohil, M., Smith, S.: N-bit parity neural networks: new solutions based on linear programming. Neurocomputing **48** (2002) 477–488
12. Torres-Moreno, J., Aguilar, J., Gordon, M.: The minimum number of errors in the n-parity and its solution with an incremental neural network. Neural Proc. Letters **16** (2002) 201–210
13. Iyoda, E., Nobuhara, H., Hirota, K.: A solution for the n-bit parity problem using a single translated multiplicative neuron. Neural Processing Letters **18** (2003) 233–238
14. Wilamowski, B., Hunter, D.: Solving parity-n problems with feedforward neural network. Int. Joint Conf. on Neural Networks (IJCNN'03), Portland, Oregon 2003, Vol I, 2546–2551
15. Duch, W.: Visualization of hidden node activity in neural networks: I. Visualization methods. II. Application to RBF networks. Springer Lecture Notes in AI 3070 (2004) 38–49
16. Duch, W.: Coloring black boxes: visualization of neural network decisions. In: Int. Joint Conf. on Neural Networks, Portland, Oregon. IEEE Press Vol I (2003) 1735–1740
17. Duch, W., Jankowski, N.: Survey of neural transfer functions. Neural Computing Surveys **2** (1999) 163-213
18. Duch, W., Jankowski, N.: Taxonomy of neural transfer functions. In: International Joint Conference on Neural Networks. Como, Italy, IEEE Press Vol III (2000) 477–484
19. Dietterich, T., Bakiri, G.: Solving multiclass learning problems via error-correcting output codes. Journal Of Artificial Intelligence Research **2** (1995) 263–286
20. Zuyev, Y.: Asymptotics of the logarithm of the number of threshold functions of the algebra of logic. Soviet Mathematics Doklady **39** (1989)
21. Duch, W., Adamczak, R., Grąbczewski, K.: A new methodology of extraction, optimization and application of crisp and fuzzy logical rules. IEEE Transactions on Neural Networks **12** (2001) 277–306
22. Ngom, A., Stojmenovic I., Zunic, J.: On the Number of Multilinear Partitions and the Computing Capacity of Multiple-Valued Multiple-Threshold Perceptrons, IEEE Transactions on Neural Networks **14** (2003) 469–477

Lazy Training of Radial Basis Neural Networks

José M. Valls, Inés M. Galván, and Pedro Isasi

Universidad Carlos III de Madrid - Departamento de Informática,
Avenida de la Universidad, 30 - 28911 Leganés (Madrid), Spain
jvalls@inf.uc3m.es

Abstract. Usually, training data are not evenly distributed in the input space. This makes non-local methods, like Neural Networks, not very accurate in those cases. On the other hand, local methods have the problem of how to know which are the best examples for each test pattern. In this work, we present a way of performing a trade off between local and non-local methods. On one hand a Radial Basis Neural Network is used like learning algorithm, on the other hand a selection of the training patterns is used for each query. Moreover, the RBNN initialization algorithm has been modified in a deterministic way to eliminate any initial condition influence. Finally, the new method has been validated in two time series domains, an artificial and a real world one.

Keywords: Lazy Learning, Local Learning, Radial Basis Neural Networks, Pattern Selection.

1 Introduction

When the training data are not evenly distributed in the input space, the non-local learning methods could be affected by decreasing their generalization capabilities. One way of resolving such problem is by using local learning methods[3,9]. Local methods use only partially the set of examples during the learning process. They select, from the whole examples set, those that consider more appropriate for the learning task. The selection is made for each new test pattern presented to the system, by means of some kind of similarity measurement to that pattern. k-NN [4] is a typical example of these systems, in which the selected learning patterns are the k closest to the test pattern by some distance metric, usually the Euclidean distance.

Those methods, usually known as lazy learning or instance-based learning algorithms [1], have the inconvenience of being computationally slow, and highly dependent on the number of examples selected and on the metric used, being frequent the situations where an Euclidean metric might not be appropriate.

Bottou and Vapnik [2] introduce a dual, local/non-local, approach to give good generalization results in non-homogeneous domains. This approach is based on the selection, for each test pattern, of the k closest examples from the training set. With these examples, a neural network is trained in order to predict the test pattern. This is a good combination between local and non-local learning. However, the neural network used is a linear classifier and the method assumes

S. Kollias et al. (Eds.): ICANN 2006, Part I, LNCS 4131, pp. 198–207, 2006.

that Euclidean distance is an appropriate metric. Besides, it considers that all test patterns have the same structure but some domains would require different behaviors when being in different regions.

In this work we introduce some modifications in the general procedure of [2], by considering the use of Radial Basis Neural Networks (RBNN)[6,5]. RBNN have some advantages when using dual techniques: they are non-linear, universal approximators [7] and therefore the metric becomes a non-critical factor; besides, their training is very fast, without increasing significatively the computational cost of standard lazy approaches.

We propose to use RBNN with a lazy learning approach, making the selection of training patterns based on a kernel function. This selection is not homogeneous, as happened in [2]; by opposite it is detected, for each testing pattern, how many training patterns would be needed, and what is the importance in the learning process of each one of them. This importance is taken into consideration, in the form of a weight, in the learning process of the network.

When a lazy approach is combined with RBNN, two important aspects must be taken into account. In one hand, the initialization of the RBNN training algorithm is a critical factor that influences their performance. This algorithm has been modified in a deterministic way to eliminate any initial condition influence. In other hand, it may occur that no training pattern is selected for certain test patterns, due to the distribution of data in the input space. In those case the system must provide some answer. We propose two different approaches to treat this problem.

The final method results to be a dual local non-local method, where the initialization of the network is deterministic and the method is able to determine the degree of locality of each region of the space, by means of a kernel function that could be considered as a parameter, and modified appropriately. In some cases a test pattern could be considered as non-local in the sense that it corresponds to more frequent situations. In this case almost the totality of the training patterns will be selected, and the method behaves like an non-local approach. This transaction between local and non-local behavior is made automatically.

2 Description of the Method

The learning method proposed in this work has been called LRBNN (Lazy RBNN method) and is based on the selection, from the whole training data, of an appropriate subset of training patterns in order to improve the answer of the network for a novel pattern. For each new pattern received or query, a new subset of training examples is selected. The main idea consists of selecting those patterns close to the new query instance, in terms of the Euclidean distance. In order to give more importance to the closest examples, a weighting measure that assigns a weight to each training example is applied. This is done by using a kernel function which depends on the Euclidean distance from the training pattern to the query. In this work, the inverse function ($K(d) = 1/d$, where d

is the distance from the training pattern to the new query) is used. A more detailed information about the use of this function can be found in [8].

To carry out this idea, a n-dimensional sphere centered at the test pattern is established, in order to select only those patterns placed into it. Its normalized radius (respect to the maximum distance from any example to the query), called r_r, will be used to select the training patterns situated into the sphere, being r_r a parameter that must be established before the application of the learning algorithm. Next, the sequential structure of LRBNN method is summarized.

Let $\mathbf{q} = (q_1, ..., q_n)$ be the query instance. Let X be the whole available training data set:$X = \{(\mathbf{x_k}, \mathbf{y_k})\ k = 1 \ldots N; \mathbf{x_k} = (x_{k1}, \ldots, x_{kn}); \mathbf{y_k} = (y_{ki}, \ldots, y_{km})\}$. For each $\mathbf{q}$,

1. The standard Euclidean distances d_k from the query to each training example are calculated. Then, the relative distance d_{rk} is calculated for each training pattern: $d_{rk} = d_k/d_{max}$, where d_{max} is the distance from the novel pattern to the furthest training pattern.
2. A kernel function $K()$ is used to calculate a weight for each training pattern from its distance to the query. This function is the inverse of the relative distance d_{rk}: $K(x_k) = 1/d_{rk}; k = 1 \ldots N$
3. These values $K(x_k)$ are normalized in such a way that the sum of them equals the number of training patterns in X. These normalized values are called normalized frequencies, f_{nk}.
4. Both d_{rk} and f_{nk} are used to decide whether the training pattern is selected and -in that case- how many times is included in the training subset. They are used to generate a natural number, n_k, following the next rule:

$$\text{if } d_{rk} < r_r \text{ then } n_k = int(f_{nk}) + 1 \text{ else } n_k = 0 \tag{1}$$

   At this point, each training pattern in X has an associated natural number, n_k, which indicates how many times the pattern (x_k, y_k) will be used to train the RBNN in order to predict the query q.
5. A new training subset associated to $\mathbf{q}$, X_q, is built up. Given a pattern (x_k, y_k) from the original training set X, it will be included in X_q if $n_k > 0$. Besides, (x_k, y_k) will be randomly placed n_k times in X_q.
6. The RBNN is trained using X_q: the neurons centers are calculated in an unsupervised way using K-means algorithm in order to cluster the input training patterns included in the subset X_q. The neurons widths are evaluated as the geometric mean of the distances from each neuron center to its two nearest centers, and the RBNN weights are estimated in a supervised way in order to minimize the mean square error measured in the training subset X_q.

In order to apply the learning method to train RBNN, two features must be taken into account: On one hand, the results would depend on the random initialization of the K-means algorithm which is used to determine the locations of the RBNN centers and must be applied for each query. On the other hand, when the test pattern is located in a region of the input space where the examples are scarce, it could happen that no training examples are selected. We present solutions to both problems, which are described next.

K-means initialization. Having the objective of achieving the best performance, a deterministic initialization, instead of the usual random ones, is proposed. The idea is to obtain a prediction of the network with a deterministic initialization of the centers whose accuracy is similar to the one obtained when several random initializations are done. The initial location of the centers will depend on the location of the closest training examples selected. The deterministic initialization is obtained as follows:

- Let $(\mathbf{x_1}, \mathbf{x_2}, \ldots, \mathbf{x_l})$ be the l selected training patterns, inversely ordered by their distance to the query instance. Let m be the number of hidden neurones of the RBNN to be trained.
- If $m \leq l$ then the center of the i_{th} neuron is initialized to the $\mathbf{x_i}$ position, for $i = 1, 2, \ldots, m$. Otherwise $(m > l)$, l neurones will be initialized to the $\mathbf{x}_i$ position, for $i = 1, 2, \ldots, l$, and the remaining $m - l$ neurones (lets call this number p) will be randomly initialized in the following way:
  - M_q, the centroid of the set X_q, is evaluated.
  - p centers $(c_{1q}, c_{2q}, \ldots. c_{pq})$ are randomly generated, such as $\|c_{jq} - M_q\| < \epsilon$, $j = 1, 2, \ldots, p$, where ϵ is a very small real number.

Empty training set. It has been observed that when the input space data is highly dimensional, in certain regions of it the data density can be so small that the sphere centered at the query instance does not include any train pattern into it if the relative radius is small. When this situation occurs, an alternative way to select the training patterns must be taken. In our work, we propose two different approaches which are experimentally evaluated.

1. If the subset X_q associated to a query $\mathbf{q}$ is empty, then we apply the method of selection to the closest training pattern, as if it was the test pattern. Thus, the selected set will have, at least, one element.
2. If X_q is empty, then the network is trained with X, the set formed by all the training patterns. In other words, the network is trained as usual, with all the available patterns.

3 Experimental Validation

We have applied LRBNN to two domains, the Mackey-Glass and the Venice Lagoon time series. As it was remarked in section 2, the relative radius r_r must be given as an external parameter of the method in order to study its influence on the performance of the model. Besides, RBNN with different architectures - i.e. different number of hidden neurons- must be trained so that the influence of the network architecture can also be studied.

The method incorporates solutions regarding to the initialization of centers and the possibility of having empty training sets. These solutions are validated in the experiments where we have applied the lazy approach with both ways of initializing the centers: the random and the deterministic one. Moreover, in the cases where some test patterns can not be predicted because the associated training subset is empty, the approaches mentioned in section 2 have been applied.

3.1 An Artificial Time Series Prediction Problem: The Mackey-Glass Time Series

The Mackey-Glass time series is a well known artificial time series widely used in the literature about RBNN, [10],[6]. The data used in this work has been generated following the studies mentioned above. The task for the RBNN is to predict the value of the time series at point $x[t+50]$ from the earlier points $(x[t], x[t-6], x[t-12], x[t-18])$. 1000 data points form the training set, corresponding to the sample time between 3500 and 4499. The test set is composed by the points corresponding to the time interval $[4500, 5000]$. Both sets have been normalized in the interval $[0, 1]$. The proposed LRBNN method has been applied to this artificial time series, where RBNN of different architectures have been trained during 500 learning cycles varying the relative radius from 0.04 to 0.24.

Table 1. Mean errors with random initialization of centers. Mackey-Glass time series

	Hidden Neurones							
r_r	7	11	15	19	23	27	NP	%PP
0.04	0.02527	0.02641	0.02683	0.02743	0.02691	0.02722	45	91
0.08	0.02005	0.01891	0.01705	0.01571	0.01716	0.01585	0	100
0.12	0.02379	0.01954	0.01792	0.01935	0.01896	0.01940	0	100
0.16	0.02752	0.02223	0.01901	0.02106	0.02228	0.02263	0	100
0.2	0.03031	0.02427	0.02432	0.02287	0.02281	0.02244	0	100

Table 2. Mean errors with deterministic initialization. Mackey-Glass time series

	Hidden Neurones							
r_r	7	11	15	19	23	27	NP	%PP
0.04	0.02904	0.03086	0.03096	0.03109	0.03231	0.03295	45	91
0.08	0.01944	0.01860	0.01666	0.01565	0.01551	0.01585	0	100
0.12	0.02131	0.01742	0.01644	0.01607	0.01628	0.01602	0	100
0.16	0.02424	0.02029	0.01812	0.01729	0.01783	0.01809	0	100
0.2	0.02837	0.02083	0.01927	0.01874	0.02006	0.02111	0	100

In order to show that the deterministic initialization lead to an appropriate performance of RBNN when they are trained following the lazy learning approach, experiments with the lazy approach where the neurons centers are randomly initialized are also made. Table 1 shows the mean performance of the method for five random initializations, when RBNN with different number of hidden neurons are trained. Each value of the error for a specific number of neurons and radius corresponds to the mean value of five different mean errors. On the other hand, when the proposed deterministic initialization is applied, the obtained results are shown in table 2. We can observe that the error values are slightly better than the ones obtained when the neurons centers were randomly located. We must emphasize the advantage of this method where a single run is needed whereas if the usual K-means algorithm is applied, several initializations must be made in order to ensure an adequate performance of the method.

The columns named "NP" displays the number of "null patterns", that is, test patterns for which the number of selected training patterns is zero. This situation might arise because of the dimensionality of the problem and the non-uniform distribution of data. The "PP" column displays the percentage of test patterns that are correctly answered ("Predicted Patterns"). As it is shown, when $r_r = 0.04$, there are 45 test patterns for which the networks can not make a prediction because the associated training sets are empty. Thus, these test patterns are discarded, corresponding the error values to the rest of patterns, that is, to the 91% of the whole test set.

We have applied the two alternative ways of treating these anomalous patterns are presented. *Method (a)*, that keeps the local approach,and *Method (b)* that renounce to the local approach and follows a global one. With the aim of studying the performance of both approaches, RBNN of different architectures are trained when a relative radius of 0.04 is taken. Both *Method (a)* and *Method (b)* have been applied and the obtained error values are shown in table 3, where we can see that method (b) behaves slightly worse than method (a) in all the cases. Thus, when a local approach is taken, the method gets better results than when all the available patterns are used to train the networks.

Table 3. Null patterns processing ($r_r = 0.04$). Mackey-Glass time series.

	Hidden Neurones							
	7	11	15	19	23	27	NP	%PP
Method (a)	0.02974	0.03043	0.03132	0.03114	0.03309	0.03373	45	100
Method (b)	0.03385	0.03641	0.03545	0.03464	0.03568	0.03408	45	100

As for the influence of the relative radius and the number of hidden neurons, it is possible to observe that the performance of the networks is scarcely influenced by the value of the relative radius when it is bigger than a certain value and the number of neurons is big enough. The mean error decreases with the radius until $r_r = 0.08$, and then it maintains its value nearly constant as the radius increases if the number of neurons is bigger than 7. Thus, the relative radius is not a critical parameter if the number of neurons is bigger than 7 and the relative radius is bigger than 0.08. When the number of neurons is small, the performance of the networks gets worse as the radius increases. This is explained because the number of training patterns selected is very big and the number of neurons of the RBNN are insufficient to fit such training set.

3.2 A Real Time Series Prediction Problem: The Venice Lagoon Time Series

The Venice lagoon time series represents the behavior of the water level at Venice lagoon. Unusually high tides result from a combination of chaotic climatic elements in conjunction with the more normal, periodic, tidal systems associated with a particular area. The most famous example of flooding in the Venice lagoon occurred in November 1966 when, driven by strong winds, the Venice Lagoon rose by nearly 2 m. above the normal water level. That phenomenon is known as "high water" and many efforts have been made in Italy to develop systems for predicting sea levels in Venice and mainly for the prediction of the high water phenomenon [11].

There is a great amount of data representing the behavior of the Venice Lagoon time series. However, the part of data associated to the stable behavior of the water is very abundant as opposed to the part associated to high water phenomena. This situation leads to the following: the RBNN trained with a complete data set is not very accurate in predictions of high water phenomena. Hence, the aim in this context is to observe whether a selection of training patterns may help to obtain better predictions. A training data set of 3000 points corresponding to the water level measured each hour has been extracted from available data in such a way that both stable situations and high water situations appear represented in the set. High-water situations are considered when the level of water is not lower than 110 cm. 20 test patterns have also been extracted from the available data and they represent a situation when the water level is higher than 110 cm.

In order to apply LRBNN, different RBNN architectures have been trained during 500 learning cycles, and the relative radius has been fixed to different values from 0.04 to 0.2. As in the previous domain, two sets of experiments have been done: the first one corresponds to the usual, random K-means initialization; in order to obtain representative results, five runs of the method have been carried out and the mean of the results is showed in table 4.

The second set of experiments, carried out only once, corresponds to the deterministic initialization of the neurons centers. The results are displayed on table 5. As it happened on the previous domains, when the deterministic initialization of the centers is done, the results are similar or slightly better than when the centers are randomly located.

Table 4. Mean errors with random initialization of centers. Venice Lagoon time series

	Hidden Neurones							
r_r	7	11	15	19	23	27	NP	%PP
0.04	0.03596	0.03866	0.04035	0.04031	0.04015	0.04169	14	30
0.08	0.03286	0.03330	0.03150	0.03547	0.03799	0.03476	2	90
0.12	0.03219	0.02693	0.02490	0.02365	0.02677	0.02738	0	100
0.16	0.02487	0.02506	0.02554	0.02783	0.02600	0.02603	0	100
0.2	0.03350	0.03035	0.03094	0.03139	0.03155	0.03179	0	100

It is important to observe that there are null patterns even when the relative radius grows to 0.08. When $r_r = 0.04$, 14 test patterns, out of 20, can not be predicted. Thus, only a 30% of the test set can be properly predicted. And still for $r_r = 0.08$ 2 patterns are not predicted. The anomalous situations are now more frequent and this is explained as follows: the dimensionality of this problem

Table 5. Mean errors with deterministic initialization. Venice Lagoon time series.

	Hidden Neurones							
r_r	7	11	15	19	23	27	NP	%PP
0.04	0.03413	0.03378	0.03328	0.03465	0.03404	0.03389	14	30
0.08	0.03181	0.03028	0.03062	0.03041	0.03148	0.03017	2	90
0.12	0.02967	0.02682	0.02269	0.02234	0.02235	0.02643	0	100
0.16	0.02869	0.02398	0.02913	0.02059	0.02514	0.02552	0	100
0.2	0.03769	0.02420	0.02411	0.02728	0.02288	0.03336	0	100

is higher than the former one because this series problem has been modeled as a six-dimension function; besides, there are regions of the input space whose data density is very low. The test set has been generated so that its data represent the high water situations, and the training examples which corresponds to these unfrequent situations are very scarce. Thus, the null patterns processing methods presented in this work are essential in the LRBNN model. Table 6 shows the errors obtained when both methods (a) and (b) are applied if null patterns are found. It is important to realize that, although it seems that the results are worse than those seen on table 5, a 100% of the test patterns are properly predicted.

Table 6. Null patterns processing. Venice Lagoon time series.

		Hidden Neurones							
r_r	Meth	7	11	15	19	23	27	NP	%PP
0.04	(a)	0.06042	0.06276	0.06292	0.06186	0.06330	0.06352	14	100
0.04	(b)	0.09542	0.08128	0.06672	0.06239	0.06333	0.06500	14	100
0.08	(a)	0.03685	0.03447	0.03011	0.03197	0.02792	0.03231	2	100
0.08	(b)	0.04497	0.04382	0.03572	0.03407	0.03266	0.03441	2	100

In this domain, the differences between both methods are significant, specially when the relative radius is 0.04. In this case, 14 null patterns are found, that is, 70% of the whole test set. We can appreciate that method (a) achieves lower errors that method (b). Thus, when a lazy learning approach is applied the result is better than when the RBNN are trained with all the available training patterns.

It is possible to observe that, as in previous cases, when the relative radius is small, mean errors are high, due to the shortage of selected training patterns, and as the relative radius increases, the mean error decreases and then it does not change significatively. Thus, as it happened with the previous domains, the relative radius is not a critical parameter if the number of neurons and the relative radius are bigger enough.

3.3 Lazy Learning Versus Global Learning

In order to compare the lazy learning strategy (LRBNN) with the traditional one, RBNN with different number of hidden neurons (from 5 to 150) have been trained,

Table 7. Lazy learning versus traditional learning

	Mackey-Glass time series	Venice Lagoon time series
LRBNN	0.01551 ($r_r = 0.08$, 23 neurons)	0.02059 ($r_r = 0.16$, 19 neurons)
Traditional Method	0.10273 (110 neurons)	0.09605 (50 neurons)

in a global way, using the whole training data set in order to build a global approximation. In this work, the traditional learning has been carried out using a training and a validation data set, stopping the iterative process when both errors become stabilized. The standard K-means algorithm has been used and several experiments with different initial centers locations are made. In both approaches, the same data sets have been used. In table 7, the best results obtained in both domains for both methods, lazy and traditional ones, are shown. As it is possible to observe, in both domains the performance of the local method is significatively better than the performance of the traditional learning approach.

4 Conclusions

In this work, we try to complement the good characteristics of local and global approaches by using a lazy learning method for selecting the training set, using RBNN for making predictions. RBNN have some advantages: they are universal approximators and therefore the assumption of local linear behavior is no longer needed; besides, their training is very fast, without increasing significatively the computational cost of standard local learning approaches. We present a method (LRBNN) that can get the locality of the input space, and then uses a non-linear method to approximate each region of the input space. In addition, the selection of patterns is made using a kernel function, taking into account the distribution of data.

When a lazy learning strategy is used, two important aspects related to RBNN training and patterns selection have been taken into account. In the first place, the initialization of the neurons centers is a critical factor that influences RBNN performance. Usually, the initial location of centers are randomly established, but in a lazy strategy, in which a network must be trained for each new query, random initialization must be avoided. For this reason, in this work, the algorithm has been modified in a deterministic way to eliminate any initial condition influence with the objective of achieving the best performance. Regarding to the selection procedure, in which the Inverse kernel function is used, it may occur that no training pattern is selected for certain test patterns, due to the distribution of data in the input space. We have proposed and validated two different approaches to treat this problem.

LRBNN has been applied to two different domains: an artificial time series (the well known Mackey-Glass time series) and a real one (representing the Venice Lagoon water level). For both domains, we present the results obtained by LRBNN when a deterministic centers initialization is made. Besides, with the aim of showing the advantages of this deterministic initialization, the same method is applied but the RBNN are trained with a random initialization of their centers. We show the mean results of several random initializations. As we said before, LRBNN provides two alternative ways of guarantying the selection of training examples for all the query instances. When the use of these alternative methods is necessary, the obtained results are also showed. Finally, LRBNN performance is compared with the performance of RBNN trained using a global approach, that is, using the whole training set.

The results obtained by LRBNN improves significatively the ones obtained by RBNN trained in a global way. Besides, the proposed deterministic initialization of the neurons centers produces similar or slightly better results than the usual random initialization, being thus preferable because only one run is necessary. Moreover, the method is able to predict 100% of the test patterns, even in those extreme cases when no train examples would be selected using the normal selection method. The experiments show that the relative radius, parameter of the method, is not a critical factor because if it reaches a minimum value and the network has a sufficient number of neurons, the error on the test set keeps its low value relatively constant.

Thus, we can conclude that the combination of lazy learning and RBNN, can produce significant improvements in some domains.

Acknowledgments. This article has been financed by the Spanish founded research MEC project OPLINK::UC3M, Ref: TIN2005-08818-C04-02

References

1. D.W. Aha, D. Kibler, and M.K. Albert. Instance-based learning algorithms. *Machine Learning*, 6:37–66, 1991.
2. L. Bottou and V. Vapnik. Local learning algorithms. *Neural Computation*, 4(6):888–900, 1992.
3. C.G. Atkenson, A.W. Moore, and S. Schaal. Locally weighted learning. *Artificial Intelligence Review*, 11:11–73, 1997.
4. B.V. Dasarathy (Editor). Nearest neighbour(NN) norms: NN pattern classification techniques. *IEEE Computer Society Press*, 1991.
5. J. Ghosh and A. Nag. *An Overview of Radial Basis Function Networks.* R.J. Howlett and L.C. Jain (Eds). Physica Verlag, 2000.
6. J.E. Moody and C. Darken. Fast learning in networks of locally tuned processing units. *Neural Computation*, 1:281–294, 1989.
7. J. Park and I.W. Sandberg. Universal approximation and radial-basis-function networks. *Neural Computation*, 5:305–316, 1993.
8. J.M. Valls, I.M. Galván, and P. Isasi. Lazy learning in radial basis neural networks: a way of achieving more accurate models. *Neural Processing Letters*, 20:105–124, 2004.
9. D. Wettschereck and T. Dietterich. Improving the perfomance of radial basis function networks by learning center locations. *Advances in Neural Information Processing Systems*, 4:1133–1140, 1992.
10. L. Yingwei, N. Sundararajan, and P. Saratchandran. A sequential learning scheme for function approximation using minimal radial basis function neural networks. *Neural Computation*, 9:461–478, 1997.
11. J.M. Zaldívar, E. Gutiérrez, I.M. Galván, F. Strozzi, and A. Tomasin. Forecasting high waters at Venice Lagoon using chaotic time series analysis and nonlinear neural networks. *Journal of Hydroinformatics*, 2:61–84, 2000.

Investigation of Topographical Stability of the Concave and Convex Self-Organizing Map Variant

Fabien Molle[1,2] and Jens Christian Claussen[2,*]

[1] Theoretical Physics, Chalmers Tekniska Högskola, Göteborg
[2] Institut für Theoretische Physik und Astrophysik
Leibnizstr. 15, 24098 Christian-Albrechts-Universität zu Kiel, Germany
claussen@theo-physik.uni-kiel.de
http://www.theo-physik.uni-kiel.de/~claussen/

Abstract. We investigate, by a systematic numerical study, the parameter dependence of the stability of the Kohonen Self-Organizing Map and the Zheng and Greenleaf concave and convex learning with respect to different input distributions, input and output dimensions.

Topical groups: Advances in Neural Network Learning Methods, Neural and hybrid architectures and learning algorithms, Self-organization.

Neural vector quantizers have become a widespreadly used tool to explore high-dimensional data sets by self-organized learning schemes. Compared to the vast literature on variants and applications that appeared the last two decades, the theoretical description proceeded more slowly. Even for the coining Self-Organizing Map (SOM) [1], still open questions remain, as a proper description of the dynamics for the case of dimension reduction and varying data dimensionality, or the question for what parameters stability of the algorithm can be guaranteed. This paper is devoted to the latter question. The stability criteria are especially interesting for modifications and variants, as the concave and convex learning [2], whose magnification behaviour has been discussed recently [3]. Especially for the variants, analytical progress becomes quite difficult, and in any case one will expect that the stability will depend on the input distribution to some —apart from special cases— unknown extent. As the invariant density in general is analytically unaccessible for input dimensions larger than one (see [4] for recent tractable cases), we expect a general theory not to be available immediately, and instead proceed with a systematic numerical exploration.

The Kohonen SOM, and the nonlinear variant of Zheng and Greenleaf. – The class of algorithms investigated here is defined by the learning rule, that for each stimulus $\mathbf{v} \in V$ each weight vector $\mathbf{w_r}$ is updated according to

$$\mathbf{w}_\mathbf{r}^{\mathsf{new}} = \mathbf{w}_\mathbf{r}^{\mathsf{old}} + \varepsilon \cdot g_{\mathbf{rs}} \cdot (\mathbf{v} - \mathbf{w}_\mathbf{r}^{\mathsf{old}})^K \tag{1}$$

* Corresponding author.

S. Kollias et al. (Eds.): ICANN 2006, Part I, LNCS 4131, pp. 208–215, 2006.

($g_{\mathbf{rs}}$ being a gaussian function (width σ) of euclidian distance $|\mathbf{r}-\mathbf{s}|$ in the neural layer, thus describing the neural topology). Herein

$$|\mathbf{w_s} - \mathbf{v}| = \min_{\mathbf{r}\in R} |\mathbf{w_r} - \mathbf{v}| \tag{2}$$

determines for each stimulus $\mathbf{v}$ the *best-matching unit* or *winner* neuron.

The case $K = 1$ is the original SOM [1], corresponding to a linear or Hebbian learning rule. The generalization to K or $1/K$ taking integer values has been proposed by Zheng and Greenleaf [2], but arbitrary nonzero real values of K can be used [3], and the choice of $K \to 0$ has been shown (for the onedimensional case) to have an invariant density with the information-theoretically optimal value of the magnification exponent one [3], i.e., the neural density is proportional to the input density and hence can be used as a density estimator.

Convergence and stability. – It is well known that for the learning rate ε, one has to fulfill the Robbins-Munro conditions (see, e.g. [4]) to ensure convergence, with all other parameters fixed. However, practically it is necessary to use a large neighborhood width at the beginning, to have the network of weight vectors ordered in input space, and decrease this width in the course of time downto a small value that ensures topology preservation during further on-line learning. Thus the situation becomes more involved when additionally also σ is made time-dependent. Here we consider the strategy where the stability border in the (ε, σ) plane always is approached from small ε with σ fixed during this final phase. An ordered state has to be generated by preceding learning phases.

Measures for Topographical Stability. – To quantify the ordered state and the topology preservation, a variety of measures is used, e.g. the topographic product [5], the Zrehen measure [6], and the average quadratic reconstruction error. To detect instable behaviour, all measures should be suitable and give similar results. For an unstable and disordered map, also the total sum over all (squared) distances between adjacent weight vectors will increase significantly; so a thresholded increase will indicate instability as well. This indicator is used below; however, for the case of a large neighborhood (of network size), the weight vectors shrink to a small volume, thus influencing the results; however, this applies to a neighborhood widths larger than that commonly used for the pre-ordering.

In addition we use here an even more simple approach than the Zrehen measure (which counts the number of neurons that lie within a circle between each pair of neurons that are adjacent in the neural layer). For a mapping from d to d dimensions, we consider the determinant of the i vectors spanned by $\boldsymbol{w}_{\boldsymbol{r}+\mathbf{e}_i} - \boldsymbol{w}_{\boldsymbol{r}}$, with $\mathbf{e}_i$ being the ith unit vector. The sign of this determinant, where $1 \leq i \leq d$, thus gives the orientation of the set of d vectors. Note that the 1-dimensional case just reads $\text{sgn}(w_{r+1} - w_r)$, which has been widely considered to detect the ordered state in the 1 to 1 dimensional case. Hence, we can define

$$\chi(\{\boldsymbol{w}_{\boldsymbol{r}}\}) := \tag{3}$$
$$1 - \frac{1}{N} \left| \sum_{\boldsymbol{r}} \text{sgn}(\det((\boldsymbol{w}_{\boldsymbol{r}+\mathbf{e}_1} - \boldsymbol{w}_{\boldsymbol{r}}), \ldots (\boldsymbol{w}_{\boldsymbol{r}+\mathbf{e}_i} - \boldsymbol{w}_{\boldsymbol{r}}), \ldots (\boldsymbol{w}_{\boldsymbol{r}+\mathbf{e}_d} - \boldsymbol{w}_{\boldsymbol{r}}))) \right| .$$

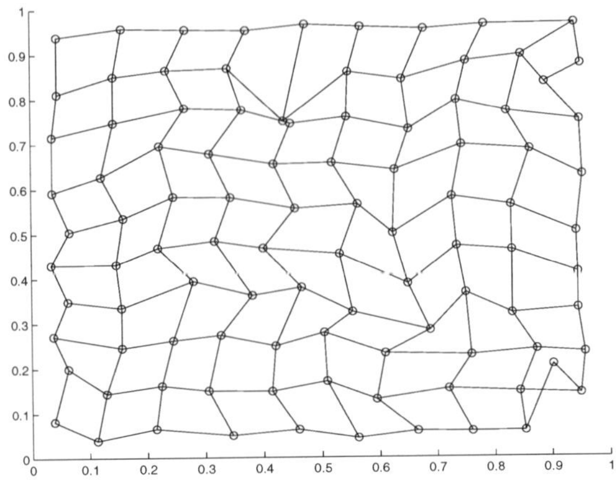

Fig. 1. Situation where one defect is detected by the crossproduct measure (3)

This evaluates the number of neurons N_+ (resp. N_-), where this sign is positive (resp. negative), hence the relative fraction of minority signs is given by $(1 - |N_+ - N_-|/N)$. A typical single defect is shown in Fig. 1. Due to its simplicity, this measure χ will be used in the remainder.

Modification of learning rules and data representation. – A classical result [7,8,9] states that the neural density for 1-D SOM (in the continuum limes) approaches not the input density itself, but a power of it, with exponent 2/3, the so-called magnification exponent. As pointed out by Linsker [10], the case of an exponent 1 would correspond to the case of maximal mutual information between input and output space. Different modifications of the winner mechanism or the learning rule, by additive or multiplicative terms, have been suggested and influence the magnification exponent [11,12,13,14]. Here we investigate the case of concave and convex learning [2,3], which defines a nonlinear generalization of the SOM.

Topographical Stability for the Self-Organizing Map. – Before investigating the case of concave and convex learning, the stability measures should be tested for the well-established SOM algorithm. Using the parameter path of Fig. 2, we first analyze the 2D $\rightarrow$ 2D case, for three input distributions: the homogeneous input density (equidistribution), an inhomogeneous input distribution $\sim \sin(\pi x_i)$ [14], and a varying-dimension dataset (Figs. 3, 4). The results are shown in Fig. 5.

Different input dimensions and varying intrinsic dimension. – As the input dimensionality is of pronounced influence on the maximal stable learning rate (Fig. 6), we also investigate an artificial dataset combining different dimensions: the box-plane-stick-loop dataset [15] (Fig. 3), or its 2D counterpart, the plane-stick-loop (Fig. 4). Here the crossproduct detection will become problematic where the input space is intrinsically 1D (stick and loop), thus the average distance criterion is used, and we restrict to the case $\sigma \leq 1$.

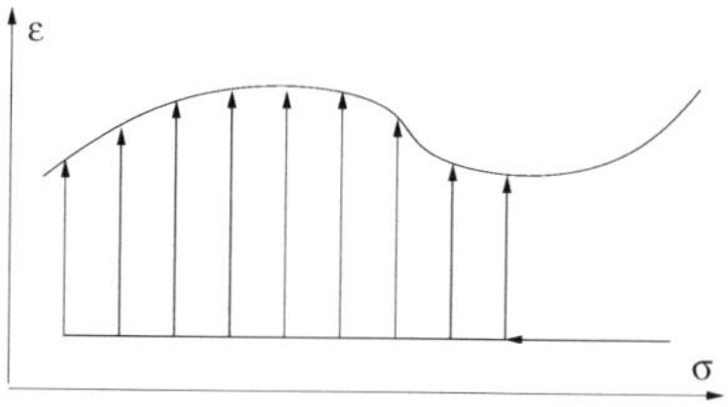

Fig. 2. Schematic diagram of the parameter path in (ε, σ) space. Starting with high values, σ is slowly decreased to the desired value, while the learning rate still is kept safely low. From there, at constant σ the learning rate is increased until instability is observed; giving an upper border to the stability area. – The same scheme is applied for the concave and convex learning, where the nonlinearity exponent is considered as a fixed parameter.

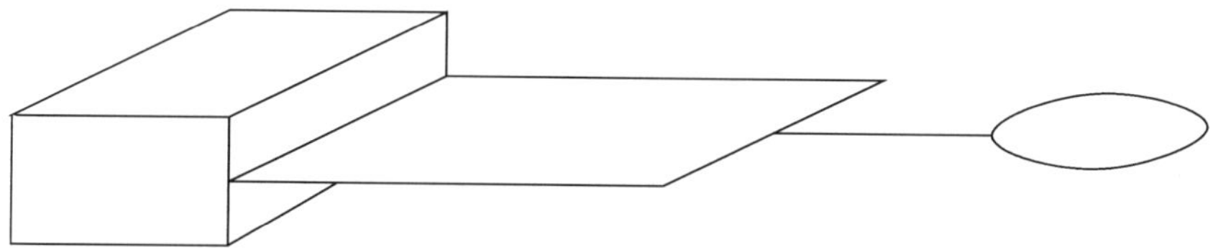

Fig. 3. Schematic view of the classical box-plane-stick-loop dataset. Its motivation is to combine locally different input data dimensions within one data set.

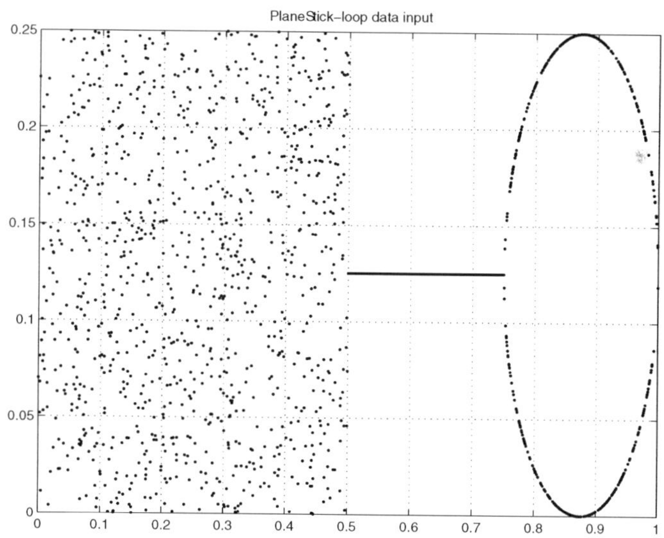

Fig. 4. Part of the input data for the 2D plane-stick-loop data set (Fig. 3)

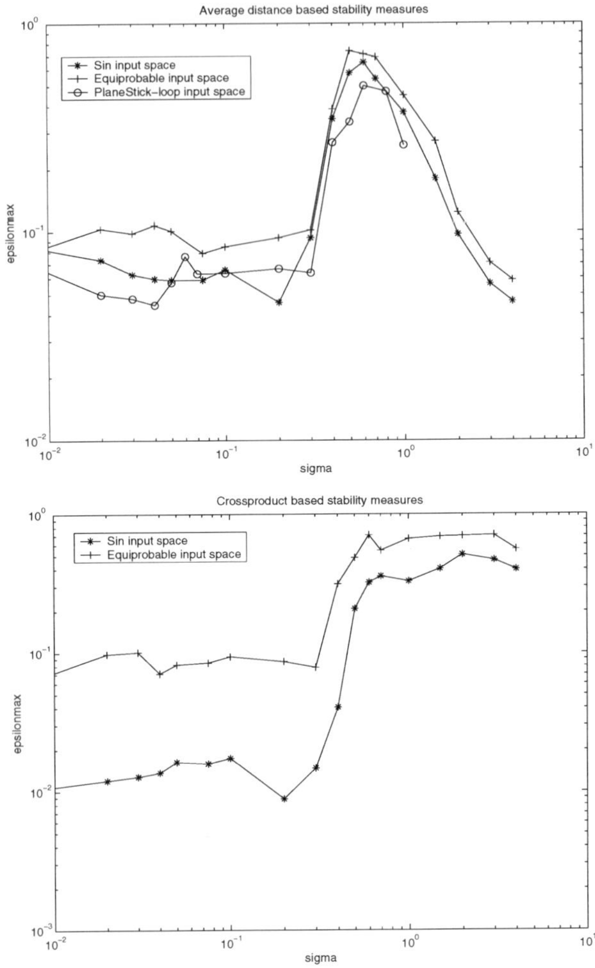

Fig. 5. Critical $\varepsilon_{\max}(\sigma)$ where (coming from small ε values, see Fig. 2) the SOM learning loses stability. Here a 2D array of 10×10 neurons was used with decay $\exp(-t/k)$ exponentially in time t, with k between 30000 and 60000 depending on ε_0 (for σ between 0.1 and 0.001, $k = 300000$). Top: Unstable ε detected from growth of the averaged distance of neurons; here a threshold of 15% was chosen. For large σ, this measure becomes less reliable due to shrinking of the network, i.e. $\forall_{\boldsymbol{r}} \boldsymbol{w_r} \to \langle \boldsymbol{v} \rangle$. Bottom: Unstable ε detected from the crossproduct measure, eq. (3), with threshold of 1 defect per 100 iterations. The ε value depends on the data distribution, but the qualitative behaviour remains similar. In all cases, below a certain σ_{crit} of about 0.3, ε has to be decreased significantly. This independently reproduces [16], here we investigate also different ε values.

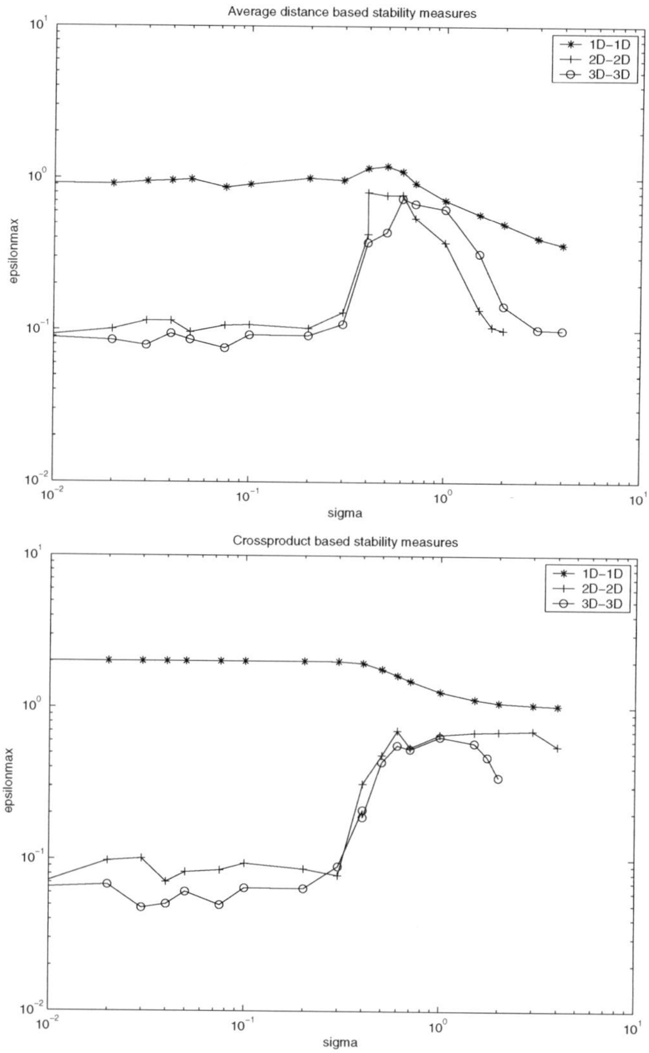

Fig. 6. Stability border dependence on input dimension (1D, 2D, 3D). The known 1D case is included for comparison. Top: Using the average length criterion (for $\sigma > 1$, the result can be misleading due to total shrinking of the network, see text). Bottom: Using the crossproduct detection for defects, similar results are obtained; for large σ instabilities are detected earlier.

Concave and convex learning: Stability of nonlinear learning. – The simulation results are given in Fig. 7: Clearly, a strong influence of the nonlinearity parameter K is observed. Especially one has to take care when decreasing σ, because for

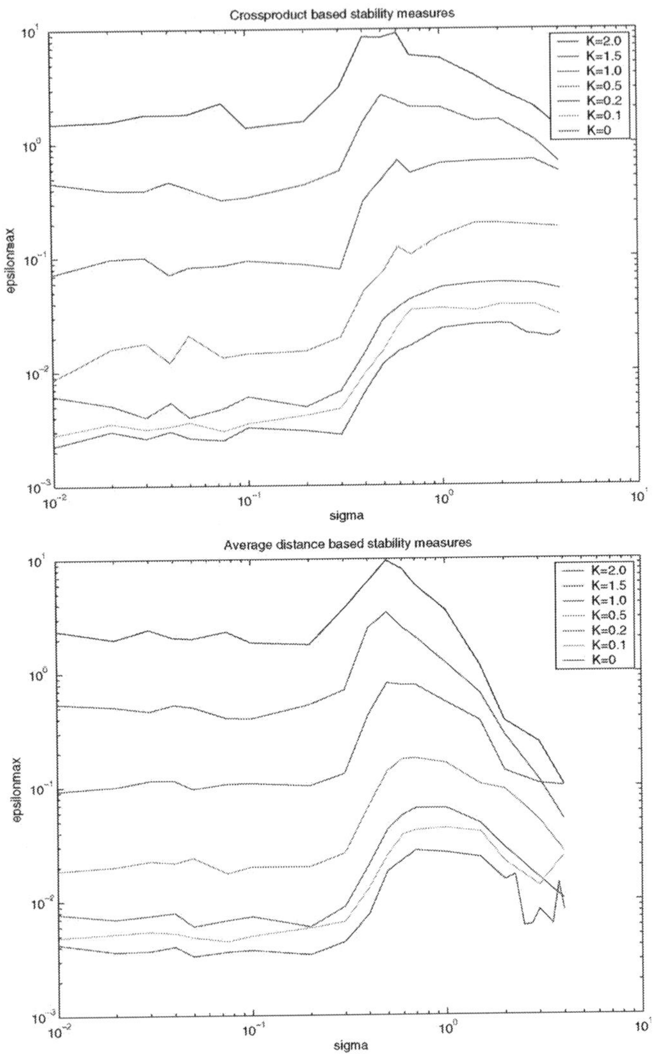

Fig. 7. Critical $\varepsilon_{\max}(\sigma)$ for different values of the nonlinearity parameter from $K = 2.0$ (top) to $K = 0$ (bottom). $K = 1$ corresponds to the SOM case.

too large ϵ the network becomes instable. For $K < 1$, much smaller values of ϵ are possible, thus considerably longer learning phases have to be taken into account compared to original SOM. For $K > 1$ the stability range becomes larger.

Discussion. – We have defined a standardized testbed for the stability analysis of SOM vector quantizers with serial pattern presentation, and compared the SOM with the recently introduced variants of concave and convex learning. The stability regions for different input distribution and dimension are of the same shape, thus qualitatively similar, but not coinciding exactly. The neighborhood width, but unfortunately also the input distribution affect the maximal stable learning rate. For the concave and convex learning, the exponent steering the nonlinear learning also crucially influences the learning rate. In all cases, a plateau for $\sigma \ll 1$ is found where the learning rate must be quite low compared to the intermediate range $0.3 \leq \sigma \leq 1$. As a too safe choice of the learning rate simply increases computational cost, an accurate knowledge of the stability range of neural vector quantizers is of direct relevance in many applications.

References

1. T. Kohonen. Self-Organized Formation of Toplogically Correct Feature Maps. Biological Cybernetics **43** (1982) 59-69
2. Y. Zheng and J. F. Greenleaf. The effect of concave and convex weight adjustments on self-organizing maps. IEEE Transactions on Neural Networks, **7** (1996) 87-96
3. T. Villmann and J. C. Claussen, Investigation of Magnification Control in Self-Organizing Maps and Neural Gas. Neural Computation **18** (2006) 449-469
4. T. Kohonen. Comparison of SOM point densities based on different criteria. Neural Computation **11** (1999) 2081-2095
5. T. Villmann, R. Der, M. Herrmann, and T. Martinetz. Topology Preservation in Self-Organizing Feature Maps: Exact Definition and Measurement. IEEE Transactions on Neural Networks, **8** (1997) 256-266
6. S. Zrehen. Analyzing Kohonen maps with geometry. In Gielen, S. and Kappen, B., editors, Proc. ICANN'93, p. 609-612, Springer, London (1993)
7. H. Ritter and K. Schulten. On the Stationary State of Kohonen's Self-Organizing Sensory Mapping. Biological Cybernetics **54** (1986) 99-106
8. H. Ritter and K. Schulten. Convergence Properties of Kohonen's Topology Conserving Maps: Fluctuations, Stability and Dimension Selection. Biological Cybernetics **60** (1988) 59-71
9. H. Ritter, T. Martinetz, and K. Schulten. Neural Computation and Self-Organizing Maps: An Introduction. Addison-Wesley (1992)
10. R. Linsker. How to generate maps by maximizing the mutual information between input and output signals. Neural Computation **1** (1989) 402-411
11. T. Heskes. Energy functions for self-organizing maps. In E. Oja and S. Kaski, editors, *Kohonen Maps*, p. 303-316. Elsevier, Amsterdam (1999)
12. E. Erwin, K. Obermayer, and K. Schulten. Self-organizing maps: Ordering, convergence properties and energy functions. Biol. Cyb. **67** (1992) 47–55
13. J. C. Claussen, Generalized Winner Relaxing Kohonen Feature Maps, Neural Computation **17** (2005) 996-1009
14. J. C. Claussen and T. Villmann, Magnification Control in Winner Relaxing Neural Gas, Neurocomputing **63** (2005) 125-137 2005.
15. T. M. Martinetz, S. G. Berkovich, and K. J. Schulten. 'Neural-gas' network for vector quantization and its application to time-series prediction. *IEEE Trans. on Neural Networks*, **4** (1993) 558–569
16. T. Villmann. PhD thesis, Leipzig, Verlag Harri Deutsch, Frankfurt (1998)

Alternatives to Parameter Selection for Kernel Methods

Alberto Muñoz[1], Isaac Martín de Diego[2], and Javier M. Moguerza[2]

[1] University Carlos III de Madrid, c/ Madrid 126, 28903 Getafe, Spain
alberto.munoz@uc3m.es
[2] University Rey Juan Carlos, c/ Tulipán s/n, 28933 Móstoles, Spain
{isaac.martin, javier.moguerza}@urjc.es

Abstract. In this paper we propose alternative methods to parameter selection techniques in order to build a kernel matrix for classification purposes using Support Vector Machines (SVMs). We describe several methods to build a unique kernel matrix from a collection of kernels built using a wide range of values for the unkown parameters. The proposed techniques have been successfully evaluated on a variety of artificial and real data sets. The new methods outperform the best individual kernel under consideration and they can be used as an alternative to the parameter selection problem in kernel methods.

1 Introduction

It is well known that the choice of kernel parameters is often critical for the good performance of Support Vector Machines (SVMs). Nevertheless, to find optimal values in terms of generalization performance for the kernel parameters is an open and hard to solve question. For instance, the effect of RBF kernels parameter within a SVM framework has been studied from a theoretical point of view [5]. Several practical proposals to choose the RBF kernel parameter have been made [14,7,2,13]. However, there is not a simple and unique technique to select the best set of parameters to build a kernel matrix. Our proposal is based on the combination of the different kernel matrices that arise with the use of a range of values for the unkown parameters. Combining kernels provides a solution that minimizes the effect of a bad parameter choice. An intuitive and usual approach to build this combination is to consider linear combinations of the matrices. This is the proposal in [6], which is based on the solution of a semi-definite programming problem to calculate the coefficients of the linear combination. The solution of this kind of optimization problem is computationally very expensive [17]. Recently, a simpler algorithm based on the same ideas for learning a linear combination of kernels has been developed [1]. The main difference is the way in which the weights within the semi-definite programming problem are found.

In this paper we propose several methods to build a kernel matrix from a collection of kernels generated from different values of the unkown parameters

S. Kollias et al. (Eds.): ICANN 2006, Part I, LNCS 4131, pp. 216–225, 2006.

in the kernel function. The functions involved in the proposed methods take advantage of class conditional probabilities and nearest neighbour techniques.

The paper is organized as follows. The general framework for the methods is presented in Section 2. The proposed methods are described in Section 3. The experimental setup and results on artificial and real data sets are described in Section 4. Section 5 concludes.

2 General Framework

As already mentioned, our proposal is based on the generation of a collection of kernel matrices using a wide range of values for the unkown kernel parameters. Once the collection has been built, we will combine the kernels in order to build a unique kernel. We will not focus on the generation step but on the combination step. Different ways to generate parameters, not only for kernels but for many other methods, can be consulted in [15] or in many simulation handbooks.

In order to combine the kernel matrices we make use of the concept of functional combination of kernel matrices. This concept was introduced originally in [10].

Let $K_1, K_2, ...K_M$ be a set of M input kernels defined on a data set X, and denote by K^* the desired output combination. Let y denote the label vector, where for simplicity $y_i \in \{-1, +1\}$ (the extension to the multiclass case is straighforward).

Consider the following (functional) weighted sum:

$$K^* = \sum_{m=1}^{M} W_m \otimes K_m \,, \tag{1}$$

where $W_m = [w_m(x_i, x_j)]$ is a matrix whose elements are nonlinear functions $w_m(x_i, x_j)$, with x_i and x_j data points in the sample, and '$\otimes$' denotes the element by element product between matrices (Hadamard product). We assume that $K_m(x_i, x_j) \in [0, 1] \quad \forall \quad i, j, m$ (otherwise the kernels can be scaled). Notice that if $w_m(x_i, x_j) = \mu_m$, where $\mu_m\,, m = 1, \ldots M$ are constants, then the method reduces to calculate a simple linear combination of matrices:

$$K^* = \sum_{m=1}^{M} \mu_m K_m \,. \tag{2}$$

Several methods have been suggested to learn the coefficients μ_m of the linear combination [1,6]. Thus, the formulation used in these papers is a particular case of the formula we use. For instance, if we take $\mu_m = \frac{1}{M}$, the average of the kernel matrices is obtained.

Regarding our proposals, consider the (i, j) element of the matrix K^* in (1):

$$K^*(x_i, x_j) = \sum_{m=1}^{M} w_m(x_i, x_j) K_m(x_i, x_j) \,. \tag{3}$$

This is the general formula of our approximation. In this way, we will generate a particular weight for each pair of elements under consideration.

An aspect that has to be treated before describing the methods is the fact that the kernel matrix arising from the combination has to be a positive semi-definite matrix. Since this can not be guaranteed in advance, we make use of some of the several solutions that have been proposed to solve this difficulty [12]. For instance, consider the spectral decomposition $K^* = Q\Lambda Q^T$, where Λ is a diagonal matrix containing (in decreasing order) the eigenvalues of K^*, and Q is the matrix of the corresponding eigenvectors. Assume that Λ has at least p positive eigenvalues. We can consider a p-dimensional representation by taking the first p columns of Q: $Q_p \Lambda_p Q_p^T$. We will refer to this technique as 'Positive Eigenvalue Transformation'. A computationally cheaper solution is to consider the definition of a new kernel matrix as K^{*2}. Notice that, in this case, the new kernel matrix is: $Q\Lambda^2 Q^T$. We call this method 'Square Eigenvalue Transformation'. In practice, there seems not to be a universally best method to solve this problem [11].

3 Some Specific Proposals

The next section is devoted to described a common aspect to the methods we will propose: The use of conditional class probabilities in order to build the weights $w_m(x_i, x_j)$ introduced in the previous section.

3.1 Conditional Class Probabilities

Consider the pair (x_i, y_i) and an unlabeled observation x_j. Given the observed value x_j, define $P(y_i|x_j)$ as the probability of x_j being in class y_i. If x_i and x_j belong to the same class this probability should be high. Unfortunately, this probability is unknown and has to be estimated. In our proposals, we will estimate it by:

$$P(y_i|x_j) = \frac{n_{ij}}{n}, \tag{4}$$

where n_{ij} is the number of the n-nearest neighbours of x_j belonging to class y_i.

Notice that each kernel induces a different type of neighborhood. In fact, there is an explicit relation between a kernel matrix and a distance matrix. For instance, consider a matrix K of inner products in an Euclidean space $\mathcal{F}$ (a kernel). Then $D^2 = ve^T + ev^T - 2K$ is a matrix of square Euclidean distances in $\mathcal{F}$ [4], where v is a vector made up of the diagonal elements of K. Hence, it is advisable to estimate this probability for each kernel representation, that is, for the kernel K_m we will estimate the conditional probabilities $P_m(y_i|x_j)$ using the induced distances matrix D_m^2. We will need the average of this conditional probabilities over the kernel matrices:

$$\bar{P}(y_i|x_j) = \frac{1}{M}\sum_{m=1}^{M} P_m(y_i|x_j), \tag{5}$$

and

$$\bar{\rho}(x_i, x_j) = \frac{\bar{P}(y_i|x_j) + \bar{P}(y_j|x_i)}{2}. \tag{6}$$

To estimate the conditional class probabilities, the appropriate size of the neighbourhood has to be determined. We propose a dynamic and automatic method: given two points x_i and x_j, we look for the first common neighbour. For each data point (x_i and x_j), the size k of the neighbourhood will be determined by the number of neighbours nearer than the common neighbour. To be more specific, let $R(x_i, n) = \{n\text{-}nearest\ neighbours\ of\ x_i\}$, then $k = \text{argmin}_n\{R(x_i, n) \cap R(x_j, n) \neq \emptyset\}$. Obviously, the size k of the neighbourhood depends on the particular pair of points under consideration.

At this point, we have the means to implement some particular proposals of combination methods.

3.2 The 'MaxMin' Method

The 'MaxMin' method (first used in [10]) produces a functional combination of two kernels, namely, the maximum and the minimum of the ordered sequence of kernels, being zero the weight assigned to the rest of the kernels. Consider the ordered sequence:

$$\min_{1 \leq m \leq M} K_m(x_i, x_j) = K_{[1]}(x_i, x_j) \leq \ldots \leq K_{[M]}(x_i, x_j) = \max_{1 \leq m \leq M} K_m(x_i, x_j) \,,$$

where the subscript $[\cdot]$ denotes the position induced by the order. This method builds each element of K^* using the formula:

$$K^*(x_i, x_j) = \bar{\rho}(x_i, x_j) K_{[M]}(x_i, x_j) + (1 - \bar{\rho}(x_i, x_j)) K_{[1]}(x_i, x_j) \,. \tag{7}$$

If x_i and x_j belong to the same class then the conditional class probabilities $\bar{\rho}(x_i, x_j)$ will be high and the method guarantees that $K^*(x_i, x_j)$ will be large. On the other hand, if x_i and x_j belong to different classes the conditional class probabilities $\bar{\rho}(x_i, x_j)$ will be low and the method will produce a value close to the minimum of the kernels. In the following, this method will be refered as **MaxMin**.

For the particular case of $K_1, \ldots, K_M$ being a collection of RBF kernels:

$$K_m(x_i, x_j) = e^{-\|x_i - x_j\|^2/(2*\sigma_m^2)} \,, \quad m = 1, \ldots, M \,, \tag{8}$$

with different σ_m values, then, the minimum and the maximum for each pair of points (x_i, x_j) correspond respectively to the highest and the lowest value of the collection of σ_m parameters.

3.3 The Percentile-in Method

Next, we propose a method whose assignment of positive weights $w_m(x_i, x_j)$ is based on the order induced by the kernels. The method builds each element of K^* using the following formulae:

$$K^*(x_i, x_j) = K_{\lceil \bar{\rho}(x_i, x_j) M \rceil} \,, \tag{9}$$

where the subscript $\lceil \cdot \rceil$ denotes the upper rounding of the argument.

We denote this method by **'Percentile-in'** method [10] . If the class probability $\bar{\rho}(x_i, x_j)$ is high, we can expect a high similarity between x_i and x_j and the method will guarantee a high $K^*(x_i, x_j)$. If the class probability $\bar{\rho}(x_i, x_j)$ is low, $K^*(x_i, x_j)$ will be also low.

3.4 The Percentile-out Method

As in the previous method, the last proposed technique is based on the order induced by the kernels. However, in this case two kernel values are considered. Each element of the K^* matrix is built as follows:

$$K^*(x_i, x_j) = \frac{1}{2}\left(K_{\lceil \bar{P}(y_i|x_j)M \rceil} + K_{\lceil \bar{P}(y_j|x_i)M \rceil}\right) , \quad (10)$$

where the subscript $\lceil \cdot \rceil$ denotes the upper rounding of the argument. We denote this method by **'Percentile-out'** method [10] .

If the conditional class probabilities $\bar{P}(y_i|x_j)$ and $\bar{P}(y_j|x_i)$ are high, we can expect a high similarity between x_i and x_j and both methods will guarantee a high $K^*(x_i, x_j)$. If the conditional class probabilities $\bar{P}(y_i|x_j)$ and $\bar{P}(y_j|x_i)$ are both low, $K^*(x_i, x_j)$ will be also low.

This method can be viewed as a smoothed MaxMin method. As in the MaxMin method, two kernels are considered for each pair of points in the sample. The conditional probabilities are used in order to obtain values not so extreme as the maximum and the minimum. Only if $\bar{P}(y_i|x_j) = \bar{P}(y_j|x_i)$ the Percentile-out method reduces to the Percentile-in method. Otherwise, this method takes into account the difference between the proportion of neighbours of x_j belonging to class y_i and the proportion of neighbours of x_i belonging to class y_j.

4 Experiments

To test the performance of the proposed methods, an SVM (with the upper bound on the dual variables fixed to 1) has been trained on several real data sets using the kernel matrix K^* constructed.

In order to classify a non-labelled data point x, $K^*(x, i)$ has to be evaluated. We calculate two different values for $K^*(x, i)$, the first one assumming x belongs to class $+1$ and the second assumming x belongs to class -1. For each assumption, we compute the distance between x and the SVM hyperplane and assign x to the class corresponding to the largest distance from the hyperplane.

Since our technique is based on the calculation of the nearest neighbours, we have compared the proposed methods with the k-Nearest Neighbour classification (k-NN, using the optimal value $k = l^{\frac{4}{d+4}}$ [16]). We have compared our methods with the linear combination of kernels method (LC) [1]. In order to evaluate the improvement provided by our proposals, we have carried out a Wilcoxon signed-rank test (see for instance [8]). This nonparametric test is used to compare the median of the results for different runs of each method. So, the null hypothesis of the test is that our methods do not improve the existing combination techniques.

4.1 Artificial Data Sets

4.2 Two Areas with Different Scattering Matrices

This data set, shown in Figure 1, is made up of 400 points in $\mathbb{R}^2$. Visually there are two areas of points (80% of the sample is in area A_1 and 20% is in area A_2). Each area A_i corresponds to a circle with radio σ_i. Here $\sigma_1 = 10^{-2}\sigma_2$, with $\sigma_2 = 1$. The first group center is $(0, 1)$ and the second group center is $(1, 1)$. Nevertheless, the areas do not coincide with the classes $\{-1, +1\}$ that are to be learned. Half of the points in each class belongs to aread A_1, and the other half to area A_2. Within each area, the classes are linearly separable. Therefore the only way to built a classifier for this data set is to take into account the area each point belongs to. We use 50% of the data for training and 50% for testing.

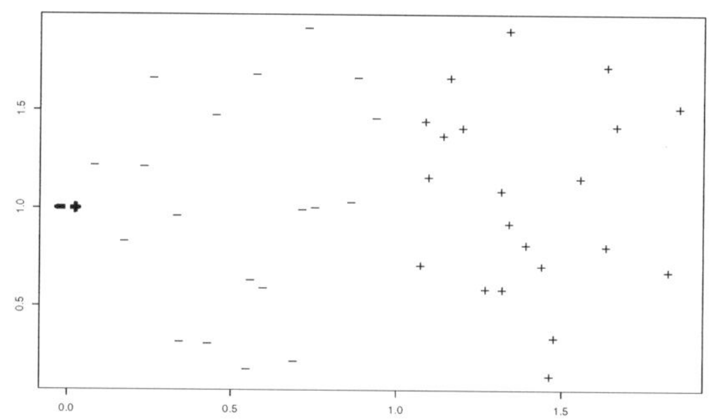

Fig. 1. Two areas with different scattering matrices. The first area center is $(0, 1)$ and the second area center is $(1, 1)$. The areas do not coincide with the classes $\{-1, +1\}$.

Let $\{K_1, \ldots, K_5\}$ be a set of five RBF kernels with parameters σ =0.5, 2.5, 5, 7.5 and 10 respectively. In order to get a positive semi-definite kernel matrix K^*, we use the Square Eigenvalue Transformation technique described in Section 2.

Table 1 shows the performance of our proposals for this data set. The results have been averaged over 10 runs. Given the geometry of the data, it is clear that is not possible to choose a unique best σ for the whole data set. As σ grows, the test error increases for the data contained in area A_1, and decreases within area A_2. The LC method seems to work fairly. Nevertheless, the MaxMin method achieves the best results on classification. Regarding the Wilcoxon signed-rank test for the comparison of our methods with the LC technique, the p-value is smaller than 0.001 for the MaxMin method.

4.3 The Three Spheres Example

The data set contains 300 data points in $\mathbb{R}^4$. We generate three different groups of observations (100 observations per group) corresponding to three spheres in $\mathbb{R}^3$. The center is the same for the three spheres $(0, 0, 0)$ and the radii are

Table 1. Percentage of missclassified data and percentage of support vectors for the two different scattering data set: A_1 stands for the less scaterring group, A_2 stands for the most dispersive one

	Train Error			**Test Error**			**Support Vectors**		
Method	Total	A_1	A_2	Total	A_1	A_2	Total	A_1	A_2
RBF$_{\sigma=0.5}$	2.1	2.6	0.0	13.5	4.1	51.0	39.6	25.1	97.5
RBF$_{\sigma=2.5}$	4.8	6.0	0.0	13.5	6.5	41.5	62.2	53.4	97.5
RBF$_{\sigma=5}$	6.6	8.2	0.0	14.0	10.1	29.5	82.8	79.2	97.0
RBF$_{\sigma=7.5}$	16.0	19.9	0.5	22.2	22.6	20.5	94.6	94.2	96.0
RBF$_{\sigma=10}$	30.7	38.2	0.5	37.3	44.1	10.0	94.2	95.4	89.5
MaxMin	0.3	0.4	0.0	4.9	0.9	21.0	27.7	9.6	100.0
Percentile-in	4.2	5.1	0.5	9.0	3.1	32.5	35.9	20.1	99.0
Percentile-out	0.7	0.9	0.0	7.7	1.1	34.0	29.0	11.4	99.5
***k*-NN**	14.5	3.5	58.5	15.5	3.5	63.5	—	—	—
LC	1.6	2.0	0.0	8.1	2.5	29.5	46.6	33.2	100.0

different (0.1,0.3, and 1 respectively). The 100 points on the sphere with radio equals to 0.3 belong to class +1, and the other 200 points belong to class −1. Finally a fourth random additional dimension is added to the data set, following a Normal distribution (centered in 0 and with 10^{-2} as standard deviation). We use 50% of the data for training and 50% for testing.

Let $\{K_1, \ldots, K_5\}$ be a set of polynomial kernels, $K(x, z) = (1 + x^T z)^d$, with parameters d = 1, 2, 3, 4, 5 respectively. In order to scale the matrices, we use the following normalization [3]: $K(x, z) = K(x, z)/(\sqrt{K(x, x)}\sqrt{K(y, y)})$. We use the Square Eigenvalue Transformation method to solve the problem of building a positive semi-definite matrix. Table 2 shows the performance of the proposed methods when combining these kernel matrices. The results have been averaged over 10 runs.

The MaxMin and Percentile methods show the best overall performance. All our combination methods provide better results than the best polynomial kernel, using usually significantly less support vectors. Regarding the Wilcoxon signed-rank test for the comparison of our methods with the LC technique, the p-values are smaller than 0.001 for all our methods. So the improvement obtained by the use of our proposals is statistically significant. Notice that the results using any of the single kernels are very poor, while the results obtained using any of our combination methods are significatively better. This example also shows that using a linear combination of the kernels may not be a good choice.

4.4 A Real Data Set

In this section we have dealt with a database from the UCI Machine Learning Repository: the Breast Cancer data set [9]. The data set consists of 683 observations with 9 features each. Let $\{K_1, \ldots, K_{12}\}$ be a set of RBF kernels with

Table 2. Percentage of missclassified data, sensitivity (Sens.), specificity (Spec.) and percentage of support vectors for the three spheres data set. Standard deviations in brackets.

Method	Train			Test			Support Vectors
	Error	Sens.	Spec.	Error	Sens.	Spec.	
Polynomial$_{d=1}$	31.8 (2.5)	0.000	1.000	34.9 (2.5)	0.000	1.000	69.5 (5.0)
Polynomial$_{d=2}$	31.8 (2.5)	0.000	1.000	34.9 (2.5)	0.000	1.000	75.7 (7.9)
Polynomial$_{d=3}$	30.6 (1.8)	0.200	0.909	36.1 (1.8)	0.200	0.891	71.7 (5.6)
Polynomial$_{d=4}$	23.7 (7.3)	0.377	0.893	31.7 (7.0)	0.293	0.816	69.5 (4.6)
Polynomial$_{d=5}$	14.7 (2.5)	0.541	0.958	24.1 (7.0)	0.436	0.798	69.5 (4.6)
MaxMin	4.0 (0.8)	0.964	0.958	5.5 (2.5)	0.921	0.958	8.4 (1.2)
Percentile-in	5.5 (1.4)	0.907	0.963	6.9 (3.2)	0.864	0.967	7.6 (1.4)
Percentile-out	4.5 (1.1)	0.941	0.959	6.9 (2.9)	0.886	0.957	8.5 (1.5)
***k*-NN**	10.9 (2.4)	0.795	0.934	15.7 (4.2)	0.725	0.904	— (—)
LC	31.8 (2.5)	0.000	1.000	34.9 (2.5)	0.000	1.000	71.5 (3.9)

parameters σ =0.1, 1, 10, 20, 30, 40, 50, 60, 70, 80, 90, 100 respectively. We use the Positive Eigenvalue Transformation to solve the problem of building a positive semi-definite matrix.

Table 3 shows the performance of the proposed methods when combining all these kernel matrices. Again, the results have been averaged over 10 runs. The MaxMin method, the Percentile-in method, and the Percentile-out method improve the best RBF kernel under consideration (test errors of 2.8% for the three methods vs. 3.1%). The results provided by all the combination methods are not degraded by the inclusion of kernels with a bad generalization performance. Our methods clearly outperform the SVM classifier using an RBF kernel with $\sigma = \sqrt{d}/2$, where d is the data dimension (see [14] for details). Regarding the

Table 3. Percentage of missclassified data, sensitivity (Sens.), specificity (Spec.) and percentage of support vectors for the cancer data using a battery of RBF kernels. Standard deviations in brackets.

Method	Train			Test			Support Vectors
	Error	Sens.	Spec.	Error	Sens.	Spec.	
Best RBF	2.3 (0.3)	0.979	0.976	3.1 (1.6)	0.976	0.966	13.6 (1.3)
Worst RBF	0.0 (0.0)	1.000	1.000	24.7 (2.3)	1.000	0.627	74.0 (2.4)
MaxMin	0.1 (0.1)	0.999	0.998	2.8 (1.6)	0.963	0.975	14.2 (1.5)
Percentile-in	2.0 (0.4)	0.982	0.979	2.8 (2.8)	0.975	0.969	7.8 (0.7)
Percentile-out	0.2 (0.1)	0.999	0.997	2.8 (1.7)	0.964	0.975	19.2 (4.5)
***k*-NN**	2.7 (0.5)	0.961	0.980	3.4 (1.5)	0.949	0.974	— (—)
LC	0.0 (0.0)	1.000	1.000	3.2 (1.6)	0.976	0.964	41.5 (4.4)
SVM	0.1 (0.1)	1.000	0.999	4.2 (1.4)	0.989	0.942	49.2 (1.0)

Wilcoxon signed-rank test for the comparison of our methods with the SVM technique, the p-values are smaller than 0.05 for the MaxMin and the Percentile-out methods, and smaller than 0.1 for the Percentile-in method. Again, the improvement obtained by the use of our proposals is statistically significant.

5 Conclusions

In this paper, we have proposed alternative methods to parameter selection techniques within a Kernel Methods framework. The proposed techniques are especially usefull when does not exist an overall and unique best parameter. The suggested kernel combination methods compare favorably to the use of one of the kernels involved in the combination. We have also shown that a linear combination of the kernels may not be a good choice. Further research will focus on the theoretical properties of the methods and extensions. In particular, the methods shown in this paper do not take full advantage of the concept of the functional weighted sum described in (1): we think that there is room for improvement and more sophisticated ways for the calculus of the weights may be designed. The method could be generalized by using more than two kernels, but then, a parameter to control the relative importance of the kernels will be needed [10].

Acknowledgments. This work was partially supported by Spanish grants TIC-2003-05982-C05-05 (MCyT) and PPR-2004-05 (URJC).

References

1. O. Bousquet and D.J.L. Herrmann. On the complexity of learning the kernel matrix. In S. Becker, S. Thurn, and K. Obermayer, editors, *Advances in Neural Information Processing Systems, 15*, pages 415–422. Cambridge, MA: The MIT Press, 2003.
2. O. Chapelle, V. Vapnik, O. Bousquet, and S. Mukherjee. Choosing multiple parameters for support vector machines. *Machine Learning*, 46(1/3):131–159, 2002.
3. N. Cristianini, J. Shawe-Taylor, A. Elisseeff, and J. Kandola. *On Kernel-Target Alignment*, pages 367–373. Cambridge, MA: MIT Press, 2002.
4. J. C. Gower and P. Legendre. Metric and euclidean properties of dissimilarity coefficients. *Journal of Classification*, 3:5–48, 1986.
5. S.S. Keerthi and C. Lin. Asymptotic behaviors of support vector machines with gaussian kernel. *Neural Computation*, 15:1667–1689, 2003.
6. G. R. G. Lanckriet, N. Cristianini, P. Barlett, L. *El Ghaoui*, and M. I. Jordan. Learning the kernel matrix with semi-definite programming. *Journal of Machine Learning Research*, 5(Jan):27–72, 2004.
7. J.-H. Lee and C.-J. Lin. Automatic model selection for support vector machines. Technical report, National Taiwan University, 2000.
8. E. L. Lehmann. *NonParametrics: Statistical Methods Based on Ranks.* McGraw-Hill, 1975.
9. O. L. Mangasarian and W. H. Wolberg. Cancer diagnosis via linear programming. *SIAM News*, 23(5):1–18, 1990.

10. J. M. Moguerza, I. Martín de Diego, and A. Muñoz. Improving support vector classificacion via the combination of multiple sources of information. In *Proc. of the IAPR International Workshops SSPR 2004 and SPR 2004, Vol. 3138 of LNCS*, pages 592–600. Berlin: Springer, 2004.
11. E. Pękalska, R. P. W. Duin, S. Günter, and H. Bunke. On not making dissimilarities euclidean. In *Proc. of the IAPR International Workshops SSPR 2004 and SPR 2004, Vol. 3138 of LNCS*, pages 1145–1154. Berlin: Springer, 2004.
12. E. Pękalska, P. Paclík, and R. P. W. Duin. A generalized kernel approach to dissimilarity-based classification. *Journal of Machine Learning Research, Special Issue on Kernel Methods*, 2(12):175–211, 2001.
13. K. Schittkowski. Optimal parameter selection in support vector machines. *Journal of Industrial and Management Optimization*, 1(4):465–476, 2005.
14. B. Schölkopf, S. Mika, C. J.C. Burges, K.-R. Müller P. Knirsch, G. Rätsch, and A. J. Smola. Input space vs. feature space in kernel-based methods. *IEEE Transactions on Neural Networks*, 1999.
15. J. Shawe-Taylor and N. Cristianini. *Kernel Methods for Pattern Analysis*. Cambridge University Press, 2004.
16. B. Silverman. *Density Estimation for Statistics and Data Analysis*. Chapman and Hall, London, 1986.
17. L. Vandenberghe and S. Boyd. Semidefinite programming. *SIAM Review*, 38(1):49–95, 1996.

Faster Learning with Overlapping Neural Assemblies

Andrei Kursin[1], Dušan Húsek[2], and Roman Neruda[3]

[1] Kharkiv Polytechnic Institute, Information Systems Department,
21 Frunze st., 61002, Kharkiv, Ukraine
ak@kpi.kharkov.ua

[2] Institute of Computer Science, Neural Networks and Nonlinear Systems Department
2 Pod Vodarenskou veži st., 18207, Prague, Czech Republic
dusan@cs.cas.cz

[3] Institute of Computer Science, Neural Theoretical Computer Science Department
2 Pod Vodarenskou veži st., 18207, Prague, Czech Republic
roman@cs.cas.cz

Abstract. Cell assemblies in neural network are often assumed as overlapping, i.e. a neuron may belong to several of them simultaneously. We argue that network structures with overlapping cell assemblies can exhibit faster learning comparing to non-overlapping ones. In such structures newly trained assemblies take advantage of their overlaps with the already trained neighbors. The assemblies learned in such manner nevertheless preserve the ability for subsequent separate firing. We discuss the implications it may have for intensification of neural network training methods and we also propose to view this learning speed-up in a broader context of inter-assembly cooperation useful for modeling concept formation in human thinking.

1 Introduction

Neural assembly is among the main concepts of connection science. It describes a set of cells which are distinguished from the rest of neuron mass due to their higher connectivity. The chief part in assembly formation is devoted to Hebbian learning process which strengthens the links among neurons that fire simultaneously.

It goes hand in hand with the very definition of neural assembly [1], [2] that the assemblies should be *overlapping*, i.e. one neuron may belong to several of them. It is evident that overlapping structure of assemblies increases storage capacity of a network but on the other hand it rises a problem of organizing connection matrix in a way to avoid palimpsest effects and to ensure separate firing of assemblies in spite of the overlaps. So far the studies of overlapping neural assemblies have concerned primarily this problem [3], [4], [5]. We want to draw the reader's attention to the fact that overlapping assembly structures may have other interesting properties. In particular, we argue here that overlapping assemblies may exhibit faster learning comparing to non-overlapping ones. We describe simulations that support our assumption and, in concluding section, we discuss what consequences such assembly cooperation may have either for neural network training methods or for modeling cognitive functions like concept formation.

S. Kollias et al. (Eds.): ICANN 2006, Part I, LNCS 4131, pp. 226–233, 2006.

2 Overview

To verify our assumption we chose to study a simple Hopfield-type neural network [6], [7], [8]. From time to time it is exposed to certain input and when input is released, the network converges to some attractor state. The configuration of the state is determined by the input pattern and the assembly structure of the network. In fact, the network tends to select a neural assembly, which is the closest to the input pattern. Assemblies are distinguished by denser connectivity of their neurons while the network as a whole is sparsely connected. Each assembly may be either trained or not. In the trained state, the most of the links among the assembly neurons are potentiated. An untrained assembly has only initially weak inter-neuron connections.

We studied the process of potentiation of intra-assembly links and found that under certain conditions, the assemblies, overlapping with already trained ones, learn their input patterns faster than assemblies without such overlaps. This can be explained by the fact that portion of their internal linkage is already potentiated. We consider this fact as the simplest form of inter-assembly interactions we hope to simulate later.

3 Network Model

3.1 Neurons

Our network consists of N excitatory neurons and an inhibitory subsystem. Each excitatory neuron i (i=1,...,N) is a simple 2-state neuron whose activation at time t is denoted as $A_i(t)$, $A_i(t) \in \{0, 1\}$. The activation is calculated as a function of the neuron input:

$$A_i(t) = \Theta(I_i(t)), \qquad (1)$$

where function Θ returns 1 for positive values of the argument and 0 otherwise.

Neuron input is a sum of the signals incoming from other excitatory neurons, inhibitory signal T_i, and external input H_i if any:

$$\tau_n \dot{I}_i(t) = -I_i + \sum_{j=1}^{k} J_{ij} A_j - T_i + H_i, \qquad (2)$$

where J_{ij} is efficacy of a link from neuron j to neuron i, k is the number of input links per neuron and τ_n is an integration time constant for excitatory neurons.

3.2 Inhibition

The inhibitory subsystem can be viewed as a single giant neuron that takes the sum of all excitatory neuron activations $I = \sum_{i=1}^{N} A_i$ as input and calculates inhibitory signal according to the transduction function

$$\varphi(I) = \begin{cases} \varphi_{\min} & I < I_c \\ \eta I & I \geq I_c \end{cases},$$

where and $\varphi_{\min} > 0$, $\eta \geq 1$ are constant values. The inhibitory signal $T_i = \varphi(I)$ is equal for every excitatory neuron. The goal of the inhibitory subsystem is to keep the number of active excitatory neurons in certain limits. Inhibitory signal is calculated each τ_h time units.

3.3 Input

The network input is described by a set E which comprises all neurons of the network activated at this time. The same value H^{ext} of external input current ensuring high activation probability is injected to all neurons in E while for all other neurons $H_i=0$ in (2). The input set E can be described as follows:

$$E = (A_q \setminus C) \cup B, \tag{3}$$

where A_q is a set of neurons belonging assemble q while C and B are the sets providing input noise. $C \subset A_q$ contains the neurons from A_q absent in current pattern and B forms "added noise" ($B \cap A_q = \emptyset$).

3.4 Link Modification Rule

As was stated above, our network is sparsely connected. Each neuron receives input links from $k= rN$ neurons, where $r \ll 1$. In the described simulations r was about 10%. A link is characterized by its conductivity J, which can be either modifiable or constant depending on the type of simulation. We make all intra-assembly links constant and potentiated when we test discriminating ability of overlapping assemblies. In learning experiments all links are modifiable.

The links are modified according to Hebbian principle, i.e. a link connecting simultaneously firing neuron is strengthened, and a link connecting a firing and a silent neuron is depressed. We use stochastic link modification rule proposed in [7], [8]. It is notable because it supports either long term potentiation (LTP) or long term depression (LTD) and has analog short term dynamics producing short term memory in attractors. The rule provides robust learning in our experiments though we believe that the main experiment results, i.e. inter-assembly interactions, may be reproduced with other learning rules as well.

A link, according to this model, has two stable conductivity values J_0 and J_1 for LTD and LTP states correspondingly. Suppose, a link is in LTD state and both neurons are active, then current conductivity gradually increases up to threshold w_{ij}, which fluctuates in some boundaries J_0' and J_1', $J_0 < J_0' < J_1' < J_1$. If the threshold is reached, the conductivity jumps to another stable value J_1. If the threshold is not reached, the conductivity quickly returns to the previous stable value in the absence of activation in the connected neurons. Depression of an LTP link occurs reversely. The rule can be formalized as follows:

$$\tau_c \dot{J}_{ij}(t) = -J_{ij}(t) + J_0 + c_{ij}(t) + (J_1 - J_0)\Theta(J_{ij}(t) - w_{ij}(t)) . \quad (4)$$

It is an integrator with time constant τ_c. The term $c_{ij}(t)$ represents Hebbian learning source; it is specified in terms of mean activation of the two neurons:

$$c_{ij}(t) = \lambda_+ \overline{A}_i(t)\overline{A}_j(t) - \lambda_-[\overline{A}_i(t) + \overline{A}_j(t)] ,$$

where $\overline{A}_i$ is the mean activation of neuron i. λ_+ and λ_- are constant coefficients selected so that transition between the stable states occur only when both values $\overline{A}$ are high or one is high and one is low. w_{ij} is the fluctuating threshold, Θ is the same as in (1) and the whole last term of (4) is the refresh source that indefinitely keeps J_i at one of the stable values in absence of activation. τ_c is taken to be sufficiently long to ensure slow stable learning. Mean activation of neurons is calculated as

$$\overline{A}_i(t) = \frac{\overline{A}_i(t - \tau_n) + A_i(t)}{2} .$$

When a neuron is active for several time steps, $\overline{A}_i$ reaches 1. In inactive state $\overline{A}_i$ quickly approaches zero.

3.5 Connection Matrix

One can study overlapping neural assemblies in several ways. For example, it is possible to choose a fully connected network and study how assemblies are formed according to correlations in input data [4], [5]. Or one may arrange a set of "innate" overlapping assemblies and train their connections from the input layer of the network [3]. We chose a structure that would probably help us to test our assumption. It is namely a sparsely connected network where cell assemblies are distinguished by denser connectivity of their neurons. To ensure existence of cell assemblies in such network the connection matrix should be organized according to proximity principle, i.e. when the probability of two neurons to be connected depends on the distance between them. It is about 1 for nearby neurons and gradually comes to zero as the distance increases. We used a square metrics where neurons are located at cross-sections of a square grid. Both dimensions of the network are assumed cycled to avoid undesirable edge effects. The assembly structure in such network is determined according to structured principle formulated in [3]: "The *minint* (minimum internal connectivity) of a set of neurons is the minimum number of innate links that any neuron in the set receives from other neurons in the set. The *maxext* (maximum external connectivity) of a set of neurons is the maximum number of innate links that any neuron outside the set receives from neurons in the set. A web [neural assembly] is a set of neurons whose minint is greater than its maxext." It was reported that there do exist overlapping assemblies in such networks and their number is usually about the number of neurons in the network [3]. Actually the number is lower if we take into account activation dynamics [2] so that each portion of an assembly can ignite the whole or be necessary for firing of the whole. Otherwise, the set of assemblies, distinguished purely structurally, contains also weakly coupled unions of smaller assemblies.

Actually, reaching maximum number of assemblies was not our purpose here, and we adopted an approximation of proximity principle to simplify control over the network during experiments. We designed our assemblies to occupy certain simple-form geometric areas on network "surface". Each neuron gets input links from all neurons of the areas to which it belongs, excluding itself. The remainder of its connections is randomly distributed over the rest of the network.

Each arrangement of assemblies achieved in such way was tested first for discrimination properties, i.e. the ability of each assembly to fire independently in the fully trained state. We arranged the connection matrix with fully potentiated intra-assembly links and tested if assemblies can fire independently in response to corresponding data. The matrix versions with insufficient discrimination characteristics did not participate in training tests.

4 Simulations and Results

The simulations were performed with a network of $N = 1024$ neurons. Mean assembly size $M = 16$. The number of input links per neuron $k = 96$. Inhibition parameters were $I_c = 14$ — 16, $\varphi_{\min} = 2$ and $\eta = 3$. $H^{ext} = 16$. Learning parameters had the following values: $J_0 = 0$, $J_1 = 1$, $J'_0 = 0.4$, $J'_1 = 0.6$, $\lambda_+ = 1.23$ and $\lambda_- = 0.41$. τ_h was taken as a minimum time unit and $\tau_n = 32\tau_h$, which means for discretely calculated network that between two successive updates of inhibitory signal T, activation states of $N/32$ randomly chosen neurons are recalculated.

During simulation, input pattern created as specified above were presented to the network for time period $t_p = 30\tau_n$, then input current was removed and the network was allowed to move to an attractor state during delay period $t_d = 60\tau_n$ whose length was chosen to be sufficient to reach an attractor in any experimental situation. In a fully trained network the attractors usually coincide with corresponding assemblies. During learning tests, the situation is not always the same. The coincidence between the reached attractor and the intended assembly is calculated as a portion of the assembly neurons present in the attractor. This number L was used as a measure of learning. As more intra-assembly links become potentiated, this number approaches 100% and remains close to this in the trained network.

The training sets contained 4-8 assemblies each. Actually, the training set sizes are not of much importance here since assemblies in a set are uncorrelated and learn independently. Sets of any size would be learned in about the same number of cycles.

Training experiments were performed according to the following scheme. The preparation stage started from unlearned connection matrix ($A_{ij} = J_0$ for every i, j). A set X of assemblies in the network were trained using corresponding sequence of input patterns. Each member of the sequence was built on the basis of certain assembly from the set X according to (3). Assemblies in X did not overlap and input patterns were uncorrelated. The sequence was repeated a number of times (S), every time with different random noise portions till L reached 100% for every assembly in the set.

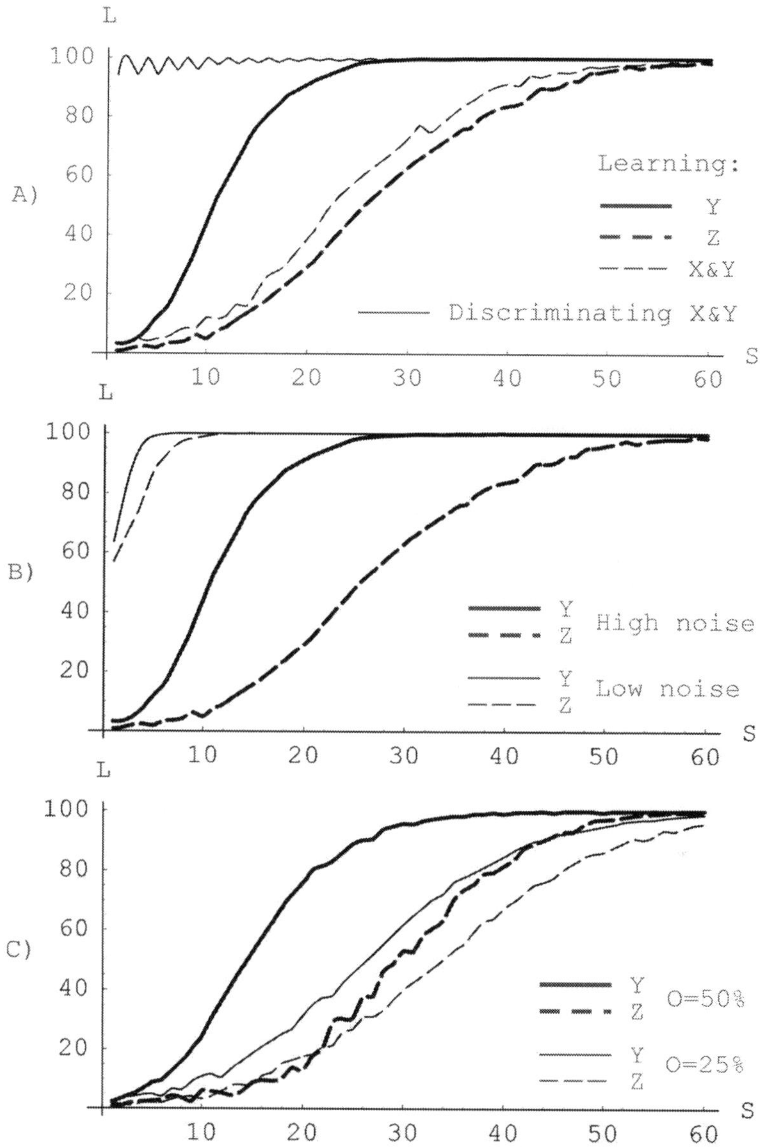

Fig. 1. Assembly-attractor matching measure L (%) versus number of input cycles S. (A) illustrates learning different pattern sets. Thin solid line corresponds to the test for X and Y discriminability after learning both sets. $|C| = 0.0625M$, $|B| = M$, overlap size O = 50%. (B) shows influence of noise: thick lines correspond to $|B| = M$, $|C| = 0.125M$, thin – to $|B| = 0.5M$, $|C| = 0.0625M$; O = 50%. (C) illustrates influence of overlap size; $|C| = 0.25M$, $|B| = M$. The data were averaged over 4–8 assemblies and 50 trials for each test.

Then sets Y and Z of assemblies were trained in the same manner in separate copies of the network obtained on the preparation stage. Members of these sets don't overlap with each other but every assembly in Y overlaps with one or more assemblies in X while assemblies of Z don't. It was found that set Y is learned faster than Z, especially on early stages. The data describing learning progress are presented in Fig. 1.

After the training stage, the networks were tested for sufficient discriminability between sets X and Y and showed good responses for either set of patterns (the thin solid line in the Fig. 1A).

A copy of each network obtained after the preparation stage was also exposed to an input sequence comprising patterns corresponding to members of both X and Y. This was done to test a network's ability to learn set Y while pertaining memory of set X. The learning rates here were slower but usually still higher than the rates for Z (the thin dashed line in Fig. 1A). However this test depends much on the ratio between quantities of X and Y members in the training set. If assemblies from X prevail, the Y learning curve after $L = 40$—50% comes close to or even below the curve for set Z.

Two factors were found to influence learning advantages of Y assemblies. First, they more significantly win in "difficult conditions", i.e. when there is sufficient noise (especially "added noise", $| B | > 0.75\ M$). For illustration see Fig. 1B. We consider it positive since sufficient noise should be expected in real tasks. Tests were performed for $| C |$ between 0 and $0.25M$, and $| B |$ between 0 and M.

Second, the overlap of assemblies should not be so great that it tends to frequently ignite a previously learned X assembly, otherwise learning abilities of assemblies in Y would be very low. Below this ignition level, increasing size of overlaps facilitates learning of set Y. Fig. 1C presents test results for overlap sizes (measured as a portion of M) O=25% and O=50%.

5 Discussion

Faster learning rates for the sets of overlapping assemblies observed in the experiments clearly follow from the ability of such assemblies to benefit from the potentiated links that they already have in their structure due to the overlaps with trained assemblies. At the same time such overlaps don't slow down learning much when the network is exposed to the sequences containing either old or new input patterns. This may be only one from the range of interesting properties that overlapping assembly structures may exhibit. But we want to draw your attention to certain consequences the results may have.

In other publications [9, 10] we argue that neural networks of certain architecture are capable of advanced type of learning resembling metaphorizing abilities of human thinking [11], i.e. when a novel concept is formed by analogy and on the basis of some already known concept. Such phenomenon is ubiquitous in human mental practice. Its cognitive gain is evident: it provides faster and better learning than purely inductive method since learning occurs not through exploration of novel concept from scratch but rather through looking for properties of an old concept in the new one.

In terms of neural networks, such learning is characteristic for the process in which an old trained assembly "helps" a new one to learn a novel piece of input data. This help can be provided only in two ways: through overlapping of neuron sets belonging to the assemblies, or via associative links between the assemblies. It seems that the former factor is more powerful since associative links of previously dormant neurons belonging to the freshly recruited assembly are probably too weak to ensure such "help". In this paper we showed that the "help" through assembly overlaps exists.

On the other hand this research has interesting implications for the field of artificial neural network training. Usage of previously obtained knowledge (when learning a new thing benefits from already knowing something similar) is clearly an effective way for intensification of training methods. This evident feature of natural neural networks is still not sufficiently accounted for in training their artificial analogs. We showed here that neural networks with overlapping assemblies have necessary properties to introduce this feature into learning. Further research may consist in application of the proposed principle to some practical task, e.g. categorization or information retrieval, like in [12].

Acknowledgements

This work is supported by INTAS grant Nr. 03-55-1661 and partly by the Institutional Research Plan AVOZ10300504 "Computer Science for the Information Society: Models, Algorithms, Applications" and grant IET100300414.

References

1. Hebb, D. O.: The Organization of Behavior. A Neuropsychological Theory. New York: John Wiley (1949).
2. Palm, G.: Neural Assemblies. Studies of Brain Function. Vol. VII. Springer, Berlin Heidelberg New York (1982).
3. Wickelgren, W.A.: Webs, Cell Assemblies, and Chunking in Neural Nets. In: Canadian Journal of Experimental psychology. Vol. 53. 1 (1999) 118-131
4. Strangman, G.: Detecting Synchronous Cell Assemblies with Limited Data and Overlapping Assemblies. In: Neural Computation. Vol. 9. (1997) 51-76
5. Huyk, C. R.: Overlapping Cell Assemblies from Correlators. In: Neurocomputing. Vol. 56. (2004) 435-439
6. Hopfield, J. Neural Nets and Physical Systems with Emergent Collective Computational Abilities. Proc. of the Nat. Academy of Sciences USA 79 (1982) 2554-2558
7. Amit, D. J., Brunel, N.: Learning Internal Representations in an Attractor Neural Network. In: Network, 6 (1995) 359-388.
8. Brunel, N.: Hebbian Learning of Context in Recurrent Neural Networks. In: Neural Computation, Vol. 8 (1996) 1677-1710
9. Kursin, A.: Neural Network: Input Anticipation May Lead To Advanced Adaptation Properties. In: Kaynak, O. et al. (eds.): Artificial Neural Networks and Neural Information Processing. Springer-Verlag, Berlin-Heidelberg, 779-785 (2003).
10. Kursin, A.: Self-Organization of Anticipatory Neural Network. In Scientific Proceedings Of Riga Technical University, series Computer Science, Information Technology and Management Science, Riga (2004) 51-59.
11. Lakoff, G., Johnson, M.: Metaphors We Live By. Univ. of Chicago Press, Chicago (1980).
12. Huyck, C., Orengo V.: Information Retrieval and Categorisation using a Cell Assembly Network. In: Neural Computing & Applications. Vol. 14. 4 (2005) 282-289.

Improved Storage Capacity of Hebbian Learning Attractor Neural Network with Bump Formations

Kostadin Koroutchev* and Elka Korutcheva**

EPS, Universidad Autónoma de Madrid,
Cantoblanco, Madrid, 28049, Spain
k.koroutchev@uam.es
Depto. de Física Fundamental,
Universidad Nacional de Educación a Distancia,
c/ Senda del Rey 9,28080 Madrid, Spain
elka@fisfun.uned.es

Abstract. Recently, bump formations in attractor neural networks with distance dependent connectivities has become of increasing interest for investigation in the field of biological and computational neuroscience. Although the distance dependent connectivity is common in biological networks, a common fault of these network is the sharp drop of the number of patterns p that can remembered, when the activity changes from global to bump-like, than effectively makes these networks low effective.

In this paper we represent a bump-based recursive network specially designed in order to increase its capacity, which is comparable with that of randomly connected sparse network. To this aim, we have tested a selection of 700 natural images on a network with $N = 64K$ neurons with connectivity per neuron C. We have shown that the capacity of the network is of order of C, that is in accordance with the capacity of highly diluted network. Preserving the number of connections per neuron, a non-trivial behavior with the radius of the connectivity has been observed. Our results show that the decrement of the capacity of the bumpy network can be avoided.

1 Introduction

Recently, the bump formations in recurrent neural networks have been analyzed in several investigations concerning linear-threshold units [1,2], binary units [3] and Small-World networks of Integrate and Fire neurons [4]. These bump formations represent geometrically localized patterns of activity and have a size proportional to the number of connections per neuron.

It has been shown that the localized retrieval is due to the short-range connectivity of the networks and it could explain the behavior in structures of biological relevance as the neocortex [5].

---

* Also with: ICCS, Bulgarian Academy of Science.

** Also with: ISSP, Bulgarian Academy of Science.

S. Kollias et al. (Eds.): ICANN 2006, Part I, LNCS 4131, pp. 234–243, 2006.

In the case of linear-threshold neural network model, the signal-to-noise analysis has been used [1,2] in the case of spatially organized networks. It has been shown that the retrieval states of the connected network have non-uniform activity profiles, when the connections are short-range enough, and that the level of localization increases by increasing the gain or the neuron's saturation level [2].

The investigation of the spontaneous activity bumps in Small-World networks (SW) [6,7] of Integrate-and Fire neurons [4], has recently shown that the network retrieves when its connectivity is close to the random and displays localized bumps of activity, when the connectivity is close to the ordered. The two regimes are mutually exclusive in the range of the parameter governing the proportion of the long-range connections on the SW topology of Integrate-and-Fire network, while the two behaviors coexist in the case of linear-threshold and smoothly saturated units.

The result related to the appearance of bump formations have been recently reported by us [3] in the case of binary model for associative network. We demonstrated that the spatially asymmetric retrieval states (SAS) can be observed when the network is constrained to have a different activity compared to that induced by the patterns.

The model we studied in Ref.[3] has a symmetric and distance dependent connectivity for all neurons within an attractor neural network (NN) of Hebbian type formed by N binary neurons $\{S_i\}, S_i \in \{-1,1\}, i = 1,...,N$, storing p binary patterns $\eta_i^\mu, \mu \in \{1...p\}$ with symmetric connectivity between the neurons $c_{ij} = c_{ji} \in \{0,1\}, c_{ii} = 0$.

The corresponding Hopfield model is [8]:

$$H = \frac{1}{N}\sum_{ij} J_{ij} S_i S_j, \tag{1}$$

with Hebbian learning rule:

$$J_{ij} = \frac{1}{c}\sum_{\mu=1}^{p} c_{ij}(\eta_i^\mu - a)(\eta_j^\mu - a). \tag{2}$$

The learned patterns are drawn from the following distribution:

$$P(\eta_i^\mu) = \frac{1+a}{2}\delta(\eta_i^\mu - 1) + \frac{1-a}{2}\delta(\eta_i^\mu + 1),$$

where the parameter a is the sparsity of the code [9]. The number of connections per neuron is $C \equiv cN$.

In order to introduce an asymmetry between the retrieval and the learning states, a condition on the mean activity of the network has to be imposed:

$$H_a = NR(\sum_i S_i/N - a).$$

Here the parameter R controls the affinity of the system toward the appearance of bump formations. This term favors states with lower total activity $\sum_i S_i$ that is equivalent to a decrease of the number of active neurons. Thus, the above term creates an asymmetry between the learning and the retrieval states. We showed in Ref.[3] that this condition is a necessary and a sufficient condition for the observation of bump formations. Similar observations have been reported in the case of linear-threshold network [1], where in order to observe bump formations, one has to constrain the activity of the network. The same is true in the case of smoothly saturating binary network [2], when the highest activity level, that can achieved by the neurons, is above the maximum activity of the units in the stored pattern.

As we have shown in Ref.[3], when the bump appears, the capacity of the network drops dramatically because the network decreases its effective size to the size of the bump.

On the other side, the spatially restricted activity means that the effective coding is restricted to very sparse coding. Usually, the capacity of the network, keeping patterns with sparsity a, is proportional to $1/a|\log a|$ and increases with the sparsity. Therefore a question arise: Is it possible to use the sparsity of the code, imposed by the bump in order to increase the capacity of the network?

In this paper we are trying to use explicitly the sparseness of the code, imposed by the bump appearance, in order to increase the capacity of a two-dimensional network that stores natural images. By means of simulations, we show that the capacity of the network can be increased to the limits typical for sparsely connected network, e.g. to achieve capacities of order of C.

2 Theoretical Analysis

The theoretical analysis of the present model has been explained in details in Ref.[3]. Here we briefly present the main points of this analysis and the most important results.

For the theoretical analysis of the spatially asymmetric states (SAS), we consider the decomposition of the connectivity matrix c_{ij} by its eigenvectors $a_i^{(k)}$:

$$c_{ij} = \sum_k b_i^{(k)} b_j^{(k)}, \tag{3}$$

with

$$b_i^k \equiv a_i^{(k)} \sqrt{\lambda_k / c} \tag{4}$$

where λ_k label the corresponding (positive) eigenvalues.

Following the classical analysis of Amit et al. [10], we study the following binary Hopfield model [8]:

$$H = -\frac{1}{cN} \sum_{ij\mu} S_i \xi_i^\mu c_{ij} \xi_j^\mu S_j - \sum_{\nu=1}^{s} h^\nu \sum_i \xi_i^\nu S_i + NR\overline{S_i b_i^0}, \tag{5}$$

where the Hamiltonian (5) has been represented in terms of the variables $\xi_i^\mu = \eta_i^\mu - a$. The second term in the Hamiltonian introduces a small external field

h, which will tend later to zero. This term is necessary to take into account the finite numbers of overlaps that condense macroscopically. The third term imposes an asymmetry in the neural network's states, which is controlled by the value of the parameter R. This term is responsible for the bump formations. In Eq. (5) the over line means a spatial averaging $\overline{(.i)} = \frac{1}{N}\sum_i(.)$.

Following the classical analysis [10], we use the variables $m^{\mu}_{\rho k}$ for each replica ρ and each eigenvalue, which are the usual overlaps between neurons and patterns. In addition, we also use two other order parameters (OP)

$$q_k^{\rho,\sigma} = \overline{(b_i^k)^2 S_i^{\rho} S_i^{\sigma}},$$

which is related to the neuron activity, and the parameter $r_k^{\rho,\sigma}$, conjugate to $q_k^{\rho,\sigma}$, that measures the noise in the system. The indexes $\rho, \sigma = 1, ..., n$ label the different replicas, while the role of the connectivity matrix is taken into account by the parameters b_i^k.

Finally, by using the replica symmetry ansatz and the saddle-point method [10], we obtain the following expression for the free energy per neuron:

$$\begin{aligned} f = &\frac{1}{2c}\alpha(1-a^2) + \frac{1}{2}\sum_k (m_k)^2 - \frac{\alpha\beta(1-a^2)}{2}\sum_k r_k q_k + \frac{\alpha\beta(1-a^2)}{2}\sum_k \mu_k r_k + \\ &+ \frac{\alpha}{2\beta}\sum_k [\ln(1-\beta(1-a^2)\mu_k + \beta(1-a^2)q_k) - \\ &- \beta(1-a^2)q_k(1-\beta(1-a^2)\mu_k + \beta(1-a^2)q_k)^{-1}] - \\ &- \frac{1}{\beta}\int \frac{dz e^{-\frac{z^2}{2}}}{\sqrt{2\pi}} \overline{\ln 2\cosh\beta\left(z\sqrt{\alpha(1-a^2)\sum_l r_l b_i^l b_i^l} + \sum_l m_l \xi_i b_i^l + R b_i^0\right)}, \end{aligned} \quad (6)$$

where $\alpha = p/N$ is the storage capacity, $\mu_k = \lambda_k/cN$ and we have used the fact that the average over a finite number of patters ξ^{ν} can be self-averaged [3], [10].

The equations for the OP r_k, m_k and q_k are respectively:

$$r_k = \frac{q_k(1-a^2)}{(1-\beta(1-a^2)(\mu_k - q_k))^2}, \quad (7)$$

$$m_k = \int \frac{dz e^{-\frac{z^2}{2}}}{\sqrt{2\pi}} \overline{\xi_i b_i^k \tanh\beta\left(z\sqrt{\alpha(1-a^2)\sum_l r_l b_i^l b_i^l} + \sum_l m_l \xi_i b_i^l + R b_i^0\right)} \quad (8)$$

and

$$q_k = \int \frac{dz e^{-\frac{z^2}{2}}}{\sqrt{2\pi}} \overline{(b_i^k)^2 \tanh^2\beta\left(z\sqrt{\alpha(1-a^2)\sum_l r_l b_i^l b_i^l} + \sum_l m_l \xi_i b_i^l + R b_i^0\right)}. \quad (9)$$

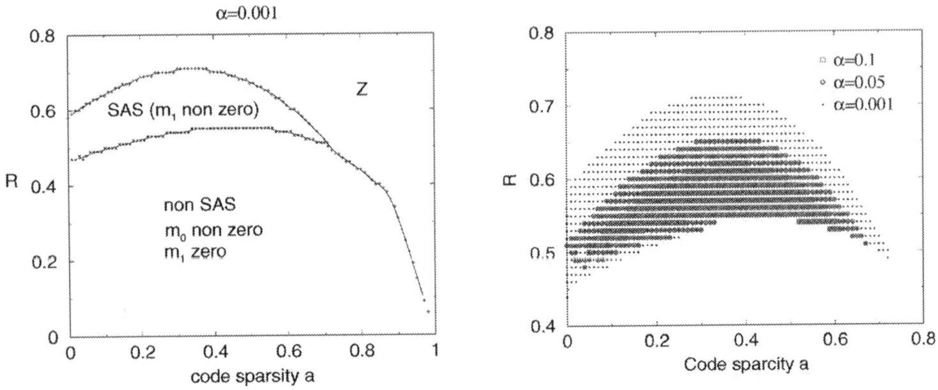

Fig. 1. Left:Phase diagram for $\alpha = 0.001$ ($\alpha/c = 0.02$). The SAS region, where local bumps are observed, is relatively large. The Z region corresponds to trivial solutions for the overlap. Right: SAS region for different values of α. High values of α limit the SAS region.

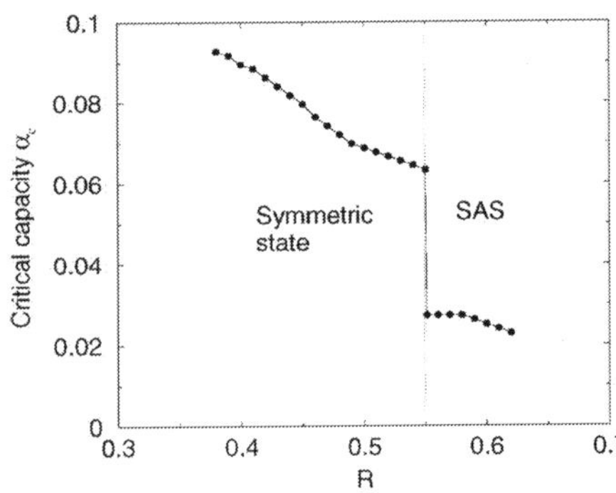

Fig. 2. The critical storage capacity α_c as a function of R for $a = 0.4$, showing a drop on the transition to a bump state

The numerical analysis of the above equations for $T = 0$ gives a stable region for the solutions corresponding to bump formations for different values of the load α, the sparsity a and the retrieval asymmetry parameter R, shown in Fig.1. As can be seen, the sparsity of the code a enhances the SAS effect, although it is also observed for $a = 0$. As we expected, the asymmetry factor R between the stored and retrieved patters is very important in order to have spatial asymmetry. The diagram in Fig.1 shows a clear phase transition with R. Only for intermediate values of R, the bump solution exists.

The dependence of the critical storage capacity α_c on the asymmetry parameter R is shown in Fig.2. The figure presents a drastic drop of α_c at the transition from homogeneous retrieval (symmetric) state to spatially localized

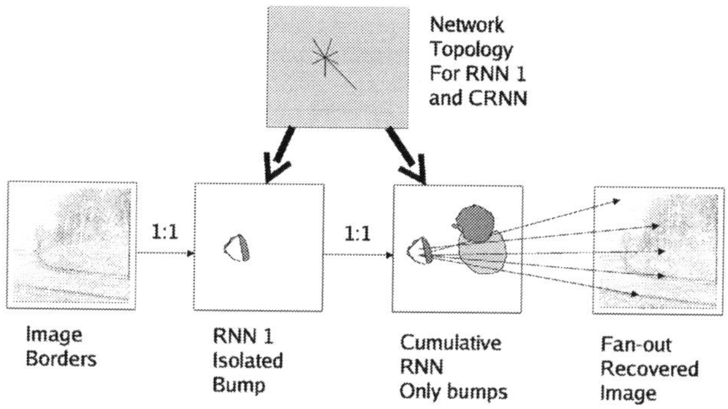

Fig. 3. Schematics of the two stage pattern storage. The first stage forms a bumps in RNN 1 that contains few patterns. The second stage records this bump in a cumulative RNN with high capacity. The fan-out recovers the original image.

(bump state). Effectively only the fraction of the network in the bump can be excited and the storage capacity drops proportionally to the size of the bump.

3 Computer Experiments

As we mentioned in the Introduction, we have performed computer experiments to show that the capacity of the network can be increased to the limits typical for sparsely connected network.

As a test we used the natural image database of van Hateren [11], as well as the Waterloo set of images. In order to make the simulation easier, we decreased the resolution of the images down to 256 by 256 points. We have used randomly selected subsets of up to 300-700 images of van Haterens database and also consecutive images of that database. No difference in the results, concerning the capacity of the network has been found. Waterloo dataset is very small and was used as a reference set together with the van Heteren database. Once again it was found that the results do not depend of the database source.

For the de-correlation of the image, we used a border detector, converting the original image into an image that represents its borders. The simulations show that the exact type of the border detector does not affect the results. Each pixel of the resulting image is represented by a neuron in the network. We tested binarised version of the images borders as well as the discrete (gray) values of the pixels in order to train the network. Both of them give approximately the same results.

We applied a two stage learning procedure Fig.3. In the first stage we trained a Hebbian network with spatially dependent connectivity and probability of connections $p(r) \propto const + e^{-r^2/2\sigma^2}$, where r is the radius of connectivity. The trained network contains few patterns, or even only the pattern in question. Then we raised the threshold in order to observe a bump with determined sparsity. The bump formation is sufficiently noise resistant and stable in order to reconstruct

Fig. 4. The contours of Lena and the bump formation with sparsity $a = 0.005$ and $r = 15$. The bump represents a compact area

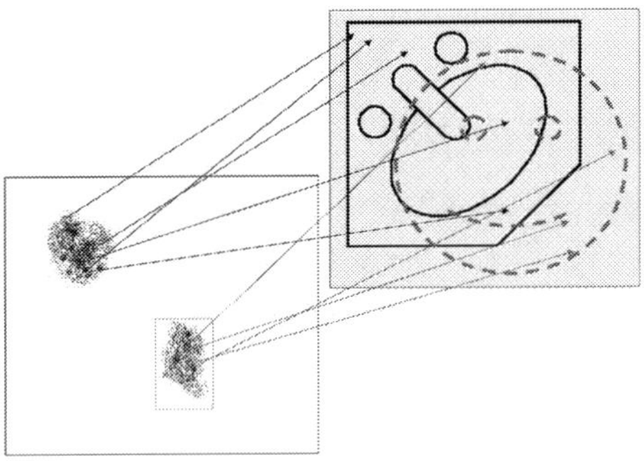

Fig. 5. Fanout to recover the image from its bump. Usually the bump is unique, although up to 4 separate bumps are easily observable.

the complete image. An example of such a bump is given in Fig. 4. The location of the bump depends exclusively on the topology of the network and on the pattern. For a typical image, usually only one of few (two to three) bumps are formed.

In the second stage, we used the bump in order to train a Hebbian network with exactly the same topology, that contains only bumps from a variety of the images. We referred this network as a Cumulative Recurrent Neural Network (CRNN).

In a parallel work [12] we have shown that the image can be restored effectively from its bump. This gives us an opportunity to save only the bump in a network and to associate another feed-forward network in order to restore the image in question.

Roughly speaking the fan-out part of the network connects a dilation of the bump/bumps to all the neurons of the fan-out network. The rest of the neurons,

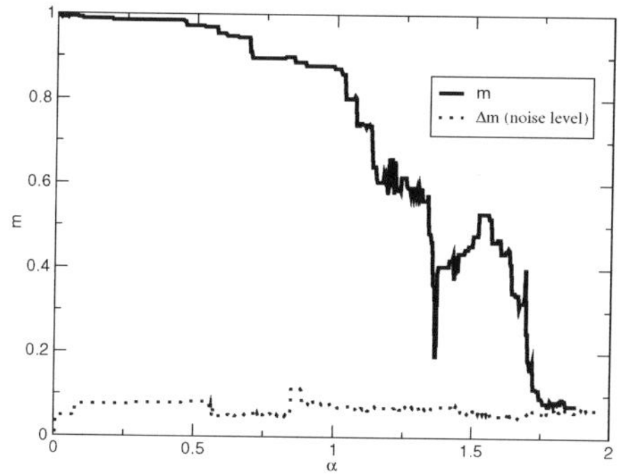

Fig. 6. Typical behavior of the order parameter m, normalized to the sparsity of the pattern as a function of the network load α. The dashed curve is the "recovering" of an arbitrary not memorized pattern, e.g. the noise level.

i.e. the neurons outside of the bump are not connected to the fan-out layer. Thus if the areas of two different bumps do not intersect, the recovered patterns clearly do no interact, Fig.5.

The intersecting bumps, when they are orthogonal enough, do not cause a problem. In other words, with capacity up to the critical capacity α_c, we have a good recovery of the initial pattern. The details of this results are presented in Ref.[12]. The noise level is usually less than the dashed line in Fig.6 that is a small quantity compared to the signal.

In order to keep the network in regime $C \propto N$, the connectivity of the system must be large enough. In these simulations we used $C = 300$, that is suitable for computer simulations. It fits well into the computer memory, it is large enough in order to be in the range $\propto N$ and shows a rich behavior. For this value of C, the correlation radius r can vary from 10 to the size of the image.

For large values of r, no bump formations can be observed. Therefore, one can expect that the capacity of the network will reach its asymptotic limit proportional to $1/a|\log a|$. However, the pattern is very unstable and susceptible to noise, which makes the capacity close to zero, Fig.7(right).

When r is very small, the network effectively converts into locally connected network. In this case the capacity is small, Fig. 7(left), but not too small because the pattern are spatially correlated (the borders are continuous curves) and therefore the bump is well formed. Thus, different images are kept in different parts of the cumulative CRNN and they are essentially orthogonal.

For intermediate levels of the correlation radius r, one can achieve capacity close to the theoretical capacity $\alpha = 1.58$ for sparse random network, Fig. 6.

The critical capacity versus the load is shown in Fig.8. One can see that the critical capacity has a maximum near $r = 20$ and in general it is very high. It is of the same order as the critical capacity of a very sparse network, up to the correlation radius where the bumps disappear and the capacity drops sharply to zero.

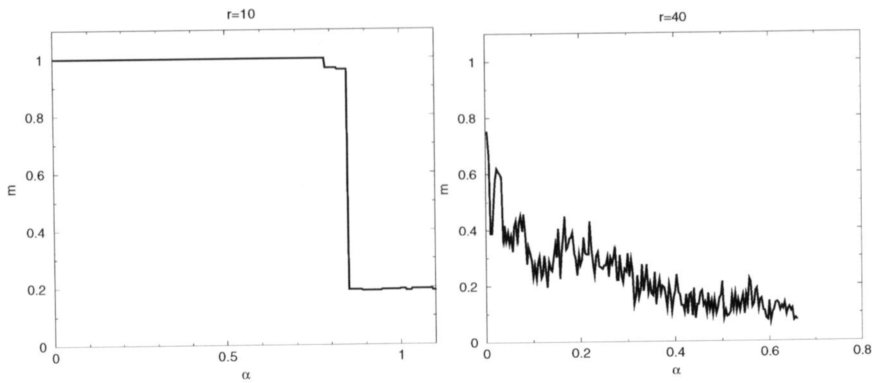

Fig. 7. The bump capacity with too small (left) and too large (right) correlation radius r

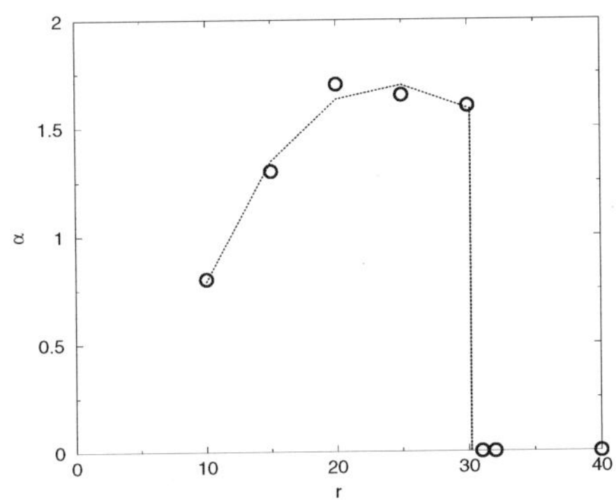

Fig. 8. Non-trivial behavior of the critical storage capacity during the bump formation phase. For $r = 31.3$, α_c drops to 0.

4 Conclusion

In this paper we have shown that by using a special two-stage learning procedure and a bumpy network, one can increase the storage capacity of this network even with Hebbian learning rule. This feature is based on the theoretical background, showing that the location of the bump is only slightly dependent on the patterns already stored in the network. Therefore, one can use the bump as an intrinsic characteristic of the image and the topology of the network . The architecture is promising when another, fan-out layer is added to it, because this allows a total recovering of the image with relatively few connections.

Acknowledgments

The authors acknowledge the financial support from the Spanish Grants DGI.M. CyT. FIS2005-1729, Plan de Promoción de la Investigación UNED and TIN

2004–07676-G01-01. We also thank David Dominguez for the fruitful discussion of the manuscript.

References

1. Y.Roudi and A.Treves, JSTAT, P07010, 1 (2004).
2. Y.Roudi and A.Treves, cond-mat/0505349.
3. K.Koroutchev and E.Korutcheva, Preprint ICTP, Trieste, Italy, IC/2004/91, 1 (2004);
   K.Koroutchev and E.Korutcheva, Central Europ. J.Phys., **3**, 409 (2005);
   K.Koroutchev and E.Korutcheva, Phys.Rev.E **73** (2006)No 2.
4. A.Anishchenko, E.Bienenstock and A.Treves, q-bio.NC/0502003.
5. V.Breitenberg and A.Schulz, *Anatomy of the Cortex*, Springer, Berlin, 1991.
6. D.J.Watts and S.H.Strogatz, Nature, **393**, 440 (1998).
7. D.J.Watts, *Small Worlds: The Dynamics of Networks Between Order and Randomness(Princeton Review in Complexity)*(Princeton University Press, Princeton, New Jersey, 1999).
8. J.Hopfield, Proc.Natl.Acad.Sci.USA, **79**, 2554 (1982).
9. M.Tsodyks and M.Feigel'man, Europhys.Lett., **6**, 101 (1988).
   Stat.Phys., **14**, 565 (1994).
10. D.Amit, H.Gutfreund and H.Sompolinsky, Ann.Phys., vol.173,(1987),pp.30-67.
11. van Hateren J. H. and van der Schaaf A, *Independent component filters of natural images compared with simple cells in primary visual cortex.* in Proc.R.Soc.Lond. B, 265:359-366, 1998.
12. K.Koroutchev and E.Korutcheva, in the Proceedings of the 9th Granada Seminar on Computational and Statistical Physics, AIP, 2006.

Error Entropy Minimization for LSTM Training*

Luís A. Alexandre[1] and J.P. Marques de Sá[2]

[1] Department of Informatics and IT-Networks and Multimedia Group, Covilhã, University of Beira Interior, Portugal,
lfbaa@di.ubi.pt

[2] Faculty of Engineering and INEB, University of Porto, Portugal,
jmsa@fe.up.pt

Abstract. In this paper we present a new training algorithm for the Long Short-Term Memory (LSTM) recurrent neural network. This algorithm uses entropy instead of the usual mean squared error as the cost function for the weight update. More precisely we use the Error Entropy Minimization approach, were the entropy of the error is minimized after each symbol is present to the network. Our experiments show that this approach enables the convergence of the LSTM more frequently than with the traditional learning algorithm. This in turn relaxes the burden of parameter tuning since learning is achieved for a wider range of parameter values. The use of EEM also reduces, in some cases, the number of epochs needed for convergence.

1 Introduction

One of the most promising machines for sequence learning is the Long Short-Term Memory (LSTM) recurrent neural network [1,2,3,4]. In fact, it has been shown that LSTM outperforms traditional recurrent neural networks (RNNs) such as Elman, Back-Propagation Through Time (BPTT) and Real-Time Recurrent Learning (RTRL) in problems where the need to retain information for long time intervals exists [1].

Traditional RNNs suffer from the problem of loosing error information pertaining to long time lapses. This occurs because the error signals tend to vanish over time [5]. LSTM is able to deal with this problem since it protects error information from decaying using gates.

Usually error backpropagation for neural network learning is made using MSE as the cost function. The authors of [6] introduced an error-entropy minimization algorithm that used Renyi's quadratic entropy. They applied their approach to problems of time-series prediction (MacKey Glass chaotic time series) and nonlinear system identification. Note that these problems do not need to retain information for long time lapses. They are usually solved using Time Delay Neural Networks (TDNNs).

In [7], the authors propose the use of the minimization of the error entropy instead of MSE as a cost function for classification purposes. In terms of the

* This work was supported by the Portuguese FCT-Fundação para a Ciência e Tecnologia (project POSC/EIA/56918/2004).

S. Kollias et al. (Eds.): ICANN 2006, Part I, LNCS 4131, pp. 244–253, 2006.

entropy measure, two approaches have been tested both with good results when compared to MSE: in [7,8] Renyi's quadratic entropy was used. In [9] the EEM algorithm was used with Shannon's entropy.

In this paper we adapt the LSTM for learning long time lapse problems using EEM and present several experiments that show the benefits that can be obtained from such an approach.

The rest of the paper is organized as follows: the next section presents LSTM, the following section shows how EEM can be incorporated into the learning algorithm of the LSTM. Section IV presents the experiments and the last section contains the conclusions.

2 LSTM

The LSTM was proposed in [1]. It results from the use of gates to keep the non-linearities (the transfer functions) in the neurons from making the error information pertaining to long time lapses vanish. Note that the use of gates is only one possibility to avoid this problem that affects traditional RNNs: other approaches can be found in [5], pp. 776.

In the following brief discussion we are referring to the original proposition in [1]: other approaches have been proposed [2,3,4].

The main element of the LSTM is the block: a block is a set of cells and two gates (see fig. 1). The gates are called input and output gates.

Each cell (see fig. 2) is composed of a central linear element called the CEC (Constant Error Carousel), two multiplicative units that are controlled by the block's input and output gates, and two non-linear functions $g()$ and $h()$.

The CEC is responsible for keeping the error unchanged for an arbitrarily long time lapse. The multiplicative units controlled by the gates decide when the error should be updated.

A LSTM network consists of the normal input and output layers. Typically hidden layers may also be used but the distinguishing characteristic is the use between the input and output layers of one or more blocks of cells. Each block may have an arbitrary number of cells.

The input layer is connected to all the gates and to all the cells and to the output layer. The gates and the cells have input connections from all cells and all gates.

For a detailed description, of the learning algorithm see [1].

3 EEM for LSTM

3.1 The EEM

The idea behind EEM is to replace the MSE, as the cost function of a learning system, with the entropy of the error.

In [6] it is shown that the minimization of the error entropy (in particular, Renyi's entropy) results in the minimization of the divergence between the joint

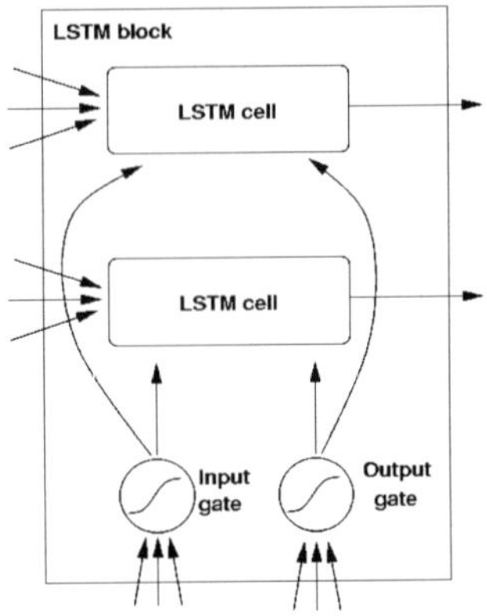

Fig. 1. An LSTM block

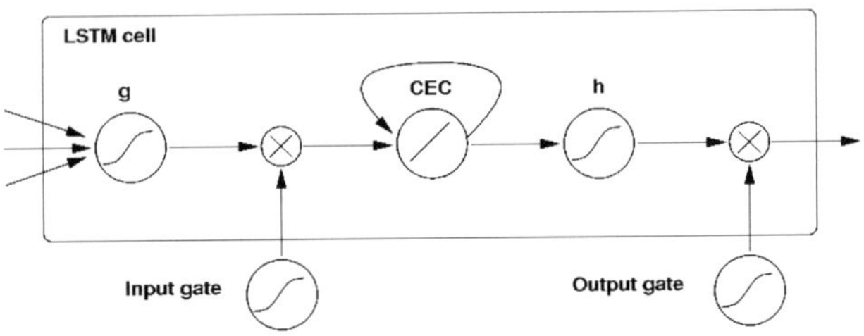

Fig. 2. An LSTM cell (also visible are the input and output gates)

pdfs of input-target and input-output signals. This suggests that the distribution of the output of the system is converging to the distribution of the targets.

Also, when the entropy is minimized, for the classification case and under certain mild conditions, implies that the error must be zero (see proof in [8]).

Let the error $e(j) = T(j) - y(j)$ represent the difference between the target T of the j output neuron and its output y, at a given time t (not given explicitly to keep the notation simple).

We will replace the MSE of the variable $e(j)$ for its EEM counterpart.

First it is necessary to estimate the pdf of the error. For this we use the Parzen window approach

$$\hat{f}(e(j)) = \frac{1}{nh} \sum_{i=1}^{n} K\left(\frac{e(j) - e(i)}{h}\right) \tag{1}$$

where h represents the bandwidth of the kernel K and n is the number of neurons in the output layer.

The kernel used is the Gaussian kernel given by

$$K(x) = \frac{1}{\sqrt{2\pi}} \exp\left(-\frac{x^2}{2}\right)$$

Renyi's quadratic entropy is given by

$$H_{R2}(x) = -\log\left(\int_C (f(x))^2 dx\right) \quad (2)$$

where C is the support of x and $f(\cdot)$ is its density function.

Note that equation (2) can be seen as the logarithm of the expected value of the pdf: $-\log E[f(x)]$. This justifies the use of the following estimator for H_{R2}

$$\hat{H}_{R2}(x) = -\log\left(\frac{1}{n}\sum_{i=1}^{n} f(x_i)\right)$$

Once we put the estimator of the pdf from expression (1) into this last expression, where x is also replaced by the error $e(j)$, we get the final practical expression of our cost function

$$\hat{H}_{R2}(e(j)) = -\log\left(\frac{1}{n^2 h}\sum_{i=1}^{n}\sum_{u=1}^{n} K\left(\frac{e(i)-e(u)}{h}\right)\right) \quad (3)$$

Note that instead of the time complexity for the MSE which is $O(n)$, the EEM approach has $O(n^2)$ complexity.

3.2 Application of EEM to LSTM Learning

In this paper we modified the gradient-based learning algorithm of LSTM presented in [1] and replace the use of the MSE as the cost function by the EEM.

The change in the derivation presented in [1] occurs in the following expression (the backpropagation error seen at the output neuron k):

$$E(k) = f'(net(k))(T(k) - y(k)) \quad (4)$$

where $f(\cdot)$ is the sigmoid transfer function, $f'(\cdot)$ represents its derivative, $net(k)$ is the activation of the output neuron k at time t, $T(k)$ is the target for the output neuron k at time t and $y(k))$ is the output of neuron k at time$t$ (there are differences in the variable names to make this expression coherent with the rest of the paper).

This expression becomes (see the appendix for the derivation)

$$E(k) = Qf'_k(net(k))\sum_{i=1}^{n} \exp\left(-\frac{a(i,k)}{2h^2}\right) a(k,i) \quad (5)$$

where Q stands for a constant term and $a(i,k)$ is

$$a(i,k) = T_i(t) - y_i(t) - T_k(t) + y_k(t) = e(i) - e(k)$$

Of course this change will affect all the backpropagated errors.

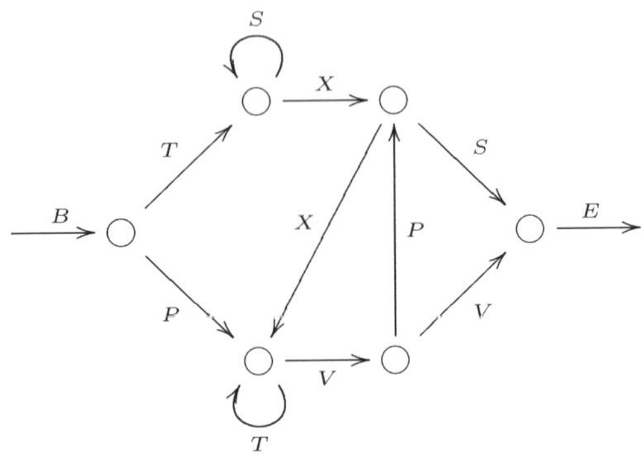

Fig. 3. Reber grammar

4 Experiments

In this section we present several experiments that compare the performance of LSTM learning with MSE versus EEM.

We use three standard data sets for this purpose: the Reber grammar problem, the embedded Reber grammar and the context-sensitive grammar $A^n B^n$.

4.1 Reber Grammar

The finite automata in figure 3 generates strings from the grammar known as the Reber grammar.

The strings are generated by starting at B, and moving from node to node, adding the symbols in the arcs to the string. The symbol E marks the end of a string. When there are two arcs leaving a node, one is chosen randomly with equal probability. This process generates strings of arbitrary length.

We conducted several experiments were the goal was the prediction of the next valid symbol of a string, after the presentation of a given symbol. For instance, if the network receives a starting symbol B it has to predict that the possible next symbols are P or T. If the network is able to correctly predict all possible symbols of all strings both from the training and test sets, using less than 250.000 sequences for learning, we say that it converged.

We used a set with 500 strings from the Reber grammar for training and a different set with 500 strings for testing. For each topology and value of the parameter h we repeated the process 100 times. The change was the random initialization of the network weights. Tables 1 and 2 present the results. They show the percentage of the trained networks that were able to learn perfectly both the training and test sets, and the average and standard deviation of the number of sequences that were used for training.

We tested the two topologies: the strings are codified in a 1-out-of-7 coding, so both input and output layers have 7 neurons. The topologies did not use any traditional neurons in the hidden layer. In the first case two blocks were used,

Table 1. Results for the experiments with the Reber grammar for the topology (7:0:2(2,1):7). ANS stands for the Average Number of Sequences necessary for convergence.

	Learning rate=0.1		Learning rate=0.2		Learning rate=0.3	
	ANS (std) [10^3]	% conv.	ANS (std) [10^3]	% conv.	ANS (std) [10^3]	% conv.
MSE	15.1 (24.5)	38	74.9 (116.5)	63	61.0 (111.5)	**56**
EEM h=1.3	81.8 (115.4)	36	42.6 (51.5)	11	113.6 (135.6)	7
EEM h=1.4	45.6 (68.2)	45	70.6 (93.8)	11	61.8 (63.6)	10
EEM h=1.5	26.0 (43.9)	54	84.1 (120.2)	29	47.2 (39.1)	13
EEM h=1.6	28.4 (43.1)	**66**	58.0 (84.2)	37	135.1 (160.1)	15
EEM h=1.7	23.0 (25.9)	64	54.9 (87.8)	40	96.9 (135.8)	30
EEM h=1.8	75.8 (51.8)	30	60.1 (96.8)	50	66.0 (111.8)	33
EEM h=1.9	78.0 (110.1)	62	53.6 (94.1)	61	48.7 (66.2)	33
EEM h=2.0	49.3 (77.6)	58	57.6 (109.0)	**67**	57.4 (83.7)	51

Table 2. Results for the experiments with the Reber grammar for the topology (7:0:2(2,2):7). ANS stands for the Average Number of Sequences necessary for convergence.

	Learning rate=0.1		Learning rate=0.2		Learning rate=0.3	
	ANS (std) [10^3]	% conv.	ANS (std) [10^3]	% conv.	ANS (std) [10^3]	% conv.
MSE	19.2 (26.1)	41	46.0 (77.2)	53	30.1 (50.7)	61
EEM h=1.3	57.2 (67.5)	43	44.3 (58.8)	18	28.4 (12.9)	5
EEM h=1.4	20.2 (27.8)	55	70.5 (112.4)	22	108.9 (130.5)	10
EEM h=1.5	33.6 (55.3)	59	68.1 (109.8)	36	31.7 (32.7)	19
EEM h=1.6	26.5 (41.9)	**65**	38.3 (47.2)	42	58.4 (82.7)	25
EEM h=1.7	32.3 (57.2)	53	48.4 (83.8)	48	55.6 (82.9)	26
EEM h=1.8	24.5 (48.8)	60	54.4 (88.3)	61	43.4 (82.0)	40
EEM h=1.9	51.8 (76.7)	49	70.6 (134.5)	66	62.1 (108.1)	**66**
EEM h=2.0	44.6 (79.3)	48	48.3 (85.8)	**68**	44.3 (70.2)	41

one with one cell and the other with two cells (table 1). In the second case, both blocks had two cells (table 2). Each table shows the results for learning rates of 0.1, 0.2 and 0.3.

The MSE line refers to the use of the original learning algorithm.

The results are discussed in section 4.4.

4.2 Embedded Reber Grammar

The second set of experiments uses the Embedded Reber Grammar (ERG): it is generated according to figure 4.

This grammar produces two types of strings: BT<reber string>TE and BP<reber string>PE. In order to recognize these strings, the learning machine has to be able to distinguish them from strings such as BP<reber string>TE and BP<reber string>TE. To do this it is essential to remember the second symbol

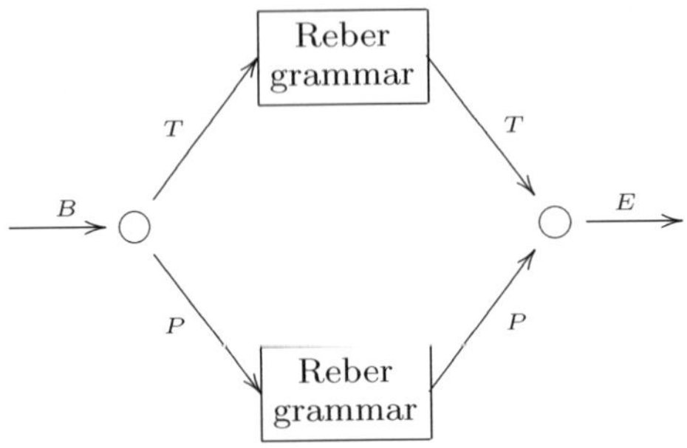

Fig. 4. Embedded Reber grammar

Table 3. Results for the experiments with the embedded Reber grammar

	Learning rate=0.1		Learning rate=0.3	
	ANS (std) [10^3]	% conv.	ANS (std) [10^3]	% conv.
MSE	44.4(48.4)	8	66.3 (40.1)	6
EEM h=1.3	79.6 (94.5)	7	-	0
EEM h=1.4	13.1 (6.5)	8	-	0
EEM h=1.5	49.9 (53.8)	12	327 (-)	1
EEM h=1.6	65.7 (46.9)	8	62.0 (74.5)	3
EEM h=1.7	41.2 (26.9)	8	183.7 (137.4)	3
EEM h=1.8	60.6 (35.8)	12	102.4 (101.4)	5
EEM h=1.9	124.3 62.4	13	76.9 (125.0)	**7**
EEM h=2.0	143.2 (111.9)	**14**	84.6 (87.6)	**7**

in the sequence such that it can be compared with the second last symbol. Notice that the length of the sequence can be arbitrarily large.

This problem is no longer learnable by an Elman net but a RTRL can learn it with some difficulty, since, as opposed to the RG problem, there is the need to retain information for long time lapses.

In this case the experiments were similar to the ones reported in the previous section but the learning rates used were 0.1 and 0.3. The train and test sets had also 500 strings each. This time only one topology was used: 7 neurons in the input and output layer (the codification is the same as in the RG example), and four blocks each with 3 cells. The experiments were also repeated 100 times. The results are in table 3.

4.3 A^nB^n Grammar

This data set consists in a series of strings from the context-sensitive grammar A^nB^n. Valid strings consist of n A symbols followed exactly by n B symbols.

The network is trained with only the correct strings for n from 1 up to 10. We consider that the network converged if it is able to correctly classify all the

Table 4. Results for the experiments with the grammar A^nB^n

	Test $n = 1, \ldots, 50$		Test $n = 1, \ldots, 100$	
	Average n. seq. (std) [10^3]	% conv.	Average n. seq. (std) [10^3]	% conv.
MSE	4.93 (2.80)	17	4.90 (2.41)	12
EEM h=1.3	10.31 (7.40)	18	8.75 (6.48)	13
EEM h=1.4	11.67 (8.11)	19	11.90 (7.94)	12
EEM h=1.5	15.31 (8.25)	28	16.08 (7.76)	18
EEM h=1.6	17.06 (8.62)	36	17.51 (8.46)	25
EEM h=1.7	18.77 (8.78)	45	18.33 (8.33)	29
EEM h=1.8	20.74 (8.74)	**50**	19.69 (7.83)	**35**
EEM h=1.9	20.80 (8.38)	48	21.09 (7.39)	**35**
EEM h=2.0	22.19 (8.20)	49	22.50 (7.52)	33

strings in both the training and test sets, using less than 50.000 sequences for learning.

In the first experiment we used for the test set the correct strings for $n = 1 \ldots 50$. In the second experiment we used the strings $n = 1 \ldots 100$. In both experiments the topology of the network was three neurons in the input layer, two blocks each with one neuron and three neurons in the output layer. Both experiments were repeated 100 times for a learning rate of 1.0. The results obtained are in table 4.

4.4 Discussion

For the Reber grammar experiments, the EEM improved the percentage of convergence in all six sets of experiments except for the first topology with learning rate 0.3. In one case (first topology and learning rate=0.2) it not only outperformed MSE but there was also a reduction in the number of sequences necessary for network converged from 74.9×10^3 to 57.6×10^3.

It can also be seen that for MSE then increase in the value of the learning rate was good in terms of percentage of convergence, specially in the case of the second topology, whereas for the EEM the best results were obtained for both topologies with a learning rate of 0.2.

In the case of the experiments with the ERG, and for the learning rate 0.1, two benefits were found from the application of the EEM: in two cases ($h = 1.4$ and 1.7) we were able to obtain the same percentage of convergence but with a smaller number of required training sequences. In four other cases, the percentage of convergence increased from 8 to 12, 13 and 14%. In these cases the number of necessary sequences also increased when compared to the use of MSE. When the learning rate was set at 0.3, there were two situations ($h = 1.3$ and 1.4) where the EEM was not able to converge. Although the best result was still obtained with the EEM for h=1.9 and 2.0. In these experiments it is apparent that the increase of the learning rate was not beneficial either for MSE nor for the EEM.

Finally, the experiments with the A^nB^n grammar confirmed that the use of the EEM is beneficial in terms of increasing the convergence percentage: from the 16 sets of 100 repetitions only one had the same performance of the MSE; all other sets improved. Again, the number of necessary training sequences increased.

5 Conclusions

In this paper a new learning algorithm for LSTM was introduced. It uses the EEM as a cost function instead of the MSE. From the experiments made we conclude that this approach is beneficial since there is a sustainable increase in the convergence percentage of the networks. This improvement comes with the cost of a longer training time since the EEM algorithm is slower than the MSE and also the networks tend to need more training epochs to achieve perfect learning.

When using pdf estimation with kernels, the problem of the value to choose for the kernel bandwidth is always present. In this paper we used fixed values during the training stage. We intend to use adaptive approaches to the setting of h so that it can adapt dynamically during training.

References

1. Hochreiter, S., Schmidhuber, J.: Long short-term memory. Neural Computation **9**(8) (1997) 1735–1780
2. Gers, F., Schmidhuber, J., Cummins, F.: Learning to forget: Continual prediction with LSTM. Neural Computation **12**(10) (2000) 2451–2471
3. Gers, F., Schmidhuber, J.: Recurrent nets that time and count. In: Proc. IJCNN'2000, Int. Joint Conf. on Neural Networks, Como, Italy (2000)
4. Pérez-Ortiz, J., Gers, F., Eck, D., Schmidhuber, J.: Kalman filters improve LSTM network performance in problems unsolvable by traditional recurrent nets. Neural Networks **16**(2) (2003) 241–250
5. Haykin, S.: Neural Networks: A Comprehensive Foundation, 2nd edition. Prentice Hall (1999)
6. Erdogmus., D., Principe, J.: An error-entropy minimization algorithm for supervised training of nonlinear adaptive systems. IEEE Trans. Signal Processing **50**(7) (2002) 1780–1786
7. Santos, J., Alexandre, L., Sereno, F., Marques de Sá, J.: Optimization of the error entropy minimization algorithm for neural network classification. In: ANNIE 2004, Intelligent Engineering Systems Through Artificial Neural Networks. Volume 14., St.Louis, USA, ASME Press Series (2004) 81–86
8. Santos, J., Alexandre, L., Marques de Sá, J.: The error entropy minimization algorithm for neural network classification. In Lofti, A., ed.: Proceedings of the 5th International Conference on Recent Advances in Soft Computing, Nottingham, United Kingdom (2004) 92–97
9. Silva, L., Marques de Sá, J., Alexandre, L.: Neural network classification using Shannon's entropy. In: 13th European Symposium on Artificial Neural Networks - ESANN 2005, Bruges, Belgium (2005) 217–222

Appendix

Here we derive expression (5). The term $(T(k) - y(k))$ in equation (4) comes from the derivative of the MSE

$$\frac{1}{n}\sum_{i=1}^{n}(T(i) - y(i))^2$$

w.r.t. the output y_k.

What we are going to do is find this same derivative, but now for expression (3). We note that since the logarithm in expression (3) is a monotonically increasing function, to minimize it is the same as to minimize its operand. So, we will find the partial derivative of the operand, which is given by

$$\frac{1}{n^2 h\sqrt{2\pi}}\sum_{i=1}^{n}\sum_{j=1}^{n}\frac{\partial}{\partial y_k}\exp\left(-\frac{(e(i) - e(j))^2}{2h^2}\right)$$

$$= \frac{1}{n^2 h\sqrt{2\pi}}\sum_{i=1}^{n}\sum_{j=1}^{n}\frac{\partial}{\partial y_k}\exp\left(-\frac{(T(i) - y(i) - T(j) + y(j))^2}{2h^2}\right) \quad (6)$$

Now, when $i = k$ the derivative of the inner term becomes

$$\exp\left(-\frac{(T(k) - y(k) - T(j) + y(j))^2}{2h^2}\right)\left(-\frac{1}{2h^2}\right)2(T(k)-y(k)-T(j)+y(j))(-1) \quad (7)$$

Likewise, if $j = k$, the derivative becomes

$$\exp\left(-\frac{(T(i) - y(i) - T(k) + y(k))^2}{2h^2}\right)\left(-\frac{1}{2h^2}\right)2(T(i)-y(i)-T(k)+y(k)) \quad (8)$$

Expressions (7) and (8) yield the same values since they only differ in sign in the term that is squared and the dummy variables i and j have both the same range (from 1 to n). This allows us to write the derivative of the operand of (3) as

$$Q\sum_{i=1}^{n}\exp\left(-\frac{(T(i) - y(i) - T(k) + y(k))^2}{2h^2}\right)(T(i) - y(i) - T(k) + y(k)) \quad (9)$$

where

$$Q = \frac{2}{n^2 h^3\sqrt{2\pi}}$$

Can AdaBoost.M1 Learn Incrementally? A Comparison to Learn^{++} Under Different Combination Rules

Hussein Syed Mohammed, James Leander, Matthew Marbach, and Robi Polikar[*]

Electrical and Computer Engineering, Rowan University, Glassboro, NJ 08028, USA
polikar@rowan.edu

Abstract. We had previously introduced Learn^{++}, inspired in part by the ensemble based AdaBoost algorithm, for incrementally learning from new data, including new concept classes, without forgetting what had been previously learned. In this effort, we compare the incremental learning performance of Learn^{++} and AdaBoost under several combination schemes, including their native, weighted majority voting. We show on several databases that changing AdaBoost's distribution update rule from hypothesis based update to ensemble based update allows significantly more efficient incremental learning ability, regardless of the combination rule used to combine the classifiers.

1 Introduction

Learning from new data without forgetting prior knowledge is known as incremental learning, and it is encountered often real world applications. This is because sufficiently dense and a representative set of training examples is usually required for satisfactory classifier performance, however, acquisition of such a representative dataset often become available in small and separate batches at different times. Under such conditions, it is necessary to incrementally update an existing classifier to accommodate new data while retaining the information from old data.

Traditionally, when new data become available, previous classifiers are discarded and retrained with the composite data obtained by combining all the data accumulated thus far. However, this approach results in loss of all previously acquired information, and it is commonly known as *catastrophic forgetting* [1]. Furthermore, this approach may not even be feasible, if the original dataset is no longer available.

Learning new data incrementally without forgetting previously acquired knowledge raises the stability-plasticity dilemma [2]: acquiring new knowledge requires plasticity, whereas retaining previously acquired knowledge requires stability. The challenge is to achieve a meaningful balance between the two conflicting properties.

Various forms of incremental learning have been studied under various conditions. In one extreme end, incremental learning is trivialized by allowing retraining with old data, while on the other end, an incremental learning algorithm is expected to learn in an online incremental setting, where learning is carried out in an instance-by-instance

[*] Corresponding author.

S. Kollias et al. (Eds.): ICANN 2006, Part I, LNCS 4131, pp. 254–263, 2006.

basis with some instances introducing new classes. Algorithms that are currently available for incremental learning, such as ARTMAP [3], typically fall somewhere in the middle of this spectrum.

2 Learn^{++} for Incremental Learning

Learn^{++} inspired in part by the ensemble structure of the AdaBoost.M1 algorithm [4], exploits the synergistic expressive power of an ensemble of classifiers to incrementally learn additional information from new data [5,6]. Specifically, for each database that becomes available, Learn^{++} generates a number of diverse classifiers, which are then combined using a suitable combination rule (originally, the weighted majority voting). The pseudocode of Learn^{++} is shown in Figure 1.

Inputs to Learn^{++} are the training data $\boldsymbol{S_k}$ of m_k samples drawn from the current database $\boldsymbol{DB_k}$, a supervised learning algorithm **BaseClassifier,** and an integer T_k specifying the number of classifiers to be generated for database $\boldsymbol{DB_k}$. Learn^{++} generates an ensemble of classifiers using different subsets of each training data, $\boldsymbol{S_k}$. This is achieved by iteratively updating a distribution $\boldsymbol{D_t}$, $t = 1,\ldots,T_k$ from which training

Inputs: For each dataset drawn from $\boldsymbol{DB_k}$ $k=1,2,\ldots,K$

- Sequence of m_k examples $\boldsymbol{S_k}=\{(\mathbf{x}_i,y_i) \mid i=1,\ldots,m_k\}$
- Supervised learning algorithm **BaseClassifier**.
- Integer T_k, specifying the number of iterations.

Do for each $k=1,2,\ldots,K$:

Initialize $w_1(i) = D_1(i) = 1/m_k,\ \forall i,\ i=1,2,\cdots,m_k$ (1)

If $k>1$, Go to Step 5, evaluate current ensemble on new $\boldsymbol{S_k}$, update weight distribution; **End If**

Do for $t = 1,2,\ldots,T_k$:

1. Set $\boldsymbol{D}_t = \mathbf{w}_t \big/ \sum_{i=1}^{m_k} w_t(i)$ so that $\boldsymbol{D_t}$ is a distribution (2)
2. Draw a training $\boldsymbol{TR_t}$ subset from the distribution $\boldsymbol{D_t}$, and train **BaseClassifier** with $\boldsymbol{TR_t}$
3. Obtain a hypothesis h_t and calculate its error on $\boldsymbol{S_k}$. If $\varepsilon_t > ½$, discard h_t, go to step 2.
$$\varepsilon_t = \sum_{i:h_t(\mathbf{x_i})\neq y_i} D_t(i) \qquad (3)$$
4. Call **CombinationRule**, and obtain the composite hypothesis H_t
5. Compute the error of the composite hypothesis
$$E_t = \sum_{i:H_t(\mathbf{x}_i)\neq y_i} D_t(i) = \sum_{i=1}^{m_k} D_t(i)[\mid H_t(\mathbf{x}_i) \neq y_i \mid] \qquad (4)$$
6. Set $B_t = E_t \big/ (1 - E_t)$, and update the weights: (5)
$$w_{t+1}(i) = w_t(i) \times B_t^{\,1-[\mid H_t(\mathbf{x_i})\neq y_i \mid]} = w_t(i) \times \begin{cases} B_t, & if\ \ H_t(\mathbf{x_i}) = y_i \\ 1\ , & otherwise \end{cases} \qquad (6)$$

Call **CombinationRule** and output the final hypothesis.

Fig. 1. Pseudocode of Algorithm Learn^{++}

subsets are chosen. The distribution itself is obtained by normalizing a set of weights assigned to each instance based on the classification performance of the classifier on that instance. In general, instances that have not yet been learned or seen are given higher weights to increase their chance of being selected into the next training data.

At each iteration t, the distribution $\boldsymbol{D}_t$ is obtained by normalizing the weights $\boldsymbol{w}_t$ of the instances updated based on their classification by the previous ensemble (step 1 of the inner **Do** loop). A new training subset $\boldsymbol{TR}_t$ is drawn according to $\boldsymbol{D}_t$ and the **Base-Classifier** is trained with $\boldsymbol{TR}_t$ to generate the hypothesis h_t (step 2). The error ε_t of this hypothesis is calculated on the current training data $\boldsymbol{S}_k$ by adding the distribution weights of the misclassified instances (step 3). If $\varepsilon_t>1/2$, current h_t is deemed too weak, and is replaced with a new h_t, generated from a fresh $\boldsymbol{TR}_t$. If $\varepsilon_t<1/2$, the current hypothesis is retained, and all hypotheses generated during the previous t iterations are combined, using an appropriate combination schemes described later, to construct the *composite hypothesis* H_t (step 4). The composite error E_t made by H_t is determined by adding the distribution weights of all instances misclassified by the ensemble (step 5). The normalized composite error, B_t is computed, and used in updating the weights $w_t(i)$, which are then used in computing the next distribution D_{t+1}, which in turn is used in selecting the next training subset $\boldsymbol{TR}_{t+1}$. Once T_k hypotheses are generated for each database $\boldsymbol{DB}_k$, the final hypothesis H_{final} is obtained by combining all hypotheses by using one of the combination rules described below.

While Learn^{++} uses similar ensemble generation structure as AdaBoost, there are several key differences: AdaBoost runs on a single database; it has no distribution re-initialization; and it stops and aborts if $\varepsilon_t > ½$ for any h_t. Most importantly, AdaBoost is designed to improve the performance of a weak classifier, for which it uses the performance of the current single hypothesis h_t to update its weight distribution [4]. Learn^{++}, however, creates a composite hypothesis H_t representing the ensemble decision, and uses the *ensemble* performance to update its weight distribution. This allows a more efficient incremental learning ability, particularly if the new database introduces instances from a previously unseen class. When instances of a new class are introduced, an existing ensemble H_t– not yet seen instances of the new class, is bound to misclassify them, forcing the algorithm to focus on these instances that carry novel information. For a weight update rule based on the performance of h_t only, the training performance of the first h_t on instances from the new class is independent of the previously generated classifiers. Therefore, the new h_t is not any more likely to misclassify new class instances, which then causes AdaBoost to focus on other *difficult to learn* instances, such as outliers, rather than the instances with novel information content. It is this claim that we investigate in this effort.

Learn^{++} was previously shown to be capable of incremental learning, however, it's incremental learning ability has not been compared to that of AdaBoost. Given that AdaBoost was not originally designed for incremental learning, one can argue whether it is fair to compare AdaBoost to an algorithm that is designed for incremental learning. However, Learn^{++} shares much of its algorithmic detail with AdaBoost. The main difference is the distribution update rule being based on ensemble decision, rather than the previous hypothesis. Therefore, a questions of particular interest is as follows: is the incremental learning ability of Learn++ primarily due to creating and combining an ensemble of classifiers, or is it due to the strategic selection of the distribution update rule? If incremental learning ability is provided primarily by combining an ensemble of classifiers, then AdaBoost should also be able to learn incrementally.

In order to answer this question, and establish the true impact of the difference in distribution update rules, the two algorithms must be made equivalent in all other aspects. Therefore, we slightly modify AdaBoost to allow it to generate additional ensembles with new data, using the same distribution re-initialization as Learn^{++} (but retaining its own single-hypothesis-based distribution update rule). We also allow AdaBoost to generate a replacement hypothesis for any h_t that does not satisfy $\varepsilon_t < ½$ requirement. Therefore, the only difference left between the modified AdaBoost and Learn^{++} is the distribution update rule.

3 Combination Rules

Properties of different combination rules for ensemble systems have been well researched [7,8]. While the best combination rule is often application dependent, certain rules, such as the sum, weighted majority, and decision templates have repeatedly shown superior performance over others, and hence are used more often. Both AdaBoost and Learn^{++} were originally designed to be used with weighted majority voting. However, in this study, we also compare each with different combination rules, to determine whether the combination rule (in addition to distribution update rule) has any effect on incremental learning performance.

Some combination rules, namely, simple and weighted majority voting (VMW), only need access to class labels. Others need the degree of support given by the classifier to each class. For the first group, let us define the decision of the tth classifier as the binary valued $d_{t,j} \in \{0,1\}$, $t=1,\ldots,T$ and $j=1,\ldots,C$, where T is the number of classifiers and C is the number of classes. If classifier h_t correctly identifies class ω_j, $d_{t,j}=1$, and zero otherwise. For other rules, we have continuous valued $d_{t,j} \in [0,1]$, which represent the degree of support given by classifier h_t to class ω_j. For any given classifier, these supports are normalized to add up to 1 over different classes, and are often interpreted as class conditional posterior probabilities, $P(\omega_j|\mathbf{x})$.

We use the *decision profile matrix* [9], to formalize all combination rules: for an instance $\mathbf{x}$, the decision profile matrix $DP(\mathbf{x})$, consists of the elements $d_{t,j}$. The tth row of $DP(\mathbf{x})$ is the support given by the tth classifier to each class, and the jth column is the support received by class ω_j from all classifiers. The total support for each class is obtained as a simple function of the supports received by individual classifiers. We represent the total support received by class ω_j as

$$\mu_j(x) = \Im[d_{1,j}(x),\cdots,d_{T,j}(x)]\cdot \tag{7}$$

where $\Im$ (.) is the combination function. We discuss the sum, product, median, simple majority, weighted majority, and decision template combination rules.

In an ensemble system, the final decision is the class that receives the largest support from the ensemble. Let ω_k be the winning class. In simple majority voting, ω_k is the class that is selected by most number of classifiers. For binary valued $d_{t,j}$,

$$\sum_{t=1}^{T} d_{t,k} = \max_{j=1}^{C} \sum_{t=1}^{T} d_{t,j} \tag{8}$$

If some classifiers are known to be more competent than others, giving higher weights to those classifiers may improve the performance. Denoting the voting weight for classifier h_t with V_t, weighted majority voting (WMV) can be obtained as

$$\sum_{t=1}^{T} V_t d_{t,k} = \max_{j=1}^{c} \sum_{t=1}^{T} V_t d_{t,j} \cdot \tag{9}$$

In original Learn^{++} and AdaBoost, these weights are inversely proportional to the training errors of h_t:

$$V_t = \log\left(\left(1-\varepsilon_t\right)/\varepsilon_t\right) \tag{10}$$

The sum, product and median rules are similar, defined by the following expressions, respectively

$$\mu_j(x) = \frac{1}{T}\sum_{t=1}^{T} d_{t,j}(x) \tag{11}$$

$$\mu_j(x) = \frac{1}{T}\prod_{t=1}^{T} d_{t,j}(x) \tag{12}$$

$$\mu_j(x) = \underset{t=1...T}{median}\{d_{t,j}(x)\} \tag{13}$$

In each case, the ensemble decision is the class ω_k for which the total support $\mu_j(\mathbf{x})$ is highest.

Perhaps the most sophisticated combination rule that uses all supports given by all classifiers to all classes is Kuncheva's decision templates [9]. For each class ω_j, the decision template DT_j is the average of all decision profiles in training data $\boldsymbol{X_j}$

$$DT_j = \frac{1}{M_j} \sum_{\mathbf{x} \in \mathbf{X}_j} DP(\mathbf{x}) \tag{14}$$

where $\boldsymbol{X_j}$ is the set of instances coming from class ω_j; and M_j is the cardinality of $\boldsymbol{X_j}$. The class ω_k whose decision template is closest to the decision profile of the current instance, e.g., using squared *Euclidean* distance, is chosen as the ensemble decision. The closest match then decides on the label of **x.**

$$\mu_j(x) = 1 - \frac{1}{T \times C}\sum_{t=1}^{T}\sum_{k=1}^{C}\left[DT_j(t,k) - d_{t,k}(x)\right]^2 \tag{15}$$

4 Simulation Results

We present and compare simulation results of Learn^{++} and AdaBoost.M1 on several real-world and benchmark databases, using six different combination rules on each. All results are given with 95% confidence interval, obtained through 10 fold cross validation. Each database was partitioned into n sets: S_1~S_n for training, where each set introduced one or more new classes, and an additional ***TEST*** set for validation, which included instances from all classes. Ensemble generalization performances after each training session $\boldsymbol{TS_1}$ ~ $\boldsymbol{TS_n}$ (trained on $\boldsymbol{S_1}$~$\boldsymbol{S_n}$ separately, and tested on ***TEST***) are presented below. Multilayer perceptrons were used as base classifiers.

4.1 Ultrasonic Weld Inspection (UWI) Dataset

The UWI dataset was obtained from ultrasonic inspection of submarine hull welding regions. The welding regions, also known as heat-affected zones, are highly susceptible to growing a variety of defects, including potentially dangerous cracks. The discontinuities within the material, such as air gaps due to cracks, cause the ultrasonic wave to be reflected back, and received by the transducer. The reflected ultrasonic wave, also called an A-scan, serves as the signature pattern of the discontinuity, which is then analyzed to determine whether it was originated from a crack. However, this analysis is hampered by the presence of other types of discontinuities, such as porosity, slag and lack of fusion (LOF), all of which generate very similar A-scans, resulting in a very challenging database with highly overlapping classes. The data distribution and percent generalization performances of both algorithms are shown in Tables 1 and 2 respectively. The number of classifiers used to train datasets S_1, S_2 and S_3 were set as 3, 5 and 7, respectively, and kept constant for all experiments. The best performance for each algorithm at the end of TS_3 is shown in bold.

Table 1. Data distribution for UWI dataset

Dataset ↓	*Crack*	*Slag*	*LOF*	*Porosity*
S_1	300	300	0	0
S_2	200	200	200	0
S_3	150	150	137	99
TEST	200	200	150	125

Table 2. Percent test performance of AdaBoost.M1 and Learn^{++} on UWI dataset

	AdaBoost.M1			Learn^{++}		
	TS_1	TS_2	TS_3	TS_1	TS_2	TS_3
SUM	49.6±1.2	59.1±1.2	57.8±1.6	52.0±0.6	65.7±0.8	70.5±0.8
PRODUCT	48.8±1.2	59.0±1.1	57.1±2.6	51.4±0.6	62.4±0.6	65.0±0.7
MEDIAN	49.1±1.3	58.7±1.2	56.6±1.6	51.5±0.7	65.1±0.8	**70.8±0.6**
M. VOTING	49.0±0.8	58.8±1.1	58.5±1.1	51.4±0.6	65.5±0.7	70.3±0.8
WMV	49.6±1.6	59.2±1.1	**59.3±1.0**	51.4±0.9	65.1±1.0	70.6±0.5
DT	49.0±0.8	59.6±1.0	58.5±1.1	52.1±0.8	65.2±0.6	68.8±0.6

Table 2 shows that both algorithms were able to learn incrementally from the new data, as indicated by the improved generalization performances from one training session to the next. Learn^{++} outperformed AdaBoost, however, with statistical significance, on all combination rules. Furthermore, the performance of AdaBoost mildly deteriorated in TS_3, compared to its performance on the previous session, TS_2. AdaBoost could not learn the new class information, at least using the number of classifiers specified. As mentioned earlier, this is attributed to the composite hypothesis based weight update rule of Learn^{++}. The performance differences among different combination rules were mostly statistically insignificant for both algorithms.

4.2 Volatile Organic Compound (VOC) Dataset

This database was generated from responses of six quartz crystal microbalances (QCMs) to various concentrations of five volatile organic compounds, including ethanol (ET), xylene (XL), octane (OC), toluene (TL), and trichloroethylene (TCE). The data distribution, indicating a new class introduced with each dataset, is shown in Table 3, and the mean generalization performances of AdaBoost.M1 and Learn++ for the VOC database are presented in Table 4. S_1 had instances from ET, OC and TL, S_2 added instances primarily from the new class TCE (and fewer instances from the previously three), and S_3 added instances from XL (and fewer instances from the previous four). The number of classifiers used to train datasets S_1, S_2 and S_3 were chosen as 2, 3 and 5, respectively, and kept constant for all experiments.

Table 3. Data distribution for VOC dataset

Dataset ↓	*ET*	*OC*	*TL*	*TCE*	*XL*
S_1	20	20	40	0	0
S_2	10	10	10	25	0
S_3	10	10	10	15	40
TEST	24	24	52	24	40

Table 4. Percent test performance of AdaBoost.M1 and Learn++ on VOC dataset

	AdaBoost.M1			**Learn++**		
	TS_1	TS_2	TS_3	TS_1	TS_2	TS_3
SUM	61.1±0.7	63.8±2.0	67. ±7.5	62.0±1.3	71.4±0.5	83.2±3.4
PRODUCT	60.3+1.4	61.2±5.3	**70.9±3.7**	60.8±0.5	62.4±3.2	71.4±4.0
MEDIAN	60.6±0.8	59.3±6.5	67.2±4.5	61.2±0.4	66.2±1.1	79.8±3.3
M.VOTING	58.9±3.4	63.9±1.8	67.7±4.5	61.4±0.4	69.9±1.3	**85.1±1.1**
WMV	60.3±1.1	59.2±6.8	69.9±2.5	61.5±0.4	71.6±0.6	**85.2±1.5**
DT	60.2±1.1	62.6±2.8	63.3±6.1	61.2±0.5	67.2±1.2	76.7± 2.4

While both algorithms achieved incremental learning, Learn++ performed significantly better than AdaBoost.M1 on all combination rules, and usually with smaller confidence intervals. As expected, majority voting, weighted majority voting and the sum rule in general performed better than others.

4.3 Wine

The wine database from the UCI repository [10] is commonly used as a benchmark dataset. The dataset describes chemical analysis of 13 constituents found in three types of Italian wines, derived from three different cultivars of the same region. The data distribution and the test performances of both algorithms are shown in Tables 5 and 6 respectively. S_1 had instances only from classes 1 and 2, whereas S_2 introduced class 3. The number of classifiers used to train datasets S_1 and S_2 were set as 2 and 4, respectively, and kept constant for all experiments.

Table 5. Data distribution for Wine dataset

Dataset ↓	$Wine_1$	$Wine_2$	$Wine_3$
S_1	30	40	0
S_2	10	10	30
TEST	71	28	21

Table 6. Percent test performance of AdaBoost.M1 and Learn^{++} on Wine dataset

	AdaBoost.M1		**Learn^{++}**	
	TS_1	TS_2	TS_3	TS_1
SUM	60.2±6.1	77.8±9.7	61.2±4.1	82.2±6.1
PRODUCT	59.1±8.7	81.4±10.2	60.5±2.6	82.1±6.2
MEDIAN	59.0±7.2	68.1±16.8	64.5±2.4	84.0±9.1
M.VOTING	58.8±6.6	77.1±14.3	60.0±2.4	82.9±8.5
WMV	54.7±8.3	76.0±12.9	62.8±3.1	82.6±6.6
DT	62.1±3.8	73.4±16.8	60.7 ±3.4	70.7±4.8

As in previous datasets, both algorithms were able to learn incrementally from the new data, as seen by the improved generalization performances from one training session to the next. Learn^{++}, however, performed significantly better than AdaBoost.M1 on all combination rules except the decision templates, and usually with smaller (though still somewhat large) confidence intervals. We should add however that due to somewhat large confidence intervals, the performance differences among different combination rules were not statistically significant.

4.4 Optical Character Recognition Database

Also obtained from the UCI repository [10], OCR database consists of handwritten numeric characters, 0 through 9, digitized on an 8x8 grid creating 64 attributes for 10 classes. The data distribution and the mean performances of the two algorithms are shown in Tables 7 and 8, respectively. Note that the data distribution was made deliberately challenging, specifically designed to test the algorithms' ability to learn multiple new classes with each dataset, while retaining the knowledge of previously learned classes. In particular, S_1 consisted of classes 0,1,2,5,6 and 9, S_2 added classes 3 and 4, and S_3 introduced classes 7 and 8, but removes instances from classes 0 and 1. The number of classifiers used to train datasets S_1, S_2 and S_3 were set as 3, 3 and 3, respectively, and kept constant for all experiments. Interesting observations can be made from the generalization performances of AdaBoost and Learn^{++}.

Table 7. Data distribution for OCR dataset

Dataset ↓	*0*	*1*	*2*	*3*	*4*	*5*	*6*	*7*	*8*	*9*
S_1	250	250	250	0	0	250	250	0	0	250
S_2	100	100	100	250	250	100	100	0	0	100
S_3	0	0	50	150	150	50	50	400	400	0
TEST	100	100	100	100	100	100	100	100	100	100

Table 8. Percent test performance of AdaBoost.M1 and Learn^{++} on OCR dataset

	AdaBoost.M1			Learn^{++}		
	TS_1	TS_2	TS_3	TS_1	TS_2	TS_3
SUM	57.4±2.0	71.1±2.6	66.2±2.2	59.2±0.2	77.0±1.2	88.5±1.5
PRODUCT	58.0±0.5	70.3±2.6	64.5±2.8	59.8±1.1	78.2±0.5	88.4±1.4
MEDIAN	55.3±4.4	70.5±2.7	**67.0±1.9**	59.0±0.2	77.9±0.2	**90.5±1.8**
M.VOTING	58.1±0.5	71.9±2.7	65.0±2.5	58.6±0.3	75.5±1.0	82.0±0.7
WMV	57.4±2.0	68.8±1.8	**67.0±3.8**	59.2±0.1	77.2±0.4	89.3±1.0
DT	58.3±0.5	72.5±1.4	65.7±3.4	59.3±0.4	79.8±0.5	82.5±1.7

Similar to the previous databases discussed above, Learn^{++} outperformed AdaBoost, with statistical significance, on all combination rules. It is interesting to observe that AdaBoost incrementally learns the second database, however it displays substantial amount of forgetting from second to third training session. This indicates that AdaBoost is having difficulty in learning new classes, and at the same time retaining the information it had previously learned, particularly if subsequently generated classifiers are no longer trained with instances of previously seen classes. Conversely, Learn^{++} was able to achieve upper 80% to lower 90% classification performance, using any of the sum, product, median and weighted majority voting combination rules. Considering that the database was designed to test the incremental learning and knowledge retaining ability of the algorithm (by leaving instances of certain classes out), we can conclude that Learn^{++} places itself more favorably along the plasticity–stability spectrum.

5 Conclusion and Discussion

Our results indicate that AdaBoost.M1 can indeed learn incrementally from new data; however, its effectiveness is limited by its single-hypothesis-based distribution update rule. We should quickly point out that this is not a short coming of AdaBoost, as the algorithm was not originally intended for incremental learning, but rather to allow weak classifiers learn in an ensemble structure. As consistently seen in all results, and in particular in hostile learning environments, where the consecutive databases may introduce instances of new classes and/or remove instances from previously seen classes, the ensemble-based distribution update rule of Learn^{++} provides substantial performance improvement.. Therefore, we conclude that the ensemble based distribution update rule is indeed crucial in achieving efficient incremental learning.

We also note that Learn^{++} achieved narrower confidence intervals in its performances. This is significant, because a narrower confidence interval indicates better stability and robustness, qualities of considerable concern in incremental learning. Improved generalization performance along with a narrower confidence interval shows that Learn^{++} can achieve a delicate balance on the stability-plasticity spectrum.

We should note that despite its relative inferior performance in incremental learning, AdaBoost is still a strong contender: it has certainly shown promise in incremental learning of certain applications, including learning new classes. We believe that AdaBoost can still be used for incremental learning applications where the learning environment is less hostile than the one we created in our simulations. Also, since we were interested in efficient incremental learning, the ensemble sizes were kept to minimum. If AdaBoost were allowed to generate additional classifiers, it could have achieved better performances. The incremental learning ability of AdaBoost under such cases is currently being investigated.

Unlike the distribution update rule, the choice of specific combination rule does not appear to be very influential in the incremental learning performance of either algorithm. While there were some differences – sometimes significant – such differences were not consistent, and we believe that the impact of a particular combination rule is relatively minor, compared to that of the distribution update rule.

References

1. R. French, "Catastrophic forgetting in connectionist networks," *Trends in Cognitive Sciences*, vol. 3, no.4, pp. 128-135, 1999.
2. S.Grossberg, "Nonlinear neural networks: principles, mechanisms and architectures," Neural Networks, vol. 1, no. 1, pp. 17-61, 1988.
3. G.A. Carpenter, S. Grossberg, N. Markuzon, J.H. Reynolds, and D.B. Rosen, "Fuzzy ARTMAP: A neural network architecture for incremental learning of analog multidimensional maps," IEEE Trans. on Neural Networks, vol. 3, no. 5, pp.698-713, 1992.
4. Y. Freund, R. Schapire, Decision-theoretic generalization of on-line learning and an application to boosting, *J. Comp. Sys. Sci.*, vol. 55, no. 1, pp. 119-13, 1997.
5. R. Polikar, L. Udpa, S. Udpa, V. Honavar., "Learn^{++}: An incremental learning algorithm for supervised neural networks," *IEEE Trans. on System, Man and Cybernetics (C),* vol. 31, no. 4, pp. 497-508, 2001.
6. R. Polikar, J. Byorick, S. Krause, A. Marino, M. Moreton, "Learn++: A classifier independent incremental learning algorithm for supervised neural networks," Proc. of Int. Joint Conference on Neural Networks (IJCNN 2002), vol. 2, pp. 1742-1747, Honolulu, HI, 12-17 May 2002.
7. J. Kittler, M. Hatef, R. P.W. Duin, and J. Matas, "On combining classifiers," IEEE Trans. on Pattern Analy. and Machine Int., vol. 20, no. 3, pp.226-239, 1998.
8. L.I. Kuncheva. Combining Pattern Classifiers: Methods and Algorithms. John Wiley & Sons, N.J., 2004.
9. L. I. Kuncheva, J. C. Bezdek, and R. P. W. Duin, "Decision templates for multiple classifier fusion: an experimental comparison," Pattern Rec., vol. 34, no. 2, pp. 299-314, 2001.
10. C.L. Blake and C.J. Merz, Univ. of California, Irvine, Repository of Machine Learning Databases at Irvine, CA.

Ensemble Learning with Local Diversity*

Ricardo Ñanculef[1], Carlos Valle[1], Héctor Allende[1], and Claudio Moraga[2,3]

[1] Universidad Técnica Federico Santa María,
Departamento de Informática, CP 110-V Valparaíso, Chile
{jnancu, cvalle, hallende}@inf.utfsm.cl
[2] European Centre for Soft Computing 33600 Mieres, Asturias, Spain
[3] Dortmund University, 44221 Dortmund, Germany
claudio.moraga@udo.edu

Abstract. The concept of *Diversity* is now recognized as a key characteristic of successful ensembles of predictors. In this paper we investigate an algorithm to generate diversity locally in regression ensembles of neural networks, which is based on the idea of imposing a neighborhood relation over the set of learners. In this algorithm each predictor iteratively improves its state considering only information about the performance of the neighbors to generate a sort of local negative correlation. We will assess our technique on two real data sets and compare this with *Negative Correlation Learning*, an effective technique to get diverse ensembles. We will demonstrate that the local approach exhibits better or comparable results than this global one.

1 Introduction

Ensemble methods offer a simple and flexible way to build powerful learning machines for a great variety of problems including classification, regression and clustering [7] [3]. An ensemble algorithm to learn a set of examples $D = \{(x_i, y_i); i = 1, \ldots, m\}$ selects a set of predictors $S = \{h_0, h_2, \ldots, h_{n-1}\}$ from some base hypothesis space H and builds a decision function f as a composition $f = \bigoplus S$, where $\bigoplus$ is an aggregation operator such as voting for categorical outputs or a linear combination for continuous outputs.

To be useful, the set S has to have some degree of heterogeneity or *diversity* that allows the group compensate individual errors and reach a better expected performance. The characterization of methods to generate diversity has matured in the last years [5] [6] and the concept is now recognised as a central element to get significant performance improvements with the ensemble. *Negative Correlation Learning* [9] [11] for example has been proved to be an effective method to get diversity in regression ensembles.

In this paper we propose to introduce a neighborhood relation over the ensemble to locally generate diversity. Each predictor iteratively improves its state

* This work was supported in part by Research Grant Fondecyt (Chile) 1040365 and 7050205, and in part by Research Grant DGIP-UTFSM (Chile). Partial support was also received from Research Grant BMBF (Germany) CHL 03-Z13.

S. Kollias et al. (Eds.): ICANN 2006, Part I, LNCS 4131, pp. 264–273, 2006.

considering only information about the performance of the neighbors to generate a sort of local negative correlation. Different neighborhood sizes control the visibility each learner has about the ensemble, and allow to cover the gap between a single learning machine and completely independent learners. We will show that a desired global behavior can be generated considering only local learning rules, with small neighborhood sizes. It is clear that the more local the learning rules the more efficient the algorithm.

The remainder of this paper is organized as follows. In the next section we introduce the concept of diversity in regression ensembles and the *Negative Correlation Learning* algorithm. Our proposal is introduced in section 3. In section 4 we provide a set of experimental results on two real data sets to assess the introduced algorithm. We also compare this with Negative Correlation Learning. Conclusions and future work close the article.

2 Diversity and Negative Correlation Learning

The problem in designing successful ensemble learning algorithms is how to select an appropriate set of predictors from the base hypothesis space H and how to aggregate them. Studies tend to show that diversity in errors of the predictors is a key characteristic to get better generalization performance, although the way to approach the concept is highly heterogeneous. Please refer to [3] for an exhaustive taxonomy of diversity creation methods in classification and regression scenarios.

In case of regression estimation it is well-known the so called *Ambiguity Decomposition* [1] for the quadratic loss of an ensemble F obtained as a convex combination of a set of n predictors $f_0, f_2, \ldots, f_{n-1}$,

$$\bar{e} = (F - y)^2 = \sum_{i=0}^{n-1} w_i (y - f_i)^2 - \sum_{i=0}^{n-1} w_i (f_i - F)^2 \tag{1}$$

where (and for the remainder of this paper)

$$F(x) = \sum_{i=0}^{n-1} w_i f_i(x) \tag{2}$$

This decomposition states that the error of the ensemble can be decomposed into two terms where the first is the aggregation of the individual errors $(y - f_i)$ and the second (called ambiguity) measures deviations of these individual predictions around the ensemble prediction. It is clear that the higher the second term, the lower the ensemble error and so this seems an appropriate approach to quantify the concept of diversity. If the ambiguity is positive the ensemble loss is guaranteed to be less than the averaged individual errors.

Ambiguity decomposition suggests that individual learners should be trained considering information about their deviations around the ensemble. For example, the i-th learner could be trained considering the objective function

$$e_i = (y - f_i)^2 - \lambda (f_i - F)^2 \tag{3}$$

where the parameter λ weights the importance of the ambiguity component. As noted in [3] if the ensemble is uniformly weighted, we obtain that

$$e_i = (y - f_i)^2 - \lambda \sum_{j \neq i} (f_i - F)(f_j - F) \tag{4}$$

which is the learning rule considered in the *Negative Correlation Learning* algorithm described in [11] [9] for neural networks ensembles and strongly related with the constructive algorithm proposed in [4]. In this approach the set of learners is trained in parallel and synchronously such that, at time t, the learner i is trained to minimize (4) with the ensemble F computed considering the states of the learners at the previous time $t-1$.

If $\lambda = 1$ it should be noted that the ensemble is trained as a great learning machine with components f_i joined using non trainable connections. It can be verified taking the derivatives of e_i and $\bar{e}$ with respect to f_i and noting that they are proportional. On the other hand, if the parameter $\lambda = 0$, each learner is trained independently. Parameter $\lambda \in [0,1]$ covers the range between the two extremes. In [3] it is said that increasing parameter λ can force a wider basin of attraction for the objective function to be minimized, increasing in this way the possible configurations of the learners necessary to get the minimum. In practice and depending on the optimization landscape induced by the data, better solutions can be achieved using λ between 0 and 1.

3 Generating Diversity Locally

In *Negative Correlation Learning* one trains each member of the ensemble introducing information about the group behavior and controlling the weight each predictor gives to this information. With $\lambda < 1$ each predictor sees ensemble performance only partially. From this observation it seems natural to think in restricting the visibility that each predictor has on the group, allowing that they learn only considering the performance of a subset of the set of learners. To accomplish this we will introduce a neighborhood relation on the ensemble.

Let $S = \{f_0, f_2, \ldots, f_{l-1}\}$ be a set of learners. A one-dimensional linear neighborhood relation of order ν on S consists of a function $\psi : S \times S \rightarrow \{0,1\}$ such that

$$\psi(f_i, f_j) = 1 \Leftrightarrow (i-j)|l \leq \nu \, \text{or} \, (j-i)|l \leq \nu \tag{5}$$

where $i|l$ denotes the modulus after division of i by l. The neighborhood V_i of a learner i is the set of learners h_j $j \neq i$ for which $\psi(h_i, h_j) = 1$. Geometrically we obtain a ring, where two learners are neighbors if they are contiguous up to ν steps.

In [8] a bi-dimensional neighborhood relation is considered to get a cellular automata of learning machines and then to select a subset of machines to build an ensemble. However learners are trained independently before the construction of the bi-dimensional arrangement and the neighborhood relation is only used to determinate which predictors will "survive" from the original pool of trained learners. We want to use the neighborhood relation to define the learning process itself for each individual learner in a way such that each predictor adapts its state based on the behavior of its neighbors. Since diversity is desirable to get successful ensembles we want to define local learning rules encouraging the diversity of the set of learners.

A way to accomplish this could be to restrict the negative correlation learning rule for the i-th learner in equation (4), such that this takes into account only deviations $y - f_j$ concerning the neighbors of f_i. The learning function for this learner would be

$$e_i^{vic} = (y - f_i)^2 - \lambda \sum_{j \in V_i} (f_i - F)(f_j - F) \tag{6}$$

It should be noted that this learning rule is not completely local because it depends on the overall ensemble function F. However, we can obtain a decomposition similar to the ambiguity decomposition where the diversity component does not depend directly on F. In fact, we have that the ambiguity component of (1) can be written as

$$\begin{aligned} \sum_{i=1}^{n} w_i (f_i - F)^2 &= \sum_{i=0}^{n-1} w_i ((f_i - y) - (F - y))^2 \\ &= \sum_{i=0}^{n-1} w_i (f_i - y)^2 - 2 \sum_{i=0}^{n-1} w_i (f_i - y)(F - y) + (F - y)^2 \end{aligned} \tag{7}$$

Replacing in (1) we obtain

$$\begin{aligned} (F - y)^2 &= 2 \sum_{i=0}^{n-1} w_i (f_i - y)(F - y) - (F - y)^2 \\ (F - y)^2 &= \sum_{i=0}^{n-1} w_i (f_i - y)(F - y) \end{aligned} \tag{8}$$

Expanding F as in equation (2) one finally obtains

$$(F - y)^2 = \sum_{i=0}^{n-1} w_i^2 (y - f_i)^2 + \sum_{i=0}^{n-1} \sum_{j \neq i} w_i w_j (f_i - y)(f_j - y) \tag{9}$$

As in the ambiguity decomposition, the first term measures the individual performance of the estimator while the second measures a sort of error correlation between the different predictors. Clearly negative error correlation is desirable to get advantages of combining the predictors.

From this decomposition and neglecting the effect of the coupling coefficients w_j, it seems natural to train each learner with the training function

$$\tilde{e}_i = (y - f_i)^2 + \lambda \sum_{j \neq i} (f_i - y)(f_j - y) \tag{10}$$

where $\lambda > 0$ controls the importance of the group information versus the information about the individual performance. Now, we can restrict the last training function to get a rule which only depends on the neighborhood of f_i,

$$l_i = (y - f_i)^2 + \lambda \sum_{j \in V_i} (f_i - y)(f_j - y) \tag{11}$$

As in the negative correlation learning algorithm we will proceed iteratively correcting the state of f_i at time t based on the state of its neighbors at time $t-1$. Using f_i^t to denote the state of the learner f_i at time t, the learning function at time t can be explicitly expressed as

$$l_i^t = (y - f_i^t)^2 + \lambda \sum_{j \in V_i} (f_i^t - y)(f_j^{t-1} - y) \tag{12}$$

This procedure can be viewed as an synchronous cellular automaton with cells corresponding to each learner, states corresponding to a function in the hypothesis space and (stochastic) transition rules $\forall i$ corresponding to

$$f_i^t = y - \lambda \sum_{j \in V_i} (f_j^{t-1} - y) + \epsilon_i^t \tag{13}$$

where ϵ_i^t is the estimation error in the minimization of (12). After calculating equation (13) for $i = 0, \ldots, n-1$ the global ensemble transition rule is obtained

$$F^t = y - \kappa(F^{t-1} - y) + \bar{\epsilon}^t \tag{14}$$

where $\kappa = 2\lambda\nu$, $\bar{\epsilon}^t = \sum_i w_i \epsilon_i^t$ is the aggregated estimation error at time t and F^t is the state of the ensemble (2) at this time. If we think equation (14) as applied to a fixed point (x, y) the gradient of $C = (y - F^{t-1})^2$ at $F^{t-1} := F^{t-1}(x)$ is simply $\nabla C = (F^{t-1} - y)$. Then, equation (14) can be written as

$$F^t = F^{t-1} + -(1 + \kappa)\nabla C + \bar{\epsilon}^t \tag{15}$$

which shows that the global transition rule is in essence a gradient descent rule with step size $1 + \kappa$. Now, let us suppose at time $t = 0$ each learner approximates y with an additive error e_i^0, such that $F^0 = y + \bar{\epsilon}^0$. Then, the expected values $S_t = E[F^t - y]$ are governed by the recurrence

$$\begin{aligned} S_t &= E[\bar{\epsilon^t}] - \kappa S_{t-1} \\ S_0 &= E[\bar{\epsilon^0}] \end{aligned} \tag{16}$$

whose solution for $|\kappa| < 1$ is

$$S_t = \sum_{i=0}^{t} (-\kappa B)^i E[\bar{\epsilon^t}] = \frac{1}{1+\kappa B} E[\bar{\epsilon^t}] \tag{17}$$

where B is the backward operator $B^s x_t = x_{t-s}$. If the sequence of expected estimation errors $E[\bar{\epsilon}^t]$ does not depend on t we have

$$F^t - y \underset{t\to\infty}{\longrightarrow} \frac{1+(-\kappa)^\infty}{1+\kappa} E[\bar{\epsilon}^0] \tag{18}$$

This property suggest to choose $\kappa \in [0,1)$ since $\kappa > 1$ yields a divergent sequence of biases around the target. Moreover, to minimize the convergence point of the biases S_t we should choose κ close to 1.

Algorithm 1. The Proposed Algorithm

1: Let n be the number of learners and ν the order of the neighborhood.
2: Let f_i^t $i = 0, \ldots, n-1$ be the function implemented by the learner f_i at time $t = 0, \ldots, T$ and $V_i = \{f_{i-\nu}, f_{i-1}, f_{i+1}, \ldots, f_{i+\nu}\}$ the neighborhood of this learner.

3: **for** $t = 1$ to T
4: Train the learner f_i a number of epochs p to achieve the state

$$f_i^t = y - \lambda \sum_{j\in V_i} (f_j^{t-1} - y)$$

which corresponds to use the learning function

$$e_i^t = (y - f_i)^2 + \kappa \sum_{j\in V_i} (f_i - y)\left(f_j^{t-1} - y\right)$$

on the set of examples $D = \{(x_k, y_k); k = 1, \ldots, m\}$.
5: **end for**
6: Set the ensemble at time t to be $F(x) = 1/n \sum_{i=0}^{n-1} f_i(x)$

The algorithm implementing the proposed approach is presented as algorithm (1). It should be noted that there are two types of learning iterations; the loop in step 4 corresponds to group iterations where the learners in the ensemble share information about its behavior, while the implicit loop at step 5 corresponds to individual iterations where each learner modifies its state based on the group behavior information. In practice it can be observed that only one individual epoch can be enough to achieve good results.

4 Experimental Results

In this section we present results of empirical studies for analyzing different aspects of the proposed approach and for comparing this with Negative Correlation

Learning. In the whole set of experiments, two real data sets were used, namely *Boston* and NO_2. A detailed description of these data sets can be obtained from [2] and [10] respectively. In addition, neural networks with five sigmoidal hidden units and trained with standard backpropagation were employed as base learners. For each experiment figures will show error-bars corresponding to t-student confidence intervals with a significance of 0.02. Mean values are plotted with a symbol 'o'.

4.1 Effect of the Neighborhood Order

In this experiment we will analyze the effect of the neighborhood order on the results obtained with the proposed algorithm. We will keep constant the number of machines in the ensemble at $M = 20$ and parameter κ at 0.9, while the neighborhood order will be varied according to $\nu = 1, 2, \ldots, 9$. Figure (1) summarizes the results in the testing set. Results in the training were omitted due to space limitations, but the behavior is analogous.

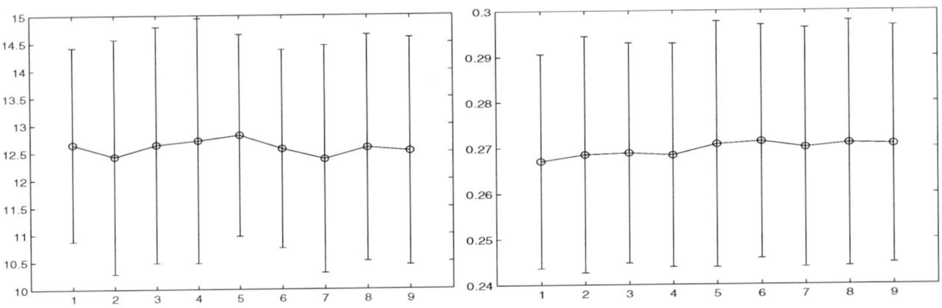

Fig. 1. MSE (vertical axis) versus neighborhood order (horizontal axis) in the Boston data set (at left) and the NO_2 data set (at right)

It can be observed that there is not a strong difference between the results obtained with different neighborhood sizes, both in terms of precision and dispersion. This difference is smaller in the testing set. Moreover, the minimum of errors occurs for both data sets at small neighborhood orders $\nu = 4$ or $\nu = 2$. This result is very attractive because it proves that a desired global behavior can be generated considering only local learning rules. In addition it should be clear that the smaller the neighborhood the more efficient the training process.

4.2 Effect of the Parameter κ

In this experiment we will analyze the effect of the κ parameter in the proposed algorithm. We will keep constant the number of machines in the ensemble at $M = 20$ and the neighborhood order at $\nu = 1$, while κ will be varied uniformly in $[0, 1.1]$ with gaps of size 0.1 and additionally in $[0.9, 1.02]$ with gaps of size 0.01. Figures (2) and (3) summarize the results for the Boston and NO2 data sets respectively.

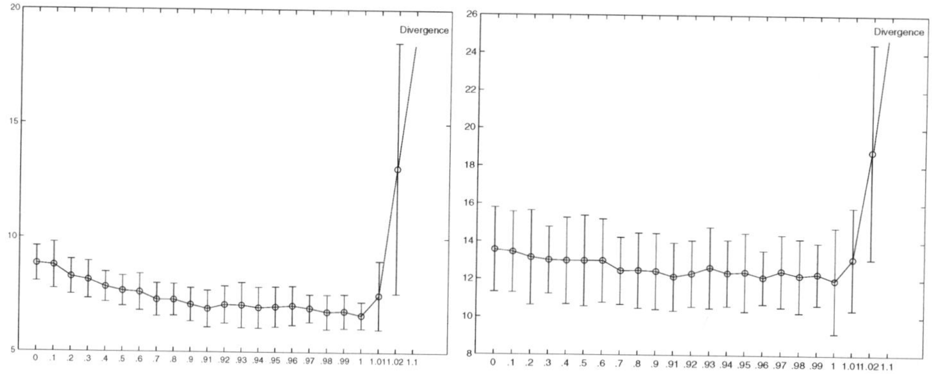

Fig. 2. MSE (vertical axis) versus value of parameter κ (horizontal axis) in the training set (at left) and the testing set (at right), corresponding to the Boston data.

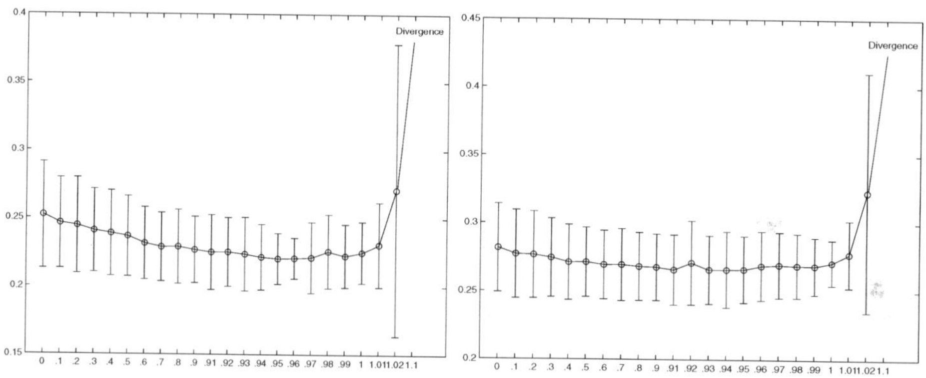

Fig. 3. MSE (vertical axis) versus value of parameter κ (horizontal axis) in the training set (at left) and the testing set (at right), corresponding to the NO_2 data.

This experiment supports the theoretical analysis of previous section for the effect of parameter κ. Both precision and variability tend to improve as κ increases from 0.0 to 1.0. Near the critical value of $\kappa = 1$ numerical instabilities appear and above this threshold the learning process becomes divergent.

4.3 Comparing with NC and Different Ensemble Sizes

In this experiment we compare our algorithm with Negative Correlation Learning (NC). Since the key characteristic in the proposed approach is the introduction of local learning rules instead of global ones we will fix the neighborhood order at the the smallest value $\nu = 1$. Based the previous experiment we will select the best parameter κ for our algorithm. Note that this "best" parameter is really the best for an ensemble of size $M = 20$, however the number of machines in the ensemble will be varied according to $M = 3, 4, 5, 6, 7, 8, 9, 20$.

Previously to this comparison we conducted experiments with NC using an ensemble of size $M = 20$ to determine the best parameter λ testing values between 0 and 1 with gaps of size 0.1. In the Boston data set the best testing result is achieved at $\lambda = 0.5$ as reported in [3]. In the NO_2 data set, the minimum testing error corresponds to $\lambda = 0.4$. Values of λ greater than 0.5 cause divergence of the algorithm in both data sets. Figures (4) and (5) summarize the results. The circle-solid curves corresponds to our algorithm and the dotted-diamond curves to Negative Correlation. To get a better visualization the last curve was horizontally (but not vertically!) shifted a bit.

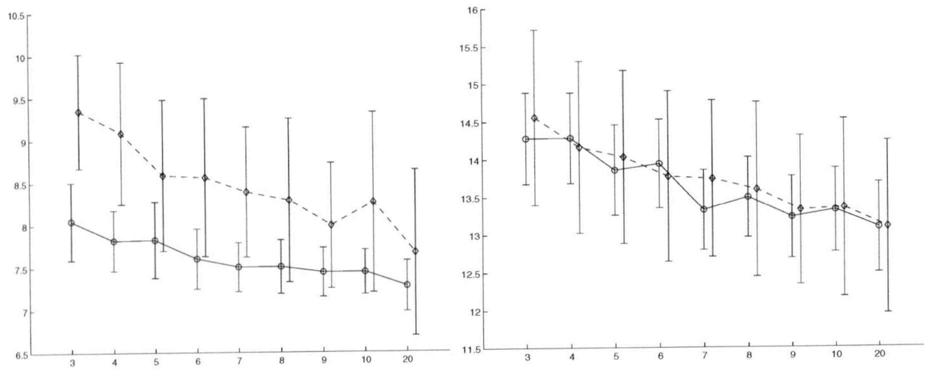

Fig. 4. MSE (vertical axis) versus number of machines (horizontal axis) in the training set (at left) and the testing set (at right), corresponding to the Boston data.

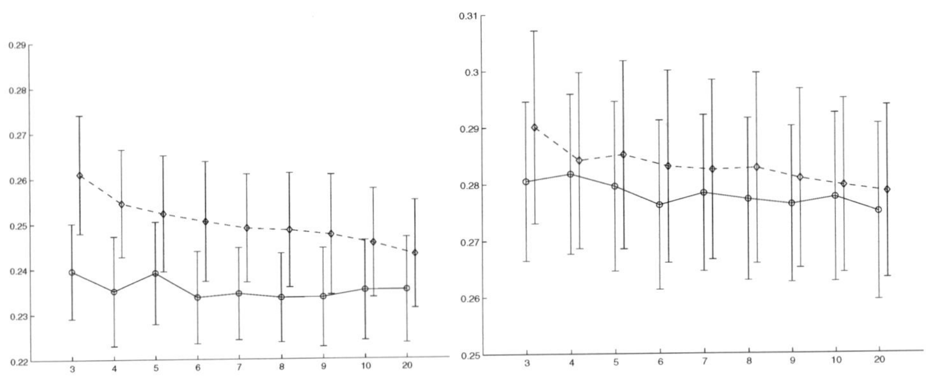

Fig. 5. MSE (vertical axis) versus number of machines (horizontal axis) in the training set (at left) and the testing set (at right), corresponding to the NO_2 data.

From this experiment we can conclude that a local approach to generate diversity exhibit better or comparable results than a global approach such as Negative Correlation (NC) both in terms of accuracy and precision. Generalization looks not so good for our algorithm but its comparable to the typical results with these data sets. It is important to remark that our algorithm has complexity proportional to $2\nu m$ in the ensemble size because each learner adapts its state based only in the behavior of 2ν fixed neighbors (we used $\nu = 1$). On the other hand, NC has complexity proportional to m^2 because at each training iteration each machine needs to know the whole ensemble performance. This advantage was verified in practice but results were omitted due to space limitations.

5 Future Work

In the future work we plan to work with more complex neighborhoods such as bi-dimensional or asymmetric arrangements. We also want to experiment with a distributed version of the algorithm taking advantage of its parallel nature. Since recent results have shown that ensembles capable to balance the influence of data points in the training can get better generalization, we are also interested in designing an algorithm capable to control the influence of an example in a machine based on information about the influence of this point in the neighborhood. Finally, experiments with pattern recognition tasks should be incorporated.

References

1. J. Vedelsby A. Krogh, *Neural network ensembles, cross-validation and active learning*, Neural Information Processing Systems **7** (1995), 231–238.
2. C.L. Blake and C.J. Merz, *UCI repository of machine learning databases*, 1998.
3. G. Brown, *Diversity in neural network ensembles*, Ph.D. thesis, School of Computer Science, University of Birmingham, 2003.
4. S. Fahlman and C. Lebiere, *The cascade-correlation learning architecture*, Advances in Neural Information Processing Systems, vol. 2, Morgan Kaufmann, San Mateo, 1990, pp. 524–532.
5. R. Harris G. Brown, J. Wyatt and X. Yao, *Diversity creation methods: A survey and categorisation*, Information Fusion Journal (Special issue on Diversity in Multiple Classifier Systems) **6** (2004), no. 1, 5–20.
6. C. Whitaker L. Kuncheva, *Measures of diversity in classifier ensembles*, Machine Learning **51** (2003), 181–207.
7. J. Kittler F. Roli N.C. Oza, R. Polikar (ed.), *Multiple classifier systems, 6th international workshop, mcs 2005, seaside, ca, usa, june 13-15, 2005, proceedings*, Lecture Notes in Computer Science, vol. 3541, Springer, 2005.
8. T.W. Druovec P. Povalej, P. Kokol and B. Stiglic, *Machine-learning with cellular automata*, Advances in Intelligent Data Analysis VI, vol. 1, Springer-Verlag, 2005, pp. 305–315.
9. B. Rosen, *Ensemble learning using decorrelated neural networks*, Connection Science **8** (1999), no. 3-4, 373–384.
10. P. Vlachos, *StatLib datasets archive*, 2005.
11. X. Yao Y. Lui, *Ensemble learning via negative correlation*, Neural Networks **12** (1999), no. 10, 1399–1404.

A Machine Learning Approach to Define Weights for Linear Combination of Forecasts

Ricardo Prudêncio[1] and Teresa Ludermir[2]

[1] Departament of Information Science, Federal University of Pernambuco, Av. dos Reitores, s/n - CEP 50670-901 - Recife (PE) - Brazil
[2] Center of Informatics, Federal University of Pernambuco, Pobox 7851 - CEP 50732-970 - Recife (PE) - Brazil

Abstract. The linear combination of forecasts is a procedure that has improved the forecasting accuracy for different time series. In this procedure, each method being combined is associated to a numerical weight that indicates the contribution of the method in the combined forecast. We present the use of machine learning techniques to define the weights for the linear combination of forecasts. In this paper, a machine learning technique uses features of the series at hand to define the adequate weights for a pre-defined number of forecasting methods. In order to evaluate this solution, we implemented a prototype that uses a MLP network to combine two widespread methods. The experiments performed revealed significantly accurate forecasts.

1 Introduction

Combining time series forecasts from different methods is a procedure commonly used to improve forecasting accuracy [1]. In the linear combination of forecasts, a weight is associated to each available method, and the combined forecast is the weighted sum of the forecasts individually provided by the methods.

An approach that uses knowledge for defining the weights in the linear combination of forecasts is based on the development of expert systems, such as the Rule-Based Forecasting system [2]. The expert rules deployed by the system use descriptive features of the time series (such as length, basic trend,...) in order to define the adequate combining weight associated to each available forecasting method. Despite its good results, developing rules in this context may be unfeasible, since good forecasting experts are not always available [3].

In this paper, we proposed the use of machine learning for predicting the linear weights for combining forecasts. In the proposed solution, each training example stores the description of a series (i.e. the series features) and the combining weights that empirically obtained the best forecasting performance for the series. A machine learning technique uses a set of such examples to relate time series features and adequate combining weights.

In order to evaluate the proposed solution, we implemented a prototype that uses MLP neural networks [4] to define the weights for two widespread methods:

S. Kollias et al. (Eds.): ICANN 2006, Part I, LNCS 4131, pp. 274–283, 2006.

the Random Walk and the Autoregressive model [5]. The prototype was evaluated in 645 yearly series and the combined forecasts were compared to the forecasts provided by benchmarking methods. The experiments revealed that the forecasts generated by the predicted weights were significantly more accurate than the benchmarking forecasts.

Section 2 presents methods for linear combination of forecasts, followed by section 3 that describes the proposed solution. Section 4 brings the implemented prototype, experiments and results. Section 5 presents some related work. Finally, section 6 presents the conclusions and future work.

2 Combining Forecasts

The combination of forecasts from different methods is a well-established procedure for improving forecasting accuracy [1]. Empirical evidence has shown that procedures that combine forecasts often outperform the individual methods that are used in the combination [1][6].

The linear combination of K methods can be described as follows. Let $Z_t (t = 1, \ldots, T)$ be the available data of a series Z and let $Z_t (t = T+1, \ldots, T+H)$ be the H future values to be forecasted. Each method k uses the available data to calibrate its parameters and then generates its individual forecasts $\widetilde{Z}_t^k (t = T+1, \ldots, T+H)$. The combined forecasts $\widetilde{Z}_t^C$ are defined as:

$$\widetilde{Z}_t^C = \sum_{k=1}^{K} w^k * \widetilde{Z}_t^k \ (t = T+1, \ldots, T+H) \tag{1}$$

The combining weights $w^k (k = 1, \ldots, K)$ are numerical values that indicate the contribution of each individual method in the combined forecasts. Eventually constraints are imposed on the weights in such a way that:

$$\sum_{k=1}^{K} w^k = 1 \ and \ w^k \geq 0 \tag{2}$$

Different approaches for defining adequate combining weights can be identified in literature [6]. A very simple approach is to define equal weights (i.e. $w^k = 1/K$) for the combined methods, which is usually referred as the Simple Average (SA) combination method. Despite its simplicity, the SA method has shown to be robust in the forecasting of different series [7].

A more sophisticated approach for defining the combining weights is to treat the problem within the regression framework [8]. In this context, the individual forecasts are viewed as explanatory variables and the actual values of the series as the response variable. In order to estimate the best weights, each combined method generates *in-sample* forecasts (i.e. forecasts for the available data Z_t). Then, a least squares method computes the weights that minimize the squared error for the combined in-sample forecasts. Empirical results revealed that the regression-based methods are almost always less accurate than the SA [9].

An alternative approach that has shown promising results in the combination of forecasts is based on the development of expert systems, such as the Rule-Based Forecasting [2]. In that work, an expert system with 99 rules is used to weight four forecasting methods. The rule base was developed based on the guidelines provided by human experts. The authors used time series features in the rules (such as basic trend, presence of outliers) to modify the weight associated to each method. Figure 1 shows an example of rule used in the system. The rule defines how to modify the weights in case of a series that presents an insignificant trend. In the experiments performed using the expert system, the improvement in accuracy over the SA method has shown to besignificant.

Rule: Insignificant Basic Trend - IF NOT a significant basic trend, THEN add 0.05 to the weight on random walk and subtract it from that on the regression.

Fig. 1. Example of rule implemented in the Rule Based Forecasting system

3 The Proposed Solution

As seen, expert systems have been successfully used to define weights for the linear combination of forecasts. Unfortunately, the knowledge acquisition in these systems depends on the availability of good human forecasting experts, who are often scarce and expensive [3]. In order to minimize this difficulty, the use of machine learning is proposed to define the weights for combining forecasts.

Figure 2 presents the architecture of the proposed solution. The system has two phases: training and use. In the training phase, the Intelligent Combiner (IC) uses a supervised algorithm to acquire knowledge from a set of examples in the Database (DB). Each training example stores the descriptive features for a particular series and the best configuration of weights used to combine K methods for that series. The learner implemented in the IC module generalizes the experience stored in the examples by associating time series features to the most adequate combining weights. In the use phase, given a time series to be forecasted, the Feature Extractor (FE) module extracts the values of the time series features. According to these values, the IC module predicts an adequate configuration of weights for combining the K methods.

The solution proposed here contributes to two different fields: (1) in time series forecasting, since we provided a new method for combining forecasts; (2) in machine learning, since we used its concepts and techniques in a problem which was not tackled yet.

In the following subsections, we provide a more formal description of each module of the proposed solution. Section 4 presents an implemented prototype and the experiments that evaluated the proposed solution.

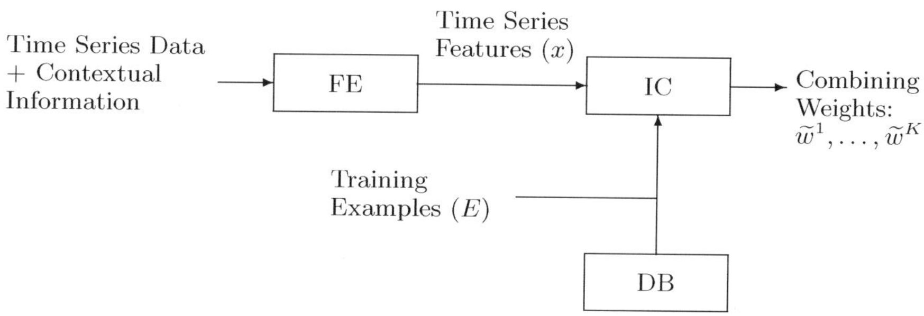

Fig. 2. System's architecture

3.1 The Feature Extractor

As said, the combining weights are predicted for a time series Z based on its description. Formally, a time series Z is described as a vector $x = (x^1, \ldots, x^p)$ where each x^j ($j = 1, \ldots, p$) corresponds to the value of a descriptive feature X_j. A time series feature can be either: (1) a contextual information directly provided by the user, such as the domain of the series, the time interval, the forecasting horizon, among others; or (2) a descriptive statistics automatically calculated from the available time series data Z_t ($t = 1, \ldots, T$).

In the proposed solution, the FE module extracts those features that are computed from the available data. These features may be for example the length, the presence of trend or seazonality, the autocorrelations, the amount of turning points, among others. Obviously the choice of appropriate features is highly dependent on the type of series at hand.

3.2 The Database

An important aspect to be considered is the generation of a set of training examples used by the IC module. Let $E = \{e_1, \ldots, e_n\}$ be a set of n examples, where each example stores: (1) the values of the p features for a particular series; and (2) the adequate weights for combining the K forecasting methods on that series. An example $e_i \in E$ is then defined as a vector $e_i = (x_i, w_i)$ where $x_i = (x_i^1, \ldots, x_i^p)$ is the description of the series and $w_i = (w_i^1, \ldots, w_i^K)$ is the best configuration of weights associated to the K methods.

In order to generate each example $e_i \in E$, we simulate the forecasting of a series Z_i by using the K methods and then we compute the configuration of weights that would obtain the best forecasting result if it was used to combine the forecasts provided by the K methods. For this, the following tasks have to be performed.

First, given a sample series Z_i ($i = 1, \ldots, n$), its data is divided into two parts: a fit period $Z_{i,t}(t = 1, \ldots, T_i)$ and a forecasting period $Z_{i,t}(t = T_i{+}1, \ldots, T_i{+}H)$. The fit period corresponds to the available data at time T_i used to calibrate the K methods. The forecasting period in turn corresponds to H observations to be forecasted by the K calibrated methods. Hence, for each series Z_i, this task results on the forecasts $\widetilde{Z}^k_{i,t}$ $(k = 1, \ldots, K)$ $(t = T_i + 1, \ldots, T_i + H)$ individually provided by the K calibrated methods for the forecasting period of the series.

In the second task, we defined the combining weights $(w_i^1, ..., w_i^K)$ that minimize a chosen forecasting error measure $\mathcal{E}$, computed for the combined forecasts $\widetilde{Z}^C_{i,t}(t = T_i + 1, ..., T_i + H)$. The measure $\mathcal{E}$ may be for example the Mean Absolute Error, or the Mean Squared Error. The task of defining the best combining weights can be formulated as an optimization problem as follows:

Minimize:

$$\mathcal{E}(\widetilde{Z}^C_{i,t}) = \mathcal{E}(\sum_{k=1}^{K} w_i^k * \widetilde{Z}^k_{i,t}) \tag{3}$$

Subject to:

$$\sum_{k=1}^{K} w_i^k = 1 \ and \ w_i^k \geq 0 \tag{4}$$

Different optimization methods may be used to solve this problem, considering the characteristics of the measure $\mathcal{E}$ to be minimized. For each series Z_i, this optimization process results on the weights $(w_i^1, ..., w_i^K)$ that would minimize the forecasting error $\mathcal{E}$ if they were used to combine the K methods on that series.

Finally, in the third task, the features $(x_i^1, \ldots, x_i^p)$ are extracted for describing the fit period of the series (as defined in the FE module). These features and the optimized weights $(w_i^1, ..., w_i^K)$ are stored in the DB as a new example.

3.3 The Intelligent Combiner

The Intelligent Combiner (IC) module implements a supervised learning algorithm that acquires knowledge from the set of training examples E. Given the set E, the algorithm is used to build a learner which is a regression model $\mathcal{L} : \mathcal{X} \rightarrow [0; 1]^K$ that receives as input a time series description and predicts the best configuration of combining weights, considering the error criteria $\mathcal{E}$. The final output of the system for a time series described as $x = (x^1, \ldots, x^p)$ is then:

$$\widetilde{w} = (\widetilde{w}^1, \ldots, \widetilde{w}^K) = \mathcal{L}(x) \tag{5}$$

The algorithms that can be used to generate the learner $\mathcal{L}$ may be for instance neural network models, algorithms for induction of regression trees, the k-nearest neighbour algorithm, among others.

4 The Implemented Prototype

In order to verify the viability of the proposal, a prototype was implemented for defining the combining weights of $K = 2$ methods: the Random Walk (RW) and the Auto-Regressive model (AR) [5]. The prototype was applied to forecast the yearly series of the M3-Competition [10], which provides a large set of time series related to economic and demographic domains. In the next subsections, we provide the implementation details as well as the experiments that evaluated the prototype. A short presentation of the implemented prototype can also be found in [11].

4.1 The Feature Extractor

In this module, the following features were used to describe the yearly series:

1. Length of the time series (L): number of observations of the series;
2. Basic Trend (BT): slope of the linear regression model;
3. Percentage of Turning Points (TP): Z_t is a turning point if $Z_{t-1} < Z_t > Z_{t+1}$ or $Z_{t-1} > Z_t < Z_{t+1}$. This feature measures the oscillation in a series;
4. First Coefficient of Autocorrelation (AC): large values of this feature suggest that the value of the series at a point influences the value at the next point;
5. Type of the time series ($TYPE$): it is represented by 5 categories, *micro, macro, industry, finances* and *demographic*.

The first four features are directly computed using the series data and TYPE in turn is a contextual information provided by the authors of M3-Competition.

4.2 The Database

In this case study, we used the 645 yearly series of the M3-Competition to generate the set of examples. Each time series was used to generate a different example as defined in the section 3.2.

First, given a series Z_i, the last $H = 6$ years of the series were defined as the forecasting period and the remaining data of the series was defined as the fit period (as suggested in the M3-Competition). After calibration, the RW and AR models provided its individual forecasts $\widetilde{Z}^k_{i,t}$ $(k = 1, 2)$ $(t = T_i + 1, \ldots, T_i + 6)$

Second, we defined the combining weights $w_i^k (k = 1, 2)$ that minimized the Mean Absolute Error (MAE) of the combined forecasts $\widetilde{Z}^C_{i,t}(t = T_i+1, ..., T_i+6)$. This task was formulated as the optimization problem:

Minimize:

$$MAE(\widetilde{Z}^C_{i,t}) = \frac{1}{6} \sum_{t=T_i+1}^{T_i+6} |Z_{i,t} - \widetilde{Z}^C_{i,t}| = \frac{1}{6} \sum_{t=T_i+1}^{T_i+6} |Z_{i,t} - \sum_{k=1}^{2} (w_i^k * \widetilde{Z}^k_{i,t})| \quad (6)$$

Subject to:

$$\sum_{k=1}^{2} w_i^k = 1 \ and \ w_i^k \geq 0 \quad (7)$$

This optimization problem was treated using a line search algorithm implemented in the Optimization toolbox for Matlab [14].

Finally, the example associated to the series Z_i is composed by the values of the five features (defined in the FE module, section 4.1) computed on the fit data and the optimum weights $w_i = (w_i^1, w_i^2)$.

4.3 The Intelligent Combiner

In the implemented prototype, the IC module uses the Multi-Layer Perceptron (MLP) network [4] (one hidden layer) as the learner. The MLP input layer has 9 units that represent the 5 time series features described in the FE module. The first four input units received the values of the numeric features (i.e. L, BT, TP, AC). The feature TYPE was represented by 5 binary attributes (either 1 or 0 value), each one associated to a different category of series (see fig. 3).

The output layer has two nodes that represented the weights associated to the RW and AR models. The output nodes used sigmoid functions which ensures that the predicted weights are non-negative. In order to ensure that the predicted weights sum to one (see eq. 2), the outputs of the MLP were normalized. Formally, let O_1 and O_2 be the values provided as output for a given time series description x. The predicted weights $\widetilde{w}^1$ and $\widetilde{w}^2$ are defined as:

$$\widetilde{w}^k = \frac{O_k}{\sum_{l=1}^{2} O_l} \quad (k = 1, 2) \tag{8}$$

The MLP training was performed by the BackPropagation (BP) algorithm [4] and followed the benchmark training rules provided in [12]. The BP algorithm was implemented by using the Neural Network Toolbox for Matlab [13].

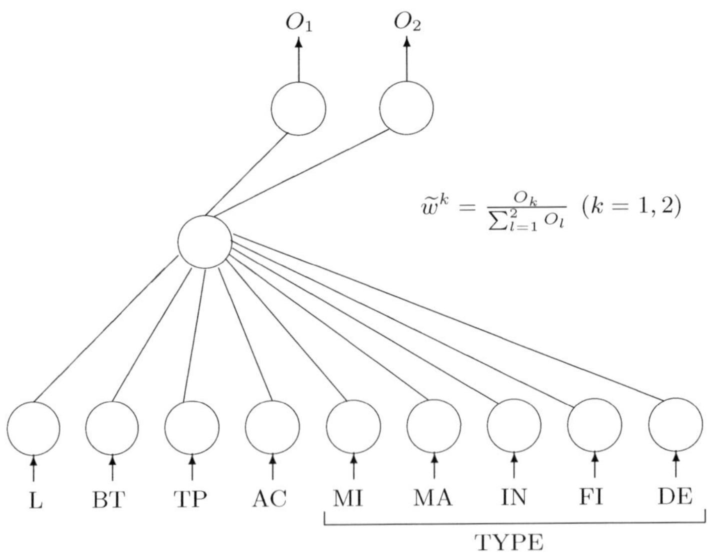

Fig. 3. MLP used to define the combining weights for the RW and AR models

4.4 Experiments and Results

In the performed experiments, the set of 645 examples in the DB was equally divided into training, validation and test sets. We trained the MLP using 2, 4, 6, 8 and 10 nodes in the hidden layer (30 runs for each value). The optimum number of hidden nodes was chosen as the value that obtained the lowest average SSE error on the validation set over the 30 runs. Table 1 summarizes the MLP results. As it can be seen, the optimum number of nodes according to the validation error was 8 nodes. The gain obtained by this value was also observed in the test set.

Table 1. Training results

Number of	SSE Training		SSE Validation		SSE Test	
Hidden Nodes	Average	Deviation	Average	Deviation	Average	Deviation
2	65.15	1.87	69.46	0.80	68.34	1.03
4	64.16	1.97	69.49	0.87	67.78	1.20
6	64.22	1.81	69.34	0.86	67.14	1.28
8	**64.13**	**2.28**	**69.29**	**0.67**	**66.74**	**1.35**
10	64.37	2.62	69.56	1.03	67.54	1.32

We further investigated the quality of the combined forecasts generated by using the weights predicted by the selected MLP (with 8 nodes). Let TEST be the set of 215 series that were used to generate the test set of the above experiments. Let $\widetilde{w}_i^1$ and $\widetilde{w}_i^2$ be the weights predicted for the series $Z_i \in$ TEST. The forecasting error produced by the combination of methods at time t is:

$$e_{i,t}^C = Z_{i,t} - \widetilde{Z}_{i,t}^C = Z_{i,t} - \sum_{k=1}^{2} (\widetilde{w}_i^k * \widetilde{Z}_{i,t}^k) \tag{9}$$

In order to evaluate the amount of these errors across all series $Z_i \in$ TEST for all forecasting points $t \in \{T_i+1, \ldots, T_i+6\}$, we considered the Percentage Better (PB) measure [15]. Given a reference method R that serves for comparison, the PB measure is computed as follows:

$$PB_R = 100 * \frac{1}{m} \sum_{Z_i \in TEST} \sum_{t=T_i+1}^{T_i+6} \delta_{i,t} \tag{10}$$

where

$$\delta_{i,t} = \begin{cases} 1, & \text{if } |e_{i,t}^R| < |e_{i,t}^C| \\ 0, & \text{otherwise} \end{cases} \tag{11}$$

In the above definition, $e_{i,t}^R$ is the forecasting error obtained by the method R in the i-th series at forecasting time t, and m is the number of times in which $|e_{i,t}^R| \neq |e_{i,t}^C|$. Hence, PB_R indicates in percentage terms, the number of times that the error obtained by the method R was lower than the error obtained using the combined forecasts. Hence, values lower than 50 indicate that the combined forecasts are more accurate than the forecasts obtained by the reference method.

The PB measure was computed for three reference methods. The first one is merely to use RW for forecasting all series and the second is to use AR for all series. The third reference method is the Simple Average (SA). Table 2 summarizes the results over the 30 runs of the best MLP. The average PB measure was lower than 50% for all reference methods which indicates that the combined

Table 2. Comparative forecasting performance measured by PB

Reference Method	PB Measure		
	Average	Deviation	Conf. Interv. (95%)
RW	42.20	0.26	[42.11; 42.29]
AR	40.20	0.44	[40.04; 40.36]
SA	43.24	1.45	[42.72; 43.76]

forecasts were more accurate. The confidence intervals suggest that the obtained gain is statistically significant.

5 Related Work

The proposed solution is closely related to previous work that used machine learning to *select* forecasting methods [3][16][17][18][19]. In the selection approach, time series features are used by a learning technique to predict the *best method* for forecasting among a set of candidates. In the solution presented here, the learner uses the features to define the *best combination of methods*. Our approach is more general since the selection problem can be seen as a special case of combination where $w^k = 1$ if k is the best method and 0 otherwise.

In a previous work [18], we adopted the selection approach and applied a MLP network to select among the RW and AR models. Experiments were performed in the same 645 yearly series and the same forecasting period adopted in the present work. Table 3 shows the *PB* value computed for the combination approach, by using the selection approach developed in [18] as reference. As it may be seen, the combination procedure obtained a performance gain when compared to the simple selection approach adopted in the previous work.

Table 3. Forecasting performance (PB) of the combined forecasts by using the selection approach as a basis for comparision

Reference Method	PB Measure		
	Average	Deviation	Conf. Interv. (95%)
Selection Approach (MLP)	48.28	0.74	[48.01; 48.54]

6 Conclusion

In this work, we proposed the use of machine learning to define linear weights for combining forecasts. In order to evaluate the proposal, we applied a MLP network to support the combination of two forecasting methodss. The performed experiments revealed a significant gain in accuracy compared to benchmarking procedures for forecasting. Modifications in the current implementation may be performed, such as augmenting the set of features, optimizing the MLP design and considering new forecasting error measures to be minimized.

Although the focus of our work is forecasting, the proposed method can also be adapted to other situations, for example, to combine classifiers. In this context, instead of time series features, characteristics of classification tasks should be considered (such as the number of examples). In this context, our solution becomes related to the Meta-Learning field [20], which studies how the performance of learning algorithms can be improved through experience. The use of the proposed solution to combine classifiers will be investigated in future work.

References

1. Hibon, M., Evgeniou, T.: To combine or not to combine: selecting among forecasts and their combinations. International Journal of Forecasting, 21(1) (2004) 15–24.
2. Adya, M., Armstrong, J. S., Collopy, F., Kennedy, M.: Automatic identification of time series features for rule-based forecasting. International Journal of Forecasting, 17(2) (2001) 143–157.
3. Arinze, B.: Selecting appropriate forecasting models using rule induction. Omega-International Journal of Management Science, 22(6) (1994) 647–658.
4. Rumelhart, D.E., Hinton, G.E., Williams, R.J.: Learning representations by back-propagation errors. Nature, 323 (1986) 533–536.
5. Harvey, A.: Time Series Models. MIT Press, Cambridge, MA (1993)
6. DeMenezes, L., Bunn, D., Taylor, J.: Review of guidelines for the use of combined forecasts. European Jour. of Operational Research, 120 (2000) 190–204.
7. Armstrong, J.: Findings from evidence-based forecasting: methods for reducing forecast error. (2005) Available at: http://www.jscottarmstrong.com/. Accessed on March 20, 2006.
8. Granger, C.W.J., Ramanathan, R.: Improved methods of combining forecasts. Journal of Forecasting, 3 (1984) 197-204.
9. Asku, C., Gunter, S.I.: An empirical analysis of the accuracy of SA, OLS, ERLS and NRLS combination of forecasts. Intern. Journal of Forecasting, 8 (1992) 27-43.
10. Makridakis, S., Hibon, M.: The M3-competition: results, conclusions and implications. International Journal of Forecasting, 16(4) (2000) 451–476.
11. Prudêncio, R., Ludermir, T.B.: Using Machine Learning Techniques to Combine Forecasting Methods. Lect. Notes in Artificial Intelligence, 3339 (2004) 1122–1127.
12. Prechelt, L.: Proben 1: a set of neural network benchmark problems and benchmarking rules, Tech. Rep. 21/94, Fakultat fur Informatik, Karlsruhe (1994).
13. Demuth, H., Beale, M.:. Neural Network Toolbox for Use with Matlab, The Mathworks Inc, (2003).
14. The Mathworks, Optimization Toolbox User's Guide, The Mathworks Inc. (2003).
15. Flores, B.E.: Use of the sign test to supplement the percentage better statistic. International Journal of Forecasting, 2 (1986) 477–489.
16. Chu, C-H., Widjaja, D.: Neural network system for forecasting method selection. Decision Support Systems, 12(1) (1994) 13–24.
17. Venkatachalan, A.R., Sohl, J.E.: An intelligent model selection and forecasting system. Journal of Forecasting, 18 (1999) 167–180.
18. Prudêncio, R., Ludermir, T.B.: Meta-learning approaches for selecting time series models. Neurocomputing Journal, 61(C) (2004) 121–137.
19. Prudêncio, R., Ludermir, T.B., DeCarvalho, F.: A modal symbolic classifier for selecting time series models. Pattern Recogn. Letters, 25(8) (2004) 911–921.
20. Giraud-Carrier, C., Vilalta, R., Brazdil, P.: Introduction to the special issue on meta-Learning. Machine Learning, 54 (2004) 187-193.

A Game-Theoretic Approach to Weighted Majority Voting for Combining SVM Classifiers

Harris Georgiou, Michael Mavroforakis, and Sergios Theodoridis

Dept. of Informatics and Telecommunications, Division of Communications and Signal Processing, University of Athens, Greece
http://www.di.uoa.gr/~dsp
157 84, Panepistimioupolis, Ilissia, Athens, Greece
Harris Georgiou, xgeorgio@di.uoa.gr

Abstract. A new approach from the game-theoretic point of view is proposed for the problem of optimally combining classifiers in dichotomous choice situations. The analysis of weighted majority voting under the viewpoint of coalition gaming, leads to the existence of *analytical solutions* to optimal weights for the classifiers based on their prior competencies. The general framework of weighted majority rules (WMR) is tested against common rank-based and simple majority models, as well as two soft-output averaging rules. Experimental results with combined support vector machine (SVM) classifiers on benchmark classification tasks have proven that WMR, employing the theoretically optimal solution for combination weights proposed in this work, outperformed all the other rank-based, simple majority and soft-output averaging methods. It also provides a very generic and theoretically well-defined framework for all hard-output (voting) combination schemes between any type of classifier architecture.

1 Introduction

1.1 Classifier Combination and Game Theory

In the discipline of collective decision-making, a group of N experts with moderate performance levels are combined in an optimal way, in order to produce a collective decision that is better than the best estimate of each individual expert in the group. According to the famous Condorcet Jury Theorem [1], if the experts' individual decisions are independent and their corresponding estimations are more likely to be correct than incorrect ($p_{correct}>0.5$), then an increase in the collective performance, as a group, is guaranteed when the individual estimations are combined. Moreover, this increase in performance continues to increase asymptotically as the size N of the group increases.

In the case where each expert selects only one out of M available options, the collective group decision can be estimated by the majority voting scheme, i.e., the choice selected is the one gathering the majority of votes. When the simple majority rule is employed, each of the N experts acts with the same common interest of reaching the

S. Kollias et al. (Eds.): ICANN 2006, Part I, LNCS 4131, pp. 284–292, 2006.

optimal collective decision. However, their individual choices place them in possibly contradicting estimations, with each expert trying *to impose* its decision to the others and to the group. This is a typical competitive situation, which can be modeled by the well-studied theory of non-zero sum competitive gaming in classic Game Theory [2]. In reality, each subgroup of consentient experts essentially represents an opposing assembly to all the other similar subgroups with different consensus of choice. It is clear that this second type of cooperative, instead of purely competitive (per expert), gaming reflects the problem of collective decision-making in the most generic way. Special sections of Game Theory, namely the Coalitions and Stable Sets in cooperative gaming [2], have studied the effects of introducing "weights" to the choice of each expert according to their competencies, in order to optimize the final decision of the group.

1.2 Weighted Majority Games and Weighted Majority Rules

The case of a dichotomous situation, where there are only two symmetrical choices for each expert (i.e., M=2) to vote for, then this restricted form is known as the weighted majority game (WMG) [2]. It has been proven by Nitzan and Paroush (1982) [3] and Shapley and Grofman (1984) [4], that the optimal decision rules, in terms of collective performance, are the weighted majority rules (WMR); this is in fact a different name for the well-known weighted majority voting schemes [5], which are often used in pattern recognition for combining hard-output classifiers. The same assertion has also been verified by Ben-Yashar and Nitzan [6] as the optimal aggregation rule for committees under the scope of informative voting in Decision Theory. Although there is in fact an exponential number of such WMR for each WMG, only a few of them can be proven to be well-defined or *qualified* combination rules and even fewer can be proven to be unique, i.e., not producing exactly the same decision profile with others [7]. For example, in the 2^{32} possible[1] voting games of five experts, there are exactly 85 qualified WMR if only positive integer weights are permitted, of which only seven are unique in terms of their decision profile [7].

In this paper, the notion of modeling dichotomous choice situations for a group of experts via the theory of WMG and WMR is for the first time applied for combining hard-output classifiers. Under the conditional independence assumption, a *closed form solution* for the voting weights in the WMR formula exists and it is *directly linked to each expert's competency*. This optimal weight profile for the voting experts is the log of the odds of their individual competencies [3], [4], [7], [8].

In this paper, this particular type of game-theoretic analytical solution for optimal expert combinations in dichotomous choice situations is tested for the first time against other popular combination schemes. The possibility of having a weighted voting scheme that is based only on the prior capabilities of the experts in the group, as well as on the theoretical assertion that this analytical solution is optimal, in terms of collective competency (at least for all non-trained, i.e., iteratively optimized, weights), is extremely attractive as an option of designing very simple yet effective combination models for an arbitrary pool of classifiers.

---

[1] For five experts with two choices each there are 2^5=32 decision profiles, each of which can be generally mapped in any of the two possible outputs of the combination rule. See [7].

2 Datasets and Methods

2.1 SVM Classifier Model

The SVM classifier was used as the base model for creating a pool of classifiers for each combination scheme. Specifically, a geometric nearest point algorithm (NPA) [9], based on the notion of reduced convex hulls (RCH) [10], was used for training a standard SVM architecture with radial-basis function (RBF) as the kernel of the non-linear mapping. In previous studies [11] have shown experimental evidence that optimal combinations of SVM classifiers can be achieved through linear combination rules, i.e., the same category of combination rules examined in this study. In the two averaging combination rules that use the soft-output of the individual classifiers, the distances from the decision boundary were used instead of the (thresholded) hard-output of the SVM classifier, as they are indicative of the corresponding classification confidence [12], [13].

2.2 Datasets and Feature Grouping

In order to assess the performance of each classifier combination method, a number of publicly available test datasets [14], with known single-classifier accuracy rates for this specific SVM training model, were used. These datasets are: 1) Diabetis, 2) Flare-Solar, 3) German, 4) Heart and 5) Waveform.

Each base dataset was randomly separated into a base training set and a validation set of samples. In order to make individually trained classifiers as "independent" as possible, the method of training them in different subspaces was employed. As it has been reported previously, e.g., [13], [15], this is an effective approach towards independence among classifiers. To this end, the training set was partitioned into K distinct segments of feature groupings, i.e., containing only some of the features (dimensions) of the initial dataset. Each group of features was created in a way that satisfied two constraints: (a) each group to be distinct, i.e., no feature is included in two or more groups, and (b) each group to contain a subset of features that can describe the classification task equally well as the other feature groups, i.e., employ a "fair" distribution of the available features into K groups. Satisfaction of the second constraint required a method for ranking all the features in terms of discrimination power against the two classes, as well as their statistical independency to all the other features in the initial training set. Thus, the MANOVA method [16] was used to assign a multivariate statistical significance value to each one of the features and then produce a sorted list based on (the log of) this value.

In order to create a "fair" partitioning of this list into equally efficient segments, features were selected in pairs from the top and bottom positions, putting the currently "best" and "worst" features in the same group. Furthermore, the efficiency of each group was measured in terms of summing the log of the statistical significance value, assigned by MANOVA, of all the features contained in this group. The log was employed in order to avoid excessive differences between the values assigned by MANOVA, thus creating more even subset sums of these values. Essentially, every such pair of features was assigned in groups sequentially, in a way that all groups contained features with approximately equal sum of the log of the values assigned by

MANOVA. In other words, the MANOVA-sorted list of features was "folded" once in the middle and then "cut" into K subsequent parts of equal sums of log-values, i.e., with every part exhibiting roughly the same sum of the log of the statistical significance values, accompanying each feature included in this part.

Each one of these K distinct feature groups was used for training an individual SVM classifier. Thus, each of these K classifiers used a different, dimensionally reduced, version of the original (full) training set and therefore learns a totally different classification task.

2.3 Classifier Combination Methods

Nine linear combination rules were examined in this study. Specifically, five hard-output combination methods were employed, namely three standard rank-based methods and two voting-based schemes. These rank-based rules are [8], [13]:

- minimum ("min")
- maximum ("max")
- median ("median")

The two majority rules, including the WMR model, are [8], [13]:

- simple majority voting ("majority")
- weighted majority voting, i.e.:

$$O_{wmr}(\vec{x}) = \sum_{i=1}^{K} w_i D_i(\vec{x}) \ . \tag{1}$$

where D_i is the hard-output of each of the K individual classifiers in the pool, w_i is its assigned weight and O_{wmr} the weighted majority sum. The final hard-output decision D_{wmr} of the WMR is taken against a fixed threshold (T) that defines the decision boundary for the combination rule [7], [8]:

$$D_{wmr}(\vec{x}) = sign\left(O_{wmr}(\vec{x}) - T\right) \ . \tag{2}$$

Specifically for the weighted majority voting scheme, three different methods for calculating the weight profile were tested for comparative results:

- "direct" weighting profile for WMR ("wmr/direct") [5], [8]:

$$w_i = p_i \quad , \quad p_i = P_i(\theta = \omega_{correct} \mid \vec{x}) \ . \tag{3}$$

- "odds" weighting profile for WMR ("wmr/odds") [7], [8]:

$$w_i = \frac{p_i}{1 - p_i} \quad , \quad p_i = P_i(\theta = \omega_{correct} \mid \vec{x}) \ . \tag{4}$$

- "logodds" weighting profile for WMR ("wmr/logodds") [7], [8]:

$$w_i = \log\left(\frac{p_i}{1 - p_i}\right) \quad , \quad p_i = P_i(\theta = \omega_{correct} \mid \vec{x}) \ . \tag{5}$$

where w_i is the combination weight assigned for the i-th classifier, p_i is its prior probability for correct classification, measured in the validation set, and θ, ω are the predicted class labels.

Additionally, two soft-output averaging models were included, a non-weighted and a weighted [8]:

- simple average ("average")
- weighted average ("lsewavg")

The weights in the weighted average rule were calculated as the optimal weighting profile of the individual classifier outputs against the correct classification tag, in terms of a least-squares error (LSE) minimization criterion [15]. Thus, this method can be considered as an example of "trained" weighting rules of soft-output classifiers. In contrast, the WMR approach employs fixed analytical weighting profile and hard-output classifications (votes) as input, that is, no further training is required.

3 Experiments and Results

The evaluation of the combination models consisted of two phases, namely: (a) the design and training of SVM classifiers, trained in distinctly different subspaces, and (b) the application of the various combination schemes to the outputs of the individual classifiers.

Each of the K classifiers was separately trained and optimized, using a different group of features from the full dataset, and subsequently evaluated using the corresponding validation set. This training/validation cycle was applied three times, for each of the five datasets, each time using a new random partitioning of the full dataset into training and validation sets. The mean values and standard deviations of the success rates of all the individual ($3K$) classifiers for each dataset, as well as the details about the size and dimensionality of each (full) training and validation sets, are presented in Table 1.

The K value, i.e., the number of feature groups for each dataset, was determined experimentally in a way that each of the corresponding K training segments would be adequate to produce a well-trained SVM classifier. Thus, the German training set was split in K=5 segments, while the Flare-Solar training set in K=4 segments.

Table 1. Single versus multiple classifier accuracy percentages per dataset and K values (number of dataset partitions)

Dataset	Train set	Vali-dat. set	Data Dim.	Single classifier accuracy	K value	Individual classifier mean acc%
diabetis	468	300	8	76.5 ± 1.7	5	68.3 ± 3.9
flare-solar	666	400	9	67.6 ± 1.8	4	55.7 ± 3.6
german	700	300	20	76.4 ± 2.1	5	68.9 ± 1.8
heart	170	100	13	84.0 ± 3.3	5	74.3 ± 2.3
waveform	400	4600	21	90.1 ± 0.4	5	81.1 ± 1.2

The classification outputs of the pool of K classifiers from each training/validation cycle were fed as input to all nine combination schemes, producing the corresponding combined classification outputs. Since the output of each of the K classifiers in the pool was calculated based on the same (dimensionally reduced) validation set, the corresponding outputs and accuracy of the combination rules also refer to this validation set.

Table 2 illustrates the mean accuracy of each combination rule (each cell corresponds to three training/validation cycles), as well as the mean value and standard deviation of the success rates of all nine combination rules, for each dataset and K value employed.

Table 2. Mean accuracy percentages of all the nine combination rules, with optimized decision threshold, per dataset and K values (number of feature groups and classifiers)

Combination Rule	Diabetis	Flare-Solar	German	Heart	Waveform
	K=5	K=4	K=5	K=5	K=5
average	71.67	66.08	70.67	85.33	88.12
lsewavg	76.11	65.58	71.56	85.00	86.79
min	68.56	55.92	70.67	69.00	72.98
max	69.11	60.42	67.33	76.67	85.95
median	69.00	58.33	69.78	80.00	81.17
majority	73.00	63.75	70.67	82.33	86.59
wmr/direct	74.00	66.58	70.67	82.33	86.59
wmr/odds	75.33	66.58	71.33	84.00	86.70
wmr/logodds	75.33	66.42	71.33	84.00	86.64
Mean	72.46	63.30	70.44	80.96	84.62
Stdev	2.99	4.06	1.28	5.25	4.77

In the sequel, the overall relative performance of each combination rule was determined in terms of ranking position for each case, i.e., according to its corresponding accuracy for each dataset and K value employed. Specifically, a weighted Borda scheme (wBorda) [17] was employed to attribute 10 points to the top-ranked combination rule, 9 points to the second, and so on. In case of a "tie" where two combination rules exhibited exactly the same classification accuracy, both got the same wBorda points for the specific ranking position. Using the results from Table 3, regarding the accuracies, Table 4 illustrates the corresponding wBorda ranking points of all nine combination rules, for each dataset and K value employed in this study.

Table 4 presents a summary of the results shown in Table 3, as well as the list of all the combination rules sorted according to their sum of wBorda points, i.e., their overall efficiency throughout the five original datasets. Tables 2 through 4 present the performance and wBorda results for all the combination rules with optimized decision threshold (T). The decision threshold employed by each combination rule was in every case optimized against final accuracy, using a typical Newton-Raphson optimization algorithm [18].

Table 3. wBorda value of all combination rules, with optimized decision threshold, per dataset and K values. Each cell value represents the ranking weight according to classification accuracies, with 10 points for top position, 9 points for the second and so on. In cases of equal accuracies, the same ranking weight was assigned to the corresponding combination rules.

Combination Rule	Diabetis	Flare-Solar	German	Heart	Waveform
	K=5	K=4	K=5	K=5	K=5
average	6	8	8	10	10
lsewavg	10	7	10	9	9
min	3	3	8	4	3
max	5	5	6	5	5
median	4	4	7	6	4
majority	7	6	8	7	6
wmr/direct	8	10	8	7	6
wmr/odds	9	10	9	8	8
wmr/logodds	9	9	9	8	7

Table 4. Overall evaluation of all the combination rules, with optimized decision threshold, using the wBorda results for all datasets and K values available. The list is sorted according to the wBorda sum and mean ranking position of each combination rule, from the best to the worst combination rule.

Combination Rule	wBorda Sum	wBorda Mean	wBorda Stdev
lsewavg	45	9.0	1.22
wmr/odds	44	8.8	0.84
average	42	8.4	1.67
wmr/logodds	42	8.4	0.89
wmr/direct	39	7.8	1.48
majority	34	6.8	0.84
max	26	5.2	0.45
median	25	5.0	1.41
min	21	4.2	2.17

4 Discussion

The results from Tables 3 and 4 clearly demonstrate the superior performance of the WMR model. Specifically, the all versions of the WMR model exhibited the best performance amongst all the other hard-output combination rules. As expected, it has been proven better than the simple majority voting, as well as all the other rank-based methods (max, min, median). The "odds" weighting profile has also been proven marginally better than the "direct"- and the "logodds"-based profiles for designing the optimal WMR formula.

Interestingly, the "odds"-based version of WMR exhibited better performance than the simple averaging rule, e.g., a soft-output combination model, losing only from the weighted averaging rule with LSE-trained weights. Thus, the WMR model, especially with the "odds" and "logodds" weighting profiles, performs equally well or better than simple soft-output averaging combination rules. All four weighted combination rules, i.e., the three WMR and the LSE-trained weighted average, have been clearly proven better than all the non-weighted hard-output combination rules.

Table 4 also demonstrates the robustness and stability of the each combination rule. For small values of standard deviation (less than unity) in the corresponding wBorda mean ranks, the relative ranking position of a combination rule against the others remains more or less the same. Thus, the maximum rule exhibits a consistently lower ranking position than the simple majority rule, while the "odds"- and the "logodds"-based versions of the WMR models perform consistently better than the simple majority and the three rank-based rules. Furthermore, the "odds"- and the "logodds"-based versions of WMR exhibit the same consistency and robustness as the simple majority rule but with higher success rates.

With respect to the overall performance of the combination rules, results from Tables 1 and 2 demonstrate that in all cases the best combination rules increased the overall success rates of the classifier pool, from +2% (German dataset) to +11% (Flare-Solar dataset), in many cases very close to or equal to the corresponding reference performance level of the single SVM classifier results.

The ensemble of these classifiers clearly demonstrates that the combination of multiple simpler models, each using a $1/K$ portion of the feature space of the dataset, instead of a single classifier for the complete feature space, can be used to reduce the overall training effort. Specifically for the SVM model, kernel evaluation employs inner product between vectors, i.e., its complexity is directly proportional to the dimensionality (number of features) in the input vectors. If this feature space reduction, from F to F/K features, results in a proportional increase in the complexity of the new (reduced) input space in terms of new class distributions, then it is expected that the training of each of the K SVM classifiers may be completed up to K times faster on average. A similar approach has also been examined in other studies [11], using an ensemble of SVM classifiers trained in small training sets, instead of one large training set for a single SVM classifier. Furthermore, there is evidence that such ensembles of kernel machines are more stable than the equivalent kernel machines [11]. This reduction in training time, of course, has to be compared to the additional overhead of calculating a combination rule for every output vector from the classifier pool. Consequently, if the optimal design of this combination rule is simple (linear) and efficient, and its weighting profile can be determined analytically with no need for iterative weight optimization, the WMR approach could prove very prominent for this role in classification tasks of high dimensionality and/or dataset sizes.

5 Conclusions

The game-theoretic modeling of combining classifiers in dichotomous choice problems leads to cooperative gaming approaches, specifically coalition gaming in the form of WMG. Theoretically optimal solutions for this type of games are the WMR schemes, often referred to as weighted majority voting. Under the conditional independence assumption for the experts, there exists a closed solution for the optimal weighting profiles for the WMR formula.

In this paper, experimental comparative results have shown that such simple combination models for ensembles of classifiers can be more efficient than all typical rank-based and simple majority schemes, as well as simple soft-output averaging schemes in some cases. Although the conditional independence assumption was moderately satisfied by using distinct partitions of the feature space, results have shown that the theoretical solution is still valid to a considerable extent. Therefore, the WMR can be asserted as a simple yet effective option for combining almost any type of classifier with others in an optimal and theoretically well-defined framework.

References

1. Condorcet, Marquis de: An Essay on the Application of Probability Theory to Plurality Decision Making: An Election Between Three Candidates. In: Sommerlad and Mclean (1989) 66-80
2. Owen, G.: Game Theory. 3rd edn. Academic Press, San Diego USA (1995)
3. Nitzan, S., Paroush, J.: Optimal Decision Rules in Uncertain Dichotomous Choice Situations. Int. Econ. Rev., 23 (1982) 289–297
4. Shapley, L., Grofman, B.: Optimizing Group Judgemental Accuracy in the Presence of Independence. Public Choice, 43 (1984) 329–343
5. Littlestone, N., Warmuth, M. K.: The Weighted Majority Algorithm. Information and Computation, 108 (1994) 212–261
6. Ben-Yashar, R., Nitzan, S.: The Optimal Decision Rule for Fixed-Size Committees in Dichotomous Choice Situations: The General Result. International Economic Review, 38 (1997) 175–186
7. Karotkin, D.: The Network of Weighted Majority Rules and Weighted Majority Games. Games and Econ. Beh., 22 (1998) 299–315
8. Kuncheva, L. I.: Combining Pattern Classifiers. John Wiley and Sons, New Jersey USA (2004)
9. Mavroforakis, M., Theodoridis, S.: Support Vector Machine Classification through Geometry. Proc. XII Eur. Sig. Proc. Conf. (EUSIPCO2005), Antalya, Turkey, Sep. 2005
10. Mavroforakis, M., Theodoridis, S.: A Geometric Approach to Support Vector Machine (SVM) Classification. IEEE Trans. NN, 2006 (in press)
11. Evgeniou, T., Pontil, M., Elisseeff, A.: Leave One Out Error, Stability, and Generalization of Voting Combinations of Classifiers. Machine Learning, 55 (2004) 71–97
12. Cristianini, N., Shawe-Taylor, J.: An Introduction to Support Vector Machines and Other Kernel-Based Learning Methods. Cambridge University Press UK (2000)
13. Jain, A., Duin, R., Mao J.: Statistical pattern recognition: a review. IEEE Trans. PAMI, 22 (2000) 4–37
14. Mavroforakis, M., Sdralis, M., Theodoridis, S.: A novel SVM Geometric Algorithm based on Reduced Convex Hulls. 18th International Conference on Pattern Recognition (ICPR 2006), Hong Kong, Aug. 2006
15. Theodoridis, S., Koutroumbas, K.: Pattern Recognition. 4th edn. Academic Press, San Diego USA (2006)
16. Cooley, W., Lohnes, P.: Multivariate data analysis. John Willey & Sons, New York USA (1971)
17. Parker, J. R.: Rank and Response Combination from Confusion Matrix Data. Information Fusion, 2 (2001) 113–120
18. Forsythe, G., Malcom, M., Moler, C.: Computer Methods for Mathematical Computations. Prentice-Hall, New Jersey USA (1977)

Improving the Expert Networks of a Modular Multi-Net System for Pattern Recognition

Mercedes Fernández-Redondo[1],
Joaquín Torres-Sospedra[1], and Carlos Hernández-Espinosa[1]

Departamento de Ingenieria y Ciencia de los Computadores, Universitat Jaume I,
Avda. Sos Baynat s/n, C.P. 12071, Castellon, Spain
{redondo, jtorres, espinosa}@icc.uji.es

Abstract. A *Modular* Multi-Net System consists on some networks which solve partially a problem. The original problem has been decomposed into subproblems and each network focuses on solving a subproblem. The *Mixture of Neural Networks* consist on some expert networks which solve the subproblems and a gating network which weights the outputs of the expert networks. The expert networks and the gating network are trained all together in order to reduce the correlation among the networks and minimize the error of the system. In this paper we present the *Mixture of Multilayer Feedforward* (*MixMF*) a method based on *MixNN* which uses *Multilayer Feedfoward* networks for the expert level. Finally, we have performed a comparison among *Simple Ensemble*, *MixNN* and *MixMF* and the results show that *MixMF* is the best performing method.

1 Introduction

The most important property of an artificial neural network is its generalization capability, which is the ability to correctly respond to inputs which were not used to adapt its weights. The use of a Multi-Net system increases this ability. We can see in [1,2] that there are some approaches to build a multi-net system for pattern recognition.

The most studied approach is the *Ensemble of Neural Networks* or *Comittee Machine* (Figure 1). It consists on training different networks and combine their output vectors in a suitable manner to give the global output or final hypothesis of the classification system. In [3,4] we can see that the use of ensembles increases the generalization capability with respect to a single neural network.

Although most of the methods to create a Multi-Net System are based on the *ensemble approach* [5,6], in this paper we focus on the *Modular Network*. The *Modular Network* is based on the idea of "*divide and conquer*". The network divides the problem into subproblems and each subproblem tends to be solved by one network. We will deal with the *Mixture of Neural Networks* (*MixNN*) because is one of the most known *modular* methods and we think it could be improved. It consists of some expert networks which solve the subproblems and a gating network which is used to combine the outputs of the expert networks. Its basic diagram is also in Figure 1.

The original *Mixture of Neural Networks* (*MixNN*) [7,8] is based on a quite simple neural network architecture. We think that *MixNN* could perform better if the method was based on *Multilayer Feedforward* networks. In this paper we present a *Mixture*

S. Kollias et al. (Eds.): ICANN 2006, Part I, LNCS 4131, pp. 293–302, 2006.

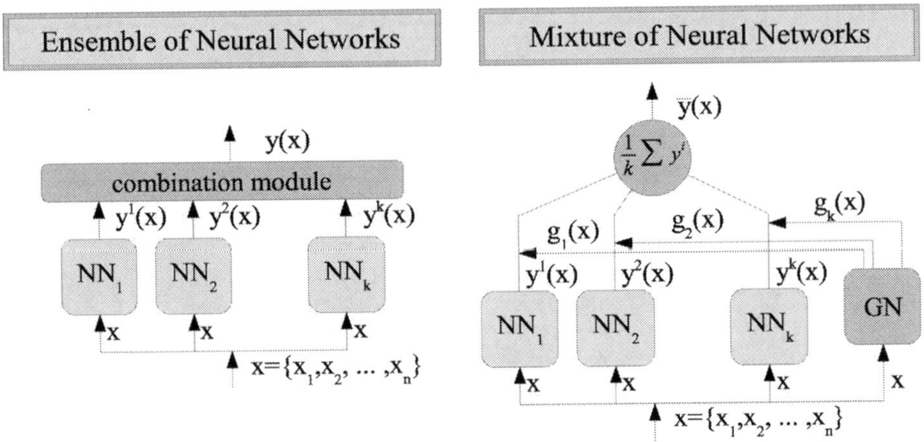

Fig. 1. Simple Ensemble and Mixture of Neural Networks diagrams

of Multilayer Feedforward Networks (*MixMF*) which is a modular approach based on Multilayer Feedforward networks trained with Backpropagation.

In section 2 we describe the *MixNN* and the *MixMF*. In subsection 3.1 we describe the databases used in our experiments. The results we have obtained are in subsection 3.2. We have also calculated the mean *Percentage of Error Reduction* to compare the methods, these results appear in subsetion 3.3.

2 Theory

In this section, we describe the *Simple Ensemble*, the *Mixture of Neural Networks* and the *Mixture of Multilayer Feedforward* Networks.

2.1 Simple Ensemble

Simple Ensemble is a method to build an ensemble of neural networks which consist on training some networks with different weights initialization. The mean of the outputs are applied to get the output of the ensemble.

$$\overline{y}_{class}(x) = \frac{1}{k} \cdot \sum_{net=1}^{k} y_{class}^{net}(x) \tag{1}$$

The output yielding the maximum of the averaged values is chosen as the correct class.

$$h_{simpleensemble}(x) = \arg \max_{class=1,\dots,q} \overline{y}_{class}(x) \tag{2}$$

Where q is the number of classes, k is the number of networks in the ensemble.

2.2 Mixture of Neural Networks

The *Mixture of Neural Networks* (*MixNN*) is a method to build a Multi-Net System based on the *Modular approach*. It consists on training *k* expert networks and a gating network. The input *x* is applied to the expert networks and the gating network. The modular network output is:

$$\overline{y}_{class} = \sum_{net=1}^{k} g_{net} \cdot y_{class}^{net} \tag{3}$$

Where the output of the expert networks is described in equation 4 and the output of the gating networks is described in equation 5

$$y_{class}^{net} = x^{T} \cdot w_{class}^{net} \tag{4}$$

$$g_{net} = \frac{\exp\left(x^{T} \cdot a_{net}\right)}{\sum_{j=1}^{k} \exp\left(x^{T} \cdot a_{j}\right)} \tag{5}$$

The output yielding the maximum of the averaged values is chosen as the correct class.

$$h_{MixNN}(x) = \arg \max_{class=1,\ldots,q} \overline{y}_{class}(x) \tag{6}$$

The neural networks used in *MixNN* are quite simple, in Figure 2 we show the diagram of an expert network.

To adapt the weights of the expert networks and the gating network, we have used the objective function described in equation 7.

$$L = \ln\left(\sum_{net=1}^{k} g_{net} \cdot \exp\left(-\frac{1}{2} \cdot \left\|d - y^{net}\right\|^{2}\right)\right) \tag{7}$$

The equations used to adapt the weights of the expert networks w and the gating network a are:

$$w_{class}^{net}(ite+1) = w_{class}^{net}(ite) + \eta \cdot h_{net} \cdot \left(d - y_{class}^{net}\right) \cdot x \tag{8}$$

$$a^{net}(ite+1) = a^{net}(ite) + \eta \cdot h_{net} \cdot \left(h_{net} - g_{net}\right) \cdot x \tag{9}$$

where:

$$h_{net} = \frac{g_{net} \cdot \exp\left(-\frac{1}{2}\left|d - y^{net}\right|^{2}\right)}{\sum_{j=1}^{k}\left(g_{j} \cdot \exp\left(-\frac{1}{2}\left|d - y^{j}\right|^{2}\right)\right)} \tag{10}$$

To train the networks of the classification system we have used the following algorithm:

Algorithm 1. Mixture of Neural Networks

Random initialization of networks
for $ite = 1$ to $Iterations$ **do**
  for each pattern from training set **do**
    for $net = 1$ to k **do**
      Adapt expert weights w^{net}
    end for
    Adapt gating weights a
  end for
  Calculate L_{ite} over Validation set
  Save weights
end for
Select iteration with maximum L (best iteration)
Set best iteration weights to network
Save final configuration

2.3 Mixture of Multilayer Feedforward Networks

Mixture of Multilayer Feedforward Networks (*MixMF*) is the new method we propose to build a modular network. *MixMF* is an approach of *MixNN* where the expert networks are *Multilayer Feedforward* networks with one hidden layer and threshold nodes. *Multilayer Feedforward* networks are more accurate than the expert networks used in *MixNN* [9], but the training process is slower. In Figure 2 we show the diagram of a *MixMF* expert network with a single hidden layer.

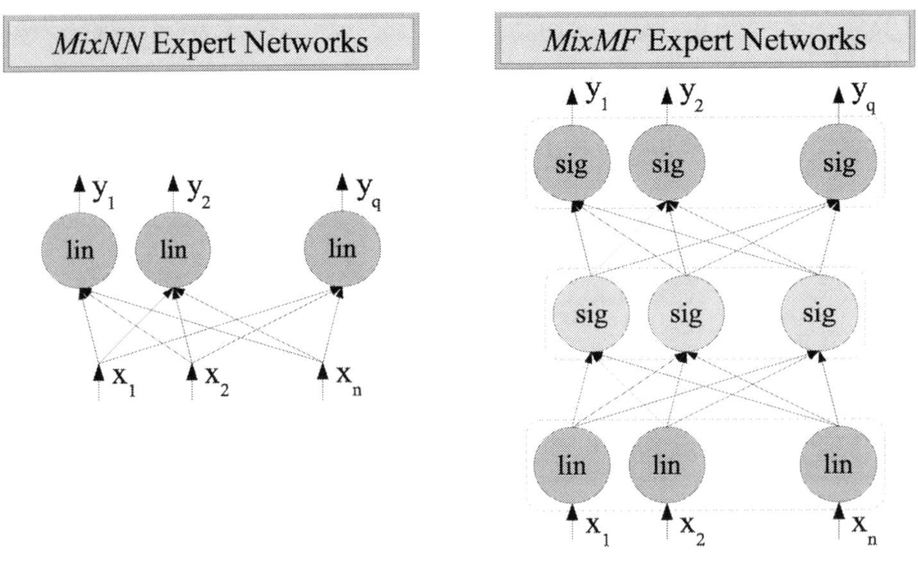

Fig. 2. Expert Network Diagram - MixNN and MixMF

In this case we have used a modified version of Algorithm 1 which take into account that the expert networks are Multilayer Feedforward Networks with one hidden layer and threshold nodes and they are trained with Backpropagation. In this subsection we describe the equations used on the *MixMF* learning process.

In order to adapt the weights of the expert networks and the gating network we have used the objective function described in equation 7. The equations used to adapt the input layer weights of the expert networks wi, the hidden layer weights of the expert networks wh and the gating network a are the following ones:

$$wh_{j,k}^{net}(ite+1) = wh_{j,k}^{net}(ite) + \eta \cdot h_{net} \cdot \delta_k^{net} \cdot ho_j^{net} \tag{11}$$

$$wi_{i,j}^{net}(ite+1) = wi_{i,j}^{net}(ite) + \eta \cdot h_{net} \cdot \delta_j^{'net} \cdot x_i \tag{12}$$

$$a^{net}(ite+1) = a^{net}(ite) + \eta \cdot h_{net} \cdot (h_{net} - g_{net}) \cdot x \tag{13}$$

where:

$$h_{net} = \frac{g_{net} \cdot \exp\left(-\frac{1}{2}\left|d - y^{net}\right|^2\right)}{\sum_{j=1}^{k}\left(g_j \cdot \exp\left(-\frac{1}{2}\left|d - y^{j}\right|^2\right)\right)} \tag{14}$$

$$\delta_k^{net} = (d_k - y_k^{net}) \cdot (1 - y_k^{net}) \cdot (y_k^{net}) \tag{15}$$

$$\delta_j^{'net} = ho_j^{net} \cdot (1 - ho_j^{net}) \cdot \sum_{h=1}^{m} \delta_h^{net} \cdot wh_{j,h}^{net} \tag{16}$$

3 Experimental Testing

In this section, we describe the experimental setup, the datasets we have used in our experiments and we show the results we have obtained. Finally, we compare the results we have obtained with *Simple Ensemble*, *MixNN* and *MixMF* on the different datasets.

For this reason we have trained multiple classification systems of 3, 9, 20 and 40 MF networks with *Simple Ensemble*, *MixNN* and *MixMF* on the eight problems described in subsection 3.1. For the expert networks of *Simple Ensemble* and *MixMF* we have used the training parameters described in table 1. In addition, we repeated ten times the whole learning process, using different partitions of data in training, validation and test sets. With this procedure we can obtain a mean performance of the ensemble for each database and an error in the performance calculated by standard error theory.

3.1 Datasets

We have used eigth different classification problems from the *UCI repository of machine learning databases* [10] to test the performance of methods. The databases we have used are: Cylinder Bands Database (band), Australian Credit Approval (cred), Solar Flare Database (flare), Glass Identification Database (glas), The monk's problem 1 (mok1), Congressional Voting Records Database (vote), Wisconsin Breast Cancer Database (wdbc).

Table 1. MF training parameters and Performance of a Single Network

Database	Hidden	Iterations	Step	Momentum	Performance
band	23	5000	0.1	0.05	72.4 ± 1.0
cred	15	8500	0.1	0.05	85.6 ± 0.5
flare	11	10000	0.6	0.05	82.1 ± 0.3
glas	3	4000	0.1	0.05	78.5 ± 0.9
mok1	6	3000	0.1	0.05	74.3 ± 1.1
survi	9	20000	0.1	0.2	74.2 ± 0.8
vote	1	2500	0.1	0.05	95.0 ± 0.4
wdbc	6	4000	0.1	0.05	97.4 ± 0.3

3.2 Results

In this subsection we present the experimental results we have obtained with the Multi-Net Systems trained with *Simple Ensemble* (Table 2), *Mixture of Neural Networks* (Table 3) and *Mixture of Multilayer Feedforward Networks* (Table 4).

Table 2. Simple Ensemble results

Database	3 Nets	9 Nets	20 Nets	40 Nets
band	73.5 ± 1.2	72.9 ± 1.5	73.8 ± 1.3	73.8 ± 1.3
cred	86.5 ± 0.7	86.4 ± 0.7	86.6 ± 0.7	86.5 ± 0.7
flare	81.8 ± 0.5	81.6 ± 0.4	81.5 ± 0.5	81.6 ± 0.5
glas	94 ± 0.8	94 ± 0.7	94 ± 0.7	94.2 ± 0.6
mok1	98.3 ± 0.9	98.8 ± 0.8	98.3 ± 0.9	98.3 ± 0.9
survi	74.3 ± 1.3	74.2 ± 1.3	74.3 ± 1.3	74.3 ± 1.3
vote	95.6 ± 0.5	95.6 ± 0.5	95.6 ± 0.5	95.6 ± 0.5
wdbc	96.9 ± 0.5	96.9 ± 0.5	96.9 ± 0.5	96.9 ± 0.5

Table 3. *Mixture of Neural Networks* results

Database	3 Nets	9 Nets	20 Nets	40 Nets
band	72.7 ± 2.2	74.4 ± 1.3	74 ± 1.9	75.5 ± 1.3
cred	86.8 ± 0.5	86.9 ± 0.5	86.5 ± 0.6	86 ± 0.5
flare	81.5 ± 0.5	81.7 ± 0.5	81.7 ± 0.6	81.8 ± 0.6
glas	89.4 ± 1	91.2 ± 1.1	90.2 ± 1.3	91 ± 1.1
mok1	87.8 ± 2.2	93.6 ± 2.6	93.6 ± 2.1	93.9 ± 2.5
survi	72.3 ± 1.2	72.6 ± 0.9	73.8 ± 0.9	73.6 ± 1.2
vote	95 ± 1.2	96.1 ± 0.6	96.1 ± 0.6	96.5 ± 0.7
wdbc	94.7 ± 0.5	94.9 ± 0.4	95.1 ± 0.6	94.6 ± 0.5

3.3 Interpretations of Results

Comparing the results showed in tables 2, 3 and 4 we can see that we have got better results with *MixMF*. We can also see that the improvement in performance using

Table 4. *Mixture of Multilayer Feedforward* results

Database	3 Nets	9 Nets	20 Nets	40 Nets
band	75.5 ± 1.9	74.2 ± 2	74.7 ± 1.7	73.8 ± 1.6
cred	85.9 ± 0.5	86.7 ± 0.7	86.5 ± 0.7	86.8 ± 0.5
flare	82.1 ± 0.6	81.9 ± 0.6	81.6 ± 0.6	81.7 ± 0.6
glas	94.6 ± 1	94.6 ± 1.2	94.2 ± 1.3	95 ± 1.2
mok1	99.3 ± 0.8	99.3 ± 0.8	98.8 ± 0.9	100 ± 0
survi	74.6 ± 1.3	74.9 ± 1.2	74.6 ± 1.1	75.1 ± 1.2
vote	96.1 ± 0.6	96.1 ± 0.6	96.1 ± 0.6	95.8 ± 0.6
wdbc	96.9 ± 0.5	96.9 ± 0.5	96.9 ± 0.5	96.9 ± 0.5

MixMF depends on the database and the number of networks used in the multinet system. For instante, in database *mok1 MixMF* has an increase of performance with respect to *MixNN* of 11.50% for the 3-network system, but only has an increase of 5.12% for the 20-network system.

In general, we can also see in these tables that the *MixMF* increase of performance with respect to *MixNN* is higher than with respect to *Simple Ensemble*, for instance in databases *mok1*, *glas* and *survi*.

The increase of performance we have shown is an *absolute* measure so we cannot see how important is the increase of performance with respect to the error. To have a *relative* measure and information about the error reduction, we have also calculated the percentage of error reduction PER of the multinet systems with respect to a *Single Network*. We have used equation (17) to calculate the PER value.

$$PER = 100 \cdot \frac{Error_{singlenetwork} - Error_{multinet}}{Error_{singlenetwork}} \tag{17}$$

The PER value ranges from 0%, where there is no improvement by the use of a particular multinet system method with respect to a single network, to 100%. There can also be negative values, which means that the performance of the multinet system is worse than the performance of the single network. This new measurement is relative and can be used to compare more clearly the different methods.

Table 5 shows PER values for the ensembles trained with *Simple Ensemble*. Tables 6 and 7 shows PER values for the modular networks trained with *MixNN* and *MixMF*.

According to the results showed in tables 5-7, we can see that, in general, *Mixture of Multilayer Feedforward Networks* is the best performing method, and *Mixture of Neural Netorks* is the worst method for 5 databases.

Furthermore, we have calculated the mean increase of performance with respect to a *Single Network* (Table 8) and the mean PER (Table 9) across all databases for each method to get a global measurement.

According to the mean PER, *MixMF* is the best performing method. The highest difference between *MixMF* and *Simple Ensemble* is in the 9-network ensemble where the mean PER increase is 3.3%. The highest difference between original *MixMF* and *MixNN* is in the 3-network ensemble where the mean PER increase is 23.90%. In Figure we can see more the diference on PER among all the methods.

Table 5. Simple Ensemble PER

Database	3 Nets	9 Nets	20 Nets	40 Nets
band	3.8	1.84	5.14	5.14
cred	6.52	5.41	7.08	6.52
flare	−1.68	−2.8	−3.36	−2.8
glas	72.09	72.09	72.09	73.02
mok1	93.19	95.13	93.19	93.19
survi	0.38	0	0.38	0.38
vote	12.59	12.59	12.59	12.59
wdbc	−19.24	−19.24	−19.24	−19.24

Table 6. *Mixture of Neural Networks* PER

Database	3 Nets	9 Nets	20 Nets	40 Nets
band	1.19	7.1	5.79	11.05
cred	8.12	8.68	5.97	2.77
flare	−3.19	−2.13	−2.13	−1.63
glas	50.69	59.06	54.41	58.13
mok1	52.33	75.21	75.21	76.18
survi	−7.37	−6.13	−1.67	−2.29
vote	0	22.59	22.59	30
wdbc	−102.7	−95.77	−88.85	−106.16

Table 7. *Mixture of Multilayer Feedforward* PER

Database	3 Nets	9 Nets	20 Nets	40 Nets
band	11.05	6.44	8.44	5.14
cred	2.22	7.56	5.97	8.12
flare	−0.28	−1.07	−2.91	−2.13
glas	74.88	74.88	73.02	76.74
mok1	97.08	97.08	95.13	100
survi	1.51	2.79	1.51	3.41
vote	22.59	22.59	22.59	15
wdbc	−18.85	−18.85	−18.85	−18.85

Table 8. Mean Increase of Performance with respect to *Single Network* across all databases

Method	3 Nets	9 Nets	20 Nets	40 Nets
S.E	5.17	5.1	5.19	5.21
MixNN	3.67	3.99	3.93	4.17
MixMF	5.62	5.63	5.48	5.69

Table 9. Mean Percentage of Error Reduction with respect to *Single Network* across all databases

Method	3 Nets	9 Nets	20 Nets	40 Nets
S.E.	20.96	20.63	20.98	21.1
MixNN	−0.12	8.58	8.92	8.51
MixMF	23.77	23.93	23.11	23.43

4 Conclusions

In this paper we have reviewed the *Mixture of Neural Networks* a modular network based on a quite simple architecture of neural networks.

Finally, we have proposed *Mixture of Multilayer Feedforward Networks*, a modular method based on *Mixture of Neural Networks* and *Multilayed Feedforward*.

Moreover, we have trained Multi-Net Systems of 3, 9, 20 and 40 networks with *Simple Ensemble*, *Mixture of Neural Networks* and *Mixture of Multilayer Feedforward* in order to test the performance of the new method and cover a wide spectrum of the number of networks in the classification system. The results showed that the performance of *Mixture of Multilayer Feedforward* was better but the improvement by the use of *Mixture of Multilayer Feedforward* depends on the database.

Futhermore, we have obtained the mean *Percentage of Error Reduction* across all databases and the mean *Increase of Performance*. According to these global measures, the method we have proposed *Mixture of Multilayer Feedforward* performs better than *Simple Ensemble* and it permorms by far better than the original *Mixture of Neural Network*. In general, the *Mixture of Multilayer Feedforward* is the best performing method.

We can conclude that the *Mixture of Neural Networks* variation we have proposed in this paper is better than the original *Mixture of Neural Networks* because it uses a better neural networks architecture to build the expert networks. Moreover, the *Mixture of Multilayer Feedforward* performs better than *Simple Ensemble* because training process and the gating network reduces the correlation among the networks.

Acknowledgments

This research was supported by project *P1·1B2004-03* of Universitat Jaume I - Bancaja in Castellón de la Plana, Spain.

References

1. Sharkey, A.J., ed.: Combining Artificial Neural Nets: Ensemble and Modular Multi-Net Systems. (1999)
2. Dara, R.A., Kamel, M.S.: Sharing training patterns among multiple classifiers. In Roli, F., Kittler, J., Windeatt, T., eds.: Multiple Classifier Systems. Volume 3077 of Lecture Notes in Computer Science., Springer (2004) 243–252
3. Tumer, K., Ghosh, J.: Error correlation and error reduction in ensemble classifiers. Connection Science **8**(3-4) (1996) 385–403

4. Raviv, Y., Intratorr, N.: Bootstrapping with noise: An effective regularization technique. Connection Science, Special issue on Combining Estimators **8** (1996) 356–372
5. Hernandez-Espinosa, C., Fernandez-Redondo, M., Torres-Sospedra, J.: Ensembles of multilayer feedforward for classification problems. In: Neural Information Processing, ICONIP 2004. Volume 3316 of Lecture Notes in Computer Science. (2005) 744–749
6. Hernandez-Espinosa, C., Torres-Sospedra, J., Fernandez-Redondo, M.: New experiments on ensembles of multilayer feedforward for classification problems. In: Proceedings of International Conference on Neural Networks, IJCNN 2005, Montreal, Canada. (2005) 1120–1124
7. Jacobs, R.A., Jordan, M.I., Nowlan, S.J., Hinton, G.E.: Adaptive mixtures of local experts. Neural Computation **3** (1991) 79–87
8. Jordan, M.I., Jacobs, R.A.: Hierarchical mixtures of experts and the EM algorithm. Technical Report AIM-1440 (1993)
9. Kuncheva, L.I.: Combining Pattern Classifiers: Methods and Algorithms. Wiley-Interscience (2004)
10. Newman, D.J., Hettich, S., Blake, C.L., Merz, C.J.: UCI repository of machine learning databases (1998)

Evaluating Users' Satisfaction in Packet Networks Using Random Neural Networks

Gerardo Rubino[1], Pierre Tirilly[1], and Martın Varela[2,⋆]

[1] Inria/Irisa, Rennes
rubino@irisa.fr, pierre.tirilly@ens.insa-rennes.fr
[2] SICS
mvarela@sics.se

Abstract. Quantifying the quality of a video or audio transmission over the Internet is usually a hard task, as based on the statistical processing of the evaluations made by a panel of humans (the corresponding and standardized area is called *subjective testing*). In this paper we describe a methodology called Pseudo-Subjective Quality Assessment (PSQA), based on Random Neural Networks, which is able to perform this task automatically, accurately and efficiently. RNN had been chosen here because of their good performances over other possibilities; this is discussed in the paper. Some new insights on PSQA's use and performance are also given. In particular we discuss new results concerning PSQA–based dynamic quality control, and conversational quality assessment.

1 Introduction

When we need to quantitatively assess the quality of an audio or video transmission over the Internet, the most accurate way to do it is to have a panel of humans perform the assessment on actual test sequences representative of the conditions studied. This is a standard procedure for which norms exist (see Subsection 2.1 for examples on voice multimedia communications). There are some methods to do an automatic quantitative assessment as well, that is, without using subjective tests, but they suffer either from poor accuracy or efficiency, or both (see Section 2). As an alternative, the Pseudo-Subjective Quality Assessment (PSQA) technology has been recently developed. It allows to automatically quantify the quality of a video or audio communication over a packet network, as perceived by the user. The PSQA technique is accurate, which means that it correlates well with the values given by panels of human observers, and efficient, because it can work, if necessary, in real time. It has been tested on video [10] and audio [13] flows. It can be applied in many areas, for instance, for the analysis of the impact of different factors on quality (see the mentioned papers, and [12] for an example of the study of Forward Error Correction techniques

⋆ M. Varela´s work was carried out during the tenure of an ERCIM fellowship.

S. Kollias et al. (Eds.): ICANN 2006, Part I, LNCS 4131, pp. 303–312, 2006.

in audio streaming), or for performance evaluation of communication networks using models [14].

PSQA is based on learning how human observers quantify the quality of a flow under standardized experimental conditions. The learning process consists of training a particular type of Neural Network, a Random Neural Network (RNN), to capture the relation between a set of factors having an *a priori* strong impact on the perceived quality and the latter. Let us briefly recall the structure of the RNN tool, before describing in some detail the technique used for quantitative quality assessment of video and audio flows. For a global presentation of PSQSA with a detailed description of the RNN tool see [11].

RNN have been developed by Erol Gelenbe in a series of papers (for starters, see [1], [3], [2]). In this paper we will use 3-layer feedforward RNN. Such a network can be seen as a parametric function $\nu()$ mapping a vector of size $I+J$, denoted here $(\boldsymbol{C},\boldsymbol{N}) = (C_1,\cdots,C_I,N_1,\cdots,N_J)$, into a real number. Let us denote by vector $\boldsymbol{W}$ the function's parameters. The input variables $C_1,\cdots,C_I$ are related to the network connection, or to the codec used (example: the bit rate), and the variables $N_1,\cdots,N_J$ correspond to the network state (example: the loss rate). The value of the function is the quality of the audio or video connection. The function's parameters are the weights in the neural network.

As a neural network, our RNN has three layers, the input one with $I+J$ variables, the hidden one with H units, and the output layer with a single node. The mapping can be explicitely written as

$$\nu(\boldsymbol{W};\boldsymbol{C},\boldsymbol{N}) = \frac{\sum_{h=1}^{H} \varrho_h w_{ho}^{+}}{r_o + \sum_{h=1}^{H} \varrho_h w_{ho}^{-}},$$

where

$$\varrho_h = \frac{\sum_{i=1}^{I} C_i r_i^{-1} w_{ih}^{+} + \sum_{j=1}^{J} N_j r_j^{-1} w_{jh}^{+}}{r_h + \sum_{i=1}^{I} C_i r_i^{-1} w_{ih}^{-} + \sum_{j=1}^{J} N_j r_j^{-1} w_{jh}^{-}}$$

is the *activity rate* of hidden neuron h. The strictly positive numbers r_o, r_h for $h = 1..H$, r_i for $i = 1..I$ and r_j for $j = 1..J$, correspond to the firing rates of the neurons in the network (respectively, for the output one, the hidden nodes, and the $I+J$ input ones). The weights are the variables tuned during the learning process. We denote by w_{uv}^{+} (resp. by w_{uv}^{-}) the weight corresponding to an exiting (resp. inhibiting) signal going from neuron u to neuron v (observe that both numbers w_{uv}^{+} and w_{uv}^{-} are ≥ 0). For the interpretation and the dynamics of a RNN see the references cited above. For our purposes here, we can just see it as a rational parametric function. Learning will thus consist of finding appropriate values of the weights capturing the mapping from $(\boldsymbol{c}^{(k)},\boldsymbol{n}^{(k)})$ to the real number $q^{(k)}$ where $q^{(k)}$ is the quality given by a panel of human observers to some audio or video sequence (depending on the application) when the source parameters had the values present in $\boldsymbol{c}^{(k)}$ and the parameters caracterizing the network had the values in vector $\boldsymbol{n}^{(k)}$, for $k = 1..K$.

To be more specific, let us describe how PSQA is used, with a simple example. Assume we have some audio or video stream whose quality, as perceived by the users, is to be evaluated. Assume we limit the application of PSQA to just $I = 1$ source parameter, the bit rate C_1, and to $J = 2$ network parameters, the end-to-end loss rate N_1 and the mean size of bursts of (consecutive) lost packets, N_2. The goal of the process is to come up with a function $\nu(\boldsymbol{W}; C_1, N_1, N_2)$ mapping the values of these 3 parameters into quality, for instance in a standard MOS range. We start by choosing K series of values for the bit rate, the loss rate and the mean loss burst size, denoted $(c_1^{(k)}, n_1^{(k)}, n_2^{(k)})$ for $k = 1..K$. Then, we select some short (audio/video) sequence (norms recommend to use sequences with a few seconds length) and we send it K times through a controllable network (using a testbed, for instance), where in the kth case the three selected parameters have values $c_1^{(k)}$, $n_1^{(k)}$ and $n_2^{(k)}$. The K resulting sequences are shown to a panel of humans and a subjective testing experiment is performed, following an appropriate norm depending on the type of flow (see below), which allows to obtain the (measured) perceived quality of each received sequence, denoted $q^{(1)}, \cdots, q^{(K)}$. Then, a RNN is trained to build such a $\nu()$ function, using the K input-output values obtained from the testing phase. The usual way to do this is to separate this data into a training part, used to build the approximation $\nu()$, and a validation one, used to test the predictive capacity of $\nu()$.

PSQA has been tested on video and on audio flows. In this paper, we will discuss only the latter, for sake of space. Next section describes previous and current work on the quantitative analyses of the perceived quality of voice communications, with some new results concerning interactive voice sessions. Section 3 will then discuss about the performance of our learning tool, RNN, as compared to other available tools such as standard Artificial Neural Networks. Section 4 concludes the paper.

2 Using PSQA to Analyze Perceived Voice Quality

2.1 On Voice Quality Assessment

When assessing voice quality, there are two very different kinds of subjective tests one can perform. The first kind, which is the one most widely used, is related to the quality of the voice itself, and so it does not take other factors inherent to conversation into account. We refer to these assessments as unidirectional, since the only point of interest is the quality of the received signal. The other kind of tests concern actual conversational quality, and in a way are a superset of the first kind. In these, we not only consider the voice quality itself, but also other factors (mostly delay-related) which affect the perceived quality of an *actual conversation.* There exist standard procedures for performing subjective assessments of both unidirectional (e.g. [6]) and interactive (e.g. [8]) speech. Interactive tests are more difficult to set up and perform than unidirectional ones.

In any case, all subjective tests are expensive in terms of time and money, so a significant research effort has been directed toward developing *objective* tests. These provide a cheaper and more practical alternative to subjective tests. Most subjective tests operate by comparing the received signal to the original one, and estimating the quality from the difference between the two. The estimation can be done by mechanisms ranging from a simple SNR measurement to very complex psychoacoustic models. Of the methods available in the literature, only the ITU E-model [4] and P.563 algorithm [5] can perform their assessment without requiring the original signal. The accuracy of objective assessment techniques is quite variable, in particular when considering VoIP. The best performance among the purely objective metrics is that of PESQ [7], which attains up to about 0.92 correlation with subjective scores (a typical "best score").

The need for being able to accurately assess VoIP quality in real–time arises in several scenarios, such as the development of dynamic quality–control mechanisms. Traditional methods of voice quality assessment tend to only cover one of the two conditions needed. Either they are adequately accurate (i.e. PESQ) but not able to perform in real–time, or they do work in real–time, but their accuracy is not as high as needed (i.e. the ITU E–model).

In this respect, PSQA offers the best of both worlds, since it is very accurate, and it can work in real–time. This is a consequence of not needing the original signal to perform the assessment, and of the simplicity of the calculations on the RNN.

2.2 Unidirectional Voice Quality Assessment and Its Applications

We have succesfully used PSQA to analize the ways in which different parameters affect the voice quality for VoIP streams [13]. To this end, we used six quality–affecting parameters, and using PSQA, we studied the relations between them and the quality as perceived by the end user. The parameters considered were the codec used, whether error correction was being used, the offset of the error correction, the loss rate and mean loss burst size found in the network, and the packetization interval (i.e. the length of the speech contained in each packet).

In this study, we obtained a 0.94 correlation coefficient between the RNN predictions and subjective scores, which is on par (slightly better, actually) with the best objective assessment techniques currently available. The results obtained allow us to understand, for instance, how the loss rate affects the perceived quality, for different codecs and with or without error correction. Figure 1(a), for example, shows how the packetization interval and the mean loss burst size affect the perceived quality.

An immediate application of these results is modifying application–level parameters to acommodate variations in the network conditions, improving, if possible, the perceived quality. We have developed two simple algorithms which allow to manipulate the codec, forward error correction, and packetization interval values to dynamically optimize the quality. The first algorithm takes a naive approach, trying to keep the quality between two thresholds at all costs. The second algorithm takes bandwidth consumption into account, and tries to keep the quality between the same bounds, but resorting to lower bit rate encodings whenever possible. Both algorithms present a similar improvement on quality over scenarios lacking dynamic control, and depending on network load, one may perform slightly better than the other. Figure 1(b) shows the performance of the simplest of both algorithms when the network conditions degrade.

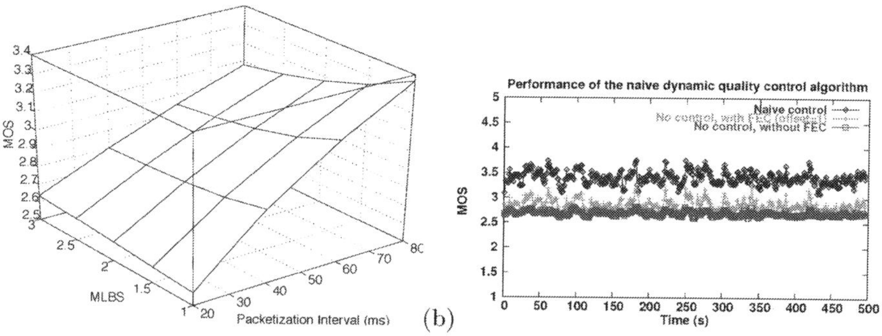

(a) (b)

Fig. 1. (a) Variation of the perceived quality as a function of the Mean Loss Burst Size (MLBS) and the Packetization interval, without error correction. Loss rate is 2%. (b) Performance of a PSQA–based control algorithm based on the same RNN. The network parameters found in these trace are as follows: Loss rate: 12.47%, Mean loss burst size: 2.13 packets. Under these network conditions, the (naive) control algorithm offers a noticeably better quality than when no control is used.

2.3 Conversational Quality Assessment

As mentioned above, conversational quality assessment is significantly more difficult to perform than the unidirectional one. To our knowledge, there is no purely objective assessment method available in the literature able to predict conversational quality, nor have there been subjective assessment campaign covering as many parameters as we used in our studies. We used PSQA to develop an understanding of the how the conversational quality evolves with the different parameters which affect it. In particular, we were interested in the relative impacts of delay and loss rate on the overall quality.

The set of parameters we considered for this study were the bit rate (using a single codec), the forward error correction settings, the packet loss rate and mean loss burst size, and the one–way delay and its variability (jitter). As the study was focused on VoIP, we used subjects with previous experience with VoIP applications. The results we obtained with the RNN present a correlation coefficient of 0.95 with subjective scores, which is very good. Given the large parameter space considered, it is worth noting that the use of RNN for the estimations allowed for excellent generalization capabilities.

Among the results we obtained by analizing the RNN behavior, the most interesting one is that VoIP users seem much more tolerant to delay than it is usually reported for PSTN systems (and by extension adopted for VoIP). Moreover, we found that the impact of the delay on the conversational quality was relatively small compared to that of the loss rate, and that one–way delays as big as 600ms made little difference on the perceived quality for loss rates of about 7% and up. This is good news, since it allows implementors to use better loss concealment or correction techniques, at the expense of delay, resulting in an improved overall quality. Figure 2 shows the impact of delay on the conversational quality for diferent loss rates.

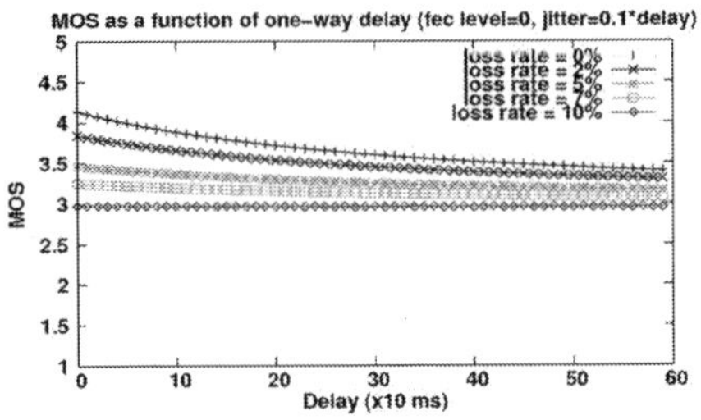

Fig. 2. Perceived quality as a function of delay for some loss rates (without FEC). Jitter was 10% of the delay. Note how the impact of the delay diminishes as the loss rate increases.

3 Comparing RNN Against Other Tools

As mentioned above, the choice of RNN over other tools for implementing PSQA is not arbitrary. In [9], [10], a first performance comparison had been made, establishing that RNN offer two advantages over traditional ANN for our applications. The first one is that of better generalization capabilities (i.e. less over-training). Since our approach only allows for relatively few learning points in a vast input space, this is a very important characteristic. The second advantage was that RNN are less sensitive to variations in the number of neurons in the hidden layer, which allows to obtain good performance without needing to optimize the neural network architecture.

In this section we discuss more in deep about the performances of RNN compared to Bayesian networks and standard ANN, the two families of alternate tools we used. We also present results of new comparison tests between ANN and RNN performance for both unidirectional and interactive VoIP applications.

3.1 Using Naive Bayesian Classifiers

We have tried naive Bayesian classifiers for providing MOS estimations for PSQA. This was meant primarily to test their accuracy more than for production use, since two reasons make them less desirable than RNN:

- they require a larger amount of training data, and
- they only provide classification into discrete values, whereas RNNs provide a continuous, differentiable function.

However, this kind of classifier is easier to implement than RNNs, and is computationally trivial, which, despite the RNN's simplicity, could be useful in certain contexts where computing power or memory are very limited.

Naive Bayesian classification is based on the observation of a sufficiently large amount of data, and the assumption that those observations are statistically independent. We perform the classification by counting, for *a large number of assessments*, the number of times each score happens for each value of each parameter considered. This allows us to estimate, for each value of the selected parameters, the probability that a given score will happen, looking at the quality as a random object. In other words, we estimate the conditional probability of the score (or quality) being equal to q given that $\boldsymbol{C} = \boldsymbol{c}$ and $\boldsymbol{N} = \boldsymbol{n}$, for any possible configuration $(\boldsymbol{c}, \boldsymbol{n})$ of the parameters' values (we are assuming a discrete range for quality as well as for the parameters). Then, we can find the score which is most likely to be associated with each configuration.

As stated above, this approach needs a large body of data for the training process. As not enough subjective assessments were available to train the classifier, we needed to generate data suitable for assessing the performance of this tool. To this end, we generated a large set of assessments (covering the whole parameter space) with a previously trained RNN, and used them instead of actual subjective scores.

We performed several tests, using over 20,000 configurations to train the classifier. Although validation results were reasonably accurate, the performance of this classifier with configurations for which we had subjective scores was consistently and significantly lower than that of the RNN.

Among the possible explanations for this bad performance, the foremost is that the quality–affecting parameters, and their relation to MOS scores are not actually independent. Given the results obtained, it is reasonable to think that they are too correlated for the assumption of statistical independence to hold.

As mentioned before, even if this approach did perform well, it does not offer the advantages of using a RNN, and in any case, it needs the RNN (or another trained estimator) in order to compensate for the usual lack of available subjective data.

3.2 RNN Compared to Standard ANN Tools

In our experiments, we also tested the performances of classical neural networks in the previously described learning tasks for automatic perceived quality assessment in VoIP applications. The global conclusion is that RNN outperforms Artificial Neural Networks in our context. In this subsection we present further results concerning the accuracy of PSQA when implemented with RNN and ANN. To this end, we used 15 different training and validation sets, with varying sizes and also by using different parts of the data for training and for predicting, for both the one-way and interactive cases in VoIP communications. We then trained RNN and an ANN with appropriate architectures and compared their performances.

In order for the comparison to be fair, we tried to optimize the performance of the neural nets by using different number of hidden layers and different sizes for those hidden layers. In the ANN case, we used commercial packages and thus we looked at the results given by different training algorithms. The best results in both types of networks were obtained with three-layer architectures. For the ANN, the optimal hidden layer sizes were between 6 and 8 neurons, depending on the input data, and for the RNN, a 10-neuron hidden layer provided the best performance. For the RNN, we also considered the simplest architecture possible, namely 6 input neurons, and one output neuron, with no hidden units, as [14] reported acceptable results even with this simple structure. The best training algorithms from the quality of the predictions were Levenberg-Marquardt for the ANN. Our implementation of RNN only uses a simple gradient descent algorithm.

Tables 1 and 2 show the sizes of the training, validation and testing sets used (all randomly chosen). As classically done in neural network methodology, the validation set was used to stop the training when the network generalized well, in order to select good candidates in each family (ANN, RNN), and the test set was used to mesure the real performance of the network (for the comparisons).

Table 1. Number of samples used to train, validate and test the neural nets (one-way streams) in order to compare RNN vs ANN performance

Training	Validation	Test	Training	Validation	Test	Training	Validation	Test
92	10	10	82	20	10	72	20	20
92	10	10	82	10	20	72	30	10
92	10	10	82	10	20	72	10	30
82	20	10	82	10	20	62	10	40
82	20	10	72	20	20	62	20	30

Table 2. Number of samples used to train, validate and test the neural nets (interactive streams) in order to compare RNN vs ANN performance

Training	Validation	Test	Training	Validation	Test	Training	Validation	Test
100	10	10	90	20	10	80	20	20
100	10	10	90	10	20	80	30	10
100	10	10	90	10	20	80	10	30
90	20	10	90	10	20	70	10	40
90	20	10	80	20	20	70	20	30

We found that, although the training error reached (calculated as MSE) is about one order of magnitude lower for ANN than for RNN, the validation results are consistently and significantly better for the RNN. Figures 3 (a) through (d) show the results obtained for both one-way and interactive streams, for the 15 validation and test data sets.

It is interesting that for the interactive scenario, the ANNs performance is noticeably better than for the one-way case. However, the MSE obtained with RNN are lower, and for the one-way case, the difference is very noticeable.

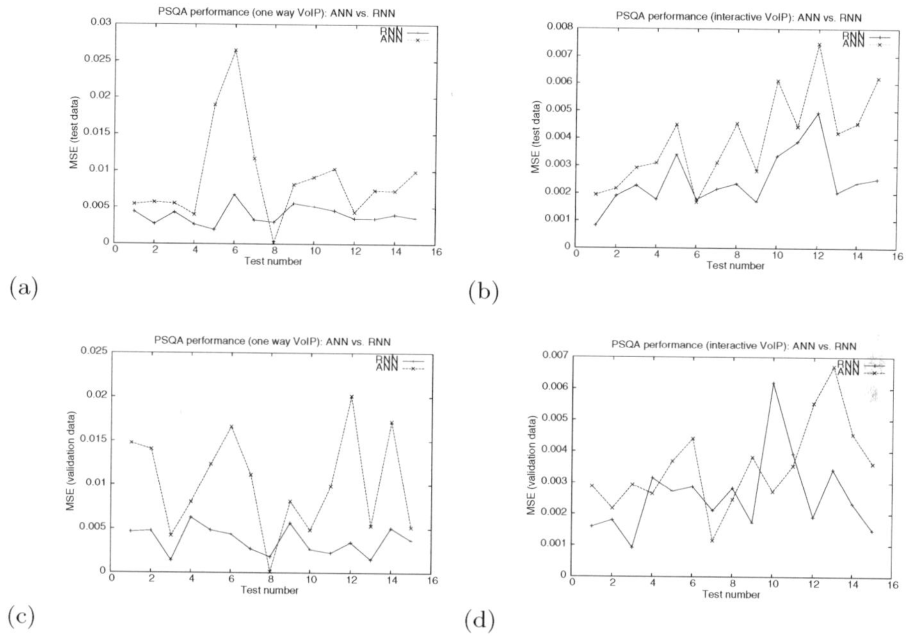

Fig. 3. MSE comparison of RNN and ANN for (a) test data, one-way streams, (b) test data, interactive streams (c) validation data, one-way streams and (d) validation data, interactive streams

4 Conclusions

The PSQA technology allows to provide an accurate and automatic quantitative evaluation of the perceived quality of an audio or video communication over a packet network, where the flow is subject to different kinds of distortions (losses by congestion, delays and jitter, etc.). It has been succesfully tested and used on video and audio streams. In this paper we report novel results on interactive speech quality, obtained by using a RNN-based PSQA tool, and about two proof–of–concept PSQA–based dynamic quality control algorithms. The method comes up with an approximation of the perceived quality as a nice function of measurable parameters, thanks to the RNN technology. It can be applied to the monitoring of an existing network or for control purposes, as well as to the analyzing the impact of specific parameters on the perceived quality.

This paper also shows some results illustrating how RNN outperforms standard ANN as well as Bayesian networks in performing the assessment task.

In particular, we have performed an in-depth performance comparison between ANN and RNN–based PSQA implementations, both for unidirectional nad interactive speech streams.

One of the topics of further research work is to extend these experiments to more cases. Another important research direction for PSQA development is the analysis of the impact of using more efficient learning techniques in the RNN tool.

References

1. E. Gelenbe. Random Neural Networks with Negative and Positive Signals and Product Form Solution. *Neural Computation*, 1(4):502–511, 1989.
2. E. Gelenbe. Stability of the Random Neural Network Model. In *Proc. of Neural Computation Workshop*, pages 56–68, Berlin, West Germany, February 1990.
3. E. Gelenbe. Learning in the Recurrent Random Neural Network. *Neural Computation*, 5(1):154–511, 1993.
4. ITU-T Recommendation G.107. The E-model, a Computational Model for Use in Transmission Planning, March 2005. http://www.itu.int/.
5. ITU-T Recommendation P.563. Single–ended Method for Objective Speech Quality Assessment in Narrow–band Telephony Applications, May 2004.
6. ITU-T Recommendation P.800. Methods for Subjective Determination of Transmission Quality, August 1996.
7. ITU-T Recommendation P.862. Perceptual Evaluation of Speech Quality (Pesq), an Objective Method for End-To-End Speech Quality Assessment of Narrowband Telephone Networks and Speech Codecs, 2001.
8. ITU-T Recommendation P.920. Interactive test methods for audiovisual communications, 2000.
9. S. Mohamed. *Automatic Evaluation of Real-Time Multimedia Quality: a Neural Network Approach.* PhD thesis, INRIA/IRISA, Univ. Rennes I, Rennes, France, jan 2003.
10. S. Mohamed and G. Rubino. A Study of Real–time Packet Video Quality Using Random Neural Networks. *IEEE Transactions On Circuits and Systems for Video Technology*, 12(12):1071–1083, December 2002.
11. G. Rubino. Quantifying the quality of audio and video transmissions over the internet: the psqa approach. In J. Barria, editor, *Design and Operations of Communication Networks: A Review of Wired and Wireless Modelling and Management Challenges.* Imperial College Press, 2005.
12. G. Rubino and M. Varela. Evaluating the utility of media–dependent FEC in VoIP flows. In *LNCS 3266: Proceedings of the Fifth International Workshop on Quality of future Internet Services (QofIS'04)*, Barcelona, Spain, September 2004.
13. G. Rubino, M. Varela, and S. Mohamed. Performance evaluation of real-time speech through a packet network: a random neural networks-based approach. *Performance Evaluation*, 57(2):141–162, May 2004.
14. Gerardo Rubino and Martn Varela. A new approach for the prediction of end-to-end performance of multimedia streams. In *In Proceedings of the Measurement of Speech and Audio Quality in Networks workshop (MESAQIN'04)*, September 2004.

Random Neural Networks for the Adaptive Control of Packet Networks

Michael Gellman and Peixiang Liu

Dept. of Electrical & Electronic Eng.,
Imperial College London
{m.gellman, p.liu}@imperial.ac.uk*

Abstract. The Random Neural Network (RNN) has been used in a wide variety of applications, including image compression, texture generation, pattern recognition, and so on. Our work focuses on the use of the RNN as a routing *decision maker* which uses Reinforcement Learning (RL) techniques to explore a search space (*i.e.* the set of all possible routes) to find the optimal route in terms of the Quality of Service metrics that are most important to the underlying traffic. We have termed this algorithm as the Cognitive Packet Network (CPN), and have shown in previous works its application to a variety of network domains. In this paper, we present a set of experiments which demonstrate how CPN performs in a realistic environment compared to *a priori*-computed optimal routes. We show that RNN with RL can autonomously *learn* the best route in the network simply through exploration in a very short time-frame. We also demonstrate the quickness with which our algorithm is able to adapt to a disruption along its current route, switching to the new optimal route in the network. These results serve as strong evidence for the benefits of the RNN Reinforcement Learning algorithm which we employ.

1 Introduction

The Cognitive Packet Network (CPN) approach has been used for routing in a number of different domains: from wire-line, to wireless Ad-hoc, and even overlay networks [1,2,3]. These have demonstrated that it is able to use the QoS goal of the user or application to selectively find the best path in the network that satisfies it.

At the core of the CPN algorithm is the Reinforcement Learning (RL) algorithm which uses a fully recurrent Random Neural Network (RNN) [4] as a decision-maker to route packets. Each output link from a CPN node (router) is represented by a neuron in the RNN, the weights of which are updated according to the quality of the routes explored by the Smart Packets (SPs) or used by the Dumb Packets (DPs). Therefore, the RL algorithm uses the observed outcome

* This work was supported by the UK Engineering and Physical Sciences Research Council under Grant GR/S52360/01.

S. Kollias et al. (Eds.): ICANN 2006, Part I, LNCS 4131, pp. 313–320, 2006.

of the current decision to either *reward* or *punish* the corresponding routing decision so that its future decisions are more likely meet the user's desired QoS goal.

In this paper, We begin by giving a brief overview of the mechanisms behind CPN, including one optimization which improves the Quality of Service given to data traffic. Following this, we describe our testbed and procedure that we have used to conduct our experiments. Finally, we present our results which demonstrate CPN's ability to *learn* the optimal route in a network, and also its ability to quickly react to a harmful change in that route.

2 CPN Overview

The CPN algorithm has already been described in detail in previous publications (*e.g.* [5]); thus, in this section, we provide a brief description of the mechanics of CPN, and focus more in-depth on an optimization that we have introduced in our experiments to improve the QoS of user traffic.

CPN is a *reactive* routing protocol – that is, it waits until a connection requests the establishment of a route to begin the routing process. For this purpose, there exists a division of labor between packets whose job it is to carry data (so-called *Dumb* packets or DPs), and those which are responsible for exploring the network and learning the best path (*Smart* packets or SPs).

Smart packets are sent at regular intervals during the connection, and are routed using a Reinforcement Learning (RL) algorithm at each hop in the network. At each router they visit, they may have some QoS information added into the packet, depending on the QoS metrics of the connection. For instance, when delay is the metric, we encode the current timestamp into the packet, in order to calculate a round-trip time estimate. When an SP reaches its destination, an acknowledgement packet (SAck) is generated which will travel along the reverse route of its SP, performing the RL algorithm at each hop to indicate the success or failure of the original SP in finding an optimal route.

In contrast to SPs, Dumb packets are source-routed using the routes brought back by SAcks. When they reach the destination, they also generate an acknowledgement (DAck) which also updates the RNN weights through the RL algorithm. This ensures that the reward information stored at each node is kept up-to-date. In the original algorithm, DPs would use the route brought back by the most recent SAck; however, in our experiments, we found that this was not always optimal, and thus we propose an optimization which we present in the following subsection.

2.1 Keeping Track of the Best Route

In the originally proposed algorithm, we proposed to use the route brought back by a SAck immediately, under the assumption that it contains the most up-to-date information about the network, and would be the best route decided upon by the network. However, this does not take into account the fact that, in order to better explore the network, each router will (with a small probability) sometimes choose a random next hop for an SP, rather than the one decided upon by the RNN. This decision may or may not result in better QoS for the route, and thus we propose to compare the quality of routes brought back by a SAck against the current route before switching.

When a SAck arrives at its destination, we compare the reward for this route to the currently active one. Only if the newly arrived SAck's reward is greater than our current route do we switch. DAcks update the reward stored with their route, but do not cause the active route to change. In the remainder of this paper we refer to this optimization as *CPN with Sorting*.

While at first this may seem to deprive us of some of the exploratory benefits of the CPN algorithm, we contend (and demonstrate in Section 3.3) that this is not the case because of DAcks. Because each DAck also triggers the RNN-RL algorithm at each router, a sudden decrease in quality along a route will punish the decision, and cause the next SP to choose an alternate next hop, which will enhance of discovering a new best route; thus, the network is still able to quickly respond to changes in network conditions, and it also ensures that data traffic always uses the best available path.

3 Experiments

We wanted to demonstrate CPN's ability to learn the optimal route in a network, as well as explore its performance for different levels of exploration overhead, and under network dynamics. Thus, we have used our network testbed facilities at Imperial College London to perform controlled experiments into each of these areas.

3.1 Testbed

Our network testbed consists of standard PCs which run Linux 2.6.15 and contain one or more quad-10/100 Ethernet cards. An important consideration for our experiments is the topology that we use; we would like it to be as realistic as possible so that we can generalize our results to other, larger networks. Towards that end, we contacted the Swiss Education and Research Network, who provided us details on their 46-router backbone, complete with bandwidth, OSPF costs, and link-level delays (Fig. 1). We have implemented this topology in its entirety in our lab[1]. We have configured IP routing using quagga 0.99.3 with the OSPF costs given to us; thus, the routes in our testbed should be exactly the same as those used in the real Swiss backbone.

[1] With the following differences: we have scaled the bandwidth down by a factor of 100, and have used artificial delays to replicate the physical layer delays in the data set (from 0-40ms).

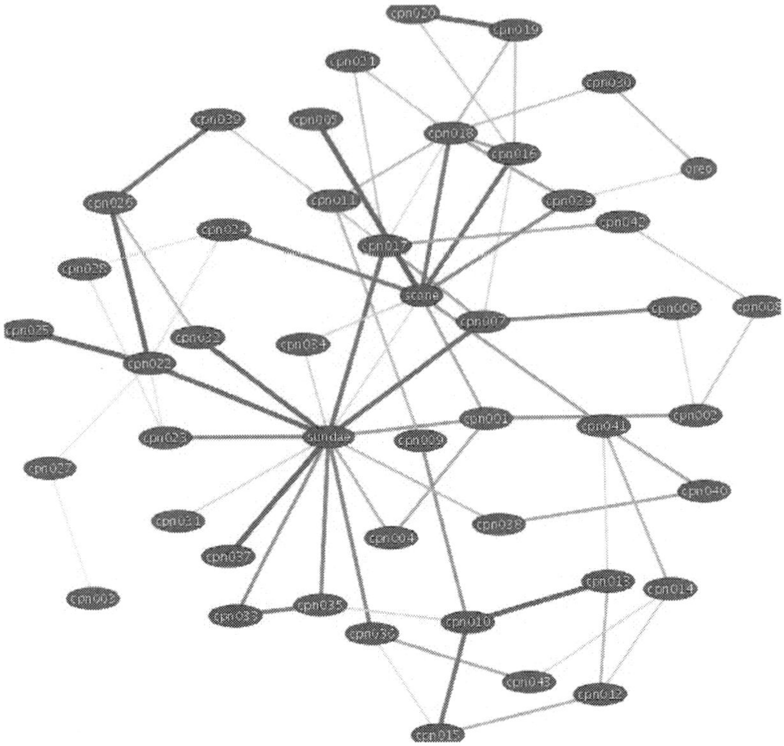

Fig. 1. The 46-node testbed topology. Links between nodes are colored and emphasized in terms of the delay of the link; lighter thinner lines are low-delay links, while darker, thicker ones denote higher delays.

3.2 Scenario and Implementation

As a traffic source, our experiments use the open-source Quake II multi-player game, which uses UDP as its transport protocol. Due to its open-source nature, we were able to modify the application to better suit it to network experimentation, including: removing all video, sound, and input code to allow clients to run without user input; and embedding *bot* code into the server so that our generated traffic resembles that of a real game complete with movement, player kills, and so forth. Because delay is critical to the performance of a first-person shooter, we configure CPN to optimize this metric, and also use it to compare our results.

Our experiments use a single server (cpn008 in the figure), and 10 clients. The delay results that we report are an average of the application-level *ping* results reported by the Quake II server, where each experiment lasts 15 minutes. Each experiment was repeated 20 times, and, unless otherwise noted, had minimal variance.

CPN is implemented as a Linux kernel module that runs at the networking layer. It receives Quake II packets through an IP-over-CPN tunnelling mechanism that hides from the application that it is using CPN; thus, no re-writing of the application is necessary. The only downside of this approach is that there is a small amount of additional packet overhead (an extra IP header), although, as we show in Section 3.3, this is minimal.

3.3 Results

In this section, we present the results from three different sets of experiments that we conducted on our testbed. The first set compares the performance of CPN with the "sorting" optimization that we described in section 2.1. Following this experiment, we investigated the impact of varying the amount of exploration overhead on CPN's ability to quickly discover the optimal path in the network. Our final set of experiments demonstrates the quickness with which CPN is able to react to a performance degradation along its optimal route.

Sorting vs No Sorting. When we first experimented with CPN's ability to find the optimal route in the network, we noticed that it would sometimes route Dumb packet traffic along non-optimal paths for short periods of time before returning to the best path (Fig. 2). Upon closer investigation, we discovered that this was due to Smart packets who, with a small probability[2] at each hop, would randomly choose a next as opposed to using the RNN-RL algorithm. While this is an integral component of the CPN algorithm because it enhances our ability to discover new paths, it was also harming the performance of data traffic that would use these new routes without considering how they compared to their current route. Thus, we implemented the optimization described in section 2.1.

The outcome of the optimization is shown in Fig. 2, where we can see it leads to a significant improvement. Once CPN has discovered the optimal route in the early stages of the experiment, it continues to use it whereas without the optimization, CPN subjects the data traffic to a number of other routes with worse delay performance. Thus, throughout the remainder of this paper, all experiments take place with this optimization enabled.

Discovering Optimal Routes. In order to demonstrate CPN's ability to quickly find the optimal route in the network, we compared its performance with that of IP using OSPF to determine its routes. Because the cost of each link is proportional to its delay, OSPF routing converges to the minimal delay path, giving us a baseline for comparison.

We conducted experiments (according to the procedure above) for different percentages of Smart packets[3], which controls the overhead of network exploration. The results can be seen in Fig. 3.

[2] In our experiments, the probability of randomly choosing a next hop is 0.05.

[3] While we refer to it as a percentage, it is more accurately described as the *probability* that, when sending a DP, that we also send an SP.

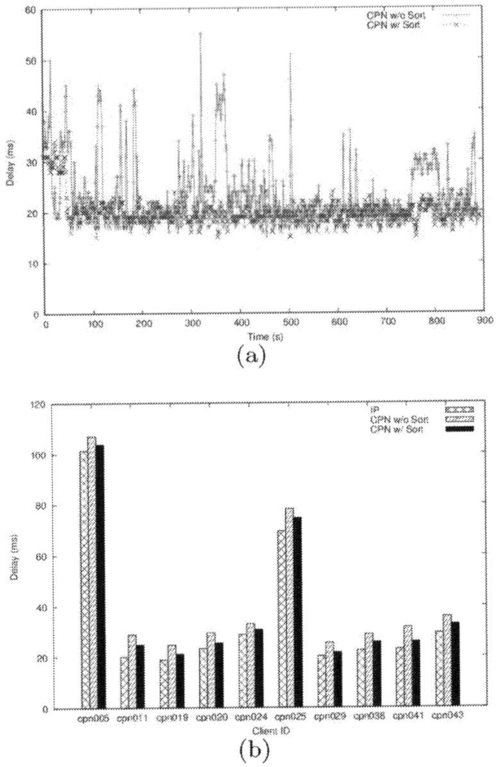

Fig. 2. Comparing the effect of sorting. In (a), where we have plotted the measured delay during the experiment for one selected client, we can see the variance in the non-sorting case is much higher, as some of the Dumb packets use a route with higher delay before switching back to the optimal one. The results for every client are summarized in (b), where we see that the average delay is lower when using the sorting optimization.

Looking at these results, we can see that CPN can always find the optimal route in the network; it just takes a longer time when the exploration overhead is low (Fig. 3.b). We can also see that the difference in delay performance between 10% and 20% SPs is minimal for many of the clients, leading us to decide to use the former for the remainder of the paper.

CPN's Response to Network Dynamics. In our final experiment, we wanted to see how quickly CPN could adapt to a sudden change along the optimal route. We decided to test this by manually increasing the delay along one of the links in the network (the link between cpn017 and cpn042 in the Fig. 1) midway through the experiment (at time $t = 450s$). The results of this experiment are shown in Fig. 4, where CPN was able to react to the change within 1 second, and find a route around the hotspot.

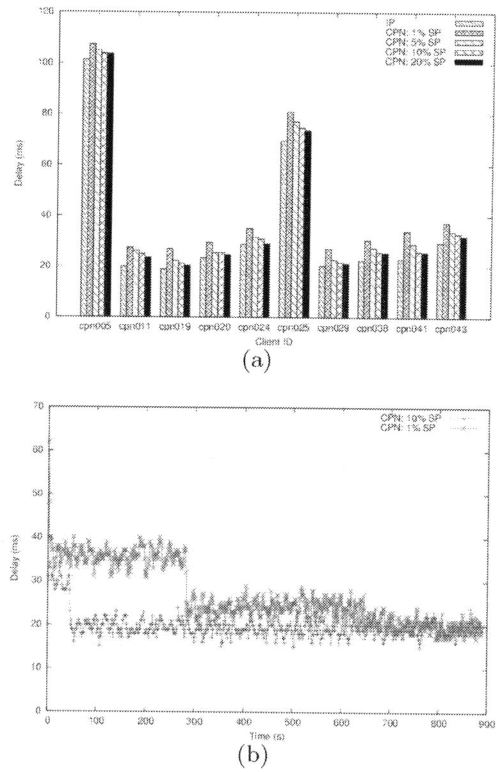

Fig. 3. IP vs. CPN for different percentages of Smart packets. In (a), we can see that the average delay for CPN is very close to optimal. Also, as the number of SPs increases, the delay decreases. The reason the average is a bit higher can be seen in (b) where at the beginning of the experiment, CPN has not yet found the optimal route, increasing the overall average. In addition, the impact of the SP percentage on the rate of discovering the optimal route is highlighted.

4 Conclusion and Future Work

In this paper, we have presented a number of experiments which demonstrate CPN's ability to find the optimal route in the network, and the speed with which it is able to do so. We showed that even with minimal exploration overhead (1% Smart Packets), CPN is still able to find the optimal route in the network, and that increasing the overhead simply accelerates the convergence process. We have also described a small optimization to the original algorithm where we only re-route data traffic when we find a better route, and have shown that this performs better than the previous method. These all point to the strength of the RNN as a decision-maker in a dynamic, on-line scenario.

Looking to the future, we would like to extend this work to a larger, more heterogeneous environment, and are currently exploring how PlanetLab can be

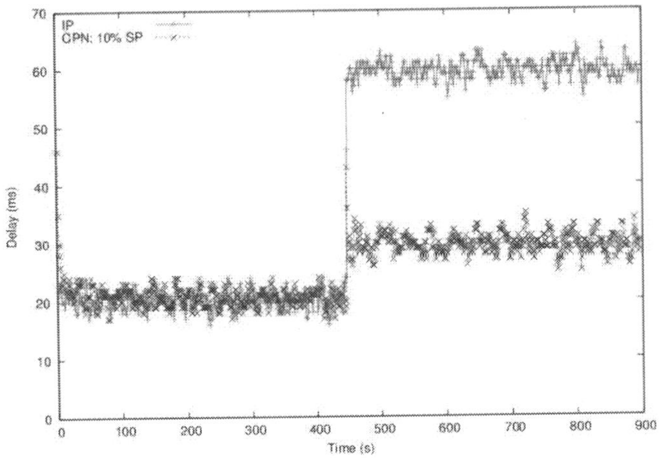

Fig. 4. How CPN reacts to a disturbance along the optimal route. At time $t = 450s$, the delay along one link in the optimal route is increased by 40ms. Within 1 second, CPN has switched to another route.

used for this purpose. We also plan to diversify the traffic types that we include in our evaluation, including TCP flows, and real-time multimedia streams.

Acknowledgements

We would like to thank our supervisor Prof. Erol Gelenbe for his support and guidance in this work. We would also like to thank the Swiss Education and Research Network for sharing their network topology data.

References

1. Gelenbe, E., Xu, Z., Seref, E.: Cognitive packet networks. In: Proceedings of the 11th International Conference on Tools with Artificial Intelligence. (1999) 47–54
2. Gelenbe, E., Gellman, M., Lent, R., Liu, P., Su, P.: Autonomous smart routing for network QoS. In: Proceedings of the First International Conference on Autonomic Computing, New York, NY (2004) 232–239
3. Gelenbe, E., Lent, R.: Power-aware ad hoc cognitive packet networks. Ad Hoc Networks Journal **2**(3) (2004) 205–216
4. Gelenbe, E.: Learning in the recurrent random neural network. Neural Computation **5** (1993) 154–164
5. Gelenbe, E., Lent, R., Xu, Z.: Measurement and performance of a cognitive packet network. Journal of Computer Networks **37**(6) (2001) 691–701

Hardware Implementation of Random Neural Networks with Reinforcement Learning

Taskin Kocak

School of Electrical Engineering and Computer Science
University of Central Florida
Orlando, FL 32816
tkocak@cpe.ucf.edu

Abstract. In this paper, we present a hardware implementation of a random neural network (RNN) model. The RNN, introduced by Gelenbe, is a spiked neural network model that possesses several mathematical properties such as the existence and uniqueness of the solution, and convergence of the learning algorithm. In particular, we discuss the implementation details for an RNN which uses a reinforcement learning algorithm. We also illustrate an example where this circuit implementation is used as a building block in a recently proposed novel network routing protocol called cognitive packet networks (CPN). CPN does not employ a routing table instead it relies on the RNN with a reinforcement algorithm to route probing packets.

1 Introduction

The random neural network (RNN) model was developed by Gelenbe in the late eighties and early nineties [1]. It has been proven to be successful in a variety of applications when used either in a feed-forward or a fully recurrent architecture [5]. In most problems, RNN yields strong generalization capabilities, even when the training data set is relatively small compared to the actual testing data. The model also achieves fast learning due to its computational simplicity for the weight updating process.

In this paper, we present hardware design and digital implementation of the RNN model. First, we illustrate the design of a single neuron and its usage in an array of neurons. Second, we describe the state machine used to implement the reinforcement learning (RL) algorithm. Then, we illustrate an example where the circuit implementation of the RL-based RNN is used as a building block in a recently proposed novel network routing protocol called cognitive packet networks (CPN) [7,9,10]. Low-level design details such as the RTL schematic and layout information are provided for the main CPN component smart packet processor which incorporates the RNN part. Further, simulations results are given to verify the execution of the learning algorithm.

S. Kollias et al. (Eds.): ICANN 2006, Part I, LNCS 4131, pp. 321–329, 2006.

2 Random Neural Networks

The RNN [1,2,3,4] is an analytically tractable spiked random neural network model. Consider a RNN with n neurons in which "positive" and "negative" signals circulate. Positive signals represent excitation, while negative signals represent inhibition. Each neuron i of the network is represented at time t by its input signal potential $k_i(t) \geq 0$, constituted only by positive signals that have accumulated. If the potential of the neuron i is positive, it may "fire" in a random sequence with rate $r(i)$, sending signals to the other neurons or to the outside of the network. Exogenous excitatory signals and inhibitory signals from outside the network to neuron i are in a Poisson stream of rate $\Lambda(i)$ and rate $\lambda(i)$ respectively. When neuron i fires, an excitatory signal may go to neuron j with probability $p^+(i,j)$, or as a negative signal with probability $p^-(i,j)$, or it may depart from the network with probability $d(i)$. The network parameters are the n by n "weight matrices" $W^+ = \{w^+(i,j)\}$ and $W^- = \{w^-(i,j)\}$ which need to be learned from input data, where $w^+(i,j)$ is the rate at which neuron i sends excitation signals to neuron j, and $w^-(i,j)$ is the rate at which neuron i sends inhibition signals to neuron j. The W^+ and W^- are learned from K input-output pairs.

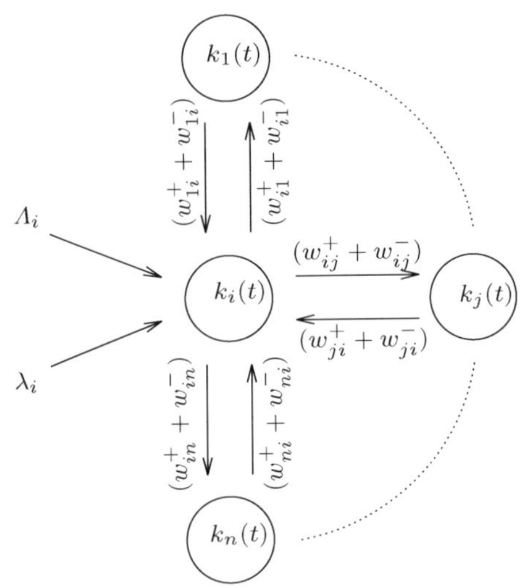

Λ_i	arrival rate of exogenous excitatory signals
λ_i	arrival rate of exogenous inhibitory signals
w_{ji}^+	arrival rate of excitatory signals from neuron j
w_{ji}^-	arrival rate of inhibitory signals from neuron j
$k_i(t)$	instantaneous potential of the neuron

Fig. 1. Representation of a neuron in the RNN

The state q_i of the ith neuron in the network is the probability that it is excited and the desired output of neuron i of the kth input is y_{ik}. The network adjusts its parameters to produce the set of desired output vectors in a manner that minimizes a cost C_k which is a function of network state $q = (q_1, \dots, q_n)$.

Figure 1 shows the representation of a neuron in the RNN using the model parameters that have been defined above. In this figure, only the transitions to and from a single neuron i are considered in a recurrent fashion. All the other neurons can be interpreted as the replicates of neuron i.

2.1 Reinforcement Learning Algorithm for RNN

There are different learning algorithms that may be applied to the RNN model. The gradient descent algorithm has been used with feed-forward topologies in many applications [5]. For the gradient descent algorithm to be implemented, the RNN output vectors need to be known a priori and provided during the training mode. In this work, we consider the RNN models that use a reinforcement learning (RL) algorithm. This is beneficial in applications such as computer networks where the desired input-output pairs are not known ahead. An RL algorithm for RNN is introduced in [6] and described here again. Let there be n neurons in the network and R denotes the reward value. If the previous decision of firing a signal from neuron i to neuron j was successful then this decision is rewarded by increasing the excitation weight value ($w^+(i,j)$) assigned to the corresponding interconnection and decreasing the inhibition weight values ($w^-(i,k), k \neq j$) for all other interconnections from neuron i as follows:

$$\begin{aligned} w^+(i,j) &= w^+(i,j) + R, \\ w^-(i,k) &= w^-(i,k) + R/(n-1), k \neq j, \end{aligned} \tag{1}$$

If the decision was not right, then it is punished by decreasing the excitation weight value between neurons i and j and increasing the inhibition weight values for all other interconnections as follows:

$$\begin{aligned} w^+(i,k) &= w^+(i,k) + R/(n-1), k \neq j, \\ w^-(i,j) &= w^-(i,j) + R. \end{aligned} \tag{2}$$

Then, weights are normalized to avoid indefinite increase in size:

$$\begin{aligned} w^+(i,j) &\leftarrow w^+(i,j) * r_i/r_i^*, \\ w^-(i,j) &\leftarrow w^-(i,j) * r_i/r_i^*. \end{aligned} \tag{3}$$

where $r_i^* = \sum[w^+(i,m) + w^-(i,m)]$.

3 Hardware Implementation of RNN

3.1 Circuit Implementation of an RNN Neuron

In RNN model, the output of a neuron i is given by

$$q_i = \frac{\sum_j q_j w_{ji}^+ + \Lambda(i)}{r(i) + \sum_j q_j w_{ji}^- + \lambda(i)} \tag{4}$$

This equation can be easily implemented in hardware as shown in Fig. 2. Note that the external inhibition signal, $\lambda(i)$, is zero in most applications; therefore it is not included in the figure.

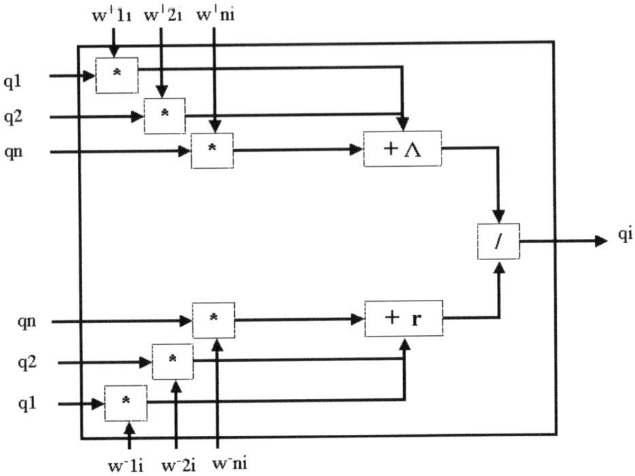

Fig. 2. Implementation of a single neuron i

Following the analysis given in [8], the number of weight terms in the RL algorithm can be reduced to $2n$ from $2n^2$. The simplified neuron output equation becomes

$$q_i = [w^+ \sum_j q_j + \Lambda(i)]/[r(i) + w^- \sum_j q_j + \lambda(i)] \tag{5}$$

This allows us to use the single neuron in an array configuration as depicted in Fig. 3 to form an n neuron RNN network.

3.2 Implementation of the Reinforcement Learning Algorithm

We implemented the RL algorithm using a finite state machine (FSM). The state diagram is illustrated in Fig. 4. In this FSM, it is assumed that the RNN weights and the previous decision are stored in a table. The table is searched first to determine if the network has given an output before or not. If so, the weight values along with the decision are retrieved from the table. Based on the previous decision's success, either the neuron is rewarded or punished by calculating new

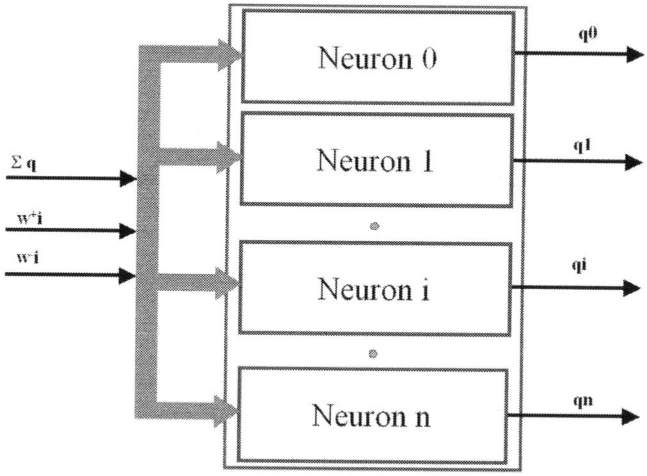

Fig. 3. Implementation of an n neuron network

weight values. However, if the search yields a miss then the default weight values are applied to the RNN.

4 Example Implementation in Networking

IP networks must evolve to meet the performance demands of todays and tomorrows users. Active networks have received a lot of attention in the last decade. Recently proposed CPN shows similarity with discrete active networks. CPN assigns a quality-of-service (QoS) goal to packets. The goal is a metric which defines the success of the outcome such as $G = \alpha * delay + \beta * loss$. The network uses the goal to provide a best-effort attempt at satisfying the QoS requirements. CPN employs intelligent based routing to pursue the goal. There are three types of packets: Smart packet (SP), Dumb packet (DP) and Acknowledgement packet (ACK). SPs probe the network for optimum routes. The next hop of the SP is determined using the output of RNN within the current router. DPs carry the payload and are source routed using best path information obtained from SP-ACK action. ACKs travel the reverse route of SP and carry the network data that is used to calculate a Reward value relevant to the goal of the packet. Reward value is used to update the status of the RNN. The function of the RNN in the CPN is to capture the effect of unpredictable network parameters and convert it into a routing decision. Each QoS-Source-Destination (QSD) combination has its own RNN model. RNN has fully recurrent topology and $2 * n^2$ weights. Each neuron corresponds to a port of the router. Neuron with highest output value (q) represents the next port.

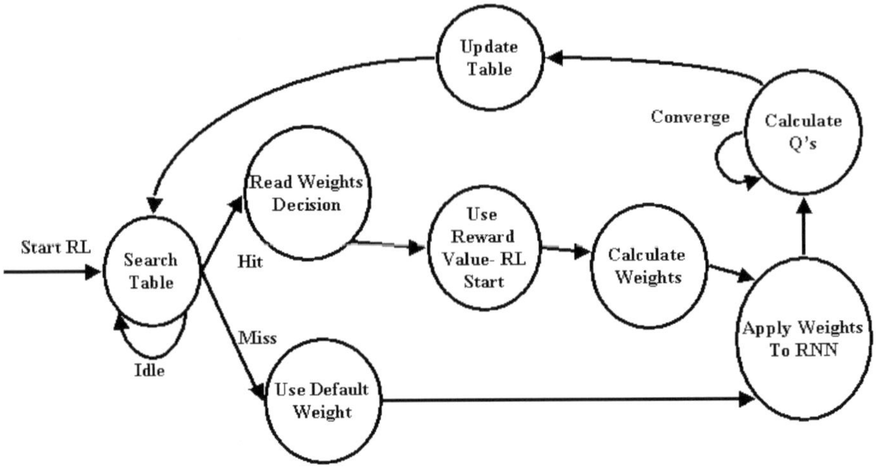

Fig. 4. State machine for the RL algorithm implementation

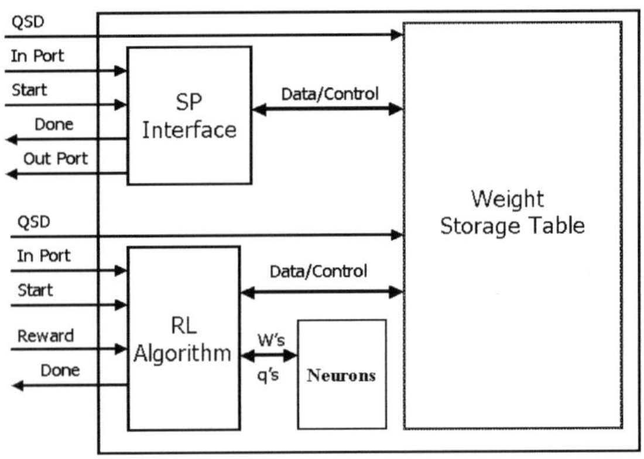

Fig. 5. SPP block diagram

4.1 Smart Packet Processor

The smart packet processor(SPP) in CPN has three main functions: 1) Determining the next hop for incoming SPs based upon related RNN status; 2) Updating the stored RNN models using data extracted from ACK; 3) Storing the RNN models indexed by QSD. The SPP block diagram is shown in Fig. 5.

The SP interface receives the smart packets and initiates the retrieval of weights, involkes the neuron array to calculate the output and directs the SP through the port whose corresponding neuron has the hisghest q. The weight storage tabel holds the neural network parameters (weights) and the network packet flow information (QSD). The SPP is implemented for a 4-port CPN test router in VHDL. The behavioral model developed in this hardware description language is synthesized to obtain hardware circuit implementation. The register-transfer-logic (RTL) schematic of the weight storage table component is shown in Fig. 6.

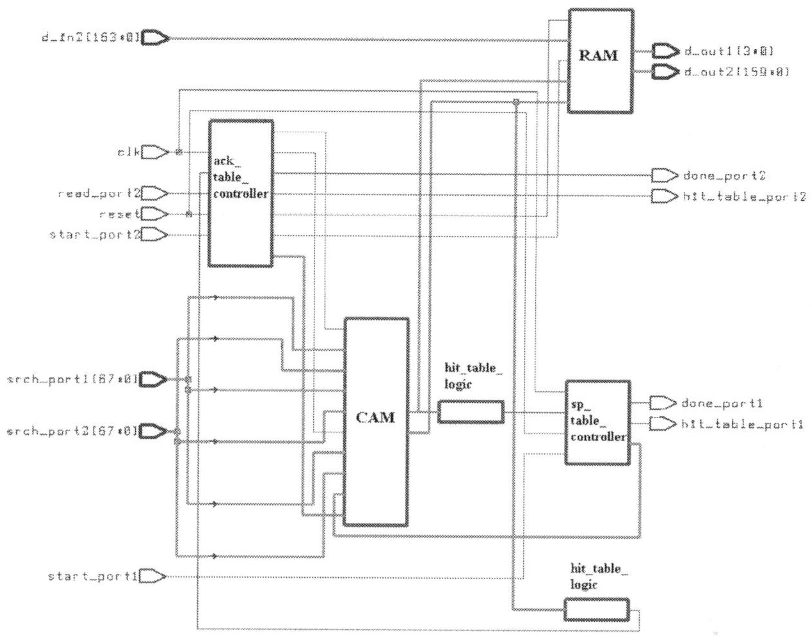

Fig. 6. RTL schematic of the weight storage table component

The synthesized gate-level netlist is imported to Cadence Silicon Ensemble, for floorplanning, placing and routing of the design. The obtained layout for the design is shown in Fig. 7. The core occupies 6.46 mm^2 in a 3-metal single-poly 0.6-μm digital CMOS process.

Simulations are run on the RTL description in Synopsys Design Analyzer to test the execution of the learning algorithm. The results shown in Fig. 8 are given for a reward scenario. The SPP is triggered with the *start_ack* signal. The QSD along with the port number are fed to the SPP. The RL module reads the threshold (the previous "smoothed" value of rewards) and the weights. The reward value is compared against the threshold, if it is greater then the RL component rewards the previous decision by increasing the excitation weight of the corresponding neuron and inhibition weigths of the other neurons. The weights are normalized and the output values are calculated until they converge. For this implementation, the weight, threshold and q terms each have 15 bits to the right of the ones position.

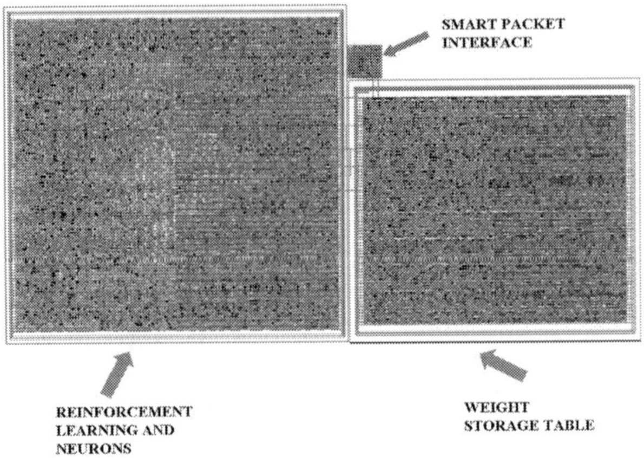

Fig. 7. Layout of the SPP design

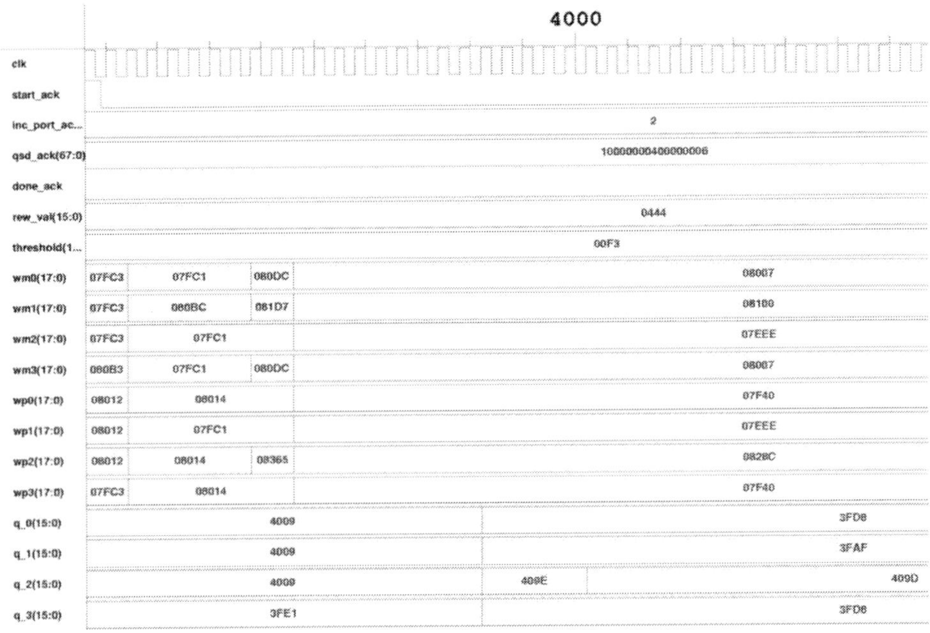

Fig. 8. Simulation results showing the learning algorithm in a reward scenario

5 Conclusions

We presented a hardware implementation for learning random neural networks (RNN) which uses reinforcement learning algorithm. Further, we illustrate an example where the RNN design used as a building block in a novel network routing protocol. Future directions for this work can be investigating whether the current digital implementation of the neurons can be replaced with an analog/mixed-signal implementation which would consume significantly less amount of space and processing time.

References

1. E. Gelenbe, "Random neural networks with negative and positive signals and product form solution," *Neural Computation,* vol. 1, no. 4, pp. 502-511, 1989.
2. E. Gelenbe, "Stability of the random neural network model," *Neural Computation,* vol. 2, no. 2, pp. 239-247, 1990.
3. E. Gelenbe, "Learning in the recurrent random neural network," *Neural Computation,* vol. 5, no. 1, pp. 154-164, 1993.
4. E. Gelenbe, Z.-H. Mao, Y.-D. Li "Function approximation with spiked random networks", *IEEE Trans. on Neural Networks*, vol. 10, No. 1, pp. 3–9, 1999.
5. H. Bakircioglu and T. Kocak, "Survey of random neural network applications", *European Journal of Operational Research*, vol. 126, pp. 319-330, 2000.
6. U. Halici, "Reinforcement learning algorithm with internal expectations for the random neural network", *European Journal of Operations Research*, vol. 126, pp.288-307, October 2000.
7. E. Gelenbe, R. Lent, and Z. Xu, "Design and analysis of cognitive packet networks", *Performance Evaluation*, vol. 46, pp. 155-176, 2001.
8. J. Seeber, "Design and implementatation of smart packet processor for the cognitive packet network router", *M.S. Thesis*, University of Central Florida, Orlando, 2002.
9. E. Gelenbe, R. Lent, A. Nunez, "Self-aware networks and QoS", *Proc. of the IEEE*, vol. 9, no. 9, pp. 1478-1489, 2004.
10. E. Gelenbe, "Cognitive Packet Network", *U.S. Patent No. 6,804,201 B1*, Oct. 12, 2004.

G-Networks and the Modeling of Adversarial Agents

Yu Wang

Dept. of Electrical & Electronic Engineering
Imperial College London, London, SW7 2BT, UK
yu.wang3@imperial.ac.uk

Abstract. As a result of the structure and content transformation of an evolving society, many large scale autonomous systems emerged in diverse areas such as biology, ecology or finance. Inspired by the desire to better understand and make the best out of these systems, we propose an approach which builds stochastic mathematical models, in particular G-networks models, that allow the efficient representation of systems of agents and offer the possibility to analyze their behavior using mathematics. This approach is capable of modelling the system at different abstraction levels, both in terms of the number of agents and the size of the geographical location. We demonstrate our approach with some urban military planning scenarios and the results suggest that this approach has tackled the problem in modelling autonomous systems at low computational cost. Apart from offering the numerical estimates of the outcome, the approach helps us identify the characteristics that impact the system most and allows us to compare alternative strategies.

Keywords: mathematical modelling, G-Networks, military strategy and planning, multi-agent systems.

1 Introduction

As a society evolves, its structure and content transform accordingly to reflect and address its needs. As a result, more and more large scale autonomous systems occur in various forms in the surrounding world, from diverse areas of study such as biology, ecology, finance or transportation. Large scale systems have been traditionally characterized by a large number of variables, nonlinearities and uncertainties. As an example taken from biology, a human body, where organs, containing billions of cells, perform different functions that contribute towards the operating of the body can be seen as a large scale system. Inspired by the desire to better understand and utilize the environment, we study such systems and hope to gain insights, predict the future and control them partially if not fully.

There have been many attempts to model large scale systems, such as building differential equations or with simulations [1-5]. However the sheer complexity and diversity of large scale systems make them difficult to be described and modelled, and it is even more difficult to provide numerical predictions of the underlying processes of such systems. To tackle these problems, we propose to use a stochastic approach, in particular G-networks [6-10], to model the individuals of the same nature collectively. In doing so, the computational complexity is greatly reduced. Another innovative aspect of our approach is that it is able to model systems with multiple geographical

S. Kollias et al. (Eds.): ICANN 2006, Part I, LNCS 4131, pp. 330 – 339, 2006.

locations at different levels of abstraction. With the approach, we aim to provide insights into systems in terms of their performance and behaviours, to identify the parameters which strongly influence them, and to evaluate how well an individual's task can be achieved and, therefore compare the effects of alternative strategies.

As presented in [13-17], our approach has many application areas, such as military planning, systems biology and computer networking. In this paper, we use strategic military planning in urban scenarios as an example to demonstrate our approach. We describe the systems of interest, the mathematical model, and the chosen scenarios. We illustrate two methods that we use in dealing with the situations where complete or incomplete world knowledge is available. To validate our model, we compare its results with those obtained from a simulator [11, 12] that was built in our group. We will briefly discuss the differences between these two approaches, and analyze the discrepancy between the results. The paper concludes with a discussion of potential extensions to the model.

2 System Descriptions

As an example of a potential application, we consider a closed system containing N distinct geographic locations and a set of C agent classes. Locations may have obstacles that an agent cannot physically enter, such as stretches of water, trees, buildings and so on. Obstacles depend of course on the physical objects that the agents represent (e.g. land and air vehicles will have different kinds of obstacles). At the moment, we do not take the local minimum problem into consideration and we assume all obstacles in the terrain are convex in shape.

In these systems, agents take actions to achieve their goals, which are reflected by their motions and behaviors. A goal, be it stationary or motion-oriented, attracts agents to move towards it. A stationary goal is a specific location, whereas a motion-oriented goal refers to the situation where an agent itself is the goal (target) of others. Agents either protect or destroy their motion-oriented goals. To achieve this, agents might need to cooperate or compete with others in the system. The difference in nature of the goals results in agents belonging to different adversary teams. Teams in the same group collaborate with each other to achieve their goals, while those in adversarial groups would exhibits competing behaviors to prevent others accomplishing their goals.

Motion in the system is a result of forces exercised by the goal and agents. There are three types of forces in our system: attractive, repulsive and long-range attractive and short-range repulsive. A straightforward example of an attractive force would be the force that a stationary goal (i.e. the destination) applies on an agent. A repulsive force can be interpreted as the tension of moving away. For example, if agent D's aim is to destroy agent B, then agent B applies an attractive force on D. On the other hand, agent D exercises a repulsive force on B. A long-range attractive and short-range repulsive force makes a group of agents stay together and keeps their distance at the same time. Thus if the agents are too close, repulsive forces are applied, otherwise, attractive forces are applied.

3 The Mathematical Model

Let $i = g(t, c, k)$ denote the location at time t of agent k belonging to team c. The location i may be a single variable or a vector, depending on the geographic representation that is being used, e.g. the (x, y, z) coordinates of a location on a map, where z may represent elevation when this is relevant. Thus in a two-dimensional coordinate space, a location will be denoted by $i=(x(i),y(i))$ and a neighboring location will be denoted by some $j=i+d$ where $d \in \{(0,0),(\pm1,0),(0,\pm1),(\pm1,\pm1)\}$. It is assumed that each agent has an initial location denoted by $S(c, k)$, and it may have (though this is not necessary) a final destination $D(c, k)$. We also assume that agents may die as a result of adversarial effects, or for other reasons, in which case they are relocated to "heaven" denoted by H. For the purpose of this model, we assume that there is just one heaven for everyone.

The mathematical model we develop aims at being able to compute, over a large number of experiments, quantities such as the probability $q(i, c, k)$ that agent k of team c is in location i. Such probabilities will also be used to compute indirectly the average time it may take certain agents to reach specific locations. The approach we take is based on constructing an ergodic model, i.e. one which has a stationary probability distribution, so that:

$$q(i,c,k) = \lim_{t\to\infty} prob[g(t,c,k) = i] \tag{1}$$

3.1 Agent Parameters

In the model, the motion of an agent is a result of two parameters:

- Its speed or rate of movement, $r(i, c, k)$ which will depend on its location (reflecting the geographic characteristics of its location)
- The direction of this movement which will be denoted by the probability $p(i, j, c, k)$ that agent (c, k) moves from i to j.

In this study, we assume that locations i and j are adjacent, in which case $j = i + d$ where d is one of the nine cardinal directions from i, including $d=(0, 0)$ which means that the agent has not moved at all.

The direction of motion for any one of the agents is determined by:

- the objectives or final destinations of the agents, when applicable, as expressed by a force of attraction
- the interaction between agents, as expressed by forces of attraction and repulsion
- the physical obstacles in the terrain, or the specific terrain related difficulties which may discourage or motion in a particular direction

In addition to motion, interaction between the agents is exhibited by their ability to destroy the others. Each agent (c, k) has a set of enemies, $E(c, k)$, that it tries to destroy, a shooting range $R(c, k)$ within which it is able to destroy an adversary, and a firing rate $f(c, k)$. Of course, these parameters may be identical in certain cases for all agents belonging to the same class or adversary team. In our model, the concept of enemy need not be reciprocal, i.e. $(c,k)\in E(c',k')$ does not necessarily imply $(c',k')\in E(c,k)$.

3.2 Forces

We use the cumulative-force exercised on an agent to determine its motion probabilities *p(i, j, c, k),* which define the direction of motion. Let *Forces(c', k', c, k)* be the force exercised on agent *(c, k)* by agent *(c', k')*. A positive coefficient implies that agent *(c, k)* is attracted to agent *(c', k')*, whereas a negative coefficient implies that agent *(c, k)* is repulsed by agent *(c', k')*. The strength of an inter-agent force varies with the distance of the two agents. The destination of an agent, *D(c, k)*, if one exists, also exercises an attractive force *G(i, d, c, k)*, which may also vary across the terrain.

The net force *v(i, d, c, k)* exerted on agent *(c, k)* at location *i* in direction *d* is computed as follows, where the function $dist(i,j)>0$ represents the way that the force component changes with the distance between agents. The set *L(i, d)* represents all locations at direction *d* from *i* in the terrain and *d* is defined as $d=\{direction(j-i)\}$.

$$v(i,d,c,k)=\sum_{all(c',k')}\sum_{j\in L(i,d)}\frac{Forces(c',k',c,k)q(j,c',k')}{dist(i,j)}+G(i,d,c,k) \tag{2}$$

Let *O(i)* be the set of neighbors of *i* which do not contain obstacles. In the process of obtaining motion probabilities, we introduce an adjusting factor to assist renormalizing *v(i, d, c, k)* to positive values. The adjusting factor is set in a way that it has a trivial impact on the accuracy of the motion probabilities. Let *V(i, c, k)* be the sum (numerical) of forces exerted on an agent from all the directions. It can be represented as:

$$V(i,c,k)=\sum_{d\in O(i)}|v(i,d,c,k)| \tag{3}$$

The motion probability, *p(i, j, c, k)*, is defined in equation (4), which also allows us to take *d= (0,0)*, i.e. the probability of staying in the current location. This of course raises the issue of certain agents getting "stuck" in a place from which they will not move away until conditions related to other agents have changed.

$$p(i,j,c,k)=\begin{cases}\dfrac{v(i,d,c,k)}{V(i,c,k)+factor} & if\ \ d\notin O(i)\\ 0 & if\ \ d\in O(i), d\neq(0,0)\\ \dfrac{factor}{V(i,c,k)+factor} & if\ \ d=(0,0)\end{cases} \tag{4}$$

3.3 Conditions Under Which the Simulation Game Ends

We consider that the simulation game ends when some subset of agents, for instance any one of the agents of some class *c*, reaches some pre-selected set of positions, which may include their destinations. Alternatively, the game may also end when some agents reach heaven (i.e. when they are killed). To formalize the terminating conditions, we define a final state set *F(c, k)* for the agent *(c, k)* as a subset:

$$F(c,k) \subseteq \{locations\ \ j\} \cup \{H\} \tag{5}$$

It is also possible that $F(c,k) = \phi$, in which case this means that this particular agent does not influence the time at which the simulation ends. The terminating condition *F* is then simply:

$$F = \cup_{all(c,k)} F(c,k) \tag{6}$$

and the interpretation we give to it is that:

$$Simulation\ \ Ends\ \ At\ \ t \Leftrightarrow if\ \ \exists(c,k),\ g(c,t,k) \in F(c,k),\ forF\ (c,k) \neq \phi \tag{7}$$

When a game attains its terminating condition, after some random time of average value 1 (this value is chosen for the purpose of normalization), each agent *(c, k)* (including the agents that made it to heaven), will move instantaneously to its initial location *S(c, k)*, and the game will start again at rate *Rstart*. For the purpose of this mathematical model, this cycle repeats itself indefinitely. This allows us to compute ensemble averages that are of interest. We assume that in the class of games, either some agent of some designated class(es) reach their destination, or all agents of designated class(es) reach heaven. Thus, we exclude situations where all agents become blocked and cannot move any further, or enter cycles of behavior which exclude the agents' demise, or that impair their ability to attain their destinations.

The terminating probability *T* is defined as the stationary probability that the model is in the terminating state:

$$T = \lim_{t \to \infty} prob\ [\vee_{all(c,k)}\, g(t,c,k) \in F(c,k)] \tag{8}$$

Similarly we define:

$$T(c,k) = \lim_{t \to \infty} prob[\vee_{all(c,k) \neq (c,k)}\, g(t,c',k') \in F(c',k')] \tag{9}$$

Thus *T(c, k)* is the stationary probability that the game is in the terminating state, given that agent *(c, k)* is already in a final state. Suppose now that we wish to compute the expected time τ it will take some specific agent *(c, k)* to reach some specific location γ. In that case we would set $F(c,k) = \{\gamma\}$, and the terminating probability, *T*, becomes $q(\gamma,c,k)$. We then have:

$$\tau = \frac{1}{q(\gamma,c,k)} \tag{10}$$

3.4 Model Equations

The equations that describe the overall long-term behavior of the system are obtained heuristically based on the equations satisfied by the stationary probability distributions of G-networks [7]. We heuristically, but plausibly, choose to relate the *q(i, c, k)* to each other and to the agents' parameters via the following equations:

$$q(i,c,k)=\begin{cases}\dfrac{Rstart+Neighbors(i,c,k)}{r(i,c,k)(1-p(i,i,c,k))+Killed(i,c,k)+MutaRate(i,c,k)}, & if\ \ i\in S(c,k)\\[2ex] \dfrac{Neighbors(i,c,k)+MutaIn(i,c,k)}{r(i,c,k)(1-p(i,i,c,k))}, & if\ \ i\in D(c,k)\\[2ex] \dfrac{\sum\limits_{(c,k)\in E(c',k')}\ \sum\limits_{i,j\neq H} q(i,c,k)Killed(i,c,k)}{Rstart}, & if\ \ i=H\\[2ex] \dfrac{Neighbors(i,c,k)+MutaIn(i,c,k)}{r(i,c,k)(1-p(i,i,c,k)+Killed(i,c,k)+MutaRate(i,c,k)}, & otherwise\end{cases} \tag{11}$$

$$Neighbors(i,c,k)=\sum_{d\notin O9i),d\neq(0,0)} q(i+d,c,k)r(i+d,c,k)p(i+d,i,c,k)$$

$$Killed\ (i,c,k)=\sum_{(c,k)\in E(c',k'),j} q(j,c',k')1[|\ j-i\ |\leq R(c',k')]f(c',k')$$

$$MutaIn(i,c,k)=\sum_{(c',k'),d\notin O(i),d\neq(0,0)} q(i,c',k')r(i+d,c',k')p(i,i+d,c',k')$$

In addition, we use the normalizing condition that the states that the sum of the probabilities that any given agent is at any one of the locations (including "heaven") is one. Thus for any *(c, k)* we have:

$$\sum_i q(i,c,k)=1 \tag{12}$$

You might have noticed that the *Rstart* rate is not defined in the above equation. Our approach is versatile in the sense that it provides insights into various aspects that are of interest based on one set of system equations. Therefore, the condition under which the process repeats itself is defined accordingly. For example, with the same model, we can examine the impact that the initial locations have on the average time that agents take to complete their tasks, the average time of a specific agent achieving its goal or the average percentage of a team achieving its goal.

3.5 Scenarios and Obtained Results

We have experimented with scenarios where agents of the same class travel towards their goals in the system. Results show that the model converges quickly and the inter-agent forces have impacts on the agents' performance depending on their strength. The detail of those scenarios and corresponding results can be found in [18].

After our initial success with a single agent class, we incorporate collaborating and competing behaviors into the model by introducing multiple agent classes, in other words, adversary teams. We demonstrate such models with a system containing three agents of different adversary teams: civilian (agent 1), terrorist (agent 2) and police (agent 3). The civilian's aim is to reach its destination alive with the help of the police. The police fights against the terrorist so that it will not kill the civilian. The terrorist attacks anybody who prevents it from killing the civilian. Both the police and the terrorist are only capable of attacking their enemies within a certain range. The collaborating and competing behaviors are not necessarily symmetrical, as illustrated in this scenario. The scenario repeats itself either when agent 1 is dead or arrives at its goal. Therefore the *Rstart* Rate has the numerical value of *Max[q(H, c, k), q(D (c, k), c, k)]*.

The detail of the scenario is as follows: in a 15×15 terrain, the agents' initial locations are (7,14) for the civilian, (5,2) for the terrorist and (13,2) for the police. The civilian has a stationary destination, location (14,14), whereas the police and the terrorist have motion-oriented goals. The nature of the terrorist attracts it towards the civilian and keeps it away from the police. The civilian travels towards its goal and avoids being killed by the terrorist. The police is attracted by the terrorist more than the civilian, simply because its task is to prevent the civilian from being killed.

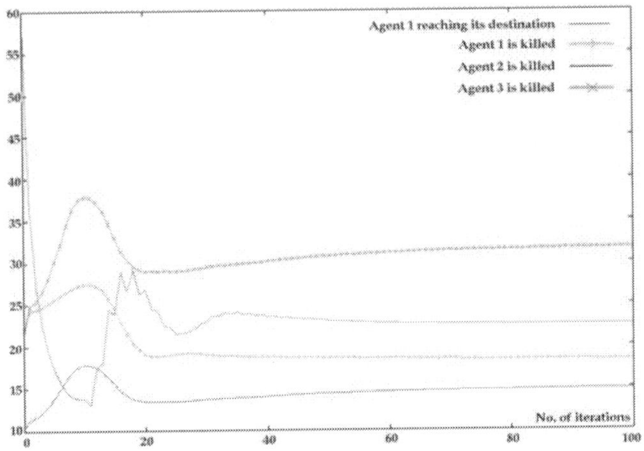

Fig. 1. Estimated time of events occurrence

Technically, the police and the terrorist are identically equipped except the police fires twice as frequently as the terrorist. Due to this nature, we expect the terrorist stands a higher chance of being killed by the police than killing the police. The result (See Fig 1) shows that, as a result of receiving help from the police, the civilian stands a higher chance of reaching the goal (55.5%) than being killed (44.5%). The reason that the police does not have a very big impact on the situation is that it takes times for the police to approach the terrorist. The result indicates that, on average, it takes 32 steps for the terrorist to kill the police and 15 steps for the police to kill the terrorist. This is inline with our predictions. As one might have noticed, for this scenario, the algorithm converges at around 50 iterations, which is shorter comparing with obtaining the average statistics from the simulator.

As mentioned before, the police and the terrorist have the same technical characteristics except their shooting rates. So if they have identical settings, it should take the same amount of effort to kill each other. We therefore assign them with the same shooting rate and see how that affects the outcome. Results confirm that, the estimated time of the police and the terrorist killing each other is the same.

In military planning, one often has to consider the trade-off between protection and agility. Weakening the protecting capacity of the police gives the terrorist more chances to launch attacks; however, it offers advantages such as agility and low consumption. Depending on the nature of the task, the police forces may have different focuses. With our approach, we can easily compare alternative strategies and select one that addresses our needs best.

Now we illustrate how mutation is modelling with an example of a system consisting of two agents. We restrain the two agents from interaction, so that mutation is the dominating factor of the outcome. The agents locate at (2,2) and (8,8) respectively. Agent 1 plans to go to (4,4) and agent 2 travels towards (14, 14). The two agents are identical apart from the fact that agent 1 has the mutating ability. With mutation rate a, agent 1 mutates and exhibits the same behaviour as agent 2. Thus after mutating, agent 1 will start pursuing agent 2's goal.

The following table (See Fig 2) shows how the estimated time of un-mutated agent 1, mutated agent 1 and agent 2 reaching their destinations. For example, with mutation rate 0.6, on average, 73% of the time agent 1 carries out an alternative task, which requires 6.35 unit time to complete, whereas 27% of the time, agent 1 preserves its nature and arrives its goal at 10.66 unit time. This feature is significant in task planning, where by altering parameters, one could "foresee" how fast multiple goals can be achieved with a reasonable overhead, if such need rises during the mission. In this case, if the team has to be split to achieve different tasks without too much overhead, mutation rate 0.4 can be a good choice.

Mutation-Rate	Non-Mutated	Mutated	A1 Goal(NM)	A2 Goal	A1 Goal(M)
0.1	0.819	0.18	2.645	6.8	50.53
0.2	0.6224	0.376	3.66	6.8	17.58
0.3	0.436	0.5624	5.47	6.8	8.68
0.4	0.33	0.663	7.5	6.8	6.62
0.5	0.296	0.703	9.05	6.8	6.47
0.6	0.2693	0.7352	10.66	6.8	6.35
0.7	0.23	0.7626	12.44	6.8	6.29
0.8	0.2143	0.7849	14.18	6.8	6.21
0.9	0.19556	0.8037	15	6.8	6.14
1	0.1812	0.8181	18.1	6.8	6.12

Fig. 2. Estimated time of an agent reaching its goal (with different mutating rate)

During the experiments, we have also discovered that our approach can be used to model systems at different levels of abstraction. For example, 1 agent in the mathematical model does not necessarily represent a single agent; instead, it can represent a certain number of agents, say 150. From another perspective, we can also estimate the outcome of a larger terrain by modelling a smaller terrain with similar natures/characteristics, as presented in [14]. The approach is also validated again a simulator that was developed in our group. Results [14] show that, despite the magnitude discrepancy, the mathematical model is inline with the statistics collected in the simulator.

So far, the features that our approach offers are desirable. However, it is assumed so far is that we can calculate the probability $q(i, c, k)$ (and therefore the estimated time for events). This is not always the case for large scale autonomous systems which are known for their uncertainties and complexities. We have also proposed a method [17,19] to overcome situations where a complete knowledge of the system is not available. This method estimates the agents' motion probability using historical observation records.

4 Conclusions

In this paper, we have presented a stochastic approach based on G-Networks, which models large scale agent systems. The approach models systems containing collaboration, competition and mutation under the condition that complete information of the system is available. We first described the systems of interest, the mathematical model and demonstrated the approach with some scenarios of military planning in urban environments. Results show that our approach identifies the parameters that strongly influence the agents' performance and allows us to compare alternative strategies at low computational cost. We then proposed an extension to the model which deals with systems where complete information is not readily available.

We plan to incorporate behaviors such as reproduction into the model so that it can be applied in fields such as system biology. After the initial success of obtaining the motion probability via observation, we are investigating how to deduce agents' intention via similar means. This undoubtedly will reduce the dependence of model to the system's knowledge.

In reality, obstacles in urban environment have an impact on an agent's behavior. For example, a building might be an obstacle for a car but not a pedestrian. Therefore we plan to change the obstacles so that they have different impact on agents' behaviors and incorporate the wall following method mentioned in [18] to deal with the local-minima problem. In doing so, we are able to model more realistic urban scenarios. Theoretical wise, we aim to study the computational constraints related to resources or time-frame, as well as conduct an extensive exploration on modelling large scale agent systems at different abstraction levels.

References

1. Amin, K.A., Mikler, A. R: Dynamic Agent population in Agent-Based Distance Vector Routing, in Proceedings of the Second international workshop on Intelligent systems design and application. (2002) 195-200
2. Burmeister, B.: Models and methodology for agent-oriented analysis and design, in K. Fischer (Eds) Working Notes of the KI'96 Workshop on Agent-Oriented Programming and Distributed Systems. (1996)
3. Cysneiros, L.M., Yu, E.: Requirements Engineering for Large-Scale Multi Agent Systems, Software Engineering for Large-Scale Multi-Agent Systems: Research Issues and Practical Applications, Vol. 2603. (2003) 39-56
4. Huang, G., Abur, A., Tsai, W.K.: A Multi-level Graded-Precision Model of Large Scale Power Systems for Fast Parallel Computation, Mathematical computing modelling, Vol. 11. (1998) 325-330
5. Kinny, D., Georgeff, M., Rao, A.: A Methodology and modelling technique for systems for BDI agents, in W. van der Velde and J. Perram (Eds). Agents Breaking Away: Proceedings of the Seventh European Workshop on Modelling Autonomous Agents in a Multi-Agent World MAAMAW'96, Vol. 1038. (1996) 56-71
6. Gelenbe, E.: Random neural networks with positive and negative signals and product form solution, Neural Computation, Vol. 1(4). (1989) 502-510
7. Gelenbe, E.: G-networks with instantaneous customer movement, Journal of Applied Probability, Vol. 30 (3). (1993) 742-748

8. Gelenbe, E.: G-Networks with signals and batch removal, Probability in the Engineering and Informational Sciences, Vol. 7. (1993) 335-342
9. Fourneau, J. M., Gelenbe, E., Suros, R.: G-networks with multiple classes of positive and negative customers, Theoretical Computer Science, Vol. 155. (1996) 141-156
10. Gelenbe, E., Labed, A.: G-networks with multiple classes of signals and positive customers, European Journal of Operations Research, Vol. 108(2). (1998) 293-305
11. Gelenbe, E., Hussain, K., Kaptan, V.: Simulating the navigation and control of autonomous agents, in Proceedings of the 7th International Conference on Information Fusion. (2004) 183-189
12. Gelenbe, E., Hussain, K., Kaptan, V.: Enabling Simulation with Augmented Reality, in Proceedings of the International Symposium on Modeling, Analysis and Simulation of Computer and Telecommunication Systems. (2004) 290-310
13. Gelenbe, E., Kaptan, V., Wang, Y.: Simulation and Modelling of Adversarial Games, in the Proceedings of the 6th European GAME-ON Conference on Simulation and AI in Computer Games. (2005) 40-44
14. Gelenbe, E., Wang, Y.: A mathematical approach for mission planning and rehearsal, to appear in the Proceedings of SPIE Defence & Security, Orlando. (2006)
15. Gelenbe, E., Wang, Y.: A Trade-off between Agility and Resilience, in Proceedings of the 13th Turkish Symposium on Artificial Intelligence and Neural Networks. (2004) 209-217
16. Gelenbe, E., Kaptan, V., Wang, Y.: Biological Metaphors for Agent Behavior, in the Proceedings of the 19th International Symposium on Computer and Information Sciences, Lecture Notes in Computer Science, Vol. 3280. (2004) 667-675
17. Gelenbe, E., Wang, Y.: Modelling Large Scale Autonomous Systems, Accepted to Fusion 2006. (2006)
18. Wang, Y.: Numerical Modelling of Autonomous Agent Movement and Conflict, to appear in Computational Management Science. (2006)

Development of a Neural Net-Based, Personalized Secure Communication Link

Dirk Neumann[1], Rolf Eckmiller[1], and Oliver Baruth[1]

Department of Computer Science, Division of Neural Computation, University of Bonn, Römerstr. 164, 53117 Bonn, Germany,
{neumann, eckmiller, baruth}@nero.uni-bonn.de,
http://www.nero.uni-bonn.de

Abstract. This paper describes a novel ultra-secure, unidirectional communication channel for use in public communication networks, which is based on

a) learning algorithms in combination with neural nets for fabrication of a unique pair of modules for encryption and decryption, and
b) in combination with decision trees for the decryption process,
c) signal transformation from spatial to temporal patterns by means of ambiguous spatial-temporal filters (ST filters),
d) absence of public- or private keys, and
e) requirement of biometric data of one of the users for both generation of the pair of hardware/software modules and for the decryption by the receiver.

To achieve these features we have implemented an encryption-unit (EU) using ST filters for encryption and a decryption unit (DU) using learning algorithms and decision trees for decryption.

1 Introduction

To establish a secure communication via a public network (e.g. www) several methods like VPN or SSH are known. All of these methods use a private key to ensure private communication [9], [13]. With the willingness of everyone to use public networks even for very private activities, e.g. information exchange with the medic institute or transactions with your credit institute via the www, the demand on easy to use encryption systems with integrated authentication arises. To enhance the acceptance of a new secure communication environment, the user should not keep an additional password in mind, but rather use his always available biometric identification [5], [15]. The proposed secure communication environment uses biometric data of an owner to establish an individually secured, encrypted point-to-point communication [1], [6]. To achieve this ultra secure communication several different modules have to be developed, some running in software on a normal PC, others being realized on special hardware (e.g. PCMCIA FPGA Card) [18], [17].

S. Kollias et al. (Eds.): ICANN 2006, Part I, LNCS 4131, pp. 340–348, 2006.

2 Generation of a Unique Pair of Hardware/Software Encryption- and Decryption Units

The secure communication environment consists of a software component, two hardware components and one biometric input device, e.g. fingertip sensor (see Fig. 1) [11]. Initally, the functionality of the system parts is not given.

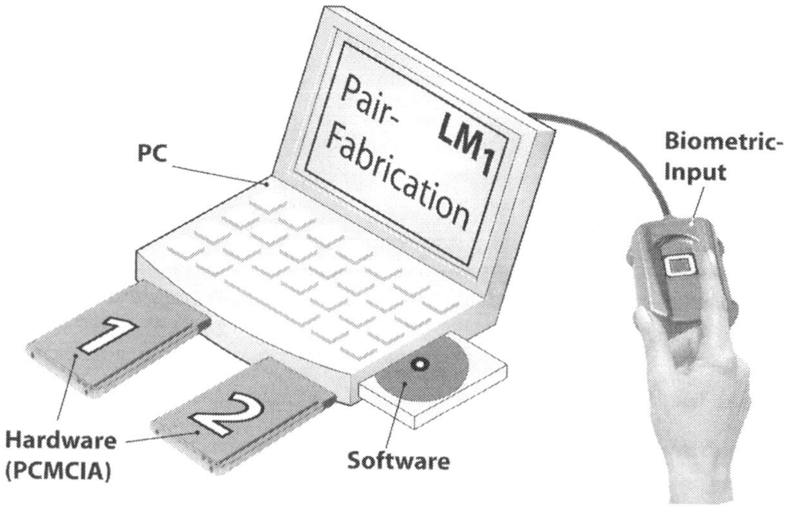

Fig. 1. System components of the secure communication environment. Two special hardware modules (preferably as programmable FPGA PCMCIA cards, depicted as 1 and 2) are programmed by a computer. A special software is used to allow biometric data to be the input to a learning module (LM) whereas the biometric data is acquired via a fingerprint sensor. The learning module LM_1 is used in a interative tuning procedure to generate two different complementary algorithms which will be embedded in the hardware modules forming a unique pair of functional complementary modules.

In a first step, the owner of the hardware kit uses a normal PC to manufacture a unique pair of hardware modules (e.g. PCMCIA FPGA Boards, intelligent smart cards) [3]. Later, these cards are essential to build the encryption unit (EU) and the decryption unit (DU). In this step the biometric data of an owner is acquired via a biometric input device. This data is used as an individual modulation vector (MV) which defines the parameter of the algorithm, of the used spatial temporal filters (ST filter) in the EU.

These filter algorithms will be implemented in hardware on a special module. Additional to the modulation vector a random part (random vector RV) exists like the salt value for hash calculations [9] so that an owner will never generate two identical encryption units. RV is not used for the ST filter algorithm, but rather to deform shape and geometrical arrangement of the original concentric symmetric detection zone (DZ) of the ST filters.

In a second step the learning module one (LM_1) is used to train iteratively a neural network (MLP) in order to define the geometrical arrangement by RV back to original concentric symmetric DZ [10], [14]. For security reasons the function of the neural net will be implemented in the second hardware module, which is used in the decryption unit.

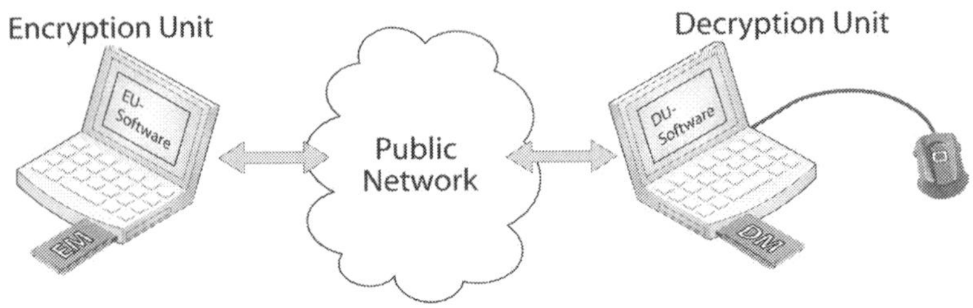

Fig. 2. Left part: A personal computer uses the first hardware module (EM) with its embedded encryption functions. On this hardware module an additional software component is placed in a flash-memory subunit which is used by the PC to act as encryption unit (EU). Right part: A second PC uses the second hardware module (DM) including another software component to form the decryption unit (DU). These two units can use a public network for communication purposes.

An other approach to generate two complementary functions for DU and EU uses only very view (e.g. only one, or defined by RV) ST filters in the encryption unit (EU) which are moved along a trajectory defined by the random vector RV. In this the case the complete input area has to be covered at least two times. In this case the learning module (LM_1) is used to find a representation of this trajectory which can be used by the decryption unit (DU).

The second hardware module also includes a special learned decision tree [10] which uses the ST filter output functions as input and reconstructs the original input data. This decision tree has to know about the used ST filter properties in EU, which are defined by MV. Whereas MV will be acquired via the biometric input device [12]. The required biometric input data (MV) during the decryption process prevent DU from unprohibited usage.

After these two steps the two hardware modules can be used in different locations. The first hardware module connected to a PC forms the encryption unit and the second hardware module connected to an other PC with a biometric input device forms the decryption unit.

3 Encryption Specifics

The encryption unit (EU) is based on the usage of different ST filters. These ST filters have their own assigned detection zone (DZ), which is used to acquire data as a binary input. This binary data can be taken from a black and white picture. Each ST filter does its own data processing (the calculation can be scheduled in parallel) on the acquired data; this computation uses two different independent calculation routines, one for the center and one for the surrounding [2], [4]. The center pixels are indicated as C and the surrounding pixels as S (see Fig. 3).

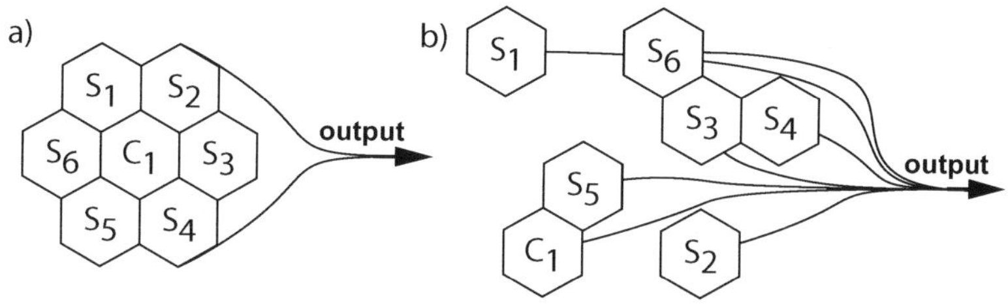

Fig. 3. (a) Example of a simple ST filter with concentric symmetric detection zone (DZ) to show the functionality. This DZ uses one input value for the center calculation (indicated as C_1) and six input values for the surrounding calculation (indicated as $S_1, \ldots, S_6$). (b) ST filter with a more spreading DZ which arises from the concentric symmetric DZ shown in (a) by a transformation depending on a random vector (RV).

Each routine uses a weighting (indicated as +, −) of the binary data and then calculates the sum over the corresponding center- and surrounding values. The temporal properties of each ST filter are implemented by means of FIR filters [7]. These ST filters are constructed in such a way that several input patterns will generate the same output value. This gives us a simple 'one way function' [9] because if we only have the output of one ST filter, we cannot calculate the original input pattern. This means that an ST filter always creates ambiguity and we cannot clearly dissolve the ambiguity of an individual ST filter. Now, if we use a special set of ST filters which have to fulfill defined conditions, we can clearly reconstruct the original input pattern [8], [16]. This reconstruction can only be done if all properties of the used ST filters are known. This means that our modulation vector (MV) is a very essential part for this calculation. These ST filters are chosen among a defined set of ST filter classes. These filter classes are defined by MV, and each ST filter is assigned to one class by MV. Each ST filter class has its own properties concerning the detection zone and the time response.

Additional to these requirements we have to define conditions for the arrangement of the detection zones used by the ST filters. The ST filters must be arranged over the whole input pattern, so that each pixel is covered of at least two detections zones of different ST filters. If we have ST filters with only one center pixel, then this center pixel needs not to be covered by an additional ST filter. For example, if we use three different classes of ST filters and each filter class has one center pixel and six surrounding pixels, we get a placement shown in Fig. 5a. We see that the surrounding pixels in the middle of the three ST filters, are each covered by two adjacent ST filters.

If we have more ST filters in a pixel plane and if we have three different ST filter classes, we could assign them to the used filters in a fashion shown in Fig. 5b (this example tiling we call basic tiling). This assignment is defined by MV. To make the encryption unit unique and independently from the used biometric data we use the random vector (RV) to deform the shape and to change the spread of each used ST filter (see Fig. 3b). Whereas these two vectors are embedded in the algorithm, which are embedded in the hardware boards and calculate the ST filter functions.

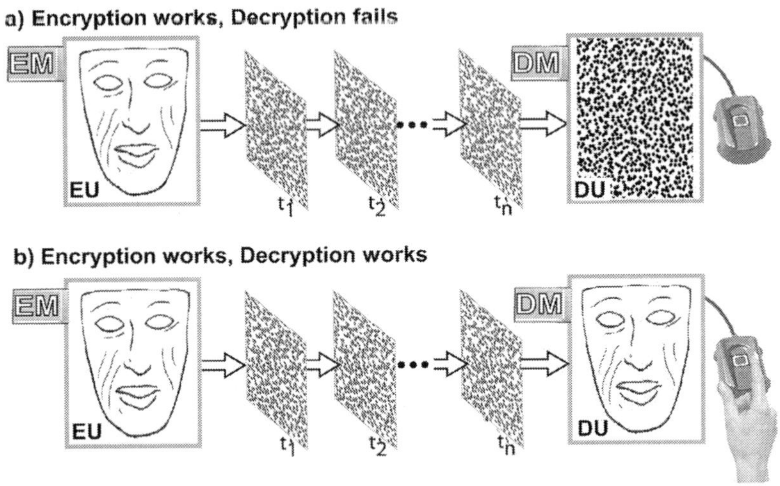

Fig. 4. The encryption unit (EU) uses the black and white picture showing a picture of a Inca mask as input. The spatial information of this picture is transformed into a pure temporal data stream by means of FIR filters, depicted by t_1, ..., t_n. This data stream is received by the corresponding decryption unit (DU) which reconstruct the original picture/spatial information by using of additional information extracted from the biometric input device. Upper part: EU uses the embedded algorithms of the encryption module (EM) to calculate the temporal data stream. The reconstruction of the original data stream fails because of the missing biometric input, and the decrypted image only shows a random image. Bottom part: The EU encrypts the data as described above. The DU is now able to reconstruct the original data, depicted as the Inca mask, with the additional biometric information acquired via the biometric input device.

4 Decryption Specifics

The decryption unit (DU) uses the second hardware board and a biometric input device to reconstruct the original data out of the encrypted data stream. Thereby the hardware board is used to reconstruct the original arrangement of each ST filter detection zone. The software component including a second learning module (LM_2) uses the biometric input device to acquire the modulation vector (MV) and uses it as additional input for a special decision tree (DT). The DT will help to reconstruct the input pattern from the beginning. The biometric input is an essential part of the decryption unit which gives us the possibility to use DU only if we are the owner and only if we permitted to do that.

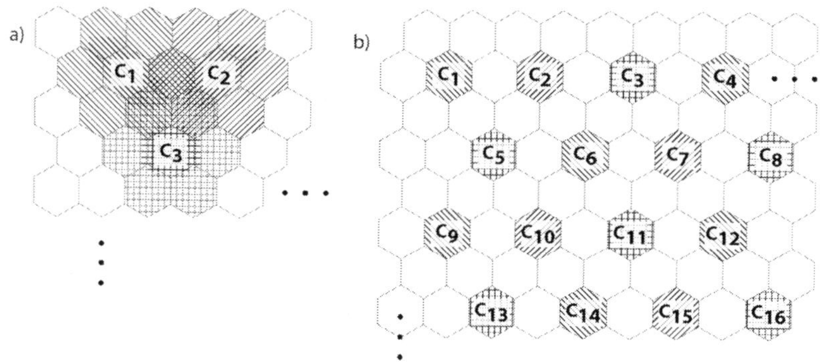

Fig. 5. (a) Example of an ST filter placement showing three different ST filter classes. Each surrounding pixel is covered by at least two different ST filters. The center values are only covered by one ST filter, thereby C_i indicates the center of ST filter i (b) Possible assignment of ST filter to three available ST filter classes over the entire input, fulfilling the properties mentioned for (a). For simplicity reasons only the center values of each ST filter are indicated.

In some cases the information of the ST filters are not sufficient to recalculate the input pattern clearly, so there is an additional part in the software solution which will request an additional set of ST filter outputs. This additional ST filter output is the result of a second calculation of the ST filter, where the position of all ST filters is shifted by one pixel (one bit) in a direction which is specified by DU. In this configuration the decryption unit with its unique second hardware board can only decrypt data which was generated by the corresponding encryption unit, using the corresponding unique first hardware board derived from the initial tuning/manufacture process described above. This means that we have authentication functionality within this unique pair of hardware boards working in the EU and the DU. Somebody can authenticate himself by using the EU to send an encrypted message to the DU, because the DU is only capable to decrypt messages from the corresponding EU.

5 First Simulation Study

To show the feasibility we have implemented a basic version of this proposed secure communication environment. This basic version (see Fig. 6) can be used to test various parameters of the used algorithms. Our implemented encryption unit uses a black and white picture with a resolution of 32x32 pixels as input data (P_{In}). For the used ST filters we developed three different classes, each with the same geometric expansion of the detection zone. The parameter of the weighting and the parameter for the time response were different. For the distribution of the ST filters we used the basic tiling (see Fig. 5b).

To cover the complete input pattern with ST filters it is necessary to use 17x17 of them, each chosen from the three ST filter classes. To transmit the

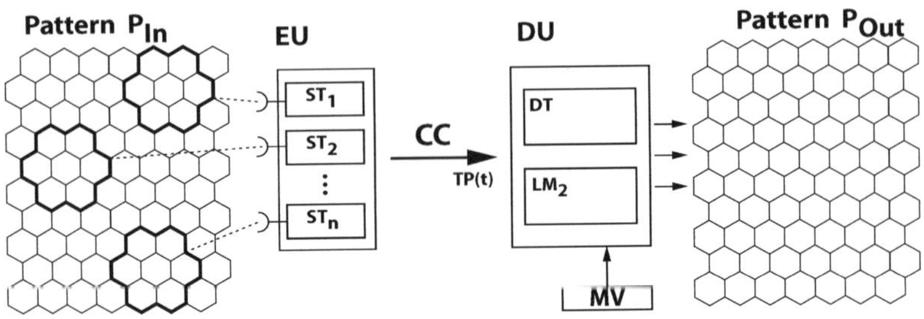

Fig. 6. Schema of the secure communication environment showing the input pattern P_1 scanned by the ST filter ensemble implemented in the encryption unit (EU); this ensemble consists of n ST filters. The calculated temporal data stream output TP(t) of the DU is transmitted via a communication channel CC to the decryption unit (DU). DU use TP(t) with external information obtained via the modulation vector (MV) to use the second learning module (LM_2) and a decision tree (DT) to reconstruct the output pattern P_2 equal to P_1.

output of these ST filters we can use the standard TCP/IP protocol or we can save the output in a file. For the decryption unit we have implemented a transceiver to gather the transmitted encrypted data. This data gets into the initially trained neural net which arranges the detection zones for the postprocessing. The following decision uses the modulation vector to get information of the used ST filters (defined ST filter classes and assignment of each ST filters to one of them). As a result of this learned decision tree we get a 32x32 pixel black and white picture P_{Out} clearly identical to the input picture (see Fig. 7). The implemented decision tree needs, only one additional output from the ST filters to reconstruct the picture, where the filters are shifted by one pixel. The runtime of the decision tree is independent of the used ST filter to ST filter class assignment and independent of the input pattern.

We have shown that single ST filters which produce ambiguity and thus are not clearly invertible can be used in ensembles to allow us to reconstruct the original input. Therefor the used detection zones must fulfill several requirements. With the currently implemented three different ST filter classes, a pair of encryption- (EU) and decryption unit (DU) could be created using the random vector (RV) to get a unique EU-DU pair. This EU-DU pair has an embedded authentication feature and can be used for authentication purposes. The modulation vector (MV) can be used to establish a usage of the DU only if the owner uses its own fingerprint to permit this. For future steps we will implement more types of ST filter classes with differently shaped detection zones. Furthermore these detection zones will overall use more pixels for their calculations.

Another meaningful extension could be a bidirectional communication between two geographically spread locations. For this purpose we have to integrate an encryption- and a decryption unit in one system or better in one hardware board, where encryption- and decryption algorithms are placed in one system. In

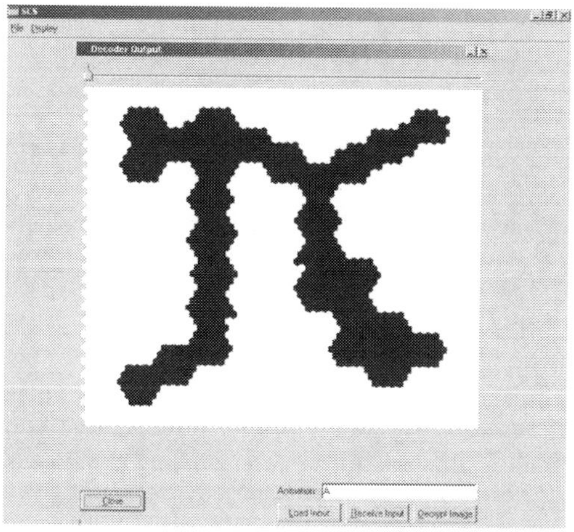

Fig. 7. Screen shot of the working decryption unit (DU). It uses the information from the modulation vector (MV) for decryption and finally shows the window of the decrypted image, in this case a black and white picture of π.

this case, encryption and decryption should have different modulation vectors. Thereby we have two different pairs of EU-DU combined in one system.

6 Conclusions

1. The application of ambiguous spatial-temporal transformations in combination with learning algorithms offers a novel mechanism for encryption and decryption.
2. The combination of unique hardware pairs for encryption- and decryption units with embedded biometric data allows an ultra secure unidirectional communication via public networks.

References

1. O. Baruth, R. Eckmiller and D. Neumann, "Retina encoder tuning and data encryption for learning retina implants," *Proc. of the Int. Joint Conf. on Neural Networks,* (IJCNN) 2003, Vol. **1**, pp. 1249-1252, Portland, Oregon, 2003.
2. E.A. Bernardete and E. Kaplan, *The dynamics of primate M retinal ganglion cells,* Visual Neuroscience, **16** pp. 355-368, 1999.
3. T.C. Clancy, N. Kiyavash and D.J. Lin, "Secure Smartcard-Based Fingerprint Authentication, " *ACM SIGMM 2003 Workshop on Biometrics Methods and Applications,* pp. 45-52, 2003.
4. D.W. Dong, "Spatiotemporal Inseparability of Natural Images and Visual Sensitivities," in *Computational, neural and ecological constraints of visual motion processing,* (page 371-380), Edited by J.M. Zanker and J. Zeil, Berlin: Springer, 2001.

5. J. Daugman, *The importance of being random: statistical principles of iris recognition,* Pattern Recognition, **36** pp. 279-291, 2003.
6. R. Eckmiller, O. Baruth and D. Neumann, "Method and Device for Decryption-Secure Transfer of Data", PCT Patent Application, PCT WO 2004021694.
7. S. Haykin, Editor, *Adaptive Filter Theory,* New Jersey: Prentice Hall, 4th Edition, 2002.
8. R.J. McEliece, Editor, *The Theory of Information and Coding,* Cambridge: Cambridge University Press, 2002.
9. A. Menezes, P. van Oorschot, and S. Vanstone, Editors, *Handbook of Applied Cryptography,* Boca Raton: CRC Press, 1997.
10. T.M. Mitchel, Editor, *Machine Learning,* New York: McGraw Hill, 1997.
11. S. Parbhakar, P. Sharath, and K. Anil, "Biometric Recognition: Security and privacy concerns," *IEEE Security and Privacy Magazine,* **1**, pp. 33-42, 2003.
12. N. Ratha and R. Bolle, Editiors, *Automatic Fingerprint Recognition Systems,* New York: Springer, 2004.
13. B. Schneier, Editor, *Applied Cryptography: Protocols, Algorithms, and Source Code in C,* New York: John Wiley and Sohns, 2nd Edition, 1996.
14. J. Si, A.G. Barto, W.B. Powell, and I.D. Wunsch, Editors, *Handook of Learning and Approximate Dynamic Programming,* Piscataway: IEEE Press and New York: Wiley-Interscience, 2004.
15. C. Soutar. D. Roberge, A. Stoianov, R. Gilroy, and K.V. Kumar, "Biometric Encryption," in *ICSA Guide to Cryptography,* (chapter 22), Edited by R.K. Nichols, McGraw-Hill, 1999.
16. K.R. Rao and P.C. Yip, Editors, *The Transform and Data Compression Handbook,* Boca Raton: CRC Press, 2001.
17. T. Wollinger, J. Guadjardo, and C. Paar, "Security on FPGAs: State-of-the-art implementations and attacks", ACM Transactions in Embedded Computing Systems, Vol. **3**, pp. 534-574, August 2004.
18. T. Wollinger and C. Paar, "Security aspects of FPGAs in cryptographic application" in *New Algorithms, Architectures, and Applications for Reconfigurable Computing,* (chapter 1), Edited by W. Rosenstiel and P. Lysaght, Dordrecht; Boston ;London: Kluwer Academic Publishers, 2004.

Exact Solutions for Recursive Principal Components Analysis of Sequences and Trees

Alessandro Sperduti

Department of Pure and Applied Mathematics, University of Padova, Italy
sperduti@math.unipd.it

Abstract. We show how a family of exact solutions to the Recursive Principal Components Analysis learning problem can be computed for sequences and tree structured inputs. These solutions are derived from eigenanalysis of extended vectorial representations of the input structures and substructures. Experimental results performed on sequences and trees generated by a context-free grammar show the effectiveness of the proposed approach.

1 Introduction

The idea to extend well known and effective mathematical tools, such as Principal Component Analysis, to the treatment of structured objects has been pursued directly (e.g. [6,5]) or indirectly (e.g. [2,3,1]) by many researchers. The aim is to devise tools for embedding discrete structures into vectorial spaces, where all the classical pattern recognition and machine learning methods can be applied.

Up to now, however, at the best of our knowledge no exact solution for Recursive Principal Components Analysis has been devised. Here we define sufficient conditions that allow us to construct a family of exact solutions to this problem. Experimental results on significantly large structures demonstrate the effectiveness of our proposal.

2 Recursive Principal Components Analysis

In [6] a connection between Recursive Principal Components Analysis and the representations developed by a simple linear recurrent neural network has been suggested. Specifically, a model with the following linear dynamics is considered:

$$\mathbf{y}_t = \mathbf{W}_\mathbf{x}\mathbf{x}_t + \sqrt{\alpha}\mathbf{W}_\mathbf{y}\mathbf{y}_{t-1} \quad (1)$$

where t is a discrete time index, $\mathbf{x}_t$ is a zero-mean input vector, $\mathbf{y}_t$ is an output vector, $\alpha \in [0, 1]$ is a gain parameter which modulates the importance of the past history, i.e. $\mathbf{y}_{t-1}$, with respect to the current input $\mathbf{x}_t$, $\mathbf{W}_\mathbf{x}$ and $\mathbf{W}_\mathbf{y}$ are the matrices of synaptic efficiencies, which correspond to feed-forward and recurrent connections, respectively. In [6] it is assumed that the time series $(\mathbf{x}_t)$ is bounded and stationary. The model is trained using an extension of the Oja's rule, however there is no proof that the proposed learning rule converges to the recursive principal components.

S. Kollias et al. (Eds.): ICANN 2006, Part I, LNCS 4131, pp. 349–356, 2006.

Here, for the sake of clearness, we consider the special case where $\alpha = 1$. The construction we are going to derive does not depend on this specific setting. We focus on the following equations

$$\mathbf{x}_t = \mathbf{W}_\mathbf{x}^\mathsf{T}\mathbf{y}_t \tag{2}$$

$$\mathbf{y}_{t-1} = \mathbf{W}_\mathbf{x}\mathbf{x}_{t-1} + \mathbf{W}_\mathbf{y}\mathbf{y}_{t-2} \tag{3}$$

$$= \mathbf{W}_\mathbf{y}^\mathsf{T}\mathbf{y}_t \tag{4}$$

which have to be satisfied in order for the network to compute recursive principal components, i.e. the sequence is first encoded using eq. (1), and then, starting from an encoding, it should be possible to reconstruct backwards the original sequence using the transposes of $\mathbf{W}_\mathbf{x}$ and $\mathbf{W}_\mathbf{y}$. In fact, the aim of recursive principal component is to find a low-dimensional representation of the input sequence (or tree) such that the expected residual error is as small as possible.

3 Exact Solutions for Sequences

For $t = 1, \ldots, T$ the following equations should be satisfied

$$\mathbf{x}_t = \mathbf{W}_\mathbf{x}^\mathsf{T}\underbrace{(\mathbf{W}_\mathbf{x}\mathbf{x}_t + \mathbf{W}_\mathbf{y}\mathbf{y}_{t-1})}_{\mathbf{y}_t} = \mathbf{W}_\mathbf{x}^\mathsf{T}\mathbf{W}_\mathbf{x}\mathbf{x}_t + \mathbf{W}_\mathbf{x}^\mathsf{T}\mathbf{W}_\mathbf{y}\mathbf{y}_{t-1} \tag{5}$$

$$\mathbf{y}_{t-1} = \mathbf{W}_\mathbf{y}^\mathsf{T}\overbrace{(\mathbf{W}_\mathbf{x}\mathbf{x}_t + \mathbf{W}_\mathbf{y}\mathbf{y}_{t-1})} = \mathbf{W}_\mathbf{y}^\mathsf{T}\mathbf{W}_\mathbf{x}\mathbf{x}_t + \mathbf{W}_\mathbf{y}^\mathsf{T}\mathbf{W}_\mathbf{y}\mathbf{y}_{t-1} \tag{6}$$

where it is usually assumed that $\mathbf{y}_0 = \mathbf{0}$. Sufficient conditions for the above equations to be satisfied for $t = 1, \ldots, T$ are as follows:

$$\mathbf{W}_\mathbf{x}^\mathsf{T}\mathbf{W}_\mathbf{x}\mathbf{x}_t = \mathbf{x}_t \tag{7}$$

$$\mathbf{W}_\mathbf{y}^\mathsf{T}\mathbf{W}_\mathbf{x}\mathbf{x}_t = \mathbf{0} \tag{8}$$

$$\mathbf{W}_\mathbf{y}^\mathsf{T}\mathbf{W}_\mathbf{y}\mathbf{y}_{t-1} = \mathbf{y}_{t-1} \tag{9}$$

$$\mathbf{W}_\mathbf{x}^\mathsf{T}\mathbf{W}_\mathbf{y}\mathbf{y}_{t-1} = \mathbf{0} \tag{10}$$

From eqs. (8) and (10) we deduce that the columns of $\mathbf{W}_\mathbf{x}$ must be orthogonal to the columns of $\mathbf{W}_\mathbf{y}$. Thus, the set of vectors $\mathbf{v}_t = \mathbf{W}_\mathbf{x}\mathbf{x}_t$ must be orthogonal to the vectors $\mathbf{z}_t = \mathbf{W}_\mathbf{y}\mathbf{y}_t$, since the vectors $\mathbf{x}_t$ and $\mathbf{y}_t$ are projected onto orthogonal subspaces of the same space $\mathbb{S}$. From this observation it is not difficult to figure out how to define a partition of $\mathbb{S}$ into two orthogonal subspaces. Let s be the dimensionality of $\mathbb{S}$. Since $\mathbf{v}_t$ represents the current input information, while $\mathbf{z}_t$ represents the "history" of the input, we can assign the first k dimensions of $\mathbb{S}$ to encode vectors $\mathbf{v}_t$, and the remaining $(s-k)$ dimensions to encode vectors $\mathbf{z}_t$. This can be done by setting to 0 the last $(s-k)$ components for vectors $\mathbf{v}_t$, while setting to 0 the first k components for vectors $\mathbf{z}_t$. Moreover, if we chose k to be equal to the dimension of $\mathbf{x}_t$ and $s = k(q+1)$, where q is the depth of the memory we want to have in our system, we can define vectors $\mathbf{v}_t \in \mathbb{S}$ as

$$\mathbf{v}_t^\mathsf{T} \equiv [\mathbf{x}_t^\mathsf{T}, \underbrace{\mathbf{0}^\mathsf{T}, \ldots, \mathbf{0}^\mathsf{T}}_{q}] \tag{11}$$

where $\mathbf{0}$ is the vector with all zeros of dimension k, and vectors $\mathbf{z}_t \in \mathbb{S}$, with $t \leq q$ to explicitly represent the history, according to the following scheme

$$\mathbf{z}_t^{\mathsf{T}} \equiv [\mathbf{0}^{\mathsf{T}}, \mathbf{x}_t^{\mathsf{T}}, \ldots, \mathbf{x}_1^{\mathsf{T}}, \underbrace{\mathbf{0}^{\mathsf{T}}, \ldots, \mathbf{0}^{\mathsf{T}}}_{(q-t)}] \tag{12}$$

Using this encoding, $\mathbf{y}_t \in \mathbb{S}$ is defined as

$$\mathbf{y}_t^{\mathsf{T}} = \mathbf{v}_t^{\mathsf{T}} + \mathbf{z}_{t-1}^{\mathsf{T}} = [\mathbf{x}_t^{\mathsf{T}}, \ldots, \mathbf{x}_1^{\mathsf{T}}, \underbrace{\mathbf{0}^{\mathsf{T}}, \ldots, \mathbf{0}^{\mathsf{T}}}_{(q-t+1)}]. \tag{13}$$

Recalling that $\mathbf{z}_t = \mathbf{W}_{\mathbf{y}}\mathbf{y}_t$, it becomes evident that the function implemented by $\mathbf{W}_{\mathbf{y}}$ is just a shift of k positions of the $\mathbf{y}_t$ vector, i.e.

$$\mathbf{W}_{\mathbf{y}} \equiv \begin{bmatrix} \mathbf{0}_{k\times kq} & \mathbf{0}_{k\times k} \\ \mathbf{I}_{kq\times kq} & \mathbf{0}_{kq\times k} \end{bmatrix}, \tag{14}$$

and recalling that $\mathbf{v}_t = \mathbf{W}_{\mathbf{x}}\mathbf{x}_t$, we have

$$\mathbf{W}_{\mathbf{x}} \equiv \begin{bmatrix} \mathbf{I}_{k\times k} \\ \mathbf{0}_{kq\times k} \end{bmatrix}. \tag{15}$$

It can be readily verified that the defined vectors and matrices satisfy eqs. (7)-(10). In fact, eq. (7) is satisfied since $\mathbf{W}_{\mathbf{x}}^{\mathsf{T}}\mathbf{W}_{\mathbf{x}} = \mathbf{I}_{k\times k}$, while eqs. (8) and (10) are satisfied because by construction the columns of $\mathbf{W}_{\mathbf{x}}$ are orthogonal to columns of $\mathbf{W}_{\mathbf{y}}$, and finally eq. (9) is satisfied since $\mathbf{W}_{\mathbf{y}}^{\mathsf{T}}\mathbf{W}_{\mathbf{y}} = \begin{bmatrix} \mathbf{I}_{kq\times kq} & \mathbf{0}_{kq\times k} \\ \mathbf{0}_{k\times kq} & \mathbf{0}_{k\times k} \end{bmatrix}$ and all $\mathbf{y}_t$, $t = 0, \ldots, T-1$, have the last k components equal to 0.

The problem with this encoding is that s is too large, and information is not compressed at all. This problem can be easily fixed by computing the principal components of vectors $\mathbf{y}_t$.

Let

$$\bar{\mathbf{y}} = \frac{1}{T}\sum_{i=1}^{T}\mathbf{y}_i \quad \text{and} \quad \mathbf{C}_{\mathbf{y}} = \frac{1}{T}\sum_{i=1}^{T}(\mathbf{y}_i - \bar{\mathbf{y}})(\mathbf{y}_i - \bar{\mathbf{y}})^{\mathsf{T}} = \mathbf{U}\mathbf{\Lambda}\mathbf{U}^{\mathsf{T}} \tag{16}$$

where $\mathbf{\Lambda}$ is a diagonal matrix with elements equal to the eigenvalues of $\mathbf{C}_{\mathbf{y}}$, and $\mathbf{U}$ is the matrix obtained by collecting by column all the corresponding eigenvectors. Let $\widetilde{\mathbf{U}} \in \mathbb{R}^{s\times p}$ be the matrix obtained by $\mathbf{U}$ removing all the eigenvectors corresponding to null eigenvalues. Notice that in some cases we can have $p \ll s$. Then, we have

$$\tilde{\mathbf{y}}_t = \widetilde{\mathbf{U}}^{\mathsf{T}}(\mathbf{y}_t - \bar{\mathbf{y}}) \quad \text{and} \quad \mathbf{y}_t = \widetilde{\mathbf{U}}\tilde{\mathbf{y}}_t + \bar{\mathbf{y}} \tag{17}$$

and using eq. (2)

$$\begin{aligned}
\tilde{\mathbf{y}}_t &= \widetilde{\mathbf{U}}^{\mathsf{T}}(\mathbf{W}_{\mathbf{x}}\mathbf{x}_t + \mathbf{W}_{\mathbf{y}}\mathbf{y}_{t-1} - \bar{\mathbf{y}}) & (18)\\
&= \widetilde{\mathbf{U}}^{\mathsf{T}}\mathbf{W}_{\mathbf{x}}\mathbf{x}_t + \widetilde{\mathbf{U}}^{\mathsf{T}}\mathbf{W}_{\mathbf{y}}\mathbf{y}_{t-1} - \widetilde{\mathbf{U}}^{\mathsf{T}}\bar{\mathbf{y}} & (19)\\
&= \widetilde{\mathbf{U}}^{\mathsf{T}}\mathbf{W}_{\mathbf{x}}\mathbf{x}_t + \widetilde{\mathbf{U}}^{\mathsf{T}}\mathbf{W}_{\mathbf{y}}(\widetilde{\mathbf{U}}\tilde{\mathbf{y}}_{t-1} + \bar{\mathbf{y}}) - \widetilde{\mathbf{U}}^{\mathsf{T}}\bar{\mathbf{y}} & (20)\\
&= \begin{bmatrix} \widetilde{\mathbf{U}}^{\mathsf{T}}\mathbf{W}_{\mathbf{x}} & \widetilde{\mathbf{U}}^{\mathsf{T}}(\mathbf{W}_{\mathbf{y}} - \mathbf{I}_{s\times s})\bar{\mathbf{y}} \end{bmatrix} \begin{bmatrix} \mathbf{x}_t \\ 1 \end{bmatrix} + \widetilde{\mathbf{U}}^{\mathsf{T}}\mathbf{W}_{\mathbf{y}}\widetilde{\mathbf{U}}\tilde{\mathbf{y}}_{t-1} & (21)\\
&= \widetilde{\mathbf{W}}_{\mathbf{x}}\tilde{\mathbf{x}}_t + \widetilde{\mathbf{W}}_{\mathbf{y}}\tilde{\mathbf{y}}_{t-1}, & (22)
\end{aligned}$$

where $\tilde{\mathbf{x}}_t \equiv \begin{bmatrix} \mathbf{x}_t \\ 1 \end{bmatrix}$, $\widetilde{\mathbf{W}}_\mathbf{x} \equiv \begin{bmatrix} \widetilde{\mathbf{U}}^\mathsf{T}\mathbf{W}_\mathbf{x} & \widetilde{\mathbf{U}}^\mathsf{T}(\mathbf{W}_\mathbf{y} - \mathbf{I}_{s\times s})\bar{\mathbf{y}} \end{bmatrix} \in \mathbb{R}^{p\times(k+1)}$ and $\widetilde{\mathbf{W}}_\mathbf{y} \equiv \widetilde{\mathbf{U}}^\mathsf{T}\mathbf{W}_\mathbf{y}\widetilde{\mathbf{U}} \in \mathbb{R}^{p\times p}$.

3.1 Trees

When considering trees, the encoding used for $\mathbf{z}$ vectors is a bit more complex. First of all, let us illustrate what happens for binary complete trees. Then, we will generalize the construction to (in)complete b-ary trees. For $b = 2$, we have the following linear model

$$\mathbf{y}_u = \mathbf{W}_\mathbf{x}\mathbf{x}_u + \mathbf{W}_\mathbf{l}\mathbf{y}_{ch_l[u]} + \mathbf{W}_\mathbf{r}\mathbf{y}_{ch_r[u]} \tag{23}$$

where u is a vertex of the tree, $ch_l[u]$ is the left child of u, $ch_r[u]$ is the right child of u, $\mathbf{W}_\mathbf{l}, \mathbf{W}_\mathbf{r} \in \mathbb{R}^{s\times s}$. In this case, the basic idea is to partition $\mathbb{S}$ according to a perfectly balanced binary tree. More precisely, each vertex u of the binary tree is associated to a binary string $id(u)$ obtained as follows: the binary string "1" is associated to the root of the tree. Any other vertex has associated the string obtained by concatenating the string of its parent with the string "0" if it is a left child, "1" otherwise. Then, all the dimensions of $\mathbb{S}$ are partitioned in s/k groups of k dimensions. The label associated to vertex v is stored into the j-th group, where j is the integer represented by the binary string $id(u)$. E.g. the label of the root is stored into group 1, since $id(root)$ ="1", the label of the vertex which can be reached by the path ll starting from the root is stored into group 4, since $id(u)$ ="100", while the label of the vertex reachable through the path rlr is stored into group 13, since $id(u)$ ="1101". Notice that, if the input tree is not complete, the components corresponding to missing vertexes are set to be equal to 0. Using this convention, vectors $\mathbf{v}_u$ maintain the definition of eq. (11), and are used to store the current input label, i.e. the label associated to the root of the (sub)tree presented up to now as input, while vectors $\mathbf{z}_u$ are defined according to the scheme described above, with the difference that the first k components (i.e., the ones storing the label of the root) are set to 0.

Matrices $\mathbf{W}_\mathbf{l}$ and $\mathbf{W}_\mathbf{r}$ are defined as follows. Both matrices are composed of two types of blocks, i.e. $\mathbf{I}_{k\times k}$ and $\mathbf{0}_{k\times k}$. Matrix $\mathbf{W}_\mathbf{l}$ has to implement a push-left operation, i.e. the tree $\mathcal{T}$ encoded by a vector $\mathbf{y}_{root(\mathcal{T})}$ has to become the left child of a new node u whose label is the current input $\mathbf{x}_u$. Thus $root(\mathcal{T})$ has to become the left child of u and also all the other vertexes in $\mathcal{T}$ have their position redefined accordingly. From a mathematical point of view, the new position of any vertex a in $\mathcal{T}$ is obtained by redefining $id(a)$ as follows: *i)* the most significative bit of $id(a)$ is set to "0", obtaining the string $id_0(a)$; *ii)* the new string $id_{new}(a)$ ="1"+$id_0(a)$ is defined, where + is the string concatenation operator. If $id_{new}(a)$ represents a number greater than s/k then this means that the vertex has been pushed outside the available memory, i.e. the vertex a is *lost*. Consequently, groups which correspond to *lost* vertexes have to be annilated. Thus, $\mathbf{W}_\mathbf{l}$ is composed of $(q+1)\times(q+1)$ blocks, all of type $\mathbf{0}_{k\times k}$, except for the blocks in row $id_{new}(a)$ and column $id(a)$, with $id_{new}(a) \le s/k$, where a block $\mathbf{I}_{k\times k}$ is placed. Matrix $\mathbf{W}_\mathbf{r}$ is defined similarly: it has to implement a push-right operation, i.e.: *i)* the most significative bit of $id(a)$ is set to "1", obtaining the string $id_1(a)$; *ii)* the new string $id_{new}(a)$ ="1"+$id_1(a)$ is defined. Matrix $\mathbf{W}_\mathbf{x}$ is defined as in eq. (15).

Generalization of the above scheme for complete b-ary trees is not difficult. The linear model becomes

$$\mathbf{y}_u = \mathbf{W_x}\mathbf{x}_u + \sum_{c=0}^{b-1} \mathbf{W_c}\mathbf{y}_{ch_c[u]} \tag{24}$$

where $ch_c[u]$ is the $c+1$-th child of u, and a matrix $\mathbf{W_c}$ is defined for each child. The string associated to each vertex is defined on the alphabet {"0","1",...,"b-1"}, since there are b children. The symbol $b-1$ is associated with the root and b push operations have to be defined. The new string associated to any vertex a in $\mathcal{T}$, after a c-push operation, is obtained by redefining $id(a)$ as follows: *i)* the most significative symbol of $id(a)$ is set to c, obtaining the string $id_c(a)$; *ii)* the new string $id_{new}(a)$ ="b-1"+$id_c(a)$ is defined. E.g., if $b=5$ and c ="3", then *i)* the most significative symbol of $id(a)$ is set to "3", obtaining the string $id_3(a)$; *ii)* the new string $id_{new}(a)$ ="b-1"+$id_3(a)$ is defined. Matrix $\mathbf{W_c}$ is defined by placing blocks $\mathbf{I}_{k\times k}$ in positions $(id_{new}(a), id(a))$ only if $id_{new}(a) \leq s/k$, where $id_{new}(a)$ is interpreted as a number represented in base b. Performing the eigenspace analysis, we obtain

$$\tilde{\mathbf{y}}_u = \widetilde{\mathbf{W}}_{\mathbf{x}}\tilde{\mathbf{x}}_u + \sum_{c=0}^{b-1} \widetilde{\mathbf{W}}_{\mathbf{c}}\tilde{\mathbf{y}}_{ch_c[u]}, \tag{25}$$

where $\tilde{\mathbf{x}}_u \equiv \begin{bmatrix}\mathbf{x}_u\\ 1\end{bmatrix}$, $\widetilde{\mathbf{W}}_{\mathbf{x}} \equiv \begin{bmatrix}\widetilde{\mathbf{U}}^{\mathsf{T}}\mathbf{W_x} & \widetilde{\mathbf{U}}^{\mathsf{T}}(\sum_{c=0}^{b-1}\mathbf{W_c} - \mathbf{I}_{s\times s})\bar{\mathbf{y}}\end{bmatrix} \in \mathbb{R}^{p\times(k+1)}$ and $\widetilde{\mathbf{W}}_{\mathbf{c}} \equiv \widetilde{\mathbf{U}}^{\mathsf{T}}\mathbf{W_c}\widetilde{\mathbf{U}} \in \mathbb{R}^{p\times p}$, $c=0,\ldots,b-1$.

A problem in dealing with complete trees is that very soon there is a combinatorial explosion of the number of paths to consider, i.e. in order for the machine to deal with moderately deep trees, a huge value for s needs to be used. In practical applications, however, the observed trees tend to follow a specific generative model, and thus there may be many topologies which are never, or very seldomly, generated. For this reason we suggest to use the following approach. Given a set of trees $\mathbf{T}$, the optimized graph $G_{\mathbf{T}}$ [4] is obtained by joining all the trees in such a way that any (sub)tree in $\mathbf{T}$ is represented only once. The optimized graph $G_{\mathbf{T}}$, which is a DAG, is then visited bottom-up, generating for each visited vertex v the set of id strings associated to the tree rooted in v, thus simulating all the different push operations which should be performed when presenting the trees in $\mathbf{T}$ to the machine. Repeated id strings are removed. The obtained set P is then used to define the state space of the machine: each string is associated to one group of k coordinates. In this way, only paths which appear in the set $\mathbf{T}$ (including all subtrees) are represented, thus drastically reducing the size of s, which will be equal to $|P|\times k$. One drawback of this approach is that if a new tree with "unknown" paths is presented to the machine, the vertexes which are reached by those paths are lost.

A final practical consideration concerns the observation that by introducing a dummy vector $\mathbf{y}_{dummy} = -\sum_i \mathbf{y}_i$, eq. (25) is simplified since $\bar{\mathbf{y}} = \mathbf{0}$, and the corresponding derived weight matrices appear to be much more effective. In the experiments reported in this paper, we have used this trick.

Sentence	*Noun Phrase*	*Verb Phrase*	*Prepositional Phrase*	*Adjectival Phrase*
s → np vp s → np V	np → D ap np → D N np → np pp	vp → V np vp → V pp	pp → P np	ap → A ap ap → A N

Fig. 1. Context-Free Grammar used in the experiments

4 Experiments

For testing our approach, we have considered the context-free grammar shown in Figure 1, and already used by Pollack [2]. Sequences and corresponding parse trees are randomly generated from the grammar, rejecting sequences longer than 30 items. In all, 177 distinct sequences (and corresponding parse trees) are generated. Among them, 101 sequences are randomly selected for training, and the remaining 76 sequences are used for testing the generalization ability of the machine. In Table 1 we have reported some statistics about the data. The dataset for trees is obtained by considering the parse trees corresponding to the selected sequences.

Sequences are composed of terminal symbols, which are encoded by 5-dimensional "one-hot" vectors (i.e. $k = 5$). Since there are up to 30 items in a sequence, $s = 150$. Trees also include nonterminal symbols. In this case, symbols are represented by 6-dimensional vectors (i.e. $k = 6$), where the first component is 0 for terminal symbols and 3 for nonterminal symbols, while the remaining 5 components follow a "one-hot" coding scheme. The state space $\mathbb{S}$ is obtained by computing the optimization graph for the training set and generating all the possible paths following the procedure described at the end of Section 3.1. In all, 351 distinct paths where generated, leading to a final dimension for the state space equal to $s = 6 \times 351 = 2106$. The computation of the optimized graph also allowed the identification of 300 unique (sub)trees, thus allowing us just to consider the same number of different non-null states. The dummy state $\mathbf{y}_{dummy}$ is used for both datasets to get zero-mean vectors.

The spectral analysis for sequences required 0.1 cpu/sec on an Athlon 1900+ based computer, while it required 377.77 cpu/sec for trees. Results are shown in Figure 2. In Figure 3 we have reported for the sequence dataset the error in label decoding (left) and the mean square error for labels (right) plotted versus the number of used components.

Table 1. Statistical properties of the datasets

Dataset/Split	# examples	Max. length (depth)	Max. number item per example	Tot. number items	Tot. number unique (sub)trees
Sequences/Training	101	30	30	1463	-
Sequences/Test	76	30	30	1158	-
Tree/Training	101	14	59	2825	300
Tree/Test	76	15	59	2240	242

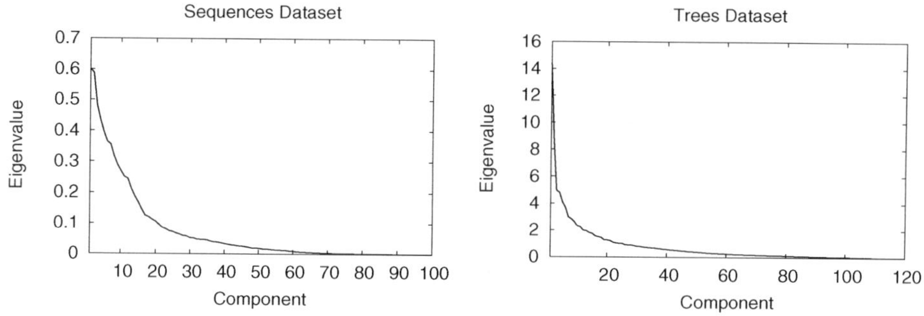

Fig. 2. Eigenvalues for sequences and trees. The most significant eigenvalue, caused by the introduction of $\mathbf{y}_{dummy}$, is not shown since it is very high (2197.62 for sequences, and 10781.63 for trees), as well as null eigenvalues beyond the shown maximum x-value (Component).

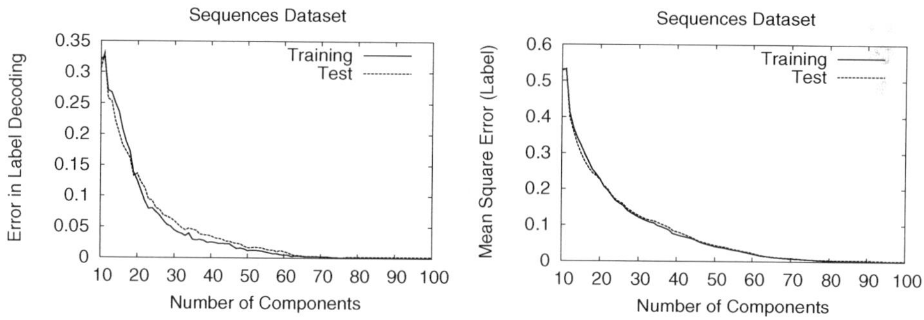

Fig. 3. Experimental results for sequences

The error in label decoding is computed as follows. Each sequence is first fed into the machine, so to get the final state for the sequence. Then the final state is decoded so to regenerate all the items (labels) of the sequence. A decoded label is considered to be correct if the position of the highest value in the decoded label matches the position of the 1 in the correct label, otherwise a loss of 1 is suffered. The final error is computed as the ratio between the total loss suffered and the total number of items (labels) in the dataset. The mean square error for labels is computed by considering the total Euclidean distance between correct and decoded labels. The final result is normalized by the number of total items. For sequences it can be seen that the machine exhibits an almost perfect generalization capability. The same result in not true for trees (see Figure 4), where in the test set there was a tree of depth 15, i.e. deeper than the deepest tree in the training set (depth 14). Thus, for this test tree the state space was not able to store all the necessary information to reconstruct it. Moreover, new paths appear in the test set which cannot as well be properly treated by the machine. Notwithstanding these difficulties, which could have been avoided by using a larger training set, the label decoding error of the machine is below 7.5% for a number of components higher than 95.

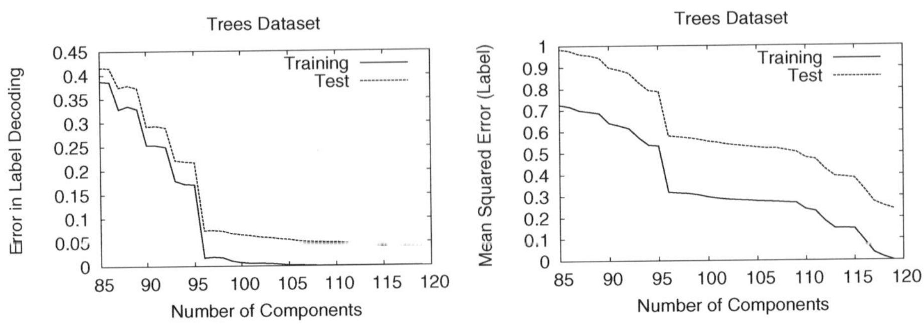

Fig. 4. Experimental results for trees

5 Conclusion

We have shown how to derive a family of exact solutions for Recursive Principal Components Analysis of sequences and trees. Basically, we have demonstrated that for these solutions there is a very close relationship with the principal components computed on explicit flat representations of the input structures, where substructures are considered as well. From a "recursive" point of view this is quite disappointing, although we have experimentally shown that in practical applications the number of parameters which a recursive solution needs is significantly lower than the number of parameters required by the (almost) equivalent flat solution.

From a mathematical point of view, the solutions are exact only if all the non-null components are used. We have still not investigated whether this property is maintained when using a subset of such components. The empirical results shown in the paper seems to indicate that using a subset of such components quite good results are obtained, even if the solution may be suboptimal. It should be pointed out that, while dealing with sequences is quite easy, the proper treatment of trees is not so trivial, due to the potential combinatorial explosion of the number of distinct paths. Thus, further study is required to devise an effective strategy for designing the explicit state space for trees.

References

1. Callan, R. E., Palmer-Brown, D.: (S)RAAM: An analytical technique for fast and reliable derivation of connectionist symbol structure representations. Connection Science, **9**(1997)139–160.
2. Pollack, J.B.: Recursive distributed representations. Artificial Intelligence **46** (1990) 77–105.
3. Sperduti, A.: Labeling RAAM. Connection Science **6** (1994) 429–459.
4. Sperduti, A., Starita, A.: Supervised neural networks for the classification of structures. IEEE Transactions on Neural Networks **8** (1997) 714–735.
5. T. Voegtlin, T., Dominey, P. F.: Linear Recursive Distributed Representations. Neural Networks **18** (2005) 878–895.
6. Voegtlin, T.: Recursive Principal Components Analysis. Neural Networks **18** (2005) 1040–50.

Active Learning with the Probabilistic RBF Classifier

Constantinos Constantinopoulos* and Aristidis Likas

Department of Computer Science, University of Ioannina, GR 45110 Ioannina, Greece
ccostas@cs.uoi.gr, arly@cs.uoi.gr

Abstract. In this work we present an active learning methodology for training the probabilistic RBF (PRBF) network. It is a special case of the RBF network, and constitutes a generalization of the Gaussian mixture model. We propose an incremental method for semi-supervised learning based on the Expectation-Maximization (EM) algorithm. Then we present an active learning method that iteratively applies the semi-supervised method for learning the labeled and unlabeled observations concurrently, and then employs a suitable criterion to select an unlabeled observation and query its label. The proposed criterion selects points near the decision boundary, and facilitates the incremental semi-supervised learning that also exploits the decision boundary. The performance of the algorithm in experiments using well-known data sets is promising.

1 Introduction

Active learning a classifier constitutes a special learning problem, where the training data are actively collected during the training. The training data are available as a stream of classified observations, but the information they carry is controlled from the classifier. The classifier determines regions of interest in the data space, and asks for training data that lie in these regions. The importance of active learning is well established, see [1] for a study on the increase of classifier's accuracy as the number of labeled data increases. Various active learning methods have been suggested; in [2] a learning method for Gaussian mixture models [3] is proposed, that selects data that minimize the variance of the learner. In [4] active learning for a committee of classifiers is proposed, which selects data for which the committee members disagree. Based on this selection method, in [5] they propose the use of available unclassified data by employing EM [6] to form a better selection criterion, that is used to train a naive Bayes classifier. In [7] they train Gaussian random fields and harmonic functions, and select data based on the estimated expected classification error.

* This research was co-funded by the European Union in the framework of the program "Heraklitos" of the "Operational Program for Education and Initial Vocational Training" of the 3rd Community Support Framework of the Hellenic Ministry of Education, funded by 25% from national sources and by 75% from the European Social Fund (ESF).

S. Kollias et al. (Eds.): ICANN 2006, Part I, LNCS 4131, pp. 357–366, 2006.

Our work concentrates on a variation of the active learning scenario called the *pool-based* active learning, also studied in [5,7]. In this case a set of labeled and unlabeled observations is available right from the start. During training we are allowed to iteratively query the label of unlabeled points, and use the acquired labels to improve the classifier. In practice this scenario is important when querying a field expert is expensive, as in medical diagnosis, or when there is a huge quantity of unlabeled data that prohibits thorough labeling, as in text classification. The intuition behind pool-based learning is that the unlabeled data can be exploited to construct a more detailed generative model for the data set. Thus this problem is closely related to *semi-supervised* learning. Algorithms for semi-supervised learning have been proposed for Gaussian mixtures in [8,9], as well as for the RBF network [10]. So it has been established that unlabeled data can reveal useful information for the distribution of the labeled data.

We concentrate here on the pool-based active learning of the probabilistic RBF (PRBF) classifier [11,12]. It is a special case of RBF network [13] that computes at each output unit the density function of a class. It adopts a cluster interpretation of the basis functions, where each cluster can generate observations of any class. This is a generalization of a Gaussian mixture model [3,13], where each cluster generates observations of only one class. In [14] an incremental learning method based on EM for supervised learning is proposed. In this work we propose an incremental learning method based on EM for semi-supervised learning. We are facilitated by the fact that each node of the PRBF describes the local distribution of potentially all the classes. For the unlabeled data we can marginalize the class labels from the update equations of EM, to use both labeled and unlabeled data in parameter estimation.

In the following section we describe an incremental algorithm for the semi-supervised training of the PRBF based on EM. In section 3 we use this algorithm to tackle the problem of active learning. Next in section 4 we present the results from our experimental study. Some discussion in section 5 concludes this work.

2 Semi-supervised Learning

Assume a set of labeled observations $X = \{(x^n, y^n)|\, n = 1, \ldots, N\}$ and a set of unlabeled observations $X_\emptyset = \{x^n|\, n = 1, \ldots, N_\emptyset\}$. The labeled observations have an "input" part $x \in \Re^d$, and an "output" part $y \in \{1, \ldots, K\}$ in the case of a classification task with K classes. This "output" part (called label) assigns an observation to one class, and in the case of unlabeled observations is missing. Let Ω be the joint set of labeled and unlabeled observations, i.e. $\Omega = X \cup X_\emptyset$. Moreover we can separate X according to the "output" labels in K disjoint sets $X_k = \{(x^n, y^n)|\, y^n = k, n = 1, \ldots, N_k\}$ one for each class, then $\Omega = \bigcup_k X_k \cup X_\emptyset$.

Adopting the Bayes decision rule, a classifier assigns a new unlabeled observation $x^\star$ to the class $k^\star$ with maximum posterior probability. If we drop the part of the posterior that depends only on $x^\star$, then

$$k^\star = \arg\max_k p(x^\star|k)p(k) \tag{1}$$

where $p(x|k)$ is the class conditional distribution of observations from class k, and $p(k)$ is the prior probability of this class. For two classes k and k' there is a *decision boundary* $p(x|k)p(k) = p(x|k')p(k')$ that divides the space of the observations.

To describe class conditional distributions we employ the PRBF network. For input x the class conditional probability $p(x|k)$ is the k-th output of a PRBF with J basis functions

$$p(x|k) = \sum_{j=1}^{J} p(j|k)\, p(x|j) \tag{2}$$

The coefficients $p(j|k)$ are non-negative and $\sum_j p(j|k) = 1$, while each basis function is a Gaussian

$$p(x|j) = \frac{1}{(2\pi\sigma_j^2)^{d/2}} \exp\{-\frac{1}{2}(x-\mu_j)^T(x-\mu_j)/\sigma_j^2\} \tag{3}$$

with mean $\mu_j \in \Re^d$ and variance σ_j^2. In order to find estimates for the parameters of the network

$$\theta = \{p(k), p(j|k), \mu_j, \sigma_j |\, j = 1, \ldots, J,\; k = 1, \ldots, K\}$$

we maximize the joint likelihood, as in [10]. Assuming i.i.d. observations, the joint log-likelihood $\mathcal{L}$ of labeled and unlabeled data is

$$\begin{aligned}\mathcal{L} = \log p(\Omega) &= \log \prod_k \prod_{x\in X_k} p(x,k) \prod_{x\in X_\emptyset} p(x)\\ &= \sum_k \sum_{x\in X_k} \log p(k) \sum_j p(j|k)p(x|j) + \sum_{x\in X_\emptyset} \log \sum_k p(k) \sum_j p(j|k)p(x|j).\end{aligned} \tag{4}$$

For the maximization of $\mathcal{L}$ we use the Expectation-Maximization (EM) algorithm [6]. The EM is an iterative algorithm that is guaranteed to converge at a local maximum of the likelihood surface. It is employed in problems where *hidden variables* exist. These variables determine the solution of the problem, although are not observable. In our case the hidden variables define the node of the network that generated an observation, and the label of an unlabeled observation. In the following we formally derive the update equations of EM.

We introduce a hidden variable $z^{(x)}$ for each $x \in \Omega$ that assigns this observation to one class and one node of the network. Each $z^{(x)}$ is a binary $J \times K$ matrix, where $z_{jk}^{(x)} = 1$ if x is assigned to the k-th class and the j-th node. This assignment is unique, so that $\sum_j \sum_k z_{jk}^{(x)} = 1$. Moreover for a labeled observation (x,k) the corresponding $z^{(x)}$ is constrained so that $z_{j\ell}^{(x)} = 0$ for all (j,ℓ) with $\ell \neq k$. Thus a hidden variable can assign a labeled observation to any node but only one class. This does not hold for the case of unlabeled observations that can be assigned to any class and any node. Given the set of hidden variables $Z = \{z^{(x)} |\, \forall x \in \Omega\}$, we define the *complete* log-likelihood

$$\mathcal{Q} = \log p(\Omega, Z) = \log \prod_{x\in\Omega} \prod_k \prod_j [p(k)p(j|k)p(x|j)]^{z_{jk}^{(x)}} \tag{5}$$

Although we can not compute $\mathcal{Q}$ directly, as it depends on the unknown values of Z, we can compute its expectation $\langle\mathcal{Q}\rangle$ w.r.t. the distribution of Z. Since the expected value of $z_{jk}^{(x)}$ is equal to the joint posterior probability $p(j,k|x)$ that x is assigned to the j-th node and the k-th class, it follows that

$$\langle\mathcal{Q}\rangle = \sum_{x\in\Omega}\sum_{k}\sum_{j} p(j,k|x)\log\{p(k)p(j|k)p(x|j)\}\,. \tag{6}$$

The EM algorithm iterates two steps until convergence. During the *E-step* it computes the expectation of the complete log-likelihood $\langle\mathcal{Q}\rangle$, given the current estimate for the parameter vector θ. During the *M-step* it provides estimates θ that maximize $\langle\mathcal{Q}\rangle$. This procedure is guaranteed to converge at a local maximum of the joint log-likelihood $\mathcal{L}$.

Explicitly described, during the E-step we compute $p(j,k|x)$ for every $x\in\Omega$, $j\in\{1,\ldots,J\}$ and $k\in\{1,\ldots,K\}$ according to

$$p(j,k|x) = p(j|k,x)p(k|x). \tag{7}$$

If x is unlabeled then we compute $p(k|x)$ and $p(j|k,x)$ for every class k using Bayes theorem

$$p(k|x) = \frac{p(x|k)p(k)}{\sum_\ell p(x|\ell)p(\ell)} \tag{8}$$

$$p(j|k,x) = \frac{p(j|k)p(x|j)}{\sum_i p(i|k)p(x|i)}. \tag{9}$$

If x is labeled, then we exploit the information of the label and set

$$p(k|x) = \begin{cases} 1 \text{ if } x\in X_k \\ 0 \text{ if } x\notin X_k \end{cases} \tag{10}$$

and we compute $p(j|k,x)$ similarly

$$p(j|k,x) = \begin{cases} \frac{p(j|k)p(x|j)}{\sum_i p(i|k)p(x|i)} & \text{if } x\in X_k \\ 0 & \text{if } x\notin X_k \end{cases} \tag{11}$$

During the M-step we maximize $\langle\mathcal{Q}\rangle$ w.r.t. θ, given the current estimation of the joint posteriors. The solution for every $j\in\{1,\ldots,J\}$ and $k\in\{1,\ldots,K\}$ is

$$\mu_j = \frac{\sum_{x\in\Omega}\sum_k p(j,k|x)\,x}{\sum_{x\in\Omega}\sum_k p(j,k|x)} \tag{12}$$

$$\sigma_j^2 = \frac{1}{d}\frac{\sum_{x\in\Omega}\sum_k p(j,k|x)\,(x-\mu_j)^T(x-\mu_j)}{\sum_{x\in\Omega}\sum_k p(j,k|x)} \tag{13}$$

$$p(j|k) = \frac{\sum_{x\in\Omega} p(j,k|x)}{N_k + \sum_j\sum_{x\in X_\emptyset} p(j,k|x)} \tag{14}$$

$$p(k) = \frac{N_k + \sum_j\sum_{x\in X_\emptyset} p(j,k|x)}{N + N_\emptyset}. \tag{15}$$

An important aspect of network training is the estimation of the number of basis functions to be used. To tackle this we adopt the incremental approach proposed in [14] for supervised learning, that we modify suitably. It is an incremental method with two stages. We start with a network having only one node, whose parameters are easily estimated from the statistics of the training data. During the first stage we iteratively add new nodes to the network, until we reach the desired complexity. Then the second stage follows, where we split all the nodes in order to increase classification performance. In the next sections we give more details for the two stages.

2.1 Addition of Nodes

Given a network with M nodes we can construct a network with $M+1$ nodes. If the given class conditional density is $p(x|k)$, then adding a Gaussian node $q(x) = \mathcal{N}(x; \mu_q, \sigma_q^2)$ results in $\hat{p}(x|k)$ as follows

$$\hat{p}(x|k) = (1 - \alpha_k)\, p(x|k) + \alpha_k\, q(x) \tag{16}$$

where α_k is the prior probability that node q generates observations from class k. However we have to estimate α_k, the mean μ_q and variance σ_q^2 of q. Thus we search for parameters such that q is near the decision boundary. Good estimation of class conditional densities near the boundary is crucial for the performance of the classifier.

According to [14] we resort to a clustering method, namely the *kd-tree* [15]. The kd-tree is a binary tree that partitions a given data set. It is constructed recursively by partioning the data of each node in two subsets. Using only the labeled points, we initially partition the data in M subsets

$$X_j = \{(x,k)|\,(x,k) \in X,\, p(j|k,x) > p(i|k,x), \forall i \neq j\}$$

one for each node. Employing the kd-tree we repartition each of X_j in six subsets. These subsets result from the construction of a kd-tree with two levels. More levels would result in a lot of small clusters. We would like to avoid that, as we want to gradually shrink the size of the clusters. Moreover, in order to add the next node, we are going to employ the kd-tree again to partition all the data in smaller clusters. The statistics of the resulting subsets are probable estimates of μ_q and σ_q^2. The corresponding estimation of prior is $\alpha_k = p(j|k)/2$. Partitioning each node we create $6M$ sets of candidates $\theta_q = \{\alpha_k, \mu_q, \sigma_q^2\}$, so we have to select the most appropriate according to a criterion.

As proposed in [14], we compute the change of the log-likelihood $\Delta\mathcal{L}_k^{(q)}$ for class k after the addition of q

$$\begin{aligned}\Delta\mathcal{L}_k^{(q)} &= \frac{1}{N_k}\left(\log \hat{p}(x|k) - \log p(x|k)\right) \\ &= \frac{1}{N_k} \sum_{x \in X_k} \log\left\{1 - \alpha_k + \alpha_k \frac{q(x)}{p(x|k)}\right\}.\end{aligned} \tag{17}$$

We retain those θ_q that increase the log-likelihood of at least two classes and discard the rest. For each retained θ_q, we add the positive $\Delta\mathcal{L}_k^q$ terms to compute the total increase of the log-likelihood $\Delta\mathcal{L}_q$. The candidate $q^\star$ whose value $\Delta\mathcal{L}_{q^\star}$ is maximum consists the parameters of the node that will be added to the current network, if this maximum value is higher than a prespecified threshold. Otherwise, we consider that the attempt to add a new node is unsuccessful. We set this threshold equal to 0.01 after experimentation, in order to avoid the addition of nodes with negligible effects on the performance of the network. So we chose a small value, to also prevent the premature termination of the procedure.

After the successful addition of a new node we apply the semi-supervised EM, as described in the previous section. This procedure can be applied iteratively, in order to add the desired number of nodes to the given network. Figure 1 illustrates the addition of the first two nodes. The initial network with only one node is illustrated in Figure 1(a). The six candidate nodes and the chosen node are illustrated in Figure 1(b) and Figure 1(c) correspondingly. Figure 1(d) illustrates the network after the application of semi-supervised EM.

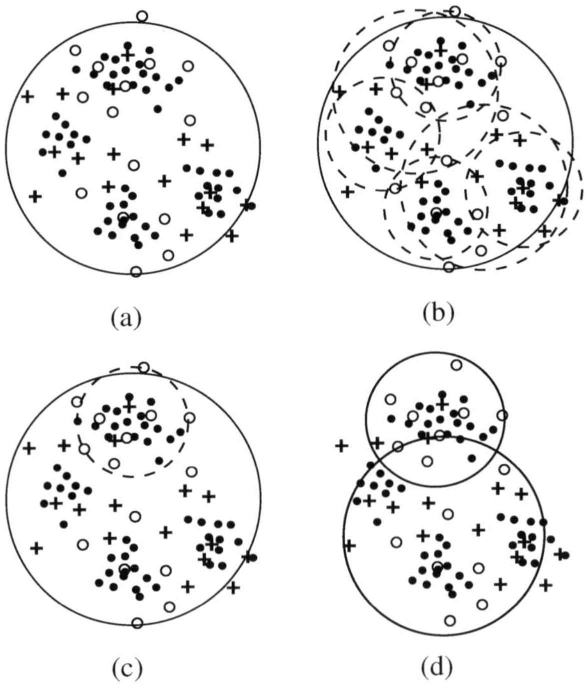

(a) (b) (c) (d)

Fig. 1. Addition of the first two nodes. The nodes of the network are drawn with solid lines, and the candidate nodes with dashed lines. The dots represent the unlabeled observations in a two-class problem.

2.2 Splitting of Nodes

After the stage of adding nodes, there may be nodes of the network located to regions with overlapping among classes. In order to increase the generalization performance of the network we follow the approach suggested in [16], and split each node. During this stage we use both supervised and unsupervised observations. We evaluate the joint posterior probabilities $p(j,k|x)$ for a node, and define if it is responsible for observations of more than one class. If $\sum_{x \in \Omega} p(j,k|x) > 0$, then we remove it from the network, and add a separate node for the k-th class. So finally each node is responsible for only one class. Splitting a node $p(x|j)$, the resulting node for class k is a Gaussian $p(x|j,k)$ with mean μ_{kj}, variance σ_{kj}^2 and mixing weight $p(j|k)$. These parameters are estimated according to:

$$\mu_{kj} = \frac{\sum_{x \in \Omega} p(j,k|x)\, x}{\sum_{x \in \Omega} p(j,k|x)} \tag{18}$$

$$\sigma_{kj}^2 = \frac{1}{d} \frac{\sum_{x \in \Omega} p(j,k|x)\, (x - \mu_{kj})^T (x - \mu_{kj})}{\sum_{x \in \Omega} p(j,k|x)} \tag{19}$$

$$p(j|k) = \frac{\sum_{x \in \Omega} p(j,k|x)}{N_k + \sum_j \sum_{x \in X_\emptyset} p(j,k|x)} . \tag{20}$$

Consequently the class conditional density is estimated as

$$p(x|k) = \sum_j p(j|k) p(x|j,k) . \tag{21}$$

In the case of a training set where all the points are labeled, the class conditional likelihood is increased for all classes after splitting as proved in [16]. However in the semi-supervised case we cannot guarantee that splitting increases the joint likelihood.

3 Active Learning

In the previous section we described an incremental algorithm for training a PRBF network using labeled and unlabeled observations. In the following we incorporate the algorithm in an active learning method, where we iteratively select an unlabeled point and query its label. After its label is given, we add the labeled point in the labeled set and train the network again. The crucial point is to pick a point that greatly benefits the training of our classifier. We propose the selection of a point that lies near the classification boundary. In this way we facilitate the iterative addition of basis functions on the classification boundary, as described in the previous section.

As a criterion of selecting a suitable point we propose the ratio of class posteriors. For each unlabeled observation $x \in X_\emptyset$ we compute the class posterior $p(k|x)$ for every class, and then find the two classes with the largest posterior values:

$$\kappa_1^{(x)} = \arg\max_k p(k|x), \ \kappa_2^{(x)} = \arg\max_{k \neq \kappa_1^{(x)}} p(k|x). \tag{22}$$

We choose to ask for the label of $\hat{x}$ that exhibits the smallest ratio of largest class posteriors:

$$\hat{x} = \arg\min_{x \in X_\emptyset} \log \frac{p(\kappa_1^{(x)}|x)}{p(\kappa_2^{(x)}|x)}. \tag{23}$$

In this way we pick the unlabeled observation that lies closer to the decision boundary of the current classifier. Note that according to (1) we classify observations to the class with the maximum class posterior. Thus for some x on the decision boundary holds that $p(\kappa_1^{(x)}|x) = p(\kappa_2^{(x)}|x)$. Consequently if an observation approaches the decision boundary between two classes, then the corresponding logarithmic ratio of class posteriors tends to zero.

Summarizing the presented methodology, we propose the following active learning algorithm:

1. Input: The set X of labeled observations, the set $X_\emptyset$ of unlabeled observations, and a degenerate network $PRBF_{J=1}$ with one basis function.
2. For $s = 0, \ldots, S-1$
   (a) Add one node to the network $PRBF_{J+s}$ to form $PRBF_{J+s+1}$.
   (b) Apply EM until convergence for semi-supervised training of $PRBF_{J+s+1}$.
3. For $s = 0, \ldots, S$
   (a) Split the nodes of $PRBF_{J+s}$ to form $PRBF_{J+s}^{split}$.
4. Select the network $PRBF_{J^\star}^{split} \in \{PRBF_J^{split}, \ldots, PRBF_{J+S}^{split}\}$ that maximizes the joint likelihood.
5. Set the current network: $PRBF_J = PRBF_{J^\star}$.
6. If $X_\emptyset$ is empty go to step 7, else
   (a) Pick an unlabeled observation $\hat{x}$ according to (23), and ask its label $\hat{y}$.
   (b) Update the sets: $X = X \cup \{(\hat{x}, \hat{y})\}$ and $X_\emptyset = X_\emptyset \setminus \{\hat{x}\}$.
   (c) Go to step 2.
7. Output: Split the nodes of $PRBF_J$ to form the output network $PRBF_J^{split}$.

In all our experiments we use $S = 1$, thus we try to add one node at each iteration of the active learning.

4 Experiments

For the experimental evaluation of our method we used three data sets, available from the UCI repository. The first is the "segmentation" set, that consists of 2310 points with 19 continuous features in 7 classes. The second is the "waveform" set, that consists of 5000 points with 21 continuous features in 3 classes. The last is the "optical digits" set, that consists of 5620 points with 62 continuous features in 10 classes. All the data sets were standardized, so that all their features exhibit zero mean and unit standard deviation. In all experiments we applied our algorithm starting with 50 uniformly selected labeled points. We treated the rest as a pool of unlabeled points, and we actively selected 400 more. Each experiment was repeated five times, and we computed the average generalization error on a separate test set that contained the 10% of the original data set. Figure 2 illustrates the average generalization error and the average number of PRBF nodes after each added label. The results are satisfactory, as the generalization error almost halved in all cases after the addition of 50 labels. After the addition of 300 labels the error had converged, and the addition of more labels offered little improvement. After the addition of 400 labels, the average error for the "segmentation" data set was 0.156, for the "waveform" data set was 0.091, and for the "optical digits" data set was 0.089. For comparison, we also applied the supervised learning method proposed in [14] using the original data sets. The generalization error of the resulting PRBF for the "segmentation" data set was 0.246, for the "waveform" data set was 0.142, and for the "optical digits" data set was 0.07. Concluding, we note that the number of nodes converged slower than the error, but eventually it also reached a plateau. The average number of nodes after the addition of 400 labels was 285.2 for the "segmentation" data set, 294.6 for the "waveform" data set, and 509 for the "optical digits" data set.

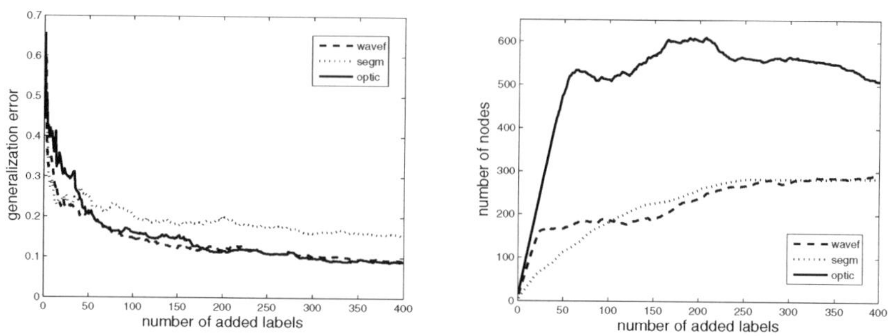

Fig. 2. The average generalization error (left), and the average number of network nodes (right) for pool-based active PRBF learning

5 Discussion

We have proposed an algorithm for the active learning of the PRBF classifier. We derived an EM algorithm for semi-supervised training of PRBF, and an incremental variation that sequentially adds nodes to the network. We use this method to estimate class conditional densities for pool-based active learning.

The experimental results are encouraging, and slight modifications of the method may further improve its performance. For example we could acquire the labels for a bunch of unlabeled observations, before we try to add a new node. The most important issue for consideration is the time complexity of the algorithm, e.g. it takes almost two hours to solve the "waveform" data set with a MATLAB implementation on a standard personal computer. A method to decrease the number of splits in each iteration would improve the execution time significantly. Another interesting issue concerns the complexity of the resulting network. We note that the interpretation of the network weights as probabilities alleviates the problem, as it forces many weights to near zero values and overfitting is avoided. However we could use a validation set for better model selection.

Our future plans include a more detailed study of the method, and elaboration on several of our choices, with the most important being the comparison with other selection methods for the active acquisition of class information. Also we plan to consider the problem of new class discovery, as a similar task that we would like to tackle.

References

1. Castelli, V., Cover, T.: On the exponential value of labeled samples. Pattern Recognition Letters **16** (1995) 105–111
2. Cohn, D., Ghahramani, Z., Jordan, M.: Active learning with statistical models. Journal of Artificial Intelligence Research **4** (1996) 129–145
3. McLachlan, G., Peel, D.: Finite Mixture Models. John Wiley & Sons (2000)
4. Freund, Y., Seung, H.S., Shamir, E., Tishby, N.: Selective sampling using the query by committee algorithm. Machine Learning **28** (1997) 133–168
5. McCallum, A.K., Nigam, K.: Employing EM in pool-based active learning for text classification. In Shavlik, J.W., ed.: Proc. 15th International Conference on Machine Learning, Morgan Kaufmann (1998)
6. Dempster, A., Laird, N., Rubin, D.: Maximum likelihood estimation from incomplete data via the EM algorithm. Journal of the Royal Statistical Society, Series B **39**(1) (1977) 1–38
7. Zhu, X., Lafferty, J., Ghahramani, Z.: Combining active learning and semi-supervised learning using Gaussian fields and harmonic functions. In: Proc. 20th International Conference on Machine Learning. (2003)
8. Ghahramani, Z., Jordan, M.: Supervised learning from incomplete data via an EM approach. In Cowan, J.D., Tesauro, G., Alspector, J., eds.: Advances in Neural Information Processing Systems 6, Morgan Kaufmann (1994)
9. Tadjudin, S., Landgrebe, A.: Robust parameter estimation for mixture model. IEEE Trans. Geoscience and Remote Sensing **38** (2000) 439–445
10. Miller, D., Uyar, H.: Combined learning and use for a mixture model equivalent to the RBF classifier. Neural Computation **10** (1998) 281–293
11. Titsias, M.K., Likas, A.: Shared kernel models for class conditional density estimation. IEEE Trans. Neural Networks **12**(5) (2001) 987–997
12. Titsias, M.K., Likas, A.: Class conditional density estimation using mixtures with constrained component sharing. IEEE Trans. Pattern Anal. and Machine Intell. **25**(7) (2003) 924–928
13. Bishop, C.M.: Neural Networks for Pattern Recognition. Oxford University Press (1995)
14. Constantinopoulos, C., Likas, A.: An incremental training method for the probabilistic RBF network. IEEE Trans. Neural Networks **to appear** (2006)
15. Bentley, J.L.: Multidimensional binary search trees used for associative searching. Communications of the ACM **18**(9) (1975) 509–517
16. Titsias, M.K., Likas, A.: Mixture of experts classification using a hierarchical mixture model. Neural Computation **14**(9) (2002) 2221–2244

Merging Echo State and Feedforward Neural Networks for Time Series Forecasting

Štefan Babinec[1] and Jiří Pospíchal[2]

[1] Department of Mathematics, Fac. of Chemical and Food Technologies
Slovak University of Technology, 812 37 Bratislava, Slovakia
Phone/Fax: +421 2 52495177
stefan.babinec@stuba.sk

[2] Institute of Applied Informatics, Fac. of Informatics and Information Technologies
Slovak University of Technology, 842 16 Bratislava, Slovakia
Phone: +421 2 60291548; Fax: +421 2 65420587
pospichal@fiit.stuba.sk

Abstract. Echo state neural networks, which are a special case of recurrent neural networks, are studied from the viewpoint of their learning ability, with a goal to achieve their greater prediction ability. A standard training of these neural networks uses pseudoinverse matrix for one-step learning of weights from hidden to output neurons. Such learning was substituted by backpropagation of error learning algorithm and output neurons were replaced by feedforward neural network. This approach was tested in temperature forecasting, and the prediction error was substantially smaller in comparison with the prediction error achieved either by a standard echo state neural network, or by a standard multi-layered perceptron with backpropagation.

1 Introduction

From the point of view of information flow, neural networks can be divided into two types: feedforward neural networks and recurrent neural networks [1]. In feedforward neural networks the input information proceeds from input neurons to output neurons. Such networks basically implement input/output mapping (functions) and they can serve as universal approximators. On the other hand, recurrent neural networks contain at least one cyclic path, where the same input information repeatedly influences the activity of neurons on the cyclic path. Such networks are more closely related to biological neural networks, which are also mostly recurrent. From the mathematical point of view the recurrent neural networks are also universal approximators, but they implement dynamic systems. These networks are less common in technical applications because of both theoretical and practical problems with learning.

Their applications are hindered by a lack of effective algorithms for super-vised training. This problem was solved by the so-called Echo state neural networks [3], where a very fast learning algorithm is used. It is based on a calculation of a pseudoinverse matrix, which is a standard numerical task.

S. Kollias et al. (Eds.): ICANN 2006, Part I, LNCS 4131, pp. 367–375, 2006.

The trouble with this approach is that it offers nearly perfect learning of the training set for the given recurrent neural network, but the predictive ability of such a network is not very good. The advantage of "one-step" learning changes into a disadvantage, when we want to improve the predictive abilities of the given trained network. The approach using the pseudoinverse matrix does not help us in any direct way in this matter. In our previous work [5,6] we explored a possibility to improve "one-step" learning by evolutionary approaches.

In this paper we have substituted the output layer by feedforward neural network and the original "one-step" learning algorithm was substituted by back-propagation of error learning algorithm. Connection between "liquid state" computing, related to Echo states, and backpropagation was mentioned previously in [4,7].

With this approach we loose the advantage of fast computation of the "one-step" optimization typical for Echo state networks, but we get flexibility and better quality of prediction.

2 Main Idea of Echo State Neural Networks

The core of echo state neural networks is a special approach to the analysis and training of recurrent neural networks. This approach leads to a fast, simple and constructive algorithm for supervised learning of recurrent neural networks. The basic idea of echo state neural networks is exploitation of a big "reservoir" of neurons recurrently connected with random weights, as a source of dynamic behavior of the neural network, from which the requested output is combined.

Under certain circumstances, the state $\mathbf{x}_n = (x_1(n), x_2(n), \ldots, x_N(n))$ of recurrent neural networks is a function of its previous inputs $\mathbf{u}(n), \mathbf{u}(n-1), \ldots$.

An output of ith hidden neuron in time n is $x_i(n)$ and N is the number of hidden neurons. An input vector is $\mathbf{u}_n = (u_1(n), u_2(n), \ldots, u_K(n))$, where $u_i(n)$ is an input to the ith input neuron in time n and K is the number of input neurons. Therefore, there exists such a function E, that

$$\mathbf{x}(n) = E(\mathbf{u}(n), \mathbf{u}(n-1), \ldots). \tag{1}$$

Metaphorically speaking, the state $\mathbf{x}(n)$ of neural network can be considered as so called "echo" recall of its previous inputs.

3 Combination of Echo State and Feedforward Neural Network

Our paper presents a combination of echo state neural network and feedforward neural network. In this approach the output layer of echo state network is substituted with feedforward neural network and "one-step" training algorithm is replaced with backpropagation of error learning algorithm (Fig.1). In original echo state neural networks we have no possibility to stop the training algorithm to avoid the overfitting problem. Therefore such neural networks have often troubles with generalization. By using backpropagation of error learning algorithm we can stop training when error on validation set does not improve.

The most complex and at the same time the most important part of echo state neural network, so called "dynamical reservoir", remained preserved. The main task of this big recurrent layer is to preprocess the input signal for the feedforward part of the whole neural network.

3.1 Description of Neural Network

The echo state part of the whole neural network consists of K input, N hidden and L output neurons. The state of input neurons in time n is characterized by a vector $\mathbf{u}_n = (u_1(n), u_2(n), \ldots, u_K(n))$, the state of output neurons in time n is given by a vector $\mathbf{y}_n = (y_1(n), y_2(n), \ldots, y_L(n))$, and similarly for hidden neurons $\mathbf{x}_n = (x_1(n), x_2(n), \ldots, x_N(n))$. The values of input – hidden synaptic weights are stored in matrix $\mathbf{W}^{in} = (w_{ij}^{in})$, hidden – hidden weights are stored in matrix $\mathbf{W} = (w_{ij})$ and hidden – to – output weights are stored in matrix $\mathbf{W}^{out} = (w_{ij}^{out})$.

The feedforward part of the whole neural network consists of L input neurons, M output neurons and of S hidden layers which may have different number of neurons in each layer. As we can see from the Fig.1, the output layer in echo state part is the same as the input layer in the feedforward part.

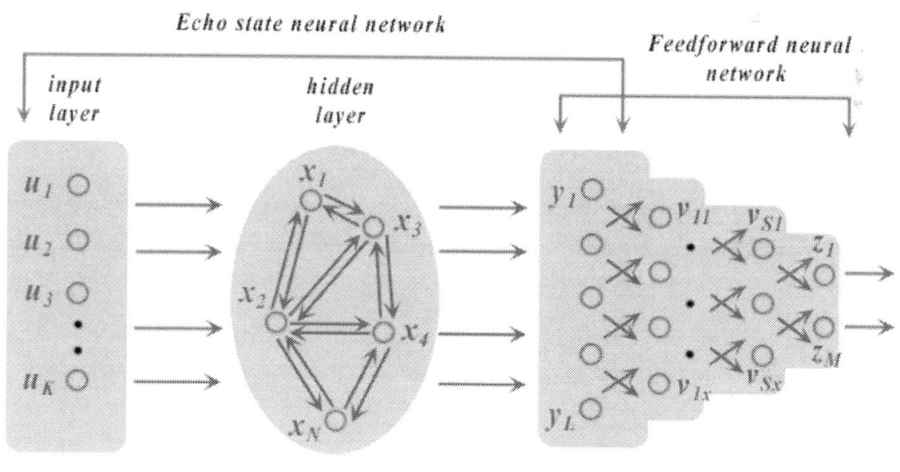

Fig. 1. The architecture used in this approach – combination of echo state neural network and feedforward neural network

3.2 The Learning Algorithm

The only weights which are trained in this combination of neural networks are the weights in the feedforward part. The whole algorithm for training and testing consists of two steps.

The First Step: The first step should create an untrained echo state neural network consisting of weights $\mathbf{W}^{in}, \mathbf{W}, \mathbf{W}^{out}$, which however can produce so called "echo" states. There exists a number of ways how to obtain such a network with the given property. We have used the following approach [3]:

- We have randomly generated an internal weight matrix $\mathbf{W}_0$.
- Then we have created a normalized matrix $\mathbf{W}_1$ with unit spectral radius from the matrix $\mathbf{W}_0$ by putting $\mathbf{W}_1 = 1/|\lambda_{max}|\mathbf{W}_0$, where $|\lambda_{max}|$ is the spectral radius of $\mathbf{W}_0$.
- After that we have scaled $\mathbf{W}_1$ to $\mathbf{W} = \alpha\mathbf{W}_1$, where $\alpha < 1$, whereby $\mathbf{W}$ obtains a spectral radius α.
- Then we have randomly generated input weights $\mathbf{W}^{in}$ and output weights $\mathbf{W}^{out}$.

Now, the untrained network is an echo state network, regardless of how $\mathbf{W}^{in}$, $\mathbf{W}^{out}$ are chosen.

The Second Step: Now we can accede to the training and testing of the whole neural network. As we mentioned before, the only weights which are trained in this approach are the weights in the feedforward part. The learning algorithm used in this part is the well known backpropagation of error learning algorithm. This algorithm is described in details in [1]. For our approach is important, how to propagate the input signal through the echo state part.

The states of hidden neurons in "dynamical reservoir" are calculated from the formula

$$\mathbf{x}(n+1) = f(\mathbf{W}^{in}\mathbf{u}(n) + \mathbf{W}\mathbf{x}(n)), \tag{2}$$

where f is the activation function of hidden neurons (we used the sigmoidal function). The states of output neurons are calculated by the formula

$$\mathbf{y}(n+1) = f^{out}(\mathbf{W}^{out}(\mathbf{u}(n)\mathbf{x}(n+1)), \tag{3}$$

where f^{out} is the activation function of output neurons (we used the sigmoidal function).

4 Prediction of Air Temperature Data

Most of the publications about prediction strive to achieve the best prediction, which is then compared with results of other prediction systems on selected data. This paper is different in this aim; its goal was to compare results achieved by original "echo state" neural network, with our new approach.

Records of average air temperature in Slovakia in years 1999 – 2002 were chosen as testing data for quality of prediction. The training set was composed of a time sequence of 1096 samples of air temperature data in years 1999, 2000, 2001 and the testing set was composed of next 31 samples – that means January of the year 2002.

This task is basically a function approximation; it is the prediction of data from previous trends in the same data.

A mean absolute percentage error (MAPE) was used for the measurement of prediction quality, where P_i^{real} and P_i^{calc} are measured, resp. predicted values, and N is the number of couples of values (the length of the predicted time series):

$$MAPE = \frac{\sum_{i=1}^{N} \left| \frac{P_i^{real} - P_i^{calc}}{P_i^{real}} \right|}{N} \times 100 \tag{4}$$

5 Experiments

Experiments were divided into two parts. The task of the first part was to find parameters of echo state neural networks, which would be optimal for the quality of prediction on the testing air temperature set. Very simple feedforward neural network was used in this first part. The reason for such simple neural network were computational demands. The network consisted of two layers with 4 neurons in first layer and 1 neuron in second layer. The results of experiments are in the following Table 1.

Table 1. Results of experiments in the first part: quality of the prediction for different parameter values

Index	*DR Size*	α	*Average MAPE*	*The best MAPE*
1	200	0.7	42.981 %	33.986 %
2	200	0.8	41.549 %	32.475 %
3	250	0.7	44.857 %	35.451 %
4	250	0.8	44.557 %	35.251 %
5	300	0.7	39.528 %	29.123 %
6	300	0.8	40.254 %	30.256 %

DR Size is dynamical reservoir, α is a parameter influencing the ability of the neural network to have echo states (the used values were chosen in agreement with the values used by Jaeger, the author of echo state networks, see [2,3]). For each value of *DR size* and parameter α, which are presented in the Table 1, values of weights in dynamical reservoir were generated randomly 50 times and for every of this initialization of weights the prediction error on the testing set was calculated. Then the average error from the whole set of attempts was calculated (attribute *Average MAPE* in Table 1). The best achieved error was recorded as well (attribute *The best MAPE* in Table 1). As we can see from Table 1, there is an evident correlation between these attributes. In the cases, where better *Average MAPE* error was achieved, better *The best MAPE* error was achieved too. The best results were achieved for *DR* which consists of 300 neurons and for the parameter α equal 0.7.

The second part of the experiments was focused on finding the best parameters of feedforward neural network and it's backpropagation of error learning algorithm. Parameters of dynamical reservoir and initial synaptic weights were chosen in accordance with the results of experiments in the first phase.

Thereafter we started with training the feedforward neural network for all samples from the training set except the last 7 samples. This last week of year 2001 was chosen as a validation set. This set was used for testing the quality of prediction on samples, which were not used during the training process.

A considerable number of experiments was carried out, the representative results of which are given in the following Table 2.

Table 2. Results of representative experiments in second part of neural network learning

Index	*Learning cycles*	*Number of neurons*	*Parameter* γ	*MAPE*
1	3854	12 – 1	0.8 %	24.547 %
2	3214	12 – 1	0.9 %	22.654 %
3	5635	14 – 8 – 1	0.8 %	15.729 %
4	4250	14 – 8 – 1	0.9 %	16.925 %
5	4411	25 – 11 – 1	0.8 %	18.521 %
6	3953	25 – 11 – 1	0.9 %	17.443 %

Attribute *Learning cycles* specifies the number of learning cycles after which the best prediction error on the testing set was achieved. Attribute *Number of neurons* specifies the number of neurons in each layer. Attribute *Parameter* γ specifies the value of learning parameter in backpropagation of error learning algorithm. Every training of feedforward part started with the same values of synaptic weights and other parameters of echo state part. Attribute *MAPE* specifies the best reached prediction error on the testing set.

In the following Table 3 we can see the comparison of best achieved errors on testing air temperature set with three different approaches. We can see graphical representation of the two most important approaches in Figures 2 and 3. It is clear from this table and figures, that the combination of echo state neural network and feedforward neural network can considerably increase the quality of prediction.

Table 3. Comparison of three different approaches.

Approach	*MAPE*
TD FFNN with BP	29.833 %
ESN	23.281 %
Comb. of ESN and FFNN with BP	15.729 %

Attribute *MAPE* in Table 3 specifies the best reached prediction error on the testing set. Attribute *Approach* specifies the approach used for prediction. *TD FFNN with BP* – time delay feedforward neural network with backpropagation of error learning algorithm. Its best error (MAPE 29.833 %) was achieved with these network's parameters: number of learning cycles: 6386, learning parameter $\gamma = 0.7$, number of neurons in each layer: 8 - 23 - 9 - 1. *ESN* – echo state neural network with "one-step" learning algorithm, *Comb. of ESN and FFNN with BP* – combination of echo state neural network and feedforward neural network with backpropagation of error learning algorithm.

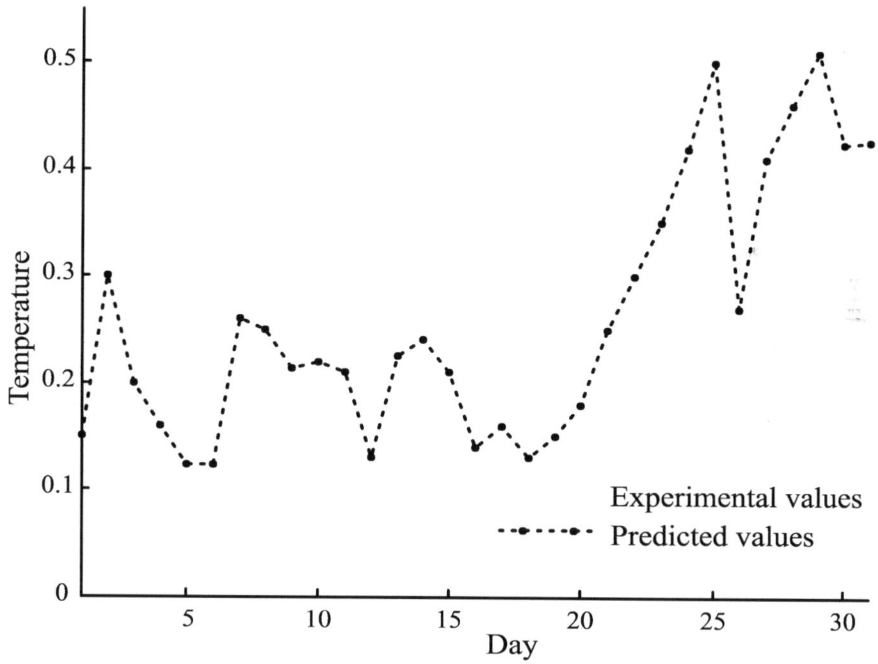

Fig. 2. Testing data: 31 records of air temperature and 31 values predicted by original echo state neural network with "one-step" learning algorithm (DR Size 250, Alpha 0.8, MAPE 23.28 %)

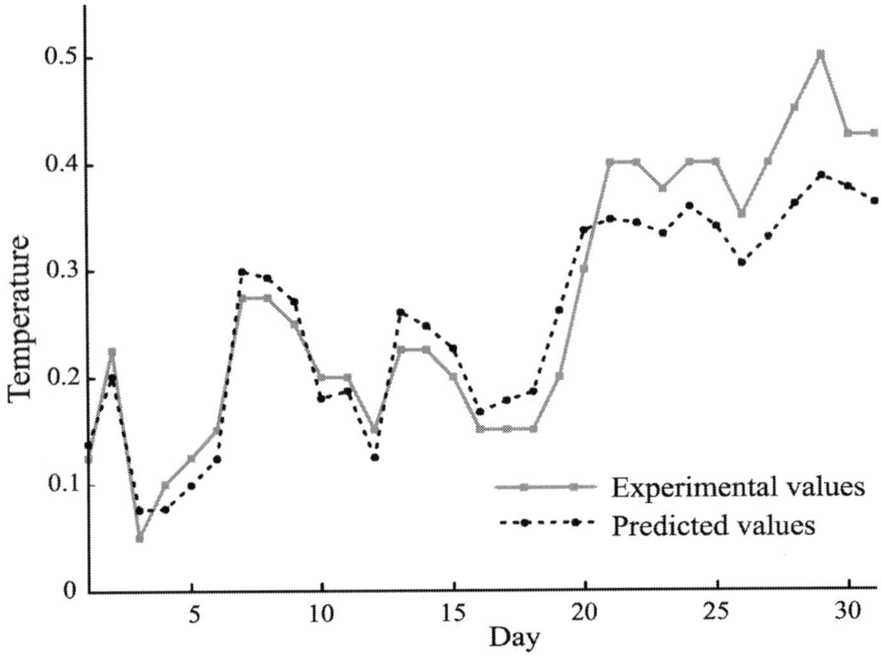

Fig. 3. Testing data: 31 records of air temperature and 31 values predicted by echo state neural network combined with feedforward neural network (Experiment No. 3 from Table 2, DR Size 300, Alpha 0.7, MAPE 15.73 %)

6 Conclusions

Echo state neural networks belong to a group of relatively new approaches in the field of neural networks. Their biggest advantage is their closer relation to biological models due to their recurrent nature and the use of the reservoir of dynamic behavior without weight setting. These networks perform extraordinarily in learning a time sequence, which is essential for example for motoric control, in human beings or in a robot, or also for language processing tasks. Compared to other types of recurrent networks, echo state networks have a major advantage in their ability of "one step learning", even though this approach is probably not very biologically plausible. Their disadvantage is a lower generalization ability and the absence of an approach, which would be able to improve a trained network.

The problem of the trained network improvement doesn't appear in common feed forward or recurrent neural networks, because in a case of need of the network's improvement the trained network can be further "trained" by another batch of iterations of the classical algorithm of back propagation of error. This however doesn't work in echo state networks, where the standard algorithm with a pseudoinverse matrix allows only the approaches "all or nothing", which means that we will, or we will not train the network, nothing in between. A network trained by this approach can't be further trained.

We have tried to solve the above mentioned problem in this work, where the output layer was substituted by feedforward neural network and the original "one-step" learning algorithm was replaced by backpropagation of error learning algorithm. Because we didn't want to work with artificially created examples, we chose real data to evaluate our algorithms. Those data represent the meteorological measurements of air temperature. Our aim was to find out if this approach is able to increase prediction quality of echo state networks. From the results shown in the paper, it is clear that this aim has been accomplished.

The combination of echo state neural network and feedforward neural network can increase the quality of the network's prediction.

Acknowledgement. This work was supported by Scientific Grant Agency Vega of Slovak Republic under grant #1/1047/04 and by Grant Agency APVT under grant APVT-20-002504.

References

1. Haykin, S.: Neural networks - A comprehensive foundation. Macmillian Publishing, 1994.
2. Jaeger, H.: The Echo State Approach to Analysing and Training Recurrent Neural Net-works. German National Research Center for Information Technology, GMD report 148, 2001.
3. Jaeger, H.: Short Term Memory in Echo State Networks. German National Research Center for Information Technology, GMD report 152, 2002.
4. Natschlager, T., Maass, W., Markram, H.: The "liquid computer":A novel strategy for real-time computing on time series. Special Issue on Foundations of Information Processing of TELEMATIK, 8(1):39-43, 2002.
5. Babinec, S., Pospichal, J.: Optimization in Echo state neural networks by Metropolis algorithm. In R. Matousek, P. Osmera (eds.): Proceedings of the 10th International Conference on Soft Copmputing, Mendel'2004. VUT Brno Publishing, 2004, pp. 155-160.
6. Babinec, S., Pospichal, J.: Two approaches to optimize echo state neural networks. In R. Matousek, P. Osmera (eds.): Proceedings of the 11th International Conference on Soft Computing, Mendel'2005. VUT Brno Publishing, 2005, pp. 39-44.
7. Goldenholz, D.: Liquid computig: A real effect. Technical report, Boston University Department of Biomedical Engineering, 2002.

Language and Cognition Integration Through Modeling Field Theory: Category Formation for Symbol Grounding

Vadim Tikhanoff[1], José F. Fontanari[2], Angelo Cangelosi[1], and Leonid I. Perlovsky[3]

[1] Adaptive Behaviour & Cognition, University of Plymouth, Plymouth PL4 8AA, UK
[2] Instituto de Física de São Carlos, Universidade de São Paulo, São Carlos, SP, Brazil
[3] Air Force Research Laboratory, Hanscom Air Force Base, MA 01731, USA
vadim.tikhanoff@plymouth.ac.uk,
fontanari@ifsc.usp.br, acangelosi@plymouth.ac.uk,
leonid.perlovsky@hanscom.af.mil

Abstract. Neural Modeling Field Theory is based on the principle of associating lower-level signals (e.g., inputs, bottom-up signals) with higher-level concept-models (e.g. internal representations, categories/concepts, top-down signals) avoiding the combinatorial complexity inherent to such a task. In this paper we present an extension of the Modeling Field Theory neural network for the classification of objects. Simulations show that (i) the system is able to dynamically adapt when an additional feature is introduced during learning, (ii) that this algorithm can be applied to the classification of action patterns in the context of cognitive robotics and (iii) that it is able to classify multi-feature objects from complex stimulus set. The use of Modeling Field Theory for studying the integration of language and cognition in robots is discussed.

1 Introduction

1.1 Grounding Language in Categorical Representations

A growing amount of research on interactive intelligent systems and cognitive robotics is focusing on the close integration of language and other cognitive capabilities [1,3,13]. One of the most important aspects in language and cognition integration is the grounding of language in perception and action. This is based on the principle that cognitive agents and robots learn to name entities, individuals and states in the external (and internal) world whilst they interact with their environment and build sensorimotor representations of it. For example, the strict relationship between language and action has been demonstrated in various empirical and theoretical studies, such as psycholinguistic experiments [10], neuroscientific studies [16] and language evolution theories [17]. This link has also been demonstrated in computational models of language [5,21].

S. Kollias et al. (Eds.): ICANN 2006, Part I, LNCS 4131, pp. 376 – 385, 2006.

Approaches based on language and cognition integration are based on the principle of grounding symbols (e.g. words) in internal meaning representations. These are normally based on categorical representations [11]. Much research has been dedicated on modeling the acquisition of categorical representation for the grounding of symbols and language. For example, Steels [19,20] has studied the emergence of shared languages in group of autonomous cognitive robotics that learn categories of objects. He uses discrimination tree techniques to represent the formation of categories of geometric shapes and colors. Cangelosi and collaborators have studied the emergence of language in multi-agent systems performing navigation and foraging tasks [2], and object manipulation tasks [6,12]. They use neural networks that acquire, through evolutionary learning, categorical representations of the objects in the world that they have to recognize and name.

1.2 Modeling Field Theory

Current grounded agent and robotic approaches have their own limitations. For example, one important issue is the scaling up of the agents' lexicon. Present models can typically deal with a few tens of words (e.g. [20]) and with a limited set of syntactic categories (e.g. nouns and verbs in [2]). This is mostly due to the use of computational intelligent techniques, the performance of which is considerably degraded by the combinatorial complexity (CC) of this problem. The issue of scaling up and combinatorial complexity in cognitive systems has been recently addressed by Perlovsky [14]. In linguistic systems, CC refers to the hierarchical combinations of bottom-up perceptual and linguistic signals and top-down internal concept-models of objects, scenes and other complex meanings. Perlovsky proposed the neural Modeling Field Theory (MFT) as a new method for overcoming the exponential growth of combinatorial complexity in the computational intelligent techniques traditionally used in cognitive systems design. Perlovsky [15] has suggested the use of MFT specifically to model linguistic abilities. By using concept-models with multiple sensorimotor modalities, a MFT system can integrate language-specific signals with other internal cognitive representations.

Modeling Field Theory is based on the principle of associating lower-level signals (e.g., inputs, bottom-up signals) with higher-level concept-models (e.g. internal representations, categories/concepts, top-down signals) avoiding the combinatorial complexity inherent to such a task. This is achieved by using measures of similarity between concept-models and input signals together with a new type of logic, so-called dynamic logic. MFT may be viewed as an unsupervised learning algorithm whereby a series of concept-models adapt to the features of the input stimuli via gradual adjustment dependent on the fuzzy similarity measures.

A MFT neural architecture was described in [14]. It combines neural architecture with models of objects. For feature-based object classification considered here, each input neuron $i = 1,\ldots,N$ encodes feature values O_i (potentially a vector of several features); each neuron i may contain a signal from a real object or from irrelevant context, clutter, or noise. We term the set $O_i, i = 1,\ldots,N$ an input neural field: it is a set of bottom-up input signals. Top-down, or priming signal-fields to these neurons are generated by models, $\mathrm{M}_k(S_k)$ where we enumerate models by index $k = 1,\ldots,M$. Each model is characterized by its parameters S_k, which may also be a vector of several features. In this contribution we will consider the simplest possible case, in which parameters model represent feature values of object, $\mathrm{M}_k(S_k) = S_k$. Interaction between bottom-up and top-down signals is determined by neural weights associating signals and models as follows. We introduce an arbitrary similarity measure $l(i\,|\,\mathrm{k})$ between bottom-up signals O_i and top-down signals S_k [see equation (2)], and define the neural weights by

$$f(k\,|\,i) = l(i\,|\,k)\Big/\sum_{k'} l(i\,|\,k'). \tag{1}$$

These weights are functions of the model parameters S_k, which in turn are dynamically adjusted so as to maximize the overall similarity between object and models. This formulation sets MFT apart from many other neural networks.

Recently, MFT has been applied to the problem of categorization and symbol grounding in language evolution models. Fontanari and Perlovsky [7] use MFT as an alternative categorization and meaning creation method to that of discrimination trees used by Steels [19]. They consider a simple world composed of few objects characterized by real-valued features. Whilst in Steels's work each object is defined by 9 features (e.g. vertical position, horizontal, R, G and B color component values), here each object consists of a real-valued number that identifies only one feature (sensor). The task of the MFT learning algorithm is to find the concept-models that best match these values. Systematic simulations with various numbers of objects, concept-models and object/model ratios, show that the algorithm can easily learn the appropriate categorical model. This MFT model has been recently extended to study the dynamic generation of concept-models to match the correct number of distinct objects in a complex environment [8]. They use the Akaike Information Criterion to gradually add concept-models until the system settles to the correct number of concepts, which corresponds to the original number of distinct objects defined by the experimenter. This method has been applied to complex classification tasks with high degree of variance and overlap between categories. Fontanari and Perlovsky [9] have also used MFT in simulations on the emergence of communication. Meanings are created through MFT categorization, and word-meaning associations are learned using two variants of the obverter procedure [18], in which the agents may, or may not, receive feedback about the success of the communication episodes. They show that optimal communication success is guaranteed in the supervised scheme, provided the size of the repertoire of signals is sufficiently large, though only a few signals are actually used in the final lexicon.

1.3 MFT for Categorization of Multi-dimensional Object Feature Representations

The above studies have demonstrated the feasibility of using MFT to model symbol grounding and fuzzy similarity-based category learning. However, the model has been applied to a very simplified definition of objects, each consisting of one feature. Simulations have also been applied to a limited number of categories (concept-models). In more realistic contexts, perceptual representations of objects consist of multiple features or complex models for each sensor, or result from the integration of different sensors. For example, in the context of interactive intelligent systems able to integrate language and cognition, their visual input would consist of objects with a high number of dimensions or complex models. These could be low-level vision features (e.g. individual pixel intensities), or some intermediate image processing features (e.g. edges and regions), or higher-level object features (color, shape, size etc.). In the context of action perception and imitation, a robot would have to integrate various input features from the posture of the teacher robot to identify the action or complex models (e.g. [6]). The same need for multiple-feature objects applies to audio stimuli related to language/speech. In addition, the interactive robot would have to deal with hundreds, or thousands, categories, and with high degrees of overlap between categories.

To address the issue of multi-feature representation of objects and that of the scaling up of the model we have extended the MFT algorithm to work with multiple-feature objects. We consider both the cases in which all features are present from the start, and the case in which the features are dynamically added during learning. For didactic purposes, first we will carry out simulations on very simple data sets, and then on data related to the problem of action recognition in interactive robots. Finally, we will present some results on the scale up of the model, using hundred of objects.

2 The Model

We consider the problem of categorizing N objects $i = 1, \cdots, N$, each of which characterized by d features $e = 1, \cdots, d$. These features are represented by real numbers $O_{ie} \in (0,1)$ - the input signals - as described before. Accordingly, we assume that there are M d-dimensional concept-models $k = 1, \cdots, M$ described by real-valued fields S_{ke}, with $e = 1, \cdots, d$ as before, that should match the object features O_{ie}. Since each feature represents a different property of the object as, for instance, color, smell, texture, height, etc. and each concept-model component is associated to a sensor sensitive to only one of those properties, we must, of course, seek for matches between the same component of objects and concept-models. Hence it is natural to define the following partial similarity measure between object i and concept k

$$l(i \mid k) = \prod_{e=1}^{d} \left(2\pi\sigma_{ke}^{2}\right)^{-1/2} \exp\left[-\left(S_{ke} - O_{ie}\right)^{2} / 2\sigma_{ke}^{2}\right] \quad (2)$$

where, at this stage, the fuzziness σ_{ke} is a parameter given *a priori*. The goal is to find an assignment between models and objects such that the global similarity

$$L = \sum_{i} \log \sum_{k} l(i \mid k) \quad (3)$$

is maximized. This maximization can be achieved using the MFT mechanism of concept formation which is based on the following dynamics for the modeling field components

$$dS_{ke}/dt = \sum_i f(k \mid i)[\partial \log l(i \mid k)/\partial S_{ke}], \tag{4}$$

which, using the similarity (1), becomes

$$dS_{ke}/dt = -\sum_i f(k \mid i)(S_{ke} - O_{ie})/\sigma_{ke}^2 . \tag{5}$$

Here the fuzzy association variables $f(k \mid i)$ are the neural weights defined in equation (1) and give a measure of the correspondence between object *i* and concept *k* relative to all other concepts *k'*. These fuzzy associations are responsible for the coupling of the equations for the different modeling fields and, even more importantly for our purposes, for the coupling of the distinct components of a same field. In this sense, the categorization of multi-dimensional objects is not a straightforward extension of the one-dimensional case because new dimensions should be associated with the appropriate models. This nontrivial interplay between the field components will become clearer in the discussion of the simulation results.

It can be shown that the dynamics (4) always converges to a (possibly local) maximum of the similarity *L* [14], but by properly adjusting the fuzziness σ_{ke} the global maximum often can be attained. A salient feature of dynamic logic is a match between parameter uncertainty and fuzziness of similarity. In what follows we decrease the fuzziness during the time evolution of the modeling fields according to the following prescription

$$\sigma_{ke}^2(t) = \sigma_a^2 \exp(-\alpha t) + \sigma_b^2 \tag{6}$$

with $\alpha = 5\times10^{-4}$, $\sigma_a = 1$ and $\sigma_b = 0.03$. Unless stated otherwise, these are the parameters we will use in the forthcoming analysis.

3 Simulations

In this section we will report results from three simulations. The first will use very simple data sets that necessitate the use of two features to correctly classify the input objects. We will demonstrate the gradual formation of appropriate concept-models though the dynamic introduction of features. In the second simulation we will demonstrate the application of the multi-feature MFT on data related to the classification of actions from interactive robotics study. Finally, in the third simulation we will consider the scaling up of the MFT to complex data sets.

To facilitate the presentation of the results, we will interpret both the object feature values and the modeling fields as d-dimensional vectors and follow the time evolution of the corresponding vector length

$$S_k = \sqrt{\sum_{e=1}^{d} (S_{ke})^2} \Big/ d \,, \tag{7}$$

which should then match the object length $O_i = \sqrt{\sum_{e=1}^{d} (O_{ie})^2} \Big/ d$.

3.1 Simulation I: Incremental Addition of Feature

Consider the case in which we have the 5 objects, initially with only one-feature information. For instance, we can consider color information only on Red, the first of the 3 RGB feature values, as used in Steels's [19] discrimination-tree implementation. The objects have the following R feature values: $O_1 = [0.1]$, $O_2 = [0.2]$, $O_3 = [0.3]$, $O_4 = [0.5]$, $O_5 = [0.5]$.

A first look at the data indicates that these 5 input stimuli belong to four color categories (concept-models) with Red values respectively 0.1, 0.2, 0.3 and 0.5. As a matter of fact, the application of the MFT algorithm to the above mono-dimensional input objects reveal the formation of 4 model fields, even when we start with the condition in which 5 fields are randomly initialized (Fig. 1).

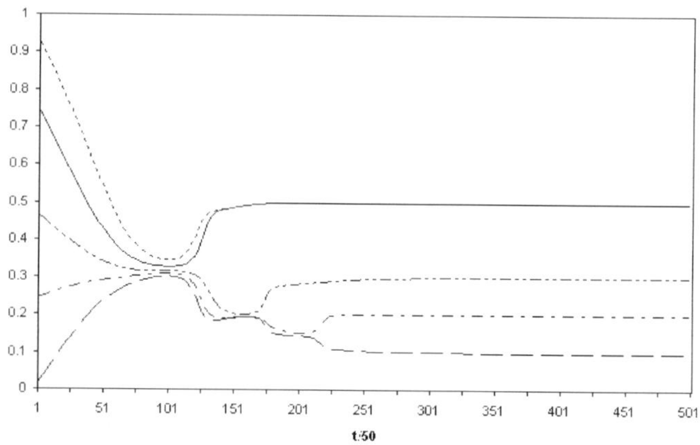

Fig. 1. Time evolution of the fields with only the first feature being used as input. Only 4 models are found, with two initial random fields converging towards the same .5 Red concept-model value.

Let us now consider the case in which we add information from the second color sensor, Green. The object input data will now look like these: $O_1 = [0.1, 0.4]$, $O_2 = [0.2, 0.5]$, $O_3 = [0.3, 0.2]$, $O_4 = [0.5, 0.3]$, $O_5 = [0.5, 0.1]$.

The same MFT algorithm is applied with 5 initial random fields. For the first 12500 training cycles (half of the previous training time), only the first feature is utilized. At timestep 12500, both features are considered when computing the fuzzy similarities. From timestep 12500, the dynamics of the σ_2 fuzziness value is initialized, following equation (7), whilst σ_1 continues[1] its decrease pattern started at

[1] We have also experimented with the alternative method of re-initializing both σ_a values, as in equation (7), whenever a new feature is added. This method produces similar results.

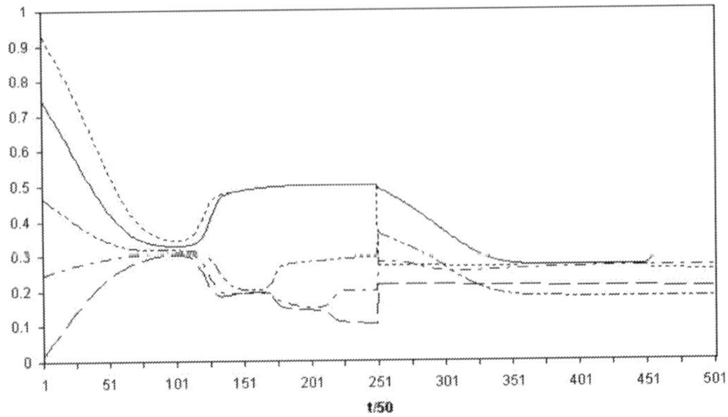

Fig. 2. Time evolution of the fields when the second feature is added at timestep 12500. The dynamic fuzziness reduction for σ_2 starts at the moment the 2nd feature is introduced, and is independent from σ_1. Note the restructuring of 4 fields initially found up to timestep 12500, and the further discovery of the model. The fields values in the first 12500 cycles is the actual mono-dimensional field value, whilst from timestep 12500 the equation in (7) is used to plot the combined fields' value.

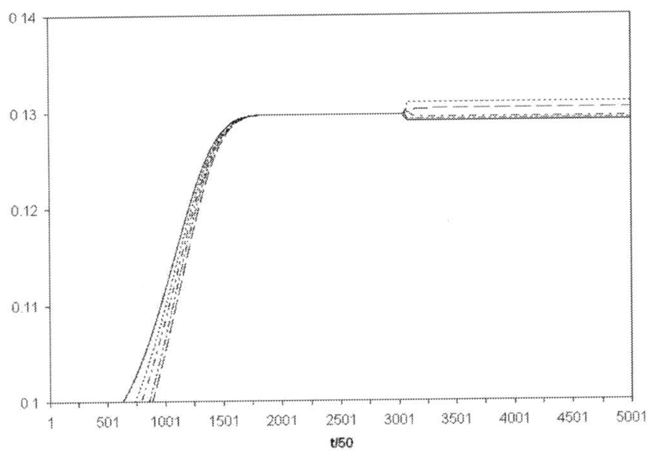

Fig. 3. Evolution of fields in the robot posture classification task. The value of the field corresponds to equation (7). Although the five fields look very close, in reality the individual field values match very well the 42 parameters of the original positions.

timestep 0. Results in Fig. 2 show that the model is now able to correctly identify 5 different fields, one per combined RG color type.

3.2 Simulation II: Categorization of Robotic Actions

In the introduction we have proposed the use of MFT for modeling the integration of language and cognition in cognitive robotic studies. This is a domain where the input to the cognitive agent (e.g. visual and auditory input) typically consists of multi-dimensional data such as images of objects/robots and speech signals. Here we apply the multi-dimensional MFT algorithm to the data on the classification of the posture of robots, as in an imitation task. We use data from a cognitive robotic model of symbol grounding [4,6]. We have collected data on the posture of robots using 42 features. This consist of the 7 main data (X, Y, Z, and rotations of joints 1, 2, 3, and 4) for each of the 6 segments of the robot's arms (right shoulder, right upperarm, right elbow, left shoulder, left upperarm, left elbow). As training set we consider 5 postures: resting position with both arms open, left arm in front, right arm in front, both arms in front, and both arms down. In this simulation, all 42 features are present from timestep 0. Fig. 3 reports the evolution of fields and the successful identification of the 5 postures.

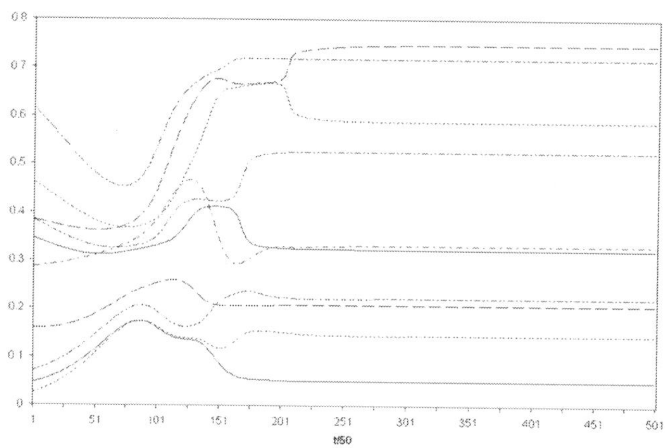

Fig. 4. Evolution of fields in the case with 1000 input objects and 10 prototypes

3.3 Simulation III: Scaling Up with Complex Stimuli Sets

Finally, we have tested the scaling-up of the multi-dimensional MFT algorithm with a complex categorization data set. The training environment is composed of 1000 objects belonging to the following 10 2-feature object prototypes: [0.1, 0.8], [0.2, 1.0], [0.3, 0.1], [0.4, 0.5], [0.5, 0.2], [0.6, 0.3], [0.7, 0.4], [0.8, 0.9], [0.9, 0.6] and [1.0, 0.7]. For each prototype, we generated 100 objects using a Gaussian distribution with standard deviation of 0.05. During training, we used 10 initial random fields.

Fig. 4 reports the time evolution of the 10 concept-models fields. The analysis of results also shows the successful identification of the 10 prototype models and the matching between the 100 stimuli generated by each object and the final values of the fields.

4 Discussion and Conclusion

In this paper we have presented an extension of the MFT algorithm for the classification of objects. In particular we have focused on the introduction of multi-dimensional features for the representation of objects. The various simulations showed that (i) the system is able to dynamically adapt when an additional feature is introduced during learning, (ii) that this algorithm can be applied to the classification of action patterns in the context of cognitive robotics and (iii) that it is able to classify multi-feature objects from complex stimulus set.

Our main interest in the adaptation of MFT to multi-dimensional objects is for its use in the integration of cognitive and linguistic abilities in cognitive robotics. MFT permits the easy integration of low-level models and objects to form higher-order concepts. This is the case of language, which is characterized by the hierarchical organization of underlying cognitive models. For example, the acquisition of the concept of "word" in a robot consists in the creation of a higher-order model that combines a semantic representation of an object model (e.g. prototype) and the phonetic representation of its lexical entry [15]. The grounding of language into categorical representation constitutes a cognitively-plausible approach to the symbol grounding problem [11]. In addition, MFT permits us to deal with the problem of combinatorial complexity, typical of models dealing with symbolic and linguistic representation. Current cognitive robotics model of language typically deal with few tens or hundred of words (e.g. [6,19]). With the integration of MFT and robotics experiments we hope to deal satisfactory with the combinatorial complexity problem.

Ongoing research is investigating the use of MFT for the acquisition of language in cognitive robotics. In particular we are currently looking at the use of multi-dimensional MFT to study the emergence of shared languages in a population of robots. Agents first develop an ability to categorize objects and actions by building concept-models of objects prototypes. Subsequently, they start to learn a lexicon to describe these objects/actions through a process of cultural learning. This is based on the acquisition of a higher-order MFT.

Acknowledgements

Effort sponsored by the Air Force Office of Scientific Research, Air Force Material Command, USAF, under grants number FA8655-05-1-3060 and FA8655-05-1-3031. The U.S. Government is authorized to reproduce and distribute reprints for Governmental purpose notwithstanding any copyright notation thereon.

References

[1] Barsalou, L. (1999), Perceptual symbol systems. *Behavioral and Brain Sciences*, 22, 577-609.

[2] Cangelosi A. (2001). Evolution of communication and language using signals, symbols and words. *IEEE Transactions on Evolutionary Computation*. 5(2), 93-101

[3] Cangelosi A., Bugmann G. & Borisyuk R. (Eds.) (2005). *Modeling Language, Cognition and Action: Proceedings of the 9th Neural Computation and Psychology Workshop*. Singapore: World Scientific.

[4] Cangelosi A., Hourdakis E. & Tikhanoff V. (2006). Language acquisition and symbol grounding transfer with neural networks and cognitive robots. *Proceedings of IJCNN2006: 2006 International Joint Conference on Neural Networks*. Vancouver, July 2006.

[5] Cangelosi, A., & Parisi, D. (2004). The processing of verbs and nouns in neural networks: Insights from synthetic brain imaging. *Brain and Language*, 89(2), 401-408.

[6] Cangelosi A, Riga T (2006). An Embodied Model for Sensorimotor Grounding and Grounding Transfer: Experiments with Epigenetic Robots, *Cognitive Science*, 30(4), 1-17.

[7] Fontanari J.F., Perlovsky L.I. (2005). Meaning creation and modeling field theory. In C. Thompson & H. Hexmoor (Eds.), *IEEE KIMAS2005: International Conference on Integration of Knowledge Intensive Multi-Agent Systems*. IEEE Press, pp. 405-410.

[8] Fontanari J.F., Perlovsky L.I. (2006a). Categorization and symbol grounding in a complex environment. *Proceedings of IJCNN2006: 2006 International Joint Conference on Neural Networks*. Vancouver, July 2006.

[9] Fontanari J.F., Perlovsky L.I. (2006b). Meaning creation and communication in a community of agents. *Proceedings of IJCNN2006: 2006 International Joint Conference on Neural Networks*. Vancouver, July 2006.

[10] Glenberg A., & Kaschak, M. (2002). Grounding language in action. *Psychonomic Bulletin & Review*, 9(3), 558-565.

[11] Harnad, S. (1990). The symbol grounding problem. *Physica D*, 42, 335-346.

[12] Marocco, D., Cangelosi, A., & Nolfi, S. (2003). The emergence of communication in evolutionary robots. *Philosophical Transactions of the Royal Society of London – A* 361, 2397-2421.

[13] Pecher, D., & Zwaan, R.A., (Eds.). (2005). Grounding cognition: The role of perception and action in memory, language, and thinking. Cambridge: Cambridge University Press.

[14] Perlovsky L. *Neural Networks and Intellect: Using Model-Based Concepts*. Oxford University Press, New York, 2001.

[15] Perlovsky L., "Integrating language and cognition," *IEEE Connections*, vol. 2, pp. 8-13, 2004

[16] Pulvermuller F. (2003) *The neuroscience of language. On brain circuits of words and serial order*. Cambridge: Cambridge University Press.

[17] Rizzolatti, G., & Arbib, M. (1998). Language within our grasp. *Trends in Neuroscience*, 21: 188-194.

[18] Smith A. D. M. (2003). Semantic generalization and the inference of meaning," In: W. Banzhaf, T. Christaller, P. Dittrich, J. T. Kim, J. Ziegler (Eds.), *Proceedings of the 7th European Conference on Artificial Life*, Lecture Notes in Artificial Intelligence, vol. 2801, pp. 499-506.

[19] Steels L (1999) *The talking heads experiment (Volume I. Words and meanings)*. Antwerpen: Laboratorium.

[20] Steels, L. (2003) Evolving grounded communication for robots. *Trends in Cognitive Sciences*, 7(7):308-312.

[21] Wermter, S., Elshaw, M., and Farrand, S., 2003, A modular approach to self-organization of robot control based on language instruction. *Connection Science*, 15(2-3): 73-94.

A Methodology for Estimating the Product Life Cycle Cost Using a Hybrid GA and ANN Model

Kwang-Kyu Seo*

Department of Industrial Information and Systems Engineering, Sangmyung University, San 98-20, Anso-Dong, Chonan, Chungnam 330-720, Korea
kwangkyu@smu.ac.kr
Tel.: +81-41-550-5371; Fax. +81-41-550-5185.

Abstract. Although the product life cycle cost (LCC) is mainly committed by early design stage, designers do not consider the costs caused in subsequent phases of life cycle at this phase. The estimation method for the product life cycle cost in early design processes has been required because of both the lack of detailed information and time for a detailed LCC for a various range of design alternatives. This paper proposes a hybrid genetic algorithm (GA) and artificial neural network (ANN) model to estimate the product LCC that allows the designer to make comparative LCC estimation between the different product concepts. In this study, GAs are employed to select feature subsets eliminated irrelevant factors and determine the number of hidden nodes and processing elements. In addition, GAs are to optimize the connection weights between layers of ANN simultaneously. Experimental results show that a hybrid GA and ANN model outperforms the conventional backpropagation neural network and verify the effectiveness of the proposed method.

1 Introduction

The ability of a company to compete effectively on the increasingly competitive global market is influenced to a large extent by the cost as well as the quality of its products and the ability to bring products onto the market in a timely manner. In order to guarantee competitive pricing of a product, cost estimates are performed repeatedly throughout the life cycle of many products. In the early phases of the product life cycle, when a new product is considered, cost estimate analyses are used to support the decision for product design. Later on when alternative designs are considered, the best alternative is selected based on its predicted life cycle cost (LCC) and its benefits. Manufacturers usually considered only how to reduce the cost the company spends for material acquisition, production, and logistics. In order to survive in the competitive market environment especially resulted from the widespread awareness of global environmental problems and legislation, manufacturers now have to consider reducing the cost of the entire life cycle of a product. The costs incurred during life cycle are mostly committed by early design decisions. Studies reported in Dowlatshahi [1] and by other researchers in design suggest that the design of the product

* Corresponding author.

S. Kollias et al. (Eds.): ICANN 2006, Part I, LNCS 4131, pp. 386–395, 2006.

influences between 70% and 85% of the total cost of a product. Therefore, designers can substantially reduce the LCC of products by giving due consideration to life cycle implications of their design decisions. The research on design methodology to minimize the LCC of a product also becomes very important and valuable [2, 3].

Generally, product designers are being asked to judge the cost of the products to be developed. Not only is this an additional task for designers, but it is also necessary something they are qualified to do. Therefore, the cost models created by cost estimators should be integrated with traditional design models, making the parametric cost results available on demand. However, the use of detailed parametric models is not well suited to early conceptual design, where ideas are diverse and numerous, details are very scarce, and the pace is swift. This is unfortunate because early phases of the design process are widely believed to be the most influential in defining the product LCC.

This paper proposes a hybrid genetic algorithm (GA) and artificial neural network (ANN) model to predict the product LCC. In this study, the GA is employed as an optimization method of relevant feature subsets selection, the determination of the number of hidden layer and processing elements. In addition, the GA globally searches and seeks optimal or near-optimal connection weights of ANN to improve the prediction accuracy.

The rests of the paper are organized as follows. Section 2 shows the overview of the proposed method. Section 3 presents a description of the hybrid GA and ANN model. Sections 4 describe the procedures of experiments and experimental results are shown in section 5. Finally, section 6 discusses the conclusions and future research issues.

2 Overview of the Proposed Method

In this paper, the possibility of the LCC prediction method based on a hybrid GA and ANN model is investigated. The proposed method provides useful LCC prediction of products in terms of product attributes and LCC factors.

The hybrid GA and ANN model based LCC prediction method of products is a different approach to other cost prediction methods. The proposed hybrid model is trained to generalize on characteristics of product concepts using product attributes and corresponding LCC factors from pre-existing LCC studies. The designer queries the hybrid GA and ANN model to predict the product LCC with product attributes to quickly obtain the LCC prediction result for a new product concept as shown in Fig. 1. Designers need to simply provide product attributes of new product concepts to gain LCC prediction. It has the flexibility to learn and grow as new information becomes available, but it does not require the creation of a new model to make as LCC prediction for a new product concept. Also, by supporting the extremely fast comparison of the cost performance of product concepts, it does not delay product development.

This study proposes a hybrid GA and ANN model to improve the prediction accuracy of the product LCC. The GA part of the hybrid model as an optimization method is proposed to select relevant product attributes and determine the number of hidden nodes and processing elements. In addition, the hybrid model simultaneously searches the connection weights between layers in ANN. In other words, the GA globally searches and seeks an optimal or near-optimal ANN topology and connection weights for the prediction of the product LCC.

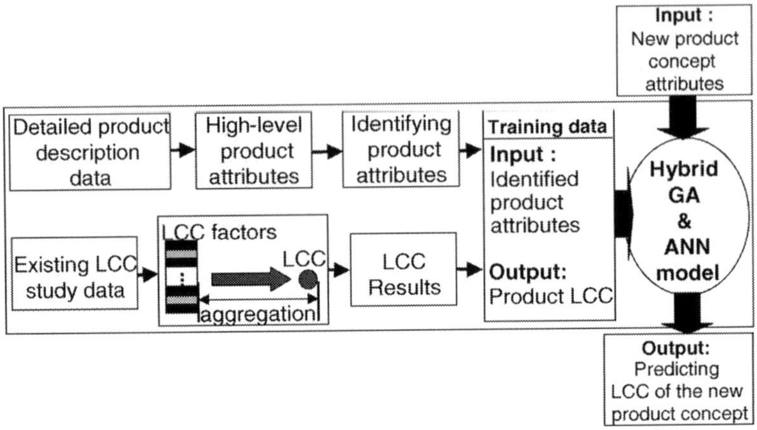

Fig. 1. The structure of the proposed method

There are two key variables that are investigated in order to evaluate the proposed LCC prediction model. Firstly, the feasible LCC factors as output to predict the product LCC are introduced. The LCC factors should be in a form useful to cost estimators and designers. Secondly, a list of reasonable product attributes as inputs is identified and LCC factors in order to make a set of meaningful attributes. The product attributes must be meaningful to designers and consist of only product attributes typically known during conceptual design.

2.1 Life Cycle Cost Factors

The first issue is to identify the feasible LCC factors as outputs for use in the hybrid model. In order to introduce the LCC factors, all the costs incurred in the product life cycle are investigated and enumerated. The LCC of a product is determined by aggregating all the LCC factors. The product LCC can be decomposed into cost factors as shown in table 1 [4, 5].

Table 1. Life cycle stage and LCC factors for use in the hybrid model

Life cycle	Cost factor
design	market recognition, development
production	materials, energy, facilities, wages, waste, pollution, health damages
usage	transportation, storage, waste, breakage, warranty/service, energy, materials, maintenance, pollution, health damages
disposal/ recycling	disposal/recycling dues, energy, waste, disposal, pollution, health damages

Table 1 provides a list of cost factors for the product life cycle that was adapted to the feasible LCC factors as useful outputs for predicting the product LCC in the the hybrid model. This decomposition is by no means the most comprehensive and representative of all products or any product for that matter.

The cost factors considered will depend on the stage in which we want to use the model, the kind of information to be extracted from the model and the product being designed. While the LCC is the aggregate of all the costs incurred in the product's life, it must be pointed out that there are differences between the cost issues that will be of interest to the designer of the product and the firm developing the product in the LCC analysis. We show the prediction of the total energy cost during life cycle and the maintenance cost in usage phase as examples in this study.

2.2 Product Attributes

2.2.1 General Product Attributes

The product attributes need to be both logically and statistically linked to LCC factors, and also be readily available during product concept design. The attributes must be sufficient to discriminate between different concepts and be compact so that the demands on the hybrid model are reasonable. Finally, they must be easily understood by designers and, as a set, span the scope of the product life cycle. These criteria were used to guide the process of systematically developing a product attribute list.

With these goals in mind, a set of candidate product attributes, based upon the literature and the experience of experts was formed [6, 7]. Experts in both product design and cost estimation discussed as candidate attributes derived from the literature. The candidate product attributes identified initially are listed in table 2. They are specified, ranked, binary or not applicable according to their properties such as an appropriate qualitative or quantitative sense or typically rank order concepts.

This study helped us identify attributes that designers could both understand and had knowledge of during the conceptual design. Furthermore, we were able to evaluate which attributes are likely to vary significantly from concept to concept.

The maintainability attributes are additionally identified because the purpose of this study is to predict the maintenance cost of usage phase in conceptual design.

Table 2. Product attribute set as feature subsets grouped for organizational properties

Group name	Associated product attributes
general design properties	durability, degradability
functional properties	mass (mass is represented as the component ratio of 9 materials), volume
manufacturing properties	assemblability, process
operational properties	lifetime, use time, energy source, mode of operation, power consumption, flexibility, upgradeability, serviceability, modularity, additional consumables
end-of-life properties	recyclability, reusability, disassemblability

2.2.2 Maintainability Attributes

Maintenance is an important aspect of life-cycle concerns and plays significant role during the usage phase of a product. It is the design attribute of a product which facilitates the performance of various maintenance activities. Design characteristics of a product which facilitate maintainability will be effective factors which support product maintenance during usage phase. Maintainability is one of the product design parameter that has a great impact in terms of ease of maintenance. Maintainability attributes for products, in general, can be identified by design, personnel and logistics support properties [8, 9]. The maintainability attributes under design property are only considered as the product attributes at the early design stage and presented in table 3. They are also estimated by an appropriate qualitative or quantitative sense.

Table 3. Maintainability attribute set as feature subsets

Property	Maintainability attributes
design	accessibility, reassemblability, simplicity, identification, standardization, diagnosability, modularity, tribo-features

3 A Hybrid GA and ANN Model for Estimating the Product LCC

The GA is a general-purpose evolutionary algorithm that can be used for optimization. When compared to traditional optimization methods, the GA provides heuristic near-optimal solutions. The GA uses a parallel search approach for locating the optimal solution. In the GA, each population member is a potential solution. Recently, the GA has been investigated and shown to be effective in exploring a complex space in an adaptive way, guided by the biological evolution mechanisms of reproduction, crossover, and mutation [10, 11].

The first step of the GA is problem representation. The problem must be represented in a suitable form to be handled by the GA. Thus, the problem is described in terms of genetic code. The GA often works with a form of binary coding. If the problems are coded as chromosomes, the populations are initialized. Each chromosome within the population gradually evolves through biological operations. There are no general rules for determining the population size. Once the population size is chosen, the initial population is randomly generated. After the initialization step, each chromosome is evaluated by a fitness function. According to the value of the fitness function, the chromosome associated with fittest individuals will be reproduced more often than those associated with unfit individuals [12].

The GA works with three operators that are iteratively used. The selection operator determines which individuals may survive. The crossover operator allows the search to fan out in diverse directions looking for attractive solutions and permits chromosomal material from different parents to be combined in a single child. In addition, the mutation operator arbitrarily alters one or more components of a selected chromosome. Mutation randomly changes a gene on a chromosome. It provides the means for introducing new information into the population. Finally, the GA tends to converge on a near optimal solution through these operators [13].

The GA is usually employed to improve the performance of artificial intelligence (AI) techniques. For ANN, the GA is popularly used to select neural network topology including optimizing relevant feature subsets, and determining the optimal number of hidden layers and processing elements. The feature subsets, the number of hidden layers, and the number of processing elements in hidden layers are the architectural factors of ANN to be determined in advance for the modeling process of ANN. However, determining these factors is still part of the art. These factors were usually determined by the trial and error approach and the subjectivity of designer. This may lead a locally optimized solution because it cannot guarantee a global optimum.

In this paper, we propose the hybrid GA and ANN model to resolve these problems of the prediction of the product LCC. In this study, the GA is used for the step of selecting relevant product attributes and optimizing the network topology of ANN. And then, the GA search near optimal connection weights in ANN. Eventually, the GA globally searches an optimal or near-optimal ANN topology and connection weights in the hybrid model.

The overall framework of the hybrid model is shown in Fig. 2.

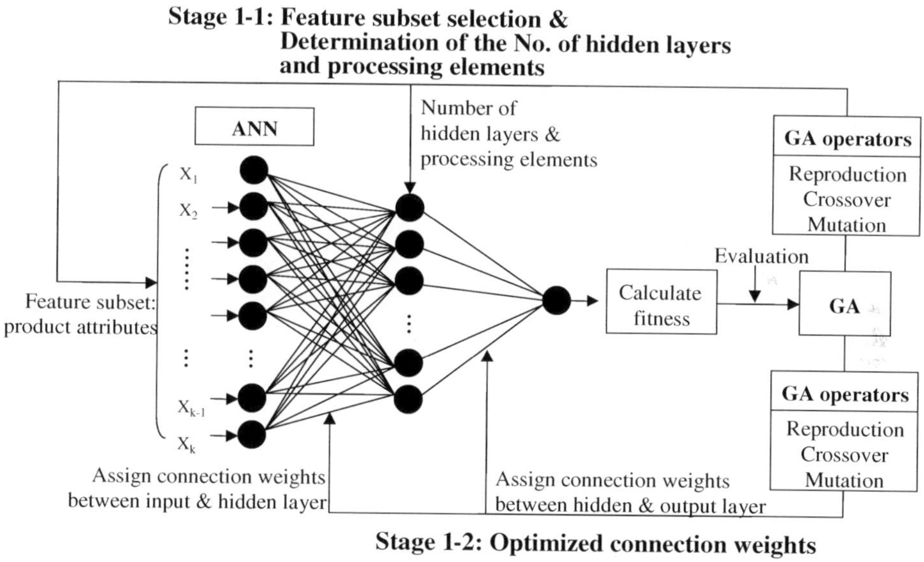

Fig. 2. The structure of the proposed hybrid model

The prediction process of the hybrid model consists of the two stages as follows:

In the first stage, the GA searches optimal or near optimal feature subset and determines the number of hidden nodes and processing elements. The GA also fines optimal or near optimal connection weights of ANNs. The first stage is divided into the following two sub-stages.

Stage 1-1 (GA for Feature Subset selection and determination of the number of Hidden layers and processing elements (GAFSH)): The populations, feature subset, the number of hidden nodes and the processing elements are initialized into random values before the search process. The feature subset, the number of hidden nodes and processing elements for searching must be encoded on chromosomes. The chromosomes for feature subset are encoded as binary strings stands for some subset of original feature set. Each bit of one chromosome represents whether the corresponding feature is selected or not. 1 in each bit means the corresponding feature is selected whereas 0 means it is not selected. In this study, the chromosomes for the feature subsets such as product attributes are encoded as 28 bit string for the energy cost and 36-bit string for the maintenance cost. The number of hidden layer is encoded 2-bit string and that of hidden nodes is encoded as 8-bit string. The encoded chromosomes are searched to optimize a fitness function. In this study, the fitness function is the average deviation between actual and predicted values of the product LCC. The parameters to be searched use only the information about the training data.

Stage 1-2 (GA for Connection Weights (GACW)): After determining the feature subset, the number of hidden nodes and processing elements, connection weights are optimized by GAs. In this study, ANN has one hidden layer and 16 hidden nodes for the energy cost at stage 1-1. In stage 1-2, GA searches optimal or near-optimal connection weights. The populations for the connection weights are initialized into random values before the search process. The parameters for searching must also be encoded on chromosomes. This study needs two sets of parameters. The first set is the set of connection weights between the input layer and the hidden layer of the network. The second set is the set of connection weights between the hidden layer and the output layer. As we know, the above two sets may mitigate the limitation of the gradient descent algorithm. The strings used in this study have the following encoding. This study uses 14 input features and employs 16 processing elements in the hidden layer for predicting the energy cost. Each processing element in the hidden layer receives 16 signals from the input layer. The first 224 bits represent the connection weights between the input layer and the hidden layer. Each processing element in the output layer receives a signal from the hidden layer. The next 16 bits indicate the connection weights between the hidden layer and the output layer. The fitness function is also used to the average deviation between actual and predicted values of product LCC. The prediction procedures for the maintenance cost are same to the process for the energy cost.

In first stage, GA operates the process of crossover and mutation on initial chromosomes and iterates until the stopping conditions are satisfied. For the controlling parameters of the GA search, the population size is set to 100 organisms and the crossover and mutation rates are varied to prevent ANN from falling into a local minimum. The range of the crossover rate is set between 0.7 and 0.9 while the mutation rate ranges from 0.05 to 0.1. As the stopping condition, only 5,000 trials are permitted.

The second stage, selected product attributes, and optimized or near-optimized ANN topology and connection weights are applied to the holdout data. This stage is indispensable to validate the generalizability because ANN has the eminent ability of learning the known data. If this stage is not carried out, the model may fall into the problem of over-fitting with the training data.

4 Research Data and Experiments

4.1 Research Data

As mentioned earlier, the feasibility test of the proposed method was conducted focusing on the total energy and maintenance cost components of the LCC factors. Research data with product attributes and corresponding energy and maintenance costs were collected for 200 different electronic products. The energy cost was obtained by total energy consumption during the life cycle of products. The maintenance cost in usage phase was calculated by equation (1). The equation is composed of labor cost, part replacement cost, and failure rate. The equation is useful in determining the maintenance cost.

$$MC = [(LC_{Fixed} + (T_L \times R_L) + C_R)] \times R_F \qquad (1)$$

Where:

LC_{Fixed} = Fixed labor cost such as the fixed cost when a maintenance representative visits a customer ($)

T_L = Labor time such as mean actual repair time (MART) or mean actual maintenance time (MAMR) (hour)

R_L = Labor rate ($/hour)

C_R = Mean replacement cost of parts or materials when maintenance occurs ($)

R_F = Failure rate

The examples of research data for the hybrid model are shown in table 4.

Table 4. Examples of research data for the hybrid model

Product	Inputs								Outputs	
	Mass (kg)	Ferrous M. (%mass)	Plastics (%mass)	Lifetime (hours)	Use time(hrs) (hours)	…	Power consump. (watt)	Modularity (0~4)	Energy cost($)*	Maintenance cost($)**
1	8.17	32.62	61.58	61320	3041	…	1064	1	596.21	6.00
2	1.04	16.19	77.65	26280	13	…	58	1	20.53	1.79
3	0.18	45.77	32.86	43800	13688	…	0	1	1.65	8.25
4	0.64	22.16	71.09	2160	45	…	13	1	11.47	1.85
5	1.93	2.85	65.54	43800	487	…	616.44	4	83.54	2.75
…	…	…	…	…	…	…	…	…	…	…
…	…	…	…	…	…	…	…	…	…	…
198	49.78	67.07	27.64	87600	87600	…	13	3	1467.29	18.00
199	40.46	8.83	25.81	87600	11680	…	616	3	2935.2	6.54
200	35.01	24.24	51.75	121764	121764	…	19	4	313.41	12.57

* The energy cost is the total cost of energy consumption during product's life cycle.

** The maintenance cost is the mean cost of product maintenance during usage phase of a product.

4.2 Experiments and Discussion

This study compares GAFSH and GACW to the backpropagation neural network (BPNN). BPNN uses the gradient descent algorithm to train the network. This is the conventional approach of previous studies. As mention earlier, GAFSH employs the GA to select feature subsets and determine the number of hidden layers and processing element of ANN. Using the selected feature subsets and the determined the number of hidden layers and processing element of ANN, GACW employs the GA to determine the connection weights of ANN. About 20% (40) of the data is used for holdout and 80% (160) is used for training. The training data is used to search the optimal or near-optimal parameters and is employed to evaluate the fitness function. The holdout data is used to test the results with the data that is not utilized to develop the model.

The GA selects 14 product attributes from 28 product attributes for the energy cost and 21-product attributes from 36 product attributes for the maintenance cost. In addition, the GA recommends one hidden layer and 16 processing elements for the energy cost and one hidden layer and 24 processing elements for the maintenance cost in ANN. Table 5 presents mean absolute percent error (MAPE) by different models. In table 5, GACW has higher prediction accuracy than BPNN and GAFSH for the holdout data. GAFSH outperforms BPNN with the gradient descent algorithm. The experimental results show that the prediction accuracy performance of ANN is sensitive not only to various feature subsets but also to different topology. Thus, this result shows that simultaneous optimization of the feature subsets and topology is need for the best prediction. In addition, we concluded that ANN with the genetically evolved connection weights provides higher prediction results than that of GAFSH.

The McNemar tests are used to examine whether the proposed model shows better performance than the conventional BPNN. This test is a nonparametric test for two related samples using the chi-square distribution. This test may be used with nominal data and is particularly useful with 'before–and-after' measurement of the same subjects [14]. The teat results show GAFSH and GACW outperforms the conventional BPNN at a 5% statistical significance level. Although GACW shows higher accuracy than that of GAFSH, GACW does not outperforms GAFSH at a 5% statistical significance level.

Table 5. Prediction accuracy by conventional ANN and a hybrid model (MAPE)

	Conventional BPNN		Hybrid GA & ANN model			
			GAFSH		GACW	
	Training	Holdout	Training	Holdout	Training	Holdout
Energy cost	19.57%	28.78%	12.74%	18.56%	11.26%	16.34%
Maintenance cost	7.38%	11.43%	4.98%	7.56%	4.52%	7.13%
Average	13.48%	20.11%	8.86%	13.06%	7.89%	11.76%

5 Conclusions

The lack of an analytic LCC prediction method motivated the development of the hybrid GA and ANN model to predict the product LCC. For the proposed method, the product attributes and LCC factors were identified and used to predict the cost performance of products. In order to resolve the local optimum problem of ANN, the hybrid GA and ANN model has been proposed. In this study, the GA was used for the step of selecting relevant feature subsets and determining the number of hidden layers and processing elements. In addition, the GA genetically evolved connection weights of ANN. The proposed model outperformed the conventional BPNN. The major advantage of the proposed model was simultaneous consideration of efficiency and effectiveness for the prediction of the product LCC.

For future work, we intend to apply the GA to other AI techniques to improve the accuracy and efficiency. We believe that there is great potential for further research with the optimization using the GA for other AI techniques.

References

1. Dowlatshahi, S.: Product design in a concurrent engineering environment: an optimisation approach. Journal of Production Research, Vol. 30(8) (1992) 1803-1818
2. .Westkämp, E., Alting, l., Arndt, G.:Life Cycle Management: Approaches and Visions Towards Sustainable Manufacturing. Annals of the CIRP, Vol. 49(2) (2000) 501-522
3. Westkämp, E., Osten-Sacken, D.v.d.: Product Life Cycle Costing Applied to Manufacturing System. Annals of the CIRP, Vol. 47(1) (1998) 353-356
4. Alting, L.: Life cycle design of products: a new opportunity for manufacturing enterprises. In Concurrent Engineering: Automation, Tools, and Techniques. A. Kusiak (Ed.) (pp. 1-17), New York: Wiley (1993)
5. Alting, L., and Legarth, J.: Life cycle Engineering and Design. Annals of the CIRP, Vol. 44(2) (1995) 569-580
6. Park, J. -H. and Seo, K. -K.: Approximate Life Cycle Assessment of Product Concepts using Multiple Regression Analysis and Artificial Neural Networks. KSME International Journal, Vol. 17(12) (2003) 1969-1976
7. Seo, K.-K, Min, S.-H. and Yoo, H.-W.: Artificial Neural Network based Life Cycle Assessment Model of Product Concepts using Product Classification Method. Lecture Notes in Computer Science, 3483 (2005) 458-466
8. Takata, S., Hiraoka, H., Asama, H., Yamoka, N., Saito, D.: Facility model for lifecycle maintenance system, Annals of the CIRP, Vol. 44(1) (1995) 117-121
9. Tarelko, W.: Control model of maintainability level, Reliability Engineering and System Safety; Vol. 47(2) (1995) 85-91
10. Goldberg, D. E.: Genetic algorithms in search, optimization, and machine learning. Reading, MA: Addison-Wesley (1989)
11. Adeli, H., and Hung, S.: Machine learning: Neural networks, genetic algorithms, and fuzzy systems. New York: Wiley (1995)
12. Davis, L.: Genetic algorithms and financial applications. In G. J. Deboeck (Ed.), Trading on the edge (pp. 133–147), New York: Wiley (1994)
13. Wong, F., and Tan, C.: Hybrid neural, genetic, and fuzzy systems. In G. J. Deboeck (Ed.), Trading on the edge (pp. 243–261), New York: Wiley (1994)
14. Cooper, D. R., & Emory, C. W.: Business research methods, Chicago, IL: Irwin (1995)

Using Self-Organizing Maps to Support Video Navigation

Thomas Bärecke[1], Ewa Kijak[1], Andreas Nürnberger[2], and Marcin Detyniecki[1]

[1] LIP6, Université Pierre et Marie Curie, Paris, France
thomas.baerecke@lip6.fr
[2] Faculty of Computer Science, Otto-von-Guericke Universität Magdeburg, Germany

Abstract. Content-based video navigation is an efficient method for browsing video information. A common approach is to cluster shots into groups and visualize them afterwards. In this paper, we present a prototype that follows in general this approach. Unlike existing systems, the clustering is based on a growing self-organizing map algorithm. We focus on studying the applicability of SOMs for video navigation support. We ignore the temporal aspect completely during the clustering, but we project the grouped data on an original time bar control afterwards. This complements our interface by providing – at the same time – an integrated view of time and content based information. The aim is to supply the user with as much information as possible on one single screen, without overwhelming him. Special attention is also given to the interaction possibilities which are hierarchically organized.

1 Introduction

Extremely large digital libraries with all types of multimedia documents are available today. Efficient methods to manage and access these archives are crucial, for instance, quick search for similar documents or effective summarization via visualization of the underlying structure.

The prototype presented in this paper implements methods to structure and visualize video content in order to support a user in navigating within a single video. We focus on the way video information is summarized in order to improve content-based navigation. Currently, a common approach is to use clustering algorithms in order to automatically group similar shots and then to visualize the discovered groups in order to provide an overview of the considered video stream [1,2]. Summarization and representation of video sequences are usually keyframe-based. A keyframe is a representative still image of a video sequence. The simplest method of extracting a keyframe for a given sequence is to choose its median frame. The keyframes can be arranged in the form of a temporal list and hierarchical browsing is then based on the clustered groups. In this paper, we use one promising unsupervised clustering approach that combines both good clustering and visualization capabilities: the self-organizing maps (SOMs). In fact, they have been successfully used for the navigation of text [3,4,5,6] and image collections [7,8].

S. Kollias et al. (Eds.): ICANN 2006, Part I, LNCS 4131, pp. 396–405, 2006.

The visualization capabilities of self-organizing maps provide an intuitive way of representing the distribution of data as well as the object similarities. As most clustering algorithms, SOMs operate on numerical feature vectors. Thus, video content has to be defined by numerical feature vectors that characterize it. A variety of significant characteristics has been defined for all types of multimedia information [9]. From video documents, colour histograms for describing the keyframes are widely used [1,2,10]. We also followed this simple approach, since our goal was to investigate the visualisation and interaction capabilities of SOMs for video structuring and navigation. For this purpose, sophisticated visual descriptors and similarity measures are not necessary. Since we use only colour features of still images as a model for the video content on which we apply the clustering algorithm, temporal information is completely ignored. However, after clustering the grouped images are projected on a time bar tool that is visualizing the temporal distribution of similar sequences.

Our system is composed of feature extraction, structuring, visualization, and user interaction components. Structuring and visualization parts are based on growing SOMs that were developed in previous works and applied to other forms of interactive retrieval [6,11]. We believe that *growing* SOMs are particularly adapted to fit video data. The visualization and user interaction components were designed with the intention to provide intuitive content-based video browsing functionalities to the user. In the following four sections we describe the system components and the processing steps. First, we discuss the video feature extraction process. Then we shortly describe how the data is structured by using growing self-organizing maps. Afterwards, a detailed description of the visualization component of the system is given. Before concluding, the last section deals with the interaction possibilities of our system.

2 Video Feature Extraction

The video feature extraction component supplies the self-organizing map with numerical vectors and therefore they form the basis of the system. This process is shown in Fig. 1. The module consists of two parts: temporal segmentation and feature extraction.

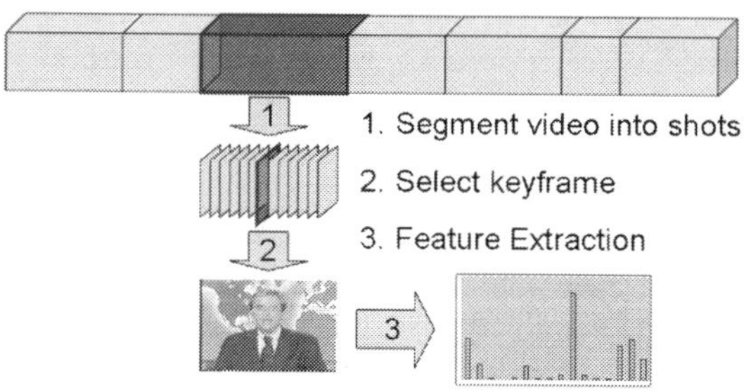

Fig. 1. Video Feature Extraction

2.1 Temporal Segmentation

The video stream is automatically segmented into shots by detecting their boundaries. A shot is a continous video sequence taken from one camera. We identify shot boundaries by searching for rapid changes of the difference between colours histograms of successive frames, using a single threshold. It was shown in [12] that this simple approach performs rather well. The colours are represented in the IHS (intensity, hue, saturation) space, because of its suitable perceptual properties and the independence between the three colourspace components. A simple filtering process allows the reduction of the number of false positives, i.e. a set of two successive frames which belong to the same shot although the difference of their colour histograms exceeds the given threshold. The shots with an insufficient number of frames (usually less than 5), are ignored. However, the number of false positives does not have a great influence on our approach, since similar shots will be assigned to the same cluster, as discussed in the following.

2.2 Feature Extraction

In order to obtain good clustering a reasonable representation of the video segments is necessary. For each shot, one keyframe is extracted (we choose the median frame of a shot) along with its colour histograms using a specified colour space. The system currently supports IHS, HSV , and RGB colour models. Apart from a global colour histogram, histograms for certain regions of the image are also extracted. Four regions are defined, the top, bottom, left, and right rectangles of the image. Each histogram is described by a numerical feature vector. In order to be able to train a self-organizing map with the resulting set of vectors, we use an early fusion strategy by merging all histogram vectors into a single virtual vector, which is then used to define each shot.

3 Structuring with Growing Self-Organizing Maps

Self-organizing maps (SOMs) [13] are artificial neural networks, well suited for clustering and visualization of high dimensional information. In fact, they map high-dimensional data into a low dimensional space (two dimensional map). The map is organized as a grid of symmetrically connected cells. During learning, similar high dimensional objects are progressively grouped together into the cells. After training, objects that are assigned to cells close to each other, in the low-dimensional space, are also close to each other in the high-dimensional space.

Our map is based on cells organized in hexagonal form, because the distances between adjacent cells are always constant on the map (see Fig. 2). In fact, in the traditional rectangular topology the distance would depend on whether the two cells are adjacent vertically (or rather horizontally) or diagonally.

The neuronal network structure of SOMs is organized in two layers (Fig. 2). The neurons in the input layer correspond to the input dimensions, here the feature vector describing the shot. The output layer (map) contains as many neurons as clusters needed. The connection weights between input and output layer of neural network encode positions in the high-dimensional feature space. Every unit in the output layer represents a prototype, i.e. here the center of a cluster of similar shots.

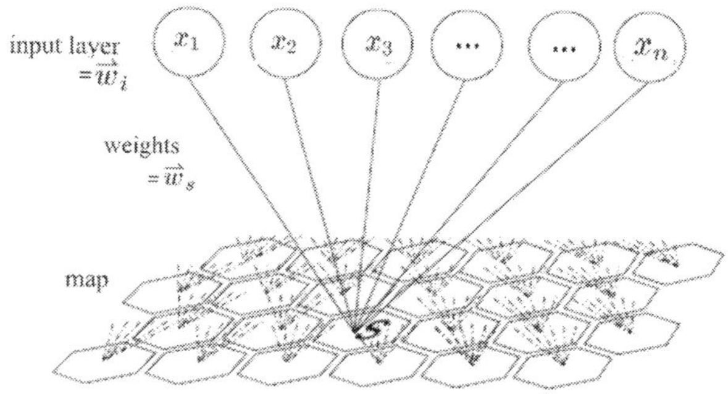

Fig. 2. Structure of a Hexagonally Organized Self-Organizing Map: The basic structure is an artificial neural network with two layers. Each element of the input layer is connected to every element of the map.

Before the learning phase of the network, the two-dimensional structure of the output units is fixed and the weights are initialized randomly. During learning, the sample vectors are repeatedly propagated through the network. The weights of the most similar prototype w_s (winner neuron) are modified such that the prototype moves towards the input vector w_i. To preserve the neighbourhood relations, prototypes that are close to the winner neuron in the two-dimensional structure are also moved in the same direction. The strength of the modification decreases with the distance from the winner neuron. Therefore, the weights w_s of the winner neuron are modified according to the following equation:

$$\forall i : w_s^{'} = w_s + v(c, i).\delta.(w_s - w_i) \quad (1)$$

where δ is a learning rate. By this learning procedure, the structure in the high-dimensional sample data is non-linearly projected to the lower-dimensional topology.

Although the application of SOMs is straightforward, a main difficulty is defining an appropriate size for the map. Indeed, the number of clusters has to be defined before starting to train the map with data. Therefore, the size of the map is usually too small or too large to map the underlying data appropriately, and the complete learning process has to be repeated several times until an appropriate size is found. Since the objective is to structure the video data, the desired size depends highly on the content. An extension of self-organizing maps that overcomes this problem is the growing self-organizing map [6].

3.1 The Growing Self-Organizing Map

The main idea is to initially start with a small map and then add during training iteratively new units, until the overall error – measured, e.g., by the inhomogeneity of objects assigned to a unit – is sufficiently small. Thus the map adapts itself to the structure of the underlying data collection. The applied method restricts the algorithm to add new units to the external units if the accumulated error of a unit exceeds a specified threshold value. This approach simplifies the growing problem (reassignment and internal-topology difficulties) and it was shown in [6] that it copes well with the introduction of data in low and high dimensional spaces. The way a new unit is inserted is illustrated in Fig. 3.

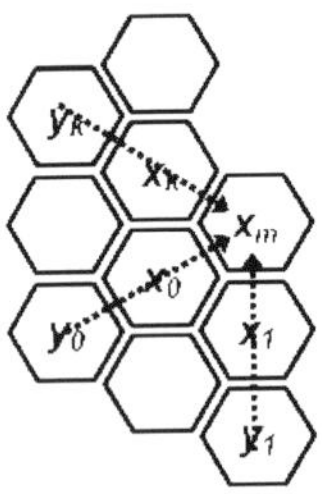

x_i, y_i: weight vectors
x_k: weight vector of unit with highest error
m: new unit
α, β: smoothness weights (defaults: $\alpha \approx 0.2, \beta \approx 0.5$)
Computation of new weight vector for x_m for m:

$$x_m = \left[x_k + \alpha * (x_k - y_k) + \sum_{i=0, i \neq k}^{n} (x_i + \beta * (x_i - y_i)) \right] * \frac{1}{n+1}$$

Fig. 3. Insertion of a new Unit: When the cumulated error of a cell exceeds a threshold, a new unit x_m is added to the map. It is placed next to the unit with the highest error at the border of the map.

3.2 Similarity Between Shots

As in all clustering algorithms the main problem is how to model the similarity between the objects that are going to be grouped into one cluster. We model the difference of two video sequences with the Euclidean distance of the two low-level feature vectors that were extracted from the video. However, this distance does not necessarily correspond to a perceived distance by a human. In addition, these features represent only a small part of the video content. In any case, there remains a semantic gap between the video content and what we see on the map. However, since for this first prototype study we are mainly interested in the capabilities of the SOMs, this approach seems sufficient, since we are not looking at grouping the shots "purely semantically", but rather at extracting a structure based on visual similarities.

4 Visualization

Our system represents a video shot by a single keyframe and constructs higher level aggregates of shots. The user has the possibility to browse the content in several ways. The basic idea is to provide as much information as possible on a single screen, without overwhelming the user. Therefore, we combined elements

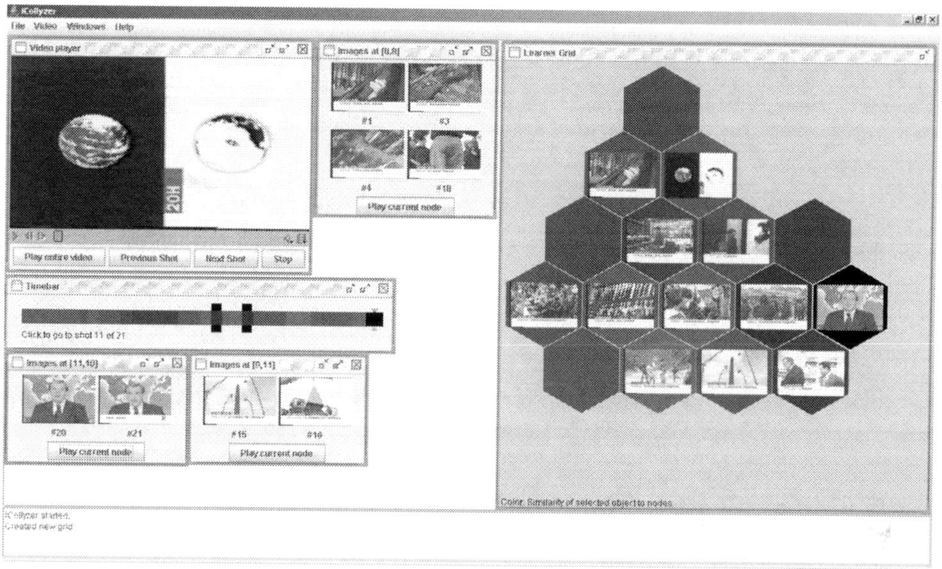

Fig. 4. Screenshot of the Interface: The player in the top left corner provides video access on the lowest interaction level. The time bar and shot list provide an intermediate level of summarized information while the growing self-organizing map on the right represents the highest abstraction level.

providing information on three abstraction levels as illustrated in Fig. 4. First, there is an overview over the whole content provided by the self-organizing map window, described in section 4.1. On each cell, the keyframe of the shot that is the nearest to the cluster centre, i.e. the most typical keyframe of a cluster, is displayed. The second level consists of a combined content-based and time-based visualization. Furthermore, a list of shots is provided for each grid cell (see Sect. 4.2) and a control (see Sect. 4.3) derived from the time-bar control helps to identify content that is similar to the currently selected shot.

4.1 Self Organizing Map Window

The self-organizing map window (see Fig. 4.1) contains the visual representation of the self organizing map where the clusters are represented by hexagonal nodes. The most typical keyframe of the cluster is displayed on each node. If there are no shots assigned to a special node no picture is displayed. The background colors of the grid cells are used to visualize different information about the clusters. After learning, shades of green indicate the distribution of keyframes: the brightness of a cell depends on the number of shots assigend to it (see Fig. 4.1a). Later, the background color indicates the similarity of the cluster to a selected shot as described below. For a thorough discussion of coloring methods for self-organizing maps we like to refer to [14].

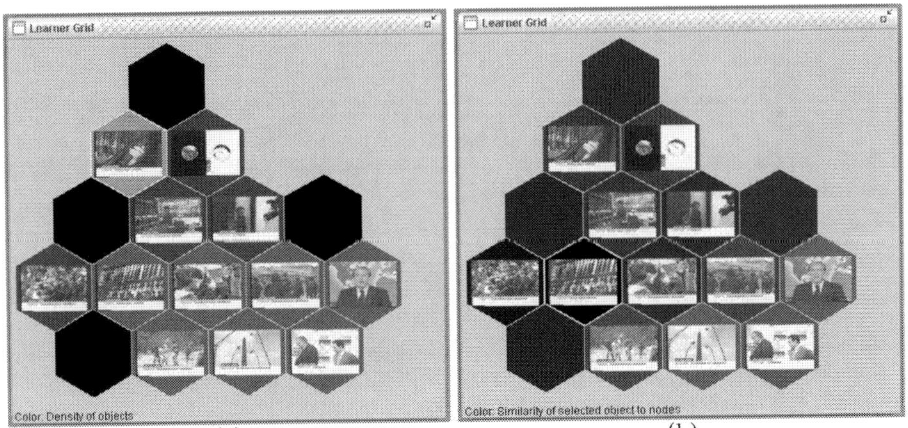

(a) (b)

Fig. 5. Growing self-organizing map: (a) After training. The brightness of a cell indicates the number of shots assigned to each node. On each node the keyframe of the shot with the smallest difference to the cluster center is displayed. (b) After a shot has been selected. The brightness of a cell indicates the distance between each cluster center and the keyframe of the chosen shot. Notice that sequences in adjacent cells are similar as intended.

After this first display, a click on a cell opens a list of shots assigned to the specific cell (see Sect. 4.2). The user can then select a specific shot from the list. As a result, the colour of the map changes to shades of red (see Fig. 4.1b). Here, the intensity of the colour depends on the distance between the cluster centres and the actually selected shot and thus is an indicator for its similarity. For instance, if we select a shot that has the visual characteristics A and B, all the nodes with these characteristics will be coloured in bright red and it will progressively change towards a darker red based on the distance. This implies in particular that the current node will be automatically coloured in bright red, since by construction all of its elements are most similar. In fact, objects that are assigned to cells close to each other, in the low-dimensional space, are also close to each other in the high-dimensional space.

But this does not mean that objects with a small distance in the high-dimensional space are necessarily assigned to cells separated by a small distance on the map. For instance, we can have on one side of the map a node with shots with the characteristics A and on another the ones with B. And then in one of both, let's say A-type, a shot with characteristics A and B. Thanks to the visualisation schema presented above, starting with a shot A&B, located in a node A, we will easily identify the nodes in which all the shots are rather of type B. This improves significantly the navigation possibilities provided by other clustering schemas.

From user interaction perspective the map is limited to the following actions: select nodes and communicate cluster assignment and colour information to the time bar. Nevertheless it is a very powerful tool which is especially useful for presenting a structured summarization of the video to the user.

4.2 Player and Shot List

The player is an essential part of every video browsing application. Since the video is segmented into shots, functionalities were added especially for the purpose of playing the previous and the next shot.

A shot list window showing all keyframes assigned to a cell (Fig. 4) is added to the interface every time a user selects a node from the map. Multiple shot lists for different nodes can be open at the same time representing each shot by a keyframe. These keyframes correspond to the actual selected node in the self-organizing map, as described in section 4.1. When clicking on one of the keyframes, the system plays the corresponding shot in the video. The button for playing the current node is a special control, which results in a consecutive play operation of all shots corresponding to the selected node, starting with the first shot. This adds another temporal visualization method to the segmented video.

4.3 Time Bar

The time bar of our prototype (Fig. 6) reintroduces the temporal aspect into the interface which we ignored in the SOM. The colours of the self organizing map are projected into the temporal axis. With this approach, it is possible to see within the same view the information about the similarity of keyframes and the corresponding temporal information. A green double arrow displays the current temporal position within the video. Additionally, there are black extensions on the time bar at the places where the corresponding shots of the selected node can be found. There are two interaction possibilities with our time bar. By clicking once on any position, the system plays the corresponding shot. Clicking twice, it forces the self organizing map to change the currently selected node to the one corresponding to the chosen frame. And therefore, the background colour schema of the map is recomputed.

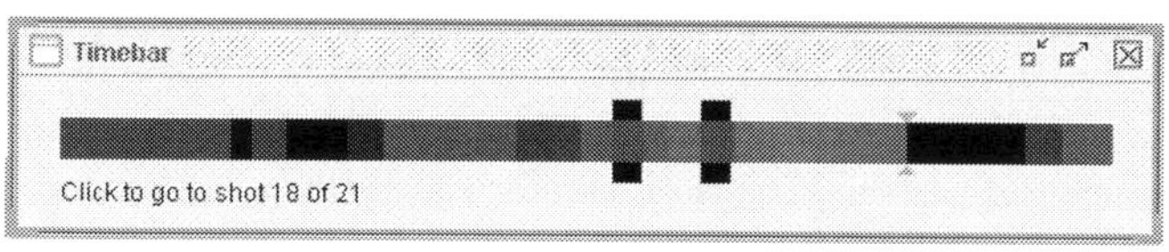

Fig. 6. Time Bar Control: The time bar control provides additional information. The brightness of the colour indicates the distribution of similar sequences on the time scale. Around the time bar, black blocks visualize the temporal positions of the shots assigned to the currently selected node. Finally, the two arrows point out the actual player position.

5 User Interaction

The four components presented above are integrated into one single screen (Fig. 4) providing a structured view of the video content. The methods for user interaction are hierarchically organized (Fig. 7). The first layer is represented by the video viewer. The shot lists and timebar visualize the data on the second layer. The self-organizing map provides the highest abstraction level.

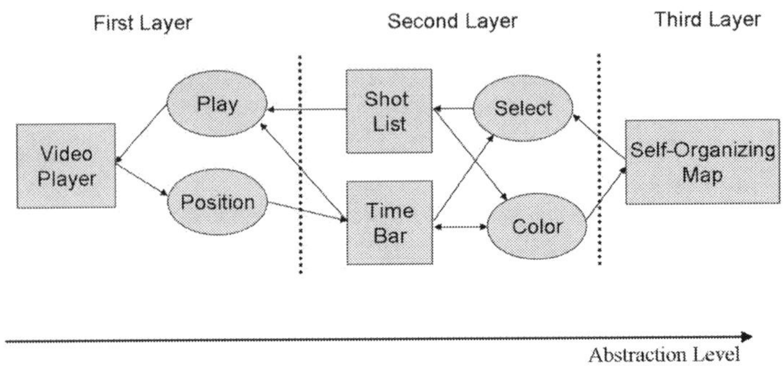

Fig. 7. User Interactions: This figure illustrates the main user interactions possible with our system. All listed elements are visible to the user on one single screen and always accessible thus providing a summarization on all layers at the same time.

The self-organizing map is situated in the third layer. The user can select nodes and retrieve their content i.e. the list of corresponding keyframes. The time bar is automatically updated by visualizing the temporal distribution of the corresponding shots when the current node is changed. Thus, a direct link from the third to the second layer is established. Furthermore the user views at the same time the temporal distribution of similar shots inside the whole video on the time bar, after a certain shot has been selected. In the other direction selecting shots using both the time bar and the list of keyframes causes the map to recompute the similarity values for its nodes and to change the selected node. The colour of the grid cells is computed based on the distance of its prototype to the selected shot. The same colours are used inside the time bar. Once the user has found a shot of interest, he can easily browse through similar shots using the colour indication on the time bar or map.

Notice that the first layer cannot be accessed directly from the third layer. Different play operations are activated by the time bar and shot lists. The player itself gives feedback about its current position to the time bar. The time bar is actualized usually when the current shot changes.

All visualization components are highly interconnected. In contrast to other multi-layer interfaces, the user can always use all provided layers simultaneously within the same view. He can select nodes from the map, keyframes from the list or from the time bar, or even nodes from the time bar by double-clicking.

6 Conclusions

The structuring and visualization of video information is a complex and challenging task. In this paper we presented a tool for content-based video navigation based on a growing self-organizing map. Our interface allows the user to browse the video content using simultaneously several perspectives, temporal as well as content-based representations of the video. Combined with the interaction possibilities between them this allows efficient searching of relevant information in video content.

References

1. Lee, H., Smeaton, A.F., Berrut, C., Murphy, N., Marlow, S., O'Connor, N.E.: Implementation and analysis of several keyframe-based browsing interfaces to digital video. In Borbinha, J., Baker, T., eds.: Lecture Notes in Computer Science. Volume 1923. (2000) 206–218
2. Girgensohn, A., Boreczky, J., Wilcox, L.: Keyframe-based user interfaces for digital video. Computer **34**(9) (2001) 61–67
3. Lin, X., Marchionini, G., Soergel, D.: A selforganizing semantic map for information retrieval. In: Proc. of the 14th Int. ACM/SIGIR Conference on Research and Development in Information Retrieval, New York, ACM Press (1991) 262–269
4. Kohonen, T., Kaski, S., Lagus, K., Salojärvi, J., Honkela, J., Paattero, V., Saarela, A.: Self organization of a massive document collection. IEEE Transactions on Neural Networks **11**(3) (2000) 574–585
5. Roussinov, D.G., Chen, H.: Information navigation on the web by clustering and summarizing query results. Inform. Proc. & Management **37**(6) (2001) 789–816
6. Nürnberger, A., Detyniecki, M.: Visualizing changes in data collections using growing self-organizing maps. In: Proc. of Int. Joint Conference on Neural Networks (IJCANN 2002), IEEE (2002) 1912–1917
7. Laaksonen, J., Koskela, M., Oja, E.: Picsom: Self-organizing maps for content-based image retrieval. In: Proc. of IEEE Int. Joint Conference on Neural Networks (IJCNN'99), Washington, DC, IEEE (1999)
8. Nürnberger, A., Klose, A.: Improving clustering and visualization of multimedia data using interactive user feedback. In: Proc. of the 9th Int. Conf. on Inform. Proc. and Management of Uncertainty in Knowledge-Based Systems. (2002) 993–999
9. Bimbo, A.D.: Visual Information Retrieval. Morgan Kaufmann (1999)
10. Miene, A., Hermes, T., Ioannidis, G.: Automatic video indexing with the advisor system. In: Proc. Int. Works. on Content-Based Multimedia Indexing, Brescia, Italy (2001)
11. Nürnberger, A., Detyniecki, M.: Adaptive multimedia retrieval: From data to user interaction. In Strackeljan, J., Leivisk, K., Gabrys, B., eds.: Do smart adaptive systems exist - Best practice for selection and combination of intelligent methods. Springer-Verlag, Berlin (2005)
12. Browne, P., Smeaton, A.F., Murphy, N., O'Connor, N., Marlow, S., Berrut, C.: Evaluating and combining digital video shot boundary detection algorithms. In: Proc. Irish Machine Vision and Image Processing Conf., Dublin, Ireland (2000)
13. Kohonen, T.: Self-Organizing Maps. Springer-Verlag, Berlin Heidelberg (1995)
14. Vesanto, J.: SOM-based data visualization methods. Intelligent Data Analysis **3** (1999) 111–126

Self-Organizing Neural Networks for Signal Recognition

Jan Koutník and Miroslav Šnorek

Department of Computer Science and Engineering, Czech Technical University, Prague, Karlovo nam. 13, Czech Republic
koutnij, snorek@fel.cvut.cz

Abstract. In this paper we introduce a self-organizing neural network that is capable of recognition of temporal signals. Conventional self-organizing neural networks like recurrent variant of Self-Organizing Map provide clustering of input sequences in space and time but the identification of the sequence itself requires supervised recognition process, when such network is used. In our network called TICALM the recognition is expressed by speed of convergence of the network while processing either learned or an unknown signal. TICALM network capabilities are shown on an experiment with handwriting recognition.

1 Introduction

Processing of sequentially dependent data became very important in modern history of artificial neural networks. An overview of neural networks for temporal data processing is given in [2] and [11]. The motivation is that many data produced by measurement of real processes involve time property. Such data we call signals. The signals are produced by dynamic systems. Our existence in spatio-temporal environment forces us to take the temporal property of signals into account. A processing system would either ignore the temporal property and process only a static form of the data, so it will not exploit all information stored within the data. Or it will use the data as a signal, exploit temporal property and then it can benefit in various data processing task such as building of the model, performing recognition of signals or perform prediction task.

Let us have several examples of input data. In handwriting recognition processing of trajectories - data sequences are more effective than processing of handwriting scanned images. Writer's hand dynamics can serve more precise data for e.g. in signature recognition task because one writer always starts the first letter of subscription from the same side. Handwriting data were used in this work as a testing data. Figure 1 contains 20 cases of four handwritten letters from all letters from *a* to *z* used in experiments.

There are more examples where signal recognition takes place instead of static data recognition. Second good example are medical data. The EEG and ECG signals have temporal property. Processing of such signals requires usage of a suitable tool. Other examples can be found in speech processing, video processing etc.

Our objective is a recognition task. Building of the signals' model only is not satisfactory. A good model can be used for recognition as it is in case of Hidden Markov Models, which are used for signal recognition with a success [12]. There are several paradigms in the field of artificial neural networks, which can be used for signal processing as well. Such networks contain recurrent connections between cells, which is one of the conditions for good signal processing.

S. Kollias et al. (Eds.): ICANN 2006, Part I, LNCS 4131, pp. 406–414, 2006.

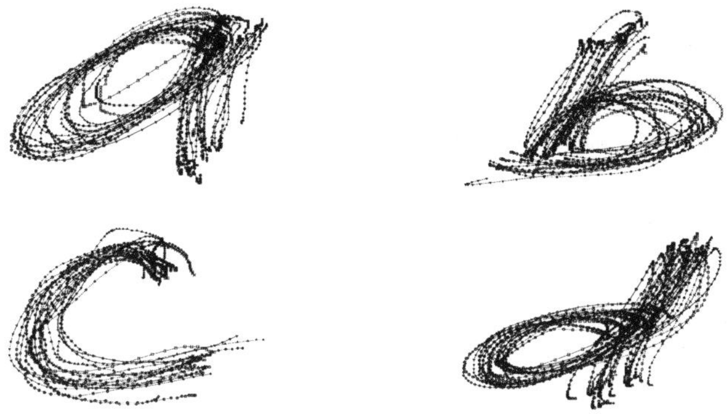

Fig. 1. Experimental data. Our objective is to teach the self-organizing neural network to recognize handwriting letters. The figure contains 4 examples of 20 cases of handwritten letters *a b c* and *d*.

Since the recurrent connections allow the network to deal with temporal information stored within a signal, the recurrent neural network models started to appear in 90's. There are supervised models like Jordan and Elman [3] networks, spiking neural networks [9], which cover the temporal property with leaky integration of an input and transmission of temporal activation pulses and recent Long Short-Term Memory described in [4]. Self-organizing approach to sequence processing can be exploited as well. One of such important neural networks is the Recurrent Self Organizing Map (RSOM) [14] because it is an obvious extension of Kohonen's Self Organizing Map (SOM) [5] with recurrent connections. Our network explained and tested in sections 2 and 3 has several properties common with RSOM, especially from the point of view of the self-organization, thus a brief overview of the RSOM is given.

1.1 Recurrent Self-Organizing Map

The Recurrent Self-Organizing Map (RSOM) [14] has been derived from feed-forward variant of Kohonen's clustering neural networks called SOM [5]. Behavior of the RSOM model can be compared to behavior of Hidden Markov Models, because the RSOM creates an automaton model for and input sequence, but as mentioned in [14] the state map cannot be competing to Hidden Markov Models since there is no unique winner (best matching unit) for a particular sequence. In other words *the RSOM itself is not capable of recognition of sequences*. But it does not mean that the RSOM could not be a part of a signal recognition system. It can be, but it plays a role of a preprocessing neural network. The state map does not provide sequence classification. The map only reduces dimension of the original signal. The RSOM can be exploited as a preprocessing tool, which reduces input space dimension and transforms input data to a sequence of code words.

One of the ideas is to create hierarchical structure of RSOM networks where the bottom network processes the input signal directly and a sequence of weight vectors (code words) of the best matching units in the bottom layer is used as a signal passed to the upper layer. The same idea was used in hierarchical SOM networks [8]. Hierarchical RSOM appeared in 2005 [1].

1.2 Self Organization and Recognition

A question that rose with RSOM – how to find the winner (best matching unit) in RSOM has to be solved more generally. The best matching unit is found for each data vector in the input signal. So we have the input signal transformed to same number of code words. Since there is normally less code words (neurons) in the RSOM map than data vectors in the input signal, the sequence of code words can be shortened by removing consequent duplicate code words. Obviously, for the recognition of a signal we have to look at the whole sequence of the code words. We could not distinguish between signals from single best matching units. We have to find another output of the model, which carries information about the recognition of the signal. We introduce a self-organized neural network that produces the desired output.

This paper is organized as follows. Section 2 briefly introduces the Categorizing and Learning Module and its proposed extension to the TICALM network. Section 3 contains experimental results, which are discussed in section 4. Section 5 concludes the contribution.

2 Temporal Information Categorizing and Learning Map

The Temporal Information Categorizing and Learning Map (TICALM) introduced in [7] solves the sequence recognition problem. In comparison with RSOM it has several differences and commons. The spatio-temporal map present in RSOM is split into two maps. The spatial map and temporal map exist separately in TICALM. The TICALM network is self-organized. The main difference is that the best matching unit is searched by lateral inhibition. This principle was used in early SOM models but in current SOM it was replaced by winner-take-all principle best matching unit search. This was because e.g. only one best matching unit has to win the competition, which the lateral inhibition in SOM did not ensure. The lateral inhibition runs in iterations in TICALM. One and only one best matching unit wins the competition. Iterative behavior of the TICALM may be taken as a performance drawback but it provides our needed feature. The iterative process does the recognition missed in RSOM model.

The idea is simple. The TICALM model represents the input sequence better, the less iterations are needed to find the best matching unit. The recognition of a signal is proportional with a speed of convergence (number of iterations) of TICALM.

2.1 Categorizing and Learning Module

The TICALM is a next generation of Categorizing and Learning Module (CALM) [10]. We briefly introduce original CALM network. The CALM sorts static input vectors into categories. It uses scalar product of input weights and input and is suitable for recognition of high dimensional input patterns. The module constructs spatial map of the input space in its input synapses. It contains a subsystem (consists of two neurons), which controls the learning rate and distinguishes between new and already learned input vector. The CALM works in iterations in comparison with conventional SOM. The best matching unit consists of a pair of neurons, which support each other and compete among other neuron pairs. Figure 2 shows a structure of the original CALM with extension to TICALM, where thick lines express the extension to TICALM.

There is a layer of R-neurons, which is connected to input vector using learned input inter-synapses. Other intra-modular synapses have constant weights, which were set up by evolutionary algorithm [10]. Each R-neuron has its matching V-neuron. R-neurons are fully connected to V-neurons layer using *cross* synapses. The *cross* synapses setup induces a topological organization of the spatial map in the CALM. The map organization can be arbitrary, it is not restricted to be rectangular or linear mesh as in SOM. Weights (*flat*) between V-neurons realize competition among V-neurons. The sensitivity subsystem consists of A-neuron and E-neuron. The system is activated during high competition activity in V-neurons layer. It means that the input vector is not learned yet. The Sensitivity subsystem starts to support new randomly chosen R-neuron, which is chosen as a new category representant. A hard competition can be recognized by longer module convergence. Practically, unknown input vector is learned in about 30 iterations. Best matching unit for already learned input vector is found within 5 iterations.

The original CALM was improved in several recent works. First modification that faced signal processing was introduced in [13]. CALM modules were connected into a network using time-delay inter-modular connections that allowed to process sequences. Addition of time-delay connections was inspired by previous models of feed-forward networks for processing of sequences.

2.2 Temporal Information Categorizing and Learning Map

The current state of improved CALM called TICALM [6] is depicted in figure 2. First, notice an added M-neurons layer. These neurons (*memory neurons*) are used to hold V-neurons activity after iterations over previous input vector has ended. The activity of V-neurons is copied to M-neurons after the end of the iteration and is diminished by a coefficient.

The M-neurons are connected back to R-neurons via *mem* synapses. These synapses are parallel to *cross* synapses that carry information from V-neurons. The *mem* synapses create the temporal map in TICALM.

The temporal map in *mem* synapses has also another important property. Besides that it drives the network to be time-dependent, it can be easily transformed to a probabilities of transitions of a probabilistic automaton, which can be *hidden* in a Hidden Markov Model [6]. The *mem* are being learned during the learning process. So not only inter-modular input synapses are being updated during networks iterations.

The TICALM uses a different input metrics than the original CALM. The CALM used a scalar product. the scalar product of an input vector and input weights could not distinguish between linear dependent input vectors. So it is not suitable for scalar input networks (one dimensional signals). We used Euclidean metrics as it is used in SOM. Other possibility could be also the Manhattan metrics used in RSOM. Other neurons use standard scalar product as defined in original CALM.

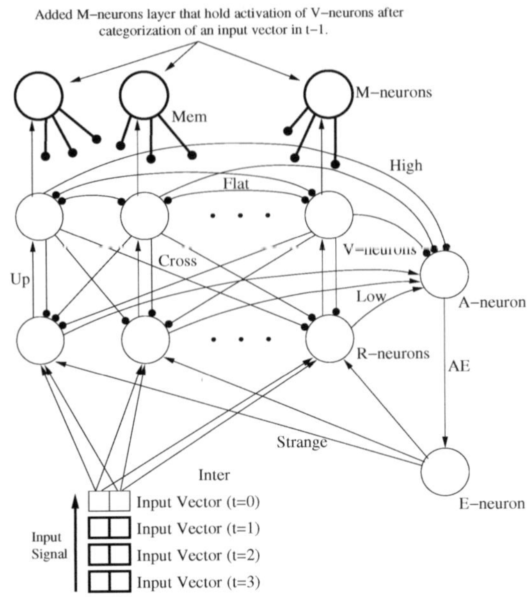

Fig. 2. Temporal Information Categorizing and Learning Map. The *mem* synapses are drawn shortened, they create full connection from M-neurons layer to R-neurons layer. The TICALM is connected to 4 dimensional input signal.

Function of all neurons in TICALM is expressed by excitation of neurons. Output neuron value is calculated using activation function a_i:

$$a_i(t+1) = \begin{cases} (1-k)a_i(t) + \frac{e_i}{1+e_i}(1-(1-k)a_i(t)), & e_i \geq 0 \\ (1-k)a_i(t) + \frac{e_i}{1-e_i}(1-k)a_i(t), & e_i < 0 \end{cases} \tag{1}$$

where $a_i(t)$ denotes neuron activation in previous iteration step, k is an activation decay factor and e_i is a neuron excitation:

$$e_i = \sum_j w_{ij} a_j(t) \tag{2}$$

where a_j is a selected neuron input and w_{ij} is a proper intra-modular weight except R-neurons where the excitation contains another part - input vector measured by Euclidean metrics:

$$e_i = l - \sum_j (a_j(t) - w_{ij})^2 \tag{3}$$

where l stands for number of input synapses (input vector dimension) in the case that input signal is transformed to $< 0, 1 >$.

Instead of modified Grossberg rule for learning inter-modular synapses used in CALM we use following rule for inter-modular synapses:

$$\Delta w_{ij}(t+1) = \mu_t a_i(a_j - w_{ij}(t)) \tag{4}$$

where $w_{ij}(t)$ is an input weight in time previous iteration and μ_t is a learning coefficient calculated as follows:

$$\mu_t = d + w_{\mu E}(1 - \frac{a_E - 0.5}{0.25}) \qquad (5)$$

where d is a constant with a small value determining base rate learning, $w_{\mu E}$ controls the influence of the E-neuron activation a_E. Afterwards, a new weight of inter-modular synapse in time $(t+1)$ is calculated:

$$w_{ij}(t+1) = max(min(w_{ij}(t) + \Delta w_{ij}(t+1), 1), 0) \qquad (6)$$

The *mem* synapses are learning according to the following rule:

$$\Delta m_{ij}(t+1) = \mu_t a_i(k_1 a_j + k_2 - m_{ij}) \qquad (7)$$

where m_{ij} is a *mem* synapse between j-th M-neuron and i-th R-neuron, k_1 and k_2 are *mem* synapses boundaries. A new *mem* weight is calculated:

$$m_{ij}(t+1) = m_{ij}(t) + \Delta m_{ij}(t+1) \qquad (8)$$

The *mem* weights are learned using Hebb's rule. Synapses between consequently winning best matching units are being forced other synapses are being diminished.

3 Experimental Results

A short experiment that reflects a temporal map creation using TICALM was described in [6]. Here, we describe another experiment, which demonstrates how the network can be applied to the recognition task.

The TICALM network recognizes a signal by the speed of it's convergence. It means that when the signal was learned into the TICALM, which means that appropriate spatial and temporal maps were created, the TICALM recognizes it by less iteration steps than if the signal in unknown for the network. So the experiment is divided into two phases. In the first phase (training) the input signal (a sequence of captured coordinates of a handwritten letter) is presented to the TICALM and the network uses the built-in learning algorithms to create spatial and temporal maps. In the second phase (recognition) the learning algorithms are switched off and the network processes the input signal in the same way.

In the recognition phase a number of iterations is measured after the best matching unit is found. If the number of the iterations is low, the signal was recognized otherwise it was not. This induces an important rule for the experiments. If we have several different input signals, we need the same number of TICALM modules to recognize them to be trained separately. After the training, in the recognition phase TICALM with the lowest number of iterations performed recognizes its matching input signal. In this way the TICALM modules are employed in recognition task in the same way as Hidden Markov Models are being used.

As a testing data a database of handwritten letters was used[1]. Figure 1 contains an example signal used in recognition. The signals were preprocessed to the length of 20 2-dimensional samples.

[1] Handwriting data (lowercase letters) are available for public at http://service.felk.cvut.cz/courses/36NAN/data/other/tablet/letters-20051104.zip

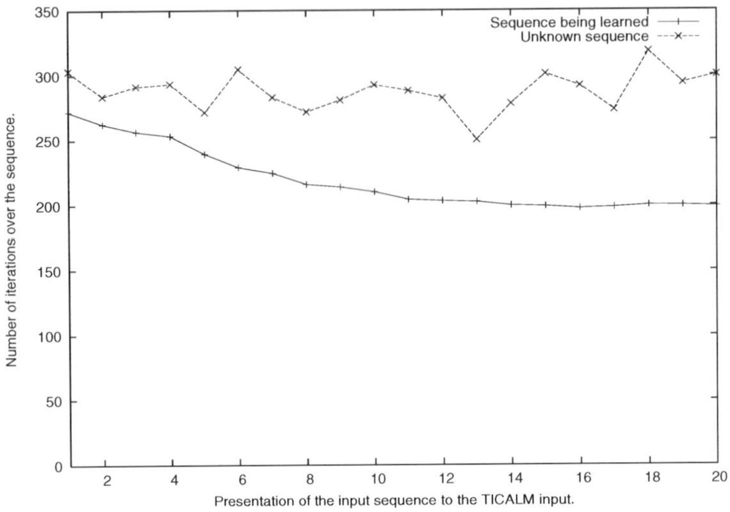

Fig. 3. Convergence of the TICALM during learning (solid curve) and presentation of an unknown input signal (dashed curve). Partially linear interpolation was used only for emphasizing purposes.

The data were captured using Wacom Tablet. We have chosen only first two dimensions of the data (x and y coordinates). Other dimensions like pressure, azimuth of the pen etc. were not used.

We trained 26 TICALM networks each to one handwritten letter. More than the overall performance of the TICALM the behavior of the module is important. Figure 3 shows average cumulative number of iterations over a signal during the learning and the recognition process. There are two curves in the graph. Solid curve shows the average convergence (number of iterations) of the network over the input signal. The average is made of 26 runs for all letters. Dashed curve express average number of iterations over an unknown input sequence. The average is calculated from presentation of rest 25 letters instead of a letter learned in the TICALM.

If the learning is switched on and the TICALM is in the initial state, it learns the input signal starting at approximately 250 iterations for 20 input samples. In the 20 consecutive presentations the number of iterations was diminished to approximately 200 iterations per the input signal. Amount of iterations over an unknown signal is approximately 1.5 times greater. A stochastic behavior of the network being presented with a non learned signal reflects the stochastic behavior of the TICALM sensitivity subsystem when an unknown input sequence is presented.

4 Discussion

Experiments demonstrate how the TICALM can recognize temporal sequences – signals. The TICALM can be directly used for recognition, so there is no need to build hierarchical structures as it is in case of original CALM network or the RSOM. Besides, the TICALM preserves several plausible features found in original CALM and adds new ones. Namely it uses different input space metrics that allows better discrimination of low dimensional input spaces. It preserves to be a self-organized map by creating a topology in *cross* synapses as used in original CALM model. It adds another map, which learns temporal property of the signal using self-organization and Hebb's rule. So the result of the training phase is not only the recognition system as for instance is the case of Back-propagation Through Time algorithm but we obtain a temporal map as a training phase side effect. The temporal map stored in *mem* synapses is important from the other point of view as well. It can be converted to a transition probability matrix of a Hidden Markov Model. The synapse weight values have meaning of transition probabilities.

Another difference to RSOM is that recurrent connections exist between all pairs of M-neurons and R-neuron in TICALM. The RSOM contains recurrent self-connections in neurons only so conversion of RSOM weights to a probabilistic model is not that obvious.

The future work deals with intensive testing if the TICALM network and search for a best setup, which leads to high speed convergence for modeled signals and long convergence for unknown signals, so the recognition process would be more stable. Another possibility is to have a layered structure of M-neurons instead of one M-neurons layer, so the short term memory in temporal map covers more than one historical steps of the TICALM. Such structure may be compared to k-order Hidden Markov Models.

5 Conclusions

We introduced a self-organized neural network called Temporal Information Categorizing and Learning Map, which is capable of recognition of signals that contain time property. The network itself constructs a self organized space map that reflects spatial property of a signal as well as temporal map that reflects temporal consequences contained in the signal. In comparison with unsupervised models like RSOM, which only constructs the spatio-temporal map, the TICALM performs a signal recognition task. The recognition of the signal or of a part of the signal is proportional to a speed of the TICALM convergence. A signal, on which the networks was trained to requires less of network steps than an unknown signal. The TICALM was tested on handwriting letters database, where it shows a capability of the letter recognition using the convergence speed measurement.

Acknowledgment

This work has been supported by the Ministry of Education of the Czech Republic under Project No. MSM6840770012.

References

1. Baier V., Motion Processing and Prediction With a Hierarchical RSOM based Model, in Proceedings of the IJCNN 2005.
2. Barreto G.A., Araujo A.F.R., Kremer S. A taxonomy for spatiotemporal connectionist networks revisited: The unsupervised case, *Neural Computation*, 15(6), pp. 1255-1320. 2003

3. Elman J.L., Finding structure in time. *Cognitive Science*, 14, pp. 179-211. 1990
4. Gers F.A., Schmidhuber J., LSTM recurrent networks learn simple context free and context sensitive languages. *IEEE Transactions on Neural Networks*, 12(6), pp. 1333-1340, 2001.
5. Kohonen T., *Self-Organizing Maps*, 3rd edition Springer-Verlag, 2001, ISBN 3-540-67921-9
6. Koutník J., Šnorek M., Neural Network Generating Hidden Markov Chain, in Adaptive and Natural Computing Algorithms - Proceedings of the International Conference in Coimbra. Wien: Springer, 2005, pp. 518-521. ISBN 3-211-24934-6.
7. Koutník J., Šnorek M., Single Categorizing and Learning Module for Temporal Sequences In Proceedings of the International Joint Conference on Neural Networks. Piscataway: IEEE, 2004, pp. 2977-2982. ISBN 0-7803-8360-5.
8. Lampinen J., Oja E., On Clustering Properties of Hierarchical Self-Organizing Maps. Journal of Mathematical Imaging and Vision, 2(2-3), pp. 261-272 1992.
9. Maass W., Networks of Spiking Neurons: The Third Generation of Neural Network Models, *Neural Networks* 10(9), pp. 1659-1671, 1997
10. Murre J.M.J., Phaf R.H., Wolters G., CALM: Categorizing and Learning Module, *Neural Networks*, Vol. 5, pp. 55-82, Pergamon Press 1992
11. Principe J., Euliano N.,Garani S., Principles and networks for self-organization in space-time, *Neural Networks*, Volume 15, Issues 8-9, pp. 1069-1083. 2002
12. Rabiner L.R., A Tutorial on Hidden Markov Models and Selected Applications in Speech Recognition, in Proceedings of the IEEE, vol.77,No.2,Feb 1989.
13. Tijsseling A.G., Sequential Information Processing using Time-Delay Connections in Ontogenic CALM networks, in IEEE Transactions on Neural Networks, Volume 16, Issue 1, Jan. 2005, pp: 145 - 159, ISSN: 1045-9227
14. Varsta M., *Self Organizing Maps in Sequence Processing*, Dissertation Thesis, Helsinki University of Technology, 2002, ISBN 951-22-6251-7

An Unsupervised Learning Rule for Class Discrimination in a Recurrent Neural Network

Juan Pablo de la Cruz Gutiérrez

Infineon Technologies AG, Am Campeon, 85579, Neubiberg, Germany
JuanPablo.delaCruz-Guiterrez@infineon.com

Abstract. A number of well-known unsupervised feature extraction neural network models are present in literature. The development of unsupervised pattern classification systems, although they share many of the principles of the aforementioned network models, has proven to be more elusive. This paper describes in detail a neural network capable of performing class separability through self-organizing Hebbian like dynamics, i.e., the network is able to autonomously find classes of patterns without the help from any external agent. The model is built around a recurrent network performing winner-takes-all competition. Automatic labelling of input data samples is based upon the induced activity pattern after presentation of the sample. Neurons compete against each other through recurrent interactions to code the input sample. Resulting active neurons update their parameters to improve the classification process. The learning dynamics are moreover absolutely stable.

1 Introduction

Recurrent neural networks are systems comprised of individual elements, the neurons, which perform some nonlinear operation based on their inputs and their recent history. When neurons are linked in a suitable fashion, in these networks, collective properties (behavior) emerge. These properties can be understood in terms of resolution of some computational task, for instance, working as a content addressable memory(CAM) [1]. The interplay between mutual interaction of neurons, and the nonlinear responsiveness to input stimuli, endows neural networks with the capability of robustly performing complex operations on large amounts of data. For example, it has been demonstrated how lateral interactions in a Hopfield network can be crafted to solve computational intensive tasks [2]. Besides, neural networks are also robust against noise, and parameter variations. A vast number of theoretical analysis of different models and applications in a wide number of fields have been developed. For the concrete case of pattern classification problems, there are a number of reviews, e.g. [3] and [4], addressing the basic questions under current research. They additionally provide references to significant contributions on the field.

Many unsupervised learning rules for neural networks, like the Principal Component Analysis (PCA) networks [11], the Bienenstock, Cooper and Munro

S. Kollias et al. (Eds.): ICANN 2006, Part I, LNCS 4131, pp. 415–424, 2006.

(BCM) rule [5], or independent component analysis (ICA) [6], exist for projection methods. Despite their differences, they share two principles which underlie most of the models in literature. The first one, and most recalled, is the Hebbian rule which has the problem of rendering the network unstable if no constraints are imposed on the dynamics. This rule states that synaptic efficacies among two neurons is enhanced, if presynaptic and postsynaptic activity are correlated in a certain fashion. The second states the dynamics of the competition among neurons, so that efficacies do not grow without control, and the neuron's response becomes tuned to some specific set of incoming stimuli. The topic is extraordinarily extensive, and the reader is referred to [7], and references therein.

The paper at hand presents a self-organizing neural network performing class separation, which incorporates those general principles. It is built around a recurrent network which adapts to statistics of input data, so that different group of neurons become selectively active to different data classes. Unlike other approaches, this model does not require the knowledge of class membership of input samples at any stage, and all elements in the network operate concurrently.

The paper is structured as follows: the next section reviews existing unsupervised learning rules. Next, the model is presented and described in detail. Finally, results are analyzed, and possible solutions for remaining open questions are outlined.

2 Neural Networks for Class Discrimination

The goal of every pattern classification system is: given a training data set $\{\mathbf{x}_i, y_i\}_{i=1}^{N}$, where N is the number of samples $\mathbf{x}_i$, and y_i are the labels indicating class membership, find the functional relationship between input samples and class labels. After training, the system must be able to correctly assign labels to new "unseen" samples, drawn from the same distribution of probability. The analysis of this formal problem has led to the development of different theories and techniques for its practical implementation, as well as applications. These techniques are based on the existence of the training set, from which they infer the statistical properties of the population from which they were drawn.

The focus of interest of this paper lies in unsupervised learning architectures, i.e., systems whose goal is to determine the statistical properties of input data without explicit feedback from a teacher. These models have been of interest in neurocomputation for the modelling of topographical maps in nervous systems [7,8,9,10]. Nevertheless, the principles employed in those models also lead to the implementation of tasks first formulated in a formal sense, like the already mentioned PCA [11] and ICA [6] networks, which implement projection methods. The attractiveness of these models is manifold: on one side no teacher signals are required, they adapt continuously to input data, and all elements operate concurrently. This last trait fostered the implementation of neural networks on silicon [2].

Feature extraction, dimensional reduction, classification and functional approximation algorithms are intimately related, as demonstrated in [12,13], where it is shown that they can be formulated in terms of a canonical generalized eigenvalue problem. This was indeed one of the facts that led to the formulation of the stochastic adaptative rules in [14] for the estimation of linear discriminant analysis (LDA). In both cases, previous knowledge about class labels and class and mixture means of data, respectively, is necessary.

The next section introduces a recurrent neural network which performs fully unsupervised class discrimation, through a self-organizing learning rules. For each sample, labels are assigned according to the resulting activity pattern after presentation. Recurrent connections develop in order to ensure that each neuron codes for one and only one class.

3 Presentation of the Model

The problem of linear class separation, sketched in figure 1, is the starting point for the formulation of the new model. In geometrical terms, linear class separation refers to the problem of finding a hyperplane such that samples from the two classes, C_+ and C_-, lie on opposite sides of the plane. In mathematical terms, this corresponds to finding the vector $\mathbf{w}$ and the parameter b, such that, $\mathbf{w}^T\mathbf{x} + b > 0, \forall \mathbf{x} \in C_+$, and $\mathbf{w}^T\mathbf{x} + b < 0, \forall \mathbf{x} \in C_-$. If d_+ and d_- are the shortest distances from the plane to the closest samples from classes C_+ and C_-, respectively, then, the optimal separating hyperplane will that which maximizes the amount $d_+ + d_-$. Given the labels for the data samples, finding optimal separating hyperplane is the core idea of Support Vector Machines (SVM) [15]. The name reflects the idea that the separating hyperplane is constructed upon a small number of vectors lying on the hyperplane itself, the so-called Support Vectors (SV). Our goal though, is to find such separating hyperplanes without knowledge of class membership.

Following this description, we can understand how to move on to the general case of an arbitrary number of classes. Given a sample vector of a given class, one would take two weight vectors and compare the projections of the sample vector on them, and then choose the winner. After carrying out the operation with all pairs of vectors, the sample should be assigned to the weight vector class who won the highest number of such competitions. All these comparisons correspond to a set of inequalities which must simultaneously hold. If we sum up them all up, the condition resulting for a neuron to be the winner is,

$$N\mathbf{w}_j^T\mathbf{x} - \sum_{l\neq j} \mathbf{w}_l^T\mathbf{x} + \sum_l w0_{jl} \begin{cases} > 0 \text{ if } j = \mathsf{argmax}_k \ \{\mathbf{w}_k^T\mathbf{x}\} \\ < 0 \text{ otherwise} \end{cases} \tag{1}$$

where $\mathbf{w}_j$ is the weight corresponding to the jth neuron, N is the total number of neurons in the network, and the matrix $\mathbf{W0} = (w0_{jl})_{N\times N}$ represents the margins of hyperplanes separating the classes. Condition 1 implies that every neuron is assumed to code for a different class. In the general case, where we have more neurons than classes, this is relaxed to allow a set of neurons win for the same class. This analysis suggests that this goal could be attained through a

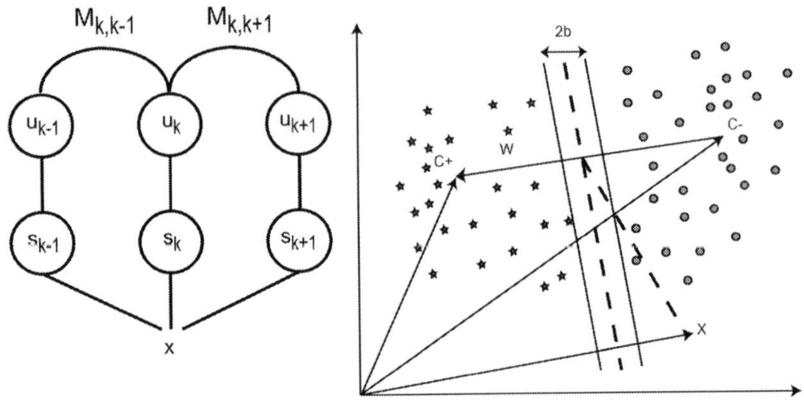

Fig. 1. *Left.* Schematic drawing of the model. Neurons in the first layer respond linearly to input pattern. Responsiveness of each neuron to input patterns is given by its weight vector and threshold, which develop according to the activity pattern that they evoke in the upper layer. See text for details. *Right.* Linear separation of two classes, whose samples are represented by stars and circles. Vectors $\mathbf{c}_+$ and $\mathbf{c}_-$ represent the mean of each class, C_+ and C_-, respectively. The parameter b represents the margin of the classifier. As we see, membership of a given sample is decided based upon the projection of the sample on the difference vector among the two class centers. If such projection exceeds the threshold (margin), b, then it will be assigned to the corresponding class.

recurrent network. The weight vector stands for the responsiveness of neuron to a given input sample. Neurons compare how well they match the input sample with the rest (recurrent interactions), and only the best tuned are chosen to represent the given class (remain active).

A recurrent neural model carrying out these ideas, is represented schematically in figure 1, and its behavior is described by the set of equations,

$$\tau_u \dot{u}_k + u_k = \sum_j M_{kj} f\left[u_j\right] + s_k(\mathbf{x}, t) - h \tag{2}$$

$$\tau_M \dot{M}_{ij} = \delta_{ij} - \sum_k M_{ik} \left(f\left[u_k\right] f\left[u_j\right] + \alpha \delta_{kj}\right) \tag{3}$$

$$\tau_w \dot{\mathbf{w}}_k = -\epsilon \mathbf{w}_k + f\left[u_k\right] \mathbf{x} \tag{4}$$

$$\tau_0 \dot{w}_{0k} = f\left[u_k\right] \left(f\left[u_k\right] - w_{0k}\right) \tag{5}$$

where $s(\mathbf{x}, t) = \mathbf{w}_k^T \mathbf{x} - w_{0k}$ is the input to the recurrent layer, h is the rest potential, f is a sigmoidal function, $f(x) = 1/\left(1 + \exp(\gamma(x - \theta))\right)$, u_k is the average membrane potential of the kth neuron, and α is a small positive constant whose role will be clarified later on. The time decay constants satisfy the relation $\tau_0 \gg \tau_w \gg \tau_M \gg \tau_u$. The rest of this section is a description of this set of equations.

The model is composed of two layers. A first layer receives a sample drawn at random from a density of probability $p(\mathbf{x})$. Every neuron responds to this pattern in a linear fashion,

$$s_k(\mathbf{x}) = \mathbf{w}_k^T \mathbf{x} + w_{0k} \tag{6}$$

where $\mathbf{w}_k$ is the weight vector which determines a neuron's responsiveness to the input vector, and w_{0k} is the threshold. The goal of the network is to tune the parameters of each neuron in such a way that the condition $\mathbf{w}_k^T \mathbf{x} + w_{0k} > 0$ holds only for samples drawn from a given class. When the sample belongs to a different class, this term becomes negative, thus inhibiting the postsynaptic neuron.

Weight vectors develop over time according to a simple Hebbian rule, equation 4, made up of two terms. A correlation term between presynaptic and postsynaptic activity, and a decay term whose goal is to keep the weights bounded. The parameter ϵ plays an important role in the dynamics, and depends on the chosen recurrent model, as it will be later explained. For this case, this value must be small, otherwise it overrides the Hebbian term. The thresholds evolve so that, only samples belonging to the corresponding class will evoke activity in the neuron. Adaptation of a neuron's weight vector and threshold occurs at the slowest time scale, since the system has to first make a decision on which neurons are representing a given input sample.

The outputs from these layers are fed into an upper layer. This layer consists of a Hopfield-like network, given by equation 2, which follows a winner-takes-all (WTA) dynamics. Functionality of recurrent networks is determined by its recurrent interactions. The study of the computational properties of a recurrent network in terms of its lateral interactions is, in general, a complex subject. Different approaches to its study, models, and open questions are reviewed in [7] and [16]. In our particular case, WTA dynamics ensures that only those neurons which best match the current input are finally active. The resulting activity in the steady state is thus understood as a labelling of data. Those neurons which remain active code for the same class, and adapt their parameters to further improve classification. Hopfield-like networks are well suited for this task, as they attain their steady state in a few neural time constant, τ_u, periods. This dynamics must necessarily take place at the fastest time scale, as the system needs to decide who the winner is, before it adapts its parameters.

The general form of the lateral interactions for such networks can be intuitively anticipated. The winners must strongly inhibit the other neurons, so that finally it is the only set still active. This is attained with a short-range excitatory and long-range inhibitory interaction kernel. The stable steady state of learning rule 3 has that kind of structure. The big differences in time scales of the different equations allow us to make use of the adiabatic approximation. In other words, we might average (equation 3) over the input ensemble yielding,

$$\tau_M \dot{\mathbf{M}} = I - \mathbf{M} \left(\mathbf{Q} + \alpha \mathbf{I} \right) \tag{7}$$

in matricial form, where $\mathbf{Q} = \langle f\left[\mathbf{u}\right] f\left[\mathbf{u}\right]^T \rangle$ is the correlation matrix of the neuron's average firing rate. Its steady state is $\left(\mathbf{Q} + \alpha \mathbf{I}\right)^{-1}$. The added term $\alpha \mathbf{I}$ has two effects on the learning dynamics. First it compensates for very low eigenvalues of the correlation matrix, which are most sensitive to noise. Second, it speeds up convergence to the steady state.

It is easy to prove that the stationary state is absolutely stable. Consider a perturbation which shifts the system away from it, $\mathbf{M} = (\mathbf{Q} + \alpha\mathbf{I})^{-1} + \delta\mathbf{M}$. Substituting back into 7, we find that the perturbation obeys the differential equation,

$$\tau_M \delta\dot{\mathbf{M}} = -\delta\mathbf{M}\left(\mathbf{Q} + \alpha\mathbf{I}\right) \tag{8}$$

Since $\mathbf{Q} + \alpha\mathbf{I}$ is a positive definite matrix, $\delta\mathbf{M}$ converges exponentially fast to the null matrix [17].

The dynamics of the recurrent weights must evolve at a faster pace than the weight dynamics, for, in order to achieve capability of discriminating classes, lateral interactions assure that only those neurons whose weight vectors are closely tuned to input patterns are active.

The adiabatic approximation can also be employed for the analysis of the development of threshold values in equation 5. Once the dynamics converges to the stationary state, samples coming from a given class activate a concrete set of neurons. Again, averaging over the input ensemble, equation 5 is approximated by,

$$\tau_0 \dot{w}_{0k} = \langle f\left[u_k\right]^2 \rangle - \langle f\left[u_k\right] \rangle w_{0k} \tag{9}$$

Since $f\left[u_k\right] \geq 0$, the dynamics is absolutely stable. Therefore, the stationary state of the total system is absolutely stable.

There is still an important caveat to point out about this sort of networks. Recurrent networks do not perform an ideal WTA dynamics, i.e., the most active neuron does not always correspond to the best tuned weight vector to the current input. First of all, its final state depends upon initial conditions, i.e., the steady state of the network after presentation of a new sample depends upon the previous state. Besides, if two close by neurons are strongly excited by input stimuli, the network will interpolate among the two, and choose one neuron in between. The first effect can be avoided during training by setting initial state to zero before presentation of a new sample. Nevertheless, for tasks like classification of time series, this trait might be advantageous, as it implicitly exploits temporal correlations present in the data. Second, one has to regard the effect of the extent of local excitatory connections upon the final solution. The shorter the excitatory range is, the higher the resolution a network can achieve.

4 Simulations

In order to first test the network toy data was created, so that possible side effects are avoided, and a good understanding of the dynamics is attainable. The data consisted of samples picked at random from three Gaussian distributions of mean $(\mu_x, \mu_y) = (2, 0), (-1, 2), (-2, -1.5)$ and variances $(\sigma_x, \sigma_y) = (0.5, 0.7), (1.3, 0.4), (0.4, 0.3)$, respectively. We chose a small network comprised of 10 neurons, so that results are easier to visualize. Simulations demonstrate that

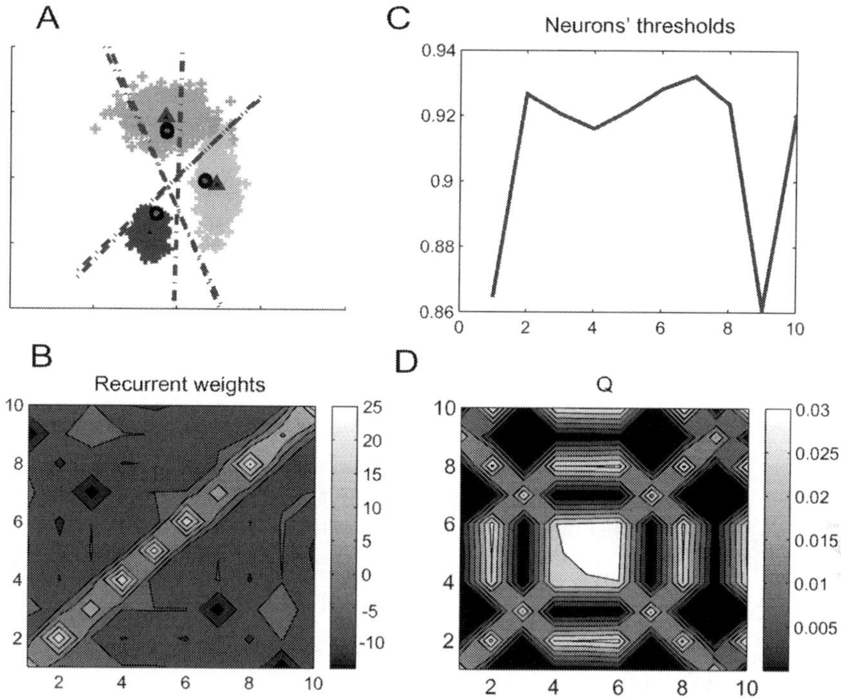

Fig. 2. Panel A shows the position of weight vectors in feature space and linear decision boundaries of each of the neurons ($\mathbf{w}_k^T\mathbf{x} + w_{0k} = 0$). See also figure 3 and caption for more details. Panel B displays the matrix representing the recurrent network connections. It corresponds to the recurrent weights of a winner-takes-all like network. Off-diagonal matrix elements are negative, which translates into inhibitory connections. Diagonal elements are positive, corresponding to a positive feedback of the neuron to itself. Panel C displays the threshold values of the neurons. Panel D shows the correlation matrix of average firing rates of neurons. The block structure of matrix corresponds to the different groupings of neurons, in terms of response selectivity to input patterns.

the system converges in a short time frame. Typically, the system requires the presentation of about 1500 presentations, to be close to the stationary state. Nevertheless, this magnitude also depends on the degree of overlap between classes. If the overlap is considerable, it can take some more time to some neurons to "decide" for one of the classes. This problem can be lessened by introducing some white noise, which does not seriously damage the network's performance since the equations 2-5 are structurally stable. Convergence takes place in two phases though. In a first fast phase, neurons become selective for a given class, and recurrent weights take display a short-range excitatory, long-range inhibition distribution. In a second phase, recurrent weights are further adapted, so that they fully decorrelate neurons responses, and threshold sharply mark the boundary class.

Figure 2 shows an state of the network close to stationary state. Plot B shows the recurrent weights. It displays the characteristics of a network performing WTA. Short range excitatory connections are around the diagonal of the matrix (positive feedback to itself) and strong long range inhibitory lateral connections. Plot C displays the activity in the recurrent network after presentation of a sample. Active neurons corresponds to those coding for its corresponding class. Plot A shows the location of weight vectors in input space. Broken lines represent the linear decision boundaries given by the relationship $\mathbf{w}_k^t\mathbf{x} + w_{0k} = 0$. Notice that the actual decision surfaces are a result of the WTA dynamics and are nonlinear.

Unlike other neural models employed for the modelling of neuron's response in early sensory stages, as retina cells [18] or cells in Lateral Geniculate Nucleus (LGN) [19], this model develops, in general, no topographical mapping in this toy example (see figure 3 and caption). The reason lies in that sensory inputs contain already correlations which are exploited by those models. For example, in the case of visual stimuli, close by spatial locations tend to have similar distribution in spatial frequency, contrast and luminosity. This fact is indeed taken into account in the mathematical formulation and analysis of those models.

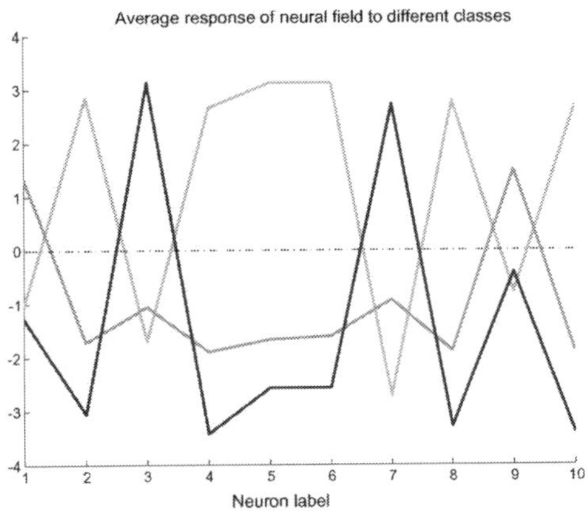

Fig. 3. Output average firing rate of every neuron in the recurrent network in response to class samples. Classes are labelled by different line colors. Classes are not mapped topographically on the neural layer, i.e. neighboring neurons do not represent, in general, classes that are spatially closest to each other. Every neuron codes samples originating from a single class. We chose a small number of neurons, so that it is easy to appreciate the tuning of every neuron in the figure.

5 Conclusions

We have presented a model built around a recurrent neural network which evolves following an unsupervised learning rule, and performs class separation. The simplicity of the combined rules makes the system amenable for mathematical analysis. Indeed, absolute stability of the stationary state of the network has been proved.

Some limitations of the system have already been commented during the description of the learning dynamics. For example, the dependency of the class discrimination capabilities on the lateral connections. When two classes overlap or are "very" close, the network tends to interpolate among the two classes. This can be compensated if we introduce some minimum separation between classes. This is however not straightforward to implement, since it introduces a baseline inhibition which specially affects the first stages of the dynamics.

This work is rooted in the model studied in [8], where it was demonstrated how, given a recurrent network with a Mexican hat like interaction kernel and fixed threshold, the system is able to develop a cortical map through self-organization. In our case, we let thresholds and recurrent weight change, so that the network adapts to differentiate among clusters of input samples with different statistical properties. We can analyze properties of resulting map, like its resolution, i.e. how well can this model tell different classes apart, and amplification, i.e. the size of the region excited by samples from each class. This analysis is left for future work.

Another aspect under current consideration is how to enforce a topographical representation of inputs, even when input patterns, as the one considered in the simulations, show no correspondence between location in input space and statistical properties, as is actually the case of natural stimuli, like visual stimuli. Different possibilities can be considered. On one side, one could introduce an additional condition in the development of recurrent weights, so that neighboring neurons have a stronger tendency to fire together. But this could, on the other hand, also be reinforced by introducing a second term in the dynamic equation for the weight vectors 4. Adding a term of the form $-\sum_l T_{kl}\mathbf{w}_l$, where $T_{kl} = T_{|k-l|}$, enforces the tendency of neighboring neurons to code for the same class.

References

1. Hopfield, J.J.: Neurons with graded response have collective computational properties like those of two-state neurons. Proc. Natl. Acad. Sci. USA **81** (1984)
2. Hopfield, J.J., Tank, D.W.: Neural computation of decisions in optimization problems. Biological Cybernetics **52** (1985) 141–152
3. Kulkarni, S.R., Lugosi, G., Venkatesh, S.S.: Learning pattern classification – a survey. IEEE Transactions on Information Theory **44**(6) (1998) 2178–2206
4. Zhang, G.P.: Neural networks for classification: A survey. IEEE Transactions on Systems, Man and Cybernetics–Part C: Applications and Reviews **30**(4) (2000) 451–462
5. Bienenstock, E.L., Cooper, L.N., Munro, P.W.: Theory for the development of neuron selectivity: orientation specificity and binocular interaction in visual cortex. The Journal of Neuroscience **2**(1) (1982) 32–48
6. Hyvärinen, A., Oja, E.: Independent component analysis by general non-linear hebbian-like learning rules. Signal Processing (1998)

7. Dayan, P., Abbott, L.F.: Theoretical Neuroscience. Computational Neuroscience. MIT press (2001)
8. Amari, S.I.: Dynamic stability for formation of cortical maps. In Arbib, M.A., Amari, S.I., eds.: Dynamic Interactions in Neural Networks: Models and Data. Volume 1 of Research notes in Neural Computing. Springer-Verlag (1989) 15–34
9. Ermentrout, B., Osan, R.: Models for pattern formation in development. In J., S., Champneys, A.R., Krauskopf, B., di Bernardo, M., Wilson, R.E., Osinga, H.M., Homer, M.E., eds.: Nonlinear Dynamics and Chaos. Where do we go from here ? Institute of Physics Publishing (2003) 321–347
10. Kohonen, T.: Self-organizing Maps. 3rd edn. Volume 30 of Information Sciences. Springer (2001)
11. Oja, E.: Principal components, minor components, and linear neural networks. Neural Networks **5** (1992) 927–935
12. Bie, T.D., Cristianini, N., Rosipal, R.: Eigenproblems in pattern recognition. In: Handbook of Computational Geometry for Pattern Recognition, Computer Vision, Neurocomputing and Robotics. E. bayro-corrochano edn. (Springer Verlag)
13. Burges, C.J.C.: Geometric methods for feature extraction and dimensional reduction. In Rokach, L., Maimon, O., eds.: Data Mining and Knowledge Discovery Handbook: A Complete Guide for Practitioners and Researchers. Kluwer Academic Publishers (2005)
14. Chatterjee, C., Roychowdhury, V.P.: On self-organizing algorithms and networks for class-separability features. IEEE Transactions on Neural Networks **8**(3) (1997) 663–678
15. Burges, C.J.C.: A tutorial on support vector machines for pattern recognition. Data Mining and Knowledge Discovery **2**(2) (1998) 121–167
16. Vogels, T.P., Rajan, K., Abbott, L.F.: Neural network dynamics. Annual Reviews Neuroscience **28** (2005)
17. Perko, L.: Differential Equations and Dynamical Systems. Texts in Applied Mathematics. Springer-Verlag (1991)
18. Atick, J.J., Redlich, A.N.: Predicting ganglion and simple cell receptive field organizations. International Journal of Neural Systems **1**(4) (1990) 305–315
19. Dong, D.W., Atick, J.J.: Temporal decorrelation: A theory of lagged and nonlagged responses in the lateral geniculate nucleus. Network: Computation in Neural Systems **6**(2) (1990) 159–178

On the Variants of the Self-Organizing Map That Are Based on Order Statistics

Vassiliki Moschou, Dimitrios Ververidis, and Constantine Kotropoulos*

Department of Informatics, Aristotle University of Thessaloniki
Box 451, Thessaloniki 54124, Greece
{vmoshou, jimver, costas}@aiia.csd.auth.gr

Abstract. Two well-known variants of the self-organizing map (SOM) that are based on order statistics are the marginal median SOM and the vector median SOM. In the past, their efficiency was demonstrated for color image quantization. In this paper, we employ the well-known IRIS data set and we assess their performance with respect to the accuracy, the average over all neurons mean squared error between the patterns that were assigned to a neuron and the neuron's weight vector, and the Rand index. All figures of merit favor the marginal median SOM and the vector median SOM against the standard SOM. Based on the aforementioned findings, the marginal median SOM and the vector median SOM are used to re-distribute emotional speech patterns from the Danish Emotional Speech database that were originally classified as being neutral to four emotional states such as hot anger, happiness, sadness, and surprise.

1 Introduction

The neural networks constitute a powerful tool in pattern recognition. They have been an active research area for the past three decades due to their wide range of applications [1]. The self-organizing map (SOM) establishes a mapping from an input data space onto a two or three dimensional lattice of nodes so that a number of topologically ordered and well defined neuron prototypes is produced. The nodes are organized on a map and they compete in order to win the input patterns [2]. The SOM is among the most popular neural networks. A number of 5384 related papers are reported in [4, 5].

We are interested in the class of SOM training algorithms that employ multivariate order statistics, such as the marginal median and the vector median [8]. These SOM variants as well as the standard SOM, that is trained with the batch algorithm (to be referred to as SOM hereafter), are applied to pattern clustering. The novel contribution of the paper is in the assessment of SOM training algorithms in clustering with respect to the accuracy, the average over all neurons mean squared error, and the Rand index. The superiority of the studied SOM variants against the SOM is demonstrated by experiments carried out using the well-known IRIS data. We also compare the studied SOM variants with the SOM in the re-distribution of emotional speech patterns from the Danish Emotional Speech (DES) database [15], that were originally classified as being

* This work has been supported by the project PYTHAGORAS II "Efficient techniques for information organization, browsing, and retrieval in multimedia", funded by the Greek Ministry of Education and the European Union.

S. Kollias et al. (Eds.): ICANN 2006, Part I, LNCS 4131, pp. 425–434, 2006.

neutral, into four emotional states such as hot anger, happiness, sadness, and surprise. The latter experiment is motivated by the following fact. There are emotional facial expression databases such as the Action-Unit coded Cohn-Kanade database [17] where the neutral emotional class is not represented adequately. Accordingly, facial expression feature vectors are not clustered to the neutral emotional class [18]. For the emotional speech databases, the utterances are regularly classified as neutral. Accordingly, when the neutral class is not represented in one modality it is difficult to develop multimodal emotion recognition algorithms (e.g. feature fusion algorithms). Frequently, the ground truth information related to emotions that is provided by the human evaluators is biased towards the neutral class. Therefore, the patterns classified as neutral might be needed to be re-distributed among the non-neutral classes to enable further experimentation.

The outline of this paper is as follows. Section 2 describes briefly the standard SOM and the batch training algorithm, as well as the SOM variants tested, namely the marginal median SOM (MMSOM) and the vector median SOM (VMSOM). In section 3, we define mathematically the evaluation measures employed, i.e. the accuracy, the average over all neurons mean squared error, and the Rand index. This section also describes the Kuhn-Munkres algorithm [13] and how it is used to calculate the SOM accuracy. In section 4, the data, we worked on, are discussed. In section 5, the experimental results for clustering the IRIS data using the SOM, the MMSOM, and the VMSOM are demonstrated. Furthermore, figures of merit are presented and discussed for the re-distribution of neutral patterns into four non-neutral emotional classes using the SOM, the MMSOM, and the VMSOM on the DES data. Finally, conclusions are drawn in section 6.

2 Self-Organizing Map and Its Variants

2.1 Self-Organizing Map (SOM)

The SOM forms a nonlinear mapping of an arbitrary D-dimensional input space onto a two or three dimensional lattice of nodes (the map). Each node is associated with a weight vector $\mathbf{w} = (w_1, w_2, \ldots, w_D)^T$ in the input space. The SOM is trained iteratively and learns the input patterns. The task of the self-organizing (unsupervised) learning lies to revealing the statistical properties of the input patterns, creating suitable representations for the features (i.e. weight vectors), and automatically creating new clusters. The map neurons compete each other in order to be activated by winning the input patterns. Only one neuron wins at each iteration and becomes the winner or the *best matching unit* (BMU) [7].

Let us denote by $\mathbf{x}_j$ the jth D-dimensional input feature vector and by $\mathbf{w}_i$ the ith D-dimensional weight vector. The first step of the algorithm is the weight vector initialization performed using the linear initialization algorithm. The weight vectors $\mathbf{w}_i$ define the *Voronoi tessellation* of the input space [1, 2]. Each Voronoi cell is represented by its centroid that corresponds to the weight vector $\mathbf{w}_i$. Each input pattern $\mathbf{x}_j$ is assigned to a Voronoi cell based on the nearest neighbor condition. That is, the BMU index, $c(j)$, of the input pattern $\mathbf{x}_j$ is defined by

$$c(j) = \arg\min_i \{\|\mathbf{x}_j - \mathbf{w}_i\|\} \tag{1}$$

where $\|.\|$ denotes the Euclidean distance. Accordingly, the SOM can be treated as a vector quantization method [6]. The most important step of the SOM algorithm is the adaptation of the neuron weight vectors. The neurons are connected to adjacent neurons by a neighborhood function dictating the structure of the map (topology). It determines how strongly the neurons are connected to each other [7]. In each training step, the neurons update depends on the neighborhood function, whose purpose is to correlate the directions of the weight updates of a large number of neurons around the BMU [20]. The larger the neighborhood, the more rigid the SOM. In our experiments, the neighborhood function used is the Gaussian. To update the winner neurons and their neighbors either a Least Mean Squared (LMS) type adaptation rule [1] or a batch algorithm can be employed. In this paper, we are interested in the latter. In the batch training algorithm, for a fixed training set $\{\mathbf{x}_j\}$, we keep record of the weight updates, but their adjustment is applied only after all samples of the set have been considered. The learning stops when a pre-determined number of iterations is reached [20]. At each training iteration, the BMU is determined. Afterwards, all the neurons that belong to the BMU's neighborhood are updated. The updating rule of the ith weight vector, $\mathbf{w}_i$, is computed as [7]

$$\mathbf{w}_i(t+1) = \frac{\sum_{j=1}^{N} \alpha(t)\ h_{ic(j)}(t)\ \mathbf{x}_j}{\sum_{j=1}^{N} h_{ic(j)}(t)} \tag{2}$$

where N defines the number of patterns $\mathbf{x}_j$ that have been assigned to the ith neuron up to the tth iteration and $h_{ic(j)}(t)$ denotes the neighborhood function around the BMU $c(j)$. The learning rate $a(t)$ is a decreasing function of time.

During training, the neighborhood function shrinks through time [2, 7]. At the first training steps, large initial learning rates and neighborhood radii are used in order to have a rigid SOM, whereas small initial learning rates and radii are used during the following training steps. Concerning the neighborhood, as its range is decreased, so does the number of neurons whose weight update direction is correlated. As a result of this correlation, neighboring neurons will be specialized for similar input patterns [20]. The topological information of the map ensures that neighboring neurons on the map possess similar attributes. It must be mentioned that, due to the neighborhood shrinking and the decreasing learning rate through time, it is usual for a SOM to have "dead" units. The "dead" units are neurons which subsequently fail to be associated with any of the input vectors, and, thus, are never organized by the input data. The "dead" neurons have zero (or very low) probability to be active [20].

2.2 SOM Variants Based on Order Statistics

The standard SOM has some disadvantages, such as lack of robustness against outliers and against erroneous choices for the winner vector due to the linear estimators [8]. In order to face these problems, the variants of the standard SOM that employ multivariate order statistics can be used. The MMSOM and the VMSOM treat efficiently the outliers, because they inherit the robustness properties of the order statistics [9].

The SOM variants under discussion differentiate in the way they update the weight vectors. The MMSOM updates the weight vectors using the marginal median, while the VMSOM applies the vector median [8, 9]. In contrast, the SOM calculates the weighted mean value of the input patterns, as can be seen in (2). Although that the MMSOM and the VMSOM, used in this paper, update only the BMU, while the SOM updates also the BMU's neighboring neurons, in principal, such update is not prohibited for MMSOM and VMSOM as well. The MMSOM and VMSOM updating rules, discussed here, can be seen as special cases of vector quantizers that employ a generalized centroid condition [6].

In subsections 2.1 and 2.2, $R_c(t)$ denotes the input patterns assigned to the BMU until the tth iteration and $\mathbf{x}(t)$ denotes the input pattern assigned to the BMU in the tth iteration.

Marginal Median SOM. The MMSOM calculates the marginal median of all patterns assigned to the winner neuron and updates only the BMU's weight vector. The MMSOM relies on the concept of marginal ordering. The marginal ordering of N input vectors $\mathbf{x}_1, \mathbf{x}_2, \ldots, \mathbf{x}_N$, where $\mathbf{x}_j = (x_{1j}, x_{2j}, \ldots, x_{Dj})^T$, is performed by ordering the winner neuron's vector components independently along each of the D dimensions [8, 9]:

$$x_{q(1)} \leq x_{q(2)} \leq \cdots \leq x_{q(N)}, \quad q = 1, 2, \ldots, D \tag{3}$$

where q denotes the index of the vector component into consideration. The new weight vector of the BMU emerges from the calculation of the marginal median of all patterns indexed by the BMU. The calculation of the marginal median is defined by [11]

$$\begin{aligned} &\text{marginal median}\ \{\mathbf{x}_1, \mathbf{x}_2, \ldots, \mathbf{x}_N\} = \\ &= \begin{cases} (x_{1(v+1)}, \ldots, x_{D(v+1)})^T, & N = 2v+1 \\ (\frac{x_{1(v)}+x_{1(v+1)}}{2}, \ldots, \frac{x_{D(v)}+x_{D(v+1)}}{2})^T, & N = 2v \end{cases} \end{aligned} \tag{4}$$

where N denotes the number of patterns assigned to the BMU, $\mathbf{w}_c$. The winner neuron is updated by

$$\mathbf{w}_c(t+1) = \text{marginal median}\ \{R_c(t) \cup \mathbf{x}(t)\}. \tag{5}$$

Vector Median SOM. The VMSOM calculates the vector median of all patterns assigned to the winner neuron and updates only the BMU's weight vector. The vector median operator is the vector that belongs to the set of input vectors indexed by the BMU, which is the closest one to all the current input vectors. The vector median of N input vectors $\mathbf{x}_1, \mathbf{x}_2, \ldots, \mathbf{x}_N$ is defined by [10]

$$\begin{aligned} &\text{vector median}\ \{\mathbf{x}_1, \mathbf{x}_2, \ldots, \mathbf{x}_N\} = \\ &= \mathbf{x}_l, \text{where} \quad l = \arg\min_k \sum_{j=1}^{N} |\mathbf{x}_j - \mathbf{x}_k| \end{aligned} \tag{6}$$

The winner neuron is updated by

$$\mathbf{w}_c(t+1) = \text{vector median}\ \{R_c(t) \cup \mathbf{x}(t)\}. \tag{7}$$

3 Clustering Evaluation Measures

Three measures are employed in order to assess the performance of the SOMs under study, namely the accuracy, the average over all neurons mean squared error (AMSE), and the Rand index.

3.1 Accuracy

Let M be the total number of patterns that compose the test set, $\mathbf{x}_j$ be the jth pattern, and $\delta(x, y)$ be the delta Kronecker which equals 1 if $x = y$ and 0 otherwise. The accuracy of the assignment performed by the SOM is defined as [12]

$$AC = \frac{1}{M} \sum_{j=1}^{M} \delta(g(\mathbf{x}_j), map(\phi(\mathbf{x}_j))) \tag{8}$$

where $g(\mathbf{x}_j)$ is the true label of the pattern, $\phi(\mathbf{x}_j)$ is the label assigned to the pattern by the SOM, and $map(v_i)$ is the *optimal matching*, which maps the label assigned to the pattern by the SOM or its variants onto the ground truth labels. The optimal matching is needed because SOM is an unsupervised training algorithm. It can be derived by the Kuhn-Munkres algorithm [13].

The problem solved by the Kuhn-Munkres algorithm is stated as follows. Consider a complete weighted bipartite graph $\Gamma = (V \bigcup U, V \times U)$. Let us denote $V = \{v_i\}$ and $U = \{u_i\}$, where $i = 1, 2, \ldots, K$ and K being the number of nodes. The weight of the edge (v_i, u_i) is denoted by $\xi(v_i, u_i)$. The goal is to find the optimal matching from V to U, that is the matching with the maximum sum of the edge weights that belong to it. Mathematically, given a $K \times K$ weight matrix Ξ, which represents the graph Γ, a permutation π of $1, 2, \ldots, K$ must be found so that the following sum

$$\sum_{i=1}^{K} \xi(v_i, u_{\pi(i)}) \tag{9}$$

is maximized. The resulted set of edges is the optimal matching. A graph that is not complete, it must be forced to become complete, by adding zeros in the weight matrix Ξ for the non-existing edges.

Let us explain the use of the Kuhn-Munkres algorithm in the calculation of the SOM clustering accuracy. The accuracy of the assignment performed by the SOM is defined by (8). Let us consider that the patterns must be clustered into K clusters. That is the number of nodes of the graph Γ equals K. The weight $\xi(v_i, u_i)$ assigned to the edge (v_i, u_i) corresponds to the profit made out, if the label assigned by the SOM is v_i and the ground truth label is u_i. The purpose is to maximize the profit. Obviously, if the two labels are the same, the profit is maximized, since the SOM has assigned the patterns to the correct ground truth class.

The input of the algorithm is a $K \times K$ weight matrix. The weights of the elements (i, i) with $i = 1, 2, \ldots, K$ are set to be $\xi(i, i) = 1$, while the weights of the elements (i, j) with $j = 1, 2, \ldots, K$ and $i \neq j$ are set to be 0 (or a very low value). The output of the algorithm is a $K \times K$ matrix. The (i, j) matrix element equals 1 if the edge (v_i, u_j) belongs to the optimal matching, otherwise it equals 0.

3.2 Average over All Neurons Mean Squared Error (AMSE)

In order to set the definition of the AMSE, we must first define the Mean Squared Error (MSE). The MSE of one neuron is the mean value of the Euclidean distances between its weight vector and all the patterns assigned to it. Mathematically, the MSE of the neuron $\mathbf{w}_i$ is calculated as follows

$$MSE_i = \frac{1}{N} \sum_{j=1}^{N} \|\mathbf{x}_{j[i]} - \mathbf{w}_i\|^2 \tag{10}$$

where N is the total number of patterns assigned to the ith neuron and $\mathbf{x}_{j[i]}$ is the j-th pattern assigned to this neuron. The average over all neurons MSE, which from now on will be referred to as AMSE, is the mean value of MSE_i for all the neurons of the map:

$$AMSE = \frac{1}{K} \sum_{i=1}^{K} MSE_i \tag{11}$$

where K is the total number of the map neurons.

3.3 Rand Index

The Rand index is a widely used evaluation measure in clustering applications. The Rand index indicates the number of input patterns that are either from the same class but are not grouped into the same cluster, or that are not from the same class but are grouped into the same cluster. The Rand index is defined as follows [3, p. 173-174]:

$$\gamma = \frac{1}{2} \sum_{i=1}^{N_c} n_{i.}^2 + \frac{1}{2} \sum_{j=1}^{N_f} n_{.j}^2 - \sum_{i=1}^{N_c} \sum_{j=1}^{N_f} n_{ij}^2 \tag{12}$$

where N_c denotes the total number of clusters that are created after training the SOMs, N_f the total number of classes that the patterns are initially grouped into according to the ground truth, $n_{i.}$ and $n_{.j}$ the total number of patterns assigned to clusters i and j, respectively, and n_{ij} the total number of patterns assigned to cluster i that belong to class j. Rand index values lie in the range $0 \leq \gamma \leq \binom{N}{2}$, with $\binom{N}{2}$ denoting the number of combinations of two patterns that can be taken out from the total set. The lower the Rand index, the better the clustering is. A perfect clustering should produce a zero Rand index [3].

4 Data

The well-known IRIS data was used in order to evaluate the performance of the algorithms for clustering. The IRIS data records information about 150 flower patterns [14]. Each pattern is characterized by 4 features namely the sepal length, the sepal width, the petal length, and the petal width. The patterns are classified into 3 classes called Setosa, Versicolor, and Virginica. The most important feature of the IRIS data is the ground truth of the patterns, i.e. the actual class each pattern is classified to. It must be noted that the IRIS data set *does* contain outliers for unsupervised learning. Accordingly, this data set is appropriate for studying the role of the outliers in clustering. This is not the case for supervised learning [19, p.346].

Motivated by the observations made on IRIS, we have compared the SOM variants against the SOM for the redistribution of neutral emotional speech feature vectors from the DES database [15] into non-neutral emotional speech patterns. We decided to work on the DES database, because it is easily accessible and well annotated. A number of 1160 emotional speech patterns are extracted. Each pattern consists of a 90-dimensional feature vector [16]. Each emotional pattern is classified into one of the five primitive emotional states, such as hot anger, happiness, neutral, sadness, and surprise. The ground truth for all patterns is also available.

5 Experimental Results

The performance of the SOM, the MMSOM, and the VMSOM on clustering are demonstrated through the accuracy, the AMSE, and the Rand index on the IRIS data. The training set consists of 120 randomly selected patterns, while the test set is composed by the 30 remaining patterns. The accuracy, the AMSE, and the Rand index were measured using 30-fold cross validation. The accuracy should increase, while the AMSE and the Rand index should decrease for a high quality clustering.

Table 1 summarizes the accuracy, the AMSE, and the Rand index of the three SOMs using different number of neurons to build the map, respectively, averaged over the 30 cross validations. The best performance concerning the accuracy, the AMSE, and the Rand index is indicated in boldface. As it can be noticed from Table 1, the MMSOM yields the best accuracy (97.33%), the VMSOM follows (97.00%), while the SOM has the worst behavior with respect to the accuracy (91.22%). Table 1 indicates that the same ordering between the three SOMs stands also with respect to the AMSE. The smallest AMSE is measured for the MMSOM (0.221). The VMSOM yields a larger AMSE than the MMSOM (0.238) and, finally, the SOM exhibits the worst performance with respect to the AMSE (0.441). In addition, the SOM yields the worst Rand index for every map size compared to both the MMSOM and the VMSOM. The best Rand index values are 33.866 for a 4×4 map, 13.233 for a 3×3 map, and 15.266 for a 3×3 map, for the SOM, the MMSOM, and the VMSOM, respectively.

As it can be noticed form Table 1 both the MMSOM and the VMSOM have similar values that do not change significantly with the number of neurons, concerning all the evaluation measures. In contrast, the SOM values change significantly with the map size

Table 1. Accuracy, AMSE, and Rand index of SOM, MMSOM, and VMSOM averaged over 30 cross validations for different map sizes on the IRIS data

Neurons	Average accuracy			Average AMSE			Average Rand index		
	SOM	MMSOM	VMSOM	SOM	MMSOM	VMSOM	SOM	MMSOM	VMSOM
3 (2×2)	60.66	89.00	89.67	1.599	0.501	0.557	76.633	51.533	52.533
4 (2×2)	82.45	90.67	88.89	1.788	0.516	0.547	98.200	49.800	58.166
5 (3×2)	90.56	97.33	95.22	1.592	0.337	0.367	56.133	22.166	25.966
6 (3×2)	91.22	96.56	95.11	1.229	0.338	0.371	45.600	23.266	23.366
7 (4×2)	90.78	95.56	94.67	0.658	0.321	0.339	48.766	21.266	23.866
8 (4×2)	82.89	96.67	94.11	1.189	0.250	0.184	78.966	17.566	16.366
9 (3×3)	84.56	97.00	96.67	1.187	0.249	0.298	71.633	17.966	**15.266**
10 (3×3)	84.56	97.11	96.33	1.175	0.266	0.337	68.533	**13.233**	23.833
11 (4×3)	88.78	96.11	94.89	0.496	0.227	0.280	46.266	19.300	18.033
12 (4×3)	90.67	97.00	94.33	0.517	0.233	0.272	41.200	18.466	22.233
16 (4×4)	**91.22**	**97.33**	**97.00**	**0.441**	**0.221**	**0.238**	**33.866**	14.933	20.100

compared to the MMSOM and the VMSOM. This fact can be explained by the number of "dead" neurons of each SOM. Let us denote by μ the mean number of patterns that a neuron wins, by σ the standard deviation, and by N the exact number of patterns a neuron wins during training. The "dead" neurons are those for which the following inequality holds: $N < \mu - \sigma$. Table 2 presents the number of "dead" neurons of each SOM for different map sizes. It is obvious that for the SOM, the number of "dead" neurons gets very large with increasing number of neurons, causing the significant difference of its performance compared to the SOM variants for different map sizes. For both the MMSOM and the VMSOM, the number of "dead" neurons is small for all map sizes, which explains their similar behavior and the small range of values they get.

Table 2. Number of "dead" neurons for different map sizes for the SOMs

Neurons	3	4	5	6	7	8	9	10	11	12	16
SOM	1	1	1	0	2	3	3	3	4	5	8
MMSOM	0	0	0	0	1	1	0	0	2	2	4
VMSOM	0	0	0	0	1	0	0	0	1	2	3

The Student t-test for unequal variances has been used to check whether the difference between the mean accuracies achieved by the following algorithm pairs (SOM, MMSOM), (SOM, VMSOM), and (MMSOM, VMSOM) is statistically significant at the 95% level of significance in a 30-fold cross validation experiment with a 4×4 map. The same assessment has also been performed for the AMSE and the Rand index. The tests have shown that the performance differences are statistically significant.

The superiority of the MMSOM was expected, because the marginal median is not affected by the outliers in contrast to the mean. Moreover, the weight vector is not constrained to be among the input vectors assigned to a neuron as the vector median does. Furthermore, the SOM contains many "dead" units and cannot represent data well. Due to the order statistic properties of the MMSOM and VMSOM, it is expected, though, that the maps created by the SOM variants are more representative, as demonstrated by Table 2. The maps created by the MMSOM and the VMSOM have less "dead" units and the classes defined on the map are well separated. However, "dead" units are inevitable for a SOM to train a non-stationary data set [20].

The SOMs were also applied to the re-distribution of emotional speech feature vectors extracted from the DES database. The primitive emotional states are anger, happiness, neutral, sadness, and surprise. Our purpose is to re-distribute the emotional speech patterns that were originally classified as neutral into the other four emotional states. That is, to find out which class is closer to the neutral one and how each training algorithm acts on the data. The training set consists of the all the non-neutral patterns and the test set consists of all the neutral patterns. The average assignment ratio was estimated using 15-fold cross validation.

Table 3 demonstrates the average assignment ratio of the neutral patterns that are labeled as angry, happy, sad, and surprised by each SOM. As can be seen, all the algorithms classify the neutral patterns as sad with a very high percentage. This means that sadness resembles the neutral state more than the other emotional states. The largest percentage is measured for the MMSOM (61.86%), the next larger percentage is provided by the VMSOM (61.51%) and, finally, the SOM yields the lowest one (58.27%). It was expected that the MMSOM would re-distribute the neutral patterns in a better manner than VMSOM and SOM. Anger is the second closer to neutrality emotion, happiness follows and, finally, surprise is the least similar to neutrality emotion. All the algorithms conform to this order.

Table 3. Average ratio of neutral emotional speech patterns assigned to non-neutral emotional classes using the SOM variants

Emotion	Average assignment ratio (%)		
	SOM	MMSOM	VMSOM
Sadness	**58.27**	**61.86**	**61.51**
Anger	13.87	14.02	15.00
Happiness	13.56	14.81	13.62
Surprise	13.16	9.59	9.82

The Student t-test for unequal variances has also found that the differences in the average assignment ratio per emotion are statistically significant for a 15-fold cross validation experiment.

Figure 1 depicts a partition of the 2D feature domain that has been resulted after selecting the five best emotional features by the Sequential Forward Selection algorithm and applying Principal Component Analysis in order to reduce the dimensionality from five dimensions (5D) to two dimensions (2D) [16]. Only the samples which belong to the interquartile range of the probability density function for each class are shown. It can be seen that the neutral emotional class does not possess any overlap with the surprise, while such an overlap is observed for sadness, anger, and happiness. Therefore, the results shown in Table 3 comply with the sample space depicted in Figure 1.

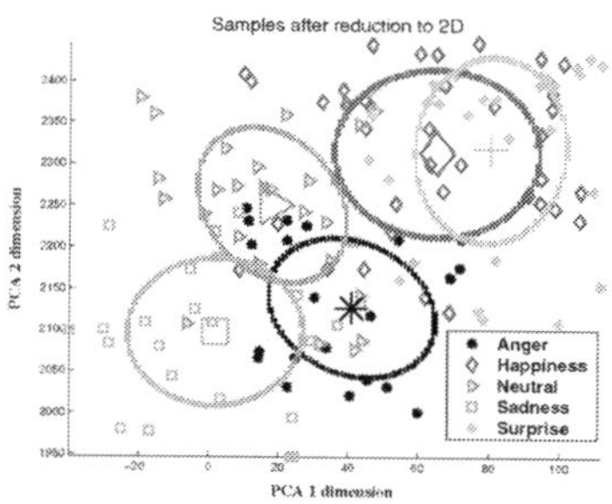

Fig. 1. Partition of the 2D domain into five emotional states derived by PCA. The samples which belong to the interquartile range of each pdf are shown. The big symbols denote the mean of each class. The ellipses denote the 60% likelihood contours for a 2-D Gauss model.

6 Conclusions

Two variants of the self organizing map, the MMSOM and the VMSOM, that are based on order statistics, have been studied. These variants have been successfully used in color quantization and document organization and retrieval. In this paper, we presented experimental evidence for their clustering quality by using the accuracy, the average over all neurons mean squared error, and the Rand index as figures of merit. The assessment was first conducted on the well-known IRIS data set. Motivated by the superiority of the SOM variants that are based on order statistics, we investigated their application in the re-distribution of emotional neutral feature vectors to non-neutral emotional states. We demonstrated that the re-distribution is consistent with the sample feature space.

References

1. S. Haykin, *Neural Networks: A Comprehensive Foundation.* Upper Saddle River, N.Y.: Prentice-Hall, 1999.
2. T. Kohonen, *Self-Organizating Maps*, 3/e. Berlin, Germany: Springer-Verlag, 2000.
3. A. K. Jain and R. C. Dubes, *Algorithms for Clustering Data.* Englewood Cliffs, N.J.: Prentice-Hall, 1988.
4. S. Kaski, J. Kangas, and T. Kohonen, "Bibliography of Self-Organizing Map (SOM) Papers: 1981-1997," *Neural Computing Surveys*, vol. 1, pp. 102-350, 1998.
5. M. Oja, S. Kaski, and T. Kohonen, "Bibliography of Self-Organizing Map (SOM) Papers: 1998-2001 Addendum," *Neural Computing Surveys*, vol. 3, pp. 1-156, 2003.
6. A. Gersho and R. M. Gray, *Vector Quantization and Signal Compression.* Boston, MA: Kluwer Academic Publishers, 1992.
7. J. Vesanto, J. Himberg, E. Alhoniemi, and J. Parhankangas, *SOM Toolbox for Matlab 5*, Finland, 2000, `www.cis.hut.fi`.
8. I. Pitas, C. Kotropoulos, N. Nikolaidis, R.Yang, and M. Gabbouj, "Order statistics learning vector quantizer", *IEEE Trans. Image Processing,* vol. 5, pp. 1048-1053, 1996.
9. C. Kotropoulos and I. Pitas,"Self-organizing maps and their applications in image processing, information organization, and retrieval," in *Nonlinear Signal and Image Processing: Theory, Methods, and Applications* (K. E. Barner and G. R. Arce, Eds.), Boca Raton, FL: CRC Press, 2004.
10. J. Astola, P. Haavisto, and Y. Neuvo, "Vector median filters," *Proceedings of the IEEE*, vol. 78, no. 4, pp. 678-689, April 1990.
11. I. Pitas and P. Tsakalides, "Multivariate ordering in color image restoration," *IEEE Trans. Circuits and Systems for Video Technology*, vol. 1, no. 3, pp. 247-259, September 1991.
12. W. Xu, X. Liu, and Y. Gong, "Document clustering based on non-negative matrix factorization," in *Proc. ACM SIGIR 03*, pp. 267-273, Toronto, Canada, 2003.
13. J. A. McHugh, *Algorithmic Graph Theory*. Prentice Hall, 1990.
14. R. A. Fisher, "The use of multiple measurements in taxonomic problems," *Ann. Eugen.*, vol. 7, pp. 179-188, 1936.
15. I. S. Engberg and A. V. Hansen, Documentation of the Danish Emotional Speech Database DES, Internal Report, Center for Person Kommunikation, Aalborg University, 1996.
16. D. Ververidis, C. Kotropoulos, and I. Pitas, "Automatic emotional speech classification," in *Proc. 2004 IEEE Int. Conf. Acoustics, Speech, and Signal Processing*, vol. I, pp. 593-596, Montreal, Canada, May 2004.
17. J. C. T. Kanade and Y. Tian, "Comprehensive database for facial expression analysis," in *Proc. IEEE Int. Conf. Face and Gesture Recognition*, pp. 46-53, March 2000.
18. I. Kotsia and I. Pitas, "Real-time facial expression recognition from image sequences using support vector machines," in *Proc. Conf. Visual Communications Image Processing*, Beijing, China, July 12-15, 2005.
19. K. V. Mardia, J. T. Kent, and J. M. Bibby, *Multivariate Analysis*. Academic Press, Harcourt Brace & Co., New York, 1979.
20. M. M. Van Hulle, *Faithful Representations and Topographic Maps. From Distortion- to Information-Base Self-Organization.* N.Y.: J. Wiley, 2000.

On the Basis Updating Rule of Adaptive-Subspace Self-Organizing Map (ASSOM)*

Huicheng Zheng, Christophe Laurent, and Grégoire Lefebvre

France Telecom R&D – DIH/HDM
4, Rue du Clos Courtel
35512 Cesson Sévigné Cedex, France

Abstract. This paper gives other views on the basis updating rule of the ASSOM proposed by Kohonen. We first show that the traditional basis vector rotation rule can be expressed as a correction to the basis vector which is a scaling of component vectors in the episode. With the latter form, some intermediate computations can be reused, leading to a computational load only linear to the input dimension and the subspace dimension, whereas a naive implementation of the traditional rotation rule has a computational load quadratic to the input dimension. We then proceed to propose a batch-mode updating of the basis vectors. We show that the correction made to each basis vector is a linear combination of component vectors in the input episode. Computations can be further saved. Experiments show that the proposed methods preserve the ability to generate topologically ordered invariant-feature filters and that the learning procedure is largely boosted.

1 Introduction

The Adaptive-Subspace Self-Organizing Map (ASSOM) [1] is basically a combination of the competitive selection and cooperative learning as in the traditional SOM [2] and a subspace method. The single weight vectors at map units in the SOM are replaced by modules of basis vectors in the ASSOM that span some linear subspaces. The ASSOM is an alternative to the standard principal component analysis (PCA) method of feature extraction. An earlier neural approach for PCA can be found in [3]. The ASSOM can generate spatially ordered feature filters thanks to spatial interactions among processing units [4]. Each module in the ASSOM can be realized as a neural network which calculates orthogonal projections of input vectors on its subspace.

The input to an ASSOM array is typically an episode, i.e. a sequence of pattern vectors supposed to approximately span some linear subspace. These vectors shall also be referred to as component vectors of the episode in this paper. By learning the episode as a whole, the ASSOM is able to capture the transformation coded in the episode. The simulation results in [1] and [4] have demonstrated that the ASSOM can induce ordered filter banks to account for translation, rotation and scaling. The relationship between the neurons in the ASSOM architecture and their biological counterparts are reported [4]. The ASSOM has been applied to speech processing [5], texture segmentation [6], image

* This work was carried out during the tenure of a MUSCLE Internal fellowship (http://www.muscle-noe.org).

S. Kollias et al. (Eds.): ICANN 2006, Part I, LNCS 4131, pp. 435–444, 2006.

retrieval [7] and image classification [7], [8], etc. in the literature. A supervised variant of ASSOM, called Supervised Adaptive-Subspace Self-Organizing Map (SASSOM), was proposed by Ruiz del Solar in [6].

The basis vector rotation rule in the traditional ASSOM implementation takes a form of matrix multiplication. This rule is hard to understand and more seriously, a naive implemenation of this rule leads to a computational load which is quadratic to the input dimension, not to mention large amount of memory required by the usually high-dimensional matrix operations. This deficiency makes the naive implementation of the basic ASSOM learning very costly for practical applications.

There were efforts in the literature to reduce the computational load associated with the basic ASSOM learning. De Ridder et al. [7] dropped topological ordering to reduce the computations involved in the cooperative learning. Furthermore, they performed an offline batch-mode updating of subspaces with PCA to avoid the time-consuming iterative updating. Similarly, López-Rubio et al. [9] proposed to combine PCA with ASSOM, but they realized an online incremental learning while retaining self-organization of generated features. The resulting algorithm is named PCASOM for Principal Component Analysis Self-Organizing Map. According to their report, under similar classification performance, their algorithm runs about twice faster than the basic ASSOM. McGlinchey et al. [10] replaced the traditional basis vector updating formula with one proposed by Oja [11]. According to their paper, the computational load is only linear to the input dimension, but quadratic to the subspace dimension.

In this paper, we first show that with the traditional basis rotation rule, the correction made to each basis vector is in fact a scaling of the component vector of the input episode. With this modified form, some intermediate computations can be reused, leading to a computational load only linear to both the input dimension and the subspace dimension. We then proceed to propose a batch-mode updating of the basis vectors, where the correction made to each basis vector is a linear combination of component vectors in the episode. This modified rule further accelerates the learning procedure by saving large amounts of computations.

This paper will be organized as follows: In Sect. 2, we review briefly the basic ASSOM learning procedure. The proposed alternative updating rules will be presented in Sect. 3. Section 4 is dedicated to experiments which demonstrate the performance of the proposed methods. This paper will be concluded by Sect. 5.

2 The Basic ASSOM Learning

An ASSOM is composed of an array of modules. Each module in the ASSOM can be realized by a two-layer neural network [4], as shown in Fig. 1. It calculates the orthogonal projection of an input vector $\mathbf{x}$ on the subspace $\mathcal{L}$ of the module. Supposing $\mathcal{L}$ is spanned by a set of basis vectors $\{\mathbf{b}_1, \mathbf{b}_2, \ldots, \mathbf{b}_H\}$, where H is the dimension of $\mathcal{L}$, the neurons in the first layer take the orthogonal projections $\mathbf{x}^{\mathrm{T}}\mathbf{b}_h$ of the input vector $\mathbf{x}$ on the individual basis vectors $\mathbf{b}_h$. The basis vectors are supposed to be orthonormalized. The only quadratic neuron of the second layer sums up the squared outputs of the first-layer neurons. The output of the module is then $\|\hat{\mathbf{x}}_{\mathcal{L}}\|^2$, with $\hat{\mathbf{x}}_{\mathcal{L}}$ being the projection of $\mathbf{x}$ on $\mathcal{L}$. This output can be regarded as a measure of the matching between the input vector $\mathbf{x}$ and the subspace $\mathcal{L}$. For an input episode $\mathbf{X} = \{\mathbf{x}(s), s \in S\}$, where S is the index set of vectors in the episode, Kohonen proposed to use the *energy* $\sum_{s \in S} \|\hat{\mathbf{x}}_{\mathcal{L}}(s)\|^2$ as the measure of matching between $\mathbf{X}$ and $\mathcal{L}$ [4].

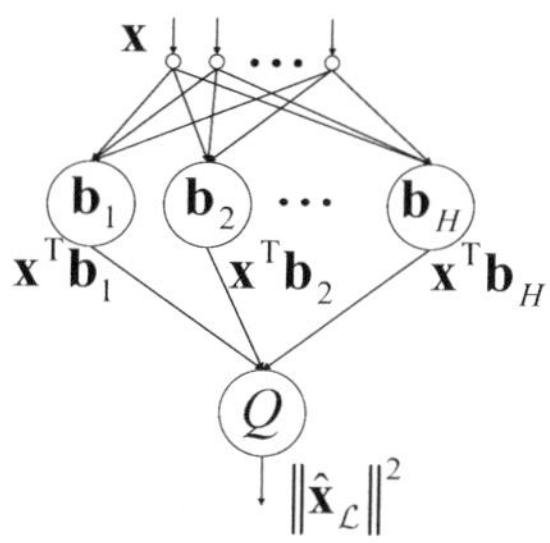

Fig. 1. A module of ASSOM realized as a neural network. $\mathbf{x}$ is an input vector. $\{\mathbf{b}_1, \mathbf{b}_2, \ldots, \mathbf{b}_H\}$ is an orthonormal basis of the linear subspace $\mathcal{L}$ of the module. Q is a quadratic neuron that sums up squares of its inputs.

The learning process of ASSOM approximately minimizes an error function in an iterative way [4]. Supposing $\mathbf{x}(s), s \in S$ is the input episode at the learning step t, then the basic ASSOM learning procedure at this step proceeds as follows:

1. Locate the winning module indexed by $c = \arg\max_{i \in I} \sum_{s \in S} \|\hat{\mathbf{x}}_{\mathcal{L}_i}(s)\|^2$, where I is the index set of modules in the ASSOM.
2. For each module i in the neighborhood of c, including c itself, update the subspace $\mathcal{L}_i$ for each component vector $\mathbf{x}(s), s \in S$, that is, update the basis vectors $\mathbf{b}_h^{(i)}$, according to the following rules:
   (a) Rotate each basis vector according to:

$$\mathbf{b}_h^{(i)} = \mathbf{P}_c^{(i)}(\mathbf{x}, t)\mathbf{b}_h^{'(i)} , \tag{1}$$

   where $\mathbf{b}_h^{(i)}$ is the new basis vector and $\mathbf{b}_h^{'(i)}$ the old one. The matrix $\mathbf{P}_c^{(i)}(\mathbf{x}, t)$ is a rotation operator defined by:

$$\mathbf{P}_c^{(i)}(\mathbf{x}, t) = \mathbf{I} + \lambda(t) h_c^{(i)}(t) \frac{\mathbf{x}(s)\mathbf{x}^{\mathrm{T}}(s)}{\|\hat{\mathbf{x}}_{\mathcal{L}_i}(s)\| \|\mathbf{x}(s)\|} , \tag{2}$$

   where $\mathbf{I}$ is the identity matrix, $\lambda(t)$ a learning-rate factor that diminishes with t. $h_c^{(i)}(t)$ is a neighborhood function defined on the ASSOM lattice.
   (b) Dissipate the basis vectors $\mathbf{b}_h^{(i)}$ to improve stability of the results [4] and then orthonormalize these basis vectors.

Through this competitive and cooperative learning procedure, the ASSOM will finally arrive at a topologically organized status, where nearby modules represent similar feature subspaces. A naive implementation of (1) requires a matrix multiplication which needs not only a large amount of memory, but also a computational load quadratic to the input dimension. It would be costly for practical applications of ASSOM.

3 On the Basis Updating Rule of ASSOM

3.1 Insight on the Basis Vector Rotation

In the first place we propose to replace the formulae (1) and (2) through a little mathematical deduction. The term $\mathbf{b}_h^{'(i)}$ in (1) can be distributed to the right side of (2), leading to the current basis vector and a correction to it:

$$\mathbf{b}_h^{(i)} = \mathbf{b}_h^{'(i)} + \Delta\mathbf{b}_h^{(i)} \ , \tag{3}$$

where

$$\Delta\mathbf{b}_h^{(i)} = \lambda(t)h_c^{(i)}(t)\frac{\mathbf{x}(s)\mathbf{x}^{\mathrm{T}}(s)\mathbf{b}_h^{'(i)}}{\|\hat{\mathbf{x}}_{\mathcal{L}_i}(s)\|\|\mathbf{x}(s)\|} \ . \tag{4}$$

$\mathbf{x}^{\mathrm{T}}(s)\mathbf{b}_h^{'(i)}$ is in fact a scalar value. The equation can be rewritten as:

$$\Delta\mathbf{b}_h^{(i)} = \alpha_{c,h}^{(i)}(s,t)\mathbf{x}(s) \ . \tag{5}$$

Here $\alpha_{c,h}^{(i)}(s,t)$ is a scalar value defined by:

$$\alpha_{c,h}^{(i)}(s,t) = \lambda(t)h_c^{(i)}(t)\frac{\mathbf{x}^{\mathrm{T}}(s)\mathbf{b}_h^{'(i)}}{\|\hat{\mathbf{x}}_{\mathcal{L}_i}(s)\|\|\mathbf{x}(s)\|} \ . \tag{6}$$

This shows that the correction $\Delta\mathbf{b}_h^{(i)}$ is in fact a scaling of the component vector $\mathbf{x}(s)$, as illustrated in Fig. 2, which seems to have been ignored by many practitioners. Equation (5) gives a clearer way to understand the basis vector rotation in ASSOM learning than the traditional rotation matrix (2). Note that in (6), $\mathbf{x}^{\mathrm{T}}(s)\mathbf{b}_h^{'(i)}$ is the projection of the component vector on the basis vectors represented by the neurons of the first layer, which we have already when computing the projection $\|\hat{\mathbf{x}}_{\mathcal{L}_i}(s)\|$ (cf. Fig. 1). If we calculate the scaling factor $\alpha_{c,h}^{(i)}(s,t)$ first, and then scale the component vector $\mathbf{x}(s)$ with this factor, the computations associated with the basis vector updating will be dramatically reduced. This implementation will be referred to as FL-ASSOM for fast-learning ASSOM. It is completely equivalent to the basic ASSOM in terms of generating topologically ordered invariant-feature filters.

Let us compare the computational loads of the basis vector updating in the basic ASSOM and the FL-ASSOM by analyzing the respective updating formulae. We assume the input dimension to be N, and the subspace dimension to be M. We first evaluate the computations required by naive implementation of the traditional basis vector updating rule (1). For each component vector $\mathbf{x}(s)$, $\mathbf{x}(s)\mathbf{x}^{\mathrm{T}}(s)$ in (2) needs N^2 multiplications, the matrix multiplication in (1) amounts to MN^2 multiplications. There are totally about $MN^2 + N^2$ multiplications. Similarly, the number of additions required by (1) can be shown to be around $MN^2 + N^2$. Finally, the computational load of naive implementation of the traditional updating rule is approximately $O(MN^2)$, i.e. quadratic to the input dimension and linear to the subspace dimension. The replacement proposed by McGlinchey et al. [10] leads to a computational load of $O(M^2N)$, as shown in their paper, i.e. linear to the input dimension but quadratic to the subspace dimension. Now

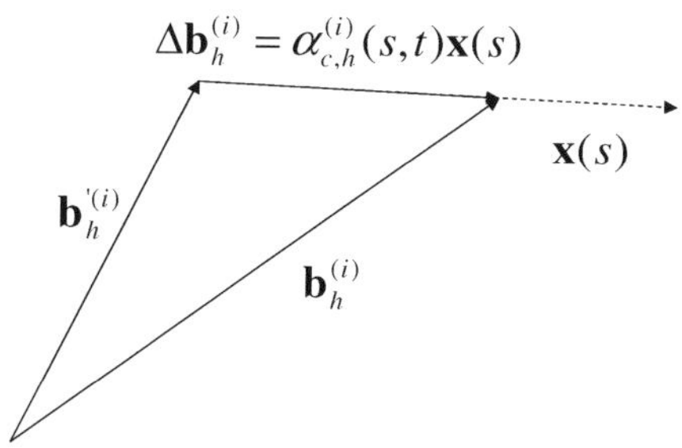

Fig. 2. An alternative view of the basis vector updating rule of ASSOM. The correction $\Delta\mathbf{b}_h^{(i)}$ is a scaling of the component vector $\mathbf{x}(s)$. After updating, $\mathbf{b}_h^{(i)}$ represents better the component vector $\mathbf{x}(s)$.

with the proposed updating rule (5), for each component vector $\mathbf{x}(s)$, the computations of $\|\hat{\mathbf{x}}_{\mathcal{L}_i}(s)\|$ and $\|\mathbf{x}(s)\|$ in (6) need about $MN+2N$ multiplications, and $\alpha_{c,h}^{(i)}(s,t)\mathbf{x}(s)$ in (5) about MN multiplications. In all (5) needs about $2MN + 2N$ multiplications. Similarly, the number of additions can be shown to be about $2MN + 2N$. So with (5), the computational load is approximately $O(MN)$, i.e. linear to both the input dimension and the subspace dimension. So there is an obvious benefit in using (5) other than a naive implementation of (1).

3.2 Further Boosting: Batch-Mode Basis Vector Updating

Basis vector updating can be further boosted by working in a batch mode. We can avoid computing the value of $\|\hat{\mathbf{x}}_{\mathcal{L}_i}(s)\|$ in (6) by using the value computed previously during module competition. However this could not be done inside the framework of FL-ASSOM since the subspaces are continuously changing in receiving each component vector of the episode. To save computation of $\|\hat{\mathbf{x}}_{\mathcal{L}_i}(s)\|$, the following batch-mode rotation operator [4] will be useful:

$$\mathbf{B}_c^{(i)}(t) = \mathbf{I} + \lambda(t)h_c^{(i)}(t)\sum_{s\in S}\frac{\mathbf{x}(s)\mathbf{x}^{\mathrm{T}}(s)}{\|\mathbf{x}(s)\|^2} \ . \tag{7}$$

With this rotation operator, each basis vector in the subspace will be rotated only once for the whole input episode.

For stability of the solution, we expect the magnitude of the correction made to the basis vectors to be monotonically decreasing with respect to $\|\hat{\mathbf{x}}_{\mathcal{L}_i}(s)\|$. We borrow the idea from the basic rotation operator $\mathbf{P}_c^{(i)}(\mathbf{x},t)$ [4] to divide the learning-rate factor $\lambda(t)$ by the scalar value $\|\hat{\mathbf{x}}_{\mathcal{L}_i}(s)\|/\|\mathbf{x}(s)\|$, which only changes the effective learning rate. The batch-mode rotation operator then becomes:

$$\mathbf{B}_c^{(i)}(t) = \mathbf{I} + \lambda(t)h_c^{(i)}(t)\sum_{s\in S}\frac{\mathbf{x}(s)\mathbf{x}^{\mathrm{T}}(s)}{\|\hat{\mathbf{x}}_{\mathcal{L}_i}(s)\|\|\mathbf{x}(s)\|} \ . \tag{8}$$

As was done for the FL-ASSOM, we distribute $\mathbf{b}_h^{'(i)}$ to terms in this operator. With similar deduction as in the FL-ASSOM, the basis vector updating rule becomes:

$$\mathbf{b}_h^{(i)} = \mathbf{b}_h^{'(i)} + \Delta \mathbf{b}_h^{(i)} \ , \tag{9}$$

where

$$\Delta \mathbf{b}_h^{(i)} = \sum_{s \in S} \left(\alpha_{c,h}^{(i)}(s,t) \mathbf{x}(s) \right) \ . \tag{10}$$

The correction made to each basis vector is thus a linear combination of the component vectors in the episode. The difference between the updating rules (3) and (9) is that the former updates the basis vectors for each component vector one by one while the latter updates the basis vectors in a batch mode for the whole episode.

The scalar parameter $\alpha_{c,h}^{(i)}(s,t)$ has the same form as (6) in the FL-ASSOM:

$$\alpha_{c,h}^{(i)}(s,t) = \lambda(t) h_c^{(i)}(t) \frac{\mathbf{x}^{\mathrm{T}}(s) \mathbf{b}_h^{'(i)}}{\|\hat{\mathbf{x}}_{\mathcal{L}_i}(s)\| \|\mathbf{x}(s)\|} \ . \tag{11}$$

The meaning of this equation is a little different from that of (6), where the subspace $\mathcal{L}_i(s)$ should be updated for each component vector $\mathbf{x}(s)$ in the episode and thus we could not reuse the computational results of module competition. Here in (11) the basis vector updating works in a batch mode, i.e. updating is performed only after the whole episode has been received. Therefore, $\|\hat{\mathbf{x}}_{\mathcal{L}_i}(s)\|$ and $\mathbf{x}^{\mathrm{T}}(s)\mathbf{b}_h^{'(i)}$ can reuse the results previously calculated during module competition. What we need to do is only store the calculated values in registers and fetch them when needed. The computational load of (11) is thus trivial. Furthermore, the dissipation as well as orthonormalization of basis vectors can be performed only once for each episode without loosing accuracy since the basis vectors are not updated during the episode. The computational load can thus be further reduced. This method will be referred to as BFL-ASSOM for batch-mode fast-learning ASSOM.

Let us estimate the computational load of BFL-ASSOM. For basis vector updating with (10), we estimate the computational load averaged on each component vector of the episode as we did for the basic ASSOM and the FL-ASSOM. As previously mentioned, the calculation of $\alpha_{c,h}^{(i)}(s,t)$ according to (11) needs only trivial computation. The majority of computation is in (10). Averaged on each vector in the episode, the computational load required by basis vector updating with BFL-ASSOM is about MN multiplications and MN additions. Furthermore, since the dissipation and orthonormalization of basis vectors can be performed only once for each episode, the whole learning time can be further reduced.

4 Experiments

We first show that the BFL-ASSOM can also generate the topologically ordered invariant-feature filters as the basic ASSOM. The results of FL-ASSOM will be shown as the ground truth since the FL-ASSOM is mathematically equivalent to the basic ASSOM. One of the most common transformations occurred to images is translation. We will show that the BFL-ASSOM permits to generate Gabor type filters from episodes subject to translation.

The input episodes are constructed from a colored noise image, which is generated by filtering a white noise image with a second-order Butterworth filter. The cut-off frequency is set to 0.6 times of the Nyquist frequency of the sampling lattice. Each episode is composed of 6 vectors, each of which is formed on a circular receptive field on the sampling lattice composed of 349 pixels. The vectors in the same episode have only random translation of no more than 5 pixels in both the horizontal and the vertical directions. The episodes are generated on random locations of the colored noise image. The mean value of components of each input vector is subtracted from each component of the vector. In order to symmetrize the filters with respect to the center of the receptive field, the input samples are weighted by a Gaussian function symmetrically placed at the center of the receptive field with a full width at half maximum (FWHM) that varies linearly with respect to the learning step t from 1 to 16 sampling lattice spacings. Each vector is normalized before entering into the ASSOM array. The ASSOM array is composed of 9×10 modules aligned in a hexagonal lattice with two basis vectors at each module. The basis vectors of all the modules are initialized randomly and orthonormalized at the beginning of the learning process. The radius of the circular neighborhood function $h_c^{(i)}(t)$ decreases linearly from 6.73 ($= 0.5 \times (9^2 + 10^2)^{1/2}$) to 0.9 ASSOM array spacings with t. The learning-rate factor has the form $\lambda(t) = 0.1 \cdot T/(T + 99t)$, where T is the total number of learning steps and set to $30,000$ for the current experiment.

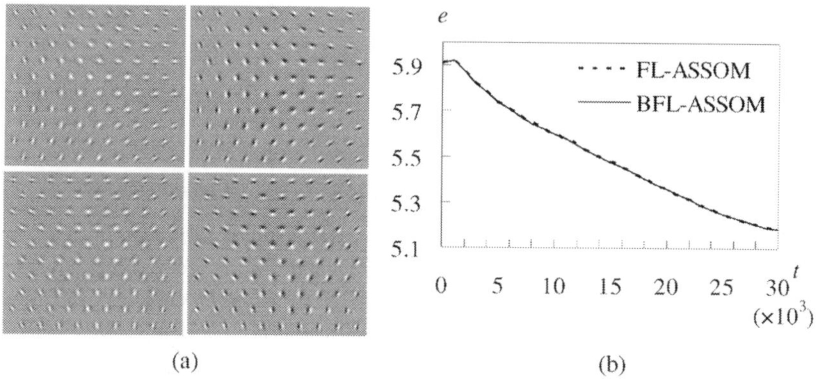

Fig. 3. (a) The Gabor type filters generated by BFL-ASSOM compared to those by FL-ASSOM on episodes subject to translation. *Top*: Filters generated by FL-ASSOM; *Bottom*: Filters generated by BFL-ASSOM. *Left*: First basis vectors. *Right*: Second basis vectors. (b) Change of the projection error e with the learning step t for the FL-ASSOM and for the BFL-ASSOM.

The translation-invariant filters generated by BFL-ASSOM compared to those by FL-ASSOM are shown in Fig. 3(a) with a gray scale. We can see that both methods generated topologically ordered Gabor-like filters. For either method, the two basis vectors at the same array locations have the same frequencies but 90 degrees of phase difference. Figure 3(b) shows how the average projection error e changes with the learning step t for the FL-ASSOM and for the BFL-ASSOM. For each input episode $\mathbf{X} = \{\mathbf{x}(s), s \in S\}$, where the component vectors $\mathbf{x}(s), s \in S$ are mean-subtracted and

Table 1. The timing results for the basic ASSOM, the FL-ASSOM and the BFL-ASSOM. VU (vector updating time) denotes the time for the basis vector updating. WL (whole learning time) denotes the time for the whole learning procedure, including module competition, basis vector dissipation and orthonormalization. All the times are given in seconds.

The basic ASSOM

	M=2		M=3		M=4	
	VU	WL	VU	WL	VU	WL
N=50	29.36	41.27	30.44	47.53	32.45	54.72
N=100	134.92	154.95	145.95	172.85	149.91	188.06
N=200	742.34	786.63	769.56	828.47	814.80	895.08
N=400	4529.69	4626.43	4956.64	5090.56	5200.78	5367.35

FL-ASSOM

	M=2		M=3		M=4	
	VU	WL	VU	WL	VU	WL
N=50	1.53	13.31	2.31	18.92	3.18	25.03
N=100	2.39	21.09	3.15	30.41	3.86	40.80
N=200	3.30	37.86	4.81	55.44	5.93	73.98
N=400	5.68	70.88	7.51	105.01	9.92	139.25

BFL-ASSOM

	M=2		M=3		M=4	
	VU	WL	VU	WL	VU	WL
N=50	0.58	4.83	1.28	6.72	1.33	8.75
N=100	0.87	6.88	1.50	10.01	1.58	13.37
N=200	1.03	11.61	1.67	17.37	2.10	22.86
N=400	1.46	21.02	2.06	31.01	2.99	41.51

normalized, the projection error is calculated according to $e(\mathbf{X}) = \sum_{s \in S} \|\mathbf{x}(s) - \hat{\mathbf{x}}(s)\|^2$, where $\hat{\mathbf{x}}(s)$ is the orthogonal projection of $\mathbf{x}(s)$ on the subspace of the winning module. e is the average of $e(\mathbf{X})$ over all the training episodes. We can see that the difference between the curves of FL-ASSOM and BFL-ASSOM is very tiny in Fig. 3(b), which reveals that their difference in terms of generating the filters is indeed very little.

In the second experiment, we compare the computational loads of the basic ASSOM, the FL-ASSOM and the BFL-ASSOM. We designed the experiment by using C++ implementations of all the methods. In this experiment, the input dimension as well as the subspace dimension vary. We count the elapsed CPU seconds for different methods. The number of iterations are fixed to $1,000$. Each episode is composed of 6 vectors. These vectors are generated randomly according to a uniform probability distribution. The rectangular ASSOM array contains 10×10 modules.

The timing results are summarized in Table 1. As was anticipated, the time of updating basis vectors with the basic ASSOM increased sharply with the input dimension and moderately with the subspace dimension. Basis vector updating is the bottleneck of the basic learning procedure, especially when the input dimension is high. With FL-ASSOM, it is clear that the time of updating basis vectors increases much more moderately with the input dimension. The response to the subspace dimension is also quite mild. Basis vector updating is no longer a bottleneck for the learning procedure. As a result, the learning time drops dramatically compared to the basic ASSOM. However the learning time outside the basis vector updating is not reduced. Now with BFL-ASSOM, we can observe that the basis vector updating time is further reduced. Moreover, the learning time outside the basis vector updating is also reduced considerably compared to the basic ASSOM and the FL-ASSOM.

The relationship between the basis vector updating time and the input dimension or the subspace dimension for the three implementations of ASSOM is visualized in Fig. 4. The basis vector updating time increases approximately linearly with respect to the input dimension for the FL-ASSOM and for the BFL-ASSOM, but apparently nonlinearly for the basic ASSOM. In all the cases, the updating time increases approximately linearly with respect to the subspace dimension.

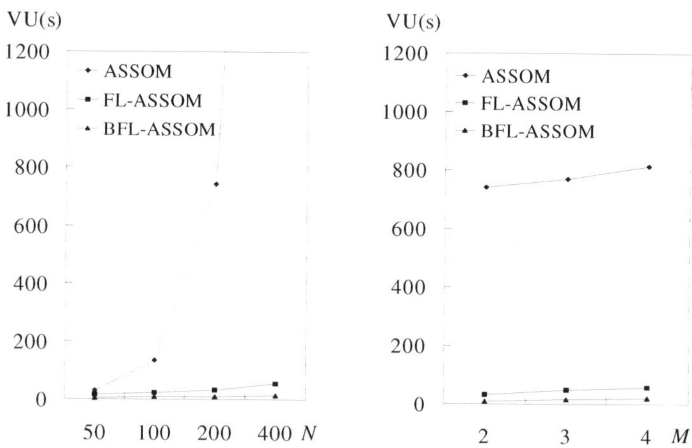

Fig. 4. *Left*: Relationship between the basis vector updating time (VU) and the input dimension N at the subspace dimension $M = 2$. *Right*: Relationship between VU and the subspace dimension M at the input dimension $N = 200$. For sake of clarity, the updating times of FL-ASSOM and BFL-ASSOM are magnified by a factor of 10

5 Conclusions

The focus of this paper is on the basis updating rule of the ASSOM learning. We first showed that the traditional basis rotation rule amounts to a correction made to the basis vectors which is a scaling of component vectors in the input episode. This gives us a better understanding of the basis updating in ASSOM learning. With this modified form of updating rule, some computations can be saved by reusing some intermediate computations. The resulting method is referred to as FL-ASSOM. A naive implementation of the traditional basis updating rule leads to a computational load linear to the subspace dimension but quadratic to the input dimension. This computational load is reduced by FL-ASSOM to be linear to both the subspace dimension and the input dimension. The ability of FL-ASSOM in generating topologically ordered invariant-feature filters is altogether preserved since the FL-ASSOM is mathematically equivalent to the basic

ASSOM. We then proceeded to present the BFL-ASSOM, where the basis vectors are updated in a batch mode. We showed that the correction made to each basis vector is a linear combination of the component vectors in the input episode. What's more, large amount of computations can be further saved by reusing more of previous computations and performing only one dissipation and orthonormalization for each episode.

Our experiments showed that the BFL-ASSOM can also generate topologically ordered Gabor-like translation-invariant filters and that the bottleneck of the basis vector updating in the learning procedure is totally removed by FL-ASSOM and BFL-ASSOM. The proposed methods can be easily adapted to the supervised ASSOM used in [6]. There is an obvious benefit in using the proposed rules instead of naive implementation of the basic learning rule.

References

1. Kohonen, T.: The Adaptive-Subspace SOM (ASSOM) and its use for the implementation of invariant feature detection. In Fogelman-Soulié, F., Gallinari, P., eds.: Proc. ICANN'95, Int. Conf. on Artificial Neural Networks. Volume 1., Paris (1995) 3–10
2. Kohonen, T.: Self-Organizing Maps. 3rd edn. Springer-Verlag, Berlin Heidelberg New York (2001)
3. Oja, E.: Principal components, minor components, and linear neural networks. Neural Networks **5** (1992) 927–935
4. Kohonen, T., Kaski, S., Lappalainen, H.: Self-organized formation of various invariant-feature filters in the Adaptive-Subspace SOM. Neural Computation **9**(6) (1997) 1321–1344
5. Hase, H., Matsuyama, H., Tokutaka, H., Kishida, S.: Speech signal processing using Adaptive Subspace SOM (ASSOM). Technical Report NC95-140, The Inst. of Electronics, Information and Communication Engineers, Tottori University, Koyama, Japan (1996)
6. Ruiz del Solar, J.: Texsom: texture segmentation using Self-Organizing Maps. Neurocomputing **21**(1–3) (1998) 7–18
7. De Ridder, D., Lemmers, O., Duin, R.P., Kittler, J.: The Adaptive Subspace Map for image description and image database retrieval. In Ferri, F., et al., eds.: SSPR&SPR 2000. Volume 1876 of LNCS., Springer-Verlag Berlin Heidelberg (2000) 94–103
8. Zhang, B., Fu, M., Yan, H., Jabri, M.: Handwritten digit recognition by Adaptive-Subspace Self-Organizing Map (ASSOM). IEEE Transactions on Neural Networks **10**(4) (1999) 939–945
9. López-Rubio, E., Muñoz Pérez, J., Gómez-Ruiz, J.: A Principal Components Analysis Self-Organizing Map. Neural Networks **17** (2004) 261–270
10. McGlinchey, S., Fyfe, C.: Fast formation of invariant feature maps. In: European Signal Processing Conference (EUSIPCO'98), Island of Rhodes, Greece (1998)
11. Oja, E.: Neural networks, principal components and subspaces. International Journal of Neural Systems **1** (1989) 61–68

Composite Algorithm for Adaptive Mesh Construction Based on Self-Organizing Maps*

Olga Nechaeva

Novosibirsk State University
Pirogova, 2, Novosibirsk, 630090, Russia
nechaeva@ssd.sscc.ru

Abstract. A neural network approach for the adaptive mesh construction based on Kohonen's Self-Organizing Maps (SOM) is considered. The approach belongs to a class of methods in which an adaptive mesh is a result of mapping of a computational domain onto a physical domain. There are some imperfections in using the SOM for mesh construction in a pure form. The composite algorithm to overcome these imperfections is proposed. The algorithm is based on the idea to alternate mesh construction on the border and inside the physical domain and includes techniques to control the consistency between boundary and interior mesh nodes and to provide an appropriate distribution of boundary nodes along the border of the domain. To increase the quality and the speed of mesh construction, a number of experiments are held to improve the learning rate. It has been shown that the quality of meshes constructed using the proposed algorithm is admissible according to the generally accepted quality criteria for finite difference meshes.

1 Introduction

Adaptive mesh methods enable us to improve the accuracy of numerical solution of problems without essential increase in the number of nodes. Nowadays, they have important applications in a variety of physical and engineering areas such as solid and fluid dynamics, combustion, heat transfer, material science, etc. [1].

Within the scope of all adaptive mesh methods, there is a class of methods in which to construct an adaptive mesh is to find a mapping of a computational domain with a given uniform mesh onto a physical domain with a desired adaptive one. This class of methods is usually used for problems with finite-difference approximation of equations. In conventional methods, e.g. the equidistribution method [2], Thompson's method [3], such a mapping is determined by solving a complicated system of nonlinear partial differential equations (PDE).

In this paper, the neural network approach for the construction of adaptive meshes resulting from the above mapping is considered [4]. The approach is based on Kohonen's Self Organizing Maps (SOM) [5]. The SOM is a neural network intended for topology preserving mapping of high-dimensional data onto a low-dimensional

* This work is supported by Grant of Rosobrazovanie, contract PHП.2.2.1.1.3653 and Program for Basic Research of RAS Presidium No. 14.15-2006.

S. Kollias et al. (Eds.): ICANN 2006, Part I, LNCS 4131, pp. 445 – 454, 2006.

space and used mainly for data analysis and visualization [6]. The property of topology preserving means that points that are near each other in the input space are mapped onto the neighboring neurons in the SOM.

The possibility of using the SOM for the adaptive mesh construction was studied earlier in [7,8]. But two main imperfections were discovered when applying the SOM in a pure form. The first consists in inaccurate mesh fitting the border of a domain, the second is associated with the construction of a mesh on non-convex domains. In this paper, the composite algorithm is proposed, which is originally based on the proposition [9] to alternate the mesh construction on the border and inside the domain. The composite algorithm enables one to overcome the first imperfection and to essentially improve the quality of meshes constructed on non-convex domains.

There are some difficulties in the above-mentioned conventional methods [2,3], which motivate the development of the new approaches to mesh construction. First, an efficient parallelization often meets overwhelming difficulties [10] conditioned by the necessity of numerical solution of nonlinear PDEs, especially, with the requirement of their compatibility with parallel implementation of a problem to be solved on the mesh. Second, for different dimensionalities of space and domain, particular nonlinear PDEs are required. Complexity of their numerical solution essentially rises with increasing dimensionalities. Finally, the above conventional methods require a preliminary construction of a good enough initial mesh and fixing mesh nodes on the border of a domain without changing their positions during the mesh construction [2].

The main advantages of the proposed approach in comparison with conventional methods are the following.

- The efficiency of parallelization [10] of the construction algorithm is high due to the internal parallelism of the SOM. Moreover, parallelization can be done according to the requirements for parallel implementation of a problem to be solved on the mesh.
- The approach uses the same algorithm of the mesh construction for different dimensionalities of both a physical domain and a space, where the domain is located.
- The algorithm of the mesh construction is able to start with arbitrary initial data and is simple to implement.

The paper is organized as follows. Section 2 describes the problem statement and the neural network approach for the adaptive mesh construction. Also, the idea of the composite algorithm is presented. In Section 3, the composite algorithm is illustrated in detail for the construction of 2D adaptive meshes on a plane. The learning rate functions selected based on experiments are proposed in Section 4. Section 5 contains examples of adaptive meshes constructed using the composite algorithm and the mesh quality evaluations. Section 6 concludes the paper.

2 Neural Network Approach

Let G be a physical domain in the Euclidean space $\mathbf{R}_{\text{ph}}$ on which an adaptive mesh G_N is to be constructed. Let Q be a computational domain in the Euclidean space $\mathbf{R}_{\text{c}}$ with a fixed mesh Q_N which is usually uniform. Also, the mesh density function $w : G \to \mathbf{R}^+$ is given such that the density of a desired adaptive mesh is to be proportional to the values of w. The problem is to find a mapping of Q onto G which transforms the mesh Q_N into the adaptive one G_N with the given mesh density. It is necessary for the mapping to ensure that the boundary nodes of the mesh Q_N are transformed into nodes distributed along the border of G.

Dimensionality of the space $\mathbf{R}_{ph}$ can be greater or equal to that of the space $\mathbf{R}_c$: $\dim(\mathbf{R}_{ph}) \geq \dim(\mathbf{R}_c)$. Also, dimensionalities of the domains G and Q are defined and considered to be equal to $\dim(\mathbf{R}_c)$. The border of the domain G has the dimensionality, which is equal to $\dim(\mathbf{R}_c) - 1$ and denoted as ∂G.

Let <dim($\mathbf{R}_{ph}$), dim($\mathbf{R}_c$)> be a configuration of dimensionalities for a given problem. The original idea of using the SOM for the mesh construction [7] is as follows. The SOM consists of two layers of neurons. In [7], it is proposed to use points selected in a random manner from G in the space $\mathbf{R}_{ph}$ as inputs for the first neuron layer of SOM. Therefore, the number of neurons in the first layer is set equal to $\dim(\mathbf{R}_{ph})$, and each neuron receives a corresponding coordinate of a random point. The second layer of the SOM is a lattice of neurons. It is convenient to associate this lattice with the mesh Q_N in such a way that the lattice structure is the same as that of the mesh Q_N, neurons of the lattice corresponding to nodes of Q_N, and hence, to the nodes of G_N. Each neuron of the second layer is connected by weighted connections to all the neurons from the first layer. The weights are adjusted during the unsupervised learning process of the SOM whereas the mesh structure is unchanged. The final weight values of the second layer neurons are the coordinates of the corresponding nodes of G_N in the space $\mathbf{R}_{ph}$. The density of the obtained mesh G_N is close to the probability distribution used for the random points generation.

The above algorithm is suitable for constructing a mesh strongly inside G, because the boundary nodes of G_N do not accurately reach the border of G during the mesh stretching over G (during the learning process). However, for solving a problem numerically, it is necessary for an adaptive mesh to be constructed both inside of the domain and on its border. To provide this condition, the composite algorithm is proposed which is based on the following.

The algorithm of the mesh construction inside a physical domain can be used for different configurations of dimensionalities <dim($\mathbf{R}_{ph}$), dim($\mathbf{R}_c$)>. For a given configuration, it is only necessary to set the number of neurons in the first layer and the structure of the second layer. Table 1 shows examples of meshes and the corresponding structures of SOM for some configurations of dimensionalities. Therefore, the algorithm can be applied for both the mesh construction inside G with the configuration <dim($\mathbf{R}_{ph}$), dim($\mathbf{R}_c$)> and the mesh construction on ∂G with the configuration <dim($\mathbf{R}_{ph}$), dim($\mathbf{R}_c$) -1>. In the second case, the border ∂G is considered to be a domain of smaller dimensionality, and random points from ∂G are used for the mesh construction on ∂G.

The composite algorithm is intended to combine, in a special way, the mesh construction for boundary and interior nodes. This algorithm is to provide the consistency between boundary and interior nodes during the mesh construction, to automatically distribute the boundary nodes along the border of G depending on its form so that those positions tend to be optimal, and to improve the quality of meshes constructed on non-convex domains.

Table 1. Examples of meshes and corresponding structures of SOM for different configurations of dimensionalities <dim($\mathbf{R}_{ph}$), dim($\mathbf{R}_c$)>

<2,1>	<2,2>	<3,2>	<3,3>
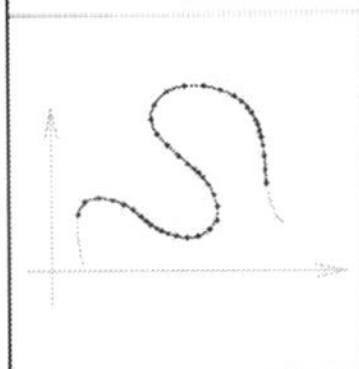	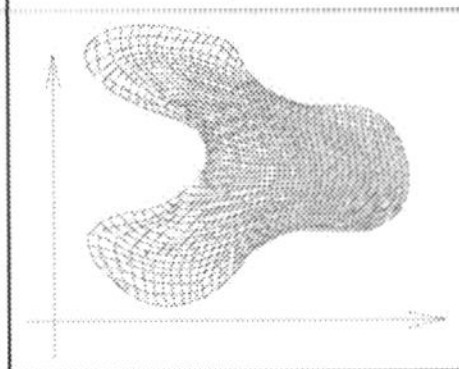	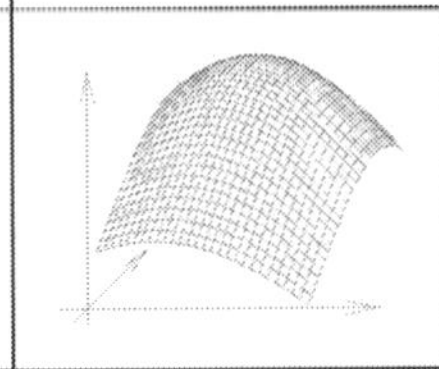	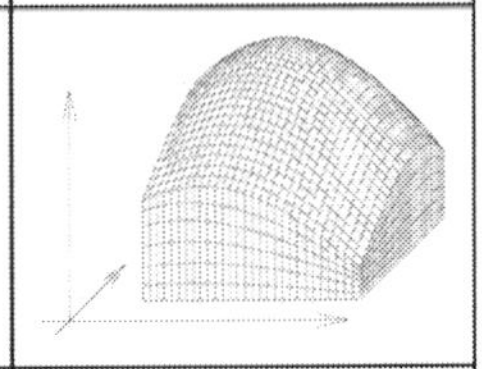
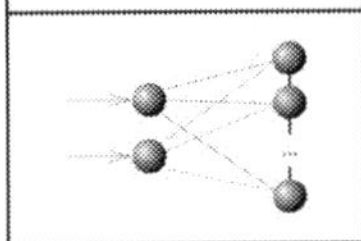	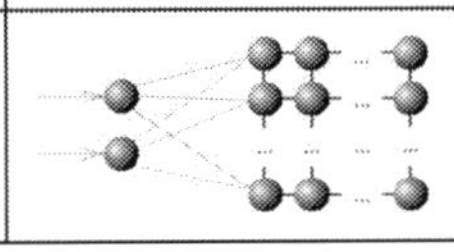	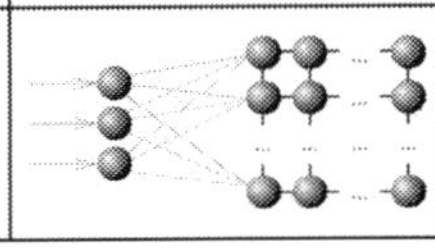	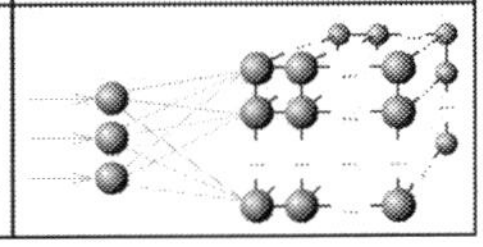

3 Composite Algorithm for 2D Adaptive Mesh Construction

In this section, the neural network approach is illustrated for $< 2, 2 >$ configuration of dimensionalities. Therefore, G is a 2D physical domain in the 2D Euclidean space $\mathbf{R}_{ph}$ with the physical coordinates $x = (x^1, x^2)$. For simplicity, we will consider the construction of adaptive meshes obtained by mapping of a uniform rectangular mesh Q_N of $N_1 \times N_2$ nodes with a computational domain being a rectangle $Q = [0, N_1-1] \times [0, N_2-1]$ and $Q_N = \{q_{ij} = (i, j), i = 0, ..., N_1-1, j = 0, ..., N_2-1\}$.

The general problem statement is to find a mapping of the computational domain Q onto the physical domain G transforming a uniform mesh Q_N into the adaptive one $G_N = \{x_{ij} = (x^1_{ij}, x^2_{ij}),\ i = 0, ..., N_1-1,\ j = 0, ..., N_2-1\}$ with mesh density distribution given by the positive density function $w : G \to \mathbf{R}^+$. For the numerical solution of problems on the mesh, it is sufficient to have only a discrete mapping Q_N onto G_N. The neural network approach enables one to obtain an inverse mapping G onto Q_N. The required mapping Q_N onto G_N is defined by weights of the second layer neurons.

Nodes coordinates of an initial mesh $G_N(0)$ can be arbitrary. For example, it is possible to set the initial positions of mesh nodes randomly or in such a way that they organize a uniform rectangular mesh. During the construction process, the initial mesh is being deformed and stretched all along the domain. The probability distribution, which is used for the random points generation, is to be proportional to values of the density function $w(x)$. The distribution can be given by the normed density function $w(x)$:

$$p(x) = \frac{w(x)}{\int_G w(x)dx}. \tag{1}$$

To present the composite algorithm for adaptive mesh construction, it is at first necessary to describe the mesh construction procedure inside G, which directly follows from the learning algorithm for the SOM. This procedure is denoted by

MeshConstruction and is listed below. Its parameters are: *Dom* – a domain, *WinnerSet* – a set of indices of the nodes among which the winner is determined, *AdjustSet* – a set of indices of the nodes changing their position at adjusting, *StartIter* – a number of the iteration, from which the procedure begins, *FinalIter* – a number of the iteration, at which the procedure terminates.

Algorithm 1. The body of the procedure *MeshConstruction* (*Dom*, *WinnerSet*, *AdjustSet*, *StartIter*, *FinalIter*).

Repeat the following operations at each iteration $t = StartIter, \ldots, FinalIter$:

1. *Point generation.* Generate a random point $y(t) = (y^1(t), y^2(t)) \in Dom$ according to the probability distribution $p(x)$ given in (1).
2. *Winner determination.* Calculate the Euclidean distances $d(\cdot,\cdot)$ between $y(t)$ and all the nodes $x_{ij}(t)$ and choose the node $x_{mn}(\mathrm{t})$ which is the closest to $y(t)$, i.e.

$$d(y(t), x_{mn}(t)) \le d(y(t), x_{ij}(t)), \tag{2}$$

   for all $(i, j) \in WinnerSet$. The node $x_{mn}(t)$ is called a *winner*.
3. *Node coordinates correction.* Adjust locations of mesh nodes using the following rule:

$$x_{ij}(t+1) = x_{ij}(t) + \theta_{mn}(t, i, j)(y(t) - x_{ij}(t)), \tag{3}$$

   for all $(i, j) \in AdjustSet$, where $\theta(t, i, j) \in [0, 1]$ is a *learning rate*.

At each iteration t, mesh nodes move towards the generated point $y(t)$. The size of displacement for each node is defined by the learning rate $\theta_{mn}(t, i, j)$. The quality and the speed of the mesh construction depend on the selection of $\theta_{mn}(t, i, j)$.

Let us denote the subsets of indices: $Interior = \{(i, j) \mid i = 1, \ldots, N_l - 2, j = 1, \ldots, N_2 - 2\}$ defines a set of interior mesh nodes, $Boundary = \{(i, j) \mid i = 0 \vee i = N_l - 1 \vee j = 0 \vee j = N_2 - 1\}$ defines a set of boundary mesh nodes.

In the composite algorithm, the mesh construction on the border and inside G alternates. Each stage of the alternation is referred to as macroiteration, whose number is denoted by s. At each macroiteration, the procedure *MeshConstruction* is performed twice – for the boundary and the interior mesh nodes. To set values of the parameters *StartIter* and *FinalIter* of the procedure *MeshConstruction*, two integer-valued functions of integer arguments are defined: $\Phi(s)$ and $\Psi(s)$.

Algorithm 2. The composite algorithm.

0. Set initial locations of mesh nodes $x_{ij}(0) = (x^1_{ij}(0), x^2_{ij}(0))$, where $(i, j) \in Interior \cup Boundary$.
1. Perform the procedure *MeshConstruction* with the following parameters: $Dom = G$, $WinnerSet = Interior \cup Boundary$, $AdjustSet = Interior \cup Boundary$, $StartIter = 1$, $FinalIter = \Phi(0)$. In other words, all neurons are learned by random points from the whole domain G during the given number of iterations $\Phi(0)$.
2. Repeat the following operations at each macroiteration $s \ge 1$:
   a) Perform the procedure *MeshConstruction* with the following parameters: $Dom = \partial G$, $WinnerSet = Boundary$, $AdjustSet = Boundary$, $StartIter =$

$\Psi(s-1)+1$, *FinalIter* = $\Psi(s)$. In other words, only the neurons corresponding to the boundary nodes are learned by random points from the border ∂G.

b) Perform the procedure *MeshConstruction* with the following parameters: *Dom* = *G*, *WinnerSet* = *Interior* ∪ *Boundary*, *AdjustSet* = *Interior*, *StartIter* = $\Phi(s-1)+1$, *FinalIter* = $\Phi(s)$. In other words, the winner is determined among all the neurons, but locations only of the interior mesh nodes are corrected.

3. The process terminates when the condition $\Phi(s) > T$ is satisfied.

Step 1 of the composite algorithm is essential because at this step the whole mesh roughly takes the form of the domain. After that, the boundary nodes tend to their optimal position on the border of *G*, e.g. the corner nodes x_{00}, x_{0,N_2-1}, $x_{N_1-1,0}$ and x_{N_1-1,N_2-1} try to be placed on the border of the most convex parts of *G*.

The consistency between the boundary and the interior mesh nodes is provided at Step 2, b). Since the winner is determined among all mesh nodes at this step, the interior nodes follow the boundary nodes that are close to them. This accelerates the movement of the interior nodes to the domain border and keeps connections between the interior and the boundary nodes that are close to each other. In particular, this technique helps one to improve the quality of meshes in the case of the non-convex domain *G* in comparison with using the SOM for the mesh construction without alternating (Fig.1.).

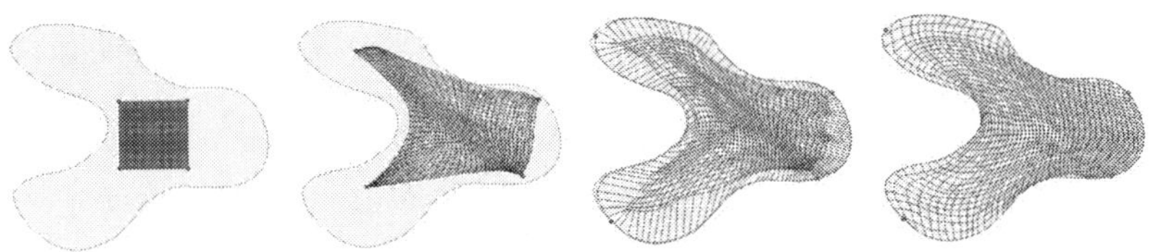

Fig. 1. The process of mesh construction using the composite algorithm

The functions $\Phi(s)$ and $\Psi(s)$ are chosen as follows.

$$\Phi(s)=\begin{cases}\Phi_0, & s=0\\ \Phi_0+\sum_{k=1}^{s}\varphi(k), & s=1,2,\dots\end{cases},\qquad \Psi(s)=\begin{cases}\Phi_0, & s=0\\ \Phi_0+\sum_{k=1}^{s}\psi(k), & s=1,2,\dots\end{cases},\tag{4}$$

where $\varphi(k)$ and $\psi(k)$ are also integer-valued functions of integer arguments.

Based on the experiments it was found that the best results are obtained when $\varphi(k)$ increases and $\psi(k)$ decreases. As a result, the boundary nodes are gradually frozen giving internal nodes an advantage until the termination of the composite algorithm. In our investigations, the following functions are used: $\varphi(k)=\lfloor 500*(1-\exp(-2k/15))\rfloor$ and $\psi(k)=\lfloor 500*\exp(-k^2/200))\rfloor$, where $\lfloor\cdot\rfloor$ is an integer part. Diagrams of these functions are shown in Fig.2.

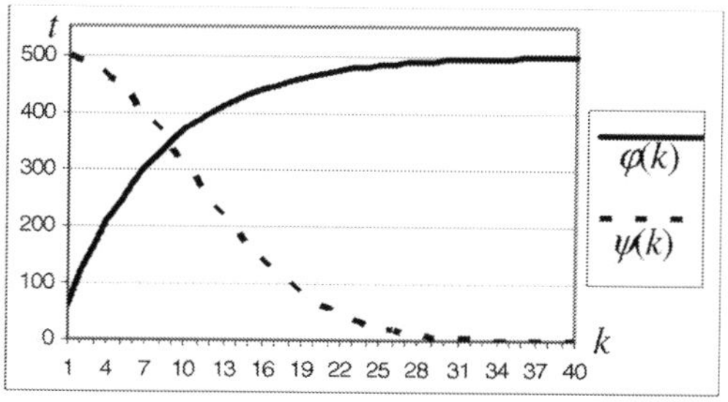

Fig. 2. Diagrams of the functions $\varphi(k)$ and $\psi(k)$

4 Learning Rate Selection

The learning rate influences the quality and the speed of mesh construction. The mesh smoothness and the absence of mesh self-crossings depend on the ratio between a learning step and a learning radius. The learning rate functions proposed below are improved in comparison with those in [7]. The strategy of fixed number of iterations has been chosen so as T is settled before the algorithm starts.

The general form of the function $\theta_{mn}(t, i, j)$ is usually given as product of two functions [11]. The first function is responsible for the learning step, the second one for the learning radius:

$$\theta_{mn}(t, i, j) = \delta(t)\eta_{mn}(t, i, j). \tag{5}$$

The function $\delta(t)$ decreases, thus, reducing all the displacements of mesh nodes. This guarantees that the iteration process of mesh construction always converges. Based on the experiments, the function $\delta(t)$ has been selected as $\delta(t) = t^{-0.2}\chi(t)$, $t = 1, ..., T$, where $\chi(t) = 1 - e^{5(t-T)/T}$, T is a maximum number of iterations. The function $\chi(t)$ is used to make a power member of $\delta(t)$ turn into zero (Fig.3.).

The function $\eta_{mn}(t, i, j)$ is responsible for the distribution of the sizes of node displacements at the iteration t. This function provides the conditions, according to which the winner receives a maximum displacement, while other nodes change their locations the less the larger the distance between the nodes q_{ij} and q_{mn} in the computational domain Q. Usually, $\eta_{mn}(t, i, j)$ is given by the following exponent [11]:

$$\eta_{mn}(t, i, j) = e^{-\frac{d(q_{ij}, q_{mn})^2}{2\sigma^2(t)}}, \tag{6}$$

where $d(q_{ij}, q_{mn})$ is the Euclidean distance between the nodes q_{ij} and q_{mn} in the space $\mathbf{R}_c$, and q_{mn} corresponds to the winner node x_{mn}.

Based on the experiments the function $\sigma(t)$ has been selected as $\sigma(t) = \sigma(T) + \chi(t)\left(\sigma(1)0{,}05^{t/T} - \sigma(T)\right)t^{-0.25}$ (Fig.3.). In our investigations, the values of $\sigma(1)$ and $\sigma(T)$ were determined in such a way that at the first and at the last iterations the given number of nodes receive the given displacements:

$\sigma(1) = d(q_{00}, q_{N_1-1,N_2-1}) / \sqrt{-2\ln(0.05)}$ and $\sigma(T) = d(q_{00}, q_{22}) / \sqrt{-2\ln(0.05)}$. These values were obtained from the equation $\eta_{mn}(t, i, j) = 0.05$ after substitution of the corresponding nodes in (6).

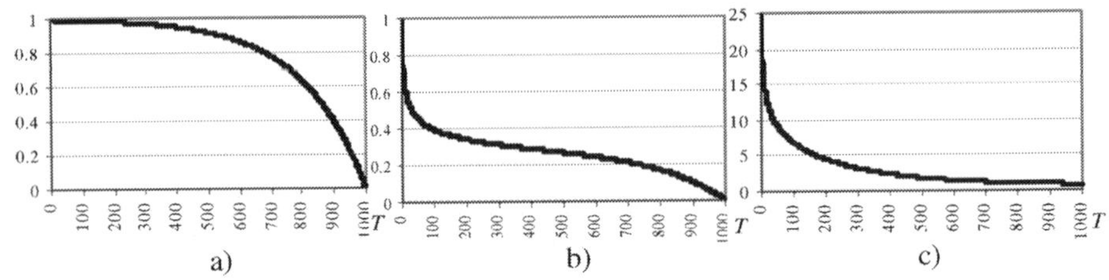

Fig.3. Diagrams of: a) function $\chi(t)$, b) function $\delta(t)$, c) function $\sigma(t)$

5 Experiments

Fig.4.a) shows an example of the adaptive mesh constructed using the composite algorithm. In the case of a mesh structure other than quadrilateral, the composite algorithm is the same. The difference can be only in enumeration of mesh nodes. Fig.4. shows an example of a hexagonal adaptive mesh constructed using the composite algorithm.

There are generally used the quality criteria for quadrilateral finite difference meshes such as the criteria of cell convexity and oblongness, the criterion of mesh lines orthogonality [12]. The criteria allow obtaining a preliminary information about a mesh before solving a problem on it and rejecting beforehand the meshes, which are unsuitable

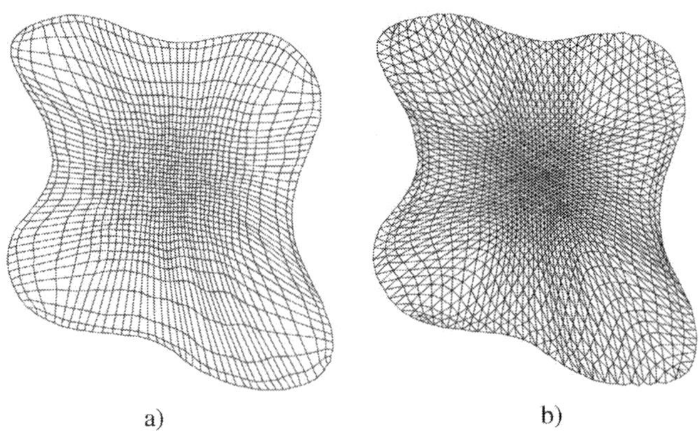

Fig. 4. Examples of adaptive meshes constructed using the composite algorithm with $T = 10000$, $N_1 = 40$, $N_2 = 40$. a) quadrilateral mesh, b) hexagonal mesh.

Table 2. Three quality criteria of finite-difference meshes with their possible and admissible values

Quality criterion	Possible values	Admissible values	Cell form for particular values
Cell convexity	$(-\infty; 1]$	$[0; 1]$	=1 – parallelogram =0 – triangle <0 – non-convex cell
Lines orthogonality	$[-1; 1]$	$[0; 1]$	=1 – rectangular =0 – triangle
Cell oblongness	$(0; 1]$	Depending on a problem	=1 – rhomb

with certainty. In Table 2 possible and admissible values of the criteria are shown. The maximum values of the criteria correspond to high-quality meshes.

The quality of quadrilateral meshes constructed using the composite algorithm has been evaluated using the above criteria. Table 3 shows the constructed meshes and the corresponding average and minimum values of the criteria among all the mesh cells. The values are in the admissible range.

Table 3. Examples of meshes constructed using the composite algorithm and the corresponding average and maximum values of the quality criteria

Mesh / Criterion	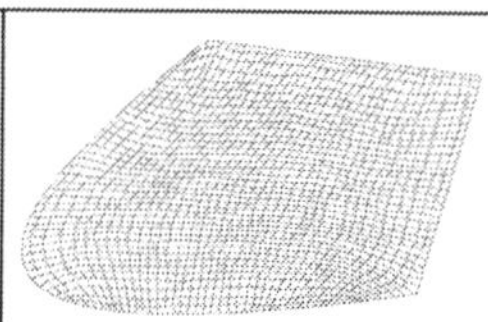	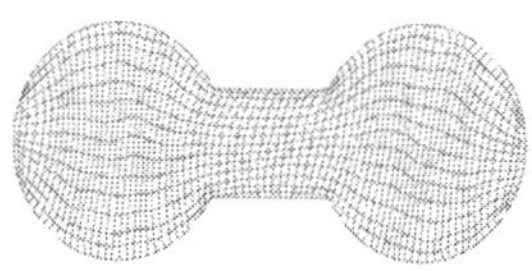	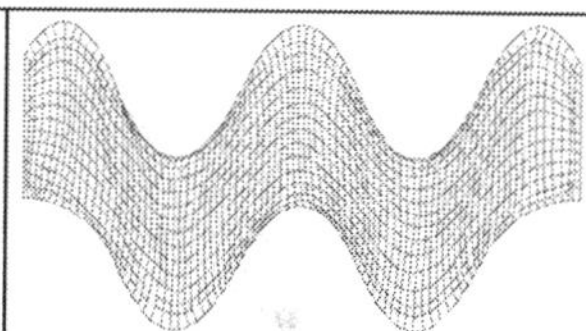
Convexity	aver.: 0.9325 min.: 0.0424	aver.: 0.9266 min.: 0.0454	aver.: 0.9280 min.: 0.1033
Ortogo-nality	aver.: 0.8882 min.: 0.0200	aver.: 0.9119 min.: 0.0587	aver.: 0.7214 min.: 0.0196
Oblongness	aver.: 0.6952 min.: 0.3155	aver.: 0.5433 min.: 0.0733	aver.: 0.6304 min.: 0.1732

6 Conclusion

The proposed composite algorithm within the neural network approach enables us to automatically construct adaptive meshes both inside and on the border of physical domains with arbitrary initial data. The composite algorithm is the same for different configurations of dimensionalities and is simple to parallelize. The learning rate, being thoroughly selected, provides the construction of qualitative adaptive meshes.

In the future, the algorithm for mesh smoothing is to be developed based on the learning of the SOM with special learning parameters. Also the comparative analysis between the neural network approach and the equidistribution method is to be made in terms of numerical solution accuracy and parallelization efficiency of both methods.

References

1. Lebedev, A.S., Liseikin, V.D., Khakimzyanov, G.S.: Development of methods for generating adaptive grids. Vychislitelnye tehnologii, Vol. 7, No. 3. (2002) 29
2. Khakimzyanov, G.S., Shokin, Yu.I., Barakhnin, V.B., Shokina, N.Y.: Numerical Modelling of Fluid Flows with Surface Waves. SB RAS, Novosibirsk (2001)
3. Thompson, J.F., Warsi Z.U.A., Mastin C.W.: Numerical grid generation, foundations and applications. North-Holland, Amsterdam (1985)
4. Nechaeva, O. I.: Neural network approach for adaptive mesh construction. Proc. of VIII National scientific conference "NeuroInformatics-2006". Part 2. MEPhI, Moscow (2006) 172-179
5. Kohonen, T.K.: Self-organization and associative memory. Springer Verlag, New York (1989)
6. Flexer, A.: On the use of self-organizing maps for clustering and visualization. Intelligent Data Analysis, Vol 5. IOS Press (2001) 373-384
7. Nechaeva, O. I.: Adaptive curvilinear mesh construction on arbitrary two-dimensional convex area with applying of Kohonen's Self Organizing Map. Neuroinformatics and its applications: The XII National Workshop. ICM SB RAS, Krasnoyarsk (2004) 101-102
8. Manevitz, L., Yousef, M., Givoli, D.: Finite Element Mesh Generation Using Self-Organizing Neural Networks. Special Issue on Machine Learning of MicroComputers in Civil Engineering, Vol. 12, No. 4. (1997) 233-250
9. Manevitz, L.: Interweaving Kohonen Maps of Different Dimensions to Handle Measure Zero Constraints on Topological Mappings. Neural Processing Letters, Vol. 5, No. 2. (1997) 83-89
10. Nechaeva, O.: Neural Network Approach for Parallel Construction of Adaptive Meshes. In: V. Malyshkin (ed.): Parallel Computing Technologies 2005. Lecture Notes in Computer Science, Vol. 3606. Springer, Berlin Heidelberg (2005) 446-451
11. Ghahramani, Z.: Unsupervised Learning. In: Bousquet, O. et al. (eds.): Machine Learning 2003. Lecture Notes in Artificial Intelligence, Vol. 3176. Springer-Verlag, Berlin Heidelberg (2004) 72–112
12. Prokopov, G. P.: About organization of comparison of algorithms and programs for 2D regular difference mesh construction. Preprint / Keldysh's Institute for Applied Mathematics AS USSA, No. 18. Moscow (1989)

A Parameter in the Learning Rule of SOM That Incorporates Activation Frequency

Antonio Neme[1,2] and Pedro Miramontes[2]

[1] Universidad Autónoma de la Ciudad de México, México, D.F., Department of Nonlinear Dynamics and Complex Systems, México
neme@nolineal.org.mx
[2] Universidad Nacional Autónoma de México, Facultad de Ciencias, México

Abstract. In the traditional self-organizing map (SOM) the best matching unit (BMU) affects other neurons, through the learning rule, as a function of distance. Here, we propose a new parameter in the learning rule so neurons are not only affected by BMU as a function of distance, but as a function of the frequency of activation from both, the BMU and input vectors, to the affected neurons. This frequency parameter allows non radial neighborhoods and the quality of the formed maps is improved with respect to those formed by traditional SOM, as we show by comparing several error measures and five data sets.

1 Introduction

Self-organizing map (SOM) is presented as a model of the self-organization of neural connections, what is translated in the ability of the algorithm to produce organization from disorder [1]. One of the main properties of SOM is its ability to preserve topographical relations present in input data in the output map [2], which is a desirable property for data visualization and clustering.

One main feature of the SOM is the ability to transform an incoming signal pattern of arbitrary dimension into a low-dimensional discrete map (usually of dimension one or two) and to adaptively transform data in a topologically ordered fashion [3, 4]. Each input data is mapped to a single neuron in the lattice, that with the closest weight vector to the input data, or best matching unit (BMU). The SOM preserves relationships during training through the neighbourhood function, which establishes the effect of the BMU to any other neuron. Weight neurons are updated accordingly to:

$$w_n(t+1) = w_n(t) + \alpha_n(t)h_n(g,t)(x_i - w_n(t)) \quad (1)$$

Where $\alpha(t)$ is the learning rate at time t and $h_n(g,t)$ is the neighbourhood function from BMU neuron g to neuron n at time t. In general, neighbourhood function decreases monotonically as a function of the distance from neuron g to neuron n. This decreasing property has been reported to be a necessary condition for convergence [5, 6]. The SOM tries to preserve relationships of input data by starting with a large neighbourhood and reducing the neighbourhood size during

S. Kollias et al. (Eds.): ICANN 2006, Part I, LNCS 4131, pp. 455–463, 2006.

the course of training [4]. It has been reported as well that the learning factor α should be a decreasing function [6].

As pointed out by Ritter [7], SOM and related algorithms share the idea of using a deformable lattice to transform data similarities into spatial relationships. The lattice is deformed by applying learning equation (1) to the neurons in the network. In this work, we propose an additional parameter that quantifies the influence of a BMU n to the neurons in the network as a function of the number of times n affects them as well as the influence of each data vector m as a function of the number of times the BMU for m affects the neurons. This frequency activation parameter allows non radial neighborhood which, as reported in the results, forms better maps, in terms of three error measures.

2 Related Work

Altough several modifications have been proposed to the SOM learning rule, they don't reflect, at least to our knowledge, the frequency of activation from other neurons. For example, Lee and Verleyen [8] propose the recursive Fisherman's rule and some hybrids from it that reflects an attenuation of the adaptation as the distance from the BMU to the affected neuron grows. The rules show a non radial neighborhood in the sense that the BMU pulls the direct neighbors and these neighbors pull farther neurons and so on, in a recursive manner.

Campoy and Vicente [9] proposed a residual activity memory for each neuron, so the SOM enlarges its temporal analysis capabilities, whereas one of the first works that incorporated a concept of memory for each neuron was in Chappell and Taylor [10], in which is defined an activation memory for each neuron, in order to define the new active neuron, and a modification of the selection mecanism is presented, so if the memory parameter is high, the previous winner neuron will win again unless another neuron matches very close the input data.

3 Frequency Function in the SOM's Learning Rule

In the traditional SOM, the BMU equally affects those neurons within its neighborhood. All neurons at the same distance (for the case of gaussian neighborhood) or inside the hypersphere (for the case of bubble neighborhood) are equally affected. We propose a modification to this scheme that includes a function of the relative frequency a given neuron n is affected by each BMU k or by each input vector m.

If during the learning process n is affected by two or more BMU (for one or more input vectors), it will not be affected the same by them (independently from the learning factor): the more a neuron n is affected by k, the larger the strenght of its influence. A new parameter, the activation frequency, $\rho_n(k, m)$, that is a function of the number of times a neuron n is affected by BMU k or by input vector m is incorporated to eq. (1). The weight modification rule is now:

$$w_n(t+1) = w_n(t) + \alpha_n(t)h_n(g,t)\rho_n(k,m)(x_i - w_n(t)) \quad (2)$$

In this model, every neuron n maintains a record of the relative frequency by which it has been affected by each BMU k, $\Omega_n(k)$, defined as the number of times BMU k has included n in its neighborhood divided by the number of times n has been affected by any BMU ($\sum_{j=1}^{|N|} \Omega_n(j)$, where N is the number of neurons in the network). Also, n has a record of the relative frequency it has been affected by each input vector m, $\beta_n(m)$, defined as the number of times vector m has affected, through any BMU, n, divided by the number of times n has been affected by any BMU (which is the same as the the number of times it has been affected by any input vector). For the gaussian neighborhood, we have defined n is influenced by k if $h_n(g,t) > 0.3$

Several frequency functions are proposed based on these two quantities and in the distance between BMU k and neuron n, $d(n,k)$. The frequency parameter $\rho_n(k,m)$ varies as a function of $\Omega_n(k)$ and $\beta_n(m)$. We have found that $\rho_n(k,m)$ should be monotonic decreasing function with respect to $\Omega_n(k)$, as it is shown in eq. (3)-(6), so the formed maps present a lower error than the formed maps by eq (1).

$$(Rule\ \ 1) \qquad \rho_n(k,m) = \Omega_n(k) \times \frac{1}{d(n,k)} \tag{3}$$

$$(Rule\ \ 2) \qquad \rho_n(k,m) = \Omega_n(k) \times \beta_n(m) \times \frac{1}{d(n,k)} \tag{4}$$

$$(Rule\ \ 3) \qquad \rho_n(k,m) = \frac{1}{1 + e^{-\psi \Omega_n(k) \times \frac{1}{d(n,k)}}} \tag{5}$$

$$(Rule\ \ 4) \qquad \rho_n(k,m) = \Omega_n(k) \tag{6}$$

As an example of the behaviour of the rules, fig. (1) shows the case for rules (1) and (3). Two neurons, i and j, such that $d(k,i) = d(k,j)$ will not be affected the same unless $\Omega_i(k) = \Omega_j(k)$. However, when $d(k,i)$ is large, the difference in the frequency activation woll be small. For low values of $d(k,i)$ the importance of $\Omega_i(k)$ becomes clear.

The proposed activation frequency functions modifies the natural neighborhood of BMU. For example, in fig. (2) it is shown the BMU and the affected neurons in four different time steps for a single data vector. It is observed that non radial and discontinuos neighborhoods are formed, which are not present in the traditiobal scheme. This discontinuity resembles the cortex activity patterns in mammals during several task processing [13].

4 Topological Preservation Quantization

To measure topological preservation, three metrics were applied. Those are the topographic error (TE) [11], error quantization (EQ) and preservation of original neighborhoods (VC) [12]. The first is defined as:

$$TE_t = \tfrac{1}{N} \textstyle\sum_{k=1}^{N} \eta(x_k), \text{ where } \eta(x_k) = \begin{cases} 1, \text{ BMU and 2nd. BMU non adjacent} \\ 0, \text{ otherwise} \end{cases}$$

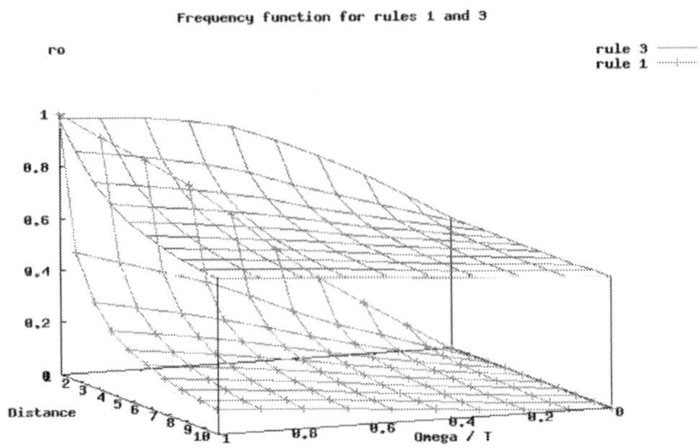

Fig. 1. Activation frequency function for rules 1, 3. For two neurons i, j situated at the same distance, $\rho_i(\Omega_i(k)) > \rho_j(\Omega_j(k))$ if $\Omega_i(k) > \Omega_j(k)$.

which is simply the proportion of data vector for which the BMU and second best matching unit are not first-neighbors. The errror quantization is:

$$EQ = \frac{1}{N}\sum_{j=1}^{N} ||x_j - w_j||^2$$

The third metric is based on the neighbohood preservation quantization, which establishes that an input data vector i has k neighbors in its neighborhood V_k in the original space and, if neighborhood preservation is complete, then, the BMU for i has as its first k active neurons those BMU for the data vectors in V_k.

$$VC = 1 - \frac{2}{Nk(2N-3k-1)}\sum_{i=1}^{N}\sum_{x_j \in V_k(x_i)}(r(x_i, x_j) - k)$$

where N is the number of data vectors and $r(x_i, x_j)$ is the rank of x_j when data vectors are ordered based on their distance from the data vector x_i after projection. The closer VC is to 1, the better the map is. As VC is a function of k, we set a value for k as $\frac{1}{10}$ of the size of each data set.

5 Results

The experiments were done in a 10x10 network. Two groups of experiments were done. In the first one, in order to test the sensivity of the proposed frequency functions to the initial values, several thousands of maps were formed(> 10000), with different initial neighborhood width h_0, learning factor α_0 and for a different number of epochs. Sensitivity results are presented in subsection 5.1. The second group of experiments establishes the ability of the proposed rules as a good alternative to form maps with a lower error that the traditional learning rule. In this group, several hundreds maps were formed, all with 1000 epochs, $\alpha_0 \in \{0.1, 0.2, 0.5, 0.7, 0.8, 0.9\}$ and $h_0 \in \{8, 9\}$, and with $\alpha_{1000} \in \{0.01, 0.001\}$, for the

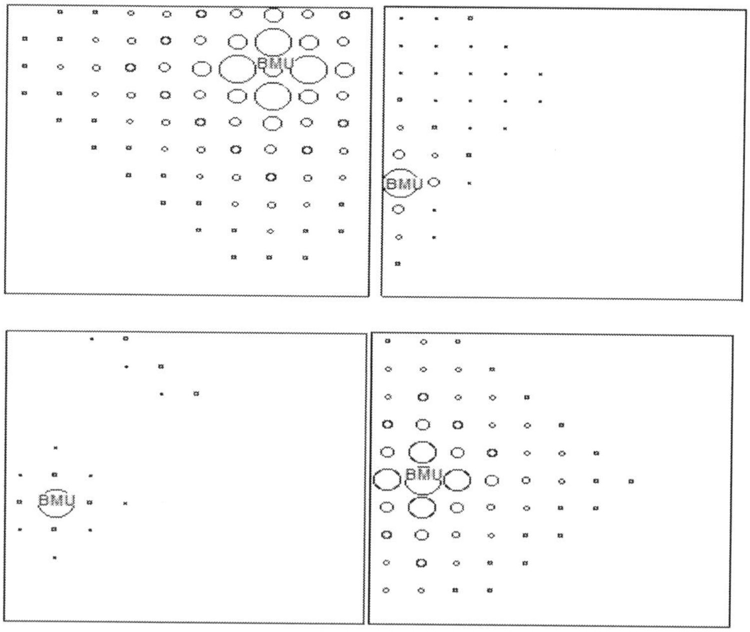

Fig. 2. BMU and affected neurons at $t = 1$, $t = 2$, $t = 3$ and $t = 9$ starting at top left for a given input vector from the spiral data set. Size of circumference is proportional to $\rho_i(k)$. In $t = 3$, there is a discontinuity in the area on influence for the BMU and for $t = 2$ there is a non radial neighborhood.

self-organizing stage, with an exponential decrease of the intermediate values for those parameters. In the second stage, convergence, the number of epochs was 10000, with initial values of $\alpha = 0.05$ and $h = 3$, exponentially decreasing until the final values $\alpha = 0.0001$ and $h = 0$. Subsection 5.2 shows convergence results.

5.1 Sensitivity

To test sensitivity to the initial conditions, SOM with several parameters were formed. Neighborhood width, h_0, was placed between 1 and 10, whereas the initial learning factor, α_0, is in the close interval $[0.001, 1.0]$ and the number of epochs is situated between 1 and 130. The final values for neighborhood width is 1 and for the learning parameter is 0.001. Altough the number of maps may be insufficient to cover the whole range of combinations for the former parameters and the choises may be arbitrary, they cover a wide range of combinations.

For each one of the five data sets and for each set of values (α_0 and h_0), a SOM were formed by the traditional SOM rule and by SOMs with the activation frequency functions described in (3) -(6), for both, bubble and gaussian neighborhood. Two data sets are bidimensional, one is five- dimensional (iris data set), one is 34-dimensional (ionosphere data set) and one is 64-dimensional (codon usage data set).

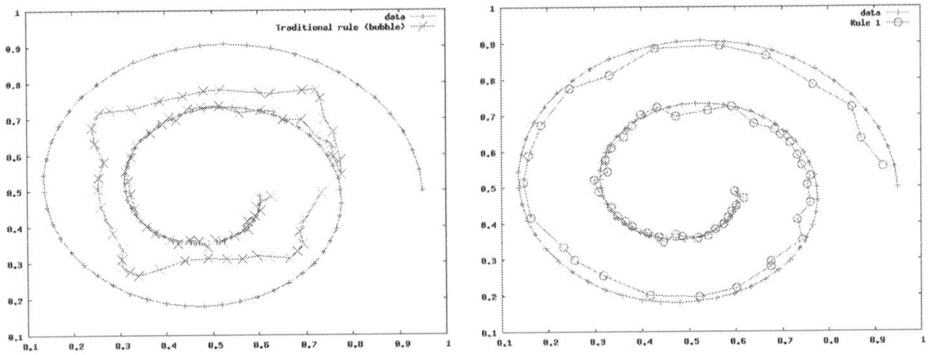

Fig. 3. Spiral data set. It is shown the SOM approximation for both, traditional rule (left) and for rule 1 (right, eq. (3)) after 10 epochs.

Table 1. Average TE, EQ and VC for the spiral and Henón data sets over all formed maps for the proposed frequency functions. Error is presented in pairs: (bubble neighborhood, gaussian neighborhood).

Rule	Spiral			Henón		
	TE	EQ	VC	TE	EQ	VC
Trad.	(0.17, 0.17)	(0.019, 0.018)	(0.71, 0.6)	(0.17, 0.29)	(0.044, 0.041)	(0.83, 0.73)
Rule 1	(0.107, 0.23)	(0.010, 0.011)	(0.64, 0.66)	(0.21, 0.22)	(0.031, 0.03)	(0.9, 0.75)
Rule 2	(0.106, .24)	(0.013, 0.013)	(0.64, 0.63)	(0.19, 0.31)	(0.028, 0.012)	(0.91, 0.89)
Rule 3	(0.104, 0.261)	(0.012, 0.011)	(0.66, 0.62)	(0.173, 0.2)	(0.025, 0.016)	(0.89, 0.89)
Rule 4	(0.102, 0.21)	(0.011, 0.015)	(0.67, 0.66)	(0.176, 0.21)	(0.026, 0.018)	(0.89, 0.89)

Table 2. Average TE, EQ and VC for the iris and ionosphere data sets over all formed maps for the proposed frequency functions. Error is presented in pairs: (bubble neighborhood, gaussian neighborhood).

Rule	Iris			Ionosphere		
	TE	EQ	VC	TE	EQ	VC
Trad.	(0.241, 0.21)	(0.0673, 0.08)	(0.68, 0.715)	(0.17, 0.196)	(0.08, 0.061)	(0.73, 0.74)
Rule 1	(0.206, 0.282)	(0.061, 0.079)	(0.83, 0.71)	(0.113, 0.195)	(0.081, 0.054)	(0.83, 0.75)
Rule 2	(0.206, 0.272)	(0.062, 0.06)	(0.82, 0.84)	(0.11, 0.195)	(0.082, 0.055)	(0.83, 0.91)
Rule 3	(0.235, 0.203)	(0.07, 0.053)	(0.8, 0.82)	(0.165, 0.145)	(0.082, 0.058)	(0.87, 0.89)
Rule 4	(0.228, 0.21)	(0.06, 0.054)	(0.82, 0.88)	(0.165, 0.169)	(0.083, 0.0.69)	(0.83, 0.85)

Table 3. Average TE, EQ and VC for the codon usage data set over all formed maps for the proposed frequency functions. Error is presented in pairs: (bubble neighborhood, gaussian neighborhood).

Rule	TE	EQ	VC
Trad.	(0.175, 0.12)	(0.24, 0.14)	(0.72, 0.72)
Rule 1	(0.103, 0.1)	(0.22, 0.17)	(0.79, 0.77)
Rule 2	(0.1, 0.098)	(0.21, 0.17)	(0.82, 0.76)
Rule 3	(0.167, 0.256)	(0.29, 0.289)	(0.83, 0.82)
Rule 4	(0.167, 0.2)	(0.19, 0.17)	(0.82, 0.71)

As the analysis of data is extensive, only some results are presented (the whole set is available from the authors). In tables (1) - (3), the average error for all maps formed by the proposed frequency functions, as well as for those generated by traditional SOM with both, bubble and gaussian neighborhood, including all initial conditions for α_0, h_0 and the number of iterations, are presented, for each one of the five data sets.

Fig. (4) shows the TE as a function of number of epochs for the traditional SOM and for SOM's with the frequency functions proposed in eqs. (3) -(6) for four of the data sets. It can be seen that the maps formed by the proposed rules are less sensitive to initial conditions that those formed by eq. (1). From fig. (3), it is clear that for the spiral data set the proposed functions form maps that folds more accurately to data than those formed by the traditional SOM.

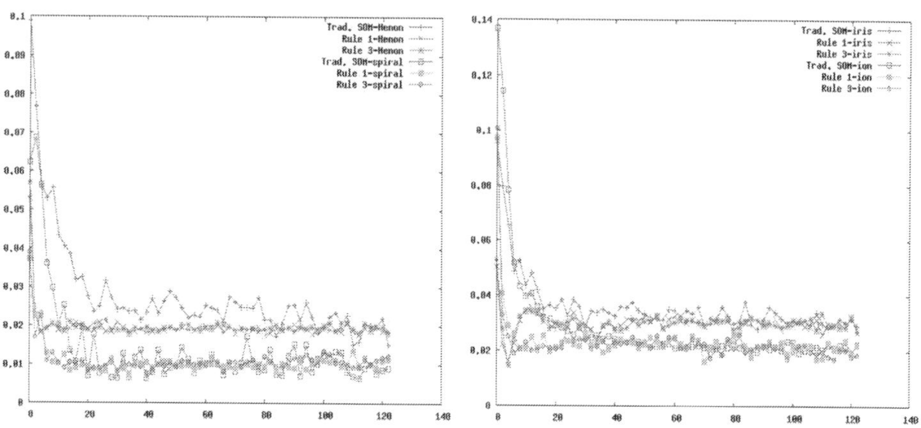

Fig. 4. TE for rules 1 and 3 and for the traditional learning rules as a function of the number of iterations for the spiral, Henón, iris and ionosphere data sets

5.2 Convergence

Once the sensitivity was analyzed, the properties of the proposed rules for maps suitable for data visualization (low error) were studied. In tables (4) and (5) it is shown the average error for several maps (> 500) for a larger number of epochs and two-stages differentiation, in contrast to what was done in the previous subsection. It is observed that the error measures are, in general, lower for those maps formed with the proposed rules. Also, a map formed by traditional rule and another formed by rule 1 are shown in fig.3.

Table 4. Average TE, EQ and VC obtained after two stages of trainning of the SOM, for the spiral and Henón data sets over all formed maps for the proposed frequency functions. Error is presented in pairs: (bubble neighborhood, gaussian neighborhood).

Rule	Spiral			Henón		
	TE	EQ	VC	TE	EQ	VC
Trad.	(0.01, 0.009)	(0.0001, 0.0001)	(0.92, 0.93)	(0.018, 0.016)	(0.0015, 0.0012)	(0.91, 0.93)
Rule 1	(0.01, 0.008)	(0.00003, 0.00002)	(0.94, 0.95)	(0.013, 0.005)	(0.001, 0.0008)	(0.9, 0.91)
Rule 2	(0.02, 0.018)	(0.00002, 0.00001)	(0.95, 0.96)	(0.012, 0.01)	(0.001, 0.0004)	(0.9, 0.92)
Rule 3	(0.015, 0.019)	(0.00001, 0.00001)	(0.95, 0.95)	(0.013, 0.014)	(0.0006, 0.0004)	(0.89, 0.92)
Rule 4	(0.018, 0.012)	(0.00002, 0.00001)	(0.96, 0.96)	(0.014, 0.012)	(0.001, 0.0003)	(0.92, 0.94)

Table 5. Average TE, EQ and VC obtained after two stages of trainning of the SOM, for the iris and codon usage data sets over all formed maps for the proposed frequency functions. Error is presented in pairs: (bubble neighborhood, gaussian neighborhood).

Rule	Iris			Codon usage		
	TE	EQ	VC	TE	EQ	VC
Trad.	(0.15, 0.105)	(0.05, 0.048)	(0.85, 0.85)	(0.137, 0.231)	(0.26, 0.25)	(0.79, 0.77)
Rule 1	(0.152, 0.1)	(0.035, 0.035)	(0.84,0.85)	(0.072, 0.045)	(0.25, 0.22)	(0.72, 0.79)
Rule 2	(0.14, 0.1)	(0.04, 0.038)	(0.91, 0.91)	(0.088, 0.069)	(0.27, 0.08)	(0.88, 0.87)
Rule 3	(0.129, 0.12)	(0.04, 0.038)	(0.89, 0.91)	(0.085, 0.12)	(0.29, 0.29)	(0.8, 0.89)
Rule 4	(0.095, 0.114)	(0.002, 0.001)	(0.93, 0.92)	(0.01, 0.012)	(0.23, 0.22)	(0.94, 0.96)

6 Discussion and Conclusions

An activation frequency parameter for the weight update equation is proposed. This parameter is a function of the activation frequency from a given BMU as well as from the relative frequency of influence of an input vector, through any BMU, to neurons within its neighborhood. Distance between BMU and neurons may also important for this parameter. This parameters add some differential influence between BMU and equally distant neurons, driven by how much those neurons are being affected by other BMUs.

The fact that the proposed rules form non radial neighborhoods gives biological plausibility, due to the fact that a neuron affects differentially other neurons not only based on their distance but in the frequency with which one of them affects the other.

Several experiments show that the error measures in the maps formed with some of the proposed activation frequency functions are lower than those formed by the traditional SOM, and, for the two-dimensional data sets, it is observed that the formed maps fold more accurately to data than those formed by the traditional SOM rule. However, we believe this results could be improved by identifying other activation frequency schemes, such as simmulated annealing. It could be of interest to study the mathematical properties of those functions as they seem to be important for the map formation.

References

1. Cottrell, M. Fort, J.C., Pagés, G. Theoretical aspects of the SOM algorithm. Neurocomputing **21** (1998) 119-138.
2. Kirk, J. Zurada, J. A two-stage algorithm for improved topography preservation in self-organizing maps. Int. Con. on Sys., Man and Cyb. **4** (2000) 2527-2532.
3. Haykin, S. Neural Networks, a comprehensive foundation. 2nd. ed. Prentice Hall. 1999.
4. Kohonen, T. Self-Organizing maps. 3rd. ed. Springer-Verlag. 2000.
5. Flanagan, J. Sufficiente conditions for self-organization in the SOM with a decreasing neighborhood function of any width. Conf. of Art. Neural Networks. Conf. pub. No. 470 (1999)
6. Erwin, Obermayer, K. Schulten, K. Self-organizing maps: Ordering, convergence properties and energy functions. Biol. Cyb. **67** (1992) 47-55
7. Ritter, H. Self-Organizing Maps on non-euclidean Spaces Kohonen Maps, 97-108, Eds.: E. Oja and S. Kaski, 1999
8. Lee, J., Verleysen, M. Self-organizing maps with recursive neighborhood adaption. Neural Networks **15** (2002) 993-1003
9. Campoy, P., Vicente, C. Residual Activity in the Neurons Allows SOMs to Learn Temporal Order LNCS 3696 (2005) 379-384
10. Chappell, G., Taylor, J. The temporal Kohonen map. Neural Networks **6** (1993) 441-445
11. Kiviluoto, K. Topology preservation in Self-Organizing maps. Proc. ICNN96, IEEE Int. Conf. on Neural Networks.
12. Venna, J., Kaski, S. Neighborhood preservation in nonlinear projection methods: An experimental study.
13. Lamm, C. Leodolter, U., Moser, E., Bauer, H. Evidence for premotor cortex activity during dynamic visuospatial imaginery. Neuroimage **14** (2001) 268-283

Nonlinear Projection Using Geodesic Distances and the Neural Gas Network

Pablo A. Estévez, Andrés M. Chong, Claudio M. Held, and Claudio A. Perez

Dept. Electrical Engineering, University of Chile,
Casilla 412-3, Santiago, Chile
pestevez@cec.uchile.cl
http://www.die.uchile.cl/~pestevez

Abstract. A nonlinear projection method that uses geodesic distances and the neural gas network is proposed. First, the neural gas algorithm is used to obtain codebook vectors, and a connectivity graph is concurrently created by using competitive Hebbian rule. A procedure is added to tear or break non-contractible cycles in the connectivity graph, in order to project efficiently 'circular' manifolds such as cylinder or torus. In the second step, the nonlinear projection is created by applying an adaptation rule for codebook positions in the projection space. The mapping quality obtained with the proposed method outperforms CDA and Isotop, in terms of the trustworthiness, continuity, and topology preservation measures.

1 Introduction

Self-organizing feature maps (SOMs) [8] have been widely used for vector quantization (VQ) and for data projection and visualization. The VQ techniques encode a manifold of data by using a finite set of reference or "codebook" vectors. An enhancement to the SOM is the curvilinear component analysis (CCA) [2], which builds a mapping that preserves well the interpoint distances. In the first step, CCA performs VQ of the data manifold in input space using SOM. In the second step, CCA makes a nonlinear projection of the quantizing vectors, which is similar to multidimensional scaling (MDS) [1], since it minimizes a cost function based on the interpoint distances. However the computational complexity of CCA is $O(N)$, while MDS is $O(N^2)$. Another difference is that in CCA the output is not a fixed grid but a continuous space that is able to take the shape of the data manifold. The codebook vectors are projected as codebook positions in output space, which are updated by a special adaptation rule. An enhanced version of CCA, called CDA (curvilinear distance analysis), makes use of geodesic or curvilinear distances instead of Euclidean distances in the input space [11, 13]. In CDA the geodesic distance is approximated by computing the sum of the Euclidean lengths of all links in the shortest path of a graph. CDA can outperform CCA in the projection of highly nonlinear databases like curved manifolds. However, only a few methods are able to project efficiently 'circular' manifolds such as a cylinder or torus. In [10] a procedure was proposed to tear or cut manifolds with essential loops to make easier their embedding in a low-dimensional space. The tearing procedure represents the manifold by a connectivity graph G. An elementary cycle is defined as a 'circular' path

S. Kollias et al. (Eds.): ICANN 2006, Part I, LNCS 4131, pp. 464–473, 2006.

in a graph with no more than C edges. A cycle is said to be non-contractible if it can not be deformed to elementary cycles. The first step of the tearing procedure consists in removing all cycles in the graph G, by computing a shortest spanning tree (SPT) [3] or alternatively, a minimum spanning tree (MST) [17]. In the second step, the removed edges are put back one at a time, but only if they do not generate any non-contractible cycle. In this way, a maximum subgraph without non-contractible cycles is created.

Another neural method for nonlinear projection is Isotop [12], which focuses on neighborhood preservation instead of distance preservation. Isotop uses competitive learning for the VQ step. The second step consists in creating a graph structure by linking the k-closest prototypes. The third step builds the mapping from the input space onto the projection space.

The neural gas (NG) is another well-known self-organizing neural network [16]. The main difference with SOM is that NG does not define an output space. The SOM is able to obtain good VQ results only if the topology of the output grid matches the topology of the data manifold. As a consequence, NG can outperform SOM when quantizing topologically arbitrary structured manifolds. Instead of a neighborhood function in the output grid, NG utilizes a neighborhood ranking of the codebook vectors within the input space. In addition, the NG adaptation rule for codebook vectors obeys a stochastic gradient descent of a global cost function, while no cost function exists for the SOM adaptation rule [15]. The NG algorithm can be combined with the competitive Hebbian rule (CHR) for forming connections among neighboring units [14]. The procedure forms induced Delaunay triangulations of the distribution of codebook vectors.

The authors have proposed elsewhere [5, 6], a nonlinear projection method, called OVI-NG, that makes use of the VQ results obtained with the NG algorithm. The codebook positions are adjusted in a continuous output space by using an adaptation rule that minimizes a cost function that favors the local distance preservation. In [4], an extension of the previous nonlinear projection method was proposed called GNLP-NG, that uses geodesic distances instead of the Euclidean ones. This method projects the codebook vectors, and it uses the competitive Hebbian rule for building a connectivity graph linking these prototypes. The proposed algorithm outperformed CDA and Isotop, in terms of several topology preserving measures.

In this paper, we present an extension to the GNLP-NG nonlinear projection method, that includes tearing or cutting graphs with non-contractible cycles. The performance of the new algorithm is compared with CDA and Isotop, using two examples of 'circular' manifolds that are not easy to project without a tearing procedure.

2 The Geodesic Nonlinear Projection Algorithm

The proposed projection method is called GNLP-NG (Geodesic Nonlinear Projection Neural Gas).

2.1 Problem Setting

Let $\{x_i : 1 \leq i \leq M\}$ and $\{w_j : 1 \leq j \leq N\}$ be D-dimensional input and codebook vectors, respectively. For a given set of input vectors, our problem is to adjust first the codebook vectors in the input space and then their respective codebook positions z_j ($j = 1, \ldots, N$) in a A-dimensional continuous output space, with $A << D$.

To obtain a distance preserving mapping a cost function is defined, which depends on the difference between the interpoint distances in both the input and the output spaces. Let $D_{j,k}$ be the Euclidean distance defined in the output space, and let $\delta_{j,k}$ be the geodesic distance between codebook vectors w_j and w_k, measured in input space.

In the Neural Gas model, the neighborhood ranking function of the codebook vectors $\mathbf{w}_j$, for $j = 1, \ldots, N$, with respect to a given input vector x_i, is defined as follows:

$$h_\lambda(x_i, w_j) = e^{-\frac{r(x_i(t), w_j(t))}{\lambda(t)}}, \qquad (1)$$

where $r_{ij} = r(x_i, w_j) \in \{0, 1, \ldots, N-1\}$ denotes the rank of the jth codebook vector, and the parameter $\lambda(t)$ controls the width of the neighborhood function,

$$\lambda(t) = \lambda_0 \left(\frac{\lambda_f}{\lambda_0}\right)^{\left(\frac{t}{t_{max}}\right)}, \qquad (2)$$

where t_{max} is the maximum number of adaptation steps.

Likewise, for projection purposes we introduce a ranking of the codebook vectors in input space using geodesic distances, with respect to a given input vector x_i. The term $\bar{r}_{ij} = \bar{r}(x_i, w_j) \in \{0, 1, \ldots, N-1\}$ denotes the rank of the jth codebook vector using geodesic distances. Here $r = 0$ and $\bar{r} = 0$ are associated with the nearest codebook vector using Euclidean and geodesic distances in input space, respectively.

2.2 Codebook Position Adaptation Rule

The following global cost function is considered:

$$E = \frac{1}{2} \sum_{j=1}^{N} \sum_{k \neq j} (D_{j,k} - \delta_{j,k})^2 F(\bar{r}_{j,k}) = \frac{1}{2} \sum_{j=1}^{N} \sum_{k \neq j} E_{j,k}, \qquad (3)$$

where the function F is defined as

$$F(f) = e^{-\left(\frac{f}{\sigma(t)}\right)^2} \qquad (4)$$

and $\sigma(t)$ is the width of the neighborhood that decreases with the number of iterations in the same way as eq. (2). The function $F(f)$ is a bounded and monotonically decreasing function, in order to favor local topology preservation.

Eq. (3) is minimized with respect to z_j, by using a simplified version of gradient descent. The codebook position associated to the winner unit, $z_j^*(t)$, is fixed, and the $N-1$ remaining positions are moved towards the winner's position, disregarding any interaction among them. The updating rule for codebook positions is given by Eq. (12).

The complexity of the proposed projection algorithm is $O(NlogN)$, due to the determination of the ranking in input space. Since the computational complexity of NG is also $O(NlogN)$ [15], the combined algorithm of NG as VQ and our GNLP visualization strategy has the same complexity.

2.3 NG Learning Algorithm

The learning algorithm combines NG for VQ, with CHR for forming connections between units. The initial topology of the network is a set of N neurons. Each neuron j has associated a D-dimensional codebook vector, w_j, $(j = 1, \ldots, N)$. Let C_{in} be a connection set, that includes the connections between units in input space. Each connection $j - k$ has an age $(\mathrm{a}_{(jk)})$ that is defined as the number of adaptation steps without being refreshed since its creation. A connection is removed if its age exceeds its lifetime $T(t)$.

1. Initialize the codebook vectors, w_j, randomly. Set the connection set to the empty set: $C_{in} = \emptyset$.
2. Present an input vector, $x_i(t)$ to the network $(i = 1, \ldots, M)$ at iteration t.
3. Find the best matching unit (BMU), j^* using:
$$j^* = \mathrm{argmin}_{j=1\ldots N} \|x_i(t) - w_j(t)\|, \tag{5}$$
and generate the ranking $r_{ij} = r(x_i(t), w_j(t)) \in \{0, 1, \ldots, N-1\}$ for each codebook vector $w_j(t)$ with respect to the input vector $x_i(t)$.
4. Update the codebook vectors:
$$w_j(t+1) = w_j(t) + \epsilon(t) h_\lambda(t)(x_i(t) - w_j(t)) \tag{6}$$
where
$$h_\lambda(t) = h_\lambda(x_i, w_j) \tag{7}$$
is the neighborhood ranking function defined in (1) and $\epsilon(t)$ is the learning rate, which depends on the number of adaptation steps t, in the same way as (2).
5. If a connection between the BMU and the second closest unit does not exist already, create a connection between units $j^* - j^{*2}$: $C_{in} = C_{in} \cup \{w_j^*, w_j^{*2}\}$, where the index of the second nearest unit is:
$$j^{*2} = \mathrm{argmin}_{j \neq j^*} \|x_i(t) - w_j(t)\|. \tag{8}$$
6. If the following condition is satisfied
$$\|w_j^{*k} - w_j^{*(k+1)}\| < \|w_j^* - w_j^{*(k+1)}\|, \tag{9}$$
for $k = 2, \ldots, K$, then create a connection between the kth-nearest unit, j^{*k}, and the (k+1)th-nearest unit, $j^{*(k+1)}$, if it does not exist already: $C_{in} = C_{in} \cup \{w_j^{*k}, w_j^{*(k+1)}\}$, else create a connection between units $j^* - j^{*(k+1)}$: $C_{in} = C_{in} \cup \{w_j^*, w_j^{*(k+1)}\}$.
Set the age of the connection $j^{*m} - j^{*n}$ to zero, for $m, n = 1 \ldots K$ ("refresh" the connection if it exists):
$$\mathrm{a}_{(j^{*m}, j^{*n})} = 0. \tag{10}$$
7. Increment the age of all edges emanating from w_j^{*k}, for $k = 1 \ldots K$:
$$\mathrm{a}_{(j^*, \ell)} = \mathrm{a}_{(j^*, \ell)} + 1, \forall \ell \in \mathcal{N}_{w_j^{*k}}, \tag{11}$$
where $\mathcal{N}_{w_j^{*k}}$ is the set of all direct topological neighbors of w_j^{*k}. Remove those connections with an age exceeding the lifetime T, where $T(t)$ has the same dependency on time t as (2).
8. If $t < t_{max}$ go back to step 2.

2.4 GNLP Algorithm

1. Apply Dijkstra's algorithm [3] to find the shortest path tree (SPT) of the neighborhood graph defined by connection matrix $\mathcal{C}_{in}$.
2. If necessary, tear or cut non-contractible cycles in the neighborhood graph, following the ad-hoc procedure described in 2.5.
3. Initialize the codebook positions, z_j, randomly.
4. Present an input vector, $x_i(t)$ to the network ($i = 1, \ldots, M$) at iteration t.
5. Find the best matching unit (BMU), j^* using eq. (5)
6. Generate the ranking using geodesic distances in input space $\bar{r}_{j^*j} = \bar{r}(w_j(t), w_j^*(t)) \in \{0, 1, \ldots, N-1\}$ for each codebook vector $w_j(t)$ with respect to the BMU.
7. Update the codebook positions:

$$z_j(t+1) = z_j(t) + \alpha(t) F(\bar{r}_{j,j^*}) \frac{(D_{j,j^*} - \delta_{j,j^*})}{D_{j,j^*}} (z_{j^*}(t) - z_j(t)), \tag{12}$$

   where $F(\bar{r}_{j,j^*})$ is a neighborhood function that depends on the ranking of units in input space according to the geodesic distance, and $\alpha(t)$ is the learning rate, which typically decreases with the number of iterations t, in the same way as eq. (2).
8. If $t < t_{max}$ go back to step 5.

2.5 Tearing Non-contractible Cycles in Graphs

The tearing procedure described below is based on Lee and Verleysen's method (2005). The only difference is that they use a simplified search procedure to compute the shortest distances in a graph, while we use Dijkstra's algorithm [3].

Let $G = (V, E)$ be the connectivity graph generated by the NG-CHR procedure described in 2.3, where V is the set of N nodes and E is the set of edges. Let $\mathcal{C}_{in}$ be the connectivity matrix associated to graph G. Let G_T be the shortest path tree (SPT). Let $\mathcal{C}_T$ be the connectivity matrix associated to graph G_T. Moreover, let G_C be a subgraph of G with no cycles including more than C edges, and $\mathcal{C}_C$ its corresponding connectivity matrix.

1. Let U be a $N \times N$ auxiliary matrix. Initialize U as the null matrix, and set $\mathcal{C}_C = \mathcal{C}_T$.
2. Sort the nodes of G_T, using a breadth-first traversal strategy, and store this order in an array p^*.
3. Let J be the set of nodes directly connected to the ith node in G. Given the ith node $v_i \in p^*$, find all its neighbors, v_j, with $j = 1, \cdots, J$, that have a direct connection to it in the connectivity graph G. Store the nodes v_j in an array p^{**}.
4. For $j = 1, \cdots, J$, consider node $v_j \in p^{**}$. If $U(i,j) = 0$ calculate the shortest distance, d_{ij}, between nodes i and j using Dijkstra's algorithm in G. Check for cycles of length less than C. If $d_{ij} + 1 < C$, set $\mathcal{C}_C(i,j) = \mathcal{C}_C(j,i) = 1$ and $U(i,j) = U(j,i) = 1$. Go back to step 4 while $j \leq J$, else if $j > J$ go to step 3. Otherwise, if $U(i,j) = 1$ then the node has already been visited.
5. Stop if $i > N$, else go back to step 4 if $j \leq J$, or to step 3 if $j > J$.

2.6 Mapping Quality

A projection onto an output space is said to be trustworthy if the set of k nearest neighbors of a point in the map are also close by in the original space. Let $U_k(i)$ be the set of samples that are in the neighborhood of the ith point in the map but not in the original space. The measure of trustworthiness of the visualization, M_1, is defined as [7, 18]:

$$M_1(k) = 1 - A(k) \sum_{i=1}^{N} \sum_{z_j \in U_k(i)} (r(w_i, w_j) - k), \tag{13}$$

where $A(k) = 2/(Nk \times (2N - 3k - 1))$, and $r(w_i, w_j)$ is the ranking in input space.

A projection onto an output space is said to be continuous if the set of k closest neighbors of a point in the original space are also close by in the output space. Let $V_k(i)$ be the set of samples that are in the neighborhood of the ith point in the original space but not in the map. The measure of continuity, M_2, is defined as:

$$M_2(k) = 1 - A(k) \sum_{i=1}^{N} \sum_{w_j \in V_k(i)} (s(z_i, z_j) - k), \tag{14}$$

where $s(z_i, z_j)$ is the ranking in output space.

The topology preservation measure q_m [9] is based on an assessment of rank order in the input and output spaces. The n nearest codebook vectors NN_{ji^w} $(i \in [1, n])$ of each codebook vector j, $(j \in [1, N])$ and the n nearest codebook positions NN_{ji^z} of each codebook position j are computed. The parameter n was set to $n = 0.4 \times k$, and k was varied from 5 to a maximum value K_q. The q_m measure takes values in $[0, 1]$, where $q_m = 1$ indicates a perfect neighborhood preservation. The global q_m measure is defined as:

$$q_m = \frac{1}{3nN} \sum_{j=1}^{N} \sum_{i=1}^{n} q_{m_{ji}}, \tag{15}$$

where

$$q_{m_{ji}} = \begin{cases} 3, \text{ if } NN_{ji^w} = NN_{ji^z} \\ 2, \text{ if } NN_{ji^w} = NN_{jl^z}, l \in [1, n], i \neq l \\ 1, \text{ if } NN_{ji^w} = NN_{jt^z}, t \in [n, k], n < k \\ 0, \text{ else.} \end{cases} \tag{16}$$

3 Simulation Results

In all simulations the parameters of the NG algorithm were set as follows: $\epsilon_0 = 0.5$, $\epsilon_f = 0.005$, $\lambda_0 = 30.0$, $\lambda_f = 0.01$, $T_0 = 20.0$, $T_f = 100.0$, $K = 2$. Likewise, the parameters of the GNLP algorithm were set as follows: $\alpha_0 = 0.3$ and $\alpha_f = 0.001$, $\sigma_0 = 0.7 \times N$, $\sigma_f = 0.1$, and $C = 6$, where N is the number of neurons.

Two artificial data sets were considered[1]: Cylinder and Torus. For each data set 300 codebook vectors were considered.The number of training epochs for NG was 20, the

[1] See http://www.dice.ucl.ac.be/~lee

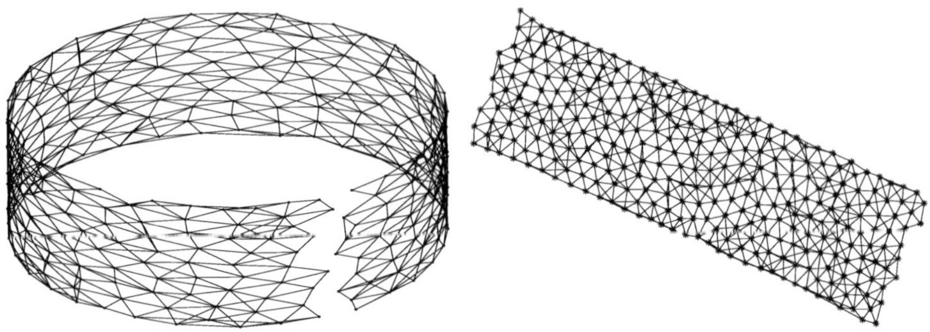

(a) (b)

Fig. 1. (a) Cylinder quantized data set using 300 prototypes and a graph tearing procedure, and (b) Cylinder data projection with GNLP-NG

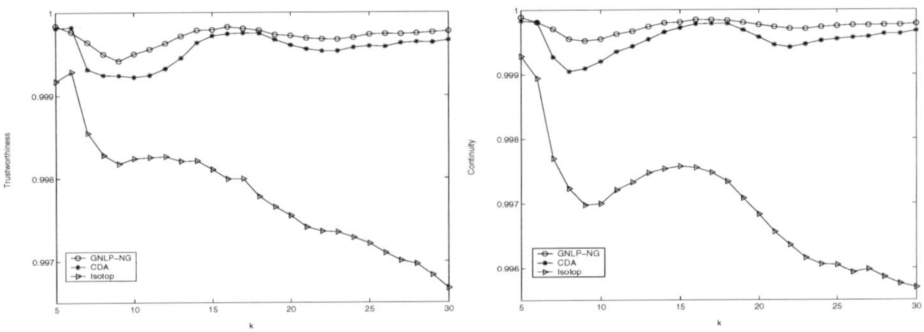

(a) (b)

Fig. 2. (a) Trustworthiness measure and (b) continuity measure as a function of the number of neighbors k, for the Cylinder data set

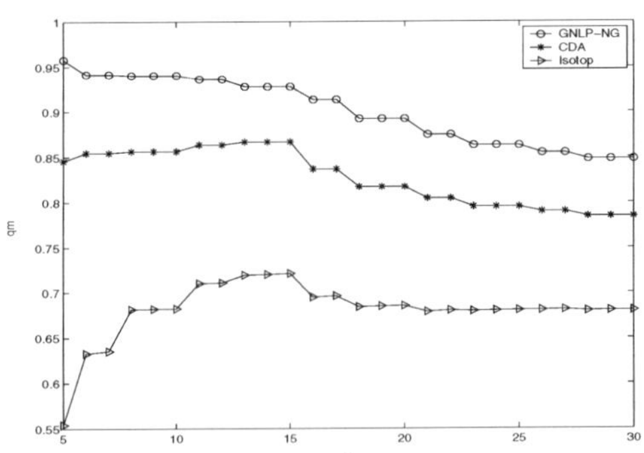

Fig. 3. Topology preservation measure q_m as a function of the number of neighbors, k, for the Cylinder data set

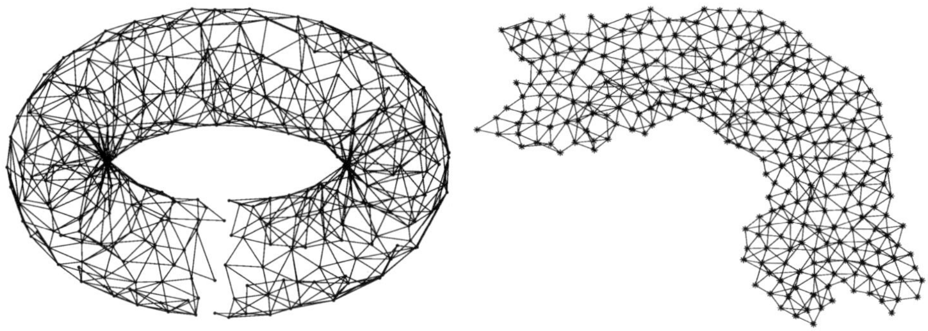

(a) (b)

Fig. 4. (a) Torus quantized data using 300 prototypes and a graph tearing procedure, and (b)Torus data projection with GNLP-NG

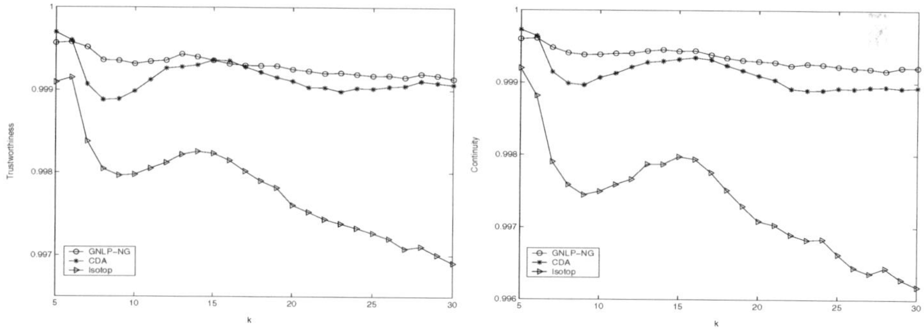

(a) (b)

Fig. 5. (a) Trustworthiness measure and (b) continuity measure as a function of the number of neighbors k, for the Torus data set

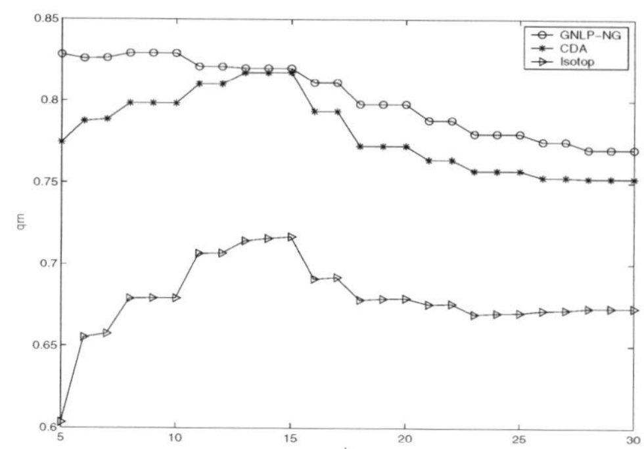

Fig. 6. Topology preservation measure q_m as a function of the number of neighbors, k, for the Torus data set

Table 1. CPU time, T, for the different steps of the GNLP-NG algorithm

	Cylinder	Torus
Algorithm	T [s]	T [s]
NG-CHR	347	188
SPT+Tearing	17	18
GNLP	472	270

same for GNLP. In the case of CDA, 30 epochs were used for training the competitive learning, 50 epochs for the nonlinear projection, and the parameter C was set to $C = 5$. The same setting was used for Isotop.

The Cylinder data set corresponds to a two-dimensional manifold embedded in a 3D space, from which 19600 samples were drawn. Due to the circularity of this manifold, its projection would get deformed or superimposed unless the corresponding connectivity graph is torn. The Torus data set corresponds to a 3D manifold, from which 10000 samples were drawn.

Fig. 1(a) shows the torn graph of the Cylinder quantized data set using NG and the tearing procedure. Fig. 1(b) illustrates the projection obtained with GNLP in output space. Fig. 2 shows (a) the trustworthiness measurement and (b) the continuity measurement, as a function of the number of nearest neighbors, k, for the Cylinder data set. It can be observed that the GNLP-NG method obtained the best trustworthiness and continuity measurements for all the k values explored (except one point), outperforming both CDA and Isotop. Fig. 3 shows the topology preservation measure q_m as a function of the number of neighbors. Here the GNLP-NG method outperformed CDA and Isotop for all k values.

Fig. 4(a) shows the torn graph of the Torus quantized data using NG and the tearing procedure. Fig. 4(b) illustrates the Torus data projection obtained with GNLP. Fig. 5 shows (a) the trustworthiness measurement and (b) the continuity measurement, as a function of the number of nearest neighbors, k, for the Torus data set. It can be observed that the best trustworthiness and continuity were obtained by GNLP-NG for $k > 6$, while CDA obtained a slightly better result for $k = 5$. Fig. 6 shows the topology preservation measure q_m as a function of the number of neighbors, for the Torus data set. Again GNLP-NG outperformed CDA and Isotop for all values of k explored.

Table 1 shows the CPU time in seconds for the different steps of the GNLP-NG algorithm, for both the Cylinder and Torus data sets. The algorithms were implemented in Matlab (except the Dijkstra's algorithm, which was implemented in C++), and executed in a Pentium IV 2.4GHz, 1GB RAM.

4 Conclusions

In the proposed GNLP-NG method the neighborhood graph is built online along with the vector quantization. This procedure allows obtaining a better graph representation than using competitive learning. A procedure for tearing or cutting graphs with non-contractible cycles was implemented, in order to make easier the nonlinear projection of manifolds with essential loops. The method is computationally efficient with

$O(N\log N)$ complexity. For two artificial data manifolds containing essential loops, the GNLP-NG method outperformed CDA and Isotop, in terms of the trustworthiness, continuity, and topology preservation measures.

Acknowledgement

This research was supported by Conicyt-Chile, under grant Fondecyt 1050751. We thank Dr. John Lee for providing us with a software implementation of CDA and Isotop.

References

1. Cox, T.F. and Cox, M.A.A.: *Multidimensional Scaling*, 2nd Edition, Boca Raton, Chapman & Hall/CRC, 2001.
2. Demartines, P., and Hérault, J.: Curvilinear component analysis: A self-organizing neural network for nonlinear mapping of data sets. *IEEE Trans. on Neural Networks*, **8** (1997) 148–154.
3. Dijkstra, F.W.: A note on two problems in connection with graphs. *Num. Math.*, **1** (1959) 269–271.
4. Estévez, P.A., and Chong, A.M.: Geodesic Nonlinear Projection using the Neural Gas Network. *Int. Joint Conference on Neural Networks*, Vancouver, Canada, 2006 (to appear).
5. Estévez, P.A., and Figueroa, C.J.: Online data visualization using the neural gas network. *Neural Networks*, (2006) (in press).
6. Estévez, P.A., and Figueroa, C.J.: Online nonlinear mapping using the neural gas network. *Proceedings of the Workshop on Self-Organizing Maps (WSOM'05)*, (2005) 299–306.
7. Kaski, S., Nikkilä, J., Oja, M., Venna, J., Törönen, P., and Castrén, E.: Trustworthiness and metrics in visualizing similarity of gene expression" *BMC Bioinformatics*, 4:48, (2003).
8. Kohonen, T.: *Self–Organizing Maps*, Berlin, Germany, Springer-Verlag, 1995.
9. König, A.: Interactive visualization and analysis of hierarchical neural projections for data mining. *IEEE Trans. on Neural Networks*, **11** (2000) 615–624.
10. Lee, J.A., and Verleysen, M.: Nonlinear dimensionality reduction of data manifolds with essential loops. *Neurocomputing*, **67** (2005) 29–53.
11. Lee, J.A., Lendasse, A., and Verleysen, M.: Nonlinear projection with curvilinear distances: Isomap versus curvilinear distance analysis, *Neurocomputing*, **57** (2004) 49–76.
12. Lee, J.A., and Verleysen, M.: Nonlinear projection with the Isotop method. *Proceedings of the International Conference on Artificial Neural Networks*, (2002) 933–938.
13. Lee, J.A., Lendasse, A., Donckers, N., and Verleysen, M.: A robust nonlinear projection method. *Proceedings of the European Symposium on Artificial Neural Networks (ESSAN'2000)*, (2000) 13–20.
14. Martinetz, T.M., and Schulten, K.J.: Topology representing networks. *Neural Networks*, **7** (1994) 507–522.
15. Martinetz, T.M., Berkovich, S.G., and Schulten, K.J.: 'Neural gas" network for vector quantization and its application to time-series prediction. *IEEE Trans. on Neural Networks*, **4** (1993) 558–569.
16. Martinetz, T.M., and Schulten, K.J.: A neural gas network learns topologies. *Artificial Neural Networks*, (1991) 397–402, Elsevier.
17. Prim, R.C.: Shortest connection networks and some generalizations. *Bell System Tech. J.*, **36** (1957) 1389–1401.
18. Venna, J., and Kaski, S.: Local multidimensional scaling with controlled tradeoff between trustworthiness and continuity. *Proceedings of the Workshop on Self-Organizing Maps (WSOM'05)*, (2005) 695–702.

Contextual Learning in the Neurosolver

Andrzej Bieszczad[1] and Kasia Bieszczad[2]

[1] Computer Science, California State University Channel Islands
One University Drive, Camarillo CA 93012
aj.bieszczad@csuci.edu

[2] Center for the Neurobiology of Learning and Memory, U.C. Irvine
320 Qureshey Research Laboratory, University of California, Irvine, CA 92697
kbies@uci.edu

Abstract. In this paper, we introduce an enhancement to the Neurosolver, a neuromorphic planner and a problem solving system. The enhanced architecture enables contextual learning. The Neurosolver was designed and tested on several problem solving and planning tasks such as re-arranging blocks and controlling a software-simulated artificial rat running in a maze. In these tasks, the Neurosolver learned temporal patterns independent of the context. However in the real world no skill is acquired in vacuum; Contextual cues are a part of every situation, and the brain can incorporate such stimuli as evidenced through experiments with live rats. Rats use cues from the environment to navigate inside mazes. The enhanced architecture of the Neurosolver accommodates similar learning.

1 Introduction

The goal of the research that led to the original introduction of Neurosolver, as reported in [1], was to design a neuromorphic device that would be able to tackle problems in the framework of the state space paradigm [2]. The research was inspired by Burnod's monograph on the workings of the human brain [3]. The class of systems that employ state spaces to present and solve problems has its roots in the early stages of AI research that derived many ideas from the studies of human information processing; e.g., on General Problem Solver [2]. This pioneering work led to very interesting problem solving (e.g. SOAR [4]) and planning systems (e.g. STRIPS [5]).

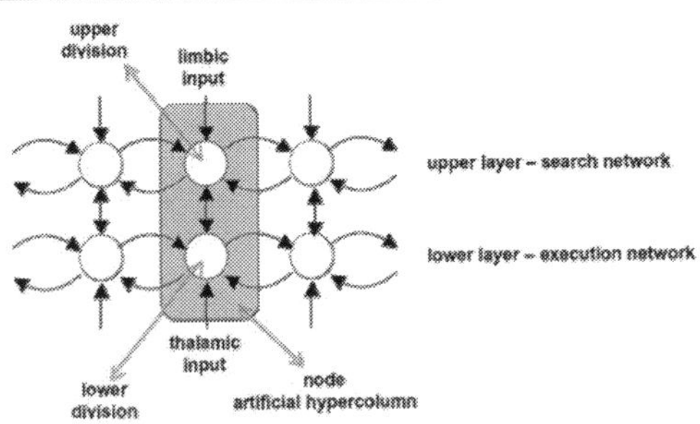

Fig. 1. An artificial cortical column

1.1 Architecture of the Neurosolver

The Neurosolver is a network of interconnected nodes. Each node is associated with a state in a problem space. A problem is presented to the Neurosolver by two signals: a

S. Kollias et al. (Eds.): ICANN 2006, Part I, LNCS 4131, pp. 474–484, 2006.

goal associated with the desired state and a sensory signal associated with the current state. The goal is delivered through the limbic input to the upper division of the node, and the percept, through the thalamic input to the lower division of the node. A sequence of firing nodes represents a trajectory in the state space. A solution to a problem is a succession of firing nodes starting with the current node and ending with the goal node.

The node used in the Neurosolver is based on a biological cortical column (references to the relevant neurobiological literature can be found in [1]). It consists of two divisions: the upper and the lower, as illustrated in Fig. 1. The upper division is a unit integrating internal signals from other upper divisions and from the control center providing the limbic input (i.e., a goal or - using more psychological terms - a drive or desire). The activity of the upper division is transmitted to the lower division where it is subsequently integrated with signals from other lower divisions and the thalamic input. The upper divisions constitute a network of units that propagate search activity from the goal, while the lower divisions constitute a network of threshold units that integrate search and sensory signals, and generate sequences of firing nodes. The output of the lower division is the output of the whole node. An inhibition mechanism prevents cycles and similar chaotic behavior. Simply, a node stays desensitized for a certain time after firing.

1.2 Learning in the Neurosolver

The Neurosolver learns by receiving instructional input. Teaching samples representing state transitions are translated into sequences of firing nodes corresponding to subsequent states in the samples. For each state transition, two connections are strengthened: one, in the direction of the transition, between the lower divisions of the two nodes, and another, in the opposite direction, between the upper divisions (Fig. 2). The strength of all inter-nodal connections is computed as a function of two probabilities: the probability that a firing source node will generate an action potential in this particular connection and the probability that the target node will fire upon receiving an action potential from the connection.

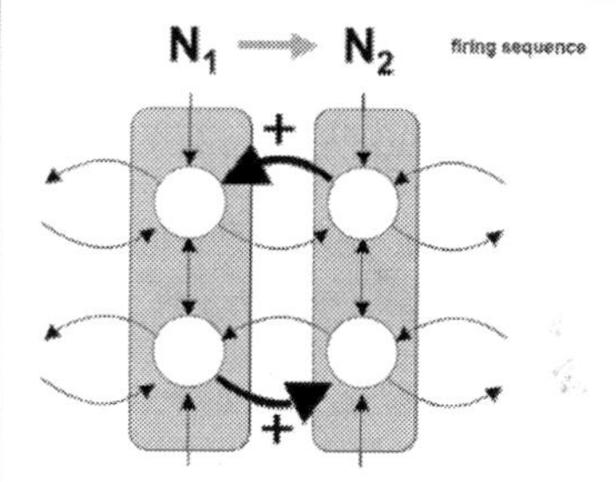

Fig. 2. Learning in the Neurosolver is probabilistic

To compute the probabilities, each division and each connection collects statistics (Fig. 3). The number of transmissions of an action potential T_{out} is recorded for each connection. The total number of cases when a division positively influenced other nodes S_{out} is collected for each division. A positive influence means that an action potential sent from a division of a firing node to another node caused that node to fire in the next cycle. In addition, we also collect

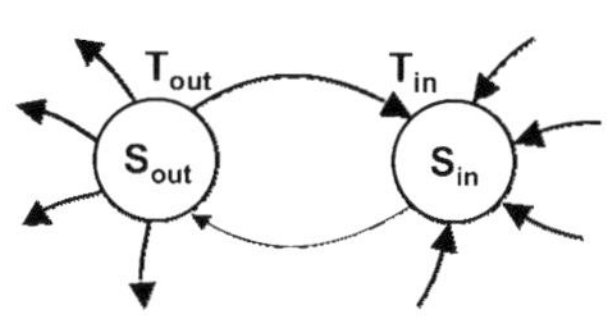

Fig. 3. Statistics collected for computing connection strengths

statistical data that relate to incoming signals. T_{in} is the number of times when an action potential transmitted over the connection contributed to the firing of the target node and is collected for each connection. S_{in}, collected for each division, is the total number of times when any node positively influenced the node. With such statistical data, we can calculate the probability that an incoming action potential will indeed cause the target node to fire. The final formula that is used for computing the strength of a connection (shown in Eq. 1) is the likelihood that a firing source node will induce an action potential in the outgoing connection, mulitplied by the likelihood that the target node will fire due to an incoming signal from the connection:

$$P = P_{out} \cdot P_{in} = (T_{out}/S_{out}) \cdot (T_{in}/ S_{in}) \quad (1)$$

1.3 Learning Sequences

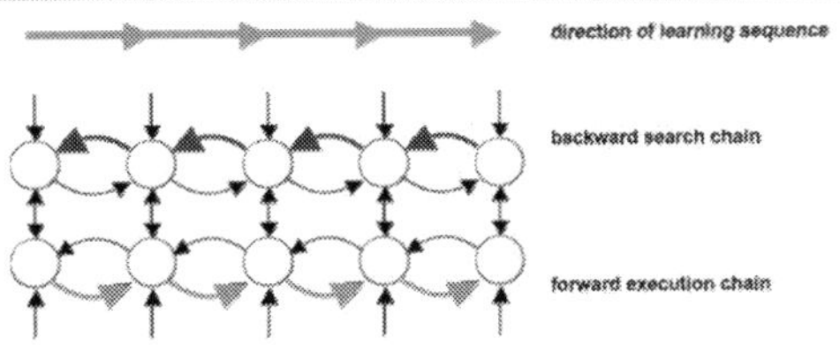

Fig. 4. The Neurosolver learn temporal patterns

As we already mentioned, the function of the network of upper divisions is to spread the search activity along upper-to-upper connections (Fig. 4) starting at the original source of activity, the node associated with the goal state that receives the limbic input,. This is a search network, because the activity spreads in hope that at some node it will be integrated within the activity of the same lower division that receives a thalamic input. If the activity exceeds the output threshold, then the node will fire triggering a resolution. The thalamic input is applied to the node corresponding to the current state. The process of spreading activity in a search tree is called goal regression [5].

The purpose of the network composed of lower divisions and their connections is to generate a sequence of output signals from firing nodes (along the connections shown in Fig. 5). Such a sequence corresponds to a path between the current state and the goal state, so it can be considered a solution to the problem. As we said, a firing of the node representing the current state triggers a solution. Each firing node sends actions potentials through the outgoing connections of its lower division. These signals may cause another node to fire if its attention (i.e., the activity in the upper division) is

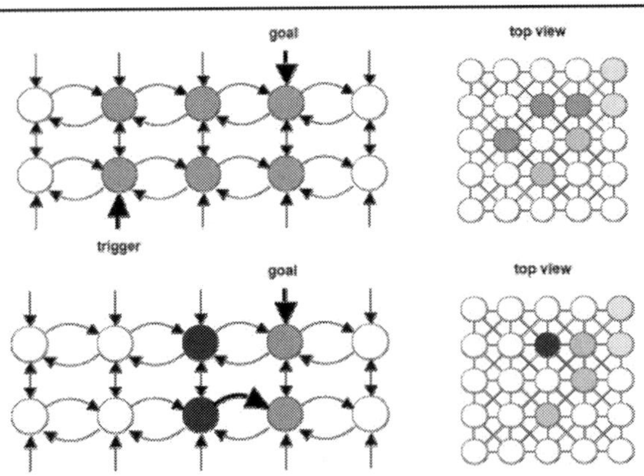

Fig. 5. A search (upper) and an execution (lower) in the Neurosolver

sufficiently high. In a way, the process of selecting the successor in the resolution path is a form of selecting the node that is activated the most. The purpose of inhibiting a node for some time after firing is to avoid oscillations. The length of the inhibition determines the length of cycles that can be prevented. The Neurosolver exhibits goal-oriented behavior similar to that introduced in [6].

2 Simulated Rat Maze

We tested the Neurosolver capabilities on several abstract data sets of various sizes and have successfully applied the Neurosolver as a planner for rearranging blocks in a block world. Our analysis of the Neurosolver capabilities in the context of symbolic planners proves that it is a promising adaptive mechanism for general planning and problem solving. We collected these and many other ideas from several earlier publications in a summary article published as [1].

Currently, to explore Neurosolver's cognitive capabilities, we apply it to control a rat running in a maze. The Neuroscience and Cognitive Science community commonly use rats to gain insight into brain processes. One area of research is concerned with rats running in several types of mazes (e.g., an excellent Web site on rat behavior [7]). Evidently, rats are good learners in this domain (e.g., [8]). In one type of maze, a rat is allowed to explore the maze with food placed in constant locations. In subsequent runs, the rat improves its ability to locate the food faster. One of the currently supported hypotheses says that the rat builds a topology of the maze in its brain (cognitive map) encoded by place cells in hippocampus ([9]).

Rats evidently use topological maps of the maze to construct best possible paths to food locations. Neurosolver has been constructed on the premise that temporary associations are made in cortex. Some researchers hypothesize that the entorhinal cortex is a link between labile episodic storage in the hippocampus to more permanent and semantic storage in the cortex ([11]).

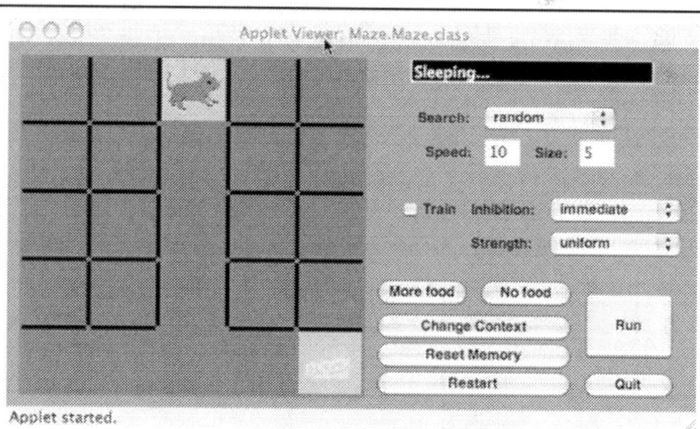

Fig. 6. The rat maze simulator

To conduct experiments, we built a simulated rat maze. The user interface is illustrated in Fig. 6. The simulator consists of the maze area (left side of the frame[1]) and the control area. The maze area consists of a number of locations separated by walls. Mazes of various shapes can be built by modifying the walls. The rat can be positioned by dragging its image to a desired location. Food (illustrated by a piece of cheese) can be placed in

[1] We will use just this area in the subsequent illustrations.

multiple locations also by dragging the image from the location in which it is posted when the More Food button is pressed. Clicking on the Run button activates the metabolic system that gradually increases the sensation of hunger. If the motivation to find food exceeds a certain threshold, the rat's limbic system is activated. The rat starts to run. The Run button changes to Pause, so the simulator can be stopped and restarted at will.

The Change Context button changes the color of the floor that is used in experiments on contextual learning. We will discuss the details later in this paper.

The rat is capable of moving in four directions: up, right, down and left. It can move only if there is no wall between two neighboring locations. The simulated rat uses an instance of Neurosolver to determine the next move in the maze. Other search modes such as random, constrained, and depth-first search are implemented to control the rat if the Neurosolver cannot find a solution .

When the rat finds food its motivation level decreases to below threshold, another parameter controlled and the rat may stop running. The threshold for this is one of the parameters that can be controlled during experiments.

Rats use cues from the environment to activate a specific place cell in the brain ([12]). A similar associative mechanism allows a rat to learn and remember where the food was found ([11]). Similarly, in the simulator, the rat remembers the locations of successful feedings, so it can use them as goals in the future runs. If the rat finds food again in the location of a goal, the strength of that goal increases, so the rat is more likely to pursue this goal in the future. Conversely, if the rat gets to a location that had food in the past, but the food is not there, the strength of the goal is lowered. If such negative experience is consistent, the goal will be completely inhibited, a behavioral process called extinction ([13]). We are planning to examine the goal management theories to a greater extent in the future.

Using our simulator, we performed a number of tests with the artificial rat running in various mazes. A T-maze is the simplest type of maze used in experiments with rats. A passage in a shape of the letter T forces the rat to choose the direction, left or right, at the crossing as shown in Fig. 6. If food is placed consistently in one arm of the T, then this is the arm that will be selected by the rats in the subsequent runs. If the rat obtained food from both arms then it will choose the one that has a better trace in memory.

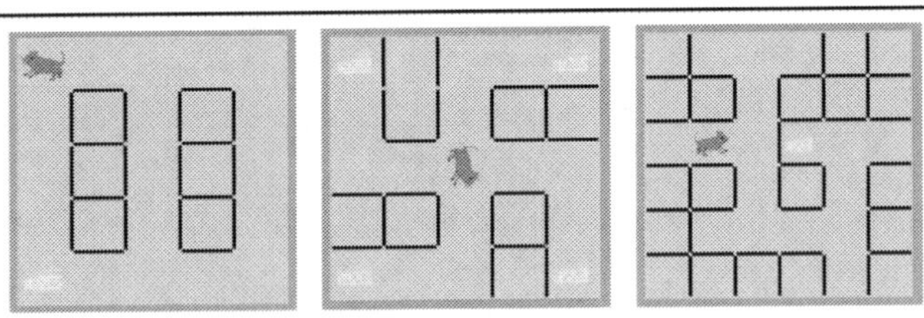

(a) (b) (c)

Fig. 7. The rat selecting best path (a), the rat pursuing multiple goals (b) and a rat faced with multiple-T maze (c)

In another experiment shown in Fig. 7 (a), there are three paths leading from the original position of the rat to the food. The rat selects the shortest path if the uniform learning is selected. If probabilistic learning is chosen, then the path most often followed in the past and therefore the most probable, will be taken. This behavior comes from the fundamental characteristics of the Neurosolver. If a wall is created along the shortest path, then the rat reconsiders the plan and selects an alternate path backtracking as necessary. The rate of degradation can be controlled from the simulator. Live rats may exhibit abberant behaviors during such stressful situations that may lower the motivation induced by hunger. As long as the threshold to search for food is met, the rat will try to get to the goal by any means putting more effort in recollections of past passages and food locations. Fig. 7 (b) shows a rat in a maze with four branches ("star-shaped"). The simulated rat trying to get all food again performs similarly to live rats. If food is removed from certain locations, then the rat will tend to move to the branches that provided consistent food-reward. There is a more complex T-maze used in tests with rats as shown in Fig. 7 (c). The rat is faced with multiple choices (T's) on their path to the food. This is a more challenging task to live rats. It also takes a longer training session for the artificial rat to build a map, and higher motivation to find a path to the food.

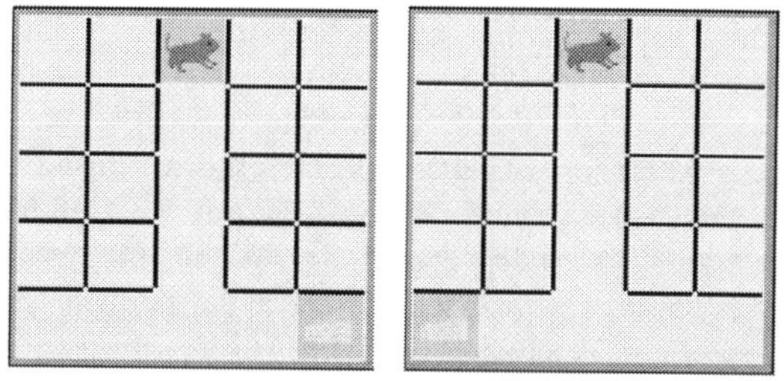

Fig. 8. The rat selecting best path (left), the rat pursuing multiple goals (center) and a rat faced with multiple-T maze

3 Applying Context

In spite of these successful experiments, a problem arises if the probability of finding food is the same in both arms of T-maze as shown in Fig. 8. For example, if the number of successful feeding in both arms of T-maze is the same, then the rat is confused. Where should the rat go left or right?

Figure 9 illustrates a top view of the Neurosolver trying to solve the rat's problem. The activity spreads from the cells corresponding to both goals (food locations) along the learned paths. The common segment of the search path is activated a bit more, because the search activity from the left goal is integrated with the activity coming from the right goal. That does not have much of an impact on the trigger in the cell corresponding to the current position of the rat in the maze. The columns fire along the common segment,

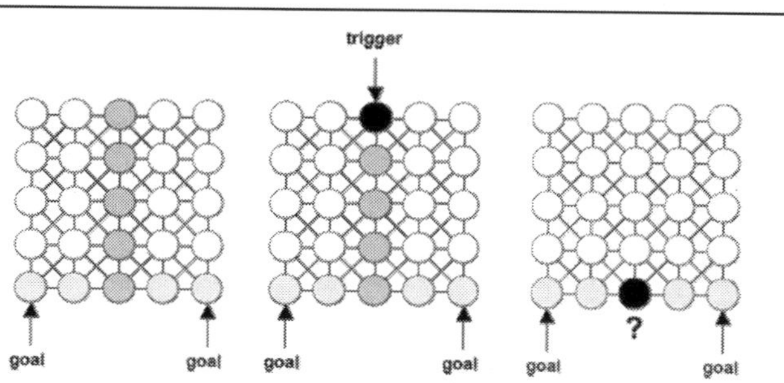

Fig. 9. The rat selecting best path (left), the rat pursuing multiple goals (center) and a rat faced with multiple-T maze

but then there is no clear-cut choice whether the next firing column should be to the left or to the right from the fork location. The Neurosolver tries to increase the activity from the goals, but it is increased by the same amount on both sides. The current version of the Neurosolver has a mechanism that arbitrarily selects one of such alternative columns and forces it to fire. The mechanism is not biologically plausible. Animals – as well as humans – make choices, and the choices are made using context.

Neuroscientists working with rats routinely incorporate context in the learning process (e.g., [7]). The Morris water maze is used as evidence for spatial contextual learning. In such a maze, the rat cannot see a hidden platform submerged in white, opaque water, but it nevertheless learns its position by using visual cues from the environment. The results (e.g., [16], [17) indicate that rats navigate using direct cues to the goal (taxon navigation) or memories based on places activated by the environment clues (praxic navigation), although the true nature of rats' maze running skills is still under discussion (e.g., [18]). When the environment changes, or obstacles like walls are used to obstruct the view, the appropriate place cells do not fire, implying that the rat is unaware of its true location and needs to re-discover the platform through more explorations. As we mentioned earlier, our simulator already includes such a discovery mechanism.

At this stage, we are interested in praxic navigation, because at the moment we do not experiment with real-time visual information that could be used to guide movements (like in hand-eye coordination). To introduce a context, the color of the maze floor or the color of the light illuminating the maze might be changed (let's say yellow and red) depending on the food location. This is of course inspired by experiments with live rats, as they tend to follow the paths leading to goals associated with the color of the floor or the color of the light.

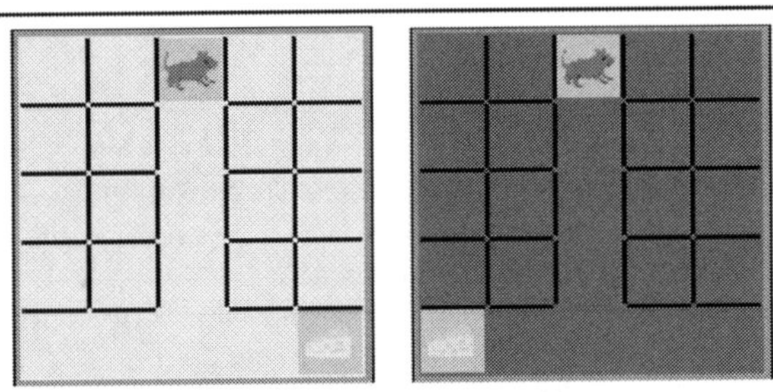

Fig. 10. Yellow floor (left; in here it's white) and red floor (right; in here it's grey) might be used for contextual learning

We have extended the maze simulator, so a colored floor can be used in the rat maze simulator as illustrated in Fig. 10. Through the addition of the color cue, we can conduct experiments involving contextual learning.

4 Architecture for Contextual Reasoning

The new capabilities of the simulated maze amount to just one element of the required modifications. More fundamentally, the Neurosolver needs new functionality to accommodate context. One way to do this would be to incorporate the contextual information as one of the dimensions that determine the place cell – in our jargon, a hypercolumn that corresponds to the location of the rat. In that way, the identical path learned with red floor would be represented in the Neurosolver separately from the same path that was learned with yellow light, because the place cells would be different[2]. That solution would be a waste of the Neurosolver resources, because each new cue would lead to yet another space. Therefore, we propose a different architecture that uses auxiliary contextual cells instead. A context cell is activated by contextual cues[3]. It is debatable which solution is more biologically plausible, but the use of contextual (or snapshot as some researchers call them) cells makes sense from a pragmatic perspective, because the storage requirements are drastically lower in the approach that we have taken.

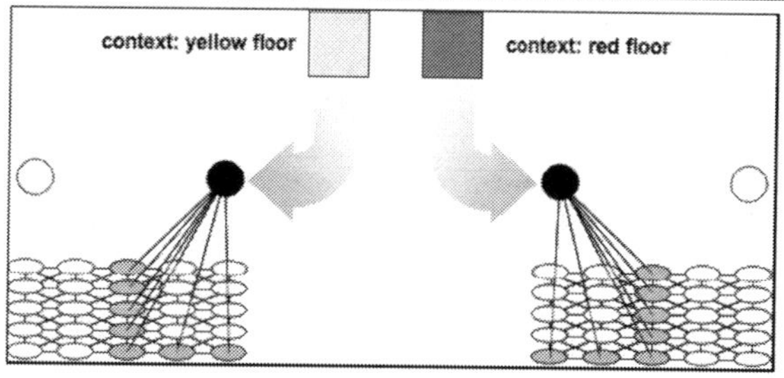

Fig. 11. Contextual association with red floor (left) and yellow floor (right)

As illustrated in Fig. 11, the color of the floor results in activation of the context cell corresponding to that color. At times, the contextual cells will be co-activated with the firing nodes of the Neurosolver corresponding to the movements of the rat. As described earlier, the Neurosolver columns fire in response to the thalamic input reflecting changes in the rat's location. Through the use of Hebbian learning rules, each node in the temporal pattern becomes gradually associated with the co-activated context cell as shown in Fig. 12. Similarly to the connection strengths in the Neurosolver, we apply statistical learning here. The strength of a connection is based on the co-activation statistics. If any given hypercolumn fired S times, and a specific context cell was co-activated T times, then the strength of the connection is the co-activation probability P_c:

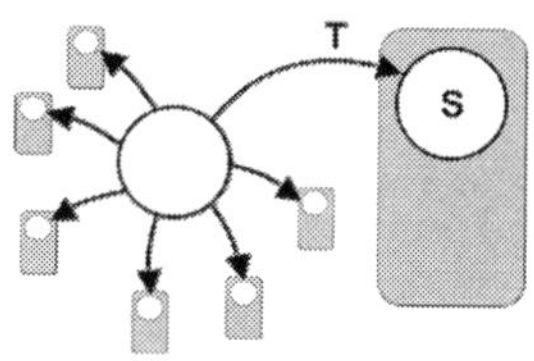

Fig. 12. Computing connection strength between the context cell and the upper division of a node

$$P_c = T/S \qquad (2)$$

After the learning, during Neurosolver's operation, any activity in the context cell is modulated by that probability and projected into the upper division of the hypercolumn.

With such associations in place, certain nodes of the Neurosolver are activated even in absence of any search activity (see Fig. 13, upper left). The nodes in the search path corresponding to a particular learned sequence will now be activated not

[2] They would be two different points in the state space that includes a color axis, and the projections on this axis ("red" or "yellow") would be different for both paths.

[3] For example, it can be selected through a competitive, winner-takes-all network. Creating such a network is not the objective of the work reported in this paper. Instead, we assume that a recognizing mechanism is in place, and red color activates one cell, while yellow activates the other one.

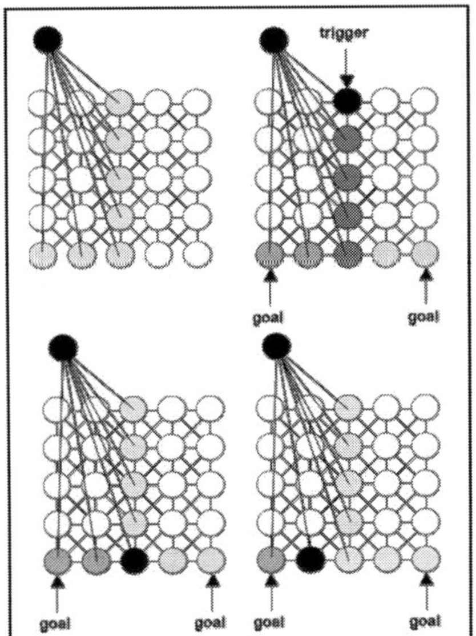

Fig. 13. Search and resolution with contextual cues

only by the activity coming from the source of the search, but also by the activity coming from the context cell. When two competing goals are applied, the search progresses as described earlier, but the nodes in the path associated with the current context are activated at a higher level, because the search activity integrates with the contextual activity. When the trigger activates a resolution path, the firing nodes will follow the highest activity in the neighborhood, so when the node corresponding to the fork in the T-maze fires, the next node to fire will more likely be the one on the path whose nodes are associated with the currently active context.

Figure 14 illustrates the corresponding behavior of the rat in the maze with two different contexts. With the yellow floor, the rat turns right and is shown just before consuming the reward after making the correct choice. In the second run, with the red floor, the rats turns to the left, and again is shown just before feeding.

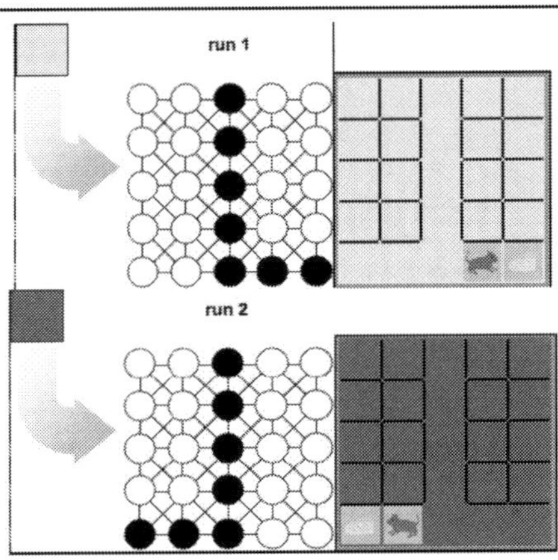

Fig. 14. Two different resolution paths in two runs. A path is selected dependent on the context – the color of the floor.

5 Conclusions and Future Work

In spite of the obvious simplifications, the Neurosolver used as a brain of the artificial rat performs analogously to live rats when tested on similar tasks. We use the analysis of any discrepancies between the behavior of the Neurosolver and live rats to evolve the architecture of the Neurosolver. One difficulty with that arises from the variety of behavioral patterns identified in lab animals. The problem of studying learning and memory in animals is that we can only infer memories from their behavior. Careful analysis of the enhanced architecture of the Neurosolver leads us to the conclusion that it indeed provides the capabilities for contextual learning. The pragmatic aspect of the new mechanism is evident, so from the engineering perspective it is a significant improvement. Nevertheless, an interesting generic question to ask is whether the behavioral patterns observed on animals can be replicated by simulated creatures controlled by Neurosolver to the same degree as the more basic alignment of behaviors reported in our previous papers.

There are plenty of other interesting venues worth pursuing in research with Neurosolver, and in more general terms in computational architectures inspired by progress in neuroscience. For example, we plan to further analyze the fundamental simplification that we consciously made in the Neurosolver that is the integration of the functionalities of the prefrontal cortex (where logical sequencing and executive function takes place) and hippocampus (where place cells are located). A related decision to map every possible location as a place cell affects the scalability of the Neurosolver, so we are exploring ideas to reduce the state space. For example, only the important places could be mapped (e.g., forks, turns, etc.), but how to modulate such a mapping is an open question.

Another idea that is very interesting for a computer scientist is "programmability" in the hippocampus. When an animal learns a new environment, he creates a map of that space that is encoded by the hippocampus. When he is out of that environment, the map, the "code", is transferred elsewhere for "storage", and the hippocampus is available for the "encoding" of a new environment. When he returns to the first environment, the "code" is "loaded" back into the hippocampus and the place cells are once again engaged. The idea that there are "grandmother cells" in the hippocampus is generally not applied in neuroscience circles.

At this moment, the searches that the Neurosolver performs are conducted off-line with locked knowledge. Allowing for real-time dynamic modifications that may alter the search or execution process is both appealing and biologically plausible. The capability to accomplish taxon navigation could be one result of such work. That in turn, could be the basis for Neurosolver-based controllers (for example, for hand-eye navigation). Another alluring idea is to explore learning by analogy that would require an association and generalization mechanism. That would allow knowledge re-use in context that engages concepts similar to those with which the current knowledge was acquired.

The path optimization capability of the Neurosolver can be viewed both positively and negatively. At its current incarnation, the Neurosolver cannot store higher order sequences. That is, for example, a sequence 1-2-3-2-4 is optimized to 1-2-4. Some applications may need such capability, so we are looking into designing an appropriate mechanism that can provide that.

Numerous technical aspects of the implementation are also worth pursuing. For example an alternative to software implementation could accommodate much larger state spaces. To provide required granularity in finding best paths and simulate analog workings of the brain, the propagation of activity is very inefficient as implemented in software. Furthermore, the Neurosolver uses statistical learning, so each of the connected elements is a little processor churning numbers. We plan to experiment with other types of weight management, perhaps more aligned with the mainstream research on Neural Networks.

Goal management is another interesting research area. How do rats choose where to go? Do they ever forget the associations between the pleasures of feeding and the location? Neuroscientists have been feeding us with new data in this area (somewhere between the amygdala, hippocampus and the cortex), and we are eager to explore the new findings.

We would like to express our gratitude to the reviewers, who raised many important and interesting issues that we tried to address in this section.

References

1. Bieszczad, A. and Pagurek, B. (1998). Neurosolver: Neuromorphic General Problem Solver, Information Sciences: An International Journal 105: 239--277, Elsevier North-Holland, New York, NY.
2. Newell, A. and Simon, H. A. (1963). GPS: A program that simulates human thought, in Feigenbaum, E. A. and Feldman, J. (Eds.), Computer and Thought. New York, NJ: McGrawHill.
3. Burnod, Y. (1988). An Adaptive Neural Network: The Cerebral Cortex. Paris, France: Masson.
4. Laird, J. E., Newell, A. and Rosenbloom, P. S. (1987). SOAR: An architecture for General Intelligence, Artificial Intelligence, 33: 1--64.
5. Nillson, N. J. (1980). Principles of Artificial Intelligence. Palo Alto, CA: Tioga Publishing Company.
6. Deutsch, M. (1960). The Effect Of Motivational Orientation Upon Trust And Suspicion, Human Relations, 13: 123--139
7. http://ratbehavior.org/
8. Hodges, H. (1996). Maze Procedures: The Radial-Arm And Water Maze Compared, Cognitive Brain Research 3 (1996) 167 – 181, Elsevier North-Holland, New York, NY.
9. Fenton, A. A. and Muller, R.U. (1998). Place Cell Discharge Is Extremely Variable During Individual Passes of The Rat Through The Firing Field, Proc. Natl. Acad. Sci. USA, Vol. 95, pp. 3182–3187.
10. Poucet, B. and Save, E. (2005). Attractors in Memory, Science, 308: 799--800, AAAS Press.
11. Fyhn, M., Molden, S. and Witter, M. P. (2004). Spatial Representation in the Entorhinal Cortex Marianne, Science, 305: 1258--1264, AAAS Press.
12. O'Keefe, J. and Dostrovsky, J. (1971). The Hippocampus as a Spatial Map. Preliminary Evidence from Unit Activity in the Freely-Moving Rat, Brain Research, 34: 171-175.
13. Pavlov, I. P. (1927). Conditioned Reflexes. Routledge and Kegan Paul, London.
14. Bitterman, M. E., Lolordo, V. M., Overmier, J. B. and Rashotte, M. E. (Eds.) (1979). Animal Learning: Survey And Analysis, New York, NJ: Plenum Press.
15. Charles C. Kemp (2001). Think Like a Rat. Paper for MIT EECS Area Exam (http://people.csail.mit.edu/cckemp/cckemp_place_cells_area_exam_2001.pdf).
16. Hartley T., Burgess N. (2002) Models Of Spatial Cognition, Encyclopaedia of Cognitive Science, MacMillan.
17. Burgess, N. and O'Keefe, J. (2002). Spatial Models of the Hippocampus in: The Handbook of Brain Theory and Neural Networks, 2nd Edition Ed: Arbib M A, MIT press, Cambridge MA.
18. Chavarriaga, R., Strösslin, T., Sheynikhovich, D. and Gerstner, W. (2005). Competition Between Cue Response And Place Response: A Model Of Rat Navigation Behavior, Connection Science, Vol. 17, Nos. 1–2, March–June 2005, 167–183.

A Computational Model for the Effect of Dopamine on Action Selection During Stroop Test

Ozkan Karabacak and N. Serap Sengor

Istanbul Technical University, Faculty of Electrical and Electronic Engineering, Electronics Engineering Department, Maslak, TR-34469, Istanbul, Turkey
{karabacak, neslihan}@ehb.itu.edu.tr

Abstract. Based on a connectionist model of cortex-basal ganglia-thalamus loop recently proposed by authors, a simple connectionist model realizing the Stroop effect is established. The connectionist model of cortex-basal ganglia-thalamus loop is a nonlinear dynamical system and the model is not only capable of revealing the action selection property of basal ganglia but also is capable of modelling the effect of dopamine on action selection. While the interpretation of action selection function is based on solutions of nonlinear dynamical system, the effect of dopamine is modelled by a parameter. The effect of dopamine in inhibiting the habitual behaviour corresponding to word reading in Stroop test and letting the novel one occur corresponding to colour naming is investigated using the model established in this work.

1 Introduction

Computational modelling of cognitive processes began to attract more attention as benefits of these models in different disciplines are recognized [1-3]. There are different approaches in computational modelling, while some consider the problem of modelling at cell level and develop models considering biophysics, others, as those using symbolic AI techniques, intend to model only behaviours. There are other models not as complex as realistic models of cell level and not as far away from the neural validity as symbolic AI techniques; these focus not only on how to get a system behaving as expected, but also aim to use the model in understanding the ongoing processes. These are models at behavioural level and while developing these models neural substrates related with the involved behaviour and their interrelations have to be considered. Thus these models are capable of explaining how the neural substrates provoke the behaviour without dealing with structures at physiological level [2, 4, 5].

Modelling at behavioural level is not only beneficial at explaining the whys of cognitive processes but also is capable in providing tools for investigating the reasons behind the abnormal behaviour. This characteristic of behavioural models makes them especially important in pharmacological studies. As animal models are not sufficient, and these models provide more flexible applications they are advantageous over models used nowadays in pharmacology [6]. Still another advantage of behavioural models is they can be used in engineering applications as designing robots and intelligent systems [3].

S. Kollias et al. (Eds.): ICANN 2006, Part I, LNCS 4131, pp. 485 – 494, 2006.

In this work behavioural approach in modelling will be considered, and previously proposed model [7] of cortex-basal ganglia-thalamus (C-BG-TH) loop for action selection will be expanded to model the Stroop task. As the model of C-BG-TH loop is capable of explaining the effect of neurotransmitter on action selection, the expansion of this model will also exploit the effect of dopamine on Stroop task.

First computational model for C-BG-TH loop will be reviewed and the model will be discussed considering the other similar models given in the literature, then expanded model of C-BG-TH loop will be established and the effect of dopamine on Stroop effect will be shown. The simulation results will reveal the effect of the dopamine system dysfunction on action selection.

2 Computational Model for Cortex-Basal Ganglia-Thalamus Loop

The basal ganglia once known for their role in motor movement control, now are also considered for their part in high order cognitive processes and motivation related acts as temporal sequence storage and generation, behavioural switching, reward evaluation, goal-directed behaviour, reinforcement learning. These behavioural functions are possible only through the interrelations of basal ganglia with cortex and thalamus and these interrelations have been investigated in [8] by means of five basal ganglia-thalamocortical circuits, which are responsible for generating different behavioural functions. Different substructures of cortex, basal ganglia and thalamus take part in each of these circuits. On the other hand the dopamine systems as mesolimbic, mesocortical and nigrostriatal systems provide regulation of information transfer through these neural circuits [9] and dysfunctions of these dopamine systems cause deficits in behavioural functions. To obtain a biologically plausible computational model of basal ganglia revealing modulating effect of dopamine, the interrelations of basal ganglia with related neural structures as cortex and thalamus has to be considered. These interrelations are considered keeping in mind that our intention is to focus on a specific cognitive process, action selection, rather than modelling all aspects of basal ganglia-thalamocortical circuits described in [8] and the model obtained will be a behavioural one.

In this section, first a simple neural structure of C-BG-TH loop will be investigated and a mathematical model of this structure will be restated. This model was inspired from [5] and proposed previously in [7]. Then how this mathematical model is used for modelling action selection will be introduced. The modification of the mathematical model for action selection was again previously given in [7] and it was inspired from [4].

2.1 Neural Substrates and Related Mathematical Model

As a cognitive process, it has been expressed that action selection is a major function of basal ganglia and is crucial in understanding human behaviour [4]. Action selection, like other cognitive processes, is initiated at cortex, where anterior cingulate system responsible for attention takes part in generating a salience signal which initiates the action selection process. Action is also terminated at cortex when the motor circuits trigger action. The salience signal causes an activation in basal ganglia, which then via thalamus initiates in cortex relevant structures for action. Thus a feedback structure, which incorporates cortex, basal ganglia and thalamus, is necessary and so a loop composed of cortex, basal ganglia and thalamus is formed. This feedback structure can be expressed as a non-linear dynamical system. The equations of the non-linear dynamical system corresponding to basic connections of cortex, basal ganglia and thalamus besides the connections within basal ganglia are given in Eq. 1 and shown in Fig. 1.

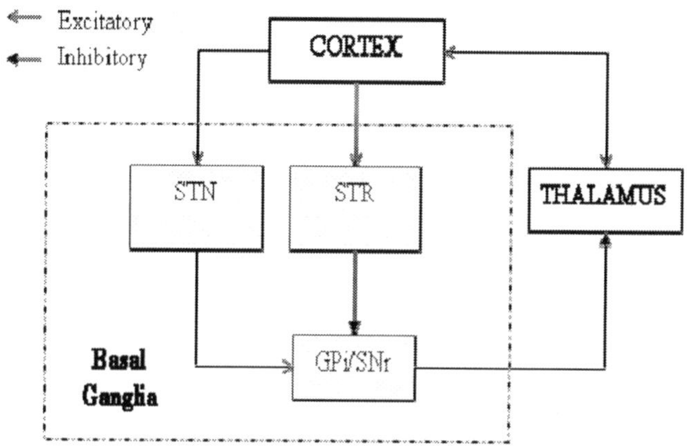

Fig. 1. Cortex-basal ganglia-thalamus (C-BG-TH) loop

$$\begin{aligned} p(k+1) &= \lambda \cdot p(k) + f(m(k)) \\ m(k+1) &= f(p(k)) - f(d(k)) \\ r(k+1) &= g(p(k), \theta_{att}) \\ n(k+1) &= f(p(k)) \\ d(k+1) &= -g(r(k), \theta_{sel}) + f(n(k)) \end{aligned} \tag{1}$$

In Eq. 1 $g(x,\theta) = 0.5\cdot(1+\tanh(3\cdot(x+\theta-1.5)))$ and $f(x) = g(x,1)$. θ_{att} models the effect of attention and is set to "1" throughout this section. The subsystem given by the Eq. (1) is rewritten in compact form as follows:

$$\Sigma_1: \ \mathbf{x}(k+1) = \mathbf{F}(\mathbf{x}(k), \theta_{att}, \theta_{sel}). \tag{2}$$

Striatum and subthalamic nucleus (STN) are initiated by cortex with excitatory glutamatergic projections, these are the inputs of basal ganglia and main outputs are globus pallidus internia (GP_i) and substantia nigra pars reticulata (SN_r). The output structures then project inhibitory GABAergic projections to thalamus. Thalamus completes the loop by excitatory glutamatergic projections to cortex. All these connections are expressed in Eq. 1, where cortex, thalamus and the substructures in BG, namely, STR, STN and GPi/SNr are denoted by $p(k)$, $m(k)$, $r(k)$, $n(k)$, and $d(k)$, respectively. While mainly GABAergic and glutamatergic projections convey this information transfer, dopamine modulates this process [10] and in Eq. 1 the parameter θ_{sel} represents the effect of dopamine while the inhibitory and excitatory effects of

GABAergic and glutamatergic projections are denoted by negative and positive coefficients in the matrix.

Even though two pathways exist in C-BG-TH loop, only the one named "direct pathway" is considered fully in Fig. 1 and the "indirect pathway " is simplified since for the modelling purpose of this work the role of basal ganglia can be expressed through this simple structure [8]. In this model, as can be followed from Eq. 1 each substructure is modelled by a single variable. These variables can be thought to denote a group of neurons that take part during the process. It is shown in [7] that this basic model of C-BG-TH loop is capable of originating two stable fixed points. These two fixed points are named "active" and "passive" points and they are interpreted as an action took place or did not take place, respectively. The attraction domains and how they depend on parameter θ_{sel}, so on neurotransmitter dopamine, are investigated in [7].

The model given by Eq. 1 is inspired from [4,5]. In action network of Taylor N.R. & Taylor J.G. [5], the motivation is to propose a model of C-BG-TH-C especially to model the temporal sequence storage and generation process, which takes part in working memory tasks. The proposed model resembles [5] as it has similar simple non-linear discrete time dynamics as the model of C-BG-TH loop. As our aim is to model action selection, we used diffusive parallel connections between loops corresponding to different actions. Thus the proposed model does not have a layered structure for action selection as in [5]. The dynamics of the proposed model also exhibits saddle point bifurcations. Furthermore the analysis of bifurcations due to a parameter has been done not for each loop separately as in [5] but for the whole system, which helps us to reveal the overall effect of dopamine on action selection.

In [4] a model of C-BG-TH loop is given for action selection and the effect of dopamine is investigated by changing the weight of connections. Control and selection loops instead of direct and indirect pathways are utilized for action selection. They also interpreted their results considering dopamine depletion and excess in relation with Parkinson's disease and Huntington's disease, and discussed that dopamine depletion gives rise to failure in selection and dopamine excess causes inability in ignoring distracts. The competition mechanism of [4], that is the use of diffusive connections from STN to GPi/SNr, is utilized in [7].

One another work that has to be mentioned is the work of Servan-Schreiber [2]. This work focuses on mesocorical dopaminergic system, but they have not concerned with C-BG-TH loop. They are more concerned with attention than action selection in Stroop test as most models deal.

2.2 Computational Model for Action Selection

Action selection depends on competition; to construct this competition using the C-BG-TH loop model given by system Σ_1 coupling of at least two Σ_1 systems is needed. Each Σ_1 system will correspond to a competing action. The coupling done is biologically plausible, as diffusive, excitatory connections from STN to GPi/SNr, which have disinhibition effect on the other loops as an overall effect, are used as shown in Fig. 2. The model for action selection is obtained connecting two Σ_1 subsystems as expressed in Eq. 3.

$$\Sigma_2: \begin{aligned} \mathbf{x}_1(k+1) &= \mathbf{F}(\mathbf{x}_1(k), \theta_{att}, \theta_{sel}) + \mathbf{G}(\mathbf{x}_2(k)) \\ \mathbf{x}_2(k+1) &= \mathbf{F}(\mathbf{x}_2(k), \theta_{att}, \theta_{sel}) + \mathbf{G}(\mathbf{x}_1(k)) \end{aligned} \quad . \tag{3}$$

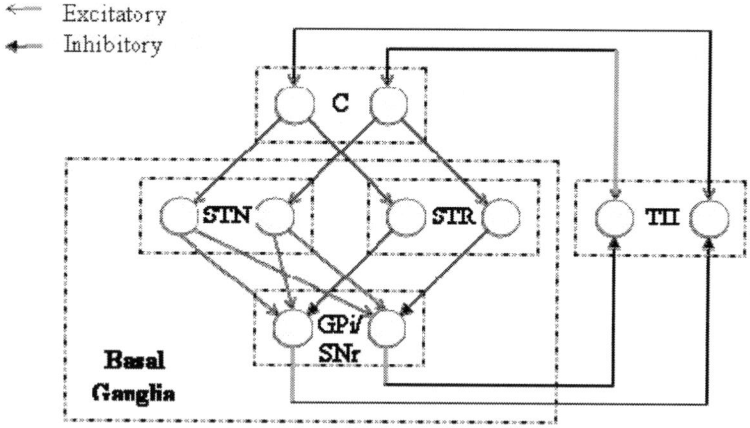

Fig. 2. Model for action selection process

Here, coupling function is $\mathbf{G}(\boldsymbol{x}) = (0\ 0\ 0\ 0\ f(x_4)/2)^T$ and stands for the abovementioned connection between Σ_1 subsystems. Due to coupling of two loops, the maximum number of the fixed points increases from two to four as shown in Fig. 3. These fixed points are denoted by x_1^*, x_2^*, x_3^* and x_4^* and these respectively correspond to the following behaviours: both subsystems are passive, only first subsystem is active, only second subsystem is active and both subsystems are active.

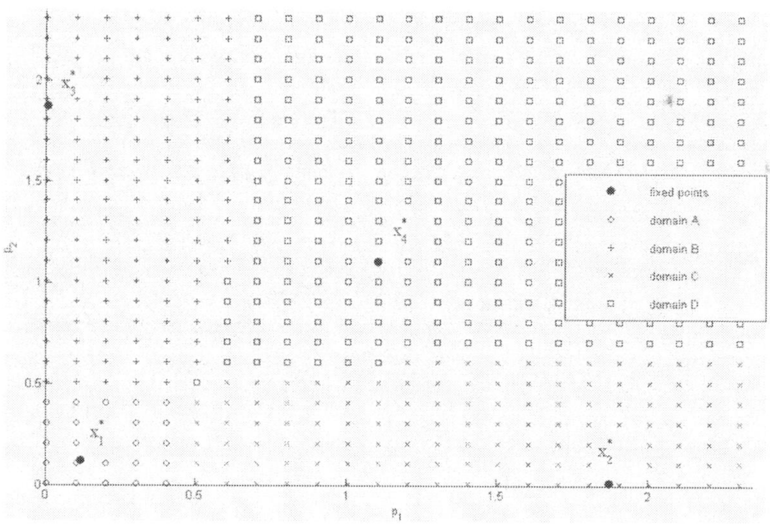

Fig. 3. Attraction domains of Σ_2 for $\theta_{sel} = 1.05$

Attraction domains of the fixed points are named A for x_1^*, B for x_2^*, C for x_3^* and D for x_4^* and they are illustrated in Fig. 3. For different initial conditions the system Σ_2 converges to different fixed points. For example, if the initial condition is $p_1 = 0.5$, $p_2 = 1$, it converges to x_3^*. Thus the second action is selected. If the initial state of the system is in region A or D, the system cannot discriminate competing actions. In the first case none of the actions and in the second case both actions are selected. If there exist all of these regions A, B, C and D like in Fig. 3, the system cannot be considered suitable for action selection. In order to realize action selection properly, Σ_2 should have larger domains of type B and C. For some values of parameter θ_{sel} this objective can be fulfilled [7]. The system with such parameter value is regarded as exploiting normal action selection behaviour, whereas occurrence of the region D and enlarging of the domain A corresponds to abnormalities in action selection. These abnormalities occur due to dopamine level. Dopamine excess/depletion causes both competing subsystems to get activated/inhibited for a large area of initial conditions. Thus action selection fails for these salience values.

3 The Effect of Dopamine on Stroop Test

Stroop test is mostly used as a measure of selective attention. During the test the subjects have to inhibit word reading tendency, which is an automatic process, fast and do not require attention and follow the novel task of colour naming, which is a controlled process, slow and requires attention [2]. So, while word reading process takes less time, colour naming takes prolonged time as some time is used to inhibit the habitual behaviour. Since Stroop test is considered as a measure of focused attention, dysfunction of attention system, i.e., anterior cingulate system is investigated in most computational models [2, 11]. Unlike previous works, in this work, effect of nigrostriatal dopamine system on Stroop test rather than mesocortical dopamine system will be investigated. This investigation has a value since behavioural consequences of nigrostriatal dopamine system in case of occurrence of salient stimuli has also been discussed [12, 14].

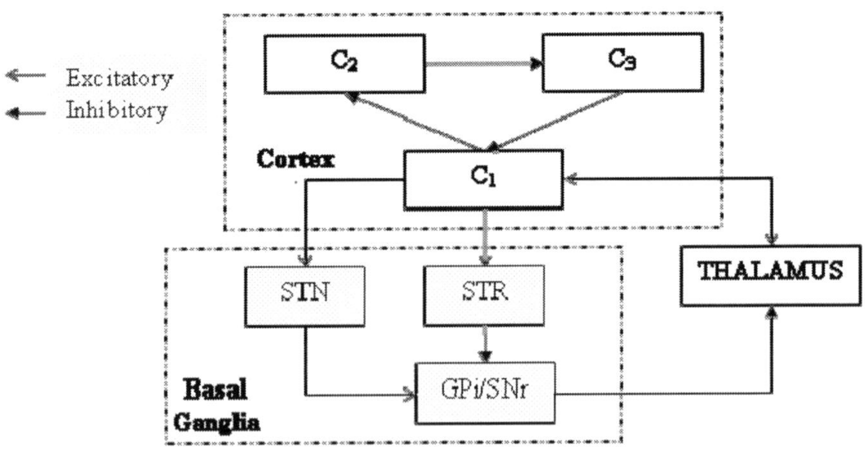

Fig. 4. Proposed model for stroop test

In order to obtain a model to investigate the effect of dopamine on Stroop test, besides the coupled C-BG-TH loops for action selection another loop, which is composed of simple representations of neural substrates responsible for attention and error detection is needed. Two structures, one the coupled C-BG-TH loops for action selection, the other cortico-cortico (C_1-C_2-C_3) loop for attention and error detection are used together but with different time scales. The model of coupled C-BG-TH loops given by system Σ_2 is utilised in this new structure after some modifications shown in Fig. 4 are done. To favour habit, one of the coupled loops in Σ_2, which is supposed to correspond to word reading task, is strengthened by changing weight of the connection from C to STR as "2", while the loop for novel task, i.e., colour naming remains same. This change of connection alters the attraction domains of Σ_2. Even for some initial values with large colour naming component, the word reading loop can win.

The error detector (C_2) and the attention block (C_3) are used to provide the inhibition of the habitual action and to support the generation of the task related action, respectively. C_1 represents the part of cortex that is in connection with other structures like BG and TH. The equations of C_1-C_2-C_3 loop are given as follows:

$$
\begin{aligned}
c_2(k) &= \begin{cases} t - f(c_1(k-5)) & , \quad k = 10, 20, 30, \dots \\ c_2(k-1) & , \quad other \end{cases} \\
c_3(k) &= \begin{cases} A \cdot f(c_2(k-1)) & , \quad k = 11, 21, 31, \dots \\ c_3(k-1) & , \quad other \end{cases} \\
\theta_{att}(k) &= \begin{cases} \theta_{att}(k-1) + 0.1 \cdot c_3(k-1) \cdot \left(2 - \min\left(\left|\theta_{att}(k-1) - \begin{pmatrix} 1 \\ 1 \end{pmatrix}\right|\right)\right) & , \quad k = 12, 22, 32, \dots \\ \theta_{att}(k-1) & , \quad other \end{cases}
\end{aligned} \tag{4}
$$

In Eq. 4, t represents the input for task and set to $(0\ 1)^T$ for colour naming task. The matrix A is a two by two matrix with "1"'s on diagonal, "-1"'s on other entries; its role is to realize both suppression of habitual behaviour and to drive attention to novel one. The initial values are taken as very small random numbers. These two loops, namely C-BG-TH and C_1-C_2-C_3, are combined at C_1 which corresponds to the variable p in Eq. 1, so C_1 is composed of p_1 and p_2. The nigrostriatal dopamine level is simulated by changing the parameter θ_{sel} in C-BG-TH loop. The activation of two cells in C_1, corresponding to word reading and colour naming salience signals are shown in Fig. 5 for a normal dopamine level. The activation of the task-irrelevant loop at the first time steps vanishes after a while and the task-relevant loop takes action. This result fits the phenomena called Stroop effect. To investigate the effect of θ_{sel} simulation results are shown in Table 1. These simulation results are obtained using the following criteria: If the value of a signal is greater than one plus the other's value during 100 time steps and during the following 100 steps it's value is never less than the other's value minus one, then the action corresponding to this signal is regarded to be generated at the end of this 200 time steps.

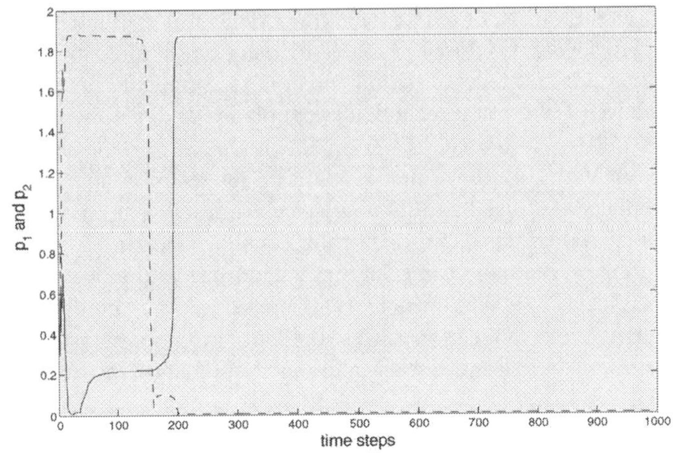

Fig. 5. Simulation result for normal dopamine level (θsel =1). Dotted line shows the activation of p_1 that corresponds to C_1 activation of word reading loop. Solid line shows the activation of p_2 that corresponds to C_1 activation of colour naming loop.

Table 1. Simulation results for different θ_{sel} values

	Dopamine Level (θ_{sel})	**Error**	**Time of the correct response**
Dopamine excess ↑	2	no	> 30000
	1.6	no	1396
	1.4	no	1261
	1.2	no	1129
	1	no	396
Dopamine depletion ↓	0.8	no	390
	0.6	yes	405
	0.55	yes	440
	0.5	yes	> 30000

In the case of dopamine excess C-BG-TH loops let both actions get activated during the first time steps. As a result, error at C_2 is less than the error in the case when C-BG-TH loop select the habitual one. The selection of both actions slows down the correction process, i.e. the work of C_1-C_2-C_3 loops. On the other hand, the effect of dopamine depletion is mainly on errors rather than on the prolonged time used for correction. In this case C_1-C_2-C_3 loops work well but the activation of the task-related action occurs late because of the low dopamine level. This causes an error according to our abovementioned criteria.

Dopamine excess corresponds to dysfunction of BG where the subject can not suppress irrelevant action and dopamine depletion corresponds to the case where subject can not perform the action, due to akinesia.

4 Discussion and Conclusion

Quite a number of computational models of basal ganglia, which evaluate the neural substrates and related neurotransmitter systems responsible for the interested behaviour have been developed [4, 5, 7]. There are also connectionist models, exploring the effect of mesocortical dopamine system on frontal cortex following similar approach [2,13] and among these [2] considers Stroop task. The objective of this work is to develop a connectionist model to investigate the role of basal ganglia during Stroop task. Computational models investigating cognitive behaviour during Stroop task focus on the dysfunction of attention system and error detection system as Stroop task is well known as a measure of selective attention. However, in this work, it is proposed that dysfunction of modulatory nigrostriatal dopamine system can impair the action selection property of basal ganglia and it is shown by simulation results that this effects the performance of subjects during Stroop task.

This investigation is based on a recently proposed model [7] exploiting the action selection property of basal ganglia which is utilized in a newly proposed neural structure, where cortico-cortico loop is considered along with C-BG-TH loop. Even though the model of C-BG-TH loop is inspired from [4, 5], it is simpler than [4] and still able to model action selection.

The effect of dopamine is modelled similar to [5] and the obtained simulation results are consistent with the observed relationship between dopamine excess/depletion and behaviour. The investigation of basal ganglia dysfunction on Stroop test is valuable, since the effect of nigrostriatal dopamine systems on behaviour other than motor actions is also investigated [12].

References

1. Churchland P.S.: Neurophilosophy. Toward a Unified Science of the Mind-Brain. A Bradford Book. The MIT Press, Cambridge, Massachusetts, London, England (1988)
2. Cohen D., Servan-Schreiber D.: Context, Cortex, and Dopamine: a Connectionist Approach to Behaviour and Biology in Schizophrenia. Psychological Review 99 (1992) 45–77
3. Doya K.: Reinforcement Learning in Continuous Time and Space. Neural Computation 12 (2000) 219-245
4. Gurney K., Prescott T.J., Redgrave P.: A Computational Model of Action Selection in the Basal Ganglia. I. A New Functional Anatomy. Biological Cybernetics 84 (2001) 401–410
5. Taylor N.R., Taylor J.G.: Hard-wired Models of Working Memory and Temporal Sequence Storage and Generation. Neural Networks 13 (2000) 201-224
6. Ashby F.G., Casale M.B.: A Model of Dopamine Modulated Cortical Activation. Nral Networks 16 (2003) 973-984
7. Karabacak O., Sengor N.S.: A Dynamical Model of a Cognitive Function: Action Selection. 16th IFAC congress (2005)
8. Alexander G.E., Crutcher M.D., DeLong M.R.: Basal Ganglia-Thalamocortical Circuits: Parallel Substrates for Motor, Oculomotor, "Prefrontal" and "Limbic" Functions. Progress in Brain Research 85 (1990) 119-146

9. Sagvolden, T., Johansen, E.B., Aase, H., Russel, V.A.: A Dynamic Developmental Theory of Attention-Deficit/Hyperactivity Disorder (ADHD) Predominantly Hyperactive/ Impulsive and Combined Subtypes. Behavioural and Brain Sciences 28 (2005) 397-468
10. Cohen J.D., Braver T.S., Brown J.W.: Computational Perspectives on Dopamine Function in Prefrontal Cortex. Current Opinion in Neurobiology 12 (2002) 223-229
11. Kaplan G.B., Sengor N.S. Gurvit H., Guzelis C.: Modelling Stroop Effect by a Connectionist Model. Proceedings of ICANN/ICONIP (2003) 457-460
12. Horvitz J.C.: Mesolimbocortical and Nigrostriatal Dopamine Responses to Salient Non-Reward Events. Neuroscience 96 (2000) 651-656
13. Servan-Schreiber D., Bruno R.M., Carter, C.S., Cohen D.: Dopamine and the Mechanisms of Cognition: Part1. A Neural Network Model Predicting Dopamine Effects on Selective Attention. Biological Psychiatry 43 (1998) 713–722
14. Djurfeldt M., Ekeberg Ö., &Graybiel A.M., "Cortex-basal ganglia interaction and attractor states". Neurocomputing, 38-40, (2001) 73-579.

A Neural Network Model of Metaphor Understanding with Dynamic Interaction Based on a Statistical Language Analysis

Asuka Terai and Masanori Nakagawa

Tokyo Institute of Technology,
2-12-1 O-okayama, Meguro-ku, Tokyo, 152-8552 Japan
{asuka, nakagawa}@nm.hum.titech.ac.jp

Abstract. The purpose of this study is to construct a human-like neural network model that represents the process of metaphor understanding with dynamic interaction, based on data obtained from statistical language analysis. In this paper, the probabilistic relationships between concepts and their attribute values are first computed from the statistical analysis of language data. Secondly, a computational model of the metaphor understanding process is constructed, including dynamic interaction among attribute values. Finally, a psychological experiment is conducted to examine the psychological validity of the model.

1 Introduction

In order to construct a computational model of metaphor understanding, we need a foundational theory that explains the same process in humans. One well-known theory in psychology is the "salience imbalance model" [1]. According to Ortony's theory, a metaphor of the form "A is like B" makes good sense metaphorically if the high-salient attribute values of a vehicle (term "B") match the low-salient attribute values of a target (term "A"). It is well known in psychology that metaphor understanding is influenced by individual differences in terms of knowledge structure. For instance, Kusumi[2] measured knowledge structure using the Semantic Differential (SD) method and statistically confirmed that individual differences in knowledge structure influence metaphor understanding.

Iwayama, Tokunaga and Tanaka[3] have proposed a computational model based on the salience imbalance model. The model is based on the assumption that each concept has a probabilistic structure of the form (concept: attribute (attribute value, attribute value probability) ...), for instance, (apple: color (red, 0.7)(green, 0.3)). In the model, metaphor understanding is represented as a process in which the high-salient attribute values of the vehicle are matched to the low-salient attribute values of the target, because the salience of an attribute value is measured in terms of entropy computed from the probabilities of attribute values. However, the model does not propose a practical method for computing the probabilistic relationships between concepts and their attribute values. Moreover, the psychological validity of Iwayama et al.'s model has not

S. Kollias et al. (Eds.): ICANN 2006, Part I, LNCS 4131, pp. 495–504, 2006.

been verified. It is necessary to investigate the psychological validity of the model because the metaphor understanding process is influenced by the structures of human knowledge and human metaphor understanding process can only be clarified by psychological experimentation.

On the other hand, there are a few studies showing that the low-salient attribute values of a target and a vehicle are emphasized within the process of metaphor understanding and that attribute values play an important part in metaphor understanding [4],[5],[6]. These attribute values have been referred to as "emergent features". However, feature emergence is not explained within the salience imbalance model. Attribute values that are closely related to the image of a vehicle and unrelated to the image of a target should be inhibited. In addition, even though an attribute value is closely related to the image of the target, in some cases, the attribute value may be inhibited due to an effect of the vehicle image. When a metaphor makes sense (e.g., "a ballerina like a butterfly"), an attribute value that is closely related to the image of the vehicle and unrelated to the image of the target (e.g., "are insects") is inhibited, while an attribute value (e.g., "athletic") that is closely related to the image of the target is also inhibited due to the effects of the vehicle image. Thus, there should be a dynamic interaction among attribute values in metaphor understanding. Utsumi[6] constructed a model that represents the mechanism of feature emergence in metaphor understanding. However, the model does not represent the dynamic interaction among attribute values.

Nakagawa, Terai and Hirose[7] constructed a neural network model of metaphor understanding with dynamic interaction. The model represents a human-like process of metaphor understanding based on a psychological experiment. However, it is not practically feasible to collect sufficient data to construct a model that adequately covers metaphorical expressions by their psychological method alone, because participants cannot respond to numerous indices within a limited time. Accordingly, a model based on only psychological experimentation cannot be extended to computational systems (e.g., a search engine).

In order to solve these problems, this study proposes a neural network model that is both based on a statistical analysis of a language corpus and validated by a psychological experiment.

2 Probabilistic Expression of Meaning Based on a Statistical Analysis of a Language Corpus

2.1 Statistical Language Analysis

For the statistical analysis of a language corpus, this study applies a statistical method with a structure similar to Pereira's method, or Probabilistic latent semantic indexing (PLSI) [8],[9],[10]. Latent semantic analysis (LSA)[11] is popular as a method of natural language analysis. In order to determine appropriate classes using LSA, weighting of terms is indispensable. Although there are various kinds of weighting methods, all of them are basically ad hoc with little mathematical support. On the other hand, the statistical analysis of language is based on probability theory and information theory. The method assumes that the co-occurrence probabilities of a term N_i and a term A_j, $P(N_i,A_j)$ is computed using formula (1):

$$P(N_i, A_j) = \sum_k P(N_i|C_k)P(A_i|C_k)P(C_k), \tag{1}$$

where $P(N_i|C_k)$ is the conditional probability of N_i, given C_k, which indicates a latent semantic class assumed in the method. Each parameter in the method, $P(C_k)$, $P(A_j|C_k)$ and $P(N_i|C_k)$, is estimated as the value that maximizes the likelihood of co-occurrence data measured from a language corpus using the EM algorithm, as follows;

E step

$$P(C_k|N_i, A_j) = \frac{P(N_i|C_k)P(A_j|C_k)P(C_k)}{\sum_k P(N_i|C_k)P(A_i|C_k)P(C_k)}, \tag{2}$$

M step

$$P(N_i|C_k) = \frac{\sum_j P(C_k|N_i, A_j)N(N_i, A_j)}{\sum_{i,j} P(C_k|N_i, A_j)N(N_i, A_j)}, \tag{3}$$

$$P(A_j|C_k) = \frac{\sum_i P(C_k|N_i, A_j)N(N_i, A_j)}{\sum_{i,j} P(C_k|N_i, A_j)N(N_i, A_j)}, \tag{4}$$

where $\sum_i P(N_i|C_k) = 1$, $\sum_j P(A_j|C_k) = 1$ and $N(N_i,A_j)$ is the co-occurrence frequency of the term N_i and the term A_j.

By using these estimated parameters to represent the probabilistic relationship between concepts and their attribute values, the model is capable of simulating the process of metaphor understanding based simply on results obtained by the statistical analysis of a language corpus.

2.2 Results of the Statistical Language Analysis

Corpus data: Data relating to adjective-noun modifications was extracted from the Japanese newspaper "MAINICHI SHINBUN" for the period 1993-2002 using a modification analysis tool called "Cabocha"[12]. The original modification data is in Japanese, and includes 3,403 adjectives and 21,671 nouns. The nouns contained in the "Word List by Semantic Principles, Revised and Enlarged Edition"[13], a Japanese thesaurus, were selected for the analysis.

In the Japanese thesaurus, nouns are classified into 4 categories at the first hierarchy level, 43 categories at the second level and 545 categories at the third rebel. Some of the nouns are also classified as an adjective. Adjectives are classified into 23 categories at the second level. In this study, the second level classes are assumed to be basic categories. The number of semantic classes in the statistical analysis was set as 70, based on the number of categories in the thesaurus.

$P(C_k)$, $P(N_i|C_k)$ and $P(A_j|C_k)$ are estimated using the statistical language analysis and $P(C_k|N_i)$ and $P(C_k|A_j)$ are computed from the Bayesian theorem, as follows:

Table 1. The meaning of the latent classes

Color Class

	$P(C_{color}\|A_i)$		$P(C_{color}\|N_i)$	
1	red	0.994	cap	0.927
2	white	0.987	flower	0.903
3	blue	0.985	helmet	0.879
4	black	0.983	jacket	0.877
5	darkish	0.978	ribbon	0.875
6	crimson	0.972	handkerchief	0.873
7	yellow	0.954	carnation	0.872
8	whitish	0.947	turban	0.869
9	snow-white	0.870	mountain range	0.859
10	pure white	0.844	powder	0.842

$$P(C_k|N_i) = \frac{P(C_k)P(N_i|C_k)}{P(N_i)} = \frac{P(C_k)P(N_i|C_k)}{\sum_h P(C_h)P(N_i|C_h)}, \quad (5)$$

$$P(C_k|A_j) = \frac{P(C_k)P(A_j|C_k)}{P(A_j)} = \frac{P(C_k)P(A_j|C_k)}{\sum_h P(C_h)P(A_j|C_h)}. \quad (6)$$

The names of the latent classes were identified from the conditional probability of the latent class C_k given adjective A_i ($P(C_k|A_j)$) and the probability of the latent class C_k given noun N_i ($P(C_k|N_i)$). For example, the class that can be labeled "Color" is shown in Table1. The listed adjectives and nouns that have relatively strong probabilities, $P(C_k|A_j)$ and $P(C_k|N_i)$, can be regarded as being more closely related to each class C_k.

3 A Model of Metaphor Understanding

3.1 Estimation of the Relationships Between Nouns and Adjectives

In this study, we computed two types of conditional probabilities. The first type is computed directly from the co-occurrence data ($P^*(A_j|N_i)$) and the other type is computed using the result of the language statistical analysis ($P(A_j|N_i)$).

The conditional probability of an adjective given a noun from the co-occurrence data, $P^*(A_j|N_i)$, is computed by function(7) using the co-occurrence frequency $N(A_j, N_i)$ and the noun frequency $N(N_i)$:

$$P^*(A_j|N_i) = \frac{N(A_j, N_i)}{N(N_i)}. \quad (7)$$

While the conditional probability from the result from the statistical language analysis, $P(A_j|N_i)$, is computed by function(8) using $P(A_j|C_k)$, $P(A_j|C_k)$ and $P(C_k)$:

$$P(A_j|N_i) = \frac{\sum_k P(N_i|C_k)P(A_j|C_k)P(C_k)}{\sum_k P(N_i|C_k)P(C_k)}. \quad (8)$$

Both of these conditional probabilities represent the relationships between nouns and adjectives. The conditional probabilities from the co-occurrence data represent more direct relationships than the probabilities from the language statistical analysis. However, these probabilities ($P^*(A_j|N_i)$) are subject to the sparseness problem. Even though the co-occurrence frequency between demon and good is 0, the co-occurrence probability would not be expected to be 0 when using a much larger scale corpus. Therefore, probabilities need to be computed based on the statistical language analysis in order to estimate the parameters of the model.

In this study, we constructed a model concerning the metaphor "a teacher like a demon". In the model, it is assumed that the high-salient attribute values of a vehicle and the attribute values of a target are related to metaphor understanding. The attribute values of the model are chosen referring to $P^*(A_j|N_i)$ and $P(A_j|N_i)$. Thus, three adjectives, "awful", "enormous" and "red" ("demon" means Japanese demon "ONI" in the metaphor "a teacher like a demon", and the color most frequently associated with Japanese demons is red), which co-occur frequently with "demon" are chosen as high-salient attribute values of "demon". Eight adjectives, "young", "gentle", "eager", "great", "horrible", "strict", "good" and "wonderful", which co-occur frequently with "teacher" are chosen as attribute values of "teacher" (see Table2). The co-occurrence probabilities of "enormous", "remarkable", "blue", "broad" and "wise" concerning "demon" have the same value. Comparing the conditional probabilities of these adjectives given "demon", which are computed based on the statistical analysis, the probability of "enormous" is the highest ($P(\text{"}enormous\text{"}|\text{"}demon\text{"}) = 0.02520$, $P(\text{"}remarkable\text{"}|\text{"}demon\text{"}) = 0.01568$, $P(\text{"}blue\text{"}|\text{"}demon\text{"}) = 0.01565$, $P(\text{"}broad\text{"}|\text{"}demon\text{"}) = 0.00623$, $P(\text{"}wise\text{"}|\text{"}demon\text{"}) = 0.00080$). Thus, "awful", "red" and "enormous" were chosen as high-salient attributes.

Table 2. The conditional probabilities of adjectives given either "teacher" or "demon" computed directly from the co-occurrence data ($P^*(A_j|N_i)$)

	$P^*(A_i\|demon)$		$P^*(A_i\|teacher)$	
1	awful	0.19355	young	0.09670
2	red	0.09677	gentle	0.04326
3	enormous	0.06452	eager	0.04199
4	remarkable	0.06452	great	0.03435
5	blue	0.06452	horrible	0.03053
6	broad	0.06452	strict	0.02926
7	wise	0.06452	good(Chinese character)	0.02672
8	cold-blooded	0.03226	good	0.02163
9	harmless	0.03226	wonderful	0.02036
10	mysterious	0.03226	favorite	0.01908

Thus, $P(A_j|\text{"}teacher\text{"})$ and $P(A_j|\text{"}demon\text{"})$ are computed in order to determine the strengths of relationships between the concepts in the metaphor and the

Table 3. The probabilities of the attribute values given either "teacher" or "demon" computed from the statistical language analysis($P(A_j|N_i)$)

		Noun	
		demon	teacher
Attribute values	awful	0.00818	0.00062
	red	0.02085	0.00088
	enormous	0.02519	0.00009
	young	0.00282	0.09679
	gentle	0.00112	0.01104
	eager	0.00095	0.00896
	great	0.00039	0.00336
	horrible	0.01129	0.00303
	strict	0.00281	0.02691
	good	0.00131	0.03182
	wonderful	0.00083	0.00938

attribute values (see Table3). The connection weights of the model are estimated using these probabilities.

3.2 The Architecture of the Model

The model consists of "teacher" and "demon" nodes (concepts nodes) as input nodes and the attribute value nodes as output nodes (Fig.1). The input weights from the "teacher" and "demon" nodes to the attribute value nodes are estimated using conditional probabilities of each attribute value given either "teacher" or "demon". The respective input weights are multiplied by either $\alpha_{teacher}$ or α_{demon}, which indicate the influences of "teacher" and "demon" on the process of metaphor understanding respectively (the values of these parameters, $\alpha_{teacher}$ or α_{demon}, vary according to relevant vehicle and target functions within the metaphor, would be different for the two metaphors of "a teacher like a demon" and "a demon like a teacher").

Every attribute value node has mutual and symmetric connections with the other attribute value nodes. The weighting of a connection between two attribute values in a metaphor may differ from the weighting of the connection between the same two attribute values in another different metaphor. For example, in the case of "a dog like a cloud", the connection between "white" and "fluffy" would be heavily weighted. On the other hand, in the case of "skin like snow", the connection between them would not so heavy. Therefore, every weight for mutual connections between the attribute value nodes is estimated using the correlation coefficient between the two attribute values in a metaphor. The correlation coefficient is computed using the conditional probabilities of each attribute value given the concept, which are classified into the same category as the target or the vehicle at the fourth level of the "Word List by Semantic Principles, Revised and Enlarged Edition"[13].

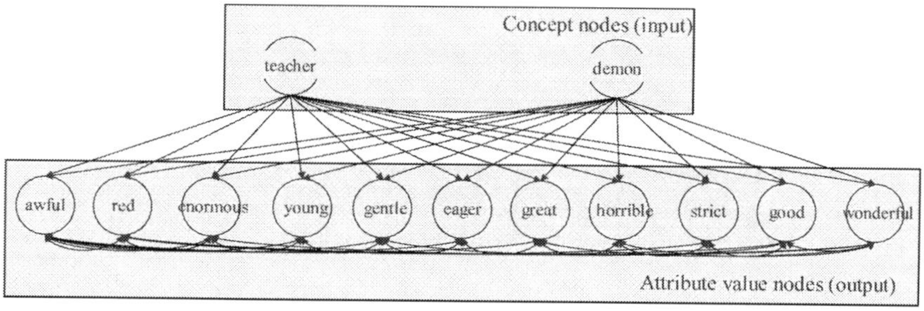

Fig. 1. The architecture of the model concerning "a teacher like a demon"

When a value of 1 is input to the "teacher" node and a value of 0 is input to the "demon" node, the attribute value nodes output the strengths of the relationships between the attribute values and "teacher." Conversely, when a value of 0 is input to "teacher" node and a value of 1 is input to the "demon" node, the attribute value nodes output the strengths of the relationships between the attribute values and "demon". When a value of 1 is input to both nodes, the attribute value nodes output the strengths of the relationships between the attribute values and the metaphorical expression "a teacher like a demon".

The dynamics of the network is based on the following system of simultaneous differential equations:

$$\frac{dx_i(t)}{dt} = -x_i(t) + f(\sum_j w_{ij}x_j(t) + \sum_k \alpha_k W_{ik} I_k) \qquad (k = \text{"teacher"}, \text{"demon"}), \tag{9}$$

where $x_i(t)$ represents the activation strength of the ith attribute value node at time t, w_{ij} denotes the weight of the connection between the ith attribute value node and the jth attribute value node, W_{ik} is the input weight from k's input node to the ith attribute value node and I_k represents the value of k's input node. In this study, the value of $\alpha_{teacher}$ is 1 and the value of α_{demon} is 0.8. The function f is the logistic function. W_{ik} is estimated using the conditional probability of each attribute value given either "teacher" or "demon" by formula(10):

$$W_{ik} = \frac{P(A_i|k)}{\sum_i P(A_i|k)}, \tag{10}$$

where A_i denotes the meaning of the ith attribute value node and k denotes a concept as a vehicle or a target.

3.3 Simulation Results of the Model

The simulation results of the model concerning "a teacher like a demon" are shown in Fig.2 (the original output values between -1 to 1 have been transformed to a 1 to 7 scale). By comparing the activation strengths of the attribute

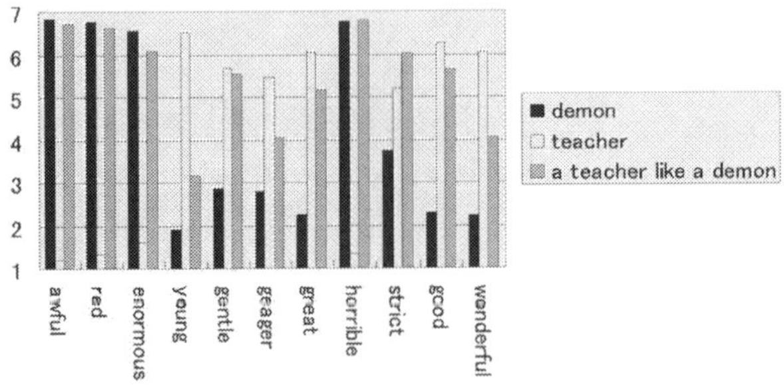

Fig. 2. The results of the simulation (a teacher likes a demon)

value nodes for "teacher" and those of the attribute value nodes for "a teacher like a demon", it is clear that the image of "demon" emphasizes the images of "awful", "red", "enormous" and "horrible", while weakening the images of "young," "gentle", "eager", "great", "good" and "wonderful" for "teacher" in the metaphorical expression of "a teacher like a demon." Furthermore, the image of "strict" for "teacher" is emphasized through the interaction among attribute values in the process of metaphor understanding.

4 Psychological Experiment

The model appears to successfully represent human metaphor understanding. However, the human process of metaphor understanding can only be clarified by psychological experimentation. In order to examine the psychological validity of the simulation of the metaphor understanding model, a psychological experiment was conducted.

4.1 Method

- Participants: 25 undergraduates.
- Metaphorical expression: "a teacher like a demon".
- Attribute values: "horrible", "strict", "strong", "big", "red", "cold", "gentle", "great", "irritable" and "noisy".
- Scale: 7-point scale, from 1 "Strongly disagree" to 7 "Strongly agree".

In pilot studies, the concepts (a target and a vehicle) and a metaphor were presented to 31 participants and they were asked to express the features of the presented concepts and the metaphor using adjectives. "Horrible", "strong", "big", "red" were chosen as features of "demon", while "gentle", "horrible", "strict", "great", "noisy" were chosen as features of "teacher". "Horrible", "strict", "strong" "cold", "irritable" were chosen as features of "a teacher like a demon".

The relationships between the attribute values and the concepts, and the relationships between the attribute values and the metaphor were measured

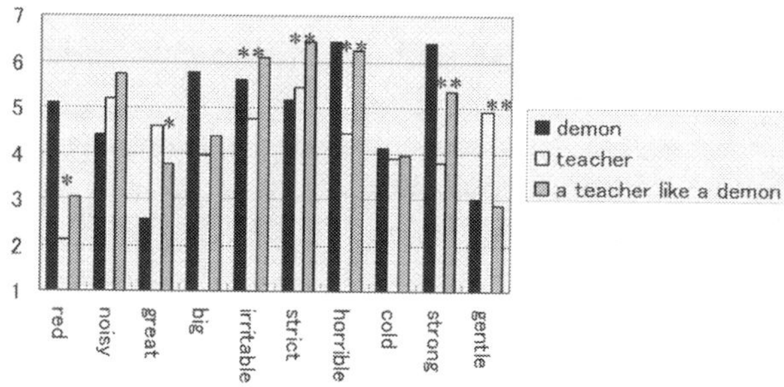

Fig. 3. The results of the psychological experiment (** $p < 0.01$, * $p < 0.05$)

by the Semantic Differential (SD) method. For example, taking the concept of "teacher" and the attribute value "horrible", the participants were asked to evaluate the extent to which the image evoked by "horrible" corresponds to the image evoked by "teacher". The participants evaluated the images of "teacher", "demon" and "a teacher like a demon".

4.2 The Results of the Psychological Experiment on Metaphor Understanding

The results of the psychological experiment indicate that for the expression" a teacher like a demon", the image of "demon" emphasizes the impressions of "red", "irritable", "horrible" and "strong" while weakening the impressions of "great", "gentle" at significant levels (Fig.3). The changes from the image of "teacher" to the image of "a teacher like a demon" are consistent with the process of metaphor understanding. The results from the simulation and from the experiment are closely matched in terms of the pattern of changes, with both sets of results showing emphasis for the impressions of "red", "horrible" and "strict" and weakening of the impressions of "great" and "gentle". In particular, the change for "strict" in the experimental results shows an effect of dynamic interaction among the attribute values in the metaphor understanding. These results support the psychological validity of the model.

5 Discussion

In this study, a computational model of metaphor understanding with dynamic interaction was constructed based on data obtained through a statistical language analysis. In addition, simulations of the dynamic process of metaphor understanding were conducted with the model. While there is still some room for improvement with the model in terms of the selection and number of attribute values and further simulations with a wider range of concepts, the simulation results are consistent with the results from the present psychological experiment;

indicating that the model can be constructed without the need for such experimentation. These findings suggest that the model can be applied to various computational systems (e.g., search engines).

Acknowledgements. This research is supported by the Tokyo Institute of Technology 21COE Program, "Framework for Systematization and Application of Large-scale Knowledge Resources".

References

1. Ortony, A.: Beyond Literal Similarity. Psychological Review. **86(3).** (1979) 161–180
2. Kusumi, T.: Hiyu no Syori Katei to Imikozo. Kazama Syobo(1995)
3. Iwayama, M., Tokunaga, T. and Tanaka, H.: The role of Salience in understanding Metaphors. Journal of the Japanese Society for Artificial Intelligence.**6(5).** (1991) 674–681
4. Nueckles, M. and Janetzko, D.: The role of semantic similarity in the comprehension of metaphor. Proceeding of the 19th Annual Conference of the Cognitive Science Society. (1997) 578–583
5. Gineste, M., Indurkhya, B. and Scart, V.: Emergence of features in metaphor comprehension. Metaphor and Symbol. **15(3).**(2000) 117–135
6. Utsumi, A.: Hiyu no ninchi / Keisan Moderu. Computer Today. **96(3).** (2000) 34–39
7. Nakagawa, M., Terai, A. and Hirose, S.: A Neural Network Model of Metaphor Understanding. Proceedings of Eighth International Conference on Cognitive and Neural Systems (2004) 32
8. Pereira, F., Tishby, N., and Lee, L.: Distributional clustering of English words. Proceedings of the 31st Meeting of the Association for Computational Linguistics. (1993) 183–190
9. Hofmann, T.: Probabilistic latent semantic indexing. Proceedings of the 22nd International Conference on Research and Development in Information Retrieval :SIGIR f99. (1999) 50–57
10. Kameya, Y., and Sato, T.: Computation of probabilistic relationship between concepts and their attributes using a statistical analysis of Japanese corpora. Proceedings of Symposium on Large-scale Knowledge Resources: LKR2005.(2005) 65–68
11. Deerwester, S., Dumais, S., Furnas, G., Landauer, T. and Harshman, R.: Indexing by Latent Semantic Analysis. Journal of the Society for Information Science. **41(6).** (1990) 391–407
12. Kudoh, T., and Matsumoto, Y.: Japanese Dependency Analysis using Cascaded Chunking. Proceedings of the 6th Conference on Natural Language Learning: CoNLL 2002. (2002) 63–69
13. The National Institute for Japanese Language: Word List by Semantic Principles, Revised and Enlarged Edition, Dainippon-Tosho. (2004)

Strong Systematicity in Sentence Processing by an Echo State Network*

Stefan L. Frank

Nijmegen Institute for Cognition and Information, Radboud University Nijmegen
P.O. Box 9104, 6500 HE Nijmegen, The Netherlands
S.Frank@nici.ru.nl

Abstract. For neural networks to be considered as realistic models of human linguistic behavior, they must be able to display the level of systematicity that is present in language. This paper investigates the systematic capacities of a sentence-processing Echo State Network. The network is trained on sentences in which particular nouns occur only as subjects and others only as objects. It is then tested on novel sentences in which these roles are reversed. Results show that the network displays so-called strong systematicity.

1 Introduction

One of the most noticeable aspects of human language is its systematicity. According to Fodor and Pylyshyn [1], this is the phenomenon that 'the ability to produce/understand some sentences is *intrinsically* connected to the ability to produce/understand certain others' (p. 37). To give an example, any speaker of English who accepts 'Capybaras eat echidnas' as a grammatical sentence, will also accept 'Echidnas eat capybaras', even without knowing what capybaras and echidnas are.[1]

The ability of neural networks to behave systematically has been fiercely debated [1, 2, 3, 4]. The importance of this discussion to cognitive science is considerable, because neural networks must be able to display systematicity in order to be considered viable models of human cognition. Moreover, it has been argued that, for connectionist systems to *explain* systematicity, they should not just be able to behave systematically but do so as a necessary consequence of their architecture [5, 6].

In a paper investigating sentence processing by neural networks, Hadley [3] defined systematicity in terms of learning and generalization: A network displays systematicity if it is trained on a subset of possible sentences and generalizes to new sentences that are structurally related to the training sentences. The degree

* This research was supported by grant 451-04-043 of the Netherlands Organization for Scientific Research (NWO).

[1] Capybaras are large South-American rodents and echidnas are egg-laying mammals that live in Australia.

S. Kollias et al. (Eds.): ICANN 2006, Part I, LNCS 4131, pp. 505–514, 2006.

of systematicity displayed by the network depends on the size of the discrepancy between training and test sentences the network can handle. In particular, Hadley distinguishes *weak* and *strong* systematicity. A network only shows weak systematicity if all words in the test sentences occur in the same 'syntactic positions' that they occupied in the training sentences. Christiansen and Chater [7] correctly note that these 'syntactic positions' are not properly defined, but for the purpose of this paper the term can be taken to refer to a noun's grammatical role (subject or object) in the sentence. Unlike weak systematicity, strong systematicity requires the network to process test sentences with words occurring in *new* syntactic positions. Also, these test sentences must have embedded clauses containing words in new syntactic positions.

An example might clarify this. Suppose a network has only been trained on sentences in which female nouns ('woman', 'girls') occur as subjects and male nouns ('man', 'boys') occur as object. Examples of such sentence are 'woman likes boys' and 'girls see man that woman likes'. The network displays weak systematicity if it can process untrained sentences that (like the training sentences) have female subject(s) and male object(s), such as 'woman likes man' and 'girls that like boys see man'. The network is strongly systematic if it can handle sentences with embedded clauses that have male subject(s) and female object(s), that is, in which the roles of males and females are reversed. Examples of such sentences are 'boys like girls that man sees' and 'man that likes woman sees girls'.

In 1994, Hadley [3] argued that the neural networks of that time showed weak systematicity at best, while strong systematicity is required for processing human language. Since then, there have been several attempts to demonstrate strong systematicity by neural networks, but these demonstrations were either restricted to a few specific test items [7] or required representations [8] or architectures [9] that were tailored specifically for displaying systematicity. In contrast, this paper demonstrates that strong systematicity on a large number of test sentences can be accomplished by using only generally applicable representations and architectures.

Instead of the common Simple Recurrent Network (SRN) [10], an adaptation to Jaeger's [11, 12] Echo State Network (ESN) shall be used because it has been shown to outperform an SRN when weak systematicity is required [13]. Presumably, this is because fewer network connections are trained in an ESN than in an SRN. Since this comes down to a smaller number of parameters to fit, generalization (and thereby systematicity) is improved. It is likely that an ESN will again be superior to an SRN when the task requires strong rather than just weak systematicity.

2 Setup of Simulations

In connectionist cognitive models, sentence processing most often comes down to performing the word-prediction task. In this task, the network processes sentences one word at a time and, after each word, should predict which word will be the next input. The network is successful if it does not predict any word that would form an ungrammatical continuation of the input sequence so far.

The word-prediction task has also been used to investigate the issue of systematicity in neural networks [7, 8, 9, 13, 14], and it shall be used here as well. Section 2.1 describes the language on which the network is trained. After training, the network, presented in Sect. 2.2, is tested on a specific set of new sentences (see Sect. 2.3) to probe its systematic capacities.

2.1 The Language

The network learns to process a language that has a 30-word lexicon, containing 9 singular and 9 plural nouns (N), 5 singular and 5 plural transitive verbs (V), the relative pronoun 'that', and an end-of-sentence marker that is denoted [end] (and also considered a word).

The training sentences are generated by the grammar in Table 1. There is no upper limit to sentence length because of relative clauses that can be recursively embedded in the sentence. Relative clauses, which are phrases beginning with the word 'that', come in two types: subject-relative clauses (SRCs) and object-relative clauses (ORCs). In a SRC, 'that' is followed by a verb phrase (VP), while in an ORC 'that' is followed by a noun. In the training sentences, 20% of the noun phrases (NP) contains a SRC and 20% contains an ORC. This results in an average sentence length of 7 words.

Table 1. Production rules for generating training sentences. Variable n denotes number: singular (sing) or plural (plu). Variable r denotes noun role: subject (subj) or object (obj).

Head		Production
S	→	$\mathrm{S}_{\mathrm{sing}} \mid \mathrm{S}_{\mathrm{plu}}$
S_n	→	$\mathrm{NP}_{n,\mathrm{subj}}$ VP_n [end]
$\mathrm{NP}_{n,r}$	→	$\mathrm{N}_{n,r} \mid \mathrm{N}_{n,r}$ $\mathrm{SRC}_n \mid \mathrm{N}_{n,r}$ ORC
VP_n	→	V_n $\mathrm{NP}_{\mathrm{sing,obj}} \mid \mathrm{V}_n$ $\mathrm{NP}_{\mathrm{plu,obj}}$
SRC_n	→	that VP_n
ORC	→	$\mathrm{ORC}_{\mathrm{sing}} \mid \mathrm{ORC}_{\mathrm{plu}}$
ORC_n	→	that $\mathrm{N}_{n,\mathrm{subj}}$ V_n
$\mathrm{N}_{\mathrm{sing,subj}}$	→	woman | girl | dog | cat | mouse | capybara | echidna
$\mathrm{N}_{\mathrm{plu,subj}}$	→	women | girls | dogs | cats | mice | capybaras | echidnas
$\mathrm{N}_{\mathrm{sing,obj}}$	→	man | boy | dog | cat | mouse | capybara | echidna
$\mathrm{N}_{\mathrm{plu,obj}}$	→	men | boys | dogs | cats | mice | capybaras | echidnas
$\mathrm{V}_{\mathrm{sing}}$	→	likes | sees | swings | loves | avoids
$\mathrm{V}_{\mathrm{plu}}$	→	like | see | swing | love | avoid

As can be seen from Table 1, nouns refer to either females, males, or animals. In training sentences, females ('woman', 'women', 'girl', and 'girls') occur only as subjects while males are always in object position. Animals can occur in either position. In test sentences, the roles of male and female nouns will be reversed (see Sect. 2.3). This means that such sentences *are* considered grammatical, they are just not training sentences, which is why they are not generated by the grammar of Table 1. To define the language *in general*, the four rewrite rules for nouns (N) in Table 1 are replaced by:

N_{sing} → woman | girl | man | boy | dog | cat | mouse | capybara | echidna
N_{plu} → women | girls | men | boys | dogs | cats | mice | capybaras | echidnas

A note is in place here about the meaning of the terms 'subject' and 'object', which can sometimes be unclear. Take, for instance, the sentence 'girls that woman likes see boys'. In the main clause of this sentence, 'girls' is the subject (because girls do the seeing) but the same word is object in the subordinate clause (because girls are being liked). To decide upon the noun's syntactic position in such cases, the method by [7] is used. For the sentences used here this comes down to: Nouns directly following a verb are objects, and all the others are subjects. Table 2 shows some examples of training sentences and indicates which nouns are subjects and which are objects.

Table 2. Examples of training sentences. Subscripts 'subj' and 'obj' indicate the sentences' subject(s) and object(s), respectively.

Type	Example sentence
Simple	girls$_{subj}$ like cat$_{obj}$ [end]
	cat$_{subj}$ likes boy$_{obj}$ [end]
SRC	girl$_{subj}$ that likes boys$_{obj}$ sees cat$_{obj}$ [end]
	girls$_{subj}$ like cats$_{obj}$ that see boy$_{obj}$ [end]
ORC	girls$_{subj}$ that cat$_{subj}$ likes see boys$_{obj}$ [end]
	girls$_{subj}$ like boy$_{obj}$ that cat$_{subj}$ sees [end]
SRC and ORC	girl$_{subj}$ that likes boys$_{obj}$ sees cats$_{obj}$ that man$_{subj}$ avoids [end]
	girl$_{subj}$ that likes boys$_{obj}$ that cats$_{subj}$ see avoids man$_{obj}$ [end]

2.2 The Network

Network Processing. Sentences are processed by an Echo State Network that has been extended with an additional hidden layer, resulting in a total of four layers. These are called the input layer, dynamical reservoir (DR), hidden layer, and output layer, respectively. The input and output layers have 30 units each, corresponding to the 30 words of the language. The DR has 1 000 linear units, and the hidden layer has 10 sigmoidal units. This network is nearly identical to Frank's [13], who showed empirically that the extra hidden layer and linear DR units are needed for successful generalization in the word-prediction task. The only difference is that the current network has a much larger DR because it needs to process longer sentences and learn long-distance dependencies between nouns and verbs.

Sentence processing by this ESN is similar to that of a four-layer SRN. At each time step (indexed by t), one word is processed. If word i forms the input at time step t, the input activation vector $\boldsymbol{a}_{\text{in}}(t) = (a_{\text{in},1}, \ldots, a_{\text{in},30})'$ has $a_{\text{in},i} = 1$ and $a_{\text{in},j} = 0$ for all $j \neq i$. The output activation after processing the word is computed from the input according to

$$\begin{aligned} \boldsymbol{a}_{\text{dr}}(t) &= \boldsymbol{W}_{\text{in}}\boldsymbol{a}_{\text{in}}(t) + \boldsymbol{W}_{\text{dr}}\boldsymbol{a}_{\text{dr}}(t-1) \\ \boldsymbol{a}_{\text{hid}}(t) &= \mathbf{f}\left(\boldsymbol{W}_{\text{hid}}\boldsymbol{a}_{\text{dr}}(t) + \boldsymbol{b}_{\text{hid}}\right) \\ \boldsymbol{a}_{\text{out}}(t) &= \mathbf{f}_{\text{sm}}\left(\boldsymbol{W}_{\text{out}}\boldsymbol{a}_{\text{hid}}(t) + \boldsymbol{b}_{\text{out}}\right) \quad , \end{aligned}$$

where $\boldsymbol{a}_{\text{in}}, \boldsymbol{a}_{\text{dr}}, \boldsymbol{a}_{\text{hid}}, \boldsymbol{a}_{\text{out}}$ are the activation vectors of the input layer, DR (with $\boldsymbol{a}_{\text{dr}}(0) = \mathbf{0}$), hidden layer, and output layer, respectively; $\boldsymbol{W}$ are the corresponding connection-weight matrices; $\boldsymbol{b}$ are bias vectors; $\mathbf{f}$ is the logistic activation function; and $\mathbf{f}_{\text{sm}}$ is the softmax activation function. As a result of applying $\mathbf{f}_{\text{sm}}$, the total output activation equals 1 and each $a_{\text{out},i}$ can be interpreted as the network's estimated probability that the next input will be word i.

Network Performance. The network's performance is defined as follows: Let G denote the set of words that can grammatically follow the current input sequence, that is, any word $i \notin G$ would be an incorrect prediction at this point. Moreover, let $a(G) = \sum_{i \in G} a_{\text{out},i}$ be the total amount of activation of output units representing words in G, that is, the total 'grammatical' activation.

Ideally, $a(G) = 1$ when there is no 'ungrammatical' output activation. In that case, the performance score equals +1. Likewise, performance is −1 if $a(G) = 0$ (there is no grammatical activation). By definition, performance equals 0 if the network learned nothing except the frequencies of words in the training set. If $fr(G)$ denotes the total frequency of the words in G, performance equals 0 if $a(G) = fr(G)$. All in all, this leads to the following definition of performance:

$$\text{performance} = \begin{cases} \frac{a(G)-fr(G)}{1-fr(G)} & \text{if } a(G) > fr(G) \\ \frac{a(G)-fr(G)}{fr(G)} & \text{otherwise} \end{cases} \quad . \tag{1}$$

Network Training. The most important difference between the network used here and an isomorphic SRN is that in ESNs, connection weight matrices $\boldsymbol{W}_{\text{in}}$ and $\boldsymbol{W}_{\text{dr}}$ are not trained but keep their initial random values. All other weights (i.e., those in $\boldsymbol{W}_{\text{hid}}, \boldsymbol{W}_{\text{dr}}, \boldsymbol{b}_{\text{hid}}$, and $\boldsymbol{b}_{\text{out}}$) were trained using the backpropagation algorithm, with a learning rate of .01, cross-entropy as error function, and without momentum. All initial weights and biases were chosen randomly from uniform distributions in the following ranges: $\boldsymbol{W}_{\text{hid}}, \boldsymbol{W}_{\text{out}}, \boldsymbol{b}_{\text{hid}}, \boldsymbol{b}_{\text{out}} \in [-0.1, +0.1]$ and $\boldsymbol{W}_{\text{in}} \in [-1, +1]$. Of the DR connections, 85% was given a zero weight. The other 15% had uniformly distributed random weights such that the spectral radius of $\boldsymbol{W}_{\text{dr}}$ equalled .7.

Ten networks were trained, differing only in their initial connection weight setting. The training sentences, generated at random, were concatenated into one input stream, so the network also had to predict the word following [end], that is, the next sentence's first word. During training, the performance score over a random but fixed set of 100 training sentences was computed after every 1 000 training sentences. As soon as the average performance exceeded .98, the network was considered sufficiently trained.

2.3 Testing for Strong Systematicity

Test Sentences. For a network to display strong systematicity, it needs to correctly process new sentences that have embedded clauses with words occurring in new syntactic positions. Four types of such sentences, listed in Table 3, constituted the test set. Each has one subject- or object-relative clause that modifies either the first or second noun. As such, the test-sentence types are labelled SRC1, SRC2, ORC1, and ORC2.

Table 3. Examples of test sentences of four types

Relative clause		Test sentence	
Type	Position	Type	Example
subject	first	SRC1	boy that likes girls sees woman [end]
	second	SRC2	boy likes girls that see woman [end]
object	first	ORC1	boys that man likes see girl [end]
	second	ORC2	boys like girl that man sees [end]

Test sentences were constructed by taking the structures of the examples in Table 3, and filling the noun and verb positions with all combinations of nouns and verbs, such that:

- Only male nouns appear in subject positions and only female nouns in object positions (note that these roles are reversed relative to the training sentences and that test sentences contain no animal names);
- The resulting sentence is grammatical (i.e., there is number agreement between a noun and the verb(s) it is bound to);
- The two verbs of SRC2, ORC1, and ORC2 sentences differ in number; In SRC1 sentences, where the verbs must have the same number, the first two nouns differ in number;
- The unbound noun (for SRC1 sentences: the third noun) was singular.

This makes a total of 2 (numbers) $\times$ 2^3 (nouns) $\times 5^2$ (verbs) $=$ 400 test sentences of each of the four types. Before processing any of these, the network was given the input [end], putting the DR-units into the right activation state for receiving the test sentence's first word.

Generalization Score. The performance measure defined in (1) assigns a score of 0 to the outputs of a network that has learned nothing but the frequencies of words in the training set. To rate the network's systematicity, a similar measure is used. Instead of using word frequencies as baseline, however, this measure assigns a score of 0 to the outputs of a hypothetical network that has learned the training set to perfection but is not systematic *at all.*

According to Hadley's [3] definition of systematicity, a complete lack of systematicity is the same as the absence of generalization. What can a non-generalizing network do when confronted with a new input sequence? By definition, it cannot use the complete sequence for predicting the next input since this would require generalization. Therefore, the best it can do is to use the most recent part of the test input that was also present in the training sentences. Assume, for instance, that the test input is the ORC1 sentence 'boys that man likes see girl'. After processing '[end] boys', the network is faced with an input sequence it was not trained on because training sentences always begin with a female noun (i.e., 'boys' never follows [end] in the training sentences). All that the network can do is to base its next-word prediction on the last input only, that is, the word 'boys'. In the training sentences, male nouns were followed by [end] in 50% of the cases. Therefore, the non-generalizing network will, by definition, result in an (incorrect) output activation $a_{\text{out},[\text{end}]} = .5$ at this point.

When the next word enters the network, the input sequence is '[end] boys that'. The two-word sequence 'boys that' has appeared in the training sentences (see Table 2). By definition, the output activations of the non-generalizing network at this point are exactly the probabilities that each of the 30 words follows 'boys that' in the training sentences. Likewise, after the word sequence '[end] boys that man', the network basis its predictions on 'man' only because 'that man' never appears in the training sentences. Again, this results in $a_{\text{out},[\text{end}]} = .5$.

The generalization score is computed by an equation identical to (1) except that $fr(G)$ is replaced by the total grammatical output activation of the hypothetical non-generalizing network. This means that positive scores indicate some degree of generalization. If the network scores positively on each word of the four types of test sentences, it displays strong systematicity.

There are several points in the test sentences were generalization is not required for making grammatical predictions. For instance, after '[end] boys that', the next word must be a noun or a plural verb. In the training sentences, 'boys that' is always followed by a plural verb. The non-generalizing network will therefore make the perfectly grammatical prediction that the next word is a plural verb. At such points, even a perfectly systematic network does no better than a non-generalizing one, so the generalization score is not defined. Note that grammatical predictions are always trivial to make at such points, because generalization is not needed. Therefore, nothing is lost by the occasional absence of a generalization score.

3 Results and Conclusion

Generalization scores, averaged over sentences of each type and over the 10 trained networks, are plotted in Fig. 1. The near-perfect performance on the first word of test sentences is a first indication of systematicity by the network. This first word is always a male noun in subject position, which it never occupied in any of the training sentences. Instead, in 50% of the cases, male nouns in training sentences occur in sentence-final position, being followed by [end]. To make the correct prediction that a sentence-initial male noun is *not* followed by [end], the network must have learned that it is the position of the noun that matters and not its particular (male) identity. As is clear from Fig. 1, the network has succeeded in doing this.

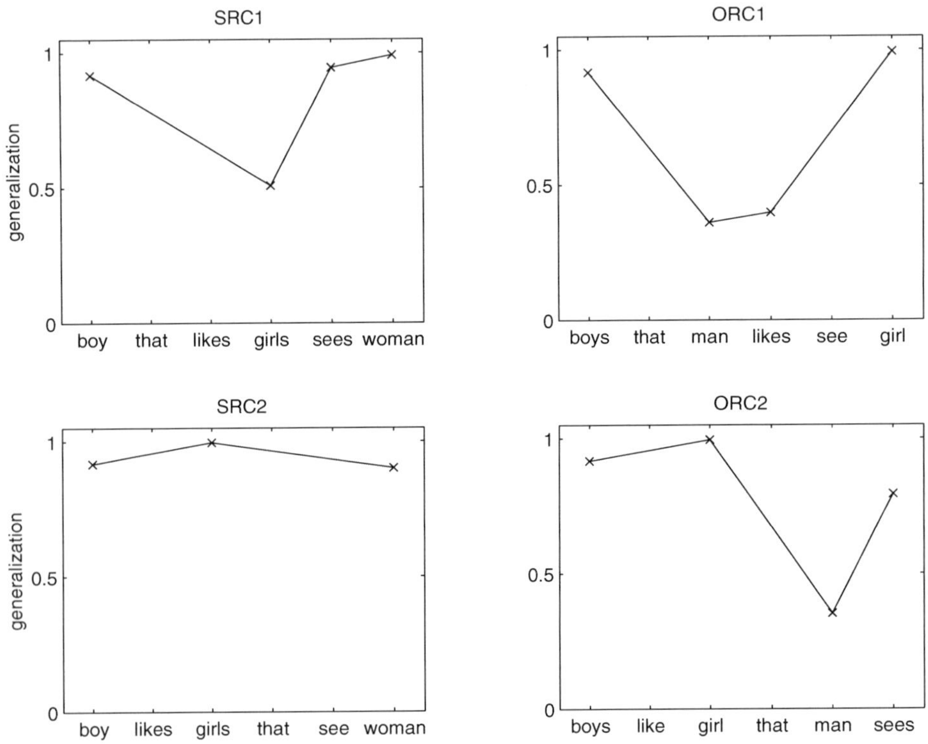

Fig. 1. Generalization scores on test sentences of four types, at all words where generalization is defined (labels on the horizontal axes are only examples of test-sentence words, the plotted performance scores are averaged over all sentences of a type).

For the systematicity to be considered *strong* in the sense of [3], it should also be demonstrated in embedded clauses. The plots in Fig. 1 show that the average performance is above 0 on all words of each test sentence type, including the words in relative clauses. Even at the point in the sentence where performance is minimal, it is still highly significantly positive, as was revealed by sign tests ($p < 10^{-16}$ for each of the four sentences types).

Strictly speaking, these results are not yet proof of systematicity because they could just be an artifact of averaging over networks and sentences. If, for instance, each individual network scores negatively at one point, but this is not the same point for the 10 trained networks, the average generalization score can be positive while none of the networks displays systematicity. The same can happen if each test sentence results in a negative score at some point, but this point differs among the sentences. However, this was clearly not the case: Of all individual generalization scores, only 0.38% was negative.

These results are a clear indication of strong systematicity by the ESN. It processed test sentences with nouns in syntactic positions they did not appear in during training, both in the main clause and in relative clauses, and performed significantly better than a non-systematic network can do. As defined in [3], successful processing of test sentences that differ this much from the training sentences requires strong systematicity.

Nevertheless, generalization decreased considerably at some points in the test sentences. As can be seen from Fig. 1, there are four points at which the generalization score is quite low (less than .6). At all of these problematic points, the network must predict a verb after having processed both a singular and a plural noun. Presumably, the difficulty lies with binding the verb to the correct noun, that is, predicting whether it is singular or plural. An investigation of the network's output vectors supported this interpretation. Nevertheless, the ESN always does better than a network that is not systematic at all and, therefore, displays some (but by no means perfect) strong systematicity.

4 Discussion

Echo State Networks can display not only weak but also strong systematicity. This is a necessary condition for a neural network that is to form a cognitive model of sentence processing. However, Hadley [15, 16] argued that human language is not only systematic with respect to word order in sentences (i.e., syntax) but also with respect to their meaning. That is, people are *semantically systematic*: They can correctly assign a semantic interpretation to sentences they have not been exposed to before. Contrary to this, the word-prediction task used in the simulations presented here is purely syntactic. Recently, however, Frank and Haselager [17] showed that semantic systematicity is not beyond the abilities of a neural network. They trained an ESN to transform sentences into distributed representations of the situations the sentences referred to, and found that the network could generalize to sentences describing novel situations.

References

1. Fodor, J.A., Pylyshyn, Z.W.: Connectionism and cognitive architecture: a critical analysis. Cognition **28** (1988) 3–71
2. Chalmers, D.J.: Connectionism and compositionality: why Fodor and Pylyshyn were wrong. Philosophical Psychology **6**(3) (1993) 305–319
3. Hadley, R.F.: Systematicity in connectionist language learning. Mind & Language **9**(3) (1994) 247–272
4. Niklasson, L.F., Van Gelder, T.: On being systematically connectionist. Mind & Language **9** (1994) 288–302

5. Fodor, J.A., McLaughlin, B.: Connectionism and the problem of systematicity: Why Smolensky's solution does not work. Cognition **35** (1990) 183–204
6. Aizawa, K.: The systematicity arguments. Dordrecht, The Netherlands: Kluwer Academic Publishers (2003)
7. Christiansen, M.H., Chater, N.: Generalization and connectionist language learning. Mind & Language **9**(3) (1994) 273–287
8. Hadley, R.F., Rotaru-Varga, A., Arnold, D.V., Cardei, V.C.: Syntactic systematicity arising from semantic predictions in a Hebbian-competetive network. Connection Science **13**(1) (2001) 73–94
9. Bodén, M.: Generalization by symbolic abstraction in cascaded recurrent networks. Neurocomputing **57** (2004) 87–104
10. Elman, J.L.: Finding structure in time. Cognitive Science **14** (1990) 179–211
11. Jaeger, H.: Adaptive nonlinear system identification with echo state networks. In Becker, S., Thrun, S., Obermayer, K., eds.: Advances in neural information processing systems. Volume 15. Cambridge, MA: MIT Press (2003)
12. Jaeger, H., Haas, H.: Harnessing nonlinearity: predicting chaotic systems and saving energy in wireless communication. Science **304** (2004) 78–80
13. Frank, S.L.: Learn more by training less: systematicity in sentence processing by recurrent networks. Connection Science (in press)
14. Van der Velde, F., Van der Voort van der Kleij, G.T., De Kamps, M.: Lack of combinatorial productivity in language processing with simple recurrent networks. Connection Science **16**(1) (2004) 21–46
15. Hadley, R.F.: Systematicity revisited: reply to Christiansen and Chater and Niklasson and van Gelder. Mind & Language **9**(4) (1994) 431–444
16. Hadley, R.F.: On the proper treatment of semantic systematicity. **14** (2004) 145–172
17. Frank, S.L., Haselager, W.F.G.: Robust semantic systematicity and distributed representations in a connectionist model of sentence comprehension. In Miyake, N., Sun, R., eds.: Proceedings of the 28th Annual Conference of the Cognitive Science Society. Mahwah, NJ: Erlbaum (in press)

Modeling Working Memory and Decision Making Using Generic Neural Microcircuits*

Prashant Joshi

Institute for Theoretical Computer Science
Technische Universität Graz
A-8010 Graz, Austria
joshi@igi.tugraz.at
http://www.igi.tugraz.at/joshi

Abstract. Classical behavioral experiments to study working memory typically involve three phases. First the subject receives a stimulus, then holds it in the working memory, and finally makes a decision by comparing it with another stimulus. A neurocomputational model using generic neural microcircuits with feedback is presented here that integrates the three computational stages into a single unified framework. The architecture is tested using the two-interval discrimination and delayed-match-to-sample experimental paradigms as benchmarks.

1 Introduction

Classical experiments in behavioral and experimental neuroscience that are employed to study the working memory typically involve three phases. First the subject receives a stimulus, then this stimulus is held in working memory, and finally a decision is made by comparing it with another incoming stimulus. Two classical experimental paradigms to test this are the two-interval discrimination [1] and the delayed-match-to-sample [2] tasks.

A recent study proposed controlled mutual inhibition of neurons in pre-frontal cortex (PFC) as the neural algorithm behind the working memory and decision making process [1]. Although the model was successful in integrating the loading, maintenance, and decision making phases, obtaining such precise tuning of mutual inhibition in cortical circuits in PFC seems biologically implausible. Also, despite existing evidence that shows synaptic learning as a responsible mechanism for working memory related tasks [2], the model used a static (no learning involved) neural circuit.

This article demonstrates that simple linear readouts from generic neural microcircuit models that send their output as a feedback signal to the circuit, can be used to model decision making process that involves the use of working memory. The neurocomputational architecture described here integrates the three crucial stages described above into a single unified framework. These sequential

* The work was partially supported by the Austrian Science Fund FWF, project $\#P17229 - N04$ and project $\#FP6 - 015879$ (FACETS) of the European Union.

S. Kollias et al. (Eds.): ICANN 2006, Part I, LNCS 4131, pp. 515–524, 2006.

stages are the initial loading (L) of stimulus into working memory, maintenance (M) of working memory, and decision (D) making (following the notation used in [1]). Essentially this model presents a unified computational framework for working memory and decision making carried out by the neurons in PFC. For comparison, the unified framework is used to build a spiking neural network model of two-interval discrimination. Additionally to show that this computational paradigm is task-independent, we employ the same paradigm to model the delayed match-to-sample task.

The core principles behind the working of this model make the assumption that the cortex can be thought of as an ultra-high dimensional dynamical system, where the afferent inputs arriving from thalamus and the recurrent cortical feedbacks are churned in a non-linear way to obtain a high-dimensional projection of the low-dimensional input space. Preceding work has demonstrated that such high dimensional transient dynamics provides the neural circuit with analog fading memory[1] that provides the circuit enough computational power for performing open-loop sensory processing tasks [3, 4].

Analog fading memory by itself is not powerful enough to render the circuits the power to hold information in working memory. The obvious reason being that analog fading memory by itself has an upper limit on the order of tens of msec, depending on the time constants of synapses and neurons in the neural circuit [4], whereas typically the working memory holds information on the order of seconds. Recent results show that feedback from trained readout neurons that are part of generic neural circuit can induce multiple co-existing "partial attractors" in the circuit dynamics [5, 6]. This result is further extended here to demonstrate that even in the presence of feedback noise, such "partial attractor" states can be held by generic neural circuits on the time-scales of several seconds, that is obviously a requirement for tasks involving working memory.

The neural microcircuit model considered in this article is described in section 2, and the results for the two-interval discrimination and delayed-match-to-sample tasks are presented in sections 3 and 4 respectively. Finally section 5 presents a discussion of the results.

2 Generic Neural Microcircuit Models

For the experiments described in this article, generic cortical microcircuit models consisting of integrate-and-fire neurons were used, with a high level of noise that reflects experimental data. Biologically realistic models of dynamic synapses were used whose individual mixture of pair-pulsed depression and facilitation

[1] A map (or filter) F from input- to output streams is defined to have *fading memory* if its current output at time t depends (up to some precision ε) only on values of the input $\mathbf{u}$ during some finite time interval $[t-T, t]$. Formally, F has fading memory if there exists for every $\varepsilon > 0$ some $\delta > 0$ and $T > 0$ so that $|(F\mathbf{u})(t) - (F\tilde{\mathbf{u}})(t)| < \varepsilon$ for any $t \in \mathbb{R}$ and any input functions $\mathbf{u}, \tilde{\mathbf{u}}$ with $\|\mathbf{u}(\tau) - \tilde{\mathbf{u}}(\tau)\| < \delta$ for all $\tau \in [t-T, t]$. This is a characteristic property of all filters that can be approximated by an integral over the input stream $\mathbf{u}$, or more generally by Volterra- or Wiener series.

(depending on the type of pre- and postsynaptic neuron) was based on experimental data [7, 8]. These circuits were not created for any specific computational task. Sparse synaptic connectivity between neurons was generated (with a biologically realistic bias towards short-range connections) by a probabilistic rule[2], and synaptic parameters were chosen randomly from distributions that depended on the type of pre- and postsynaptic neurons (in accordance with empirical data from [7, 8])[3]. The neurons in the generic cortical microcircuit models were placed on the integer-points of a 3-D grid[4], and 20% of these neurons were randomly chosen to be inhibitory.

Each readout neuron was trained by linear regression to output at any time t, a particular target value $f(t)$. Linear regression was applied to a set of data points of the form $\langle y(t), f(t)\rangle$, for many time points t, where $y(t)$ is the output of low-pass filters applied to the spike-trains of pre-synaptic neurons, and $f(t)$ is the target output. During training, the feedback from readouts performing diverse computational tasks was replaced by a *noisy* version of their target output ("teacher forcing") [5]. Note that teacher forcing with noisy versions of target feedback values trains these readouts to correct errors resulting from imprecision in their preceding feedback (rather than amplifying errors)[9].

The generic neural microcircuit model received analog input streams from 4 sources for the experiment modeling two-interval discrimination (from 5 sources in the experiment modeling delayed-match-to-sample). The outcomes of the experiments discussed in this article were all negative if these analog input streams were directly fed into the circuit (as input current for selected neurons in the circuit). Apparently the variance of the resulting spike trains were too large to

---

[2] The probability of a synaptic connection from neuron a to neuron b (as well as that of a synaptic connection from neuron b to neuron a) was defined as $C \cdot \exp(-D^2(a,b)/\lambda^2)$, where $D(a,b)$ is the Euclidean distance between neurons a and b, and λ is a parameter which controls both the average number of connections and the average distance between neurons that are synaptically connected. Depending on whether the pre- or postsynaptic neuron were excitatory (E) or inhibitory (I), the value of C was set according to [8] to 0.3 (EE), 0.2 (EI), 0.4 (IE), 0.1 (II). For the experiment modeling *a)* the two-interval discrimination task, $C = 1$, $\lambda = 1.5$; *b)* the delayed-match-to-sample task, $C = 1$, $\lambda = 1.2$.

[3] *Neuron Parameters*: membrane time constant 30 ms, absolute refractory period 3 ms (excitatory neurons), 2 ms (inhibitory neurons), threshold 15 mV (for a resting membrane potential assumed to be 0), reset voltage drawn uniformly from the interval [13.8, 14.5] mV, constant non-specific background current I_b uniformly drawn from the interval [13.5, 14.5] nA for each neuron, noise at each time-step I_{noise} drawn from a gaussian distribution with mean 0 and SD chosen for each neuron randomly from a Gaussian distribution over the interval [4.0, 5.0] nA, input resistance 1 $M\Omega$, the initial condition for each neuron, i.e. its membrane potential at time $t = 0$, was drawn randomly (uniform distribution) from the interval [13.5, 14.9] mV.

[4] 400(500) neurons, arranged on a $20 \times 5 \times 4$ ($20 \times 5 \times 5$) grid for the circuit modeling the two-interval discrimination (delayed-match-to-sample) task.

[5] At each time-step t, a different noise value of $0.0001 \times \rho \times f(t)$ was added, where ρ is a random number drawn from a gaussian distribution with mean 0 and SD 1, and $f(t)$ is the current value of the input signal (signal-dependent noise).

make the information about the slowly varying values of these input streams readily accessible to the circuit. Therefore the input streams were instead fed into the circuit with a simple form of spatial coding, where the location of neurons that were activated encoded the current values of the input variables. More precisely, each input variable is first scaled into the range $[0, 1]$. This range is linearly mapped onto an array of 50 symbolic input neurons. At each time step, one of these 50 neurons, whose number $n(t) \in \{1, \ldots, 50\}$ reflects the current value $i_n(t) \in [0, 1]$ which is the normalized value of input variable $i(t)$ (e.g. $n(t) = 1$ if $i_n(t) = 0$, $n(t) = 50$ if $i_n(t) = 1$). The neuron $n(t)$ then outputs at time step t, the value of $i(t)$. In addition the 3 closest neighbors on both sides of neuron $n(t)$ in this linear array get activated at time t by a scaled down amount according to a gaussian function (the neuron number n outputs at time step t the value $i(t)\frac{1}{\sigma\sqrt{2\pi}} \exp^{\frac{-(n-n(t))^2}{2\sigma^2}}$, where $\sigma = 0.8$). Thus the value of each of the 4(5)[6] input variables is encoded at any time by the output values of the associated 50 symbolic input neurons (of which at least 43 neurons output at any time the value 0). The neuron in each of these 4(5) linear arrays are connected with one of the 4(5) layers consisting of 100 neurons in the previously described circuit of $((20 \times 5) \times 4)$ $(((20 \times 5) \times 5))$ integrate-and-fire neurons.

The coding scheme used here is similar to the population coding [10], i.e., the neurons in the circuit fire maximally for their own preferred stimuli, and different neurons have different preferred stimuli. This coding scheme is suitable in context of this article as previous studies demonstrate that neurons in PFC process information via a labelled-line code [11, 12].

3 Two-Interval Discrimination

In the actual experimental protocol for two-interval discrimination task, a pre-trained subject is presented with two frequencies f_1 and f_2, separated by a certain delay interval. Initially the frequency f_1 is loaded into the working memory, and its value is maintained during the delay phase, and on presenting the f_2 frequency, the subject is required to decide whether "$f_1 > f_2$?". Two kind of neurons have been observed in PFC which show opposite activities in response to the above question [1]. The first type of neurons (called "+" neurons from now), show an increase in their activity during the D phase, when the answer to the above question is "yes", and the second type of neurons (called "-" neurons from now), show an increase in their neural activity when the answer to the above question is "no". The information required to carry out the task is present in the firing activity of "+" and "-" neurons independently, and the reason for the simultaneous presence of both these sets is till now unknown.

In this setup[7], the "+" and "-" neurons have been modeled as simple linear readouts that send feedback of their activity to the neural circuit (see Figure 1).

[6] 4 input variables for the two-interval discrimination task, and 5 input variables for the delayed-match-to-sample task.

[7] Total simulation time for one trial 3.5 s, simulation time-step 10 ms, f_1 and f_2 are presented for 0.5 s each, during the L and D phases respectively.

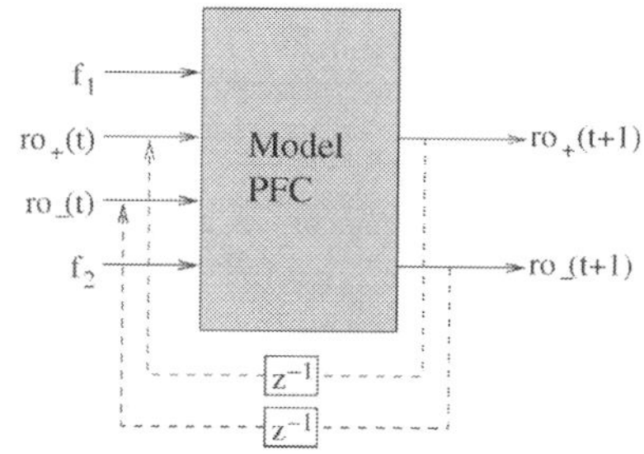

Fig. 1. Closed-loop setup for the two-interval discrimination task. The model PFC circuit is made of 400 integrate-and-fire neurons arranged on the integer points of a $20 \times 5 \times 4$ cube. The circuit receives 2 frequencies (f_1, f_2) as external inputs, and two feedback signals ($ro_+(t)$, $ro_-(t)$)from the "+" and "-" readouts. The task is to answer the question if "$f_1 > f_2$?" The "+"("-") neurons show an increase in their activity when the answer to the above question is "yes"("no"). The notation z^{-1} denotes a unit time-step delay in feedback signal.

In addition to the feedback, the circuit receives 2 external inputs (f_1 and f_2). The input signals f_1 and f_2 are presented during the L and D phases respectively (see Figure 2 A). The training data consisted of 2 noisy versions of each of the 10 input pairs (f_1, f_2) (see Figure 2 B). The target functions of "+" and "-" readouts are as described above. Figure 2 E shows a 200 ms blowup of the circuit response of 100 randomly chosen neurons (activity of excitatory neurons shown in black) during one of the closed-loop validation runs (f_1 = 18 Hz, f_2 = 26 Hz). The panels C and D of figure 2 show the desired (black) and observed values of the "+" and "-" readouts during this run. Panel F and G show the response of the "+" and "-" readouts for the 10 pairs of input frequencies, (f_1, f_2) (note the similarity to Figure 1, panels C and D in [1], which show the actual data from "+" and "-" neurons in PFC during the two-interval discrimination task).

To test the robustness of the neural model, experiments were done where after the readouts have been trained, a subset κ_n (κ_n progressively increased from 0.5% to 5% in 10 trials such that $\kappa_n \subset \kappa_{n+1}$) of synapses converging onto the "+" readouts were randomly chosen and pruned (synaptic weight set to 0). The resulting setup was tested with the 10 frequency pairs (f_1, f_2). Panels H and I of figure 2 show the result of these robustness tests. Panel H shows the mean and standard error of correlation values for progressively higher levels of pruned synapses. Panel I shows the resulting matrix of correlation values, where each square shows the correlation value for a particular validation run, for a particular pruning percentage. The control correlation values[8] (no pruning) are shown in the row on the top. Results indicate that the model shows graceful degradation in presence of suboptimal inputs. This is quite interesting, as traditional models of attractor based neural computation fail to demonstrate robustness [13].

[8] Over 10 validation runs, mean = 0.9556, SD = 0.0206.

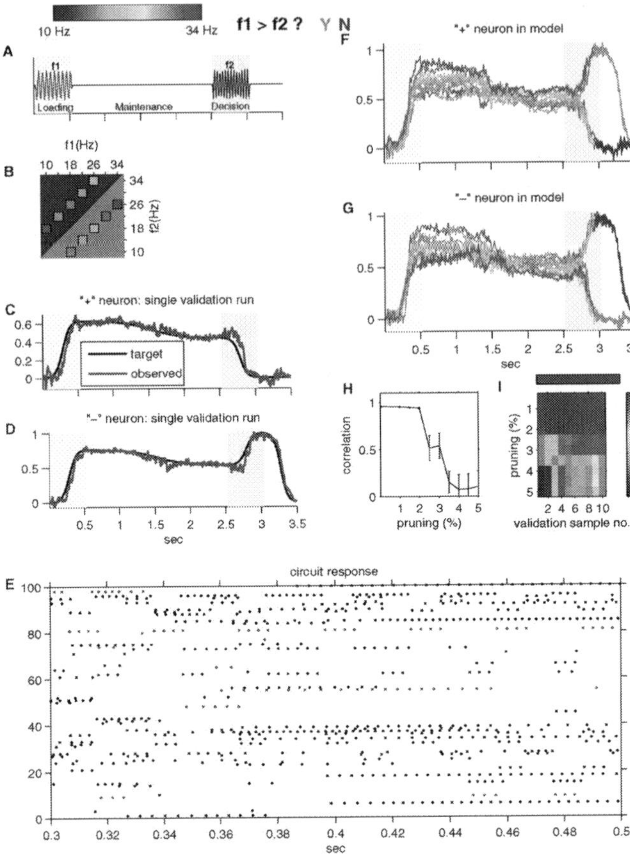

Fig. 2. **(A)** The external frequencies f_1 (18 Hz), and f_2 (26 Hz) presented during the L and D phase for one of the close-loop validation trials. **(B)** Stimulus frequencies used in this study. The target (black) and observed values for the **(C)** "+" readout, and **(D)** the "-" readout. **(E)** A blowup of 200 ms of resulting firing activity of 100 randomly chosen neurons in the circuit. Excitatory neurons are shown in black. Responses of the **(F)** "+" and **(G)** "-" readouts for each of the frequency pairs. The colorbar at upper left indicates the value of f_1 used in each trial. **(H)** Mean and standard error of correlation values for trials with progressively higher pruning percentage. **(I)** The resulting matrix of correlation values where each square shows the correlation value for a particular validation run, for a particular pruning percentage. The control correlation values (no pruning) are shown in the row on the top.

4 Delayed-Match-to-Sample

In this task, a pre-trained subject is presented with a cue visual stimulus during the L phase (for example a small colored square on a predetermined region of the screen), which is followed by a delay period (M phase), and subsequently in the D phase two simultaneous probe stimuli are presented at two different

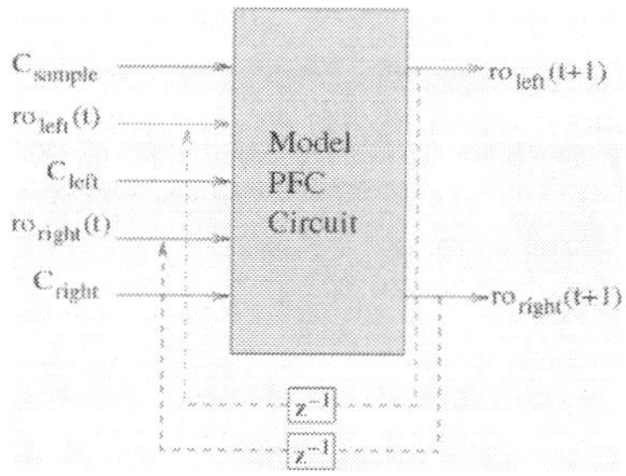

Fig. 3. Closed-loop setup for the delayed-match-to-sample task. The model PFC circuit is made of 500 integrate-and-fire neurons arranged on the integer points of a $20 \times 5 \times 5$ cube. The circuit received 3 color stimulus (C_{cue}, C_{left}, C_{right}) as external inputs, and 2 feedback signals ($ro_{left}(t)$, $ro_{right}(t)$) from the "left" and "right" readouts. The "left"("right") neurons show an increase in their activity during D phase when $C_{cue} = C_{left}(C_{right})$.

places on the screen. The task is to decide which of the probe stimuli has the same color as the sample stimulus[2].

Figure 3 shows the setup[9] used in this experiment. The model PFC circuit received 3 external inputs (C_{cue}, the cue color; C_{left}, the left probe color; C_{right}, the right probe color), and 2 feedback signals (from the "left" and "right" readouts). The cue stimulus was presented during the L phase and the probe stimuli were presented during the D phase. The neurons in the model PFC circuit made projections to two linear readouts (called "left" and "right" readouts from now on) which had similar functionality as the "+" and "-" readouts in the two-interval discrimination task. The "left" readout showed high activity if the answer to the question "$C_{cue} = C_{left}$?" was "yes". The "right" readout behaved exactly opposite to this and showed high amount of activity if the probe stimulus shown on the right matched the sample stimulus.

For this experiment the training data consisted of 2 noisy versions of each of the 5 input color triplets $(C_{cue}, C_{left}, C_{right})$ (see Figure 4 B). Figure 4 shows the result of a closed loop validation run. Panel A shows one such external triplet. The target (black) and observed response of the "left" and "right" readout are shown in panel C and D respectively. The panels E and F of figure 4 show the response of the "left" and "right" readouts for the 5 input stimuli (each line drawn in the color of corresponding cue stimulus).

To test the performance of the setup, we tested the setup in 100 different validation runs. Panels G and H of figure 4 present the histogram of correlation values for the "left"[10] and "right"[11] readouts over the set of these 100 validation runs.

[9] Total simulation time for one trial 1.95 s, simulation time-step 10 ms, C_{cue} is presented for 0.32 s during the L phase, and C_{left} and C_{right} are presented simultaneously for 0.6 s, during the D phase.

[10] mean = 0.9962, SD = 0.0018.

[11] mean = 0.9942, SD = 0.0018.

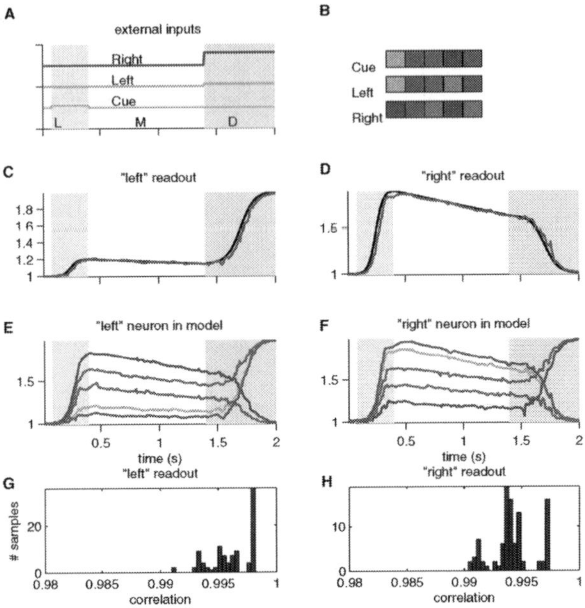

Fig. 4. (A) The external stimuli to the model PFC circuit during one of the closed loop validation runs. The cue stimulus is presented during the L phase and the left-and-right probe stimuli are presented during the D phase. **(B)** The colors of the cue, left, and right stimuli used in trials. The target (black) and observed values for the **(C)** "left" readout, and **(D)** the "right" readout. Responses of the **(E)** "left" and **(F)** "right" readouts for each of the color triplets. The performance of the setup was tested using 100 different closed loop validation runs. Histogram of correlation values for the **(G)** "left" and the **(H)** "right" readout.

5 Discussion

A new neurocomputational paradigm is described that uses synaptic learning mechanisms to present a unified model for working memory and decision making using biologically realistic neural microcircuit models composed of spiking neurons and dynamic synapses. Additionally results for the two-interval-discrimination task show that the neural algorithm is task independent. It is to be noted however that although spiking neural circuits were used to model the interval-discrimination tasks for added biological realism, it is not a requirement, and any recurrent circuit would give similar results, as long as it has the kernel property.

Readouts make a binary decision by reaching one of the states corresponding to the decision made by them. The actual point of time when the readout makes a decision can be thought of as a threshold crossing event, i.e. the first time when the readout crosses a threshold after the presentation of the probe stimulus in the D phase.

It was found that using population coding to encode the inputs projecting on to model PFC circuits was essential to obtain the demonstrated results. Using a population of neurons to encode each analog input stream is not unrealistic, as sufficient evidence exists in current literature of its existence. It is however not claimed here that the precise mechanism of population coding used in this article is the one used in cortical circuitry of PFC.

The feedback from the trained readouts played an apparently important role in the neural model described above. Although the generic neural microcircuits used to simulate the model PFC circuit are endowed with fading memory due to the kernel property of the circuits, this is not sufficient for holding information in working memory for longer timespans ranging in the order of seconds. Apparently the feedback from trained readouts provides the circuit with additional needed information that falls outside the window of fading memory, hence enhancing the information present in the circuit dynamics.

Obviously closed-loop applications of generic neural microcircuit models like the ones discussed in this article present a harder computational challenge than open-loop sensory processing tasks, since small imprecisions in their output are likely to be amplified by the plant (e.g. the arm model) to yield even larger deviations in the feedback, which is likely to further enhance the imprecision of subsequent outputs. This problem can be solved by teaching the readouts from the neural microcircuit during training to ignore smaller recent deviations reported by feedback, thereby making the target trajectory of output torques an attractor in the resulting closed-loop dynamical system.

This study also demonstrates the ability of generic neural microcircuit models to hold "partial attractor" states in their circuit dynamics for significantly longer and biologically relevant time-scales ranging in the order of a couple of seconds, in presence of noise. Also a point of interest is the robustness of this neurocomputational model to factors such as synaptic pruning, and feedback noise.

According to the "internal model" hypothesis [14, 15], there exists an internal model for each of the tasks that we have learned throughout our lives. One outcome of such a hypothesis would be that a given neuron may participate with a different synaptic weight in a number of neural assemblies, each supporting a different internal model. Interestingly, this is reflected in our setup too, as neurons in the generic neural circuit make synaptic projections to the set of readouts with different synaptic weights assigned for each task.

The results presented in this article demonstrate the role of feedback in enhancing the inherent fading memory of a neural circuit. Further it also shows the ability of generic neural circuits to model working memory and decision making, which happens at significantly longer time-scales. Further work is needed to explore if this neurocomputational architecture is extensible across diverse cognitive modalities, e.g. decision making, and motor control.

Acknowledgements

The author is thankful to Wolfgang Maass for critical comments on the draft version of the manuscript, and anonymous reviewers for helpful suggestions.

References

1. C. K. Machens, R. Romo, and C. D. Brody. Flexible control of mutual inhibition: A neural model of two-interval discrimination. *Science*, 307:1121–1124, 2005.
2. G. Rainer, H. Lee, and N. K. Logothetis. The effect of learning on the function of monkey extrastriate visual cortex. *PLoS Biology*, 2(2):275–284, 2004.
3. D. V. Buonomano and M. M. Merzenich. Temporal information transformed into a spatial code by a neural network with realistic properties. *Science*, 267:1028–1030, Feb. 1995.
4. W. Maass, T. Natschläger, and H. Markram. Real-time computing without stable states: A new framework for neural computation based on perturbations. *Neural Computation*, 14(11):2531–2560, 2002.
5. W. Maass, P. Joshi, and E. D. Sontag. Principles of real-time computing with feedback applied to cortical microcircuit models. *NIPS*, 2005.
6. W. Maass, P. Joshi, and E. D. Sontag. Computational aspects of feedback in neural circuits. *submitted for publication*, 2005.
7. H. Markram, Y. Wang, and M. Tsodyks. Differential signaling via the same axon of neocortical pyramidal neurons. *PNAS*, 95:5323–5328, 1998.
8. A. Gupta, Y. Wang, and H. Markram. Organizing principles for a diversity of GABAergic interneurons and synapses in the neocortex. *Science*, 287:273–278, 2000.
9. P. Joshi and W. Maass. Movement generation with circuits of spiking neurons. *Neural Computation*, 17(8):1715–1738, 2005.
10. A. Pouget and P. E. Latham. Population codes. In M. A. Arbib, editor, *The handbook of brain theory and neural networks*, pages 893–897. MIT Press, 2003.
11. J. M. Fuster. Unit activity in prefrontal cortex during delayed-response performance: neuronal correlates of transient memory. *J. Neurophysiol.*, 36:61–78, 1973.
12. J. M. Fuster, M. Bodner, and J. K. Kroger. Cross-modal and cross-temporal association in neurons of frontal cortex. *Nature*, 405:347–351, 2000.
13. H. S. Seung, D. D. Lee, B. Y. Reis, and D. W. Tank. Stability of the memory of eye position in a recurrent network of conductance-based model neurons. *Neuron*, 26(1):259–271, 2000.
14. Kenneth J. W. Craik. *The Nature of Explanation*. Cambridge University Press, 1943.
15. E. Bizzi and F. A. Mussa-Ivaldi. Neural basis of motor control and its cognitive implications. *Trends in Cognitive Sciences*, 2(3):97–102, 1998.

A Virtual Machine for Neural Computers

João Pedro Neto

Faculty of Sciences, University of Lisbon, Portugal

Abstract. Neural Networks are mainly seen as algorithmic solutions for optimization and learning tasks where the ability to spread the acquired knowledge into several neurons, i.e., the use of sub-symbolic computation, is the key. We have shown in previous works that neural networks can perform other types of computation, namely symbolic and chaotic computations. Here in, we show how these nets can be decomposed into tuples which can be efficient calculated by software or hardware simpler than previous neural solutions.

1 Introduction

The initial works of McCulloch and Pitts in the 1940's presented neural networks as computational models for logic operations considering that with some associated model of memory they could calculate the same computable functions as Turing Machines [1]. The computational equivalence of a linear model of neural net to Turing Machines was achieved only in the 1990's by Siegelmann and Sontag [2], [3].

What happened between these fifty years? Neural Networks at the 1950's where seen as models for approximation and learning. Ideas like the Hebb Law, the Perceptron and, years later, the backpropagation algorithm or Kohonen's competitive learning (just to mention a few) imprinted to the scientific community what neural networks were.

The common neural architecture by layers is able to split feature space with a high degree of approximation (if provided good enough algorithms) of the sample set. If a neuron is deactivated, the represented feature space lose accuracy but maintains some performance, since the gathered knowledge is spread by the net (the information kept within a neuron is sub-symbolic, to use Smolensky's term [4]). This and other features opened a wealth of possibilities that explains the righteous success of neural nets.

However, neural networks can also compute symbolic computation, i.e., computation where information has a defined and well specified type (like integers or floats). If provided a high-level description of an algorithm A, is it possible to automatically create a neural network that computes the function described by A?

2 Neural Symbolic Computation

In [5], [6] we gave an affirmative answer to the previous question This section briefly presents the structure of this answer.

S. Kollias et al. (Eds.): ICANN 2006, Part I, LNCS 4131, pp. 525–534, 2006.

The chosen analog recurrent neural net model is a discrete time dynamic system, $x(t+1) = \phi(x(t), u(t))$, with initial state $x(0) = x_0$, where t denotes time, $x_i(t)$ denotes the activity (firing frequency) of neuron i at time t, within a population of N interconnected neurons, and $u_k(t)$ denotes the value of input channel k at time t, within a set of M input channels. The application map ϕ is taken as a composition of an affine map with a piecewise linear map of the interval [0,1], known as the piecewise linear function σ:

$$\sigma(x) = \begin{cases} 1, \text{ if } x \geq 1 \\ x, \text{ if } 0 < x < 1 \\ 0, \text{ if } x \leq 0 \end{cases} \tag{1}$$

The dynamic system becomes,

$$x_j(t+1) = \sigma\Big(\sum_{i=1}^{N} a_{ji} \cdot x_i(t) + \sum_{k=1}^{M} b_{jk} \cdot u_k(t) + c_j\Big) \tag{2}$$

where a_{ji}, b_{jk} and c_j are rational weights. Fig. 1 displays a graphical representation of equation (2), used throughout this paper. When a_{ji} (or b_{jk} or a_{jj}) takes value 1, it is not displayed in the graph.

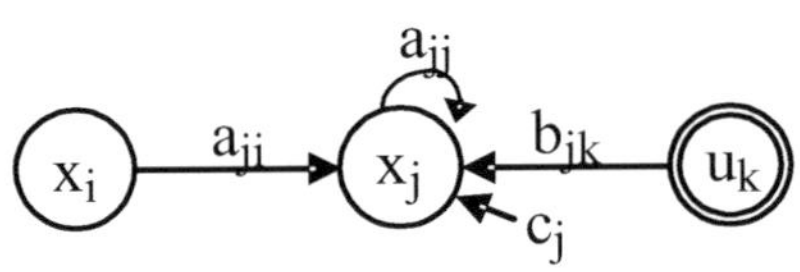

Fig. 1. Graphical notation for neurons, input channels and their interconnections

Using this model, we designed a high-level programming language, called NETDEF, to hard-wire the neural network model in order to perform symbolic computation. Programs written in NETDEF can be converted into neural nets through a compiler available at *www.di.fc.ul.pt/ jpn/netdef/netdef.htm.*

NETDEF is an imperative language and its main concepts are processes and channels. A program can be described as a collection of processes executing concurrently, and communicating with each other through channels or shared memory. The language has assignment, conditional and loop control structures (fig. 2 presents a recursive and modular construction of a process), and it supports several data types, variable and function declarations, and many other processes. It uses a modular synchronization mechanism based on handshaking for process ordering (the IN/OUT interface in fig. 2). A detailed description of NETDEF may be found at *www.di.fc.ul.pt/biblioteca/tech-reports* (report 99-5).

The information flow between neurons, due to the activation function σ, is preserved only within [0, 1], implying that data types must be coded in this interval. The real coding for values within [-a, a], where 'a' is a positive integer, is given by $\alpha(x) = (x+a)/2a$, which is an one to one mapping of [-a, a] into set [0, 1].

Input channels u_i are the interface between the system and the environment. They act as typical NETDEF blocking one-to-one channels. There is also a FIFO data structure for each u_i to keep unprocessed information (this happens whenever the incoming information rate is higher than the system processing capacity).

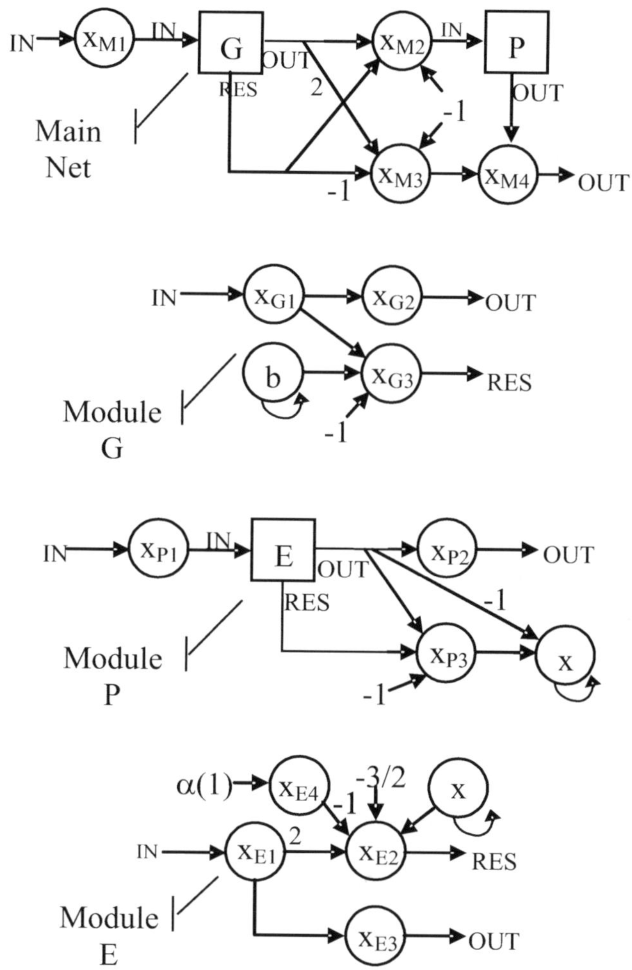

Fig. 2. Process construction of: IF b THEN x := x-1

The compiler takes a NETDEF program and translates it into a text description defining the neural network. Given a neural hardware, an interface would translate the final description into suitable syntax, so that the neural system may execute. The use of neural networks to implement arbitrary complex algorithms can be then handled through compilers like NETDEF.

As illustration of a symbolic module, fig. 2 shows the process construction for IF b THEN x := x-1. Synapse IN sends value 1 (by some neuron x_{IN}) into x_{M1} neuron, starting the computation. Module G (denoted by a square) computes the value of boolean variable 'b' and sends the 0/1 result through synapse RES. This module accesses the value 'b' and outputs it through neuron x_{G3}. This is achieved because x_{G3} bias 1.0 is compensated by value 1 sent by x_{G1}, allowing value 'b' to be the activation of x_{G3}. This result is synchronized with an output of 1 through synapse OUT. The next two neurons (on the Main Net) decide between entering module P (if 'b' is true) or stopping the process (if 'b' is false). Module P makes an assignment to the real variable 'x' with the value computed by module E. Before neuron x receives the activation value of x_{P3}, the module uses the output signal of E to erase its previous value. In module E the decrement of 'x' is computed (using $\alpha(1)$ for the code of real 1). The 1/2 bias of neuron x_{E2} for subtraction is necessary due to coding α.

The dynamics of neuron x is given by (3). However, if neuron x is used in other modules, the compiler will add more synaptic links to its equation.

$$x(t+1) = \sigma(x(t) + x_{P3}(t) - x_{E3}(t)) \qquad (3)$$

This resulting neural network is homogenous (all neurons have the same activation function) and the system is composed only by linear, i.e., first-order neurons. The network is also an independent module, which can be used in some other context. Regarding time and space complexity, the compiled nets are proportional to the respective algorithm complexity.

3 Neural Computers

Most work with neural nets uses software applications. Computers are fast, cheap and, for most tasks, the flexibility of these applications is more than enough. But this is also a consequence of the traditional vision of neural nets: fixed architecture schemes performing specialized algorithms. This vision also reflects on most hardware neural networks [7], [8]. Hardware for sub-symbolic computation is often constructed to execute (with much more speed and reliability) a set of algorithms on a set of topologies (e.g., layered networks), however flexible the hardware turns out to be. Solutions usually provide only for the implementation of a defined set of techniques, like backpropagation or competitive learning, to be run on dedicated hardware. But a typical neural hardware will not be able to compute a list with the first hundred prime numbers just by using these tools.

But is this necessary? Can we just use our standard von-Neumann computer to execute symbolic computations, and use a neural software or hardware to perform sub-symbolic computations? Yes, but then we are one step shorter for a fully definition of a *neural computer*. A neural computer is a system able to perform computations, any computations, in a stand alone mode. It is not just an extra component for specialized learning computations. A neural computer should be able to perform any algorithm, not just a defined set of classic sub-symbolic algorithms. A neural computer is more than an attachable module for an Intel-like CPU, it is a replacement of that CPU.

4 Vector and Tuple Representation

We noted, in section 2, the existence of a compiler which translates high-level algorithmic descriptions into neural nets.

It is easy to simulate the execution of these nets using a regular computer (our software can compile and execute its resulting nets). But, what kind of hardware is needed in order to calculate them? The NETDEF language produces modules which communicate via a small number of channels, but nonetheless these resulting nets are highly non-planar with very complex topologies. It would not be feasible to translate this into a 3-D hardware of neurons and synapses. Besides, every algorithm produces a different network, so a fixed architecture would be useful just for a specific problem. It is theoretically possible to implement a universal algorithm, i.e., to implement a neural network that codes and executes any algorithm, but there are easier solutions.

Neural nets are massive parallel models of computation. In this model, every neuron computes its output value in a synchronized way but independently from each other's future value. This feature is used at NETDEF in two ways: (a) it exists a parallel block constructor, where several processes are started at the same time, and (b) primitive type operations are automatically executed in parallel (e.g., to calculate 'x+y' the network evaluates 'x' and 'y' at the same time). So, our system should use the parallel characteristic of neural models.

Neural networks can be seen as vector machines. Assume a neural network Ψ with n neurons. The neurons activation at time t can be represented by vector $x_t = (x_1(t), x_2(t) \dots x_n(t), 1)$. This vector includes all the relevant data to define Ψ's state. The topology of Ψ is given by a $(n+1) \times (n+1)$ matrix M containing all synaptic weight's (the extra row/column is for biases). So, the network dynamics is given by:

$$\begin{aligned} x_0 &= (x_1(0), x_2(0) \dots x_n(0), 1) \\ x_{t+1} &= M_\Psi \cdot x_t \end{aligned} \tag{4}$$

which, afterwards, apply function σ to every element of the resulting vector (see figure 3).

The implementation of this type of machine is direct and it uses just sums, products and the σ function. However there are disadvantages. The typical

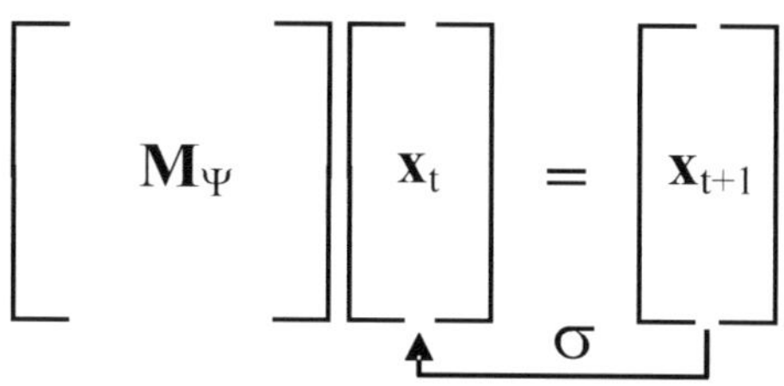

Fig. 3. Updating the network state

NETDEF networks produce sparse matrixes M_Ψ and even using known results to reduce their calculation (cf. [9], [10]) it results on unnecessary space quadratic complexity. Also, when optimized it is difficult to change the matrix values. If M_Ψ has fixed values, no learning is possible (since the synaptic values are not able to change). If M_Ψ is variable, the network may adapt its own structure and be able to perform learning and adaptation, which is possible with NETDEF networks, thus mixing symbolic and sub-symbolic computations (read [11] for proper details).

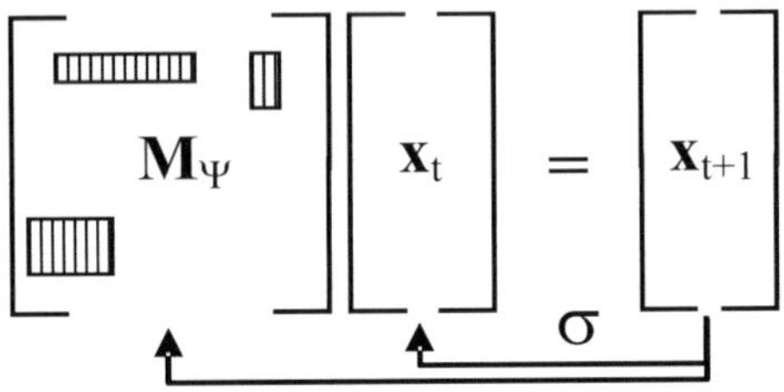

Fig. 4. Updating the network state with a changeable topology

Our proposed solution is to split the matrix into smaller tokens of information, namely triples, looking at a neural network as a list of synapses. A classic synapse has three attributes: (a) the reference to the output neuron (or 1 if it is a bias synapse), (b) its synaptic value, and (c) the reference to the input neuron.

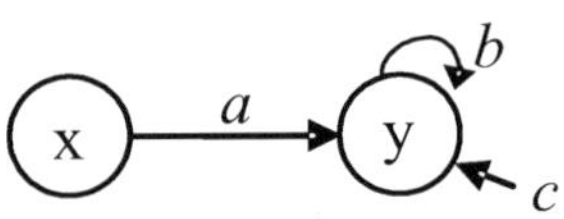

Fig. 5. This neural net translates to [(x,a,y), (y,b,y), (1,c,y)]

As stated in [11], an extended model of these networks (with no greater computing power but easier to describe network adaptation) is able to change dynamically the values of some special synapses, called synaptic-synapses. Although we are not concerned with biological plausibility, the existence of neuron-synaptic connections in the brain is known to exist in the complex dendritic trees of genuine neural nets (see [16] for details). In our model their main task is to convey values and use them to update and change other synaptic weights. Fig. 6 displays a diagram of a neuron-synaptic connection, linking neuron z to the connection between neurons x and y. Semantically, synapse of weight w_{xy} receives the previous activation value of z. The dynamics of neuron z is defined by the high order dynamic rule

$$y(t+1) = \sigma(2a.z(t).x(t) - a(z(t) + x(t)) + 0.5a + 0.5) \qquad (5)$$

Expression (5) is the result of $\alpha(\alpha^{-1}(z(t)) \times \alpha^{-1}(x(t)))$. This calculation is necessary because the data flow values are encoded through coding α. To avoid ambiguities, the first argument refers to the neuron synapse connection, and the second, to the input neuron. To identify this synaptic type, the triple consists of (a) the output neuron reference, (b) the neuron reference keeping the actual synaptic value, and (c) the input neuron reference.

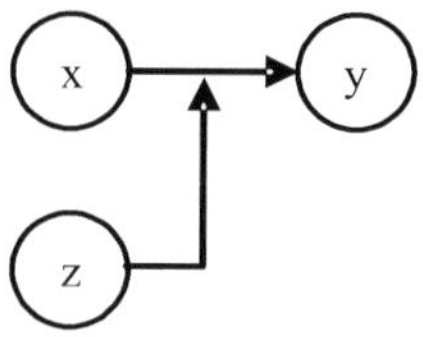

Fig. 6. This neural net translates to [(x,z,y)]

It is the type of the terms that defines the type of synapse. For typical synapses, there is a synaptic weight in the middle term. For biases, the number 1 is in the first term. And for synaptic-synapses, the middle term is a neuron id.

We may translate M_Ψ into a list of tuples, L_Ψ. The list L_Ψ size is proportional to the number of synapses. On the worst case (a totally connected network) it has space quadratic complexity (the same as the matrix approach). But the usual NETDEF network is highly sparse, making it, in practice, proportional to the number of neurons.

Notice there is no need to keep detailed information about each neuron; they are implicitly defined at L_Ψ. This list, herein, has a fixed size: it is possible to change the synaptic values dynamically but is not possible to create new neurons or delete existing ones. There is, however, the possibility of deactivating a neuron by assigning zero values to its input and output synapses. With some external process of neuron creation/destruction, it would be straightforward to insert/delete the proper triples at L_Ψ.

5 Executing the Network

In the last section we showed how to represent a complex neural network with a list of triples. These lists can be executed by software, by sequential or by parallel hardware and yet the list textual representation is independent of this choice. This description can be computed by very different hardware via virtual machines. We now present what that virtual machine for a real system should perform. A system able to process this model of neural network consists of:

1. A vector $\mathbf{x}$ with the current system state.
2. A triple list L_Ψ with the current net topology.
3. A vector x^+ with the next system state.

By convention, when x_1 (i.e., the first component of $\mathbf{x}$) is 1 the system starts processing, and stops processing when x_2 is 1. These do not represent neurons from the actual network, but are used, respectively, to activate the input signal of the main neural module and to receive its output signal. The system initializes with the following steps:

1. Download the net topology into L_Ψ.
2. Initialize $\mathbf{x}$ and x^+ to $\mathbf{0}$.
3. Partition L_Ψ into the available P processors.
4. Update x_1 to 1.

The third step focuses on parallelization. The computation of each triple depends only at the access of $\mathbf{x}$, which has two simple solutions: the vector is kept on some shared memory available to all processors, or before each execution, $\mathbf{x}$ is copied into the local memory of every processor. There is a similar writing problem with x^+. In this case, using a single shared memory would imply mutual exclusion for writing, creating traffic problems for neurons with high fan-ins. On the other hand, using local copies would mean that all copies should be added before the execution of the next cycle. Abstracting this problem, the typical step of execution at each processor is:

1. Get $\mathbf{x}$.
2. For each triple (a,b,c), if it is a:
   (a) synapse: $x^+[c] = x^+[c] + b * x[a]$
   (b) bias: $x^+[c] = x^+[c] + b$
   (c) synaptic-synapse: $x^+[c] = x^+[c] + x[b] * x[a]$
3. Update x^+.

The list L_Ψ is never explicitly changed. The system dynamics adapts through the neuron's value changes associated with synaptic-synapses. These special neurons do not directly alter other neurons activations but indirectly change them by the synaptic changes they promote. So, while the implicit network may suffer changes due to learning, the triples representing it do not suffer any alteration. A fixed L_Ψ helps the task of optimizing the initial distribution of tuples to the available processors.

When all processors end, the system updates $\mathbf{x}$ using x^+:

1. For each i, $x[i] = \sigma(x^+[i])$
2. If $x_2 \neq 1$ then $x^+ = \mathbf{0}$; activate processors else halt

Notice that $\mathbf{x}$ keeps the current state, i.e., all the current neural activations, and thus acts as the system memory. L_Ψ is the algorithm specified by the neural network dynamics. Since these neural networks are (discrete) dynamical systems, the algorithmic flow is the orbit of that system through time. This connection between computation and dynamic systems is discussed by Chris Moore at [12]. If dynamical systems of sufficient complexity perform computational tasks, the results from theory of computation (like the Halting Problem) are inherited by dynamical system. On the other hand, a computational system with chaotic features (like neural networks) can use them to perform new types of computation (cf. [12]–[15]).

6 Conclusion

Neural networks are more than nice data structures to apply optimization and learning algorithms. They can be used to perform symbolic computations. It has already been proven that every computable function can be exactly calculated by a discrete analog neural network with a piece-wise linear activation function.

Neural networks can, this way, be used to perform sub-symbolic and symbolic algorithms, and mix both within a homogenous architecture. In [11] it was shown how to use symbolic modules to control sub-symbolic tasks. In this way, neural nets can act as stand alone computers without the need of a von-Neumann computer.

This paper focused on how to translate neural networks into an abstract description based on tuples making it possible to execute them with simpler software or hardware solutions.

References

1. McCulloch, W., Pitts, W.: A logical calculus of the ideas immanent in nervous activity Bulletin of Mathematical Biophysics. **5** (1943) 115–133.
2. Siegelmann, H., Sontag, E.: Analog Computation via Neural Networks", Theoretical Computer Science, Elsevier. **131** (1994) 331–360
3. Siegelmann, H.: Neural Networks and Analog Computation, Beyond the Turing Limit, Birkhuser. (1999)
4. Smolensky, P.: "On the proper treatment of connectionism", Behavioral and Brain Sciences, Cambridge Univ. Press. **11** (1988) 1–74
5. Neto, J., Siegelmann, H., Costa, J.: On the Implementation of Programming Languages with Neural Nets, First International Conference on Computing Anticipatory Systems, CHAOS. **1** (1998) 201–208
6. Neto, J., Costa, J., Siegelmann, H.: Symbolic Processing in Neural Networks, Journal of Brazilian Computer Society. **8-3** (2003) 58–70
7. Lindsey, C.: Neural Networks in Hardware, Architectures, Products and Applications, *www.particle.kth.se/ lindsey/HardwareNNW/Course*, (2002)
8. Duong, T., Eberhardt, S., Daud, T., Thakoor, A.: Learning in Neural Networks: VLSI Implementation Strategies, Fuzzy Logic and Neural Network Handbook, C. Chen (ed.), McGraw-Hill. (1996)
9. Sangiovanni-Venticelli, A.: Optimization for Sparse Systems, Sparse Matrix Computations, D. Rose (ed.), Academic Press. (1976) 97–110
10. Pan, V.: How to Multiply Matrices Faster, Lecture Notes in Computer Science 179, Springer Verlag. (1982) 31–45
11. Neto, J., Costa, J., Ferreira, A.: Merging Sub-symbolic and Symbolic Computation in H. Bothe and R. Rojas (eds), Proceedings of the Second International ICSC Symposium on Neural Computation (NC'2000), ICSC Academic Press. (2000) 329–335
12. Moore, C.: Unpredictability and Undecidability in Dynamical Systems, Phys. Rev. Lett. **64-20** (1990) 2354–2357
13. Sinha, S., Ditto, W.: Computing with distributed chaos, Phys. Rev. E. **60-1** (1998) 363–377

14. Munakata, T., Sinha, S., Ditto, W.: Computing: Implementations of Fundamental Logical Gates by Chaotic Elements, IEEE Trans. Circuits Syst. **49-11** (2002) 1629–1633
15. Neto, J., Ferreira, A., Coelho, H.: On Computation over Chaos using Neural Networks Application to Blind Search and Random Number Generation, Journal of Bifurcation and Chaos in Applied Sciences and Engineering. **16-1** (2006) *to be published.*
16. Shepherd, G. M.: Neurobiology, 3rd ed., Oxford University Press. (1994)

Machine Cognition and the EC Cognitive Systems Projects: Now and in the Future

John G Taylor

King's College London, Department of Mathematics, The Strand, London, U.K. WC2R 2LS
john.g.taylor@kcl.ac.uk

Abstract. The strong support for the development of cognitive machines by the EC (under INFSO E5 - Cognition) will be reviewed, covering the main ideas of the 23 projects in this unit funded under FP6. The variety of approaches to cognition contained in these will be summarized, and future developments in FP7 considered. Conclusions on the future of the development of cognitive machine seen from this European perspective will conclude the paper.

1 Introduction

Machine cognition is now becoming a research area in which important new developments are expected to occur. This is because of a number of coinciding developments: the growth of ever larger computing systems, able to handle teraflops of data; the increased sensitivity and subtlety of robotic embodiment platforms; the increased sophistication of methods of machine intelligence for handling the various components suspected to be present in cognition; and last but not least the advances being made in brain science in understanding the neural components involved in cognitive processes and the dynamic flow of activity during such cognitive processing. This helps explain why the EC has funded 23 new projects on machine cognition in the latest FP6 round of projects. It has also set up the Unit IFSO E5 for cognition, under the acronym of ACS, denoting artificial cognitive systems (with the same underlying concepts as 'machine cognition') [1].

In this paper we will briefly survey these projects, and in the process extract what are regarded in the European scene as the fundamental components of a cognitive machine or ACS. The brief survey of the projects is given in the next section, and possible principles underpinning human cognition in the following section. How this relates to those involved in the EC projects is considered in section four. Section five considers possible futures (as in FP7) as well as possible further questions as to how cognitive machines could develop. We conclude the paper with a discussion.

2 The FP6 ACS Machine Cognition Projects

From the EC Unit E5 website [1] the objectives and focus of the projects are stated as:

'To develop artificial systems that can interpret data arising from real-world events and processes (mainly in the form of data-streams from sensors of all types and in

S. Kollias et al. (Eds.): ICANN 2006, Part I, LNCS 4131, pp. 535 – 542, 2006.

particular from visual and/or audio sources); acquire situated knowledge of their environment; act, make or suggest decisions and communicate with people on human terms, thereby supporting them in performing complex tasks.

Focus is on research into ways of endowing artificial systems with high-level cognitive capabilities, typically perception, understanding, learning, knowledge representation and deliberation, thus advancing enabling technologies for scene interpretation, natural language understanding, automated reasoning and problem-solving, robotics and automation, that are relevant for dealing with complex real-world systems. It aims at systems that develop their reasoning, planning and communication faculties through grounding in interactive and collaborative environments, which are part of, or connected to the real world.

These systems are expected to exhibit appropriate degrees of autonomy and also to learn through "social" interaction among themselves and/or through human-agent cooperation; in a longer term perspective, research will explore models for cognitive traits such as affect, consciousness or theory of mind.'

The content of this program is therefore ambitious, with even the mention of consciousness and theory of mind. We note that similarly ambitious aims have been expressed as the foundation of several new adventures in cognitive research, such as at the Brain Sciences Institute (BSI) in Tokyo. This latter has made enormous progress towards its aims, so the fact that these may be highly ambitious need not cause a negative reaction; it should make for careful assessment of the overall project and results being obtained, however, with realisation that some of the goals are harder than initially thought.

To be able to assess the realism of the aims of the EC Cognitive Projects then, let us consider the projects further.

The 23 ACS projects are listed below: (in alphabetical order):

1) BACS: Bayesian Approach to Cognitive Systems
2) CASBLIP: Cognitive Aid System for Blind People
3) CLASS: Cognitive-Level Annotation using Latent Statistical Structure
4) COSPAL: Cognitive Systems using Perception-Action Learning
5) COSY: Cognitive Systems for Cognitive Assistants
6) DECISIONS-IN-MOTION: Neural Decision-Making in Motion
7) DIRAC: Detection and Identification of Rare Audio-visual Cues
8) eTRIMS: eTraining for Interpreting Images of Man Made Scenes
9) euCOGNITION: European Network for the Advancement of Artificial Cognitive Systems
10) GNOSYS: An Abstraction Architecture for Cognitive Agents
11) HERMES: Human-Expressive Representations of Motion and their Evaluation in Sequence
12) ICEA: Integrating Cognition, Emotion and Autonomy
13) JAST: Joint-Action Science and Technology
14) MACS: Multisensory Autonomous Cognitive Systems
15) MATHESIS: Observational Learning in Cognitive Agents
16) MindRaces: Mind RACES: from Reactive to Anticipatory Cognitive Embodied Systems
17) PACO-PLUS: Perception, Action and Cognition through Learning of Object-Action Complexes

18) PASCAL: Pattern Analysis, Statistical Modelling and Computational Learning
19) POP: Perception On Purpose
20) RASCALLI: Responsive Artificial Situated Cognitive Agents Living and Learning on the Internet
21) ROBOT-CUB: Robotic Open-architecture Technology for Cognition, Understanding and Behaviours
22) SENSOPAC: SENSOrimotor structuring of Perception and Action for emerging Cognition
23) SPARK:Spatial-temporal patterns for action-oriented perception in roving robots

How can we begin to appreciate this vast range of ideas and approaches to cognition? To begin with, we can gather the projects together under several headings, emphasisng various approaches to cognition::

a) Embodiment driven (# 13, 21),
b) Applications-driven (# 2, 5, 7, 8, 20)
c) Machine-intelligence driven (# 1, 3, 18)
d) Neural-based (# 15)
e) Cognitive science based (symbolic: #5)
f) Hybrid (# 10)
g) Dynamic systems (# 23)

The above clustering of the projects is broad, but sufficient for the analysis to follow; we will not go into further details of the projects, but refer the reader to the appropriate web-site. These various approaches need not, however, be the most effective to achieve a breakthrough in the creation of an autonomous cognitive machine. In order to appreciate better the task facing the creation of a cognitive machine, and the necessary criteria, we develop a general model of human cognition and then relate the model to the projects.

3 The Nature of Human Cognition

There are numerous definitions of human cognition, such as: 1) 'The conscious process of knowing or being aware of thoughts or perceptions, including understanding and reasoning'; 2) 'The mental process or faculty of knowing, including aspects such as awareness, perception, reasoning, and judgment'; 3) 'High level functions carried out by the human brain, including comprehension and use of speech, visual perception and construction, calculation ability, attention (information processing), memory, and executive functions such as planning, problem-solving, and self-monitoring.' These definitions cover a multitude of functions. Here the basic components of cognition will be taken to be: awareness, thinking, knowing, reasoning and executive functions.

Awareness presents an enormous difficulty, since there is no universally accepted model of its nature or any accepted model of its possible creation by the brain. A model of awareness, developed from a control model of attention using engineering control ideas [1, 6] has some claim for reality, with support coming from a range of paradigms in brain science. The claim that awareness is properly included in such a model has been argued in detail elsewhere [6, 7]; some such indication is needed for any self-respecting model of cognition to be worth its salt.

At last year's ICANN05 a set of sub-component processes and resulting principles of cognitive processing were suggested [2]. These arose from looking more closely at the sub-components of cognitive processing being carried out by specific regions in the brain: 1) Storage and retrieval of memories in hippocampus (HC) and related areas; 2) Rehearsal of desired inputs in working memory; 3) Comparison of goals with new posterior activity;.4) Transformation of buffered material into a new, goal-directed form (such as spatial rotation of an image held in the mind); 5) Inhibition of pre-potent responses; 6) The development of forward maps of attention in both sensory and motor modalities, so that possibly consequences of attended actions on the world can be imagined; 7) Determination of the value of elements of sequences of sensory-motor states as they are being activated in forward model recurrence; 8) Learning of automatic sequences (chunks) so as to speed up the cognitive process

The resulting principles of cognition deduced from these sub-components were [2]:

P1: There is overall control by attention of the cognitive process, using attention-based control signals to achieve suitable transformations to solve cognitive tasks;

P2: Fusion of attention control (in parietal lobe and PFC) and long-term learning in HC occurs to achieve an expanding state space of stimuli and actions, and of corresponding attention control;

P3: The creation of a valuation of goals occurs in PFC to handle reward prediction biasing of the processes 1) to 6) above;

P4: Transformations on buffered stimuli is achieved by creation of suitable PFC goals associated with the required transformations being carried out on existing buffered activities, under the control of attention;

P5: Forward models (along the lines of equation (1) in section 2)) are created under attention control so that, by their recurrence, sequential passage through sequences of attended sensory-motor states is achieved (as in thinking), possibly to reach a desired goal (as in reasoning) or valued states that may even correspond to new ways of looking at a problem (as in creativity);

P6: There is creation of, and ability to access, automated 'chunks' of knowledge, so they can be inserted into forward model sequences under P4. These automated chunks are initially created by effortful attention at an earlier time (using error-based learning in cerebellum) but are then gradually transferred to automatic mode by suitably long rehearsal (through reinforcement training by a reward signal based on the neuromodulator dopamine);

P7: Attention is employed as a gateway to consciousness, with consciousness providing an overarching control function of speed-up of attention, thereby giving consciousness overall guidance over ongoing processes (plus other features still to be defined).

How much overlap occurs between theses principles and those being explicated in the EC ACS Cognitive Machine projects listed above? We turn to that in the next section.

4 The Relation of Human Cognition to EC-Based Cognition

The possible approaches to cognition extracted from the set of EC Cognitive Systems projects at the end of section 2 are broad, and certainly cover the range of principles presented in the previous section in a general manner. As expected the brain-based projects, such as GNOSYS and MATHESIS are closest to the thrust of these principles, the former project being heavily guided by the nature of brain processing, the latter involving both brain imaging experiments and experiments on children to create a model of motor action learning by example by means of neural network systems. Similarly the projects involving embodiment also are well related to some of the aspects of human cognitive principles, although not necessarily following them through to the higher level especially of consciousness. Such relationship is unclear in some of the other projects, particularly those involved with symbolic processing styles, so these need to be considered further. But also GNOSYS possesses a certain degree of ambivalence to symbols, since it does not use them heavily but does have a symbolic goal tree structure, and similar symbolic components in drive and emotion driven responses.

The main problem these features lead us to is the existence of a symbolic-sub-symbolic divide in approaches to cognition. That does not mean that the more theoretical approaches (Bayesian or dynamical systems theory, for example) should be neglected (nor are they in the EC Cognitive Systems projects list), but it would appear that there is still a problem in how to bridge this symbolic-sub-symbolic gap which is still to be properly explored. The other theoretical approaches are undoubtedly important in any analysis of human cognition: the dynamical system properties of neural networks in the brain are being analyzed ever more successfully by numerous tools, and are leading to increased insights into such aspects as the binding problem, capacity limits on memory systems, and so on. Similarly Bayesian approaches have been found to be of great value in understanding and modeling pattern analysis and decision making in the brain, as well as various aspects of motor control. However it is the gap between such projects as COSY and GNOSYS that are at issue. There is an enormous power in the symbolic approach, although CYC and related projects have run into severe overload problems when trying to make progress in common-sense reasoning by collecting millions of examples of facts about the world.

Returning to the problem of the existence of a gap between symbolic processing and neural network-based sub-symbolic processing, it is clear that the extreme of universal grammar of Chomsky, and the associated LAD (for Language acquisition device he proposed to be genetically based in the brain of the infant) are not the mechanisms by which language is learnt in the brain. This learning process takes several years, and is developmentally complex. The codes for nouns and for actions are inextricably intertwined in the brain with representations for stimuli. So is it necessary to understand the details of brain-based language abilities, and especially attempt to implement them in a neural network framework, in order to build a proper basis for understanding human cognition?

Much can be done by means of linguistic reasoning, as is well known from recent researches into different forms of reasoning that humans use to solve problems. But there is always the symbol grounding problem: how do the symbols in the brain gain their basis of meaning so as to refer to objects in the world? It is that a purely symbolic system cannot achieve without addition of the sub-symbolic representations of objects and actions which are described by words at the symbolic level.

The mechanism by which this language grounding is achieved in the brain is being explored assiduously, especially by the use of brain imaging systems to enable dissociations between various brain areas involved in language processing to be made, as well as to descry their functionality in such processing. At the same time developmental aspects are being carefully explored so as to be able to track the learning processes involved in learning a language.

It would seem that there are at least two main aspects of this learning process, which develop sequentially one after the other. The first is the creation of a vocabulary of semantically well-defined words, related to the various stimuli an infant may see as it grows up. This semantic definition needs a set of teachers, so that through joint attention of the child and the teacher to a given stimulus the word for it can be heard and ultimately learnt by the child; this allows ever increasingly precise classification of the world in terms first of basic categories (dog, cup,…) and then increasingly of refined super-ordinate categories, such as terrier, Yorkshire terrier, and so on. This semantic process undoubtedly goes on throughout life but has a special spurt at about 2 – 4 years of age.

The second, much harder aspect of language learning is that of syntax and the related agent semantics. Verbs are learnt in far fewer in number in the first few years, and the concept of agent (subject) acting on an object is thereby more difficult to comprehend for the developing child. There may be a number of reasons for this slower growth of syntactic power: the frontal lobes where action goals are stored takes time to myelinate in the growing child, the working memory sites able to store increasingly long sequences of words involved in syntactically complex sentences may not also be available for young children, a model of the causal nature of actions on objects does not develop till late in children, and so on. But for whatever reason there is a difficulty in getting syntax right and the associated agent semantic model of the world.

It is correspondingly difficult to create an artificial model of the processes involved in semantic analysis, especially of complex sentences. However this problem needs to be solved before a machine system can be created to be able to be said to 'understand' language inputted to it; the problems of encoding and outputting a stream of words are not of the same level of difficulty. It is how the string of words is analyzed, not only by the semantic net assumedly learnt by the methods of joint attention of teacher and pupil, but more especially by the syntactic analysis system of the brain, that will determine the level of understanding of the system to the word string.

There are numerous recurrent net systems that have been developed to model various more detailed aspects of syntactic analysis. However there still has to be developed a more complete approach to language understanding and learning which can hopefully lead us forward to a machine capable of understanding language at the level of a 5-year old.

The present gap between sub-symbolic and symbolic processing, and especially in the lack of suitably powerful models able to bridge that gap, indicates a serious lacuna in modern cognitive systems research. The sub-symbolic approaches will thereby have to proceed without language, so that they are cut off from those projects using symbolic techniques.

The only level of reasoning for the sub-symbolic systems will be that of non-human animals. Their reasoning powers are now being realized as more powerful than appreciated heretofore. So it is possible to attempt to create a reasoning or thinking system that is solely based on sub-symbolic processing. In this way cognition can be analyzed up to a certain depth.

5 The Future of Machine Cognition

The future of machine cognition in general will depend on how successful are the various projects of section 2 above, as well as of the many projects around the world of a similar nature. Thus it is early days yet to speak to this question with any certainty. It may be that it is necessary, on order to obtain cognition at the level of humans, to include three aspects of the program in developing machine cognition which so far I have neglected or left to one side:

1) Providing suitable computing power. There are now a variety of large parallel computers throughout the world; one of these (Deep Blue of IBM) is being applied at EPFL, Lausanne to build a mini-column of ten thousand neurons by means of computing the dynamics of each neuron on a dedicated computing unit. The present record of such machines is of 250,000 such nodes at Los Alamos (apparently dedicated to defense work). It is clear that the computing demands are high for learning suitably precise lower-level representations, say as in the brain's V1 or V2, that are able to be used as a basis for more developed higher level representations of objects. It is these representations to which object words are to be attached by attention; that gives a solid basis for semantics. Syntax needs models of prefrontal cortices to enable sequencing and chunking to occur, so as to allow causal models of the world to be created (as internal control models) and thereby provide a firm syntactic basis for the analysis of word strings. But here again a large amount of computing power is needed. Such provision of a dedicated parallel computer for such learning processes should be considered as a serious priority in future many-project work.

2) Properly understand the nature of consciousness and its presence in the cognitive process. If the CODAM model of consciousness is correct (as arising from a control signal equal to the efference copy of the attention movement signal generated in parietal areas to move attention) then consciousness will play an important role in any cognitive processing. It will enable faster and more error-free processing to occur, as well as allow the scaling problem to be handled of how to live effectively in a complex world (by removing distracters through careful attention control, as in the CODAM approach).

3) The development of a language system able to learn continuously by means of increasingly complex sub-symbolic modules guided by human speech development. To solve this ambitious task will require a multi-disciplinary team working across brain science, linguistics, neural network modeling, parallel computing and ancillary disciplines. Much is being learnt in all these fields; it is increasingly appropriate to attempt to fuse this knowledge by the production of a truly brain-based LAD.

It is these two aspects and their future exploration which need to be developed as part of future EC support.

6 Discussion

In this paper I have briefly surveyed the various projects involved in creating cognitive systems in the EC Cognitive Systems Unit. It has been recognized that these cover a broad area, and are making important progress in attacking basic problems in the area, both at a fundamental and at an applied level. The relevant members of the

EC should be congratulated on their forward-looking approach to the topic. At the same time a set of principles were presented in this paper on which cognitive systems appear to be based in the normal usage of 'cognitive'. These principles were deduced form analysis of some of the fundamental components of cognitive processing. The relation between the two sets of approaches - between the EC Cognitive Systems projects and the cognitive principles –are then seen to be somewhat consistent.

However several problems were thrown up by this analysis:

1) The need to bridge the gap between sub-symbolic and symbolic processing styles;
2) The need to provide a much greater level of processing power;
3) The need to bring in consciousness as a working component of cognition, as is recognized in all the main discussions on the subject.

There are obvious (although highly non-trivial) solutions to the above problems:

1) To the first problem, it is necessary to attack the sub-symbolic basis of language by creating linguistically defined object and action representations, so that the semantic aspects of language are correctly sub-symbolically founded. This structure then needs to be extended to a similar founding of syntax and agent semantics. Small parts of this program have been attempted, but only a concerted and interdisciplinary attempt will be successful.
2) Create a central parallel large-scale computer (upwards of 100,000 nodes) so as to enable the very large-scale computations to be achieved for learning connection weights in multi-modular models of components of the human brain;
3) Finally attack the problem of consciousness using developed models of attention (such as CODAM [3], although other models would be relevant here). These would enable a more powerful system, endowed with some of the powers carried by consciousness, to allow the system to move beyond lower-level attention control

All of these routes to the future could be followed; the more assiduously they are developed and solved the sooner will we begin to move towards a future that is mind!

Acknowledgements

I would like to thank the EC, under the GNOSYS project and more recently MATHESIS, for support to carry out this work.

References

[1] http://cordis.europa.eu.int/ist/cognition/index.html
[2] Taylor JG (2005) The Principles of Cognitive Systems. Proc ICANN2005, Warsaw.
[3] Taylor JG (2005) Mind and Consciousness: Towards a Final Answer? Physics of Life Reviews 2:1-45

A Complex Neural Network Model for Memory Functioning in Psychopathology

Roseli S. Wedemann[1], Luís Alfredo V. de Carvalho[2], and Raul Donangelo[3]

[1] Instituto de Matemática e Estatística, Universidade do Estado do Rio de Janeiro
Rua São Francisco Xavier, 524, 20550-013, Rio de Janeiro, RJ, Brazil
roseli@ime.uerj.br
[2] Eng. Sistemas e Computação, COPPE - Universidade Federal do Rio de Janeiro,
Caixa Postal 68511, 21945-970, Rio de Janeiro, RJ, Brazil
LuisAlfredo@ufrj.br
[3] Instituto de Física, Universidade Federal do Rio de Janeiro
Caixa Postal 68528, 21941-972, Rio de Janeiro, RJ, Brazil
donangel@if.ufrj.br

Abstract. In an earlier paper [1], we described the mental pathology known as neurosis in terms of its relation to memory function. We proposed a mechanism whereby neurotic behavior may be understood as an associative memory process in the brain, and the symbolic associative process involved in psychoanalytic working-through can be mapped onto a process of reconfiguration of the neuronal network. Memory was modeled by a Boltzmann machine represented by a complete graph. However, it is known that brain neuronal topology is selectively structured. Here, we further develop the memory model, by including known mechanisms that control synaptic properties, showing that the network self organizes to a hierarchical, clustered structure. Two modules corresponding to sensorial and declarative memory interact, producing sensorial and symbolic activity, representing unconscious and conscious mental processes. This extension of the model allows an evaluation of the idea of working-through in a hierarchical network structure.

1 Introduction

Psychoanalytic research regarding the *transference neuroses* has found that traumatic and repressed memories are knowledge which is present in the subject, but which is symbolically inaccessible to him. It is therefore considered *unconscious* knowledge [2]. Freud observed that neurotic patients systematically repeated symptoms in the form of ideas and impulses, and called this tendency a *compulsion to repeat* [3]. He related the compulsion to repeat to repressed or traumatic memory traces, caused by a conflict associated with libidinal fixation and frustration [2].

Neurotic analysands have been able to obtain relief and cure of painful symptoms through a mechanism called *working-through*. This procedure aims at developing knowledge regarding the causes of symptoms by accessing unconscious memories, and understanding and changing the way in which the analysand obtains satisfaction, *i.e.*, changing his compulsion to repeat [3].

S. Kollias et al. (Eds.): ICANN 2006, Part I, LNCS 4131, pp. 543–552, 2006.

Although inconclusive, psychodynamic theories seem to suggest correlations between creativity, psychopathology and unconsciousness [2,3,4,5]. We explored these commonalities and proposed, in a previous paper [1], a schematic functional model for some concepts associated with neurotic mental processes, as described by Sigmund Freud [2,3]. Our description is based on the current view that the brain is a cognitive system composed of neurons, interconnected by a network of synapses, that cooperate locally among themselves to process information in a distributed fashion. Mental states thus appear as the result of the global cooperation of the distributed neural cell activity in the brain [6,7]. We also consider that the emergence of a global state of the neural network of the brain generates a bodily response which we call an *act*.

As a first approximation, in [1] memory was modeled by a Boltzmann machine represented by a complete graph. It is known, however, that brain neuronal topology is selectively structured. Neurons interact mainly with spatially close neighbors, having fewer long-range synapses connecting them to other neurons farther away [8,6]. In this paper, we further develop the memory model, by including some known microscopic mechanisms that control synaptic properties, and show that the network self organizes to a hierarchical, clustered structure. We propose a memory organization, where two hierarchically structured modules corresponding to sensorial and declarative memories interact, producing sensorial and symbolic activity, representing unconscious and conscious mental processes. In proposing this organization, we followed Freud's idea that unconscious memories are those that we cannot access symbolically, *i.e.* cannot talk about [2,3]. An organization involving language processing brain areas, derived from a neurophysiological point of view, can be found in [9]. A model emphasizing brain mechanisms for attention involved in conscious activity can be found in [10]. Our main contribution is to propose a neuronal mechanism, based on known microscopical, biological brain mechanisms, to describe conscious and unconscious mental activity involved in neurotic behavior and psychoanalytic therapy by memory functioning, as described by Freud. We thus represent brain mechanisms involved in neurosis, as a complex system, based on a neural network model and analyse it according to recent methods developed for the study of complex networks.

In the next section, we review our functional and computational model for neurosis. In Section 3, we describe a self-organizing model for generating hierarchically clustered memory modules, representing sensorial and declarative memories. We then show results from computer simulations with some mathematical properties of these complex networks. In the last section, we draw some conclusions and perspectives for future work.

2 Functional and Computational Model for the Neuroses

In this section, we review the model described in [1]. There we proposed that the neuroses manifest themselves as an associative memory process, a mechanism where the network returns a stored pattern when it is shown another input pattern sufficiently similar to the stored one [11]. We modeled the compulsion to repeat neurotic symptoms by supposing that such a symptom is acted when the subject is presented with a stimulus which resembles, at least partially, a repressed or traumatic memory trace. The stimulus causes a stabilization of the neural net onto a minimal energy state, corresponding to the memory trace that synthesizes the original repressed experience, which in turn generates a neurotic response (an act). The neurotic act is not a result of the stimulus as a new situation but a response to the repressed memory trace. Repression can be accounted for in part by a mechanism which inhibits formation of certain synaptic connections, in the cortical map formation process, thus inhibiting symbolic representation.

We mapped the linguistic, symbolic associative process involved in psychoanalytic working-through into a corresponding process of reinforcing synapses among memory traces in the brain. These connections should involve declarative memory, leading to at least partial transformation of repressed memory to consciousness. This has a relation to the importance of language in psychoanalytic sessions and the idea that unconscious memories are those that cannot be expressed symbolically. We propose that as the analysand symbolically elaborates manifestations of unconscious material through transference in psychoanalytic sessions, he is reconfiguring the topology of his neural net, by creating new connections and reinforcing or inhibiting older ones. The network topology which results from this reconfiguration process will stabilize onto new energy minima, associated with new acts. The process of slowly and repeatedly reconfiguring synaptic connections to elaborate knowledge accounts for the long durations of psychoanalytic processes, where repetition is specially important.

Memory functioning is modeled by a Boltzmann machine, where node states take binary values [11]. Pattern retrieval on the net is achieved by a standard simulated annealing process, in which the network temperature parameter is gradually lowered by a factor α. We considered that the network initially had random symmetric connection weights, and was divided into two weakly linked subsets, representing conscious and unconscious parts of memory. These parts should correspond to a map formation described in [12], representing neurotic memory states.

To simulate the working-through process, we stimulate the net by means of a change in a randomly chosen node n_i belonging to the "unconscious" section of a neurotic memory pattern. This stimulus is then presented to the network and, if the Boltzmann machine retrieves a pattern with conscious configuration different than that of the neurotic pattern, we interpret this as a new conscious association, and enhance all weights from n_i to the changed nodes in the conscious subset. The increment values are proportional to the products of the states of the neurons connected by the synapse and the learning parameter β. New knowledge is learned only when the stimulus from the analyst is not similar to the neurotic memory trace. This procedure must be repeated for various reinforcement iterations in an adaptive learning process, and also each set of reinforcement iterations must be repeated for various initial annealing temperature values.

3 Hierarchical Memory Model

In a further refinement of our model, we now consider that neurons belong to two different subsets, corresponding to *sensorial* and *declarative memories*. Memory traces stored in sensorial memory represent mental images of stimuli received by sensory receptors (located in eyes, ears, skin and other parts of the body). Sensorial memory represents brain structures such as the amygdala, cerebellum, reflex pathways, hippocampus, and prefrontal, limbic and parieto-occipital-temporal cortices that synthesize visual, auditory and somatic information. A trace in sensorial memory may "become conscious" if associated to a pattern in declarative memory. Declarative memory stores representations of memory traces stored in sensorial memory, i. e. *symbols*, and represents brain structures such as areas of the medial temporal lobe, hippocampus, Broca's and Wernicke's areas and areas of the frontal cortex. These latter areas are associated with language and, because of them, we can associate a word such as "red" to the visual sensation of seeing a red object. With declarative memory we represent Freud's concept of conscious memory traces (in some of his works this is referred to as the preconscious). Attentional mechanisms, which we did not model, select stimuli to sensorial memory and allow them to become conscious if associated to a trace in declarative memory. Here we concentrate on the relations between sensorial and declarative memories and the possibility of becoming conscious. We thus consider that, when a stimulus S that retrieves a pattern in sensorial memory stimulates also retrieval of an associated pattern in declarative memory, it becomes conscious. Stimuli of sensorial memory which do not cause activation of declarative memory remain unconscious. This mechanism is similar to ideas proposed by Edelman in [9], and strongly reflects Freud's concepts of conscious and unconscious mental processes and the role of language in psychoanalysis.

In order to model the structure of the topology of each of the two memories, we consider the following microscopic biological mechanisms. Brain cells in many animals have a structure called *on-center/off-surround*, in which a neuron is in cooperation, through excitatory synapses, with other neurons in its immediate neighborhood, whereas it is in competition with neurons that lay outside these surroundings. Competition and cooperation are found statically hardwired, and also as part of many neuronal dynamical processes, where neurons compete for certain chemicals [6]. In synaptogenesis, for example, substances generically called neural growth factors are released by stimulated neurons and, spreading through diffusion, reach neighboring cells, promoting synaptic growth. Cells that receive neural growth factors make synapses and live, while cells that have no contact with these substances die [6]. A neuron that releases neural growth factor guides the process of synaptic formation in its tri-dimensional neighborhood, becoming a center of synaptic convergence. When neighboring neurons release different neural growth factors in different amounts, many synaptic convergence centers are generated and a competition is established between them through the synapses of their surroundings. A signaling network is thus established to control development and plasticity of neuronal circuits. Since this competition is started and controlled by environmental stimulation, it is possible to have an idea of the way environment represents itself in the brain.

Based on these microscopic mechanisms, we developed the following *clustering algorithm* to model the self-organizing process which controls synaptic plasticity, resulting in a structured topology of each of the two memories.

Step 1. Neurons are uniformly distributed in a square bi-dimensional sheet.

Step 2. To avoid the unnecessary and time-consuming numerical solution of the diffusion equation of the neural growth factors for simulation of synaptic growth, we assume a Gaussian solution. Therefore, a synapse is allocated to connect a neuron n_i to a neuron n_j according to a Gaussian probability, given by eq. 1

$$P_{ij} = \exp(-(\boldsymbol{r_j} - \boldsymbol{r_i})^2/(2\sigma^2))/\sqrt{2\pi\sigma^2}\,, \tag{1}$$

where $\boldsymbol{r_j}$ and $\boldsymbol{r_i}$ are the positions of n_j and n_i in the bi-dimensional sheet and σ is the standard deviation of the distribution and is considered here a model parameter. If a synapse is allocated to connect n_i and n_j, its strength is proportional to P_{ij}.

Step 3. We verified in [12] that cortical maps representing different stimuli are formed such that each stimulus activates a group of neurons spatially close to each other, and that these groups are uniformly distributed along the sheet of neurons representing memory. We thus now randomly choose m neurons which will each be a center of the representation of a stimulus. The value of m should be chosen considering the storage capacity of the Boltzmann machine [11].

Step 4. For each of the m centers chosen in Step 3, reinforce adjacent synapses according to the following criteria. If n_i is a center, define $sum_{n_i} = \sum_j |w_{ij}|$, where w_{ij} is the weight of the synapse connecting n_j to n_i. For each n_j adjacent to n_i, increase $|w_{ij}|$ by Δw_{ij}, with probability $Prob_{n_j} = |w_{ij}|/sum_{n_i}$, where $\Delta w_{ij} = \eta Prob_{n_j}$ and $\eta \in \Re$ is a model parameter chosen in $[0, 1]$. After incrementing $|w_{ij}|$, decrement Δw_{ij} from the weights of all the other neighbors of n_i, according to: $\forall k \neq j, |w_{ik}| = |w_{ik}| - \Delta w_{ik}$, where $\Delta w_{ik} = (1 - |w_{ik}|/\sum_{k \neq j} |w_{ik}|)\Delta w_{ij}$.

Step 5. Repeat step 4 until a clustering criterion is met.

In the above clustering algorithm, steps 2 and 3 are justified in the algorithm's description. Step 4 regulates synaptic intensities, i. e. *plasticity*, by strengthening synapses within a cluster and reducing synaptic strength between clusters (disconnects clusters). By cluster, we mean a group of neurons that are spatially close, with higher probability of being adjacent by stronger synapses. This step represents a kind of preferential attachment criterion with some conservation of energy (neurosubstances) among neurons, controlling synaptic plasticity. Neurons that have received stronger sensorial stimulation and are therefore more strongly connected, will stimulate their neighborhoods and promote still stronger connections. This is in agreement with the microscopic biological mechanisms we mentioned above.

The growth of long-range synapses is energetically more costly than short-range synaptic growth, and therefore the former are less frequent in the brain than the latter. For allocating long-range synapses which connect clusters, we should consider the basic learning mechanism proposed by Hebb [6,9,11], based on the fact that synaptic growth among two neurons is promoted by simultaneous stimulation of the pair. This also suggests a mechanism through which the external world, culture and language, reflects itself in the brain. Memory traces stored by configurations of certain states of neuronal groups, which receive simultaneous stimuli, should enhance synaptic growth among the neuronal groups representing these traces, allowing association among traces. Since memory traces represent both sensorial information and concepts (declarative memories), we are also representing association of ideas or symbols by long-range synapses. This may suggest a way in which basic language structure (and maybe Chomsky's concept of Universal Grammar [13]) is mapped onto brain topology.

We are studying these processes and, since we are still not aware of the synaptic distributions that result in such topologies, as a first approximation, we allocate synapses randomly among clusters. Within a cluster C, a neuron n_i is chosen to receive a connection with probability $P_i = \sum_j |w_{ij}| / \sum_{n_j \in C} \sum_k |w_{jk}|$. If the synapse connects clusters in different memory sheets (sensorial and declarative memories), its randomly chosen weight is multiplied by a real number κ in the interval $[0, 1]$, reflecting the fact that, in neurotic patterns, sensorial information is weakly accessible to consciousness, *i.e.* repressed.

Mechanisms of memory storage and retrieval by the Boltzmann machine and simulation of the working-through psychoanalytical process are then carried on as reviewed in Section 2 and described in [1].

4 Simulation Illustration and Network Properties

We illustrate the model with a preliminary simulation experiment for a network with $N = 32$ neurons, such that $N_{sens} = 16$ of them belong to the sensorial memory subset. We are aware that this number of neurons is extremely small for realistic brain studies. However, in this first approach, the purpose of our simulations is to illustrate these basic concepts and mechanisms at a semantic level. The short range microscopic mechanisms are scalable and, since our algorithms are based on microscopic biological mechanisms, we do expect that this level should be amenable to mapping to a biological substratum. Some simulations of system configurations resistant to learning have taken more than one day, on a sequential processor, even for this small system. In the future, we plan to parallelize these algorithms in order to simulate significantly larger systems. Synapses connecting different memories are multiplied by $\kappa = 0.5$, square memory sheets have size 1.5×1.5, $\sigma = 0.58$, $\beta = 0.3$, $\eta = 0.1$ and the other parameters for the annealing process in the Boltzmann Machine are attributed the same values we have presented in [1].

In Fig. 1, we show one of the topologies generated after executing only the clustering algorithm and, in Fig. 2, the corresponding topology after long-range synaptic generation. Although the number of neurons is still quite small, it suffices to illustrate the fact that the algorithm self organizes the network in a clustered and hierarchical manner. Fig. 3 gives a more quantitative view.

We generated 1000 different topologies from the same initial system parameter values mentioned above and measured the average node degree (k) distribution, to evaluate the effects of competitive and cooperative mechanisms on network structure. In Fig. 3, the cross symbols represent the values found in our simulations and the curve a fit of a Poisson distribution $k^\lambda \exp(-k)/\lambda!$. It is known that random graphs follow the Poisson distribution of node degrees [14]. Our graphs are not random, but the spatially homogeneous allocation of synapses of the clustering algorithm may explain the close fit of the distribution for smaller values of k. The deviation from Poisson distribution for higher values of k may be attributed to the competitive associative biological mechanisms we described in the previous section, which introduce some structure to the topology.

The average clustering coefficient as defined in [15,14] for the topology in Fig. 2 is 0.43. This is higher than the value of 0.28 which was measured for the neural network

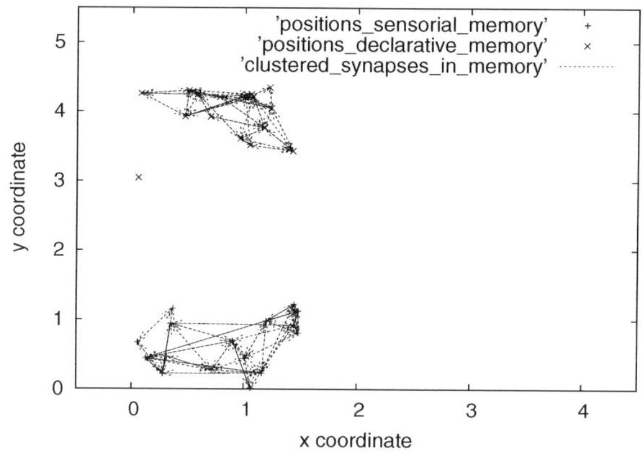

Fig. 1. Network topology after clustering

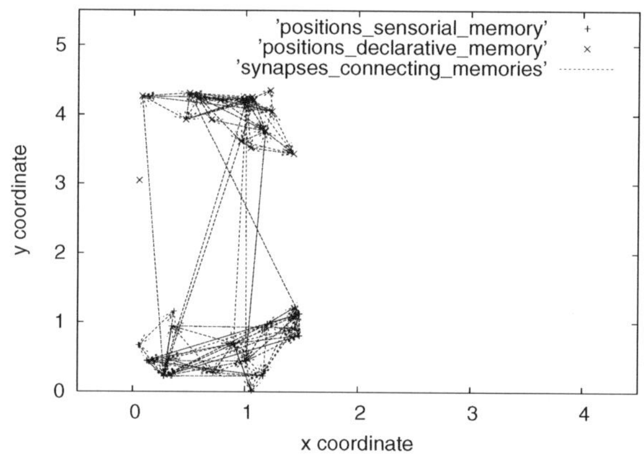

Fig. 2. Network topology with long range synapses

of the worm *C. Elegans* (the clustering coefficient of this worm's network would be 0.049, if it were a random graph) [14].

For the initial topology shown in Fig. 2, the network stored 13 memory patterns before working-through. Of these, 38.4% were still stored after working through, which shows that the network does adapt with the simulation of working-through, freeing itself from some of the "neurotic" states. For smaller values of κ the network has learning difficulties, which suggests a difficulty in associating unconscious memory traces among themselves and with declarative memory.

In neural network modeling, temperature is inspired by the fact that real neurons fire with variable strength, and there are delays in synapses, random fluctuations from the release of neurotransmitters, and so on. These are effects that we can loosely think of as noise [11,6]. So temperature in Boltzmann machines controls noise. In our model, temperature allows associativity among memory configurations, lowering synaptic

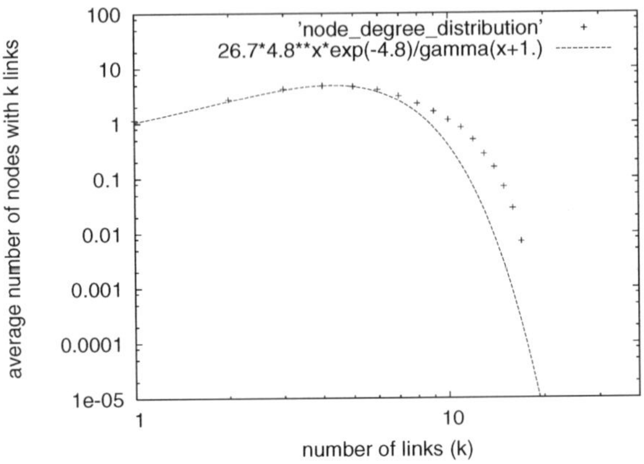

Fig. 3. Node degree distribution in logarithmic scale. Cross symbols represent values found in our simulations and the curve a fit of a Poisson distribution.

inhibition, in an analogy with the idea that freely talking in analytic sessions, and stimulation from the analyst lower resistances and allow greater associativity.

It is also possible to relate Boltzmann machine functioning with the model described in [12], where the signal-to-noise ratio at the neuronal level, induced by the catecholamines (dopamine, norepinephrine, etc.), alters the way the brain performs associations of ideas. Increases or decreases in the catecholaminergic levels have behavioral consequences in arousal, attention, learning, memory, and motor responses [4]. It is still not clearly verified, but it seems plausible to assume that catecholamines affect the neuronal ability to discern what is information from what is noise in a signal. According to some of these hypotheses, an increase of dopamine (hyperdopaminergy) promotes an enhancement of the signal and therefore of the signal-to-noise ratio [4,5,12]. Since noise is responsible for more free association of ideas, its reduction weakens associativity resulting in fixation to certain ideas. On the other hand, a decrease in dopamine increases noise which permits a broader association of ideas. Excessive noise may disorganize thought by permitting the association of distantly correlated ideas and breaking the logic of thought. We can thus relate temperature and noise in our Boltzmann machine to associativity of thought as in [12].

We have not yet thoroughly explored the parameter space of the model. This task will be continued in future work. We show some results from preliminary simulations, for different values of σ (see Equation 1), in Table 1. In our algorithms, associativity is sensitive to values of σ in cluster formation. If σ is too low, the network is more resistant to associativity and learning. For larger values of σ, associativity, represented by interactions among neurons, is higher and the network associates more loosely. In Table 1, T_{learn} is the temperature value for the first learning event during working-through, and the third column shows the percentage of original (neurotic) patterns that are still stored after working-through. These results are still not conclusive because they also depend on other model parameters, which we are still studying. In these cases, lower values of σ require larger temperatures for associativity that leads to learning.

Table 1. System behavior for different values of σ

σ	T_{learn}	% Original Patterns	Clustering Coefficient
0.465	0.530	100	0.452
0.520	0.031	41.4	0.439
0.578	0.001	38.4	0.430

Psychoanalysis promotes new thought associations and learning, and this cannot be carried through if catacholamine levels are either too high or too low. We suggest that psychoanalytic working-through explores the neurophysiologic potentialities of the subject, inducing network reconfiguration. When these potentialities are limited by chemical alterations, such as in strong psychotic cases, working-through should be limited or even impossible.

5 Conclusions

We have further developed the memory model presented in [1] to include microscopic biological neuronal mechanisms, and verified that the memory complex network self organizes into a hierarchical clustered structure. We have proposed a memory organization, where two hierarchically structured modules corresponding to sensorial and declarative memories interact, producing sensorial and symbolic activity, representing unconscious and conscious mental processes. This memory structure and functioning along with an adaptive learning process is used to explain a possible mechanism for neurotic behavior and psychoanalytical working-through.

The model is in agreement with psychoanalytic experience that working-through is a slow process, where the individual slowly elaborates knowledge by re-associating weakly connected memory traces and new experiences. This repetitive self-reconfiguration process, which we represent in the model by a change in network connectivity, will correspond to new outcomes in the subjects life history.

We are proceeding in further model refinement and analysis. It is still necessary to test the dependence of model behavior on various parameters such as temperature and κ. We are very interested in trying to map language structure and processing into network topology and dynamics, although we are not yet sure whether this is possible. Our main contribution is to propose a neuronal mechanism, based on known microscopical, biological brain mechanisms, that describes conscious and unconscious memory activity involved in neurotic behavior, as described by Freud. However, we do not sustain or prove that this is the actual mechanism that occurs in the human brain. Although biologically plausible, in accordance with many aspects described by psychodynamic and psychoanalytic clinical experience, and experimentally based on simulations, the model is very schematic. It nevertheless seems to be a good metaphorical view of facets of mental phenomena, for which we seek a neuronal substratum, and suggests directions of search.

We finish with a quote from the Introduction of [16], Freud's early work from late 1890's, first published only in 1950: "The intention is to furnish a psychology that shall be a natural science: that is, to represent psychical processes as quantitatively determinate states of specifiable material particles, thus making those processes perspicuous and free from contradiction." Although Freud stopped work on his model for lack of instruments at the time, these ideas pervaded all of his work, and it impresses one to see how his ideas regarding the unconscious give strong insight into contemporary models of consciousness, such as those in [9] arrived at by neurophysiological investigation.

This research was developed with grants from the Brazilian National Research Council (CNPq), the Rio de Janeiro State Research Foundation (FAPERJ) and the Brazilian agency which funds graduate studies (CAPES).

References

1. Wedemann, R.S., Donangelo, R., Carvalho, L.A.V., Martins, I.H.: Memory Functioning in Psychopathology. In: Sloot, P.M.A. et al. (Eds.): Lecture Notes in Computer Science **2329**. Springer-Verlag, Berlin Heidelberg (2002) 236–245
2. Freud, S.: Introductory Lectures on Psycho-Analysis. Standard Edition. W. W. Norton and Company, New York - London (1966). First German edition (1917)
3. Freud, S.: Beyond the Pleasure Principle. Standard Edition. The Hogarth Press, London (1974). First German edition (1920)
4. Spitzer, M.: A Cognitive Neuroscience View of Schizophrenic Thought Disorder. Schizophrenia Bulletin, **23** no. 1 (1997) 29–50
5. Servan-Schreiber, D., Printz, H., Cohen, J.: A Network Model of Catecholamine Effects: Gain, Signal-to-Noise Ratio, and Behavior. Science, **249** (1990) 892–895
6. Kandel, E.R., Schwartz, J.H., Jessel, T.M. (Eds.): Principles of Neural Science. MacGraw Hill, USA (2000)
7. Varela, F.J., Thompson, E., Rosch, E.: The Embodied Mind. The MIT Press, Cambridge, MA (1997)
8. Ganong, W.F.: Review of Medical Physiology. MacGraw Hill, USA (2003)
9. Edelman, G.M.: Wider than the Sky, a Revolutionary View of Consciousness. Penguin Books, London (2005)
10. Taylor, J.G.: The Codam Model and Deficits of Consciousness I & II. In: Palade, V. et al. (Eds.): Lecture Notes in Artificial Intelligence **2774**. Springer-Verlag, Berlin Heidelberg (2003) 1130–1148
11. Hertz, J.A., Krogh, A., Palmer, R.G. (ed.): Introduction to the Theory of Neural Computation. Lecture Notes, Vol. I, Santa Fe Institute, Studies in the Science of Complexity. Perseus Books, Cambridge, MA (1991)
12. Carvalho, L.A.V., Mendes, D.Q., Wedemann, R.S.: Creativity and Delusions: The Dopaminergic Modulation of Cortical Maps. In: Sloot, P.M.A. et al. (Eds.): Proceedings of ICCS 2003. Lecture Notes in Computer Science **2657**. Springer-Verlag, Berlin Heidelberg (2003) 511–520
13. Chomsky, N.: On Nature and Language. Cambridge University Press, UK (2002)
14. Newman, M.E.J.: Random Graphs as Models of Networks. In: Bornholdt, S. et al. (Eds.): Proceedings of the International Conference on Dynamical Networks in Complex Systems, Handbook of Graphs and Networks, from the Genome to the Internet. Wiley-VCH, USA (2003)
15. Watts, D.J., Strogatz, S.H.: Collective dynamics of 'small world' networks. Nature **393** (1998) 440–442
16. Freud, S.: Project for a Scientific Psychology. Standard Edition. Volume I, The Hogarth Press, London (1966). First German edition (1950)

Modelling Working Memory Through Attentional Mechanisms

John Taylor, Nickolaos Fragopanagos, and Nienke Korsten

[1] Department of Mathematics, King's College London, Strand, London, WC2R 2LS, United Kingdom
{john.g.taylor, nickolaos.fragopanagos, nienke.korsten}@kcl.ac.uk

Abstract. Recent studies of working memory have shown that the network of brain areas that supports working memory function overlaps heavily with the well studied network of selective attention. It has thus been suggested that working memory may operate by means of a repeated focusing of attention on the internal representations of the items that need to be maintained. We have employed our CODAM model of attention to simulate a specific working memory paradigm based on precisely this concept of 'refreshing' internal representations using attention. We propose here that the well known capacity limit of working memory can be attributed to the 'scarceness' of attentional resources. The specific mechanism of CODAM for modelling such scarceness is used in the paradigm to explain the behavioural and brain imaging data. This and related paradigms allow us to extend the specification of CODAM sites and functions to more detailed executive functions under executive control.

Keywords: Working memory, Attention, Refreshing, Computational model.

1 Introduction

Traditionally working memory has been viewed as being distinct from other cognitive functions as well as being mostly supported by prefrontal cortical areas [1,2,3]. Recent working memory research however is building up a body of evidence that challenges this view. On the one hand, posterior and parietal cortical areas appear to be heavily involved in the maintenance of information in working memory tasks [4,5,6,7] and, on the other, considerable overlap is observed between areas activated under spatial working memory and those under spatial attention [8,9,10]. The latter finding suggests that there may also be a functional overlap of working memory and attention. This has led Awh and colleagues [8,9,11,12] to suggest that, in fact, it is spatial attention that supports rehearsal in spatial working memory. Drawing from this concept of attention-based rehearsal we present here an extension of our CODAM attention model[13] that is able to model working memory as a repeated (but limited) focusing of attention on items to be remembered.

Numerous studies of attention have shown that, when attention is focused on a specific item from the external world, then the activity of the neuronal populations coding for this item is modulated by a signal arising from a network of frontoparietal

S. Kollias et al. (Eds.): ICANN 2006, Part I, LNCS 4131, pp. 553 – 562, 2006.

cortical areas [14,15,16]. More recent studies have shown that attention can be applied not only to 'online' representations but also to 'offline' representations -held in working memory- according to task goals. This has been shown to occur, for instance, using cues presented during the maintenance period of a working memory task (retro-cues) that direct attention to specific locations [17,18,19] or objects [20] held in working memory. Retro-cues have been shown to improve performance considerably in a working memory task very similar to the one being modelled below [21] although the maintenance period in this paradigm is quite short (<1500ms) and thus is unlikely to involve rehearsal mechanisms. Nevertheless, it offers further support to the ability of attention to modulate representations even when these are no longer externally supported.

Computational models of working memory so far have not employed any attention-based maintenance mechanisms. Most of them use self-excitatory PFC units or recurrent loops between PFC and basal ganglia structures to achieve stable delay-period states by means of bifurcation [22,23,24,25,26,27,28]. Deco and Rolls [29] attempt to merge a working memory model with an attention model but this is only within the PFC and so it fails to account for the more recent evidence of posterior-parietal-sited working memory maintenance. Moreover, it does not involve attentional modulation as a mechanism of refreshing working memory representations but only as a biasing signal of the competition for further processing. The limitation of the existing computational models of working memory mentioned above is that they usually include an ad-hoc signal (typically linked to dopamine) that initiates (and terminates) the required maintenance. On the contrary, our maintenance mechanism is based on the attention system reboosting representations held in a buffer as soon as a monitor detects that these representations have decayed beyond a critical level. In that sense our proposed mechanism is more 'natural' and benefits from using the same system as the one used to attend to the external world.

2 Simulations

In order to clarify this issue, we have modeled a WM task as presented by Pessoa et al [30] and Pessoa & Ungerleider [10], where subjects maintain a visual stimulus consisting of eight simple, rectangular bars at four possible different orientations. The subjects are asked to remember these orientations over a delay period of 6 seconds, then to indicate if the subsequent test stimulus, also consisting of eight bars, is the same or different. Their fMRI results indicate differential, preformance-related activations for different brain areas during different phases of the experiment (encoding, delay, test). We have modelled the neural (fMRI) activation during the delay phase, as well as the behavioral results of this paradigm.

This model is based on the CODAM (COrollary Discharge of Attention Model) [31] which has been successfully used to model the AB in [13]. This is a model of attentional processing, including WM functionality, with such features as graded neurons and intra-module mutual inhibition in object representing sites.

The CODAM model is based on the well-attested fact that attention is created by a signal sited in parietal lobe which amplifies target stimulus representations in lower

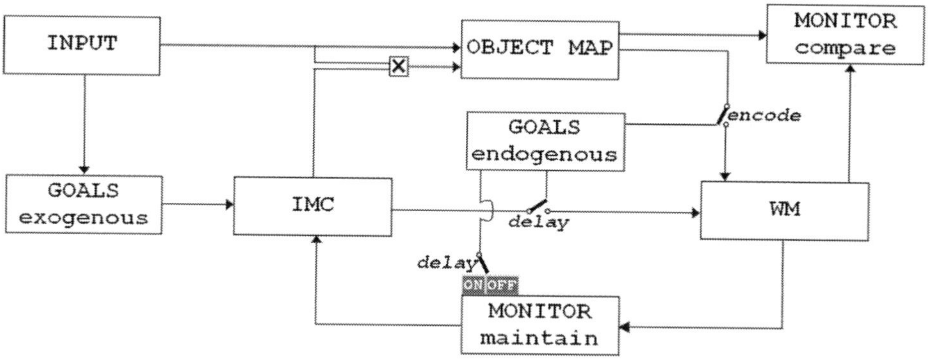

Fig. 1. An outline of the model

cortex at the same time reducing the activation of stimuli representing distracters. This attention movement control signal, generated by the inverse model controller (IMC) in engineering control terms, sends its amplification/inhibition effects to lower level cortices by what is very likely some form of contrast gain, although that is controversial [32]. This attention control signal is itself guided by signals from a prefrontal goal site, such as in ventro-lateral or dorso-lateral prefrontal cortex: this is the origin of the Goals site in figure 1. In the original CODAM model [31,13] it was suggested that there was a corollary discharge of the attention movement signal (from whence the acronym CODAM arose) which was used, as in many engineering control models and in control models of motor action in the brain, to speed up the movement of attention as well as to help it reduce errors in guidance.

We have adapted this model to the paradigm of Pessoa & Ungeleider [30,10] as outlined in figure 1. As shown in this figure, the corollary discharge module has been discarded from the original model, since its functionality – balancing endogenous and exogenous movement of attention – was obsolete in the current paradigm. Instead, a monitor module (MONITOR maintain) has been added to guide the movement of the attentional signal and thereby create long-term (>1.5 sec) maintenance. Another addition is the comparator module (MONITOR compare) which was added for the purpose of comparing the remembered output of WM to the new, probe input of the object map. These new modules were added on the basis of their functional necessity (which is further elaborated below) and an attempt to match their activities to the activity of specific areas observed experimentally was also made.

Each node in any module is coding for a particular representation. The connections between the modules are one on one, i.e. a node dedicated to one particular stimulus is connected to nodes in other modules that represent this same stimulus. All weights between modules are 1, except for the INPUT – OBJECT MAP, which is 0.5. Several modules contain lateral inhibition between all neurons in the module, as described below. Single neurons are graded with a sigmoid response function (unless otherwise indicated for separate modules) as in the original CODAM model (see appendix of [13] for equations). We present an outline of the function of and structure within the various modules, with their possible associated brain sites. Modules contain eight nodes, one for each initially activated combination of location and orientation, unless otherwise indicated. We have chosen to only include eight nodes and eliminate the nodes representing the alternative locations x orientations since they would not be activated at any point throughout the simulation.

GOALS endogenous. This module contains three nodes, coding for the task rules which are 'encode sample set of stimuli' (0.5 s.) then 'remember sample set of stimuli until test set appears' (6 s.) then 'compare sample set to test' (0.5 s.); these three signals are multiplied with appropriate WM inputs (OBJECT MAP or IMC) at the encode, delay and test phases, where they allow or deny encoding/maintenance activity, as well as prevent maintenance activity in the maintenance MONITOR during the encode and test phases.

GOALS exogenous. This module codes for the bias-weights associated with the reactive (or exogenous) attentional modulation of the stimuli representations.

IMC. This module is the controller of attention involved in all three phases of the paradigm; during the encode and test phases it amplifies (sigma-pi) representations in OBJ for the sample set of stimuli; during the delay phase it amplifies (sigma-pi) recurrent connections in WM for the sample set of stimuli (as attention-based rehearsal). Within this module, each node receives excitatory input from MONITOR maintain and GOALS exogenous, and inhibitory input from all other nodes in the module (with a 0.2 gain).

WM. This is the sensory working memory for the sample and the test stimuli; the conjunctions of features coded in object maps and locations coded in spatial maps are coded and maintained in this module for the sample stimuli to be later compared to the test stimuli. This module contains an internal maintenance system with a decay time of approximately 3.5 s. (from an activation of 1.7 to 0.2). This is reached through bifurcation with an additional 'parasite' node to each node in the module, which can be modulated by the IMC (see figure 2). Weights to and from the parasite nodes are 0.9 and 0.8, respectively. IMC input is multiplied by 2.3.

OBJ. Formally, this module should consist of a separate feature and a spatial map for the stimuli but for puposes of simplicity we will take this to be a conjoined map; this module is responsible for the perceptual/categorical processing of the stimuli that then activates the appropriate WM nodes for further maintenance. Each node in the module is inhibited by the other nodes, with a 0.1 gain. Since, strictly speaking, this function should be represented by two modules, there are two associated brain areas.

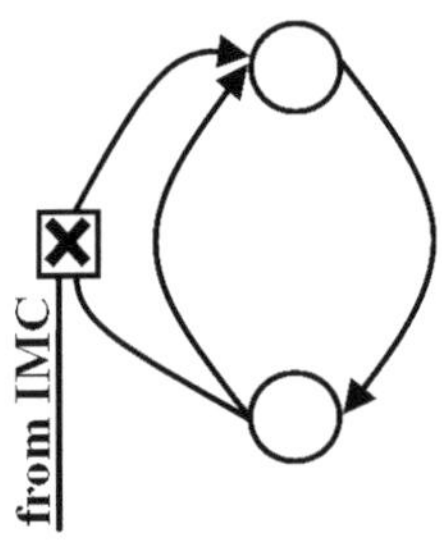

Fig. 2. Sigma-pi mod ulation within WM

MONITOR maintain. This modules function is to monitor the level of activations in WM and trigger a change of attentional focus to a less activated node when refreshing of the attended node is complete. The nodes in this module do not contain the standard sigmoidal transfer functions, but instead one node at a time is activated with a permanent output of 1. When this nodes WM membrane potential reaches its upper threshold of 1.6, a separate 'trigger' system within the module assigns attention to a different node, which is then activated if its WM membrane potential is between 0.2 (above noise level) and 0.6. See below for a more extensive explanation of the maintenance system.

MONITOR compare. This monitor module is used for comparison between the sample set of stimuli (in WM) and the test set (in OBJ) and generate an output (change/no change). For each node, if the activation of the corresponding WM node is nonzero, this activation is multiplied with the activation of the corresponding OBJ node. It will therefore give a nonzero output if both the encode and test stimulus correspond to the orientation x location represented by this node, and the encode stimulus is remembered (the WM has remained active). If the WM input is zero (i.e. if the encode stimulus is not remembered), a random output will be generated for this node. In order to generate a 'no-change' response, all outputs need to be 1. Otherwise a 'change' response is generated.

We now present a table showing our proposed module-to-brain area associations. For further elaboration on these associations and experimental support see [13, 31].

Table 1. Overview of modules with their function and proposed corresponding brain sites

Module	Function	Suggested brain site
GOALS endogenous	maintain task goals	DLPFC
GOALS exogenous	attentional boosting of input	MFG
IMC	attention control	SPL/IPS
WM	stim maintenance during delay	SPL/IPS
OBJECT MAP	perceptual processing of input	DOC/IT
MONITOR maintain	monitor WM activations	FEF
MONITOR compare	compare WM & probe	Cb/pulv

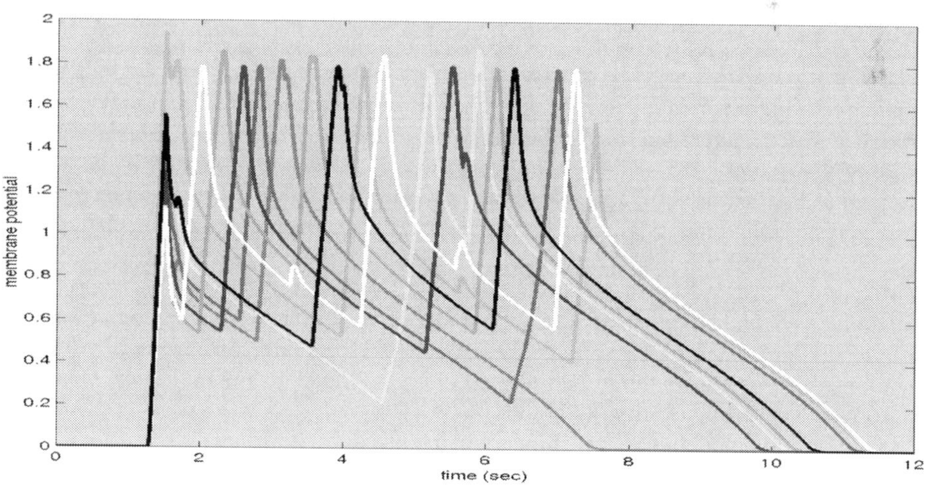

Fig. 3. Membrane potentials of the WM nodes for a typical trial illustrating the sequential attentional reboosting

We finish off this section by elaborating somewhat further on the maintenance system: The maintenance MONITOR module contains a control system for the boosting of the WM by the IMC, through sigma-pi modulation of the recurrent connections inside WM by the IMC. This monitor system only allows for one of its eight nodes to be active at a particular time, thereby allowing one IMC node to increase activation in one WM node. This 'refreshing' continues until this one attended WM nodes activation reaches its upper threshold (1.6), at which point the attentional focus switches, i.e. the monitor activates a different IMC node to boost WM activation. This boosting signal can only switch to a node with a WM activation lower than 0.6 and higher than 0.2. This nodes activation is then increased until it reaches threshold etc. etc, until the end of the delay phase. This sequential attentional reboosting of the WM representations is very clearly illustrated in the figure above (Fig. 3) where the WM nodes' membrane potentials are plotted for a typical trial.

Let us close this section by stating that the values of the various parameters were chosen so that the best fit to the experimental data was obtained. The 'refreshing' mechanism specifically is particularly sensitive to the values of the parameters that control it such as the thresholds used in the maintenance MONITOR module and the strength of the sigma-pi modulation of the WM representations by the IMC.

3 Results

The behavioural results of Pessoa et al. [30] report a mean performance across subjects of 71.4% correct for high-confidence trials, which dropped to 60.8% correct for the low-confidence trials. Distinguishing between high- and low-confidence trials is beyond the scope of this paper, so we have averaged over the two types of trials and taken a performance grand average of about 65%. We have been able to reproduce this result with our model by running as many simulated trials as experimental ones (160). We have assumed that a 'correct' trial is one where all the nodes of the MONITOR/compare have correctly identified a change or a non-change (as applicable). For each node of the MONITOR/compare a correct outcome can be produced when either the corresponding WM representation of the sample set was successfully maintained or when maintenance failed but the randomly generated orientation for this node matched the one that had decayed. To obtain an average of correct responses similar to the experimental data, according to this mechanism of deciding the overall change/non-change response, the weights that control the modulation of the WM pair recurrence (i.e. the 'refreshing' strength) were set to a high value. This results in an average of about 2 nodes of the WM decaying before the test set is presented. Or, conversely, about 6 items are on average perfectly maintained in WM throughout the required delay period. This result, in turn, could be interpreted as a WM capacity of about 6 items.

In order to model the neuroimaging results of the paper, we followed a methodology of relating synaptic activity to fMRI BOLD responses that is becoming more and more popular in the computational neuroscience literature. Logothetis and colleagues [33] analyzed simultaneously recorded neuronal and fMRI responses from the visual cortex of monkeys and found it was local field potentials (LPFs) that correlated most highly with the BOLD signals. This suggests that the BOLD signal reflects the synaptic input activity of the observed area and not its firing output. Thus, to obtain the simulated BOLD signal for a given module we convolve the total synaptic input to its neurons (excitatory and inhibitory) with a suitable model of a hemodynamic response function (HRF) such as the one from [34].

The main thrust of the Pessoa et al. [30] study is that one can predict the outcome of a trial (correct/incorrect) just by looking at the levels of fMRI activations of specific areas. In other words, the activations of certain areas (as measured by fMRI) were found to correlate with behavioural performance in the task. Our aim is to firstly show that once we have assigned a certain function to each area for which BOLD signals where obtained and, by implication, a certain module of our model to that area, we can reproduce the BOLD signals recorded experimentally using the synaptic activity of the module that corresponds to each area. Secondly, we aim to show that the deterioration or the limited operation of certain critical modules of our model (that would manifest as a reduced BOLD signal) leads to behavioural performance impairment (calculated by the MONITOR/compare described above).

The correspondence of functional components of our model with brain areas was presented in the previous section. So we present our first results of the simulated BOLD signals from the IMC as they relate to the experimentally recorded signals from the right IPS in Fig. 4.

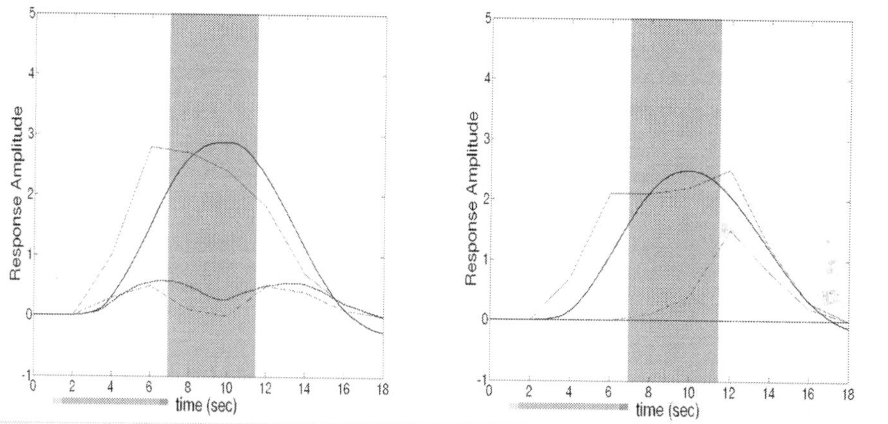

Fig. 4. Left-hand side: simulated BOLD responses (bold lines) from the model's IMC and the MONITOR/Maintain plotted against the experimental BOLD signals (light lines) from the right IPS. Right-hand side: simulated BOLD responses (bold lines) from the model's MONITOR/ Maintain plotted against the experimental BOLD signals (light lines) from the right FEF. Solid lines in both cases correspond to correct trials and dashed lines to incorrect ones.

As we can see on the left hand side of figure 4 we have managed to reproduce the timecourse of the activation of the right IPS (model's IMC) satisfactorily both for the correct and the incorrect trials. The difference between correct and incorrect trials in our model corresponds to having the MONITOR/maintain on or off. When the MONITOR/maintain is off no WM maintenance takes place and so all decisions at the test phase are at chance level. In the same way we have modelled the correct/incorrect activation difference in the right FEF which corresponds to our model's MONITOR maintain. The resulting BOLD signals are shown on the right hand side of figure 4.

Finally we present the simulated BOLD responses for the model's Object Map which corresponds to the experimental DOC/IT areas (figure 5, LHS). Note that our simulated BOLD signals do not differ between correct and incorrect trials as the Object Map does not participate in the delay period where we set the MONITOR/maintain on or off accordingly. Instead the difference observed during the encode phase experimentally may have to do with some attentional or perceptual failure during encoding which causes erroneous representations to be encoded and maintained thus leading to incorrect responses at test.

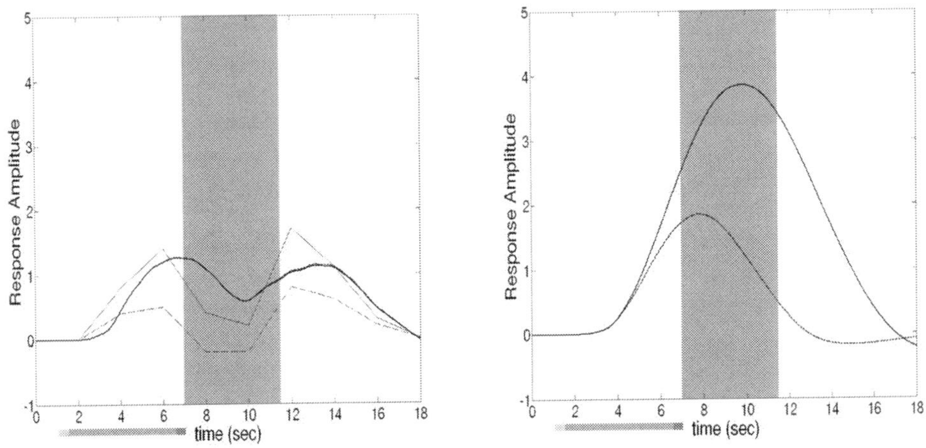

Fig. 5. Left-hand side: simulated BOLD responses (bold lines) from the model's Object Map plotted against the experimental BOLD signals (light lines) from the DOC/IT. Right-hand side: simulated BOLD responses (bold lines) from the model's Working Memory. Solid lines in both cases correspond to incorrect trials and dashed lines to correct ones.

The figure plotted on the right hand side of figure 5 shows the simulated BOLD responses (bold lines) from the model's Working Memory. We couldn't match these BOLD signals to any of the experimental ones from the Pessoa et al. paper as not all of the areas that exhibited behavioural performance predictiveness were plotted.

4 Conclusions

In the paper we have presented simulation results, based on the CODAM engineering control model of attention [13, 31] to explain recent behavioural and brain imaging results on maintaining information in subjects in working memory sites from [10, 30]. Our model used a simplified form of buffer working memory site (by recurrence of activity between two neurons), and used a comparator to determine when refreshing of the buffer activity should occur (using contrast gain feedback of attention). We suggested identification of the functional modules in our simulation with various of those observed in the experiment [10, 30] with considerable success. These identifications support those arrived at earlier [30, 31, 13], and in particular through the similarity of the temporal flow of activity reported in [10, 30] in comparison with that reported here. Finally we note there are numerous questions we have not been able to discuss in any depths here: the detailed nature of working memory buffer refreshment, the predictions of our model for single cell activity, the nature of attention feedback itself. We hope to turn to these elsewhere.

Acknowledgements. This work has been partially supported by the HUMAINE EC Network of Excellence and the Biotechnology and Biological Sciences Research Council (BBSRC), UK; one of us (JGT) would like to thank the EC for financial support under the GNOSYS cognitive robot project (FP6-IST-2-003835) to allow him to pursue this work.

References

1. Goldman-Rakic, P.S.: Development of Cortical Circuitry and Cognitive Function. Child Dev. 58 (1987) 601-622
2. Smith, E.E., Jonides, J.: Storage and Executive Processes in the Frontal Lobes. Science 283 (1999) 1657-1661
3. Courtney, S.M.: Attention and Cognitive Control As Emergent Properties of Information Representation in Working Memory. Cogn Affect. Behav. Neurosci. 4 (2004) 501-516
4. Fuster, J.M., Alexander, G.E.: Neuron Activity Related to Short-Term Memory. Science 173 (1971) 652-654
5. D'Esposito, M., Postle, B.R., Rypma, B.: Prefrontal Cortical Contributions to Working Memory: Evidence From Event-Related FMRI Studies. Exp. Brain Res. 133 (2000) 3-11
6. Petrides, M.: Dissociable Roles of Mid-Dorsolateral Prefrontal and Anterior Inferotemporal Cortex in Visual Working Memory. J. Neurosci. 20 (2000) 7496-7503
7. Passingham, D., Sakai, K.: The Prefrontal Cortex and Working Memory: Physiology and Brain Imaging. Curr. Opin. Neurobiol. 14 (2004) 163-168
8. Awh, E., Jonides, J., Reuter-Lorenz, P.A.: Rehearsal in Spatial Working Memory. J. Exp. Psychol. Hum. Percept. Perform. 24 (1998) 780-790
9. Awh, E., Gehring, W.J.: The Anterior Cingulate Cortex Lends a Hand in Response Selection. Nat. Neurosci. 2 (1999) 853-854
10. Pessoa, L., Ungerleider, L.G.: Neural Correlates of Change Detection and Change Blindness in a Working Memory Task. Cereb. Cortex 14 (2004) 511-520
11. Awh, E., Jonides, J.: Overlapping Mechanisms of Attention and Spatial Working Memory. Trends Cogn Sci. 5 (2001) 119-126
12. Postle, B.R., Awh, E., Jonides, J., Smith, E.E., D'Esposito, M.: The Where and How of Attention-Based Rehearsal in Spatial Working Memory. Brain Res. Cogn Brain Res. 20 (2004) 194-205
13. Fragopanagos, N., Kockelkoren, S., Taylor, J.G.: A Neurodynamic Model of the Attentional Blink. Brain Res. Cogn Brain Res. 24 (2005) 568-586
14. Kastner, S., Pinsk, M.A., De Weerd, P., Desimone, R., Ungerleider, L.G.: Increased Activity in Human Visual Cortex During Directed Attention in the Absence of Visual Stimulation. Neuron 22 (1999) 751-761
15. Kastner, S., Ungerleider, L.G.: The Neural Basis of Biased Competition in Human Visual Cortex. Neuropsychologia 39 (2001) 1263-1276
16. Corbetta, M., Shulman, G.L.: Control of Goal-Directed and Stimulus-Driven Attention in the Brain. Nat. Rev. Neurosci. 3 (2002) 201-215

17. Griffin, I.C., Nobre, A.C.: Orienting Attention to Locations in Internal Representations. J. Cogn Neurosci. 15 (2003) 1176-1194
18. Lepsien, J., Griffin, I.C., Devlin, J.T., Nobre, A.C.: Directing Spatial Attention in Mental Representations: Interactions Between Attentional Orienting and Working-Memory Load. Neuroimage. 26 (2005) 733-743
19. Nobre, A.C., Coull, J.T., Maquet, P., Frith, C.D., Vandenberghe, R., Mesulam, M.M.: Orienting Attention to Locations in Perceptual Versus Mental Representations. J. Cogn Neurosci. 16 (2004) 363-373
20. Lepsien, J., Nobre, A.C.: Attentional Modulation of Object Representations in Working Memory . in preparation (2006)
21. Lamme, V.A.: Separate Neural Definitions of Visual Consciousness and Visual Attention; a Case for Phenomenal Awareness. Neural Netw. 17 (2004) 861-872
22. Durstewitz, D., Kelc, M., Gunturkun, O.: A Neurocomputational Theory of the Dopaminergic Modulation of Working Memory Functions. J. Neurosci. 19 (1999) 2807-2822
23. Tagamets, M.A., Horwitz, B.: A Model of Working Memory: Bridging the Gap Between Electrophysiology and Human Brain Imaging. Neural Netw. 13 (2000) 941-952
24. Frank, M.J., Loughry, B., O'Reilly, R.C.: Interactions Between Frontal Cortex and Basal Ganglia in Working Memory: a Computational Model. Cogn Affect. Behav. Neurosci. 1 (2001) 137-160
25. Brunel, N., Wang, X.J.: Effects of Neuromodulation in a Cortical Network Model of Object Working Memory Dominated by Recurrent Inhibition. J. Comput. Neurosci. 11 (2001) 63-85
26. Deco, G., Rolls, E.T., Horwitz, B.: "What" and "Where" in Visual Working Memory: a Computational Neurodynamical Perspective for Integrating FMRI and Single-Neuron Data. J. Cogn Neurosci. 16 (2004) 683-701
27. Ashby, F.G., Ell, S.W., Valentin, V.V., Casale, M.B.: FROST: a Distributed Neurocomputational Model of Working Memory Maintenance. J. Cogn Neurosci. 17 (2005) 1728-1743
28. Chadderdon, G.L., Sporns, O.: A Large-Scale Neurocomputational Model of Task-Oriented Behavior Selection and Working Memory in Prefrontal Cortex. J. Cogn Neurosci. 18 (2006) 242-257
29. Deco, G., Rolls, E.T.: Attention and Working Memory: a Dynamical Model of Neuronal Activity in the Prefrontal Cortex. Eur. J. Neurosci. 18 (2003) 2374-2390
30. Pessoa, L., Gutierrez, E., Bandettini, P., Ungerleider, L.: Neural Correlates of Visual Working Memory: FMRI Amplitude Predicts Task Performance. Neuron 35 (2002) 975-987
31. Taylor, J.G.: Mind and Consciousness: Towards a Final Answer? Physics of Life Reviews 2 (2005) 1-45
32. Taylor, N., Hartley, M., Taylor, J.G.: Analysing Attention at Neuron Level. BICS 2006 (2006)
33. Logothetis, N.K., Pauls, J., Augath, M., Trinath, T., Oeltermann, A.: Neurophysiological Investigation of the Basis of the FMRI Signal. Nature 412 (2001) 150-157
34. Glover, G.H.: Deconvolution of Impulse Response in Event-Related BOLD FMRI. Neuroimage. 9 (1999) 416-429

A Cognitive Model of Multi-objective Multi-concept Formation

Toshihiko Matsuka[1], Yasuaki Sakamoto[1],
Jeffrey V. Nickerson[1], and Arieta Chouchourelou[2]

[1] Stevens Institute of Technology, Hoboken, NJ 07030, USA
[2] Rutgers University, Newark, NJ 07102, USA

Abstract. The majority of previous computational models of high-order human cognition incorporate gradient descent algorithms for their learning mechanisms and strict error minimization as the sole objective of learning. Recently, however, the validity of gradient descent as a descriptive model of real human cognitive processes has been criticized. In the present paper, we introduce a new framework for descriptive models of human learning that offers qualitatively plausible interpretations of cognitive behaviors. Specifically, we apply a simple multi-objective evolutionary algorithm as a learning method for modeling human category learning, where the definition of the learning objective is not based solely on the accuracy of knowledge, but also on the subjectively and contextually determined utility of knowledge being acquired. In addition, unlike gradient descent, our model assumes that humans entertain multiple hypotheses and learn not only by modifying a single existing hypothesis but also by combining a set of hypotheses. This learning-by-combination has been empirically supported, but largely overlooked in computational modeling research. Simulation studies show that our new modeling framework successfully replicated observed phenomena.

1 Introduction

The ability to categorize plays a central role in high-order human cognition. By compressing the vast amount of available information, categorization allows humans (and other intelligent agents) to process, understand, and communicate complex thoughts and ideas by efficiently utilizing salient information while ignoring other types. Given the importance of categories as the building units of human knowledge [1], researchers in cognitive science and related fields have shown a great deal of interest in understanding and explaining human category learning.

Human category learning research has been strongly associated with computational modeling as a means to test various theories of how humans encode, organize, and use knowledge. Previous modeling efforts, however, have limited emphases because their principal focus was on describing categorization processes (i.e., forward algorithm specifying how various types of information are integrated to classify items to proper categories), largely overlooking the learning processes (i.e., backward algorithm specifying how category representations are modified based on experiences). In fact, a central debate in the categorization discourse is on how categories are represented (e.g., [2] [3]). The majority of modeling work employed a single learning method, namely

S. Kollias et al. (Eds.): ICANN 2006, Part I, LNCS 4131, pp. 563–572, 2006.

gradient descent (e.g., [2] [3]) with strict (categorization) error minimization as the sole objective of learning.

Recently, Matsuka [4] raised a concern that the lack of emphasis on learning processes will lead to a limited understanding of the nature of human concept formation. One limitation of using the traditional learning algorithms in human category learning research is that it assumes that every human learner modifies his or her category knowledge by local optimization to reduce misclassification irrespective of the contextual factors (e.g., the learner's goals). The result is little variability in learning among the simulated learners. A typical gradient descent optimization method can be considered a normative rather than descriptive account of human learning processes [4].

Unlike a typical gradient descent method, human learning is variable, not based exclusively on error minimization. For example, humans tend to develop simpler category representations instead of complex representations that are slightly more diagnostic in discriminating members of different categories [5]. Moreover, given two stimulus dimensions or features (e.g., size and color) that are equally diagnostic in distinguishing across categories, humans learn to attend to a single, randomly selected diagnostic dimension when making a classification judgment. They ignore the other dimension that is redundant but equally diagnostic [5], leading to individual differences in learning. Unlike human learning, which is often arbitrary and tends to be biased toward simplicity, the gradient descent algorithm always "correctly" updates the category representations, resulting in allocating attention to the two diagnostic dimensions.

The purpose of this work is to introduce a new framework for descriptive models of human learning that offers qualitatively plausible interpretations of cognitive behaviors. We apply a simple evolutionary algorithm as a learning method for modeling human category learning. Here, we show that the evolutionary algorithm can account for variations in human attention allocation during category learning that existing gradient descent models cannot. We conclude with a discussion of how our new framework might be extended to incorporate sensitivity to various contextual factors, such as learners' goals, that humans display.

2 Modeling Category Learning with Evolutionary Algorithm

Our new framework is called CLEAR for Category Learning with Evolutionary AlgoRithm. CLEAR models human learning processes by evolutionary processes where multiple chromosomes compete with one another for survival (i.e., optimization). Each chromosome, consisting of a set of genes, is a particular solution to a problem. The interpretation of each gene is model specific. For example, as in the current modeling approach, a gene may represent a coefficient that corresponds to an attention weight allocated to a stimulus dimension.

CLEAR assumes that human learning involves consideration of multiple solutions (i.e., chromosomes) according to their usefulness in a given task. Each of these solutions determines which aspects of the categories are psychologically salient and which can be ignored. The use of a population of solutions in CLEAR is an innovative contribution to cognitive modeling as virtually all previous categorization models have optimized a single solution using gradient descent on prediction errors (e.g. [2]).

The assumption that humans entertain a range of solutions is consistent with the results from human laboratory experiments [6]. For example, Anderson and Pichert [6] asked people to read a story about a house from the perspective of either a burglar or home-buyer. The story contained pieces of information relevant to one perspective but irrelevant to the other. For example, a color television set was relevant to the burglar but not to the home-buyer. Alternatively, a leaking roof was relevant to the home-buyer but not the burglar. In a free recall task conducted after learning about the story, people recalled more information relevant to the perspective they took than irrelevant information. More important, shifting people's perspective after the initial recall allowed people to recall information that they could not recall previously, suggesting that people indeed have multiple solutions that activate different pieces of information.

Although CLEAR always has multiple solutions in its mind, CLEAR, like the participants in Anderson and Pichert [6], opts for and applies a single solution with the highest predicted utility (e.g., accuracy, score, etc.) to make one response at a time (e.g., categorize an input instance). The functions for estimating the utility for each solution is described in a later section. In the next section, we describe the processes that evolve into solutions in CLEAR.

2.1 Learning Via Evolutionary Algorithm

CLEAR utilizes the Evolution Strategy (ES) method (e.g. [7]) for its learning processes. CLEAR, as in a typical ES application, assumes three key processes in learning: *crossover*, *mutation*, and (survivor) *selection*. In the *crossover* process, the randomly selected chromosomes form a pair and exchange gene information, creating a new pair of chromosomes. In human cognition, the crossover process can be interpreted as conceptual combination, in which new solutions are created based on merging ideas from existing solutions that are useful (e.g., creative discovery). In the *mutation* process, each gene (i.e., coefficient) is randomly altered. A mutation can be considered as a modification of a solution by randomly creating a new hypothesis. In the *selection* process, a certain number of solutions are deterministically selected on the basis of their fitness in relation to the environment for survival. Those selected solutions (i.e., chromosomes) will be kept in CLEAR's memory trace (i.e., population space), while non-selected solutions become obsolete or are forgotten.

Unlike previous modeling approaches to category learning research which modify a single solution (i.e., a single set of coefficients), CLEAR maintains, modifies, and combines a set of solutions. The idea of having a population of solutions (as opposed to having an individual solution) is important because it allows not only the selection and concept combination (i.e., crossover processes) in learning, but also the creation of diverse solutions, making learning more robust. Thus, unlike previous models, CLEAR assumes that humans have the potential to maintain a range of solutions and are able to apply a solution most suitable for a particular set of situational characteristics. To our knowledge, these capabilities (i.e., using diverse solutions and learning-by-combination) have not been addressed in the category learning modeling discourse. This may warrant future research. The utility of having homogeneous versus heterogeneous solutions likely depends on situational factors (e.g., a motivation to test a range of strategies) that will vary from one context to another [8].

Another important feature of CLEAR is that it allows the hypothetical error surface to be non-smooth or discontinuous. This characteristic has not been successfully incorporated in cognitive modeling using the gradient descent optimization method. CLEAR, because of the stochastic nature of its optimization method, can incorporate multi-objective functions in learning that are consistent with the complexity of human learning and the possibly discontinuous nature of knowledge utility hypersurface.

Finally, another significant contribution of CLEAR to cognitive modeling is adaptive search strategies, which can be interpreted as analogous to learning rate. As in many recent ES, CLEAR incorporates a self-adoption mechanism for modifying solutions (i.e., coefficient mutation), dynamically altering range of search areas. This mechanism allows CLEAR to be sensitive to the topology of knowledge utility hypersurface (e.g., searching within a smaller area if solutions are close to an optimum). This, in turn, makes CLEAR sensitive to changes in learning objectives. Virtually all previous cognitive models incorporate either static or time-decreasing learning rate.

2.2 Categorization Processes in CLEAR

Rather than introducing new forward algorithms, we apply CLEAR's learning processes to ALCOVE's [2] categorization processes. ALCOVE is a computational model of category learning that assumes that humans store every studied instance or exemplar in memory. We chose ALCOVE because of its popularity and demonstrated predictive capability using relatively simple mechanisms in perceptual classification research.

In ALCOVE, categorization decision is based on the activations of stored exemplars. As shown in Equation 1, each exemplar's activation in ALCOVE, scaled by specificity, c (which determines generalization gradient), is based on the inverse distance between an input, x, and a stored exemplar, ψ_j, in multi-dimensional representational space, scaled by dimensional selective attention weights, α. The exemplar activations are then fed forward to the k-th output node (e.g., output for category k), O_k, weighted by w, which determines the strength of association between each exemplar j and each output node k:

$$O_k^m(x) = \sum_j w_{kj}^m \left[\exp\left(-c \cdot \sum_i \alpha_i^m |\psi_{ji} - x_i|\right)\right] \tag{1}$$

where superscript m indicates m-th solution being utilized.

The probability of categorizing input instance x to category C is based on the activation of output node C relative to the activations of all output nodes:

$$P(C) = \frac{\exp(\phi \cdot O_c^o(x))}{\sum_k \exp(\phi \cdot O_k^o(x))}. \tag{2}$$

where ϕ controls decisiveness of classification response, and superscript o indicates the solution adopted to make a categorization response.

In the traditional ALCOVE model, attention (α) and association weights (w) are updated by gradient descent that minimizes the actual and the predicted value (i.e., Eq. 2). In the current work, a solution consists of these coefficients (i.e., attention and association weights). CLEAR optimizes a set of these solutions. We now describe the functions for estimating the utility of a solution (i.e., attention and association weights).

2.3 Learning Process in CLEAR

Since CLEAR is based on Evolutionary Strategy, its genotype space is identical to its phenotype space, except the genes (coefficients) for self-adapting strategy, σ. In CLEAR, there are two coefficients for self-adaptation; σ_w for defining search area for *w*, and σ_α for α. For the sake of simplicity, we use the following notation $(\mathbf{w}^m, \boldsymbol{\alpha}^m) \in \boldsymbol{\theta}^m$.

Hypotheses Combinations. In CLEAR, randomly selected pairs of solutions are exchanging information to combine knowledge. In particular, CLEAR utilizes discrete recombination of ALCOVE coefficients and intermediary recombination of the coefficient for self-adaptation. Thus, parent solutions $\boldsymbol{\theta}^{p1}$ and $\boldsymbol{\theta}^{p2}$ would produce a child solution $\boldsymbol{\theta}^c$, where $\theta_i^c = \theta_i^{p1}$ if UNI ≤ 0.5 or θ_i^{p2} otherwise, where UNI is a random number drawn from the Uniform distribution. For self-adapting strategy, $\sigma_i^c = 0.5 \cdot (\sigma_i^{p1} + \sigma_i^{p2})$. This combination process continues until the number of children solutions produced reaches the memory capacity of CLEAR.

Hypotheses Modifications. After the recombination process, CLEAR randomly modifies its solutions, using a self-adapting strategy. Thus,

$$\sigma_w^m(t+1) = \sigma_w^m(t) \cdot \exp(N(0,\gamma)) \quad w_{kj}^m(t+1) = w_{kj}^m(t) + N(0, \sigma_w^m(t+1)) \tag{3}$$

$$\sigma_\alpha^m(t+1) = \sigma_\alpha^m(t) \cdot \exp(N(0,\gamma)) \quad \alpha_i^m(t+1) = \alpha_i^m(t) + N(0, \sigma_\alpha^m(t+1)) \tag{4}$$

where *t* indicates time, γ defines search width (via σ's), and N(0.σ) is a random number drawn from the Normal distribution with the corresponding parameters.

Selection of Surviving Hypotheses. After creating new sets of solutions, CLEAR selects a limited number of solutions to be maintained in its memory. In CLEAR, the survivor selection is done deterministically, selecting best solutions on the basis of estimated utility of concepts or knowledge. The function defining utility of knowledge is described in the next section.

2.4 Estimating Utility

The utility of each solution or a set of coefficients determines the selection process in CLEAR, which takes place twice. During categorization, CLEAR selects a single solution with the highest predicted utility to make a categorization response (referred to as concept utility for response or UR hereafter). During learning, CLEAR selects best fit solutions to update its knowledge (utility for learning or UL hereafter). In both selection processes, the solution utility is subjectively and contextually defined, and a general function is given as:

$$U(\boldsymbol{\theta}^m) = \Upsilon\left(E(\boldsymbol{\theta}^m), Q_1(\boldsymbol{\theta}^m), ..., Q_L(\boldsymbol{\theta}^m)\right) \tag{5}$$

where Υ is a function that takes concept inaccuracy (i.e., *E*) and *L* contextual factors (i.e., *Q*) and returns an estimated solution utility value (Note that CLEAR's learning is framed as a minimization problem). There are virtually infinite contextual functions appropriately defined for Eq. 5 (e.g. concept abstraction, domain expertise and knowledge commonality). For example, in ordinary situations, humans prefer simpler solutions (e.g. requiring a smaller amount of diagnostic information to be processed) to complex ones, as long as both are sufficiently accurate, whereas in highly critical tasks (e.g. medical diagnosis), many might choose a solution with the highest accuracy disregarding complexity.

Note that functions for UR and UL do not have to be the same. For example, domain experts often know multiple approaches to categorize objects and such an ability appears to be a very important characteristic and thus be a part of their UL. However, "knowledge diversity" is only relevant for selecting a population of solutions (for survival), but not for selection of a particular solution to make a categorization response. Thus, knowledge diversity should not be considered for UR.

In CLEAR, the predicted (in)accuracy of a solution during categorization is estimated based on a retrospective verification function [9], which assumes that humans estimate the accuracies of the solutions by applying the current solutions to previously encountered instances with a memory decay mechanism. Thus,

$$E(\boldsymbol{\theta}^m) = \sum_{g=1}^{G} \left[\left(\frac{\sum_{\forall i | x^{(i)} = x^{(g)}} (\tau^{(i)} + 1)^{-D}}{\sum_{g} \sum_{\forall i | x^{(i)} = x^{(g)}} (\tau^{(i)} + 1)^{-D}} \right) \sum_{k}^{K} \left[d_k^{(g)} - O_k^m \left(x^{(g)} \right) \right]^2 \right] \tag{6}$$

where g indicates particular training exemplars, G is the number of unique training exemplars, the last term is the sum of squared error with d being the desired output, and the middle term within a parenthesis is the (training) exemplar retention function defining the strength of the retaining training exemplar $x^{(g)}$. Memory decay parameter, D, in the exemplar retention function controls speed of memory decay, and τ indicates how many instances were presented since $x^{(g)}$ appeared, with the current training being represented with "0." Thus, $\tau = 1$ indicates $x^{(g)}$ appeared one instance before the current trial. The denominator in the exemplar retaining function normalizes retention strengths, and thus it controls the relative effect of training exemplar, $x^{(g)}$, in evaluating the accuracy of knowledge or concept. $E(\boldsymbol{\theta})$ is strongly influenced by more recently encountered training exemplars in early training trials, but it evenly accounts for various exemplars in later training trials, simultaneously accounting for the Power Law of Forgetting and the Power Law of Learning.

3 Simulations

In order to investigate the capability of CLEAR to replicate observed empirical data, a simulation study was conducted. In particular, we simulated an empirical study that suggests that human concept formation might be driven by more than minimization of classification error [5]. Table 1 shows the schematic representation of the stimulus used in the empirical study and the present simulation study. (Note: Dimensions 1 and 2 are redundant and are also perfectly correlated with the category membership, each being a necessary and sufficient diagnostic dimension). The left column of Fig. 1 shows the observed data. The majority of subjects in the empirical study [5] chose to pay attention primarily if not exclusively to either one of them, as shown in Fig.1 (bottom

Table 1. Schematic representation of stimulus set used in Simulation Study

Stimulus Set				
Category	Dim1	Dim2	Dim3	Dim4
A	1	1	3	4
A	1	1	4	1
A	1	1	1	2
B	2	2	2	1
B	2	2	3	2
B	2	2	4	3
C	3	3	1	3
C	3	3	2	4
C	3	3	3	1
D	4	4	4	2
D	4	4	2	3
D	4	4	1	4

left), where the amount of attention allocated to each feature was operationally defined by the feature viewing time. Thus, the majority of subjects seemed to have at least two learning objectives: (1) to minimize categorization error and (2) to minimize knowledge complexity. In the present study we simulated this multi-objective learning phenomenon with CLEAR.

Another interesting phenomenon reported in that study is that some subjects who learned to pay attention to either dimension exclusively reported that they realized that there was another diagnostic dimension, indicating the possibility of possessing multiple concepts or solutions. Although it is uncertain whether these subjects were deliberately acquiring multiple solutions or not, we simulated this type of learners as well.

Methods: There were two types of CLEAR learners involved in the present simulation study, namely SA who tries to acquire simple accurate classes of knowledge, and MSA whose learning objective is to acquire multiple simple accurate class of knowledge. Equations 7 and 8 describe knowledge utility functions for learning (UL) for SA and MSA, respectively. For both models, Eq.7 is used for their UR, as knowledge diversity should have no effect in selecting a solution to make a response. The knowledge utility for SA (and UR of MSA) is given as:

$$U_{a+s}(\boldsymbol{\theta}^m) = E(\boldsymbol{\theta}^m) + \lambda_w \sum_k \sum_j w_{kj}^2 + \lambda_\alpha \sum_i \left[1 + \alpha_i^{-2} \cdot \sum_l^I \alpha_l^2\right]^{-1} \quad (7)$$

where the first term is defined as in Eq.6, the second term is a weight decay function regularizing w, the third term is relative attention elimination function reducing the number of dimensions attended, and λ's are scalars weighting different contextual factors. The knolwedge utility for learning (UL) for MSA is given as:

$$U_{a+s+d}(\boldsymbol{\theta}^m) = U_{a+s}(\boldsymbol{\theta}^m) \cdot \left[1 + \sum_n \xi(\delta(m,n))\right] \quad (8)$$

where the last term controls diversity (penalizing having "similar" concepts), δ indicates the distance between solution m and n, and ξ is defined as

$$\xi(\delta(m,n)) = \begin{cases} 1 - \left(r^{-1} \cdot \sqrt{\sum_i (\theta_i^m - \theta_i^n)^2}\right)^2, & \text{if } \sqrt{\sum_i (\theta_i^m - \theta_i^n)^2} \leq r \\ 0, & \text{otherwise.} \end{cases} \quad (9)$$

where r is a parameter defining the penalizing radius. The diversity controlling function is an application of a modified version of the fitness sharing method [7].

Both models were run in a simulated training procedure to learn the correct classification responses for the stimuli with corrective feedback. The basic training procedures followed that of the original study [5]. There were a total of 8 training blocks, each of which was organized as a random presentation of 12 unique exemplars. The model parameters were selected arbitrarily; $c = 1.25$, $\phi = 5$,$D = 1$,$\gamma = 0.5$, $\lambda_w = 0.01$, $\lambda_\alpha = 0.75$, $r = 1.5$. Note that the same parameter values were used for SA and MSA, except r, which was only applicable for MSA. The memory sizes for the two models were 10 (i.e., possessing 10 solutions at a time). There were a total of 500 simulated subjects for both models.

Results & Discussion: Figure 1 shows the results of the simulation study. Both SA and MSA successfully replicated the observed learning curve and aggregated attention learning curves (Fig. 1 top row). In replicating individual attention allocation patterns, SA seemed more consistent with the empirical data than MSA, indicating that (the majority of) human subjects in that study might have had simplicity and accuracy as their learning objectives, but not concept diversity.

To test if MSA had acquired diverse solutions, we investigated all solutions in MSA's memory trace and compared them with those of SA's. Let us define the diagnostic feature dimension selected to be attended by the manifesting concept (i.e., concept or solution whose UR value was the lowest) as a *dominant* dimension and the other as a *recessive* dimension. For MSA, the average difference in the relative amount of attention *to be* allocated to a recessive and dominant dimensions was 0.24 in the latent concepts (i.e., solutions whose UR values are not lowest), whereas that for SA was 0.01. The maximum differences were 0.67 and 0.21 for MSA and SA, respectively. While the differences may not be great, MSA successfully acquired more diverse solutions than SA. Although with different parameter configurations, we observed greater degrees of differences, there were too much attentional shifts, exhibiting too much alternation of the dominant and recessive dimensions than empirical data.

Possible reasons for the smaller degree of difference was that the solution diversity in MSA took into account both association and attention weights. We might have achieved a greater degree of diversity in attention allocation, if we only controlled attention distribution patterns. But, there was no clear empirical or theoretical justification for embedding such a cognitive mechanism, and thus we did not test the possibility. Furthermore, more explicit diversity control learning mechanisms, such as Pareto-optimal search methods, would find more diverse effective solutions, we are uncertain about their empirical or theoretical justification for such learning mechanisms as cognitive processes.

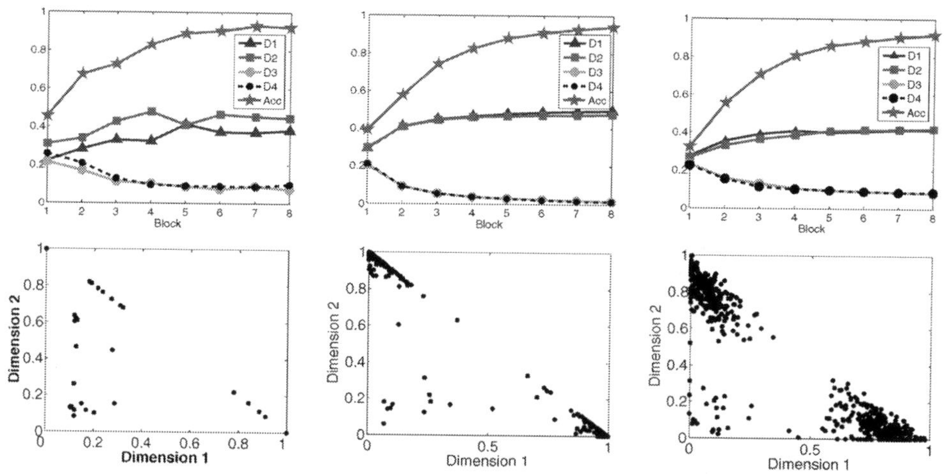

Fig. 1. Left column: Observed empirical results. Middle column: SA, Right Column MSA. The graphs in the top row show classification accuracies and the amounts of relative attention allocated to the four feature dimensions. The scatter plots compare relative attention allocated to Dimensions 1 and 2 for the last 3 blocks for empirical data and the last block for SA and MSA.

4 Conclusion

We introduced a CLEAR framework, which used an evolutionary algorithm as a learning method for computational models of human category learning. CLEAR, unlike gradient descent, uses multiple solutions (e.g., attention and association weights). Each solution can be regarded as a filtering device that determines which aspects of the stored information become salient and which should be ignored. The use of multiple solutions allows CLEAR to have a diverse set of solutions. Because different solutions filter different pieces of information, CLEAR can account for variability across learners and tasks that gradient descent method cannot.

As an initial step, we applied CLEAR to ALCOVE's categorization processes. ALCOVE is an exemplar model of category learning that assumes that humans store every previously encountered example in memory. CLEAR optimized multiple solutions, each of which highlighted different aspects of the stored exemplars. CLEAR was able to correctly replicate observed category learning phenomena that were characterized as multi-objective optimization. In addition, CLEAR was able to acquire diverse solutions during such multi-objective learning processes.

Future Directions. Future research might address how different tasks and goals are linked to different solutions in CLEAR. Research on this issue involves specifying the contextual factor in CLEAR's utility calculation. Human cognitive processing is greatly affected by situational factors, such as goals and motivations, and thus the contextual factor in CLEAR likely plays an important role in modeling human cognition. For example, even when people are learning about the same categories, their knowledge about the categories can vary when they engage in different tasks and have different goals (e.g., [8]). Specifying the contextual factor in CLEAR to incorporate sensitivity to different tasks and goals may allow CLEAR to form ad hoc categories to achieve temporary goals, such as things to sell at a garage sale to dispose of unwanted possessions, like humans do [10].

Final Note. Most previous category learning models have used the gradient descent optimization method, which can be regarded as a normative rather than a descriptive account of human learning processes [4]. Unlike the gradient descent method, human learning varies across individuals and contexts, and is based on factors other than misclassification. The present work is an initial attempt to develop a more descriptive account of human learning by applying a simple evolutionary algorithm. More rigorous research in descriptive models of human learning is indispensable in the structural elucidation of cognitive information flow and, more generally, in the advancement in cognitive science.

Acknowledgements

This research was supported in part by the Office of Naval Research, Grant # N00014-05-1-00632.

References

1. Solomon, K. O., Medin, D. L., Lynch, E.: Concepts Do More Than Categorize. Trends in Cognitive Sciences 3 (1999) 99–105
2. Kruschke, J. K.: ALCOVE: An Exemplar-Based Connectionist Model of Category Learning. Psychological Review 99 (1992) 22–44
3. Love, B. C., Medin, D. L., Gureckis, T. M.: SUSTAIN: A Network Model of Human Category Learning. Psychological Review 111 (2004) 309–332
4. Matsuka, T.: Simple, Individually Unique, and Context-Dependent Learning Method for Models of Human Category Learning. Behavior Research Methods 37 (2005) 240–255
5. Matsuka, T., Corter, J. E.: Process Tracing of Attention Allocation in Category Learning. (2006) Under review.
6. Anderson, R. C., Pichert, J. W: Recall of Previously Unrecallable Information Following a Shift in Perspective. Journal of Verbal Learning and Verbal Behavior 17 (1978) 1–12
7. Eiben, A. E., Smith, J. E: Introduction to Evolutionary Computing, Berlin: Springer-Verlag (2003)
8. Markman, A. B., Baldwin, G. C., Maddox, W. T.: The Interaction of Payoff Structure and Regulatory Focus in Classification. Psychological Science 16 (2005) 852–855
9. Matsuka T., Chouchourelou, A.: A Model of Human Category Learning with Dynamic Multi-Objective Hypotheses Testing with Retrospective Verifications. In *Proc. IJCNN 2006*. (2006) Forthcoming.
10. Barsalou, L. W.: Ad Hoc Categories. Memory and Cognition 11 (1983) 211–227

A Basis for Cognitive Machines

J.G. Taylor, S. Kasderidis, P. Trahanias, and M. Hartley

Dept of Mathematics, King's College London UK
Institute of Computer Science, FORTH, Heraklion, Crete

Abstract. We propose a general attention-based approach to thinking and cognition (more specifically reasoning and planning) in cognitive machines as based on the ability to manipulate neural activity in a virtual manner so as to achieve certain goals; this can then lead to decisions to make movements or to no actions whatever. The basic components are proposed to consist of forward/inverse model motor control pairs in an attention-control architecture, in which buffers are used to achieve sequencing by recurrence of virtual actions and attended states. How this model can apply to various reasoning paradigm will be described and first simulations presented using a virtual robot environment.

1 Introduction

We will consider in this paper reasoning, and planning as advanced components of cognition. We leave out speech since non-linguistic reasoning occurs in non-human animals. We also leave out consciousness/awareness as an important but subtle component of cognition. We note here, however, that consciousness has been proposed as a feature arising out of working memory activations controlled by attention, for example as through the CODAM approach [1], [2]. However although we consider the role of attention we will not pursue it up to consciousness.

Various approaches have been developed for reasoning and planning. Among them we single out three that have recently proved important:

a) Symbolic, using logical inference embedded in linguistic/symbolic structures;
b) Probabilistic, using cognition as defined as probabilistic inference;
c) Connectionist, with numerous discussions of how inference can be obtained from low-level neural network structures. We will discuss here only the third of the above approaches since it fits in most naturally to the neural systems relevant to the consideration of animal cognition, more specifically animal reasoning. Cues on higher cognition from brain-based systems possessed by animals could be important is helping us better understand how such feats are achieved in humans.

The most important components of our approach to cognition will be based on (i) forward models, to be used as a basis of encoding the causality of the world; (ii) working memory modules, for imagining future events when combined with forward models (this process is termed 'prospection' in [3]), (iii) attention control, enabling the selection of only one of numerous distracters in lower level cortices to be evaluated or transformed to enable certain goals to be attained. These components have already been included in the CODAM model [1], [2], although we must be careful to differentiate between a state estimator and a forward model. Here we need to do so: a state estimator is one that estimates the state of the plant being controlled at a given

S. Kollias et al. (Eds.): ICANN 2006, Part I, LNCS 4131, pp. 573 – 582, 2006.

time, whereas a forward model makes a prediction of a future state of the plant. Such a prediction can arise from building a state estimator not for the state now but for that one second (or whenever, after recurrent running of the predictor) time ahead. This would therefore require an efference copy of the control signal to update the state estimate to that for the next time step.

That such forward models occur in the brain in the parallel case of the motor control system has been demonstrated experimentally by numerous authors. For example in [4] it has been shown how adaptation to novel force fields by humans is only explicable in terms of both an inverse controller and a learnable forward model. More recent work has proposed methods by which such forward models can be used in planning (where motor action is inhibited during the running of the forward model) or in developing a model of the actions of another person [5]. Such planning has been analysed in those references and numerous other publications for motor control and actions, but not for more general thinking, especially including reasoning. Nor has the increasingly extensive literature on imagining motor actions been appealed to: it is important to incorporate how motor actions are imagined as taking place on imagined objects, so as to 'reason' as to what will be optimally rewarded possible actions. In order to proceed with this aspect of reasoning we need to extend various aspects of attention control:

1) To create and use forward models under sensory attention (as in the CODAM model);
2) To fuse a sensory attention with a motor attention architecture, as already constructed in one way in [6];
3) To determine how the 'virtual' running of motor actions on stimuli (through imagination) can be used to achieve specific goals (as already suggested in [5]);
4) To extend this to include the influence of reward-value maps on the biasing of actions to determine the optimally rewarded set of actions (as in the actor-critic type of reward model), but now with attention included to handle distracters in complex domains;
5) To attempt to insert some low-level creativity by means of activation of a previously learnt pair of forward and inverse models by associative connections so as to achieve the desired goal, as would occur in arguing by analogy.

We will consider further aspects of these points now.

Under point 1), we have only defined the working memory, the proposed crucial site for imagination, as composed of a buffer site, which functions as an estimate of the attended state of the world inside the head. Where would a forward model for such an estimated state reside in the brain? It produces an estimate of the attended state from that previously obtained and a copy of the action control signal. Functionally it will be expected to be a component of the overall working memory system, so reside in the prefrontal cortex, in the parietal cortex, or in both sites. We have already considered the PFC, as in the ACTION net model [7], as a storage system for recurrent states of the system, so as a TSSG (temporal sequence storage and generation unit). Sequences of states can thus be glued together by the ACTION net, as 'chunks' or schemata (involving sensory/response sequences in general, sewn together possibly by the hippocampus). How can these structures be used to help achieve reasoning/thinking powers?

The schemata of the PFC are intrinsically forward models of sensory/response sequences: starting at the initial state of the sequence, the further ones will be generated sequentially, using if necessary further buffer capacity to become aware of the various states of the sequence, or the final one. Thinking through the consequences of an action on a stimulus could thus be achieved in that way, by running through a given schemata from an initial state. Planning would require some final goal state being active, and comparison with the various states in a sequence generated in the above manner made to check if the goal was yet reached. In the process all response patterns would be inhibited from making any real actions (as observed in over-activity of striatum when imagining motor actions). Various strategies for good planning (such as back-tracking, etc) would need to be used; however these appear only to be second order features of the basic architecture, so will be presently neglected. This overall architecture provides, then, a principled approach to reasoning and planning in neural systems. The corresponding architecture for achieving this is shown schematically in figure 1. We add finally that the forward models provided by the ACTION-network style of the PFC may be an extension of more time-limited forward models in parietal cortex, as proposed in terms of parietal deficits.

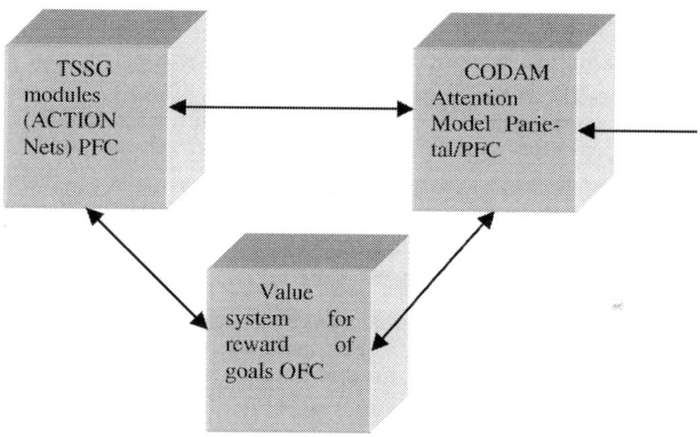

Fig. 1. The ACTION networks contain a system of forward models (fused sensory-and-action based) for running a possible stimulus input forward to a later possible goal state. This is fused with the CODAM attention system, being checked by comparing the produced state to the goal state in a CODAM architecture (with a suitable monitor module), at the same time determining its reward value (by the limbic value reward system). TSSG module = Temporal Sequence Storage and Generation Module.

2 Low-Level Reasoning in Animals

Reasoning is considered by some to be the sole prerogative of humans, but this is now clearly negated by observations on animal reasoning processes. These processes are increasingly well known in chimpanzees, such as in tool use or in imitation [8], [9]. However an ever larger circle of animals appears to be involved in some form of reasoning, such as the corvids (including crows, etc) [3]). Components of intelligent response are observed in the powers of pigeons, for example, to develop classificatory powers for sets of stimuli, such as houses versus trees, or in apes in terms of the results of language learning experiments, such as those involving the chimpanzee Washoe. Even more powerful appears to be the reasoning powers of crows, such as Betty, who can fashion tools appropriate for extracting food from receptacles by means of bending pieces of wire into suitable hooked shapes so as to be able to retrieve the food. We note that corvids (crows, jays, ravens, jackdaws) have excessively large brains for their body weight, at about the same ratio of brain volume/body weight as primates. They also possess similar reasoning powers, especially of a social nature (such as reburying food in a different site if they see that they have been observed by another of their species when caching the food in the first instance). As noted in [3] these reasoning powers are most likely involved in handling complex social interactions, as in the re-caching powers mentioned above or in the ability to construct dominance hierarchies with inheritance powers.

The detailed nature of the mental sub-processes involved in animal reasoning was teased out by [10]. They ran a set of experiments on orang-utans and gorillas. One was to ascertain if they could choose the most appropriate length of stick, placed initially out of their reach, to draw food to themselves using only one of two sticks of different lengths placed in front of them; both species were successful at this task A second experiment used a barrier between the food table and the table for the sticks, so that direct perception could not alone be used to solve the task; again there was success.

To prevent the animals from only selecting the longer tool at every instance, a third experiment was run in which the tools were presented sequentially and not simultaneously, so as to avoid the simple rule of 'always choose the longer tool by direct perceptual comparison', regardless of its appropriateness (in some cases the shorter tool would have done just as well as the longer one in retrieving the food). However the animals could still have used the simple rule 'always choose the longer tool', using a form of working memory to hold the first tool in memory and then comparing with the percept of the second tool, so a fourth experiment added a cost to the longer tool. This was achieved by placing the longer tool on the table flush with the mesh, but at a distance from it so that it could only be retrieved by use of the shorter tool. Finally sequential presentation of the tools was combined with separate views of the reward and the tool table. In that case both species tended to use the safer strategy of going for the longer tool (by using the short one to draw the longer one to them) rather than choosing the length of tool most appropriate for the task. However when offered a tool that was too short to be successful they refused to make any attempt to retrieve the food significantly more times than when the tool was long enough, in conditions when both tool and reward were visible or when they were only available sequentially. We conclude that in this case only a limited understanding of the tools (sticks) was available to the animals.

From [9] and the references therein, there is a development of causal reasoning powers as one proceeds from capuchin monkey -> chimpanzee -> human child. The capuchin monkey is only able to create a simple rule for the 'food in tube' task (where sticks are available to retrieve from form inside a transparent tube, open at each end): R1 'push stick into end of tube farthest from food'. Chimpanzees can develop a more general rule R2: 'push stick into end closer to the trapping tube (created buy joining a vertical tube to the horizontal one) than food'. R2 is more complex since it requires the need to evaluate the relation between the trapping tube and the food before the end to push the stick into can be deduced. From the comments in [9] it would appear that children have a deeper understanding of the nature of a stick and its causal relation to the food and the trapping tube than either monkey or chimpanzee. They may well be able to develop a predictive model of the effect of the stick pushing the food and the effect of the trap on the food. The most complete reasoning approach would be to activate a visual imagery model of the tube, the food and the stick, and use manipulations on that by imagined actions on the stick to imagine where the food will go under various sorts of pushes by the stick. Rules could be formulated from such imagining, although it is more likely, especially in the child's case, that the flexibility of the imagining model does not need a rule-based intermediate step to work out how to gain the food by use of the stick.

3 An Architecture for Reasoning

One form of a neural network model needed to achieve the above causal reasoning by imagery was already given in figure 1. A more specific form (extending [5]) is given in figure 2: This shows the visual input (from parietal lobe) proceeding to an inverse model controller to generate a suitable action to achieve the desired (goal) from the visual state arising from the visual input. There is a corollary discharge of the IMC motor control signal (whose main output will generate motor responses in motor cortex/Cerebellum/spinal chord) fed to a forward model, and so allow a prediction to be made of the next sensory (visual) input. This process can be speeded up by use of prefrontal 'chunking' of sequences of actions (such as from the initial position of the bent stick at the top of the tube to its touching the handle of the food bucket at the bottom of the tube).

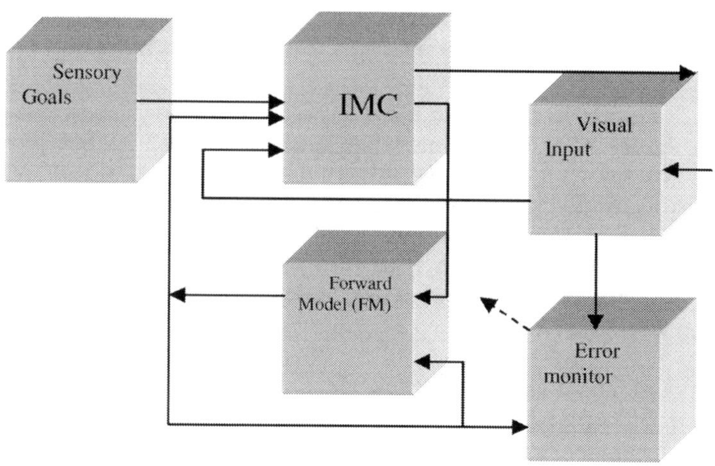

Fig. 2. Basic Architecture for Reasoning. This is a standard motor reasoning system, where the input to the Inverse Model Controller (IMC) in the recurrent mode is solely that arising from the FM from a previous step, with no input being used from the Visual Input module. The output (the dashed arrow) of the error monitor (which compares visual input to that predicted by the FM) can be used to train the FM and IMC. The output from the IMC is sent not only to the FM (as an efference copy) but mainly to the motor system (motor cortex, Cerebellum, basal ganglia, spinal chord) to move the muscles, and thence cause action on the stimulus generating the visual input.

There is no attention control system present in the system of figure 2, so leaving the system vulnerable to distracters. This can be remedied by including a CODAM form of attention control, as shown in figure 3.

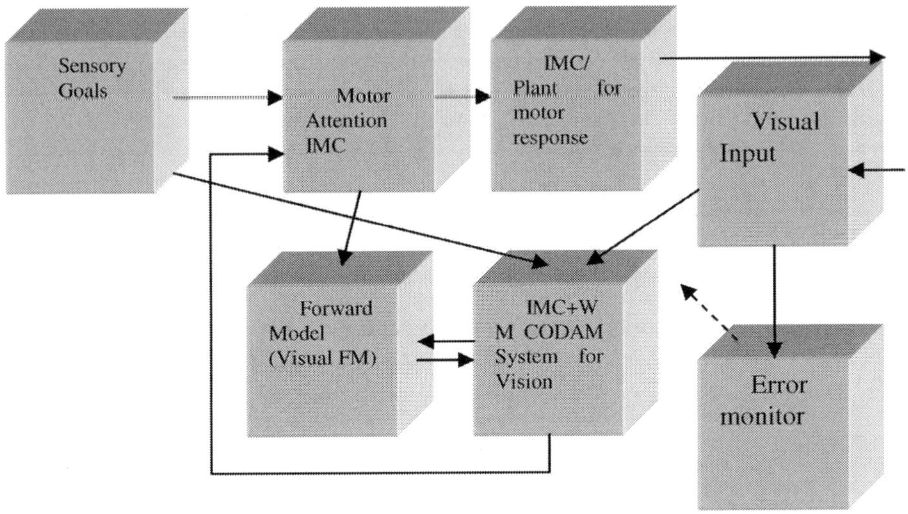

Fig. 3. Attention-controlled Forward/Inverse Model Paris for Reasoning. Note that the visual input to the forward model is assumed to be under the control of the attended visual output, as from the visual working memory (WM), and the FM feeds back to update the WM, as occurs in mental imagery involving transformations of imagined objects. The goals module has been labelled 'visual', to emphasise that it involves the final visual states of objects to be manipulated. See text for further explanation.

The extension from fig 2 to fig 3 is by the addition of two further modules. One of these is a motor attention IMC module, applying an attention signal to the motor 'plant' (consisting of the IMC for motor responses, supposedly sited in motor CX, Cb, BG, etc) to pick out the desired action signal; this motor attention IMC is known present in the left angular gyrus [11], [12]. It uses as input the goal position and the actual visual input from the WM (visual) module in the Visual CODAM system. Thus only the attended visual state is used to guide the motor attention IMC, and is fed back from it to update it.

The other extra module is a set of CODAM-like modules for visual attention, consisting of a visual attention IMC, a WM(visual), a WM(corollary discharge) and an error monitor; this latter is different from the explicit monitor in fig 3 imported from fig 2, which is purely for training the sensory FM and the motor attention IMC. Thus we have a full CODAM-type model for visual processing (so as to avoid visual distracters by the full use of attention, and for which there is experimental evidence for each component) whilst we have taken only a ballistic attention model for motor attention (although there may also be the CODAM-type extensions, these only complicate an already complicated figure).

We note that the goal module in figures 2 and 3 have been taken as that of the final sensory state to be achieved by the movement. This is different from suggestions in [6] for example, where actual motor actions were taken to be represented in prefrontal goal modules. However there is considerable simplification achieved by considering prefrontal goal states only as desired states of external stimuli, such as objects or components of the body. This choice is consistent with the usage of goal states considered in motor planning in [5] and in motor control in [13], who define the goal as a particular desired state of a given external stimulus. This is also consistent with more general remarks on goals in motor control as corresponding to effector final states [14] There is also direct experimental evidence of this assumption [15]. Thus goal states in general are not specific actions themselves, but are in the sensory domain. They generate the requisite actions by use of the control apparatus (motor attention IMC and lower level motor IMC). This is also consistent with coding in the highest areas of PFC which are sensory in content, as compared to motor codes in the lower non-primary motor areas (PMC, SMA, and related areas).

The crucial mode of action of the visual FM is to accept input from the motor attention planner and from the attended visual output of the WM(visual) site, possibly accompanied by the lower level codes for object and feature maps if necessary (these also arise from the CODAM-type visual module in figure 3). These two sets of inputs provide an update of the visual activation by the FM, as a new sensory state arrived at by the action input to the FM. The new sensory state then can lead, in combination with the desired state goal in the goal module to an action produced by the motor IMC that will cause the new visual state to be transformed to the desired goal state if the specific action is taken. The resulting motor action is then, in the reasoning mode, not taken in actuality but is fed back to the FM (visual), to be either used for further updating of the FM (as in continued planning or imagining of a sequence of visual stimuli) or (in the acting mode) to provide a specific external motor action by updating the IMC (motor attention). Also, in the reasoning mode there is associated inhibition of the lower level motor apparatus when the overall motor attention IMC and the visual FM are used in reasoning (imaging sequences of actions on sequences of visual stimuli).

From several lines of evidence the visual FM of figure 3 may be regarded as a point of contact between the two CODAM networks, one for motor attention and the other for visual attention, that have been proposed in [6], [16]. It has been suggested as being in the posterior parietal lobe and/or cerebellum [13].

4 Learning the FM/IMC Pairs

The main question we have not yet discussed is as to the creation of the FM/IMC pairs used above. We take to heart the criticism of the standard use of feedback-error-learning, and the related training of the IMC so that it becomes so accurate that the IMC is all that is needed in control and the feedback used by the FM becomes superfluous [17], [18]. In particular the results of [17] show that this is not the case in human tracking data, where loss of sensory feedback shows up the errors in the IMC, which are especially considerable at low frequency, whereas it would be expected by other approaches that there would be no such error present.

The method used by [17] was to construct a good model of the FM by training on the error it produces, by comparing sensory prediction with actual sensory feedback, whilst using an approximate version of the actual IMC as a linear IMC (as a PID controller, initially with high gain). The same approach can be used in the construction of the FM/IMC pairs needed above. Thus the IMC for moving a stick to a point would use a linear IMC, with initial and final end points of the stick leading to an approximate motor control to move there. It would have a non-linear FM, trained on the error in the final stick end-point. The FM in a number of reasoning cases can be created by using the IMC to generate sensory effects of the response and then using this sensory feedback to create the error in reaching the goal set in the IMC, and thence to use error- based learning to achieve a suitably trained FM (with no resulting error) . This is shown in figure 2, this being a general architecture for achieving effective learning of the FM, with some parameter modification of the IMC after the FM has been created, as suggested in [17] is possible to relate the architecture of figure 2 to that of the Cerebellum, which is supposedly the most-well studied error learning system in the brain [19].

There is presently data that supports the existence of the FM/IMC pair in motor attention. In particular we can make the identification that the motor attention IMC is in the supramarginal gyrus (SMG) in the inferior parietal lobe, especially in the left hemisphere, and composed of areas BA7b and AIP (in the intrapareital sulcus); the motor attention FM is in the Parietal Reach region (PRR) in the SPL and IPS. The evidence for the placing of the IMC(am) as above is given in [11], [12], and arises from data on the deficit caused by TMS applied to SMG(L) for incorrectly cued stimuli in the motor Posner paradigm. The evidence for FM in the PRR is of a variety of sorts, but the strongest seems to be that for the updating of the body map by tool use, and resulting difficulties with tools in the case of damage to the relevant PRR area [12]. More generally the presence of forward (state updating) models in the IPS for both eye movements, attention orienting and motor movement preparations are supported by predictive updating of stimulus responses in single neurons in these regions.

At the same time there are good connections observed between SMG and PMCx, relevant for the movement selection processing known to be carried out by the PMCx, and thereby helped in the plan selection by SMG (as required for an IMC-type of module), whereas the PRR is better connected to visual and somato-sensory inputs, as well as those from the IMC in SMG [20] Finally we conclude that the IMC(am) acts on the PMC, itself identified with the IMC(m) used in more general unattended motor control, but guided for by the specific motor attention control signal from the IMC(am) to which the PMC is well-connected [20].

We note that the FM carries a map of the body of the animal/system it represents. This is now known to be used in tool use by animals, there being an extension of the body map when a tool is learnt to be used; thus on using a hammer to drive in a nail, the user does not still hit the nail with their hand/extremity, but with the hammer as an extension of their hand. This extension (or addition of a new FM) to the old FM needs careful modeling for GNOSYS.

5 The Simulation

We have developed a virtual robot, through the webots simulator [21 so as to enable a software simulation of the above architectures (from fig 2 through ultimately to fig 4). We have designed a software version of the forward and backward models for the control of a gripper designed for the webot. This gripper has 3 DoFs, denoted by θ1, θ2, θ3. The relation between these and the 3-d co-ordinates of the end of the gripper r are known algebraically, and the corresponding forward and inverse model controller in fig 2 can be designated as

$$r' = F(r, \theta) \tag{1a}$$

$$\theta = G(r', r) \tag{1b}$$

where r and θ in (1a) are the start 3-d position and the angle of change of the gripper to determine the new 3-d position r', and r, r' are the original and desired 3-d position of the end of the gripper to be achieved by the turn angle θ in (1b).

The pair of forward and inverse models of equation (1a), (1b), together with information on the visual input (such as from a hierarchical architecture like that of GNOSYS) fills in all of the modules of figure 2 except for the monitor (which would be training an artificial neural network version of the pair of models). The overall architecture can be extended by adding a working memory buffer, as in figure 3, to allow for recurrence in the sequences INPUT -> IMC ->FM -> WMbuffer ->IMC ->.& so recurrently until the goal state in the IMC is attained and the sequence stops (where the WMbuffer is needed to hold the previous output of the forward model till it is updated by the next output from the forward model). This system can be used for reasoning about use of a stick, for example (in the case of chimpanzees or crows) if there is a set of (2L) nodes, one for each length L observed for a stick, where L runs over a suitable discrete set. The action of the node 2L, when active, is: GOTO stick of length L; grasp stick; change forward/inverse model pairs to hose for a gripper with extra length L. This process allows reasoning about use of the extended gripper in dragging food reward towards itself. Thus it can run through mentally such manipulations to determine if a particular stick would be successful to help it obtain the food.

The purpose of this simulation is to be able to test the paradigm described earlier [10] in which primates must sequentially use different lengths of stick to be able to attain a food reward. The system may then also be extended to further reasoning paradigms.

6 Conclusions

It is seen that recent ideas in animal reasoning are very relevant to guide how we may develop primitive reasoning powers in the GNOSYS software. The four basic components in animal reasoning proposed in [3] of causality, imagination, prospection and flexibility are all developed as part of the CODAM attention control system and used in the above suggested paradigms: causality in terms of forward models (for example used to run through how a stick of a suitable length would or would not be able to retrieve the distant reward), imagination (in terms of buffer working memory and its activation by imagined situations as part of running through the forward model), prospection (in terms of using the forward model in conjunction with the buffer, as just mentioned) and flexibility (in terms of rapid re-activation of past memories or rapid encoding for use of new ones relevant to the situation).

The approach to training the basic FM/IMC pairs of internal models is developed using the approach of [17] were the FM is non-linear and trained using an approximate IMC; at a later stage of usage the parameters in the IMC can be modified to give more accurate response.

References

1. Taylor JG (2003) Paying Attention to Consciousness. Progress in Neurobiology 71:305-335
2. Taylor JG (2005) From Matter to Consciousness: Towards a Final Solution? Physics of Life Reviews 2:1-44
3. Emery NJ & Clayton NS (2004) The Mentality of Crows: Convergent Evolution of Intelligence in Corvids and Apes. Science 306: 1903-1907
4. Bhushan N & Shadmehr R (1999) Computational nature of human adaptive control during learning of reaching movements in force fields. Biol Cybern, 81:39
5. Oztop et al (2005) Mental state inference using visual control parameters. Brain Res Cogn Brain Res 22:129
6. Taylor JG & Fragopanagos N (2003) Simulations of Attention Control Models in Sensory and Motor Paradigms.Proc ICANN03, Istanbul
7. Taylor NR & Taylor JG (2000) Hard-wired models of working memory and temporal sequence storage and generation. Neural Netw. 12:201
8. McGrew WC (1992) Chimpanzee Material Culture. Cambridge: Cambridge University Press.
9. Boysen ST & Himes GT (1999) Current Issues and Emerging Theories in Animal Cognition. Annual Reviews of Psychology 50:683-705
10. Mulcahy NJ, Call J & Dunbar RIM (2005) Gorillas and Orang Utans Encode Relevant Problem Features in a Tool Using Task. Journal of Comparative Psychology 119:23-32
11. Rushworth MFS, Ellison A & Walsh V (2001) Complementary localization and lateralization of orienting and motor attention. Nature Neuroscience 4(6):656-661
12. Rushworth MFS, Johansen-Berg H, Gobel SM & Devlin JT (2003): The left parietal and premotor cortices: motor attention and selection. NeuroImage 20:S89-S100
13. Desmurget M, Grafton S (2000): Forward modeling allows feedback control for fast reaching movements. Trends Cogn Sci 4:423
14. Wise SP & Shadmehr R (2002) Motor Control. Vol 3:1-21 in Encyclopedia of the Brain: USA: Elsevier.
15. Morasso, P (1981) Spatial control of arm movements. Experimental Brain Research, 42, 223-227.
16. Taylor JG & Fragopanagos N (2004) Modelling Human Attention and Emotions. Proc IJCNN04, Budapest
17. Davidson PR, Jones RD, Andreae JH & Sirisena HR (2002) Simulating Closed and Open-Loop Voluntary Movement: A Nonlinear Control-Systems Approach. IEEE Trans Biomedical Engineering 49:1242-1252
18. Neilson PD & Neilson MD (1999) A neuroengineering solution to the optimal tracking problem. Human Movement Science 18:155-183
19. Ohyama T, Nores WL, Murphy M & Mauk MD (2003) What the cerebellum computes. Trends in Neuroscience 26(4):222-6
20. Rozzi S, Calzavara R, Belmalih A, Borra E, Gregoriou GG, Matelli M & Luppino G (2005) Cortical Connections of the Parietal Cortical Convexity of the Macaque Monkey. Cerebral Cortex (Nov 23, 2005)
21. Webots. http://www.cyberbotics.com. Commercial Mobile Robot Simulation Software

Neural Model of Dopaminergic Control of Arm Movements in Parkinson's Disease Bradykinesia

Vassilis Cutsuridis

Computational Intelligence Laboratory, Institute of Informatics and Telecommunications, National Center for Scientific Research "Demokritos", Agia Paraskevi, Athens GR-15310
vcut@iit.demokritos.gr

Abstract. Patients suffering from Parkinson's disease display a number of symptoms such a resting tremor, bradykinesia, etc. Bradykinesia is the hallmark and most disabling symptom of Parkinson's disease (PD). Herein, a basal ganglia-cortico-spinal circuit for the control of voluntary arm movements in PD bradykinesia is extended by incorporating DAergic innervation of cells in the cortical and spinal components of the circuit. The resultant model simulates successfully several of the main reported effects of DA depletion on neuronal, electromyographic and movement parameters of PD bradykinesia.

1 Introduction

The most severe symptom of Parkinson's disease is bradykinesia (i.e. slowness of movement). It is not known what causes bradykinesia because there are many pathways from the sites of neuronal degeneration to the muscles (see Figure 1). The most important pathways are: (1) the pathway from the substantia nigra pars compacta (SNc) and the ventral tegmental area (VTA) to the striatum and from the striatum to the substantia nigra pars reticulata (SNr) and the globus pallidus internal segment (GPi) and from there to the thalamus and the frontal cortex, (2) the pathway from the SNc and the VTA to the striatum and from the striatum to the SNr and the GPi and from there to the brainstem, and (3) the pathway from the SNc/VTA to cortical areas such as the supplementary motor area (SMA), the parietal cortex, and the primary motor cortex (M1), and from there to the spinal cord. [12]

The currently accepted view of what causes bradykinesia is that cortical motor centers are not activated sufficiently the basal ganglia (BG) circuits. As a result, inadequate facilitation is provided to motor cortical and spinal neuron pools and hence movements are small and weak [1]. The implication of this view is that cells in the cortex and spinal cord are functioning normally. This paper suggests otherwise.

In this paper, I integrate experimental data on the anatomy and neurophysiology of the globus pallidus internal segment [19], the cortex [10] and the spinal cord structures, as well as data on PD psychophysics [16, 17, 18, 20] to extend a neural model of basal ganglia–cortex–spinal cord interactions during movement production [2, 3, 4, 5, 6, 7, 8]. Computer simulations show that disruptions of the BG output and of the SNc's DA input to frontal and parietal cortices and spinal cord may be responsible for delayed movement initiation. The main hypothesis of the model is that

S. Kollias et al. (Eds.): ICANN 2006, Part I, LNCS 4131, pp. 583–591, 2006.

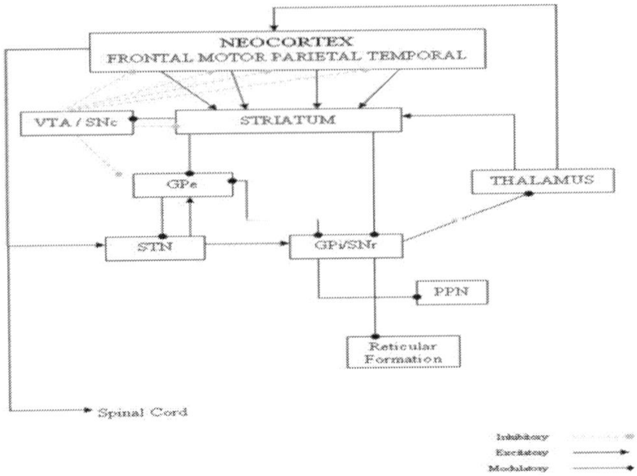

Fig. 1. Schematic diagram of dopaminergic innervation of basal ganglia and sensory-motor cortex. Arrow-ending solid lines, excitatory projections; Dot-ending solid lines, inhibitory projections; Diamond-ending dotted lines, dopamine (DA) modulatory projections; STN, subthalamic nucleus; GPi, globus pallidus internal segment; GPe, globus pallidus external segment; SNr, substantia nigra pars reticulata; SNc, substantia nigra pars compacta; VTA, ventral tegmental area; PPN, pedunculopontine nucleus.

elimination of DA modulation from the SNc disrupts, via several pathways, the build-up of the pattern of movement-related responses in the primary motor and parietal cortex, and results in a loss of directional specificity of reciprocal and bidirectional cells in the motor cortex as well as in a reduction in their activities and their rates of change. These changes result in delays in recruiting the appropriate level of muscle force sufficiently fast and in an inappropriate scaling of the dynamic muscle force to the movement parameters. A repetitive triphasic pattern of muscle activation is sometimes needed to complete the movement. All of these result in an increase of mean reaction time and a slowness of movement (i.e. bradykinesia). This work has been published in [12, 13, 14, 15].

2 Materials and Methods

2.1 Basis of the Model

Figure 2 schematizes the basal ganglio-cortico-spinal network model. As a basal ganglio-cortical network, the VITE (Vector Integration To-End point) model of [3] was chosen, which is here extended. In my proposed version of the VITE model, the types and properties of the cortically identified neurons are extended and the effects of dopamine depletion on key cortical cellular sites are studied. Briefly in the model, an arm movement difference vector (DV) is computed in parietal area 5 from a comparison of a target position vector (TPV) with a representation of the current position called perceived position vector (PPV). The DV signal then projects to area

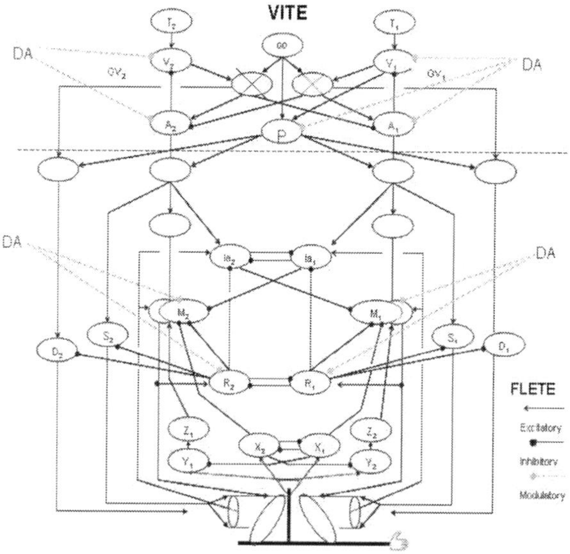

Fig. 2. Neural network representation of the cortico-spinal control system. (Top) The VITE model for variable-speed trajectory generation. (Bottom) the FLETE model of the opponent processing spinomuscular system. Arrow lines, excitatory projections; solid-dot lines, inhibitory projections; diamond dashed lines, dopamine modulatory inputs; dotted arrow lines, feedback pathways from sensors embedded in muscles; DA, dopamine modulatory signal; GO, basal ganglia output signal; P, bi-directional co-contractive signal; T, target position command; V, DV activity; GV, DVV activity; A, current position command; M, alpha motoneuronal (MN) activity; R, renshaw cell activity; X, Y, Z, spinal inhibitory interneuron (IN) activities; Ia, spinal type a inhibitory IN activity; S, static gamma MN activity; D, dynamic gamma MN activity; 1,2, antagonist cell pair (adapted from [12]).

4, where a desired velocity vector (DVV) and a non-specific co-contractive signal (P) [9] are formed. A voluntarily scalable GO signal multiplies (i.e. gates) the DV input to both the DVV and P in area 4, and thus volitional-sensitive velocity and non-specific co-contractive commands are generated, which activate the lower spinal centers. In my model, the DVV signal represents the activity of reciprocal neurons [10], and it is organized for the reciprocal activation of antagonist muscles, whereas the P signal represents the activity of bidirectional neurons (i.e. neurons whose activity decreases or increases for both directions of movement [10]), and it is organized for the co-contraction of antagonist muscles.

The spinal recipient of my model is the FLETE (Factorization of LEngth and Tension) model [2, 3, 4, 5, 6]. Briefly, the FLETE model is an opponent processing muscle control model of how spinal circuits afford independent voluntary control of joint stiffness and joint position. It incorporates second-order dynamics, which play a large role in realistic limb movements. I extended the original FLETE model by incorporating the effect of the now cortically controlled co-contractive signal (in the original FLETE model, the co-contraction signal was simply a parameter) onto its spinal elements. Finally, I studied the effects that dopamine depletion on key spinal centers has on voluntary movements.

2.2 Architecture

The mathematical formalism of the basal ganglio-cortico-spinal model has been described elsewhere [12].

3 Results

3.1 Dopamine Depletion Effects on the Discharge of Globus Pallidus Internal Segment Neurons

Figure 3 shows a qualitative comparison of abnormal cellular responses of GPi neurons to striatal stimulation in MPTP-treated monkeys [19] and simulated long duration of late inhibitions (B) and oscillatory (D) GPi neuronal responses. I propose that GPi responses similar to figure 3B are used to complete small amplitude

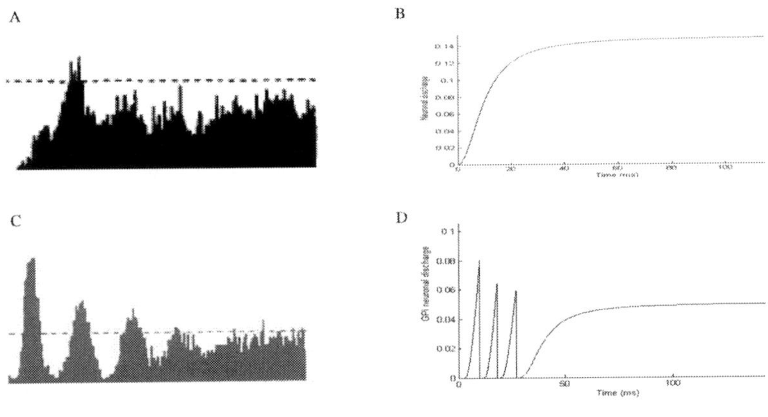

Fig. 3. Comparison of (A) peristimulus histograms (PSTH) of neuronal responses of GPi cells to striatal stimulation in MPTP-treated monkeys (adapted from Tremblay et al., 1989, Fig. 2, p. 23), (B) simulated dopamine depleted GPi neuronal response, (C) peristimulus histograms (PSTH) of abnormal oscillatory responses of GPi neurons to striatal stimulation in MPTP-treated monkeys (adapted from Tremblay et al., 1989, Fig. 2, p. 23), and (D) simulated oscillatory disrupted GPi responses. Time units in ms.

movements, whereas GPi responses similar to figure 3D are used to complete large amplitude movements.

3.2 Dopamine Depletion Effects on the Discharge of Cells in the Primary Motor and Posterior Parietal Cortices

Figure 4 depicts a composite schematic of simulated discharges of neurons in the primary motor and posterior parietal cortices in normal and dopamine depleted conditions. An increase in baseline activity of area's 4 bidirectional and reciprocal cells is evident as it has been observed experimentally [10]. Also, a significant reduction of peak activity of the bidirectional and reciprocal cells in the dopamine depleted case [10]. Finally, a disinhibition of reciprocally activated cells is observed in column 2 of figure 4D.

3.3 Dopamine Depletion Effects on the Discharge of Cells in the Spinal Cord

Figure 5 depicts a qualitative comparison of alpha-MN activity in normal and dopamine depleted conditions for small and large amplitude movements. In the dopamine depleted case, a significant decrease in the peak alpha-MN activity of both agonist and antagonist muscles is observed [20]. What is also evident is the absence of a co-contractive activation of antagonist muscles as it has been observed in monkey stimulation studies [21], but not in human studies [20].

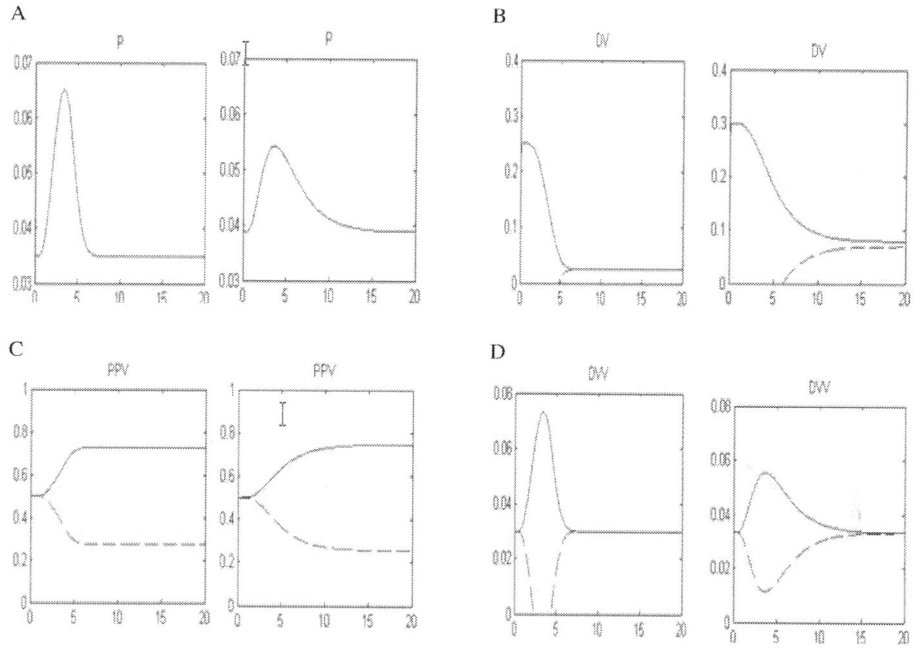

Fig. 4. Model dopamine depleted cell responses of primary motor and posterior parietal cortices in normal (column 1 of A, B, C, and D) and dopamine depleted (column 2 of A, B, C, and D) conditions. P: bidirectional neuronal response; DV: posterior area 5 phasic cell response; PPV: area 4 tonic cell response; DVV: area 4 reciprocal (phasic) cell response. Time (x-axis) in ms.

Also, a repetitive biphasic agonist-antagonist muscle activation can be observed as in [22]. The GO signal used to simulate such repetitive muscle activation is the one from figure 3B.

3.4 Dopamine Depletion Effects on Movement Variables

Figure 6 depicts the position, velocity and force profiles produced in the dopamine depleted condition of a large-amplitude movement. The GO signal used for these

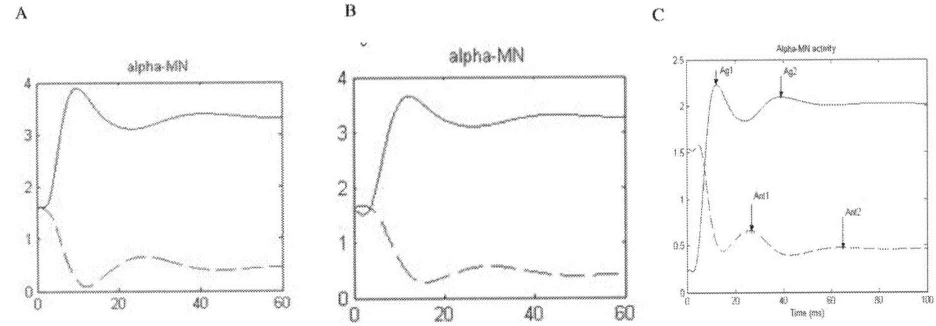

Fig. 5. Comparison of (A) simulated triphasic alpha motorneuronal (MN) activation under normal conditions, (B) simulated disrupted a-MN activation under dopamine depleted conditions, and (C) simulated repetitive biphasic a-MN activity in a dopamine depleted movement. The agonist a-MN activity is scaled down by 1.6 marks, so that a clearer repetitive biphasic pattern of muscle activation is shown. The GO signal used in (B) is the same as in figure 3B, whereas the GO signal used in (C) is the same as in figure 6A. Time (x-axis) in ms.

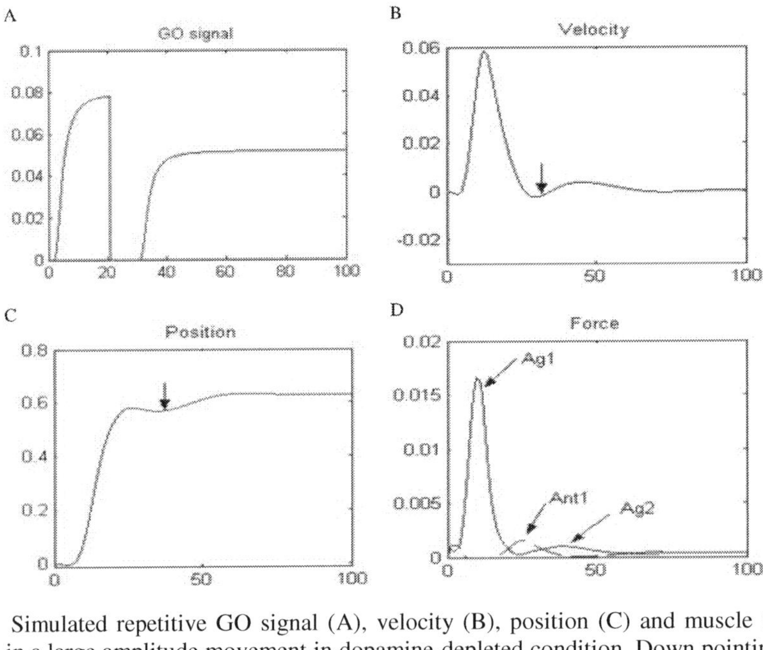

Fig. 6. Simulated repetitive GO signal (A), velocity (B), position (C) and muscle force (D) profiles in a large amplitude movement in dopamine-depleted condition. Down pointing arrows indicate sub-movements needed to complete the movement.

simulated is shown in figure 6A. They are evident the two sub-movements required to complete the movement. Figure 7 depicts the position, velocity and force profiles produced in the dopamine depleted condition of a small-amplitude movement. The GO signal used for these simulated is shown in figure 7A. For this movement a single motor command produced by the DVV cells is needed to complete it.

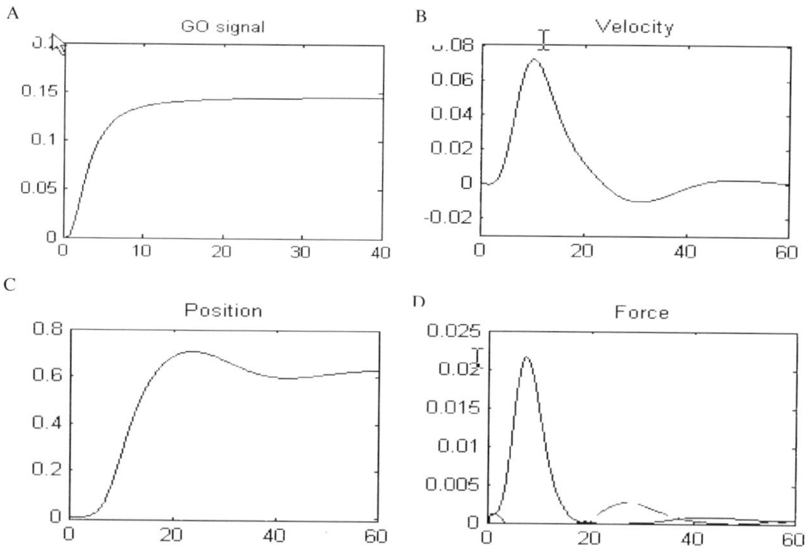

Fig. 7. Simulated GO signal (A), velocity (B), position (C) and muscle force (D) profiles in a small amplitude movement in dopamine-depleted condition

4 Conclusion

The present model is a model of voluntary movement and proprioception that offers an integrated interpretation of the functional roles of the diverse cell types in movement related areas of the primate cortex. The model is based on known cortico-spinal neuroanatomical connectivity (see Tables 1 and 2 of [8]). The model is successful at providing an integrative perspective on cortico-spinal control of parkinsonian voluntary movement by studying the effects of dopamine depletion on the output of the basal ganglia, cortex and spinal cord. It can account for the many known empirical signatures of Parkinsonian willful action such as

- Reduction of firing intensity and firing rate of cells in primary motor cortex Abnormal oscillatory GPi response
- Disinhibition of reciprocally tuned cells
- Repetitive bursts of muscle activation
- Reduction in the size and rate of development of the first agonist burst of EMG activity
- Asymmetric increase in the time-to-peak and deceleration time
- Decrease in the peak value of the velocity trace
- Increase in movement duration
- Substantial reduction in the size and rate of development of muscle production

These findings provide enough evidence to support the main hypothesis of the model reported earlier in the paper. A much larger set of experimental evidence that the model successfully simulated are shown in [12].

References

1. Albin, R. L., Young, A. B., & Penney, J. B.: The functional anatomy of basal ganglia disorders. Trends in Neurosciences. 12 (1989) 366–375
2. Bullock, D., & Contreras-Vidal, J. L.: How spinal neural networks reduce discrepancies between motor intention and motor realization. In K. Newel, & D. Corcos (Eds.), Variability and motor control (pp. 183–221). Champaign, IL: Human Kinetics Press, 1993.
3. Bullock, D., Grossberg, S.: Neural dynamics of planned arm movements: Emergent invariants and speed-accuracy properties during trajectory formation. Psychological Review. 95 (1988) 49–90.
4. Bullock, D., Grossberg, S.: VITE and FLETE: Neural modules for trajectory formation and tension control. In W. Hershberger (Ed.), Volitional action (pp. 253–297). Amsterdam, The Netherlands: North-Holland, 1992.
5. Bullock, D., Grossberg, S.: Adaptive neural networks for control of movement trajectories invariant under speed and force rescaling. Human Movement Science. 10 (1991) 3–53.
6. Bullock, D., Grossberg, S.: Emergence of triphasic muscle activation from the nonlinear interactions of central and spinal neural networks circuits. Human Movement Science. 11 (1992) 157–167.
7. Bullock, D., Cisek, P., Grossberg, S.: Cortical networks for control of voluntary arm movements under variable force conditions. Cerebral Cortex. 8 (1998) 48–62.
8. Contreras-Vidal, J. L., Grossberg, S., Bullock, D.: A neural model of cerebellar learning for arm movement control: Cortico-spino-cerebellar dynamics. Learning and Memory. 3(6) (1997) 475–502.
9. Humphrey, D. R., & Reed, D. J.: Separate cortical systems for control of joint movement and joint stiffness: Reciprocal activation and coactivation of antagonist muscles. In J. E. Desmedt (Ed.), Motor control mechanisms in health and disease. New York: Raven Press, 1983.
10. Doudet, D. J., Gross, C., Arluison, M., Bioulac, B.: Modifications of precentral cortex discharge and EMG activity in monkeys with MPTP induced lesions of DA nigral lesions. Experimental Brain Research. 80 (1990) 177–188.
11. Dormand, J. R., & Prince, P. J.: A family of embedded Runge-Kutta formulae. Journal of Computational and Applied Mathematics. 6 (1980) 19–26.
12. Cutsuridis, V., Perantonis, S. (in press): A Neural Model of Parkinson's Disease Bradykinesia. Neural Networks.
13. Cutsuridis, V.: A neural network model of normal and Parkinsonian EMG activity of fast arm movements. Book of abstracts of the 18 Conference of Hellenic Society for Neuroscience, Athens, Greece, October 17-19, 2003.
14. Cutsuridis, V., Bullock, D.: A Neural Circuit Model of the Effects of Cortical Dopamine Depletion on Task-Related Discharge Patterns of Cells in the Primary Motor Cortex. Rethymnon, Crete, Book of abstracts of the 17th Conference of Hellenic Society for Neuroscience, Poster 3, p. 39, October 4-6, 2002.
15. Cutsuridis, V., Bullock, D.: A Neural Circuit Model of the Effects of Cortical Dopamine Depletion on Task-Related Discharge Patterns of Cells in the Primary Motor Cortex. Poster Session II: Sensory-Motor Control and Robotics, Book of abstracts of the 6th International Neural Network Conference, Boston, MA, May 30 - June 1, 2002.
16. Stelmach, G.E., Teasdale, N., Phillips, J., Worringham, C.J.: Force production characteristics in Parkinson's disease. Exp Brain Res. 76 (1989) 165-172.
17. Rand, M.K., Stelmach, G.E., Bloedel, J.R.: Movement Accuracy Constraints in Parkinson's Disease Patients. Neuropsychologia. 38 (2000) 203-212.

18. Camarata, P.J., Parker, P.G., Park, S.K., Haines, S.J., Turner, D.A., Chae, H., Ebner, T.J.: Effects of MPTP induced hemiparkinsonism on the kinematics of a two-dimensional, multi-joint arm movement in the rhesus monkey. Neuroscience. 48(3) (1992) 607-619.
19. Tremblay, L., Filion, M., & Bedard, P. J.: Responses of pallidal neurons to striatal stimulation in monkeys with MPTP-induced parkinsonism. Brain Research. 498(1) (1989) 17–33.
20. Godaux, E., Koulischer, D., & Jacquy, J.: Parkinsonian bradykinesia is due to depression in the rate of rise of muscle activity. Annals of Neurology. 31(1) (1992) 93–100.
21. Benazzouz, A., Gross, C., Dupont, J., Bioulac, B.: MPTP induced hemiparkinsonism in monkeys: Behavioral, mechanographic, electromyographic and immunohistochemical studies. Experimental Brain Research. 90 (1992) 116–120.
22. Hallett, M., Khoshbin, S.: A physiological mechanism of bradykinesia. Brain. 103 (1980) 301–314.

Occlusion, Attention and Object Representations

Neill R. Taylor[1], Christo Panchev[2], Matthew Hartley[1],
Stathis Kasderidis[3], and John G. Taylor[1]

[1] King's College London, Department of Mathematics, The Strand,
London, WC2R 2LS, U.K.
{neill.taylor, john.g.taylor}@kcl.ac.uk,
mhartley@mth.kcl.ac.uk
[2] Sunderland University, School of Computing and Technology, St. Peter's Campus,
Sunderland SR6 0DD, U.K.
christo.panchev@sunderland.ac.uk
[3] Foundation for Research & Technology Hellas, Institute of Computer Science,
Vassilika Vouton, 71110 Heraklion, Greece
stathis@ics.forth.gr

Abstract. Occlusion is currently at the centre of analysis in machine vision. We present an approach to it that uses attention feedback to an occluded object to obtain its correct recognition. Various simulations are performed using a hierarchical visual attention feedback system, based on contrast gain (which we discuss as to its relation to possible hallucinations that could be caused by feedback). We then discuss implications of our results for object representations per se.

1 Introduction

In any complex visual scene there will be many occluded objects. It has proved a challenging problem for machine vision to see how occluded shapes can be segmented and identified in such scenes [1], though has been dealt with by a number of models notably the Neocognitron [2]. Here we consider the use of attention as a technique to help the recognition problem. We do that on the basis of a hierarchical set of neural modules, as is known to occur in the brain. Brain guidance of this form has been much used in the past to deduce pyramidal vision architectures. We add to that hierarchy an additional processing system in the brain, that of attention. This is also well known to act as a filter on inputs, with those stimuli succeeding to get through the filter being singled out for further high-level processing involving prefrontal cortical executive functions.

In this paper we propose to employ both hints from the brain: hierarchy and attention feedback, to help resolve the problem of occlusion. In particular we will show that attention helps to make an object representation of the occluded stimulus more separate from that of the occluder, so helping improve the recognition of the occluded object. At the same time the position of the occluded object can be better specified. We also show that the process of attending to an occluded object does not cause it to become a hallucination, so that it regains all of its components in the internal image of the stimulus in the brain. This is achieved partly by an approach using attention feedback defined as contrast gain on the inputs to a neuron from the attended stimulus.

S. Kollias et al. (Eds.): ICANN 2006, Part I, LNCS 4131, pp. 592–601, 2006.

In that way inputs that are zero (due to occlusion) will not become increased at all, whereas special constraints would have to be imposed to prevent other sorts of feedback from eliciting posterior activation of the whole of the encoded occluded stimulus.

The paper starts in the next section with a description of the hierarchical neural architecture. It continues in section 3 with a description of the results obtained on simulating an occluded shape by another different one and the activations in the network arising from attention to the stimulus or elsewhere. In section 4 we use these results to characterize object representations in the brain. Section 5 is a concluding discussion section.

2 Visual Architecture

The visual system in the brain is hierarchical in nature. We model that by a sequence of neural network modules forming two hierarchies, one dorsal for spatial representations and one ventral for object representations. Both of these are well known to occur in the brain. We take a simplified model composed of leaky-integrate-and-fire neurons, with the hierarchies as shown in fig. 1, table 1 shows the sizes of all modules.

The model is composed of the ventral 'what' stream and the dorsal 'where' stream. The ventral stream is trained from V2 upwards in a cumulative process, whilst the dorsal stream is hard-wired. The retinal input undergoes edge detection and orientation detection (4 orientations 0°, 45°, 90°, 135°) to provide the input to the ventral stream lateral geniculate nucleus (LGN). Above LGN all regions are composed of a topographically related excitatory layer and an inhibitory layer of neurons that are reciprocally connected, both layers receive excitatory input from other regions; and there are lateral excitatory connections in the excitatory layer.

The ventral 'what' stream via hard-wired and trained connections has a progression of combining object features from oriented bars, to angles formed by 2 bars, to arc segments composed of 3 and 4 bars and finally to objects. At each level there is a loss of spatial information due to the reduction in layer size until at the modelled anterior part of the inferotemporal cortex (area TE) and above representations are spatially invariant. The ventral primary visual cortex (V1 ventral) is composed of 4 excitatory layers, one for each orientation, which are interdigitated in 2-dimensions to provide a pin-wheel structure such that neurons responding to the 4 orientations for the same spatial position are grouped together. V2 is known to preferentially respond to angles formed by pairs of oriented bars [3], so the model V2 is trained using guided spike-time dependent plasticity (STDP) to give similar responses. More specifically, V2 is trained on pairs of bars forming single angles that are present in the objects (square, triangle and circle), and for each single pattern at the input only the neurons assigned to represent that angle are allowed to fire and thereby adapt to the stimulus. V4 receives input from V1 and V2; those to the excitatory neurons are trained using STDP.

Experimental results [4] indicate that V4 responds preferentially to arc segments composed of 3-4 lines; we train V4 on a random selection of length 5 arcs from our object group (square, triangle and circle). TEO (within posterior inferotemporal cortex) receives inputs from V2 and V4; only those connections to excitatory TEO

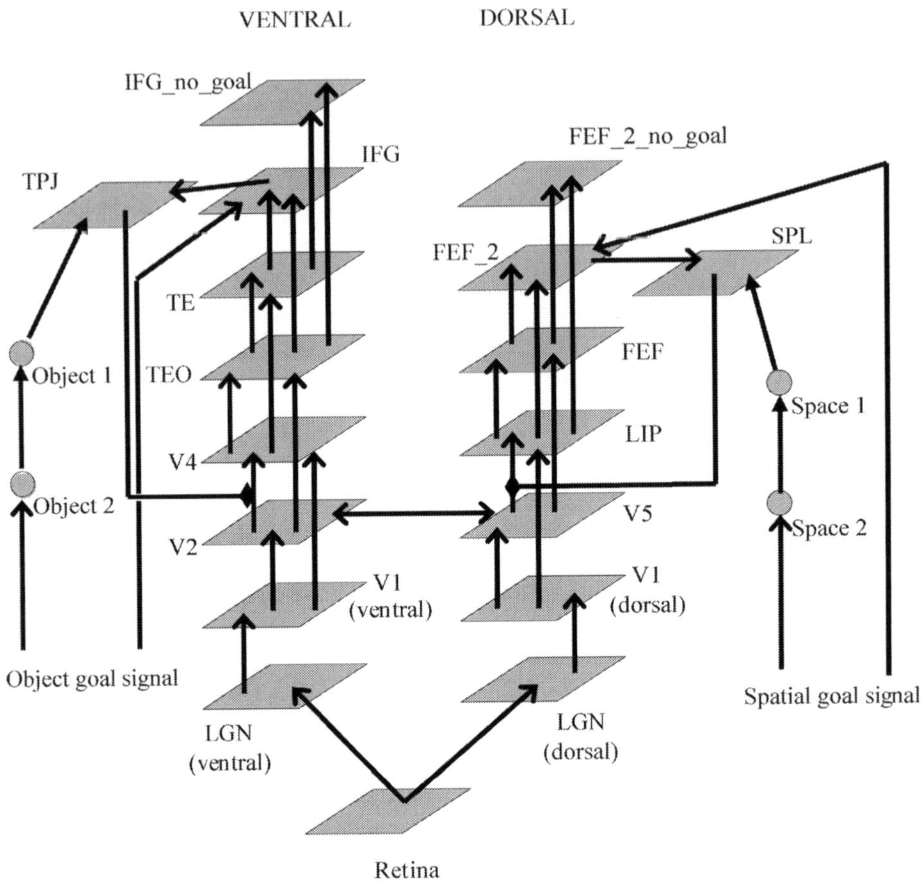

Fig. 1. Hierarchical Neural Model Architecture. Open arrow heads indicate excitatory connections, closed-arrow heads are inhibitory connections, and closed-diamond connections indicate sigma-pi weights. We only show the general hierarchical structure, the internal structure of each region is described in the text.

neurons undergo learning. Inputs are complete objects (square, triangle and circle) and use the previously trained V4. TE is also trained on complete objects, with inputs from V4 and TEO. The inferior frontal gyrus (IFG) modules, IFG and IFG_no_goal, are exactly the same and composed of 3 neurons, each one of which represents one of our objects and is hardwired to the TE neurons that preferentially respond to that object. The IFG module can receive inputs from an object goal site such that the system can have its attention directed to a particular object group, where the goal is currently externally determined. This attentional process occurs through sigma-pi weights from temporal–parietal junction (TPJ), the generator of attention movement control signals, onto the V2→V4 (excitatory nodes) connections. Normally when there is no object goal TPJ is kept inhibited by high spontaneous firing of a node termed Object 2; when a goal is setup, this node is itself inhibited by the Object 1 node, which allows TPJ nodes to become active. The sigma-pi weights connect TPJ object nodes to V2→V4 connections that have been identified as being important for developing representations for that object at the higher levels (TE, IFG modules). Hence when the goal 'square' is set-up, say, we see an increase in the firing rate of the neurons that represent arc segments of squares in V4 (only for neurons that have V2 'square' inputs since this attentional process is modulatory not additive), and via this increased activity to TEO, TE and IFG_no_goal site. The IFG_no_goal module indicates the results of the attentional process; the IFG module does not show the results of attention since it is where the object goal is formed and hence the majority of its activation is goal related.

The dorsal 'where' stream is hard-wired to refine spatial representations as information is passed upwards via V1 dorsal, V5, and frontal eye field (FEF) modules: FEF_1 and FEF_2. Spatial goals can be set-up in the dorsal stream with a particular location being excited in FEF_2. Superior parietal lobule (SPL) is the dorsal equivalent of the ventral TPJ as the generator of the movement of spatial attention signals to posterior sites; SPL receives input from FEF_2 but can only fire if a spatial goal is set allowing for disinhibition via the Space 1 and Space 2 nodes. The spatial goal position is currently determined externally. Sigma-pi weights connect SPL to the V5→lateral intraparietal area (LIP) connections, only the weights between excitatory nodes are affected. A spatial goal can, via the sigma-pi connections, modulate the inputs to LIP nodes from V5; an increased firing rate for nodes in this region results for LIP and higher dorsal modules if an input exists at the same spatial location as the goal. We use the FEF_2_no_goal module to view the affects of spatial attention. We have previously shown, using a similar dorsal route architecture, that attention can highlight a particular spatial location [5].

Table 1. The sizes of modules

Module	Size
LGN	38*28
V1 ventral	76*56
V2	76*56
V4	30*20
TEO	15*10
TE	5*5
IFG, IFG_no_goal, TPJ	3*1
V1 dorsal, V5. LIP, FEF_1, FEF_2, FEF_2_no_goal, SPL	19*14

Lateral connections between V4 and LIP allow for the passage of information between the 2 streams [6, 7]. When an object is attended to (object goal set) these connections lead to increased firing in the location of the attended object in LIP which then through processing at higher dorsal levels indicates the position of the object. Alternatively spatial attention in the dorsal stream increases firing rates of V4 nodes at that location; via TEO and TE the activations in IFG_no_goal indicate which object is located at that spatial location.

Parameter searches are performed at each level such that useful partial object representations and object representations are found at TEO and TE, respectively.

A more detailed look at the way that the three figures are encoded into the various modules is shown in fig. 2, where the preference maps are shown in a) for V4 and in b) for TE. These maps assign the colours: grey, black and white for square, triangle or circle, respectively, to each of the neurons in the relevant module according to the stimulus shape to which it is most responsive. As we see in fig. 2, in V4 the preference map has a semi-topographic form, with clusters of nodes preferring a given stimulus in their nearby area; in TE there is no topography and the representations are spatially invariant.

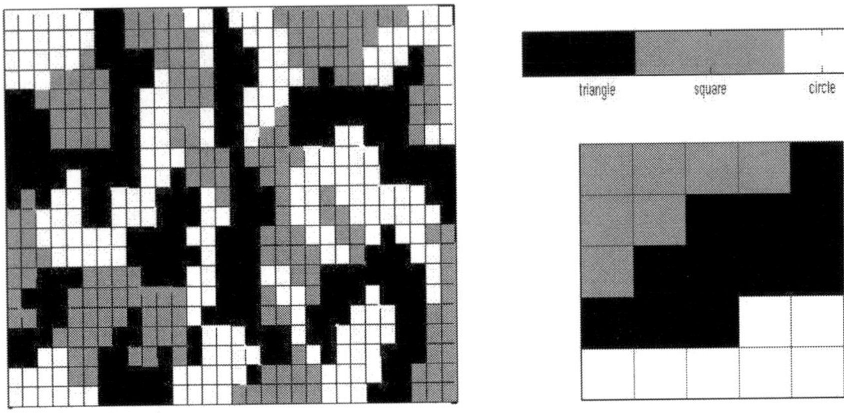

a) V4 preference map b) TE preference map

Fig. 2. Preference maps in V4 and TE for the shapes: triangle, square and circle

We now turn to consider in the next section what happens when we present an occluded stimulus.

3 Occlusion with Attention

The total image we investigate here is shown in fig. 3. It consists of a square and a triangle. Of course it is problematic which of the two figures is occluding the other in fig. 3, since it is two dimensional with no hint of three-dimensionality (to which the resolution of occlusion can contribute). We will discuss that question later, but consider here that we are looking at either figure as occluding the other, and ask the question as to how one might extract a clearer recognition of each figure in the presence of the distorting second figure. This task is thus more general and difficult, we suspect, than occlusion, where one figure is to be taken as the occluding figure in the foreground and the other (the occluded figure) as background or ground (as detected by the three-dimensional clues).

The activities in several of the modules are shown in figures 4, 5 and 6. Of these some entries have negative values, these being where the change between two different conditions has been calculated. In particular the figure 4a, denoted 'V4 Square-Away' shows the firing rates for V4 neurons where the results for attend triangle have had the attend-away results subtracted from them to show the overall difference that attending the square causes. A similar result holds for fig. 4b.

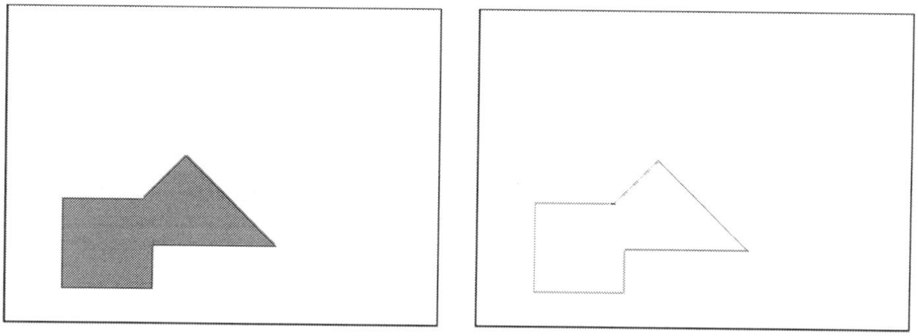

Fig. 3. Occluded input, composed of a square and a triangle. Left figure shows the complete input, right the edge detected input.

In particular we see from fig. 4a that when attention is turned to the square, in comparison with the case when no attention feedback is used at all, then there is considerable increase in the firing rates of V4 neurons preferring the square, as seen from the V4 preference map of fig. 3a. Symmetrically, the triangle-representing neurons increase their firing considerably when attention is directed towards the triangle, to the detriment to those neurons preferring the square.

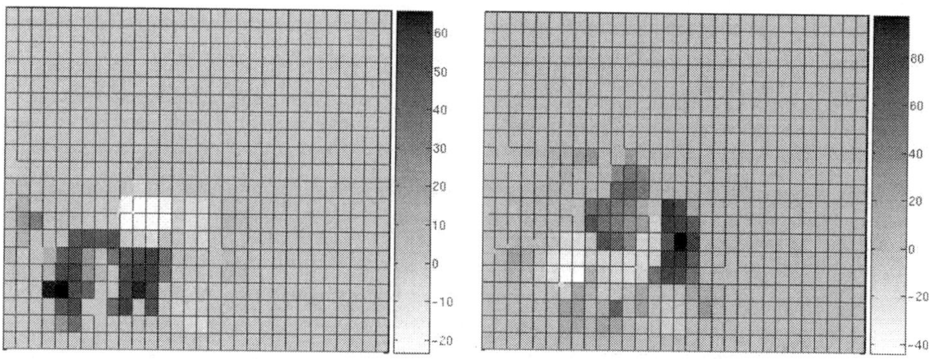

a) V4 square-away b) V4 triangle-square

Fig. 4. V4 responses. Scale in Hz. We plot a) V4 response to attend square of composite object minus response to composite image without attention.; b) is the response when the triangle of the composite object is attended to minus the attend away firing rate.

A similar situation occurs for the neurons in TE as seen in fig. 5. In fig. 5a it is exactly those neurons preferring the square that are boosted, with an average firing rate increase of about 64 Hz. In the same manner, the attention feedback to the triangle-causes the triangle-preferring nodes in TE to increase their firing rates by above 100Hz on average.

These results indicate that either figure becomes more strongly represented in V4 and TE when attention is turned to it. This is in general to the detriment of the other figure, as is to be expected.

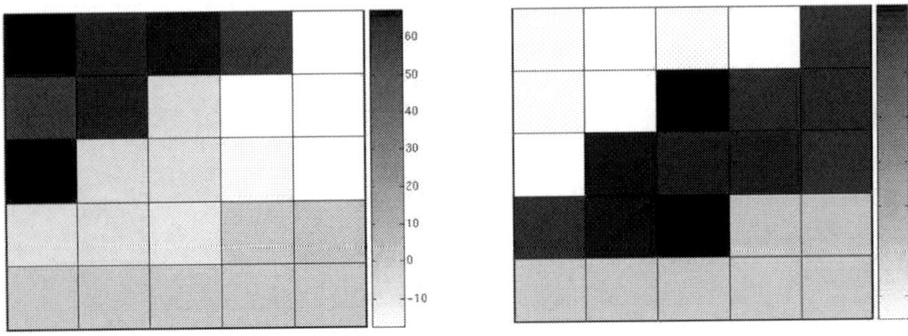

a) TE Square-Away b) TE Triangle Away

Fig. 5. TE responses. Scale in Hz. Where a) is the TE firing rate response when the square is attended to minus the response in the no attention case; b) is the TE response to attend triangle minus the attend away response.

The IFG_no_goal module shows the affects of object attention. When there is no attention the square representation is the most highly activated having a stable firing rate of 49Hz, whilst the triangle representation has a firing rate of only 9Hz. With attention directed to the square, the firing rate for the square increases to 118Hz, and the triangle reduces to 1Hz, but when attention is focused on the triangle the firing rates become 12Hz for the square representing node and 123Hz for the triangle. These results agree with the previous ones described for activities residing in lower-level modules under the affect of attention to one or other of the shapes - square or triangle (as shown in figures 4 and 5).

4 Object Representation

As previously mentioned the lateral connections between the ventral and dorsal streams allow for activations modulated by attention to be transferred from stream to stream. We see from fig. 6 the change in FEF responses when attention is directed to the square or triangle (within the ventral stream) versus the no attention condition. In both cases firing rates increase by up to 40-50Hz, highlighting the centres of the objects as well as parts of the overlap region, there are also decreases near the centre of the non-attended object in range of 20-30Hz. The highest firing rates occur near the overlap regions, but overall the firing rates are higher for the spatial location of the attended object.

Such results indicate that a neural 'object representation' in our architecture is dependent on where attention is directed. If it is directed at the stimulus providing activation of the learnt object representation in the brain then there are two aspects of the representation which are different from the set of neurons activated when there is no attention to the object or attention is directed to another object in the visual field. These effects are

1) An extension of the representation into the other stream (from ventral to dorsal stream activations if objects are being attended to, or from spatial position to object, so from dorsal to ventral streams, if positions are being attended to. This extension is

not as strong if there is no attention to one or other of space or object, with attention in the ventral stream there are increases in LIP firing rates at the location of the attended object of up to 100Hz, in FEF_no_goal module the increases are up to 50Hz in attention cases as against the no attention cases.

2) A lateral extension of the representation in each module when attention is involved. Such a lateral extension, for example in V4, is due to the lateral excitatory connections in the module itself, which allow for the spread of activity due to the increased firing of central neurons to the object representation when attention is being used. Whilst there is little change in the numbers and positions of active V4 excitatory neurons, an increase of ~8% in the number of neurons with most of these having low firing rates (<10Hz), there is a large increase in the firing rates of neurons that are active and showing a preference to the attended object (up to 80Hz), and a decrease in firing rates of active nodes showing a preference for the unattended object caused by increased inhibition. The spreading of activity at the excitatory layer is prevented by an increase in numbers and firing rates of the V4 inhibitory neurons, from 3 neurons in the non-attentive case to 16 for the attentive case and firing rates increasing by up to 20Hz.

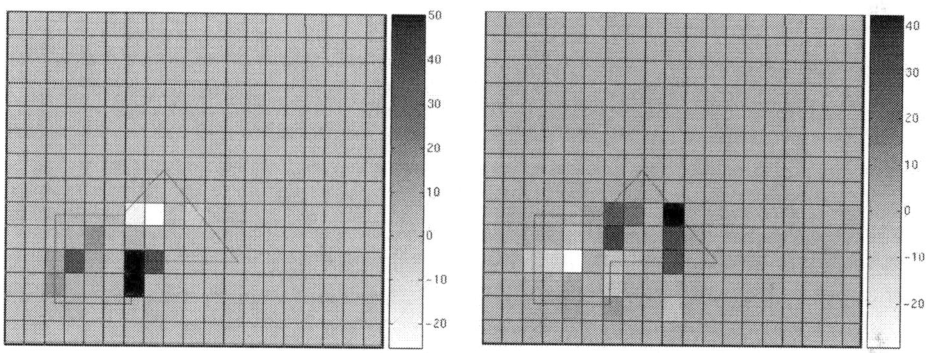

a) FEF Square-Away b) FEF Triangle-Away

Fig. 6. FEF responses, with the perimeter of the input scaled and superimposed on the neuron firing rate changes. Scale in Hz. The plots are a) FEF response to attend square minus response to attend away; and b) is the response to attend triangle minus the attend away result.

We conclude that an object representation can only be specified when the attention state of the system is itself known. This corresponds to the statement that when feedback is present then not only do the afferent synaptic weights to the neurons of a module specify the unattended representations it carries, but the feedback weights are also needed to specify how these representations are modified by the attention state.

It can be commented that this feature of attention-dependent object representations is well known. However we are here indicating, beyond the feature itself of attention-dependence, that this feature is important to help resolve the question of occlusion, and to separate out components of an occluded/occluding set of objects which the objects are. Such a process needs, as a base to start from, the object representations of the unattended figures. The resolution of the occluded figures into its component objects can then be achieved by using the slightly activated classifier neurons in IFG (without attention) so as to return to their component attended object representations sequentially to check that these objects are indeed present. Without attention the firing rates in the IFG_no_goal module are: triangle 4Hz, square 64Hz, circle 0Hz for our occluded input (fig. 2).

5 Discussion

We have shown using a hierarchical visual model how attention, modelled as contrast gain, can be used to recognise occluded objects. By activating object goals in the ventral visual stream the components of an occluded object can be recognised, in this case a square and a triangle. Additionally the increased activation due to this attentional process can be transferred into the dorsal stream to cause increased activations in the spatial location of the attended object. Indeed, though not shown here, results have shown that with lateral connectivity at the V4 – LIP level spatial attention can lead in the ventral to the identification of the object at the attended location. Contrast gain attention, as modelled here, does not cause hallucination of the occluded part of an object, since the occluded parts have zero activation and cannot be increased by the attentive modulation. Additive attention could lead to the occluded parts of the object becoming active, as the feedback from higher levels travels down the visual stream. We are not necessarily talking here about 'filling-in' since it is clear that looking at figure 2 we can attend to the part of the object that resembles a square without 'seeing' the occluded vertex. We have yet to include in a comprehensive manner the known feedback connections in the visual cortex. Limited additions of these weights have shown that small weight values help attention as contrast gain to distinguish between attended and unattended objects, as well as helping to refine representations at higher levels in the attend away cases.

Certain attention results [8] have so far only successfully been modelled using attention as multiplicative for graded and spiking neurons in a variety of architectures [8, 9, 10, 11]. Visual models using additive attention include [12] and recent models [13] have suggested that a multiplicative component of attention can be the result of additive attention, though they did not investigate whether the model gave similar results to the experimental studies [8].

There is the question of what is occlusion in a 2-dimensional image in a monocular system. Here, we have both objects the same colour and perhaps this is not occlusion but a composite object that attention can be moved around via goals to find which parts resemble the system's learnt objects (square, triangle). This could be a harder problem than dealing with occlusion in 3-dimensions with a binocular system and the aid of extra information such as depth, and our results on the 2-dimensional problem give good grounds for claiming that attention will be crucial in resolving 3-dimensional occlusion.

Finally, attention-dependent object representations are important in resolving occlusion, by the separation of the composite object into component objects, using the unattended object representations.

Acknowledgments

One of us (NRT) would like to thank EPSRC, others (JGT, CP, & SK) would like to thank the EC, under the GNOSYS (FP6-003835) project, and another (MH) would like to thank the EC, under the MATHESIS project for support to carry out this work.

References

1. Zitnick C. L. & Kanade T.: 'A Cooperative Algorithm for Stereo Matching and Occlusion Detection', IEEE Trans. Patt. Anal. Mach. Intel. (2000) 22: 675-684
2. Fukushima K.: "Recognition of partly occluded patterns: a neural network model", Biol Cybern (2001) 84: 251-259.
3. Ito M. & Komatsu H.: Representation of Angles Embedded within Contour Stimuli in Area V2 of Macaque Monkeys. J Neuroscience (2004) 24: 3313-3324.
4. Pasupathy A. & Connor C.E.: Shape Representation in Area V4: Position-Specific Tuning for Boundary Configuration. J Neurophysiol (2001) 86:2505-2519.
5. Taylor J.G., Hartley M., & Taylor N.R. 'Attention as Sigma-Pi controlled ACh-based feedback', Proc. of IJCNN'05 (2005).
6. Lanyon L. J. & Denham S.L.: 'A model of active visual search with object-based attention guiding scan paths', Neural Netw. (2004) 17: 873-97.
7. van der Velde F. & de Kamps M.: 'From Knowing What to Knowing Where: Modeling Object-Based Attention with Feedback Disinhibition of Activation', J. Cog. Neurosci. (2001) 13: 479-491.
8. Reynolds J.H., Chelazzi L. & Desimone R.: 'Competitive mechanisms subserve attention in Macaque areas V2 and V4', J. Neurosci. (1999) 19: 1736-53.
9. Taylor J.G. & Rogers M.: 'A control model of the movement of attention' Neural Netw. (2002) 15:309-326.
10. Taylor N.R., Hartley M. & Taylor J.G. 'The Micro-Structure of Attention', accepted for CNS'06 (2006).
11. Taylor N.R., Hartley M. & Taylor J.G. 'Analysing Attention at Neuron level' accepted for BICS'06 (2006).
12. Grossberg S. & Raizada R. D. 'Contrast-sensitive perceptual grouping and object-based attention in the laminar circuits of primary visual cortex', Vision Res. (2000) 40: 1413-32.
13. Deco G. & Rolls E.T.: 'Neurodynamics of biased competition and cooperation for attention: a model with spiking neurons', J. Neurophysiol. (2005) 94: 295-313.

A Forward / Inverse Motor Controller for Cognitive Robotics*

Vishwanathan Mohan[1,2] and Pietro Morasso[1]

[1] Neurolab, DIST, University of Genova,Via Opera Pia 13, Genova 16145 Italy.
morasso@dist.unige.it
www.biomedica.laboratorium.dist.unige.it/morasso.html
[2] Doctoral School on Humanoid Technologies, Italian Institute of Technology, Genova, Italy
vishwanathan.mohan@unige.it

Abstract. Before making a movement aimed at achieving a task, human beings either run a mental process that attempts to find a feasible course of action (at the same time, it must be compatible with a number of internal and external constraints and near-optimal according to some criterion) or select it from a repertoire of previously learned actions, according to the parameters of the task. If neither reasoning process succeeds, a typical backup strategy is to look for a tool that might allow the operator to match all the task constraints. A cognitive robot should support a similar reasoning system. A central element of this architecture is a coupled pair of controllers: FMC (forward motor controller: it maps tentative trajectories in the joint space into the corresponding trajectories of the end-effector variables in the workspace) and IMC (inverse motor controller: it maps desired trajectories of the end-effector into feasible trajectories in the joint space). The proposed FMC/IMC architecture operates with any degree of redundancy and can deal with geometric constraints (range of motion in the joint space, internal and external constraints in the workspace) and effort-related constraints (range of torque of the actuators, etc.). It operates by alternating two basic operations: 1) relaxation in the configuration space (for reaching a target pose); 2) relaxation in the null space of the kinematic transformation (for producing the required interaction force). The failure of either relaxation can trigger a higher level of reasoning. For both elements of the architecture we propose a closed-form solution and a solution based on ANNs.

1 Introduction

Humans exhibit an enormous repertoire of motor behaviour which enables us to effectively interact with many different objects under a variety of environmental conditions. Modelling the way in which humans learn to coordinate their movements in daily life or in more demanding activities is an important scientific topic from many points of view, such as medical, psychological, kinesiological, and cybernetic. The ability to perform in such a varying and often uncertain environment is also a feature which is conspicuously absent from most robotic control, as robots tend to be

* This research was partly supported by the EU FP6 project GNOSYS.

S. Kollias et al. (Eds.): ICANN 2006, Part I, LNCS 4131, pp. 602–611, 2006.

designed to operate within rather limited environmental situations. Cognitive problems have been explored in recent years from the point of view of the design of robots or the development of humanoid technologies. The research has clustered around broad and related concepts: developmental robotics, epigenetic robotics, bio robotics [5, 6]. In this context, the idea that perception is not a passive mechanism for receiving and interpreting sensory data but is the active process of anticipating the sensory consequences of an action provides a computational alternative to the conventional view based on a segregation of perceptual, motor and cognitive processes in different parts of the brain, according to some kind of hierarchical organization. This implies that adaptive behaviour can best be understood within the context of the (biomechanics of the) body, the (structure of the organism's) environment, and the continuous exchange of signals/energy between the nervous system, the body and the environment. In other words the appropriate question to ask is not what the neural basis of adaptive behaviour is but what the contributions of all components of the coupled system to adaptive behaviour and their mutual interactions are. The brain has a body and the body has its constraints and affordances.

Our ability to execute movements using different end effector systems and on different scales (writing on a paper and a black board for example), makes it plausible that a level of representation of movement may exist independent of how the movement will be performed [7]. In order to complete the specification of motor plan, the task description must be bound to the specific end effector, which on its turn, must be translated into the proximal space, thus completing the desired kinematic picture of movement. The transformation form distal space to body space requires inversion of the forward kinematic mapping x=f(q), which is usually redundant both in terms of structure and task [1]. Consider the 'stick paradigm' requiring a use of an appropriate tool for obtaining a goal that is not achievable by direct use of end effector: the robot's body imposes specific constraints on the range of attributes (physical dimensions, weight, moment of inertia, surface characteristics etc) of sticks that can be grasped and effectively used by the robot. This description is independent of the specific task, but the task may impose additional constraints on the stick attributes that must be taken into account when reasoning about the task and eventually planning a course of actions. In general, it is very difficult to formalize a priori the constraints that qualify an effective "stick" in the context of a given task and thus only a very approximate schema can be formulated in terms of explicit rules and structural descriptions: the crucial part is best represented in terms of experiments, either "mental experiments" (carried out by means of an internal model of the body and the environment) or "actual experiments", which require a corresponding sequence of virtual or actual movements for achieving the task. In addition, it is essential to reach a reasonable solution which also takes into account internal and external constraints and the fact that in general a "target position" is associated with a "target effort".

In this framework, we propose a general biomimetic system for the coordination of body/arm/tool movements that operates with any degree of redundancy, for any configuration of limbs, and can deal with geometric constraints (range of motion in the joint space, internal and external constraints in the workspace) and effort-related constraints (range of torque of the actuators, etc.).The crux of this architecture implements the Passive Motion Paradigm: the "virtual stiffness" determines an attractive force field to the target. However, this field can be distorted in order to take into account external constraints or obstacles. The pair of transformations, determined by the Jacobian and the transpose Jacobian, solve "the transfer from the end-point trajectory to a requisite motor plan" or also the inverse kinematic transformation in the general redundant case. In other words, this is a method to generate end-effector movements according to a plan. At the end of the simulation if the residual error is null, we can say that the target is reachable. If it is not, the residual error is an estimate of the size of the required tool. However, we should also consider that the final position at the end of the relaxation to x_{des} is a local minimum of an energy function and given the redundancy of the system it may well be that another better equilibrium configuration exists, for example a configuration with a smaller residual error that then requires a tool of smaller size. How to search for such better configurations? One possibility is to carry out null-space movements i.e. movements that exploit the redundancy of the transformation (the fact that for each position of the end effector there is an infinity of arm configurations that are consistent with it). The equilibrium configurations can be memorized and rewarded/punished according to performance measures. In sum, the proposed FMC/IMC pair integrates end-effector movements and null-space movements in the same computational mechanism that becomes a generator of potential solutions or a reporter of general impossibility to achieve the goal.

2 The FMC/IMC Pair

Let x be the vector that identifies the pose of the end-effector of a robot in the workspace and q the vector that identifies the configuration of the robot in the joint space: $x = f(q)$ is the kinematic transformation or, for each time instant of a planning process,

$$\dot{x} = J(q) \cdot \dot{q} \tag{1}$$

where J is the Jacobian matrix of the kinematic transformation. In general we shall consider a redundant robot, i.e. a robot in which the dimensionality of q is (much) greater than the dimensionality of x. Even a 5 or 6 DOFs manipulator becomes redundant if you mount it on a mobile platform. An inverse motor controller is one that, given a desired law of motion to a target in the workspace, computes a coordinated joint rotation pattern that implements it. We build the IMC by using the passive motion paradigm (PMP: Mussa Ivaldi et al 1988) that consists of the following steps:

1) Define a virtual attractive force field to a designated target

$$x_T : \; F = K_x (x_T - x) \tag{2}$$

2) Map the force vector into an equivalent torque vector (principle of virtual works):

$$T = J^T F \tag{3}$$

3) Relax the arm configuration in the applied field:

$$\dot{q} = B \cdot T \tag{4}$$

4) Map the arm movement into the workspace:

$$\dot{x} = J \cdot \dot{q} \tag{5}$$

where K_x is a stiffness matrix and B is a viscosity matrix. Fig. 1 illustrates the method:

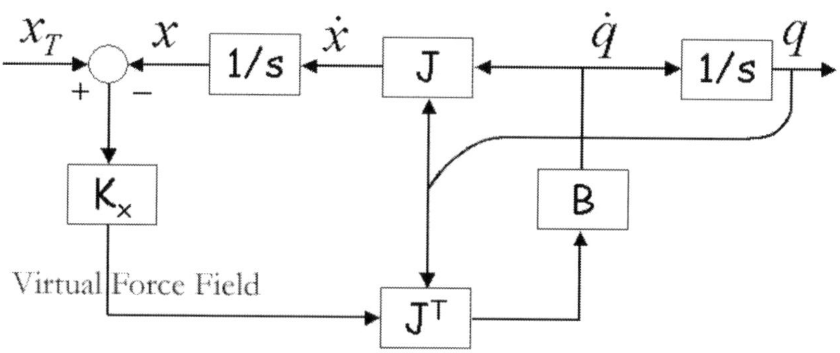

Fig. 1. FMC/IMC Pair

The algorithm always converges to a "reasonable" equilibrium state, whatever the degree of redundancy of the robot: if the target is within the workspace of the robot, it is reached; if it is not reachable, the robot fully extends the arm to a position that is at a minimum distance from the target. Moreover, the timing of the relaxation process can be controlled by using a time base generator (Tsuji et al 2002) and the concept of terminal attractor dynamics (Zak 1991): this can be simply implemented by substituting the relaxation equation (4) with the following one:

$$\dot{q} = \Gamma(t) \cdot B \cdot T \tag{6}$$

where a possible form of the time-varying gain that implements the terminal attractor dynamics is the following one (it uses a minimum-jerk time base generator with duration τ):

$$\begin{cases} \xi(t) = 6 \cdot (t/\tau)^5 - 15(t/\tau)^4 + 10(t/\tau)^3 \\ \Gamma(t) = \dot{\xi} \big/ (1-\xi) \end{cases} \tag{7}$$

The model above can be further extended in order to deal with a variety of internal and external constraints and the fact that in general a "target position" is associated with a "target effort", i.e. the fact that after the target pose x_T has been reached (and possibly an object has been grasped, but the analysis of grasping is outside the scope of this paper) a force vector F_T has to be applied to some object. Figure 2 shows how we can incorporate in the model the following two requirements that both exploit the redundancy of the system:

1) Keep the robot away from the joint limits (an internal constraint): this can be expressed by adding an attractive force field in the joint space $K_q(q_{ref} - q)$, where q_{ref} is the set of joint rotation values in the middle of the range of motion (an alternative implementation would be a repulsive field from the joint limits);

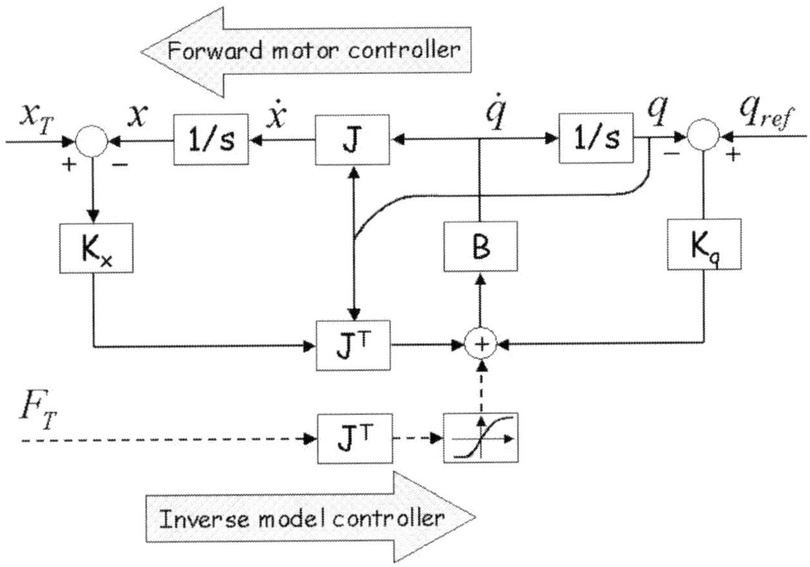

Fig. 2. First relaxation: to the target pose in agreement with the joint limits. Second relaxation (dashed line): in the null space in order to produce the target effort in agreement with the actuator limits.

2) After reaching the target pose, perform a movement in the null space of the kinematic transformation[1] in such a way that for a given desired effort F_T the torque required of each actuator is within the allowed limits: this is implemented by introducing F_T as an additional force drive in the IMC and saturating the computed torque vector according to the actuator constraints.

[1] The null space of the kinematic transformation $x = f(q)$ is characterized by the equation: $J(q) \cdot \dot{q} = 0$.

The scheme of fig. 2 can be implemented in a neural way by training a multi-layer feed forward network to learn the kinematic transformation $x = f(q)$. From this trained network it is possible to extract the Jacobian matrix in a simple way.

Suppose for example that we use a three-layer network (fig. 3):

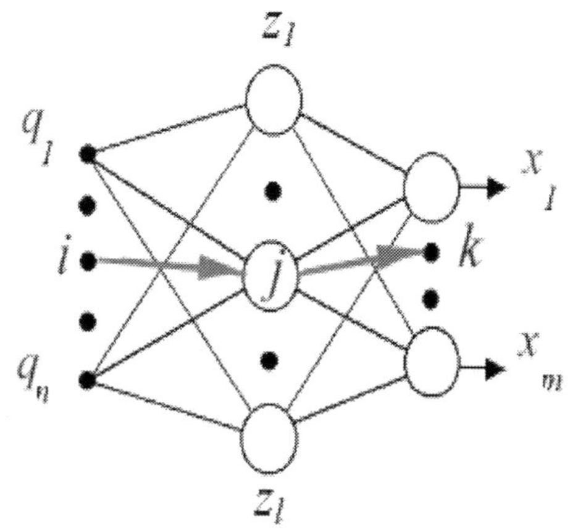

Fig. 3. Three-layer feed forward neural network trained to learn the kinematic transformation

$$x = f(q) \Rightarrow \begin{cases} h_j = \sum_i w_{ij} q_i \\ z_j = g'(h_j) \\ x_k = \sum_j w_{jk} z_j \end{cases} \tag{8}$$

After training, we can extract the Jacobian matrix from the network in the following way:

$$J_{ki} = \frac{\partial x_k}{\partial q_i} = \sum_j \frac{\partial x_k}{\partial z_j} \frac{\partial z_j}{\partial h_j} \frac{\partial h_j}{\partial q_i} = \sum_j w_{jk} g'(h_j) w_{ij} \tag{9}$$

The procedure can be easily generalized to a network with more than 1 hidden layer.

3 Implementation and Simulation Results

The FMC/IMC pair was implemented with respect to a planar robot with three revolute joints and link lengths of one for each link. Assuming for each joint a range of motion of $\pm 90^{\circ}$ we sampled the configuration space with 60K samples and after a few trials we chose a (3, 30, 15, 2) network which tracked the targets reasonably well to an error of 10^{-3} . The MLN was trained using the Levenberg-Marquardt algorithm designed to approach second-order training speed without having to compute the

Hessian matrix [Hagan et al, 1994].The Jacobians and torques can be computed using the trained two layer MLP as follows :

$$\dot{x} = J \cdot \dot{q} \Rightarrow \sum_i J_{ki}\dot{q}_i = \sum_i \left(\sum_l w_{lk}.g'(p_l).\sum_j w_{jl}.g'(h_j).w_{ij} \right)\dot{q}_i \qquad (10)$$

$$T = J^T \cdot F \Rightarrow T_i = \sum_k J_{ik} F_k$$

Where g(.) is the tansig function for the hidden layers, j and l are the first and second hidden layers respectively.

Figure 4 illustrates the reaching trajectories obtained by the FMC/IMC pair without application of Joint reference fields and the external force drive (for Null Space movements), starting form an initial configuration of (π/4, π/4, 0). Of the four different targets are presented, three are within the workspace and one (-3, 4) was not reachable. Note that in this case, the robot fully extends the arm to a position that is at a minimum distance from the target. The timing of the relaxation process is controlled by using a neural time base generator.

The next step was to perform a coordinated body/arm/gripper virtual movement that brings the gripper to the target according to some optimality criterion which takes

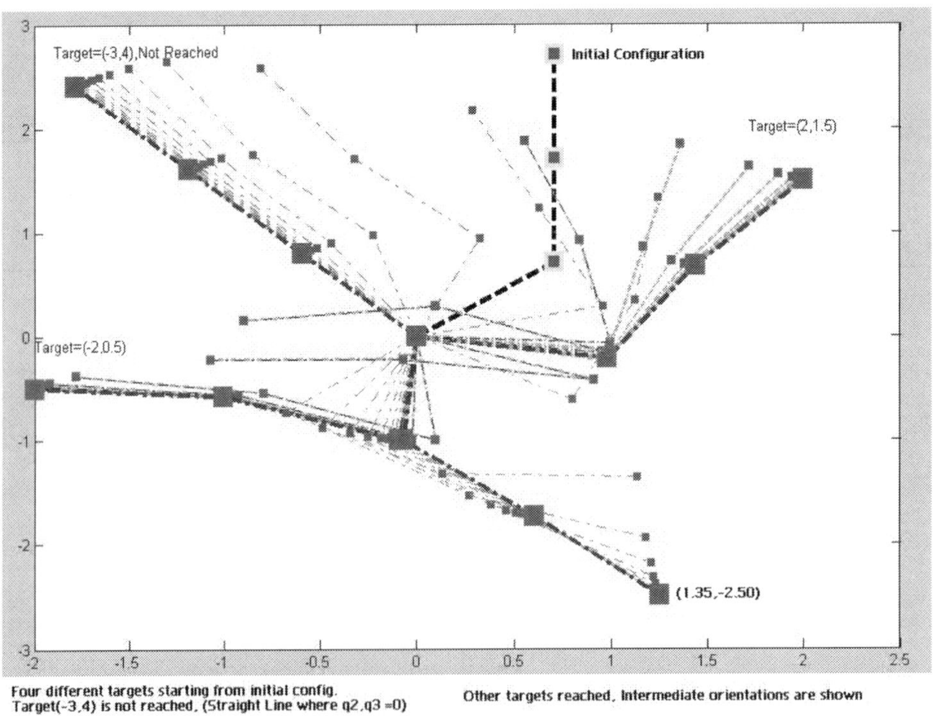

Fig. 4. Relaxation to target pose using FMC/IMC pair

into account internal constraints. We considered the case of joint limits as an imposed internal constraint, which were set to rotation values in the middle of the range of motion. Figure 5 shows the solutions reached after attractive force field in the joint space was added, target to be reached is (-1,1).

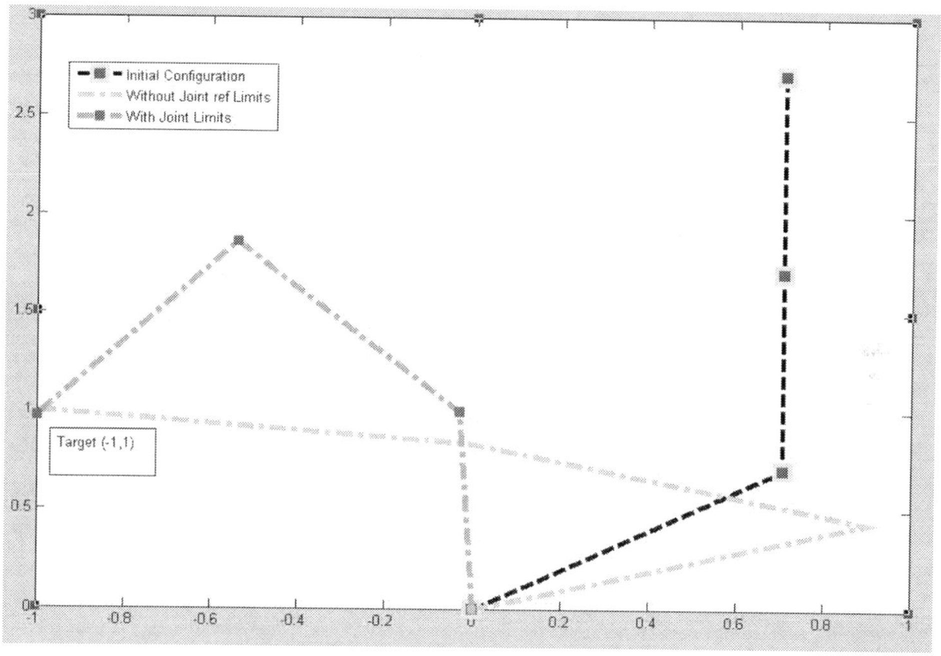

Fig. 5. Relaxation to target pose after application of attractive force field in joint space

Finally, we simulated virtual movements in the null-space in such a way to minimize the overall effort required of all the actuators while satisfying the required target effort. These are coordinated movements of the body/arm/gripper that keep the same pose and are characterized by equation $J(q) \cdot \dot{q} = 0$. Note that this is only possible for a redundant system and the effort constraint Ft is turned on only after target Xt has been reached. Figure 6 and 7 show the final trajectories reached by the FMC/ IMC pair after performing null space movements.

Hence, for an arm that is mounted on a mobile robot which includes in its geometric/kinematic description, the position/orientation of the mobile platform (3 dofs) as well as the 5 dofs of the arm and the 2 dofs of the gripper, the proposed FMC/ IMC pair provides a dynamic approach to integrate motor redundancy, internal constraints (as regards to geometry, self-interference, and range of forces/torques) and external constraints (obstacles). If the forward simulation is successful then the movement is executed, otherwise the residual "error" or measure of inconsistency can be taken as a starting point for breaking the action plan into a sequence of sub actions.

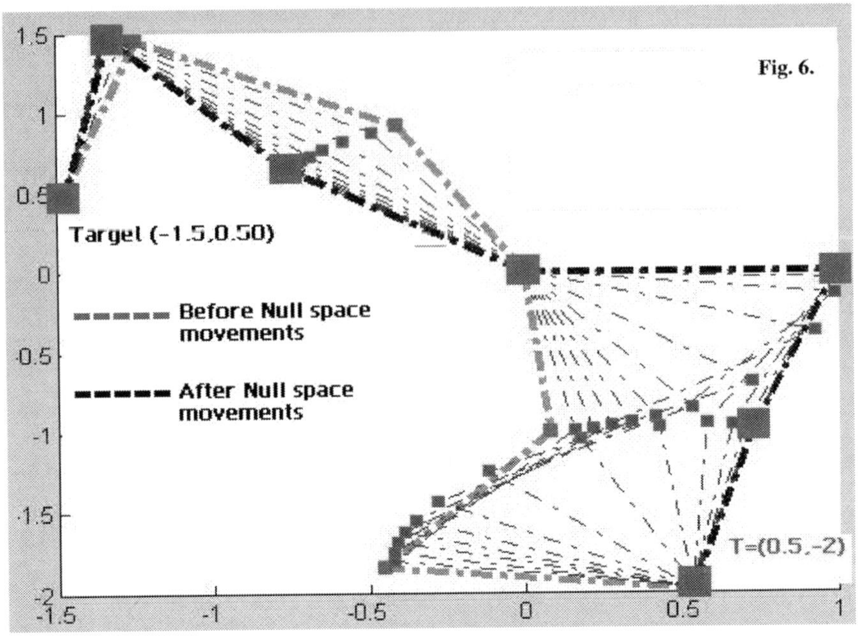

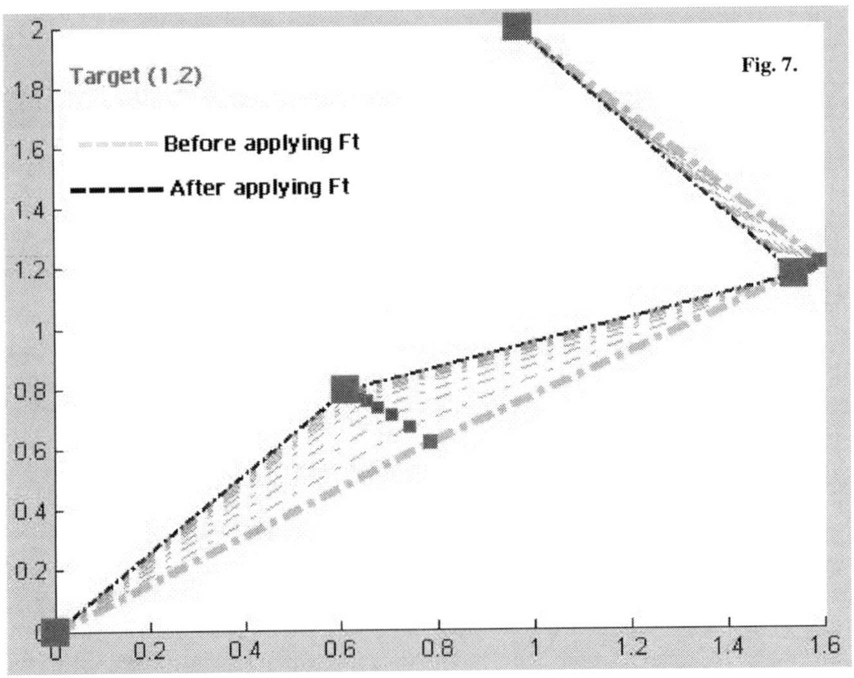

Fig. 6 & 7. Relaxation to target pose after virtual movement in the null-space

4 Concluding Remarks

The reasoning system of a cognitive robot must incorporate a scheme for imagination of motor actions, taking place on imagined objects, so as to 'reason' possible optimally rewarded actions, at the same time satisfying a range of internal constraints (joint limits), external constraints (obstacles for the body/arm/gripper) and effort related constraints (range of torques of actuators etc). Using a coupled pair of controllers FMC/IMC, we provide a computational frame work to perform the motor cognitive functions of trajectory determination, coordinate transformations and generation of motor commands implicitly knowledgeable of geometric constraints as well as effort-related constraints. While the FMC predicts trajectories of the end-effector variables in the workspace based on trajectories in joint space, the IMC calculates necessary motor commands from desired trajectories in workspace.

The mental simulation using the proposed framework can run as follows: a) Reaching the positional target according to some optimality criterion b) Performing movements in the null field to reach the effort target. In the case that the goal is satisfied, the selected configuration is memorized and actual movement is initiated. A failure of either relaxation triggers a higher level of reasoning based on the residual "mismatches" in order to decide on two possible outcomes: a) abort the task because it cannot be possibly executed; b) evaluate the geometric/physical parameters of a tool (within a given repertoire) that might be useful for solving the task; In the case of outcome 'b', a sub goal can be instantiated so as to fetch the tool, grasp it and, if appropriate, perform some babbling movements to update the arm/gripper/tool kinematics. In this case, a given task can then be broken down to a sequence of target positions (assembled in correct order), to be attained independently by the use of the proposed relaxation dynamics, with the especially important components of the memory, recalibration and influence of rewards.

References

1. Mussa Ivaldi FA, Morasso P, Zaccaria R (1988) Kinematic Networks. A Distributed Model for Representing and Regularizing Motor Redundancy. Biological Cybernetics, 60, 1-16.
2. Tsuji T, Tanaka Y, Morasso P, Sanguineti V, Kaneko M (2002) Bio-Mimetic Trajectory Generation of Robots via Artificial Potential Field with Time Base Generator. IEEE Transactions on Systems, Man, and Cybernetics, Part C - Applications, 88, 4, 426-439.
3. Zak M (1991) Terminal chaos for information processing in neurodynamics. Biological Cybernetics 64, 343-351.
4. Z.: Hagan, M. T., and M. Menhaj, (1994) "Training feed forward networks with the Marquardt algorithm," IEEE Transactions on Neural Networks, vol. 5, no. 6, pp. 989-993.
5. Beer, R.D., Quinn, R.D., Chiel, H.J., and Ritzmann, R.E. (1997). Biologically- inspired approaches to robotics. Communications of the ACM 40(3):30-38.
6. Weng J (2004). Developmental Robotics: Theory and Experiments. International Journal of Humanoid Robotics.

A Computational Model for Multiple Goals

Stathis Kasderidis

Foundation for Research and Technology - Hellas
Institute of Computer Science, Computational Vision and Robotics Laboratory
Science and Technology Park of Crete
Vassilika Vouton, P.O. Box 1385, 71110 Heraklion, Greece
stathis@ics.forth.gr

Abstract. The paper discusses a computational model suitable for the monitoring and execution of multiple co-existing goals inside an autonomous agent. The model uses a number of mechanisms to calculate dynamically goal priority. We provide an overview of the model, a discussion of a Cognitive Agent architecture that includes it and we provide results that support the current design. We conclude with a discussion of the results, points of interest and future work.

1 Introduction

There has been in the past decade great interest and advancements for planning algorithms in the AI literature [1] (and references therein). The main focus of such algorithms is the development of a suitable plan that will achieve a target state. For autonomous agents, with arbitrary long lifetimes, situated in a dynamic environment the above approach was extended with execution monitoring and re-planning facilities. Usually the framing of the problem took the form of Hierarchical Task Networks [1], or that of Abstraction and Hierarchical Planning [2, 3].

However, while the above approaches clearly attack an important aspect of the overall problem, they do not attack the full problem. For this reason we suggest that the whole planning problem be decomposed into distinct but interwoven subproblems: i. The problem of 'reasoning', ii. The problem of 'execution'. The first problem deals mainly with the development of a suitable plan which will achieve a target state, while the second one is related to the monitoring of the plan and the provision of re-planning requests to the 'reasoning' component dynamically.

This paper will discuss the 'execution' component of the planning process inside the agent. The paper is structured as follows: in section 2 we will provide a concise overview of the Cognitive Architecture and of the links of the execution component (which we call the Computational Model) with the Reasoning, Motivation, Concept System and Goal Generation modules of the architecture. In section 3 we present the Computational Model and justify our motivation. In section 4 we provide results that support the current work. Section 5 concludes with a discussion on various points of interest and future work.

S. Kollias et al. (Eds.): ICANN 2006, Part I, LNCS 4131, pp. 612–622, 2006.

2 A Cognitive Architecture

The GNOSYS Architecture aims to control robotic agents in arbitrary complex, novel and unstructured environments. It consists of a number of high-level modules that provide the basic cognitive capabilities of the agent. For the purposes of this paper, figure 1 provides a reduced version, including modules only necessary for the discussion of our Computational Model.

The information flow starts with the request for the achievement of a target state. This request comes from two primary sources. It is either a *User* request or a request from the *Motivational System*. The realisation of the request takes the form of *Goal* object, which is generated by the *Goal Generation System*. The Goal includes a plan and a number of parameters (for definition see section 2.1). The necessary plan is provided by the Reasoning System, while initial values of the parameters are given by the Concept System and corresponding Value Maps. After the Goal object is created and appropriately parameterised inside the Goal Generation module it is sent to the Goal Space. There it becomes active and starts its life as a self-contained execution entity. In effect the Goal Space corresponds to the notion of the 'core execution loop' for our architecture.

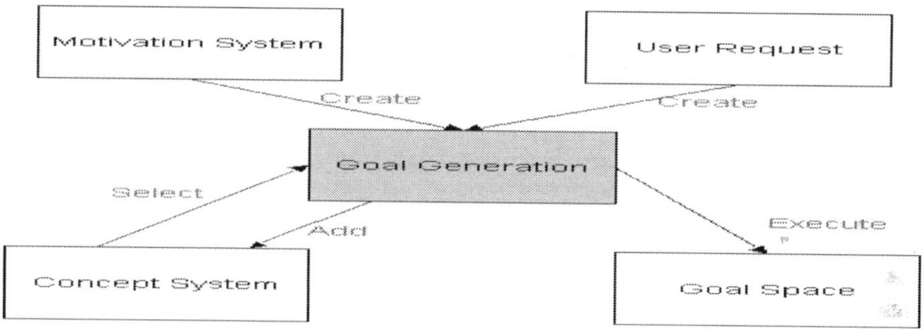

Fig. 1. Components of the GNOSYS Architecture and relations to Goal Space (which implements the Computational Model)

The Goal object and the Goal Space constitute our structural decomposition of the Computational Model while the functional decomposition corresponds to the actual mechanisms that control the lifetime and priority of a given goal; they will be discussed in sections 2 and 3 respectively.

2.1 Definition of a Goal

A Goal object is a self-contained entity (an execution scope). It encapsulates the following components: A goal state; A plan which will provide the sequence of actions needed in order to achieve the goal state; Various parameters that control the lifetime and priority of the Goal. This type of encapsulation might seem strange at first but the justification behind this definition follows a number of ideas:

i. Partition of Output Space in disjoint action domains;
ii. To keep statistics about the utility of a plan for reaching a target state
iii. To allow for the case of Hierarchical Plans for the Reasoning System
iv. To allow the representation of a Goal as a concept.

2.1.1 Partition of the Output Space

The *Output Space* is the space of actions produced by the agent. These actions might be directed to the agent's environment or internally to the agent itself. The main idea for handling and monitoring the complexity of generated action is to use a divide-and-conquer strategy. This leads naturally to the division of the output space in mutually exclusive domains of 'action'. Goals in a given domain compete with a WTA strategy for the right of realising their actions in the current processing step. They are organised as a set of competing families. Each family represents a plan with a specific structure. This will be discussed further in section 3.2. This approach allows the overall architecture to scale efficiently with the dimensionality of the problem and to accept distributed implementation so as to bring in more computational resources as needed.

2.1.2 Goals and Reinforcement Learning

So far we have not explicitly mentioned any Learning process that takes place inside our agent. Indeed such functionality is present. The Value Maps module stores maps that associate states with actions using a RL learning paradigm. However, while the RL framework is very clear on how to build these associations when one uses primitive actions and states, complications arise when we have 'abstract actions' (and abstract states), where we use a hierarchical planner to decompose a high-level task to finer and finer plans up to the level of primitive actions. Care is needed when abstract actions and states are involved. This is due to the context that initiated the related primitive actions. The same primitive action can be evoked by a number of more abstract plans, so the action should not be treated only as a simple action but as primitive Goal; in this way it can be linked to information regarding the actual hierarchical plan that is member of. The standard RL framework can be extended in this way to include the effects of internal context (created by the processing of multiple co-existing Goals) The utility of a plan (schema) depends not only on the sum of utilities of its primitive actions, but also on the contextual information related to the utilities of other co-executing plans.

2.1.3 Reasoning and Hierarchical Plans

While it is not the purpose of this paper to discuss the actual problem of development of a Reasoning system, we have hinted so far that such a system is based partially in a hierarchical planner. Our primary concern is to treat a plan that as a whole that must compete and evaluate its priority. For this reason it is more natural, to our opinion, to call the plan and its goal state a Goal and then take the view that a Reasoning system needs to define a 'meta-plan' (a *decomposition pattern*) that decomposes a high-level goal into a sequence of sub-goals. Having said that we do not want to be more specific as to the nature of the 'meta-plan'. A brief discussion will be given in section 3.2. Independent of the form of the actual meta-plan we can design mechanisms that are mostly independent of the decomposition pattern used and are based on general ideas.

2.1.4 Goals as Concepts

In the GNOSYS Architecture all internal representations are treated as concepts at a suitable level of abstraction. For example Object, Schema & Belief concepts exists. A schema is an abstract plan that is parameterised with appropriate concrete target states and a set of allowable primitive actions, constraints, etc. These aspects constitute features of the concept, which take specific values with a given instantiation. To treat schemas as concepts the Concept System represents them as feature sets and creates links from one schema to another due to shared sub-plans, value maps, or other such features. Thus one can easily imagine semantic-style networks of concepts that relate Goals with each other in the same or different abstraction levels. These semantic networks are built through agent experience and are updated dynamically. The end result is that when we search appropriate initial values for the parameterisation of a Goal (by the Goal Generation System), we can retrieve appropriate information by the use of stored information in the Concept System.

2.1.5 Goal Attributes

Attributes of a Goal can be separated into a number of classes according to their use. A high-level classification is as follows:

- Identification Flags (includes ID, Parent ID, Goal class (i.e. concept), Owner, Domain ID)
- Type Flags (includes Mode and Type (Primitive or Complex))
- Importance Flags (includes Weight, Emotional Weight)
- Action-Attention Flags (Action Index, Sensory Attention Index, Motor Attention index, Boundary Attention Index)
- Drive Flags (includes Commitment, Engagement)
- Termination Flags (Termination Probability, Elimination Probability, Execution Count, User Termination Flag)
- Resource Flags (includes the IDs of the corresponding action domains that the Goal executes in as well as IDs of input modalities)

Most of the above classes are self-explanatory. It suffices to say that Identification Flags cover also the Goal class (as a concept), the Owner (which could be the 'Self' concept or the agent's user) and the Domain ID. The latter is used to denote the domain that the Goal competes into. A Goal can exist and compete independently into multiple domains. For example a high-level Goal of 'fetch the cup from the table' can be translated into a plan of moving from an initial point to a point closer to the table (one action domain – BodyMove) and start a combined sub-plan of movement and grasping actions such as to position the robot body in a suitable pose so as the robotic Arm to grasp the target object. In this sub-plan we have the appearance of a second action domain (that of the Robotic Arm). In the case of the sub-plan the actual high-level command of Fetch (CupOnTheTable), exists concurrently in both domains and acts as a family leader that spawns children sub-goals. For the BodyMove domain these might be Goals such as Move, Rotate, etc, and for the Arm domain these might be Goals such as MoveArm, Grasp, Open/Close Fingers, etc. The importance, drive, action-attention and termination flags are used by the various mechanisms that constitute the functional decomposition of the Computational Model. They will be discussed in section 3. Finally the Resource Flags provide the IDs of the Action Domains that the Goal will be registered with to compete into and the IDs of the input modalities. The notion of the input modalities corresponds to the partition of the input space (an analogous case to the Output space discussed above) into separate and mutually exclusive sources of input information. The set of input modalities provide the State Vector of the agent in a given time. The input modalities include both environmental and internal sources of information.

2.1.6 Goal Structure

Figure 2 shows the internal structure of goal as a set of modules. Each module corresponds to a sub-process that executes in the scope of the Goal.

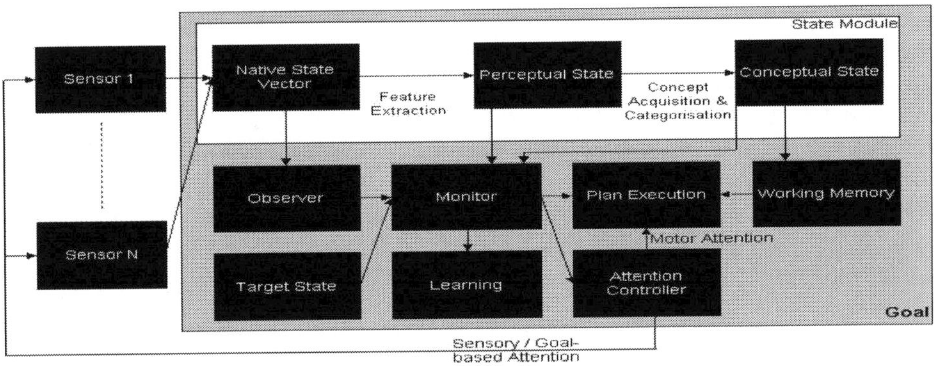

Fig. 2. Components of a Goal Object. See text for description.

Sensors correspond to input modalities. The State Module transforms the native (raw signals) state into perceptual representations (i.e. set of features). The perceptual state is transformed to a conceptual one by recognition of present object concepts in the sensory input. If novel objects exist new concepts are formed in the concept system. In either case, instances of the concepts are activated in the working memory module. This information is used next by the plan execution module. The plan is provided by the reasoning system and it is executed by the plan execution component. It is the module that implements the selected action in the current processing step. After an action is executed a new cycle begins. Erroneous plan execution is indicated by motor attention events. The target state module provides a representation of the goal state. The observer module builds a model of the environment / internal state as needed. The monitor module provides a comparison of the current state with the target state so as to signal the termination of the plan. It also measures discrepancies between the expected state (coming from the observer) and the actual state so as to raise sensory attention events. The attention events include sensory, motor and boundary attention events. These are stored in the attention controller module, which implements a conceptual queue. Attention events might request re-planning through the reasoning system. Part of the plan may be the change of the resolution of the state information through the input modalities (this aims to either enhance information about particular objects, *goal-based* attention, or to better sample novel signals, *sensory attention*). If there are discrepancies in the prediction of the observer the monitor module initiates an on-line learning phase of the current observer models. The learning module implements an RL mechanism, which learns utilities of state/action pairs. These are stored in value maps in the concept system. At the termination of the Goal the various models are stored back to the concept system as features of the current Goal and are associated with the Goal's parents as an indication of the computational context that produced them.

The internal operation of a Goal is shown schematically in figure 2. A Goal object terminates in two ways:

i. By achieving its Target State;
ii. By elimination from the Goal Space.

The actual definition of reaching the target state is included in the Monitor module of fig. 2 and obviously depends on the state representation. In our case the states are considered to live in continuous spaces and we use a Euclidean distance between the current state and the target one. In practice we compare this distance with another attribute (not described in section 2.1.5) to determine if the target state has been approached within an ε-neighbourhood.

2.2 Definition of the Goal Space

The Goal Space acts as the highest-level container of the Goal objects in our architecture. It is divided in a number of action-domains that correspond to the actual output modalities. However the active number of action domains at any given time depends on the classes of Goals that are currently executed. Figure 3 provides a schematic representation of the Goal Space of our agent and the legend provides an explanation of the concept.

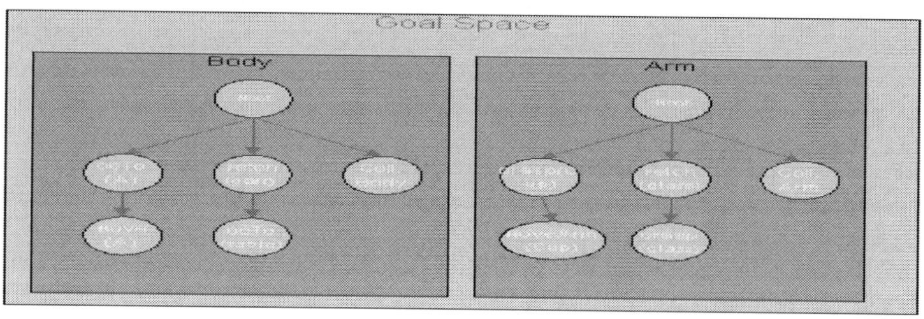

Fig. 3. Goal Space of the GNOSYS Agent architecture. Two action domains are currently active. In both domains three family trees compete for access to actual output. This is called *Global Attention Control* and the winning family will produce actual results. The outputs from both domains are executed in parallel as they represent mutually exclusive sub-sets of the action space.

2.2.1 Action Domains

The concept of the Action Domain has already been explained above. It suffices to add that the actual grouping of the competing Goals takes place as a tree of Goals (see fig. 3), which we call GoalTree. The actual process of calculating the priorities is by the use of the *Action Index* concept. It will be discussed in section 3 more fully. It is a real-valued variable bounded in the interval [0,1]. The actual process of priority calculation has the following high-level steps: Termination Check; Process; Action Index Calculation; Elimination; Output.

The first step checks the termination conditions of a Goal object and if they are satisfied, it is deleted from the current GoalTree. The remaining Goals (if any) continue to the Process phase where they calculate their responses for the current state input. Their responses are added in a list of virtual actions that is maintained by the domain. In the next step the Action Index Calculation takes place in a recursive manner from the Family Parent to its children and so forth. In the fourth step, the winning family is identified and the suggested virtual actions of the losing families are eliminated from the virtual actions list. The determination of the winning family is based on the value of the highest action index. Finally the virtual actions are sent to the actuators to be realised as output. Another processing step starts again. Steps should be taken so as to terminate Goal families that do not seem to converge to their target state. This issue will be discussed next in section 3.

3 A Computational Model

In this section we provide a description of the actual mechanisms that underlie our computational model. Two distinct issues must be discussed. The first one relates to the calculation of the goal priority and termination conditions. The second one relates to the Goal/sub-Goal dependencies and how these affect the computation of scheduling priorities. These issues are discussed next in their corresponding sections.

3.1 Goal Priority and Termination

A Goal's priority is effectively given by the value of its Action Index. The actual formula is given by (1):

$$\text{Action Index}=(W + EW + \text{TermProb} + \text{S-AI} + \text{M-AI} + \text{B-AI} + \Sigma_j\, \delta(j, \text{contribut.})) / (6+\#\ \text{Contributing Children}) \quad (1)$$

Let us explain the main components of formula (1). We assume that each Goal has an *extrinsic value* (W) that comes either by the User or by experience (built by RL and it is thus stored in the value maps). This weight, as all other terms in (1), is suitably scaled in the interval [0, 1]. The term EW represents the *intrinsic value* of the Goal. This value is updated in every processing step by the Motivation System and captures the influence of the Goal to the agent's well being. This idea allows us to assign different value on a Goal even when the agent faces the same external state. In this case the value of a Goal will depend critically on the importance that the Motivation places on the Goal in comparison with other co-existing Goals. The Termination Probability is discussed below. The terms S-AI, M-AI and B-AI correspond to three different (local) attention mechanisms that capture complimentary aspects. S-AI corresponds to *sensory attention*, which captures novelty in the environment. M-AI captures related *motor attention* events. This idea roughly corresponds to the fact that in a real system, actions might not be performed perfectly, or that they will not terminate in the expected time. A motor attention event does not necessarily imply a need for re-planning; it might simply indicate the need for a second attempt to achieve the current plan action. However after a number of unsuccessful re-trials this will indicate a need for re-planning. The B-AI captures attention events coming from the internal agent environment; more specifically the deviation of one or more *homeostatic variables* from their equilibrium state. It is called *boundary attention*. Suitable definitions and concrete example definitions of these concepts can be found in [4, 5]. The last contribution comes from a Goal's children and corresponds to the idea that an attention event, which might be raised in any level of a Goal hierarchy, might influence the execution of the high-level Goals as well. For this reason it is prudent to raise the whole family's priority. The actual mechanism in (1) uses the number of *contributing children* for a given goal. A goal is called contributing if it produces an attention event or it has a child that produces such an event.

Normally a Goal is terminated if it converges to its target state, if it is executed a set number of times, or it is stopped by the User. However there is the pathological case where Goals might not converge to their targets, due to environment changes, interference from other Goals' plans or otherwise. In such case the actual Termination Check in 2.2.1 evaluates first the *formal termination conditions* for a Goal. If these are not satisfied, it generates a random number in the range [0, Elimination Probability] and compares this against the Termination Probability. If the Termination Probability is less than the random number, the Goal is eliminated from the GoalTree. In our current implementation the Termination Probability is a function of the distances between the current and the target states when sampled in appropriate intervals, e.g. every 5 secs. The Elimination Prob. is a more complex concept. In every processing step a Goal receives from the Motivational System updates for the *Commitment* and *Engagement* variables. Commitment in our architecture captures the persistence that the agent places to the Goal. Engagement captures higher-order effects that indicate agent withdrawal due to a prolonged period of low emotional importance (EW) or of suspended execution. Thus it is a function of EW and Termination Probability. We do not provide here formulas for Commitment and Engagement as these are actually components of the Motivational System and will be discussed there in a different paper. For the current paper it suffices to say that the Elimination Probability has been defined as in (2).

$$\text{ElimProb}=1.0-\text{Commitment}*\text{Engagement} \quad (2)$$

During Goal initialisation the corresponding values of Commitment and Engagement are set to 0.9. They are both bounded in [0, 1].

3.2 Goals, Plans and Priorities

So far we have mentioned that the Reasoning System uses partially a hierarchical / abstraction approach to planning. Without actually going to the details of how this might be achieved it is useful to identify the decomposition classes that might exist in our architecture. These include the following:

A. Independent sub-Goals (a. Sequential steps, b. Concurrent Cooperative, c. Concurrent Competitive)
B. Dependent sub-Goals (a. Serialisable, b. Block Serialisable, c. High-order dependence)

Let us offer some examples of the above patterns. Case A.a corresponds to a set of movement commands to a robot to visit points of interest in a room. Case A.b corresponds to the case of concurrent robot movements to achieve a pose against a table in order to lift an occluded object, while at the same time moves its robotic arm. Case A.c corresponds to an example of movement where we take care of collision avoidance with other objects in the environment. Serialisable Goals are those which a specific sequence of commands must be carried out in a given order, e.g. as in cooking a meal (case B.a). Block serialisable, B.b, is a sub-set of Goals that are not pair- wise independent either with goals in the same sub-set or with any other Goal. However, as a set are serialisable with other Goals that might exist, but themselves form a non-decomposable problem. The general non-decomposable problem is shown in B.c. We have not yet established how best to handle the priorities of a set of non-decomposable Goals. However, for the case of concurrent cooperative Goals the current model will be augmented to increase the priority of all goals in the set if any of the goals is a winner in its respective action-domain. This fact comes from the recognition that it does not make sense to try to approach a table in a suitable pose while the corresponding Grasp Goal in the Arm domain might be suppressed by another domain Goal, for example WaveHand. The Movement and Grasping commands, even though independent in a pair-wise manner, still form a group with dependencies in order to achieve the final Goal of *Grasping an object from the table*. In this case this Goal might be considered as an emergent Goal.

4 Results

The Computational Model in this paper has been tested both in a simulated and in a real robotic agent. Results on the priority of competing Goal families are shown in figure 4. The scenario used is the following: We provide the real robotic agent with a set of spatial goals in a grid [-1,1] x [-1,1] using consecutive commands of the type MoveTo (x,y) (SP1-4). We have used four spatial commands corresponding roughly to the four corners of the grid. After the second MoveTo command a MoveArm (AP1) command was issued asking the robotic arm to be extended in a forward direction as the robot starts executing the movement towards the third goal. While the robot was in transit a series of obstacles were placed in front of the third spatial goal so as to make this inaccessible. All goals of the scenario had equal Commitment, Engagement, EW and W values, so as the Motivation system did not affect them in the observed time scale. We see that goals SP1 and SP2 are executed and terminate sequentially as expected. When SP1 terminates, the Global Attentional competition assumes SP2 as the winner. The interest in this experiment lies in the behaviour

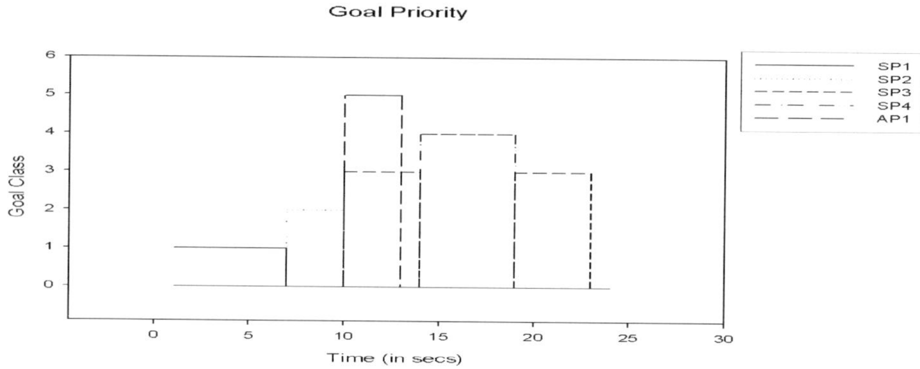

Fig. 4. Goal priorities for spatial goal SP1-4 and arm goal AP1. Duration of experiment is 25 secs for all goals to complete. The duration of each goal reflects the time that takes for the robotic agent to other move into space or extend the arm.

of goal SP3. After the termination of SP2, SP3 assumes execution and in parallel AP1 is executed as well. While the robot is in transit we place some obstacles in front of the SP3 location. The robot remains in the vicinity of SP3 for a while. After some seconds have passed the SP3 termination probability is lowered, thus resulting in SP4 becoming active. SP4 is now normally executed. After termination of SP4, SP3 is resumed again at which point we remove the obstacles and allow the Goal to be achieved and terminate. The interest in this case is the ability of the system to 're-member' given goals and react according to a changing environment. Goals may be suspended and re-activated again without being terminated (in a sensible time-scale). Repeating the experiment a second time with obstacles' removed after a longer time (360 secs) resulted in the constancy of the rate of target convergence and the corre-sponding lowering of the Termination Probability. Eventually the Termination Prob-ability became smaller then the Elimination Probability and the Goal was deleted from the GoalSpace as it should be expected for a goal that does not achieve any progress.

5 Discussion

We have presented a Computational Model that corresponds to the execution compo-nent of our Cognitive Architecture. We made a distinction between a plan coming from the Reasoning System and its actual run-time monitoring which raises re-plan and learning requests among other things. We have discussed in detail the concept of Goal for our architecture and we provided our justification for this approach. We also discussed numerous Goal attributes that influence Goal priority and termination. These attributes provide the necessary integration links with other components of the Cognitive Architecture. Finally we discussed the problem of priorities for classes of plan decompositions. There is still future work in this direction so as to cover all iden-tifiable classes. Finally there is one more issue regarding priority. One can imagine a case where a losing goal can very speedily be satisfied; thus to allow *opportunistic execution*. In this way we will maximise the throughput of the agent. However, one should investigate further if this increased processing throughput will lead actually to a higher motivational level, which is the ultimate cost function that must be optimised. This idea is already implemented in (1) but it needs further work to test its value. One has to make more concrete the expected gain of the approach by studying in concert the computational model together with the motivational system.

Acknowledgements

We would like to acknowledge the support of the European Union through the IST GNOSYS project (FP6-003835) of the Cognitive Systems Initiative to our work.

References

1. Rusell, S., Norving, P. *Artificial Intelligence: A Modern Approach,* Prentice Hall, 2nd Ed., (2003).
2. Korf, R. 'Planning as Search: A Quantitative approach', *Artificial Intelligence* 33, 65-88, (1987).
3. Knoblock, C. 'Learning Abstraction Hierarchies for Problem Solving', In *Proc. of the Eighth National Conf. on Artificial Intelligence*, Boston, MA, (1990).
4. Kasderidis S. & Taylor J. G., 'Attentional Agents and Robot Control', *International Journal of Knowledge-based and Intelligent Systems* 8, 69-89, (2004).
5. Kasderidis, S., Taylor, J.G. 'Combining Attention & Value Maps', *In Proc. of 15th Int. Conf. on Artificial Neural Networks (ICANN 2005),* Warsaw, Poland, Vol. I, Ed. W. Duch et al, Lect. Notes in Comp. Sci, Vol 3696, pp. 79-84.

Detection of a Dynamical System Attractor from Spike Train Analysis

Yoshiyuki Asai[1,2], Takashi Yokoi[1], and Alessandro E.P. Villa[2,3,4]

[1] National Institute of Advanced Industrial Science and Technology (AIST), Tsukuba, Japan
{yoshiyuki.asai, takashi-yokoi}@aist.go.jp,
[2] NeuroHeuristic Research Group, INFORGE Institute of Computer Science and Organization, University of Lausanne, Switzerland
{oyasai, avilla}@neuroheuristic.org
[3] INSERM U318, Laboratoire db Neurosciences Précliniques, Grenoble, France
[4] Laboratoire de Neurobiophysique, University Joseph Fourier, Grenoble, France
alessandro.villa@ujf-grenoble.fr
http://openadap.net:8080/oan/

Abstract. Dynamics of the activity of neuronal networks have been intensively studied from the view point of the nonlinear dynamical system. The neuronal activities are recorded as multivariate time series of the epochs of spike occurrences–the spike trains–which are often effected by intrinsic and measuring noise. The spike trains can be considered as a mixture of a realization of deterministic and stochastic processes. Within this framework we considered several simulated spike trains derived from the Zaslavskii map with additive noise. The ensemble of all preferred firing sequences detected by the pattern grouping algorithm (PGA) in the noisy spike trains form a new time series that retains the dynamics of the original mapping.

1 Introduction

Neurons can be considered as nonlinear gating elements, and the activity of each neuron propagates through the entire neural network. The electrophysiological recordings of neural activity usually generate multivariate time series corresponding to the timing of spike discharges. These time series, often referred to as spike trains, can be considered as observed realizations of dynamics with high degree of freedom embedded in a small subspace. The dynamical system theory provides powerful tools to exploit such time series, to infer the underlying system. The recorded time series are often affected by random noise such that some realizations are missed by the observer and some observations are not associated to the dynamical system but correspond to the realizations of a stochastic process that depends on the method of measurement. The possibility to filter out the noisy components from the time series observed in nature may be a clue to ascertain the deterministic feature of the underlying dynamical process and to study the topological characteristics of the attractor.

S. Kollias et al. (Eds.): ICANN 2006, Part I, LNCS 4131, pp. 623–631, 2006.

In a previous study we showed the main idea of a noise filtering procedure based on the pattern grouping algorithm (PGA) (Villa and Tetko, 1999; Tetko and Villa, 2001). The algorithm was able to detect temporal patterns of events that repeated more frequently than expected by chance. It was developed to detect preferred firing sequences from simultaneously recorded multivariate time series of neural activities (Villa and Tetko, 1999). The detection of patterns of firing within a noisy time series may be exploited to reconstruct a time series related to the generating attractors, if any. In this article we investigate the performance of denoising by PGA depending on two parameters, *i.e.*, the maximum duration of a pattern and time jitter in a simulated spike train generated from the Zaslavskii map.

2 Methods

2.1 Simulated Spike Train

We consider the Zaslavskii map (Zaslavskii, 1978), which is well-known as a discrete dynamical system exhibiting chaotic behavior, to produce the simulated spike train. Let us consider the following equations

$$\begin{cases} x_{n+1} = x_n + v(1 + \mu y_n) + \varepsilon v \mu \cos x_n & (mod.\ 2\pi) \\ y_{n+1} = e^{-\gamma}(y_n + \varepsilon \cos x_n) \ , \end{cases} \tag{1}$$

where $x,\ y \in \mathbf{R}$, the parameters are real numbers with $\mu = \frac{1-e^{-\gamma}}{\gamma}$, $v = \frac{4}{3} \cdot 100$.

With this parameter set, the system exhibits a chaotic behavior. The initial conditions were set to $x_0 = 0.3$ and $y_0 = 0.3$. By iterative calculation, we get the time series $\{x_n\}$. A new time series $\{w_n\}$ is derived by taking the difference between two adjacent values of $\{x_n\}$, and adding a constant C to make all values positive, *i.e.* , $w_n = x_{n+1} - x_n + C$, where $C = \min\{(x_{n+1} - x_n)\} + 0.1$. The time series $\{w_n\}$ are assumed to correspond to the sequence of the inter-spike-intervals. In order to let the time series be comparable to the observed dynamics in usual neurophysiological experiments, the simulated time series was rescaled in time, and had 3 *events/s* (*i.e.*, 3 *spikes/s* in neurophysiological terms) on average, and the unit time corresponded to "*ms*". Ten thousand points ($N = 10,000$) were generated in each series. The plot of w_{n+1} against w_n–return map–of the simulated spike train, without noise, is shown in Fig. 1(a).

2.2 Noise

Let us consider two types of the observational noises: (i), an additive noise corresponding to the random insertion and deletion of points in the original time series; (ii), a jitter noise corresponding to a slight shift in time of the points of the time series. Let us consider an original time series of 10,000 points. The procedure to introduce the noise in the time series is the following. At first, P_d % of the total amount of points was chosen stochastically according to a uniform distribution and deleted from $\{w_n\}$. As a result, the time series $(\{w'_n\})$

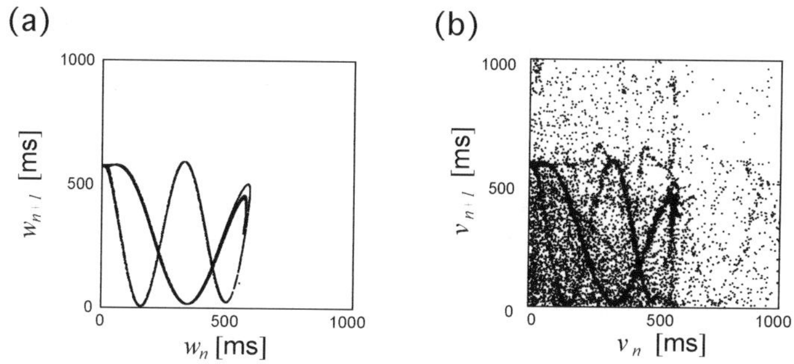

Fig. 1. (a) Return maps of original time series, and (b) one with additive noise (with $P_d = 20\%$, $P_a = 20\%$) and jitter noise ($J = 5$)

includes $(10,000 - 100 \times P_d)$ events. Secondly, the jitter noise was applied to $\{w'_n\}$. The time of occurrence of each point in $\{w'_n\}$ was shifted Δt in time, where the amount of jitter Δt was a value in $[-J, J]$ uniformly distributed. The total counts of points in the noisy time series $\{w''_n\}$ remained the same of $\{w'_n\}$. Thirdly, we added new points to $\{w''_n\}$ following a uniform distribution without any overlapping to any existing point. The percentage of newly inserted points is P_a % of the total number of points of $\{w_n\}$. This overall procedure produces a noisy simulated spike train, referred to as $\{v_n\}$, with $(10,000 + 100 \times (P_a - P_d))$ points. The return map of a noisy simulated spike train is shown in Fig. 1(b).

2.3 Detection of Temporal Patterns

Preferred firing sequences are temporal patterns defined as sequences of spikes (formed by three or more) with high temporal precision of the order of milliseconds. The pattern detection algorithm begins with finding all single or multineuron sequences of intervals that repeat two or more times within a record. Secondly, the "pattern grouping algorithm (PGA)" (Villa and Tetko, 1999; Tetko and Villa, 2001) computes how many of such sequences of intervals can be expected by chance, clusterizes into one group those sequences of intervals with slight difference in spike timing, and provides confidence limits for this estimation. The general notation describes the timing features of a triplet as follows: $< a, b, c\ ;\ t_1 \pm \Delta t_1, t_2 \pm \Delta t_2 >$. This means that the pattern starts with a spike of unit #a, then $t_1 \pm \Delta t_1$ ms later a spike of unit #b and a third spike of unit #c $t_2 \pm \Delta t_2$ ms from pattern start. In this article, since we considered univariate time series, all events had the same label "1" (*i.e.*, $a = b = c = 1$).

The algorithm is controlled by two main parameters. The *window duration* determines the maximal duration of the temporal pattern, *i.e.*, the interval between the first and the last event in a pattern, to be considered in the analysis. The *jitter* determines the time precision of the events within the temporal pattern. In the current study the jitter of detection was set ± 5 ms for all analyses.

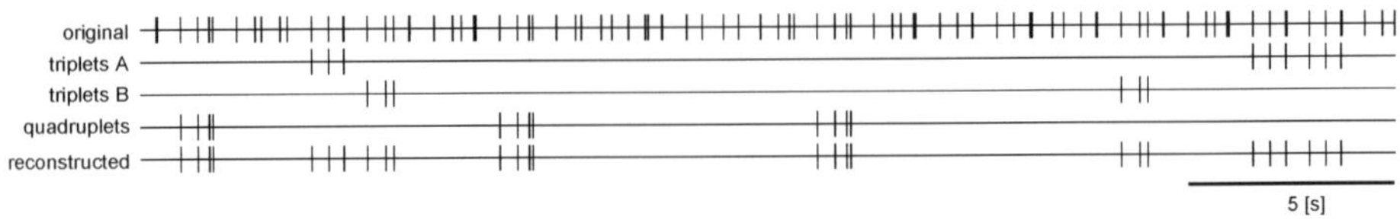

Fig. 2. The top line shows the original time series. The second line shows a triplet occurring 3 times in the original series. The third line shows a triplet occurring 2 times and the fourth line shows a quadruplet occurring 3 times. If these patterns were the only ones that were detected, then the reconstructed time series would correspond to the ensemble of points that belonged to the patterns, as shown in the bottom line.

2.4 Reconstruction of Time Series

The procedure of time series reconstruction by using the PGA algorithm as a denoising filter can be easily illustrated as follows. Firstly, the PGA algorithm detects a number of temporal patterns that repeated above the chance level. Secondly, the points belonging to all repetitions of those patterns are sorted out from the original series and grouped into a new series, which is actually a subset of the original (Fig. 2).

3 Results

In the absence of noise $\{w_n\}$ and with window parameter 800 ms PGA found 105 types of quadruplet patterns and 154 types of triplets patterns The overall amount of points that belonged to the distinct 259 temporal patterns was 7,321 (about 73% of the points of $\{w_n\}$). Such reconstructed time series, in the absence of noise, is labelled R_0. Fig. 3a shows a quadruplet of R_0.

The quadruplet is denoted $< 1, 1, 1, 1; 423 \pm 1.0, 705 \pm 4.5, 789 \pm 5.5 >$, meaning that the first event at time 0, the second 423 ms later with ± 1.0 ms time precision, the third 705 ± 4.5 ms later of the first event, and the last one occurred 789 ± 5.5 ms after the start of the pattern. We found 33 repetitions of this specific temporal pattern. It is interesting to notice a fifth occurrence in Fig. 3a, with a latency of 1,359 ms from pattern start. Because of the limitation of the window parameter to 800 ms the pentaplet was not detected as such. The same procedure is applied to different conditions of noise. Fig. 3b shows a triplet pattern detected in noisy time series $\{v_n\}$ (with $P_d = 20\%, P_a = 20\%$ and J = 5), which is described as $< 1, 1, 1; 423 \pm 5.5, 786 \pm 5.5 >$ and recurred 26 times. The triplet shown in the figure matches one of the possible subpatterns produced by the quadruplet of Fig. 3a. It is of interest to notice that two events, the third and the fifth, belonging to the temporal pattern detected in the absence of noise were less preserved, but still visible by naked eye in the raster display of Fig. 3b.

The performance of PGA as a function of window duration was studied with fixed jitter of detection of $\pm 5ms$. The reconstructed time series R_0 in absence of noise with window durations of 800, 1200 and 1600 ms, from $\{w_n\}$ are illustrated by Fig. 4a-c. The oblique line is a bias due to the size of window duration, which

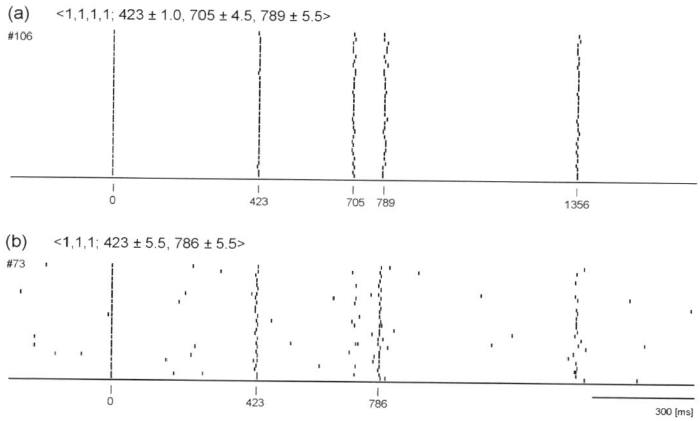

Fig. 3. Raster display of patterns aligned by their first event at time 0 which were found by PGA with window parameter 800 *ms* and fixed time accuracy at ± 5 *ms*. (a) Quadruplets (n=33) denoted as $< 1,1,1,1; 423\pm 1.0, 705\pm 4.5, 789\pm 5.5 >$. (b) Triplets ($n = 26$) denoted as $< 1,1,1; 423 \pm 5.5, 786 \pm 5.5 >$.

is expressed as $r_{n+1} = -r_n + D$, where $D \in [800, 1200, 1600]$. The window duration effect is also illustrated by the return maps at Fig. 4d-f in noisy time series $\{v_n\}$ with parameters $P_d = 20\%, P_a = 20\%$ and $J = 5$.

The points included in any reconstructed time series R as well as in the original time series $\{w_n\}$, are denoted using the logical expression as $W \cap R$. The points belonging to this intersection are interesting as they are related to the original deterministic process found in $\{v_n\}$. The number of points in R_0, *i.e.*, the reconstructed time series in the absence of noise can be considered as the maximum number of detectable points in W given fixed parameters of PGA. The reconstruction index (RI) is defined as a ratio of the number of points in $W \cap R$ to the number of points of R_0. A larger RI means better detection of the original dynamics masked by noise. $R \cap W^c$, where W^c represents a complementary set of W, is a set of events in R but not included in W, *i.e.*, elements in this set are events either added as noise to $\{w_n\}$ or events which were in $\{w_n\}$ and shifted by the jitter noise. The size of $R \cap W^c$ also provides an indication of the ability of the reconstruction (smaller set, better reconstruction).

Table 1 summarizes the sample sizes of the reconstructed series and corresponding RI as a function of the window duration. Notice that RI increased as the window duration became longer but the efficacy of the increment of RI decreased for larger windows. For example, the difference between RIs for window parameter 800 and 1200 *ms* was 10.3 (=43.7 - 33.4). This difference was 5.8 between RIs with window parameter 1400 and 1600 *ms*.

The robustness of the denoising property of PGA to reconstruct the original time series was investigated with nine types of noisy time series determined by the combination of three additive noise $(P_d, P_a) \in [(20,20),(10,20),(20,10)]$ and three levels of jitter noise $J \in [2,3,5]$. The return maps of the nine noisy

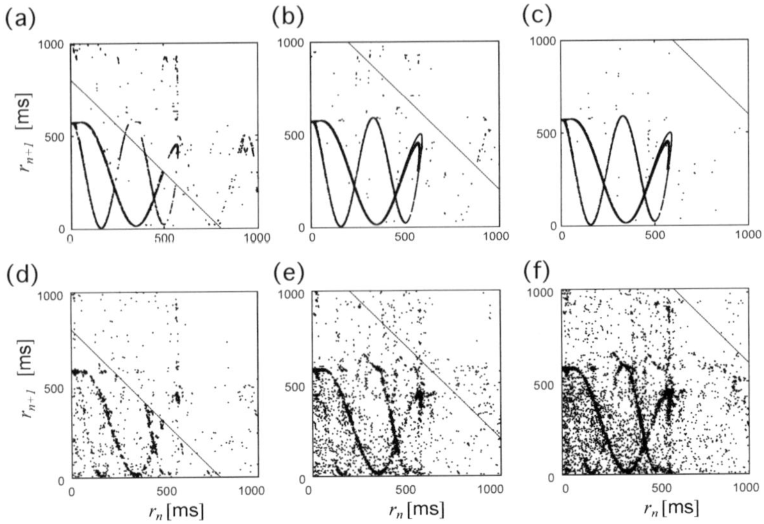

Fig. 4. (a-c) The return maps of the reconstructed time series from $\{w_n\}$ shown in Fig. 1(a) by PGA with window parameter = 800, 1200 and 1600 ms, respectively. (d-e) The return map of the reconstructed time series from $\{v_n\}$ with $P_d = 20\%, P_a = 20\%$ and $J = 5$ shown in Fig. 1b.

Table 1. Number of events in reconstructed time series by PGA with various window parameters and fixed time accuracy of 5 ms. P_3: triplets; P_4:quadruplets. See text for definition of the other symbols.

Window	Points in the time series							N	
Duration [ms]	W, V	R_0	R	$V \cap R^c$	$W \cap R$	$R \cap W^c$	RI	P_4	P_3
800	10000	7321	2711	7289	2444	267	33.4	17	48
1000	10000	9336	4556	5444	4078	478	43.7	39	76
1200	10000	9723	5614	4386	5002	612	51.5	84	99
1400	10000	9907	6496	3504	5722	774	57.8	178	137
1600	10000	9966	7256	2744	6330	926	63.6	292	173

time series are shown at Fig. 5. The reconstruction of the noisy time series was performed by applying a 1,000 ms window duration. The corresponding return maps of the reconstructed time series are shown at Fig. 6. For example, a panel at left top corner in Fig. 6 shows the reconstructed time series from the time series with noise of $P_d = 20\%, P_a = 20\%$ and $J = 2$ shown in panel at left top corner in Fig. 5.

Table 2 summarizes the sample sizes of the reconstructed series and corresponding RI as a function of the parameters (P_d, P_a) and J used to generate the noisy series. For any type of additive noise, larger level of jitter noise provoked a decrease in performance as measured by RI, in particular with jitter noise $J = 8$, which is larger than the jitter accuracy of PGA ($\pm 5ms$) used in this

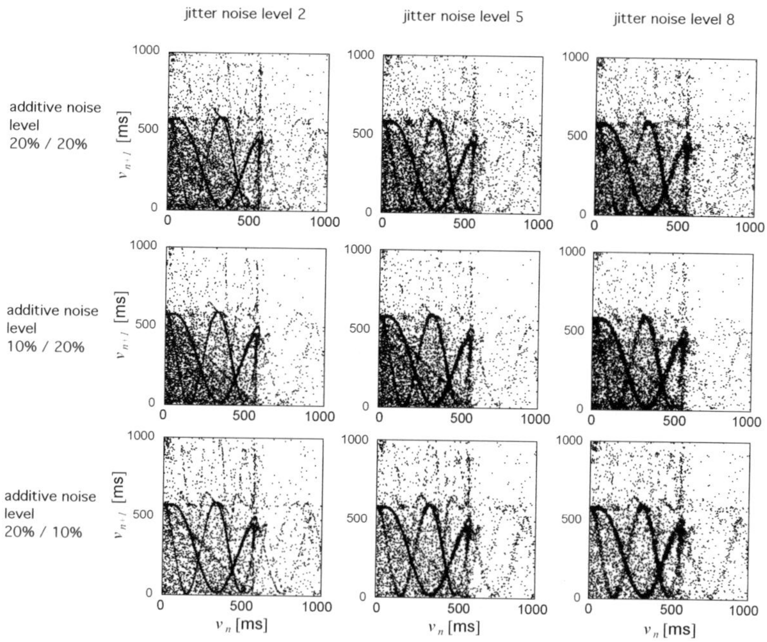

Fig. 5. The return maps of the original time series after insertion of various noise. We tested three combination of additive noise, as noted left side of figure in order $P_d\%$ / $P_a\%$. For each additive noise, three levels of jitter noise were considered as shown top of figure.

Table 2. Number of events in the reconstructed time series including various noise by PGA with a fixed window duration at 1,000 ms and time precision of 5 ms. P_3: triplets; P_4:quadruplets. See text for definitions of the other symbols.

Noise		Points in the time series							N	
P_d, P_a	J	W, V	R_0	R	$V \cap R^c$	$W \cap R$	$R \cap W^c$	RI	P_4	P_3
20%, 20%	2	10000		5831	4169	5271	560	56.5	87	109
	5	10000	9336	4556	5444	4078	478	43.7	39	76
	8	10000		3887	6113	2252	1635	24.1	27	68
10%, 20%	2	11000		7589	3411	6851	738	67.0	153	123
	5	11000	10218	6307	4693	5725	581	56.0	88	92
	8	11000		5318	5682	3199	2119	31.3	60	74
20%, 10%	2	9000		5812	3188	4982	830	59.2	90	144
	5	9000	8420	4442	4558	3809	628	45.2	45	94
	8	9000		3623	5377	2070	1550	24.6	24	81

analysis. It interesting to notice that the sample size of $R \cap W^c$ is minimal with jitter noise $J = 5$ regardless the type of additive noise. This is due to the fact that the jitter noise matches the jitter accuracy of PGA. It is also interesting to notice that adding 10 or 20% points at random had little influence on the final performance with regard to an identical rate of 20% of deletion.

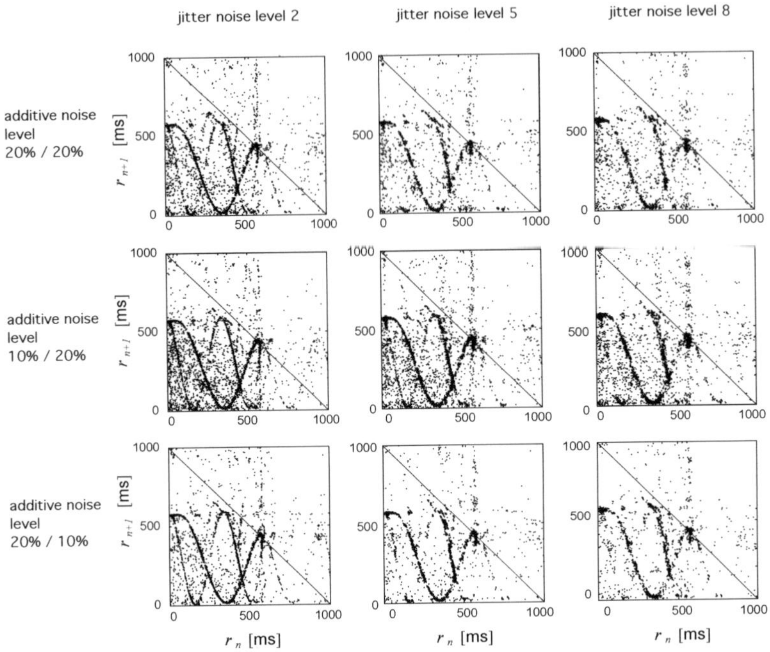

Fig. 6. The return maps of the reconstructed time series from the time series inserted various noise shown in Fig. 5 by PGA with window duration = 1,000 *ms* and time precision of 5 *ms*. The positions of panels correspond to those of Fig. 5.

4 Discussion

The result presented here show that the PGA algorithm could efficiently detect a significant amount of events belonging to an attractor of a deterministic process out of a noisy spike train characterized by deterministic and stochastic processes. The efficiency of the reconstruction depended on the parameter related to the algorithm as well as that of noise. The dependency of denoising by PGA on the parameter of the window duration (Fig. 4) suggested that the longer the duration of the window the better is the performance. However, the amount of noisy points ($R \cap W^c$) also increased with an increase in window duration, thus suggesting that there might be a compromise to select an optimal window duration. The goal of this study was to determine the suitability of this technique to search for dynamical system attractors embedded in the spike trains and not at seeking the most appropriate parameters of PGA. It is clear from Fig. 6 that the denoising performance of PGA is depending on the nature of noise, and, as reported previously, on the kind of deterministic dynamics. This means that there might be no fixed suitable parameters for PGA to provide the best performance in all cases.

We have been trying to apply PGA algorithm to detect recurrent patterns from the electrophysiologically obtained time series (Villa and Tetko, 1999). It has been proposed that sequences of synchronous activity ("synfire chain") propagate through the cortical network assuming diverging/converging feed-forward links with high temporal fidelity (Abeles and Gerstein, 1988). According to this theory, the reactivation of the same chain would produce the repetitions of precise spatio-temporal firing patterns. If such chains exist, it is supposed that PGA algorithm can detect those patterns. In this manuscript we demonstrated that PGA can successfully detect patterns from the chaotic process generated by discrete deterministic dynamical system with presence of noise. This may open the way not merely to the detection of dynamical systems (Celletti and Villa, 1996; Segundo, 2003) in neural activity but even to the reconstruction of the generating attractor. The presence of preferred firing sequences and the presence of dynamical systems may represent two faces of the same coin as recently suggested (Villa, 2000).

References

Abeles, M. and Gerstein, G. (1988). Detecting spatiotemporal firing patterns among simultaneously recorded single neurons, *J. Neurophysiol.* **60**: 909–924.

Celletti, A. and Villa, A. E. P. (1996). Low dimensional chaotic attractors in the rat brain, *Biological Cybernetics* **74**: 387–394.

Segundo, J. P. (2003). Nonlinear dynamics of point process systems and data, *International Journal of Bifurcation and Chaos* **13**: 2035–2116.

Tetko, I. V. and Villa, A. E. P. (2001). A pattern grouping algorithm for analysis of spatiotemporal patterns in neuronal spike trains. 1. detection of repeated patterns, *J. Neurosci. Meth.* **105**: 1–14.

Villa, A. E. P. (2000). Empirical evidence about temporal structure in multi-unit recordings, *in* R. Miller (ed.), *Time and the Brain*, Harwood Academic Publishers, chapter 1, pp. 1–51.

Villa, A. E. P. and Tetko, I. V. (1999). Spatiotemporal activity patterns detected from single cell measurements from behaving animals, *Proceedings SPIE* **3728**: 20–34.

Zaslavskii, G. M. (1978). The simplest case of a strange attractor, *Phys. Let.* **69A**: 145–147.

Recurrent Neural Networks Are Universal Approximators

Anton Maximilian Schäfer[1,2] and Hans Georg Zimmermann[1]

[1] Information & Communications, Learning Systems
Siemens AG, Corporate Technology, 81739 Munich, Germany
{Schaefer.Anton.ext, Hans_Georg.Zimmermann}@siemens.com
[2] Department Optimisation and Operations Research, University of Ulm, 89069 Ulm, Germany

Abstract. Neural networks represent a class of functions for the efficient identification and forecasting of dynamical systems. It has been shown that feedforward networks are able to approximate any (Borel-)measurable function on a compact domain [1,2,3]. Recurrent neural networks (RNNs) have been developed for a better understanding and analysis of open dynamical systems. Compared to feedforward networks they have several advantages which have been discussed extensively in several papers and books, e.g. [4]. Still the question often arises if RNNs are able to map every open dynamical system, which would be desirable for a broad spectrum of applications. In this paper we give a proof for the universal approximation ability of RNNs in state space model form. The proof is based on the work of Hornik, Stinchcombe, and White about feedforward neural networks [1].

1 Introduction

Recurrent neural networks (RNNs) allow the identification of dynamical systems in form of high dimensional, nonlinear state space models. They offer an explicit modeling of time and memory [5].

In previous papers, e.g. [6], we discussed the modeling of open dynamical systems based on time-delay recurrent neural networks which can be represented in a state space model form. We solved the system identification task by finite unfolding in time, i.e., we transferred the temporal problem into a spatial architecture [6], which can be handled by error backpropagation through time [7]. Further we enforced the learning of the autonomous dynamics in an open system by overshooting [6]. Consequently our RNNs not only learn from data but also integrate prior knowledge and first principles into the modeling in form of architectural concepts. However, the question arises if the outlined RNNs are able to identify and approximate any open dynamical system, i.e., if they hold an universal approximation ability.

In 1989 Hornik, Stinchcombe, and White [1] could show that any Borel-measurable function on a compact domain can be approximated by a three-layered feedforward network, i.e., a feedforward network with one hidden layer, with an arbitrary accuracy. In the same year Cybenko [2] and Funahashi [3] found similar results, each with different methods. Whereas the proof of Hornik, Stinchcombe, and White [1] is based on the Stone-Weierstrass theorem, Cybenko [2] makes in principle use of the Hahn-Banach

S. Kollias et al. (Eds.): ICANN 2006, Part I, LNCS 4131, pp. 632–640, 2006.

und Riesz theorem. Funahashi [3] mainly applies the Irie-Miyake and the Kolmogorov-Arnold-Sprecher theorem.

Some work has already been done on the capability of RNN to approximate measurable functions, e.g. [8]. In this paper we focus on open dynamical systems and prove that those can be approximated by RNNs in state space model form with an arbitrary accuracy. We start with a short introduction on open dynamical systems and RNNs in state space model form (sec. 2). We further recall the basic results of the universal approximation theorem of Hornik, Stinchcombe, and White [1] (sec. 3). Subsequent we show that these results can be extended to RNNs in state space model form and we consequently give a proof for their universal approximation ability (sec. 4). We conclude with a short summary and an outlook on further research (sec. 5).

2 Open Dynamical Systems and Recurrent Neural Networks

Figure 1 illustrates an open dynamical system in discrete time which can be described as a set of equations, consisting of a state transition and an output equation [5,6]:

$$\begin{aligned} s_{t+1} &= g(s_t, u_t) && \text{state transition} \\ y_t &= h(s_t) && \text{output equation} \end{aligned} \tag{1}$$

The state transition is a mapping from the present internal hidden state of the system s_t and the influence of external inputs u_t to the new state s_{t+1}. The output equation computes the observable output y_t.

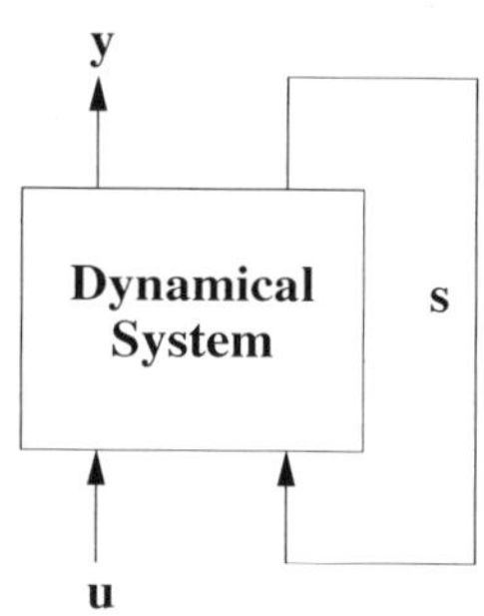

Fig. 1. Open dynamical system with input u, hidden state s and output y

The system can be viewed as a partially observable autoregressiv dynamic state transition $s_t \rightarrow s_{t+1}$ that is also driven by external forces u_t. Without the external inputs the system is called an autonomous system [5]. However, most real world systems are driven by a superposition of an autonomous development and external influences.

If we assume that the state transition does not depend on s_t, i.e., $y_t = h(s_t) = h(g(u_{t-1}))$, we are back in the framework of feedforward neural networks [6]. However, the inclusion of the internal hidden dynamics makes the modeling task much harder, because it allows varying inter-temporal dependencies. Theoretically, in the recurrent framework an event s_{t+1} is explained by a superposition of external inputs $u_t, u_{t-1}, \ldots$ from all the previous time steps [5].

In previous papers, e.g. [6], we proposed to map open dynamical systems (eq. 1) by a recurrent neural network (RNN) in state space model form

$$\begin{aligned} s_{t+1} &= f(As_t + Bu_t + \theta) \quad \text{state transition} \\ y_t &= Cs_t \quad \text{output equation} \end{aligned} \tag{2}$$

where A, B, and C are weight matrices of appropriate dimensions and θ is a bias, which handles offsets in the input variables u_t [5,6]. f is the so called activation function of the network which is typically sigmoidal (def. 3) like e.g., the hyperbolic tangent.

A major advantage of RNNs written in form of a state space model (eq. 2) is the explicit correspondence between equations and architecture. It is easy to see, that the set of equations (2) can be directly transferred into a spatial neural network architecture using so called finite unfolding in time and shared weight matrices A, B, and C [5,7]. Figure 2 depicts the resulting model [6].

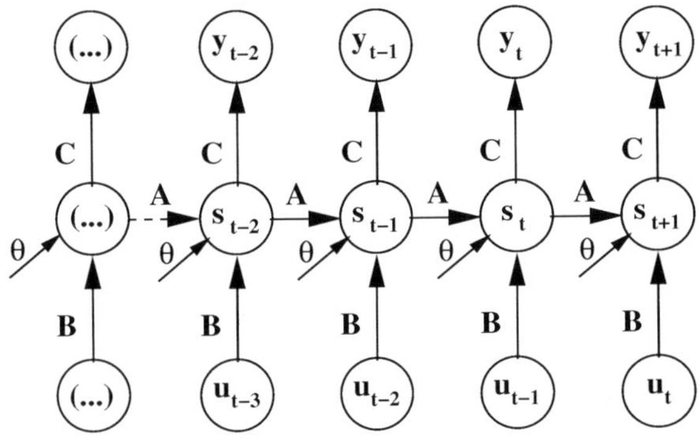

Fig. 2. Recurrent neural network unfolded in time

A more detailed description about RNNs in state space model form can be found in [4] or [6].

3 Universal Approximation Theorem for Feedforward Neural Networks

Our proof for RNNs in state space model form (sec. 4) is based on the work of Hornik, Stinchcombe und White [1]. In the following we therefore recall their definitions and main results:

Definition 1. *Let* A^I *with* $I \in \mathbb{N}$ *be the set of all affine mappings* $A(x) = w \cdot x - \theta$ *from* $\mathbb{R}^I$ *to* $\mathbb{R}$ *with* $w, x \in \mathbb{R}^I$ *and* $\theta \in \mathbb{R}$. *'·' denotes the scalar product.*

Transferred to neural networks x corresponds to the input, w to the network weights and θ to the bias.

Definition 2. *For any (Borel-)measurable function $f(\cdot):\mathbb{R}\rightarrow\mathbb{R}$ and $I\in\mathbb{N}$ be $\sum^I(f)$ the class of functions*

$$\{NN:\mathbb{R}^I\rightarrow\mathbb{R}:NN(x)=\sum_{j=1}^{J}v_jf(A_j(x)),x\in\mathbb{R}^I,v_j\in\mathbb{R},A_j\in\mathrm{A}^I,J\in\mathbb{N}\}. \tag{3}$$

Here NN stands for a three-layered feedforward neural network, i.e., a feedforward network with one hidden layer, with I input-neurons, J hidden-neurons and one output-neuron. v_j denotes the weights between hidden- and output-neurons. f is an arbitrary activation function (sec. 2).

Remark 1. *The function class $\sum^I(f)$ can also be written in matrix form*

$$NN(x)=vf(Wx-\theta) \tag{4}$$

where $x\in\mathbb{R}^I$, $v,\theta\in\mathbb{R}^J$ and $W\in\mathbb{R}^{J\times I}$.

In this context the computation of the function $f(\cdot):\mathbb{R}^J\rightarrow\mathbb{R}^J$ be defined component-wise, i.e.,

$$f(Wx-\theta):=\begin{pmatrix} f(W_1\cdot x-\theta_1)\\ \vdots\\ f(W_j\cdot x-\theta_j)\\ \vdots\\ f(W_J\cdot x-\theta_J)\end{pmatrix} \tag{5}$$

where W_j $(j=1,\ldots,J)$ denotes the $j-th$ row of the matrix W.

Definition 3. *A function f is called a sigmoid function, if f is monotonically increasing and bounded, i.e.,*

$$f(a)\in[\alpha,\beta],\ \textit{whereas}\ \lim_{a\rightarrow-\infty}f(a)=\alpha\ \textit{and}\ \lim_{a\rightarrow\infty}f(a)=\beta \tag{6}$$

with $\alpha,\beta\in\mathbb{R}$ and $\alpha<\beta$. In the following we define $\alpha=0$ and $\beta=1$ which bounds the sigmoid function on the interval $[0,1]$.

Definition 4. *Let $\mathcal{C}^I$ and $\mathcal{M}^I$ be the sets of all continuous and respectively all Borel-measurable functions from $\mathbb{R}^I$ to $\mathbb{R}$. Further denote $\mathbb{B}^I$ the Borel-σ-algebra of $\mathbb{R}^I$ and $(\mathbb{R}^I,\mathbb{B}^I)$ the I-dimensional Borel-measurable space.*

$\mathcal{M}^I$ contains all functions relevant for applications. $\mathcal{C}^I$ is a subset of it. Consequently, for every Borel-measurable function f the class $\sum^I(f)$ belongs to the set $\mathcal{M}^I$ and for every continuous f to its subset $\mathcal{C}^I$.

Definition 5. *A subset S of a metric space (X,ρ) is ρ-dense in a subset T, if there exists, for any $\varepsilon>0$ and any $t\in T$, $s\in S$, such that $\rho(s,t)<\varepsilon$.*

This means that every element of S can approximate any element of T with an arbitrary accuracy. In the following we replace T and X by $\mathcal{C}^I$ and $\mathcal{M}^I$ respectively and S by $\sum^I(f)$ with an arbitrary but fixed f. The metric ρ is chosen accordingly.

Definition 6. *A subset S of $\mathcal{C}^I$ is uniformly dense on a compact domain in $\mathcal{C}^I$, if, for any compact subset $K \subset \mathbb{R}^I$, S is ρ_K-dense in $\mathcal{C}^I$, where for $f, g \in \mathcal{C}^I$ $\rho_K(f,g) \equiv \sup_{x \in K} |f(x) - g(x)|$.*

Definition 7. *Given a probability measure μ on $(\mathbb{R}^I, \mathbb{B}^I)$, the metric $\rho_\mu : \mathcal{M}^I \times \mathcal{M}^I \to \mathbb{R}^+$ be defined as follows*

$$\rho_\mu(f,g) = \inf\{\varepsilon > 0 : \mu\{x : |f(x) - g(x)| > \varepsilon\} < \varepsilon\}. \tag{7}$$

Theorem 1. ***(Universal Approximation Theorem for Feedforward Networks)***
For any sigmoid activation function f, any dimension I and any probability measure μ on $(\mathbb{R}^I, \mathbb{B}^I)$, $\sum^I(f)$ is uniformly dense on a compact domain in $\mathcal{C}^I$ and ρ_μ-dense in $\mathcal{M}^I$.

This theorem states that a three-layered feedforward neural network, i.e., a feedforward neural network with one hidden layer, is able to approximate any continuous function uniformly on a compact domain and any measurable function in the ρ_μ-metric with an arbitrary accuracy. The proposition is independent of the applied sigmoid activation function f (def. 3), the dimension of the input space I, and the underlying probability measure μ. Consequently three-layered feedforward neural networks are universal approximators.

Theorem 1 is only valid for feedforward neural networks with I input-, J hidden- and a single output-neuron. Accordingly, only functions from $\mathbb{R}^I$ to $\mathbb{R}$ can be approximated. However with a simple extension it can be shown that the theorem holds for networks with a multiple output (cor. 1).

For this, the set of all continuous functions from $\mathbb{R}^I$ to $\mathbb{R}^n$, $I, n \in \mathbb{N}$, be denoted by $\mathcal{C}^{I,n}$ and the one of (Borel-)measurable functions from $\mathbb{R}^I$ to $\mathbb{R}^n$ by $\mathcal{M}^{I,n}$ respectively. The function class $\sum^I$ gets extended to $\sum^{I,n}$ by (re-)defining the weights v_j ($j = 1, \ldots, J$) in definition 2 as $n \times 1$ vectors. In matrix-form the class $\sum^{I,n}$ is then given by

$$NN(x) = Vf(Wx - \theta) \tag{8}$$

with $x \in \mathbb{R}^I, \theta \in \mathbb{R}^J, W \in \mathbb{R}^{J \times I}$ and $V \in \mathbb{R}^{n \times J}$. The computation of the function $f(\cdot) : \mathbb{R}^J \to \mathbb{R}^J$ be once more defined component-wise (rem. 1).

In the following, function $g : \mathbb{R}^I \to \mathbb{R}^n$ has got the elements g_k, $k = 1, \ldots, n$.

Corollary 1. *Theorem 1 holds for the approximation of functions in $\mathcal{C}^{I,n}$ and $\mathcal{M}^{I,n}$ by the extended function class $\sum^{I,n}$. Thereby the metric ρ_μ is replaced by $\rho_\mu^n := \sum_{k=1}^{n} \rho_\mu(f_k, g_k)$.*

Consequently three-layered multi-output feedforward networks are universal approximators for vector-valued functions.

4 Universal Approximation Theorem for RNNs

The universal approximation theorem for feedforward neural networks (theo. 1) proves, that any (Borel-)measurable function can be approximated by a three-layered feedforward neural network. We now show, that RNNs in state space model form (eq. 2) are also universal approximators and able to approximate every open dynamical system (eq. 1) with an arbitrary accuracy.

Definition 8. *For any (Borel-)measurable function $f(\cdot) : \mathbb{R}^J \to \mathbb{R}^J$ and $I, n \in \mathbb{N}$ be $RNN^{I,n}(f)$ the class of functions*

$$\begin{aligned} s_{t+1} &= f(As_t + Bu_t - \theta) \\ y_t &= Cs_t \quad . \end{aligned} \tag{9}$$

Thereby be $u_t \in \mathbb{R}^I$, $s_t \in \mathbb{R}^J$ and $y_t \in \mathbb{R}^n$, with $t = 1, \ldots, T$. Further be the matrices $A \in \mathbb{R}^{J \times J}$, $B \in \mathbb{R}^{J \times I}$, and $C \in \mathbb{R}^{n \times J}$ and the bias $\theta \in \mathbb{R}^J$. In the following, analogue to remark 1, the calculation of the function f be defined component-wise, i.e.,

$$s_{t+1_j} = f(A_j s_t + B_j u_t - \theta_j), \tag{10}$$

where A_j and B_j $(j = 1, \ldots, J)$ denote the $j - th$ row of the matrices A and B respectively.

It is obvious, that the class $RNN^{I,n}(f)$ is equivalent to the RNN in state space model form (eq. 2). Analogue to its description in section 2 as well as definition 2, I stands for the number of input-neurons, J for the number of hidden-neurons and n for the number of output-neurons. u_t denotes the external inputs, s_t the inner states and y_t the outputs of the neural network. The matrices A, B, and C correspond to the weight-matrices between hidden- and hidden-, input- and hidden- and hidden- and output-neurons respectively. f is an arbitrary activation function.

Theorem 2. ***(Universal Approximation Theorem for Recurrent Neural Networks)*** *Let $g : \mathbb{R}^J \times \mathbb{R}^I \to \mathbb{R}^J$ be measurable and $h : \mathbb{R}^J \to \mathbb{R}^n$ be continuous, the external inputs $u_t \in \mathbb{R}^I$, the inner states $s_t \in \mathbb{R}^J$, and the outputs $y_t \in \mathbb{R}^n$ $(t = 1, \ldots, T)$. Then, any open dynamical system of the form*

$$\begin{aligned} s_{t+1} &= g(s_t, u_t) \\ y_t &= h(s_t) \end{aligned} \tag{11}$$

can be approximated by an element of the function class $RNN^{I,n}(f)$ (def. 8) with an arbitrary accuracy, where f is a continuous sigmoide activation function (def. 3).

Proof. The proof is given in two steps. Thereby the equations of the dynamical system are traced back to the representation by a three-layered feedforward network.

In the first step, we conclude that the state space equation of the open dynamical system, $s_{t+1} = g(s_t, u_t)$, can be approximated by a neural network of the form $\bar{s}_{t+1} = f(A\bar{s}_t + Bu_t - \theta)$ for all $t = 1, \ldots, T$.

Let now be $\varepsilon > 0$ and $f : \mathbb{R}^{\bar{J}} \to \mathbb{R}^{\bar{J}}$ be a continuous sigmoid activation function. Further let $K \in \mathbb{R}^J \times \mathbb{R}^I$ be a compact set, which contains $s_t, \bar{s}_t$ and u_t for all $t = 1, \ldots, T$. From the universal approximation theorem for feedforward networks (theo. 1) and the subsequent corollary (cor. 1) we know, that for any measurable function $g(s_t, u_t) : \mathbb{R}^J \times \mathbb{R}^I \to \mathbb{R}^J$ and for an arbitrary $\delta > 0$, a function

$$NN(s_t, u_t) = Vf(Ws_t + Bu_t - \bar{\theta}), \tag{12}$$

with weight matrices $V \in \mathbb{R}^{J \times \bar{J}}$, $W \in \mathbb{R}^{\bar{J} \times J}$ and $B \in \mathbb{R}^{\bar{J} \times I}$ and a bias $\bar{\theta} \in \mathbb{R}^{\bar{J}}$ exists, such that

$$\sup_{s_t, u_t \in K} |g(s_t, u_t) - NN(s_t, u_t)| < \delta \quad \forall\ t = 1, \ldots, T. \tag{13}$$

As f is continuous and T finite, there exists a $\delta > 0$, such that according to the ε-δ-criterion we get out of equation (13), that for the dynamics

$$\bar{s}_{t+1} = Vf(W\bar{s}_t + Bu_t - \bar{\theta}) \tag{14}$$

the following condition holds

$$|s_t - \bar{s}_t| < \varepsilon \quad \forall\ t = 1, \ldots, T. \tag{15}$$

Further let

$$s'_{t+1} := f(W\bar{s}_t + Bu_t - \bar{\theta}) \tag{16}$$

which gives us, that

$$\bar{s}_t = Vs'_t. \tag{17}$$

With the help of a variable transformation from $\bar{s}$ to s'_t and the replacement $A := WV (\in \mathbb{R}^{\bar{J} \times \bar{J}})$, we get the desired function on state s':

$$s'_{t+1} = f(As'_t + Bu_t - \bar{\theta}) \tag{18}$$

Remark 2. *The transformation from s to s' might involve an enlargement of the internal state space dimension.*

In the second step we show, that the output equation $y_t = h(s_t)$ can be approximated by a neural network of the form $\bar{y}_t = C\bar{s}_t$. Thereby we have to cope with the additional challenge, to approach the nonlinear function $h(s_t)$ of the open dynamical system by a linear equation $C\bar{s}_t$.

Let $\tilde{\varepsilon} > 0$. As h is continuous per definition, there exist an $\varepsilon > 0$, such that (according to the ε-δ-criterion) out of $|s_t - \bar{s}_t| < \varepsilon$ (eq. 15) follows, that $|h(s_t) - h(\bar{s}_t)| < \tilde{\varepsilon}$. Consequently it is sufficient to show, that $\hat{y}_t = h(\bar{s}_t)$ can be approximated by a function of the form $\bar{y}_t = C\bar{s}_t$ with an arbitrary accuracy. The proposition then follows out of the triangle inequality.

Once more we use the universal approximation theorem for feedforward networks (theo. 1) and the subsequent corollary (cor. 1), which gives us that equation

$$\hat{y}_t = h(\bar{s}_t) \tag{19}$$

can be approximated by a feedforward neural network of the form

$$\bar{y}_t = Nf(M\bar{s}_t - \hat{\theta}) \tag{20}$$

where $N \in \mathbb{R}^{n \times \hat{J}}$ and $M \in \mathbb{R}^{\hat{J} \times J}$ be suitable weight matrices, $f : \mathbb{R}^{\hat{J}} \to \mathbb{R}^{\hat{J}}$ a sigmoid activation function, and $\hat{\theta} \in \mathbb{R}^{\hat{J}}$ a bias. According to equation (17) and equation (18) we know that $\bar{s}_t = Vs'_t$ and $s'_{t+1} = f(As'_t + Bu_t - \bar{\theta})$. By insertion we get

$$\begin{aligned} \bar{y}_t &= Nf(M\bar{s}_t - \hat{\theta}) \\ &= Nf(MVs'_t - \hat{\theta}) \\ &= Nf(MVf(As'_{t-1} + Bu_{t-1} - \bar{\theta}) - \hat{\theta}) \quad . \end{aligned} \tag{21}$$

Using again theorem 1 equation (21) can be approximated by

$$\tilde{y}_t = Df(Es'_{t-1} + Fu_{t-1} - \tilde{\theta}) \quad , \tag{22}$$

with suitable weight matrices $D \in \mathbb{R}^{n\times\bar{\bar{J}}}$, $E \in \mathbb{R}^{\bar{\bar{J}}\times\bar{J}}$, and $F \in \mathbb{R}^{\bar{\bar{J}}\times I}$, a bias $\tilde{\theta} \in \mathbb{R}^{\bar{\bar{J}}}$, and a (continuous) sigmoid activation function $f : \mathbb{R}^{\bar{\bar{J}}} \to \mathbb{R}^{\bar{\bar{J}}}$.

If we further set

$$r_{t+1} := f(Es'_t + Fu_t - \tilde{\theta}) \quad (\in \mathbb{R}^{\bar{\bar{J}}}) \tag{23}$$

and enlarge the system equations (18) and (22) about this additional component, we achieve the following form

$$\begin{aligned} \begin{pmatrix} s'_{t+1} \\ r_{t+1} \end{pmatrix} &= f\left(\begin{pmatrix} A & 0 \\ E & 0 \end{pmatrix}\begin{pmatrix} s'_t \\ r_t \end{pmatrix} + \begin{pmatrix} B \\ F \end{pmatrix} u_t - \begin{pmatrix} \bar{\theta} \\ \tilde{\theta} \end{pmatrix}\right) \\ \tilde{y}_t &= (0 \quad D)\begin{pmatrix} s'_t \\ r_t \end{pmatrix} . \end{aligned} \tag{24}$$

Their equivalence to the original equations (18) and (22) is easy to see by a componentwise computation.

Finally out of

$$\begin{aligned} \tilde{J} := \bar{J} + \bar{\bar{J}}, \tilde{s}_t &:= \begin{pmatrix} s'_t \\ r_t \end{pmatrix} \in \mathbb{R}^{\tilde{J}}, \\ \tilde{A} := \begin{pmatrix} A & 0 \\ E & 0 \end{pmatrix} \in \mathbb{R}^{\tilde{J}\times\tilde{J}}, \tilde{B} &:= \begin{pmatrix} B \\ F \end{pmatrix} \in \mathbb{R}^{\tilde{J}\times I}, \\ \tilde{C} := (0 \quad D) \in \mathbb{R}^{n\times\tilde{J}} \text{ and } \theta &:= \begin{pmatrix} \bar{\theta} \\ \tilde{\theta} \end{pmatrix} \in \mathbb{R}^{\tilde{J}}, \end{aligned}$$

follows

$$\begin{aligned} \tilde{s}_{t+1} &= f(\tilde{A}\tilde{s}_t + \tilde{B}u_t - \theta) \\ \tilde{y}_t &= \tilde{C}\tilde{s}_t \quad . \end{aligned} \tag{25}$$

Equation (25) is apparently an element of the function class $RNN^{I,n}(f)$. Thus the theorem is proven.

q. e. d.

5 Conclusion

In this paper we gave a proof for the universal approximation ability of RNNs in state space model form. After a short introduction into open dynamical systems and RNNs in state space model form we recalled the universal approximation theorem for feedforward neural networks. Based on this result we proofed that RNNs in state space model form are able to approximate any open dynamical system with an arbitrary accuracy.

The proof can be seen as a basis for future work on RNNs in state space model form as well as a justification for their use in many real-world applications. It also underlines the good results we achieved by applying RNNs to various time-series problems.

Nevertheless further research is done on a constant enhancement of RNNs for a more efficient use in different practical questions and problems. In this context it is important to note that for the application of RNNs to real-world problems an adaption of the model to the respective task is advantageous as it improves its quality. Besides that we will continue our work on high-dimensional and dynamical consistent neural networks [4].

References

1. Hornik, K., Stinchcombe, M., White, H.: Multi-layer feedforward networks are universal approximators. Neural Networks **2** (1989) 359–366
2. Cybenko, G.: Approximation by superpositions of a sigmoidal function. In: Mathematics of Control, Signals and Systems. Springer, New York (1989) 303–314
3. Funahashi, K.I.: On the approximate realization of continuous mappings by neural networks. Neural Networks **2** (1989) 183–192
4. Zimmermann, H.G., Grothmann, R., Schaefer, A.M., Tietz, C.: Identification and forecasting of large dynamical systems by dynamical consistent neural networks. In Haykin, S., J. Principe, T.S., McWhirter, J., eds.: New Directions in Statistical Signal Processing: From Systems to Brain. MIT Press (2006)
5. Haykin, S.: Neural Networks: A Comprehensive Foundation. Macmillan, New York (1994)
6. Zimmermann, H.G., Neuneier, R.: Neural network architectures for the modeling of dynamical systems. In Kolen, J.F., Kremer, S., eds.: A Field Guide to Dynamical Recurrent Networks. IEEE Press (2001) 311–350
7. Rumelhart, D.E., Hinton, G.E., Williams, R.J.: Learning internal representations by error propagation. In Rumelhart, D.E., et al., J.L.M., eds.: Parallel Distributed Processing: Explorations in The Microstructure of Cognition. Volume 1. MIT Press, Cambridge (1986) 318–362
8. Hammer, B.: On the approximation capability of recurrent neural networks. In: International Symposium on Neural Computation. (1998)

A Discrete Adaptive Stochastic Neural Model for Constrained Optimization

Giuliano Grossi

Dipartimento di Scienze dell'Informazione
Università degli Studi di Milano
Via Comelico 39, I-20135 Milano, Italy
grossi@dsi.unimi.it

Abstract. The ability to map and solve combinatorial optimization problems with constraints on neural networks has frequently motivated a proposal for using such a model of computation.

We introduce a new stochastic neural model, working out for a specific class of constraints, which is able to choose adaptively its weights in order to find solutions into a proper subspace (feasible region) of the search space.

We show its asymptotic convergence properties and give evidence of its ability to find hight quality solution on benchmark and randomly generated instances of a specific problem.

1 Introduction

The reputation of neural networks for combinatorial optimization, widely documented in over two decades of research, has known varies degrees of success. In some cases they showed they were not competitive with ad-hoc heuristics designed for a specific problem, nevertheless almost every type of combinatorial problem has been tackled by neural networks and many approaches result in behaviour comparable to alternative techniques in terms of solution quality (for a review see [1]).

With the aim to deal with a wide class of problems, we propose here a neural computing model which is promising when applied to the optimization of a particular class of NP-hard constrained combinatorial optimization problems ([2,3]). More precisely, we consider problems having quadratic pseudo-boolean constraint functions, i.e., mappings from the family of subsets of a finite ground set to the set of integers. We focus on such functions because there is a large number of combinatorial optimization problems which arise naturally or can be easily formulated as pseudo-boolean optimization problems [4].

The idea in our architecture, based on a stochastic state transition mechanism, is to adaptively choose network connections in such a way that the energy function associated with the network is optimized for a set of desired network states corresponding to the admissible solutions. Unfortunately, because of its nonlinear character, the network can also exhibit non-desirable, local minima.

S. Kollias et al. (Eds.): ICANN 2006, Part I, LNCS 4131, pp. 641–650, 2006.

A common approach to handle constraints in neural optimization is to apply a penalty function to bias the search toward a feasible solution. Among many examples [1], one was introduced in [5], in which a discrete-time Hopfield model [6] was proposed for solving the maximum clique problem on undirected graphs. The method builds a finite sequence of discrete Hopfield networks in which the energy function of the $(t+1)$-th network is the energy function of the t-th network augmented by a penalty factor depending on the violations.

The model proposed here is based on a similar network evolution strategy, but adopts stochastic units. A penalization strategy is applied to adaptively modify the weights in such a way that the probability to find non-admissible solutions continuously decreases up to negligible values. The network performs a constraint-satisfaction search process that begins with "weak" constraints and then proceeds by gradually strengthening them until a feasible state is found.

The algorithm has been analyzed and tested on benchmark and random graphs of different size and density, generally showing better results than that proposed in [5].

2 Constrained Pseudo-boolean Optimization

A family of functions that often plays an important role in optimization models are the so-called *pseudo-boolean functions* [4]. Their polynomial representation, with particular regard to the quadratic and symmetric one, corresponds in a natural way to the objective (or cost) function of many optimization problems. Our aim here is twofold: to use these functions as cost but also to discriminate admissible solutions from non-admissible ones introducing suitable pseudo-boolean penalty functions. To this end, we will use equality quadratic constraints based on two independent variables and expressed as boolean functions.

Let $\mathbf{B}^n$ be a vector space of n-tuples of elements from $\mathbf{B} = \{0,1\}$. We shall consider functions in n boolean (binary) variables $x_1, \ldots, x_n$, and denote with $\boldsymbol{x} = (x_1, \ldots, x_n) \in \mathbf{B}^n$ binary vectors. The variables x_i $(1 \leq i \leq n)$ together their complements $\bar{x}_i = 1 - x_1$ form the set of literals $L = \{x_1, \bar{x}_1, \ldots, x_n, \bar{x}_n\}$. Let $V = \{1, \ldots, n\}$ and let S denote an arbitrary subset of V, then there is a unique vector $\boldsymbol{x}^S \in \mathbf{B}^n$ representing the characteristic vector of S.

Mappings $f : \mathbf{B}^n \to \mathbf{R}$ are called pseudo-boolean functions. Since there is a one-to-one correspondence between the subsets of V and the binary vectors of $\mathbf{B}^n$, these functions are *set functions*, that is, mappings that associate a real value to every subset S of V.

In the following, we will represent such pseudo-boolean functions by means of multi-linear polynomials having the form:

$$f(x_1, \ldots, x_n) = \sum_{S \subseteq V} c_S \prod_{k \in S} x_k \ ,$$

where c_S are integer coefficients and, by convention, $\prod_{k \in \varnothing} x_k = 1$. The size of the largest subset $S \subseteq V$ for which $c_S \neq 0$ is called the *degree* of f, and is denoted by deg (f). Naturally, a pseudo-boolean function f is *quadratic* if deg $(f) \leq 2$.

2.1 Constraint Definition and Objective Function

For many optimization problems P the set of admissible solutions Sol_P is a proper subset of the search space $\mathbf{B}^n$ on which the pseudo-boolean functions are defined, i.e., $Sol_P \subset \mathbf{B}^n$. In order to use our space-independent general technique to find optima, we need to introduce constraints as base ingredients to succeed in finding an admissible solution.

Frequently the constraints are given in terms of boolean functions $g : \mathbf{B}^2 \to \mathbf{B}$ and a constraint g is satisfied by an input $b \in \mathbf{B}^2$ if $g(b) = 1$. An admissible solution $S \in Sol_P$ must then satisfy a finite collection $\{g_1, \ldots, g_m\}$ of pseudo-boolean functions, i.e., $\bigwedge_{k=1}^{m} g_k = 1$, which is equivalent to satisfy the equality $\sum_{k=1}^{m} g_k(l_i, l_j) = m$, with $l_i, l_j \in L$.

In order to illustrate the feasibility of mapping a broad class of combinatorial problems into a discrete neural model, we formalize a *minimization* problem with the following general form

$$\begin{aligned} \text{minimize} \quad & \sum_{i=1}^{n} x_i \\ \text{subject to} \quad & \sum_{k=1}^{m} g_k(l_i, l_j) = m \ . \end{aligned} \tag{1}$$

The same can be easily derived for *maximization* problems.

To derive an energy function whose minimum value corresponds to the "best" solution for P, we combine both the cost and the penalty function in the same objective function, obtaining the general form

$$\begin{aligned} f(x_1, \ldots, x_n) &= \alpha \sum_{i=1}^{n} x_i - \sum_{k=1}^{m} \gamma_k g_k(l_i, l_j) \\ &= \sum_{i=1}^{n} \lambda_i x_i - \sum_{i<j} w_{ij} x_i x_j + C \ , \end{aligned} \tag{2}$$

where $\alpha, \gamma_1, \ldots, \gamma_m$ are positive integer parameters useful in finding better solutions (we will explain their role later) while preserving optima, $(w_{ij})_{n \times n}$ is an integer symmetric matrix with null diagonal, $\boldsymbol{\lambda}$ an integer vector and C an integer constant. We will interpret $(w_{ij})_{n \times n}$ and $\boldsymbol{\lambda}$ as weigths and tresholds respectively of the the neural network used to find optimal solutions.

2.2 Examples

In this section we recall some combinatorial optimization problems, which arise naturally in the area of algorithmic graph theory, and can be easily formulated as pseudo-boolean optimization problems.

All graphs $G = \langle V, E \rangle$ considered are arbitrary undirected graphs, where $V = \{1, \ldots, n\}$ is the set of vertices and $E \subseteq V \oplus V$ (not ordered pairs) is the set of edges. Two distinct vertices i and j are called *adjacent* if they are connected by an edge; the set $\mathcal{N}(i)$ denotes the *neighbours* of vertex i, i.e., the vertices that are adjacent to i.

Max Clique (Max Independent Set). It consists in finding the largest cardinality subset of vertices which are pairwise connected, i.e., adjacent. This problem can be formulated as

$$\text{maximize } \alpha \sum_{i \in V} x_i, \qquad \text{subject to} \sum_{\{i,j\} \in E^c} (\bar{x}_i \vee \bar{x}_j) = |E^c| \ , \tag{3}$$

where E^c denotes the complement of E.

Min Independent Dominating Set. It is to find the minimum cardinality subset of independent vertices (all pairwise not connected) such that every vertex non in the subset is adjacent to one in the subset. It is equivalent to

$$\text{minimize } \alpha \sum_{i \in V} x_i, \qquad \text{subject to} \sum_{\{i,j\} \in E} (x_i \wedge x_j) = 0 \ . \tag{4}$$

Min Vertex Cover. A vertex cover for a graph G is a subset of vertices such that each edge has at least one end-point in the subset. The objective function of this problem is:

$$\text{minimize } \alpha \sum_{i \in V} x_i, \qquad \text{subject to} \sum_{\{i,j\} \in E} (x_i \vee x_j) = |E| \ . \tag{5}$$

Max k-colorable Induced Subgraph. This problem can be translated into the search of a maximum independent set (problem (3)) on an expanded graph $\tilde{G} = \langle \tilde{V}, \tilde{E} \rangle$ in which for each $i \in V$ the set $\{i_1, \ldots, i_k\} \subseteq \tilde{V}$ and for each $\{i,j\} \in E$ the sets $\{\{i_p, j_p\} \mid 1 \leq p \leq k\}$ and $\{\{i_p, i_q\} \mid 1 \leq p, q \leq k \ \wedge \ p \neq q\}$ belong to $\tilde{E}$.

3 A Stochastic Neural Model

In this section we introduce and briefly analyze a stochastic recurrent network model to find admissible (approximate) solutions of the problem whose objective function has the form given in (2).

The network architecture is derived from the problem (for instance, the graph topology), with the set of neurons (or *units*) isomorphic to the set of binary variables representing the solutions and the connections between neurons derived according the logical dependencies between variable pairs. The absence of such a connection between a variable pair, implies neither strength in the synaptic connection between neurons nor constraint between the variables themselves.

Each unit i ($1 \leq i \leq n$) is stochastic, assuming the state $x_i = 1$ with probability $\phi(h_i)$ and $x_i = 0$ with probability $1 - \phi(h_i)$, where

$$h_i = \lambda_i - \sum_{j \in \mathcal{N}(i)} w_{ij} x_j \qquad \text{and} \qquad \phi(h) = \frac{1}{1+e^{-h}} \ , \tag{6}$$

which is the usual *Glauber* [7] or *logistic* sigmoid-shaped function.

Note that we do not introduce into the function the parameter representing an analogue of the temperature as in the annealing algorithms [8,9] and used to control the gain (or slope) of the activation function. Instead, we will simply let the system adaptively choose its weights to lead the search toward a feasible solution.

Solving a combinatorial optimization problem on a neural net requires a mapping of the problem onto this system in such a way that one can decipher a solution from the output of the neurons. In particular the energy function of such a network must be related to the objective function of the problem. Under this preamble, we adopt as energy the quadratic function expressed in (2), we use the stochastic activation function described above in order to derive a stochastic dynamic able to lead the system status in the saturation region of the sigmoid function.

We can therefore interpret the pseudo-boolean quadratic function

$$E(x_1, \ldots, x_n) = \sum_{i \in V} \lambda_i x_i - \sum_{i<j} w_{ij} x_i x_j \tag{7}$$

as energy of the network. Moreover, the energy has to satisfy two fundamental properties:

1. by multiplying the linear term $\sum_{i \in V} x_i$ by a positive constant α, the optima are preserved. In this way, since the linear term contrasts the quadratic one (penalty factor) during the network's evolution, lower values of the linear term should lead towards stable states with lower energy (better solutions).
2. the quadratic form $-\frac{1}{2}\boldsymbol{x} \cdot W \cdot \boldsymbol{x}^T + (\boldsymbol{\lambda} - \alpha) \cdot \boldsymbol{x}^T$ appearing in (2) is easy to maximize (in polynomial time) because the coefficients are all positive or all negative (they are obtained from the set of identical quadratic function g_k, with $1 \leq k \leq m$, as showed in Section 2).

By adopting the saturated-nonlinear activation function (6) and letting the network evolve with asynchronous dynamics, the resulting dynamic system locally minimizes the network energy with high probability. However, we are interested in a process that, while minimizing the energy, also guarantees that the final stable state correspond to a valid solution. For this reason, the proposed process alternates the following two phases called *units updating* and *weights strengthening*:

Units updating *(Phase 1.)*. Asynchronously update the units with the following stochastic dynamics:

$$x_i = \begin{cases} 1, & \text{with probability } \phi(h_i) \\ 0, & \text{with probability } 1 - \phi(h_i) \end{cases} \qquad \text{for each unit } i \in [1, \ldots, n] \ ;$$

Weights strengthening *(Phase 2.)*. Increase the weights w_{ij} and the thresholds λ_i, λ_j of the connections between the units i and j that violate the problem constraints $g(l_i, l_j)$. Many different penalization strategies can be defined: here we update only those connections that cause violations in the current state. This strategy can be formalized as follows:

$$\forall k \quad \text{if} \quad g_k(l_i, l_j) = 0 \quad \Rightarrow \quad \gamma_k = \gamma_k + 1 \ . \tag{8}$$

Since for each i, the relationship between the (6) and the (7) is expressed by

$$\begin{aligned} h_i(x_1, \ldots, x_n) = \Delta E_i = \frac{\partial}{\partial x_i} E(x_1, \ldots, x_n) \\ = E(x_1, \ldots, x_{i-1}, 1, x_{i+1}, \ldots, x_n) \\ - E(x_1, \ldots, x_{i-1}, 0, x_{i+1}, \ldots, x_n) \ . \end{aligned}$$

thus, flipping the state of the unit i produces a variation of energy $\Delta E_i = h_i(x_1, \ldots, x_n)$, which does not depend on x_i. This energy dependency from h_i is just used to force the convergence toward feasible solution, because h_i is a function of those constraints g_k containing the variable x_i. Thus, changing the absolute value of γ_k (as in (8)) result in a changing of the sign of h_i.

The algorithm ASNM (Adaptive Stochastic Neural Model) that iterates the simulation of the neural network behaviour and the selective penalization is sketched in the Algorithm 1. The algorithm takes in input the adjacency matrix

Algorithm 1. ASNM

Require: An initial matrix W, a vector $\boldsymbol{\lambda}$ and an integer $\alpha > 0$
  while (there are violations) **do**
    for all $i = 1$ to n **do**
      $h_i = \lambda_i - \sum_{k \in \mathcal{N}(i)} w_{ik}\, x_k$;
      $x_i = \begin{cases} 1, & \text{with probability } \phi(h_i) \\ 0, & \text{with probability } 1 - \phi(h_i) \end{cases}$
    end for
    for all $k = 1$ to m **do**
      if (g_k not satisfied) **then**
        $\gamma_k = \gamma_k + 1$
      end if
    end for
  end while
Ensure: admissible solution (characteristic vector) $\boldsymbol{x}$

of a graph and the threshold α. It has to be proved that this algorithm stops when a stable state of a suitable network is reached, where the related subset of vertices is a maximal independent one therefore representing a solution of the maximum independent set problem.

This statement is summarized in the following

Theorem 1. *Let $\boldsymbol{x}(t)$ be the state found by the algorithm at time t and let $Z(t)$ be the random variable denoting the number of violations at the same time. Then,*

$$\lim_{t \to +\infty} \mathbf{E}[Z(t)] = 0 \ ,$$

where $\mathbf{E}[\cdot]$ denotes the expectation of a random variable.

Moreover, regarding the optimality of the feasible solution found it can be stated that:

Corollary 1. *Let $\boldsymbol{x}$ be the stable state of the network found by the ASNM algorithm. Then $\boldsymbol{x}$ represents, with high probability, an optimal solution, i.e., a suboptimal solution but not subset of another one.*

Contrarily to neural systems based on the Ising model with Glauber dynamics [7], like the Boltzmann machines ([10,11]) and the stochastic Hopfield networks [12], the proposed model does not provide a control parameter like temperature. Nevertheless, for minimization (maximization) problems, this system naturally moves in the direction of decreasing (increasing) energy but it allows to move in the opposite direction. The probability of such a move is initially high, but it decreases during the system evolution (the argument of the activation function gets far away from zero, in the regions of saturation). In other words, the probability of making a contrary move depends on the absolute value of (6), which is proportional to the number of violations.

4 Experimental Results

To give an idea of the performances (solution quality and time) of the ASNM heuristic we chose the MAX CLIQUE problem and did two kinds of experiments: the first based on a widely used reference benchmark and the second on randomly generated graphs of various size.

First experiment. The instances examined here consist on graphs of different sizes (from 16 to 1000 vertices) selected among those collected in the benchmark of the second DIMACS challenge (held in 1993 with the purpose of finding efficient approximation algorithms for MAX CLIQUE, GRAPH COLORING, and SATISFIABILITY) and widely used subsequently by various researchers for heuristic testing purpose.

For this test we set the parameter $\alpha = 10$ (sufficiently little but without a particular meaning), we run the algorithm 10 times and take the best value (in many cases the variance is zero), The results obtained by ASNM on DIMACS graphs are presented in Table 4.

The first column (**Graph**) contains the name of the instances; the second ($|\mathbf{V}|$) reports the number of vertices; the third (**ASNM**) gives the size of the

Table 1. Results on the DIMACS benchmark instances. The second column reports the size of the graph, the third column gives the clique size found by ASNM, while the last column lists the best results of all participants (with star if optimality is proved). We write in bold the values of the second column if they coincide with the best achieved in the challenge.

Graph $\langle \mathbf{V}, \mathbf{E} \rangle$	**$\|\mathbf{V}\|$**	ASNM	**Best**	**Graph $\langle \mathbf{V}, \mathbf{E} \rangle$**	**$\|\mathbf{V}\|$**	ASNM	**Best**
MANN9	16	**16**	16^*	Johnson8-2-4	28	**4**	4^*
MANN27	378	**126**	126^*	Johnson8-4-4	70	**14**	14^*
MANN45	1035	343	345^*	Johnson16-2-4	120	**8**	8^*
Hamming6-2	64	**32**	32^*	Johnson32-2-4	496	**16**	16
Hamming6-4	64	**4**	4^*	c-fat200-1	200	**12**	12^*
Hamming8-2	256	**128**	128^*	c-fat200-2	200	**24**	24^*
Hamming8-4	256	**16**	16^*	c-fat200-5	200	**58**	58^*
Hamming10-2	1024	**512**	512^*	c-fat500-1	500	**14**	14^*
Hamming10-4	1024	**40**	40	c-fat500-2	500	**26**	26^*
p.hat300-1	300	**8**	8^*	c-fat500-5	500	**64**	64^*
p.hat300-2	300	**25**	25^*	san200-0.7.1	200	**30**	30^*
p.hat300-3	300	**36**	36^*	san200-0.7.2	200	**18**	18^*
p.hat500-1	500	**9**	9^*	san200-0.9.1	200	**70**	70^*
p.hat500-2	500	**36**	36^*	san200-0.9.2	200	**60**	60^*
p.hat500-3	500	**50**	50	san200-0.9.3	200	**44**	44^*
p.hat700-1	700	**11**	11^*	san400-0.7.1	400	**40**	40^*
p.hat700-2	700	**44**	44^*	san400-0.7.2	400	**30**	30^*
p.hat700-3	700	**61**	62^*	san400-0.9.1	400	**100**	100^*
p.hat1000-1	1000	**10**	10	sanr200-0.7	200	**18**	18^*
p.hat1000-2	1000	**46**	46	sanr200-0.9	200	**42**	42^*
p.hat1000-3	1000	**68**	66	sanr400-0.5	400	**13**	13^*
keller4	171	**11**	11^*	sanr400-0.7	400	**21**	21
keller5	776	**27**	27^*	san1000	1000	**10**	10
r100-0.5	100	**9**	9^*	r200-0.5	200	**11**	11^*
r300-0.5	300	**12**	12^*	r400-0.5	400	**13**	13^*

cliques found by ASNM, and the fourth (**Best**) gives the best results obtained in the challenge over all heuristics.

Second experiment. In this experiment we take the so-called p-random graphs into account. They are represented by the pair $\langle V, E \rangle$, where $V = \{1, \ldots, n\}$ and E is obtained selecting $\{i, j\}$ as edge with probability p $(1 \leq i < j \leq n)$. To show the behaviour of the algorithm on this kind of instances we give a direct

Table 2. Clique average size with relative standard deviation at confidence level 95% obtained by ASNM and IHN (columns 4 and 5) on p-random graphs for various values of n and p (column 1). Column 2 gives the expected size $C(n,p)$ of the maximum clique while column 3 reports the average time (in seconds) spent by ASNM to converge.

			Average ± stdev	
n-p	$\mathbf{C(n,p)}$	Av. time	ASNM	IHN
200-0.2	6.2	2.15	5.76 ± 0.07	5.53 ± 0.09
200-0.5	11.6	0.96	10.73 ± 0.09	10.23 ± 0.11
200-0.8	26.6	0.25	24.13 ± 0.12	22.96 ± 0.12
400-0.2	7.0	29.14	6.20 ± 0.07	6.06 ± 0.04
400-0.5	13.3	13.84	12.30 ± 0.08	11.83 ± 0.10
400-0.8	31.7	3.48	28.70 ± 0.14	26.80 ± 0.17
600-0.2	7.4	135.7	6.96 ± 0.03	6.76 ± 0.07
600-0.5	14.2	69.48	13.10 ± 0.05	12.53 ± 0.10
600-0.8	34.6	17.11	31.40 ± 0.10	29.83 ± 0.14

comparison with the meta-heuristic IHN presented and tested in [5], which is in some sense the deterministic version of ASNM. It builds a finite sequence of discrete Hopfield networks in which the energy function of the $(t+1)$-th network is the energy function of the t-th network augmented by a penalty factor depending on the violations. We choose such a meta-heuristic because its performances has been already shown good in [5,13], in which it was compared with many other well known heuristics for MAX CLIQUE presented at DIMACS. Also in this case we set the parameter $\alpha = 10$, run the algorithm 10 times for each instance graph and take the best value obtained.

In Table 2 (column 4) we report, for some values of n and p, the clique average size (on 30 graphs randomly generated for each pair n and p) found by ASNM at confidence level 95%. Observe that the average values found by ASNM are always better than those found by IHN (column 5). These results are also compared with the theoretical evaluation of the expected maximum clique for p-random graphs of size n, obtained according to an asymptotic result (column 2)[1].

Summarizing, both Table 4 and Table 2 show that the quality of the solutions found by ASNM are good. On all but one (MANN45) instances of the DIMACS challenge, ASNM found the best value. On p-random graphs the clique sizes found by ASNM are always better than those found by IHN and they are at least 90% of the optimal estimate $C(n,p)$.

[1] The expected number of cliques of size k in a p-random graph of size n is $\binom{n}{k}p^{\binom{k}{2}}$. For p fixed $(0 < p < 1)$ and sufficiently large n, let $C(n,p)$ be the real number k such that $\binom{n}{k}p^{\binom{k}{2}} = 1$; it was shown in [14] that, for $n \to \infty$, the probability that a p-random graph of size n has a clique of size $C(n,p)$ approaches 0. Therefore, $C(n,p)$ estimates quite well the expected size of the maximum clique.

References

1. Smith, K.: Neural networks for combinatorial optimization: A review of more than a decade of research (1999)
2. Garey, M.R., Johnson, D.S.: Computers and Intractability. A Guide to the Theory of NP-Completeness. W. H. Freeman & Co., San Francisco, CA (1979)
3. Karp, R.M. Complexity of Computer Computations. In: Reducibility among Combinatorial Problems. Plenum Press, New York (1972) 85–103
4. Boros, Hammer: Pseudo-boolean optimization. DAMATH: Discrete Applied Mathematics and Combinatorial Operations Research and Computer Science **123** (2002)
5. Bertoni, A., Campadelli, P., Grossi, G.: A neural algorithm for the maximum clique problem: Analysis, experiments and circuit implementation. Algoritmica **33**(1) (2002) 71–88
6. Hopfield, J.J.: Neurons with graded response have collective computational properties like those of two-state neurons. In: Proceedings of the National Academy of Sciences. Number 81, NAS (1984) 3088–3092
7. Glauber, R.J.: Time-dependent statistics of the Ising model. Journal of Mathematical Physics **4**(2) (1963) 294–307
8. Kirkpatrick, S., Gelatt, C., Vecchi, M.: Optimization by simulated annealing. Science **220** (1983) 671–680
9. Laarhoven, P.J.M., Aarts, E.H.L.: Simulated Annealing: Theory and Applications. Mathematics and its Applications. Reidel Publisching Company (1987)
10. Ackley, D., Hinton, G., Sejnowski, T.: A learning algorithm for boltzmann machines. Cognitive Science **9** (1985)
11. Aarts, E., Korst, J.: Simulated annealing and Boltzmann machines: a stochastic approach to combinatorial optimization and neural computing. John Wiley & Sons, Inc., New York, NY, USA (1989)
12. Hertz, J., Krogh, A., Palmer, R.G.: Introduction to the Theory of Neural Computation. Addison-Wesley (1991)
13. Grossi, G., Posenato, R.: A distributed algorithm for max independent set problem based on Hopfield networks. In Marinaro, M., Tagliaferri, R., eds.: Neural Nets: 13th Italian Workshop on Neural Nets (WIRN 2002). LNCS 2486, Springer-Verlag (2002) 64–74
14. Matula, D.: On the complete subgraph of a random graph. Combinatory Mathematics and its Applications (1970) 356–369

Quantum Perceptron Network

Rigui Zhou[1,2], Ling Qin[1], and Nan Jiang[1]

[1]Department of Computer Science and Technology,Nanjing University of Aeronautics and Astronautics, Nanjing, Jiangsu, 210016, China
riguizhou@nuaa.edu.cn
[2]Nanchang Institute of Aeronautical Technology, Nanchang, Jiangxi, 330034, China

Abstract. A novel neural network,quantum perceptron network(QPN),is presented built upon the combination of classical perceptron network and quantum computing.This quantum perceptron network utilizing quantum phase adequately has the computing power that the conventional perceptron is unable to realize. Through case、performance analysis and simulation,a quantum perceptron with only one neuron can realize XOR function unrealizable with a classical perceptron having a neuron. Simple network structure can achieve comparatively complicated network function,which will throw heavy influence on the field of artificial intelligence and control engineering.

1 Introduction

In 1995,Kak firstly presented the concept of Quantum Neural Computation[1] that is one of the young and outlying science. In 1997,Tohru Nitta extended the Back-Propagation Algorithm to Complex Numbers fields[2]. Recently,researchers from around the world have also begun considering the implications of quantum computation for the fields such as neural networks (both biological and artificial), so the studies in quantum computation have been enriched by new works addressing the idea of developing quantum neural networks. Quantum Artificial Neural Network (QANN) produced from this is a new paradigm based on the combination of classical neural computation and quantum computing and it has the high value for theoretic study and the potential application. These QANNs have many promising characteristics, both in the case of supervised and unsupervised learning. In 1998, A first systematic and deep examination of QANN was done by T Menneer in his Ph.D.thesis[3]. At the same time,many QANN models were presented. For example,in 1995, Quantum Inspired Neural Nets[4] and Entangled Neural Nerworks [5]were proposed by Narayanan et al;In particular,an associative memory[6,7,8] based on the use of Grover's quantum search algorithm has been introduced by Ventura and Martinez. This associative memory network can solve the completion problem, that is, it can restore the full pattern when initially presented with just a part of that pattern. Besides, one of the most attractive properties of it is its exponential capacity,which may have distinct adventages over its classical cousin.

In 2001, M.V.Altaisky[9] developed a simple quantum perceptron that realizes perceptron function by selecting different operators, but he did not at all make use of quantum phase.This paper presented quantum perceptron network utilizing quantum phase adequately, it, having the computing power unrealizable with a classical perceptron, is very different from M.V.Altaisky's network.

S. Kollias et al. (Eds.): ICANN 2006, Part I, LNCS 4131, pp. 651 – 657, 2006.

The remainder of this paper is organized as follows:In the section 2, quantum theory is briefly introduced; Section 3 presents quantum perceptron and its learning;In the section 4, the case analysis is discussed; In the last, this paper concludes with final remarks and future work.

2 Quantum Theory

Quantum compution is based upon physical principles from the theory of quantum mechanics(QM), which in many ways is counterintuitive. In quantum compute- r, "qubits" are the counterparts of "bits" in classical computers. Qubit has two basis quantum states, $|0\rangle$ and $|1\rangle$. $|0\rangle$ corresponds to the bit 0 of classical computers, and $|1\rangle$ to the bit 1. Several necessary quantum concepts that form the basis for the study of QPN are briefly reviewed here.

2.1 Linear Superposition[10]

Quantum systems are described by a wave function $|\varphi\rangle$ that exists in a Hilbert space.The state $|\varphi\rangle$ of a general quantum system can be described by the linear superposition of the basis states $|\phi_i\rangle$ as $|\varphi\rangle = \sum_i c_i |\phi_i\rangle$,where c_i is a probability amplitudes and meets $\sum_i |c_i|^2 = 1$. and $|c_i|^2$ gives the probability of $|\varphi\rangle$ collapsing into state $|\phi_i\rangle$ if it decoheres.Note that the wave function $|\varphi\rangle$ describes a real physical system that must collapse to exactly one basis state.Use is made here of the Dirac bracket notation, where the ket $|\bullet\rangle$ is analogous to a column vector, and the bra $\langle\bullet|$ is analogous to the complex conjugate transpose of the ket $|\bullet\rangle$. In the same,the probability that a quantum state $|\varphi\rangle$ will collapse into an eigenstate $|\phi_i\rangle$ can be written $|\langle\phi_i|\varphi\rangle|^2$ too.

2.2 Quantum Gates and Their Representations

In two state quantum system, a quantum state can be expressed as $|\varphi\rangle = c_1|0\rangle + c_2|1\rangle$ (see 2.1 section), where c_1,c_2 are complex numbers, this shows quantum state can be in $|1\rangle$ state, $|0\rangle$ state or simultaneously in both (superposition). That is,when qubit state $|\varphi\rangle$ collapses into either the $|0\rangle$ state or the $|1\rangle$ state, it does so with probabilities $|c_1|^2$ and $|c_2|^2$, respectively, where $|c_1|^2 + |c_2|^2 = 1$.

It is possible to construct any arbitrary quantum logic gate using the 1-bit rotation gate and the 2-bit controlled NOT gate as primitive elements [10].Based on the Bloch sphere, formula (2) can be act as another way to express the formula (1).

$$|\varphi\rangle = c_1|0\rangle + c_2|1\rangle \tag{1}$$

$$|\varphi\rangle = \cos\frac{\theta}{2}|0\rangle + e^{i\phi}\sin\frac{\theta}{2}|1\rangle \tag{2}$$

$$f(\varphi) = e^{i\phi} = \cos\phi + i\sin\phi \tag{3}$$

Where i is the imaginary unit and ϕ is the phase that describes the quantum state. Formula(3) is used to simplify formula (2) for the convenience of operating quantum state.

3 Quantum Perceptron(QP)

3.1 Structure and Learning of QP

On the basis of creationary work done by Kouda[11-14] et al, we design a quantum perceptron showed in the Fig.1. (For simpleness,we only draw two input nodes P_1, P_2 and a output node). Quantum computer is still used only in the lab, therefore the range of input of quantum perceptron is (0,1), which can be simulated in the classical computer.

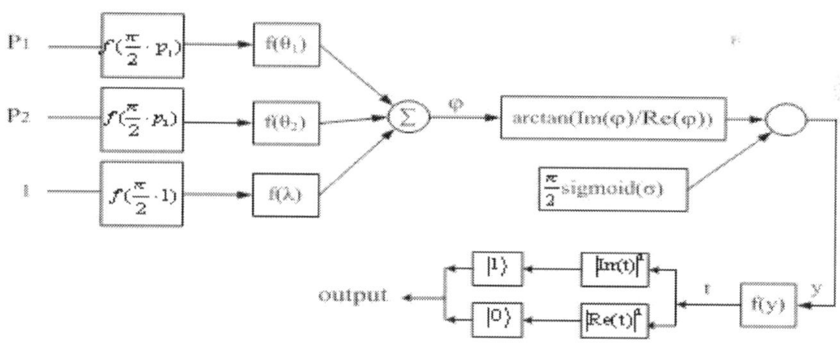

Fig. 1. quantum perceptron

In Fig.1., the output of quantum perceptron equals to $|1\rangle$ or $|0\rangle$ with probabilities $|\mathrm{Im}(t)|^2$ or $|\mathrm{Re}(t)|^2$, respectively.

$$t = f(y) \quad sigmoid(x) = \frac{1}{1+e^{-x}} \tag{4}$$

$$y = \frac{\pi}{2} sigmoid(\sigma) - \arctan(\mathrm{Im}(\varphi) / \mathrm{Re}(\varphi)) \tag{5}$$

$$\varphi = f(\frac{\pi}{2} P_1) f(\theta_1) + f(\frac{\pi}{2} P_2) f(\theta_2) - f(\frac{\pi}{2} \cdot 1) f(\lambda) \tag{6}$$

Where θ_1、θ_2 and λ are phase parameters in the form of weight connection and threshold, respectively, and σ is the phase control gene, P_1, P_2 are input data.

How does the network learn in the work process? We define a quantum version of the well-known perceptron algorithm.This rule is represented by the following equations:

$$\begin{aligned} \theta_l^{new} &= \theta_l^{old} + \alpha e \ input \\ \lambda^{new} &= \lambda^{old} + \alpha e \ input \\ \sigma^{new} &= \sigma^{old} + \alpha e \ input \end{aligned} \tag{7}$$

Where $e = T - output$, T is a target output(it is a real number, not complex number); $l \in (1,2)$, α is a learning rate and within the range 0.0—1.0, in usual. Of course α can adopt $\alpha = \frac{1}{k}$, here k is a iteration numbers, thus α can decrease step by step with iteration number increases, which quickens convergence process of network.

3.2 Convergence Prove[9]

For the sake of convergence prove of quantum perceptron,we still adopt the form of quantum state, such as $|\varphi\rangle = c_1 |0\rangle + c_2 |1\rangle$ which is entirely equal to formula (2), (3):

$$\left\| |T\rangle - |x_{output}(k+1)\rangle \right\|^2 = \left\| |T\rangle - \sum_{k=1}^{L} W_k(k+1) |x(k)\rangle \right\|^2 = \left\| |T\rangle - \sum_{k=1}^{L} W_k(k) |x_k\rangle + \alpha(|T\rangle - x(k)) \right\|^2$$

$\langle x_k | x_k \rangle = 1$ is used in the formula, namely, $|x_k\rangle$ is a normalized input states:

$$\left\| |T\rangle - |x_{output}(k+1)\rangle \right\|^2 = \left\| |T\rangle - |x_{output}(k)\rangle - \sum_{k}^{L} (\alpha |T\rangle - |x(k)\rangle) \right\|^2 = (1 - L\alpha)^2 \left\| |T\rangle - |x(k)\rangle \right\|^2$$

Where L is the input number of quantum neuron,i.e. P_1、P_2 . For small α (0< α <1/L), the result of iteration converges to the desired state $|T\rangle$, especially, quantum perceptron converges the quickest when $\alpha = \frac{1}{L}$.

4 Case Analysis

There is a case called XOR function whose input/output is:

$$\{x_1 = \begin{bmatrix} 1 \\ 1 \end{bmatrix}, T_1 = 0\}\{x_2 = \begin{bmatrix} 0 \\ 1 \end{bmatrix}, T_2 = 1\}\{x_3 = \begin{bmatrix} 1 \\ 0 \end{bmatrix}, T_3 = 1\}\{x_4 = \begin{bmatrix} 0 \\ 0 \end{bmatrix}, T_4 = 0\},$$

From this, we can meet input/output condition with single-layer quantum perceptron with two inputs(see Fig.1.).But classical one-neuron perceptron unable to accomplish this line-impartibility problem. Can quantum perceptron with a neuron accomplish this? Supposing connection weight θ_k (k=1,2) 、threshold λ and phase control gene σ are all equal to zero,i.e.f(θ_k)=f(λ)=sigmoid(σ)=1. The network carry through iteration computing for the first time as follows:

1. The quantum perceptron computes with the first input/output pair:f($\frac{\pi}{2}$ P$_1$)=i, f($\frac{\pi}{2}$ P$_2$)=i, evolving according to (4),(5),(6), φ=i*1+i*1-i*1=i, $y=\frac{\pi}{2}-arctan(\frac{1}{0})=0$, $t=1$, so the probability of output equaling to $|0\rangle$ is 100%, which corresponds to classical "o" and is consistent with the goal $T_1 = 0$, this network realizes the goal and need not to adjust parameter θ_k, λ, σ.

2. As above, network computes with the second input/output:f($\frac{\pi}{2}$ P$_1$)=1,f($\frac{\pi}{2}$ P$_2$)=i, φ=1*1+i*1-i*1=1, $y = \frac{\pi}{2} - \arctan(0) = \frac{\pi}{2}$, $t = i$, the probability of output= $|1\rangle$ is 100% and is consistent with the goal $T_1 = 1$, need not to adjust parameter.

3. The network computes with the third input/output: f($\frac{\pi}{2}$ P$_1$)=i,f($\frac{\pi}{2}$ P$_2$)=1, φ = 1*1+i*1-i*1=1, $y = \frac{\pi}{2} - \arctan(0) = \frac{\pi}{2}$, $t = i$, the output is consistent with the goal $f(T_3 \frac{\pi}{2}) = f(\frac{\pi}{2}) = i$,not adjusting parameter.

4. At last, network computes with the fourth input/output: f($\frac{\pi}{2}$ P$_1$)=1,f($\frac{\pi}{2}$ P$_2$)=1, φ=1*1+1*1-i*1=2-i, $y = \frac{\pi}{2} - \arctan(-0.5) = 2.0344$, $t = -0.4472 + i0.8944$,

the output collapses in the state $|0\rangle$ with the probability of $output = |\mathrm{Re}(t)|^2 = 20\%$, which isn't entirely consistent with the goal $T_4 = 0$,namely,the output is identical with the goal only with the probability of 20%.Network completes the first iteration compution after adjusting the parameter θ_k, λ, σ following formula (7).

The quantum perceptron with one neuron goes along the second iteration compution as the same fashion.From the result of simulation,the network can realize XOR function only by 16 iterations.

5 Conclusion and Expectation

quantum perceptron network with only one neuron can realize XOR function unrealizable with a classical perceptron having a neuron.Through case analysis land simulation,it proves that quantum perceptron network designed can solve line-impartibility problem which must be performed by the two-layer classical artificial neural network(CANN).This shows ulteriorly that the size of quantum perceptron may be less or simpler than that of its conventional counterpart when they encounter the same task. The results suggest that the excellent learning performance of our quantum perceptron is not simply due to the complex-valued neuron parameters or the introduction of the restricted polar radius, but to the quantum characteristics, i.e., the quantum superposition, quantum phsae and the probability interpretation.But, it is left for future study to clarify this efficiency in mathematical terms. Other future work will focus on the use of the quantum perceptron in more practical applications.

In a word,the research of QPN includes not only investigating a kind of new algorithm, but also exploring a novel computing architecture and model, science problems containing in this network and research into these problems will drive the research of neural network、quantum computing and other correlative field to a new development.

Acknowledgement

This work is supported by the Grand Fundamental Advanced Research of Chinese National Defense under Grant No.S0500A001 and the Science and Technology Foundation of the Education Department of Jiangxi Province under Grant No.2005169

References

1. Kak S. C.: On Quantum Neural Computing.Information Sciences, 1995, 83:143-160
2. Tohru Nitta.:An Extension of the Back-Propagation Algorithm to Complex Numbers. Neural Networks, Vol.10, No.8, pp.1391-1415, 1997.

3. Menneer T.: Quantum Artificial Neural Networks. Ph. D. thesis of The Univ. of Exeter, UK, 1998.
4. Menneer T,Narayanan A.:Quantum-inspired Neural Networks.Tech. Rep. R329, Univ. of Exeter, 1995
5. Li Wei-gang: Entangled Neural Networks.http://www.cic.unb.br/~weifang/qc/enn2000.pdf
6. Ventura D, Martinez T R.: Quantum Associative Memory.Information Sciences,2000, 124:273-296.
7. A.A. Ezhov a, A.V. Nifanova, Dan Ventura:Quantum associative memory with distributed queries, Information Sciences,128 (2000),271-293
8. Ventura D,Martinez T R.:Quantum Associative Memory.Information Sciences, 2000, 124:273-29
9. M.V. Altaisky.Quantum neural network.Technical report, 2001.http://xxx.lanl.gov/ quantph/0107012
10. M.Nielsen and I.Chuang:Quantum Computation and Quantum Information,Cambridge University Press, Cambridge,2000.
11. Kouda,N., Matsui, N., Nishimura, H. and Peper, F:Qubit Neural Network and Its Efficiency In:Proceedings of Knowledge- Based Intelligent Information Engineering Systems (KES2003), LNAI-2774, pp. 304–310, Springer-Verlag, 2003.
12. Kouda, N., Matsui, N., Nishimura, H. and Peper, F.: Qubit Neural Network and Its Learning Efficiency, Neural Computing and Applications,Springer-Verlag,DOI:10.1007/ s00521-004 -0446-8, 2005
13. Kouda, N., Matsui, N., Nishimura, H. and Peper, F.:An Examination of Qubit Neural Network in Controlling an Inverted Pendulum.Neural Processing Letters (2005) 22:277–290 . Springer 2005
14. Kouda, N.,Matsui, N. and Nishimura, H.: Image compression by layered quantum neural networks, Neural Processing Letters, **16** (1) (2002), 67–80.
15. BobRicks, Dan Ventura:Training a Quantum Neural network, http://books.nips.cc/ papers /files/nips16/NIPS2003_ET05.pdf
16. Li Fei, Zhao Shengmei, Zheng Baoyu:Performance of a single quantum neuron.Chinese Journal of Electronics, Vol.14, No.1, Jan.2005:111-114.
17. Xie Guangjun,Zhang Zhenquan:A Quantum Computitive Learning Algorithm.Chinese Journal Of Quantum Electronics, Vol.20, No.1, Feb., 2003 (in Chinese).

Critical Echo State Networks

Márton Albert Hajnal and András Lőrincz

Eötvös Loránd University, Pázmány P. sétány 1/C, Budapest, Hungary, H-1117,
ouraborous@ludens.elte.hu, andras.lorincz@elte.hu
http://nipg.inf.elte.hu/

Abstract. We are interested in the optimization of the recurrent connection structure of Echo State Networks (ESNs), because their topology can strongly influence performance. We study ESN predictive capacity by numerical simulations on Mackey-Glass time series, and find that a particular small subset of ESNs is much better than ordinary ESNs provided that the topology of the recurrent feedback connections satisfies certain conditions. We argue that the small subset separates two large sets of ESNs and this separation can be characterized in terms of phase transitions. With regard to the criticality of this phase transition, we introduce the notion of Critical Echo State Networks (CESN). We discuss why CESNs perform better than other ESNs.

Keywords: time series, prediction, echo state network, phase transition, critical point.

1 Introduction

Motivation: We are interested in learning the dynamics of deterministic nonlinear systems with artificial neural networks. It is relevant for us that (i) the network captures and represents the dynamical properties, (ii) learning should be fast, and (iii) learning has a neural form.

The Echo State Network (ESN) is an important candidate for such efforts. Despite of its simplicity, it shows immense representation capacity for nonlinear-dynamical systems. Further, the speed of learning is unique amongst Recurrent Neural Networks (RNNs) due to its fast Linear Mean Squared Error (LMSE) tuning algorithm. Finally, the on-line form of any LMSE algorithm corresponds to the well known local Delta-rule that can be implemented in neural networks.

We shall show by numerical experiments that under certain conditions, ESN can gain more than an order of magnitude for Mackey-Glass (MG) time series in terms of prediction length. We provide a set of conditions that achieve this gain. The basic finding is that such ESNs correspond to a critical condition. We describe a framework to measure if an ESN exhibits phase transition and critical behavior. The framework also helps us provide an interpretation.

The paper is built as follows. First, we briefly review background information about ESNs (Section 2.1) and about critical phenomena (Sect. 2.2). We describe our methods in Sect. 3. We study 'macroscopic behavior', optimize the topology,

S. Kollias et al. (Eds.): ICANN 2006, Part I, LNCS 4131, pp. 658–667, 2006.

and test prediction capacities. Section 4 is about our results on the critical point of ESN phase transition and about the predictive potential of some critical ESNs (CESNs). Discussion can be found in Sect. 5. We close with a short summary.

2 Preliminaries

2.1 Echo State Networks

Echo State Network was first introduced by Jaeger [1,2]. We study simple ESNs that contain all necessary components. The ESN has a hidden layer that holds the hidden representation $\mathbf{a} \in \mathbb{R}^l$. It receives input $\mathbf{x} \in \mathbb{R}^k$ and provides output $\mathbf{y} \in \mathbb{R}^m$ (Fig. 1). Network dynamics is governed by the following equations:

$$\mathbf{a}_t = (1-\mu)\mathbf{a}_{t-1} + \sigma(\mathbf{F}\mathbf{a}_{t-1} + \mathbf{W}\mathbf{x}_{t-1}) \quad (1)$$

$$\mathbf{y}_t = \mathbf{H}\mathbf{a}_t \quad (2)$$

where $\mathbf{W}$ and $\mathbf{H}$ are the input and output mappings, respectively, $\mathbf{F}$ represents the recurrent feedback connections of the hidden layer, $\sigma(\cdot)$ is a component-wise non-linearity that we set to $\tanh(\cdot)$, and μ is the parameter of leaky integration. In the ESN approach, a large number of neurons is used with random recurrent connections at the hidden layer ($l \gg k$). They seem to play the role of a 'dynamic reservoir'. We shall consider the configuration when the output of network is an estimation of the next input, that is, $\mathbf{x}_{t+1} = \mathbf{y}_t$ (and $k = m$) at every time step $t > t_0$. In this mode, and upon tuning, the network is capable of approximating the continuation of the experienced time series in the absence of further inputs.

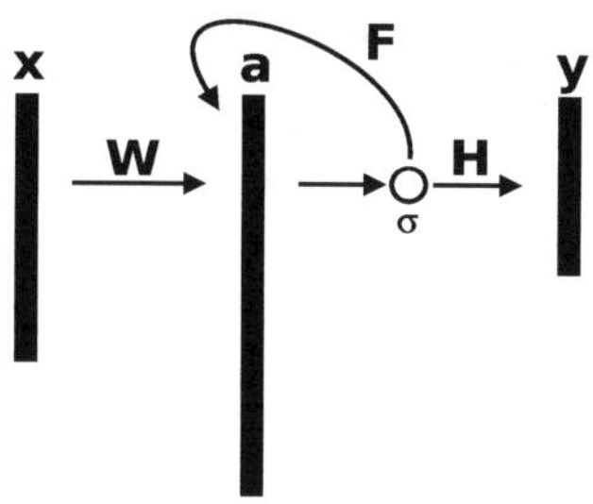

Fig. 1. Structure of the Echo State Network. $\mathbf{x}$: input, $\mathbf{a}$: hidden representation, $\mathbf{y}$: output, $\mathbf{W}$, $\mathbf{F}$, $\mathbf{H}$ linear transformations, σ: nonlinearity.

ESNs are special RNNs: only the hidden-to-output connections (matrix $\mathbf{H}$) are trained. Training is a simple linear regression task that minimizes the time averaged mean squared error between the output and training signal, which is the input itself in our case :

$$J = \frac{1}{2}\sum_t \varepsilon(t) = \frac{1}{2}\sum_t |\mathbf{y}_t - \mathbf{x}_t|^2. \quad (3)$$

Usually, random initialization is used for matrices **W** and **F**. It has been found that the so called echo states may not appear, unless the ESN satisfies the following constraints [3]: Matrix **F** should be *sparse*; only a few percent of its elements is non-zero and thus *connectivity* p is low. Also, matrix **F** should be *contractive*: the magnitude of singular values should not exceed 1. More details on the operation and tuning of ESNs can be found in the literature, see, e.g. [1,3,4,5]. There are studies about performance and alteration of ESNs [6,7,8]. It has been noted that the nature of the *dynamical reservoir* is not understood yet [9]. Our work aims to shed light on this issue.

2.2 Critical Phenomena

Critical phenomena are notable concepts in physics. The notion refers to many-body interactions, where 'body' is meant in a very general sense. Critical phenomena appear in second order phase transitions and percolation processes, among others. In general, critical phenomena may occur in the transition region that separates 'phases', which may differ in their symmetry properties, in the macroscopic parameters, in their structure, i.e., in their long range order. It is typical to define the *order parameter* of the transition that appears or disappears in one of the phases. The transition between the phases can be a function of the size of the system. The change of the order parameter becomes infinitely sharp in the limit of infinite size. This singular value of the parameter is called the *transition point* of the phase transition. Chaotic behavior is typical for temporal changes at the transition point. These concepts are sufficient for us to proceed. For further details about critical phenomena and for a review of the vast literature of the subject, see, e.g., [10] and references therein.

Below, we define an order parameter for ESNs and present computer simulations. They show that the transition of the network becomes sharp by increasing the size of the network. We also find that at around the transition point predictive capabilities of ESNs can be much better than those of ordinary ESNs.

3 Methods

In this section we describe how the long term behavior of the hidden layer was studied. We establish conditions for finding ESNs with better hidden layer recurrent connections. Our efforts lead to an order parameter and a test that captures the essence of chaotic time series.

3.1 Time Evolving Properties

We are to describe and quantify the special condition mentioned in Sect. 2. Consider the long term behavior of the components of the hidden layer, $a_{i,t}$, $i = 1, \ldots, l$. We would like to eliminate the effects of the input $\mathbf{x}$ and we set $\mathbf{W} \equiv \mathbf{0}$. Under this condition, qualitative description of the time evolution of the components $a_{j,t}$ can be provided, because activity propagation that starts at time 0 and ends at time t is determined solely by $\mathbf{F}^t$ apart from non-linearities.

The proportion of non-zero elements in $\mathbf{F}^t$ will be called *time evolving connectivity* and we denote it by p_t. Similarly, let q_t denote the number of non-zero matrix elements of $\mathbf{F}^t$. Thus $q_t = l^2 p_t$, where l is the number of neurons at the hidden layer. Quantities p_t and q_t are *macroscopic* measures of the connectivity structure that – in a broad sense – characterize information transfer from $\mathbf{a}_0$ to $\mathbf{a}_t$ through the non-zero elements of matrix $\mathbf{F}^t$. In the limit $t \to \infty$, p_t may converge. In this case, p_t may increase, decrease, or even vanish. Alternatively, it is easy to find cases, when p_t may keep changing for all times around its average value. In this case, we take this average as p_∞. In line with this note, we shall see that $o = \frac{p_\infty}{p_0}$ is an appropriate order parameter for us.

The activities of the hidden layer are also subject to temporal changes. For $\mu = 1$, Eq. (1) can be rewritten as $\mathbf{a}_t \approx \sigma((\mathbf{F} + \mathbf{WH})\mathbf{a}_{t-1})$. Upon optimizing matrix $\mathbf{H}$ for objective (3), the largest eigenvalue of $\hat{\mathbf{F}} = \mathbf{F} + \mathbf{WH}$ will approximate 1 for non-vanishing deterministic processes.

3.2 Prediction Test

We tested ESNs on Mackey-Glass (MG) [11] time series, derived by means of the delayed parameter differential equation:

$$\dot{x}(t) = -\gamma x(t) + \frac{\alpha x(t-\tau)}{1 + x(t-\tau)^\beta}, \tag{4}$$

where parameter β influences bifurcation, whereas delay parameter τ influences the complexity of the time series. We used $\alpha = 0.2$ and $\gamma = 0.1$, which are widespread in the literature.

Mean squared error is the typical measure of accuracy in the ESN literature. However, if networks are tested on MG time series that may exhibit chaotic patterns depending on the delay parameter, a peculiar effect occurs: prediction estimates usually follow the original trajectory accurately for some time, but – apparently – the network looses the dynamics suddenly. This phenomenon is a general property of chaotic systems, because the divergence of individual trajectories can be exponential. Predictive capacity for chaotic systems is thus better described by the exponent of the divergence of trajectories or by thresholds.

For the comparison of different networks, we introduce a measure of predictive capacity: *successful prediction length*, ζ. Prediction is called 'θ-successful' for time τ with parameters t_p and T, or 'successful', for short, if starting to predict at time t_p and predicting for time durations $t \leq \tau$, the average of the squared prediction error $\varepsilon(t)$ over time interval T does not exceed θ, but it does if $t > \tau$:

$$\zeta(t_\mathrm{p}) = \arg\max_\tau \left(\langle \varepsilon(t_\mathrm{p} + \tau + i) \rangle_{i=1,\ldots,T} < \theta \right), \tag{5}$$

where τ is the growing length of attempted predictions, $\langle \cdot \rangle$ denotes averaging, and i is the running index of averaging.

Should $\langle \varepsilon \rangle$ exceed θ, we consider that the system can not keep the predicted output close to the true input trajectory $\mathbf{x}(t)$ any further. Measure ζ captures the essence of chaotic dynamics [12].

In numerical experiments, different transformations **F** and **W**, starting point t_p, and training lengths were used to learn the distributions of $\zeta(t_p)$ for one-dimensional MG time series. Parameters of this study are provided in Table 1. It may be worth noting that training length and network sizes are much smaller than those of [1]. Now, we describe our experimental findings.

Table 1. Experimental parameters

size of hidden layer	l in the range $20 - 400$
value of elements of **W**	randomly chosen; $\approx \pm 0.07$
value of non-zero elements of **F**	*equal* and positive
max. eigenvalue of **F** (scale factor)	0.9
value of leaky integrator, μ	0.7
Mackey-Glass parameters as in [1]	$\alpha = 0.2, \gamma = 0.1, \beta = 10$
Mackey-Glass delay parameters	$\tau = 17$ & 30
training length	1500 (with sub sampling 10)
threshold and averaging window in Eq. (5)	$\theta = 0.2, T = 10$

4 Results

First, we shall show that a sharp transition appears in time evolving connectivity and describe how the final phase depends on the initialization. After measuring the value of the critical point we shall conclude that permutation matrices, or orthogonal matrices in general, satisfy the critical condition. We shall demonstrate the superior performance of permutation matrices.

4.1 ESN Phase Transition and the Critical Point

We have created a large number of networks of various sizes. The connectivity structure of the hidden layer was set randomly. We have determined p_∞ for all of them. We have plotted p_∞ against p_0 (Fig. 2a) for different inner layer sizes. Two phases emerged with a transition interval between them. By increasing the size of the network, the position of the interval underwent a monotone shift towards lower values and the width of the interval became narrower (Fig. 2b).

According to our original *critical point conjecture* sharp phase transition emerges with a critical point that separates the two phases at around $p_t \approx p_0$. We define the critical point of ESNs as $p_c = p_0 = p_\infty$ (but see also Sect. 5). Figure 2b shows that a $p_c \approx 1/l$ relation is apparent for larger network sizes with a few percent relative standard error. Thus, according to Section 3.1, we have $q_c \approx l$, because $q_t = l^2 p_t$ and $p_c \approx 1/l$. For a *critical network* subject to our choices detailed in Table 1, the number of equal and non-zero elements in **F** is equal to the dimension of the hidden layer.

In the next section we introduce *exact critical structures* for the hidden matrix. We shall see that critical structure often exhibits superior performance.

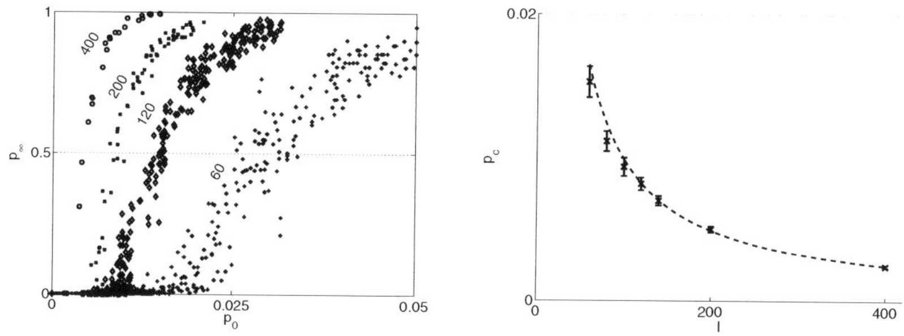

(a) Connectivity in the limit p_∞ versus initial connectivity p_0

(b) Critical 'point' p_c versus network size l

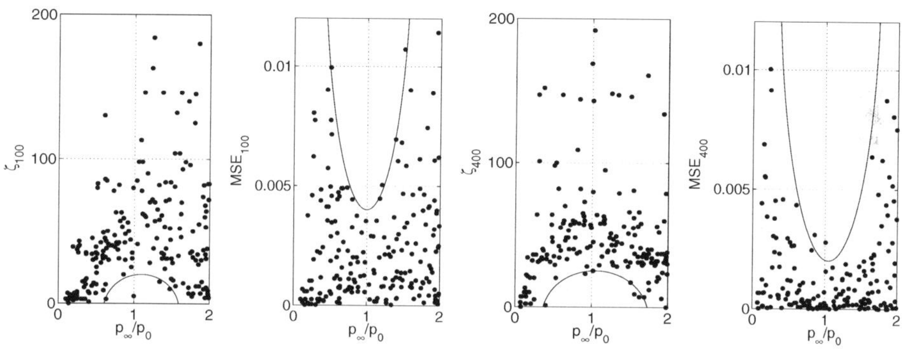

(c) **(d)** **(e)** **(f)**

Fig. 2. Phase transition and improved performance around the critical point. **(a):** Phase transition in time evolving connectivity. *Zero phase*: connections of the hidden layer disappear for sufficiently large, but finite times. *Saturated phase*: (almost) all connections contribute after sufficiently large times. Transition between the phases becomes sharp for larger hidden layers. **(b):** Position of p_c shifts to lower values as hidden layer size l increases. Dashed line: fit by assuming $p_c = 1/l$. **(c)** and **(e):** estimated successful prediction length ζ, **(d)** and **(f):** MSE of ζ, **(c)** and **(d):** size of hidden network is 100, **(e)** and **(f):** size of hidden network is 400. Solid lines: approximate (indicative) 'boundaries' that show improvements around the critical point $p_\infty/p_0 = 1$.

4.2 Critical Echo State Networks

Condition $q_c = l$ for matrix $\mathbf{F}^t$ is satisfied e.g., if every row and every column of $\mathbf{F}^t$ contains one non-zero element in the limit. Such structure will be called *exact critical structure*. For example, ESNs with permutation matrices in the hidden layer (PESNs) have exact critical structure.

Before proceeding, we conclude for the general case: according to our numerical studies, there is a critical region for networks. In this region, the time evolving hidden layer connectivity may not loose all connections (may not enter the *zero phase*) or may not get close to full connectivity (the *saturation phase*). See also

Fig. 2a and the caption of Fig. 2. Such networks will be called *Critical Echo State Networks* (CESNs).

There are special cases that belong to CESNs. For example, if the eigenvalues of matrix $\mathbf{F}$ are bounded by the unit sphere, two of them are on this sphere, and these two do not form a diagonal sub-matrix, that is they mix elements of the internal representation, then $\mathbf{F}^t$ will not belong to the zero phase nor to the saturation phase. Also, ESNs with hidden orthogonal matrices are CESNs, because their connectivity structure neither vanishes nor saturates in the limit.

In our investigations we shall turn to hidden permutation matrices, because otherwise the relative number of critical structures generated randomly may be very low, especially for large hidden layers. A particular $l \times l$ permutation matrix contains $1 \leq l_\nu \leq l$ number of cycles. A cycle of length l_ν exchanges the corresponding elements of a vector in l_ν steps. Similarly, orthogonal matrices mix subspaces.

Now, we present results for hidden permutation matrices, i.e., for PESNs and we set

$$\mathbf{F} = \mathbf{P} \ , \tag{6}$$

where $\mathbf{P}$ is a permutation matrix.

4.3 Prediction Gain over Ordinary ESNs

In this section we compare the performance of ESNs with PESNs on Mackey-Glass time series with delay parameters 17 (MG17) and 30 (MG30).

Figures 3a and 3b show distributions of successful prediction length ζ for 4,000 ordinary ESNs. The distributions are compact. The same figures depict the distributions for 4,000 PESNs with randomly generated input matrix, hidden permutation matrix, and optimized output matrix. PESNs show more asymmetric distributions for ζs. The average and the median are about the same for the two distributions, but the ESN distributions are much narrower. A large proportion of PESNs are very successful, whereas we have barely encountered significantly better than average randomly initialized ESNs, in agreement with the results reported in the literature. In Figs. 3a and 3b, the decrease of the PESN distribution is slower than that of the ESN distribution; the PESN distribution seems to have a long tail. For the more difficult MG30 time series prediction length is shorter for both networks.

Figures 3c and 3d compare the number of occasions that a particular successful prediction length was achieved by ESNs and PESNs for MG17 (Fig. 3c) and for MG30 (Fig. 3d). Performances were evaluated over 2500 different starting points and two comparisons were made. The best PESN out of 4000 randomly chosen PESN networks was compared to (a) the average ESN out of 4000 randomly chosen ESN networks and (b) the best ESN out of the same 4000 randomly chosen ESN networks. For high ratios, i.e., when the performance of the PESN is much better than that of the ESN, the curves become similar for both MG17 and MG30. Results indicate that for large successful prediction lengths, performances of the average and the best ESNs out of 4000 randomly generated networks are poor and are very similar. Thus, high performance ESNs are rare compared to

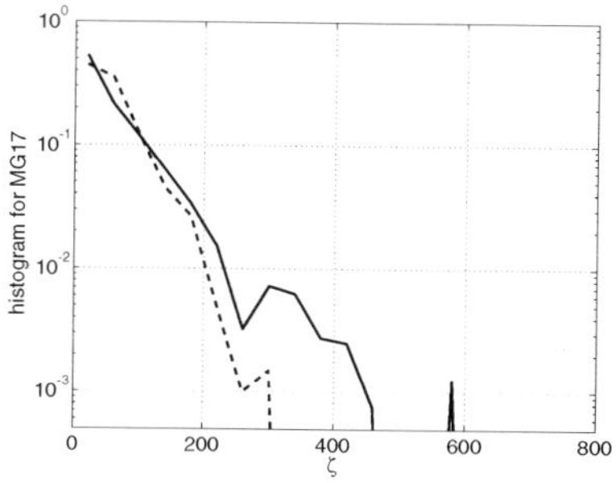

(a) Performance distributions over 2500 starting points for randomly generated ESN and PESN networks for MG17.

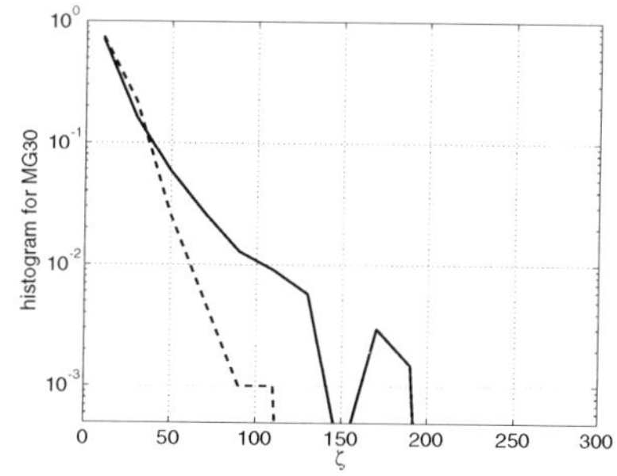

(b) Performance distributions over 2500 starting points for randomly generated ESN and PESN networks for MG30.

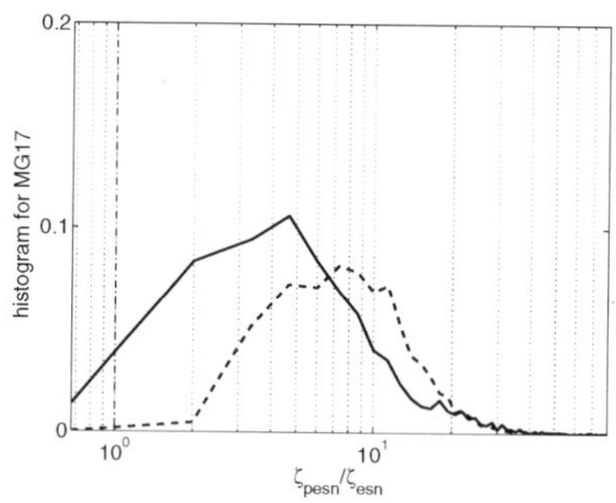

(c) Ratios of numbers of occasions of successful predictions length of PESNs and ESNs for 2500 starting points for MG17.

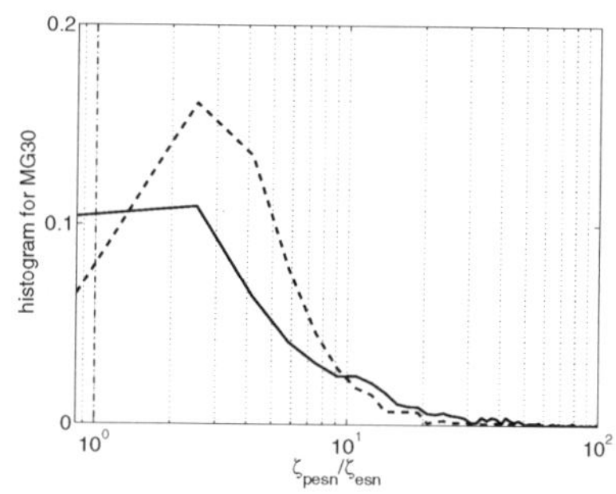

(d) Ratios of numbers of occasions of successful predictions length of PESNs and ESNs for 2500 starting points for MG30.

Fig. 3. Comparisons of ESNs and PESNs for MG17 and MG30 time series. **(a)** and **(b)**: Dashed lines: ESN, solid lines: PESN, network size: $l = 60$, **(c)** and **(d)**: Dashed lines: average ESN (out of 4000) and best PESN (out of 4000), solid lines: best ESN (out of 4000) and best PESN (out of 4000), network size: $l = 60$, vertical line at value 1 (at 10^0): 'curve' for identical distributions.

high performance PESNs: PESNs form a highly efficient subgroup within ESN networks – at least for MG chaotic time series.

We found that matrix $\mathbf{W}$ had an effect on the performance of PESNs. For example, $\mathbf{W}$ with similar elements had a negative effect. Uneven averages and variances for elements of $\mathbf{W}$ belonging to *different* cycles improved performance.

5 Discussion

We studied critical ESNs with single inputs. PESN performances have broader distributions than ESN ones. For PESNs, the *probability* that extremely good

ESN is found is dramatically increased. One can quickly find extremely good PESNs, whereas good ESNs are rare amongst ordinarily initialized ESNs.

Why do we find high performance PESNs significantly more often? Consider the permutation matrix in the hidden layer of the PESN. In general, it connects disjoint sets of elements, that is, we have disjoint cycles. We found that neither the single cycle case, nor the case of a large number but small cycles exhibited good performances. This was expected because of the following reason. For a single cycle of size n, identical representation arises after n steps. However, if there are more cycles, the identical representation appears after m_{LCM} steps, where m_{LCM} is the least common multiple of the sizes of the cycles. LCM is small if all cycles are equal, if cycles are small, or if there are single cycles. Such PESNs, show poor performances, but form only a small subset of randomly generated PESNs.

Now, consider general orthogonal matrices in the hidden layer. They belong to the class of CESNs, because their time evolving connectivity can neither saturate nor disappear for large times. It is possible that connectivity structure does not converge: periodic or never repeating structures may occur. We have studied CESNs starting from good PESNs. For example, we changed the sign of one or more non-zero elements of a permutation matrix. In all cases, the good predictive performance dropped to average. We also tried to modify different non-diagonal 2×2 sub-matrices defined by two non-zero elements of the permutation matrix to a rotation matrix. Performance decreased in most cases unless the angle of rotation was small. Note that permutation corresponds to rotation by 90^0 *and* a reflection, whereas 180^0 rotation corresponds to the change of the sign of one of the components. Combinations of these changes also spoiled performance in an overwhelming majority of the experiments.

The hidden permutation matrix is able to approximate non-periodic dynamical systems, because the hidden layer is embedded by the input matrix $\mathbf{W}$ and the output matrix $\mathbf{H}$, and they can modify finite cycles; matrix $\hat{\mathbf{F}} = \mathbf{F} + \mathbf{WH}$ counts in this respect. Note however, that matrix $\mathbf{WH}$, which can modify the permutation matrix, has limited capabilities, because the rank of this matrix is 1. Chances are high that changes of permutation matrices of good PESNs destroy performance, thus such changes seem to be out of reach for the optimization procedure of matrix $\mathbf{H}$ of the PESN.

Identification capabilities of *general* CESNs for dynamical systems *beyond* MG time series deserve further studies. A rich repertoire of phenomena may appear for input dimensions larger than 1.

6 Summary

We have shown that ESNs undergo sharp phase transition depending on the connectivity properties at the hidden layer. We have introduced and studied *critical echo state networks*. We found – by means of a large number of numerical simulations – that a large proportion of exact critical structures exhibit highly superior performance as opposed to ordinary ESNs, at least for MG time series.

We have argued that CESNs with permutation matrices in the hidden layer can identify both periodic and aperiodic time series, because the permutation matrix is complemented by other structures of the ESN.

References

1. Jaeger, H.: The Echo State Approach to Analysing and Training Recurrent Neural Networks. Technical Report 148, Fraunhofer Institute for Autonomous Intelligent Systems (2001)
2. Maas, W., Natschläger, T., Markram, H.: Real-Time Computing Without Stable States: A New Framework for Neural Computation Based on Perturbations. Neural Computation **14** (2002) 2531–2560
3. Jaeger, H.: Short Term Memory in Echo State Networks. Technical report, German National Research Center for Information Technology (2002)
4. Jaeger, H., Haas, H.: Harnessing Nonlinearity: Predicting Chaotic Systems and Saving Energy in Wireless Communication. Science **304** (2004) 78–80
5. Ishii, K., Zant, T., Becanovic, V., Plöger, P.: Identification of Motion with Echo State Network. In: Proc. Oceans. (2004) 1205–1230
6. Baier, N., De Feo, O.: Chaotic Model Identification Using a Biologically Inspired Algorithm. Aperest, Universidad Complutense de Madrid (2004)
7. Mayer, N., Browne, M.: Echo State Networks and Self-Prediction. In: Lecture Notes in Computer Science. Volume 3141., Springer Berlin / Heidelberg (2004) 40
8. Fette, G., Eggert, J.: Short Term Memory and Pattern Matching with Simple Echo State Networks. In: Lecture Notes in Computer Science. Volume 3696., Springer Berlin / Heidelberg (2005) 13
9. Jaeger, H.: Reservoir Riddles: Suggestions for Echo State Network Research (Extended Abstract). In: Proceedings of International Joint Conference on Neural Networks, Montreal, Canada. (2005)
10. Sornette, D.: Critical Phenomena in Natural Sciences. Springer Series in Synergetics. Springer, Berlin, Germany (2003)
11. Mackey, M., Glass, L.: Oscillation and chaos in physiological control systems. Science **197** (1977) 287–289
12. Cvitanovic, P., Artuso, R., Mainieri, R., Tanner, G., Vattay, G., Whelan, N., Wirzba, A.: Chaos: Classical and Quantum. Niels Bohr Institue, Copenhagen, chaosbook.org version 11 (2004)

Rapid Correspondence Finding in Networks of Cortical Columns

Jörg Lücke[1,2] and Christoph von der Malsburg[2,3,4]

[1] Gatsby Computational Neuroscience Unit, UCL, London WC1N 3AR, UK
[2] Institut für Neuroinformatik, Ruhr-Universität, 44780 Bochum, Germany
[3] Frankfurt Institute for Advanced Studies, 60438 Frankfurt a. M., Germany
[4] Computer Science Dept., University of Southern California, LA 90089-2520, USA

Abstract. We describe a neural network able to rapidly establish correspondence between neural fields. The network is based on a cortical columnar model described earlier. It realizes dynamic links with the help of specialized columns that evaluate similarities between the activity distributions of local feature cell populations, are subject to a topology constraint, and gate the transfer of feature information between the neural fields. Correspondence finding requires little time (estimated to 10-40 ms in physiological terms) and is robust to noise in feature signals.

1 Introduction

For various purposes it is necessary for the brain to find point-to-point correspondences between structured neural arrays. Among these is stereo vision and visual motion extraction. There are good reasons to assume that the brain also performs *correspondence-based invariant object recognition* [1]. In the technical domain, this represents state-of-the-art object and face recognition technology [2,3].

Objects can be recognized in less then 100 ms, see e.g. [4], and after each saccade a new system of correspondences needs to be established, making it clear that the mechanism must be very fast. A previous neural model of correspondence-based recognition [5] had problems with the evaluation of feature similarity and with speed. The model [6] is fast but did not attempt to cope with different feature types.

2 Columnar Network Model

The central element of our model is the *cortical column*. As discussed in [7] our column corresponds approximately to what in neuroscience is called a hypercolumn or macrocolumn and in primary visual cortex comprises all neurons that are activated from one point in visual space. A column contains sub-units called *minicolumns* or simply *units* that comprise on the order of one hundred neurons which are connected by mutual excitation. The activity of a unit is described

S. Kollias et al. (Eds.): ICANN 2006, Part I, LNCS 4131, pp. 668–677, 2006.

collectively by a variable p, and the units of one column mutually inhibit each other. The coupling coefficient ν of this inhibition is cyclically driven (*ν-cycle*), such that when ν is low all units are on, and when ν approaches a critical value some units switch off in sequence, thus reflecting the relative strengths of their afferent input. The dynamics of columns is described in the next section.

A simple model setting for the process of correspondence finding (see Fig. 1B) consists of an input domain $\mathcal{I}$, left column of large shaded ellipses, and a model domain $\mathcal{M}$, right column. Both domains consist of neural sheets that represent images by the activity distribution of fields of local feature detectors. (Correspondingly they should be two-dimensional, but for simplicity we limit ourselves here to one-dimensional chains. And the model domain should contain many such sheets to represent objects in memory, but for the time being we focus on just one.)

In each point of the two domains there are two columns, one to represent local features (horizontal ellipses within the shaded regions of Fig. 1B) and one to control links between the two domains (vertical ellipses). This double column (shaded ellipse in Fig. 1B) is called a *node*. *Feature columns* represent, with their activity, the local texture of the image or model, usually represented by units that are excited by different local spatial frequencies and orientations of the image's gray-level distribution. Typical feature distributions as used on our simulations are shown in Fig. 1A, where each row corresponds to one feature column, index i, and each column to one feature type, index α.

The domains communicate through *links*, which connect feature columns by as many fibers as there are feature types. In a *link control column* each unit stands for one link entering the node and does three things. One, it compares the activity distributions of the feature columns at the two ends of the link, two, it tries to be consistent with activities of units controlling parallel links, and three, by its activity it keeps open its link. The situation is shown in more detail in Fig. 2. At the end of a ν-cycle, when only one control unit is left active, all but one of the links into the node are switched off. This unit or link is selected by a combination of two criteria. One is feature similarity, the other is a topology constraint. The latter is to favor those link arrangements that connect neighbors in one domain with neighbors in the other domain, and is implemented by connections between control units in neighboring nodes (symbolized at the extreme right of Fig. 1B).

3 System Dynamics

The dynamics of the system is described by a set of coupled stochastic differential equations. We first introduce some notation. Let $\mathcal{L} \in \{\mathcal{I}, \mathcal{M}\}$ and $\mathcal{L}' \in \{\mathcal{I}, \mathcal{M}\}\backslash\{\mathcal{L}\}$ be indices for the two domains, i.e., $(\mathcal{L}, \mathcal{L}') = (\mathcal{I}, \mathcal{M})$ or $(\mathcal{M}, \mathcal{I})$. Further, let $p_\alpha^{\mathcal{L}i}$ stand for the activity of the feature unit α in node i of domain $\mathcal{L}$. We assume α runs from 1 to k and i from 1 to N. Let us designate by $W^{\mathcal{L}i,\mathcal{L}'j}$ the activity of the control unit with index j in node i of domain $\mathcal{L}$ (each

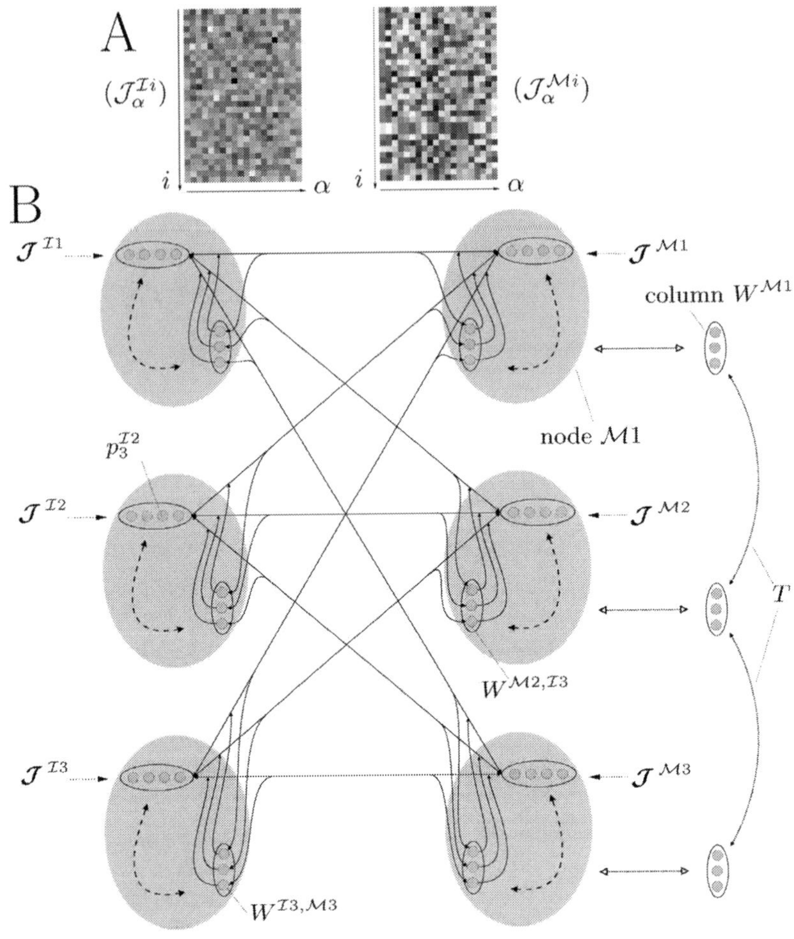

Fig. 1. A A collection of feature vectors (rows) with $k = 20$ entries. Model feature vectors on the right-hand-side are iid and uniformly distributed in $[0, 1]$. The input feature vectors on the lhs are noisy copies of the vectors on the rhs. **B** Network of columns for correspondence finding. The network consists of an input domain and a model domain with nodes $\mathcal{I}1$ to $\mathcal{I}3$ and $\mathcal{M}1$ to $\mathcal{M}3$, respectively. Each node consists of a feature column with $k = 4$ minicolumns and of a control column with $N = 3$ minicolumns. Each node in the input layer receives input from each node in the model layer, and *vice versa*. The inputs to a node are modulated by its control column according to the interconnectivity as displayed in Fig. 2. The control columns receive input from the units of feature columns of both layers and from neighboring control columns.

control column must contain as many control units as there are nodes in the other domain, in order to control as many links). As introduced and discussed in [7], the dynamics of the feature columns is then described by

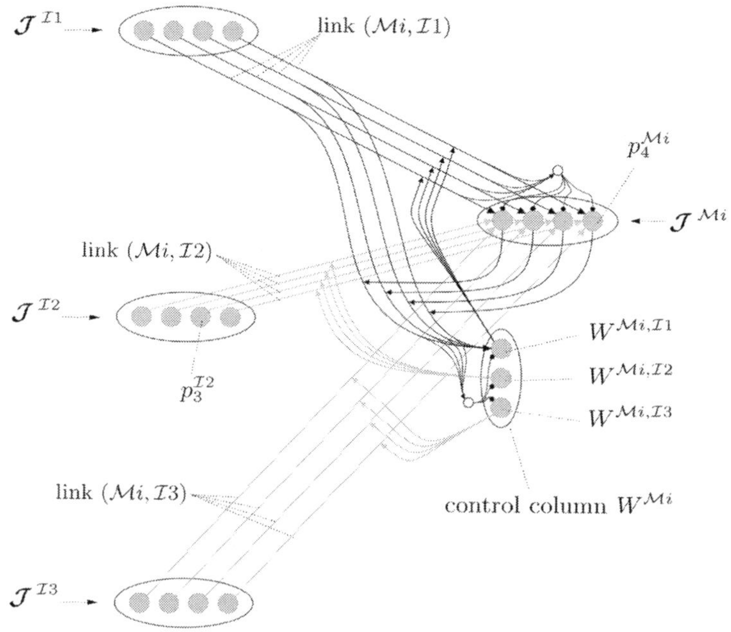

Fig. 2. Detailed connectivity of one node. On the input side three feature columns $\mathcal{I}1$, $\mathcal{I}2$ and $\mathcal{I}3$ are shown, which by their activity distribution represent three input feature vectors. On the model side, one node $\mathcal{M}i$ consisting of feature and control column is shown. For one control unit, $W^{\mathcal{M}i,\mathcal{I}j}$, the connection details are indicated by bold lines. On its input side, the unit evaluates the similarity of feature vectors in terms of their scalar product (multiplicative interactions indicated by arrowheads touching connecting fibers) and with its output gates the incoming link it stands for. Connectivity with neighboring control columns is not shown. Small circles represent neurons which by their inhibition subtract the mean of incoming feature vectors.

$$\frac{d}{dt} p_\alpha^{\mathcal{L}i} = f(p_\alpha^{\mathcal{L}i}, \nu \max_{\beta=1,\ldots,k} \{p_\beta^{\mathcal{L}i}\}) + \kappa E_\alpha^{\mathcal{L}i}, \tag{1}$$

where $E_\alpha^{\mathcal{L}i}$ is input to the unit, controlled in its strength by parameter κ, and where

$$f(p, h) = a\, p\, (p - h - p^2) + \sigma\eta_t \tag{2}$$

is a control function in the form of a polynomial of third degree, including the Gaussian noise term $\sigma\eta_t$ with variance σ^2. The inhibition is determined by the most active unit in the column, modulated by the inhibitory coefficient ν, which, as stated above, is controlled cyclically:

$$\nu(t) = \begin{cases} 0 & \text{if } \tilde{t} < T_{\text{init}} \\ (\nu_{\max} - \nu_{\min}) \frac{\tilde{t} - T_{\text{init}}}{T - T_{\text{init}}} + \nu_{\min} & \text{if } \tilde{t} \geq T_{\text{init}} \end{cases}, \tag{3}$$

where $\tilde{t} = t \bmod T$, which is $t - nT$, with n the greatest integer satisfying $t - nT \geq 0$.

To specify the input $E_\alpha^{\mathcal{L}i}$ in (1) we first have to define a few quantities. The feature input to feature unit α in node $\mathcal{L}i$ is designated as $\tilde{\mathcal{J}}_\alpha^{\mathcal{L}i} = \mathcal{J}_\alpha^{\mathcal{L}i} - \frac{1}{k}\sum_{\beta=1}^{k} \mathcal{J}_\beta^{\mathcal{L}i}$, where the mean is subtracted from the raw feature inputs. The momentary coupling strength from node j in domain $\mathcal{L}'$ to node i in domain $\mathcal{L}$ is set equal to the mean-free activity of the control unit of that link, $\tilde{W}^{\mathcal{L}i,\mathcal{L}'j} = W^{\mathcal{L}i,\mathcal{L}'j} - \frac{1}{N}\sum_{l=1}^{N} W^{\mathcal{L}i,\mathcal{L}'l}$. We then define the input into feature unit $p_\alpha^{\mathcal{L}}i$ in (1) as

$$E_\alpha^{\mathcal{L}i} = C_E\, \tilde{\mathcal{J}}_\alpha^{\mathcal{L}i} + (1 - C_E) \sum_{j=1}^{N} \sum_{\beta=1}^{k} \tilde{W}^{\mathcal{L}i,\mathcal{L}'j}\, R_{\alpha\beta}^{\mathcal{L}i,\mathcal{L}'j}\, p_\beta^{\mathcal{L}'j}, \tag{4}$$

where the parameter $C_E \in [0,1]$ controls the relative strength of the two sources of input. The matrix $R_{\alpha\beta}^{\mathcal{L}i,\mathcal{L}'j}$ defines feature-preserving interconnections between feature columns in the two domains: $R_{\alpha\beta}^{\mathcal{L}i,\mathcal{L}'j} = \delta_{\alpha\beta} - \frac{1}{k}$. Here finally are the dynamic equations of the control units:

$$\frac{d}{dt} W^{\mathcal{L}i,\mathcal{L}'j} = f(W^{\mathcal{L}i,\mathcal{L}'j}, \nu \max_{l=1,\ldots,N} \{W^{\mathcal{L}i,\mathcal{L}'l}\}) + \kappa\, I^{\mathcal{L}i,\mathcal{L}'j}, \tag{5}$$

$$I^{\mathcal{L}i,\mathcal{L}'j} = C_I \sum_{\alpha,\beta=1}^{k} p_\alpha^{\mathcal{L}i} R_{\alpha\beta}^{\mathcal{L}i,\mathcal{L}'j} p_\beta^{\mathcal{L}'j} + \overbrace{(1 - C_I) \sum_{a,b=1}^{N} T_{ab}^{\mathcal{L}i,\mathcal{L}'j} \tilde{W}^{\mathcal{L}a,\mathcal{L}'b}}^{\text{topology term}}, \tag{6}$$

where $C_I \in [0,1]$ controls the relative influence of the two terms in (6). The first term evaluates feature similarity. It resembles a scalar product with Euclidean metric between the activity vectors $\boldsymbol{p}^{\mathcal{L}i}$ and $\boldsymbol{p}^{\mathcal{L}'j}$ (other choices of $R^{\mathcal{L}i,\mathcal{L}'j}$ would correspond to other metrics). However, the situation is somewhat more complicated, as the activities of feature units do reflect feature values more in terms of the timing of their switching off in the course of the ν-cycle (later for stronger values) than by their firing strength at any moment.

The topology term in (6) implements link-to-link interactions. With vanishing topology term, $C_I = 1$, dynamics (1) to (6) would converge to a one-to-one connectivity that connected the most similar feature vectors in the model and input domains. Unfortunately it turns out that if there are non-trivial differences between model and image of the same object many nodes find their most similar feature vector in non-corresponding points of the other domain [8]. To remedy this problem, system dynamics should favor link arrangements that preserve neighborhood relationships. Accordingly, we structure the intra-layer connections ($T_{ab}^{\mathcal{L}i,\mathcal{L}'j}$) such that parallel links excite each other:

$$T_{ab}^{\mathcal{L}i,\mathcal{L}'j} = \sum_{c,d=-L}^{L} A_{c,d}\, \delta_{a,i+c}\, \delta_{b,i+d} - \tfrac{1}{N}\,, \qquad (A_{c,d}) = \begin{pmatrix} 0 & 0 & & & & & & & \\ 0.3 & 0.4 & 0.3 & & & & & & \\ & 0.1 & 0.8 & 0.1 & & & & & \\ & & 0 & 1 & 0 & & & & \\ & & & 0 & 0 & 0 & & & \\ & & & & 0 & 1 & 0 & & \\ & & & & & 0.1 & 0.8 & 0.1 & \\ & & & & & & 0.3 & 0.4 & 0.3 \\ & & & & & & & 0 & 0 \end{pmatrix} \qquad (7)$$

Here, all empty entries are meant to be zero. To understand the interactions of control columns consider the simpler limiting case $A_{c,d} = \delta_{c,d} - \delta_{c,0}\delta_{d,0}$ and $N = \infty$. If we inserted the resulting $(T_{ab}^{\mathcal{L}i,\mathcal{L}'j})$ into (6), the topology term would take the form $(1 - C_I) \sum_{c=-L, c\neq 0}^{L} \tilde{W}^{\mathcal{L}(i+c),\mathcal{L}'(j+c)}$, and only exactly parallel links in the range $[-L, +L]$ would excite each other. We use instead (7) in the following because we want links to excite each other also if they are only approximately parallel. For the two-dimensional case, $(T_{ab}^{\mathcal{L}i,\mathcal{L}'j})$ will have to be generalized appropriately.

4 Simulations

For numerical simulations of the differential equations we use the Euler method with time steps $\Delta t = \frac{1}{100}$ms. As domains we use chains of $N = 30$ nodes and cyclic boundary conditions, so that the last and the first nodes of the one-dimensional chain are neighbors. We choose the parameters $\kappa = 1.0\,\text{ms}^{-1}$ and $\sigma_{\text{no.}} = 0.01\,\text{ms}^{-1}$. The parameter a of the function f in (5) is chosen as in [9], $a = 200\,\text{ms}^{-1}$. The system is operated with oscillating inhibition coefficient ν, *cf.* (3), with period length $T = 25$ ms, $T_{\text{init}} = 2$ ms, $\nu_{\text{min}} = 0.4$ and $\nu_{\text{max}} = 0.52$ (a value slightly above the critical value $\nu_c = 0.5$, see [7]).

The influence of the topology term in (6) is best studied by setting C_I to zero, which lets the system ignore feature similarities and consider only the topology interactions within each layer. This results in decoupling the feature column dynamics, (1) and (4), from that of the link control columns, (5) and (6), which therefore can be simulated in isolation. A typical time course of the minicolumn activities $(W^{\mathcal{L}i,\mathcal{L}'j})$ during a ν-cycle is shown in Fig. 3. As can be seen, the system converges to a shifted diagonal connectivity matrix $(W^{\mathcal{L}i,\mathcal{L}'j})$, i.e., to a neighborhood-preserving one-to-one connectivity pattern. To which diagonal the system converges is decided by spontaneous symmetry breaking induced by noise when ν approaches a critical value.

If we choose an intermediate value for C_I, link dynamics is influenced by both neighborhood relationships and feature similarities. Both influences are essential to find the right correspondences and their relative strengths can be chosen using C_I. We simulate dynamics (1) to (7) with $C_E = 0.6$, so that feature columns are slightly more sensitive to their own feature vector than to input from the other layer, and with $C_I = 0.5$, giving equal weight to feature similarities and

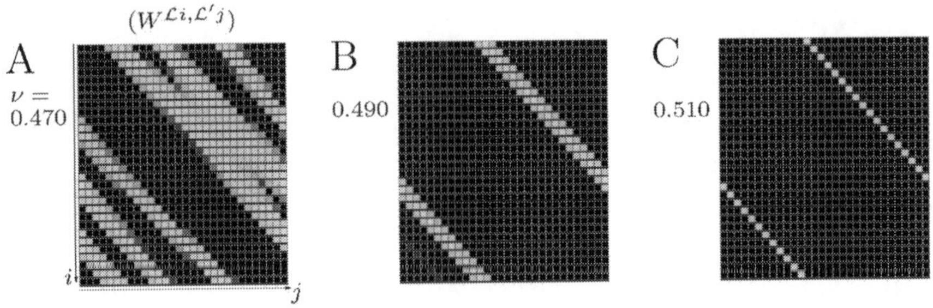

Fig. 3. Time course of network activities $(W^{\mathcal{L}i,\mathcal{L}'j})$ during a ν-cycle for $C_I = 0$. In a ν-cycle, ν increases from 0.4 to 0.52 in about 25ms. In the beginning all minicolumns $(W^{\mathcal{L}i,\mathcal{L}'j})$ of all columns are equally active (light grey). **A** For $0.45 < \nu \leq 0.47$ minicolumns start to be deactivated. **B** Because of the special choice of $(T_{ab}^{\mathcal{L}i,\mathcal{L}'j})$, diagonally arranged minicolumns are exciting each other and survive the increasing inhibition longer. Note that diagonals in $(W^{\mathcal{L}i,\mathcal{L}'j})$ correspond to neighborhood-preserving connectivity patterns between input and model domain. **C** Finally just one minicolumn per control column survives and the connectivity matrix $(W^{\mathcal{L}i,\mathcal{L}'j})$ is a shifted diagonal.

neighborhood relationships in the control of links. As input and model we use feature vectors $(\mathcal{J}^{\mathcal{I}i})$ and $(\mathcal{J}^{\mathcal{M}i})$ as given in Fig. 1A. The input feature vectors are noisy versions of the model feature vectors[1]. In Fig. 4 the result of a simulation with these feature vectors and parameters is shown for one ν-cycle. As can be observed, the dynamics converges to a symmetric one-to-one connectivity pattern between input and model layer. For visualization purposes we have chosen input feature vectors that were not translated w.r.t. the model feature vectors. For translated input feature vectors $(\mathcal{J}'^{\mathcal{I}i}) = (\mathcal{J}^{\mathcal{I}(i+\text{const})})$ (respecting the cyclic boundary conditions) the system converges the corresponding shifted diagonal. For input generated as above the system reliably finds the right correspondences for noise levels up to about $\sigma = 0.6$ which shows a remarkably high noise tolerance. For $k > 20$ results improve and for $k < 20$ the error rate increases. Note in this context that in our technical applications feature vectors typically have $k \geq 40$ entries [2,3] and that cortical columns in primary sensory areas are estimated to contain about $k = 80$ minicolumns [10].

[1] We use $N = 30$ model and $N = 30$ feature vectors, each with $k = 20$ entries. Correspondingly, the numbers of nodes per layer is N and the number of minicolumns per feature column is k. The model feature vectors consist of randomly ordered copies of 10 different feature vectors whose entries contain equally, identically, and independently distributed random values between zero and one. An input feature vector $(\mathcal{J}^{\mathcal{I}i})$ is generated from the model vector by adding Gaussian white noise with $\sigma = 0.6$ to the values $(\mathcal{J}^{\mathcal{M}i})$. Subsequently, the set of all values $(\mathcal{J}^{\mathcal{I}i})$ is rescaled such that all feature vector entries lie in the interval $[0, 1]$ again. The resulting image $(\mathcal{J}^{\mathcal{I}i})$ has on average smaller component deviations from the mean, due to the rescaling after adding noise.

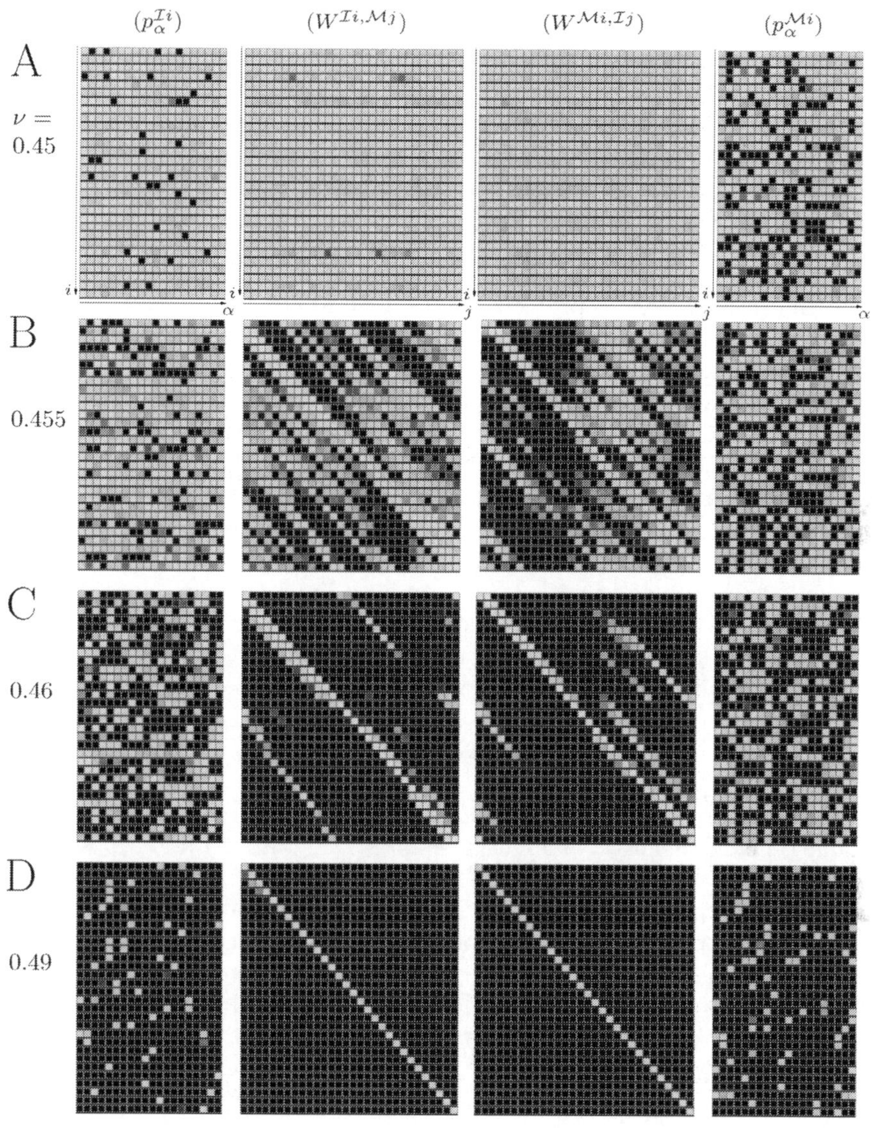

Fig. 4. Time course of the dynamical variables of the system displayed in Fig. 1B if feature vectors Fig. 1A are used. During a ν-cycle, ν increases from 0.4 to 0.52 in about 25ms. In the beginning of the ν-cycle all minicolumns are active (high gray values) and start deactivating at about $\nu = 0.45$ (see **A**). Note that the feature columns start first to deactivate their minicolumns because the input they get from their feature vectors is patterned. Feature vectors on the model side have inputs of higher variance and deactivate their minicolumns earlier. **B-C** Minicolumns of the control columns are deactivated according to similarities in feature columns and activities of other control columns. **D** Finally, control minicolumns remain active that correspond to diagonals in $(W^{\mathcal{I}i,\mathcal{M}j})$ and $(W^{\mathcal{M}i,\mathcal{I}j})$. The system has found the right correspondences as a neighborhood preserving mapping between similar features.

On the basis of feature similarities alone, $C_I = 1$, a system with otherwise the same parameters and noise level $\sigma = 0.6$ converges to one-to-one connectivities which are not neighborhood preserving and in which $80 - 90\%$ of the surviving links connect non-corresponding points.

5 Conclusion

Finding homomorphic, that is, structure-preserving, mappings between neural fields —the correspondence problem— is a capability of fundamental importance for the brain, not only for the visual system (stereo matching, motion field extraction) or perceptual systems in general (invariant pattern recognition), but more fundamentally for the application of abstract schemas to concrete situations and analogical thinking, and thus for intelligence on all levels. By its very nature, correspondence requires for its establishment and expression neural implementation media for the formulation of structural relationships and for the expression of dynamic links.

Both roles are played in our system by control columns, whose implementation turned out to be possible with fairly standard neurons. Our model describes minicolumn activity by abstract continuous variables, but as shown in previous work [9] this is capturing the essential properties of a more direct modeling of a system of spiking neurons [11]. Our model makes essential use of sigma-pi neurons, requiring sums of products of signals, *cf.* the second term in (4) and the first term in (6). Both cases involve control neurons, on the input side in one case, the output side in the other.

The activity of control columns and of feature columns is described here by the same type of stochastic differential equation (equations 1 and 5), but feature neurons and control neurons are probably of a different nature, the two types of columns playing very different roles. Control columns evaluate the similarity of local structure expressed by feature columns (this similarity being defined by the R-matrices in equations 4 and 6) and define, by interactions with each other, the homomorphy aspect of the correspondence (topology term). Feature columns, on the other hand, express local structure and are able to transmit it over distance. The handling of feature columns as integrated entity in the evaluation of similarities makes it possible to represent whole feature spaces, instead of single sample points in such spaces, as are represented by the combination-coding neurons that are conventionally used to represent higher features.

Our system solves several problems with previous models. One of them is the evaluation of feature similarities, which was a problem for [1], [6] and [5]. Another is excessive time requirement in [5]. As we demonstrate here, neural correspondence finding is possible in time-scales well below 100 ms because of the use of population rates. In fact, our simulations show that convergence to the right correspondences is possible within a critical period of a single ν-cycle of a few tens of ms (25 ms for our simulations) which would correspond to gamma range oscillations. During the critical phase (see Fig. 4) neurons typically spike only few times (<10) as discussed in [11]. In the limit of short period lengths with still reliable convergences ($\approx$10 ms) neurons have time to spike only 1 to 2 times in this period.

There are some challenges ahead of us. A full visual object recognition system will need a two-dimensional version and a model domain with many dozens of thousands of models. This threatens to require excessive numbers of control units. However, by using the maplet idea [1] and intermediate layers between the image domain and the model domain [6] this dragon could likely be tamed. Another, more formidable challenge concerns the ontogenetic development of the highly specific network structures involved in our model.

Acknowledgments. We thank J. D. Bouecke in helping us aquiring simulation results and gratefully acknowledge funding by the Gatsby Charitable Foundation, the Hertie Foundation, and by the EU project FP6-2005-015803.

References

1. J. Zhu and C. von der Malsburg. Maplets for correspondence-based object recognition. *Neural Networks, Special Issue 'New Developments in Self-Organizing Systems'*, 17/8–9:1311 – 1326, 2004.
2. P. Jonathan Philips, Hyeonjoon Moon, Syed A. Rizvi, and Patrick J. Rauss. The FERET evaluation methodology for face-recognition algorithms. *IEEE Transactions on Pattern Analysis and Machine Intelligence*, 22(10):1090–1104, 2000.
3. Kieron Messer et al. Face authentication test on the BANCA database. In *Proceedings of ICPR 2004, Cambridge*, volume 4, pages 523 – 532, 2004.
4. S. Thorpe, D. Fize, and C. Marlot. Speed of processing in the human visual system. *Nature*, 381:520–522, June 1996.
5. Laurenz Wiskott and Christoph von der Malsburg. Face recognition by dynamic link matching. In J. Sirosh, R. Miikkulainen, and Y. Choe, editors, *Lateral Interactions in the Cortex: Structure and Function*, chapter 4. WorldWideWeb, www.cs.utexas.edu/users/nn/book/bwdraft.html, 1995. ISBN 0-9647060-0-8.
6. B. A. Olshausen, C. H. Anderson, and D. C. Van Essen. A neurobiological model of visual attention and invariant pattern recognition based on dynamic routing of information. *The Journal of Neuroscience*, 13(11):4700–4719, 1993.
7. J. Lücke. Dynamics of cortical columns – sensitive decision making. In *Proc. ICANN*, LNCS 3696, pages 25 – 30. Springer, 2005.
8. L. Wiskott. The role of topographical constraints in face recognition. *Pattern Recognition Letters*, 20(1):89–96, 1999.
9. J. Lücke and J. D. Bouecke. Dynamics of cortical columns – self-organization of receptive fields. In *Proc. ICANN*, LNCS 3696, pages 31 – 37. Springer, 2005.
10. V. B. Mountcastle. The columnar organization of the neocortex. *Brain*, 120:701 – 722, 1997.
11. J. Lücke and C. von der Malsburg. Rapid processing and unsupervised learning in a model of the cortical macrocolumn. *Neural Computation*, 16:501 – 533, 2004.

Adaptive Thresholds for Layered Neural Networks with Synaptic Noise

D. Bollé and R. Heylen

Institute for Theoretical Physics
Katholieke Universiteit Leuven, Celestijnenlaan 200 D
B-3001, Leuven, Belgium

Abstract. The inclusion of a macroscopic adaptive threshold is studied for the retrieval dynamics of layered feedforward neural network models with synaptic noise. It is shown that if the threshold is chosen appropriately as a function of the cross-talk noise and of the activity of the stored patterns, adapting itself automatically in the course of the recall process, an autonomous functioning of the network is guaranteed. This self-control mechanism considerably improves the quality of retrieval, in particular the storage capacity, the basins of attraction and the mutual information content.

1 Introduction

As is common knowledge by now, layered feedforward neural network models are the workhorses in many practical applications in several areas of research and, therefore, any new insight in their capabilities and limitations should thus be welcome. In view of the fact that in many of these applications, e.g., pattern recognition in general, information is mostly encoded by a small fraction of bits and that also in neurophysiological studies the activity level of real neurons is found to be low, any reasonable network model has to allow variable activity of the neurons. The limit of low activity, i.e., sparse coding is then especially interesting. Indeed, sparsely coded models have a very large storage capacity behaving as $1/(a \ln a)$ for small a, where a is the activity (see, e.g., [1,2,3,4] and references therein). However, for low activity the basins of attraction might become very small and the information content in a single pattern is reduced [4]. Therefore, the necessity of a control of the activity of the neurons has been emphasized such that the latter stays the same as the activity of the stored patterns during the recall process. This has led to several discussions imposing external constraints on the dynamics of the network. However, the enforcement of such a constraint at every time step destroys part of the autonomous functioning of the network, i.e., a functioning that has to be independent precisely from such external constraints or control mechanisms. To solve this problem, quite recently a self-control mechanism has been introduced in the dynamics of networks for so-called diluted architectures [5]. This self-control mechanism introduces a time-dependent threshold in the transfer function [5,6]. It is determined as a function of both the cross-talk noise and the activity of the stored patterns in the network,

S. Kollias et al. (Eds.): ICANN 2006, Part I, LNCS 4131, pp. 678–687, 2006.

and adapts itself in the course of the recall process. It furthermore allows to reach optimal retrieval performance both in the absence and in the presence of synaptic noise [5,6,7,8]. These diluted architectures contain no common ancestors nodes, in contrast with feedforward architectures. It has then been shown that a similar mechanism can be introduced succesfully for layered feedforward architectures but, without synaptic noise [9].

The purpose of the present contribution is to generalise this self-control mechanism for layered architectures when synaptic noise is allowed, and to show that it leads to a substantial improvement of the quality of retrieval, in particular the storage capacity, the basins of attraction and the mutual information content.

2 The Model

Consider a neural network composed of binary neurons arranged in layers, each layer containing N neurons. A neuron can take values $\sigma_i(t) \in \{0,1\}$ where $t = 1,\dots,L$ is the layer index and $i = 1,\dots,N$ labels the neurons. Each neuron on layer t is unidirectionally connected to all neurons on layer $t+1$. We want to memorize p patterns $\{\xi_i^\mu(t)\}, i = 1,\dots,N,\ \mu = 1,\dots,p$ on each layer t, taking the values $\{0,1\}$. They are assumed to be independent identically distributed random variables (i.i.d.r.v.) with respect to i, μ and t, determined by the probability distribution: $p(\xi_i^\mu(t)) = a\delta(\xi_i^\mu(t)-1)+(1-a)\delta(\xi_i^\mu(t))$. From this form we find that the expectation value and the variance of the patterns are given by $E[\xi_i^\mu(t)] = E[\xi_i^\mu(t)^2] = a$. Moreover, no statistical correlations occur, in fact for $\mu \neq \nu$ the covariance vanishes.

The state $\sigma_i(t+1)$ of neuron i on layer $t+1$ is determined by the state of the neurons on the previous layer t according to the stochastic rule

$$P(\sigma_i(t+1) \mid \sigma_1(t),\dots,\sigma_N(t)) = \{1+\exp[2(2\sigma_i(t+1)-1)\beta h_i(t)]\}^{-1}. \quad (1)$$

The right hand side is the logistic function. The "temperature" $T = 1/\beta$ controls the stochasticity of the network dynamics, it measures the synaptic noise level [10]. Given the network state $\{\sigma_i(t)\}; i = 1,\dots,N$ on layer t, the so-called "local field" $h_i(t)$ of neuron i on the next layer $t+1$ is given by

$$h_i(t) = \sum_{j=1}^{N} J_{ij}(t)(\sigma_j(t)-a) - \theta(t) \quad (2)$$

with $\theta(t)$ the threshold to be specified later. The couplings $J_{ij}(t)$ are the synaptic strengths of the interaction between neuron j on layer t and neuron i on layer $t+1$. They depend on the stored patterns at different layers according to the covariance rule

$$J_{ij}(t) = \frac{1}{Na(1-a)} \sum_{\mu=1}^{N} (\xi_i^\mu(t+1)-a)(\xi_j^\mu(t)-a)\ . \quad (3)$$

These couplings then permit to store sets of patterns to be retrieved by the layered network.

The dynamics of this network is defined as follows (see [11]). Initially the first layer (the input) is externally set in some fixed state. In response to that, all neurons of the second layer update synchronously at the next time step, according to the stochastic rule (1), and so on.

At this point we remark that the couplings (3) are of infinite range (each neuron interacts with infinitely many others) such that our model allows a so-called mean-field theory approximation. This essentially means that we focus on the dynamics of a single neuron while replacing all the other neurons by an average background local field. In other words, no fluctuations of the other neurons are taken into account. In our case this approximation becomes exact because, crudely speaking, $h_i(t)$ is the sum of very many terms and a central limit theorem can be applied [10].

It is standard knowledge by now that mean-field theory dynamics can be solved exactly for these layered architectures (e.g., [11,12]). By exact analytic treatment we mean that, given the state of the first layer as initial state, the state on layer t that results from the dynamics is predicted by recursion formulas. This is essentially due to the fact that the representations of the patterns on different layers are chosen independently. Hence, the big advantage is that this will allow us to determine the effects from self-control in an exact way.

The relevant parameters describing the solution of this dynamics are the *main overlap* of the state of the network and the μ-th pattern, and the *neural activity* of the neurons

$$M^{\mu}(t) = \frac{1}{Na(1-a)} \sum_{i=1}^{N} (\xi_i^{\mu}(t) - a)(\sigma_i(t) - a), \qquad q(t) = \frac{1}{N} \sum_{i=1}^{N} \sigma_i(t) \,. \tag{4}$$

In order to measure the retrieval quality of the recall process, we use the mutual information function [5,6,13,14]. In general, it measures the average amount of information that can be received by the user by observing the signal at the output of a channel [15,16]. For the recall process of stored patterns that we are discussing here, at each layer the process can be regarded as a channel with input $\xi_i^{\mu}(t)$ and output $\sigma_i(t)$ such that this mutual information function can be defined as [5,15]

$$I(\sigma_i(t); \xi_i^{\mu}(t)) = S(\sigma_i(t)) - \langle S(\sigma_i(t)|\xi_i^{\mu}(t)) \rangle_{\xi^{\mu}(t)} \tag{5}$$

where $S(\sigma_i(t))$ and $S(\sigma_i(t)|\xi_i^{\mu}(t))$ are the entropy and the conditional entropy of the output, respectively

$$S(\sigma_i(t)) = -\sum_{\sigma_i} p(\sigma_i(t)) \ln[p(\sigma_i(t))] \tag{6}$$

$$S(\sigma_i(t)|\xi_i^{\mu}(t)) = -\sum_{\sigma_i} p(\sigma_i(t)|\xi_i^{\mu}(t)) \ln[p(\sigma_i(t)|\xi_i^{\mu}(t))] \,. \tag{7}$$

These information entropies are peculiar to the probability distributions of the output. The quantity $p(\sigma_i(t))$ denotes the probability distribution for the neurons at layer t and $p(\sigma_i(t)|\xi_i^{\mu}(t))$ indicates the conditional probability that the

i-th neuron is in a state $\sigma_i(t)$ at layer t given that the i-th site of the pattern to be retrieved is $\xi_i^\mu(t)$. Hereby, we have assumed that the conditional probability of all the neurons factorizes, i.e., $p(\{\sigma_i(t)\}|\{\xi_i(t)\}) = \prod_j p(\sigma_j(t)|\xi_j(t))$, which is a consequence of the mean-field theory character of our model explained above. We remark that a similar factorization has also been used in Schwenker et al. [17].

The calculation of the different terms in the expression (5) proceeds as follows. Because of the mean-field character of our model the following formula hold for every neuron i on each layer t. Formally writing (forgetting about the pattern index μ) $\langle O\rangle \equiv \langle\langle O\rangle_{\sigma|\xi}\rangle_\xi = \sum_\xi p(\xi)\sum_\sigma p(\sigma|\xi)O$ for an arbitrary quantity O the conditional probability can be obtained in a rather straightforward way by using the complete knowledge about the system: $\langle\xi\rangle = a$, $\langle\sigma\rangle = q$, $\langle(\sigma-a)(\xi-a)\rangle = M$, $\langle 1\rangle = 1$.

The result reads

$$p(\sigma|\xi) = [\gamma_0\xi + (\gamma_1-\gamma_0)\xi]\delta(\sigma-1) + [1-\gamma_0-(\gamma_1-\gamma_0)\xi]\delta(\sigma) \tag{8}$$

where $\gamma_0 = q - aM$ and $\gamma_1 = (1-a)M + q$, and where the M and q are precisely the relevant parameters (4) for large N. Using the probability distribution of the patterns we obtain

$$p(\sigma) = q\delta(\sigma-1) + (1-q)\delta(\sigma)\ . \tag{9}$$

Hence the entropy (6) and the conditional entropy (7) become

$$S(\sigma) = -\ q\ln q - (1-q)\ln(1-q) \tag{10}$$

$$\begin{aligned} S(\sigma|\xi) = &-\ [\gamma_0 + (\gamma_1-\gamma_0)\xi]\ln[\gamma_0 + (\gamma_1-\gamma_0)\xi] \\ &-[1-\gamma_0-(\gamma_1-\gamma_0)\xi]\ln[1-\gamma_0-(\gamma_1-\gamma_0)\xi]\ . \end{aligned} \tag{11}$$

By averaging the conditional entropy over the pattern ξ we finally get for the mutual information function (5) for the layered model

$$\begin{aligned} I(\sigma;\xi) = -q\ln q - (1-q)\ln(1-q) + a[\gamma_1\ln\gamma_1 + (1-\gamma_1)\ln(1-\gamma_1)] \\ +(1-a)[\gamma_0\ln\gamma_0 + (1-\gamma_0)\ln(1-\gamma_0)]\ . \end{aligned} \tag{12}$$

3 Adaptive Thresholds

It is standard knowledge (e.g., [11]) that the synchronous dynamics for layered architectures can be solved exactly following the method based upon a signal-to-noise analysis of the local field (2) (e.g., [4,12,18,19] and references therein). Without loss of generality we focus on the recall of one pattern, say $\mu = 1$, meaning that only $M^1(t)$ is macroscopic, i.e., of order 1 and the rest of the patterns causes a cross-talk noise at each step of the dynamics.

We suppose that the initial state of the network model $\{\sigma_i(1)\}$ is a collection of i.i.d.r.v. with average and variance given by $E[\sigma_i(1)] = E[(\sigma_i(1))^2] = q_0$. We furthermore assume that this state is correlated with only one stored pattern, say pattern $\mu = 1$, such that $\mathrm{Cov}(\xi_i^\mu(1), \sigma_i(1)) = \delta_{\mu,1}\ M_0^1\ a(1-a)$.

Then the full recall proces is described by [11,12]

$$M^1(t+1) = \frac{1}{2}\left\{\int \mathcal{D}x \tanh\left[\beta((1-a)M^1(t) - \theta(t) + \sqrt{\alpha D(t)}\,x)\right]\right.$$
$$\left. + \int \mathcal{D}x \tanh\left[\beta(-aM^1(t) - \theta(t) + \sqrt{\alpha D(t)}\,x)\right]\right\} \quad (13)$$
$$q(t+1) = aM^1(t+1)$$
$$+\frac{1}{2}\left\{1 + \int \mathcal{D}x \tanh\left[\beta(-aM^1(t) - \theta(t) + \sqrt{\alpha D(t)}\,x)\right]\right\} \quad (14)$$
$$D(t+1) = Q(t+1)$$
$$+\frac{\beta}{2}\left\{1 - a\int \mathcal{D}x \tanh^2 \beta\left[(1-a)M^1(t) - \theta(t) + \sqrt{\alpha D(t)}\,x\right]\right.$$
$$\left. - (1-a)\int \mathcal{D}x \tanh^2 \beta\left[-aM^1(t) - \theta(t) + \sqrt{\alpha D(t)}\,x\right]\right\}^2 D(t) \quad (15)$$

where $\alpha = p/N$, $\mathcal{D}x$ is the Gaussian measure $\mathcal{D}x = dx(2\pi)^{-1/2}\exp(-x^2/2)$, where $Q(t) = [(1-2a)q(t) + a^2]$ and where $D(t)$ contains the influence of the cross-talk noise caused by the patterns $\mu > 1$. As mentioned before, $\theta(t)$ is an adaptive threshold that has to be chosen.

In the sequel we discuss two different choices and both will be compared for networks with synaptic noise and various activities. Of course, it is known that the quality of the recall process is influenced by the cross-talk noise. An idea is then to introduce a threshold that adapts itself autonomously in the course of the recall process and that counters, at each layer, the cross-talk noise. This is the self-control method proposed in [5]. This has been studied for layered neural network models without synaptic noise, i.e., at $T = 0$, where the rule (1) reduces to the deterministic form $\sigma_i(t+1) = \Theta(h_i(t))$ with $\Theta(x)$ the Heaviside function taking the value $\{0, 1\}$. For sparsely coded models, meaning that the pattern activity a is very small and tends to zero for N large, it has been found [9] that

$$\theta(t)_{sc} = c(a)\sqrt{\alpha D(t)}, \quad c(a) = \sqrt{-2\ln a} \quad (16)$$

makes the second term on the r.h.s of Eq.(14) at $T = 0$, asymptotically vanish faster than a such that $q \sim a$. It turns out that the inclusion of this self-control threshold considerably improves the quality of retrieval, in particular the storage capacity, the basins of attraction and the information content.

The second approach chooses a threshold by maximizing the information content, $i = \alpha I$ of the network (recall Eq. (12)). This function depends on $M^1(t)$, $q(t)$, a, α and β. The evolution of $M^1(t)$ and of $q(t)$ (13), (14) depends on the specific choice of the threshold through the local field (2). We consider a layer independent threshold $\theta(t) = \theta$ and calculate the value of (12) for fixed a, α, M_0^1, q_0 and β. The optimal threshold, $\theta = \theta_{opt}$, is then the one for which the mutual information function is maximal. The latter is non-trivial because it is even rather difficult, especially in the limit of sparse coding, to choose a threshold interval by hand such that i is non-zero. The computational cost will thus be larger compared to the one of the self-control approach. To illustrate this we plot in Figure 1 the information content i as a function of θ without self-control or a priori optimization, for $a = 0.005$ and different values of α. For every value of

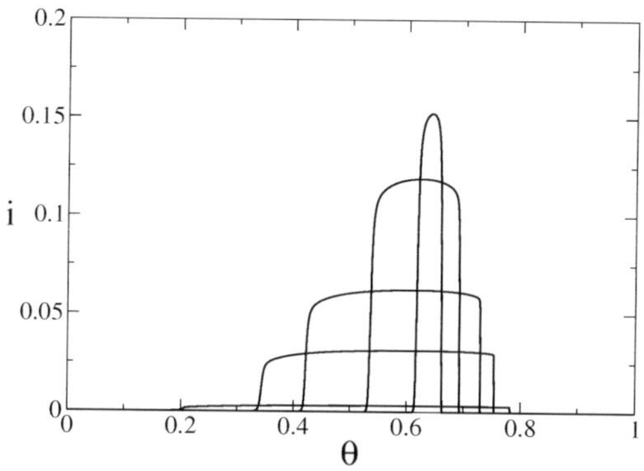

Fig. 1. The information $i = \alpha I$ as a function of θ for $a = 0.005$, $T = 0.1$ and several values of the load parameter $\alpha = 0.1, 1, 2, 4, 6$ (bottom to top)

α, below its critical value, there is a range for the threshold where the information content is different from zero and hence, retrieval is possible. This retrieval range becomes very small when the storage capacity approaches its critical value $\alpha_c = 6.4$.

Concerning then the self-control approach, the next problem to be posed in analogy with the case without synaptic noise is the following one. Can one determine a form for the threshold $\theta(t)$ such that the integral in the second term on the r.h.s of Eq.(14) at $T \neq 0$ vanishes asymptotically faster than a?

In contrast with the case at zero temperature where due to the simple form of the transfer function, this threshold could be determined analytically (recall Eq. (16)), a detailed study of the asymptotics of the integral in Eq. (14) gives no satisfactory analytic solution. Therefore, we have designed a systematic numerical procedure through the following steps:

- Choose a small value for the activity a'.
- Determine through numerical integration the threshold θ' such that

$$\int_{-\infty}^{\infty} \frac{dx\; e^{-x^2/2\sigma^2}}{\sigma\sqrt{2\pi}} \Theta(x-\theta) \leq a' \quad \text{for} \quad \theta > \theta' \tag{17}$$

  for different values of the variance $\sigma^2 = \alpha D(t)$.
- Determine as a function of $T = 1/\beta$, the value for θ'_T such that

$$\int_{-\infty}^{\infty} \frac{dx\; e^{-y^2/\sigma^2}}{2\sigma\sqrt{2\pi}} [1 + \tanh[\beta(x-\theta)]] \leq a' \quad \text{for} \quad \theta > \theta' + \theta'_T. \tag{18}$$

The second step leads precisely to a threshold having the form of Eq. (16). The third step determining the temperature-dependent part θ'_T leads to the final proposal

$$\theta_t(a,T) = \sqrt{-2\ln(a)\alpha D(t)} - \frac{1}{2}\ln(a)T^2. \tag{19}$$

This dynamical threshold is again a macroscopic parameter, thus no average must be taken over the microscopic random variables at each step t of the recall process.

We have solved these self-controlled dynamics, Eqs.(13)-(15) and (19), for our model with synaptic noise, in the limit of sparse coding, numerically. In particular, we have studied in detail the influence of the T-dependent part of the threshold. Of course, we are only interested in the retrieval solutions with $M > 0$ (we forget about the index 1) and carrying a non-zero information $i = \alpha I$. The important features of the solution are illustrated, for a typical value of a in Figures 2-4. In Figure 2 we show the basin of attraction for the whole retrieval

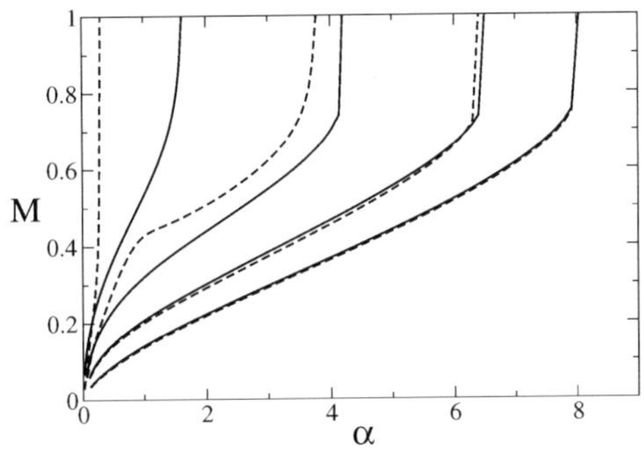

Fig. 2. The basin of attraction as a function of α for $a = 0.005$ and $T = 0.2, 0.15, 0.1, 0.05$ (from left to right) with (full lines) and without (dashed lines) the T-dependent part in the threshold (19)

phase for the model with threshold (16) (dashed curves) compared to the model with the noise-dependent threshold (19) (full curves). We see that there is no clear improvement for low T but there is a substantial one for higher T. Even near the border of critical storage the results are still improved such that also the storage capacity itself is larger.

This is further illustrated in Figure 3 where we compare the evolution of the retrieval overlap $M(t)$ starting from several initial values, M_0, for the model with (Figure 3 (a)) and without (Figure 3 (b)) the T-correction in the threshold and for the optimal threshold model (Figure 3 (c)). Here this temperature correction is absolutely crucial to guarantee retrieval, i.e., $M \approx 1$. It really makes the difference between retrieval and non-retrieval in the model. Furthermore, the model with the self-control threshold with noise-correction has even a wider basin of attraction than the model with optimal threshold.

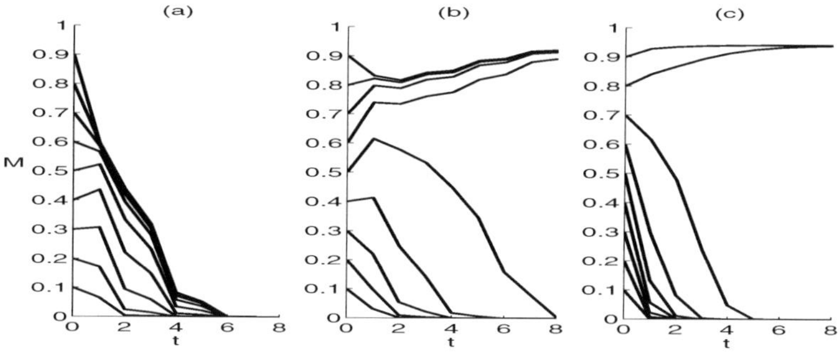

Fig. 3. The evolution of the main overlap $M(t)$ for several initial values M_0 with $T = 0.2$, $q_0 = a = 0.005$, $\alpha = 1$ for the self-control model (19) without (a) and with T-dependent part (b) and for the optimal threshold model (c)

In Figure 4 we plot the information content i as a function of the temperature for the self-control dynamics with the threshold (19) (full curves), respectively (16) (dashed curves). We see that a substantial improvement of the information content is obtained.

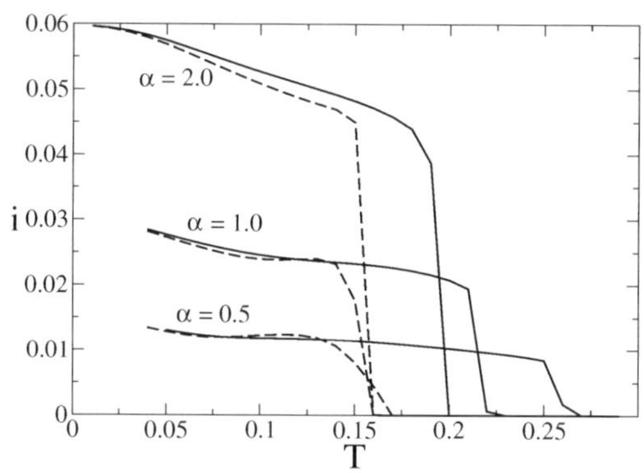

Fig. 4. The information content $i = \alpha I$ as a function of T for several values of the loading α and $a = 0.005$ with (full lines) and without (dashed lines) the T-correction in the threshold

Finally we show in Figure 5 a $T - \alpha$ plot for $a = 0.005$ (a) and $a = 0.02$ (b) with (full line) and without (dashed line) noise-correction in the self-control threshold and with optimal threshold (dotted line). These lines indicate two phases of the layered model: below the lines our model allows recall, above the lines it does not. For $a = 0.005$ we see that the T-dependent term in the self-control threshold leads to a big improvement in the region for large noise and small loading and in the region of critical loading. For $a = 0.02$ the results for the self-control threshold with and without noise-correction and those for the optimal

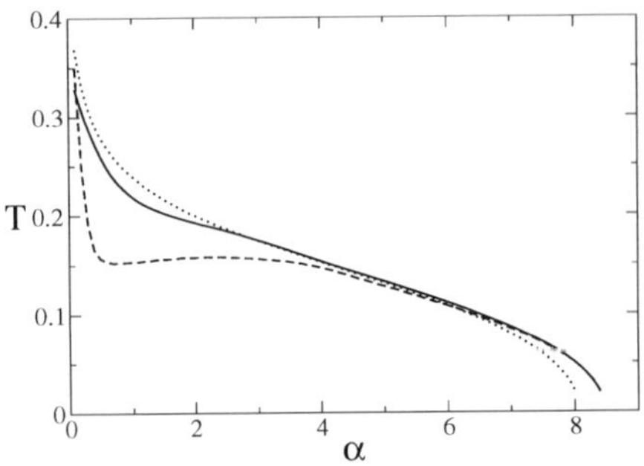

(a)

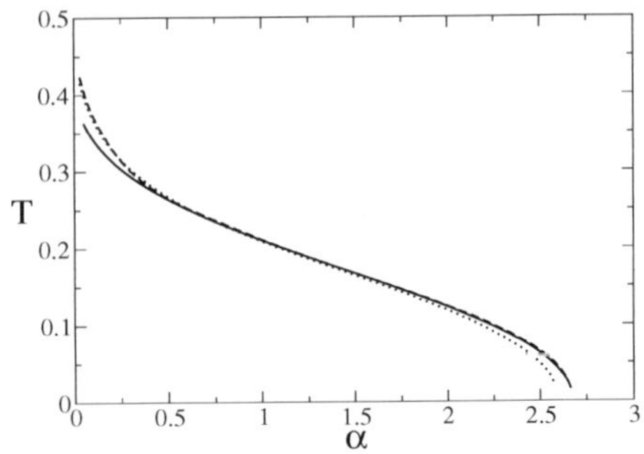

(b)

Fig. 5. Phases in the $T - \alpha$ plane for $a = 0.005$ (a) and $a = 0.02$ (b) with (full line) and without (dashed line) the temperature correction in the self-control threshold and with optimal threshold (dotted line)

thresholds almost coincide, but we recall that the calculation with self-control is autonomously done by the network and less demanding computationally.

4 Conclusions

In this work we have studied the inclusion of an adaptive threshold in sparsely coded layered neural networks with synaptic noise. We have presented an analytic form for a self-control threshold, allowing an autonomous functioning of the network, and compared it with an optimal threshold obtained by maximizing the mutual information which has to be calculated externally each time one of the network parameters (activity, loading, temperature) is changed. The consequences of this self-control mechanism on the quality of the recall process have been studied.

We find that the basins of attraction of the retrieval solutions as well as the storage capacity are enlarged. For some activities the self-control threshold even sets the border between retrieval and non-retrieval. This confirms the considerable improvement of the quality of recall by self-control, also for layered network models with synaptic noise.

This allows us to conjecture that self-control might be relevant for other architectures in the presence of synaptic noise, and even for dynamical systems in general, when trying to improve, e.g., basins of attraction .

Acknowledgment

This work has been supported by the Fund for Scientific Research- Flanders (Belgium).

References

1. Willshaw D J, Buneman O P, and Longuet-Higgins H C, Nonholographic associative memory, *Nature* **222** (1969) 960.
2. Palm G, On the storage capacity of an associative memory with random distributed storage elements, *Biol. Cyber.* **39** (1981) 125.
3. Gardner E, The space of interactions in neural network models, *J. Phys. A: Math. Gen.* **21** (1988) 257.
4. Okada M, Notions of associative memory and sparse coding, *Neural Networks* **9** (1996) 1429.
5. Dominguez D R C and Bollé D, Self-control in sparsely coded networks, *Phys. Rev. Lett.* **80** (1998) 2961.
6. Bollé D, Dominguez D R C and Amari S, Mutual information of sparsely coded associative memory with self-control and ternary neurons, *Neural Networks* **13**(2000) 455.
7. Bollé D and Heylen R, Self-control dynamics for sparsely coded networks with synaptic noise, in *2004 Proceedings of the IEEE International Joint Conference on Neural Networks*, p.3195
8. Dominguez D R C, Korutcheva E, Theumann W K and Erichsen Jr. R, Flow diagrams of the quadratic neural network, *Lecture Notes in Computer Science*, **2415**, (2002) 129.
9. Bollé D and Massolo G, Thresholds in layered neural networks with variable activity, *J. Phys. A: Math. Gen.* **33** (2000) 2597.
10. Hertz J, Krogh A and Palmer R G, *Introduction to the Theory of Neural Computation*, Addison-Wesley, Redwood City (1991).
11. Domany E, Kinzel W and Meir R, Layered Neural Networks, *J.Phys. A: Math. Gen.* **22** (1989) 2081.
12. Bollé D, Multi-state neural networks based upon spin-glasses: a biased overview, in *Advances in Condensed Matter and Statistical Mechanics* eds. Korutcheva E and Cuerno R., Nova Science Publishers, New-York,(2004 p. 321-349.
13. Nadal J-P, Brunel N and Parga N, Nonlinear feedforward networks with stochastic outputs: infomax implies redundancy reduction, *Network: Computation in Neural Systems* **9** (1998) 207.
14. Schultz S and Treves A, Stability of the replica-symmetric solution for the information conveyed by a neural network. *Phys. Rev. E* **57** (1998) 3302.
15. Blahut R E, *Principles and Practice of Information Theory*, Reading, MA: Addison-Wesley (1990).
16. Shannon C E, A mathematical theory for communication, *Bell Systems Technical Journal* **27** (1948) 379.
17. Schwenker F, Sommer F T and Palm G, Iterative retrieval of sparsely coded associative memory patterns, *Neural Networks* **9** (1996) 445.
18. Amari S, Neural theory and association of concept information, *Biol. Cyber.* **26** (1977) 175.
19. Amari S and Maginu K, Statistical neurodynamics of associative memory, *Neural Networks* **1** (1988) 63.

Backbone Structure of Hairy Memory

Cheng-Yuan Liou

Department of Computer Science and Information Engineering
National Taiwan University
cyliou@csie.ntu.edu.tw

Abstract. This paper presents a new memory of the Hopfield model that fixes many drawbacks of the model, such as loading capacity, limit cycle and error tolerance. This memory is derived from the hairy model [15]. This paper also constructs a training process to further balance the vulnerable memory parts and improve the memory.

Keywords: hairy model, neural network, associative memory, Hopfield network, modular biology.

1 Introduction

This paper presents a new memory derived from the hairy model [15]. This memory is the backbone of the Hopfield memory [7] that can balance the vulnerable parts of memory patterns. It enlarges the memory basins in order to resist corruption. We briefly review associative memory (AM) and notations. The hairy model and memory basin are also reviewed.

Associative memory
Auto-AM is used to store patterns: when a reasonable subset of a certain pattern is received with another corrupted part, it has the ability to recover that pattern. There are many models for training auto-AM that can improve its accuracy, efficiency and capacity. One of these is the famous model proposed by Hopfield (1982). It applies Hebb's postulate to generate weights collectively. This model carries very rich meaning in both biology and physics. But it has drawbacks in loading capacity, limit cycles and error tolerance with respect to noisy patterns. Many designs aimed at remedying these drawbacks have achieved varying degrees of success [5][13][15].

Auto-AM is a connected network with N neurons. Each neuron i has N neural weights connecting it to all the neurons j, including itself, a threshold θ_i and a state value v_i. The state value is updated according to the formula

$$v_i(t+1) = \operatorname{sgn}\left[\sum_{j=1}^{N} w_{ij} v_j(t) - \theta_i\right], \tag{1}$$

or in matrix form,

$$V(t+1) = \operatorname{sgn}\left[W^T V(t) - \theta\right], \tag{2}$$

S. Kollias et al. (Eds.): ICANN 2006, Part I, LNCS 4131, pp. 688–697, 2006.

where W is an $N \times N$ neural weight matrix, θ is an $N \times 1$ threshold vector, $V(t)$ is an $N \times 1$ state vector representing the state at iteration (or evolution) time t, and sgn() is the signum function returning 1 with input greater than or equal to zero and -1 with negative input. During the learning phase, the network is trained by P memory patterns $X^k, k = 1, \ldots, P$, using any training algorithm. During the retrieving phase, the input is presented to the network as $V(0)$. The network operates repeatedly according to Eq.(1) or Eq.(2) until evolution converges to a stable state or falls into a limit cycle. A stable state (memory pattern) meets the requirement

$$V(t) = \text{sgn}\left[W^T V(t) - \theta\right] \tag{3}$$

no matter whether the network is operating in the synchronous mode or asynchronous mode.

Each neuron has a bipolar state (a bit), and there are 2^N states in total. Therefore, we view the network as an N-dimensional (N-D) hypercube with each state located at a cube corner [10]. The current state is located at a corner and serves as the next input to the network. After evolution proceeds according to Eq.(1) or Eq.(2), the current state either moves to another corner or stays at the original corner. Corners that remain unchanged are stable states. The memory patterns we intend to save are located at certain stable corners. The goal of AM is to evolve an arbitrary initial state to a nearby stable corner where a pattern is stored. Figure 1 shows a 3-D cube corresponding to a network with three neurons (three bits). In this figure, a neuron represents a plane (the shaded plane) dividing the cube into positive and negative regions (sides). In Fig. 1(a), states $(1,1,1)^T$, $(1,-1,1)^T$, $(1,1,-1)^T$ and $(-1,1,1)^T$ are in the positive region, while the other corners are in the negative region. The memory patterns denoted by black dots are: (a). $(1,1,1)^T$, $(-1,-1,-1)^T$; (b). $(1,1,1)^T$, $(-1,-1,-1)^T$, $(1,-1,-1)^T$, $(-1,1,-1)^T$; (c). $(-1,1,1)^T$, $(1,-1,1)^T$, $(1,1,-1)^T$, $(-1,-1,-1)^T$; (d). $(1,1,1)^T$, $(-1,1,1)^T$, $(1,-1,-1)^T$, $(-1,-1,-1)^T$; (e). $(-1,1,1)^T$, $(-1,1,-1)^T$, $(1,1,-1)^T$.

The state of each neuron i is determined by an $(N-1)$-D decision hyperplane (the shaded surfaces in Fig.1) with the following equation:

$$w_{i1}v_1 + w_{i2}v_2 + w_{i3}v_3 + \cdots + w_{iN}v_N - \theta_i = 0, \quad i = 1, \ldots, N. \tag{4}$$

The $N \times 1$ weight vector $W_i = (w_{i1}, w_{i2}, \ldots, w_{iN})^T$ of neuron i is the normal vector of the corresponding hyperplane, and this hyperplane divides the hypercube into a positive region (side) to which the normal vector points and a negative region. The length of vector W_i, $|W_i|$, is normalized to one. The hairy learning adjusts the hyperplane to make all the P patterns stable. That is, when the ith bit of a pattern is equal to 1, this stable pattern should be located in the positive region of the ith hyperplane; on the other hand, if its ith bit is equal to -1, it should be located in the negative region.

Since each decision hyperplane can be trained separately in the hairy model, we will discuss the hyperplane for neuron i only. For neuron i, we have P equations, $k = 1, \ldots, P$:

$$\begin{cases} w_{i1}X_1^k + w_{i2}X_2^k + \cdots + w_{iN}X_N^k - \theta_i = W_i^T X^k - \theta_i > 0, & if \quad X_i^k = 1 \\ w_{i1}X_1^k + w_{i2}X_2^k + \cdots + w_{iN}X_N^k - \theta_i = W_i^T X^k - \theta_i < 0, & if \quad X_i^k = -1 \end{cases} . \tag{5}$$

We discard the case in Eq.(5), where $W_i^T X^k - \theta_i = 0$, because it rarely happens in analog operation. For computational convenience, we multiply every element in Eq.(5) by X_i^k and obtain a compact formula:

$$\begin{cases} s_i^k = & \{w_{i2}X_2^k + \cdots + w_{iN}X_N^k - \theta_i\}X_i^k \\ & = \{W_i^T X^k - \theta_i\}X_i^k > 0, \text{ for both cases } X_i^k = 1 \text{ and } X_i^k = -1, \end{cases} \quad k = 1, \ldots, P. \tag{6}$$

Whenever we say that the ith hyperplane is stable, we mean that all P patterns in the correct (stable) regions of the hyperplane satisfy Eq.(6). This hyperplane is not stable whenever there is a pattern in the wrong region.

The hairy model provides biologically plausible (Hebb's postulate) solutions to increase the memory basin without using any hidden neurons or annealing process. Note that this basin (basin-λ) is defined for perfect recalls in the model. It successfully remedies the drawbacks of the Hopfield model. It explores the flexible space in between the two pattern sets, $U_{i,1} = \{X^k; k = 1, \ldots, P,$ and $X_i^k = 1\}$ and $U_{i,-1} = \{X^k; k = 1, \ldots, P,$ and $X_i^k = -1\}$, in order to locate the decision hyperplane under the stability conditions in Eq.(5). It is *consistent* with Hopfield model and Hebb's postulate. It somehow reveals the mysterious underlying mechanics of the postulate. This means that the postulate increases the distance, s_i^k, indirectly.

The idea of the backbone memory

The idea of the proposed new memory comes from a condition in (Eq.(9) in [13]; Eq.(14) in [15]). This condition says that one selects a nearest pair of patterns $\{c_{i,p}, c_{i,n}\}$ from each of the two sets, $U_{i,1}$ and $U_{i,-1}$, where $c_{i,p}$ is a pattern in $U_{i,1}$ and $c_{i,n}$ is a pattern in $U_{i,-1}$. We suppose that both sets, $U_{i,1}$ and $U_{i,-1}$, are not empty. This pair is the nearest pair (in terms of the Euclidean distance) among all the pairs across the two sets (see Fig. 1(a,e)). These two nearest patterns are most vulnerable to each other's corrupted patterns. The weights of the hyperplane (Eq. (14) in [15]) are set as

$$w_{ij} = c_{i,p,j} - c_{i,n,j}, \text{ normalize } W_i \ , \tag{7}$$

$$\theta_i = \sum_{j=1}^{N} w_{ij} \left(\frac{c_{i,p,j} + c_{i,n,j}}{2} \right) .$$

This hyperplane is right in the middle between the two nearest patterns. It is perpendicular to the line section connecting the two patterns $\{c_{i,p}, c_{i,n}\}$ and passes through the center of this section. $c_{i,p} - c_{i,n}$ is the direction of the normal vector of this hyperplane. From our evidence, this condition may not be useful when there are multiple minima (see Fig. 1(b,c,d)). In other words, many patterns across the two sets will share the same minimum distance. There are many equally vulnerable pair patterns. To keep the balance of such multiple minima, we devised a memory to balance the hyperplane among all the minimal pair

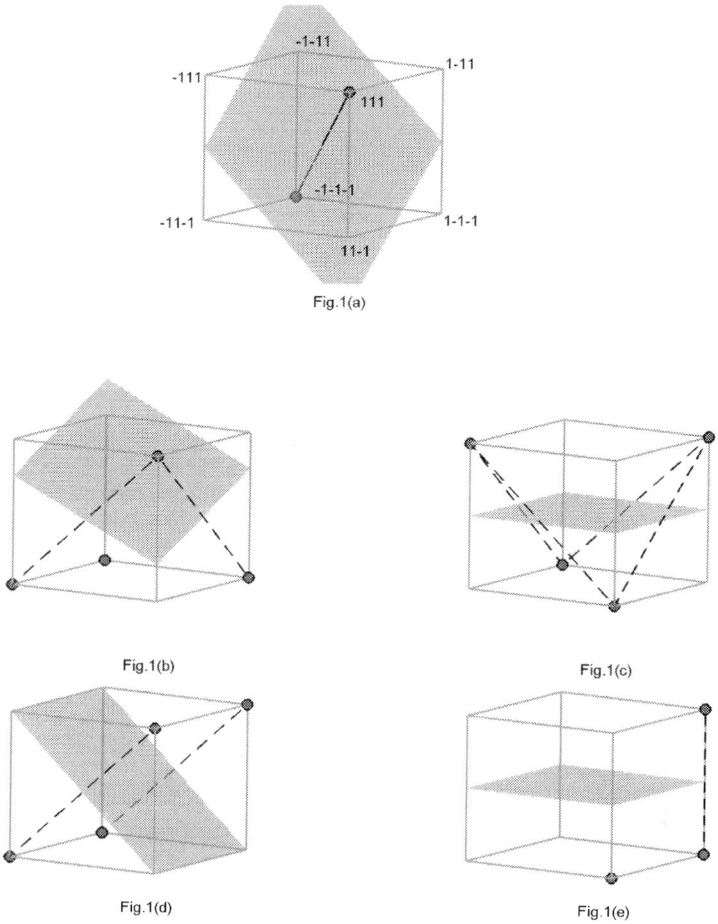

Fig. 1. 3-D cube network with three neurons. Black dots denote memory patterns. The shaded plane is the decision plane corresponding to the third neuron (bit) studied in this paper. This plane separates the memory patterns on the two sides (regions) according to the third bit of each pattern. A stable pattern on the positive (negative) side will have a +1 (-1) value in its third bit. The patterns connected by dashed lines are used to determine the position of the decision plane.

patterns. We expect that this memory can effectively resolve the drawbacks of the Hopfield model.

2 Backbone Memory

For the multiple minima, we save those patterns which have the same minimum pair distance in two sets, $U_{i,p}$ and $U_{i,n}$, where $U_{i,p} = \{X^k$; pattern X^k with the minimum distance, $k = 1, \ldots, P$, and $X_i^k = 1\}$ and $U_{i,n} = \{X^k$; pattern

X^k with the minimum distance, $k = 1, \ldots, P$, and $X_i^k = -1\}$. Suppose there are $N_{i,p}$ patterns in set $U_{i,p}$ and $N_{i,n}$ patterns in $U_{i,n}$. In Fig. 1(b), there are two nearest pairs: $N_{3,p} = 1$, $N_{3,n} = 2$, $U_{3,p} = U_{3,1} = \{(1,1,1)^T\}$, $U_{3,n} = \{(1,-1,-1)^T, (-1,1,-1)^T\}$, $U_{3,-1} = \{(1,-1,-1)^T, (-1,1,-1)^T, (-1,-1,-1)^T\}$; in Fig. 1(c), there are four nearest pairs: $N_{3,p} = 2$, $N_{3,n} = 2$, $U_{3,p} = U_{3,1} = \{(-1,1,1)^T, (1,-1,1)^T\}$, $U_{3,n} = U_{3,-1} = \{(1,1,-1)^T, (-1,-1,-1)^T\}$; in Fig 1(d), there are two pairs: $N_{3,p} = 2$, $N_{3,n} = 2$, $U_{3,p} = U_{3,1} = \{(1,1,1)^T, (-1,1,1)^T\}$, $U_{3,n} = U_{3,-1} = \{(1,-1,-1)^T, (-1,-1,-1)^T\}$.

We prefer a hyperplane which evenly balances these minimum distance patterns. We place the hyperplane right in the middle between the center of the patterns in $U_{i,p}$ and the center of the patterns in $U_{i,n}$. That is,

$$
\begin{aligned}
C_{i,p,j} &= \frac{1}{N_{i,p}} \sum_{X^k \in U_{i,p}} X_j^k , \\
C_{i,n,j} &= \frac{1}{N_{i,n}} \sum_{X^k \in U_{i,n}} X_j^k , \\
w_{ij} &= C_{i,p,j} - C_{i,n,j} \text{ , normalize } W_i, \\
\theta_i &= \sum_{j=1}^{N} w_{ij} \left(\frac{C_{i,p,j} + C_{i,n,j}}{2} \right),
\end{aligned}
\tag{8}
$$

where $C_{i,p}$ and $C_{i,n}$ are the centers of the patterns in $U_{i,p}$ and the patterns in $U_{i,n}$, respectively. Accordingly, we rewrite the above weights as that in Hopfield model

$$
w_{ij} = \frac{1}{N_{i,p}} \sum_{X^k \in U_{i,p}} X_i^k X_j^k + \frac{1}{N_{i,n}} \sum_{X^k \in U_{i,n}} X_i^k X_j^k , \tag{9}
$$

where the summation is over all the minimal distance patterns in each set. These patterns sustain the Hopfield memory and are backbones (or supports) of the memory. Eq.(9) is exactly the same as Eq.(8) with zero threshold $\theta_i = 0$. The planes in Fig. 1(a,c,d,e) are also those obtained by Eq.(9).

Experiments on associative memory

We performed experiments (also see those in [13] and [15]), to compare the performance of the Little model (LM) [16], the error-correction rule (ECR) [19], et-AM [13] and e-AM [15].

Initial condition

The initial $w_{ij}(0)$ and $\theta_i(0)$ are set so as to make all the patterns stable [10] in order to ensure positivity. They are set as follows:

$$
w_{ij} = \begin{cases} 1, & if \quad i = j \\ 0, & if \quad i \neq j \end{cases}, \tag{10}
$$
$$
\theta_i = 0.
$$

These initial weights can store stable patterns up to the network limit, 2^N, without any error tolerance. Below, we briefly review the Hopfield model for reference purposes.

Hopfield model

The Hopfield Model is constructed by using the outer product rule to compute the weights as follows:

$$w_{ij} = \begin{cases} \frac{1}{N}\sum_{k=1}^{P} X_i^k X_j^k \\ \quad = \frac{1}{N}\sum_{X^k \in U_{i,1}} X_i^k X_j^k + \frac{1}{N}\sum_{X^k \in U_{i,-1}} X_i^k X_j^k \,, & if \quad i \neq j; \\ 0, & if \quad i = j. \end{cases} \tag{11}$$

Experimental Results

Table 1 lists experimental results for (N=10, P=5). In this case the storage parameter $\alpha = \frac{P}{N} = 0.5$. In each experiment, we presented 10 sets of randomly produced patterns to the networks and then got the averaged results. The networks in Eq.(8) and Eq.(9) are labeled as 'b_θ-AM' and '$b_{\theta=0}$-AM' in Tabel 1, respectively. We explain each row item below:

SP (# of Stored Patterns ($/P$)):	given P patterns, the number of patterns successfully stored;
SS (# of Stable States ($/2^N$)):	the number of stable states;
TS (# of States to Stable ($/2^N$)):	the number of transient states converging to all stable states;
C (# of Cycles):	the number of limit cycles;
IC (# of States in Cycles ($/2^N$)):	the number of states involved in all limit cycles. ($\geq$ 2C);
TC (# of States to Cycles ($/2^N$)):	the number of transient states falling into limit cycles;
R (Recovery ($/NP$)):	given NP 1-bit-error patterns, the number of patterns converging to the original stored patterns.

Table 1. Comparison among LM, ECR, et-AM, e-AM, $b_{\theta=0}$-AM, b_θ-AM, $g_{\theta=0}$-AM and g_θ-AM

$N = 10$, $P = 5$, $\alpha = (5/10) = 0.5$

	LM	ECR	et-AM	e-AM	$b_{\theta=0}$-AM	b_θ-AM	$g_{\theta=0}$-AM	g_θ-AM
SP (/5)	1.9	5	5	5	2.9	4.4	5.0	5.0
SS (/1024)	5.0	43.9	60.0	52.7	29.9	15.8	371.4	40.8
TS (/1024)	744.8	978.4	964.0	971.3	994.1	1008.2	652.6	983.2
C	44.3	0.2	0	0	0	0	0	0
IC (/1024)	88.6	0.4	0	0	0	0	0	0
TC(/1024)	185.6	1.3	0	0	0	0	0	0
R (/50)	12.1	13.5	38.9	39.6	15.7	36.2	20.8	40.1

From Table 1, LM has rather limited capacity; it could not even store five patterns in a ten-neuron network, and neither could the Hopfield model. ECR produces cycles. We will not discuss LM and ECR with regard to further experiments. $b_{\theta=0}$-AM and b_θ-AM have better performance than the LM does. Their spurious stable states have very small basins and cannot withstand thermal pertubation. Note that 100 hyperplanes were trained in this experiment: $100 = 10$(hyperplanes/set)$\times 10$(sets). Among them, 46 hyperplanes had multiple minimal distance pairs.

$b_{\theta=0}$-AM and b_θ-AM produce large basins for the stored patterns and generate no cycle. Their cycle performances are better than those of LM and ECR. They cannot store five patterns. To fix their storage capacity, we construct a simple training to tune their weights. In the construction, $b_{\theta=0}$-AM is used as a reference to train the all-stable hyperplane Eq.(10) linearly toward its reference hyperplane in Eq.(9). This means that the initial weights and reference weights are set according to Eq.(10) and Eq.(9) separately, that is, $w_{ij}(0) =$ Eq.(10) and w_{ij}^{ref} =Eq.(9). Calculate $\triangle w_{ij} = \varepsilon[w_{ij}^{ref} - w_{ij}(0)]$. Update all weights according to $w_{ij}(t) = w_{ij}(0) + t\triangle w_{ij}$. ε is a small number. We set $\varepsilon = 0.001$ in all experiments. Note that this updation increases s_i^k indirectly and does not follow the steepest ascent direction and its dynamics does not follow the hairy model. The training is stopped just at iteration $t = T$. In this iteration, $SP_{t=T} = P$ and after this iteration, $SP_{t=T+1} < P$. The performances are listed in Table 1 under the label '$g_{\theta=0}$-AM'. It can store all patterns, tolerate noisy patterns, and generate no cycle. This training remedies the storage capacity of $b_{\theta=0}$-AM and still keeps Hebb's postulate.

The experiment in this table under the label 'g_θ-AM' is obtained by training the all-stable hyperplane toward the reference hyperplane Eq.(8). This means that the initial weights and reference weights are set according to Eq.(10) and Eq.(8) separately, that is, $w_{ij}(0) =$ Eq.(10) and w_{ij}^{ref} =Eq.(8). Calculate $\triangle w_{ij} = \varepsilon[w_{ij}^{ref} - w_{ij}(0)]$. Update all weights according to $w_{ij}(t) = w_{ij}(0) + t\triangle w_{ij}$. Both methods '$g_{\theta=0}$-AM' and '$g_\theta$-AM' have similar performance. All methods et-AM, e-AM, b-AM and g-AM produced no limit cycles, as we expected. In almost all of our experiments, the evolution of noisy states converged in a single iteration during recall after learning. This is very different from the evolutionary recall process in many other models.

In Table 2 we did experiments for a large network with twenty neurons $N = 20$ and twenty patterns $P = 20$. In this case $\alpha = 1$.

In Tables 2 and 3, we presented 30 sets of randomly produced patterns to the network and then got the averaged results. We also did experiments on the pseudo-inverse learning rule under the label 'PsdoInv' [17][8]. This rule can store N linearly independent patterns and cannot generate any basins for patterns. Table 3 shows two experimentals with more patterns, $P = 25$ and $P = 45$. In Table 3, $\alpha = \frac{25}{20} = 1.25 > 1$ and $\alpha = \frac{45}{20} = 2.25 > 2$, respectively.

Note that the loading capacity α under various conditions has been studied [3], $\alpha = 0.14$ [7]; $\alpha < (1/[2(lnN)])$ [18]; $\alpha = 0.16$ [2]; $\alpha = 1$ [8]. The bound $\alpha < 2$ has been derived in [4] based on thermal statistic average. All of them are

Table 2. Simulations for $N = 20$, $P = 20$ using PsdoInv, $g_{\theta=0}$-AM, and g_θ-AM

$P = 20$	PsdoInv	$g_{\theta=0}$-AM	g_θ-AM
SP (/20)	20	20	20
SS ($/2^{20}$)	1048576.0	559983.7	324250.8
TS ($/2^{20}$)	0	488592.3	724325.2
C	0	0	0
IC ($/2^{20}$)	0	0	0
TC($/2^{20}$)	0	0	0
R (/400)	0	68.2	93.4

Table 3. Simulations for $N = 20$, $P = 25$ and $P = 45$ using $g_{\theta=0}$-AM, and g_θ-AM

$P = 25$	g_θ-AM	$g_{\theta-0}$-AM	$P = 45$	g_θ-AM	$g_{\theta=0}$-AM
SP (/25)	25	25	SP (/45)	45	45
SS ($/2^{20}$)	542919.1	730262.3	SS ($/2^{20}$)	823705.8	864453.8
TS ($/2^{20}$)	505656.9	318313.7	TS ($/2^{20}$)	224870.2	184122.2
C	0	0	C	0	0
IC ($/2^{20}$)	0	0	IC ($/2^{20}$)	0	0
TC($/2^{20}$)	0	0	TC($/2^{20}$)	0	0
R (/500)	72.5	51.0	R (/900)	38.3	35.3

derived for imperfect recalls. The methods, et-AM, e-AM and g-AM are designed for perfect recalls.

Next, we tested the error tolerance of the method g_θ-AM by using the $P = 25$ case in Table 3. For each pattern, we randomly generated 1000 noisy patterns with 2, 4, 6 and 8 error bits, respectively. We used g_θ-AM to recover them. The numbers shown in Table 4 are the total numbers of patterns in the 75000 noisy patterns, 75000 = 1000(noisy patterns/pattern)×25(patterns/set)×30(sets), that can be restored by the method according to restoration iterations in both synchronous and asynchronous modes. These results show that many noisy patterns with 2 or 4 errors can be recovered by g_θ-AM. Also, we found that for all 75000 patterns, none of them fell into a limit cycle.

Table 4. Error Tolerance using g_θ-AM in different recall modes. Label 'Iteration t' denotes the number of iterations for a noisy state evolving to its memory pattern.

N=20,P=25	g_θ-AM, Synchronous Eq.(2)					g_θ-AM, Asynchronous Eq.(1)				
Iteration t	$t = 1$	$t = 2$	$t = 3$	$t = 4$	$t = 5$	$t = 1$	$t = 2$	$t = 3$	$t = 4$	$t = 5$
2 error bits	13897	7944	0	0	0	29529	8357	0	0	0
4 error bits	170	459	282	60	0	376	841	169	3	0
6 error bits	3	16	20	11	5	0	20	24	4	1
8 error bits	0	0	0	1	1	0	0	0	0	0

3 Discussion

The idea of the backbone memory leads us to design the g-AM and b-AM. They can serve as a foundation for exploring physiological implications [9][12], such as cell-cell-adhesion [14] and modular biology [6]. The patterns in $U_{i,p}$ and $U_{i,n}$ resemble the selective adhesion proteins with adhesion features.

The computational cost is linearly proportional to the network size, N, and the number of patterns, P. In g_θ-AM and $g_{\theta=0}$-AM, each hyperplane is independent of all others during tuning. This is beneficial for learning in asynchronous mode. This is very different from correlation matrix designs, such as the pseudo-inverse method. All methods et-AM, e-AM and g-AM can accommodate stable patterns up to the network limit, 2^N.

Finally, the hidden neurons in the Boltzmann machine [1][11] can be designed according to et-AM, e-AM, or g_θ-AM. One can add additional neurons (bits) as hidden neurons to the memory patterns, which are states of visible neurons. These added neurons serve as hidden neurons and can further stabilize the visible neurons. There is no need to use the noise clamping technique or annealing process. With such designed hidden neurons, the visible memory patterns can tolerate catastrophic corruption, when all N bits of a memory pattern could be corrupted. The extra neurons provide hidden coupling, hologram-like contents, for restoring the memory. We think that the apparently non-working sections of DNA may have a function that is somehow similar to that of hidden neurons, in that they maintain the integrity of the working DNA. We are still working in this issue.

Acknowledgement

This work was supported by the National Science Council under projects NSC 93-2213-E-002-034&-105.

References

1. Ackley, D.H., Hinton, G.E. and Sejnowski, T.J.: A learning algorithm for Boltzmann machine. Cognitive Science **9** (1985) 147-169
2. Amari, S.I. and Maginu, K.: Statistical Neurodynamics of Associative Memory. Neural Networks, **1(1)** (1988) 63-73
3. Amit, D.J.: Modeling brain function: The world of attractor neural networks. Cambridge University Press (1989)
4. Gardner, E. and Derrida, B.: Optimal storages properties of neural network models. Journal of Physics A **21** (1988) 271-284
5. Gardner, E.: Optimal basins of attraction in randomly sparse neural network models. Journal of Physics A **22(12)** (1989) 1969-1974
6. Hartwell, L.H., Hopfield, J.J., Leibler, S. and Murray, A.W.: From molecular to modular cell biology. Nature, Suppl. **402** (1999) C47-C52
7. Hopfield, J.J.: Neural networks and physical systems with emergent collective computational ability. Proceedings of the National Academy of Sciences of the United States of America **79** (1982) 2554-2558

8. Kanter, I. and Sompolinsky, H.: Associative recall of memory without errors. Physical Review A **35(1)** (1987) 380-392
9. Kauffman, S.A.: Antichaos and adaptation. Scientific American, August (1991) 64-70
10. Li, J., Michel, A.N. and Porod, W.: Analysis and synthesis of a class of neural networks: linear systems operating on a closed hypercube. IEEE Transactions on Circuits and Systems **36(11)** (1989) 1405-1422
11. Liou, C.-Y. and Lin, S.-L.: The other variant Boltzmann machine. Proceedings of International Joint Conference on Neural Networks, Washington DC (1989) 449-454
12. Liou, C.-Y. and Wu, J.-M.: Self-organization using Potts models. Neural Networks **9(4)** (1996) 671-684
13. Liou, C.-Y. and Yuan, S.-K.: Error tolerant associative memory. Biological Cybernetics **81** (1999) 331-342
14. Liou, C.-Y. and Yang, H.-C.: Selective feature-to-feature adhesion for recognition of cursive handprinted characters. IEEE Transactions on Pattern Analysis and Machine Intelligence **21(2)** (1999) 184-191
15. Liou, C.-Y. and Lin, S.-L.: Finite memory loading in hairy neurons. Natural Computing **5(1)** (2006) 15-42
16. Little, W.A.: The existence of persistent states in the brain. Mathematical Biosciences **19** (1974) 101-120
17. Personnaz L., Guyon, I. and Dreyfus, G.: Information storage and retrieval in spin-glass like neural networks. Journal Physique Lett. **46** (1985) 359-365
18. Weisbuch, G. and Fogelman-Soulie, F.: Scaling laws for the attractors of Hopfield networks. Journal De Physique Lett. **46** (1985) 623-630
19. Widrow, B. and Hoff, M.E. Jr.: Adaptive switching circuits. IRE WESCON Convention Record (1960) 96-104

Dynamics of Citation Networks

Gábor Csárdi[1,2]

[1] Department of Biophysics, KFKI Research Institute for Particle and Nuclear Physics of the Hungarian Academy of Sciences, Budapest, Hungary
[2] Center for Complex Systems Studies, Kalamazoo College, Kalamazoo, MI 49006, USA
csardi@rmki.kfki.hu

Abstract. The aim of this paper is to give theoretical and experimental tools for measuring the *driving force* in evolving complex networks. First a discrete-time stochastic model framework is introduced to state the question of how the dynamics of these networks depend on the properties of the parts of the system. Then a method is presented to determine this dependence in the possession of the required data about the system. This measurement method is applied to the citation network of high energy physics papers to extract the in-degree and age dependence of the dynamics. It is shown that the method yields close to "optimal" results.

1 Introduction

The network concept is an abstract representation. A simple network (or graph, the two are the same for our purposes) is simply a homogeneous relation over a set. The relation can be symmetric (undirected networks) or asymmetric (directed networks). While this is an adequate definition of a network, usually we imagine a network as an interconnected set of vertices (also called nodes), while the connections are called edges or arcs. In a neural network the vertices represent neurons and the edges the synapses between them; this network is clearly asymmetric, the synapse 'leads' from the presynaptic cell to the postsynaptic one. In a citation network, the vertices represent ('are') scientific papers published in journals and the edges are citations from one paper to another, forming again a directed network. In a collaboration network two vertices representing researchers are connected by an edge if they have published at least (say) one joint paper in a journal, this network is an undirected one.

There has been an upsurge in the field of complex networks recently; networked representations of various complex systems have shed light to a number of structural and dynamical phenomena. The main advantage of the network schema is that its simplicity makes it universal: every large enough system consists of many – structurally, dynamically and/or functionally interconnected parts. For recent reviews written by researchers in different fields see [1,2,3,4].

In this work we address evolving networks, and study the dynamical process of adding and removing vertices and edges to/from the the network. Particularly we

S. Kollias et al. (Eds.): ICANN 2006, Part I, LNCS 4131, pp. 698–709, 2006.

are interested in the question of how the structural and non-structural properties on the vertices determine the place of the next edge addition.

There have been a number of network evolution models in the literature recently, the most successful being the preferential attachment model proposed by Barabási and Albert [5]. They suggest a simple mechanism in which the rate for attaching new edges to a node is proportional to its number of adjacent edges at each time step. This model is thought to be valid for very different kinds of networks (showing the ease of the universal network representation) based on indirect evidence: the scale-free degree distribution. It is observed that in many networks the distribution of the vertex degree (which is simply the number of adjacent edges for a vertex) is a power-law distribution; and the BA-model is known to generate power-law degree distributions [6], so it is likely (or at least possible) that this simple mechanism is at work in many networks. It is shown however that the scale-free degree distribution can be obtained without preferential attachment, by assuming vertex intrinsic fitness, see [7]. It is also true that there may be several *underlying* causes producing preferential attachment [8,9]. Only a few studies addressed the direct observation, ie. somehow measuring the actual attachment probabilities in the evolving network as a function of the vertex degree or vertex fitness, see [10,11,12] for examples.

This neglect is partially caused by the lack of data. For calculating the actual degree distribution of a network we only need to know the *current* structure of it, ie. the binary relation defining which vertices a given vertex connects to. For studying the process of vertex and edge addition and deletion however we need to know the structure of the network at any time in the past. (At least this would be the ideal case.) It comes not as a surprise that we usually don't have this data, except in a few cases. This indicates that the rare dynamical data is very important and can be used to validate various network evolution models. Our work discussed here serves as an example for such a study.

This paper is organized as follows. In Sect. 2 we introduce a model framework and a measuring method for extracting the dependence of the network dynamics on the hypothetical dynamical parameters. In Sect. 3 we show two applications for the model and method: measuring the dynamics of scientific citation networks, and predicting the number of future citations for scientific citation networks. Finally in Sect. 4 we discuss our results and other possible applications.

2 Methods

The networked representation of a dynamic complex system is an evolving graph: vertices join to the system, they form new connections, some old connections break and perhaps some vertices are removed from the network. In each time step the network has a configuration in which the vertices and edges exhibit various structural properties. Further on, the vertices and edges may also exhibit some intrinsic properties we don't intend to ignore: in a neural network some neurons are pyramidal cells, others are interneurons and this distinction is important for most purposes.

A very natural question is the following: what structural and/or intrinsic properties determine the evolution of a given network? Another question coming hand in hand with this: how is it possible to describe the form of the dependence? (If it is possible at all.) In the rest of this section we will give a model framework and method for answering these questions in some special cases.

Let us focus on the simplest kind of evolving networks first: citation networks. We do this for two reasons. First, citation networks are simple in the sense that all outgoing edges of a vertex are added to the network right after adding the vertex itself, in the same time step. Second, there is data available for citation networks of scientific papers.

A number of important structural properties may play significant roles in the evolution of a particular citation network: the in-degree of the vertices, their transitivity (ie. if every vertex citing vertex v also cites vertex w so far, then it is likely that this will happen in the future as well). Some intrinsic properties of the vertices are also thought to be important: the topic of a paper, since it is likely that two topically close papers will cite each other; or the age of the papers since it is a reasonable assumption that out-of-date (or common knowledge) papers are not or only rarely cited.

2.1 Preferential Attachment

Let us now define the framework in which our questions can be stated formally.

The first structural property we will address is the in-degree of the vertices. Let us assume that the probability that at time step t an outgoing edge (e) of a newly added v vertex will cite a given w vertex depends on the in-degree of w, and the in-degree of other vertices in the network:

$$P[e \text{ cites } w](t) = \frac{A(d_w(t))}{\sum_{i \in V(t)} A(d_i(t))} . \tag{1}$$

Here $d_w(t)$ is the in-degree of vertex w in time step t and $V(t)$ is the set of all vertices in time step t. The $A(\cdot)$ attachment kernel function defines the dependence of the network dynamics on the in-degree of the vertices. In this simple framework this function stochastically governs the network evolution. The preferential attachment model suggests that for many networks this function is simply $A(k) = k + 1$. There are also other models which fit into this framework, see [13,14,15].

Similarly, the probability that in time step t an e edge of a newly added v vertex cites *any* other vertex with in-degree k is given by:

$$P[e \text{ cites a } k \text{ in-degree vertex}](t) = P_e(k) = \frac{N_k(t)A(k)}{\sum_{i \in V(t)} A(d_i(t))} . \tag{2}$$

$N_k(t)$ is the number of k-degree nodes in the network at time step t.

From this formula we can extract $A(k)$:

$$A(k) = \frac{P_e(k)S(t)}{N_k(t)} \tag{3}$$

by using the notation $S(t) = \sum_{i \in V(t)} A(d_i(t))$. From the data we can estimate $P_e(k)$, so if we manage to determine $S(t)$ then $A(k)$ can be estimated as well. For $S(t)$ we can use the following simple iterative approach: first we assume that $S(t)$ is constant and estimate $A(k)$ for each k. Then by using this estimation we calculate the next approximation of $S(t)$ which in turn allows us to better estimate $A(k)$. While the convergence of this iteration is hard to prove, in practice it converges fast.

In Sect.3 we show applications for the in-degree dependence of the network dynamics in scientific citation networks.

2.2 Preferential Attachment and Aging

Let us now assume that an additional intrinsic vertex property, the *age* of the vertex, also contributes to the network evolution. For simplicity from now on we measure "time" by the addition of the new vertices, ie. in each time step a single vertex is added to the network; we denote vertices by the time step of their addition, ie. vertex 1 is added in the first time step, vertex 2 in the second, etc. This implies that in time step t the age of vertex i is simply $t - i$.

Similarly to the previous section the probability that edge e of vertex v added in time step t cites vertex w is given by

$$P[e \text{ cites } w] = \frac{A(d_w(t), l_w(t))}{\sum_{i \in V(t)} A(d_i(t), l_i(t))} \ . \tag{4}$$

The probability that edge e of vertex v added in time step t cites some vertex with in-degree k and age l is

$$P[e \text{ cites a } k \text{ in-degree, } l \text{ age vertex}] = \frac{A(k,l)N_{k,l}(t)}{S(t)} \ . \tag{5}$$

Using the data for estimating $P_e(k,l)$ and the iteration technique introduced in the previous section we can extract $A(k,l)$, the function governing the evolution of the network.

2.3 Validating the Method

For validating this measurement method and software, we've applied it to various toy networks generated by different attachment rules, ie. different built-in $A(k)$ and $A(k,l)$ functions.

To validate the in-degree based method we've generated networks by the Barabási-Albert model and compared the measured $A(k)$ function to the expected linear dependence. These test networks had 300,000 nodes each having

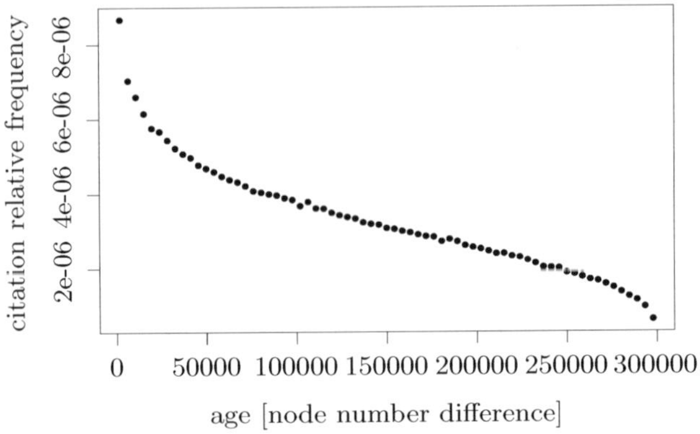

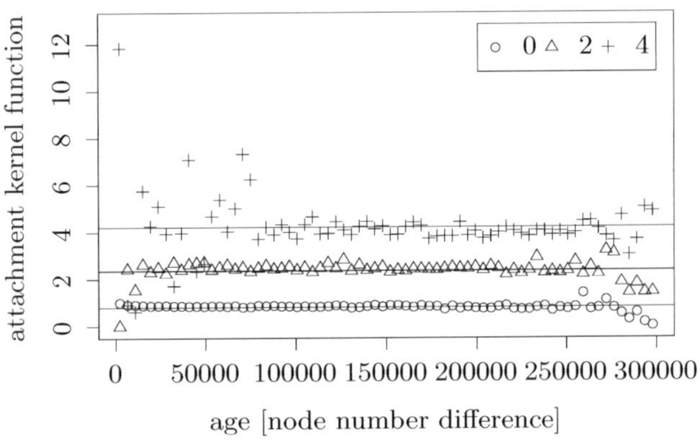

Fig. 1. The naive (upper) and non-naive (lower) methods for measuring the age dependence of the $A(k,l)$ function. The network was generated according to the Barabási-model, has 300,000 nodes and the out-degree of each node is 2. The age axes are binned into 70 units. The lower plot shows the measured $A(k,l)$ functions for various k values. The horizontal lines were added by least square fitting the data points, they can be considered as the "correct" values of the $A(k,l)$ function.

out-degree two. The measurement yielded the expected 1.0 exponent with minimal error (± 0.05).

Next we've checked the in-degree and age based method and software by similar toy networks. The measurement method very well reproduced the expected attachment rules. These experiments however have also shown that the method cannot predict the "rare" events in the evolution. As there are almost never any young nodes with high in-degree in the network, the $A(k,l)$ function for large k and small l values cannot be estimated well.

Although one might argue that for the age dependence of the $A(k,l)$ function a simpler approach could be used, we show here that this is not the case. A naive approach would simply consider the distribution of the age differences (citation lags) between the citing and the cited node as the age dependent component of $A(k,l)$, however this is clearly biased: small citation lags are overrepresented in the network because of two reasons. The first is that young nodes are more likely to be cited when the network is still small because there is less competition in the network. If older nodes are also present then the competition in higher as the network is also bigger. Second, young nodes have simply more chance to get cited, as they are present in small and big networks as well.

Figure 1 shows the two types of measurement of the age dependence of $A(k,l)$ for a simple Barabási network. While it is clear that there is no age dependence in this model, the histogram of the citation lags does not show a horizontal line. Our proposed measurement method correctly finds that $A(k,l)$ is independent of the age of the nodes.

3 Applications

3.1 The Pace of Science

In this section we apply the method described in the previous one to a scientific citation network, consisting of 28632 high energy physics papers with 367790 directed edges among them. This data is available online from the homepage of the 2003 KDD Cup (`http://www.cs.cornell.edu/projects/kddcup/datasets.html`).

First we've cleared up the dataset by removing forward citations. A forward citation means that a paper cites a more recent one. This is possible either because of errors in the database or because some papers were updated (with new citations) after their first submission without changing their original submission date.

Then the dynamics of the network (ie. the $A(\cdot)$ function) was measured in terms of the in-degree and the age of the nodes. The age of the papers was simply defined by assigning numbers to them in the order their first submission date and binning these numbers into 70 units.

After the extraction of the $A(k,l)$ function the measured data has shown that the effects of k and l can be separated, and $A(k,l)$ can be written in the form

$$A(k,l) = A_k(k) \cdot A_l(l) \ . \qquad (6)$$

This separation supports the assumptions made by various network models with aging, see works by [16] and [17]. The measured $A_k(k)$ and $A_l(l)$ functions can be seen in Figs. 2 and 3. They can be well fitted by $A_l(l) = l^{-\beta}$ and $A_k(k) = k^{\alpha}+1$, with $\alpha = 1.11$ and $\beta = 1.13$. This α value is close to the celebrated linear preferential attachment phenomenon, thought to be universal, although rarely measured directly.

The fact that the β exponent is close to one shows that ceteris paribus the "importance" of a paper is inversely proportional to its age. This defines the "pace" of science.

3.2 Citation Prediction

The ACM Special Interest Group on Knowledge Discovery and Data Mining organizes a conference each year and together with the conference they also

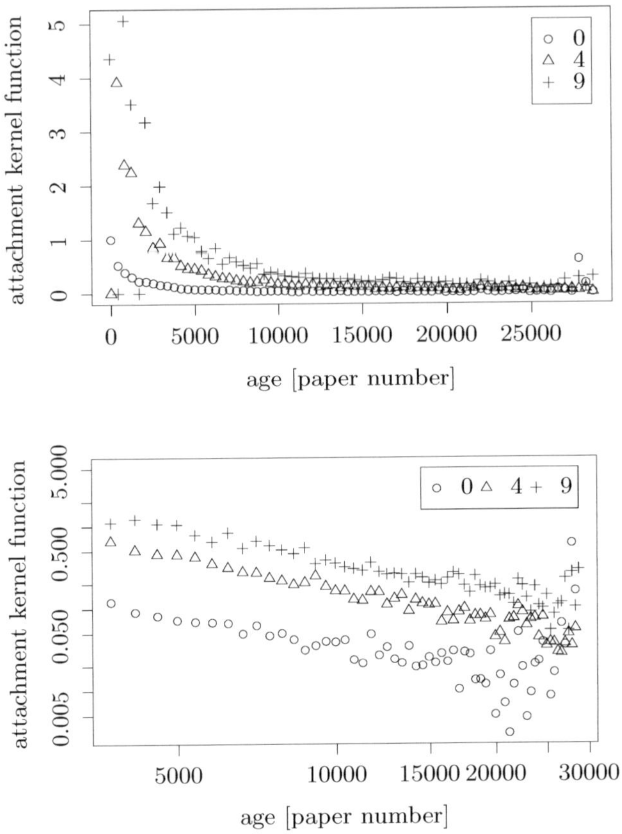

Fig. 2. The age dependence of the attachment kernel function of the high energy physics network for various in-degrees. The upper plot has linear, the lower one logarithmic axes. The lower plot clearly shows that the aging is well described by a power-law decrease independently of the degree.

host a data mining competition called KDD Cup. In 2003 the first task of the KDD Cup was to predict the citations to the papers in the high energy physics database. This database contains high energy papers submitted to the arXiv e-print archive between 1992 and July 31, 2003. The deadline for the KDD Cup submission was before April 30, 2003 and the citations made by papers in the next three months had to be predicted.

The evaluation of the prediction algorithms was done by considering only papers receiving at least six citations during the period February 1, 2003 – April 30, 2003. For these papers first the target vector, the difference between the citations received between May and August and between February and May were calculated. The specific task was the prediction of this vector. The error of the prediction was simply defined by the sum of the absolute value of the difference of the prediction and the target vector.

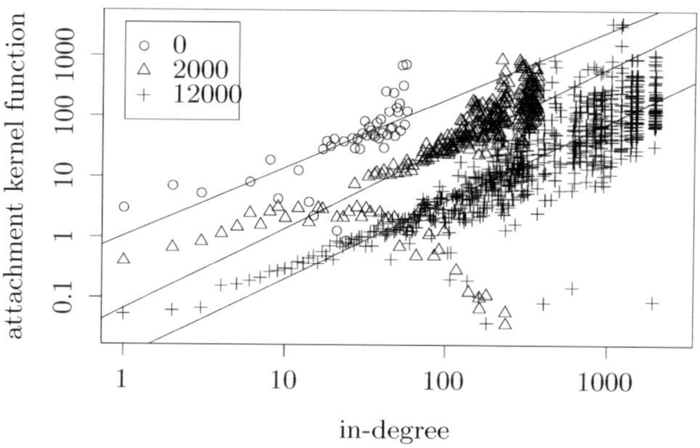

Fig. 3. The degree dependence of the attachment kernel function for various node ages. The lines are simple least square fits for the data points. The axes are logarithmic. The plot shows that the in-degree dependent part of the $A(k, l)$ function can be reasonable well estimated by an increasing power-law function, independently of the age of the nodes.

While the method in the previous section is not developed for citation prediction, is can be used for that in the following way. We can measure the dynamics (ie. the $A(\cdot)$ function) of the network up to *now* and assuming that this function will be the same in the future we can simulate the growth of the network according to the measured dynamics and see a possible realization of how the network will look like (say) three months later. By generating many realizations and taking the average number of citations a node received in these realizations we can predict the "average" expected evolution of the network.

Another important reason to do the prediction task with our proposed method is that we can compare the error of the measured $A(\cdot)$ function to other $A(\cdot)$ functions to evaluate it. If a given $A_1(\cdot)$ function proves to be a better predictor than another $A_2(\cdot)$ attachment kernel that would mean that the former one is based on more relevant properties than the latter.

At the 2003 KDD Cup, the error of the winner algorithm was 1329. The totally random network evolution, when each new node connects to a number of randomly selected nodes yields on the average an error of 3463. This value was obtained by averaging hundred totally random realizations. These error values can be used as baselines to place the error of the predictions of our method.

First we measured the $A(\cdot)$ function based on the in-degree of the nodes solely and found that the

$$A(k) = k^{\alpha} + 1 \tag{7}$$

form gives a reasonable good fit with the measured data. We fitted this form by a simple weighted least square method and got $\alpha \approx 0.85$. The prediction with this $A(k)$ function yielded an error about $2473.51(\pm 4.39)$. These values were obtained by generating 100 realizations five times, the error is simply the standard deviation of the five predictions.

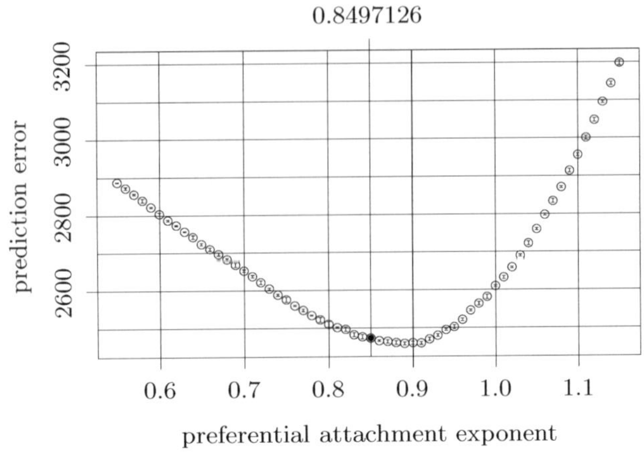

Fig. 4. Prediction error for different α values in (7). The plot was obtained by running five times 100 realizations for each α value, the error bars show the standard deviation of the five predictions. The measured 0.85 exponent is close to the optimal 0.89 value.

To evaluate our dynamics measurement method we've calculated predictions with other α exponents as well, and found that the $\alpha = 0.85$ value is very close to the "optimal" exponent, optimal in terms of the error of this prediction.

Instead if using solely the in-degree as the predictor, now we will also add the age of the nodes, and by applying the measurement method described in the previous section we measure the $A(k,l)$ function (as before k being the in-degree and l being the age of a node) governing the dynamics of the network. The measured $A(k,l)$ function can be reasonably well fitted by the following form:

$$A(k,l) = (k^\alpha + 1)\, l^{-\beta} \ . \tag{8}$$

This form assumes that the effect of in-degree and age can be separated, our data supports this assumption. By fitting this form using weighted least square fits we arrive to the exponents: $\alpha \approx 1.14$ and $\beta \approx 1.14$. By using these values in generating possible realizations of the HEP network for the prediction we get a prediction error 1732.76 ± 6.19. The fact that this prediction is much better than the "in-degree-only" one, indicates that the age of the nodes makes an important contribution to the edge-dynamics of the evolving network.

Note that the exponent of the preferential attachment is lower if we don't use the age of the papers as a property, $\alpha_k \approx 0.85$ versus $\alpha_{k,l} \approx 1.14$. This is clearly because in the former the effect of the aging is "built in" into the preferential exponent and since aging works *against* preferential attachment it makes the exponent smaller. Some works suggest that the preferential attachment mechanism can be present even in network not showing the scale-free degree distribution because there is another, opposite effect working in the system, such as limits for the number of edges a node can acquire or because the nodes lose their "attractiveness" by getting older, ie. aging, see [18]. To our knowledge the work

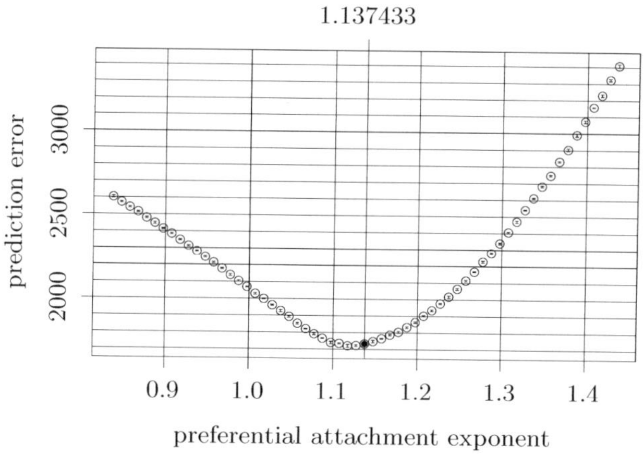

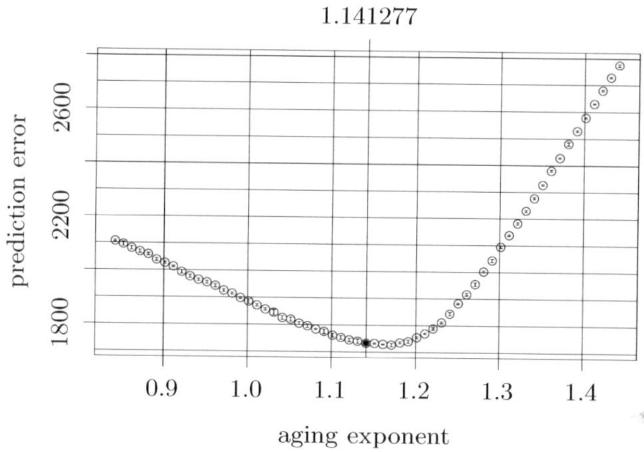

Fig. 5. Prediction error for different preferential attachment exponents (α, upper plot) and aging exponents (β, lower plot). For both exponents the dynamics measurement method gives solutions close to the optimal ones.

presented in this paper is the first one giving experimental evidence for this assumption.

4 Discussion

We have presented a model framework and a measurement method for defining and determining the dynamics of citation networks based on the properties of their nodes.

We've applied this method to a network of high energy physics papers and extracted the $A(k,l)$ function which stochastically governs the evolution of the network in terms of the in-degree and age of the nodes. Without assuming any favored form for this function we found that it can be estimated as the product of the in-degree dependent $A_k(k)$ and the age-dependent $A_l(l)$ function. The in-degree dependent part shows slightly superlinear preferential attachment while the age-dependent part shows power-law decrease.

We've evaluated the results given by the measurement method by predicting the citations received by important papers in the last three months of the high energy physics papers database and found that the measured preferential attachment and aging exponents are close to the "optimal".

We believe that the framework and method presented in this paper is a useful tool for researchers of any field interested in the evolution of complex systems. Also, it can be generalized for general evolving networks with node and edge additions and deletions, our experiments show promising results in this direction.

The citation prediction study presented here can be a general way for evaluating the description of a system based on various properties, just like we've shown that adding the age of the nodes to the considered properties resulted a much better citation prediction.

Acknowledgement

The authors would like to thank Jan Tobochnik, Katherine J. Strandburg, Péter Érdi, László Zalányi and Tamás Kiss for their cooperation. This work was funded in part by the EU FP6 Programme under grant numbers IST-4-027173-STP and IST-4-027819-IP and by the Henry R. Luce Foundation.

References

1. Newman, M.E.J.: The structure and function of complex networks. SIAM Review **45** (2003) 167–256
2. Watts, D.J.: The "new" science of networks. Annual Review of Sociology **30** (2004) 243–270
3. Barabási, A.L., Oltvai, Z.N.: Network biology: Understanding the cells's functional organization. Nature Reviews Genetics **5** (2004) 101–113
4. Boccaletti, S., Latora, V., Moreno, Y., Chavez, M., Hwang, D.U.: Complex networks: Structure and dynamics. Physics Reports **424** (2006) 175–308
5. Barabási, A.L., Albert, R.: Emergence of scaling in random networks. Science **286**(5439) (1999) 509–512
6. Mitzenmacher, M.: A brief history of generative models for power law and lognormal distributions. Internet Mathematics **1** (2004) 226–251
7. Caldarelli, G., Capocci, A., Rios, P., Muñoz, M.: Scale-free networks from varying vertex intrinsic fitness. Physical Review Letters **89** (2002) 258702
8. Kleinberg, J.M., R., K.S., Raghavan, P., Rajagopalan, S., Tomkins, A.: The web as a graph: Measurements, models and methods. In: Proceedings of the International Conference on Combinatorics and Computing, no. 1627 in Lecture Notes in Computer Science, Springer (1999)
9. Berger, N., Borgs, C., Chayes, J.T., D'Souza, R.M., Kleinberg, R.D.: Competition-induced preferential attachment. In: Proceedings of the 31st International Colloquium on Automata, Languages and Programming. (2004) 208–221

10. Jeong, H., Néda, Z., Barabási, A.L.: Measuring preferential attachment for evolving networks. Europhys. Lett. **61** (2003) 567–572
11. Redner, S.: Citation statistics from 110 years of physical review. Physics Today **58** (2005) 49
12. Roth, C.: Measuring generalized preferential attachment in dynamic social networks. arxiv:nlin.AO/0507021 (2005)
13. Krapivsky, P.L., Redner, S.: Organization of growing random networks. Phyisical Review E **63** (2001) 066123
14. Ergun, G., Rodgers, G.J.: Growing random networks with fitness. Physica A **303** (2002) 261–272
15. G., B., Barabási, A.L.: Competition and multiscaling in evolving networks. Europhysics Letters **54** (2001) 436–442
16. Dorogovtsev, S.N., Mendes, J.F.F.: Evolution of networks with aging of sites. Phys. Rev. E **62**(2) (2000) 1842–1845
17. Zhu, H., Wang, X., Zhu, J.Y.: Effect of aging on network structure. Phys. Rev. E **68** (2003) 056121
18. Amaral, L.A.N., Scala, A., Barhélémy, M., Stanley, H.E.: Classes of small-world networks. Proc. Natl. Acad. Sci. USA **97**(21) (2000) 11149–11152

Processing of Information in Synchroneously Firing Chains in Networks of Neurons

Jens Christian Claussen

Institut für Theoretische Physik und Astrophysik
Christian-Albrechts-Universität zu Kiel, Leibnizstr.15,
D-24098 Kiel, Germany
claussen@theo-physik.uni-kiel.de
http://www.theo-physik.uni-kiel.de/~claussen/

Abstract. The Abeles model of cortical activity assumes that in absence of stimulation neural activity in zero order can be described by a Poisson process. Here the model is extended to describe information processing by synfire chains within a network of activity uncorrelated to the synfire chain. A quantitative derivation of the transfer function from this concept is given.

Two seminal concepts were introduced by Abeles [1,2,3]: A quantitative model for uncorrelated activity in the cortex in absence of external stimulation, and the concept of the *synfire chain*, a spatiotemporal pattern of synchroneous activity of neurons being active in the same cortical task.

Synchroneous spiking, as a refinement of averaged firing rates, has been used as an equivalent mathematical basis for neural models [4,5]. The experimental and theoretical aspects of synfire chains remain a field of active research [6,7] and also provide a conceptual basis for neural computing architectures [8]. This paper analyzes the extension to formulate processing and propagation of information in such a network.

1 The Abeles Model of Cortical Activity

The model of uncorrelated cortical activity given by Abeles [1], here referred to as Abeles Model, is a direct approach to understand why randomly firing by self-excitation can be a stationary and robust firing mode in a neural network. The underlying experiments are interpreted in the following way: Even if the cortex is not excited by sensory input, the neurons are firing randomly (Poisson process) and excite each other. Obviously, this is to be interpreted as a "ground state" of the cortical network. An interesting question is whether random firing is a stable mode of a network or not. Because 99% of the inputs to the cortex are coming from the same or other cortical areas [9], we shall at first neglect the 1% (sensory) input and therefore consider a network with 100% feedback.

S. Kollias et al. (Eds.): ICANN 2006, Part I, LNCS 4131, pp. 710–717, 2006.

2 Definition of the Abeles Model

The defining assumptions to the Abeles Model are [1]:

(i) Each postsynaptic potential has the shape of a falling exponential (for $t \geq 0$):

$$(+A)\mathrm{e}^{-t/\tau} \qquad \text{for excitatory inputs,}$$

or,

$$(-A)\mathrm{e}^{-t/\tau} \qquad \text{for inhibitory inputs.}$$

This assumption does not only include an idealization of the waveform, it also includes that the values of synaptic strength A and the time constants τ are the same for all neurons.

(ii) All postsynaptic potentials sum up in a linear fashion giving the intracellular potential; the neuron generates a spike if the intracellular potential reaches a threshold T.

(iii) All neurons are firing independently. This, however, is eqivalent to: No information is processed.

(iv) Each neuron has N synaptic inputs which can be excitatory or inhibitory in any proportion.

(v) The neurons fire at an average rate of λ spikes per second.

3 The Self-consistence Equation for the Average Firing Rate

For high rates of inputs, the input spikes add up to nearly random fluctuations of the intracellular potential; the probability density of the intracellular potential therefore is Gaussian. This means: The firing rate of a cell is proportional to the probability for the intracellular potential to be above threshold:

$$\lambda = \frac{1}{\sigma} \cdot \frac{K}{\sqrt{2\pi}} \cdot \int_T^\infty \mathrm{e}^{-\frac{x^2}{\sigma^2}} \mathrm{d}x = \frac{K}{\sqrt{2\pi}} \cdot \int_{\frac{T}{\sigma}}^\infty \mathrm{e}^{-y^2} \mathrm{d}y \tag{1}$$

where K is an unknown constant and σ^2 is the variance of the intracellular potential, which can be calculated as follows:

Each postsynaptic potential contributes a variance of

$$\int_0^\infty ((\pm A)\mathrm{e}^{-t/\tau})^2 \mathrm{d}t = A^2 \frac{\tau}{2}.$$

Nota bene, excitatory and inhibitory connections here contribute equally.

The independent linear superposition of $N \cdot \lambda$ spikes (per second) gives the total variance

$$\sigma^2 = (N\lambda) \cdot (A^2 \frac{\tau}{2}),$$

or

$$\frac{\sigma}{A} = \sqrt{N\lambda \frac{\tau}{2}}. \tag{2}$$

This means: For random firing at a *constant* average firing rate we have to satisfy a self-consistence-equation

$$\lambda = \frac{K}{\sqrt{2\pi}} \cdot \int_{\frac{T}{A\sqrt{N\lambda\frac{\tau}{2}}}}^{\infty} \mathrm{e}^{-y^2} \mathrm{d}y \tag{3}$$

which still has K and T/A as free parameters to be fitted to the experimental data. Abeles' estimation for T/σ is as follows: If the neuron fires at a rate λ and each spike is generated if the membrane potential is approximately 1ms ($\approx 0.4\tau$) above threshold, the probability of the intracellular potential for being above threshold is approximately $\lambda \cdot 1\text{ms} = 0.005$, which is numerically equivalent to T/σ being 2.58. Therefore only one parameter (K) is free, it can be evaluated by solving equation (3) for K.

The main results of the Abeles Model are quantitative estimations of network parameters from realistic neurophysiological properties. Using $N = 20000, \lambda = 5s^{-1}, \tau = 2.5\text{ms}, K = 1000\text{s}^{-1}$, one obtains [2]:

(i) $\sigma/A = \sqrt{125} \cong 11$: The variance of the intracellular potential is 11 times bigger than the amplitude of a single spike.
(ii) $T/A = T/\sigma \cdot \sigma/A \cong 2.58 \cdot 11 \cong 29$: Only 29 *synchroneous* excitatory spikes will lift the membrane potential to threshold. (This is a small value compared with $N\lambda = 100000$ spikes that every cell receives per second.)
(iii) A *single* spike has no detectable effect on the output rate: The firing rate increases from 5 per second to 6.4 per second, but relaxes back to 5 per second with the time constant τ. This causes only 0.003 extra output spikes.

To conclude, synaptic strength seems weak for detecting a single spike, but fairly strong for detecting coincidence inputs. This is a consequence of the highly nonlinear error function, which determines by (1) the firing rate λ. As analyzed in the appendix, below a critical firing rate λ_c random firing is unstable, so that a certain level of activity is required to transmit information.

4 How Can We Describe Processing of Information? – Extension of the Abeles Model

As one of the fundamental assumptions of the Abeles model is the randomly firing of all neurons, which means that all spikes are completely uncorrelated, it is *a priori* unable to describe information transfer.

If the number of spikes carrying the information is much less than the number of random spikes ($N \cdot \lambda \approx 100000$), the probability density of the intracellular potential can be assumed to be approximately Gaussian, so that the mechanism is still the same: The fluctuations converging to each neuron raise the intracellular potential to threshold. Remarkably this condition does not explicitely restrict the correlated activity of a *single* neuron, so it can be involved constantly in information processing.

How can we understand simple processing of information in a real network, whose 'ground state' is randomly firing at a rather low rate? The concept given by Abeles is the *'synfire chain'*: Groups of synchroneously firing neurons are carrying the information; their number must be sufficiently high (at least 10–20) to excite the following neurons.

A possible quantitative description of processing of information within this concept is given by the model described in the remainder of this paper. The basic properties of the extended model [10] are defined as follows:

(i) All input spikes –same as in the Abeles model– are assumed to share the common waveform of a falling exponential,
$x_i(t) = A_i \mathrm{e}^{-(t-t_0)/\tau}$(for $t \geq 0$), which may idealize the signal through the axon.

(ii) The synaptic strength, which was a constant A in the Abeles model, may be inhibitory ($A_i < 0$) or excitatory ($A_i > 0$), and is assumed to have different values for each neuron. In general, we may assume the synaptic weights also to be time-dependent, so that synaptic plasticity can be described. However, this time-dependence takes place on a much larger time-scale than the spike dynamic.

(iii) All postsynaptic (episynaptic) potentials are assumed to sum up to the intracellular potential:

$$I(t) = \sum_i A_i x_i(t). \tag{4}$$

(iv) In addition to the Abeles Model we consider synchroneous and random inputs seperately:

$$I(t) = \sum_{i(\mathrm{sync})} A_i x_i(t) + \sum_{i(\mathrm{async})} A_i x_i(t). \tag{5}$$

As the number of randomly firing inputs is large, the difference in synaptic strength will not disturb the Gaussian distribution, and we can write for the second sum

$$\bar{A} \sum_{i(\mathrm{async})} x_i(t). \tag{6}$$

This is the same property as in the Abeles Model, although the firing rate may have a slightly different value.

For the synchroneous inputs, we now only consider one group of firing neurons, so all these inputs have the same t_0, so we can assume $t_0 = 0$, and we have, writing $x_i(t) = X_i \cdot \mathrm{e}^{-t/\tau}$:

$$\sum_{i(\mathrm{sync})} A_i x_i(t) = \sum_{i(\mathrm{sync})} A_i X_i \cdot \mathrm{e}^{-t/\tau} = \mathrm{e}^{-(t-t_0)/\tau} \sum_{i(\mathrm{sync})} A_i X_i, \tag{7}$$

where the X_i are 'digital' values (0 for no spike, 1 for a spike correlated with the synfire chain). Hence we can interprete the synchroneous inputs converging to the cell as a time-dependent lowering of the potential threshold T :

$$I(t) - \mathrm{e}^{-t/\tau} \sum_{i(\text{sync})} A_i X_i = \bar{A} \sum_{i(\text{async})} x_i(t). \tag{8}$$

Therefore, the firing rate is given by (all sums in the following text are sums only over the synfire chain inputs):

$$\lambda(t) = \frac{K}{\sqrt{2\pi}} \int_{\frac{(T-\sum A_i x_i(t))}{\sigma}}^{\infty} \mathrm{e}^{-y^2} \mathrm{d}y, \tag{9}$$

where $x_i(t) = X_i \cdot \mathrm{e}^{-t/\tau}$. We shall write for the input sum:

$$\mathcal{X} := \sum A_i X_i. \tag{10}$$

If we ask: What is the total number of extra spikes, generated by an input $\mathcal{X} \neq 0$, i.e., $\Delta\lambda(t) := \lambda_{\mathcal{X}\neq 0}(t) - \lambda_{\mathcal{X}=0}(t)$? – We have to integrate the firing rate,

$$\int_0^{\infty} \Delta\lambda(t) \mathrm{d}t = \frac{K}{\sqrt{2\pi}} \int_0^{\infty} \mathrm{d}t \left[\int_{\frac{(T-\sum A_i x_i(t))}{\sigma}}^{\infty} \mathrm{d}y \mathrm{e}^{-y^2} - \int_{\frac{T}{\sigma}}^{\infty} \mathrm{d}y \mathrm{e}^{-y^2} \right], \tag{11}$$

but this expression counts all extra spikes from $t = 0$ to $t = \infty$. However, if the output shows too much time delay, it will not be correlated to the synfire chain any more. As the time constant of the exponential is τ, we only take into account the outputs between $t = 0$ and $t = \Delta t$, where Δt is a time constant which may have a similar or smaller value than τ.So the average number of correlated output spikes $\langle \mathcal{Y} \rangle$ to a given input $\mathcal{X}$ is given by:

$$\langle \mathcal{Y}(\mathcal{X}) \rangle = \frac{K}{\sqrt{2\pi}} \int_0^{\Delta t} \mathrm{d}t \int_{\frac{T-\mathcal{X}\mathrm{e}^{-t/\tau}}{\sigma}}^{\infty} \mathrm{d}y \mathrm{e}^{-y^2}. \tag{12}$$

Here we have *not* subtracted the accidental output spikes, for their value is finite and rather small in this short time interval. For $\mathcal{X} \to (-\infty)$ the average output vanishes, which is the limit of strong inhibitory inputs. For $\mathcal{X} \to (+\infty)$, which is equivalent to strong excitation, we obtain:

$$\begin{aligned} \lim_{\mathcal{X}\to\infty} \langle \mathcal{Y}(\mathcal{X}) \rangle &= \frac{K}{\sqrt{2\pi}} \int_0^{\Delta t} \mathrm{d}t \int_{-\infty}^{\infty} \mathrm{d}y \mathrm{e}^{-y^2} \\ &= \frac{K}{\sqrt{2\pi}} \int_0^{\Delta t} \mathrm{d}t \sqrt{\pi} = \frac{K}{\sqrt{2}} \cdot \Delta t. \end{aligned} \tag{13}$$

If we choose our free parameter $\Delta t := \sqrt{2}/K$, the function $f(x) := \langle \mathcal{Y}(\mathcal{X}) \rangle$, as defined by equation (12), is a function of sigmoid type and describes the probability that an output spike is generated. For $\mathcal{X} = 0$ we have the probability of 0.005, which is the probability of accidental output spikes.

We recognize this result as the McCulloch-and-Pitts [11] Neuron Model, but in a fairly new light: Patterns of synchroneously firing neurons can be transferred and processed in a quasi-digital manner even in a randomly firing network, and the fluctuations are necessary to understand the sigmoidal character of the response function.

5 Conclusions and Outlook

Within the framework based on the activity model [2] and the concept of synfire chains, it has been shown how processing of information can be described quantitatively. Considering correlated and uncorrelated neural activity seperately, it is possible to describe information processing by synfire chains through a network of (in ground state) randomly firing neurons in a quantitative manner.

The crudest idealizations concern the waveform of the spikes. The stochastic description of the firing process and the representation of 'one bit' by more than one neuron are essential in the network for error-tolerance and the ability to generalization.

For synchroneously firing groups of neurons the 'quasi-digital' McCulloch-and-Pitts neuron Model is valid; the fluctuations of the other neurons determine the input-output characteristic to be sigmoidal.

The extended model can be generalized in a straightforward manner to describe also inhibitiory synapses and spatio-temporal aspects of real networks by use of (on larger time-scales) time-dependent values $A_i(t)$ of synaptic strength.

Acknowledgment. The author gratefully acknowledges partial financial support by Deutsche Forschungsgemeinschaft (DFG) within SFB 654.

References

1. M. Abeles. The role of the cortical neuron: Integrator or coincidence detector? Israel Journal of Medical Sciences **18** (1982) 83-92
2. M. Abeles, Local Cortical Circuits, Springer, Berlin (1982)
3. M. Abeles. *Corticonics*, Cambridge University Press (1991)
4. W. Gerstner, R. Ritz, J. L. van Hemmen. Why spikes? Hebbian learning and retrieval of time-resolved excitation patterns, Biological Cybernetics **69** (1993) 503-515
5. M. Herrmann, J. A. Hertz, A. Prügel-Bennett. Analysis of synfire chains, Network **6** (1995) 403-414
6. Kazushi Ikeda. A synfire chain in layered coincidence detectors with random synaptic delays, Neural Networks **16** (2003) 39-46
7. J. P. Sougne, A learning algorithm for synfire chains. In R. M. Franch and J. P. Sougne (eds.): Connectionist Models of Learning, Development and Evolution, p. 23-32, Springer, London (2001)
8. S. Wermter, C. Panchev. Hybrid preference machines based on inspiration from neuroscience. Cognitive Systems Research **3** (2002) 255-270

9. V. Braitenberg. Cortical architectonics: general and areal. In: M. A. B. Brazier, H. Petche (eds.), Architectonics of the cerebral cortex. Raven Press, New York, p. 443-465 (1978)
10. J. C. Claussen (born Gruel). p. 89–92 in: Diploma thesis, Kiel, Germany (1992)
11. W. S. McCulloch and W. H. Pitts. A Logical Calculus of the Ideas Immanent in Nervous Activity. Bulletin of Mathematical Biophysics **5** (1943) 115-133

Appendix: Stability Analysis of the Abeles Model

We now investigate whether the fixed point satisfying the self-consistence-equation (3) is stable or instable. Although we do not know the exact dynamical properties of the network, we can answer this question. A small change in λ will lead to a change in σ, the variance of the intracellular potential, where $\sigma(\lambda(t), t)$ is given by (2). The changed variance of the intracellular potential will cause a change in the firing rate, given by (1). However, this will need a certain delay Δt, so that the stationary equation (1) has to be modified to the iterative expression

$$\lambda(\sigma(t), t + \Delta t) = \frac{K}{\sqrt{2\pi}} \cdot \int_{\frac{T}{\sigma(t)}}^{\infty} \mathrm{e}^{-y^2} dy. \tag{14}$$

Therefore we can approximate the real dynamics by the iteration

$$\lambda(\lambda(t), t + \Delta t) = \frac{K}{\sqrt{2\pi}} \cdot \int_{\frac{T}{A\sqrt{N\lambda(t)\tau/2}}}^{\infty} \mathrm{e}^{-y^2} dy \tag{15}$$

and we obtain the answer to an increase of λ by the amount of $\Delta\lambda$:

$$\begin{aligned}\Delta\lambda(t + \Delta t) &= \lambda(\sigma(\lambda(t) + \Delta\lambda), t + \Delta t) - \lambda(\sigma(\lambda(t)), t) \\ &= \frac{\partial\lambda(\sigma(\lambda(t)), t)}{\partial\sigma(t)} \cdot \frac{\partial\sigma(\lambda(t), t)}{\partial\lambda(t)} \cdot \Delta\lambda(t),\end{aligned}$$

which means that every iteration stretches $\Delta\lambda$ by the factor

$$\alpha(\lambda) = \frac{\Delta\lambda(t + \Delta t)}{\Delta\lambda(t)} = \frac{\partial\lambda(\sigma(\lambda(t)), t)}{\partial\sigma(t)} \cdot \frac{\partial\sigma(\lambda(t), t)}{\partial\lambda(t)}. \tag{16}$$

Since
$\frac{\partial}{\partial\lambda}(A\sqrt{N\lambda\tau/2}) = \frac{\sigma}{2\lambda}$, and

$$\frac{\partial}{\partial\sigma}\left(\frac{K}{\sqrt{2\pi}} \int_{\frac{T}{\sigma(t)}}^{\infty} \mathrm{e}^{-y^2} dy\right) = \frac{K}{\sqrt{2\pi}}(-\mathrm{e}^{-(T/\sigma)^2}) \cdot \left(-\frac{T}{\sigma^2}\right),$$

we obtain

$$\alpha = \frac{K}{\lambda} \frac{1}{2\sqrt{2\pi}} \frac{T}{\sigma} \mathrm{e}^{-(T/\sigma)^2}. \tag{17}$$

For sufficiently small $\Delta\lambda$ the iteration values λ_i are close to the start value λ_0, so that the Liapunov exponent of the iteration is given by $L = \ln|\alpha(\lambda_0)|$. Obviously the fixed point, which is assumed to represent a 'ground state' of randomly firing, is a stable one if and only if the Liapunov exponent is negative, which means that $|\alpha| < 1$. Using the experimental values given by Abeles for the cortex of the cat, $T/\sigma = 2.58$, $K = 1000s^{-1}$, and $\lambda = 5s^{-1}$, we obtain the Liapunov exponent $L = -2.02$ or $\alpha = 0.13$, which is much less than 1. In this fixed point the network gives strong damping to both fluctuations and external stimulus. This includes also sufficient stability of the 'Randomly Firing Mode': Neither a fade-out nor a collective 'explosion' of the firing can be generated by small perturbations. To understand the effects of strong perturbations, we will take a short view on the stability function $\alpha(T/\sigma)$. Using equation (17), we have to remember that λ/K is a function of T/σ, so that we can use the expression $(x := T/\sigma)$:

$$\alpha(x) = \frac{\frac{x}{2}\mathrm{e}^{-x^2}}{\int_x^\infty \mathrm{e}^{-y^2}\mathrm{d}y}. \tag{18}$$

Two limiting cases can be considered: For $x \to 0$, which is the limes of very high firing rates, $\alpha(x)$ is asymptotic to $x/\sqrt{\pi}$, so that $\alpha(x)$ decreases to zero. This expresses the damping of avalanche effects. For $x \to \infty$, which is the limes of very low firing rates, the integral is asymptotic to $\frac{1}{2x}\mathrm{e}^{-x^2}$, therefore $\alpha(x)$ is asymptotic to x^2. As $\alpha(x)$ is continuous, there must exist a critical firing rate λ_c, where $\alpha(\lambda_c) = 1$. It is the point where the Liapunov exponent changes its sign. If the firing rate is higher than λ_c, we still have damping, same as in the ground state itself. If the firing rate is lower than the critical value, the cortical feedback amplifies any fluctuations of the firing rate, so that the fluctuations lead to a fade-out of the network. To conclude, if the firing rate is lower than a critical firing rate λ_c, randomly firing cannot be a stable mode of a neural network.

Phase Precession and Recession with STDP and Anti-STDP

Răzvan V. Florian[1,2,4] and Raul C. Mureşan[1,3]

[1] Center for Cognitive and Neural Studies (Coneural), Str. Saturn nr. 24, 400504 Cluj-Napoca, Romania
[2] Babeş-Bolyai University, Institute for Interdisciplinary Experimental Research, Str. T. Laurian nr. 42, 400271 Cluj-Napoca, Romania
[3] Frankfurt Institute for Advanced Studies, Johann Wolfgang Goethe University Max-von-Laue-Strasse 1, 60438 Frankfurt am Main, Germany
[4] florian@coneural.org, http://www.coneural.org/florian

Abstract. We show that standard, Hebbian spike-timing dependent plasticity (STDP) induces the precession of the firing phase of neurons in oscillatory networks, while anti-Hebbian STDP induces phase recession. In networks that are subject to oscillatory inhibition, the intensity of excitatory input relative to the inhibitory one determines whether the phase can precess due to STDP or whether the phase is fixed. This phenomenon can give a very simple explanation to the experimentally-observed hippocampal phase precession. Modulation of STDP can lead, through precession and recession, to the synchronization of the firing of a trained neuron to a target phase.

1 Introduction

Spike-timing dependent plasticity (STDP) is the dependence of synaptic changes on the relative timing of pre- and postsynaptic action potentials, a phenomenon that has been experimentally observed in biological neural systems [1,2,3]. The type of STDP that has been mostly studied is characterized by the potentiation of a synapse when the postsynaptic spike follows the presynaptic spike within a time window of a few tens of milliseconds, and the depression of the synapse when the order of the spikes is reversed. This type of STDP is sometimes called Hebbian, because it is consistent with the original postulate of Hebb that predicted the strengthening of a synapse when the presynaptic neuron causes the postsynaptic neuron to fire. Experiments have also found synapses with anti-Hebbian STDP (also called anti-STDP), where the sign of the changes is reversed, in comparison to Hebbian STDP [4,5,6,7].

Many studies have investigated the computational properties of Hebbian STDP, and have shown its function in neural homeostasis, unsupervised and supervised learning [8,9,10,11,12,13,14,15,16,17,18]. Anti-Hebbian STDP is, at a first glance, not as interesting as the Hebbian mechanism, as it leads, by itself, to an overall depression of the synapses towards zero efficacy [19]. We have recently

S. Kollias et al. (Eds.): ICANN 2006, Part I, LNCS 4131, pp. 718–727, 2006.

shown that modulating STDP with a reward signal (i.e., having both Hebbian and anti-Hebbian STDP) leads to reinforcement learning [20,21]. None of these studies have specifically investigated the consequences of STDP in oscillatory networks.

Here we study through computer simulations the effects of Hebbian and anti-Hebbian STDP in networks of neurons that fire periodically with a common period. This has biological relevance because there is such rhythmical activity in the brain, for example the hippocampal theta rhythm [22,23]. We first describe our model (Section 2) and then demonstrate some general effects induced by STDP in oscillatory networks (Section 3). Afterwards we study the interplay between these effects and oscillatory inhibition (Section 4) and how the effects can be used to teach a neuron to fire at a given phase (Section 5).

2 Methods

We study an integrate-and-fire neuron driven by N_e excitatory and N_i inhibitory input neurons. The excitatory synapses are plastic, while the inhibitory ones are static. This setup is similar to the one in [10]. We model the network's rhythmic activity by considering that input neurons fire periodically with a common period $T = 125$ ms (corresponding to the 8 Hz theta hippocampal rhythm). In the brain, neurons sometimes skip cycles, while still firing at a constant phase, but we ignore this possibility here, for the sake of simplicity, and consider that each of the input neurons fires once per period, at a predetermined phase ϕ_k. These phases are generated randomly at the beginning of the experiments. The phases of excitatory neurons are generated uniformly between 0 and 2π. In experiments where we use inhibitory neurons, the total inhibition is considered to be modulated by the global oscillation, as in other models of the hippocampal theta rhythm [26,27], and thus their phases are generated with a probability density $p(\phi_k) = [\cos(\phi_k) + 1]/(2\pi)$ (see also Fig. 2e-g).

The dynamics of the postsynaptic integrate-and-fire neuron is given by the following equation:

$$\tau_m \frac{\mathrm{d}V(t)}{\mathrm{d}t} = -(V - V_0) + \sum_{k=1}^{N_e+N_i} g_k(t)\,[E_k - V(t)]\,, \tag{1}$$

where V is the membrane potential, V_0 =-70 mV is the resting potential, τ_m =20 ms is the decay time constant, g_k are synaptic conductances and E_k are reversal potentials. When the membrane potential reaches a threshold of -54 mV, the neuron fires and V is reset to -60 mV. We consider E_k=0 mV for excitatory synapses and E_k=-70 mV for inhibitory ones (parameters from [10,24]).

Each presynaptic spike determines an instantaneous rise in the synaptic conductance, which decays then exponentially. Thus, the dynamics of the synaptic conductances is given by

$$\frac{\mathrm{d}g_k(t)}{\mathrm{d}t} = -\frac{g_k(t)}{\tau_g} + g_k^s(t)\,\Phi_k(t), \tag{2}$$

where $\tau_g = 5$ ms, g_k^s are the peak synaptic conductances, and $\Phi_k(t)$ represents the firing train of input neuron k as a sum of Dirac functions:

$$\Phi_k(t) = \sum_{n=0}^{\infty} \delta\left(t - (n\,T + \phi_k)\right). \tag{3}$$

For inhibitory synapses, g_k^s is constant and is generated randomly at the beginning of the experiment, with an uniform distribution, between 0 and g_{max}^s. For excitatory synapses, g_k^s is also initialized randomly between 0 and g_{max}^s, but varies in time due to STDP. As in previous studies [10,25], we use an exponential dependence of plasticity on the relative spike timings, we consider that the effect of different spike pairs is additive, and we limit the range of possible synaptic strengths with hard bounds, between 0 and g_{max}^s. To model Hebbian as well as anti-Hebbian STDP, we consider that plasticity is modulated by a variable $r(t)$ that can be positive as well as negative. Hence, the dynamics of the excitatory synaptic conductances is given by

$$\frac{\mathrm{d}g_k^0(t)}{\mathrm{d}t} = r(t)\left[\Phi_0(t)\,A_+ \sum_{\mathcal{F}_k^t} \exp\left(-\frac{t-t_k^f}{\tau_+}\right) + \Phi_k(t)\,A_- \sum_{\mathcal{F}_0^t} \exp\left(-\frac{t-t_0^f}{\tau_-}\right)\right], \tag{4}$$

with the additional hard bounds. We noted with $\mathcal{F}_k^t$ the set of firing times t_k^f previous to t of input neuron k, and $\mathcal{F}_0^t$ is the analogue for the postsynaptic neuron. $\Phi_0(t)$ is the spike train of the postsynaptic neuron; $A_\pm$ are constant parameters that determine the magnitude of synaptic changes, $A_+ = 0.005\,g_{max}^s$, $A_- = -A_+$; $\tau_\pm$ are the decay time constants of the exponential STDP windows, $\tau_+ = \tau_- = 20$ ms.

Following [10], we use a set of variables P_k^+ that track the influence of presynaptic spikes and P_0^- that tracks the influence of postsynaptic spikes on the synapses. These variables simplify the simulation and may also have biochemical counterparts in biological neurons. We then have:

$$\frac{\mathrm{d}P_k^+}{\mathrm{d}t} = -\frac{P_k^+}{\tau_+} + A_+\,\Phi_k(t) \tag{5}$$

$$\frac{\mathrm{d}P_0^-}{\mathrm{d}t} = -\frac{P_0^-}{\tau_-} + A_-\,\Phi_0(t) \tag{6}$$

$$\frac{\mathrm{d}g_k^0(t)}{\mathrm{d}t} = r(t)\,\left[\Phi_0(t)\,P_k^+ + \Phi_k(t)\,P_0^-\right] \tag{7}$$

In some of the experiments, we use a homeostatic mechanism [28] that scales up or down the synapses in order to keep the postsynaptic firing rate constant, at one spike per oscillation period T. We estimate the postsynaptic firing rate ν by using a leaky accumulator (equivalent to an integration with an exponential kernel),

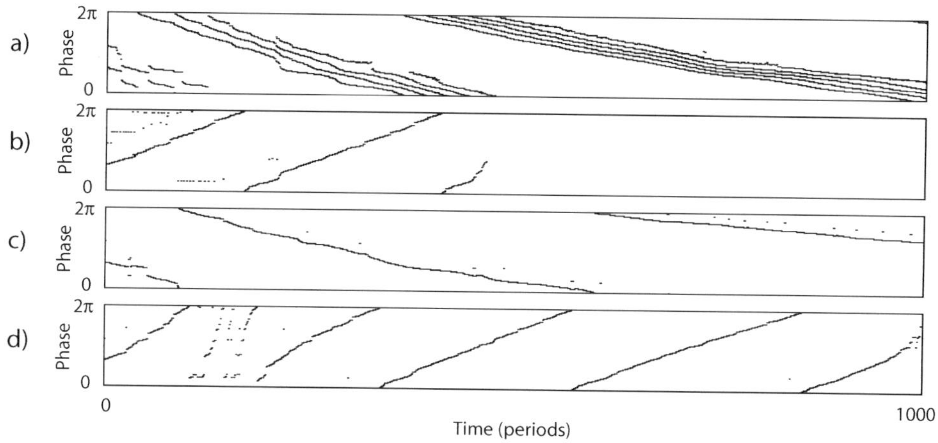

Fig. 1. The evolution in time of the phase of postsynaptic spikes relative to the input oscillation. The graphs illustrate the first 1000 periods of the experiments. All experiments start with identical conditions. a), c): Hebbian STDP. b), d): Anti-Hebbian STDP. a), b): Without homeostasis. c), d): With homeostasis.

$$\frac{\mathrm{d}\nu(t)}{\mathrm{d}t} = -\frac{\nu(t)}{\tau_\nu} + \frac{1}{\tau_\nu}\,\Phi_0(t), \tag{8}$$

with $\tau_\nu = 500$ ms. We then scale all excitatory synapses according to

$$\frac{\mathrm{d}g_k^0}{\mathrm{d}t} = \alpha\; g_k^0 \left[\frac{1}{T} - \nu(t)\right], \tag{9}$$

with $\alpha = 0.04$. This mechanism is applied additionally to the plasticity mechanisms already mentioned.

The network is simulated with a timestep of 0.5 ms.

3 Precession and Recession

We first consider a setup with $N_e = 1000$ excitatory input neurons and no inhibitory input. We use $g^s_{max} = 0.014$ and no homeostasis. If we set $r(t) = 1$, i.e. Hebbian STDP, and let the network run, we observe that the phase of the postsynaptic spikes relative to the input oscillation precesses, i.e. has a tendency to occur earlier in the cycle (Fig. 1a). This is consistent with previous observations that STDP tends to reduce the latency of postsynaptic firing in response to the same stimulus (input) [10,29] and that STDP allows the postsynaptic neuron to predict its input [12]. These properties of STDP also make inputs that fire before the postsynaptic neuron to become more and more effective in causing the postsynaptic neuron to fire, and eventually increase the total excitation that this neuron receives. This means that the neuron may start to fire more spikes per period, a phenomenon that can be seen in Fig. 1a.

If we set $r(t) = -1$, i.e. we have anti-Hebbian STDP, we observe the opposite, namely that the phase of the postsynaptic spikes recesses (has a tendency to occur later in the cycle), and that the excitation that the neuron receives diminishes, eventually leading the neuron to stop firing (Fig. 1b). This is consistent with the previous observation that anti-Hebbian STDP leads to a global weakening of the synapses [19].

However, if we also introduce the previously mentioned homeostatic mechanism that keeps the postsynaptic neuron firing once per period, we observe that the precession/recession corresponding to Hebbian/anti-Hebbian STDP becomes a stable behavior of the neuron (Fig. 1c,d).

4 Precession Control Through Oscillatory Inhibition

We now add to the previously described setup $N_i = 1000$ inhibitory input neurons, that provide an oscillatory inhibitory input current to the postsynaptic neuron (each inhibitory neuron fires once per cycle, and the number of neurons that fire at a particular phase oscillates as a function of phase). We use $r(t) = 1$, $g^s_{max} = 0.015$ and homeostasis. The phase precession is not disturbed by the oscillatory inhibition (Fig. 2a). Precession is stopped, however, by a much stronger inhibition, for example if we reduce the number of excitatory inputs from 1000 to 500 (Fig. 2b), as the neuron can fire only at phases where excitation overcomes inhibition.

This means that by modulating the ratio of excitation versus oscillatory inhibition, in conjunction with STDP, we may switch from precession to a state of constant phase firing. This is illustrated in Fig. 2c-d, where, after the firing phase stabilizes because oscillating inhibition dominates excitation, we increase the excitation received by the output neuron, by adding extra excitatory inputs. Until $t_1 = 1200\,T$, the postsynaptic neuron is driven by 500 excitatory neurons. From t_1 to $t_2 = 1500\,T$, we constantly add new excitatory inputs to the postsynaptic neuron until their number reaches 1000 at t_2. With greater excitation, the phase starts to precess. From t_2 to $t_3 = 1800\,T$, we gradually remove the newly added excitatory inputs; excitation decreases and then phase stabilizes again to a value close to the one previous to the increase in excitation.

This very simple model is thus capable to explain the basic features of hippocampal phase precession. It has been observed that when a rat moves through the receptive field of a place cell, the firing rate of the neuron correlates with the position in the place field, and the firing phase of the neuron precesses as the animal traverses the place field. The initial phase at which the neuron starts firing when the animal enters the place field is constant for every traversal of the field [30]. The simple model presented here is consistent with these observations: as the excitation of the place cell increases because the animal enters into its receptive field, its firing phase precesses simply because of STDP and because excitation overcomes the phase locking by the oscillatory inhibition.

Among the many computational models that tried to explain phase precession, only two others used STDP. The first one used STDP to explain the skewness

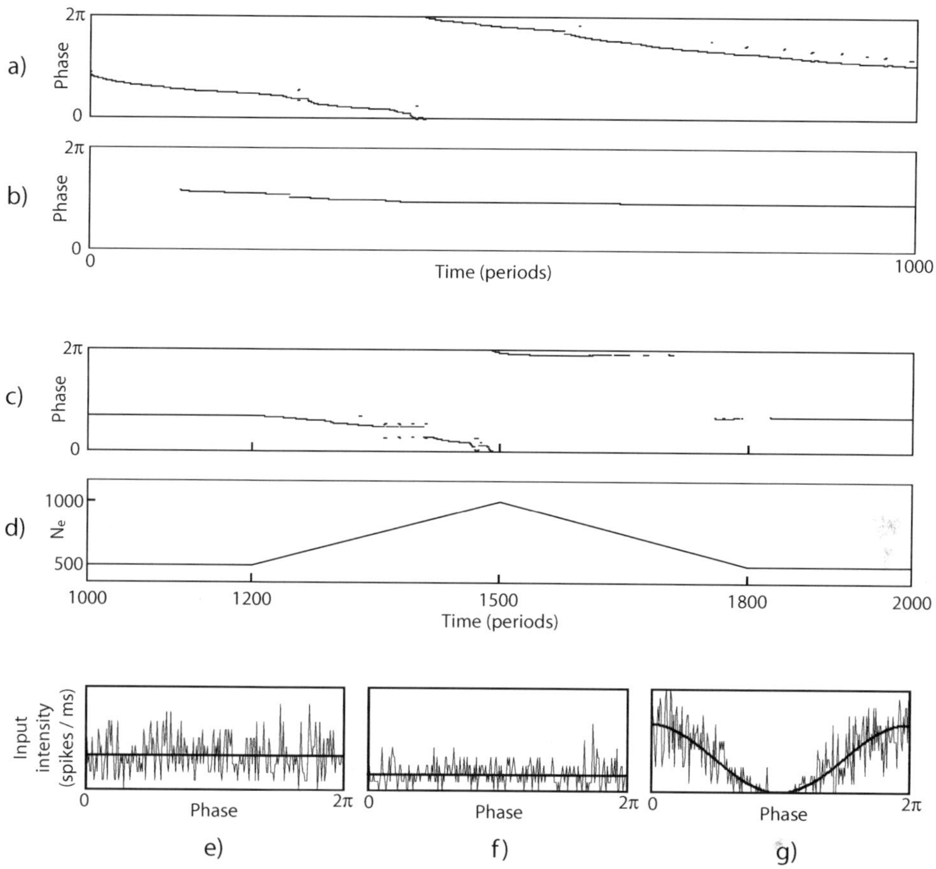

Fig. 2. a)-c) Effect of oscillatory inhibition on the dynamics of the phase of postsynaptic spikes relative to the input rhythm. All experiments start with identical conditions. a) 1000 excitatory inputs, 1000 inhibitory inputs. b) 500 excitatory inputs, 1000 inhibitory inputs. c) 500–1000 excitatory inputs, 1000 inhibitory inputs. d) The evolution in time of the number of the excitatory inputs for the experiment presented in c). e)-f) Intensity of the total excitatory and, respectively, inhibitory inputs (number of spikes per time unit) as a function of phase. The smooth line represents the average number of input spikes per timestep corresponding to the probability with which they were generated, the rugged line represents the actual histogram of the input spikes as a function of phase, corresponding to the experiments illustrated here. e) Excitatory input intensity for the experiment presented in a). f) Excitatory input intensity for the experiment presented in b). g) Inhibitory input intensity for all experiments. In the experiment presented in c), the input intensity varies between the one presented in f) and the one presented in e).

of the place fields, which, at its turn, explained phase precession, through interaction with the inhibitory oscillation [31]. The second one takes from STDP only the idea of temporally asymmetric interactions between neurons, as it uses

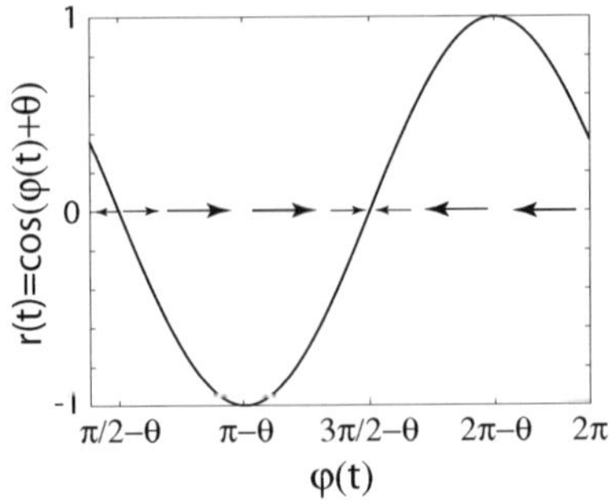

Fig. 3. The phase ϕ_0 precesses if $r(t) > 0$ and recesses if $r(t) < 0$, as indicated by the arrows. If $r(t) = \cos(\varphi(t) + \theta)$, the phase converges to $\phi_0 = 3\pi/2 - \theta$ because this is a stable point for the dynamics of ϕ_0; $\phi_0 = \pi/2 - \theta$ is an unstable equilibrium point

neurons with continuous activations instead of spiking neurons [32]. The model presented here is much simpler than these previous models, yet it captures the essential features of hippocampal phase precession.

5 Controlling the Firing Phase by Modulating STDP

Since the firing phase can be manipulated by STDP and anti-STDP, it is straightforward to devise a mechanism for moving it to a target phase, by modulating STDP. Modulation of STDP by a global reward signal proved to be a robust reinforcement learning mechanism for generic spiking neural networks and could be implemented in the brain by a neuromodulator [20,21]. Here we may use a similar modulation, but with a form that depends on the target phase at which we want the postsynaptic neuron to fire, instead of an external reward.

For example, if the variable $r(t)$ that modulates STDP oscillates as a function of the input oscillation phase, with the same period, and the STDP temporal constants $\tau_\pm$ are smaller than the oscillation period, the output neuron will always decrease its phase if the phase is in certain intervals, and increase it in others. If an oscillatory $r(t)$ is a continuous function of input oscillation phase $\varphi = 2\pi\ t/T$, and has both positive and negative values, the phase of the postsynaptic neuron will have at least two points of equilibrium ($r = 0$), among which one will be stable and one unstable. For example, if $r(t) = \cos(\varphi(t) + \theta)$, the equilibrium point will be $\phi_0 = 3\pi/2 - \theta$ (see Fig. 3). The firing phase of a postsynaptic neuron with synapses featuring STDP modulated by an $r(t)$ of this form will thus move to the phase of stable equilibrium. This means that we can train a neuron or a population of (independent) neurons to fire at a particular phase by using STDP in conjunction to an appropriate form of $r(t)$. The efficacy of this approach is illustrated in Fig. 4. The neuron learns the target firing phase within 200 periods (25 s). The same signal $r(t)$ may train an arbitrary number of neurons to fire at the same phase, thus synchronizing them.

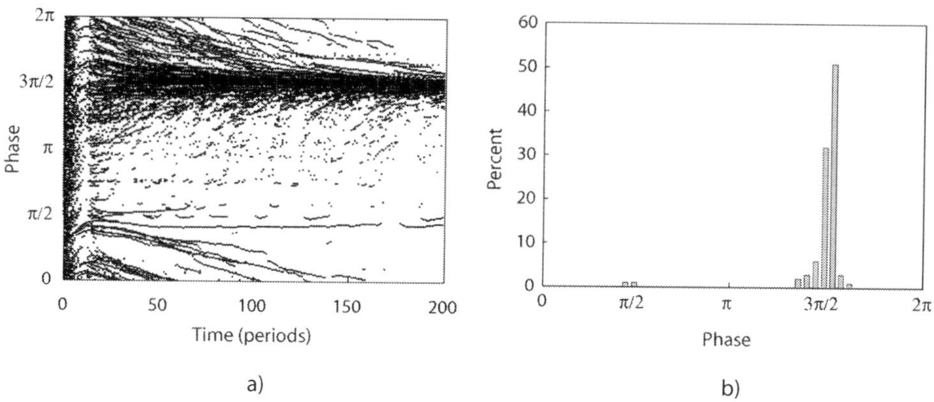

Fig. 4. Learning to fire at a target phase ($\phi_0 = 3\pi/2$) with modulated STDP. a) Superposition of the spikes emitted by the postsynaptic neuron in 100 trials, each having different, random, initial conditions. b) The distribution of the phases of the postsynaptic neuron after 200 periods of training (data from the same trials). It can be seen that in most cases the phase converges towards the stable equilibrium point $3\pi/2$, while in very few cases it converges towards the unstable equilibrium point $\pi/2$.

Alternatively, $r(t)$ need not to vary continuously during a cycle, as a function of the global oscillation phase, but may vary when the postsynaptic neuron fires, as a function of the difference between its phase $\phi_0(t)$ at which it last fired and the target phase ϕ_0^{target}. For example, if $r(t) = sin(\phi_0(t) - \phi_0^{target})$, we achieve a convergence of the postsynaptic phase to the target phase similar to the one obtained with the previous method (data not shown).

Unfortunately none of these two methods of synchronizing neurons seems to be biologically plausible. The first method requires a neuromodulator with a fast temporal variation, and the second requires the neuromodulator to represent the error between the actual phase and the target phase. However, these methods may be useful in some applications for training artificial spiking neural networks.

6 Conclusion

We have shown that, if a neuron is driven by periodically-firing excitatory inputs, its phase will precess if the synapses connecting it to the input feature Hebbian STDP, and will recess if the synapses feature anti-Hebbian STDP. If there exists a homeostatic mechanism that enforces the neuron to fire once per cycle, the precession or recession will be stable. A strong oscillatory inhibition may stop the precession caused by STDP and lock the firing phase. If the ratio of excitation versus inhibition varies in time, firing may alternate between periods of precession and of stable phase. This phenomenon can explain in a very simple fashion the hippocampal phase precession. Precession and recession may be combined by modulating STDP to train neurons to fire at a given target phase.

Acknowledgements

R.V.F. was partly supported by Arxia SRL and by a grant of the Romanian Government (MEdC-ANCS). R.C.M. acknowledges the partial support of Hertie Foundation.

References

1. Markram, H., Lübke, J., Frotscher, M., Sakmann, B.: Regulation of synaptic efficacy by coincidence of postsynaptic APs and EPSPs. Science **275** (1997) 213–215
2. Bi, G.Q., Poo, M.M.: Synaptic modifications in cultured hippocampal neurons: Dependence on spike timing, synaptic strength, and postsynaptic cell type. Journal of Neuroscience **18** (1998) 10464–10472
3. Dan, Y., Poo, M.M.: Spike timing-dependent plasticity of neural circuits. Neuron **44** (2004) 23–30
4. Dan, Y., Poo, M.M.: Hebbian depression of isolated neuromuscular synapses in vitro. Science **256** (1992) 1570–1573
5. Bell, C.C., Han, V.Z., Sugawara, Y., Grant, K.: Synaptic plasticity in a cerebellum-like structure depends on temporal order. Nature **387** (1997) 278–281
6. Egger, V., Feldmeyer, D., Sakmann, B.: Coincidence detection and changes of synaptic efficacy in spiny stellate neurons in rat barrel cortex. Nature Neuroscience **2** (1999) 1098–1105
7. Roberts, P.D., Bell, C.C.: Spike timing dependent synaptic plasticity in biological systems. Biological Cybernetics **87** (2002) 392–403
8. Kempter, R., Gerstner, W., van Hemmen, J.L.: Hebbian learning and spiking neurons. Physical Review E **59** (1999) 4498–4514
9. Kempter, R., Gerstner, W., van Hemmen, J.L.: Intrinsic stabilization of output rates by spike-based Hebbian learning. Neural Computation **13** (2001) 2709–2742
10. Song, S., Miller, K.D., Abbott, L.F.: Competitive hebbian learning through spike-timing-dependent synaptic plasticity. Nature Neuroscience **3** (2000) 919–926
11. Roberts, P.: Computational consequences of temporally asymmetric learning rules: I. Differential Hebbian learning. Journal of Computational Neuroscience **7** (1999) 235–246
12. Rao, R.P.N., Sejnowski, T.J.: Spike-timing-dependent Hebbian plasticity as temporal difference learning. Neural Computation **13** (2001) 2221–2237
13. Toyoizumi, T., Pfister, J.P., Aihara, K., Gerstner, W.: Spike-timing dependent plasticity and mutual information maximization for a spiking neuron model. In Saul, L., Weiss, Y., Bottou, L., eds.: Advances in Neural Information Processing Systems. Volume 17., MIT Press (2005) 1409–1416
14. Bell, A.J., Parrara, L.C.: Maximising sensitivity in a spiking network. In Saul, L.K., Weiss, Y., Bottou, L., eds.: Advances in Neural Information Processing Systems. Volume 17., Cambridge, MA, MIT Press (2004)
15. Chechik, G.: Spike time dependent plasticity and information maximization. Neural Computation **15** (2003) 1481–1510
16. Bohte, S.M., Mozer, C.: Reducing spike train variability: A computational theory of spike-timing dependent plasticity. In Saul, L.K., Weiss, Y., Bottou, L., eds.: Advances in Neural Information Processing Systems. Volume 17., Cambridge, MA, MIT Press (2004)

17. Hopfield, J.J., Brody, C.D.: Learning rules and network repair in spike-timing-based computation networks. Proceedings of the National Academy of Sciences **101** (2004) 337–342
18. Legenstein, R., Naeger, C., Maass, W.: What can a neuron learn with spike-timing-dependent plasticity? Neural Computation **17** (2005) 2337–2382
19. Abbott, L.F., Gerstner, W.: Homeostasis and learning through spike-timing dependent plasticity. In Gutkin, B., Hansel, D., Meunier, C., Dalibard, J., Chow, C., eds.: Methods and Models in Neurophysics: Proceedings of the Les Houches Summer School 2003. Elsevier Science (2005)
20. Florian, R.V.: A reinforcement learning algorithm for spiking neural networks. In Zaharie, D., Petcu, D., Negru, V., Jebelean, T., Ciobanu, G., Cicortaş, A., Abraham, A., Paprzycki, M., eds.: Proceedings of the Seventh International Symposium on Symbolic and Numeric Algorithms for Scientific Computing (SYNASC 2005), IEEE Computer Society (2005) 299–306
21. Florian, R.V.: Reinforcement learning through modulation of spike-timing dependent plasticity. Neural Computation (2006) In press.
22. Buzsaki, G.: Theta oscillations in the hippocampus. Neuron **33** (2002) 325–340
23. Buzsaki, G., Draguhn, A.: Neuronal oscillations in cortical networks. Science **304** (2004) 1926–1929
24. Troyer, T.W., Miller, K.D.: Physiological gain leads to high ISI variability in a simple model of a cortical regular spiking cell. Neural Computation **9** (1997) 971–983
25. Abbott, L.F., Nelson, S.B.: Synaptic plasticity: taming the beast. Nature Neuroscience **3** (2000) 1178–1183
26. Tsodyks, M.V., Skaggs, W.E., Sejnowski, T.J., McNaughton, B.L.: Population dynamics and theta rhythm phase precession of hippocampal place cell firing: A spiking neuron model. Hippocampus **6** (1996) 271–280
27. Mehta, M.R., Lee, A.K., Wilson, M.A.: Role of experience and oscillations in transforming a rate code into a temporal code. Nature **417** (2002) 741–746
28. Turrigiano, G.G., Nelson, S.B.: Homeostatic plasticity in the developing nervous system. Nature Reviews Neuroscience **5** (2004) 97–107
29. Gerstner, W., Kistler, W.M.: Spiking neuron models. Cambridge University Press, Cambridge, UK (2002)
30. O'Keefe, J., Recce, M.L.: Phase relationship between hippocampal place units and the EEG theta rhythm. Hippocampus **3** (1993) 317–330
31. Mehta, M.R., Quirk, M.C., Wilson, M.A.: Experience-dependent asymmetric shape of hippocampal receptive fields. Neuron **25** (2000) 707–715
32. Scarpetta, S., Marinaro, M.: A learning rule for place fields in a cortical model: Theta phase precession as a network effect. Hippocampus **15** (2005) 979–989

Visual Pathways for Detection of Landmark Points

Konstantinos Raftopoulos, Nikolaos Papadakis, and Klimis Ntalianis

Dept. of Electrical and Computer Engineering, National Technical University of Athens,
9 Heroon Polytechneiou Str., 157 73 Zografou, Athens, Greece
raftop@image.ntua.gr

Abstract. We describe a neuron multi-layered architecture that extracts landmark points of high curvature from 2d shapes and resembles the visual pathway of primates. We demonstrate how the rotated orientation specific receptive fields of the simple neurons that were discovered by Hubel and Wiesel can perform landmark point detection on the 2d contour of the shape that is projected on the retina of the eye. Detection of landmark points of high curvature is a trivial task with sophisticated machine equipment but we demonstrate how such a task can be accomplished by only using the hardware of the visual cortex of primates abiding to the discoveries of Hubel and Wiesel regarding the rotated arrangements of orientation specific simple neurons. The proposed layered architecture first extracts the 2dimensional shape from the projection on the retina then it rotates the extracted shape in multiple layers in order to detect the landmark points. Since rotating the image about the focal origin is equivalent to the rotation of the simple cells orientation field, our model offers an explanation regarding the mystery of the arrangement of the cortical cells in the areas of layer 2 and 3 on the basis of shape cognition from its landmark points.

Keywords: Visual cortex, Landmark points, Shape encoding, Curvature detection.

1 Introduction

Our first knowledge about cortical neurons and their receptive fields we owe to Hubel and Wiesel. The Nobel Prize winners made a remarkable progress during their 25 years of collaboration in elucidating the responses of cortical neurons by using stimuli of great relevance to vision. In their papers they define the ways in which area VI receptive fields differ from LGN receptive fields. The qualitative methods they used for studying the cortex continue to dominate experimental physiology (Hubel and Wiesel, 1959, 1962, 1968, 1977; Hubel, 1982).

Hubel and Wiesel recorded the activity of cortical neurons while displaying patterned stimuli, mainly line segments and spots, on a screen that was imaged through the animal's cornea and lens onto the retina. As the microelectrode penetrated the visual cortex, they presented line segments whose width and length could be adjusted. First, they varied the position of the stimulus on the screen, searching for the neuron's receptive field. Once the receptive-field position was established, they

S. Kollias et al. (Eds.): ICANN 2006, Part I, LNCS 4131, pp. 728 – 739, 2006.

measured the response of the neuron to lines, bars and spots presented individually. From Hubel's Nobel lecture we quote:

"Our first indication of the beauty of the arrangements of cell groupings came in 1961 in one of our first recordings from striate cortex of monkey, a spider monkey named George. In one penetration, which went into the cortex at an angle of about 45° and was 2.5 mm long, we were struck right away by something we had only seen hints of before: as the electrode advanced the orientations of successively recorded cells progressed in small steps, of about 10° for every advance of 50 μm. We began the penetration around 8:00 p.m.; five hours later we had recorded 53 successive orientations without a single large jump in orientation. During the entire time, in which I wielded the slide projector and Torsten mapped the fields, neither of us moved from our seats. Fortunately our fluid intake that day had not been excessive!" Hubel, Nobel Lecture, December 1981.

In this paper we show how these specific arrangements of the cortical cell groupings can lead to the extraction of landmark points of high curvature on a 2d shape contour that is projected on the retina of the eye. We describe an artificial neural architecture that abides to the described by Hubel arrangements of cell groupings and achieves landmark points extraction, we therefore imply that the specific arrangements of the cell groupings in the visual cortex of primates perform landmark point extraction in a way similar to our artificial model. Our purpose in this paper is to describe how what we know about the cortical cells of the primates can be used to perform landmark extraction from shapes.

The rest of the paper is organized as follows: First we present related work from both fields of neuropsychology and computer science that support the concept of landmark points in shape perception. Our specific layered architecture is presented in the next section where first we properly define the term "landmark point' and "landmark region" that will be used hereinafter and then we explain how our proposed architecture simulates the way the specific arrangements of the cortical cells are likely used to extract landmark points and regions.

2 Related Work

The concept of landmark points for shape summarization has been appreciated by many researchers in many different areas of neuropsychology and computer science. The importance of landmark points of high curvature in the way that humans perceive shapes is apparent in the work of Goodale et al [Goodale, 1994], [Goodale 1991]. In this study patients with unilateral or bilateral lesions of the visual cortex are "*unable to calibrate their grasp according to the best "grasp lines" made up from points of maximum convexity or concavity along the boundary of an object where the most stable grip should be expected*". Other patients with damages in the visual cortex are unable to discriminate from different shapes and orientations. Goodale concludes that "*The brain damage that the patient suffered as a consequence of anoxia appears to have interrupted the normal flow of shape and contour information into her perceptual systems*".

Leslie G. Ungerleider et al [Ungerleider, 1998] report that the visual processing pathways in primates:

"*....appear to be organized hierarchically, in the sense that low-level inputs are transformed into progressively more integrated representations through successive stages of processing. Within the ventral stream, for example, the processing of object features begins with simple spatial filtering by cells in V1, but by the time the inferior temporal cortex (area TE) is activated, the cells respond to global object features, such as shape .... Thus, much of the neural mechanism for both object vision and spatial vision can be viewed as a bottom-up process subserved by feed-forward projections within a pathway*".

A. Dobbins, S.W. Zucker and M.S. Cynader presented evidence that the curvature detection is related to *end-stopping neurons*, they also presented a supporting mathematical model [Dobbins, 1987]. Our work is very similar to theirs but our model is faster since it calculates the curvature in terms of simplest parallel calculations.

In the field of human cognition systematic work has been done in systems of human psychology and learning [Drigas, 2005]. Further research has revealed that the extraction of landmark points is a critical process in human perception and the basis for potential mechanisms of shape identification and recognition [Biederman, 1987], [Kayaert, Biederman 2003].

In biology biometrics Bookstein was among the first to define landmark points on various biological shape for species classification [Bookstein 1996].

At the same time many researchers in various fields of computer science have been studying shape representation techniques and many have appreciated the use of landmark points as the most compatible to the human cognition method of representing and encoding shape information. Berreti introduces a decomposition of the shape into primitives based on the curvature [Berreti, 2000]. Attneave et al and Pomerantz et al notice that the curvature of a curve has salient perceptual characteristics [Attneave, 1954], [Pomerantz, 1977] and has proven to be useful for shape recognition [Pavlidis, 1980]–[Wang, 1999]. Asada and Brandy have developed the "curvature primal sketch" descriptor [Asada, 1986], a multiscale structure based on the extraction of changes in curvature. From curvature features, a description of the contour in terms of structural primitives (e.g., ends, cranks, etc.) is constructed. Mokhtarian and Mackworth [Mokhtarian, 1986], [Mokharian, 1992] showed that curvature inflection points extracted using a Gaussian scale space can be used to recognize curved objects. Dudek and Tsotsos [Dudek, 1997] presented a technique for shape representation and recognition of objects based on multiscale curvature information. Another technique based on the landmark points of high curvature, is also introduced by [Super, 2004].

In this paper we realize an inherit advantage of the similar to [Shams,1997] and [Fukusmima, 1982] neuron based architectures in implementing the allocation of landmarks on shapes by proposing a hierarchical model that incorporates visual acquisition and landmark allocation in distinct layers emulating the visual pathway of primates.

3 Detection of Landmark Points

David H. Hubel mentions in his Nobel lecture:

"Orientation-specific simple or complex cells "detect" or are specific for the direction of a short line segment. The cells are thus best not thought of as "line detectors": they arc no more line detectors than they are curve detectors. If our perception of a certain line or curve depends on simple or complex cells it presumably depends on a whole set of them, and how the information from such sets of cells is assembled at subsequent stages in the path, to build up what we call "percepts" of lines or curves (if indeed anything like that happens at all), is still a complete mystery." Hubel, Nobel Lecture, December 1981.

Simple cells have oriented receptive fields, and hence they respond to stimuli in some orientations better than others. This receptive field property is called orientation selectivity. The orientation of the stimulus that evokes the most powerful response is called the cell's preferred orientation. Orientation selectivity of cortical neurons is a critical receptive-field property. LGN and retinal neurons have circularly symmetric receptive fields, and they respond almost equally well to all stimulus orientations. Orientation-selective neurons are found throughout layers 2 and 3, though they are relatively rare in the primary inputs within layer 4C.

In the rest of this paper we will present our model for shape encoding from landmark points. We will show that continuous successive orientations by 10 degrees of an orientation selective filter, like the ones discovered by Hubel and Wiesel in the visual cortex of primates, can be a mechanism of landmark point detection. We will describe the mechanism and the neuron connectivity model that under the above assumptions encodes the shape of a 2d contour on the basis of connected landmark points of high curvature.

3.1 The Proposed Landmark Points

Let X be a planar curve parameterized on the scalar t, the parametric representation of X is then $\vec{c}(t) = (x(t), y(t))$, where x(t) and y(t) the coordinate functions. We need a way to identify landmark points of high curvature on the curve X and be consistent to the functionality of the cortical cells. We know from Hubel and Wiesel that the cortical cells in layers 2 and 3 have orientation selective receptive fields and that this orientation changes direction continuously in successive layers. We show now that with successive rotations of an orientation selective filter we can indeed measure the curvature at every point on the curve. The idea is to use the orientation selectivity to locate the direction which is tangent to the curve at a specific point and at the same time measure the curvature at this point by accumulating the firings of the successive layers in which the rotated field keeps being close to the direction of the tangent. This way we use the rotation operation that we know happens in the visual cortex cells in successive layers and the orientation selectivity to perform curvature detection in a way that is compatible to our best knowledge regarding the functionality of the visual cortex cells of layers 2 and 3. Let $\vec{c}(t) = (x(t), y(t))$ a point on the curve.

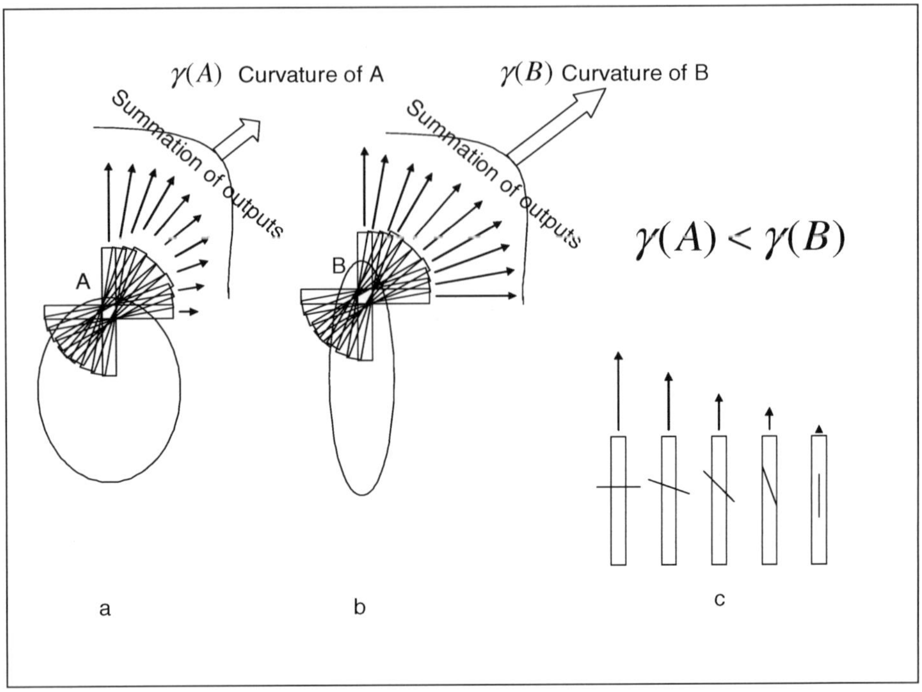

Fig. 1. Curvature detection by rotating an orientation selective receptive field. The curvature is proportional to the sum of the outputs of the rotated receptive fields. In (c) we see the orientation selective receptive field as a rectangular area and the respective outputs as arrows for several orientations from 0 to 90 degrees. The length of the arrow is proportional to the output strength for the given orientation. In (a) and (b) we illustrate the calculation of the curvature for two points A and B. In both cases the orientation selective receptive field is rotated form 0 to 90 degrees with step 10 degrees and the output for each step of rotation is proportional to the degree of the approximation of the curve on the direction permitted by the receptive field for this rotation. The outputs in case (b) will be stronger than the outputs in case (a) for most of the rotations since the change of the orientation of the receptive field approximates better the change of the direction of the tangent of the curve at the neighborhood of B.

If we rotate the axis this point will become $\vec{c}_R(t) = (x_R(t), y_R(t))$ where $x_R(t) = y(t)\cos(\theta) - x(t)\sin(\theta)$, $y_R(t) = y(t)\sin(\theta) + x(t)\cos(\theta)$ and θ the angle of rotation.

Let us look now at the second coordinate function $y_R(t)$. Its first derivative is $\dot{y}_R(t) = \dot{y}(t)\cos(\theta) - \dot{x}(t)\sin(\theta)$ and when it becomes zero we get

$$\dot{y}_R(t) = 0 \Leftrightarrow \dot{y}(t)\cos(\theta) - \dot{x}(t)\sin(\theta) = 0 \Leftrightarrow \dot{y}(t)\cos(\theta) = \dot{x}(t)\sin(\theta) \Leftrightarrow \tan(\theta) = \frac{\dot{y}(t)}{\dot{x}(t)} \quad (1)$$

This result is consistent to our intuition that the derivative of the second coordinate function at some point t becomes zero when the axis are rotated at an angle equal to the angle of the tangent to the curve at the point $\vec{c}(t)$. An orientation selective receptive field will emit the strongest output at this angle since the direction of the tangent best approximates the curve at this point. Now recall that the curvature at this point is defined as $\frac{d\theta}{ds}$, the rate of change of the angle of the tangent to the curve at the specific point per unit arc length. A rotation of the orientation selective receptive field corresponds to a step in tangent direction and arc length. If therefore we rotate the orientation selective receptive field and the respective neurons keep firing it means that we are still approximating the curve by being on the direction of the tangent even if we have moved by a unit of arc length. The more we keep approximating the curve by rotating the tangent direction and moving on the arc by a unit of length, the more is the curvature at the specific point and we can measure this curvature by accumulating the firings of all the layers, each layer corresponding to a rotation of the receptive field by 10 degrees. Just to illustrate the simplicity of the just described method we recall that in analytical terms the curvature is calculated as: $\frac{d\theta}{ds} = \frac{\frac{d\theta}{dt}}{\frac{ds}{dt}}$ where $\frac{d\theta}{dt}$ can be calculated by (1) where:

$$\tan(\theta(t)) = \frac{\dot{y}(t)}{\dot{x}(t)} \Leftrightarrow \theta(t) = \arctan\frac{\dot{y}(t)}{\dot{x}(t)} \Leftrightarrow \dot{\theta}(t) = \frac{\dot{x}(t)\ddot{y}(t) - \dot{y}(t)\ddot{x}(t)}{\left\|\dot{\vec{c}}(t)\right\|^2}$$

and $\frac{ds}{dt} = \left\|\dot{\vec{c}}(t)\right\|$ thus the curvature of the curve at the point $\vec{c}(t) = (x(t), y(t))$ is given by:

$$\gamma(t) = \frac{\dot{x}(t)\ddot{y}(t) - \ddot{x}(t)\dot{y}(t)}{\left\|\dot{\vec{c}}(t)\right\|^3} \quad (2)$$

We see that the analytical calculation of the curvature at each point on the curve involves the first and second derivatives of the coordinate functions but we manage to describe the calculation of the curvature of a planar curve through a series of operations that are consistent to our knowledge regarding the functionality of the cortical cells. In fact we explain that rotating an orientation selective two dimensional receptive field is the nature's suggestion for measuring the curvature at each of the points of a planar curve. In Fig. 1 we can see an orientation selective receptive field and its successive rotations in two scenarios of different curvatures. In the case of high curvature the field approximates better the direction of the tangent to the curve for several successive rotations.

We call each one of these rotations, a *view* from that angle of rotation. We call *interesting point*, a point on the contour at which the first derivative of $y_R(t)$ becomes zero. We call *strength* of an *interesting* point the curvature at this point. We call *landmark point*, an *interesting* point on the contour that has strength more than a given threshold S.

A *Landmark region* is made out of *interesting* points (on the contour) that fail to qualify as landmark points according to the definition above thus a landmark region is a collection of neighboring points on the contour that have strength less then S. The term landmark for the region can be justified if we think that quantity can compensate for quality. We call *width* of the landmark region the number of points included in the region.

Landmark points and regions partition the contour in a morphological meaningful way while at the same time they are defined through native quantitative methods. Notice that according to our definitions above the property of a point in being a landmark for the contours shape is defined through local measures. A higher layer in which landmark points and regions are combined to provide a global descriptor that depends on the contours whole shape is described in the extended version of this paper.

3.2 The Proposed Architecture

We propose a layered architecture as a visual pathway model for detecting landmark points of high curvature. The idea that we are going to implement here is the successive rotations of the orientation selective receptive field as was explained in the previous section.

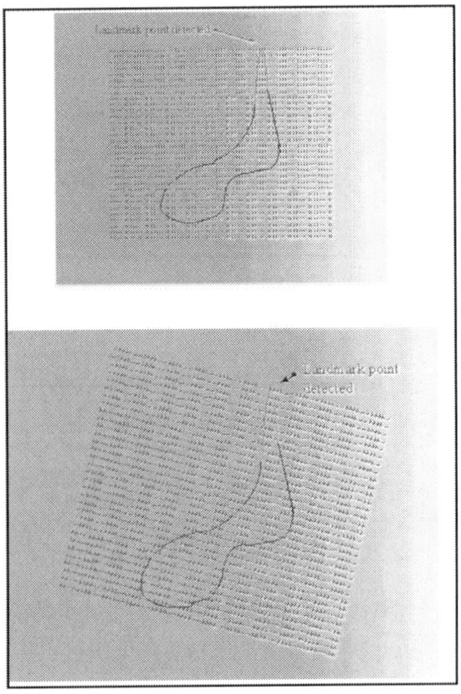

Fig. 2. The pictures above demonstrate landmark detection for the presented shape in two L1-sublayers. The same receptive field is rotated by 10° in the second image. The landmark point is detected in both the sub-layers, despite the rotation of the receptive field.

The lowest layer L0 is made out of perceptual units that we call L0-neurons and will play the role of the retinal photosensitive cells. These neurons receive the image intensity input and they become active on large intensity changes. We use these L0-neurons to detect the contour that outlines the shape of the presented object. The L0-neurons feed the inputs of the L1-neurons in layer 1.

The L1-layer is consisted of a pack of sub-layers. Every L0-neuron sends its output to the corresponding L1-neuron in each L1-sub-layer. We can imagine all the layers staggered and aligned and the connections are only across neurons that correspond to the same coordinates in their respective layers. An image containing a given random shape is captured through the sensitivity of the L0-neurons to the image's intensity differences. When a L0 neuron is adequately stimulated sends an active output to all the corresponding staggered neurons in layer L1, this way once an image is presented to layer L0 after a while all of the L1 sub-layers receive an active input from the L0 neurons that happened to be located on the boundary contour of the shape.

The L1-sublayers in our architecture act as directional oriented *interesting* point detectors. Recall that *landmark* points are *interesting* points above a threshold and they are defined through local shape characteristics of curvature. In this layer, we will detect the contour's landmark points of high curvature. Each L1-sublayer consists of neurons with the necessary receptive field to detect *interesting* points for a given orientation. The orientation bias of the receptive field of the neurons varies smoothly form L1-sublayer to the next and spans a complete angle of at least 90 degrees of rotation. Our L1-sublayers are the analogous to the hyper-columns of the human visual cortex [Hubel, 1981].

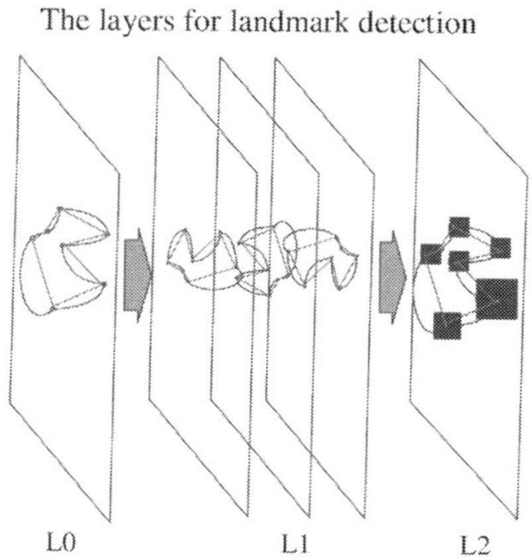

Fig. 3. The proposed layered architecture for extracting landmark points. The layers together with a graphical high level description are shown.

L1-neurons fire whenever an *interesting* point is present at their receptive field. The receptive field necessary to cause firing under *interesting* point conditions can be constructed by combining simple orientation specific receptive fields. The computational equivalent of this is a two dimensional filter with adjusted weights so as to penalize non horizontal straight segments. For the other directions the same receptive field rotated to the appropriate angle can be assumed. In figure 2 you can see a landmark-receptive field for two different L1-sublayers detecting *interesting* points on a given shape.

The same *interesting* point will cause firing in some L1 sub-layers but not in others, depending on the rotation of the receptive field of the layer's L1-neurons, but the bigger the strength of the *interesting* point is, the more are the L1-sublayers it will be visible from, where visibility of point A form layer B is defined as the fact of at least one neuron firing at sub-layer B because of the *interesting* edge point A.

The L2-cells receive the firings that were caused from contour's *interesting* points at layer L1. A strong *interesting* point will be *visible* after many consecutive rotation steps and will therefore cause a series of L1-firings in subsequent L1-sublayers that will drive the same L2-neuron. The magnitude, therefore of the input of an L2-neuron resembles the strength of the respective *interesting* point. Since every point is an *interesting* point from one view, the L2-layer will receive a *complete copy* of the contour but with different input magnitude for each point depending on the point's landmark potential. Morphological properties of the closed contour have been therefore transformed to equivalent differences in the stimulation of the L2-neurons, without loosing the spatial correspondence of the respective L0-neurons.

The functionality of the L2-neurons just described is consistent to the curvature detection observed in higher layer cortical neurons since it is clear that the magnitude of the input signal of the L2-neurons indicate *interesting* points of the analogous strength while the location of the landmark neurons describe the relative location of the landmark points on the contour.

Table 1. A landmark filter. It detects horizontal high curvatures on the boundary. The amount of locality that we accept for the formation depends on the size of the filter.

0	0	0	0	0
−4	−4	−4	−4	−4
−1	1	1	1	−1
−1	1	1	1	−1
1	1	−4	1	1

A landmark on the contour will cause firings in several consecutive L1-sublayers at the same location and therefore will accumulate a stronger stimulus to the respective L2-neuron. On the other hand a landmark region made of weaker points will cause many neighboring L2-neurons to receive weaker input. The wider the landmark region is the more neighboring L2-neurons will be stimulated because to its increased visibility at different neighboring locations across several rotations.

The task for the L2-layer is to collect the firings caused by the landmark points and regions on the contour and encode the shape from the strength of its landmark features. In figure 3 the layered architecture for curvature detection is illustrated graphically.

4 Experimental Results

We have described a biological inspired neural architecture that performs landmark point detection on the closed contour of an arbitrary shape. Here we demonstrate experimental results by using the above described methodology to detect points of high curvature.

The L0-layer that performs capturing and edge detection is trivial to implement with a capturing device and a direction-less edge detector, applied on the captured raster image. The resulting edge image plays the role of the signals sent from the L0-layer to all the L1-sublayers. The directional oriented, *interesting* and landmark point detection for each L1-sublayer corresponds to a filtering with the appropriately rotated landmark filter. A receptive field for the L2-neurons suitable to detect *interesting* points on the shape's boundary can be implemented by an actual digital filter that we call *landmark filter*. Instead of rotating the landmark filter, we equivalently use the same filter but rotate the edge image for each step. The result of the landmark detection for each step of rotation above corresponds to the task of a single L1-sublayer. These results, if superimposed, correspond to the Layer-L2 input coming from the L1-layer. In Figure 4 we use the landmark filter defined by the stencil shown in Table 1 to demonstrate the detection of a landmark point and its morphological strength, on a simple ellipsoidal shape.

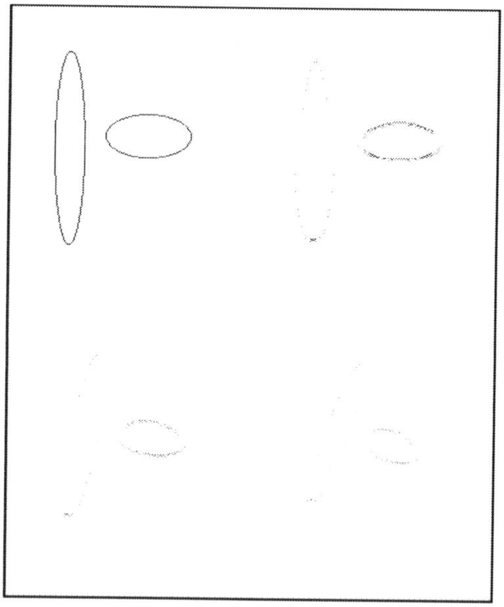

Fig. 4. In this picture we demonstrate the persistence of a landmark point to be detected by the landmark filter despite successive rotations of the image. In the upper left corner we have the original image of two ellipses that we want to detect the landmark points of their shape. The rest of the pictures correspond to the original image rotated by 0°,10°,20° respectively (left to right first) and filtered by the landmark filter that is defined in Table 1. We see that for this range of rotation the same strong landmark points of the left ellipsis are detected by the same filter in all the rotated images. On the other hand the detected points on the right ellipsis depend strongly on the rotation of the image a fact that indicates their low landmark potential.

By super-positioning the interesting point images we accumulate the strength of these points in the intensities of a single image. Points with high intensity values correspond to morphological significant (landmark) points on the contour where clusters of points of low intensity correspond to regions that connect the landmark points. A ranking therefore of the contour points according to their intensity at this stage, corresponds to their morphological significance on the contour.

A significant contribution in our approach of measuring and detecting curvature is that it can be performed with rotated filters on a single image simultaneously and in parallel, therefore the curvature detection can be performed in just the time it takes to filter a single curve image, assuming that the superposition of the filtering results happens instantly.

Further treatment regarding shape recognition and classification is not the intention of this paper but it is apparent that our representation not only abides to the physical and psychological discoveries regarding the perceptional tasks in primates but also presents a feature extraction technique that combined with standard computational techniques leads to state of the art performance and recognition capability.

5 Conclusion

We have offered an explanation regarding the rotated orientation selective receptive fields of the cortical neurons in layers 2 and 3 of the visual cortex of primates that were first discovered by Hubel and Wiesel. We have shown that by means of this specific arrangement of neurons it is possible to perform detection of landmark points of high curvature. We appreciated the completeness and performance superiority of our approach compared to other curvature detection methods since our model provides the fastest possible mechanism for curvature measurement of planar curves.

References

[Asada, 1986] H. Asada and M. Brady, "The curvature primal sketch," IEEE Trans. Pattern Anal. Mach. Intell., vol. PAMI-8, no. 1, pp. 2–14, Jan. 1986.

[Attneave, 1954] F. Attneave, "Some informational aspects of visual perception," Psychol. Rev., vol. 61, pp. 183–193, 1954.

[Berreti, 2000] Stefano Berretti, Alberto Del Bimbo and Pietro Pala, "Retrieval by Shape Similarity with Perceptual Distance and Effective Indexing", IEEE Trans on Multimedia, vol. 2, no. 4, pp. 225-239, Dec. 2000.

[Biederman, 1987] Biederman, I. (1987). Recognition-by-Components: A Theory of Human Image Understanding. Psychological Review, 94, 115-147.

[Bookstein, 1996] F.L. Bookstein. Landmark methods for forms without landmarks: morphometrics of group differences in outline shape. Med. Im. Anal., 1(3):225–243, 1996.

[Brenner, 1996] Brenner, E. & Smeets, J.B.J. (1996). Size illusion influences how we lift but not how we grasp an object. Experimental Brain Research, 111, 473-476.

[Dudek, 1997] G. Dudek and J. K. Tsotsos, "Shape representation and recognition from multiscale curvature," Comput. Vis. Image Understand., vol. 68, pp. 170–189, 1997.

[Drigas, 2005] S. Drigas, G. Koukianakis and V. Papagerasimou, "A System For Hybrid Learning And Hybrid Psychology", 2nd International Conference on Cybernetics and Information Technologies, Systems and Applications: CITSA 2005, Orlando, Florida..

[Fukushima, 1982] K. Fukushima, S. Miyake: "Neocognitron: A self-organizing neural network model for a mechanism of visual pattern recognition", Competition and Cooperation in Neural Nets, Lecture Notes in Biomathematics 45, eds.: S. Amari, M. A. Arbib, pp. 267-285, Berlin, Heidelberg, New York: Springer-Verlag (1982).

[Goodale, 1994] Goodale, M.A., Meenan, J.P., Buelthoff, H.H., Nicolle, D.A., Murphy, K.J., & Racicot, C.I. (1994b). Separate neural pathways for the visual analysis of object shape in perception and prehension. Current Biol., 4, 604-610.

[Goodale 1991] Goodale, M.A., Milner, A.D., Jakobson, L.S., & Carey, D.P. (1991). A neurological dissociation between perceiving objects and grasping them. Nature, 349, 154-156.

[Hubel, 1981] David H. Hubel, Evolution of ideas on the primary visual cortex, 1955-1978: A biased historical account, Nobel lecture, 8 December 1981, Harvard Medical School, Department of Neurobiology, Boston, Massachusetts, U.S.A., Nature, 299:515-524.

[Hubel D.H, Wiesel T., 1990] Brain mechanisms of vision, In I. Rock (ed.) The Perceptual world , pp. 3-24. W. H. Freeman NY.

[Hubel D.H, Wiesel T., 1968] Receptive fields and functional architecture of monkey striate cortex, J. Physiol., 195:215-243.

[Hubel D.H, Wiesel T.N., Stryker M. 1978] Anatomical demonstration of orientation columns in macaque monkey, J. Comp. Neurol., 177:361-380.

[Jalba, 2006] Andrei C. Jalba, Michael H. F. Wilkinson, Jos B. T. M. Roerdink, "Shape Representation and Recognition Through Morphological Curvature Scale Spaces" IEEE Trans. Image Proc., vol. 15, no. 2,pp. 331-341 ,Feb. 2006.

[Kayaert, 2003] Greet Kayaert, Irving Biederman and Rufin Vogels, "Shape Tuning in Macaque Inferior Temporal Cortex", the Journal of Neuroscience, April 1, 2003 • 23(7), pp. 3016 –3027.

[Milner, 1993] Milner, A.D. & Goodale, M.A. (1993). Visual pathways to perception and action. In T.P. Hicks, S. Molotchnikoff & T. Ono (Eds.) Progress in Brain Research, Vol. 95 (pp. 317-337). Amsterdam: Elsevier.

[Mokhtarian, 1986] F. Mokhtarian and A. K. Mackworth, "Scale-based description and recognition of planar curves and two-dimensional shapes," IEEE Trans. Pattern Anal. Mach. Intell., vol. PAMI-8, no. 1, pp. 34–43, Jan. 1986.

[Mokharian, 1992] , "A theory of multiscale, curvature-based shape representation for planar curves," IEEE Trans. Pattern Anal. Mach. Intell., vol. 14, no. 6, pp. 789–805, Jun. 1992.

[Pomerantz, 1977] J. R. Pomerantz, L. C. Sager, and R. J. Stoever, "Perception of wholes and their component parts: Some configural superiority effects," J. Exp. Psychol., vol. 3, pp. 422–435, 1977.

[Pavlidis, 1980] T. Pavlidis, "Algorithms for shape analysis of contours and waveforms," IEEE Trans. Pattern Anal. Mach. Intell., vol. PAMI-2, no. 3, pp. 301–312, Mar. 1980.

[Shams, 1997] Soheil Shams, "Affine Invariant Object Recognition through Self Organization", Hughes Research Laboratories, 1997.

[Super, 2004] B.J Super. Fast correspondence-based system for shape-retrieval. Patt. Recog. Lett., 25:217–225, 2004.

[Ungerleider, 1998] L. G. Ungerleider, S. M. Courtney and J. V. Haxby, "A neural system for human visual working memory", Proc. Natl. Acad. Sci. USA, vol. 95, pp. 883–890, Feb. 1998.

[Wang, 1999] S. L. Y.-P. Wang and K. T. Lee, "Multiscale curvature based shape representtation using b-spline wavelets," IEEE Trans. Image Process., vol. 8, no. 10, pp. 1586–1592, Oct. 1999.

A Model of Grid Cells Based on a Path Integration Mechanism

Alexis Guanella[1,⋆] and Paul F.M.J. Verschure[1,2]

[1] Institute of Neuroinformatics, University and ETH Zürich, CH-8057 Zürich, Switzerland
guanella@ini.phys.ethz.ch
[2] ICREA and Technology Department, University Pompeu Fabra E-08002 Barcelona, Spain

Abstract. The grid cells of the dorsocaudal medial entorhinal cortex (dMEC) in rats show higher firing rates when the position of the animal correlates with the vertices of regular triangular tessellations covering the environment. Strong evidence indicates that these neurons are part of a path integration system. This raises the question, how such a system could be implemented in the brain. Here, we present a cyclically connected artificial neural network based on a path integration mechanism, implementing grid cells on a simulated mobile agent. Our results show that the synaptic connectivity of the network, which can be represented by a twisted torus, allows the generation of regular triangular grids across the environment. These tessellations share same spacing and orientation, as neighboring grid cells in the dMEC. A simple gain and bias mechanism allows to control the spacing and the orientation of the grids, which suggests that these different characteristics can be generated by a unique algorithm in the brain.

Keywords: grid cells, entorhinal cortex, path integration, twisted torus.

1 Introduction

Found in the dorsocaudal medial entorhinal cortex (dMEC) of rats, grid cells [1,2] show increased firing frequency when the animal visits regularly distributed regions in an environment. It has been shown, using auto-correlative maps, that these regions (so-called subfields) form regular triangular tessellations, or grids [1]. It is possible to describe these tessellations, and, thus, the characteristics of a grid cell, with only a few parameters: the orientation and the phase of the grid, and the spacing (minimal inter-subfields distance, i.e. d in this study) and the size of its subfields. Using these parameters, it was shown that grid cells are topographically organized in the dMEC: first, neighboring cells share common orientation and spacing. Second, the spacing of the grid increases isometrically along the dorsoventral axis (in [1], d varies between 39 to 73 cm). Third, the similitude of neighboring grid cells of different layers of the entorhinal cortex, sharing common orientations and spacing, suggests that they are organized in cortical columns [3].

⋆ Corresponding author.

S. Kollias et al. (Eds.): ICANN 2006, Part I, LNCS 4131, pp. 740–749, 2006.

In spite of these observations, the role of grid cells is still poorly understood. Briefly, it has been proposed that grid cells may be part of a generalized path integration system [1,4], and could be the basis of a metric of spatial relationships [5]. Many arguments verify this hypothesis. First, the grid cell activity and its regular patterns persist after the deprivation of external landmarks (e.g. in the dark [1]). Second, entorhinal lesions disrupt the return path of rats [6]. Third and fourth, suggesting also hard wired mechanisms, the grid structure is expressed instantaneously in novel environments and the spacing parameter seems to be universal (the grid spacing remains constant when increasing the size of the arena) [1]. Fifth, the periodicity of the grid implies a covering of arbitrary big environments. These arguments raise thus the question, how grid cells could be incorporated into a path integration system.

The goal of this article is to describe an artificial neural network implementing grid cells based on a path integration mechanism. In our model, the activity of rate coded neurons is shifted by asymmetric synaptic connections. These connections are modulated by the velocity of the animal, represented by a simulated mobile agent exploring randomly a square arena. The neurons of the network represent a population of neighboring grid cells of the dMEC, whose grids share thus same orientation and spacing, but have different phases. A simple gain and bias mechanism allows the control of the spacing and the orientation of the grid (suggesting that exactly the same algorithm may be used to generate grid cells along the dorsoventral locations of the dMEC). The synaptic connectivity of the network is organized cyclically, and can be represented by a twisted torus. This topology is shown to exactly generate the same regular triangular tessellations of space as grid cells. Stability and robust activity is ensured by attractor dynamics and normalization mechanisms.

2 Methods

2.1 Neurons

We construct a population of N neurons organized in a matrix covering the repetitive rectangular structure of the subfields of grid cells (Fig. 1a). In order to conserve the ratio between the height and the side of an equilateral triangle (which is the core element of a regular triangular tessellation) and in order to have the same density of cells along both x- and y-axes, the number of cells in each row is approximately $2/\sqrt{3}$ times bigger than the number of cells in each column (Fig. 1b).

Activity and Stabilization. The neurons of the network are initialized with a random activity uniformly distributed between 0 and $1/\sqrt{N}$. The activity of a cell i at time $t+1$, i.e. $A_i(t+1)$ is defined using a linear transfer function $B_i(t+1)$ given by

$$B_i(t+1) = A_i(t) + \sum_{j=1}^{N} A_j(t)\, w_{ji} \ , \tag{1}$$

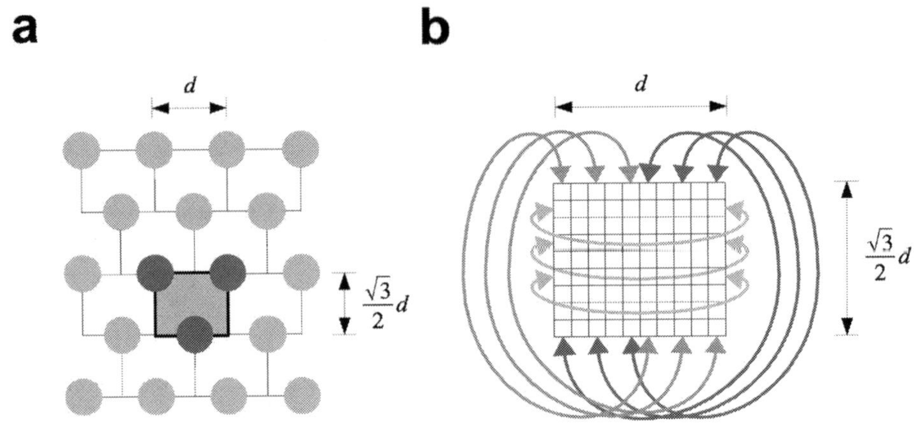

Fig. 1. (a) Repetitive rectangular structure (gray filled rectangle) of the subfields (gray circles) of grid cells defining a regular triangular tessellation of space. **(b)** Matrix of a population of 10×9 grid cells. Neighboring relationships between cells on the side of the structure are represented by gray arrows. For instance, neurons at two opposite vertical sides are neighbors.

where N is the number of cells in the network, w_{ji} is the synaptic weight connecting cell j to cell i, with $i, j \in \{1, 2, ..., N\}$. A floating average normalization mechanism over the cells activity ensures the stability of the network. Thus, $A_i(t+1)$ is defined by

$$A_i(t+1) = B_i(t+1) + \tau \left(\frac{B_i(t+1)}{< B_j(t) >_{j=1}^{N}} - B_i(t+1) \right) , \tag{2}$$

where the function $< \, . \, >_{j=1}^{N}$ is the mean over the cells of the network, and the parameter τ determines the stabilization strength. To implement this mechanism locally, we use an additional cell (cell $N+1$, which is not a grid cell), that computes the sum of all activities of the grid cells. Normalized by N, the activity of this external cell $< B_j(t) >_{j=1}^{N}$ is transferred back to the cells of the network. In order to prevent negative cell activities, we set $A_i(t+1) = 0$ when $A_i(t+1)$ is smaller than zero.

2.2 Synapses

The synapses of the network can be divided into two distinct populations. The first population is formed by the synapses which are used to compute the mean activity of the network and to stabilize the cells activity. They connect in both directions all the cells of the network and the external cell $N+1$ (Fig. 2a). Their synaptic weights are all constant and set to 1.

The second population is formed by the synapses implementing the attractor dynamics of the network. These synapses connect each cell i to each cell j, with $i, j \in \{1, 2, ..., N\}$. As we will see, their synaptic weights are computed as a Gaussian function of the distance between cells (exciting neighboring and inhibiting distal cells (Fig. 2b and 2c)). They are furthermore modulated by the input of the network (i.e. the speed of the mobile agent), which allows to shift the activity packet of the grid cells when the mobile agent is moving.

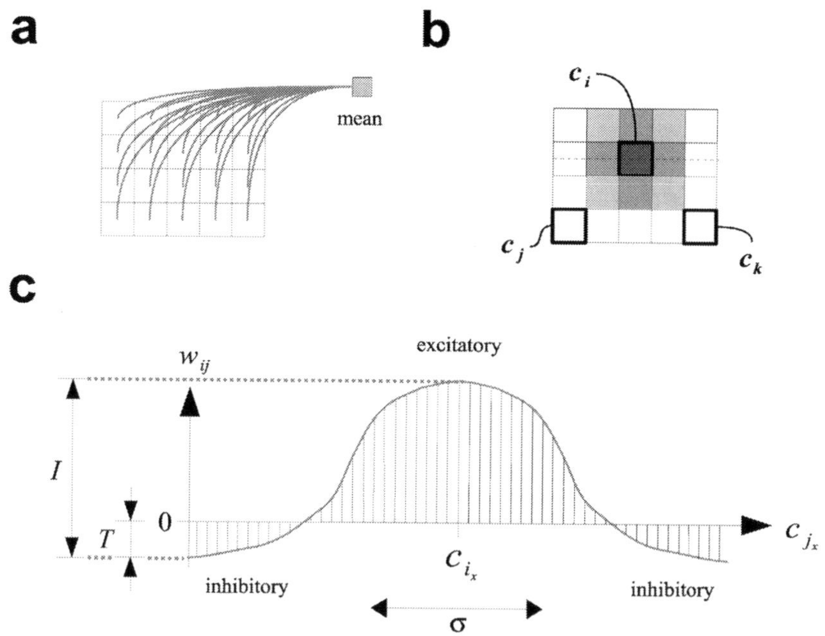

Fig. 2. (a) First population of synaptic weights connecting in both directions all the cells of the network with an external cell used to compute the mean activity. **(b)** Example of synaptic weights of the second population, connecting the neuron $\boldsymbol{c_i}$ to all the cells of the network (including itself). The dark gray color represents a high synaptic weight (e.g. the connection from $\boldsymbol{c_i}$ to itself) and a light gray color a low synaptic weight. The topology of the network implies that $\boldsymbol{c_j}$ and $\boldsymbol{c_k}$ are neighbors. **(c)** Synaptic weights of the cell $\boldsymbol{c_i}$ along an horizontal axis (represented by the dashed line in (b)). Their intensity, shift and width are parametrized respectively by I, T and σ, defining excitatory and inhibitory connections.

Attractor Dynamics. The synaptic patterns connecting grid cells in the network are defined by a Gaussian weight function. We have

$$w_{ij} = I \exp\left(-\frac{\| \boldsymbol{c_i} - \boldsymbol{c_j} \|_{tri}^2}{\sigma^2}\right) - T \ , \tag{3}$$

where $\boldsymbol{c_i} = (c_{i_x}, c_{i_y})$ is the position of the cell i, defined respectively by $c_{i_x} = (i_x - 0.5)/N_x$ and by $c_{i_y} = \frac{\sqrt{3}}{2}(i_y - 0.5)/N_y$ (with $i_x \in \{1, 2, ..., N_x\}$ and $i_y \in \{1, 2, ..., N_y\}$), and where N_x and N_y are the number of columns and rows in the cells matrix (Fig. 1) and i_x and i_y the column and the row numbers of cell i. I is the intensity parameter, defining the overall strength of the synapses,

σ regulates the size of the Gaussian and T is the shift parameter determining excitatory and inhibitory zones (Fig. 2c). The norm $||\,.\,||_{tri}$ defines the induced metric $\text{dist}_{tri}(.\,,.)$ of the network. To obtain the repetitive rectangular structure of the grid subfields, the cells at the border of the layer have to be neighbors of the cells at the opposite border (as an example, the two grid cells j and k on Fig. 2a should be neighbors). This can be seen as a torus topology. However, this is not sufficient to form a triangular grid, since a simple torus would lead, in our model, to a rectangular tessellation of space. The regular triangular tessellation is generated by twisting the torus. This is represented in the definition of the distance $\text{dist}_{tri}(.\,,.)$ or the norm $||\,.\,||_{tri}$ which permits to obtain regular triangular tessellations:

$$\text{dist}_{tri}(\boldsymbol{c_i},\boldsymbol{c_j}) := ||\,\boldsymbol{c_1}-\boldsymbol{c_2}\,||_{tri} = \min_{j=1}^{7}\,||\,\boldsymbol{c_1}-\boldsymbol{c_2}+\boldsymbol{s_j}\,||\ , \tag{4}$$

where

$$\boldsymbol{s_1} := (0,0)\ , \tag{5}$$
$$\boldsymbol{s_2} := (-.5,\frac{\sqrt{3}}{2})\ , \tag{6}$$
$$\boldsymbol{s_3} := (-.5,-\frac{\sqrt{3}}{2})\ , \tag{7}$$
$$\boldsymbol{s_4} := (.5,\frac{\sqrt{3}}{2})\ , \tag{8}$$
$$\boldsymbol{s_5} := (.5,-\frac{\sqrt{3}}{2})\ , \tag{9}$$
$$\boldsymbol{s_6} := (-1,0)\ , \tag{10}$$
$$\boldsymbol{s_7} := (1,0)\ , \tag{11}$$

and where $||\,.\,||$ is the Euclidean norm.

Modulation. The input of the network is the speed vector $\boldsymbol{v} := (v_x, v_y)$, which represents the speed of the mobile agent. This input doesn't depend on any absolute information about location. The maximum velocity of the mobile agent is given by the parameter v_{max} such as $||\,v\,||$ is always smaller than v_{max}.

It is possible to increase or decrease the size and the spacing of the subfields, as well as changing the orientation of the grid by changing only two parameters in the model, i.e. the gain $\alpha \in \mathbb{R}^+$ and bias $\beta \in [0,\pi/3]$. The input of the network is thus modulated and biased by the gain and the bias parameters, with

$$\boldsymbol{v} \longmapsto \alpha\,\boldsymbol{R_\beta v}\ , \tag{12}$$

where $\boldsymbol{R_\beta}$ is the rotation matrix of angle β defined by

$$\boldsymbol{R_\beta} = \begin{pmatrix} \cos(\beta) & -\sin(\beta) \\ \sin(\beta) & \cos(\beta) \end{pmatrix}\ . \tag{13}$$

The activity pattern is stable when the mobile agent stays immobile. However, when the agent moves, the synaptic connections of the network shift in the direction of the speed vector of the robot (Fig. 3). When expressing the synaptic weight as a function of time, we have

$$w_{ij}(t+1) = I \exp\left(-\frac{\|\,\boldsymbol{c_i} - \boldsymbol{c_j} + \alpha\, \boldsymbol{R_\beta}\, \boldsymbol{v}(t)\,\|_{tri}^2}{\sigma^2}\right) - T \ . \tag{14}$$

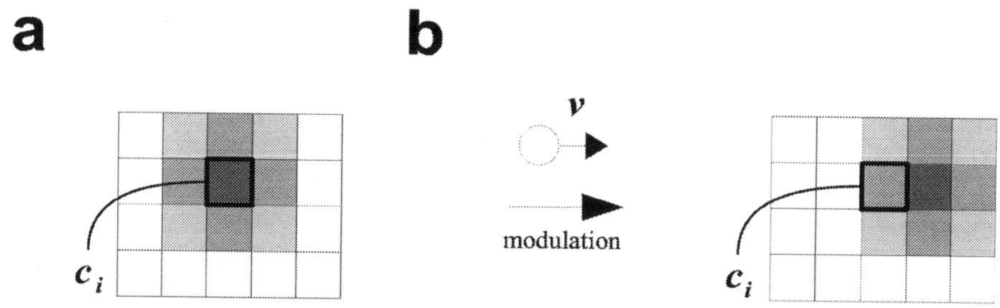

Fig. 3. Modulation of the synaptic connections of cell i. **(a)** Before modulation, the synaptic pattern of the cell i is centered around $\boldsymbol{c_i}$. **(b)** After modulation, the synaptic pattern is shifted proportionally in the direction of the speed $\boldsymbol{v}$ of the mobile agent.

2.3 Mobile Agent and Environment

The experiments are performed using a simulated Khepera robot (K-Team, Yverdon, Switzerland), which randomly explores a square arena. When approaching a wall, a Braitenberg control algorithm [7] is activated to avoid collisions. The size of the square arena is 1×1 meter, such as one arena side is approximately 18 times bigger than a robot diameter (.055 meter).

2.4 Parameters

The values of the parameters used in this study are given in table 1. These values have to satisfy two criteria. First, they have to ensure the stability of the cells of the network. This means for instance that the cells activity should not be growing endlessly. Second, they must induce the attractor dynamics of the network, and

Table 1. Values of the parameters used in this study

Parameter		Value	Unit
N	=	90	[cell]
N_x	=	10	[cell]
N_y	=	9	[cell]
τ	=	0.8	[no unit]
I	=	0.3	[no unit]
σ	=	0.24	[meter]
T	=	0.05	[no unit]
v_{max}	=	0.0275	[meter/time step]

therefore a single and stable activity packet should be continuously observed in the cell population. These criteria have to be tested over a large number of time steps. Since no objective or cost function is given in this study, no parameter search for optimization is computed.

3 Results

To analyze the activity of the neurons of the network, we first computed their mean activity maps, i.e. the mean activity of a cell as a function of the position of the mobile agent (Fig. 4). These maps show the coherent and stable activity of the multiple grid subfields. To determine if these subfields were organized in a regular triangular tessellation, we fitted the mean activity maps to regular triangular tessellations composed of Gaussian subfields. We computed the mean square residuals over all the network cells, and found the value of 0.0028 +/- 0.0004 (mean +/- standard deviation (std.)). The mean square residuals for each cell were always smaller than 0.005. These results strongly suggest that our model generates regular triangular tessellating subfields. Note that in order to get coherent results for this computation, we normalized the mean activity maps such as the maximum and the minimum intensity of these maps were respectively 1 and 0. The stability of the network was assessed by running experiments over an extended number of time steps (repetitively stable when tested over 1 million time steps).

Gain and Bias. An interesting feature of the network is the possibility to vary the spacing and the orientation of the grids by just varying the gain and bias parameters. As shown in Figs. 4 and 5, higher gain values lead to denser grids (and therefore smaller spacing between subfields) whereas higher bias values rotate the grids. We made a regression analysis to model and determined the relationship between the gain and the bias parameters and respectively the spacing and the orientation of the grids. For the gain, we found $y = a + b\log_2(x)$ with $a = -0.90$ and $b = -0.39$, with mean least square residuals of 0.00016, and where x and y are respectively gain and grid spacing. For the bias, we found $y = a + bx$ with $a = 0.00$ and $b = 1.00$, with mean least square residuals of 0.00004, and where x and y are respectively bias and grid orientation.

4 Discussion

In this article, we have presented a model of grid cells based on path a integration mechanism, embedded on a simulated mobile agent. We have shown that the neural activity of the network generates regular triangular tessellations of the environment, as grid cells in the dMEC. In our model, the grids cells share same orientation and spacing, as neighboring grid cells in the dMEC, and, as suggested in [3], as grid cells in cortical columns of the dMEC. A simple gain and bias mechanism on the input allows to vary, respectively in a log-linear and a linear relationship, the spacing and the orientation of the grids. Our model

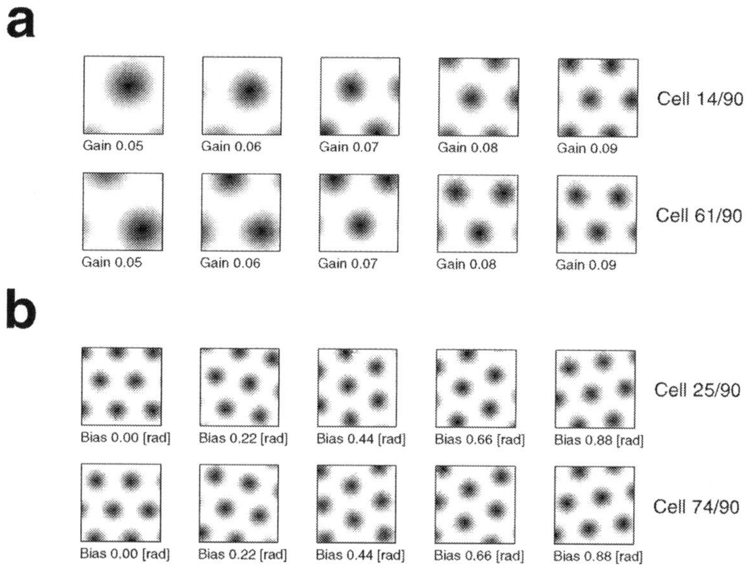

Fig. 4. **(a)** Mean activity maps of two grid cells with different gain values (here, the bias value is set to zero). Dark regions and light regions represent respectively high and low mean activity. These maps were computed over 50000 time steps. The discretization of the arena correspond to 40×40 bins. **(b)** Mean activity maps of two grid cells with different bias values (here, the gain value is set to 0.11).

gives thus a concrete example of a cortical circuit which can implement in a same algorithm grid cells with different spacings and orientations. In the dMEC, the spacing of the grid isometrically increases along the dorsoventral axis, which could thus suggest an exponential increase of the rat velocity gain along this axis.

Many studies present the implementation of path integration mechanisms based on attractor dynamics [8,9,10,11]. The idea to apply these methods for grid cells was first presented in [1] and described further in [4]. It has been implemented first in [12], as a symmetric locally connected neural network. Here, it is the first time that an implementation of such a system is explicitly described and implemented on a cyclically connected map. This synaptic architecture, which can be represented by a twisted torus, is new, and was shown in this study to be able to generate effectively grid cells with regular triangular tessellating subfields. The advantages of such a system is that it allows to implement in a relatively small population of cells a representation of space covering arbitrary big environments. Moreover, because of this particular synaptic connectivity, all the network cells have regular triangular tessellating subfields.

Our model of grid cells may be used as the proprioceptive element of a robust, modulatory and biologically based navigational system combining idiothetic (internal) and allocentric (external) sensory inputs. On the first hand, one of the classical problem of path integration mechanisms is the accumulation of

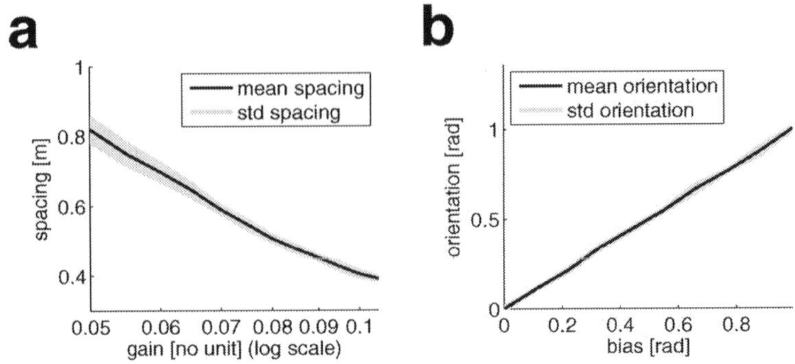

Fig. 5. **(a)** Spacing of the grid as a function of the gain parameter. **(b)** Orientation of the grid as a function of the bias parameter.

proprioceptive errors over time. For instance, in our model, the mean activity maps of grid cells would collapse if we would introduce noise in the speed input. On the other hand, one of classical problem of allocentric systems is their inability to disambiguate between two similar inputs. For instance, realistic models of place cells [13,14] of the hippocampus, based on visual inputs (e.g. [15]) are not able to distinguish between two visually similar places. A combination of these two approaches would be useful to deal with their respective weaknesses. Allocentric place cells (e.g. based on vision) would be used to recalibrate the activity of the grid cells in case of path integration errors, and, in turn, the activity of grid cells could be used to generate place cells (using simple a simple Hebbian mechanism as proposed in [4]), able to disambiguate between two visually similar places. The location of the dMEC, upstream the hippocampus, which is, in turn, an afferent of the entorhinal cortex, provides an anatomical basis for such a modulatory system.

Acknowledgments

This research was supported by the Swiss National Science Foundation (grant nr. 205321-100604). The authors thank Reto Wyss for providing the artificial neural networks simulation software *wSim* and Paul Rogister for useful comments.

References

1. Hafting, T., Fyhn, M., Molden, S., Moser, M.B., Moser, E.I.: Microstructure of a spatial map in the entorhinal cortex. Nature **436**(7052) (2005) 801–6
2. Fyhn, M., Molden, S., Witter, M.P., Moser, E.I., Moser, M.B.: Spatial representation in the entorhinal cortex. Science **305**(5688) (2004) 1258–64
3. Moser, M., Sargolini, F., Fyhn, M., Hafting, T., Witer, M., Moser, E.: Grid cells in medial entorhinal cortex: indications of columnar organization. Soc Neurosci Abstr **198**(4) (2005)

4. O'Keefe, J., Burgess, N.: Dual phase and rate coding in hippocampal place cells: theoretical significance and relationship to entorhinal grid cells. Hippocampus **15**(7) (2005) 853–66
5. Jeffery, K.J., Burgess, N.: A metric for the cognitive map: found at last? Trends Cogn Sci **10**(1) (2006) 1–3
6. Parron, C., Save, E.: Evidence for entorhinal and parietal cortices involvement in path integration in the rat. Exp Brain Res **159**(3) (2004) 349–59
7. Braitenberg, V.: Vehicles, experiments in synthetic psychology. MIT Press (1984)
8. McNaughton, B., Barnes, C., Gerrard, J., Gothard, K., Jung, M., Knierim, J., Kudrimoti, H., Qin, Y., Skaggs, W., Suster, M., Weaver, K.: Deciphering the hippocampal polyglot: the hippocampus as a path integration system. J Exp Biol **199** (1996) 173–85
9. Samsonovich, A., McNaughton, B.: Path integration and cognitive mapping in a continuous attractor neural network model. J Neurosci **17**(15) (1997) 5900–20
10. Stringer, S., Rolls, E., Trappenberg, T., de Araujo, I.: Self-organizing continuous attractor networks and path integration: two-dimensional models of place cells. Comput Neural Syst **13**(4) (2002) 429–46
11. Conklin, J., Eliasmith, C.: A controlled attractor network model of path integration in the rat. J Comput Neurosci **18**(2) (2005) 183–203
12. Fuhs, M.C., Touretzky, D.S.: A spin glass model of path integration in rat medial entorhinal cortex. J Neurosci **26**(16) (2006) 4266–76
13. O'Keefe, J., Dostrovsky, J.: The hippocampus as a spatial map. Preliminary evidence from unit activity in the freely-moving rat. Brain Res **34**(1) (1971) 171–5
14. O'Keefe, J., Nadel, L.: The hippocampus as a cognitive map. Clarendon Press, Oxford (1978)
15. Wyss, R., König, P., Verschure, P.F.M.J.: A model of the ventral visual system based on temporal stability and local memory. PLoS Biol **4**(5) (2006) e120

Temporal Processing in a Spiking Model of the Visual System

Christo Panchev

School of Computing and Technology
St. Peter's Campus, University of Sunderland
Sunderland SR6 0DD, United Kingdom
christo.panchev@sunderland.ac.uk

Abstract. Increasing amount of evidence suggests that the brain has the necessary mechanisms to and indeed does generate and process temporal information from the very early stages of sensory pathways. This paper presents a novel biologically motivated model of the visual system based on temporal encoding of the visual stimuli and temporally precise lateral geniculate nucleus (LGN) spikes. The work investigates whether such a network could be developed using an extended type of integrate-and-fire neurons (ADDS) and trained to recognise objects of different shapes using STDP learning. The experimental results contribute further support to the argument that temporal encoding can provide a mechanism for representing information in the visual system and has the potential to complement firing-rate-based architectures toward building more realistic and powerful models.

1 Introduction

Recent experiments have shown that in response to stimuli, populations of neurons in the primary visual areas could keep their mean firing rate constant while increasing the correlation of the single spikes [1]. This suggest that beginning from the primary areas, the brain could be using the timing of the individual spikes to represent external or internal stimuli, while preserving the mean firing rate of the neural populations [2, 3]. Theoretical analysis of spike trains has also shown that the individual spike timings are much more reliable than those in a random spike train with the same mean and variance [4], suggesting that the brain does have the necessary mechanisms to generate and operate with temporally precise encoding schemes.

The processing of visual information involves several stages with increasing complexity from the eye to the visual cortex along two main routes, the ventral and dorsal streams. Most experimental and modelling work suggest that the information in the visual pathways is carried in the firing rate of the neurons [5, 6]. However, there is an increasing amount of evidence that the temporal structure of the neural responses in the visual system is much more precise and reliable than previously thought [3, 7, 8, 9, 10, 1, 11], opening the possibilities for a new range of paradigms and models of the visual cortex.

S. Kollias et al. (Eds.): ICANN 2006, Part I, LNCS 4131, pp. 750–759, 2006.

Different aspects of the visual system have been studied in a number of computational models. The LISSOM model [12, 13, 14] implements the organisation and functionality of the primary visual cortex V1. Their work shows that the organisation and connectivity of V1 can be achieved through the combination of lateral competition and activity driven modification of afferent connections.

The model of visual processing presented by Thorpe et. al. [15, 16, 17] is based on the fast response times observed in the visual system, e.g. selective IT neuron responses to faces at about 80-100 ms after stimulus onset [18], and points that in order to achieve such fast response timings, the visual system has time for no more that a few spikes per neuron at each stage of a forward visual processing pathway [19]. The models are developed using large scale networks of rank-coding spiking neurons and one-shot learning [20].

The VisNet and VisNet2 models presented in [21] have been built as four hierarchically organised maps of competitive layers with firing rate neurons. These networks model the functionality, organisation and connectivity of the ventral stream of the visual cortex. Forward connections between the consecutive layers are assigned within receptive fields with radial Gaussian distribution of connection probability. The sizes of the receptive fields are such that neurons on the top layer can potentially be influenced by activity in the most distant parts of the input. The model achieves shape recognition via gradual bottom-up integration of features.

The study presented in this paper extends the contributions of those models by incorporating a network of spiking neurons, Spike-Timing-Dependent Plasticity (STPD) and processing of temporally encoded visual stimuli in a task of recognising objects of different shapes. The developed network is based on the integrate-and-fire spiking neuron with active dendrites and dynamic synapses (ADDS) and its STDP learning protocol [22, 23, 24, 25]. Following the successful simulation results, the visual model was also tested in a robot control system for local navigation in tasks like finding, approaching and manipulating target objects.

2 Motivation and Basis for Modelling Temporal Coding in the Visual System

Neural activity from the retinal cells reaches the lateral geniculate nucleus (LGN) of the thalamus. The LGN cells have narrow isotropic (i.e. without orientation selectivity) receptive fields and perform a type of edge detection by responding primarily at the borders between areas with different colours and shades. LGN responses are non-linearly proportional to the contrast between the areas and have firing patterns which are highly reproducible between trials and temporally precise - with the timing of the spikes considered meaningful with a precision as high as 1 msec [11, 10].

The visual information enters the cerebral cortex at the area of the primary visual cortex (V1). Neurons in V1 selectively respond to local features of the object in the visual field, such as spatial disparity or edges with particular orientation

[26, 27]. V1 responses to contrast are highly non-linear, with the mean firing rate of the neurons having a sigmoid form [28]. While most of the past studies of V1 responses to contrast have concentrated on the firing rate of the neurons, more recent studies have indicated that important part of the information about the contrast may be encoded in the temporal structure of neurons' responses [29, 30, 31]. There can be as much or more contrast-related information encoded in the temporal structure of V1 neural activity as in the firing rate [1]. Most of this information is encoded in the latency of the onset of spike train responses relative to the stimulus onset, which can vary independently of the overall firing rate.

Several stages further in the ventral pathway, the visual information reaches V4 - an area of neurons with larger receptive fields than those in V1, that respond to complex shapes in the visual scene [21, 32]. Here, objects in the visual scene are believed to be represented by distributed assemblies where individual neurons encode specific shapes constituting parts of the overall figure [33].

The final stage of visual processing considered here is the inferotemporal cortex (IT). Neurons in this area perform object recognition by selectively responding to objects in the visual scene independent of their size and position [34]. IT neurons have been found to interpret visual information beyond the simple grouping of perceptual input, exhibiting activity which also reflects the internal representations of objects [35, 36].

The work presented in this paper is based on the above experimental evidence and aims at building a neural network model of ADDS integrate-and-fire neurons, and train it to recognise objects in a realistic environment using STDP type learning.

3 Biologically Motivated Model of the Visual System

3.1 ADDS Neuron and STDP Learning Algorithm

This section briefly describes the model of the Active-Dendrites-and-Dynamic-Synapses (ADDS) neuron and implementation of the Spike-Timing-Dependent Plasticity algorithm. More detailed description and analysis of the models has been presented in [24]. Experimental work showed that spiking neurons incorporating the active properties of the dendrites and the dynamics of the synapse provide the computational mechanisms for processing temporal codes in a number of tasks and different time scales, including the precise selective responses in the millisecond range that is relevant for the processing tasks considered in this paper. Furthermore, the learning algorithm used here goes beyond the explicit simple dependency on the relative timing between a pair if single pre- and post-synaptic spikes following much closer the experimental evidence, and has been shown to effectively train the ADDS spiking neurons toward responses selective to the spatio-temporal distribution of input spike patterns [22, 23, 24, 25].

The ADDS spiking neuron is described as follows: For a synapse $i \in \mathcal{D}$ with strength w^i attached to an active dendrite, the total post-synaptic current I_d^i is defined by:

$$\tau_d^i \frac{d}{dt} I_d^i(t) = -I_d^i(t) + R_d^i w^i \delta(t - t_{(f)}^i) \tag{1}$$

where $t_{(f)}^i \in \mathcal{F}^i$ is the set of pre-synaptic spike times filtered as Dirac δ-pulses. Here, the time constant τ_d^i and resistance R_d^i define the active properties of the artificial dendrite as a function of the synaptic strength.

The neuron also has a set $\mathcal{S}$ of somatic synapses feeding close to or directly to the soma. The post-synaptic current generated by those synapses is described by:

$$\tau_s \frac{d}{dt} I_s(t) = -I_s(t) + \sum_{i \in \mathcal{S}} w^i \delta(t - t_{(f)}^i) \tag{2}$$

where τ_s is a fixed time constant for all such synapses, and synapse i has synaptic strength w^i and pre-synaptic spike times $t_{(f)}^i \in \mathcal{F}^i$.

Finally, the soma membrane potential u_m is:

$$\tau_m \frac{d}{dt} u_m(t) = -u_m(t) + R_m(I_d(t) + I_s(t)) \tag{3}$$

where $I_d(t) = \sum_i I_d^i(t)$ is the total current from the dendritic tree, τ_m is the membrane time constant at the soma, and R_m is the somatic resistance. Upon crossing the firing threshold θ from below, the neuron emits a spike and its potential is reset to u_{reset}. Following that event, the potential is clamped to u_{reset} for an absolute refractory period t_{refr} and is then released into a relative refractory period during which it recovers to the resting potential u_{rest}.

The current from dendrite i produces a membrane potential change at the soma, which is referred to as *partial membrane potential* and annotated as u_m^i. The *total partial membrane potential* $u_m^d = \sum_i u_m^i$ is the somatic membrane potential change generated from all dendrites.

The STDP learning is governed by the following algorithm: Immediately following a post-synaptic spike at time $\hat{t}$, synapse i, which has received a recent pre-synaptic spike, is sent a weight correction signal:

$$\Delta w^i = \begin{cases} \Delta w_d^i & \text{if } \frac{d}{dt} u_m^d(\hat{t}) \leq \varepsilon \\ \Delta w_m & \text{otherwise} \end{cases}$$
$$\text{where}$$
$$\Delta w_d^i = R_d^i I_d^i(\hat{t}) - u_m^i(\hat{t}) \quad \text{and} \quad \Delta w_m = u_m^d(\hat{t}) - R_m I_d(\hat{t}) \tag{4}$$

The learning rule has two main elements: Δw_d^i which depends on the state of the dendrite at the time of post-synaptic spike and Δw_m which depends on the state of the soma potential and total dendritic input at the time of post-synaptic spike. Depending on the gradient of the partial somatic membrane potential at the time of the post-synaptic spike (relative to a predefined constant ε) one of these two elements dominates in the determination of direction and magnitude of change in the synaptic strength.

3.2 Network Architecture

The task of the model is to recognise objects with 3 different shapes (*ball, box* and *cone*). 320×220 pixel images are collected from a robot's camera and preprocessed on a dedicated workstation. The preprocessing involves identifying areas of interest in the visual scene, segmenting and extracting information for each area and sending it to the input (LGN) layer of the vision module network. An area of interest is defined as the bounding box of a part of the visual scene and includes either an object that can be recognised by the robot or another feature of the the environment which has a different colour from the wall and floor. For each area of the visual input, the following information is extracted: a resized grey scale image of the segment (30×20 pixels), horizontal position of the segment relative to the centre of the current camera image and the colour of the central point of the segment. The extracted segments are sent to the vision module in a sequence. The architecture models parts of the ventral

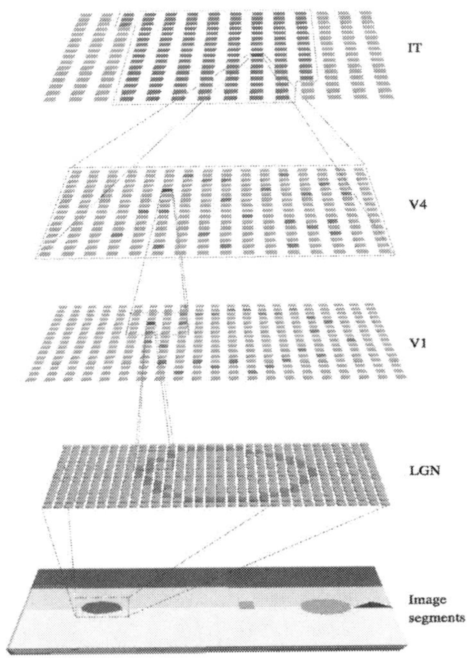

Fig. 1. Network architecture of the vision module

stream: LGN-V1-V4-IT. In order to reduce the computational requirements of the visual system, colour processing has been removed from the ventral stream, which receives as input grey-scale images with the task of recognising the shape of the object in the image.

The input layer of the network, LGN (figure 1), contains 30×20 spiking neurons driven by a LoG type ON-centre OFF-surround edge detection filter with a 3×3 mask. Each LGN neuron responds with a single spike with latency relative to the stimulus onset and proportional to the contrast difference within the area covered by the mask. Lower contrast causes higher latency of the neuron's response.

The next layer, V1, contains a map of 28×36 map simple orientation cells which receive input from LGN with a receptive field 3×3 over the neurons in LGN. Each V1 neuron selectively responds to the firing of a combination of LGN neurons forming a line with orientation of either 0, 45, 90 or 135 degrees. Such orientations are sufficient for the V1 neurons to respond to edges generated by the three different shapes of objects which the robot has to recognise: the lines from boxes and cones, and arcs from the balls.

The neurons in the next layer, V4, formed a 20×12 map and are trained to respond to a combination of spikes from the orientation cells in V1. Each V4 neuron receives input connections from V1 with a receptive field of 9×7 cells. Lateral inhibitory connections are also created between the V4 neurons with overlapping receptive fields.

The IT layer contains 200 neurons receiving excitatory input from all V4 neurons and lateral inhibition from all other IT neurons. IT neuron(s) who have learnt a spatio-temporal pattern closest to the current activity in V4 will respond first and inhibit other IT neurons. Thereby, only IT neuron(s) presenting the shape of the figure at the input respond.

All afferent connections are implemented as synapses attached to active dendrites and are subject to synaptic plasticity, and all lateral connections are implemented as synapses attached to the soma and have fixed strength.

3.3 Training and Results

V1 was trained over a set of collected and preprocessed images containing the three possible shapes of objects in the environment. For all layers, the afferent connections were trained using the ADDS neuron's STDP synaptic plasticity algorithm. The V1 neurons were competitively trained off-line with single lines plotted at various positions on LGN. The competitive learning was supported by lateral inhibitory connections between the V1 neurons with radius equal to their receptive field. The V4 neurons were trained using collected and preprocessed images of the 9 objects. IT neurons were trained using a form of guided learning - that is, for each training pattern only the neurons representing the figure at the input were allowed to fire and therefore learn the spatio-temporal patter generated by that figure. As with the previous layers, the competition between the neurons plays a critical role in the correct response of the neurons. Each IT neuron is trained to recognise a particular shape (ball, box or cone) by selectively responding to a combination of features detected in V4. 120 neurons are trained to recognise ball, 20 neurons are trained on box and 60 on cone. The number of neurons necessary for each figure was dictated by the noise resulting from the preprocessing of the images and was determined empirically.

The network was able to learn and later recognise the objects under different view angles and viewing conditions. The successful results confirmed the thesis that temporal encoding alone can provide sufficient information for the training of a spiking neural network and recognising simple objects in a realistic robot environment.

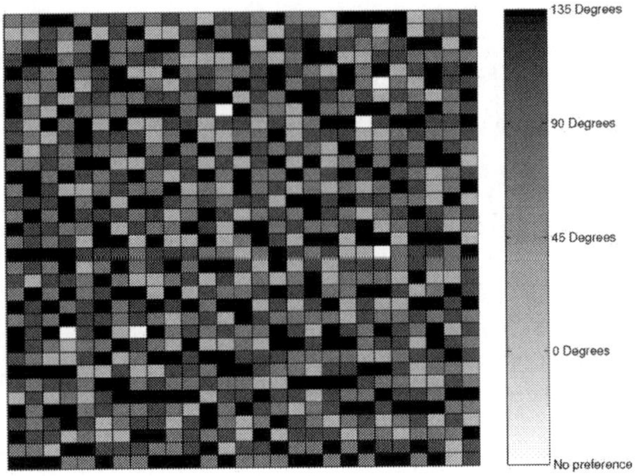

Fig. 2. Orientation selectivity map of the V1 cells

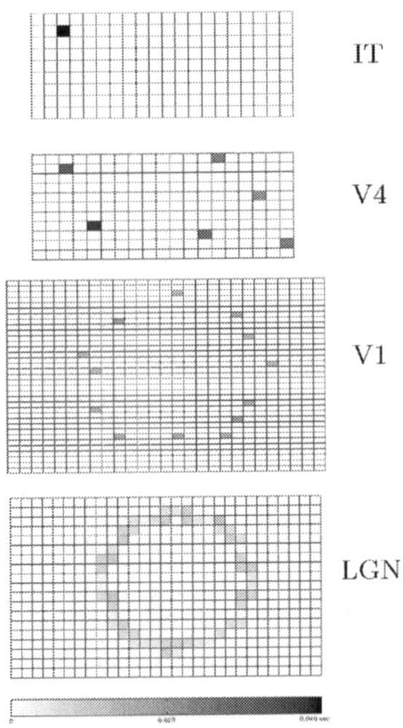

Fig. 3. Visual areas activity in processing an image of a ball. The grey scale indicates the response times relative to the onset of the image.

Figure 2 presents the orientation selectivity map of V1 after training. The target of the design and training of this layer was to achieve neurons responding to all four orientations within each small area of the image. As a result, the orientation selectivity map of V1 achieved here is organised in singularities/pinwheel structures distributed across the layer.

Figure 3 presents the activation of the LGN, V1 and V4 layers for the processing of an image containing a ball. The time taken to recognise an object in IT is within 25-50 ms with V1 responding in the interval of 10-20 ms, and V4 responding in 20-40 ms.

4 Conclusion

The paper presented a novel biologically motivated computational model of the visual system. The model is build of integrate-and-fire ADDS spiking neurons and trained using an STDP learning algorithm and a combination of competitive and guided training strategies. The network implemented parts of the ventral stream of the visual system (LGN-V1-V4-IT) and its input comprises of temporally encoded patterns extracted from the input stimuli - proportional to local contrast latency of the LGN responses. The neurons are trained to recognise the shape of the objects in the visual field based on the temporally encoded information at LGN (figure 3). After training, V1 developed a self-organised orientation selectivity map (figure 2). Considering the relatively low resolution of this layer, continuous and periodic organisation of the V1 neurons could not have been computationally affordable for the performance of the visual module required for testing it on a mobile robot. The target of the design and training of this layer was to achieve neurons responding to all four orientations within each small area of the image. As a result, the orientation selectivity map of V1 achieved here is organised in singularities distributed across the layer. Finally, the network model can recognise the 9 real objects in a robot's environment, and in further experiments was successfully tested as part of a robot's control architecture for navigation and object manipulation [24].

Complimentary to recent experimental neurobiological evidence on spatio-temporal codes used by the brain in the visual areas, the results from the model further confirm that the temporal information derived from input sensory stimuli can provide sufficient information for the discrimination and recognition of real visual patterns. Of course, the real brain also employs firing rate codes in processing information in these modalities. Future development in the work presented here will focus on integrating and taking advantages from both firing rate and temporal encoding schemes in the processing of visual information.

References

1. Reich, D.S., Mechler, F., Victor, J.D.: Temporal coding of contrast in primary visual cortex: When, what, and why. Journal of Neurophysiology **85**(3) (2001) 1039 – 1050
2. Vaadia, E., Haalman, I., Abeles, M., Bergman, H., Prut, Y., Slovin, H., Aertsen, A.: Dynamics of neuronal interaction in monkey cortex in relation to behavioural events. Nature **373** (1995) 515–518

3. Mainen, Z.F., Sejnowski, T.: Reliability of spike timing in neocortical neurons. Science **268** (1995) 1503–1506
4. de Ruyter van Steveninck, R.R., Lewen, G., Strong, S., Koberle, R., Bialek, W.: Reproducibility and variability in neural spike trains. Science **275** (1997) 1805–1808
5. Adorján, P., Levitt, J.B., Lund, J.S., Obermayer, K.: A model for the intra-cortical origin of orientation preference and tuning in macaque striate cortex. Visual Neuroscience **16** (1999) 303–318
6. Hirsch, J.A., Alonso, J.M., Reid, R.C., Martinez, L.M.: Synaptic integration in striate cortical simple cells. Journal of Neuroscience **18**(22) (1998) 9517–9528
7. Bair, W., Koch, C.: Temporal precision of spike trains in extrastriate cortex of the behaving macaque monkey. Neural Computation **8**(6) (1996) 1185–1202
8. Buracas, G., Zador, A., DeWeese, M., Albright, T.: Efficient discrimination of temporal patterns by motion-sensitive neurons in primate visual cortex. Neuron **20**(5) (1998) 959–969
9. Bair, W.: Spike timing in the mammalian visual system. Current Opinion in Neurobiology **9** (1999) 447–453
10. Reinagel, P., Reid, R.C.: Temporal coding of visual information in the thalamus. Journal of Neuroscience **20**(14) (2000) 5392–5400
11. Reinagel, P., Reid, R.: Precise firing events are conserved across neurons. Journal of Neuroscience **22**(16) (2002) 6837–6841
12. Miikkulainen, R., Bednar, J.A., Choe, Y., Sirosh, J.: Self-organisation, plasticity, and low-level visual phenomena in a laterally connected map model of the primary visual cortex. In Goldstine, R., Schyns, P., Medin, D., eds.: Perceptual Learning. Volume 36 of Psychology of Learning and Motivation. Academic Press, San Diego, CA (1997) 257–308
13. Sirosh, J., Miikkulainen, R.: Cooperative self-organisation of afferent and lateral connections in cortical maps. Biological Cybernetics **71**(1) (1994) 65–78
14. Sirosh, J.: A Self-Organizing Neural Network Model of the Primary Visual Cortex. PhD thesis, The University of Texas, Austin, TX (1995)
15. Delorme, A., Thorpe, S.: Face identification using one spike per neuron: resistance to image degradations. Neural Networks **14**(6-7) (2000) 795–803
16. VanRullen, R., Delorme, A., S.J., T.: Feed-forward contour integration in primary visual cortex based on asynchronous spike propagation. Neurocomputing **38-40**(1-4) (2001) 1003–1009
17. Delorme, A.: Early cortical orientation selectivity: How fast shunting inhibition decodes the order of spike latencies. Journal of Computational Neuroscience **15** (2003) 357–365
18. Jeffreys, D.: Evoked potential studies of face and object processing. Visual Cognition **3** (1996) 1–38
19. Perrinet, L., Delorme, A., Thorpe, S.: Network of integrate-and-fire neurons using rank order coding a: how to implement spike timing dependant plasticity. Neurocomputing **38-40**(1-4) (2001) 817–822
20. Delorme, A., Thorpe, S.: Spikenet: An event-driven simulation package for modelling large networks of spiking neurons. Network: Computation. Neural Systems **14** (2003) 613:627
21. Rolls, E.T., Deco, G.: Computational neuroscience of vision. Oxford University Press, Oxford (2002)
22. Panchev, C., Wermter, S.: Spike-timing-dependent synaptic plasticity from single spikes to spike trains. Neurocomputing **58-60** (2004) 365–371

23. Panchev, C., Wermter, S., Chen, H.: Spike-timing dependent competitive learning of integrate-and-fire neurons with active dendrites. In: Lecture Notes in Computer Science. Proceedings of the International Conference on Artificial Neural Networks, Madrid, Spain, Springer (2002) 896–901
24. Panchev, C.: Spatio-Temporal and Multimodal Processing in a Spiking Neural Mind of a Robot. PhD thesis, University of Sunderland (2005)
25. Panchev, C., Wermter, S.: Temporal sequence detection with spiking neurons: towards recognizing robot language instructions. Connection Science **18**(1) (2006) 1–22
26. Shapley, R.: A new view of the primary visual cortex. Neural Networks **17** (2004) 615 623
27. Ringach, D.L., Hawken, M.J., Shapley, R.: Dynamics of orientation tuning in macaque primary visual cortex. Nature **387** (1997) 281 – 284
28. Albreght, D.G., Hamilton, D.B.: Striate cortex of monkey and cat: contrast response function. Journal of Neurophysiology **48** (1982) 217–237
29. Gawne, T.J., Kjaer, T.W., Richmond, B.J.: Latency: another potential code for feature binding in striate cortex. Journal of Neurophysiology **76** (1996) 1356–1360
30. Mechler, F., Victor, J.D., Purpura, K.P., Shapley, R.: Robust temporal coding of contrast by V1 neurons for transient but not steady-state stimuli. Journal of Neuroscience **18** (1998) 6583–6598
31. Purpura, K., Chee-Orts, M.N., Optican, L.M.: Temporal encoding of texture properties in visual cortex of awake monkey. Society of Neuroscience Abstracts **19** (1993) 315
32. Gallant, J.L., Shoup, R.E., Mazer, J.A.: A human extrastriate area functionally homologous to macaque V4. Neuron **27** (2000) 227–235
33. Pasupathy, A., Connor, C.E.: Shape representation in area V4: Position-specific tuning for boundary conformation. Journal of Neurophysiology **86**(5) (2001) 2505–2519
34. Ito, M., Tamura, H., Fujita, I., Tanaka, K.: Size and position invariance of neuronal responses in monkey inferotemporal cortex. Journal of Neurophysiology **173**(1) (1995) 218–226
35. Bruce, C., Desimone, R., Gross, C.G.: Visual properties of neurons in a polysensory area in superior temporal sulcus of the macaque. Journal of Neurophysiology **46**(2) (1981) 369–384
36. Kobatake, E., Wang, G., Tanaka, K.: Effects of shape-discrimination training on the selectivity of inferotemporal cells in adult monkeys. Journal of Neurophysiology **80**(1) (1998) 324–330

Accelerating Event Based Simulation for Multi-synapse Spiking Neural Networks

Michiel D'Haene*, Benjamin Schrauwen**, and Dirk Stroobandt

Ghent University, Department of Electronics and Information Systems, Ghent, Belgium
{Michiel.DHaene, Benjamin.Schrauwen, Dirk.Stroobandt}@ugent.be
http://www.elis.ugent.be/SNN

Abstract. The simulation of large spiking neural networks (SNN) is still a very time consuming task. Therefore most simulations are limited to rather unrealistic small or medium sized networks (typically hundreds of neurons). In this paper, some methods for the fast simulation of large SNN are discussed. Our results equally amongst others show that event based simulation is an efficient way of simulating SNN, although not all neuron models are suited for an event based approach. We compare some models and discuss several techniques for accelerating the simulation of more complex models. Finally we present an algorithm that is able to handle multi-synapse models efficiently.

1 Introduction

Despite the ever increasing computational power in computer systems, there is still a huge demand for more computational power, especially in the field of spiking neural networks (SNN). Spiking neurons are biologically inspired neurons that communicate by using spikes. Because here the timing of the spikes is considered, SNN are able to handle temporal problems more efficiently (for example speech recognition [18]) and have more computational power than artificial neural networks [9] which use the average firing rate of neurons as inputs. Furthermore, they communicate through discrete spikes instead of analog values which significantly reduces the communication costs between neurons. This makes them particularly better suited for hardware implementations [13][14].

The behaviour of a spiking neuron can be represented by an internal membrane potential which is influenced by incoming spikes. When the potential of the membrane reaches a certain threshold value, the membrane potential will be reset to a lower value and a spike is emitted. It is important to note that each neuron operates independently, except when a spike is communicated between neurons.

An obvious way of implementing SNN in hardware is a one to one placement of the neurons into physical components. This approach benefits from the inherent parallel nature of spiking neural networks and allows extremely fast simulations (orders of

* Michiel D'Haene is sponsored by the Institute for the Promotion of Innovation through Science and Technology in Flanders (IWT-Vlaanderen)

** Benjamin Schrauwen is funded by FWO Flanders project G.0317.05

S. Kollias et al. (Eds.): ICANN 2006, Part I, LNCS 4131, pp. 760–769, 2006.

magnitude faster than realtime). Unfortunately it has to deal with some important drawbacks. First, the size of the networks is limited by the amount of hardware available on the chip. In current FPGA's[1], one can implement a couple of thousand simplified spiking neurons in a direct manner [15]. By combining several chips, one should be able to create larger networks, however this is a very cost inefficient method. Another important drawback of this approach lies in the low activity, which is for a typical spiking neural network less than 1% ([5]). This means that generally most of the synapses and neurons are inactive (i.e. potential on its restpotential) and thus occupy costly space in hardware. Direct hardware implementations are thus very space inefficient.

Therefore, in practice, large SNN are simulated using more conventional architectures. There are essentially two ways of simulating SNN: time-step driven and event based simulation. The first one divides the simulation into fixed time-steps. At each time-step, the complete network is evaluated and the new state of each neuron is calculated. The precision of the simulation depends on the size of each time-step, which also affects the simulation time. Although this is a very simple approach, the asynchronous nature of the spikes requires small time-steps ($\leq$ 1 ms) in order to achieve accurate simulation results [16].

Usually we are only interested in the external behaviour of a neuron (i.e. emitted spikes) due to incoming spikes instead of the internal membrane changes. Event based simulation takes advantage of this. Instead of evaluating the whole network on regular time intervals, the membrane potential of each neuron is evaluated only when necessary, i.e. when a neuron receives a spike, or when it will fire.

In the next section, we compare time-step driven simulation with event based simulation using two accurate simulation environments. In Section 3 we briefly discuss different models of spiking neurons and show why some of them can be simulated efficiently in an event driven manner while very biological realistic models are much more difficult to simulate. We show the results of some accelaration techniques for more complex models. Finally, because existing simulators do not support multi-synapse models efficiently, we present an algorithm that handle these models efficiently.

2 Event Based Simulation Versus Time-Step Driven Simulation

In order to compare event based simulation with time-step driven simulation, we created a random network consisting of 500 neurons with an interconnection fraction of 0.1 (i.e. each neuron has on average 50 input connections). Input spikes are generated through one input neuron. The neuron model used is developed by Olaf Booij (personal communication). It is a special case of a leaky integrate and fire neuron with exponential synapses where $\tau_m = 2\tau_s$ (this will be explained in Section 3.1).

For the time-step driven simulator, we used CSIM[2] which is a well known open source C++ simulator, optimized for speed by calculating only the active synapses. The resolution (time-step) was set to 0.1 ms.

The event-based simulator used is ESSpiNN, a simulator developed at our research group. The core of this simulator is based on MVASpike, a general event based C++

[1] Field Programmable Gate Array.

[2] http://www.lsm.tugraz.at/csim

simulator for SNN from Olivier Rochel [12]. We have adapted this simulator to our goals, and extended it in order to allow more general neuron models (e.g. SRM_0) to be simulated. Besides a broad range of neuron models, it can simulate different types of (delayed) connections such as STDP, dynamic synapses, etc. The time resolution for the neuron model used in this example is only limited by the resolution of the double precision floating point numbers, which results in very accurate simulations compared to time-step driven methods.

In Table 1 we measured the execution time of both simulators for different numbers of input spikes (on the same platform, i.e. AMD Athlon 64 3400+). We can see that the average speedup of the event driven simulator is 60 times for a relatively high neuron activity (on average 100 spikes per neuron per second). For 100 and 1000 input spikes, we used a simulation time of 1s, which means that the incoming spike activity is a factor 10 higher for the second case. When we look at CSIM, we can see that despite the 10 times lower activity, its execution speed is only a factor 3.75 faster. This is due to its time-step driven nature: time-step based simulation scales in the first place with the simulated time, while event-based simulation scales mainly with the number of spikes.

Table 1. Comparison between CSIM and ESSpiNN for a network of 500 neurons, interconnection fraction of 0.1 and 1 input neuron. The neuron model is Booij's integrate and fire neuron where $\tau_m = 2\tau_s$.

Number of spikes	Simulated time	Execution time CSIM	Execution time ESSpiNN
100	1 sec	14.6 sec	0.10 sec
1 000	1 sec	55.0 sec	0.80 sec
10 000	10 sec	451 sec	7.13 sec
100 000	100 sec	/	72.8 sec

3 Event Simulation of Different Spiking Neuron Models

In event based simulation, a new calculation will be performed only when an event occurs. An event-simulator has to keep track of all events in the system. This can be done with a queue that keeps all generated events in chronological order (the event-queue). The simulator takes the event with the smallest time stamp from this event queue, processes it and adds new events to the queue if necessary. Then it takes the next event with the smallest time stamp, etc.

An obvious requirement for event simulation of SNN is that incoming and outgoing pulses can be considered as discrete events. However, not all neural models fulfill this condition. Some models are very easy to simulate in an event driven fashion, while others are much harder. Below, we discuss some important models and their applicability to event simulation. Then we give some results measured on our event simulator for different neuron models.

3.1 Some Common Neuron Models

Hodgkin Huxley model. The model of Hodgkin and Huxley [7] is a very good imitation of a certain type of biological neuron (Fig. 1a). It consists of a collection of

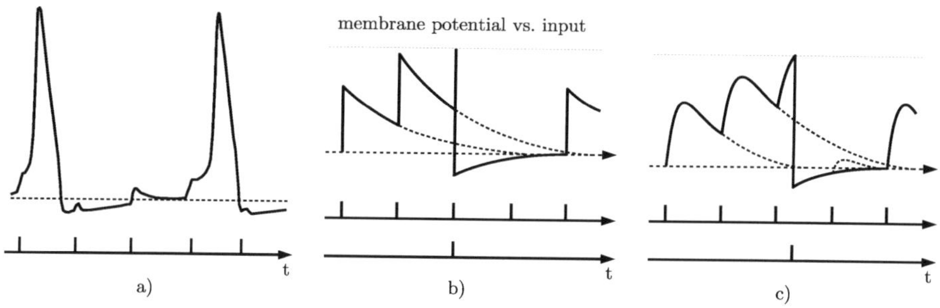

Fig. 1. Some popular spiking neural models a) example of a Hodgkin-Huxley-model, b) the Leaky IF-neuron and c) an approximation with SRM (0th order) model

differential equations that model the internal processing. It is an important model for the detailed study of the biological behaviour of neurons that deals with ion channels, different types of synapses and geometry of individual neurons. This model is in essence a continuous model, which makes it very difficult to simulate in an event driven fashion without making certain concessions [8]. However it is an important reference model for more simple models.

(Leaky) Integrate and Fire model (IF model). IF models are one of the best known examples of the so-called threshold model neurons, which means that a pulse is generated as soon as the membrane potential reaches a certain threshold. Pulses are stereotypical events which are completely characterized by their firing time stamp. After firing, the neuron optionally has a refractory period during which it stays inactive for a certain amount of time. In Fig. 1b, an example of a Leaky IF model is shown (without a synapse model). Because firing happens only when an incoming spike arrives, the event based simulation of this model is simple. However, this property also drastically limits the expression power of this model [10].

Spike Response Model (SRM). The SRM [4] can be seen as a generalization of the Leaky IF model. One of the differences however is that the membrane potential is expressed as a function of the time passed since the last fire event, instead of a function of itself. A special case of the SRM is the 0th order SRM or SRM_0. All incoming pulses now have the same shape, independent of the time since the last incoming event and the state of the membrane. A simple example is shown in Fig. 1c. The model shown is an exponential Leaky IF model with exponential synapses. For most applications, the SRM_0 is still sufficient and has been used for example for the analysis of the computational power of neurons [10], studies of collective phenomena in local coupled networks [4] and in the Liquid State Machine (LSM) [6].

There exist a number of event based simulators specially built for a specific type of IF neurons (most without synapse model) and networks. These simulators allow fast simulations but –given the simplicity of the neuron model– they have limited applicabilities and are often built with a specific application in mind. A totally different approach is to try to create a biological very realistic simulator [8]. However the aim of our work is not to create a biological very realistic simulator, but to allow efficient emulations of huge networks of powerful spiking neurons. The LIF with a synapse model is a good starting point for such a simulator, but it involves some difficulties when we try to simulate it in an event-based manner. In the next subsection, we briefly explain these difficulties and show some techniques and optimisations to solve them efficiently.

3.2 A Simple SRM$_0$ Neuron

To show the problems of simulating a more general IF model, we consider the exponential Leaky IF model with exponential synapses (Fig. 1c). The postsynaptic potential $u_i(t)$ of neuron i with n inputs at time t can be written as [4]

$$u_i(t) = \sum_f \eta_i\left(t - t_i^{(f)}\right) + \sum_j w_{ij} \sum_f \epsilon_{i,j}\left(t - t_i^{(f)}\right) \tag{1}$$

with kernels (after convolution of $\epsilon_0(s)$)

$$\eta_i(s) = -\left(\theta - u_{rest}\right)\exp\left(-\frac{s}{\tau_{m,i}}\right)\mathcal{H}(s) \tag{2}$$

$$\epsilon_{i,j}(s) = \frac{1}{\max_{i,j}}\left[\exp\left(-\frac{s}{\tau_{m,i}}\right) - \exp\left(-\frac{s}{\tau_{s,i,j}}\right)\right]\mathcal{H}(s) \tag{3}$$

with θ the threshold, u_{rest} the restpotential and τ_m and $\tau_{s,j}$ the decay constants of the membrane and the synapse. $\max_{i,j}$ is a rescaling constant to assure that a weight of 1 generates a pulse of the same height. To schedule the next update-time of a neuron, the event-simulator must be able to predict the next fire-time stamp of a neuron, i.e. the time stamp when ϵ_0 reaches the threshold θ.

Therefore equation (1) with $u_i(t) = \theta$ must be solved for t. However for non-natural values of τ_m and $\tau_{s,j}$ this can not be solved analytically. There exists different solutions for this problem which we divide roughly into tree classes: using a restricted model which can be solved very efficiently, using look-up tables, or using iterative techniques to approximate the fire-time stamp.

A good example of the first solution is the neuron model developed by Olaf Booij (personal communication). He assumes that each input synapse has the same τ_s, and that $\tau_m = 2\tau_s$. Now, the solution for t is reduced to a quadratic equation which can be solved analytically very quickly.

The use of lookup tables is a well known method to estimate the fire-time stamp which offers high speed with (depending on the size of the table) sufficient accuracy [1]. Some researchers even replaced all calculations with precalculated lookup tables [2]. An important drawback of lookup tables however is the memory size that grows more than exponentially with the desired accuracy. Also, separate lookup tables are required for each τ_s. Therefore, most researchers limit their models to a single synapse model.

An example of the last type of solution is developed by Makino [11]. His simulator is able to estimate the fire time stamp of an arbitrary continuous membrane function by dividing the function into linear envelopes where each envelope containes at most one threshold crossing. Inside each envelope, a Newton-Rhapson based method is used to estimate the fire-time stamp.

It is clear that most techniques are developed to be used with one or a very restricted number of synapse constants. When extending these techniques to neurons with several synaptic time constants (multi-synapse neurons, in the most common case, each input has its own synaptic time constant), they have to deal with the calculation of each synapse for each update of their state, which quickly decreases their efficienty. Therefore, we developed an efficient technique that can handle several synaptic time constants, which we will describe briefly below.

3.3 Efficient Processing of Multi-synapse Neurons

An obvious optimisation for the simulation of multi-synapse models that has also been used in e.g. CSIM is to separate inactive synapses from active synapses and to consider only the active synapses of a neuron in the computations. Due to the typical low activity of spiking neural networks, this results already in a significant speedup (a factor 2).

Another optimisation follows from the observation that a neuron generally has to receive several spikes before it will emit a spike itself. However, estimating if a the membrane potential will reach the threshold, is a time consuming process itself when using several synaptic constants.

Therefore, we developed a simple but fast membrane-value estimation function. Whenever the neuron receives a spike, we consider the (updated) maximum influence of the current synapse on the membrane potential by adding it to the total approximated membrane maximum (Fig. 2). When a spike enters, we only update the synapse potential of the input that receives the spike. When there is a possibility for the membrane potential to reach its threshold (i.e. the approximated membrane maximum reaches the threshold), we start updating the total membrane potential by calculating the exact potential of each synapse at the current time stamp.

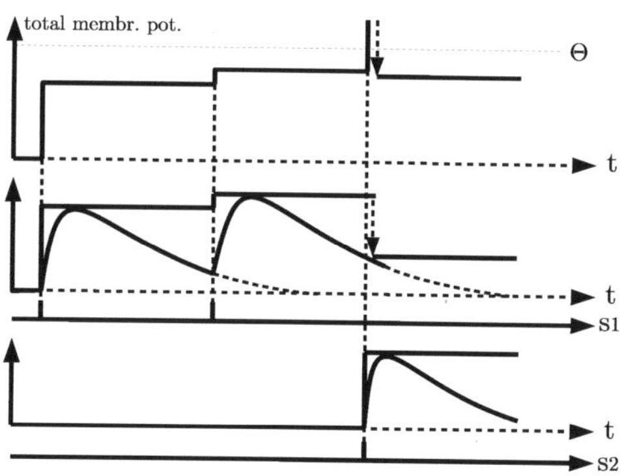

Fig. 2. Maximum approximation of the membrane potential. On time 3, the approximated maximum reaches the threshold. After updating synapse 2, we see already that the threshold will not be reached.

In many cases, after updating just a couple of synapses, the approximated membrane potential will drop below the threshold, so we can stop updating and wait for the next spike to enter. By updating the synapses with the oldest update-time first, we have the highest likelihood that only a couple of updates are sufficient to see that the threshold will not be reached because the maximum drops below the threshold. Therefore, we keep the synapses ordered by their update-time. We implemented this in an efficient way by placing them in a circular doubly linked list. Due to the algorithm, when the oldest synapse needs to be updated (to the current time stamp), only the start-pointer of the list needs to be updated, which involves no sorting actions (Fig. 3). As soon as the influence of a synapse on the membrane potential becomes negligable, we remove the synapse from the active synapse list.

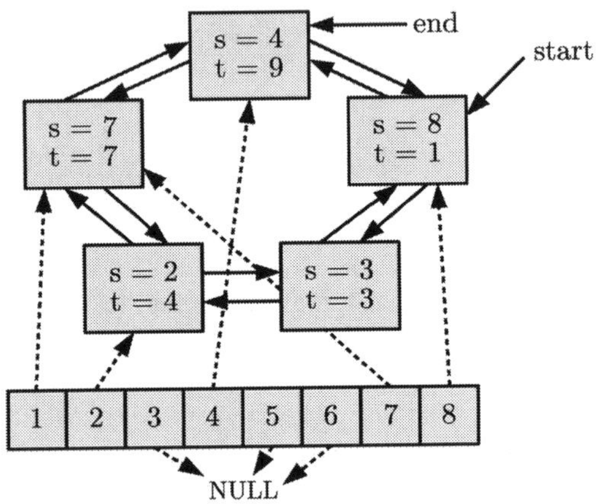

Fig. 3. A circular doubly linked list is used to keep synapses ordered by their update time, thus eliminating a sorting action. Also an index structure allows fast random access of synapses.

When the threshold is still reached after updating all synapses, we apply a Newton-Rhapson (NR) method to estimate the exact fire-time stamp. Measurements show that we have a high accuracy after only a few NR iteration steps. A drawback of our approach however is that for very high activity (spike-frequency ~kHz, which is very unrealistic), the maximum membrane approximation becomes almost useless and introduces an overhead to the simulation.

3.4 Results

In order to be able to compare the simulation speed of our model with other models, we used Booij's restrictions to the multi-synapse neurons (i.e. each synapse $\tau_{s,i}$ was set to $\tau_{m,i}/2$) and compared the execution-times for several implementations. We used multi-synapse neurons with a separate synapse constant for each input and neglected of course the fact that the synapses can be actually combined to one synapse (because they all have the same time constant, and can thus be calculated more efficiently).

In Figure 4 (top), we compare the execution time of several implementations of the general LIF model for different network activities. The first optimization (calculating

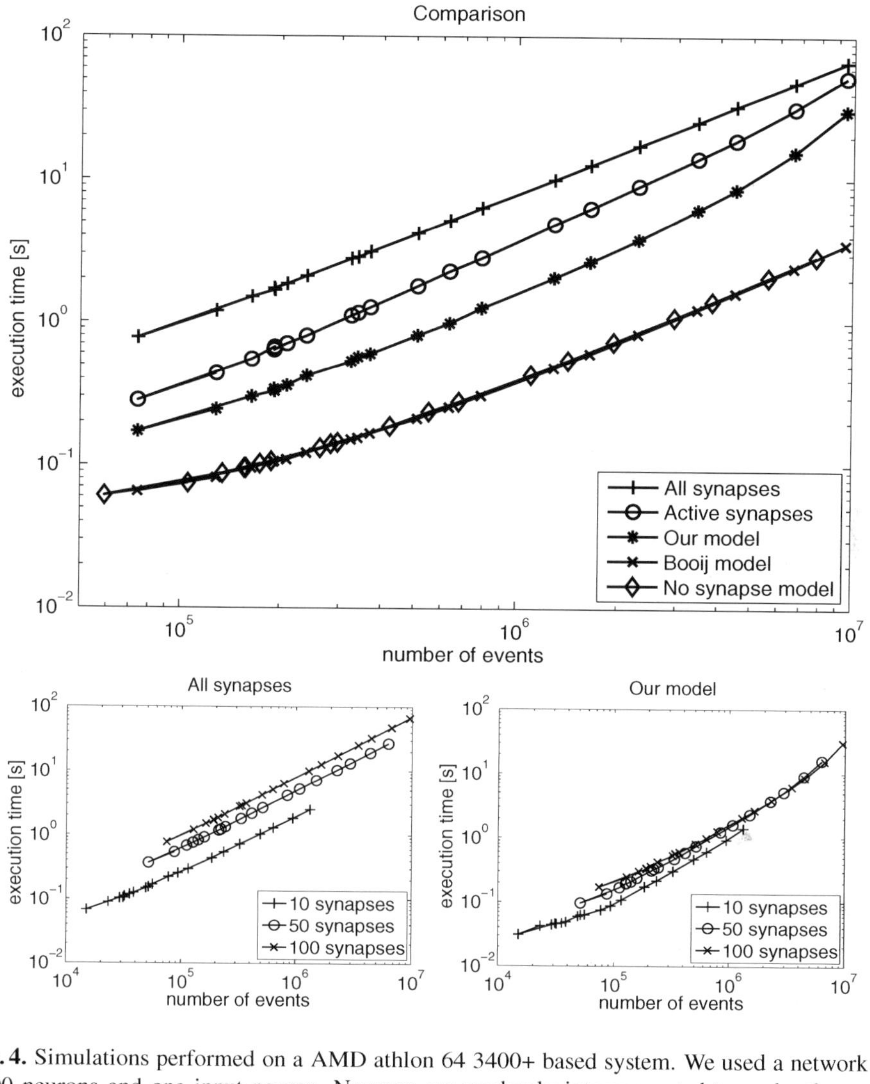

Fig. 4. Simulations performed on a AMD athlon 64 3400+ based system. We used a network of 1000 neurons and one input neuron. Neurons are randomly interconnected to each other with a spectral radius of 0.9 (this is the absolute value of the highest eigen-value of the connection matrix). The input neuron is randomly connected to the network with an interconnection fraction of 10% and fixed weights of 0.9. Each internal connection has a random delay between 0 and 10 ms and a random weight which is rescaled afterwards according to the spectral radius of the network [17]. Inputs are generated by a Poisson process within the simulated interval [0,10] seconds. The horizontal axis shows the total number of spikes (external and internal) processed by the simulator in the simulated interval ($\sim$ the network activity).

Figure top: comparison between Booij's model and several optimizations of our common LIF model with exponential synapses. All simulations use exact the same network with on average 100 synapse inputs per neuron.

Figure left: scaling of the not optimized common LIF model according to the average number of synapses per neuron. Figure right: the same but for the optimized model.

only active synapses) improves the performance of the simulator with approximately a factor 2, dependent on the activity of the network. The membrane-value estimation function further improves the performance with another factor 2. To show the efficienty of Booij's simulation model, we also plotted the execution time of a simple no-synapse LIF model. We see that the performance of both models is almost the same, although the computation power of Booij's model is much higher [10]!

A second advantage of our algorithm is that the simulation of multi-synapse neurons becomes less dependent of the interconnection density of the network, i.e. the average number of input connections of each neuron. This is shown in Figures 4 (left) and (right). We can see that for the unoptimized version of the algorithm, the execution time is highly dependent of the number of inputs, because for each incoming spike, all synapses have to be updated. In the optimized version however, it appears that the influence of the average number of synapses per neuron has decreased significantly. We did not plot the results for Booij's model, but we found that the time to calculate a number of events does not depend on the number of inputs of each neuron, as was expected (because each input shares the same synapse).

4 Future Work

The event based principle discussed in this paper is a sequential process: events must be handled one by one in the correct time-order. Although this allows yet for much faster simulations compared to time-step based simulation, it is still much to slow to be interesting for many applications (e.g. realtime simulations on modern architectures are still limited to networks of order of magnitude 10.000 neurons [3]).

An obvious way to accelerate this sequential process is the use of more processing units in parallel or through pipelining. Unfortunately the intense memory interaction of event simulation creates an important memory bottleneck. Also the inherent parallelism of SNN remains unused.

Our ultimate goal is to implement an event based simulator in parallel hardware. Therefore, we are building a SystemC framework for parallel discrete event simulation of SNN. It will allow us to investigate and optimize different synchronization mechanisms for parallel event based simulation (discribed in [3]) in order to build an efficient parallel SNN emulator. An important aspect is a hardware-friendly design: in a later stadium, the simulator will be implemented in digital hardware (FPGA) in order to benefit optimally from the inherent parallelism of SNN.

5 Conclusions

In this paper, we have shown that event based simulation is a good candidate for efficient simulation of SNN, characterized by discrete pulses and very low activity. However, not all neuron models are suited for such an event driven approach. A good compromise between complexity and possibilities is offered by the SRM_0 model. We discussed several techniques to efficiently implement a simple SRM_0 with exponential membrane function and synapse model. Because these techniques do not support multi-synapse models efficiently, we presented an algorithm that handles multi-synapse models much more efficient. It provides a significant speedup of the simulations and moreover it improves the scalability of the simulator with regard to the number of synapses.

References

1. R. Brette. Exact simulation of integrate-and-fire models with synaptic conductances. Submitted, 2005.
2. R. Carrillo, E. Ros, E. Ortigosa, B. Barbour, and R. Agis. Lookup table powered neural event-driven simulator. In *Proceedings of the 8th International Work-Conference on Artificial neural Networks, IWANN 2005*, pages 168–175, 2005.
3. M. D'Haene. Parallelle event-gebaseerde simulatietechnieken en hun toepassing binnen gepulste neurale netwerken. Technical report, Universiteit Gent, 2005.
4. W. Gerstner and W. M. Kistler. *Spiking Neuron Models*. Cambridge University Press, 2002.
5. C. Grassmann and J. Anlauf. Distributed, event driven simulation of spiking neural networks. In *Proceedings of the International ICSC /IFAC Symposium on Neural Computation*, pages 100–105, 1998.
6. A. Graves, D. Eck, N. Beringer, and J. Schmidhuber. Biologically plausible speech recognition with LSTM neural nets. In *Proceedings of Bio-ADIT*, pages 127–136, 2004.
7. A. L. Hodgkin and A. F. Huxley. A quantitative description of ion currents and its applications to conduction and excitation in nerve membranes. *J. Physiol. (London)*, 117:500–544, 1952.
8. C. Lobb, Z. Chao, R. Fujimoto, and S. Potter. Parallel event-driven neural network simulation using the Hodgkin-Huxley neuron model. In *Proceedings of the 19th Workshop on Principles of Advanced and Distributed Simulation*, 2005.
9. W. Maass. Lower bounds for the computational power of networks of spiking neurons. *Neural Computation*, 8(1):1–40, 1996.
10. W. Maass. Computation with spiking neurons. In *The Handbook of Brain Theory and Neural Networks*. The MIT Press, 2001.
11. T. Makino. A discrete-event neural network simulator for general neural models. *Neural Computing & Applications*, 11:210–223, 2003.
12. O. Rochel and D. Martinez. An event-driven framework for the simulation of networks of spiking neurons. In *Proceedings of the 11th European Symposium on Artifical Neural Networks, ESANN 2003*, pages 295–300, 2003.
13. B. Schrauwen. Embedded spiking neural networks. In *Doctoraatssymposium Faculteit Toegepaste Wetenschappen*, pages on CD–ROM. Universiteit Gent, Gent, December 2002.
14. B. Schrauwen and M. D'Haene. Compact digital hardware implementations of spiking neural networks. In J. Van Campenhout, editor, *Sixth FirW PhD Symposium*, page on CD, 1 2005.
15. B. Schrauwen and J. Van Campenhout. Parallel hardware implementation of a broad class of spiking neurons using serial arithmetic. In *Proceedings of ESANN'06*, 2006. To be published.
16. W. R. Softky. Simple codes versus efficient codes. *Current opinion in neurobiology*, 5:239–247, 1995.
17. D. Verstraeten, B. Schrauwen, and D. Stroobandt. Reservoir-based techniques for speech recognition. 2006. Accepted for publication at WCCI'06.
18. D. Verstraeten, B. Schrauwen, D. Stroobandt, and J. Van Campenhout. Isolated word recognition with the liquid state machine: a case study. *Information Processing Letters*, 95(6):521–528, 2005.

A Neurocomputational Model of an Imitation Deficit Following Brain Lesion

Biljana Petreska and Aude G. Billard

Learning Algorithms and Systems Laboratory (LASA)
Ecole Polytechnique Fédérale de Lausanne (EPFL)
Station 9, CH-1015 Lausanne, Switzerland
biljana.petreska@epfl.ch, aude.billard@epfl.ch

Abstract. This paper investigates the neural mechanisms of *visuo-motor imitation* in humans through convergent evidence from neuroscience. In particular, we consider a deficit in imitation following *callosal brain lesion*, based on the rational that looking at how imitation is impaired can unveil its underlying neural principles. We ground the functional architecture and information flow of our model in brain imaging studies and use findings from monkey brain neurophysiological studies to drive the choice of implementation of our processing modules. Our neural model of visuo-motor imitation is based on self-organizing maps with associated activities. Patterns of impairment of the model, realized by adding uncertainty in the transfer of information between the networks, account for the scores found in a clinical examination of imitation [1]. The model also allows several interesting predictions.

1 Introduction

Apraxia is generally defined as the inability to perform voluntary movements that cannot be explained by elementary motor, sensory or cognitive deficits (not caused by weakness, ataxia, akinesia, deafferentation, inattention to commands or poor comprehension). A standard test for clinical examinations of apraxia is *imitation of meaningless gestures* which is believed to test the integrity of a direct route from visual perception to motor control, not mediated by semantic representations or verbal concepts [2]. Goldenberg has shown that knowledge about body parts is also relevant, as apraxic patients were unable to map body configurations to their own body nor to a mannikin [3]. Kinematic studies of apraxia show spatial parapraxias (i.e., normal kinematic profiles with abnormal final positions) that seem to arise from a basic deficit that concerns the mental representation of the target position [4].

Goldenberg's study. A seminal study of imitation of meaningless gestures examines a patient with callosal brain lesion (disconnected hemispheres) [1]. The patient was asked to imitate a set of visual stimuli that present different positions of the hand relative to the head (see Fig. 1). To disentangle the contribution of each hemisphere the patient was tested tachistoscopically (i. e., the stimulus was

S. Kollias et al. (Eds.): ICANN 2006, Part I, LNCS 4131, pp. 770–779, 2006.

presented either to the left or right visual field) in a left- or right-hand imitation condition. As shown on the figure (upper right) the pattern of errors varies as a function of the visual field to which the stimuli were displayed and the hand used to execute the imitative movement. The schema in Figure 1 shows the hypothesized non-uniform information flow across the two hemispheres in the different conditions, related to regions in the brain based on brain imaging and lesion studies [5, 6, 7]. The stimulus is visually processed in the hemisphere contralateral to the visual field (due to optic chiasm) and the motor command is prepared in the hemisphere contralateral to the hand. The arrows show the necessary transfer of information between the two hemispheres, thus a possible source of spatial errors in the imitation (as the patient suffers from disconnected hemispheres). Imitation was perfect only in the right visual field - right hand condition, indicating a lateralization of the processing to the left hemisphere and a necessary computational process in the brain area shown in dark grey.

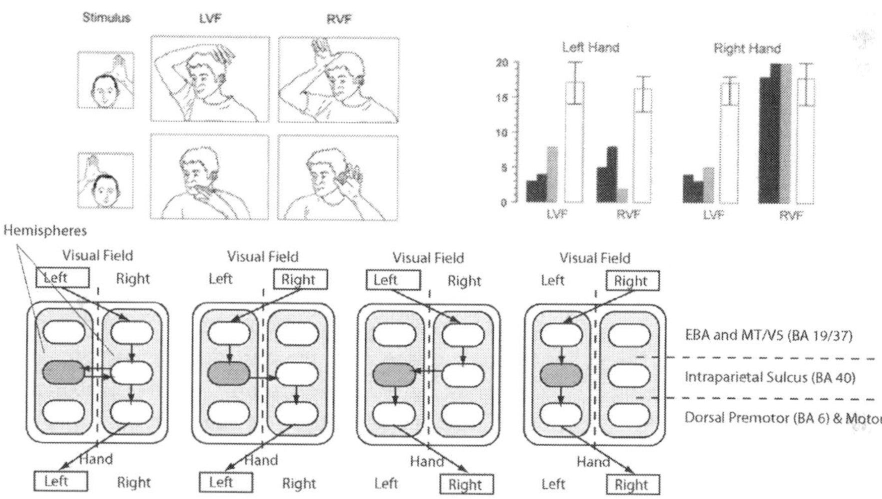

Fig. 1. Upper left *Goldenberg's experiment* of imitation of meaningless gestures, an example of a visual stimulus to imitate and the errors made by the patient. Upper right the patient's score of success in the four conditions (several trials, in white control data), taken from [1]. In the lower part, *schema of information flow* through the left and right hemispheres of the brain in the four conditions, see the text for explanation.

2 Neurocomputational Model of Imitation

In this paper, we investigate impaired imitation of meaningless gestures, namely hand postures relative to the head as the one shown in Figure 1. This work follows from a general effort in our group to decipher the neural mechanisms of visuo-motor imitation [8, 9]. In order to model the behavioral data reported in Goldenberg's study, we developed a neural network architecture that accounts for the transformations required to translate the observation of the visual stimulus to imitate to the corresponding tactile and proprioceptive information that will guide the imitative gesture. We simulate a callosal lesion by impairing the transfer of information between the networks and observe the occurrence of spatial parapraxias. Next, we describe the model.

2.1 Description of the Model

The model is composed of three neural networks, see Fig. 2: a *face visual network* in Brodmann Area BA 19/37 at the level of the occipito-temporal junction, a *face somatic network* in area BA 40 in the parietal cortex and a *hand position network* probably in dorsal premotor area BA 6. As it is the case in imitation of meaningless gestures we have implemented a visuo-motor route mediated by somatic knowledge of body parts. The face visual network receives geometrical properties of the visual stimulus to imitate (such as the position and angle of the hand relative to the nose, see Fig. 2. The face somatic network receives input from the face visual network and somatic input from tactile sensors of the face. The hand position network receives visuo-somatic input from the face somatic network and proprioceptive input from the arm. The neurons in our model are leaky integrator neurons in order to account for variations of the membrane potential in time and to have integrating properties.

Face visual network. The face visual network encodes geometrical properties of the stimulus to imitate. The network receives the two-dimensional input $\boldsymbol{x^H}$ composed of the distance $d^H \in \mathbb{R}[0,9]$ and angle $\phi^H \in \mathbb{R}[0,2\pi]$ of the hand relative the nose (shown on Fig. 2). We decided to use these two properties as they univocally define the stimulus to imitate and are quantities easy to process visually. It is certain that the brain uses also other quantities when imitating a hand posture relative to the head (position relative to the eye may be more appropriate in some cases), however we decided to limit the number of visual properties for simplicity. It was important that the visual and somatic networks rely on completely different representations.

The membrane potential m_i of the visual neuron i is governed by a first order differential equation modulated by a gaussian input:

$$\tau^V \frac{d}{dt} m_i^V = -m_i^V + \mathrm{e}^{-(\frac{|\boldsymbol{w_i^H} - \boldsymbol{x^H}|}{2\sigma_V^2})} \tag{1}$$

where τ^V is a time constant, $\boldsymbol{w_i^H}$ are the synaptic weights that connect the neuron i to the input $\boldsymbol{x^H} = \{d^H, \phi^H\}$ and σ_V corresponds to the "sensitivity" of the neuron to the input (a neuron with a large σ_V responds to a larger interval of inputs values).

The firing rate is a sigmoid function of the membrane potential with slope a and offset b:

$$g(m_i^V) = \frac{1}{(1 + \mathrm{e}^{a(-m_i^V + b)})} \tag{2}$$

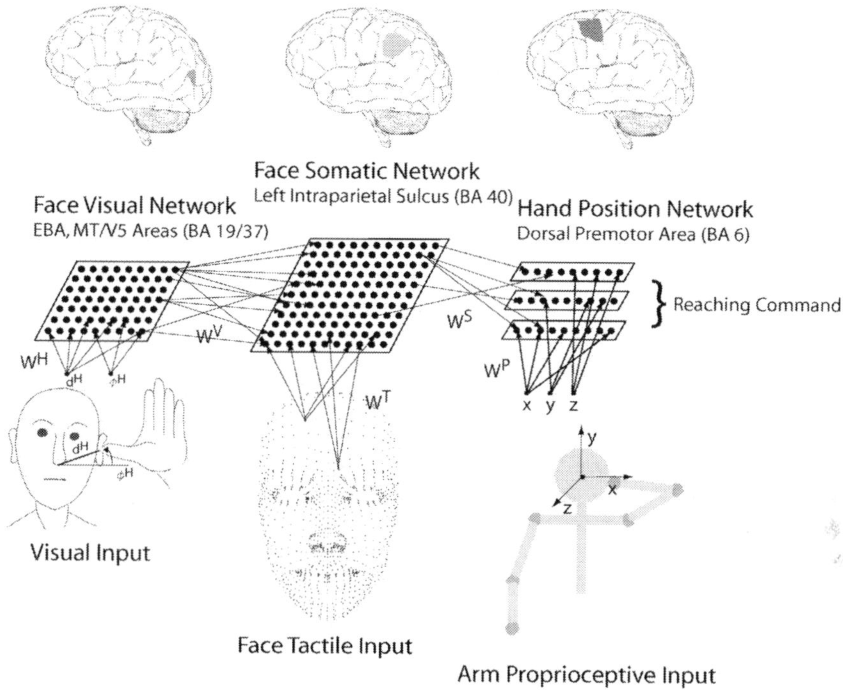

Fig. 2. Schema of the neurocomputational model. The model is composed of three neural networks that receive visual, tactile and proprioceptive input: a *face visual network* that corresponds to Brodmann Area BA 19/37 at the level of the occipito-temporal junction, a *face somatic network* that corresponds to area BA 40 in the parietal cortex and a *hand position network* in dorsal premotor area BA 6.

Face somatic Network. The face somatic network is a somatotopically organized network principally processing tactile information from the face. It receives input $\boldsymbol{x^T} \in \mathrm{IR}^{N^T}[0,1]$ from $N^T = 1500$ tactile sensors non-uniformly distributed on the face (with a preponderant number of sensors around the eyes, nose and mouth). It also receives visual input from the face visual network described previously. The membrane potential m_j^S of a somatic neuron with index j is equal to:

$$\tau^S \frac{d}{dt} m_j^S = -m_j^S + \sum_{k=1}^{N^T} w_{jk}^T e^{\left(-\frac{|\boldsymbol{x^P} - \boldsymbol{r_k^T}|}{2\sigma_T^2}\right)} + \sum_{i=1}^{N^V} w_{ji}^V g(m_i^V) \tag{3}$$

where τ^S is a time constant, w_{ik}^T is the synaptic weight of the neuron to the tactile sensor with index k and w_{ij}^V is the synaptic weight to the visual neuron with index i, N^T and N^V are the numbers of tactile sensors and visual neurons respectively, $\boldsymbol{r_k^T} \in \mathrm{IR}^3$ is the position of the tactile sensor k in space, $\boldsymbol{x^P} \in \mathrm{IR}^3$ is the center position of the hand-face contact and σ_T is the width of the contact. Note that the face somatic network integrates inputs of different types, namely somatic input from the tactile sensors and visual input preprocessed by the face visual network.

Three layers of the *hand position network* encode proprioceptive information from the arm. Each layer encodes a different coordinate of the position of contact $\boldsymbol{x^P} \in \mathbb{R}^3$ of the hand and the face, expressed in head-centered cartesian coordinates. Our motivations were the following: there is no "real" proprioceptive information from the face and we hypothesized that this information could be learned from correlations between the face tactile sensory activity and arm proprioceptive activity during reaching movements toward the face. A "positional code" may well be used in the brain where different coordinates are processed in segregated neural substrates, possibly in Cartesian coordinates [10]. The frame of reference is centered in the head to maximize the invariance of the positions of the tactile sensors (which would not be the case in a body-centered frame of reference because of the rotation of the head).

The neurons in the hand position network each have a preferred coordinate value c_k, preferred values were uniformly distributed in a volume that contains the head $\mathbb{R}^3$[-8,8]. The membrane potential m_k^P integrates over the proprioceptive input $\boldsymbol{x^P}$ and the visuo-somatic input $g(m_j^S)$ from the face somatic network (the vectorial notation expresses the three layers of the hand position network):

$$\tau^P \frac{d}{dt}\boldsymbol{m_k^P} = -\boldsymbol{m_k^P} + \mathrm{e}^{-(\frac{(\boldsymbol{x^P}-\boldsymbol{c_k})^2}{2\sigma_P^2})} + \sum_{j=1}^{N^S} \boldsymbol{w_{kj}^S} g(m_j^S) \tag{4}$$

where w_{kj}^S are the weights between a somatic neuron j and a position neuron k, σ_P is the width of the receptive field of the position neuron and N^S is the number of neurons in the face somatic network. The activation function is the same as in equation 2. The output of the hand position network is decoded using a weighted average of N^P (number of position neurons) firing rates which corresponds to the position $\boldsymbol{p}$ on the face:

$$\boldsymbol{p} = \frac{\sum_{k=1}^{N^P} \boldsymbol{c_k} g(\boldsymbol{m_k^P})}{\sum_{k=1}^{N^P} g(\boldsymbol{m_k^P})} \tag{5}$$

The decoded activity of the hand position network is used as a target for the imitation of a visual stimulus.

2.2 Training the Weights

The synaptic weights between the networks and their sensory inputs (i. e., weights W^H between the face visual network and the extracted visual parameters and weights W^T between the face somatic network and the face tactile input) have been trained with Kohonen's algorithm [11]. Thus our networks are self-organizing maps (SOM) whose weights preserve the topology of the input. The unsupervised learning algorithm consists of randomly choosing a sensory input $\boldsymbol{x}$ and determining the "winning neuron" with index j^* whose weights are closest to the input. It then updates the synaptic weights of the "winning" neuron and neurons in its neighborhood by the following rule:

$$\Delta \boldsymbol{w_i}(j^*) = \epsilon \cdot \mathrm{e}^{-\frac{|i-j^*|}{2\sigma_K^2}} [\boldsymbol{x} - \boldsymbol{w_i}] \tag{6}$$

where ϵ is the learning rate, $\boldsymbol{w_i}$ are the synaptic weights of the neuron with index i and σ_K corresponds to the size of the neighborhood[1]. In the end stimuli close in the input space are also close in the 2D neural space and more frequent inputs yield larger neural activities.

The synaptic weights between the networks (i. e., weights W^V between the visual and the face somatic network and weights W^S between the face somatic and hand position network) were trained with a presynaptic gating anti-hebbian learning rule:

$$\Delta w_{i,j} = \eta \cdot x_j [2 \sum w_{i,k} x_k - m_i] \tag{7}$$

where $w_{i,j}$ is the synaptic weight between a presynaptic neuron x_j and a post-synaptic neuron with membrane potential m_i and η is the learning rate. The learning process associates correlated activities of two networks. The connecting weights learn a mapping between the neural activity of one input and one output network for a given stimulus. In other words the weights organize in order to have the sensory activity in the input network represent the sensory activity in the output network. Both W^V and W^S were trained during the same process of self-observation, which simulates sensory input during reaching movements toward the face in front of a mirror. For example, the activity in the face visual network is associated to the somatic activity due to touching the face and is

Table 1. Parameter values

$N^V = 400$	$\tau^V = 35ms$	$\sigma_V = 0.6$	$\sigma_{VK} = 8$	$l^V = 0.97$	$\epsilon^V = 1$	$n^V = 0.98$
$N^S = 1225$	$\tau^S = 35ms$	$\sigma_S = 0.3$	$\sigma_{SK} = 22$	$l^S = 0.9996$	$\epsilon^S = 1$	$n^S = 0.99999$
$N^P = 3\text{x}100$	$\tau^P = 35ms$	$\sigma_P = 0.3$				
$a = 15$	$b = 0.5$		$\sigma = l\sigma$		$\epsilon = n\epsilon$	$\eta = -0.02$

associated with a position in space through proprioceptive information from the arm. In the end presentation of the visual stimulus to imitate alone yields the corresponding neural activities in the face somatic and position networks thus guiding a correct imitative action. The values used for the parameters of the model were selected by trial and error and are shown in Table 1[2].

[1] The weights were initialized with random values between 0 and 1 and the parameters ϵ and σ_K were decreased at each step according to the functions in Table 1.

[2] The inputs selected in the learning processes form a random uniform distribution in the input space. For a faster convergence all the time constants were set to 1. The Kohonen algorithm was run 100 times for the face visual network, 9000 times for the face somatic network and the anti-hebbian learning process was iterated 5000 times.

2.3 Simulation of the Lesion

For simulating the lesion of the corpus callosum (i. e., impaired transfer of information across the two hemispheres) we have taken into account two observations. First, some of the visual information must cross the callosum since the patient succeeds to imitate some hand positions when he/she visually processes the stimulus in one hemisphere and prepares the motor command in the other hemisphere. Second, interestingly enough, time is a very important variable. If the patient was given "unlimited time" he/she imitated correctly [12]. To model the observation that some of the information crosses, we introduce a probability of information transfer ρ . The impairing function is either applied at the level of the connection (model 1) or at the level of the input of the neuron (model 2). To model the improvement of the patient's performance with time we hypothesized an integrating factor greater than the decay factor. We added a constant $\lambda \in \mathbb{R}[0,1]$, which slows down the membrane decay. The dynamics of the membrane potential m of one neuron for the two models is then expressed by:

$$1)\ \tau\frac{d}{dt}m = -\lambda m + Wf(I) \quad 2)\ \tau\frac{d}{dt}m = -\lambda m + f(WI) \quad m < f(WI) \Rightarrow \lambda = 0 \tag{8}$$

where W is the weights matrix, I is the membrane input and f is the impairment function such as $f(x) = x$ with probability ρ and $f(x) = 0$ otherwise, see Figure 3. Therefore even if the neuron receives bits of information from time to time, the membrane potential is no more precisely tuned to the input but continues to integrate. As the face somatotopic network is situated in the left parietal cortex,we impair the connecting weights W^V in condition "left visual field" and the weights W^S in condition "left hand".

3 Results

To analyze the performance of our impairment models we have trained the weights once, then quantified the spatial parapraxias as the distance E between the desired end-target position $\boldsymbol{r}$ and the position $\boldsymbol{p}$ computed from the hand position network under different patterns of impairment[3].

A property of the model is to always converge to the right response given unlimited time no matter how impaired the transfer of information is, as long as some information does transfer ($\rho > 0$) and λ is small. As you can see in Figure 4 even for a probability of information transfer as small as $\rho = 0.1$ at the level of a single connection, the model converges to the correct position over time given a sufficiently small λ (0.1 in model 1 and 0.03 in model 2). The presence of λ deteriorates the performance in the unimpaired situation ($\rho = 1$) in model 1 (see Fig. 4) as the neuron membrane "overintegrates" in the first model, as shown

[3] For a simpler analysis of the results we have impaired all the connections equally, but our implementation allows variations of the percentage of impairment or location and size of the lesion.

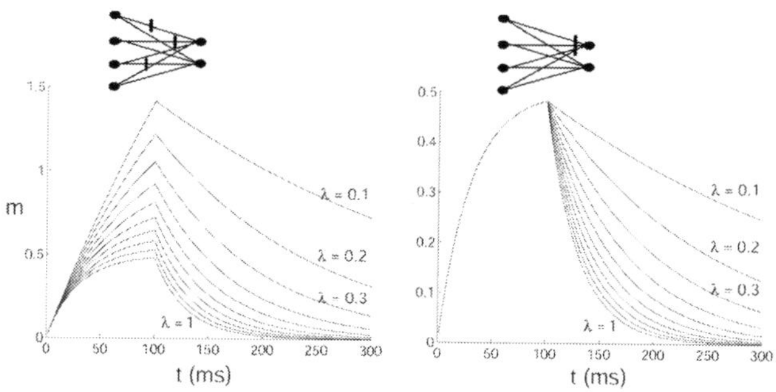

Fig. 3. The dynamics of the membrane potential of one neuron in model 1 which impairs the connection (on the left) and in model 2 where the impairment occurs at the neuron's input (on the right). Input $I = 0.5$ was applied during 100ms and τ was set to 30ms.

on Figure 3. Another drawback is that small values of λ render adaptation to a novel stimulus slower. However a longer decay time presents the advantage of having a "fading memory" of the stimulus, the stimulus remains represented in the brain after the presentation time, which is compatible with the occurrence of perseveration errors observed in experimental studies. Several predictions can be made on the basis of these models. With severe lesions, the patient needs more time to do a correct imitation, shown in Figure 4. It suggests that it is possible to obtain a measure of severity of the lesion based on the time needed by the patient to do the imitation. Small λ values would enable a correct processing even at very high impairment rates, but would depreciate the reaction time.

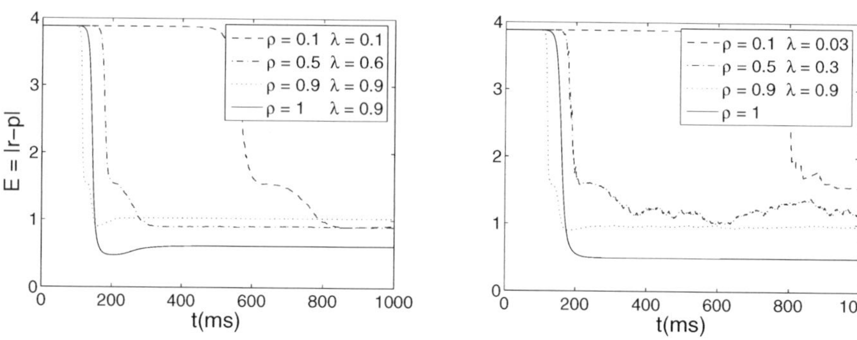

Fig. 4. The error in imitation computed as the distance between the desired and simulated end-target position for different values of ρ and λ according to model 1 (on the left) and model 2 (on the right). The observation that the patient required 180ms of visual stimulus presentation time to be able to imitate motivated the choice for the parameters of the activation function and of τ ($\tau = 35$). The starting position is the same throughout the trials and corresponds to the origin of the head-centered coordinate axis. We observe that more severe lesions necessitate longer processing time.

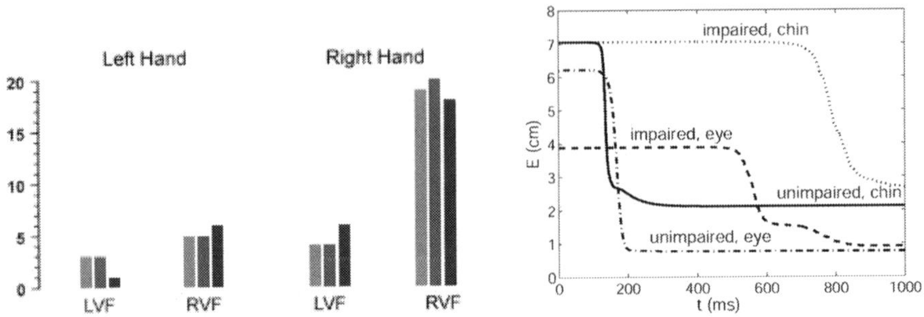

Fig. 5. Left, comparison of the results of the Goldenberg's study (light grey histograms) to the results of the simulations using the impairment model 2 (dark grey, $\tau = 30ms$, $\rho = 0.5$, and $\lambda = 0.3$) and model 1 (black, $\tau = 30ms$, $\rho = 0.4$, and $\lambda = 0.4$) respectively. The imitation was considered correct if the error distance was lower than 2.5/1.3. Right, inhomogeneities in the precision and processing time of imitation gestures toward different parts of the face, dependent on how well represented they are in the face somatic network (in our case the eye has a larger representation than the chin).

We compared the results of the simulations to the scores in Goldenberg's study with some adaptation. As we consider only the end-target spatial errors and not errors in the hand posture (such as orientation of the hand or finger configuration), we took the upper bound of the score used in the study (2 points for a correct imitation). We replicated the same experimental conditions (i. e., same visual stimuli, 180ms of stimulus presentation and weights impairment coherent with the four conditions as described in the study). A set of values could explain the scores in the Goldenberg study, as shown in Figure 5. The second model shows slightly better results, however this was not significant.

The representation of parts of the face in the "face somatic" network is non-uniform, some face parts such as the eyes or the mouth are overrepresented in contrast with the cheek or the chin. This is due to the non-uniform distribution of the tactile sensors. Therefore we observe inhomogeneities in the precision of the imitation task and in the processing times (shown in Fig. 5). Interesting predictions can be made from focal rather than diffuse lesions (i. e., stroke vs degenerative lesions). If only one part of the information transfer in weights W^V connecting the visual and somatic networks is impaired, then one should observe deficits in imitation only in some parts of the face and not in others. Specific local impairment of the weights W^S connecting the somatic and position networks could provoke errors in only one coordinate. For example, if the brain really uses a Cartesian representation in a head-centered frame of reference, then the position of the hand when reaching for the final target would be shifted only along one coordinate axis around the head. Spatial errors made by stroke patients should be used to test the plausibility of the model. However, because of brain reorganization, one should look at the impairment in imitation immediately after the lesion. As our model has learning properties, the model could possibly account for some of the effects of brain organization.

3.1 Conclusion

We presented a neural network architecture that could reproduce the deficits in visuo-motor imitation of meaningless gestures, reported in Goldenberg's seminal study [1]. We modelled two types of lesions that would affect either the integrative computation of the neuron or the connectivity across the neurons, leading to different predictions. Further, the model makes hypotheses on the type of representation used for the stimuli, for which there is as yet no neurological evidence. Further behavioral studies will be required to validate or invalidate the model's hypotheses and predictions.

References

1. Goldenberg, G., Laimgruber, K., Hermsdörfer, J.: Imitation of gestures by disconnected hemispheres. Neuropsychologia **39** (2001) 1432–43
2. Poeck, K., Kerschensteiner, M.: Ideomotor apraxia following right-sided cerebral lesion in a left-handed subject. Neuropsychologia **9** (1971) 359–361
3. Goldenberg, G.: Imitating gestures and manipulating a mannikin – the representation of the human body in ideomotor apraxia. Neuropsychologia **33**(1) (1995) 63–72
4. Hermsdörfer, J., Mai, N., Spatt, J., Marquardt, C., Veltkamp, R., Goldenberg, G.: Kinematic analysis of movement imitation in apraxia. Brain **119** (1996) 1575–1586
5. Decety, J., Grèzes, J., Costes, N., Jeannerod, M., Procyk, E., Grassi, E., Fazio, F.: Brain activity during observation of actions. Brain **120** (1997) 1763–1777
6. Mühlau, M., Hermsdörfer, J., Goldenberg, G., Wohlschläger, A.M., Castrop, F., Stahl, R., Röttinger, M., Erhard, P., Haslinger, B., Ceballos-Baumann, A.O., Conrad, B., Boecker, H.: Left inferior parietal dominance in gesture imitation: an fMRI study. Neuropsychologia **43** (2005) 1086–1098
7. Haaland, K.Y., Harrington, D.L., Knight, R.T.: Spatial deficits in ideomotor limb apraxia. A kinematic analysis of aiming movements. Brain **122** (1999) 1169–1182
8. Sauser, E., Billard, A.: Parallel and distributed neural models of the ideomotor principle: An investigation of imitative cortical pathways. Neural Networks, Special Issue on The Brain Mechanisms of Imitation Learning **19(3)** (2006)
9. Billard, A.: Imitation: In M. A. Arbib (ed.). Handbook of Brain Theory and Neural Networks (2002) 566–569
10. Lacquaniti, F., Guignon, E., Bianchi, L., Ferraina, S., Caminiti, R.: Representing spatial information for limb movement: role of area 5 in the monkey. Cerebral Cortex **5** (1995) 391–409
11. Kohonen, T.: Self-Organizing Maps. 3. ed., Springer-Verlag (2001)
12. Zaidel, D., Sperry, R.W.: Some long term motor effects of cerebral commissurotomy in man. Neuropsychologia **15** (1977) 193–204

Temporal Data Encoding and SequenceLearning with Spiking Neural Networks

Robert H. Fujii and Kenjyu Oozeki

School of Computer Science and Engineering
University of Aizu
Fukushima Prefecture, Aizu-Wakamatsu
fujii@u-aizu.ac.jp

Abstract. Sequence Learning using a Spiking Neural Network (SNN) was performed. An SNN is a type of Artificial Neural Network (ANN) that uses input signal arrival time information to process temporal data. An SNN can learn not only combinational inputs but also sequential inputs over some limited amount of time without using a recurrent network. Music melodies were encoded using unit amplitude spikes having various inter-spike interval times. These spikes were then fed into an SNN learning system. The SNN learning system was able to recognize various melodies after learning. The SNN could identify the original and noise-added melody versions properly in most cases.

Keywords: Spiking Neural Network, sequence learning, temporal data processing.

1 Introduction

Intensive research in the neuro-computational area is currently taking place in many parts of the world with the goal of understanding how the brain processes complex information quickly and seemingly without effort. Brain research has uncovered biophysical evidence that various types of neurons with specialized capabilities are involved in the information processing that occurs in the brain and the spinal cord. Artificial Neural Networks (ANNs) are simplified models of the kinds of neural networks used in the brain. An artificial neuron model is shown in Figure 1. The input, the output, the response function, and the weight of the ANN model correspond respectively to the biological neuron's dendrite, axon, soma response function, and synapse weight. A dendrite receives inputs from other neurons through its synapses. An axon corresponds to the output terminal of a neuron. A soma is the main body of a neuron. A synapse ties one neuron with other neurons; the transmission efficiency of a synapse is dependent on how much the input signal is attenuated at the synapse.

Traditional neural networks [2] constructed using neurons based on the ANN model, process input data (analog amplitude or digital 1 or 0 data) at discrete times and it is generally assumed that all neurons in a large network work synchronously (i.e. a global clock is needed). Such neural networks have been used mainly to process combinational inputs (rate-code processing). For sequential input processing using the

S. Kollias et al. (Eds.): ICANN 2006, Part I, LNCS 4131, pp. 780 – 789, 2006.

traditional neural networks, a feedback connection from the neuron output to its input is used; neural networks that use feedback connections are called recurrent neural networks [4]. The problem with recurrent networks for sequence recognition has been that past sequence information gradually gets "forgotten" as new information is input.

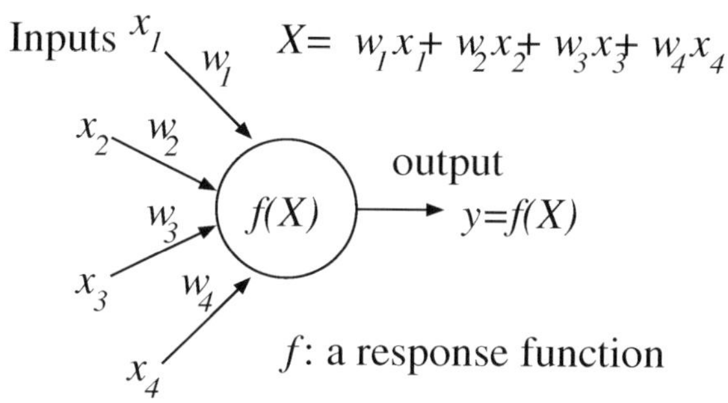

Fig. 1. ANN Model

A different type of artificial neuron is the spiking neuron model shown in Figure 2. Although a spiking neuron with only a single synaptic input is shown in Figure 2, a spiking neuron can have multiple synapses.

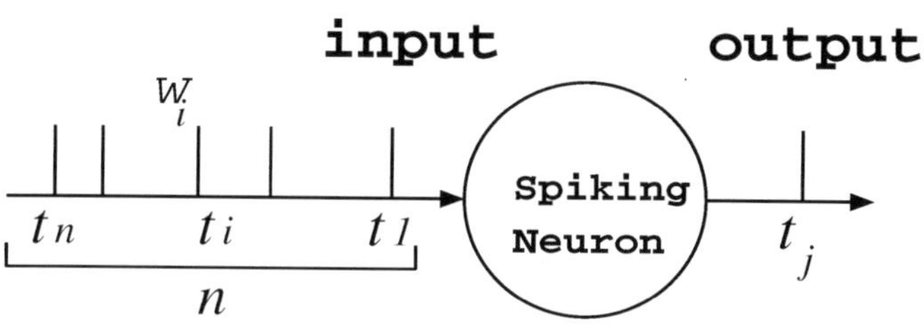

Fig. 2. Spiking Neuron with Spike Train Input

The inputs into a spiking neuron synapse consist of a sequence of unit amplitude digital spikes of very short duration. A sequence of spike inputs entering a synapse is called a spike train. The inter-spike times in a spike train are real valued and can take an infinite number of values within some specified range. Thus, information can be carried by the inter-spike times as well as by their sequence. This neuron's response function is based on the Leaky Integrator Fire Neuron model [3].

A spiking neuron has the following characteristics that differentiate it from the traditional ANN model: a) although it can process combinational inputs, its response function makes it easier to process inter-spike temporal information; and b) only a local timing reference is needed. There is a neurological basis to the spiking neuron model; recent neurological research has shown that in addition to the average firing frequency (rate code), biological neurons utilize the temporal information contained in the sequence of inter-spike times. The inter-spike interval (ISI) time information encoding space can be much larger than the one provided by the rate code used in traditional neural networks as was shown in [7]. Thus, the rich bandwidth for ISI information coding can lead to efficient neural networks. For further background on SNNs and their computational capabilities the reader is referred to [1, 3]. Spiking neural networks have been used for applications such as sound recognition and function approximation [5, 6].

In this paper, learning sequential data using an SNN is proposed. Past approaches for learning sequences using neural net like structures have included recurrent networks such as those described in [4]. The sequence learning method proposed in this paper uses spiking neurons connected in a feed-forward neural network structure and does not have the "forgetting" problem associated with recurrent networks. To test the efficacy of the proposed sequence learning method, short sequences (six to eight measures) of musical notes forming sub-sections of musical melodies were used as sequential data to be learned by an SNN learning system. The melody data consisting of note pitches and note time duration information was encoded into spike trains; these spike trains were then converted into a simpler form that could be more easily learned by the SNN learning stage. The SNN learning system was able to distinguish various melodies with and without added noise. An advanced version of this system could be used as a music database search machine.

2 Sequence Learning Using a Spiking Neural Network

The learning algorithm described in this paper utilizes two models of the spiking neuron: 1) the Leaky Integrate and Fire Neuron (LIFN) model [3] used for the coincidence detection neuron in the Learning Unit; and 2) a simplified version of the LIFN model used for the Mapping and Learning Units. The Mapping and Learning Units are described in section 3 of this paper. The structure of the learning system is shown in Figure 5.

The internal potential function of an LIFN spiking neuron [1] is defined as follows:

$$x_j(t) = \sum_{i=1}^{n} \beta\, \mathbf{W}_i\, \alpha(t - t_i) \tag{1}$$

Wi is the ith synaptic weight variable. $x_j(t)$ is the jth neuron's internal potential at time t. β is a weight scaling parameter. α (t) is the spike neuron's internal potential function

$$\alpha(t) = \frac{t}{\tau} e^{1 - t/\tau}. \tag{2}$$

τ is a time constant parameter that determines the rate of decay.

For the simplified LIFN model spiking neuron, the internal neuron potential function (2) was changed from an exponential function with decay to a linear integrator function with no decay as shown in (3). The exponential function in the LIFN model was changed to a linear $\alpha(\mathbf{t})$ function by letting $\tau \gg \mathbf{t}$. The simplified LIFN model makes the theoretical analysis of a spiking neural network more manageable. The simplified LIFN model based spiking neuron has the following $\alpha(t)$ and $x_j(t)$ functions:

$$\alpha(t) = t\frac{e}{\tau}; \quad x_j(t) = \Sigma_{i=1}^{n} \beta \frac{t_i}{\tau} e. \tag{3}$$

In the melody learning experiments, it was assumed that the inter- spike time could be any real value ≥ 3 ms. The 3 ms minimum inter-spike time value was selected to reflect the fact that a biological brain does not appear to be able to distinguish inter-spike times that are smaller than approximately 3ms. A much smaller minimum inter-spike time can be selected if the spike train processing is carried out using a spiking neural network implemented in hardware (e.g. as an integrated circuit) or in the case of software-based computer simulations.

2.1 Mapping of Spike Train Data

Twenty -five music note pitches (from C4 to C6, two octaves) and the time duration of the notes were used to represent the notes of a melody. Note duration time encoding examples are shown in Figure 3. A sixteenth note was represented by two spikes spaced 6 ms apart (first row of Figure 3, spikes at time 0 and 6 ms). The end of a note was represented by a spike 3ms after the last spike representing a given note duration (first row of Figure 3, spike at time 9 ms).

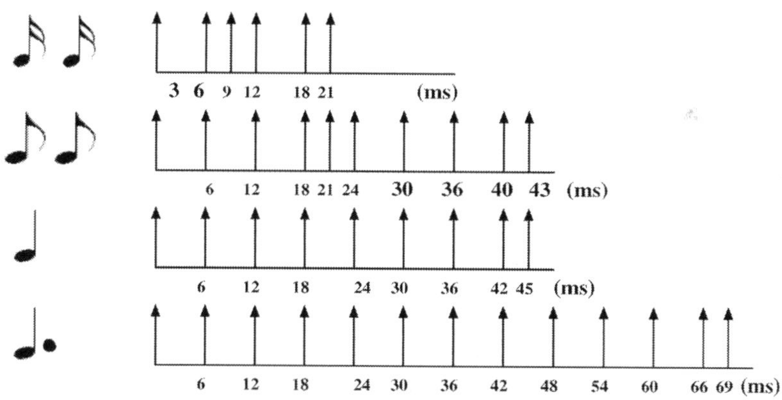

Fig. 3. Spike Train Encoding of Note Duration

For the second sixteenth note, spikes occurred at times 12 ms and 18 ms followed by a spike at time 21ms (first row of Figure 3). The separation between two notes that were to be heard as two distinct sounds was encoded using: a 3 ms delay after the end of a note spike. Two sixteenth notes connected by a slur (i.e. the two sixteenth notes are to be sounded continuously for a one eighth time duration), were represented with four spikes having inter-spike times of 6 ms (refer to row 2 of Figure 3, spikes at times 0, 6 ms, 12 ms, and 18 ms followed by a note separation spike at time 21 ms) .

The eighth note, fourth note, and the dotted fourth note representations are also shown in Figure 3.

Instead of trying to directly learn the complex characteristics of the spike trains that represented a musical melody, spike train data was first converted (mapped) into a simpler form using mapping units. A mapping unit receiving a spike train and producing two output firing times from the spiking neurons labeled ISI1 and ISI2 is shown in Figure 4. A particular aspect of the melody (e.g. a certain note pitch and its time duration) was assigned its own mapping unit. A mapping unit received the spike train data and then mapped it into the output firing times of the ISI1 and ISI2 neurons. Thus, an almost ideal one-to-one mapping between a spike train and the output firing times of two neurons could be achieved. However, realizing a unique one-to one mapping was not possible because the inter-spike times were real valued numbers ≥ 3 ms. An attempt was hence made to use a spike train mapping scheme that would have a very low probability of producing the same output firing times for different spike trains. Two kinds of Inter-Spike Interval (ISI) spiking neurons having different weight assignment functions were selected to perform the spike train mapping: ISI1 and ISI2 neurons. The ISI neurons are based on the simplified LIFN model.

The ISI1 neuron dynamically assigns a synaptic weight to an incoming spike input depending on the spike's arrival time with respect to a local reference time. The synaptic weight $Wi = \frac{\beta_1}{\tau} t_i$ (i.e. proportional to the spike arrival time) when $\mathbf{i} \neq 1$ and W1 when i =1 in (4) for the ISI1 neuron.

For the ISI2 neuron the synaptic weight $Wi = \beta_2 / \tau$ (i.e. a constant weight) when $\mathbf{i} \neq 1$ and W1 when i =1 in (4). It should be noted that the weight assignment for ISI2 does not depend on the input time as in ISI1 but the ISI2 neuron output firing time is still dependent on the input time of the spikes in a spike train.

$$x(t) = t_{out} w_1 + (t_{out} - t_2)\beta \frac{w_2}{\tau} + \cdots + (t_{out} - t_n)\beta \frac{w_n}{\tau} = t_{out}\left(w_1 + \frac{\beta}{\tau}\sum_{i=2}^{n} w_i\right) - \frac{\beta}{\tau}\sum_{i=2}^{n} w_i t_i . \tag{4}$$

x (t) = Internal Potential of neuron; w1 = constant weight for fist spike
wi = Weight for ith spike; ti = ith spike time; tout = output spike time
τ = time constant parameter; n = nth spike; β = Weight scaling parameter

The ISI1 or ISI2 neuron internal potential function is shown in (4). When x (t) in (4) reaches a specified threshold voltage v from below, the neuron fires a spike at time tout. W1 is the weight assigned for the first spike at the input of the ISI neuron. For the ISI1 neuron, the Wi weights are proportional to the ith spike input time. For the ISI2 neuron, the Wi weight is a constant.

The threshold voltage of an ISI neuron can be derived from (4) as follows:

$$ISI1: v1 = t_{out}(W_1 + (\beta_1/\tau)\sum_{i=2}^{n} t_i) - (\beta_1/\tau)\sum_{i=2}^{n} t_i^2. \tag{5}$$

$$ISI2: v_2 = t_{out} W_1 + (\beta_2/\tau)\sum_{i=2}^{n} (t_{out} - t_i). \tag{6}$$

v1 and v2 are the threshold voltages for the ISI1 and ISI2 neurons respectively. The neuron fires at time tout when the threshold voltage is reached. The W1 weight value used for $t_1 = 0$, was assigned a constant value of 0.7 based on experimental data. n is the number of spikes in a spike train.

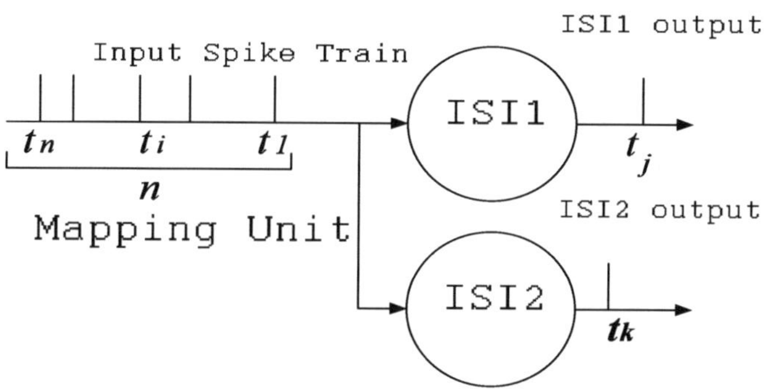

Fig. 4. Spike Train Data Mapped by ISI1 and ISI2 into output times tj and tk

2.2 Determining the β Parameter Value

The β parameter value used in equations (1, 3 – 6) for the ISI units in the mapping and learning units has to be determined appropriately in order for the SNN learning system to work accurately and quickly. When β is set too small, the neuron firing time becomes long and thus the overall processing speed slows down; when β is set too high, not all input spikes may be accounted for because some of the spiking neurons will fire before having received all the spikes. The β values for the mapping and learning units have to be set separately because the number of spikes entering these units is different. Some assumptions were made: 1) the threshold voltages (v1 and v2 in (5) and (6)) for the ISI1 and ISI2 neurons were set to 0.8V assuming that the learning system would be implemented in some VLSI technology with transistors that have a threshold voltage in the 0.8 V range; and 2) in order to account for all spike inputs within a spike train, the spiking neuron's firing time was made to occur after a pre-specified input time window (e.g. one second). To account for the worst case inputs that cause the neuron to fire at the earliest time, the β_1 parameter for the ISI1 neuron was computed by assuming that all spikes in a spike train arrive at $\frac{tinp}{2}$, where t_{inp} was either the mapping or the learning unit's input time window + 3ms (the minimum inter-spike time δt). A spike train in which spikes arrive at the earliest possible input times makes the ISI2 neuron fire the earliest. β_1 and β_2 were computed as follows:

$$\mathrm{ISI1}: \beta_1 = \frac{\tau(v_1 - (t_{inp} + \delta t)W_1)}{(t_{inp} + \delta t)(n-1)\frac{t_{inp} + \delta t}{2} - (n-1)\frac{(t_{inp} + \delta t)^2}{2}}. \tag{7}$$

$$ISI2: \beta_2 = \frac{\tau(v_2 - (t_{inp} + \delta t)W_1)}{\sum_{i=2}^{n}(t_{inp} + \delta t) - \frac{1}{2}\delta t\,(n-1)n}. \tag{8}$$

Using the β values computed using (7) and (8), the ISI1 and ISI2 neurons fired only after receiving all input spikes in a spike train. The value of n, the maximum number of spikes in a spike train, was set arbitrarily to 100 for the mapping unit and to 50 for the learning unit. For the mapping unit, six to eight measures of a musical melody line were assumed to have on the average up to 32 notes and that not all the notes in the melody were identical; hence, a spike train representing the note durations for one note pitch would most likely have less than 100 spikes. For the learning unit, the maximum number of spikes was set to 50 because there were 25 mapping units (one mapping unit for each note pitch) and two neuron output firings times for each mapping unit. For the mapping and learning units, the actual number of spikes in a spike train could vary from 0 to 100 and from 2 to 50 respectively depending on the melody. The maximum number n of spikes can be set to a different value provided various parameters such as β are adjusted properly.

It should be noted that there is no guarantee of a one-to-one input-output mapping result despite the fact that two types of ISI neurons are used to map one spike train. However, the choice of these two particular ISI1 and ISI2 neurons can be shown to have a low probability of generating identical ISI1 and ISI2 output times for two different input spike trains.

3 Melody Learning System

The structure of the learning system is shown in Figure 5. The system is comprised of the Mapping stage and the Learning stage. The Mapping stage and the Learning stage consist respectively of Mapping Units (MUs) and Learning Units (LUs). The learning system was simulated using MATLAB [8].

3.1 Mapping Stage

The Mapping stage is composed of MUs as shown in Figure 5. Each MU is composed of one ISI1neuron and one ISI2 neuron as described in Section 2 and shown in Figure 4. 25 music note pitches (from C4 to C6, two octaves) and the time duration of the notes were used to represent a melody. One MU unit was assigned to one note pitch (e.g. note C4 was assigned to MU1). Therefore, altogether 25 MUs were used in the Mapping stage. The melody data consisting of notes of various pitches and of various durations were encoded into spike trains (as described in section 2) and were then fed into the mapping stage where each spike train data was converted to two neuron output times (ISI1 and ISI2 output times) in each MU. Thus, if 25 spike train inputs are fed into the 25 Mapping units, 50 output spikes are output by the ISI neurons. .

The 50 output spikes from all the MUs formed the encoding (mapping) of the melody's 25 spike trains. All 50 output spikes from the Mapping stage were fed into the Leaning stage LUs as shown in Figure 5.

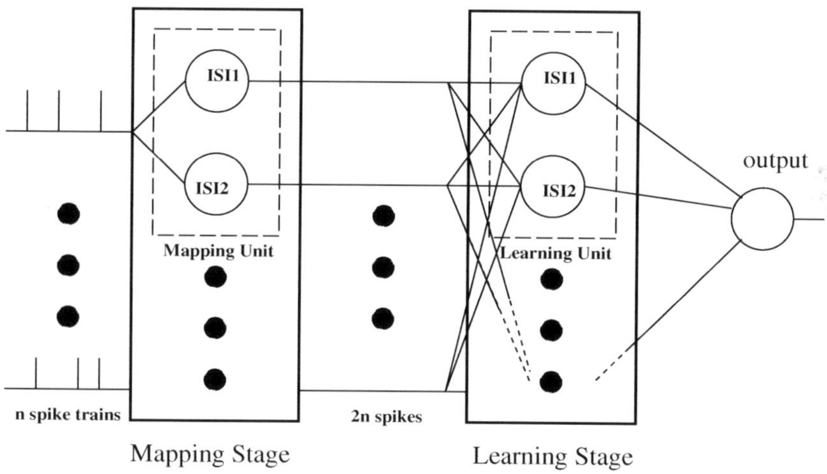

Fig. 5. Mapping and Learning System

3.2 Learning Stage

The Learning stage is composed of Learning Units (LUs) as shown in Figure 5. Each LU is composed of one ISI1 and one ISI2 neuron. Each ISI neuron receives 50 inputs from the mapping stage. Supervised learning is performed.

Initially there are no LUs. When learning starts, the first LU is generated; the synaptic weights for this LU are determined by the output firing times produced by the Mapping stage for a particular melody. In an LU, once the synaptic weights have been determined, the weights remain fixed so that a particular melody can be recognized by the output firing times of the ISI1 and ISI2 neurons in the LU. Thus, unlike the MU's ISI1 synaptic weights that dynamically change every time a spike is input, the LU's ISI1 synaptic weights remain fixed after learning. The supervised learning algorithm is as follows:

1) Select one original noise-free melody (e.g. melody 1) and assign weights to the LU (e.g. LU1) that is to represent this melody. The weights may be re-adjusted later in order to accommodate noisy versions of the same melody. 2) Select a noisy version of the original noise-free melody chosen in step 1. Apply the mapping stage outputs for this noisy version to the LU for which weights were assigned in step 1. 3) Compare

the output firing times of the LU for the original noise-free melody and the noisy version. If the output times differ by less than some ε time, the noisy version of the melody can be recognized using the same LU assigned for the original noise-free melody. If the output firing times differ by more than some ε time, a new LU is assigned for the noisy version of the melody. 4) Repeat steps 2 and 3 until all noisy versions of the original noise-free melody have been selected. 5) Repeat from step 1 for the remaining original noise-free melodies. At least one LU is needed for each melody. Additional LU units were sometimes needed to recognize a noisy version of the original noise-free melody.

3.3 Music Melody Data

Five different melodies and their noisy versions were used. Noise consisted of adding/ deleting a note, changing the time duration of a note, changing the pitch of a note, and shifting the times of the spikes slightly. Altogether fifty melody patterns were used for the learning simulations. Melody patterns 1 – 5 were the five original noise-free melodies. Melody patterns 6 and 11were made by adding one unused note pitch to the original melody 1. Melody pattern 7 was made by interchanging two spike trains representing two different pitches in melody 2. Melody pattern 8 was made by prolonging the duration of one note in melody 3. Melody pattern 9 was made by shortening the duration value of one note in melody 4. Melody pattern 10 was melody pattern 5 with one of the notes removed. Random spike time shifting noise was added to these eleven patterns in order to make 39 additional patterns. Ideally, five LUs should have been sufficient to distinguish the five melodies and their noisy versions as shown in Table 1. However, nine LUs were needed because some of the noisy versions could not be grouped together into the same LU that had been assigned for the original noise-free melody.

Table 1. Ideal Melody Learning

	Learning Unit				
	LU1	LU2	LU3	LU4	LU5
MELODY	1	2	3	4	5
	6	7	8	9	10
	11	12	13	14	15
	16	17	18	19	20
	21	22	23	24	25
	26	27	28	29	30
	31	32	33	34	35
	36	37	38	39	40
	41	42	43	44	45
	46	47	48	49	50

Table 2. Actual Learning Results

	Learning Unit								
	LU1	LU2	LU3	LU4	LU5	LU6	LU7	LU8	LU9
MELODY	1	2	3	4	5	6	7	10	11
	16	12	8	9	15	31	32	35	36
	21	17	13	14	20				
	26	22	18	19	25				
	41	27	23	24	30				
	46	37	28	29	40				
		42	33	34	45				
		47	38	39	50				
			43	44					
			48	49					

Nine LUs were needed to group the 50 melody patterns. Depending on the type of noise, more than one LU was needed to recognize a melody; for example, LU1, LU6, and LU9 were needed for pattern 1 and its noisy versions. After learning, a test of the

SNN system was performed to ascertain that the system did not cause more than one LU unit to fire at the same time for one melody.

Two problems occurred during the learning of melodies:

1) Noisy versions of the original melodies required additional LUs. One way to solve this problem is to re-adjust the already "fixed" weights of the LUs so that additional noisy versions of the melodies can also be included in order to reduce the number of LUs.
2) An LU that was supposed to respond to only one melody group can incorrectly respond to a melody belonging to a different group with nearly the same LU output times. There are several ways in which this problem can be solved: a) coincidence detection neurons, based on the Leaky Integrate and Fire spiking neuron model described in the introduction, can be used to distinguish slightly different LU output times; b) delay one of the output times with respect to the other by modifying β ; c) sub-divide the 50 inputs of the offending LU into groups of smaller number of inputs by using two or more LUs.

4 Conclusions

The recognition of temporal sequential data using a spiking neural network was carried out. Simple musical melodies were used as examples of temporal sequential data

The proposed Mapping-Learning SNN was able to recognize various melodies and their noise added versions properly. However, areas that remain to be further developed include: a) longer sequence learning; b) good algorithm for the minimization of Learning Units; and c) an unsupervised learning scheme.

References

[1] W. Maass and C. M. Bishop, "Pulsed Neural Networks," MIT Press, Cambridge, Massachusetts, 1999.
[2] S. Haykin, "Neural Networks," Prentice-Hall Publishers, 1999.
[3] W. Gerstner and W. Kistler, "Spiking Neuron Models," Cambridge University Press, 2002.
[4] Deliang Wang and Michael A. Arbib, "Complex Temporal Sequence Learning Based on Short-term Memory," Proceedings of the IEEE, Vol. 78, No. 9, Sept. 1990, pp. 1536-1543.
[5] H. H. Amin and R. H. Fujii, "Spike Train Decoding Scheme for a Spiking Neural Network," Proceedings of the 2004 International Joint Conference on Neural Networks (IJCNN '04), IEEE, pp.477-482, 2004.
[6] H. H. Amin and R. H. Fujii, "Spike Train Learning Algorithm, Applications, and Analysis," 48th IEEE Int'l Midwest Symposium on Circuits and Systems, Ohio, 2005.
[7] W. Maass, "Lower Bounds for the Computational Power of Networks of Spiking Neurons," Neural Computation, Vol. 8, No. 1, pp. 1-40, 1996.
[8] MATLAB,< http://www.cybernet.co.jp/matlab/>.

Optimal Tuning of Continual Online Exploration in Reinforcement Learning

Youssef Achbany, Francois Fouss, Luh Yen, Alain Pirotte, and Marco Saerens

Information Systems Research Unit (ISYS)
Place des Doyens 1, Université de Louvain, Belgium
{youssef.achbany, francois.fouss, luh.yen, alain.pirotte, marco.saerens}@ucLouvain.be

Abstract. This paper presents a framework allowing to tune continual exploration in an optimal way. It first quantifies the rate of exploration by defining the **degree of exploration** of a state as the probability-distribution entropy for choosing an admissible action. Then, the exploration/exploitation tradeoff is stated as a **global optimization problem**: find the exploration strategy that minimizes the expected cumulated cost, while maintaining fixed degrees of exploration at same nodes. In other words, "exploitation" is maximized for constant "exploration". This formulation leads to a set of nonlinear updating rules reminiscent of the value-iteration algorithm. Convergence of these rules to a local minimum can be proved for a stationary environment. Interestingly, in the deterministic case, when there is no exploration, these equations reduce to the Bellman equations for finding the shortest path while, when it is maximum, a full "blind" exploration is performed.

1 Introduction

One of the specific challenges of reinforcement learning is the tradeoff between exploration and exploitation. Exploration aims to continually try new ways of solving the problem, while exploitation aims to capitalize on already well-established solutions. Exploration is especially relevant when the environment is changing, i.e. nonstationary. In this case, good solutions can deteriorate over time and better solutions can appear. Without exploration, the system sends agents only along the up-to-now best path without exploring alternative paths. The system is therefore unaware of the changes and its performance inevitably deteriorates with time. One of the key features of reinforcement learning is that it explicitly addresses the exploration/exploitation issue as well as the online estimation of the probability distributions in an integrated way [18].

This work makes a clear distinction between "*preliminary*" or "*initial exploration*", and "*continual online exploration*". The objective of *preliminary exploration* is to discover relevant goals, or destination states, and to estimate a first optimal policy for exploiting them. On the other hand, *continual online exploration* aims to continually explore the environment, after the preliminary exploration stage, in order to adjust the policy to changes in the environment.

S. Kollias et al. (Eds.): ICANN 2006, Part I, LNCS 4131, pp. 790–800, 2006.

In the case of *preliminary exploration*, two further distinctions are often made [19,20,21,22]. A first group of strategies uses randomness for exploration and is often referred to as *undirected exploration*. Control actions are selected with a probability distribution, taking the expected cost into account. The second group, referred to as *directed exploration*, uses domain-specific knowledge for guiding exploration [19,20,21,22]. Usually, *directed exploration* provides better results in terms of learning time and cost.

On the other hand, "*continual online exploration*" can be performed by, for instance, re-exploring the environment either periodically or continually [6,15] by using a ϵ-greedy or a Boltzmann exploration strategy. For instance, joint estimation of the exploration strategy and the state-transition probabilities for continual online exploration can be performed within the SARSA framework [14,16,18].

This work presents a unified framework integrating exploitation and exploration for *undirected, continual, exploration*. Exploration is formally defined as the association of a probability distribution to the set of admissible control actions in each state (choice randomization). The *rate of exploration* is quantified with the concept of **degree of exploration**, defined as the (Shannon) entropy [10] of the probability distribution for the set of admissible actions in a given state. If no exploration is performed, the agents are routed on the best path with probability one – they just exploit the solution. With exploration, the agents continually explore a possibly changing environment to keep current with it. When the entropy is zero in a state, no exploration is performed from this state, while, when the entropy is maximal, a full, blind exploration with equal probability of choosing any action is performed.

The online exploration/exploitation issue is then stated as a **global optimization problem**: learn the exploration strategy that minimizes the expected cumulated cost from the initial state to the goal while maintaining a fixed degree of exploration. In other words, "exploitation" is maximized for constant "exploration". This problem leads to a set of nonlinear equations defining the optimal solution. These equations can be solved by iterating them until convergence, which is proved for a stationary environment and a particular initialization strategy. They provide the action policy (the probability distribution of choosing an action in a given state) that minimizes the average cost from the initial state to the destination states, given the degree of exploration in each state. Interestingly, when the degree of exploration is zero in all states, which corresponds to the deterministic case, the nonlinear equations reduce to the Bellman equations for finding the shortest path from the initial state to the destination states. The main drawback of this method is that it is computationally demanding since it relies on iterative algorithms like the value-iteration algorithm.

For the sake of simplicity, we first concentrate here on "*deterministic shortest-path problem*", as defined for instance in [5], where any chosen control action deterministically drives the agent to a unique successor state. On the other hand, if the actions have uncertain effects, the resulting state is given by a probability distribution and one speaks of "*stochastic shortest-path problems*".

In this case, a probability distribution on the successor states is introduced and it must be estimated by the agents; stochastic shortest-path problems are studied in Section 4.

Section 2 introduces the notations, the standard deterministic shortest-path problem, and the management of continual exploration. Section 3 describes our procedure for solving the deterministic shortest-path problem with continual exploration, while the stochastic shortest-path problem is discussed in Section 4. Section 5 is the conclusion.

2 Statement of the Problem and Notations

2.1 Statement of the Problem

During every state transition, a finite cost $c(k,u)$ is incurred when leaving state $k \in \{1, 2, \ldots, n\}$ while executing a control action u selected from a set $U(k)$ of admissible actions, or choices, available in state k. The cost can be positive (penalty), negative (reward), or zero provided that no cycle exists whose total cost is negative. This is a standard requirement in shortest-path problems [8]; indeed, if such a cycle exists, then traversing it an arbitrary large number of times would result in a path with an arbitrary small cost so that a best path could not be defined. In particular, this implies that, if the graph of the states is nondirected, all costs are nonnegative.

The **control action** u is chosen according to a **policy** Π that maps every state k to the set $U(k)$ of admissible actions with a certain probability distribution, $\pi_k(u)$, with $u \in U(k)$. Thus the policy associates to each state k a probability distribution on the set of admissible actions $U(k)$: $\Pi \equiv \{\pi_k(u), k = 1, 2, \ldots, n\}$. For instance, if the admissible actions in state k are $U(k) = \{u_1, u_2, u_3\}$, the distribution $\pi_k(u)$ specifies three probabilities $\pi_k(u_1)$, $\pi_k(u_2)$, and $\pi_k(u_3)$. The **degree of exploration** is quantified as the entropy of this probability distribution (see next section). Randomized choices are common in a variety of fields, for instance decision sciences [13] or game theory, where they are called mixed strategies (see, e.g., [12]). Thus, the problem tackled in this section corresponds to a **randomized shortest-path problem**.

Moreover, we assume that once the action has been chosen, the next state k' is known deterministically, $k' = f_k(u)$ where f is a one-to-one mapping between (states, actions) and resulting state. We assume that different actions lead to different states. This framework corresponds to a deterministic shortest-path problem. A simple modeling of this problem would do without actions and directly defined state-transition probabilities. The more general formalism fits full stochastic problems for which both the choice of actions and the state transitions are governed by probability distributions (see Section 4).

We assume, as in [5], that there is a special cost-free **destination** or **goal state**; once the system has reached that state, it remains there at no further cost. The goal is to minimize the **total expected cost** $V_\Pi(k_0)$ (Equation (2.1)) accumulated over a path $k_0, k_1, ...$ in the graph starting from an initial (or source) state k_0:

$$V_\Pi(k_0) = E_\Pi\left[\sum_{i=0}^{\infty} c(k_i, u_i)\right] \tag{2.1}$$

The expectation E_Π is taken on the policy Π that is, on all the random choices of action u_i in state k_i.

Moreover, we consider a problem structure such that termination is guaranteed, at least under an optimal policy. Thus, the horizon is finite, but its length is random and it depends on the policy. The conditions for which termination holds are equivalent to establishing that the destination state can be reached in a finite number of steps from any potential initial state; for a rigorous treatment, see [3,5].

2.2 Controling Exploration by Defining Entropy at Each State

The **degree of exploration** E_k at each state k is defined by

$$E_k = -\sum_{i \in U(k)} \pi_k(i) \log \pi_k(i) \tag{2.2}$$

which is simply the entropy of the probability distribution of the control actions in state k [9,10]. E_k characterizes the uncertainty about the choice at state k. It is equal to zero when there is no uncertainty at all ($\pi_k(i)$ reduces to a Kronecker delta); it is equal to $\log(n_k)$, where n_k is the number of admissible choices at node k, in the case of maximum uncertainty, $\pi_k(i) = 1/n_k$ (a uniform distribution).

The **exploration rate** $E_k^r = E_k / \log(n_k)$ is the ratio between the actual value of E_k and its maximum value. It takes its values in the interval $[0, 1]$. Fixing the entropy at a state sets the exploration level out of this state; increasing the entropy increases exploration up to the maximal value, in which case there is no more exploitation since the next action is chosen completely at random, with a uniform distribution, without taking the costs into account. This way, the agents can easily control their exploration by adjusting the exploration rates.

3 Optimal Policy Under Exploration Constraints for Deterministic Shortest-Path Problems

3.1 Optimal Policy and Expected Cost

We turn to the determination of the optimal policy under exploration constraints. More precisely, we will seek the policy $\Pi \equiv \{\pi_k(u), k = 1, 2, \ldots, n\}$, for which the expected cost $V_\Pi(k_0)$ from initial state k_0 is minimal while maintaining a given degree of exploration at each state k. The destination state is an absorbing state, i.e., with no outgoing link. Computing the expected cost (2.1) from any state k is similar to computing the average first-passage time in the associated Markov chain [11]. The problem is thus to find the transition probabilities leading to the minimal expected cost, $V^*(k_0) = \min\limits_{\Pi}(V_\Pi(k_0))$. It can be formulated as a constrained optimization problem involving a Lagrange function.

In [1], we derive the optimal probability distribution of control actions in state k, which is a logit distribution:

$$\pi_k^*(i) = \frac{\exp\left[-\theta_k\left(c(k,i) + V^*(k_i')\right)\right]}{\sum\limits_{j \in U(k)} \exp\left[-\theta_k\left(c(k,j) + V^*(k_j')\right)\right]}, \tag{3.1}$$

where $k_i' = f_k(i)$ is a following state and V^* is the optimal (minimum) expected cost given by

$$\begin{cases} V^*(k) = \sum\limits_{i \in U(k)} \pi_k^*(i)\left[c(k,i) + V^*(k_i')\right], \text{ with } k_i' = f_k(i) \text{ and } k \neq d \\ V^*(d) = 0, \text{ for the destination state } d \end{cases} \tag{3.2}$$

The control actions probability distribution (3.1) is often called "Boltzmann distributed exploration". In Equation (3.1), θ_k must be chosen in order to satisfy

$$\sum_{i \in U(k)} \pi_k(i) \log \pi_k(i) = -E_k \tag{3.3}$$

for each state k and given E_k. It takes its values in $[0, \infty]$. Of course if, for some state, the number of possible control actions reduces to one (no choice), no entropy constraint is introduced. Since Equation (3.3) has no analytical solution, θ_k must be computed numerically in terms of E_k. This is in fact quite easy since it can be shown that the function $\theta_k(E_k)$ is strictly monotonic decreasing, so that a line search algorithm (such as the bisection method, see [2]) or a simple binary search can efficiently find the θ_k value corresponding to a given E_k value.

Equation (3.1) has a simple appealing interpretation: choose preferably (with highest probability) action i leading to state k_i' of lowest expected cost, including the cost of performing the action, $c(k,i) + V^*(k_i')$. Thus, the agent is routed preferably to the state which is nearest (on average) to the destination state.

The same necessary optimality conditions can also be expressed in terms of the Q-values coming from the popular Q-learning framework [18,23,24]. Indeed, in the deterministic case, the Q-value represents the expected cost from state k when choosing action i, $Q(k,i) = c(k,i) + V(k_i')$. The relationship between Q and V is thus simply $V(k) = \sum_{i \in U(k)} \pi_k(i)\, Q(k,i)$; we thus easily obtain

$$\begin{cases} Q^*(k,i) = c(k,i) + \sum\limits_{i \in U(k_i')} \pi_{k_i'}^*(i)\, Q^*(k_i', i), \text{ with } k_i' = f_k(i) \text{ and } k \neq d \\ Q^*(d,i) = 0, \text{ for the destination state } d \end{cases} \tag{3.4}$$

and the $\pi_k^*(i)$ are given by

$$\pi_k^*(i) = \frac{\exp\left[-\theta_k Q^*(k,i)\right]}{\sum\limits_{j \in U(k)} \exp\left[-\theta_k Q^*(k,j)\right]} \tag{3.5}$$

which corresponds to a Boltzmann exploration involving the Q-value. Thus, a Boltzmann exploration involving the Q-value may be considered as "optimal" since it provides the best expected performances for fixed degrees of exploration.

3.2 Computation of the Optimal Policy

Equations (3.1) and (3.2) suggest an iterative procedure very similar to the well-known value-iteration algorithm for the computation of both the expected cost and the policy.

More concretely, we consider that agents are sent from the initial state and that they choose an action i in each state k with probability distribution $\pi_k(u = i)$. The agent then performs the chosen action, say action i, and incurs the associated cost, $c(k,i)$ (which, in a non-stationary environment, may vary over time), together with the new state, k'. This allows the agent to update the estimates of the cost, of the policy, and of the average cost until destination; these estimates will be denoted by $\widehat{c}(k,i)$, $\widehat{\pi}_k(i)$ and $\widehat{V}(k)$ and are known (shared) by all the agents.

1. **Initialization phase**
   - Choose an initial policy, $\widehat{\pi}_k(i)$, $\forall i, k$, satisfying the exploration rate constraints (3.3) and
   - Compute the corresponding expected cost until destination $\widehat{V}(k)$ by using any procedure for solving the set of linear equations (3.2) where we substitue $V^*(k)$, $\pi_k^*(i)$ by $\widehat{V}(k)$, $\widehat{\pi}_k(i)$. The $\widehat{\pi}_k(i)$ are kept fixed in the initialization phase. Any standard iterative procedure (for instance, a Gauss-Seidel like algorithm) for computing the expected cost until absorption in a Markov chain could be used (see [11]).
2. **Computation of the policy and the expected cost under exploration constraints**
   For each visited state k, do until convergence:
   - Choose an action i with current probability estimate $\widehat{\pi}_k(i)$, observe the current cost $c(k,i)$ for performing this action, update its estimate $\widehat{c}(k,i)$, and jump to the next state, k'_i

   $$\widehat{c}(k,i) \leftarrow c(k,i) \tag{3.6}$$

   - Update the probability distribution for state k as:

   $$\widehat{\pi}_k(i) \leftarrow \frac{\exp\left[-\widehat{\theta}_k\left(\widehat{c}(k,i) + \widehat{V}(k'_i)\right)\right]}{\sum\limits_{j \in U(k)} \exp\left[-\widehat{\theta}_k\left(\widehat{c}(k,j) + \widehat{V}(k'_j)\right)\right]}, \tag{3.7}$$

   where $k'_i = f_k(i)$ and $\widehat{\theta}_k$ is set in order to respect the given degree of entropy (see Equation (3.3)).

– Update the expected cost of state k:

$$\begin{cases} \widehat{V}(k) \leftarrow \sum\limits_{i \in U(k)} \widehat{\pi}_k(i) \left[\widehat{c}(k,i) + \widehat{V}(k'_i) \right], \text{ with } k'_i = f_k(i) \text{ and } k \neq d \\ \widehat{V}(d) \leftarrow 0, \text{ where } d \text{ is the destination state} \end{cases} \tag{3.8}$$

The convergence of these updating equations is proved for a stationary environment in [1]. However, the described procedure is computationally demanding since it relies on iterative procedures like the value-iteration algorithm in Markov decision processes.

Thus, the above procedure allows to optimize the expected cost $V(k_0)$ and to obtain a local minimum of this criterion. It does not guarantee to converge to a global minimum, however. Whether $V(k_0)$ has only one global minimum or many local minima remains an open question.

Notice also that, while the initialization phase is necessary in our convergence proof, other simpler initialization schemes could also be applied. For instance, set initially $\widehat{c}(k,i) = 0$, $\widehat{\pi}_k(i) = 1/n_k$, $\widehat{V}(k) = 0$, where n_k is the number of admissible actions in state k; then proceed by directly applying updating rules (3.7) and (3.8). While convergence is not proved in this case, we observed that this updating rule works well in practice; in particular, we did not observe any convergence problem. This rule is used in the experiments presented in [1].

3.3 Some Limit Cases

We will now show that when the degree of exploration is zero for all states, the nonlinear equations reduce to Bellman's equations for finding the shortest path from the initial state to the destination state.

Indeed, from Equations (3.7)-(3.8), if the parameter $\widehat{\theta}_k$ is very large, which corresponds to a near-zero entropy, the probability of choosing the action with the lowest value of $(\widehat{c}(k,i) + \widehat{V}(k'_i))$ dominates all the others. In other words, $\widehat{\pi}_k(j) \simeq 1$ for the action j corresponding to the lowest average cost (including the action cost), while $\widehat{\pi}_k(i) \simeq 0$ for the other alternatives $i \neq j$. Equations (3.8) can therefore be rewritten as

$$\begin{cases} \widehat{V}(k) \leftarrow \min\limits_{i \in U(k)} \left[\widehat{c}(k,i) + \widehat{V}(k'_i) \right], \text{ with } k'_i = f_k(i) \text{ and } k \neq d \\ \widehat{V}(d) \leftarrow 0, \text{ where } d \text{ is the destination state} \end{cases} \tag{3.9}$$

which are Bellman's updating equations for finding the shortest path to the destination state [4,5]. In terms of Q-values, the optimality conditions reduce to

$$\begin{cases} Q^*(k,i) = c(k,i) + \min\limits_{i \in U(k)} Q^*(k'_i, i), \text{ with } k'_i = f_k(i) \text{ and } k \neq d \\ Q^*(d,i) = 0, \text{ for the destination state } d \end{cases} \tag{3.10}$$

On the other hand, when $\widehat{\theta}_k = 0$, the choice probability distribution reduces to $\widehat{\pi}_k(i) = 1/n_k$, and the degree of exploration is maximum for all states. In this case, the nonlinear equations reduce to the linear equations allowing to compute the average cost for reaching the destination state from the initial state in a Markov chain with transition probabilities equal to $1/n_k$. In other words, we then perform a "blind" random exploration, for the choice probability distribution.

Any intermediary setting $0 < E_k < \log(n_k)$ leads to an optimal exploration vs. exploitation strategy minimizing the expected cost, and favoring short paths to the solution. In [1], we further show that, if the graph of states is directed and acyclic, the nonlinear equations can easily be solved by performing a single backward pass from the destination state.

Experimental simulations illustrating the behaviour of the algorithm, as well as comparisons with a naive Boltzmann and a ϵ-greedy exploration strategy, are provided in [1].

4 Optimal Policy Under Exploration Constraints for Stochastic Shortest Path Problems

We now consider **stochastic shortest path problems** for which, once an action has been performed, the transition to the next state is no longer deterministic but stochastic [5]. More precisely, when an agent chooses action i in state k, it jumps to state k' with a probability $\mathrm{P}(s = k'|u = i, s = k) = p_{kk'}(i)$ (transition probabilities). Notice that there are now two different probability distributions associated to the system: the probability of choosing an action i within the state k, $\pi_k(i)$, and the probability of jumping to a state $s = k'$ after having chosen the action i within the state k, $p_{kk'}(i)$.

By first-step analysis (see [1]), the recurence relations allowing to compute the expected cost $V_\Pi(k)$, given policy Π are easily found:

$$\begin{cases} V_\Pi(k) = \sum\limits_{i\in U(k)} \pi_k(i)\,[c(k,i) + \sum\limits_{k'=1}^{n} p_{kk'}(i)\, V_\Pi(k')], \\ V_\Pi(d) = 0, \text{ where } d \text{ is the destination state} \end{cases} \tag{4.1}$$

Furthermore, by defining the average cost when having chosen control action i in state k by $\overline{V}_\Pi(k,i) = \sum_{k'} p_{kk'}(i) V_\Pi(k')$, Equation (4.1) can be rewritten as

$$\begin{cases} V_\Pi(k) = \sum\limits_{i\in U(k)} \pi_k(i)\,[c(k,i) + \overline{V}_\Pi(k,i)], \\ V_\Pi(d) = 0, \text{ where } d \text{ is the destination state} \end{cases} \tag{4.2}$$

Thus, the optimal policy is obtained by substituting $V_\Pi(k'_i)$ by $\overline{V}^*(k,i)$ in both (3.1) and (3.2):

$$\pi_k^*(i) = \frac{\exp\left[-\theta_k\left(c(k,i) + \overline{V}^*(k,i)\right)\right]}{\sum\limits_{j\in U(k)} \exp\left[-\theta_k\left(c(k,j) + \overline{V}^*(k,j)\right)\right]} \tag{4.3}$$

The details are provided in [1]. The additional difficulty here, in comparison with a deterministic problem, is that the probability distributions $p_{kk'}(i)$, if unknown, have to be estimated on-line, together with the costs and the distribution of the randomized control actions [18].

4.1 On-Line Estimation of the Transition Probabilities

The transition probabilities $p_{kk'}(i)$ might be unknown and, consequently, should be estimated on-line, together with the costs and the distribution of the randomized control actions [18]. An alternative solution is to directly estimate the average cost $\overline{V}_{\Pi}(k,i) = \sum_{k'} p_{kk'}(i) V_{\Pi}(k')$ based on the observation of the value of V_{Π} in the next state k'. There is a large range of potential techniques for solving this issue, depending on the problem at hand (see for example [7]). One could simply use an exponential smoothing, leading to $\widehat{\overline{V}}(k,i) \leftarrow \alpha\widehat{V}(k') + (1-\alpha)\widehat{\overline{V}}(k,i)$, or a stochastic approximation scheme, $\widehat{\overline{V}}(k,i) \leftarrow \widehat{\overline{V}}(k,i) + \alpha\left[\widehat{V}(k') - \widehat{\overline{V}}(k,i)\right]$, which converges for a suitable decreasing policy of α [17].

This leads to the following updating rules:
For each visited state k, do until convergence:

- Choose an action i with current probability estimate $\widehat{\pi}_k(i)$, observe the current cost $c(k,i)$ for performing this action, update its estimate $\widehat{c}(k,i)$ by

$$\widehat{c}(k,i) \leftarrow c(k,i) \tag{4.4}$$

- Perform the action i and observe the current value $\widehat{V}(k')$ of the next state k'. Update $\widehat{\overline{V}}(k,i)$ accordingly (here, we choose the exponential smoothing scheme),

$$\widehat{\overline{V}}(k,i) \leftarrow \alpha\overline{V}(k') + (1-\alpha)\widehat{\overline{V}}(k,i) \tag{4.5}$$

- Update the probability distribution for state k as:

$$\widehat{\pi}_k(i) \leftarrow \frac{\exp\left[-\widehat{\theta}_k\left(\widehat{c}(k,i) + \widehat{\overline{V}}(k,i)\right)\right]}{\sum\limits_{j\in U(k)} \exp\left[-\widehat{\theta}_k\left(\widehat{c}(k,j) + \widehat{\overline{V}}(k,j)\right)\right]}, \tag{4.6}$$

  where $\widehat{\theta}_k$ is set in order to respect the prescribed degree of entropy (see Equation (3.3)).
- Update the expected cost of state k asynchronously:

$$\begin{cases} \widehat{V}(k) = \sum\limits_{i\in U(k)} \pi_k(i)\,[\widehat{c}(k,i) + \widehat{\overline{V}}(k,i)], \\ \widehat{V}(d) = 0, \text{ where } d \text{ is the destination state} \end{cases} \tag{4.7}$$

This iterative scheme is closely linked to the SARSA on-policy control algorithm [14,16,18]; a discussion of these relationships is provided in [1].

5 Conclusions

We have presented a model integrating continual exploration and exploitation in a common framework. The exploration rate is controlled by the entropy of the choice probability distribution defined on the states of the system. When no exploration is performed (zero entropy on each node), the model reduces to the common value-iteration algorithm computing the minimum cost policy. On the other hand, when full exploration is performed (maximum entropy on each node), the model reduces to a "blind" exploration, without considering the costs. The main drawback of the present approach is that it is computationally demanding since it relies on iterative procedures such as the value-iteration algorithm.

Further work will investigate the relationships with SARSA, as well as alternative cost formulations, such as the "average cost per step". We also plan to exploit the proposed exploration framework in Markov games.

Acknowledgments

Part of this work has been funded by projects with the "Région wallonne" and the "Région de Bruxelles-Capitale".

References

1. Y. Achbany, F. Fouss, L. Yen, A. Pirotte, and M. Saerens. Tuning continual exploration in reinforcement learning. *Technical report, http://www.isys.ucl.ac.be/staff/francois/Articles/Achbany2005a.pdf*, 2005.
2. M. S. Bazaraa, H. D. Sherali, and C. M. Shetty. *Nonlinear programming: Theory and algorithms.* John Wiley and Sons, 1993.
3. D. P. Bertsekas. *Neuro-dynamic programming.* Athena Scientific, 1996.
4. D. P. Bertsekas. *Network optimization: continuous and discrete models.* Athena Scientific, 1998.
5. D. P. Bertsekas. *Dynamic programming and optimal control.* Athena sientific, 2000.
6. J. A. Boyan and M. L. Littman. Packet routing in dynamically changing networks: A reinforcement learning approach. *Advances in Neural Information Processing Systems 6 (NIPS6)*, pages 671–678, 1994.
7. R. G. Brown. *Smoothing, forecasting and prediction of discrete time series.* Prentice-hall, 1962.
8. N. Christofides. *Graph theory: An algorithmic approach.* Academic Press, 1975.
9. T. M. Cover and J. A. Thomas. *Elements of information theory.* John Wiley and Sons, 1991.
10. J. N. Kapur and H. K. Kesavan. *Entropy optimization principles with applications.* Academic Press, 1992.
11. J. G. Kemeny and J. L. Snell. *Finite markov chains.* Springer-Verlag, 1976.
12. M. J. Osborne. *An introduction to game theory.* Oxford University Press, 2004.
13. H. Raiffa. *Decision analysis.* Addison-Wesley, 1970.
14. G. Rummery and M. Niranjan. On-line q-learning using connectionist systems. *Technical Report CUED/F-INFENG/TR 166, Cambridge University Engineering Departement*, 1994.

15. G. Shani, R. Brafman, and S. Shimony. Adaptation for changing stochastic environments through online pomdp policy learning. In *Workshop on Reinforcement Learning in Non-Stationary Environments , ECML 2005*, pages 61–70, 2005.
16. S. Singh and R. Sutton. Reinforcement learning with replacing eligibility traces. *Machine Learning*, 22:123–158, 1996.
17. J. C. Spall. *Introduction to stochastic search and optimization.* Wiley, 2003.
18. R. S. Sutton and A. G. Barto. *Reinforcement learning: an introduction.* The MIT Press, 1998.
19. S. Thrun. Efficient exploration in reinforcement learning. *Technical report, School of Computer Science, Carnegie Mellon University*, 1992.
20. S. Thrun. The role of exploration in learning control. In D. White and D. Sofge, editors, *Handbook for Intelligent Control: Neural, Fuzzy and Adaptive Approaches.* Van Nostrand Reinhold, Florence, Kentucky 41022, 1992.
21. S. Thrun, W. Burgard, and D. Fox. *Probabilistic Robotics.* MIT Press, 2005.
22. K. Verbeeck. Coordinated exploration in multi-agent reinforcement learning. PhD thesis, Vrije Universiteit Brussel, Belgium, 2004.
23. J. C. Watkins. Learning from delayed rewards. PhD thesis, King's College of Cambridge, UK, 1989.
24. J. C. Watkins and P. Dayan. Q-learning. *Machine Learning*, 8(3-4):279–292, 1992.

Vague Neural Network Controller and Its Applications

Yibiao Zhao[1], Rui Fang[2], Shun Zhang[1], and Siwei Luo[2]

[1] Beijing Jiaotong University, School of Traffic and Transportation
100044 Beijing, China
Yibiao_zhao@ieee.org
[2] Beijing Jiaotong University, School of Computer and Information Technology
100044 Beijing, China

Abstract. Fuzzy neural network is a promising intelligence system that combines the artificial neural network and fuzzy logic. However fuzzy neural network has its shortages: fuzzy membership function has only one single value, it cannot get more reasonable classified and cognizable results. While, vague sets theory is a generalization of fuzzy sets theory, its' distinguishing feature is having a truth-membership function and a false-membership function. It presents both of the opposite factors to deal with nonlinearities and uncertain of control system in the control fields., this paper has been accomplished the controller of vague neural networks based on the vague set theory, this controller combines the advantage of vague set in handling uncertain information and the capability of artificial neural networks in learning process. Moreover, in the application of inverted pendulum, the character of vague neural network controller was expressed.

Keywords: Vague set, vague neural network, fuzzy neural network, inverted pendulum.

1 Introduction

On the basis of fuzzy theory, Fuzzy Neural Network—FNN is the combination of fuzzy logic system and Artificial Neural Network (ANN), whose essential is the dual-simulation of the structure and thought function of human's brain, namely, the simulation of topological structure of human's brain neural network and the function of dealing with ambiguous data. FNN has the advantages of both fuzzy logic system and ANN, moreover, it also supplies the gap of them. Neural networks have extensively been used as associative memory and optimization. They have many well known advantages such as error tolerant and self-learning capacity. The base idea of the composition method of the fuzzy neural network is to realize the process of fuzzy reasoning by the structure of neural network and to make the parameters of fuzzy reasoning be expressed by the connection weights of neural network. However, Fuzzy theory has it's shortages as pointed out by Daniel J.Buehrer and Wen-Lung Gun in [5]: fuzzy neural network has its shortages: fuzzy membership function has only one single value, it cannot get more reasonable classified and cognizable results. The

S. Kollias et al. (Eds.): ICANN 2006, Part I, LNCS 4131, pp. 801 – 810, 2006.

vague set theory is a new extended form fuzzy set: it's truth and false membership function present both of the opposite factors to deal with nonlinearities and uncertain of control system in the control fields. It overcomes the disadvantage of membership function in fuzzy set which cannot characterize both the similarity and dissimilarity between pairs of objects.

Therefore, it has been conceived as a new effective tool to deal with ambiguous data and applied successfully in different fields. In this paper, vague neural network controller is proposed. It is a new extended form of FNNC. On the basis of that, we present an optimal scheme for the design of a vague neural network as a controller. And then we discussed the application of inverted pendulum.

2 Vague Set

2.1 Fuzzy Set

A fuzzy set A of the universe of discourse $U=\{x_1,x_2,...x_n,\}$ can be represented by:

$$A=\mu_A(u_1)/u_1+\mu_A(u_2)/u_2+...+\mu_A(u_n)/u_n \tag{1}$$

Where μ_A is the membership function of the fuzzy set $\mu_A:U\rightarrow[0,1]$, $\mu_A(u_i)$ is a single value and it indicates the grade of membership of u_i to the fuzzy set A

2.2 Vague Set

Let U be the universe of discourse, $U=\{x_1,x_2,...x_n\}$ and let, $t_v(x)$ and, $f_v(x)$ be the truth-membership function and false-membership function of the Vague set V. $t_v(x)$ is a lower bound on the membership degree of x derived from the evidence for x, and $f_v(x)$ is a lower bound on the negation of x derived from the evidence against x. $f_v(x)$ and $f_v(x)$ both associate a real number in the interval [0,1]with each x in U, where $t_v(x)+f_v(x)\leq 1$. This approach bounds the membership degree of x to a subinterval [$t_v(x)$, 1-$f_v(x)$] of [0, 1] A vague set in the universe of discourse U is illustrated in Fig.1

The precision of our knowledge about x is immediately clear, with our uncertainty characterized by the difference 1-$t_v(x)$-$f_v(x)$ If this is small, our knowledge about x is relatively precise; If it is large, we know correspondingly little; if1-$f_v(x)$ is equal to $t_v(x)$,our knowledge about x is exact, and the theory reverts back to that of fuzzy sets; If 1-$f_v(x)$ and $t_v(x)$ are both equal to 1 or 0,our knowledge about x is very exact and the theory revert back to that of ordinary set.

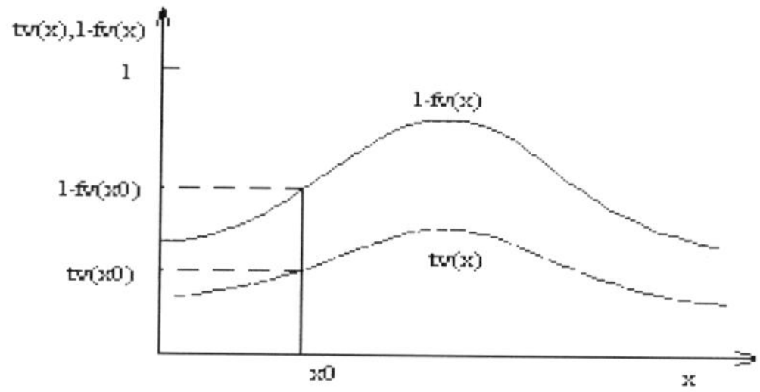

Fig. 1. Vague set

With the relationship between vague sets intuitionistic fuzzy sets, and interval-valued fuzzy sets that, vague neural network can be extended to interval-valued vague neural network by replacing the point membership degree to interval-valued membership degree. We can see that the membership degree of an element $u_i \in U$ in a fuzzy set is represented a by single value which combines the evidence for $u_i \in U$ and the evidence against $u_i \in U$, and without indicating how much there is of each. Furthermore, it tells us nothing about its accuracy. Thus, the concept of vague sets is proposed to solve the problem effectively. Vague set contains precise information of a pattern and can characterize the pattern more accurately. And, the author proposed a new neural network based on vague set theory.

3 Fuzzy Neural Networks

Fig.2 shows the architecture of the fuzzy neural network, it composes of input, fuzzification, inference, defuzzification layer and the output layer.

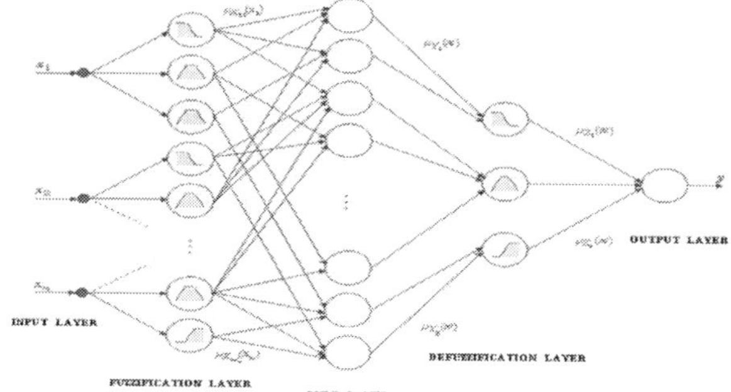

Fig. 2. The structure of FNN

Layer 1, the input layer, consists of x_i, $i=1,2...N$. Each neuron in the fuzzification layer represents a fuzzy membership function for one of the input variables.

Layer 2 comprises the term neurons which correspond to the linguistic variables such as NM, NS, NZ, Z, PZ, PS, PM, etc. each neuron in this layer performs the computation of a membership function. A bell-shaped function is usually used to compute the membership degree as:

$$\mu_{ji} = \exp\left(\frac{-(x_j - c_{ji})^2}{\delta_{ji}^2}\right) \tag{2}$$

Where c_{ij}, and δ_{ij}^2 denotes the centre and the width of the membership function corresponding to the jth linguistic variable of xi and Ni is the number of linguistic variable terms.

Layer 3 is the rule layer. each neuron in this layer corresponds to a knowledge based. rule ,Generally, a fuzzy rule can be described as:

IF x_1 is A_{1i} ...and x_m is A_{mi} THEN y is B_i

Where $x_i(i = 1, 2 \ldots)$ are input variables, "and" is logic operator "$\wedge$". A_{ji} and B_i are linguistic terms characterized by two fuzzy sets on domain of x_i and y, respectively. The total number of neurons in this layer depends on the number of match neuron constituting in layer 2.the neuron performs the fuzzy AND operation, which is a minimum operation to obtain the firing strength of rules.

$$\alpha_r = \min\{\mu_{1i}, \mu_{1j}, \ldots, \mu_{nk}\} \tag{3}$$

$i = 1, 2 \ldots N_1$, $j = 1, 2 \ldots N_2$, $k = 1, 2 \ldots N_n$, $r = 1, 2 \ldots N_A$, $N_A = \prod_{i=1}^{n} N_i$

α_r degree of match between the fuzzy linguistic variables occurring as one of the antecedents of the rule;

$N_1 N_2 \ldots N_n$ number of the linguistic variables corresponding to input x_i

N_A number of the fuzzy rules

Generally, only the linguistic variables near the input value have significant memberships. Only a small number of the neurons have significant outputs and most of the neuron's outputs are 0 or almost 0.

Layer 4 and Layer 5 are the defuzzification layer and output layer of the FNN, which perform the normalization computation of the firing strengths.

$$\overline{\alpha_r} = \alpha_r \Big/ \sum_{i=1}^{N_A} \alpha_i \quad r = 1, 2 \ldots N_A \tag{4}$$

$\overline{\alpha_r}$ denotes the normalized firing string strength for the r th rule.

4 Vague Neural Network Controller

The fuzzy neural networks are achieved by adding a fuzzification layer to a conventional feed forward neural network. The difference between fuzzy neural networks and conventional neural networks lies in how they estimate sampled input-output relationships [10]. They differ in the kind of samples used, how they represent and store those samples, and how they associatively "inference" or map inputs to outputs. Here VNN extends FNN and estimates functions with vague set samples, it can handle real inputs as well as fuzzy inputs.

4.1 Vague IF-THEN Control Rules

In VNN, The fuzzy if-then rules are extended to vague if-then rules described as:

IF x_t is $X_t(1)$ and x_f is $X_f(1)$ THEN y_t is $Y_t(1)$ and y_f is $Y_f(1)$;

IF x_t is $X_t(2)$ and x_f is $X_f(2)$ THEN y_t is $Y_t(2)$ and y_f is $Y_f(2)$;

……

IF x_t is $X_t(n)$ and x_f is $X_f(n)$ THEN y_t is $Y_t(n)$ and y_f is $Y_f(n)$.

There are n rules in all. Vague logical is represented as a rule relation matrix R:

$$R = \bigcup_{i=1}^{n} [(X_t(i) \times X_f(i)) \rightarrow (Y_t(i) \times Y_f(i))] \tag{5}$$

After the structure analysis and approximate process, above-mentioned dual-input and dual-output vague controller could be represented as:

$$Y = \bigcap_{i=1}^{n} Y_i \tag{6}$$

$$Y_k = Y_t(k) \cap Y_f(k) = \left[X_t(k) \times X_f(k) \right] \circ R_k \tag{7}$$

Where $x_t(k)$, $x_f(k)$ are input variables in the *k*th second, "and" is logic operator"$\wedge$". $X_t(i)$, $X_f(i)$, $Y_t(i)$ and $Y_f(i)$ are linguistic terms characterized by two vague sets on domain of x_i and y respectively.

4.2 "Vaguefication" in Controller

Compared with the term of "fuzzification" in fuzzy theory, The process of changing precise value to vague value is call" vaguefication" in this paper, and it is the key of internal value controller design. The easy and efficient vaguefication method will be given next.

universe of discourse E={e1, e2, ···, en}, "A" is a vague set{the biggest}, then, A can be described as a true membership function t_A and a false membership function f_A:

$$t_A(e_i) = \frac{1}{1+(e_i - a)^2/a} \tag{8}$$

$$f_A(e_i) = \frac{(e_i - a)^2/(a+1)}{1+(e_i - a)^2/(a+1)} \tag{9}$$

Where, $t_A(e_i) \in [0,1], f_A(e_i) \in [0,1], 0 \le t_A(e_i) + f_A(e_i) \le 1, a = \max\{e_i \mid (i = 1, 2, \cdots, n)\}$

In these expressions, $t_A(e_i)$ is the lower of the vague membership degree, which expresses the degree for e_i's belonging to A; $1 - f_A(e_i)$ is the upper of the vague membership degree; $1 - t_A(e_i) - f_A(e_i)$ denotes uncertain information, then, the vague set A can be denoted as:

$$A = \sum_{i=1}^{n} ([t_A(e_i), 1 - f_A(e_i)]) \tag{10}$$

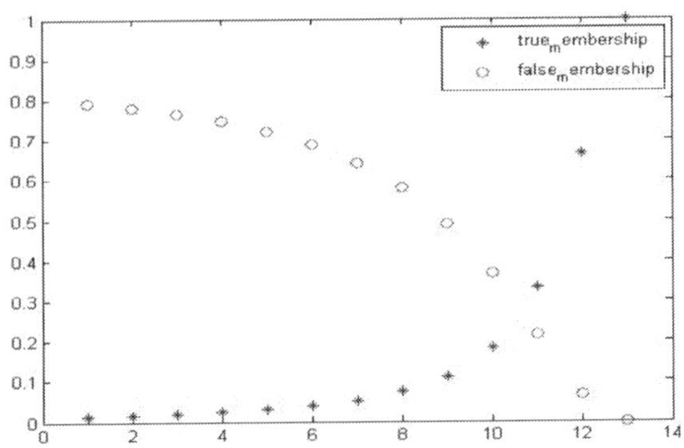

Fig. 3. True and false membership degree

5 Applications

Inverted pendulum is recognized as typical equipment in automatic control theory, also a rare typical physical model in control theory teaching and researches. Inverted pendulum is a natural unstable object, it reflects many pivotal questions in controlling process such as nonlinear issues, system's robustness, tracking problems, grave issue and etc. Its terminal aim is to make inverted pendulum object become a stable system

by introducing a proper control method. Usually, the fuzzy control rule of fuzzy logic control algorithm in multivariable input is hard to complete for complex calculation. Comparing with this, the combination of ANN and Vague Set has many merits when multivariable input is needed, like high precision, quick convergence, easy calculation, good robustness etc. which meet system's need.
When applying VNN as a controller, the overall structure of our control system may be chosen as shown in Fig.4. In two cases, there are two input signals and one output. They are cart position x, axis angle phi, and system output power y.

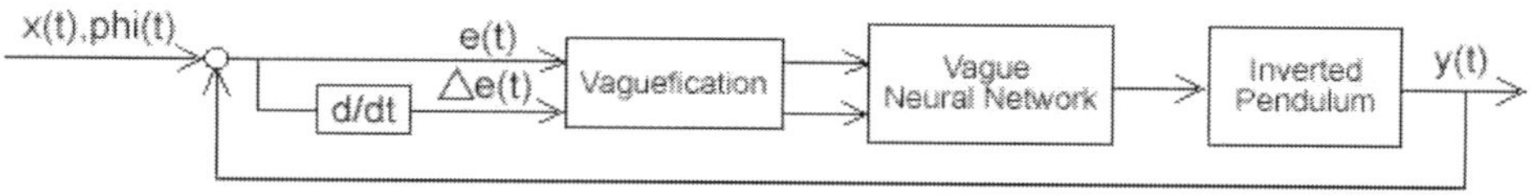

Fig. 4. Vague neural network controller

In order to compare simulation result with conventional fuzzy neural networks, we just change the biggest linguistic variable into vague set with true and false membership degree .With the proposed method, we set the Linguistic variable membership function and control rules as shown in Table.1 and Table.2 respectively.

Table 1. Linguistic variable membership function

		-6	-5	-4	-3	-2	-1	0	1	2	3	4	5	6
NB		1.0	0.8	0.7	0.4	0.1	0.0	0.0	0.0	0.0	0.0	0.0	0.0	0.0
NM		0.2	0.7	1.0	0.7	0.3	0.0	0.0	0.0	0.0	0.0	0.0	0.0	0.0
NS		0.0	0.1	0.3	0.7	1.0	0.7	0.2	0.0	0.0	0.0	0.0	0.0	0.0
NZ		0.0	0.0	0.0	0.0	0.1	0.6	1.0	0.6	0.1	0.0	0.0	0.0	0.0
PZ		0.0	0.0	0.0	0.0	0.0	0.0	1.0	0.6	0.1	0.0	0.0	0.0	0.0
PS		0.0	0.0	0.0	0.0	0.0	0.0	0.2	0.7	0.1	0.7	0.3	0.1	0.0
PM		0.0	0.0	0.0	0.0	0.0	0.0	0.0	0.0	0.2	0.7	1.0	0.7	0.3
PB	t	0.11	0.12	0.13	0.14	0.15	0.18	0.21	0.25	0.31	0.41	0.56	0.80	1.0
	f	0.79	0.78	0.77	0.75	0.72	0.69	0.64	0.58	0.49	0.37	0.33	0.07	0.0

Table 2. VNN control rules

	NB	NM	NS	ZE	PS	PM	PB
NB	NB	NB	NB	NB	NM	ZE	ZE
NM	NB	NB	NB	NB	NM	ZE	ZE
NS	NM	NM	NM	NM	ZE	PS	PS
NZ	NM	NM	NS	ZE	PS	PM	PM
PZ	NM	NM	NS	ZE	PS	PM	PM
PS	NS	NS	ZE	PM	PM	PM	PM
PM	ZE	ZE	PM	PB	PB	PB	PB
PB	ZE	ZE	PM	PB	PB	PB	PB

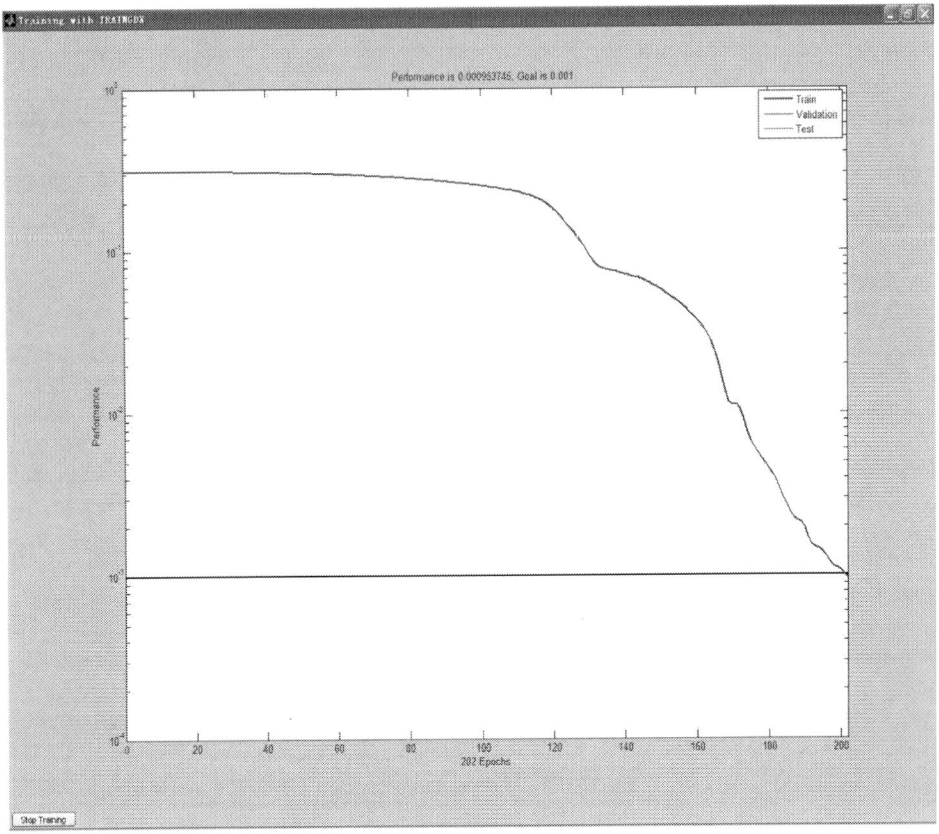

Fig. 5. Curve of training error

After 202 epochs, the network converged with training error reduced to 10^{-3}. And the training result is shown in Fig.6 and Fig.7.

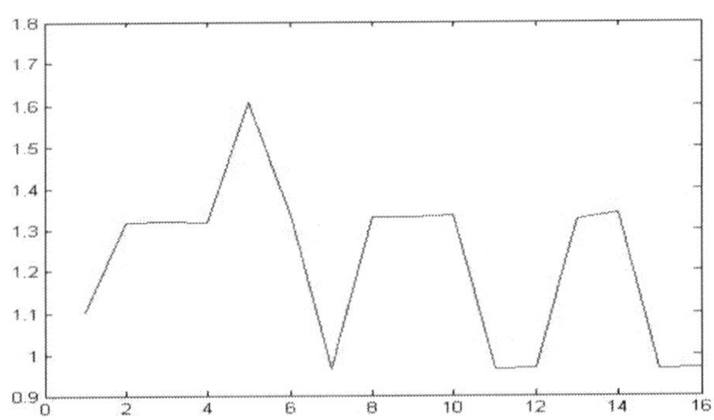

Fig. 6. Training result of VNNC

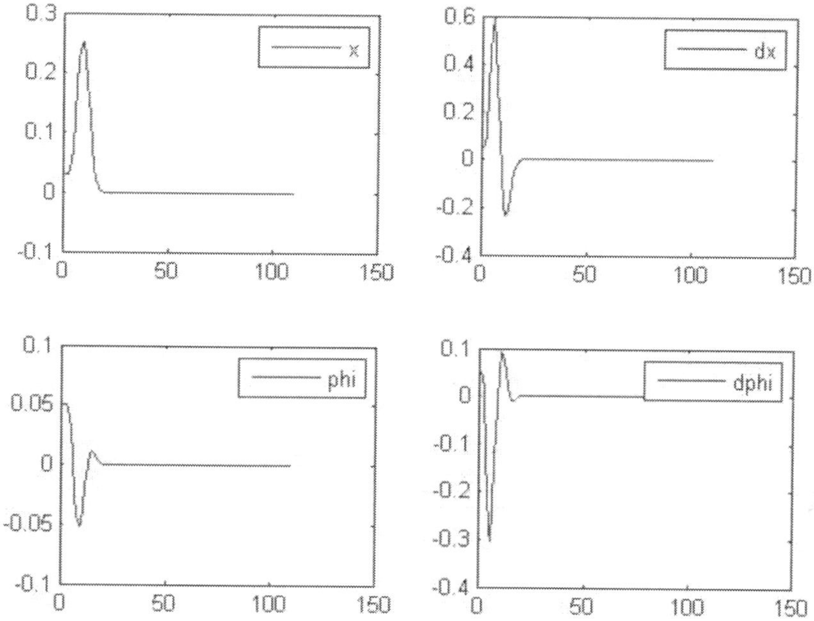

Fig. 7. Results of simulink

Through the figures above, we can see that the result meets the need of inverted pendulum control. In contract with traditional fuzzy neural network, the vague neural network considers more aspects of the input data before forwarding them into the inner of the network.

6 Conclusion

Based on vague set theory, in this paper, intelligent control system based on Vague Neural Network is put forward. According to the advantages of VNN, we utilize genetic algorithm and adapted controlling to experiment. Actually, VNN can have the same structure as a specified FNN, it can also adopt FNN's training and learning algorithm. For some real applications, FNN can be easily extended to VNN by replacing the single membership function of fuzzy set and the fuzzy if-then rule with vague set. As the discussion mentioned above, intelligent control system based on Vague Neural Network is tried and true. Its performance is better, and the result is efficient and valid.

References

1. HORIKAWA,S. FURUHASHI,T., and UCHIKAWA, Y: On fuzzy modeling using fuzzy neural network with the back-propagation algorithm, IEEE Trans., 1992, NN-3,(5),pp.801-806;

2. HASEGAWA,T.,HORIKAWA,S.,FURUHASHI,T.,UCHIKAWA,Y.,SHIMAMURA,S.,Y AMADA,T,KUNITAKE,O., AND OTSUKA,S.: An application of fuzzy neural network to fuzzy modeling of a basic oxygen furnace. Proceedings of IEEE International workshop on neural-fuzzy control, 1993. (Muroran , Japan),pp.133-138;
3. Mori,H. Fuzzy neural network applications to power systems. Power Engineering Society Winter Meeting,2000 IEEE:2,23-27;
4. Hon Keung, Kwan,Yaling Cai, A Fuzzy Neural Network and its Application to Pattern Recognition, IEEE Trans on Fuzzy System.Vol.2,NO.3,1994;
5. Gau W L, Buehrer D J.Vague sets[J].IEEE Transactions on Systems, Man, and Cybernetics,1993,23(2):610-614;
6. Bustince H, Burillo P. Vague sets are intuitionistic fuzzy sets. Fuzzy Sets and Systems 1996, 79(1): 403~405;
7. Hong D H,Choi C H. Multicriteria fuzzy decision making problems based on Vague set theory Fuzzy Set and System,2000, 114:103-113;
8. Chen,S.M. Fuzzy system reliability analysis based on vague set theory. 1997 IEEE International Conference on Computational Cybernetics and Simulation.1997,2:1650-1655.
9. Hong D H,Choi CH.Multi-criteria fuzzy decision-making problems based on vague set theory.Fuzzy Sets and Systems, 2000,114:103-113;
10. B.Kosko,Neural Networks and Fuzzy Systems-A Dynamic System Approach to Machine Intelligence,Englewood Cliffs, NJ:Prentice Hall, 1992;
11. R.P.Brent, Fast training algorithm for multilayer neural network, IEEE Trans.Neural Networks,vol.2,pp.346-354,May,1991;
12. R.Fletcher,C.M.Reeves, Function Minimization by Conjugate Gadients, Computer Journal,7,1964,p149-154.;
13. Holland,J,Adaptation in Natural and Artificial systems, MIT Press, 1975;
14. Goldberg,E,Genetic Algorithms in Search Optimization and Machine Learning Addison-Wesley. 1989;
15. K.Atanassov. Intuitionistic Fuzzy set[J].Fuzzy Set and Systems,1986,20(2);87-96;
16. L.A.Zadeh, The concept of a linguistic variable and its application to approximate reasoning-I.Inform.Sci. (1975)199-249;
17. Madan M Gupta, Jerzy B Kiszka, G M Trojan.Multicaziable Straucture of Fuzzy Control System. IEEE Transations on System, Man and Cybernetics. 1993.23(4);
18. Ahmed Rubaai Daniel Ricketts,Development and Implementation of an Adaptive Fuzzy-Neural-Network Controller for Brushless Drives Industry Applications, IEEE Transactions on Volume 38, Issue 2, March-April 2002 Page(s):441 – 447;
19. Cheng-Zhi Cao, Guang-Hua Wei, Qedong Zhang, Xin Wang, Optimization Design Of Fuzzy Neural Network Controller In Direct Torque Control System, Proceedings of the Third International Conference on Machine Learning and Cybemetics,2004,378-382;
20. Wu jin-pei, Xiao jian-hua, Intelligence fault diagnose and Expert system, science press, 1997;
21. Xuezhong Guan, Xiaoyu Zhao, Yong Guan, Liang Dong, A Controller Design Based on Vague marching reasoning. Control Engineering of China, Vol.13, No.1, Jan 2006
22. Rong-Jong Wai, and Li-Jung Chang, Stabilizing and Tracking Control of Nonlinear Dual-Axis Inverted-Pendulum System Using Fuzzy Neural Network, IEEE TRANSACTIONS ON FUZZY SYSTEMS, VOL. 14, NO. 1, FEBRUARY 2006

Parallel Distributed Profit Sharing for PC Cluster

Takuya Fujishiro, Hidehiro Nakano, and Arata Miyauchi

Musashi Institute of Technology
1-28-1 Tamazutsumi,
Setagayaku, Tokyo, 158-8557 Japan
{fujisiro, nakano, miyauchi}@ic.cs.musashi-tech.ac.jp

Abstract. This paper presents a parallel reinforcement learning method considered communication cost. In our method, each agent communicates only action sequences with a constant episode interval. As the communication interval is longer, communication cost is smaller, but parallelism is lower. Implementing our method on PC cluster, we investigate such trade-off characteristics. We show that computation time to learning can be reduced by properly adjusting the communication interval.

1 Introduction

Reinforcement learning is one of the machine learning methods which do not need supervisors. A learning agent repeats an interaction to target environments, and updates own value function. The agent can learn without prior knowledge on the target environments. Therefore, reinforcement learning can be applied easily to various tasks. However, reinforcement learning requires long computation time. In order to overcome such a drawback, parallel distributed processing is an efficient method to accelerate reinforcement learning. There have been some parallel reinforcement learning methods by using multiple agents [1],[2]. Since multiple agents learn independently, the learning can be accelerated by properly sharing and combining information which each agent has. However, as these methods are implemented on actual PC cluster, long computation time tends to be required. This is due to large communication cost.

This paper presents a parallel reinforcement learning method considered communication cost. In the conventional method, each agent communicates value functions as shared and combined information. Considering communication cost, in our method, each agent communicates only action sequences with a constant episode interval. As the communication interval is longer, communication cost is smaller, but parallelism is lower. Implementing our method on PC cluster, we investigate such trade-off characteristics. We show that computation time to learning can be reduced by properly adjusting the communication interval.

S. Kollias et al. (Eds.): ICANN 2006, Part I, LNCS 4131, pp. 811–819, 2006.

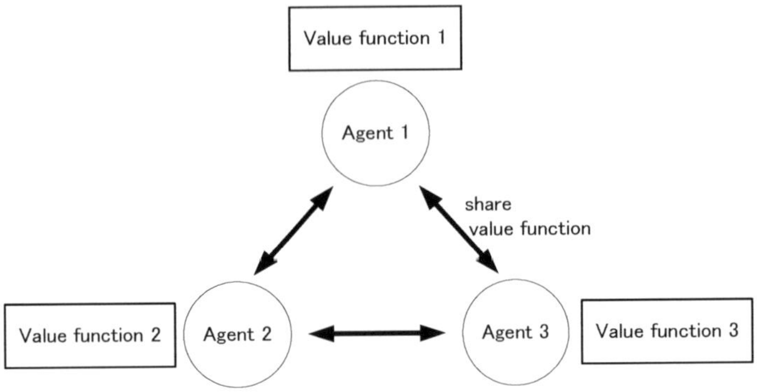

Fig. 1. Parallel Reinforcement Learning

2 Parallel Reinforcement Learning

Ref. [1] has proposed the method which can accelerate reinforcement learning by using multiple agents. Each agent learns the same task independently, and shares and combines each own value function (see Fig. 1).

In Ref. [1], the learning algorithm is based on Q-Learning [3]. Each agent updates own value function in parallel, and shares the value function to each other in the learning process. The shared value functions are combined by the following equation.

$$Q_1 = \frac{Q_1 * k_1 + Q_2 * k_2}{k_1 + k_2}, \tag{1}$$

where Q_1 is own value function, k_1 is own learning times, Q_2 is other agent's value function and k_2 is other agent's learning times.

Ref. [1] has shown the advantage of the above method by using multiple agents, and Ref. [2] has presented experimental results for various tasks. These papers have shown effectiveness of parallel reinforcement learning. However, they have not been implemented on actual parallel computer systems and/or multiprocessor systems. As we implement these methods on actual PC cluster, long computation time is required. This is due to communication cost between each cluster node. In general, communication cost is a problem in implementing on parallel computer systems.

3 Parallel Distributed Reinforcement Learning Method Considered Communication Cost

This paper proposes a parallel distributed reinforcement learning method considered the communication cost for implementing PC cluster.

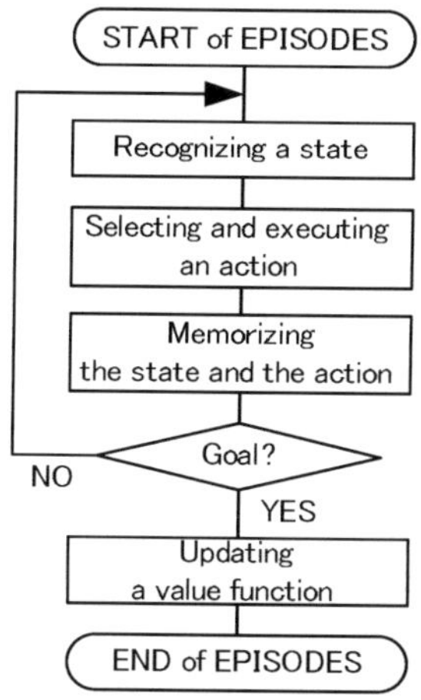

Fig. 2. Flowchart of Profit Sharing

3.1 Algorithm

1. Multiple agents learn the same task independently.
2. Each agent updates own value function in parallel.
3. Each agent shares the value function to each other in learning process.
4. Shared value functions are combined by adding to each value function.

We use Profit Sharing [4] as the reinforcement learning method. Fig. 2 shows the flowchart of Profit Sharing. Generally, in Profit Sharing, the following geometric decreasing function is used as the reinforcement function.

$$f_i = \frac{1}{S} f_{i-1}, \quad i = 1, 2, \cdots, W-1, \tag{2}$$

where W is the length of an episode, and i is the number of steps from a goal state to the ith state. f_i is reward value to the ith rule, and $1/S$ is a decreasing ratio in the reinforcement function. If the parameter S is set to the number of executable actions, Profit Sharing can realize rational learning such that an agent must achieve a goal state.

The combining method is to simply add other agent's value function to own value function. In Profit Sharing, experiences which each agent obtains are accumulated by repeating episode. It should be noted that there are many parameters in Q-Learning. It is necessary to keep those parameters in combining the value function based on Eq. (1). Accordingly, Parallel Distributed Profit Sharing can keep rational learning by simply adding experiences. Therefore, Profit Sharing is more suitable for Parallel Distributed Reinforcement Learning than Q-Learning.

3.2 Implementation

Fig. 3 shows an implementation example of Parallel Distributed Profit Sharing on PC cluster. Each agent is assigned to each node in the PC cluster. Each agent

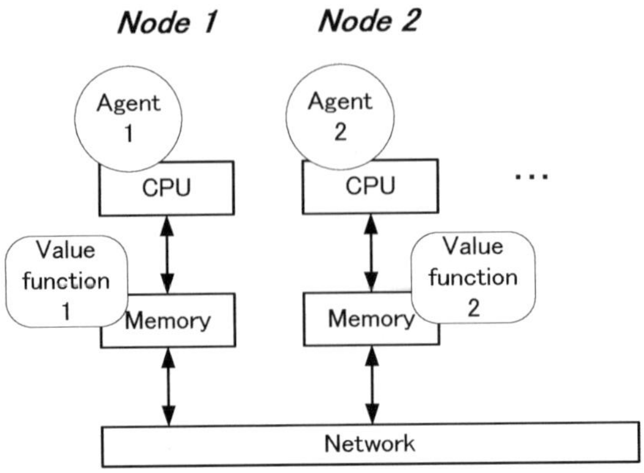

Fig. 3. Implementing PC cluster

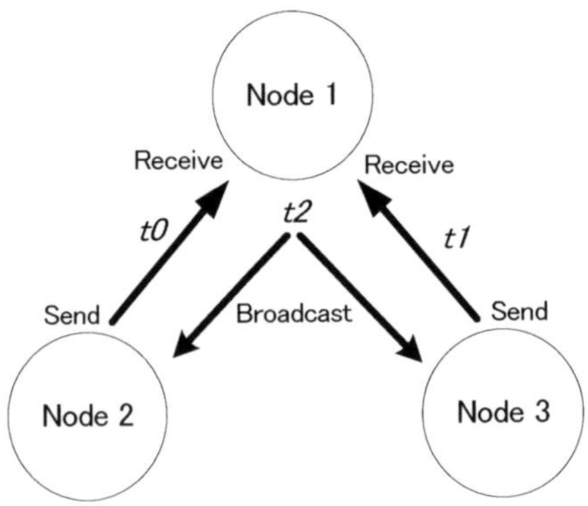

Fig. 4. Communication between nodes

updates own value function stored in the memory on own node. Synchronization and communication are repeated with a constant episode interval. Share of experience data is realized by communicating as follows (see Fig. 4):

$t0, t1$: Nodes 2 and 3 (slave node) send experience data to Node 1 (master node). Node 1 receives those experience data.
$t2$: Node 1 broadcasts the combined new value function data to all slave nodes.

3.3 Reduction of the Communication Cost

In order to reduce the communication cost, the communication data size and the communication frequency should be reduced.

For reduction of the communication data size, we propose the method in which only action sequences are sent. In the conventional methods the value functions are sent as shared information. However, the data size of a value function is large

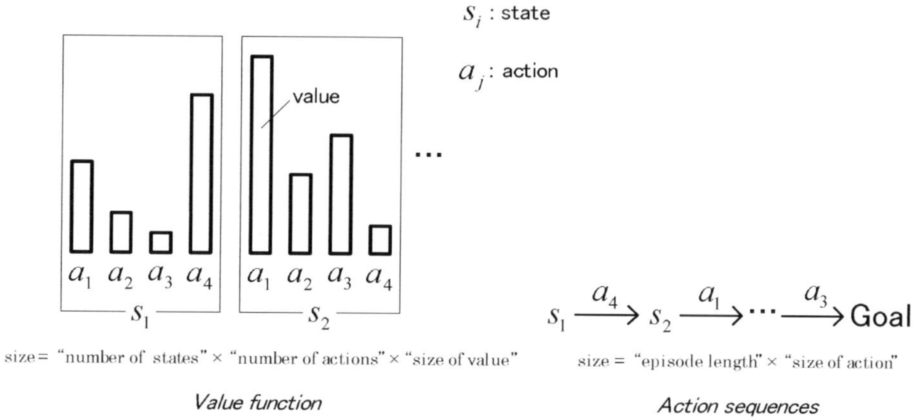

Fig. 5. The value function and the action sequences

as communication data. Therefore, as the conventional method are implemented on actual PC cluster, long computation time tends to be required. In contrast, the action sequences from the initial state to the goal state can be smaller data size than value functions (see Fig. 5). From all slave nodes, master node receives action sequences from the initial state to the goal state. The value function is combined when reward is distributed based on received action sequences.

The communication between each node has a large overhead. Therefore, the communication frequency should be reduced. In the proposed method each node communicates to each other with a constant episode interval. Then, the communication frequency can be reduced, and the calculation load of a master node can be reduced. Because, action sequences are sent only at the communication timing. A master node combines a value function based on the received action sequences, applying a weight parameter ω. We consider that the last received action sequence is repeated until the next communication timing. The value of the weight parameter ω is set to be equal to the communication interval. As the communication interval is longer, communication cost is smaller, but parallelism is lower. Implementing our method on PC cluster, we investigate such trade-off characteristics.

4 Numerical Experiments

A maze problem as shown in Fig. 6 is used for the task of an experiment. The agents learn actions from the initial state S to the goal state G. There exist plural effective rules to achieve a goal state. Each state has four executable actions: UP, DOWN, LEFT, RIGHT. The minimum steps to the goal state is 4. In order to guarantee rational learning, the parameter of reinforcement function, S, is set to 4 which is equal to the number of executable actions. The action selection rule is roulette selection. The initial reward of each rule is 10^7 The reward obtained at each episode is 10^7 Each experimental result represents the average values

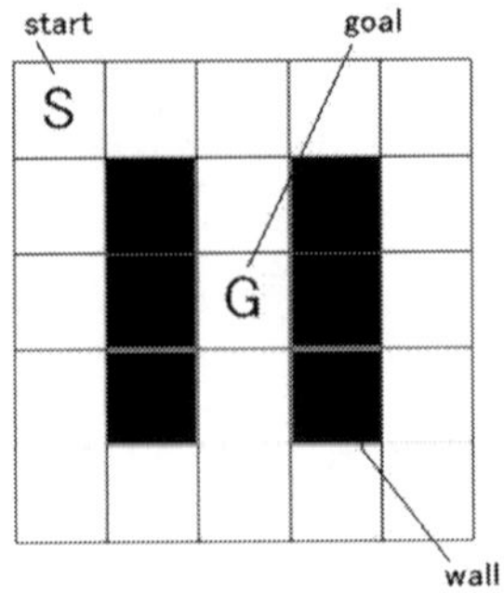

Fig. 6. A maze problem with plural effective rules

Table 1. Specification of PC cluster

CPU	Pentium III 866MHz
Memory	512MByte
LAN	100Base-TX
OS	Linux 2.4.21
MPI	MPICH 1.2.5

for 5000 trials with diffrent random sequences. The random sequences for each agent are also different. Table 1 shows specification of PC cluster.

Table 2 and Fig. 7 show typical experimental results for the methods with every episode communication. Here, convergence episodes mean the number of episodes at which the learning curve converge. We regard that learning curve converges if slope of learning curve is sufficiently small ($|slope|<10^{-5}$). The experience number is given by the following equation.

$$Experience\ \ number = \sum_{x=1}^{n} f(x), \tag{3}$$

n: Convergence Episodes $f(x)$: Learning Curve x: Episodes.

That is, experience number represents the total number of executed actions until the learning curve converges. Computation time is the time to convergence episodes. These values represent summation of computation time for 100 trials with different random sequences. 1 node means to use common Profit Sharing without parallelization. As shown in Table 2 and Fig. 7, convergence episodes and experience number decreases as the number of nodes increases.

Fig. 8 shows the improvement factor of number of experiences. As the number of nodes increases, this factor increases linearly. These results show that Parallel Distributed Profit Sharing can accelerate the learning. However, computation time is longer than common Profit Sharing without parallelization as shown in Table 2. This is due to the communication cost between each node.

Table 2. Experimental results for the method with every episode communication

Number of nodes	Convergence episodes	Experience number	Computation time [sec]
1	15398	85443.59	7.003
2	6800	38854.21	274.955
3	4644	26393.34	257.066
4	3528	19982.51	264.282
5	2842	16056.47	255.712

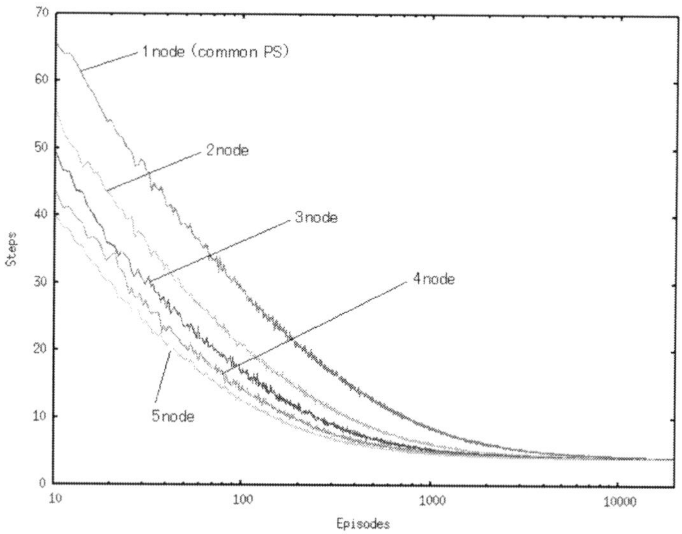

Fig. 7. Learning curves for the method with every episode communication

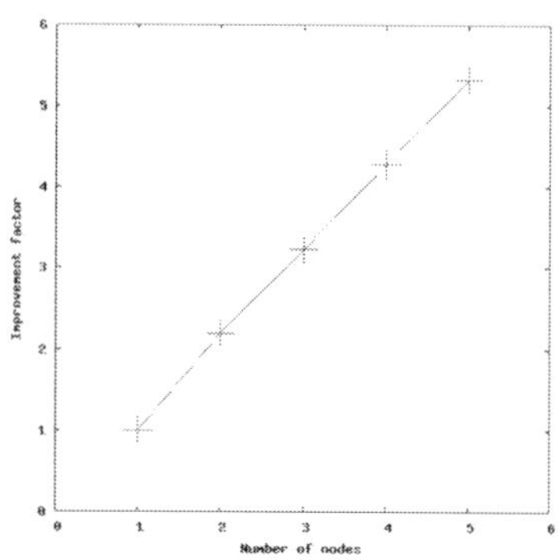

Fig. 8. Improvement factor of experience number

Fig. 9 shows the learning curves for the method with a communication interval. The learning curve for the method with 400 episode communication interval stepwise decreases at every communication timing. Also, this learning curve

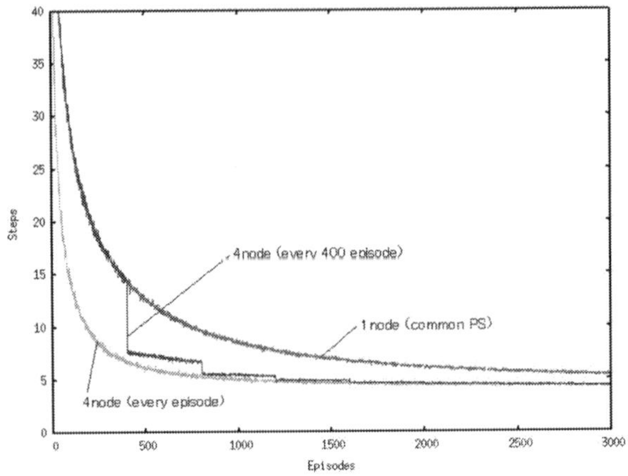

Fig. 9. Learning curves with a communication interval

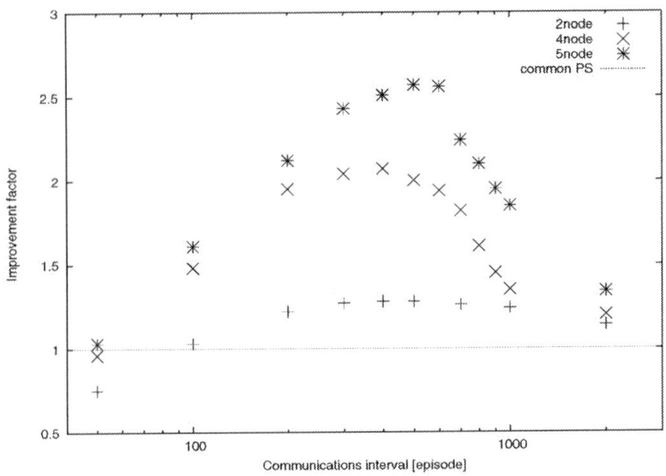

Fig. 10. Improvement factor of computation time

Table 3. The best computation time in the experimental results

	Convergence episodes	Computation time [sec]	Speed-up rate
2 node every 400 episode	9640	5.475	1.28
4 node every 400 episode	4588	3.379	2.07
5 node every 500 episode	3453	2.723	2.57
common PS	15398	7.003	1.00

approaches to the learning curve for the method with every episode communication. Fig. 10 shows the improvement factor of computation time. As the communication interval is longer, communication cost is smaller, but parallelism is lower. We can confirm such trade-off characteristics. Table 3 shows the best computation time in the experimental results. We can confirm that our method can reduce computation time to reinforcement learning by properly adjusting communication interval.

5 Conclusions

In this paper, we have proposed the parallelization method of Profit Sharing considered communication cost. In our method, each agent communicates only action sequences with a constant episode interval. We have investigated trade-off characteristics between parallelism and communication cost. We have confirmed that our method can reduce computation time to reinforcement learning by properly adjusting communication interval.

Future works include application to large-scale problems, detailed examination of the calculation load in a master node, and consideration of the optimal communication interval and its task dependency.

References

1. R. Matthew Kretchmar, "Parallel Reinforcement Learning", Proc. of the 6th World Conference on Systemics, Cybernetics and Informatics, vol.6, pp.114-118 (2002).
2. Daria Antonova, "Parallel Reinforcement Learning - Extending the Concept to Continuous Multi-State Tasks", thesis, Denison University (2003).
3. Watkins, C. J. H., and Dayan, P. Technical note: Q-learning. Machine Learning, Vol.8: 55-68, (1992).
4. Grefenstette J. J. Credit Assignment in Rule Discovery Systems Based on Genetic Algorithms, Machine Learning, Vol.3, pages 225-245 (1988).

Feature Extraction for Decision-Theoretic Planning in Partially Observable Environments

Hajime Fujita, Yutaka Nakamura, and Shin Ishii

Graduate School of Information Science,
Nara Institute of Science and Technology (NAIST)
8916-5 Takayama, Ikoma, 630-0192, Japan
{hajime-f, yutak-na, ishii}@is.naist.jp

Abstract. In this article, we propose a feature extraction technique for decision-theoretic planning problems in partially observable stochastic domains and show a novel approach for solving them. To maximize an expected future reward, all the agent has to do is to estimate a Markov chain over a statistic variable related to rewards. In our approach, an auxiliary state variable whose stochastic process satisfies the Markov property, called internal state, is introduced to the model with the assumption that the rewards are dependent on the pair of an internal state and an action. The agent then estimates the dynamics of an internal state model based on the maximum likelihood inference made while acquiring its policy; the internal state model represents an essential feature necessary to decision-making. Computer simulation results show that our technique can find an appropriate feature for acquiring a good policy, and can achieve faster learning with fewer policy parameters than a conventional algorithm, in a reasonably sized partially observable problem.

1 Introduction

A Markov decision process (MDP) is a framework for describing agents that act and learn in a stochastic environment, and decision-making problems formulated as MDPs can be solved by algorithms based on dynamic programming (DP) [1] or reinforcement learning (RL) [2]. These learning models have been studied as an approach to obtain an optimal solution of decision-theoretic planning problems, and have widespread applications in machine learning problems; for example, real robot control [3, 4], game playing [5, 6] and multi-agent control [7, 8] have been achieved as realistic applications. Since the environments often have partial observability (the agents cannot directly access real states of the environment but can get only observations which contain partial information about the state), RL researchers are now devoting much attention to partially observable problems, which can be formulated as partially observable Markov decision processes (POMDPs) [9]. Many approaches to acquire a good policy in a partially observable world, therefore, have been studied recently [10, 11], but several long-standing issues still remain.

First, learning of the value function over the belief space is difficult even with an effective approximation [12]; optimization of the value function requires

S. Kollias et al. (Eds.): ICANN 2006, Part I, LNCS 4131, pp. 820–829, 2006.

heavy computation, because its input is a continuous probability distribution and usually has high dimensionality. Second, an environmental model and the number of real states are required for explicit estimation of unobservable states; environmental information is used to calculate the belief state. These difficulties thus make dealing with various POMDP problems infeasible.

In this article, we propose a feature extraction technique for decision-theoretic planning problems in partially observable stochastic domains and show a novel approach for solving such problems. The main idea of our method is that all the agent has to do for solving the problems is to estimate the Markov chain over a statistic variable related to rewards. Our model includes an auxiliary variable, called *internal state* [13], whose stochastic process has the Markov property, and we assume that the rewards are dependent not only on the action but also on the internal state. The agent, therefore, estimates the dynamics of the internal state based on the maximum likelihood inference, given reward sequences obtained by interactions with the environment while acquiring its policy. An essential feature necessary to decision-making can be found with implicit learning of the environmental model and estimation of unobservable states; the environmental model of the target system is not required for the decision-making. The agent selects its action based on a current internal state, and learns its policy which maps an internal state into an action; the agent can then learn its policy without having to make an optimization over a continuous belief space. We applied our method to a reasonably sized problem with partial observability. Computer simulation results show that our technique can estimate an appropriate internal model for acquiring a good policy and achieve faster learning with fewer policy parameters than a conventional POMDP algorithm, in a reasonably sized partially observable problem.

2 Preliminary

A POMDP [9] is a framework to make an agent learn and act in a partially observable environment, and consists of (1) a set of real states $\mathcal{S} = \{s_1, s_2, \cdots, s_{|\mathcal{S}|}\}$, which cannot be determined with complete certainty, (2) a set of observation states $\mathcal{O} = \{o_1, o_2, \cdots, o_{|\mathcal{O}|}\}$, which can be perceived by the agents, (3) a set of actions $\mathcal{A} = \{a_1, a_2, \cdots, a_{|\mathcal{A}|}\}$, which can be executed by the agents, and (4) a reward function $R : \mathcal{S} \times \mathcal{A} \rightarrow \mathcal{R}$, which maps the pair of a state and an action into a numerical reward. The dynamics of the model is represented as transition probability $P(s_{t+1}|s_t, a_t)$ and observation probability $P(o_t|s_t, a_t)$. The objective of each agent is to acquire the policy which maximizes an expected future reward in the partially observable world. The state s_t is not observable for each agent; only the observation o_t, which contains partial information about the state, is available. Figure 1(left) represents the dynamic Bayesian network of the conventional POMDP framework. The circles denote unobservable variables and the squares denote observable ones to the agent. If the policy is determined only from an immediate observation, without estimating an unobservable state explicitly or implicitly, the policy does not usually converge to a global optimum [14], because the observation

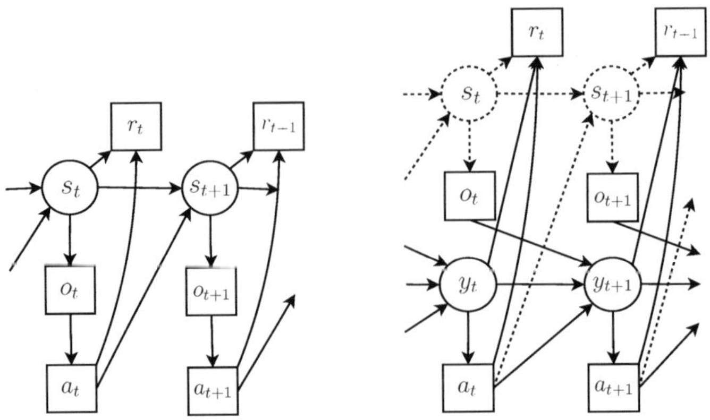

Fig. 1. The dynamic Bayesian networks which represent the two models
left: The conventional belief state MDP model. According to the incremental Bayes inference with the environmental model, the agent calculates a belief state $b(s_t)$ in every time step, and its policy is defined over the continuous belief space, $\pi \equiv P(a_t|b_t)$.
right: The proposed model with internal states. The real state s_t can be removed by the assumption that rewards are dependent on a pair of an internal state y_t and an action a_t; the agent does not have to make an explicit state estimation. It estimates an internal state model, $P(y_{t+1}|y_t, a_t, o_t)$ and $P(r_t|y_t, a_t)$, and its policy is defined over the discrete internal state space, $\pi \equiv P(a_t|y_t)$.

does not satisfy the Markov property. The simplest way to overcome this problem is to use a history of the agent's experience: $H_t = \{(o_t, -), (o_{t-1}, a_{t-1}), \cdots, (o_1, a_1)\}$. Because it is difficult to maintain such a naive history with a limited memory, however, a belief state $b(s_t) \equiv P(s_t|H_t)$ is often used, which summarizes the history as a probability distribution over $\mathcal{S}$. Since the belief state is a sufficient statistic for the history and has the Markov property, the optimal policy that maps a belief state into an action $P(a_t|b_t)$ becomes a solution of a continuous space "belief state MDP". Although this formulation can solve a POMDP, an exact solution is hard to achieve because of the requirement for computing a policy over the entire belief space, the cost of which increases exponentially with the increase in state number of the underlying MDP. In addition, to compute a belief state in every time step, the environmental model, $P(s_{t+1}|s_t, a_t)$ and $P(o_t|s_t, a_t)$, is required by the agent. These limitations make it difficult for realistic agents to deal with general POMDP problems.

In this study, we use internal state representation; our model consists of (5) a set of internal states $\mathcal{Y} = \{y_1, y_2, \cdots, y_{|\mathcal{Y}|}\}$, in addition to the four constituents in the conventional POMDP model described above. The stochastic processes over the internal states are represented as transition probability $P(y_{t+1}|y_t, a_t, o_t)$ and reward (observation) probability $P(r_t|y_t, a_t)$. Figure 1(right) represents the dynamic Bayesian network of our proposed model. The agent based on our method estimates these processes of the internal state based on the maximum likelihood inference, given reward sequences; the reward sequence is regarded as the series of observations, and the expectation-maximization (EM) inference is carried out to estimate their dynamics. By assuming that all variables take discrete values, we use the multinomial model; the EM inference is then done by the input/output hidden Markov model (HMM) [15]. Note that, in this study, the number of possible internal states is assumed to be given to the agent in advance.

3 Method

In this article, we use the following notations: sequence of internal states, actions taken by the agent, rewards and observations obtained from the environment are denoted by $\boldsymbol{y} = \{y_1, \dots, y_{T+1}\}$, $\boldsymbol{a} = \{a_1, \dots, a_T\}$, $\boldsymbol{r} = \{r_1, \dots, r_T\}$ and $\boldsymbol{o} = \{o_1, \dots, o_T\}$, respectively, where T is the termination time. The likelihood function is here given as an HMM:

$$P(\boldsymbol{r}|\boldsymbol{a}, \boldsymbol{o}; \boldsymbol{\theta}, \boldsymbol{\sigma}) = \sum_{\boldsymbol{y}} P(\boldsymbol{y}, \boldsymbol{r}|\boldsymbol{a}, \boldsymbol{o}; \boldsymbol{\theta}, \boldsymbol{\sigma})$$

$$= \sum_{\boldsymbol{y}} P(y_1) \prod_{t=1}^{T} P(r_t|y_t, a_t) P(y_{t+1}|y_t, a_t, o_t). \quad (1)$$

Our aim is to obtain the parameters $\boldsymbol{\theta}$ and $\boldsymbol{\sigma}$, related to the internal state transition and observation processes, respectively, based on the maximum likelihood inference so that the likelihood function in equation (1) is maximized. More concretely, $\boldsymbol{\theta} \equiv \{\theta_{ij} = P(y_{t+1} = j|y_t = i, a_t, o_t)|i = 1, \dots, |\mathcal{Y}|, j = 1, \dots, |\mathcal{Y}|\}, \boldsymbol{\sigma} \equiv \{\sigma_{ik} = P(r_t = k|y_t = i, a_t)|i = 1, \dots, |\mathcal{Y}|, k = 1, \dots, |\mathcal{R}|\}$. Note that the initial internal state y_1 is fixed; $P(y_1)$ is 1 for a particular internal state, otherwise 0. Since our model is based on a discrete-space HMM (that is, it is a multinomial model), the following maximum likelihood estimates can be obtained:

$$\theta_{ij} = \frac{\langle N_{ij} \rangle}{\sum_j \langle N_{ij} \rangle}, \qquad \sigma_{ik} = \frac{\langle N_{ik} \rangle}{\sum_k \langle N_{ik} \rangle}, \quad (2)$$

where N_{ij} and N_{ik} denote the number of appearances of $\{y_t = i, y_{t+1} = j, r_t = k\}$ in the sequences, and $\langle \cdot \rangle$ denotes an expectation with respect to the posterior probability $P(\boldsymbol{y}|\boldsymbol{r}, \boldsymbol{a}, \boldsymbol{o})$. Note that we use the forward-backward algorithm for the input/output HMM model to calculate the expectation, because direct summation over all sequences of the internal state $\sum_{\boldsymbol{y}}$ requires heavy computation.

Once an internal state model is established according to the above method, the remaining processes for solving a problem are to select an action and to learn a policy with the model. These are achieved by the following three steps: the first step is to determine a current internal state y_t according to the probability $P(y_{t+1}|y_t, a_t, o_t)$; the second step is to select an action based on a current policy $P(a_t|y_t)$; and the last step is to learn the policy based on an RL algorithm. Since a POMDP can become a quasi-equivalent MDP by our technique, various algorithms are available to acquire a good policy in a simple MDP environment [16]; our method thus requires many fewer policy parameters than other POMDP algorithms.

4 Computer Simulation

To evaluate the performance of our feature extraction technique, we applied our method to a partially observable problem, "The Tiger Problem", introduced by Kaelbling et al. [9]. An agent is standing in front of two closed doors; behind one of the doors is a treasure box (reward) and behind the other is a tiger (penalty). The agent cannot perceive the real position of the tiger, but can listen to sounds from behind the closed doors with incomplete accuracy; there is a chance that the agent hears the tiger's roar from the wrong door. $\mathcal{A} = \{\text{LEFT}, \text{RIGHT}, \text{LISTEN}\}$ are executable actions for the agent, and $\mathcal{O} = \{\text{LEFT}, \text{RIGHT}\}$ are observations obtained by taking the LISTEN action. The rewards r_t for opening the door with the treasure box, for opening the door with the tiger and for taking the LISTEN action are $+10$, -100 and -1, respectively. The LISTEN action does not change the tiger's position, but the LEFT and RIGHT actions, that is, opening either door, are followed by a state transition with a uniform probability; in other words, the episode is terminated by opening a door, and then another episode is initialized randomly. The probability that the agent can get a correct observation is $p = 0.85$. In this experiment, the agent estimated an internal state model using the data generated from 500 episodes in an environment where an action was selected randomly at each time step. To improve the policy, the REINFORCE algorithm [17], which is a type of policy-gradient-based RL algorithm, was used. We restricted the maximum number of actions to ten so that the agent could not take the LISTEN action for all time, and assumed that the number of internal states was five: $|\mathcal{Y}| = 5$.

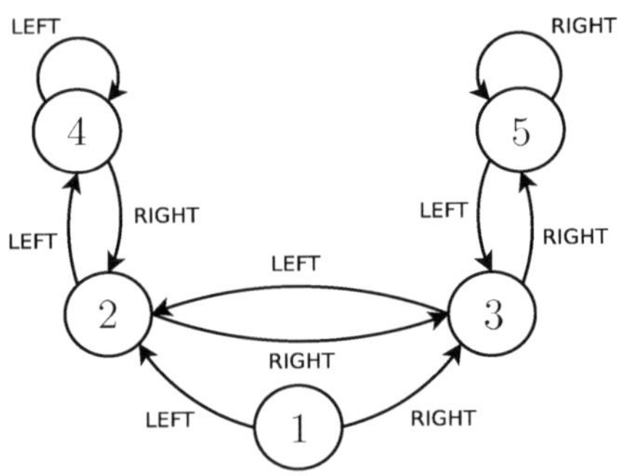

Fig. 2. The state transition when the agent takes the LISTEN action, $y_{t+1} = \text{argmax}_{y_{t+1}} P(y_{t+1}|y_t, a_t = \text{LISTEN}, o_t)$, in the Tiger Problem with $p = 0.85$

Figure 2 represents an internal state model estimated by the agent. Arrows represent transitions with a maximum probability when the agent selects the LISTEN action; each arrow is pointing toward $y_{t+1} = \text{argmax}_{y_{t+1}} P(y_{t+1}|y_t, a_t = \text{LISTEN}, o_t)$. LEFT and RIGHT in the figure represent observations obtained by taking the LISTEN action. Table 1 shows action selection probabilities acquired by the RL algorithm for the estimated model. Note that the performance of the

Table 1. Action selection probability acquired by the agent in the Tiger Problem with $p = 0.85$

y_t \ a_t	LISTEN	LEFT	RIGHT
1	0.999992	0.000003	0.000005
2	0.997126	0.000003	0.002871
3	0.999746	0.000000	0.000254
4	0.009600	0.000000	0.990400
5	0.003291	0.996709	0.000000

agent has two factors: validity of the internal state model and performance of the policy acquired for the model. Figure 3 represents learning curves for one agent based on our method and for three types of agents based on history-based methods, which learn a policy based on a k-length history defined as an observation sequence consisting of the past k observations: $h_t^k = \{o_t, \cdots, o_{t-k+1}\}$; the agent based on k-length history solves the MDP problem with $\sum_{i=0}^{k} |\mathcal{O}|^i$ states. Note that this Tiger Problem with $p = 0.85$ can be optimally solved by using 4-length history. Since the transition in Fig. 2 has the same representation ability as 2-length history, the agent based on our method could not achieve the exact optimal solution but only a sub-optimal one. The averaged reward of this

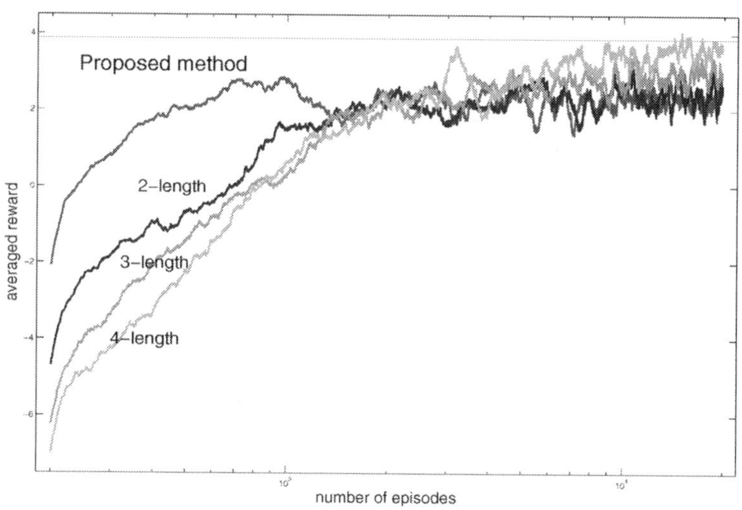

Fig. 3. Learning curves of four agents based on our method and three types of history-based methods with 2, 3 and 4-length observation history. The abscissa denotes the number of episodes in log scale and the ordinate denotes the averaged reward. We executed 20 learning runs, each consisting of 20,000 episodes. Each point represents the moving average for 400 episodes over the 20 runs. The horizontal line represents the optimal averaged reward.

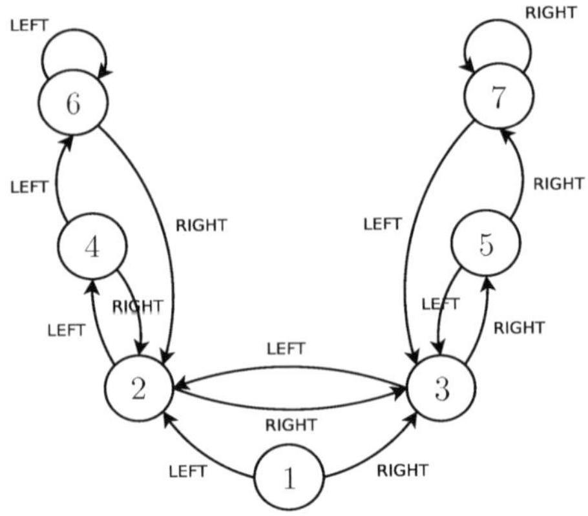

Fig. 4. The state transition when the agent takes the LISTEN action, $y_{t+1} = \text{argmax}_{y_{t+1}} P(y_{t+1}|y_t, a_t = \text{LISTEN}, o_t)$, in the Tiger Problem with $p = 0.80$

agent, however, increased at the fastest speed, because it has the fewest policy parameters. In the history-based method, the number of policy parameters increased exponentially as the history length increased; the exponential growth of the computational cost for acquiring a policy has been essentially inevitable in most conventional POMDP algorithms, even those with effective approximations, and this intractability can make problem-solving infeasible. In our method, on the other hand, the number of parameters increases linearly as the number of internal states increases; our feature extraction technique allows us to reduce the number of parameters by finding an essential structure of the reward delivery in POMDPs. The simulation result shows that our method can solve the intractability in the conventional POMDP formulation and provide us with possibilities to deal with various partially observable problems.

We also applied our technique to the Tiger Problem with $p = 0.80$, after assuming that the number of internal states is seven: $|\mathcal{Y}| = 7$. Note that, in this setting, the problem can be optimally solved by using 7-length history. Figure 4 represents an internal state model estimated by the agent, which has the same

Table 2. Action selection probability acquired by the agent in the Tiger Problem with $p = 0.80$

y_t \ a_t	LISTEN	LEFT	RIGHT
1	0.999896	0.000002	0.000102
2	0.999417	0.000568	0.000015
3	0.999985	0.000011	0.000004
4	1.000000	0.000000	0.000000
5	0.999992	0.000008	0.000000
6	0.000539	0.000000	0.999461
7	0.006082	0.993917	0.000000

representation ability as 4-length history. Table 2 shows action selection probabilities acquired by the RL algorithm for the estimated model. The agent based on our method could achieve many faster learning than a conventional algorithm, which is similar to Fig. 3, because the agent has much fewer parameters; it requires only 1/4 of the parameters of the 4-length history-based agent.

5 Discussion

Kaelbling et al. proposed an exact algorithm for solving POMDP problems offline and articulated a limitation for acquiring an optimal solution [9]. Many studies have since devised methods to reduce the computational effort by using approximate approaches. Although various applications in diverse fields have been demonstrated, with effective methods, few studies can solve the exponential growth of the computational cost due to the increase of policy parameters over the entire belief space; this fact implies that the formulation based on the belief-state MDP has a serious limitation, because the exponential increase is essentially inevitable and few approximations can follow this intractable growth. New approaches based on other formulations, therefore, are required. In this article, we proposed a novel method for solving partially observable problems which is not based on the conventional POMDP formulation.

Chrisman [18] proposed an HMM-based algorithm; it estimated unobservable states using an HMM, which is similar to our method because the maximum likelihood inference over a discrete space is used for solving partially observable problems. The algorithm, however, was formulated as a belief state MDP, namely, the agent based on the HMM-based method made an explicit estimation by calculating a belief state using the forward-backward algorithm and learned a policy over the belief space. In contrast, the agent in our method does not require any information about the real state space $\mathcal{S}$, because it does not estimate unobservable states but instead an internal state model and can learn a policy over the internal feature space. Once an appropriate feature space is found, the computation for acquiring a policy can be achieved in polynomial time, because the original POMDP problem is converted to a quasi-equivalent MDP in the feature space. Our new approach can solve the complexity of partially observable problems and should be able to make many difficult problems tractable.

Boutilier et al. proposed a factored representation technique for making a model structure tractable by introducing dynamic Bayesian networks [19], and McAllester and Singh extended this method to be applicable to POMDPs [20]. In these methods, when an action was executed, the resulting value of a state variable usually depended only on a subset of state variables; the agent could avoid exhaustive state enumeration by taking advantage of local dependencies among stochastic variables. Kitakoshi et al. used a mixture model consisting of multiple Bayesian networks to represent prior knowledge of an environment and showed that it had a better performance than a naive approach [21]. The framework of Bayesian networks is available to the RL field; we also used it to consider the variable structure in Fig. 1.

Two issues still remain for our future work. First, the number of internal states should be estimated; in this work, the number of internal states $|\mathcal{Y}|$ is given to the agent in advance, but it should be possible to estimate with the model estimation process. An appropriate number of internal states can be gradually estimated by increases or decreases in the state number with decision-making processes, or alternatively mixture models such as a Dirichlet process, non-parametric Bayesian inference, may be beneficial for this issue; application of such techniques is our future work. Second, the two separate phases, the model estimation phase by the maximum likelihood inference and the policy acquisition phase by an RL algorithm, should be unified; there are two objective functions optimized separately. Since this separation may prevent acquisition of an optimal solution, the internal model should be estimated with reinforcement learning of a policy on-line. We also plan to explore this method in our future work.

6 Concluding Remarks

In this article, we proposed a feature extraction technique for decision-theoretic planning problems in partially observable stochastic domains and showed a novel approach for solving the problems. To acquire a good policy, all the agent has to do is to estimate a Markov chain over a statistic variable related to rewards. We added an auxiliary state variable, called *internal state*, whose stochastic process has the Markov property, and assumed that the rewards are determined by a pair of an internal state and an action. The agent, therefore, estimated the process based on the maximum likelihood inference, given reward sequences obtained by interactions with the environment. An essential feature necessary to decision-making could be found with implicit learning of the environmental model and estimation of unobservable states. Computer simulation results showed that our technique could estimate an appropriate internal model for acquiring a good policy and achieve faster learning with fewer policy parameters than conventional POMDP algorithms.

References

1. Bertsekas, D.P., Tsitsiklis, J.N.: Neuro-Dynamic Programming. Athena Scientific (1996)
2. Sutton, R.S., Barto, A.G.: Reinforcement Learning: An Introduction. MIT Press (1998)
3. Morimoto, J., Doya, K.: Acquisition of stand-up behavior by a real robot using hierarchical reinforcement learning. Robotics and Autonomous Systems **36** (2001) 37–51
4. Collins, S., Ruina, A., Tedrake, R., Wisse, M.: Efficient bipedal robots based on passive dynamic walkers. Science Magazine **307** (2005) 1082–1085
5. Tesauro, G.: TD-gammon, a self-teaching backgammon program, achieves master-level play. Neural Computation **6** (1994) 215–219
6. Ishii, S., Fujita, H., Mitsutake, M., Yamazaki, T., Matsuda, J., Matsuno, Y.: A reinforcement learning scheme for a partially-observable multi-agent game. Machine Learning **59** (2005) 31–54

7. Singh, S., Bertsekas, D.: Reinforcement learning for dynamic channel allocation in cellular telephone systems. In: Advances in Neural Information Processing Systems. Volume 9. (1996) 974–980
8. Claus, C., Boutilier, C.: The dynamics of reinforcement learning in cooperative multiagent systems. In: Proceedings of the 15th National Conference on Artificial Intelligence. (1998) 746–752
9. Kaelbling, L.P., Littman, M.L., Cassandra, A.R.: Planning and acting in partially observable stochastic domains. Artificial Intelligence **101** (1998) 99–134
10. Thrun, S.: Monte Carlo POMDPs. In: Advances in Neural Information Processing Systems. Volume 12. (2000) 1064–1070
11. Yoshimoto, J., Ishii, S., Sato, M.: System identification based on on-line variational Bayes method and its application to reinforcement learning. In: Proceedings of the International Conference on Artificial Neural Networks and Neural Information Processing. Volume 2714. (2003) 123–131
12. Hauskrecht, M.: Value-function approximations for partially observable Markov decision processes. Journal of Artificial Intelligence Research **13** (2000) 33–94
13. Nakamura, Y., Mori, T., Ishii, S.: An off-policy natural gradient method for a partially observable Markov decision process. In: Proceedings of the International Conference on Artificial Neural Networks. Volume 3697. (2005) 431–436
14. Chrisman, L., Littman, M.L.: Hidden state and short-term memory. Presentation at Reinforcement learning workshop, Machine Learning Conference (1993)
15. Bengio, Y., Frasconi, P.: Input-output HMM's for sequence processing. IEEE Transactions on Neural Networks **7** (1996) 1231–1249
16. Kaelbling, L.P., Littman, M.L., Moore, A.W.: Reinforcement learning: A survey. Journal of Artificial Intelligence Research **4** (1996) 237–285
17. Williams, R.J.: Simple statistical gradient-following algorithms for connectionist reinforcement learning. Machine Learning **8** (1992) 229–256
18. Chrisman, L.: Reinforcement learning with perceptual aliasing: The perceptual distinctions approach. In: Proceedings of the 10th National Conference on Artificial Intelligence. (1992) 183–188
19. Boutilier, C., Dearden, R., Goldszmidt, M.: Exploiting structure in policy construction. In: Proceedings of the 14th International Joint Conference on Artificial Intelligence. (1995) 1104–1111
20. McAllester, D., Singh, S.: Approximate planning for factored POMDPs using belief state simplification. In: Proceedings of the 15th Annual Conference on Uncertainty in Artificial Intelligence. (1999) 409–416
21. Kitakoshi, D., Shioya, H., Nakano, R.: Analysis for adaptability of policy-improving system with a mixture model of bayesian networks to dynamic environments. In: Proceedings of the 9th International Conference on Knowledge-based Intelligent Information & Engineering Systems (KES'05). (2005) 730–737

Reinforcement Learning with Echo State Networks

István Szita, Viktor Gyenes, and András Lőrincz*

Eötvös Loránd University, Pázmány P. sétány 1/C, Budapest, Hungary, H-1117
{szityu@eotvos.elte.hu, gyenesvi@inf.elte.hu, andras.lorincz@elte.hu}
http://nipg.inf.elte.hu/

Abstract. Function approximators are often used in reinforcement learning tasks with large or continuous state spaces. Artificial neural networks, among them recurrent neural networks are popular function approximators, especially in tasks where some kind of of memory is needed, like in real-world partially observable scenarios. However, convergence guarantees for such methods are rarely available. Here, we propose a method using a class of novel RNNs, the echo state networks. Proof of convergence to a bounded region is provided for k-order Markov decision processes. Runs on POMDPs were performed to test and illustrate the working of the architecture.

1 Introduction

Reinforcement learning (RL) has become a popular method for training computer agents how to behave in complex scenarios, based on rewards and penalties received from the environment. As opposed to supervised learning, no explicit training examples are needed, instead, the agent is let to explore its environment. The trainer only needs to provide feedback on whether the actions had a positive or negative outcome in the form of possibly delayed rewards.

Traditional RL frameworks focus on Markov decision problems (MDPs) [1]; here, the state representation that forms the basis of an agent's decisions has the Markov property. The Markov property means that state signals retain all information relevant for decision making. An ideal state signal would summarize the past sensations compactly and yet, it would keep all relevant information.

Most learning methods in MDPs are concerned with learning a mapping from state-action pairs to values [1], which reflect the expected long term reward gained by choosing specific actions. In case of large state spaces that are intractable for tabular methods, function approximators are used to maintain the value mapping. Artificial neural networks are popular function approximators, and recurrent neural networks are spreading to be used in tasks which require memory, since they naturally retain information about previous states. However, proofs of convergence are only available in some special cases, like linear function approximators. Gordon's results about linear function approximators [2] ensure

* Corresponding author.

S. Kollias et al. (Eds.): ICANN 2006, Part I, LNCS 4131, pp. 830–839, 2006.

convergence to a bounded region when the Sarsa(λ) method is used. As far as we know, no positive result about ANNs have been reported so far.

In this paper, we propose a novel method utilizing echo state networks (ESN) [3] as function approximators in reinforcement learning. ESNs are novel kind of recurrent neural networks proposed by Jaeger. These networks were found to be particularly well suited for learning and predicting time series [4]. This way, ESNs are promising candidates for partially observable problems where information about the past may improve performance (e.g. k-order Markov processes).

An important aspect of these networks compared to other RNNs is that they are relatively easy to train. ESNs have random recurrent connections, and only feedforward output connections are trained in a supervised, least squares manner. Since ESNs are effectively linear function approximators acting on the *internal state* representation built from the previous observations, Gordon's results about linear function approximators [2] can be transferred to the ESN architecture. Building on this, we provide proof of convergence to a bounded region for ESN training in the case of k-order Markov decision processes. To the best of our knowledge, this is the first positive result about the convergence of artificial neural networks used for value prediction in reinforcement learning tasks.

2 Related Work

Artificial neural networks have widely been used as function approximators in RL for maintaining the value function of an agent [5, 6]. On the contrary, only limited work has already been done using recurrent neural networks, probably because of difficulties in training such networks. RNNs have the ability to retain state over time, because of their recurrent connections, and they are promising candidates for compactly storing moments of series of observations.

One of the first results with RNNs used for RL was achieved with Elman-style recurrent networks [7]. An Elman network [8] differs from a multi-layer feedforward neural network in that it has *context units*, which hold copies of the hidden unit activations of the previous time step. Because the hidden unit activations are partly determined by the context unit activations, the context units can, in principle, retain information from many time steps ago. Elman networks have also been used for RL-like tasks by Glickman et. al. [9]. They utilized an evolutionary algorithm to train the connection weights of the networks.

Perhaps the most similar work to ours is the work of Bakker [10, 11], who used two types of RNNs for RL tasks that require memory, focusing on tasks that are not fully observable, and investigated tasks with long term dependencies between events. He emphasizes the difficulty in discovering the correlation between a piece of information and the moment at which that information becomes relevant. As a solution, he introduced long short-term memory networks [10].

Various other recurrent neural network approaches have also been proposed. The interested reader is referred to a detailed review of Schmidhuber [12], whose work in the field precedes Bakker's work considerably. However, it must be noted, that none of these works provide convergence guarantees.

3 Short Introduction to ESNs

Recurrent neural networks are efficient black-box models of nonlinear systems. Their feedback connections are able to maintain an ongoing activation even in the absence of inputs. In principle, general RNNs can learn to mimic a target system with arbitrary accuracy. However, training methods available for RNNs (back-propagation through time, real-time recurrent learning, extended Kalman-filtering, etc., [3]) have not been widely employed in technical applications because of their slow convergence.

In the ESN approach, a larger-than-normal layer of neurons is used with random recurrent connections. They are not modified during training, they serve as a dynamic 'reservoir'. The network has an output layer, and may also have an input layer. Input-to-hidden layer connections are randomly generated and are not modified during training. The hidden-to-output connections are trained and training becomes a simple linear regression task. Upon tuning, the output weights minimize the squared error on the training sequence.

The idea behind the ESN approach is that the activations in the hidden layer will mimic systematic variations of the teacher sequence: traces of the hidden layer activations echo spatial and temporal combinations of the previous inputs. It is important that the 'echo' signals be richly varied. This is ensured by a sparse connectivity in the hidden layer, the reservoir. For more details, see [3].

4 Reinforcement Learning with ESNs

A large family of RL algorithms uses value function estimation: they estimate how good it is for the agent to perform a given action in a given state. The notion of 'how good' is defined in terms of expected cumulated future rewards. Value functions are defined with respect to particular policies. The value of taking action $a \in \mathcal{A}$ in state $s \in \mathcal{S}$ under policy $\pi = \pi(s, a)$ is the expected return starting from s, taking action a, and thereafter following policy π and it is denoted by $Q^\pi(s, a)$:

$$Q^\pi(s, a) = E_\pi\{R_t | s_t = s, a_t = a\}, \tag{1}$$

where policy $\pi(s, a)$ is a probability distribution function $\pi : \mathcal{S} \times \mathcal{A} \to \mathbb{R}$ that determines the probability of each action in each state, r_t is the immediate reward received in time step t, $R_t = \sum_{k=0}^{\infty} \gamma^k r_{t+k+1}$ is the cumulated discounted reward received after time step t, and γ is the discount factor.

Agents can either use value tables, or parameterized function approximators to maintain the value function. A popular update method in RL is the so called SARSA update [1], in which the value function is updated by bootstrapping, using the immediate reward and our estimate for the next step:

$$Q(s_t, a_t) \leftarrow (1 - \alpha)Q(s_t, a_t) + \alpha[r_{t+1} + \gamma Q(s_{t+1}, a_{t+1})], \tag{2}$$

The temporally extended version of the update is the SARSA(λ) update, which virtually looks ahead more than one steps to incorporate future rewards discounted by λ into the immediate reward. For $\lambda = 0$, it yields the SARSA update, whereas for $\lambda = 1$, it yields the Monte Carlo update, which takes a whole episode into account. For details, see [1]. Equations for training linear function approximators of the form $\hat{V}_t(s) = \boldsymbol{\theta_t}^T \boldsymbol{\phi_s}$ by gradient descent are also available:

$$\boldsymbol{\theta_{t+1}} \leftarrow \boldsymbol{\theta_t} + \alpha_t (q_t - Q_t(s_t, a_t))\boldsymbol{\phi_{s_t}}, \tag{3}$$

where $\boldsymbol{\theta_t}$ is the weight vector to be tuned, α_t is a learning rate, q_t is the tth training example, $Q_t(s_t, a_t)$ is our estimated value as a function of $\boldsymbol{\theta_t}$ and $\boldsymbol{\phi_{s_t}}$ is the vector of features corresponding to state s_t.

ESNs can be viewed as linear function approximators acting on an *internal state* developed from a series of previous inputs, thus incorporating the past into the state representation. The observation at time step t alone is not sufficient to choose an optimal action, but the internal representation should be more adequate since it is more likely to have the Markov property.

For a formal description, let x_t be the input to the ESN, and let u_t be the internal representation of the network at time step t. Let G be the input weight matrix, and F be the recurrent matrix of the ESN, which are defined to be random sparse matrices with a spectral radius less than one [3]. Then the internal state is calculated by the following equation:

$$u_t = \sigma(F u_{t-1} + G x_t), \tag{4}$$

where σ is some nonlinearity, usually the tanh function. Suppose we have k actions $a_1, ..., a_k$ to choose from. Then the network will have k outputs, one corresponding to each of the actions. We want to train an output matrix H, so that the ith output yields the Q value of the ith action in state u:

$$Q(u, A_i) = H_i u \tag{5}$$

where H_i denotes the ith row of H. To achieve this, from (3) we can derive the following update rule:

$$H_i^{t+1} \leftarrow H_i^t + \alpha_t(\mathit{training\_value}^t - \mathit{esn\_output}_i^t)u_t. \tag{6}$$

5 Theoretical Results

Although proofs of convergence are available for many tabular RL algorithms, the case of function approximators seems to be somewhat more problematic. Even in the simplest linear function approximator case, learning may diverge [13]. The use of neural networks seems even more difficult.

There is a positive result about linear function approximators: the TD(λ) method can be used in a convergent manner to evaluate a fixed strategy. If the strategy is adapted (instead of evaluating a fixed strategy), then care must be taken, because even Q-learning, the simplest extension of the TD method, may diverge [14]. On the other hand, Gordon had shown [15] that using the SARSA algorithm, the value function converges to a bounded region. At the same time,

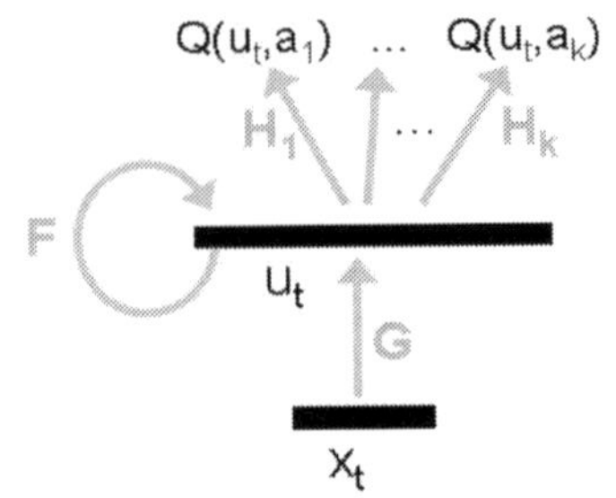

Fig. 1. Value learning with ESN. G is the input matrix, F is the matrix of recurrent weights, which are random sparse matrices. The output matrix H is tuned by the least square method to yield the Q-values.

he showed [2], that using a linear function approximator with SARSA, the value function may oscillate, and also gave a sufficient condition for the algorithm to converge.

Building on these results, we show that using an ESN as a nonlinear function approximator with the SARSA(λ) algorithm, the value function also converges to a bounded region, if the task to learn is an MDP. What is more, we also show that this result holds for k-order MDPs, too. This extension is made available by the memory present in the ESN representation. As far as we know, this is the first such result for (recurrent) neural networks.

Theorem 1 (Gordon). *Assume that a finite MDP is given and SARSA(λ) learning is being used with a linear function approximator. If the learning rates satisfy the Robbins-Monro conditions ($\alpha_t > 0$, $\sum_{t=0}^{\infty} \alpha_t = \infty$, $\sum_{t=0}^{\infty} \alpha_t^2 < \infty$), then the weight vector sequence generated by the algorithm converges to some bounded region R with probability 1.*

We note that the proof for finite MDPs can trivially be extended for continuous state space. Now, let us consider what the ESN-SARSA algorithm does: (1) the observations x_t are nonlinearly mapped to the continuous internal representation u_t, (2) on this representation a linear function approximator is trained using the SARSA method. This means that if the process u_t has the Markov property, Gordon's theorem can be applied.

Let us suppose that the input vectors x_t only contain ± 1 entries.

Definition 2 (k-step unambiguous ESN). *Given is an ESN with initial state u_0. Let us suppose, that the input sequence $x_0, \ldots, x_t$ results in an internal state u, and the input sequence $x'_0, \ldots, x'_{t'}$ results in an internal state u'. We say, that the ESN is k-step unambiguous, if $u = u'$ implies that $x_{t-i} = x'_{t'-i}$ for all $i = 0 \ldots k-1$.*

Definition 3 (unambiguous input matrix). *The matrix G of size $n \times m$ is said to be an unambiguous input matrix, if for any nonzero vector $z \in \mathbb{R}^m$ such that $z_i \in \{-1, 0, 1\}$ $(i \in \{1, \ldots, m\})$, Gz is also nonzero.*

Lemma 4. *Let G be a matrix of size $n \times m$ ($n > m$), whose entries are uniform random values from the set $\{-C, 0, C\}$. The probability that G is an unambiguous input matrix, is at least $1 - (1/3)^{n-m}$.*

Proof. Let z be a fixed vector with entries 0 or ± 1, and let us suppose that its first entry is nonzero. Let the ith row of G be denoted by G_i. The first entry of the row can have three possible values (0 or $\pm C$), from which at most one makes $G_i z = 0$. This means that at most one third of the row's possible settings make $G_i z = 0$. Having n rows, the probability that for all i, $G_i z = 0$ is at most $(1/3)^n$. Finally z can take $3^m - 1$ different values, thus the probability that for any of these values, $Gz = 0$ for a randomly selected matrix G, is at most $(1/3)^n \cdot (3^m - 1) < (1/3)^{n-m}$.

The following lemma states that if the input weights are significantly greater than the recurrent weights, then the $x \to u$ mapping is unambiguous.

Lemma 5. *Let the entries of the G matrix of the ESN be randomly chosen uniformly from the set $\{0, \pm C\}$, where $C > \sqrt{n}$. Let the recurrent matrix F be a sparse random matrix with $\|F\| < 1$. Then the ESN is 1-step unambiguous with probability $1 - (1/3)^{n-m}$.*

Proof. The input matrix is unambiguous with probability $1 - (1/3)^{n-m}$, so let us suppose for now that G is unambiguous. Indirectly, let us suppose that there exists two input series whose last elements are different, but they transfer the ESN to the same internal state. This boils down to the existence of such $x_1 \neq x_2$ and u_1, u_2 that

$$u_t = \sigma(Fu_1 + Gx_1) = \sigma(Fu_2 + Gx_2).$$

Then

$$F(u_1 - u_2) = G(x_2 - x_1),$$

and by denoting $z = (x_2 - x_1)/2$

$$F\frac{u_1 - u_2}{2} = Gz.$$

G is an unambiguous input matrix, thus at least one component of Gz is nonzero. However, the nonzero terms of Gz have the form $(\pm C) \cdot (\pm 1)$, so $\|Gz\| \geq C$. On the other hand, the entries of u_i are real values between -1 and 1, thus

$$\|F\frac{u_1 - u_2}{2}\| \leq \|F\|\|\frac{u_1 - u_2}{2}\| < 1 \cdot \sqrt{n},$$

which is a contradiction.

Lemma 6. *If the ESN is 1-step unambiguous, then it is k-step unambiguous for all $k \geq 1$.*

Proof. We proceed by induction on k: for $k = 1$ the lemma holds, and by supposing that it holds for all $n < k$, we show that it also holds for $k + 1$. For $j = 0, 1, \ldots, t$, let

$$u_{t-j} = \sigma(Fu_{t-j-1} + Gx_{t-j}),$$

and similarly, for $j = 0, 1, \ldots, t'$, let

$$u'_{t'-j} = \sigma(Fu_{t'-j-1} + Gx_{t'-j}),$$

and let us suppose that $u_t = u'_{t'}$. Since the ESN is 1-step unambiguous, then $x_t = x'_{t'}$, and this implies $u_{t-1} = u'_{t'-1}$. Using our supposition that the ESN is k-step unambiguous, it must hold for the previous k observations that $x_{t-1} = x'_{t'-1}, \ldots, x_{t-k} = x'_{t'-k}$, which means that the ESN is $k+1$-step unambiguous.

Definition 7 (induced ESN decision process). *Given is the following decision process, denoted by M: X is the finite state space, A is the finite action space, $(X \times A)^*$ denotes the space of series made of elements of $X \times A$, $R : X \to \mathbb{R}$ is the reward function, $P : (X \times A)^* \times X \to [0,1]$ is the transition probability function, that gives the probabilities of the next states based on the trajectory of states travelled and actions applied so far. The ESN-decision process induced by the decision process M is the following: $U = \mathbb{R}^n \cup Z$ is the state space, where Z is an abstract terminal state, A is the action space; and $P' : U \times U \to [0,1]$, $R' : U \to \mathbb{R}$. P' and R' are defined as follows: for any $u \in \mathbb{R}^n$ let us observe the sequence set*

$$S(u) = \{s = (x_0, a_0, x_1, \ldots, a_{t-1}, x_t, a_t) \,|\, \textit{ESN input } s \textit{ results in internal state } u\}.$$

The following three cases must be treated separately:

- *$S(u) = \emptyset$: then let $P'(Z|u,a) := 1$, $P'(v|u,a) := 0$ and $R'(u) := 0$ for all a and $v \neq Z$.*
- *for all $s \in S(u)$ the probabilities $P(.|s)$ are all the same. Then*

$$P'(v|u,a_t) := \begin{cases} P(x|s), & \text{if } \exists x : v = \sigma(Fu + Gx) \\ 0, & \text{otherwise.} \end{cases}$$

- *if the value of P is ambiguous on the elements of $S(u)$, then let $P'(Z|u,a) := 1$, $P'(v|u,a) := 0$ and $R'(u) := 0$ for all a and $v \neq Z$.*

If the last case does not occur for any $u \in \mathbb{R}^n$, then the induced decision process is said to be well defined.

The previous lemmas imply our main theorem:

Theorem 8. *Given is a k-step unambiguous ESN, and given is an $M := (X, A, P, R, x_0)$ k-order MDP, then the decision process induced by the ESN is well defined, furthermore, the decision process is an MDP, on which the value function sequence generated by the ESN-SARSA algorithm converges to a bounded region.*

Note: The theorem gives the important result that the ESN-SARSA algorithm is convergent to a region (that is, the algorithm cannot diverge) for *any* k. This is because it only states that the algorithm is convergent, it does not tell anything about the speed of convergence and how good the resulting value function will be. As k is increased, the effect of earlier steps decreases exponentially, which means that the temporal resolution of the function approximator about the far past will become poorer.

6 Test Runs

There are several benchmark tasks to test RL algorithms for partially observed environments. These tasks typically require some amount of memory. We have tested the ESN-RL architecture on some of these tasks. In these tests, the ESN was always trained 'on-line'.

The first problem was described by Littman et al. [16]. You stand in front of two doors: behind one door is a tiger and behind the other is a vast reward. You may open either door, receiving a large penalty if you chose the one with the tiger and a large reward if you chose the other. You have the additional option of simply listening. If the tiger is on the left, then with probability $1 \geq p > 0.5$ you will hear the tiger on your left and with probability $(1 - p)$ you will hear it on your right; symmetrically for the case in which the tiger is on your right. If you listen, you will pay a small penalty. The problem is this: How long should you stand and listen before you choose a door?

By varying the value of p the difficulty of the task varies; as p decreases, the task gets more difficult, more listening is needed to safely determine the position of the tiger (note, that at $p = 0.5$, one can not do better than to guess randomly). Figure 2 shows our results on the tiger problem. It can be seen, that as p increases, the ESN learns that less listening is needed, and its performance also increases up to 1, when the task becomes deterministic. In this case, the ESN learns to answer after 1 round of listening. The number of listenings learned by the ESN is similar to the results in [16]. At $p = 0.85$, their system learned to listen until it hears two more roars from one side then the other. This is comparable to the ESN's result of listening 2.37 times on average.

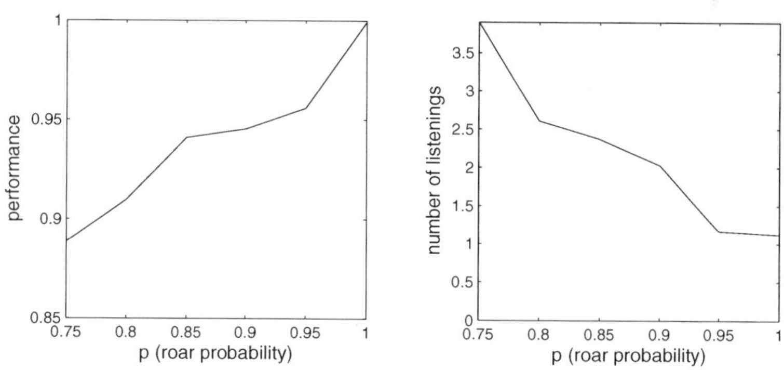

Fig. 2. Results on the tiger problem. Left: performance, right: number of listenings needed to safely open the right door. Results are averages of 100 runs

The second problem was a simple 4x3 maze example proposed by Russell and Norvig [17]. The maze has an obstacle in the middle, and has two special states, one which gives a reward of +1, one which gives a penalty of -1. The actions, N, S, E, W, have the expected result 80% of the time, and transition in a direction perpendicular to the intended with a 10% probability for each direction. Observation is limited to two wall detectors that can detect when a wall is to the left or right. The task is to find the +1 state repeatedly and avoid the -1 state, starting from a random position. As Figure 3 shows, the ESN was also able to learn this navigation problem. It must be noted that adding the agent's own previous action to the observations in the next time step increased the stability of the ESN, which might be because of the heavy partially observable nature of the task, probably being compensated by considering the previous moves.

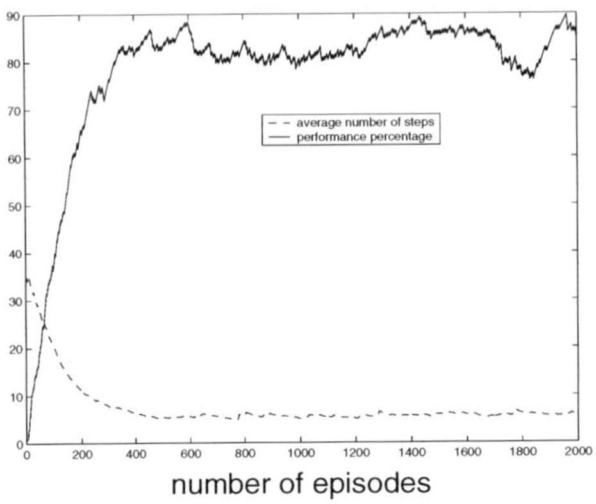

Fig. 3. An example run on the 4x3 maze problem. The performance goes up to 90 percent, and the average number of steps of the agent is around 5: the agent has learned the routes to the goal

We also tested the architecture on a T-maze problem well suited to test the state retaining properties in memory required tasks [10]. At the beginning of a T-shaped maze, the agent is shown a sign indicating whether it should to turn left or right at the end. By varying the length of the T-shape, an algorithm can be tested how long it is able to retain previous observations. Whether the ESN approach was able to learn the task seemed to depend on the randomly generated G and F matrices. With proper matrices, the agent easily learned the task. The matrices are thought to be proper, if due to the random connections, the activation resulting from observing the sign is maintained long enough to effect the decision at the end. The larger the k in the Markov process, the less probable that the random matrices will be proper. For $k < 10$, almost 100% of the matrices were proper; training was always successful. As k was increased, this percentage slowly fell, and reached 0 at $k = 25$, being around 50% at $k = 20$.

7 Conclusions

An echo state network approach for maintaining the value function of an agent in reinforcement learning tasks was proposed. In principle, the ESN is able to compactly summarize recent observations in an internal state, upon which the networks output weights can be trained to act as a linear function approximator. Theoretical results ensure convergence to a bounded region when the task is a k-order MDP. Tests of the architecture on artificial problems predict that the ESN approach may be a simple method for partially observable MDP tasks.

References

[1] Sutton, R., Barto, A.G.: Reinforcement Learning: An Introduction. MIT Press, Cambridge, MA (1998)

[2] Gordon, G.J.: Chattering in SARSA(lambda) - a CMU Learning Lab Internal Report (1996)

[3] Jaeger, H.: Tutorial on training recurrent neural networks, covering BPTT, RTRL, EKF and the 'echo state network' approach. Technical Report GMD Report 159, German National Research Center for Information Technology (2002)

[4] Jaeger, H., Haas, H.: Harnessing nonlinearity: predicting chaotic systems and saving energy in wireless telecommunication. Science (2004) 78–80

[5] Tesauro, G., Sejnowski, T.J.: A parallel network that learns to play backgammon. Artificial Intelligence **39** (1989) 357–390

[6] Bertsekas, D.P., Tsitsiklis, J.N.: Neuro-Dynamic Programming. Athena Scientific, Belmont, MA (1996)

[7] Lin, L.J., Mitchell, T.M.: Memory approaches to reinforcement learning in non-markovian domains. Technical Report CMU-CS-92-138, Carnegie Mellon University, Pittsburgh, PA (1992)

[8] Elman, J.L.: Finding structure in time. Cognitive Science **14** (1990) 179–211

[9] Glickman, M.R., Sycara, K.: Evolution of goal-directed behavior from limited information in a complex environment. In: Proc. of the Genetic and Evol. Comp. Conf., Orlando, Florida, USA, Morgan Kaufmann (1999) 1281–1288

[10] Bakker, B.: Reinforcement learning with long short-term memory. Advances in Neural Information Processing Systems **14** (2002) 1475–1482

[11] Bakker, P.B.: The State of Mind - Reinforcement Learning with Recurrent Neural Networks. PhD thesis, Universiteit Leiden (2004)

[12] Schmidhuber, J.: Making the world differentiable. Technical Report TR-FKI-126-90, Institut für Informatik, Technische Universität München (1990)

[13] Baird, L.C.: Residual algorithms: Reinforcement learning with function approximation. In: International Conference on Machine Learning. (1995) 30–37

[14] Watkins, C.J.C.H.: Learning from Delayed Rewards. PhD thesis, Cambridge University, Cambridge, UK (1989)

[15] Gordon, G.J.: Reinforcement learning with function approximation converges to a region. In: Advances in Neural Information Processing Systems. Volume 13., Cambridge, MA, MIT Press (2001) 1040–1046

[16] L. P. Kaelbling, A.R.C., Littman, M.L.: Acting optimally in partially observable stochastic domains. In: Proc. of the 12th Nat'l Conf. on Artif. Intell. (1994)

[17] Russell, S.J., Norvig, P.: Artificial Intelligence: a Modern Approach. Prentice-Hall, Englewood Cliffs, New Jersey (1994)

Reward Function and Initial Values: Better Choices for Accelerated Goal-Directed Reinforcement Learning

Laëtitia Matignon, Guillaume J. Laurent, and Nadine Le Fort-Piat

Laboratoire d'Automatique de Besançon UMR CNRS 6596,
24 rue Alain Savary, 25000 Besançon, France
{laetitia.matignon, guillaume.laurent, nadine.piat}@ens2m.fr

Abstract. An important issue in Reinforcement Learning (RL) is to accelerate or improve the learning process. In this paper, we study the influence of some RL parameters over the learning speed. Indeed, although RL convergence properties have been widely studied, no precise rules exist to correctly choose the reward function and initial Q-values. Our method helps the choice of these RL parameters within the context of reaching a goal in a minimal time. We develop a theoretical study and also provide experimental justifications for choosing on the one hand the reward function, and on the other hand particular initial Q-values based on a goal bias function.

1 Introduction

The reinforcement learning (RL) paradigm [1] provides techniques in which an agent can optimize environmental payoff for the autonomous resolution of tasks. A RL agent tries to *learn a policy, i.e.* it learns by trial-and-error to select actions that maximize its expected discounted future rewards for state-action pairs, represented by the action values. Q-learning [2] is a commonly form of RL where the optimal policy is learned implicitly in the form of a Q-function.

One of the main limitations of RL is the *slowness in convergence*. Thus, several methods have been proposed to speed up RL. They involve the incorporation of prior knowledge or bias into RL. [3] proposed a methodology for *designing reward functions* that take advantage of implicit domain knowledge. It involves the use of continuous reward functions and *progress estimators*. Likewise, with *reward shaping*, the rewards from the environment are augmented with additional rewards [4]. However, reward shaping can lead the agent into learning suboptimal policies and so, traps the system. [5] completed the reward shaping study and moreover, proved certain similarities between potential-based shaping and *initial Q-values*. Indeed, the most elementary method for biasing learning is to choose the initial Q-values [6]. So [7] studied various representations of reward functions and the complexity of Q-learning methods depending on the choice of RL representation. Finally, concerning *imitative reinforcement*, [8] proposed to give the learning agent access to the Q-values of the experienced agent.

S. Kollias et al. (Eds.): ICANN 2006, Part I, LNCS 4131, pp. 840–849, 2006.

Thus, reward function and initial Q-values play an important part in RL. Nevertheless, although RL has been studied extensively and its convergence properties are well known, in practice, people often choose reward function on one's intuition and initial Q-values arbitrarily [1]. In this paper, we discuss the effects of RL parameters on the policy in order to suggest a generic analysis. We validate our analysis with Q-learning algorithm. The main issue is to shed light on *how to correctly initialize RL parameters in order to obtain the desired optimal behavior in a minimal time within the context of a goal directed task.*

2 Reinforcement Learning

The framework of most of RL algorithms is a *Markov Decision Process* (MDP), defined as a finite set of states, S, a finite set of actions, A, and a transition function $T : S \times A \times S \rightarrow [0;1]$ giving for each state and action a probability distribution over states. $R : S \times A \times S \rightarrow \mathbb{R}$ is a reward function giving the expected immediate reward or *reinforcement* received under each transition. The goal is to learn a mapping from states to actions, called a *policy*, π.

In this work, we have validated our analysis with *Q-learning* [2] algorithm. In Q-learning, an *action-value function* $Q^{\pi}(s,a)$ is estimated over the learning process and stored in a tabular representation. An *action-value* represents the expected sum of rewards [1] the agent expects to receive by executing the action a from state s and following the policy π . The optimal action-value function Q^* is known to be the unique solution to the Bellman equation,

$$Q^*(s,a) = \sum_{\forall s' \in S} T(s,a,s') \left[R(s,a,s') + \gamma \max_{a'} Q^*(s',a') \right] . \quad (1)$$

Q-learning is an *off-policy* method and its updating rule is :

$$Q(s,a) \leftarrow Q(s,a) + \alpha \left[r + \gamma \max_{a'} Q(s',a') - Q(s,a) \right] \quad (2)$$

where r is the reward received for the transition from the state s to the new state s' by executing the action a. $\alpha \in]0;1]$ is the learning-rate parameter and $\gamma \in [0;1[$ the discount factor. Under some cases [9], Q-learning algorithm is guaranteed to converge to the optimal value function. The Q-Learning algorithm chooses the action according to an exploration/exploitation criteria. We used the *ϵ-greedy method*, in which the probability of taking a random action is ϵ and, otherwise, the selected action is the one with the largest Q-value in the current state [2].

3 Choice of Uniform Initial Q-Values with Binary Rewards

We make the assumption that two general trends stand out : *the global policy* and *a specific behavior at the beginning* of the learning process. In this section,

[1] The rewards are discounted by a discount factor γ that controls the balance between the significance of immediate rewards and future rewards.

[2] If several values are identical, the choice will be random among greedy actions.

we are going to set out that these both tendencies depend on the shape of the reward function and on the initialization of the action-value function. We first considered *a binary reward function* which has the advantage to include a lot of cases and it is possible to extrapolate.

3.1 Optimal Policy

The binary reward function is such as the reward received is always r_∞ except if the new state is the goal state and then, the reward is r_g. It's given by :

$$\forall s \in S\ \forall a \in A,\ R(s,a,s') = \begin{cases} r_g \text{ if } s' = s_g \\ r_\infty \text{ else} \end{cases} \tag{3}$$

where s' is the state obtained by executing the action a from s, and s_g the goal state. In case of all rewards are identical ($r_g = r_\infty$), the solution of the Bellman equation (1) is a constant noted Q_∞,

$$\forall s\ \forall a\ Q^*(s,a) = Q_\infty = \frac{r_\infty}{1-\gamma}\ . \tag{4}$$

That is to say that during the learning process, Q-values for all state-action values converge to Q_∞. Nevertheless, if $r_g \neq r_\infty$, Q_∞ is the limit of the action-value function $Q^*(s,a)$ when the distance between s and s_g tends toward infinity. So, toward the goal, states are *more and more or less and less attractive* depending on r_g and Q_∞. On the one hand, if $r_g > Q_\infty$, the Q-value of state-action pairs moving to the goal state will be more and more attractive than Q_∞. So the global optimal policy is *the shortest way* toward s_g. On the other hand, if $r_g < Q_\infty$, the optimal policy is random everywhere except *a local repulsion of* s_g.

As a matter of course, the shortest way toward the goal is the sought optimal policy concerning goal-directed tasks. *So r_g must always be superior to Q_∞.*

3.2 Behavior at the Beginning of the Learning Process

As well as the reward function, the initial value Q_i of the action-value function has an effect on the policy, but only at the beginning of the learning process. We believe that a global trend can be underscored during the first trials of the learning process.

Let's examine what the Q-values are expected to be at the beginning. If we calculate the first updating of a state-action pair (s,a) thanks to (2), such as the next state s' is not the goal state and has not been updated, we have :

$$\begin{aligned} Q(s,a) &\leftarrow Q_i + \alpha\left[r_\infty + (\gamma-1)Q_i\right] \\ &\leftarrow Q_i + \alpha(1-\gamma)(Q_\infty - Q_i)\ . \end{aligned} \tag{5}$$

So the discriminating value of Q_i is also Q_∞. According to the value of Q_i compared to Q_∞, *states already visited will be more or less attractive* as long as the agent has not reached plenty of times the goal state.

- **If** $Q_i > Q_\infty$ **:** states which have already been visited will have a value lower than the value of states which haven't yet been visited ($Q(s,a) < Q_i$). In other words, states which haven't yet been visited will be more attractive. It induces the agent *to explore* more systematically at the beginning of the learning than a random exploration. We called it *systematic exploration behavior*.
- **If** $Q_i < Q_\infty$ **:** states which have already been visited will have a value superior to the value of states which haven't yet been visited ($Q(s,a) > Q_i$). That's to say states which have already been visited will be more attractive. It leads the agent into less exploration at the beginning, which considerably slows down the learning. We named this behavior "*moving round in circles*". Of course, it's better to avoid it.
- **If** $Q_i = Q_\infty$ **:** states which have already been visited will have the same value than states which haven't yet been visited ($Q(s,a) = Q_i$). So we obtain at the beginning *a pure random behavior*.

3.3 Gridworld Experiments

We have discussed how traditional reward functions and arbitrary initial Q-values may slow down the learning of an interesting policy. At this point, we are going *to validate this previous analysis* and we have chosen for simplicity and clarity to use at first a non-deterministic gridworld domain to demonstrate how the behavior is influenced by using binary rewards and different initial Q-values.

Benchmark. The system is represented by a mouse going around a maze (Fig. 1a). Each mouse's position is a discrete state. When the mouse reaches the goal state, the trial ends. The mouse chooses from four actions, representing an intention to move in one of the four cardinal directions (N,E,S,W). An action that would move the mouse in a wall instead leaves the mouse in its current position. Any movement moves the mouse in the intended direction with probability 0.6, and otherwise in a random state of the four neighboring states of the expected state. All trials use Q-learning with a learning-rate α of 0.1, a discount factor γ of 0.9, a tabular Q-table initialized uniformly Q_i and follow a policy where the greedy action is taken with a probability 0.9 ($\epsilon = 0.1$).

Binary Reward Function. Our reward function replicates the function given in (3), with $r_g = 1$ and $r_\infty = 0$. With such a reward function, the optimal policy is *the shortest way toward the goal* and the discriminating value of Q_i is 0 (Q_∞=0). Fig. 1b illustrates our previous analysis. As can be readily seen, using *the systematic exploration behavior* helped speed up learning during the first trials. The agent visited every nook and cranny of the complex maze and the goal state s_g was discovered faster than in case of a random exploration. But given that the agent was always spurred on to explore, it always took more steps to reach the goal after some trials. Besides, we used different values of Q_i for the systematic exploration and we notice the more Q_i was superior to Q_∞,

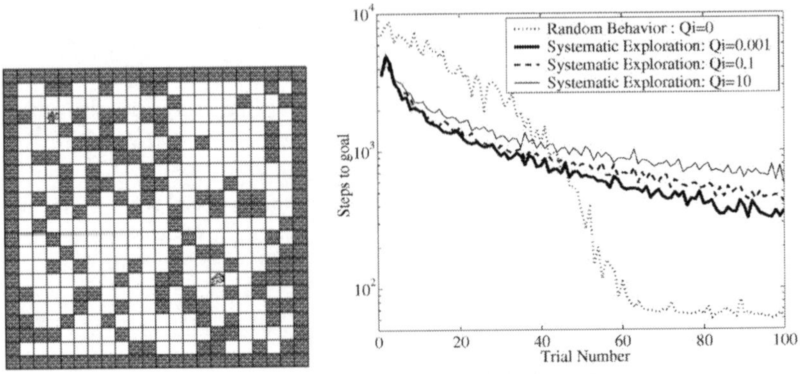

(a) Gridworld (b) Experiments

Fig. 1. On the left, non-deterministic 20×20 gridworld with a single start state (state [2, 2]) and a goal state (cheese) (state [14, 14]). On the right, gridworld experiments with different Q_i averaged over 50 independent runs. Steps to goal *vs.* trial number.

the more the agent explored. So, *the more the difference between* Q_i *and* Q_∞ *is important, the more the general behavior is underlined.*

Concerning the *moving round in circles behavior* ($Q_i < 0$), we do not submit any experiments because the agent took too much time to reach the goal. Anyway, *this behavior has to be avoided.*

3.4 Conclusion

This shed light on *the importance of initial Q-values.* The choice of Q_i is not trivial and must be done according to the desired behavior. When the system naturally goes away from the goal, a systematic exploration should be preferred in order to speed up learning at the beginning. Indeed, systematic exploration forces the controller to explore unvisited states, and so to approach the goal.

4 Choice of Continuous Reward Function and Heterogeneous Initial Q-Values

In order to broaden the scope of our study, we propose henceforth to use first two different *continuous reward functions* with uniform initial Q-values, and secondly to initialize the action value function with *a goal bias function.*

4.1 Reward Function Using Progress Estimators

First, we propose to study the use of a continuous reward function instead of binary rewards. Thus, some authors introduce reward functions by using *progress estimators* [3] or potential-based shaping [5]. Progress estimators provide a measure of improvement relative to an objective. They do not supply a complete information but only partial, goal-specific "advice". For instance, concerning the gridworld, the progress estimator can be an assessment of the expected number of steps needed to get to the goal from the new state s', defined as $\varphi(s', a) = d(s', s_g)$. d is the manhattan distance between s' and s_g. The aim of the agent in the gridworld is to minimize this function and the parameters could be then :

$$\begin{cases} \mathrm{r}(s, a, s') = -\varphi^2 = -d^2(s', s_g) \\ Q_i = 0 \end{cases} \tag{6}$$

This way, the agent is less and less punished by approaching the goal. The global policy is the shortest way toward the goal. Given our maze, this reinforcement is spurious as there are plenty of walls between the initial state and the goal. In particular, it entails an *unlearning phenomenon* after few trials. Indeed, if the agent was taken off toward a dead end (that moves the agent closer to the goal), it would get out of the trap only thanks to exploration because states are more and more attractive toward the goal. At the beginning, Q-values are near 0 so systematic exploration is strong : going out of the trap is possible. But after few trials, turning back is tantamount to choosing a less attractive Q-value and will happen only if several exploration actions follow one another, *i.e.* seldom.

So progress estimators and potential-based shaping are risky. It's better to use cautiously these approaches insofar as they may lead to a pernicious behavior.

4.2 Continuous Reward Function Inspired by Gaussian Function

Consequently, we propose a continuous reward function such that on the one hand, r *is uniform* for some states far from the goal in order to avoid the unlearning phenomenon, and on the other hand, there is *a reward gradient* in a zone around the goal. We suggest the reward function inspired by *the gaussian function* :

$$\mathrm{r}(s, a, s') = \beta e^{-\frac{d(s', s_g)^2}{2\sigma^2}} . \tag{7}$$

Q_i values are uniform, β adjusts the amplitude of the function and σ, the standard deviation, specifies the *reward gradient influence area*. As a matter of course, "moving round in circles" behavior must be avoided, *i.e.* $Q_i \geq \frac{\beta}{1-\gamma}$.

For the gridworld task, we have chosen $Q_i = 100$ and $\beta = 10$. Fig. 2a shows the unlearning phenomenon as from 80 trials with $\sigma = 3.5$ [3]. Indeed, the reward gradient influence area is too large and includes few dead ends. On the contrary, if the reward gradient influence area is only 6 steps around the goal ($\sigma = 2$), there won't be any unlearning phenomenons and the learning process is accelerated.

Such a continuous reward function is *adjustable* so as to avoid a harmful behavior. Anyway, the best approach would be to influence fleetingly the learning.

4.3 Goal Bias

In view of the importance of the action-value function initialization, we propose to be inspired by *progress estimators* in order to *initialize the action-value*

[3] *i.e.* states 10 steps away from the goal have uniform r.

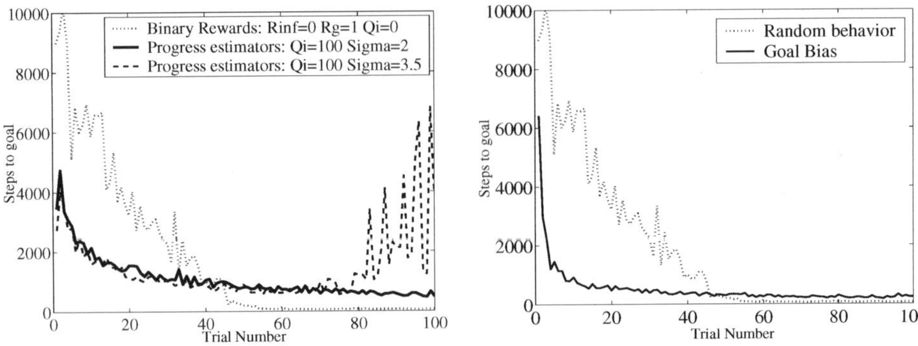

(a) Continuous reward (b) Goal bias

Fig. 2. Gridworld experiments averaged over 20 independent runs. Steps to goal *vs.* trial number. On the left, reward function inspired by gaussian function . $Q_i = 100$; $\mathrm{r}(s,a,s') = 10e^{-\frac{d(s',s_g)^2}{2\sigma^2}}$. On the right, goal bias function with binary rewards. Random test is $Q_i = 0$. Goal bias function is $Q_i(s,a) = 0.001e^{-\frac{d(s,s_g)^2}{2\times 13^2}}$.

function with more precise information. In this section, the reward function is the binary one given by (3) with $r_\infty = 0$ and $r_g = 1$.

We are going to settle a correct goal bias function thanks to our previous analysis. An interesting bias shall achieve *an adjustable state gradient* and in addition, must avoid the "moving round in circles" behavior. We suggest for instance *a gaussian goal bias function* :

$$Q_i(s,a) = \beta e^{-\frac{d(s,s_g)^2}{2\sigma^2}} + \delta + Q_\infty \,. \tag{8}$$

δ fixes the level of systematic exploration far away from the goal, β the amplitude of the bias and σ the bias influence area.

Concerning the gridworld, the bias is such that states near the goal are more and more interesting *a priori* than states far away from the goal. So δ and β must be chosen very small compared to one (in order to avoid too much systematic exploration). Fig. 2b presents goal bias results on the previous gridworld which are unambiguous. *The goal bias leads to a much faster learning process.* It is worth noticing that there is *no problem concerning dead ends even if the bias is wrong.* Contrary to Sect. 4.2, the effect of the goal bias function is transient. It advises the agent *only at the beginning of the learning process.*

4.4 Conclusion

Both progress estimators and potential-based shaping methods must be used cautiously to design a continuous reward function. Consequently, we have proposed a continuous reward function inspired by a gaussian function and whose reward gradient influence area is adjustable in order to deal with risks. *Anyway, the best solution is to choose a suitable goal bias function that does not lead to any problems. Our analysis helps the choice of a correct goal bias function.*

5 Experiments with the Pendulum Swing-Up Problem

Last of all, we validate our results on the continuous space control task of *a pendulum swinging upwards with limited torque* [10] (Fig. 3a). The control of this one degree of freedom system is non-trivial if the maximal output torque u^{max} is smaller than the maximal load torque mgl. The controller has to swing the pendulum several times to build up enough momentum to bring it upright and has to decelerate the pendulum early enough to prevent it from falling over.

We have chosen a two-dimensional state space $\mathbf{x} = (\theta, \omega)$. $30 \times 30 \times 9$ bases were used for the state-action space (θ, ω, u). Each trial was started from an initial state $\mathbf{x}(0) = (\pi, 0.1)$ and lasted 20 seconds. The sample time is 0.03 seconds. As a measure of the swing-up performance, we have chosen t_{up} as the time in which the pendulum stayed up ($|\theta| < \pi/4$). A trial was regarded as "successful" when t_{up} is superior to the t_{up} average of the 1000 last trials.

We tested the performance of the Q-Learning algorithm depending on the shape of the reward function and initial Q-values (Fig. 3b). In all cases, the t_{up} average after 10000 trials is around 14 seconds.

We tested first the binary reward function

$$R(\mathbf{x}, u, \mathbf{x}') = \begin{cases} 1 \text{ if } |\theta'| < \pi/4 \\ 0 \text{ otherwise} \end{cases} \tag{9}$$

with different uniform initial Q-values. With $Q_i = 0$, the task was really difficult to learn (`bar1`) because the behavior far from the goal is random. A better performance concerning the binary reward and uniform Q_i was observed with $Q_i > 0$ (`bar2`), *i.e.* the followed behavior when $|\theta| > \pi/4$ is *systematic exploration*. The policy drives the agent to unexplored areas which are assigned higher Q-values, *i.e.* the controller is spurred on to swing the pendulum upwards. Systematic exploration is the best strategy in this specific case.

Then, we tested goal bias with binary rewards : $Q_i(\mathbf{x}) = \beta e^{-\frac{\theta^2}{2\sigma^2}} + \delta$. Given our previous results, it is obvious that the goal bias function shall favor systematic exploration when $|\theta| > \pi/4$, so we chose $\delta = 0.1$, $\beta = 1$ and $\sigma = 0.25$ (`bar3`). Goal bias does not improve obviously the learning.

The system is a continuous space control task so the classical reward [10] is given by the height of the tip of the pendulum, *i.e.* $R(\mathbf{x}, u, \mathbf{x}') = \cos(\theta')$ and the arbitrary Q_i value is 0 (`bar4`). Thus, the pendulum moves round in circles when $|\theta| < \pi/2$ and explores systematically when $|\theta| > \pi/2$ at the beginning. The result is disappointing.

We applied our gaussian reward function (7) with $Q_i = 10$ and $\beta = 1$. The distance is defined as $d(\mathbf{x'}, \mathbf{x}_{up}) = \theta'$ with $\mathbf{x'}$ the new state and $\mathbf{x}_{up}$ the goal state. So the continuous reward function is

$$R(\mathbf{x}, u, \mathbf{x}') = e^{-\frac{\theta'^2}{2\sigma^2}} \ . \tag{10}$$

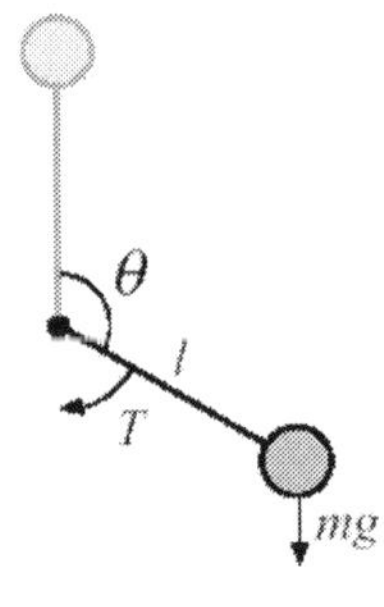

(a) A pendulum

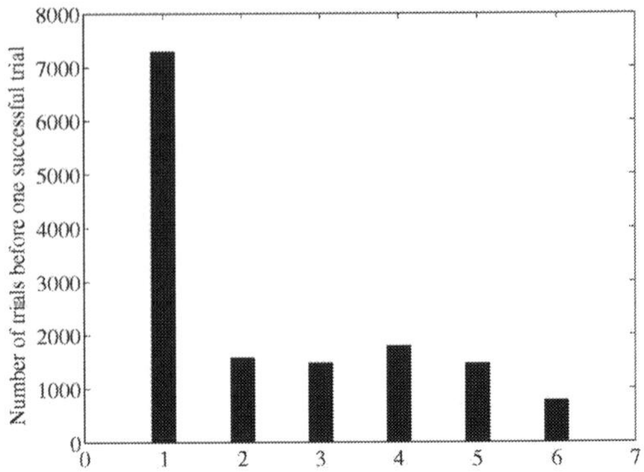

(b) Experiments

Fig. 3. On the right, comparison of number of trials before one successful trial. The simulation lasted 10000 trials averaged over 10 independent runs.
`bar1`: {binary reward ; $Q_i(\mathbf{x}) = 0$} `bar2`: {binary reward ; $Q_i(\mathbf{x}) = 0.1$}
`bar3`: {binary reward and goal bias } `bar4`: {$R(\mathbf{x}, u, \mathbf{x}') = \cos(\theta')$; $Q_i(\mathbf{x}) = 0$}
`bar5`: {continuous reward and $Q_i(\mathbf{x}) = 10$ } `bar6`: {continuous reward and goal bias}

$\sigma = 0.25$ so that the reward gradient influence area is only around $|\theta| < \pi/4$. Results (`bar5`) are near the case of a binary reward and systematic exploration.

Lastly, we tried goal bias with continuous reward function. The reward function (10) is the continuous equivalent to the binary one so we have kept this one. The reward function is continuous, it is the same for Q_i which must be higher than $\frac{R(\mathbf{x})}{1-\gamma}$. So goal bias is $Q_i(\mathbf{x}) = \beta(1 + \frac{1}{1-\gamma})e^{-\frac{d^2}{2\sigma^2}} + \delta$. We have kept previous choices: $\delta = 0.1$, $\beta = 1$ and $\sigma = 0.25$. This last simulation (`bar6`) is the better performance concerning the pendulum.

6 Conclusion

In this paper, we have called into question the arbitrary choice of the reward function and initial Q-values within the context of *goal directed tasks*. Our analysis has resulted in *rules to correctly evaluate rewards and initial Q-values* according to the desired behavior. Notably, some values of Q_i lead to a detrimental behavior that must be avoided. Thanks to our experiments, we have confirmed the presence of bounds which mark out diverse behaviors. It is worth noticing that the farther Q_i is from the bounds, the more the characteristic behaviors are distinguished.

Moreover, we advise to be wary of progressive estimators or potential-based shaping that may entail pernicious behavior. A *safer adjustable continuous reward function* is also suggested. At last, thanks to our conditions on the initial Q-values, we developed a *generic goal bias function*, whose main feature is to be transient. This method turns out to be *an effective way to improve the learning performance of goal-directed tasks.* Table 1 recapitulates the better choices.

Table 1. Better choices of reward function and initial Q-values for goal-directed RL

Binary reward function for discrete state space	Continuous reward function for continuous state space
$r(s,a,s') = \begin{cases} r_g & \text{if } s' = s_g \\ r_\infty & \text{else} \end{cases}$ Choice of r_g and r_∞ : $r_g \geq \frac{r_\infty}{1-\gamma}$	$r(s,a,s') = \beta e^{-\frac{d(s',s_g)^2}{2\sigma^2}}$
Choice of uniform initial Q-values : $Q_i = \frac{r_\infty}{1-\gamma} + \delta$	Choice of uniform initial Q-values : $Q_i = \frac{\beta}{1-\gamma}$
Choice of goal biased initial Q-values : $Q_i(s) = \beta e^{-\frac{d(s,s_g)^2}{2\sigma^2}} + \delta + \frac{r_\infty}{1-\gamma}$	Choice of goal biased initial Q-values : $Q_i(s) = \beta(1 + \frac{1}{1-\gamma})e^{-\frac{d(s,s_g)^2}{2\sigma^2}} + \delta$
$\delta \geq 0$ fixes the level of systematic exploration far away from the goal, $\beta > 0$ adjusts the amplitude of the gradient, σ specifies the gradient influence area and γ is the discount factor s is the previous state, s' the new state, s_g the goal state	

References

1. Sutton, R.S., Barto, A.G.: Reinforcement Learning: An Introduction. The MIT Press, Cambridge (1998)
2. Watkins, C.: Learning from Delayed Rewards. PhD thesis, Cambridge University, Cambridge, England (1989)
3. Mataric, M.J.: Reward functions for accelerated learning. In: Proc. of the 11th ICML. (1994) 181–189
4. Ng, A.Y., Harada, D., Russell, S.: Policy invariance under reward transformations: theory and application to reward shaping. In: Proc. of the 16th ICML. (1999) 278–287
5. Wiewiora, E.: Potential-based shaping and Q-value initialization are equivalent. Journal of Artificial Intelligence Research **19** (2003) 205–208
6. Hailu, G., Sommer, G.: On amount and quality of bias in reinforcement learning. In: Proc. of the IEEE International Conference on Systems, Man and Cybernetics, Tokyo (1999) 1491–1495
7. Koenig, S., Simmons, R.G.: The effect of representation and knowledge on goal-directed exploration with reinforcement-learning algorithms. Machine Learning **22**(1-3) (1996) 227–250
8. Behnke, S., Bennewitz, M.: Learning to play soccer using imitative reinforcement. In: Proc. of the ICRA Workshop on Social Aspects of Robot Programming through Demonstration, Barcelona (2005)
9. Watkins, C., Dayan, P.: Technical note: Q-learning. Machine Learning **8** (1992) 279–292
10. Doya, K.: Reinforcement learning in continuous time and space. Neural Computation **12**(1) (2000) 219–245

Nearly Optimal Exploration-Exploitation Decision Thresholds

Christos Dimitrakakis *

IDIAP Research Institute, 4 Rue de Simplon, Martigny CH 1920, Switzerland
dimitrak@idiap.ch

Abstract. While in general trading off exploration and exploitation in reinforcement learning is hard, under some formulations relatively simple solutions exist. Optimal decision thresholds for the multi-armed bandit problem, one for the infinite horizon discounted reward case and one for the finite horizon undiscounted reward case are derived, which make the link between the reward horizon, uncertainty and the need for exploration explicit. From this result follow two practical approximate algorithms, which are illustrated experimentally.

1 Introduction

In reinforcement learning, the dilemma between selecting actions to maximise the expected return according to the current world model and to improve the world model such as to *potentially* be able to achieve a higher expected return is referred to as the *exploration-exploitation trade-off*. This has been the subject of much interest before, one of the earliest developments being the theory of sequential sampling in statistics, as developed by [1]. This dealt mostly with making sequential decisions for accepting one among a set of particular hypotheses, with a view towards applying it to jointly decide the termination of an experiment and the acceptance of a hypothesis. A more general overview of sequential decision problems from a Bayesian viewpoint is offered in [2].

The optimal, but intractable, Bayesian solution for bandit problems was given in [3], while recently tight bounds on the sample complexity of exploration have been found [4]. An approximation to the full Bayesian case for the general reinforcement learning problem is given in [5], while an alternative technique based on eliminating actions which are confidently estimated as low-value is given in [6].

The following section formulates the intuitive concept of trading exploration and exploitation as a natural consequence of the *definition* of the problem of reinforcement learning. After the problem definitions which correspond to either extreme are identified, Sec. 3 derives a threshold for switching from exploratory to greedy behaviour in bandit problems. This threshold is found to depend on the

* Thanks to M. Keller and R. Chavarriaga, for comments and interesting discussions. This work has received financial support from the Swiss NSF under the MULTI project (2000-068231.021/1) and from IDIAP.

S. Kollias et al. (Eds.): ICANN 2006, Part I, LNCS 4131, pp. 850–859, 2006.

effective reward horizon of the optimal policy and on our current belief distribution of the expected rewards of each action. A sketch of the extension to MDPs is presented in Sec. 4. Section 5 uses an upper bound on the value of exploration to derive practical algorithms, which are then illustrated experimentally in Sec. 6. We conclude with a discussion on the relations with other methods.

2 Exploration Versus Exploitation

Let us assume a standard multi-armed bandit setting, where a reward distribution $p(r_{t+1}|a_t)$ is conditioned on actions in $a_t \in \mathcal{A}$, with $r_t \in \mathbb{R}$. The aim is to discover a policy $\pi = \{P(a_t = i)|i \in \mathcal{A}\}$, where $P(a_t = i)$ is the probability that action i is chosen at time t, which maximises $E[r_{t+1}|\pi]$, the expected value of the reward at the following time-step under the distribution defined by the policy π. It follows that the optimal gambler, or oracle, for this problem would be a policy which always chooses $i \in \mathcal{A}$ such that $E[r_{t+1}|a_t = i] \geq E[r_{t+1}|a_t = j]$ for all $j \in \mathcal{A}$. Given the conditional expectations, implementing the oracle is trivial. However this tells us little about the optimal way to select actions when the expectations are unknown. As it turns out, the optimal action selection mechanism will depend upon the problem formulation. We initially consider the two simplest cases in order to illustrate that the exploration/exploitation tradeoff is and should be viewed in terms of problem and model definition.

In the first problem formulation the objective is to discover a parameterized probabilistic policy $\pi = \big\{P(a_t|\theta_t) \bigm| a_t \in \mathcal{A}\big\}$, with parameters θ_t, for selecting actions such that $E[r_{t+1}|\pi]$ is maximised. If we consider a model whose parameters are the set of estimates $\theta_t = \big\{q_i = \hat{E}_t[r_{t+1}|a_t = i] \bigm| i \in \mathcal{A}\big\}$, then the optimal choice is to select a_t for which the estimated expected value of the reward is highest, because according to our current belief any other choice will necessarily lead to a lower expectation. Thus, stating the bandit problem in this way does not allow the exploration of seemingly lower, but potentially higher value actions and it results in a *greedy* policy.

In the second formulation, we wish to minimise the discrepancy between our estimate q_i and the true expectation. This could be written as the following minimisation problem:

$$\sum_{i \in \mathcal{A}} E\big[\|r_{t+1} - q_i\|^2 \bigm| a_t = i\big].$$

For point estimates of the expected reward, this requires sampling *uniformly* from all actions and thus represents a purely exploratory policy. If the problem is stated as simply minimising the discrepancy asymptotically, then uniformity is not required and it is only necessary to sample from all actions infinitely often. This condition holds when $P(a_t = i) > 0 \; \forall i \in \mathcal{A}, \; t > 0$ and can be satisfied by mixing the optimal policies for the two formulations, with a probability ϵ of using the uniform action selection and a probability $1 - \epsilon$ of using the greedy action selection. This results in the well-known ϵ-greedy policy (see for example [7]), with the parameter $\epsilon \in [0, 1]$ used to control exploration.

This formulation of the exploration-exploitation problem, though leading to an intuitive result, does not lead to an obvious way to optimally select actions. In the following section we shall consider bandit problems for which the functional to be maximised is

$$E\left[\sum_{k=0}^{N} g(k) r_{t+k+1} \middle| \pi\right], \quad g(k) \in [0,1], \ N \geq 0,$$

with $\sum_{k=0}^{\infty} g(k) < \infty$. In this formulation of the problem we are not only interested in maximising the expected reward at the next time step, but in the subsequent N steps, with the $g(\cdot)$ function providing another convenient way to weigh our preference among short and long-term rewards. Intuitively it is expected that the optimal policy for this problem will be different depending on how long-term are the rewards that we are interested in. As will be shown later, by lengthening the effective reward horizon through manipulation of g and N, i.e. by changing the definition of the problem that we wish to solve, the exploration bias is increased automatically.

3 Optimal Exploration Threshold for Bandit Problems

We want to know when it is a better decision to take action i rather than some other action j, with $i, j \in \mathcal{A}$, given that we have estimates q_i, q_j for $E[r_{t+1}|a_t = i]$ and $E[r_{t+1}|a_t = j]$ respectively[1]. We shall attempt to see under which conditions it is better to take an action different than the one whose expected reward is greatest. For this we shall need the following assumption:

Assumption 1 (Expected rewards are bounded from below). *There exists $b \in \mathbb{R}$ such that*

$$E[r_{t+1}|a_t = i] \geq b \quad \forall\, i \in \mathcal{A}, \tag{1}$$

The above assumption is necessary for imposing a lower bound on the expected return of exploratory actions: no matter what action is taken, we are guaranteed that $E[r_t] > b$. Without this condition, exploratory actions would be too risky to be taken at all.

Given two possible actions to take, where one action is currently estimated to have a lower expected reward than the other, then it might be worthwhile to pursue the lower-valued action if the following conditions are true: (a) there is a degree of uncertainty such that the lower-valued action can potentially be better than the higher-valued one, (b) we are interested in maximising more than just the expectation of the next reward, but the expectation of a weighted sum of future rewards, (c) we will be able to accurately determine whether one action is better than the other quickly enough, so that not a lot of resources will be wasted in exploration.

[1] For bandit problems with states in a state space $\mathcal{S}$, similar arguments can be made by considering $i, j \in \mathcal{S} \times \mathcal{A}$.

We now start viewing q_i as random variables for which we hold belief distributions $p(q_i)$, with $\bar{q_i} = E[q_i] = \hat{E}[r_{t+1}|a_t = i]$. The problem can be defined as deciding when action i, is better than taking action j, under the condition that doing so allows us to determine whether $q_i > q_j + \delta$ with high probability after $T \geq 1$ exploratory actions. For this reason we will need the following bound on the expected return of exploration.

Lemma 1 (Exploration bound). *For any return of the form $R_t = \sum_{k=0}^{N} g(k)r_{t+k+1}$, with $g(k) \geq 0$, assuming (1) holds, the expected return of taking action i for T time-steps and following a greedy policy thereafter, when $\bar{q_i} > \bar{q_j}$, is bounded below by*

$$U(i,j,T,\delta,b) = \sum_{k=T}^{N} g(k)\big((\bar{q_j}+\delta)P(q_i > q_j+\delta) + \bar{q_j}P(q_i \leq q_j+\delta)\big) + \sum_{k=0}^{T-1} g(k)\big((\bar{q_j}+\delta)P(q_i > q_j+\delta) + bP(q_i \leq q_j+\delta)\big) \quad (2)$$

for some $\delta > 0$.

This follows immediately from Assumption 1. The greedy behaviour supposes we are following a policy where we continue to perform i if we know that $P(q_i > q_j + \delta) \approx 1$ after T steps and switch back to j otherwise.

Without loss of generality, in the sequel we will assume that $b = 0$ (If expected rewards are bounded by some $b \neq 0$, we can always subtract b from all rewards and obtain the same). For further convenience, we set $p_i = P(q_i \geq q_j + \delta)$. Then we may write that we must take action i if the expected return of simply taking action j is smaller than the expected return of taking action i for T steps and then behaving greedily, i.e. if the following holds:

$$\sum_{k=0}^{N} g(k)\bar{q_j} < \sum_{k=T}^{N} g(k)\big((\bar{q_j}+\delta)p_i + \bar{q_j}(1-p_i)\big) + \sum_{k=0}^{T-1} g(k)(\bar{q_j}+\delta)p_i \quad (3)$$

$$\sum_{k=0}^{T-1} g(k)\big(\bar{q_j} - (\bar{q_j}+\delta)p_i\big) < \sum_{k=T}^{N} g(k)\big(\delta p_i\big) \quad (4)$$

Let $g(k) = \gamma^k$, with $\gamma \in [0,1]$. In this case, any choice of T can be made equivalent to $T = 1$ by dividing everything with $\sum_{k=0}^{T-1}\gamma^k$. We explore two cases: $\gamma < 1$, $N \to \infty$ and $\gamma = 1$, $N < \infty$. In the first case, which corresponds to infinite horizon exponentially discounted reward maximisation problems, we obtain the following:

$$\bar{q_j} - (\bar{q_j}+\delta)p_i < \sum_{k=1}^{\infty} \gamma^k \delta p_i \quad (5)$$

$$\frac{\bar{q_j} - (\bar{q_j}+\delta)p_i}{(1-p_i)\bar{q_j}} < \gamma. \quad (6)$$

It is possible to simplify this expression considerably. When $P(q_i \geq \bar{q_j}+\delta) = 1/2$, it follows from (6) that

$$\gamma > \frac{\bar{q_j} - (\bar{q_j} + \delta)/2}{\bar{q_j}/2} = \frac{\bar{q_j} - \delta}{\bar{q_j}}. \tag{7}$$

Thus, for infinite horizon discounted reward maximisation problems, when it is known that the all expected rewards are non-negative, all we need to do is find δ such that $P(q_i \geq q_j + \delta) = 1/2$. Then (7) can be used to make a decision on whether it is worthwhile to perform exploration. Although it might seem strange that q_i is omitted from this expression, its distribution is implicitly expressed through the value of δ.

In the second case, finite horizon cumulative reward maximisation problems, exploration should be performed when the following condition is satisfied:

$$N\delta p_i > \bar{q_j} - (\bar{q_j} + \delta)p_i \tag{8}$$

Here the decision making function is of a different nature, since it depends on both estimates. However, in both cases, the longer the effective horizon becomes and the larger the uncertainty is, the more the bias towards exploration is increased. We furthermore note that in the finite horizon case, the backward induction procedure can be used to make optimal decisions (see [2] Sec. 12.4).

3.1 Solutions for Specific Distributions

If we have a specific form for the distribution $P(q_i > q_j + \delta)$ it may be possible to obtain analytical solutions. To see how this can be achieved, consider that from (6), we have:

$$\begin{aligned} \gamma\bar{q_j} &> \bar{q_j} - \delta\frac{p_i}{1-p_i} \\ 0 &< \delta\frac{P(q_i > q_j + \delta)}{1 - P(q_j > q_j + \delta)} - (1-\gamma)\bar{q_j}, \end{aligned} \tag{9}$$

recalling that all mean rewards are non-negative.

If this condition is satisfied for some δ then exploration must be performed. We observe that if the first term is maximised for some δ^* for which the inequality is not satisfied, then there is no $\delta \neq \delta^*$ that can satisfy it. Thus, we can attempt to examine some distributions for which this δ^* can be determined. We shall restrict ourselves to distributions that are bounded below, due to Assumption 1.

3.2 Solutions for the Exponential Distribution

One such distribution is the exponential distribution, defined as

$$P(X > \delta) = \int_\delta^\infty \beta e^{-\beta(x-\mu)} dx = e^{-\beta(\delta-\mu)}$$

if $\delta > \mu$, 1 otherwise. We may plug this into (9) as follows

$$f(\delta) = \delta \frac{P(q_i > q_j + \delta)}{1 - P(q_i > q_j + \delta)} = \delta \frac{e^{-\beta_i(\mu_j + \delta - \mu_i)}}{1 - e^{-\beta_i(\mu_j + \delta - \mu_i)}} = \frac{\delta}{e^{\beta_i(\mu_j + \delta - \mu_i)} - 1}$$

Now we should attempt to find $\delta^* = \arg\max_\delta f(\delta)$. We begin by taking the derivative with respect to δ. Set $g(\delta) = e^{h(\delta)} - 1$, $h(\delta) = \beta_i(\bar{q_j} + \delta - \mu_i)$

$$\nabla f(\delta) = \frac{g(\delta) - \delta \nabla g(\delta)}{g(\delta)^2} = \frac{g(\delta) - \delta \beta_i \nabla_h g(\delta)}{g(\delta)^2} = \frac{e^{h(\delta)}(1 - \delta\beta_i) - 1}{(e^{h(\delta)} - 1)^2}$$

Necessary and sufficient conditions for some point δ^* to be a local maximum for a continuous differentiable function $f(\delta)$ are that $\nabla_\delta f(\delta^*) = 0$ and $\nabla^2_\delta f(\delta^*) < 0$. The necessary condition for δ results in

$$e^{\beta_i(q_k + \delta - \mu_i)}(1 - \delta\beta_i) = 1. \tag{10}$$

Unfortunately (10) has no closed form solution, but it is related to the Lambert W function for which iterative solutions do exist [8]. The found solution can then be plugged into (9) to see whether the conditions for exploration are satisfied.

4 Extension to the General Case

In the general reinforcement learning setting, the reward distribution does not only depend on the action taken but additionally on a state variable. The state transition distribution is conditioned on actions and has the Markov property. Each particular task within this framework can be summarised as a Markov decision process:

Definition 1 (Markov decision process). *A Markov decision process is defined by a set of states $\mathcal{S}$, a set of actions $\mathcal{A}$, a transition distribution $\mathfrak{T}(s', s, a) = P(s'_{t+1}|s_t = s, a_t = a)$ and a reward distribution $\mathfrak{R}(s', s, a) = p(r_{t+1}|s_{t+1} = s', s_t = s, a_t = a)$.*

The simplest way to extend the bandit case to the more general one of MDPs is to find conditions under which the latter reduces to the former. This can be done for example by considering choices not between simple actions but between *temporally extended actions*, which we will refer to as *options* following [9]. We shall only need a simplified version of this framework, where each possible *option* x corresponds to some policy $\pi^x : \mathcal{S} \times \mathcal{A} \to [0, 1]$. This is sufficient for sketching the conditions under which the equivalence arises.

In particular, we examine the case where we have two options. The first option is to always select actions according to some exploratory principle, such picking them from a uniform distribution. The second is to always select actions greedily, i.e. by picking the action with the highest expected return.

We assume that each option will last for time T. One further necessary component for this framework is the notion of mixing time

Definition 2 (Exploration mixing time). *We define the exploration mixing time for a particular MDP $\mathcal{M}$ and a policy π $T_\epsilon(\mathcal{M}, \pi)$ as the expected number*

of time steps after which the state distribution is close to the stationary state distribution of π after we have taken an exploratory action i at time step t, i.e. the expected number of steps T such that the following condition holds:

$$\frac{1}{\|\mathbb{S}\|}\sum_s \|P(s_{t+T}=s|s_t,\pi) - P(s_{t+T}=s|a_t=i,s_t,\pi)\| < \epsilon$$

It is of course necessary for the MDP to be ergodic for this to be finite. If we only consider switching between options at time periods greater than $T_\epsilon(\mathcal{M},\pi)$, then the option framework's roughly corresponds to the bandit framework, and T_ϵ in the former to T in the latter. This means that whenever we take an exploratory action i (one that does not correspond to the action that would have been selected by the greedy policy π), the distribution of states would remain to be significantly different from that under π for $T_\epsilon(\mathcal{M},\pi)$ time steps. Thus we could consider the exploration to be taking place during all of T_ϵ, after which we would be free to continue exploration or not. Although there is no direct correspondence between the two cases, this limited equivalence could be sufficient for motivating the use of similar techniques for determining the optimal exploration exploitation threshold in full MDPs.

5 Optimistic Evaluation

In order to utilise Lemma 1 in a practical setting we must define T in some sense. The simplest solution is to set $T = 1$, which results in an optimistic estimate for exploratory actions as will be shown below. By rearranging (2) we have

$$U(i,j,T,\delta,b) = \sum_{k=0}^{N} g(k)\bar{q_j} + \sum_{k=0}^{N} g(k)\delta p_i + (1-p_i)\left(\sum_{k=0}^{T-1} g(k)(b-\bar{q_j}))\right) \quad (11)$$

from which it is evident, since $q_j \geq b$ and $g(k) \geq 0$, that $U(i,j,T_1,\delta,b) \geq U(i,j,T_2,\delta,b)$ when $T_1 < T_2$, thus $U(i,j,1,\delta,b) \geq U(i,j,T,\delta,b)$ for any $T \geq 1$. This can now be used to obtain Alg. 1 for optimistic exploration.

Nevertheless, testing for the existence of a suitable δ can be costly since, barring an analytic procedure it requires an exhaustive search. On the other hand, it may be possible to achieve a similar result through sampling for different values of δ. Herein, the following sampling method is considered: Firstly, we determine the action j with the greatest $\bar{q_j}$. Then, for each action i we take a sample x from the distribution $p(q_i)$ and set $\delta = x - \bar{q_j}$. This is quite an arbitrary sampling method, but we may expect to obtain a $\delta > 0$ with high probability if i has a high probability to be significantly better than j. This method is summarised in Alg. 2.

An alternative exploration method is given by Alg. 3, which samples each action with probability equal to the probability that its expected reward is the highest. It can perhaps be viewed as a crude approximation to Alg. 2 when $\gamma \to 1$ and has the advantage that it is extremely simple.

Algorithm 1. Optimistic exploration

if $\exists \delta \; : \; U(i, j, 1, \delta, b) > \sum_{k=0}^{N} g(k)\bar{q}_j$ **then**
  $a \Leftarrow i$
else
  $a \Leftarrow j$
end if

Algorithm 2. Optimistic stochastic exploration

$j \Leftarrow \arg\max_i \bar{q}_i$.
$u_j = \sum_{k=0}^{N} g(k)\bar{q}_j$.
for all $i \neq j$ **do**
  $\delta \Leftarrow x - \bar{q}_j, \quad x \sim p(q_i)$
  $u_i \Leftarrow U(i, j, 1, \delta, b)$
end for
$a \Leftarrow \arg\max_i u_i$

6 Experiments

A small experiment was performed on a n-armed bandit problem with rewards $r_t \in \{0, 1\}$ drawn from a Bernoulli distribution. Alg. 2 was used with $g(k) = \gamma^k$ and $b = 0$, which is in agreement with the distribution. This was compared with Alg. 3, which can be perhaps viewed as a crude approximation to Alg. 2 when $\gamma \to 1$. The performance of ϵ-greedy action selection with $\epsilon = 0.01$ was evaluated for reference.

The ϵ-greedy algorithm used point estimates for $\bar{q}_i$, which were updated with gradient descent with a step size of $\alpha = 0.01$, such that for each action-reward observation tuple $(a_t = i, r_{t+1})$, $\bar{q}_i \Leftarrow \alpha(r_{t+1} - \bar{q}_i)$, with initial estimates being uniformly distributed in $[0, 1]$. In the other two cases, the complete distribution of q_i was maintained via a population $\{p_i^k\}_{k=0}^{K}$ of point estimates, with $K = 16$. Each point estimate in the population was maintained in the same manner as the single point estimates in the ϵ-greedy approach. Sampling actions was performed by sampling uniformly from the members of the population for each action.

The results for two different bandit tasks, one with 16 and the other with 128 arms, averaged over 1,000 runs, are summarised in Fig. 1. For each run, the expected reward of each bandit was sampled uniformly from $[0, 1]$.

As can be seen from the figure, the ϵ-greedy approach performs relatively well when used with reasonable first initial estimates. The sampling greedy approach, while having the same complexity, appears to perform better asymptotically. More importantly, Alg. 2 exhibits better long-term versus short-term performance when the effective reward horizon is increased as $\gamma \to 1$.

Algorithm 3. Sampling-greedy

$a \Leftarrow i$ with probability $P(a = i) = P(q_i > q_j) \; \forall j \neq i$

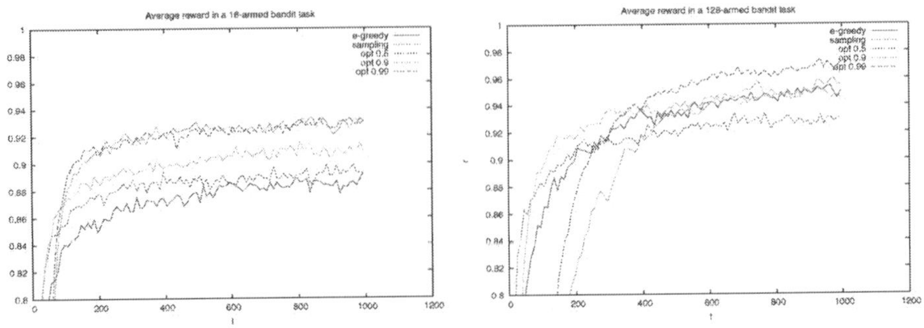

Fig. 1. Average reward in an multi-armed bandit task averaged over 1,000 experiments, smoothed with a moving average over 10 time-steps. Results are shown for ϵ-greedy (**e-greedy**), sampling-greedy (**sampling**) and Alg. 2 (**opt**) with $\gamma \in \{0.5, 0.9, 0.99\}$.

7 Discussion and Conclusion

This paper has presented a formulation of an optimal exploration-exploitation threshold for in a n-armed bandit task, which links the need for exploration to the effective reward horizon and model uncertainty. Additionally, a practical algorithm, based on an optimistic bound on the value of exploration, is introduced. Experimental results show that this algorithm exhibits the expected long-term versus short-term performance trade-off when the effective reward horizon is increased.

While the above formulation fits well within a reinforcement learning framework, other useful formulations may exist. In *budgeted learning*, any exploratory action results in a fixed cost. Such a formulation is used in [10] for the bandit problem (see also [11] for the active learning case). Then the problem essentially becomes that of how to best sample from actions in the next T moves such that the expected return of the optimal policy *after* T moves is maximised and corresponds to $g(k) = 0 \; \forall k < T$ in the framework presented in this paper. A further alternative, described in [6], is to stop exploring those parts of the state-action space which lead to sub-optimal returns with high probability.

When a distribution or a confidence interval is available for expected returns, it is common to use the optimistic side of the confidence interval for action selection [12]. This practice can be partially justified through the framework presented herein, or alternatively, through considering maximising the expected information to be gained by exploration, as proposed by [13]. In a similar manner, other methods which represent uncertainty as a simple additive factor to the normal expected reward estimates, acquire further meaning when viewed through a statistical decision making framework. For example the Dyna-Q+ algorithm (see [7] chap. 9) includes a slowly increasing *exploration bonus* for state-action pairs which have not been recently explored. From a statistical viewpoint, the exploration bonus corresponds to a model of a non-stationary world, where uncertainty about past experiences increases with elapsed time.

In general, the conditions defined in Sec. 3 require maintaining some type of belief distribution over the expected return of actions. A natural choice for this would be to use a fully analytical Bayesian framework. Unfortunately this makes it more difficult to calculate $P(q_i > d)$, so it might be better to consider simple numerical approaches from the outset. We have previously considered some simple such estimates in [14], where we relied on estimating the gradient of the expected return with respect to the parameters. The estimated gradient was then used as a measure of uncertainty. Further research on the use of population-based methods for explicitly representing a distribution of estimates is currently under way.

References

1. Wald, A.: Sequential Analysis. John Wiley & Sons (1947) Republished by Dover in 2004.
2. DeGroot, M.H.: Optimal Statistical Decisions. John Wiley & Sons (1970) Republished in 2004.
3. Bellman, R.E.: A problem in the sequential design of experiments. Sankhya **16** (1957) 221–229
4. Mannor, S., Tsitsiklis, J.N.: The sample complexity of exploration in the multi-armed bandit problem. Journal of Machine Learning Research **5** (2004) 623–648
5. Dearden, R., Friedman, N., Russell, S.J.: Bayesian Q-learning. In: AAAI/IAAI. (1998) 761–768
6. Even-Dar, E., Mannor, S., Mansour, Y.: Action elimination and stopping conditions for the multi-armed and reinforcement learning problems. Journal of Machine Learning Research (2006) to appear.
7. Sutton, R.S., Barto, A.G.: Reinforcement Learning: An Introduction. MIT Press (1998)
8. Corless, R.M., Gonnet, G.H., Hare, D.E.G., Jeffrey, D.J., Knuth, D.E.: On the lambert W function. Advances in Computational Mathematics **5** (1996) 329–359
9. Sutton, R.S., Precup, D., Singh, S.P.: Between MDPs and semi-MDPs: A framework for temporal abstraction in reinforcement learning. Artificial Intelligence **112**(1-2) (1999) 181–211
10. Madani, O., Lizotte, D.J., Greiner, R.: The budgeted multi-armed bandit problem. In: Learning Theory: 17th Annual Conference on earning Theory, COLT 2004. Volume 3120 of Lecture Notes in Computer Science., Springer-Verlag (2004) 643–645
11. Madani, O., Lizotte, D.J., Greiner, R.: Active model selection. In: Proceedings of the 20th Conference on Uncertainty in Artificial Intelligence, Banff, Canada, AUAI Press, Arlington, Virginia (2004) 357–365
12. Auer, P.: Models for trading exploration and exploitation using upper confidence bounds. In: PASCAL workshop on principled methods of trading exploration and exploitation, PASCAL Network (2005)
13. Bernardo, J.M.: Expected information as expected utility. In: The Annals of Statistics. Volume 7., Institute of Mathematical Statistics (1979) 686–690
14. Dimitrakakis, C., Bengio, S.: Gradient estimates of return. IDIAP-RR 05-29, IDIAP (2005)

Dual Adaptive ANN Controllers Based on Wiener Models for Controlling Stable Nonlinear Systems

D. Sbarbaro*

Department of Electrical Engineering, Universidad de Concepción, Chile
dsbarbar@udec.cl

Abstract. This paper presents two nonlinear adaptive predictive algorithms based on Artificial Neural Network (ANN) and a Wiener structure for controlling asymptotically stable nonlinear plants. The first algorithm is based on the minimization of a cost function taking into account the future tracking error and the Certainty Equivalence (CE) principle, under which the estimated parameters are used as if they were the true parameters. In order to improve the performance of the adaptive algorithm, we propose to use a cost function, considering not only the future tracking error, but also the effect of the control signal over the estimated parameters. A simulated chemical reactor example illustrates the performance and feasibility of both approaches.

1 Introduction

Adaptive control of discrete nonlinear systems, using flexible nonlinear parameterization like Artificial Neural Networks, have received a great deal of attention [1]. Most of these works have relied on the use of inverse model approach assuming that the system has a stable inverse and is affine in the control signal. These assumptions have limited their range of applications. The adaptive algorithms can be classified, in general, as *indirect* or *direct* ones. The former adapts the parameter of the controller with respect to some performance index, while the latter calculates the parameter of the controller based on an identified model of the plant. Adaptive controllers that are based on the CE principle completely ignore the uncertainty associated to the parameters. This may lead to inadequate transient and poor parameter convergence. Some authors have addressed these problems, in the context of inverse control, by modelling the parameters as random variables and taking into account their uncertainty in the control law [2] [3]. An algorithm that takes into account not only the control objective, but also the effect of the control signal on the convergence of the estimation algorithm is called adaptive *dual* control system [4].

In order to overcome the limitation of inverse control approaches, nonlinear predictive strategies have been proposed. As any control design tool, it requires

* This work was supported by Fondecyt project 1040486.

S. Kollias et al. (Eds.): ICANN 2006, Part I, LNCS 4131, pp. 860–867, 2006.

dynamical models of the nonlinear systems to be controlled; if these models are not available, then some empirical ones, such as: Neural Networks, NARMAX, Volterra [5], and Wiener [6], can be considered. The latter are particularly useful in representing open loop stable nonlinear processes; without introducing the stability problems associated to general recursive nonlinear models.

Adaptive Predictive controllers based on Wiener type of models and the Certainty Equivalence principle, have been proposed by several authors [6][7]. In addition, Wiener structure can be combined with ANN to provide a powerful modelling framework [8].

In general, if the parameters are modelled as random variables, then the problem posed by the predictive controllers cannot even be solved numerically, because it requires the prediction of the posterior densities. These densities can not be evaluated, since the estimated mean depends on the future output. It has been suggested, in [9], to approximate these densities by only propagating the covariance matrix. This approximation has given good results in the linear context. Hence, in this work, we explore the use of a Wiener structure combined with an ANN model, to design a dual adaptive non-linear predictive controller for stable unknown linear system, with a stochastic additive disturbance acting at the output. The dual characteristic means that the controller is able to *cautiously* track the desired reference signal; and at the same time, it *excites* the system to improve its identification, so that the performance of the overall controller can be improved in future time intervals.

The paper is organized as follows: section 2 describes the characteristic of the nonlinear model to be considered in the design of the predictive controller. Section 3 introduces the adaptive model based on ANNs, and section 4 the design of the adaptive controllers based on the CE principle and the prediction of the posterior densities. Section 5 illustrates the performance of both algorithms. Finally, in section 6 some conclusions are drawn.

2 The Nonlinear Model

The general model considered in this work can be represented by the following state space representation

$$\mathbf{x}(k+1) = \mathbf{A}\mathbf{x}(k) + \mathbf{b}u(k) \tag{1}$$

$$y(k) = h(\mathbf{x}(k)) \tag{2}$$

where $y(k)$ is the measured variable, $u(k)$ is the input variable, $\mathbf{x}(k)$ is a vector of dimension N and $h(\mathbf{x}(k))$ a continuous function. The matrix $\mathbf{A}$ and vector $\mathbf{b}$ define the dynamic of the models. It has been demonstrated that this type of structure can approximate any stable nonlinear system with any degree of accuracy [10][6]. There are two popular choices for the matrix $\mathbf{A}$ and vector $\mathbf{b}$: orthonormal filters, which are suitable for process modelling [10], and gamma filters more suitable for signal processing [11]. This approach leads naturally to a Wiener model, as it was originally proposed by Wiener [12]. The general structure of the Wiener model, using Laguerre filters, can be described in state space equations, as eq. (1) with :

$$\mathbf{A}_c = -\mu \begin{bmatrix} 1 & 0 & \dots & 0 \\ 2 & 1 & \dots & 0 \\ \vdots & \vdots & \vdots & \vdots \\ 2 & \dots & 2 & 1 \end{bmatrix}, \mathbf{b}_c = \sqrt{(2\mu)} \begin{bmatrix} 1 \\ 1 \\ \vdots \\ 1] \end{bmatrix} \tag{3}$$

where μ is the scale factor. The discrete parameters are:

$$\mathbf{A} = e^{\mathbf{A}_c T_m}, \mathbf{b} = (\mathbf{A} - \mathbf{I})\mathbf{A}_c^{-1}\mathbf{b}_c, \tag{4}$$

where T_m is the sampling time.

3 An ANN Model

Let's consider that the unknown nonlinear function $h(\mathbf{x}(k))$ can be represented as a parameterized nonlinear function:

$$h(\mathbf{x}(k)) = n(\mathbf{x}(k), \theta) \tag{5}$$

where θ represents a vector of unknown parameters. The nonlinear function is approximated by a two layer neural network as follows :

$$y(k) = \mathbf{W}_1 \sigma(\mathbf{V}_1 \mathbf{x}(k) + \mathbf{v}_0) + w_0. \tag{6}$$

where σ is a sigmoidal function. The parameters are collected in a vector of parameters θ.

Thus, the system (1) can be described by:

$$\mathbf{x}(k+1) = \mathbf{A}\mathbf{x}(k) + \mathbf{b}u(k) \tag{7}$$

$$y(k) = n(\mathbf{x}(k), \theta) + \eta(k) \tag{8}$$

where θ is the unknown vector of parameter, and $\eta(k)$ is an independent, identically distributed Gaussian random variable, with a distribution given by $N(0, \sigma^2)$. We will assume that the parameters are modelled as a random variable with a normal prior distribution given by $N(\theta(0), \mathbf{P}(0))$, where $\theta(0)$ and $\mathbf{P}(0)$ define the initial mean and covariance matrix respectively. The on-line computation of the conditional mean and covariance of θ can be carried out by the following Extended-Kalman filter:

$$\theta(k+1) = \theta(k) + \frac{\mathbf{m}(k)'\mathbf{P}(k)e(k)}{1 + \mathbf{m}(k)'\mathbf{P}(k)\mathbf{m}(k)} \tag{9}$$

$$\mathbf{P}(k+1) = \mathbf{P}(k) - \frac{\mathbf{P}\mathbf{m}(k)\mathbf{m}(k)'\mathbf{P}(k)}{1 + \mathbf{m}(k)'\mathbf{P}(k)\mathbf{m}(k)} \tag{10}$$

where $\mathbf{m}(k) = \nabla_\theta n(\mathbf{x}(k), \theta)$. Note that the covariance matrix depends only on the input signal, $u(k)$, through the variables $\mathbf{m}(k)$.

4 The Predictive Control Strategies

Predictive control can be seen as a dynamic programming problem [13]. Thus, the general predictive control in the stochastic dynamic programming setting, finds a set of future control signals $\mathbf{u}(k) = [u(k) \ldots u(k+T)]'$, so that, Bellman's equations are satisfied:

$$J^{k,k+T}(\Im_k) = min_{\mathbf{u}(k)} E\left[q_0(w(k+1) - n(\mathbf{x}(k), \theta))^2 + r_0 u(k)^2 + J^{k,k+T}(\Im_{k+1})|\Im_k\right] \tag{11}$$

where $\Im_k = \{y(0), \ldots, y(k), n(\mathbf{x}(0)), \ldots, n(\mathbf{x}(k))\}$ defines the information vector at time k, $J_k^{k+T}(\Im_k)$ is the optimal cost to go at step k, with $J_{k+T}^{k+T+1}(\Im_T) = 0$. Once the solution is obtained, only the first value is applied to the process, and the minimization is carried out each sampling time. The problem posed by (11) can not be solved, because it requires the knowledge about the future values of the posterior densities, which depend on the future control signals and future output of the process. The solution of (11) gives as a result a control signal that not only takes into account the control objective, but also the effect of the input over the parameter estimation algorithm. Several approximations to (11) can be formulated in order to obtain some practical solutions.

4.1 Certainty Equivalence Controller

The predictive controller based on the CE assumption calculates a set of future control signals without considering the uncertainty in the parameters; so that:

$$min_{\mathbf{u}(k)} J_{CE}^{k,k+T}(\Im_k) = E\left[\sum_{i=0}^{T} q_i(w(k+i+1) - n(\mathbf{x}(k+i+1), \theta))^2 + r_i u(k+i)^2\right], \tag{12}$$

where q_i and r_i are weighting factors, and $w(k)$ is the reference signal. Under the receding horizon principle, only the first control is applied to the system. As all signals are deterministic, the expectation is just:

$$min_{\mathbf{u}(k)} J_{CE}^{k,k+T}(\Im_k) = \sum_{i=0}^{T} q_i(w(k+i+1) - n(\mathbf{x}(k+i+1), \theta))^2 + r_i u(k+i)^2 \tag{13}$$

subject to the system equations. To reduce the dimension of the optimization problem several approaches can be applied. For instance, if the future control signal is assumed constant [14]; i.e. $u(k) = u(k+1) \ldots = u(k+T)$, then the problem is reduced to one dimensional optimization problem by considering the predictions as follows:

$$n(k+j) = n(\mathbf{A}^j \mathbf{x}(k) + \sum_{i=0}^{j-1} \mathbf{A}^{j-i} \mathbf{b} u(k), \theta(k)) \tag{14}$$

In addition, if we consider $r_i = q_i = 0$, $i = 0, .., T-1$ and $q_T = 1$, $r_T = 1$, the final index will be:

$$min_{u(k)} J_{CE}^{k,k+T}(\Im_k) = (w(k+1+T) - n(\mathbf{A}^j \mathbf{x}(k) + \sum_{i=0}^{j-1} \mathbf{A}^{j-i} \mathbf{b} u(k), \theta(k)))^2 \quad (15)$$

In general, a simple line search algorithm can be used to obtain the solution to this optimization problem.

4.2 An Approximate Dual Controller

The dual strategy has associated the cost function (11), which in order to be optimized in terms of $\mathbf{u}(k)$ requires the knowledge of the posterior densities $N(\theta(k+i), \mathbf{P}(k+i))$. Unfortunately, these densities can not be evaluated since the estimated mean depends on the future output. In [9] has been suggested to approximate these densities by $N(\theta(k), \mathbf{P}(k+i))$. Thus, taking the expectation, the following approximation can be found:

$$\begin{aligned} min_{\mathbf{u}(k)} J_{CE}^{k,k+T}(\Im_k) &= \textstyle\sum_{i=0}^{T} q_i (w(k+i+1) - n(\mathbf{x}(k+i+1), \theta))^2 + ... \\ & r_i u(k+i)^2 + q_i \mathbf{m}(k+1+i)' \mathbf{P}(k+1+i) \mathbf{m}(k+1+i) \end{aligned} \quad (16)$$

The above cost function subject to the system model equations (8) and the covariance equation (10) can be minimized with respect to the future control signal $\mathbf{u}(k)$. An active strategy is obtained, since the future values of the covariance matrix, are included in the index. In this way, the control signal will also try to bring the uncertainty of the parameters to some low level.

5 Simulation Results

In this section, the algorithms are applied to control two reactions in series ($A \rightarrow B \rightarrow C$) in a Continuous Stirred Tank Reactor [10]. The desired product is the intermediate product B. The differential equations describing the system are given by :

$$\dot{x}_1 = 1 - x_1 - E_3 e^{-E_1/x_3} x_1 + E_4 e^{-E_2/x_3} x_2 \quad (17)$$
$$\dot{x}_2 = -x_2 + E_3 e^{-E_1/x_3} x_1 - E_4 e^{-E_2/x_3} x_2 \quad (18)$$
$$\dot{x}_3 = u - x_3 + .005(E_3 e^{-E_1/x_3} x_1 - E_4 e^{-E_2/x_3} x_2) \quad (19)$$

with $E_1 = 50$, $E_2 = 70$, $E_3 = 300000$, and $E_4 = 60 \cdot 10^6$; where x_1 and x_2 are dimensionless concentrations of A and B, x_3 is the dimensionless temperature of the jacket surrounding the reactor. In order to model the relationship between the concentration of the desired product; i.e. x_2, and the control signal, we have considered a scale factor $\mu = .6$, sampling time $T_m = 1$, and six Laguerre filters. A neural network with 6 inputs and 4 hidden units, was trained on line. The measured concentration,$y(kT)$, considered a zero mean noise signal, $\eta(kT)$,

$$y(kT) = x_2(kT) + \eta(kT). \quad (20)$$

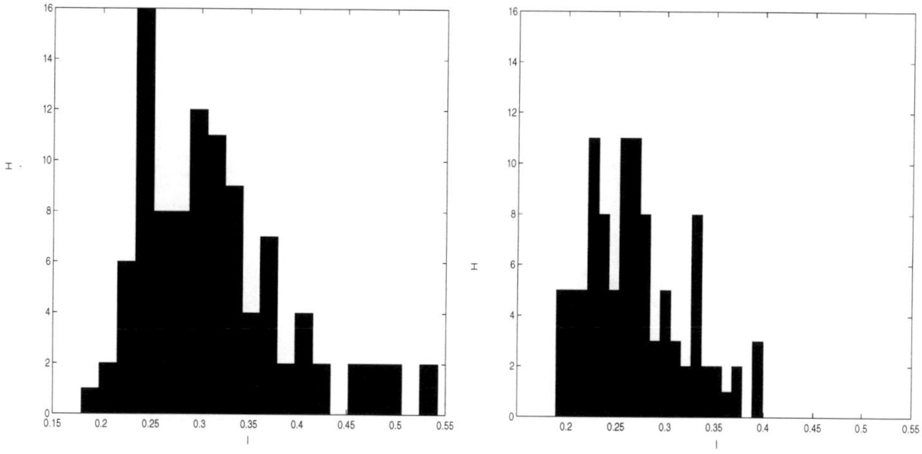

Fig. 1. Distribution of the cost function: CE controller and Dual controller

As a measure of the control performance the following index is estimated for different realizations of the noise signal:

$$I = \sqrt{\frac{1}{N}\sum_{k=0}^{N}(y_d(kT_m) - y(kT_m))^2}, \tag{21}$$

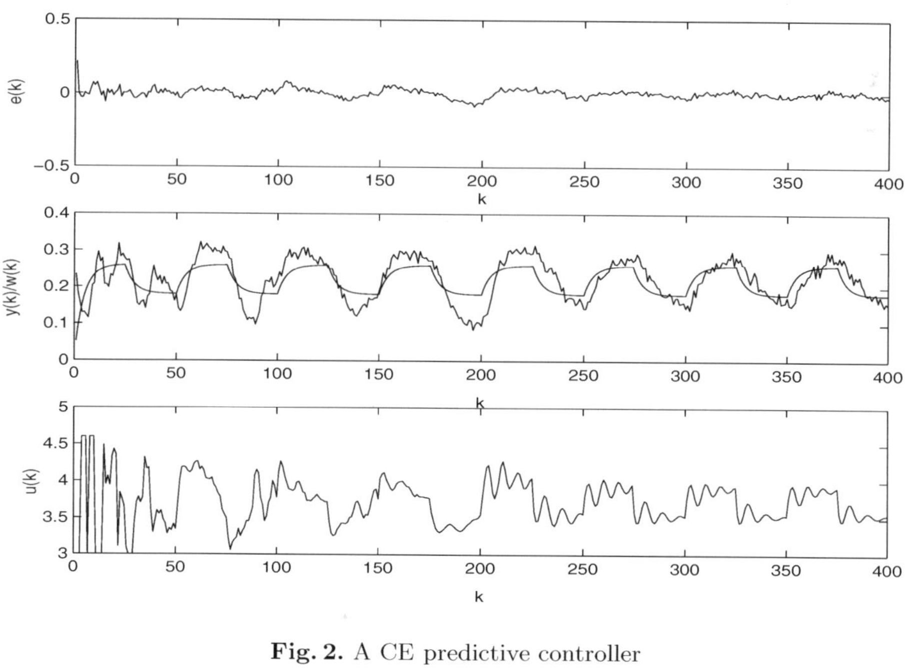

Fig. 2. A CE predictive controller

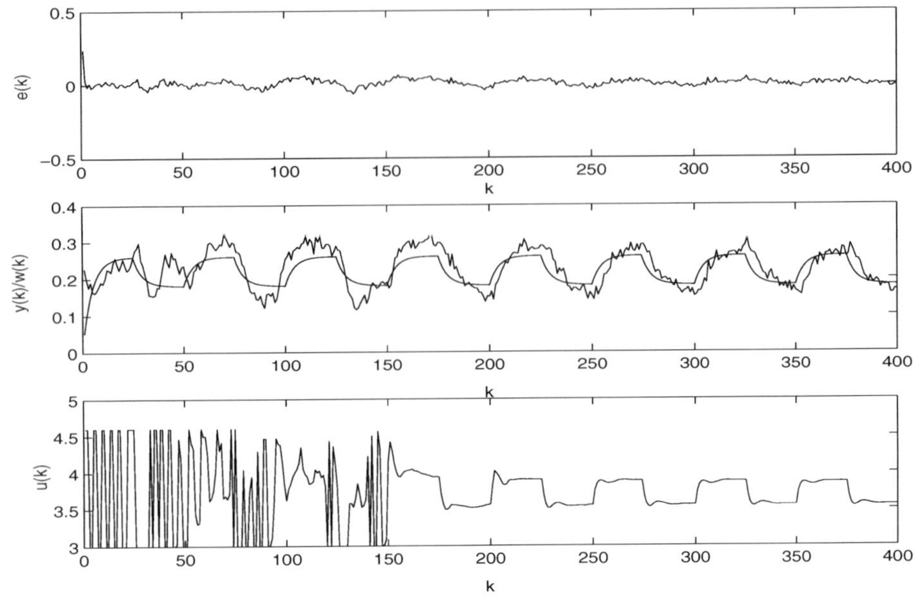

Fig. 3. A Dual predictive controller

where N is the number of samples considered in the cost function, $y_d(kT)$ and $y(kT)$ are the desired set-point and measured concentration respectively. The controller prediction horizon was set to $T = 5$ and the same initial conditions were used for all the simulations. Performing Monte Carlo simulations, the index I was evaluated for 100 realizations of the measurement noise. Figure 1 shows the distribution of the cost function for both controllers, clearly the dual controller shifts the distribution toward smaller values of the cost function. Figure 2 shows the behavior of the controller based on the CE principle, and figure 3 the one of the dual controller, both figures were obtained for the same noise realization. By comparing both figures, we can see that the latter provides larger excitation signals, but without compromising tracking performance. This extra excitation, at initial stages of the adaptive process, means smaller identification errors and better tracking performance in future time instants.

6 Final Remarks

We have presented a methodology to design an adaptive predictive controller based on an Artificial Neural Network model, considering the minimization of a cost function taking into account the future tracking errors and the effect of the control signal on the parameter estimation algorithm. The parameters of the ANN are considered as random variables and the network training algorithm is based on an Extended Kalman-filter. The obtained result shows that this controller provides more excitation to the system at initial stages than a controller based on the CE principle. This key feature gives as a result a better control performance in future time intervals. Future works consider the real-time implementation of this type of controllers.

References

1. Special issue on neural network feedback control. *Automatica* **37**(8), 2001.
2. S. Fabri and V. Kadirkamanathan, "Dual Adaptive Control of Nonlinear Stochastic Systems using Neural Networks", Automatica, Vol. 34, No. 2, February 1998, pp. 245-253.
3. D. Sbarbaro, N. Filatov and H. Unbehauen, "Adaptive dual controller for a class of nonlinear systems", Preprints of the 1998 IFAC Workshop on Adaptive Systems in Control and Signal Processing, pp. 28-33, Scotland, August 1998.
4. N. Filatov and H. Hunbeauen,"Adaptive dual control", Springer, New York, 2004.
5. Y. Chikkula, J.H. Lee, and B. A. Ogunnaike, "Robust model predictive control of nonlinear systems using input-output models", Proceedings of the ACC, Seattle, pp. 2205-2209, 1995.
6. G. Sentoni, O. Agamennoni, A. Desages, and J. Romagnoli, "Approximate models for nonlinear process control", AIChE Journal, Vol 42, no 8, pp. 2240-2250, 1996.
7. B.R. Maner, F.J. Doyle III, B.A. Ogunnaike, and R. K. Pearson, "Nonlinear predictive control of a simulated multivariable polymerization reactor using second-order Volterra models", Automatica, Vol 32, no 9, pp. 1285-1301, 1996.
8. G. Sentoni, L. Biegler, J.Guiver, and H. Zaho, "State Space Nonlinear modeling: Identification and Universality", AIChE Journal, Vol 44, no 10, pp. 2229-2239, 1998.
9. C. Kulcsar, L. Pronzato and E. Walter, "Dual Control of linearly parametrised models via prediction of posterior densities", European Journal of Control, no.2, pp 135-143., 1996.
10. Q. Zheng and E. Zafiriou, " Nonlinear Identification for control using volterra-Laguerre expansion ", Proceedings of the ACC, Seattle, pp. 2195-2199, 1995.
11. B. deVries and J. Principe. The gamma model - a new neural model for temporal processing. Neural Networks, 5(4):565-576, 1992.
12. M Schetzen, "The Volterra and Wiener theory of non-linear systems", Wiley, New York, 1980.
13. N. Filatov and H. Hunbeauen, "Survey of Adaptive dual control methods", IEE Proc.-Control Theory Appl., Vol 147, no 1, pp 118-128, 2000.
14. G. Dumont, and Y. Fu, "Non-linear adaptive control via laguerre expansion of volterra kernels", International Journal of Adaptive Control and Signal Processing, vol. 7, 367-382. 1993.

Online Stabilization of Chaotic Maps Via Support Vector Machines Based Generalized Predictive Control

Serdar Iplikci

Pamukkale University, Department of Electrical and Electronics Engineering,
Kinikli Campus, 20040, Denizli, Turkey
iplikci@pamukkale.edu.tr

Abstract. In this study, the previously proposed Online Support Vector Machines Based Generalized Predictive Control method [1] is applied to the problem of stabilizing discrete-time chaotic systems with small parameter perturbations. The method combines the Accurate Online Support Vector Regression (AOSVR) algorithm [2] with the Support Vector Machines Based Generalized Predictive Control (SVM-Based GPC) approach [3] and thus provides a powerful scheme for controlling chaotic maps in an adaptive manner. The simulation results on chaotic maps have revealed that Online SVM-Based GPC provides an excellent online stabilization performance and maintains it when some measurement noise is added to output of the underlying map.

1 Introduction

Existence of strange attractors and sensitive dependence on initial conditions and parameter perturbations are some peculiar properties of chaotic behavior, which are undesirable from the control engineering point of view. However, these properties can be exploited to achieve a desirable motion [4], e.g. stabilized motion to one of the unstable equilibrium points (UEPs). In order to stabilize the chaotic maps to their UEPs by applying judiciously chosen tiny perturbations, we employed in this study an advanced control algorithm that relies on a well-known control approach: Generalized Predictive Control (GPC) [5,6]. GPC belongs to the class of Model-Based Predictive Control (MPC) or Receding Horizon Control (RHC) techniques that have been applied to a wide spectrum of areas [7]. Since the model of the unknown plant plays very crucial role in the GPC architecture, many linear and nonlinear modeling techniques have been proposed in the literature. Moreover, in the last decade, some computationally intelligent tools such as neural networks [8], fuzzy systems [9], neuro-fuzzy hybrids [10] and, most recently, Support Vector Machines [11] have been utilized to obtain accurate models of the chaotic systems in the GPC loop. However, obtained models in these approaches are fixed and they cannot adapt to the changes in the plant dynamics and/or operating conditions, which may become as a major disadvantage when the changes are not negligible. The aim of this study is to

S. Kollias et al. (Eds.): ICANN 2006, Part I, LNCS 4131, pp. 868–877, 2006.

show the applicability of the previously proposed adaptive online method [1] to discrete-time chaotic systems. This paper is organized as follows. In the next section, the GPC approach is reviewed by introducing its components. Then, in the first subsection of Section 3, SVMs and AOSVR have been discussed very briefly. In the next subsections, the SVM-Based GPC scheme is formulated for the RBF kernels, and then the Online SVM-Based GPC algorithm is explained. Finally, investigated chaotic maps and the simulation results are given in Section 4.

2 Generalized Predictive Control

By following [3,11], consider a chaotic map represented by the NARX model

$$y_n = f(p_n, ..., p_{n-n_p}, y_{n-1}, ..., y_{n-n_y}), \tag{1}$$

where p_n is the control parameter applied to the plant at the time index n, y_n is corresponding output of the plant, and n_p and n_y denote the number of lags in the control and the output signals involved in the model, respectively.

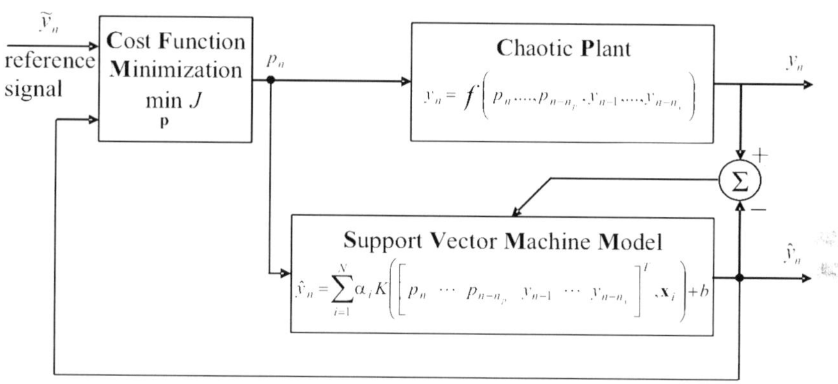

Fig. 1. The GPC architecture

Fig. 1 illustrates the architecture of the GPC scheme, where $\hat{y}_n$ is the output of the model at the time index n and $\tilde{y}$ is the reference trajectory which is desired to be followed by the plant. In the GPC scheme, model of the plant accounts for the prediction of future behavior of the plant in response to the candidate control vector $\mathbf{p}$, and also it is used to obtain some gradient information that will be required in the CFM block, which is the other component of GPC. The aim of CFM is to minimize the performance index J in (2) with respect to the candidate control vector $\mathbf{p}$.

$$J = \sum_{j=N_1}^{N_2} (\tilde{y}_{n+j} - \hat{y}_{n+j})^2 + \sum_{j=1}^{N_u} \lambda_j (\Delta p_{n+j})^2 + \sum_{j=1}^{N_u} \left(\frac{\mu}{p_{n+j} + \frac{\rho}{2} - \vartheta} + \frac{\mu}{\frac{\rho}{2} + \vartheta - p_{n+j}} - \frac{4}{\rho} \right), \quad (2)$$

where N_1 is the minimum costing horizon, N_2 is the maximum costing horizon, N_u is the control horizon, λ is the weighting factor, μ is the sharpness of the constraint function for the control signal, ρ is the range of the control signal, ϑ is the offset to the range and Δp_{n+j} is $\Delta p_{n+j} = p_{n+j} - p_{n+j-1}$ [8]. In the CFM algorithm, the entries of the candidate control vector **p**, where $\mathbf{p} = \left[p_{n+1}\, p_{n+2} \cdots p_{n+N_u}\right]^T$, are altered within the allowable range according to the general update rule, $\mathbf{p} \leftarrow \mathbf{p} + s\mathbf{z}$, where **z** is the search direction and s is the step-length. In this work, we have adopted the Modified Newton's method $\left(\mathbf{z} = -\mathbf{H}^{-1}\mathbf{g}\right)$ for computation of **z** and the Golden Section algorithm to obtain the best possible step-length s, where **g** is the gradient vector as in (3),

$$\mathbf{g} = \frac{\partial J}{\partial \mathbf{p}} = \left[\frac{\partial J}{\partial p_{n+1}} \; \frac{\partial J}{\partial p_{n+2}} \cdots \frac{\partial J}{\partial p_{n+N_u}} \right]^T, \quad (3)$$

and **H** is the Hessian matrix given as

$$\mathbf{H} = \frac{\partial^2 J}{\partial \mathbf{p}^2} = \begin{bmatrix} \frac{\partial^2 J}{\partial p_{n+1} p_{n+1}} & \frac{\partial^2 J}{\partial p_{n+1} p_{n+2}} & \cdots & \frac{\partial^2 J}{\partial p_{n+1} p_{n+N_u}} \\ \frac{\partial^2 J}{\partial p_{n+2} p_{n+1}} & \frac{\partial^2 J}{\partial p_{n+2} p_{n+2}} & \cdots & \frac{\partial^2 J}{\partial p_{n+2} p_{n+N_u}} \\ \vdots & \vdots & \cdots & \vdots \\ \frac{\partial^2 J}{\partial p_{n+N_u} p_{n+1}} & \frac{\partial^2 J}{\partial p_{n+N_u} p_{n+2}} & \cdots & \frac{\partial^2 J}{\partial p_{n+N_u} p_{n+N_u}} \end{bmatrix}. \quad (4)$$

In order to obtain the hth element $(\frac{\partial J}{\partial p_{n+h}})$ of the gradient vector **g**, we have to calculate the first order terms $(\frac{\partial \hat{y}_{n+j}}{\partial p_{n+h}})$'s, while the mth, h^{th} element $(\frac{\partial^2 J}{\partial p_{n+m} p_{n+h}})$ of the Hessian matrix **H** necessitates the calculation of the second order terms $(\frac{\partial^2 \hat{y}_{n+j}}{\partial p_{n+m} p_{n+h}})$'s. Once the optimum search direction is determined, to find the optimum step-length becomes a one-dimensional problem that can be solved by one of the line-search techniques [12,13].

3 The Online SVM-Based GPC Method

3.1 Support Vector Machines

The SVM algorithms [14,15,16] achieve global solution by transforming the regression problem into a quadratic programming (QP) problem and then solving it by a QP solver. Finding global solution and possessing higher generalization capability constitute the major advantages of the SVM algorithms over other regression techniques. In the last decade, SVM-Based algorithms have been developed very rapidly and have been applied to many areas [17,18]. Consider a training set $T = \{\mathbf{x}_k, y_k\}_{k=1}^{k=N}$, where $\mathbf{x}_k \in X \subseteq \mathbb{R}^n$ is the kth input data point in input space and $y_k \in Y \subseteq \mathbb{R}$ is corresponding output value. It is desired to model the relationship between the input and output data points by a Support Vector Machines regression model (5), which is linear in a higher dimensional feature space $\boldsymbol{F}$.

$$\hat{y}(\mathbf{x}) = \langle \mathbf{w}, \boldsymbol{\Phi}(\mathbf{x}) \rangle + b, \tag{5}$$

where $\mathbf{w}$ is a vector in feature space $\boldsymbol{F}$, $\boldsymbol{\Phi}(\mathbf{x})$ is a mapping from input space to the feature space, b is the bias term and $\langle \cdot, \cdot \rangle$ stands for the inner product operation in $\boldsymbol{F}$. When a linear ε-insensitive loss function $L(\varepsilon, y, \hat{y})$ is employed, one can obtain the dual form of the regression model as

$$\hat{y}(\mathbf{x}) = \sum_{\substack{j=1 \\ j \in SV}}^{\#SV} \alpha_i K(\mathbf{x}, \mathbf{x}_j) + b, \tag{6}$$

where $\#SV$ denotes the number of support vectors in the model and $K(\mathbf{x}_i, \mathbf{x}_j)$ is a kernel function given by $K(\mathbf{x}_i, \mathbf{x}_j) = \boldsymbol{\Phi}(\mathbf{x}_i)^T \boldsymbol{\Phi}(\mathbf{x}_j) = K_{ij}$. For more details, refer to [17,18,19].

The AOSVR algorithm [2] proceeds with optimization of the Lagrangian of dual form, and then puts the Karush-Kuhn-Tucker (KKT) optimality conditions to another form by introducing an error function. In the incremental algorithm, when a new training point $\mathbf{x}_c$ is received, the largest possible α_c value is calculated providing that the system is at the equilibrium with respect to the KKT conditions. The process is repeated until the KKT conditions are satisfied for all data points including the most recent one. When any previously used training point is desired be removed from the training set, one can use the decremental algorithm, which is totally opposite of the incremental one.

3.2 The SVM-Based GPC Formulation

In this subsection, formulation of the SVM-Based GPC approach is given as in [11]. If the current state vector is formed as

$$\mathbf{c}_n = \left[p_n \; p_{n-1} \cdots p_{n-n_p} \; y_{n-1} \; y_{n-2} \cdots y_{n-n_y}\right]^T, \tag{7}$$

then the corresponding output of the model becomes, $\hat{y}_n = \sum_{j=1}^{\#SV} \alpha_j K(\mathbf{c}_n, \mathbf{x}_j) + b$. In SVM-Based GPC, the Radial Basis Function-RBF (8) is adopted as the kernel function.

$$K_{ij} = K(\mathbf{x}_i, \mathbf{x}_j) = \exp\left(- \frac{(\mathbf{x}_i - \mathbf{x}_j)^T (\mathbf{x}_i - \mathbf{x}_j)}{2\sigma^2} \right), \tag{8}$$

where σ is the width parameter. If d_{jn} is defined as the Euclidean distance between the jth support vector $\mathbf{x}_j$ and the current state vector $\mathbf{c}_n$ as in (9),

$$d_{jn} = (\mathbf{c}_n - \mathbf{x}_j)^T (\mathbf{c}_n - \mathbf{x}_j) = \sum_{i=0}^{n_p} (x_{j,i+1} - p_{n-i})^2 + \sum_{i=1}^{n_y} (x_{j,n_p+i+1} - y_{n-i})^2, \tag{9}$$

then the kernel function can be rewritten as $K(\mathbf{c}_n, \mathbf{x}_j) = \exp\left(-\frac{d_{jn}}{2\sigma^2}\right)$ and the regression model becomes $\hat{y}_n = \sum_{j=1}^{\#SV} \alpha_j \exp\left(-\frac{d_{jn}}{2\sigma^2}\right) + b$. Now, the SVM regression model can be used to predict future trajectory of the plant as in (10).

$$\hat{y}_{n+k} = \sum_{j=1}^{\#SV} \alpha_j \exp\left(-\frac{d_{j,n+k}}{2\sigma^2}\right) + b, k = N_1, N_1+1, \ldots, N_2, \tag{10}$$

where

$$\begin{aligned} d_{j,n+k} = & \sum_{i=1}^{min(k,n_y)} (x_{j,n_p+i+1} - \hat{y}_{n+k-i})^2 + \sum_{i=k+1}^{n_y} (x_{j,n_p+i+1} - \hat{y}_{n+k-i})^2 \\ & + \sum_{i=0}^{n_p} \begin{cases} (x_{j,i+1} - p_{n+k-i})^2, & k - N_u < i \\ (x_{j,i+1} - p_{n+N_u})^2, & k - N_u \geq i \end{cases} \end{aligned} \tag{11}$$

Thus, the first order partial derivatives can be written as

$$\frac{\partial \hat{y}_{n+k}}{\partial p_{n+h}} = \sum_{j=1}^{\#SV} \alpha_j \frac{\partial \exp(-\frac{d_{j,n+k}}{2\sigma^2})}{\partial p_{n+h}}, \tag{12}$$

where

$$\frac{\partial \exp(-\frac{d_{j,n+k}}{2\sigma^2})}{\partial p_{n+h}} = \frac{\partial \exp(-\frac{d_{j,n+k}}{2\sigma^2})}{\partial d_{j,n+k}} \frac{\partial d_{j,n+k}}{\partial p_{n+h}} = -\frac{1}{2\sigma^2} \exp(\frac{-d_{j,n+k}}{2\sigma^2}) \frac{\partial d_{j,n+k}}{\partial p_{n+h}} \tag{13}$$

and

$$\begin{aligned} \frac{\partial d_{j,n+k}}{\partial p_{n+h}} = & \sum_{i=1}^{min(k,n_y)} (-2)(x_{j,n_p+i+1} - \hat{y}_{n+k-i}) \frac{\partial \hat{y}_{n+k-i}}{\partial p_{n+h}} \delta_1(k-i-1) \\ & + \sum_{i=0}^{n_p} \begin{cases} (-2)(x_{j,i+1} - p_{n+k-i})\delta_{k-i,h}, & k - N_u < i \\ (-2)(x_{j,i+1} - p_{n+N_u})\delta_{N_u,h}, & k - N_u \geq i \end{cases} \end{aligned} \tag{14}$$

where $\delta_{i,j}$ is the Kronecker Delta function and $\delta_1(\cdot)$ stands for the unit step function. The second order partial derivatives are

$$\frac{\partial^2 \hat{y}_{n+k}}{\partial p_{n+h} \partial p_{n+m}} = \sum_{j=1}^{\#SV} \alpha_j \frac{\partial^2 \exp(-\frac{d_{j,n+k}}{2\sigma^2})}{\partial p_{n+h} \partial p_{n+m}}, \tag{15}$$

where

$$\begin{aligned} \frac{\partial^2 \exp(-\frac{d_{j,n+k}}{2\sigma^2})}{\partial p_{n+h} \partial p_{n+m}} = & \frac{\partial \exp(-\frac{d_{j,n+k}}{2\sigma^2})}{\partial d_{j,n+k}} \frac{\partial^2 d_{j,n+k}}{\partial p_{n+h} \partial p_{n+m}} \\ & + \frac{\partial^2 \exp(-\frac{d_{j,n+k}}{2\sigma^2})}{\partial d_{j,n+k}^2} \frac{\partial d_{j,n+k}}{\partial p_{n+h}} \frac{\partial d_{j,n+k}}{\partial p_{n+m}} \end{aligned} \tag{16}$$

and

$$\frac{\partial^2 d_{j,n+k}}{\partial p_{n+h} \partial p_{n+m}} = \sum_{i=1}^{min(k,n_y)} (-2) \frac{\partial \hat{y}_{n+k-i}}{\partial p_{n+h}} \frac{\partial \hat{y}_{n+k-i}}{\partial p_{n+m}} \delta_1(k-i-1)$$
$$+ \sum_{i=1}^{min(k,n_y)} (-2) \left((x_{j,n_p+i+1} - \hat{y}_{n+k-i}) \frac{\partial^2 \hat{y}_{n+k-i}}{\partial p_{n+h} \partial p_{n+m}} \right) \delta_1(k-i-1). \tag{17}$$

3.3 The Online SVM-Based GPC Algorithm

The algorithm starts with an empty SVM model as can be seen from the flowchart in Fig. 2. The accuracy of the model is determined by checking whether the one-step-ahead prediction error (OSAPE) of the model at the previous iteration is less than ε or not, where OSAPE=$|y_n - \hat{y}_n|$. At every iteration, if the model is accepted as accurate, then a parameter value is produced by the SVM-Based GPC mechanism and then it is applied to the plant, otherwise a control parameter of random magnitude within allowable range is generated and then applied to the plant. After the response of the plant is obtained, the accuracy of the model is determined for the next iteration. If the model is accurate, then no update action is performed. Otherwise, if the model is not accurate, then the output of the plant is used to form a new training point and then the model is updated by Incremental Learning Algorithm. If the number of training data exceeds L, then Decremental Unlearning Algorithm is activated to remove the oldest training point from training data in order to keep the memory usage of the algorithm as low as possible.

4 Example Systems and the Simulation Results

In the simulations, the kernel parameter is fixed as $\sigma = 2$, C is set to 1000, the maximum number of training points used by AOSVR is chosen as L=100 and some GPC parameters appeared in the performance index (2) are chosen as N_1=1 and μ=10^{-20}. For the sake of robustness, the prediction horizon and the control horizon are chosen as N_2=3, N_u=3, respectively. Furthermore, in order to determine the robustness of the method with respect to measurement noise, the measured output of the underlying system is contaminated by additive zero mean Gaussian noise for which the signal-to-noise ratio SNR is given by $SNR = 10\log_{10}(\sigma_y^2/\sigma_v^2)$ dB, where σ_y^2 and σ_v^2 are the variances of the measured output and the additive noise, respectively.

In order to test the efficiency of the proposed method in a statistically convincing sense, we have repeated the simulations for 1000 times starting from different initial conditions. For each initial condition, we recorded the number of iterations beyond which the controller is able to keep the chaotic plant at the equilibrium point forever. Thus, we have obtained the average number of iterations necessary for the proposed controller to be able to stabilize the map under investigation.

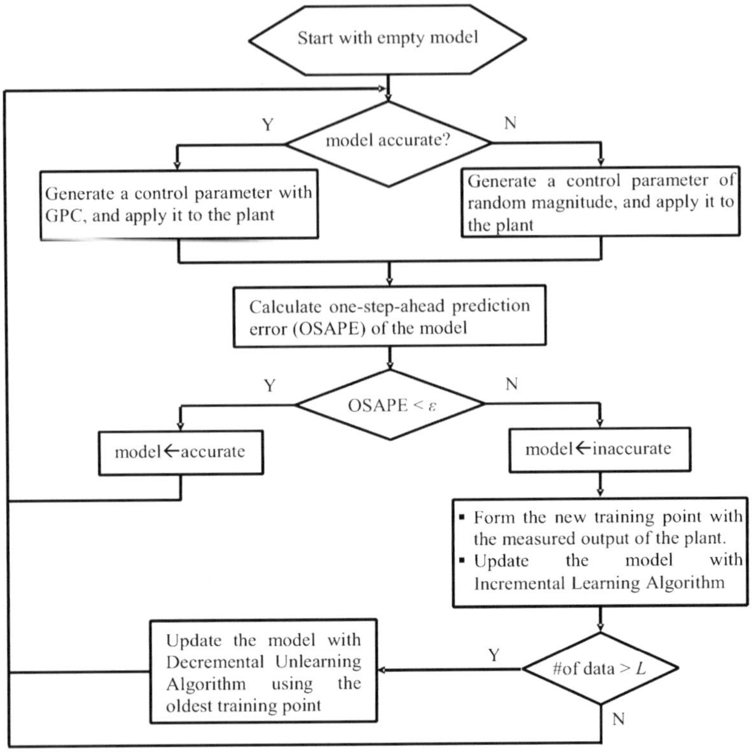

Fig. 2. Flowchart of the online control algorithm

4.1 The Logistic Map

The Logistic map is a well-known one-dimensional chaotic system governed by $y_{n+1} = p_n y_n(1 - y_n)$, where p_n denotes the value of the parameter at the time index n. The map exhibits chaotic behavior for $p_n \in [3.8, 4.0]$, which is the allowable parameter range. For the nominal parameter value $p_{nom} = 3.9$, the map has an unstable equilibrium point at $y^* = 0.7436$, which is chosen as the target. The NARX parameters are determined as n_p=1 and n_y=1, whereas the SVM parameter is chosen as ε=0.002 for the noiseless case and ε=0.007 for the noisy case. The simulation results for the map are illustrated in Fig. 3a and Fig. 3b for the noiseless and noisy conditions, respectively. For the Logistic map, the average number of iterations necessary for the proposed controller to be able to stabilize the map is 61 and 63 for noiseless and noisy conditions, respectively.

4.2 The Henon Map

The Henon map is a second-order chaotic map and governed by $y_{n+2} = p_n - y_{n+1}^2 + 0.3y_n$, where p_n is the value of the parameter at the time index n. It

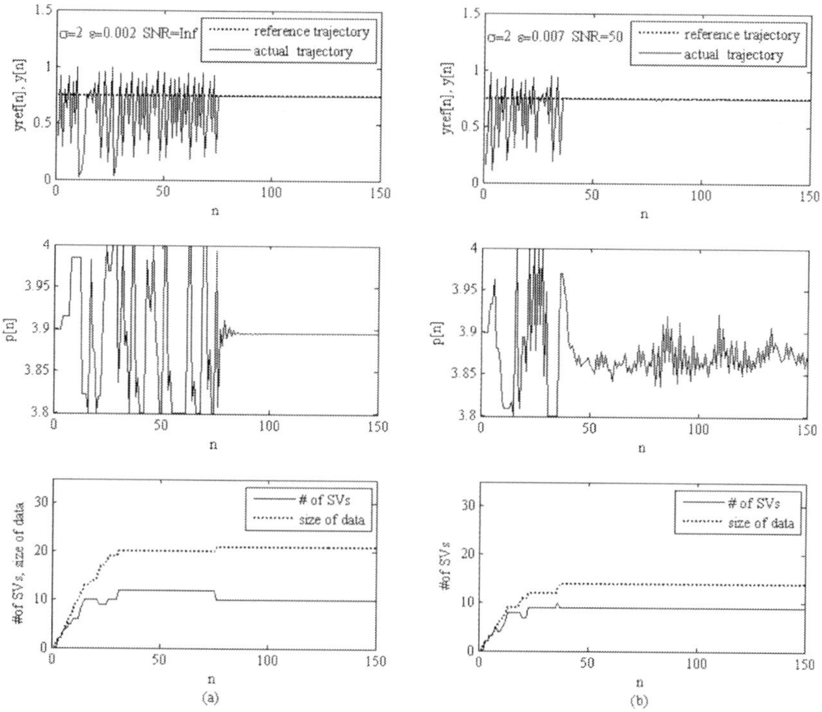

Fig. 3. Simulation results for the Logistic map: (a) noiseless and (b) noisy conditions

exhibits chaotic behavior for $p_n \in [1.34, 1.40]$, which is the allowable parameter range. For the nominal parameter value $p_{nom} = 1.37$, one of the unstable equilibrium points is $y^* = 0.8717$, which is chosen as the target. While the SVM parameter is chosen as ε=0.003 for the noiseless case and ε=0.012 for the noisy case, the NARX parameters are determined as n_p=2 and n_y=2. The simulation results can be seen Fig. 4a and in Fig. 4b for the noiseless and noisy conditions, respectively. The proposed controller requires averagely 180 iterations to stabilize the Henon map in the absence of noise, while averagely 189 iterations are required in the existence of measurement noise.

For both chaotic systems, it is observed from the simulation results that the stabilization performance of Online SVM-Based GPC is very poor at the beginning since the SVM models are initially empty. However, as the model of the underlying map is updated with the training data obtained when the model is inaccurate, the accuracy of the model is gradually improved and, after a short transient, the map can be stabilized to its UEP with very small steady-state error. Similar observations can be made in the presence of additive measurement noise. The proposed method maintain its performance when the output of the system is contaminated by zero mean Gaussian noise up to a certain level. As can be seen from the results, the investigated chaotic maps can be stabilized to their UEPs successfully by Online SVM-Based GPC applying tiny parameter

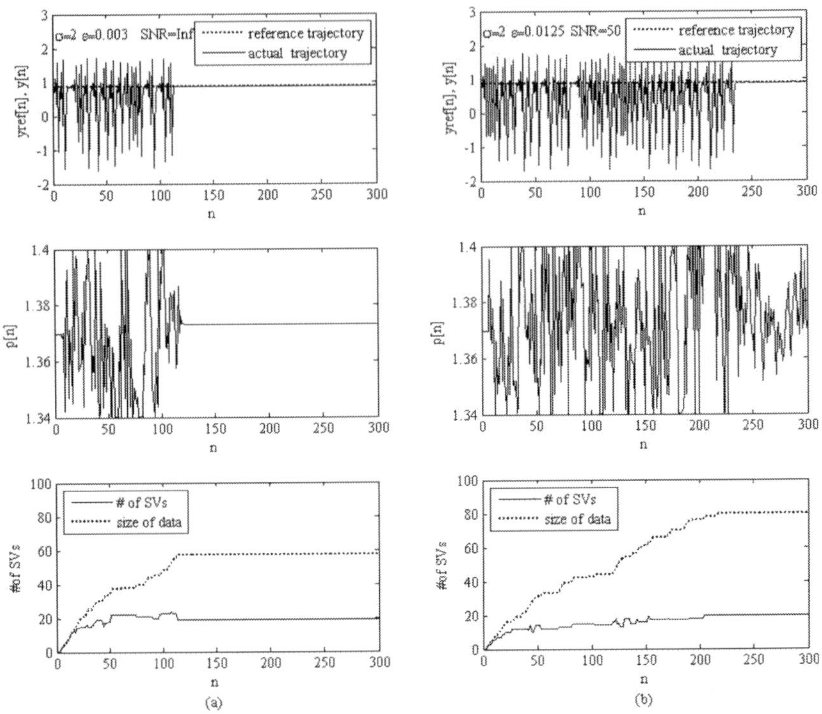

Fig. 4. Simulation results for the Henon map: (a) noiseless and (b) noisy conditions

perturbations within the allowable range. In the long run, if we proceed with the simulations for a long period, we observe that the method keeps the output of the underlying map very close to its UEP with very small steady-state error.

5 Conclusions

This study presents the applicability of the previously proposed control approach, Online SVM-Based GPC, to online stabilization of discrete-time chaotic systems. The controller performs modeling and stabilization tasks simultaneously and provides very efficient and robust online control performance by exploiting outstanding advantages of GPC and SVM techniques. In the method, the SVM model of the map is adaptive to the variations in operating conditions in the sense that the model is updated whenever it is regarded as inaccurate. Moreover, stabilization of the investigated chaotic maps is carried out by only tiny parameter perturbations within the allowable parameter range for which the maps are in their chaotic regimes. The simulation results have confirmed that the Online SVM-Based GPC structure can provide an acceptable stabilization quality for both noiseless and noisy conditions. In other words, the unknown map stabilized by online trained SVM-Based GPC can be kept at its UEP with very small transient and steady-state errors. In conclusion, the Online SVM-Based GPC scheme carries out the desired stabilization task with an acceptable performance for both noiseless and noisy conditions. Moreover, the proposed controller can further be developed as the SVM techniques are improved.

References

1. Iplikci, S.: Online trained support vector machines based generalized predictive control of nonlinear systems. *under review.*
2. Ma J., Theiler J., Perkins S.: Accurate online support vector regression. Neural Computation **15** (2003) 2683–2703.
3. Iplikci S.: Support vector machines based generalized predictive control, *under review.*
4. Ott E., Grebogi C., Yorke J.A.: Controlling chaos. Phy. Rev. Lett. **64** (1990) 1196–1199.
5. Clarke D.W., Mohtadi C., Tuffs P.C.: Generalized predictive control - part 1: the basic algorithm. Automatica 23 (1987) 137–148.
6. Clarke D.W., Mohtadi C., Tuffs P.C.: Generalized predictive control - part 2: the basic algorithm. Automatica 23 (1987) 149–163.
7. Qin S.J., Badgwell T.A.: A survey of industrial model predictive control technology. Control Engineering Practice 11 (2003) 733–764.
8. Soloway D., Haley P.J.: Neural generalized predictive control: a Newton-Raphson algorithm. Proc. of the IEEE Int'l Sym. on Int. Cont., MI, (1996) 277–282.
9. Chen L., Chen G.: Fuzzy modeling, prediction, and control of uncertain chaotic systems based on time series. IEEE Transactions on Circuits and Systems I: Fundamental Theory and Applications 47 (2000) 1527–1531.
10. Choi J.T., Choi Y.H.: Fuzzy neural network based predictive control of chaotic nonlinear systems. IEICE Transactions on Fundamentals of Electronics Communications and Computer Sciences E87A(5), (2004) 1270–1279.
11. Iplikci S.: Support Vector Machines based generalized predictive control of chaotic systems. *accepted for publication.*
12. Nocedal J., Wright S.J.: Numerical Optimization. Springer Series in Op. Res., Springer-Verlag, New York, 1999.
13. Venkataraman P.: Applied optimization with MATLAB programming. John Wiley and Sons, New York, 2002.
14. Vapnik V.: The Nature of Statistical Learning Theory. Springer-Verlag, 1995.
15. Vapnik V.: Statistical Learning Theory. John Wiley, New York, 1998.
16. Vapnik V.: The Support Vector Method of Function Estimation. Nonlinear Modeling Advanced Black Box Techniques, Kluwer Academic Publishers, Boston, 1998.
17. Cristianini N., Taylor J.S.: An Introduction to Support Vector Machines and other kernel-based learning methods. Cambridge University Press, New York, 2000.
18. Schölkopf B., Burges C.J.C., Smola A.J.: Advances in kernel methods: Support Vector Learning. The MIT Press, Cambridge MA, 1999.
19. Smola A.J., Schölkopf B.: A tutorial on support vector regression. NeuroCOLT Tech. Rep. No. NC-TR-98-030. Royal Holloway College, Univ. of London, 1998.

Morphological Neural Networks and Vision Based Mobile Robot Navigation

I. Villaverde, M. Graña, and A. d'Anjou*

Dept. CCIA, UPV/EHU, Apdo. 649, 20080 San Sebastian, Spain
ccpgrrom@si.ehu.es

Abstract. Morphological Associative Memories (MAM) have been proposed for image denoising and pattern recognition. We have shown that they can be applied to other domains, like image retrieval and hyperspectral image unsupervised segmentation. In both cases the key idea is that Morphological Autoassociative Memories (MAAM) selective sensitivity to erosive and dilative noise can be applied to detect the morphological independence between patterns. The convex coordinates obtained by linear unmixing based on the sets of morphological independent patterns define a feature extraction process. These features may be useful either for pattern classification. We present some results on the task of visual landmark recognition for a mobile robot self-localization task.

1 Introduction

Navigation is the ability of an agent to move around its environment with a specific purpose [2]. It implies some knowledge of the environment, be it topological or not. A needed ability for navigation is self-localization: the capacity of the robot to ascertain, more or less accurately, where it is from the information provided by its sensors. This knowledge makes possible other navigation related tasks like planning. The basic self-localization procedure is odometry: self-sensing and keeping track of motion commands. However the uncertainties related to the environment and the robot internal status call for sensor based external confirmation of the position internal estimation. Self-localization based on low range external sensors has been formulated in a probabilistic framework [4]. Vision based [3] and mixed [17,28] systems are proposed to increase the sensing range and self-localization robustness. Visual self-localization methods usually are based on landmark recognition [18,1,27,13,14,16,10,29,22]. For instance, the system described in [1] computes for each stored view a graph model representation of salient points in the image as measured by the information content of a neighborhood of the point in a gradient image. To recognize the view, the salient points in the current image are compared to the stored models. Model matching as performed in [1] is not invariant or robust against translations and rotations. Therefore each view is only recognized when the robot is within a small neighborhood of the physical position and orientation where the landmark was detected originally. Recognition in this case is not continuous, unless the stored views built up a dense map. Our approach based on

* The Spanish Ministerio de Educacion y Ciencia supports this work through grants DPI2003-06972 and VIMS-2003-20088-c04-04.

S. Kollias et al. (Eds.): ICANN 2006, Part I, LNCS 4131, pp. 878–887, 2006.

Morphological Autoassociative Memories (MAAM) [24,23], falls within the class of holistic approaches. The stated goal is to recognize, with some degree of robustness, several predetermined robot placements and orientations based on the recognition of the visual information captured by the robot. We can associate to each position an area of the physical environment where the robot recognizes this position, like in [1],[14]. The robot is supposed to wander looking forward, taking images at a steady rate. Each image corresponds to a view of the world, characterized by a physical position and orientation. Images are analyzed continuously and when a scene is recognized a certain spatial position is assumed for the robot. Robustness implies that the recognition must cope with some variations in lighting and small rotations and translations of the camera due to the uncertainty of the robot position, which, in its turn, is due to the uncertainties in the motion of the robot. This paper departs from previous work [19,20,32] where we first proposed Morphological Associative Memories to solve the self-localization problem, although we continue working on the same platform: a Pioneer 2DX (ActivMedia, Ma.). The approach we follow is to characterize the data by a convex region that encloses them or most of them. The features extracted are the relative coordinates of the data points in this region, the convex coordinates. These convex coordinates are the result of the linear unmixing relative to the vertices of this convex region. Therefore the dimensionality reduction depends on the degree of detail of the definition of this convex region: the number of vertices that describe it.

The Morphological Associative Memories (MAM) [24], [23], [25] were thought of as the morphological counterpart of the well known Hopfield Associative Memories [11]. AMM's are constructed as correlation matrices computed by either Min or Max matrix product. Dual constructions can be made using the dual Min and Max operators. The AMM selective sensitivity to specific types of noise (erosive and dilative noise) is of special interest to us. It was established that AMM are able to store and recall morphologically strongly independent sets of patterns. The notion of morphological independence and morphological strong independence was introduced in [25] to study the construction of AMM robust to general noise. When the input pattern is morphologically independent of the stored patterns, the result of recall is a morphological polynomial on the stored patterns [30]. We construct the erosive and dilative memories to store the patterns. Any input patters whose recalled output corresponds to one of the stored patterns in both kind of memories lies inside the convex region already defined by the stored patterns. Otherwise the pattern is a new vertex of the convex region enclosing the data. The data patterns are filtered dilatively and erosively before being binarized to construct and test the MAM's.

The early application focus of our approach was the unsupervised analysis of hyperspectral images [6,7] to obtain salient image regions that may deserve further analysis and search for labeled data. The approach we favored is that of linear filtering for target detection [15] and the "spectral unmixing" [12] model. In [8,9] we have done the next step assuming that the convex coordinates of the pixel spectra can be used as feature vectors for classification, with surprising good results. Here we try to extend this approach to the visual recognition of landmark views for mobile robot self-localization. We compare our approach with a well known linear feature extraction algorithms, Principal Component Analysis (PCA) [5].

In the following sections we will review the definition of the linear mixing model, the basics on Morphological Associative Memories, some experimental results and, finally, some conclusions and directions for further work.

2 Linear Mixing Model and Spectral Unmixing

The linear mixing model [12] can be expressed as follows:

$$\mathbf{x} = \sum_{i=1}^{M} a_i \mathbf{s}_i + \mathbf{w} = \mathbf{S}\mathbf{a} + \mathbf{w}, \tag{1}$$

where $\mathbf{x}$ is the d-dimension pattern vector, $\mathbf{S}$ is the $d \times M$ matrix whose columns are the d-dimension vertices of the convex region covering the data $\mathbf{s}_i, i = 1, .., M$, $\mathbf{a}$ is the M-dimension fractional abundance vector, and $\mathbf{w}$ is the d-dimension additive observation noise vector. The linear mixing model is subjected to two constraints on the abundance coefficients. First, to be physically meaningful, all abundance coefficients must be non-negative $a_i \geq 0, i = 1, .., M$. Second, to account for the entire composition, they must be fully additive $\sum_{i=1}^{M} a_i = 1$. Once the convex region vertices have been determined the unmixing is the computation of the matrix inversion that gives the coordinates of the point inside the convex region. The simplest approach is the unconstrained least squared error estimation given by:

$$\widehat{\mathbf{a}} = \left(\mathbf{S}^T \mathbf{S}\right)^{-1} \mathbf{S}^T \mathbf{x}. \tag{2}$$

The coefficients that result from this computation do not necessarily fulfill the non-negativity and full additivity conditions. It is possible to enforce each condition separately, but rather difficult to enforce both simultaneously [12]. The added complexity may render the whole approach impractical, therefore we will use unconstrained estimation to compute them. These coefficients are our convex coordinates which we will use as features for pattern classification.

3 Morphological Associative Memories

The work on Morphological Associative Memories stems from the consideration of an algebraic lattice structure $(\mathbb{R}, \vee, \wedge, +)$ as the alternative to the algebraic $(\mathbb{R}, +, \cdot)$ framework for the definition of Neural Networks computation [24] [23]. The operators $\vee$ and $\wedge$ denote, respectively, the discrete $\max$ and $\min$ operators (resp. sup and inf in a continuous setting), which correspond to the morphological dilation and erosion operators, respectively. Given a set of input/output pairs of pattern $(X, Y) = \left\{\left(\mathbf{x}^\xi, \mathbf{y}^\xi\right); \xi = 1, .., k\right\}$, an heteroassociative neural network based on the pattern's cross correlation [11] is built up as $W = \sum_\xi \mathbf{y}^\xi \cdot \left(\mathbf{x}^\xi\right)'$. Mimicking this construction procedure [24], [23] propose the following constructions of Heteroassociative Morphological Memories (HMM's):

$$W_{XY} = \bigwedge_{\xi=1}^{k} \left[\mathbf{y}^\xi \times \left(-\mathbf{x}^\xi\right)'\right] \text{ and } M_{XY} = \bigvee_{\xi=1}^{k} \left[\mathbf{y}^\xi \times \left(-\mathbf{x}^\xi\right)'\right], \tag{3}$$

where $\times$ is any of the $\boxdot$ or $\boxtimes$ operators. Here $\boxdot$ and $\boxtimes$ denote the max and min matrix product, respectively defined as follows:

$$C = A \boxdot B = [c_{ij}] \Leftrightarrow c_{ij} = \bigvee_{k=1..n} \{a_{ik} + b_{kj}\}, \tag{4}$$

$$C = A \boxtimes B = [c_{ij}] \Leftrightarrow c_{ij} = \bigwedge_{k=1..n} \{a_{ik} + b_{kj}\}. \tag{5}$$

If $X = Y$ then the HMM memories are Autoassociative Morphological Memories (AMM). Conditions of perfect recall by the HMM's and AMM's of the stored patterns are proved in [24],[23]. In the continuous case, the AMM's are able to store and recall any set of patterns:

$$W_{XX} \boxdot X = X = M_{XX} \boxtimes X, \tag{6}$$

for any X.

These results hold when we try to recover the output patterns from the noise-free input pattern. Let it be $\widetilde{\mathbf{x}}^\gamma$ a noisy version of $\mathbf{x}^\gamma$. If $\widetilde{\mathbf{x}}^\gamma \leq \mathbf{x}^\gamma$ then $\widetilde{\mathbf{x}}^\gamma$ is an eroded version of $\mathbf{x}^\gamma$, alternatively we say that $\widetilde{\mathbf{x}}^\gamma$ is corrupted by erosive noise. If $\widetilde{\mathbf{x}}^\gamma \geq \mathbf{x}^\gamma$ then $\widetilde{\mathbf{x}}^\gamma$ is a dilated version of $\mathbf{x}^\gamma$, alternatively we say that $\widetilde{\mathbf{x}}^\gamma$ is corrupted by dilative noise.

Morphological memories are selectively sensitive to these kinds of noise. The conditions of *robust* perfect recall are proven in [24], [23]. Here we will remember them for the sake of the reader, because they are on the basis of the proposed algorithm. Given patterns X, the equality

$$W_{XX} \boxdot \widetilde{\mathbf{x}}^\gamma = \mathbf{x}^\gamma \tag{7}$$

holds when the noise affecting the pattern is erosive $\widetilde{\mathbf{x}}^\gamma \leq \mathbf{x}^\gamma$ and the following relation holds:

$$\forall i \exists j_i ; \widetilde{x}^\gamma_{j_i} = x^\gamma_{j_i} \vee \left(\bigvee_{\xi \neq \gamma} \left(x^\gamma_i - x^\xi_i + x^\xi_{j_i} \right) \right). \tag{8}$$

Similarly, the equality

$$M_{XY} \boxtimes \widetilde{\mathbf{x}}^\gamma = \mathbf{x}^\gamma \tag{9}$$

holds when the noise affecting the pattern is dilative $\widetilde{\mathbf{x}}^\gamma \geq \mathbf{x}^\gamma$ and the following relation holds:

$$\forall i \exists j_i ; \widetilde{x}^\gamma_{j_i} = x^\gamma_{j_i} \wedge \left(\bigwedge_{\xi \neq \gamma} \left(x^\gamma_i - x^\xi_i + x^\xi_{j_i} \right) \right). \tag{10}$$

Therefore, the AMM will fail to recall the pattern if the noise is a mixture of erosive and dilative noise. In [21] we have proposed a morphological scale space method to increase the robustness of AMM.

Other approach to obtain general noise robustness is based on the so-called kernel patterns [23], [25], [30] . Related to the construction of the kernels, [25] introduced the notion of morphological independence. Here we distinguish erosive and dilative versions of this definition: Given a set of pattern vectors $X = \left(\mathbf{x}^1, ..., \mathbf{x}^k\right)$, a pattern vector $\mathbf{y}$ is said to be morphologically independent of X in the erosive sense if $\mathbf{y} \nleq \mathbf{x}^\gamma; \gamma = \{1, .., k\}$, and morphologically independent of X in the dilative sense if $\mathbf{y} \ngeq \mathbf{x}^\gamma; \gamma = \{1, .., k\}$. The set of pattern vectors X is said to be morphologically independent in either sense when all the patterns are morphologically independent of the remaining patterns in the set. For the current application we want to use AMM as detectors of the set extreme points, to obtain a rough approximation of the minimal simplex that covers the data points. We note that given a set of pattern vectors $X = \left(\mathbf{x}^1, ..., \mathbf{x}^k\right)$, and the erosive W_{XX} and dilative M_{XX} memories constructed from it, and a test pattern $\mathbf{y} \notin X$, if $\mathbf{y}$ is morphologically independent of X in the erosive sense, then $W_{XX} \boxvee \mathbf{y} \notin X$. Also, if $\mathbf{y}$ is morphologically independent of X in the dilative sense, then $M_{XX} \boxwedge \mathbf{y} \notin X$. Therefore the AMM's can be used as detectors of morphological independence.

The vector patterns that we are searching for define a high dimensional box centered at the origin of the high dimensional space (the data mean is shifted to the origin). They are morphologically independent vectors both in the erosive and dilative senses, and they enclose the remaining vectors. Working with integer valued vectors, given a set of pattern vectors $X = \left(\mathbf{x}^1, ..., \mathbf{x}^k\right)$ and the erosive W_{XX} and dilative M_{XX} memories constructed from it, if a test pattern $\mathbf{y} < \mathbf{x}^\gamma$ for some $\gamma \in \{1, .., k\}$ then $W_{XX} \boxvee \mathbf{y} \notin X$. Also, if the test pattern $\mathbf{y} > \mathbf{x}^\gamma$ for some $\gamma \in \{1, .., k\}$ then $M_{XX} \boxwedge \mathbf{y} \geq\notin X$. Therefore, working with integer valued patterns the AMM will be useless for the detection of morphologically independent patterns. However, if we consider the binary vectors obtained as the sign of the vector components, then morphological independence would be detected as suggested above: The already detected endmembers are used to build the erosive and dilative AMM. If the output recalled by a new pattern does not coincide with any of the endmembers, then the new pattern is a new endmember. Those endmembers are selected from the data following the algorithm presented in [33].

4 Experimental Results

The experimental setup is as follows. First a path was defined from our laboratory to the stairs hall on 3d floor of our building. The mobile robot platform was guided manually seven times following this path. In each of those trips, the odometry was recorded and the images taken from the camera, at an average of 10 frames per second, were also recorded.

In the computational experiments that follow, the first trip was used to train the system parameters, and the six other trips were used as test sequences, simulating a real trip. The task to perform is to recognize a given set of map positions from their landmark views. The positions were selected on the floor plane, selecting places of practical relevancy, like doors to other laboratories, and the corresponding landmark views were selected from the first trip image sequence based on odometry readings. Figure 1 shows the map position landmark views as extracted from the image sequence.

Classes are composed for each of the selected landmark position, assigning the images in the sequences to the closest map position according to its odometry reading. Therefore the task becomes the classification of the images into one of the map classes, where each class is composed of images taken in robot path positions before and after the map position for which it is the closest map position. The clasification was done using k-NN. Each of the images of the training trip was assigned to a class, using them as a cluster of images representing each selected position. For the test trips, each image was classified on those classes using 3-NN.

¿From the image sequence of the first trip a PCA transformation consisting of 230 eigenvectors was computed. All the following computations were done on the PCA coefficients of the images. We apply the algorithm described in [33] several times to the different image sequences, after their transformation by the 230 eigenvector PCA. The noise tolerance parameter with best results was set, after some tuning, to $\alpha = 5$ An instance of the extrema selected with this value is given in figure 2. Despite their similarity to the collection of images shown in figure 1, these are obtained from a completely unsupervised process, while those in figure 1 correspond to a human made selection. An interesting question is whether our approach could be used as an automatic map position determination whose landmark views correspond to the extrema found by our algorithm.

Fig. 1. The landmark views corresponding to the positions selected to build up the map

Fig. 2. The views corresponding to the extrema selected in one instance of the execution of the algorithm described in [33]

As the algorithm described in [33] has a random start, different runs of it may give different results. In tables 2 and 3 we present the classification success ratios for each of the sequences on the convex coordinates computed from the extrema found by the algorithm after each run, with different values of α. The last column shows the number of extrema found at each run. Notice that the success ratio decreases as the image sequence is farther in time from the initial one used for training. Notice also that the number of extrema is in the order of 10 for the best results, meaning a dimension reduction from 230 to O(10). The Av. column shows the average success for each image sequence, including the training image sequence. For comparison we perform some further dimension reduction on the PCA coefficient vectors selecting the most significant eigenvectors for the transformation. The average results of the image classification are given in table 1, the average being computed as in the last column of the other tables.

The comparison of the results in tables 2, 3 and 1 shows that there is at least an instance of the convex coordinates features which improves the best results of the PCA, and that the convex coordinates are comparable or improve the results of the PCA with a the samo or much stronger dimensionality reduction. Although these results are much less spectacular than the ones reported in [8,9], they confirm the trend that points to the usefulness of convex coordinates as a feature selection algorithm. We are trying to recompute the results using the original images as inputs to the extrema selection algorithm, in the hope of improving the results of our algorithm relative to the PCA. When applied to the PCA coefficient vectors, our algorithm may be greatly restricted by the data transformation already performed.

Table 1. Landmark recognition success rate based on the PCA representation of the navigation images for several sets of eigenvectors selected, using 3-nn

Set	Train	Pass 1	Pass 2	Pass 3	Pass 4	Pass 5	Pass 6	Av.
PCA 197	0,95	0,95	0,84	0,76	0,65	0,79	0,77	0,82
PCA 150	0,95	0,95	0,84	0,76	0,65	0,78	0,76	0,81
PCA 100	0,95	0,95	0,85	0,76	0,65	0,79	0,76	0,82
PCA 50	0,95	0,94	0,85	0,76	0,65	0,78	0,78	0,81
PCA 30	0,96	0,94	0,87	0,77	0,64	0,78	0,78	0,82
PCA 10	0,96	0,94	0,86	0,78	0,66	0,76	0,73	0,81
Av.	0,96	0,94	0,85	0,76	0,65	0,78	0,76	0,82

Table 2. Landmark recognition success rate based on the convex coordinates representation of the navigation images for several runs of the extrema extraction algorithm with $\alpha = 6$ and using 3-nn

Run	Train	Pass 1	Pass 2	Pass 3	Pass 4	Pass 5	Pass 6	Av.	#extrema
1	0,92	0,91	0,78	0,74	0,66	0,68	0,62	0,76	8
2	0,95	0,93	0,77	0,73	0,74	0,72	0,62	0,78	7
3	0,95	0,93	0,83	0,73	0,67	0,72	0,64	0,78	8
4	0,94	0,93	0,78	0,71	0,67	0,69	0,64	0,77	7
5	0,93	0,90	0,76	0,71	0,65	0,67	0,62	0,75	7
6	0,93	0,91	0,77	0,72	0,69	0,68	0,57	0,75	8
7	0,95	0,93	0,78	0,70	0,61	0,66	0,62	0,75	8
8	0,95	0,93	0,78	0,69	0,58	0,69	0,62	0,75	8
9	0,93	0,92	0,80	0,73	0,70	0,73	0,63	0,78	8
10	0,94	0,94	0,80	0,73	0,65	0,69	0,67	0,77	9
Av.	0,94	0,92	0,79	0,72	0,66	0,69	0,62	0,76	

Table 3. Landmark recognition success rate based on the convex coordinates representation of the navigation images for several runs of the extrema extraction algorithm with $\alpha = 5$ and using 3-nn

Run	Train	Pass 1	Pass 2	Pass 3	Pass 4	Pass 5	Pass 6	Av.	#extrema
1	0,94	0,93	0,81	0,76	0,72	0,73	0,67	0,79	13
2	0,94	0,93	0,85	0,77	0,69	0,78	0,71	0,81	14
3	0,94	0,93	0,84	0,75	0,70	0,75	0,74	0,81	13
4	0,94	0,93	0,83	0,71	0,63	0,73	0,67	0,78	14
5	0,94	0,93	0,85	0,79	0,69	0,78	0,72	0,81	12
6	0,93	0,93	0,80	0,70	0,67	0,69	0,70	0,77	12
7	0,94	0,93	0,83	0,71	0,59	0,70	0,66	0,77	12
8	0,93	0,93	0,82	0,76	0,69	0,74	0,66	0,79	12
9	0,94	0,93	0,79	0,73	0,64	0,70	0,63	0,77	14
10	0,92	0,92	0,79	0,70	0,63	0,65	0,60	0,75	12
Av.	0,94	0,93	0,82	0,74	0,67	0,73	0,68	0,78	

5 Conclusions and Further Work

We claim that we can use the convex coordinates of the data points based on the vertices of a convex region covering the data as features for pattern classification. The convex region vertices are selected using MAM to detect the morphologically independent sets. Mobile robot self-localization is stated as a classification of images taken from the camera. We use the PCA and the convex coordinates based on the MAM detected endmembers as the features for classification. We found that our approach improves the PCA approach in some runs, while on average performs similar than most PCA eigenvector selections tried. One possible future line of work could be to use the detection of endmembers as an automatic landmark selection tool, applying it to the SLAM problem.

References

1. C. Balkenius, L. Kopp, (1997) Robust Self-Localization Using Elastic Templates, in T. Lindberg (ed.), Proceedings of Swedish Symposium on Image Analysis,.
2. R. Chatila, (1995) Deliberation and Reactivity in Autonomous Mobile Robots, Robotics and Autonomous Systems, 16, pp. 197-211
3. DeSouza, G. N., Kak, A. C. (2002). Vision for Mobile Robot Navigation: A Survey. IEEE Transactions on Pattern Analysis and Machine Intelligence 24(2), pp. 237-267.
4. D. Fox, Markov Localization: A Probabilistic Framework for Mobile Robot Localization and Navigdation, Ph. D. Thesis, University of Bonn, Germany, December 1998
5. Fukunaga K., *Introduction to statistical pattern recognition*, Academic Press, Boston, MA 1990
6. Graña M., J. Gallego, "Associative Mophological Memories for endmember induction", Proc. IGARSS'2003, Tolouse, France.
7. Graña M., P. Sussner, G. Ritter, "Associative Morphological Memories for Endmember Determination in Spectral Unmixing", Proc. FUZZ-IEEE'03
8. M. Graña, A. d'Anjou (2004) "Feature Extraction by Linear Spectral Unmixing" in M. Negoita, R.J. Howlett, L.C. Jain (eds) Knowledge-Based Intelligent Information and Engineering Systems. Part I, Springer Verlag 2004, LNAI 3213, pp:692-697
9. Manuel Graña, Alicia d'Anjou, Xabier Albizuri (2005) Morphological memories for feature extraction in hyperspectral imagesMichael Verleysen, (ed.) ESANN 2005, dFacto press, 2005, pp.497-502
10. H. M. Gross, A. Koening, H. J. Boehme and Ch. Schroeter, (2002) Vision-based Monte Carlo Self-localization for a Mobile Service Robot Acting as Shopping Assistant in a Home Store, Proceedings of the IEEE Intl. Conference on Intelligent Robots and Systems.
11. Hopfield J.J., (1982) "Neural networks and physical systems with emergent collective computational abilities", Proc. Nat. Acad. Sciences, vol. 79, pp.2554-2558,
12. Keshava N., J.F. Mustard "Spectral unimixing", IEEE Signal Proc. Mag. 19(1) pp:44-57 (2002)
13. S. Livatino, C. Madsen, (1999) Optimization of Robot Self-Localization Accuracy by Automatic Visual-Landmark Selection, Proceedings of 11th Scabdinavian Conference on Image Analysis (SCIA), pp. 501-506
14. S. Livatino, C. Madsen, (1999) Autonomous Robot Navigation with Automatic Learning of Visual Landmarks, International Symposium of Intelligent Robotic Systems (SIRS99), 1999.
15. Manolakis D., G. Shaw "Detection algorithms for hyperspectral imaging applications", IEEE Signal Proc. Mag. 19(1) pp:29-43 (2002)

16. F. Marando, M. Piaggio and A. Scalzo, (2001) Real Time Self Localization Using a Single Frontal Camera, International Symposium of Intelligent Robotic Systems (SIRS01)
17. Ohya, A., Kosaka, A., Kak, A. C. (1998) Vision-Based Navigation by a Mobile Robot with Obstacle Avoidance Using Single-Camera Vision and Ultrasonic Sensing. IEEE Journal of Robotics and Automation, 14(6), pp. 969-978.
18. C.F. Olson, (2000) Landmark Selection for Terrain Matching, Proceedings ICRA2000.
19. Raducanu B., M. Graa, P. Sussner, Morphological Neural Networks for vision based self-localization, Proc. ICRA2001
20. Raducanu B., M. Graa, P. Sussner, Steps towards one-shot vision-based self-localization pp 265-294 in Biologically inspired robot behavior engineering; Richard Duro, Jose Santos, Manuel Graa (eds) Springer Verlag, 2002
21. Raducanu B., M. Graña, X. Albizuri (2003) "Morphological scale spaces and associative morphological memories: results on robustness and practical applications", J. Math. Imaging and Vision 19(2):113-122
22. J. Reuter, (2000) Mobile Robot Self-Localization Using PDAB, Proceedings of International Conference on Robotics and Automation (ICRA)
23. Ritter G. X., J. L. Diaz-de-Leon, P. Sussner. (1999) "Morphological bidirectional associative memories". Neural Networks, Volume 12, pages 851-867,
24. Ritter G. X., P. Sussner, J. L. Diaz-de-Leon. (1998) "Morphological associative memories". IEEE Trans. on Neural Networks, 9(2):281-292,
25. Ritter G.X., G. Urcid, L. Iancu, (2003) "Reconstruction of patterns from moisy inputs using morphological associative memories", J. Math. Imaging and Vision 19(2):95-112
26. Ritter GX, G Urcid (2003) "Lattice algebra approach to single-neuron computation". IEEE Trans Neural Networks 14(2): 282-295.
27. A. Rizzi, D. Duina, S. Inelli, R. Cassinis, (2002) Unsupervised Matching of Visual Landmarks for Robotic Homing using Fourier-Mellin Transform, Robotics and Autonomous Systems, 40, pp. 131-138.
28. A. Saffiotti and L. P. Wesley, (1996) Perception-Based Self-Localization Using Fuzzy Location, in L. Dorst, M. Van Lambalgen and F. Voorbraak (eds.), Lecture Notes in Artificial Intelligence 1093, Springer-Verlag, pp. 368-385
29. D. Sekimori, T. Usui, Y. Masutani and F. Miyazaki, (2002) High-Speed Obstacle Avoidance and Self-Localization for Mobile Robots Based on Omni-Directional Imaging of Floor Region, IPSJ Transactions on Computer Vision and Image Media, 42 NoSIG13-012.
30. Sussner P., (2001) "Observations on Morphological Associative Memories and the Kernel Method", Proc. IJCNN'2001, Washington DC, July
31. Sussner P., (2003) "Generalizing operations of binary autoassociative morphological memories using fuzzy set theory", J. Math. Imaging and Vision 19(2):81-94
32. I. Villaverde, S. Ibañez, F. X. Albizuri, M. Graña, (2005)Morphological neural networks for real-time vision based self-localizationAjith Abrham, Y. Dote, T. Furuhashi, M. Köpen, A. Ohuchi, Y. Ohsawa (eds) Soft Computing as transdisciplinary Science and Techonology, Proc. WSTST'05 Springer Verlag, Advances in Soft Computing, 2005, pp.70-79
33. I. Villaverde, M. Graña and A. D'Anjou, "Morphological Neural Networks for Localization and Mapping". Proceedings of the IEEE International Conference on Computational Intelligence for Measurement Systems and Applications (CIMSA'06), July 12-14, 2006, La Coruña (Spain), On Print.

Position Control Based on Static Neural Networks of Anthropomorphic Robotic Fingers

Juan Ignacio Mulero-Martínez, Francisco García-Córdova, and Juan López-Coronado

Department of System Engineering and Automatic
Polytechnic University of Cartagena
Campus Muralla del Mar, 30202, Cartagena, Murcia, Spain
{juan.mulero, francisco.garcia, jl.coronado}@upct.es

Abstract. A dynamic neurocontroller for positioning robot manipulators with a tendon-driven transmission system has been developed allowing to track desired trajectories and reject external disturbances. The controller is characterised as providing motor torques rather than joint torques. In this sense, the redundant problem regarded with the tendon-driven transmission systems is solved using neural networks that are able to learned the linear transformation that maps motor torques into joint torques. The neurocontroller not only learn the dynamics associated with the robot manipulator but also the parameters attached to the transmission system such as pulley radii. A theorem relying on the Lyapunov theory has been developed, guaranteeing the uniformly ultimately bounded stability of the whole system and providing both the control laws and weight updating laws.

1 Introduction

The calculation of tendon forces is an indeterminate problem, see [1], because of the different dimension of the space of joint torques (n-dimensional) and the dimension of the space of tendon forces (m-dimensional). The inputs of the dynamic model are the motor torques whereas the controller provides control actions based on the joint torques. Thus, a redundant system (m>n) is managed in order to get a total control according to the Morecki's property and this implies to solve a system of equations where the number of unknown variables is greater than the number of equations. In this sense, the relationship between the joint space and the tendon space is stated by a linear transformation, called structural matrix with more columns than rows. Therefore, the solution of forces for a given torque is not unique. In fact, a solution for this equation in terms of the tendon force can be expressed as the summation of an homogeneous solution and a particular solution. The particular solution minimizes the norm of the solution whereas the homogeneous is physically associated with the tendon forces that do not work but provide an increasing of tension for the tendon wrapping the

S. Kollias et al. (Eds.): ICANN 2006, Part I, LNCS 4131, pp. 888–897, 2006.

pulley. In contrast, the particular solution generates work onto the system and as a result becomes the most important component for being cause of motion. All the proposed methods in the literature try to minimize the homogeneous solution since implies an energetic consume which is not used for the motion. A solution for the problem exists if and only if the structure matrix is full rank, see [2], [3].

The objective of the control of a tendon-driven transmission system is to track the position and force of the end-effector by means of a set of input variables such as motor torques from the actuators. The control of tendon-driven manipulators has been investigated by a few researchers. The main contribution of this paper is the design and implementation of an adaptive neural network controller that tracks desired trajectories and reject disturbances with abilities of picking up parametric structured uncertainties in a tendon-driven transmission system. The neurocontroller learns not only the non-linear terms regarding to the dynamics, such as inertial, coriolis/centripetal and gravitational terms, but also the tendon routing and dimensions of pulleys in the transmission system. The use of tendons adds challenges to the controller design because the tendons can exert force in only one direction. A functional link neural network (FLNN) is developed to estimate the non-linear terms in the filtered error dynamics. The touchstone of this work is a theorem relying on the Lyapunov theory guaranteeing the uniformly ultimately bounded stability of the whole system and providing both the control laws and weight updating laws. In this scheme no preliminary off-line learning stage is required, so that the weights are initialised to zero with the exception of those regarding to the structural matrix that are initialised according to the tendon routing matrix wich items adopting discrete values of -1, 0 or 1.

The paper is organized as follows. In section II, the closed-loop error dynamics is derived using functional-link neural networks (FLNN). Section III is devoted to present the design of the neural network controller in terms of a control law and an usupervised version of the continuous-time backpropagation for weight updating law. This section concludes with a theorem which plays a central role in this theory and guarantees the ultimately uniformly boundedness for the whole system and is based in the Lyapunov energy. In section IV, the control law has been applied to an anthropomorfic finger that looks like the Standford/JPL finger with a tendon-driven transmission system. The concluding remarks are discussed in section V.Closed-loop error dynamics.

2 Closed-Loop Error Dynamics

The dynamics of an n-link robot manipulator with a tendon-driven transmission system may be expressed from the Lagrange or Hamilton formulation, see [4]

$$\left(M\left(\Theta\right)+\tilde{M}\right)\ddot{\Theta}+\tilde{B}\dot{\Theta}+C\left(\Theta,\dot{\Theta}\right)\dot{\Theta}+G\left(\Theta\right)=R^{T}B^{T}R_{m}^{-1}\tau_{m}=\tau \qquad (1)$$

where B stands for the tendon-routing matrix, R is a diagonal matrix for the pulley radii, R_m a matrix of radii of driver pulleys, $M(\Theta)$ a matrix of inertias and τ_m the motor torques, $\tilde{M}$ and $\tilde{B}$ represent respectively the rotor inertia and the motor damping matrix reflected onto the joint space. By using a filtered error variable $r = \dot{e} + \Lambda e$, the error dynamic equation is derived.

$$-\bar{M}(\Theta)\dot{r} - \bar{C}\left(\Theta, \dot{\Theta}\right) r + R^T B^T R_m^{-1} f(x) + \tau_d = R^T B^T R_m^{-1} \tau_m \qquad (2)$$

where the function $f(x)$ collects all the non-linear terms of the system and $\bar{M}(\Theta) = \left(M(\Theta) + \tilde{M}\right)$ and $\bar{C}\left(\Theta, \dot{\Theta}\right) = \left(C\left(\Theta, \dot{\Theta}\right) + \tilde{B}\right)$. The operator $()^\dagger$ stands for the Pseudo-inverse of Penrose-Moore, see [5]. The linear transformation $R^T B^T R_m^{-1}$ mapping motor torques into joint torques, is approximated by a neural network providing nominal weights $V \in \mathbb{R}^{m \times n}$ and reconstruction error ε_V,

$$R^T B^T R_m^{-1} f(x) = V^T f(x) + \varepsilon_V \qquad (3)$$

Instead of using joint torques, $\tau = R^T B^T R_m^{-1} \tau_m$ approximations $\hat{\tau} = \hat{V}^T \tau_m$ are proposed so that the torque deviations ε_τ are defined as

$$\varepsilon_\tau = \tau - \hat{\tau} = \left(R^T B^T R_m^{-1} - \hat{V}^T\right) \tau_m \qquad (4)$$

Using the FLNN as defined in (3), the filtered error dynamics turns into

$$-\bar{M}(\Theta)\dot{r} - \bar{C}\left(\Theta, \dot{\Theta}\right) r + R^T B^T R_m^{-1} f(x) + \tau_d = \hat{V}^T \tau_m \qquad (5)$$

3 Controller Structure, Control Law and Updtating Law

The controller consists of three terms: a PD-controller to guarantee a good tracking performance, $\tau_{pd} = \frac{K_v \hat{V} r}{\|\hat{V} r\|} = \frac{K_v \hat{V}}{\|\hat{V} r\|}(\dot{e} + \Lambda e)$, a compensator of non-linearities, $\tau_{nl} = \hat{f}$ and a robust controller τ_r to absorbe no modelling dynamics,

$$\tau_m = \frac{K_v \hat{V} r}{\left\|\hat{V} r\right\|} + \hat{f} + \tau_r \qquad (6)$$

where $K_v = K_v^T > 0$ is a gain matrix and $\hat{f}$ is an approximation of the non-linear function $f(x)$. An important remark is that the exact control action is expressed as motor torques instead of joint torques. Replacing the control law given by equation (6) into the plant described by (5) the closed-loop error dynamics becomes

$$\left(M(\Theta) + \tilde{M}\right)\dot{r} = -\left(C\left(\Theta, \dot{\Theta}\right) + \tilde{B}\right) r + V^T f(x) - \hat{V}^T K_v \hat{V} r - \hat{V}^T \hat{f} - \hat{V}^T \tau_r + \tau_d \qquad (7)$$

In the next proposition the functional estimation error $\tilde{f}(x) = f(x) - \hat{f}(x)$ and the weight estimation error $\tilde{V} = V - \hat{V}$ are related to the term $V^T f(x) - \hat{V}^T \hat{f}(x)$ in the filtered error dynamic equation (7).

Proposition 1. *Let define the continous function $f : \Omega_x \to \mathbb{R}^m$ where $\Omega_x \subseteq \mathbb{R}^{5n}$ is a compact set and smooth enough to a certain degree α over Ω_x, i.e. $f \in C^\alpha(\Omega_x)$. The term $V^T f(x) - \hat{V}^T \hat{f}(x)$ can be written as follows*

$$V^T f(x) - \hat{V}^T \hat{f}(x) = \tilde{V}^T \hat{W}^T \gamma(x) + \hat{V}^T \tilde{W}^T \gamma(x) + w(t) \tag{8}$$

where $w(t)$ stands for a modelling disturbance given as

$$w(t) = V^T \varepsilon + \tilde{V}^T \tilde{W}^T \gamma(x) \tag{9}$$

$V \in \mathbb{R}^{m \times n}$ are the nominal weights for the linear transformation $R^T B^T R_m^{-1} f(x) = V^T f(x) + \varepsilon_V$, $W \in R^{L \times m}$ are the ideal weights in the approximation of the nonlinear function $f(x)$ and $\tilde{f}(x) = f(x) - \hat{f}(x)$ represents the functional estimation error.

Using the proposition (1) the closed-loop system becomes

$$\left(M(\Theta) + \tilde{M}\right)\dot{r} = -\left(C\left(\Theta, \dot{\Theta}\right) + \tilde{B}\right) r + \tilde{V}^T \hat{W}^T \gamma(x) + \hat{V}^T \tilde{W}^T \gamma(x) + w(t) - \frac{\hat{V}^T K_v \hat{V} r}{\|r\|} - \hat{V}^T \tau_r + \tau_d \tag{10}$$

The next lemma is very important in order to overbound the disturbance term $w_1(t)$ at each time by a known computable function. Previously some assumptions on the boundedness of the weights and the activation functions must be considered

Lemma 1 (Bounds on the disturbance term $w(t)$). *The disturbance term w as given by the equation (9) is bounded according to the expression*

$$\|w(t)\| \leq C_0 + C_1 \left\|\hat{W}\right\|_F + C_2 \left\|\hat{W}\right\|_F \left\|\hat{V}\right\|_F + C_3 \left\|\hat{V}\right\|_F \tag{11}$$

where C_0, C_1, C_2 and C_3 are computable positive constants defined as

$$\begin{aligned} C_0 &= V_B (W_B \gamma_B + \varepsilon_B) \\ C_1 &= V_B \gamma_B \\ C_2 &= \gamma_B \\ C_3 &= W_B \gamma_B \end{aligned} \tag{12}$$

The complete control scheme appeared in the figure 1. The next theorem relies on the Lyapunov theory guaranteeing the uniformly ultimately bounded of the equilibrium point, where the uniform property means that the time interval T does not depend on t_0. A practical notion of stability is the ultimately uniformly

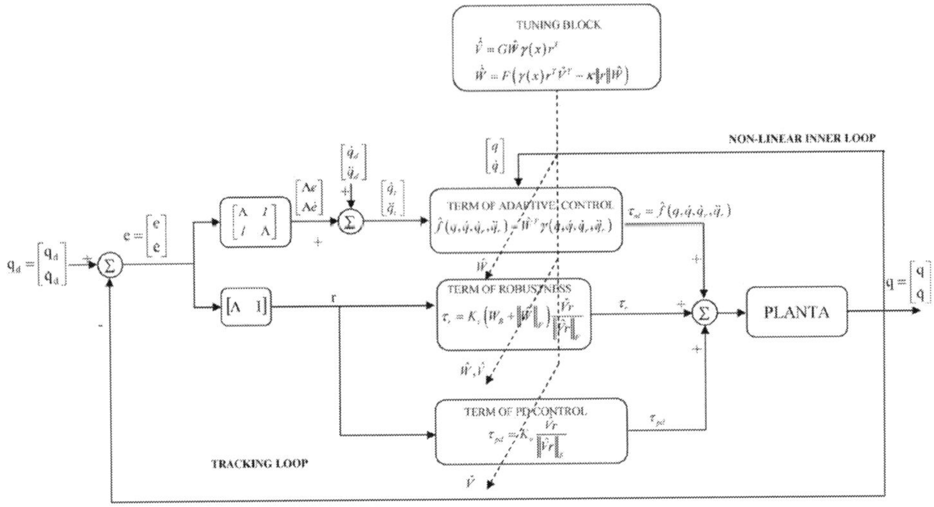

Fig. 1. Neural net control structure

boundedness (UUB) who allows to guarantee the boundedness of the signals for the closed system. The UUB is a pragmatic criterium to be applied to closed-loop systems containing unknown disturbunces, which are often be bounded by some account or modelling errors. As it is claimed in [6], the UUB is guaranteed if the Lyapunov derivative is negative outside some bounded region or $\mathbb{R}^n$.

Theorem 1. *Let the desired trajectory* $\Theta_d(t)$ *bounded by* Θ_B. *and the initial condition for* $r(t)$ *satisfying* $\|r(0)\| < \frac{b_x - \Theta_B}{c_0 + c_2}$ *where* b_x *is an upper bound for* x, $c_0 = \frac{1+\sigma_{\max}(\Lambda)}{\sigma_{\min}(\Lambda)}$ *and* $c_1 = (1 + \sigma_{\max}(\Lambda)) \|e_0\| + q_B$. *Suppose that the approximation error* ε *and unmodeled disturbances* $\tau_d(t)$ *are upper bounded by* ε_N *and* d_B *respectively. In the ideal case,* $\varepsilon_N = d_B = 0$. *Let the control law given by equation (6) with a robust term*

$$\tau_r = K_z \left(\left\| \hat{W} \right\|_F + W_B \right) \frac{\hat{V} r}{\left\| \hat{V} \right\|_F \|r\|} \tag{13}$$

where $K_z = diag(k_{z_{ii}}) > 0$ *with* $k_{z_{ii}} \geq \gamma_B$. *Let the next weight updating laws be provided by*

$$\dot{\hat{V}} = G\hat{W}^T \gamma(x) r^T \tag{14}$$

$$\dot{\hat{W}} = F\gamma(x) r^T \hat{V}^T - F\kappa \|r\| \hat{W} \tag{15}$$

with $\Gamma_m = \Gamma_m^T > 0$, $\Gamma_g = \Gamma_g^T > 0$ *and* $\Gamma_f = \Gamma_f^T > 0$ *symmetric positive-definite constant matrices. The filtered error* $r(t)$ *and the weight estimates* $\hat{W}_m$, $\hat{W}_c$ *and* $\hat{W}_g$ *are UUB. The bounds are*

$$\|r\| > \frac{\frac{1}{4}\kappa\left(W_B + \frac{V_B\gamma_B}{\kappa}\right)^2 + V_B\left(\varepsilon_B + W_B\gamma_B\right)}{B_{\min}} \equiv b_r$$

$$\left\|\hat{V}\right\|_F > \frac{\frac{1}{4}\kappa\left(W_B + \frac{V_B\gamma_B}{\kappa}\right)^2 + V_B\left(\varepsilon_B + W_B\gamma_B\right)}{K_{v_{\min}}} \equiv b_V$$

$$\left\|\hat{W}\right\|_F > \frac{1}{2}\left(W_B + \frac{V_B\gamma_B}{\kappa}\right) + \sqrt{\frac{1}{4}\left(W_B + \frac{V_B\gamma_B}{\kappa}\right)^2 + \frac{V_B\left(\varepsilon_B + W_B\gamma_B\right)}{\kappa}} \equiv b_W$$

The filtered error can be as small as desired increasing the gain κ.

The weight tuning is carried out on-line causing an unsupervised version of the continuous-time backpropagation algorithm and the weights are tuned using the filtered error instead of the output of the plant. The weights $\hat{W}$ are initialised to zero and the weights $\hat{V}$ to the routing values, i.e. a matrix consisting of elements -1, 0 or 1. As a result, there is no preliminary off-line phase and in the first steps the controller behaves just as a PD-controller. The main benefit of this approach is that it is not necessary to choose initial weights to make stable the system.

4 Ilustrative Design and Simulation

The effectiveness of the neurocontroller is tested on a three-link anthropomorphic arm widely used in literature for illustration purposes. The dynamic equations are given in [1] and it is assumed that the links have uniform density so that the center of mass is located in the midpoint of the link. The dynamic parameters of the anthropomorfic robot are $m_1 = 90\ 10^{-3}\ Kg$, $m_2 = m_3 = 100\ 10^{-3}\ Kg$, $l_1 = 30\ mm$, $l_2 = l_3 = 40\ mm$, $\mathrm{I}_{zz} = 2.7\ 10^{-5} Kg\ m^2$, $r_m = 8\ mm$, $B = \begin{pmatrix} -1 & 1 & -1 \\ -1 & -1 & 1 \\ 1 & -1 & -1 \\ 1 & 1 & 1 \end{pmatrix}$ $8\ mm$ and a reduction ratio=1/12.

In order to design the FLNN controller, previously it is necessary to choose a basis set of activation functions $\phi(x)$. Generally, it is difficult to select a set of activation functions providing a basis, however it is possible to accomplish this problem by selecting scaling and shift parameters randomly. This kind of nets are called random vector functional link (RVFL) nets and were introduced by [7]. In this case the activation functions take the form $\phi(x) = \sigma\left(V^T x\right)$ where V is an arbitrary matrix of parameters. In our work, the system's own functions have been selected as basis functions providing a network referred to as parametric network, see [8], [9] and [10], or functional layer network,see [11].

The matrix of weights $W_m \in \mathbb{R}^{nL\times m}$ is upper bounded. The Frobenius norm depends on the numerical radius of the inertia matrix, the number of degrees of freedom and the lower bound of the activation functions $\phi_m(x)$. If all the pulleys have the same radius and the isotropy property is assumed for the transmission system, the next bound is obtained

$$\|W_m\|_F \leq 2n^2 r\,(M\,(q)) = 18\ 0.0011 = 0.0198 \tag{16}$$

For the adaptive parametric controller the adaptation gains were selected as $K_v = 10^{-3}\ I_{4\times4}, G_m = G_c = G_g = \Lambda = 10\ I_{4\times4},\ \kappa = 20,\ K_z = 10^{-2}\ I_{4\times4},\ W_B = 0.1,\ W_y = 10\ I_{4\times6},\ W_c = 10^{-4}\ I_{4\times17},\ W_m = 10^{-3}\ I_{4\times9}$

The desired trajectory is chosen as a periodic sinusoid with amplitude 1 and frequency 2π rad/seg. Thus, all the joints has been excited with sinusoid signals providing phase displacements of 0, $\frac{\pi}{2}$ and π rad. This provides bounded signals for position, velocity and acceleration. It is assumed that initially there is no knowledged about the system so that all the weights are initialized to zero. The system was simulated in SIMULINK using Runge-Kutta method with an integration fixed step of $\Delta t = 10^{-3}$ sec and a simulation range from 0 to 3 sec. The response of the controller with weight tuning (e.g. Theorem 1) appears in

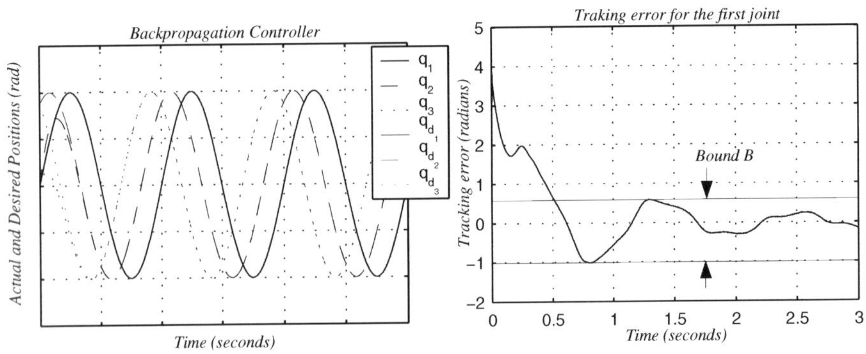

(a) (b)

Fig. 2. (a) Response of NN controller with backprop weight tuning. Actual and desired joint angles. (b) Representation of the tracking error for the first joint with $\kappa = 20$.

figure 2 (a). The figure 3 (a) shows the tracking errors for the three joints. The tracking error remains bounded for all time as it can be observed in the figure 2 (b), where the signal is swinging upwards and downwards inside a ball of radius less than 10^{-3} *rad*. It is possible to improve the tracking performance by increasing the gains of the controller. In 3 (b) the tracking error for the first joint has been plotted with $k = 20, 23, 26, 29, 33$. As it can be observed, varying the parameter k the error is bounded for all time. Therefore, it is concluded that the designed neurocontroller provides a good tracking of desired trajectories.The dynamic evolution of the weight estimations have been plotted in the figure 4, in a time range of $[0, 10]$ sec. The induced euclidean norm for matrices defined as $\|A\|_2 = \max_i \sqrt{\sigma_i\,(A)}$ where $\sigma_i\,(A)$ is a singular value of A has been used. As it can be observed all the estimations are bounded for all time. Indeed as shown

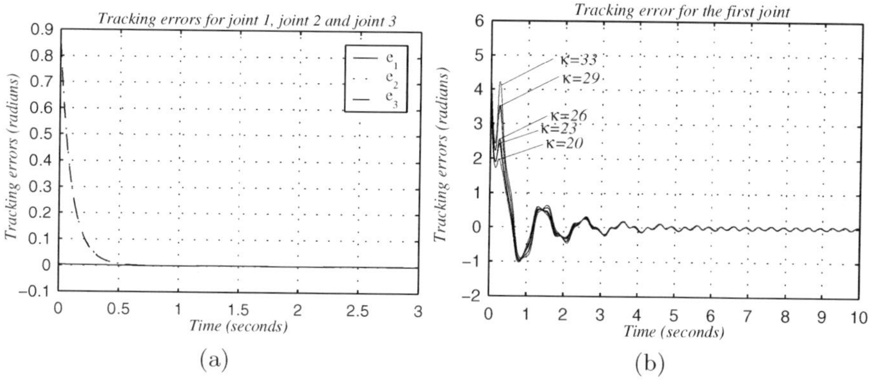

Fig. 3. (a) Response of NN controller with backprop weight tuning. Representation of tracking errors (b) Tracking error for the first joint with κ=20,23,26,29,33.

in the figure 4, the larger the design gain κ the larger the norm of the weights W. Yet, increasing the design parameter κ the convergence radius b_W decreases. This is a consequence of the dependence between the convergence radius b_W and κ. Indeed, the radius b_W tends asymptotically to W_B since $\lim_{\kappa \to \infty} b_W(\kappa) = \frac{1}{2}W_B + \frac{1}{2}W_B$.

5 Conclusion

The main advantage of this adaptive control law is that provides adequate robot manipulator performance as compared to other non-adaptive control schemes. Indeed, the tracking errors are much smaller than those obtained from non-adaptive control laws because of the learning mechanism. Another benefit of this neural network controller is that the models are valid for all the robot manipulators belonging to the same configuration class, i.e. having the same sequence of joint types and the same number of degrees of freedom.

A neural network controller based on the passivity properties of the robot manipulators has been developed to properly drive a mechanical system with a tendon-driven transmission system. The main problem of these tendon transmissions, according to the Morecki's property, is that of the different dimension of the tendon space and torque space resulting in a redundant system. A new control strategy has been developed in order to afford the problem of redundancy. Not only the dynamic parameters of the mechanical system such as masses or inertias are learned by the adaptive scheme but also all the dimensions of the spooler and drving pulleys are learned.

Both tuning and control laws were derived from a theorem that plays a central role in this paper guaranteeing good tracking and boundedness of the weights and signals. Lastly, the control law has been implemented and applied to an

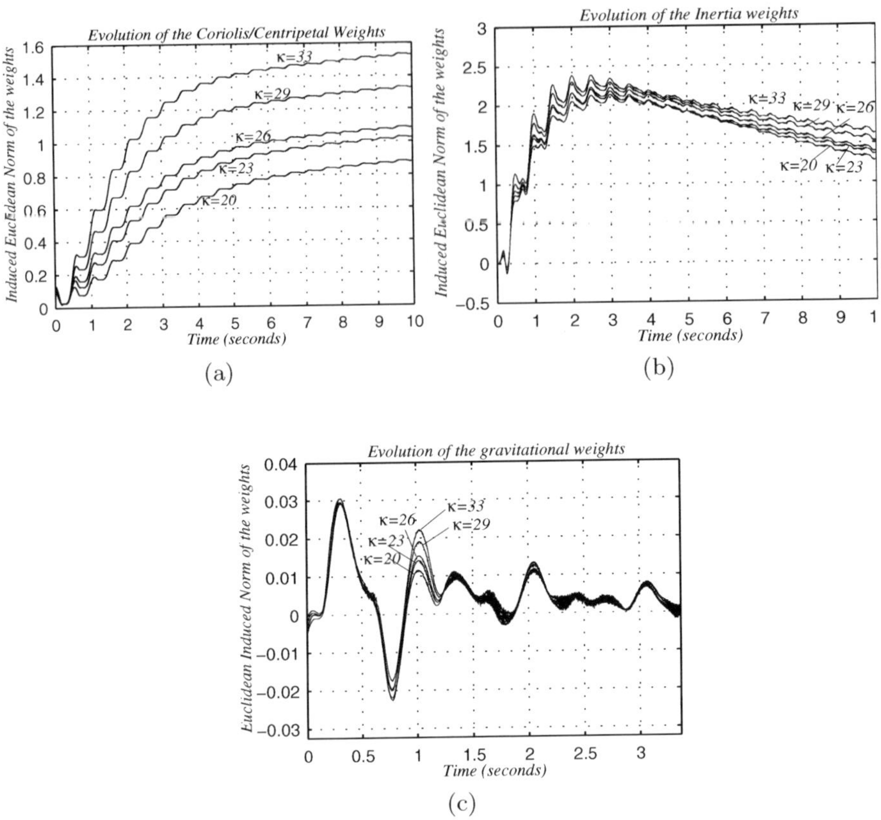

Fig. 4. (a) Evolution of the coriolis/centripetal weight estimations (b) Evolution of the inertia weight estimations. (c) Evolution of the gravitational weight estimations. The parameter κ varies in the range κ=20,23,26,29,33.

anthropomorfic finger that looks like the Standford/JPL finger with a tendon-driven transmission system. The reference signals were selected as sinoidal exciting signals to get experimental proof of the stability of the system. A sensibility analysis was made to show the influence of some design parameters in the convergence raidus of the weights and filtered error signal and then, getting proof of the assumptions and validity conditions of the theorem.

Acknowledgement

This work is supported by the project TIC2003-08164-C03-03. The authors would like to thank the Spanish Research Foundation (CICYT) by the support given to this project.

References

1. Lee, J.J.: Tendon-Driven Manipulators: Analysis, Sysnthesis and Control. PhD thesis, Harvard University and Systems Research Center (1991)
2. Morecki, A., Busko, Z., Gasztold, H., Jaworek, K.: Synthesis and control of the anthropomorphic two-handed manipulator, Proceedings of the 10th International Symposium on Industrial Robots (1980) 461–474
3. Lee, J.J., Tsai, L.W.: On the structural synthesis of tendon-driven manipulators having pseudo-triangular matrix. International Journal of Robotics Research **10** (1991) 255–262
4. Lee, J.J., Tsai, L.W.: Dynamic simulation of tendon-driven manipulators. Technical Report TR 91-53, Harvard (1991)
5. Horn, R.A., Johnson, C.R.: Topics in Matrix Analysis. Cambridge University Press (1999)
6. Lewis, F.L., Jagannathan, S., Yesildirek, A.: Neural Network Control of Robot Manipulators and Nonlinear Systems. Taylor and Francis Ltd. (1999)
7. Igelnik, B., Pao, Y.H.: Stochastic choice of basis functions in adaptive funtion approximation and the functional-link net. IEEE Trans. Neural Networks **6** (1995) 1320–1329
8. Ge, S.S., Hang, C.C., Woon, L.C.: Experimental studies of network-based adaptive control of robot manipulators,. Journal of the Institution of Engineers Singapore **37** (1997) 40–48
9. Ge, S.S., Hang, C.C.: Network modelling of rigid body robots. Proceedings of the 2nd Asian Control Conference (AsCC), Seoul, South Korea **3** (1997.) 251–254
10. Ge, S.S., Lee, T.H., Harris, C.J.: Adaptive Neural Network Control of Robotic Manipulators. World Scientific, London (1998)
11. Lewis, F.L., Liu, K., Yesildirek, A.: Neural net robot controller with guaranteed tracking performance. IEEE Transactions on Neural Networks **6** (1995) 703–716

Learning Multiple Models of Non-linear Dynamics for Control Under Varying Contexts

Georgios Petkos*, Marc Toussaint, and Sethu Vijayakumar

Institute of Perception, Action and Behaviour, School of Informatics
University of Edinburgh EH9 3JZ

Abstract. For stationary systems, efficient techniques for adaptive motor control exist which learn the system's inverse dynamics online and use this single model for control. However, in realistic domains the system dynamics often change depending on an external unobserved context, for instance the work load of the system or contact conditions with other objects. A solution to context-dependent control is to learn multiple inverse models for different contexts and to infer the current context by analyzing the experienced dynamics. Previous multiple model approaches have only been tested on linear systems. This paper presents an efficient multiple model approach for non-linear dynamics, which can bootstrap context separation from context-unlabeled data and realizes simultaneous online context estimation, control, and training of multiple inverse models. The approach formulates a consistent probabilistic model used to infer the unobserved context and uses Locally Weighted Projection Regression as an efficient online regressor which provides local confidence bounds estimates used for inference.

1 Introduction

Learning dynamics for control is essential in situations where analytical derivation of the plant dynamics is not feasible. This can be either due to the complexity of the system or due to lack of or inaccurate knowledge of the physical properties of the system being controlled. Adaptive control is an established research area that has offered a multitude of methods that can be used in such cases. However, the dynamics of the environment that the system has to interact with or even of the system itself are often changing in a rapid or discontinuous fashion. For example, a robot arm may be required to manipulate objects of different weights – an instantiation of control under multiple contexts. In these cases, classic adaptive control methods are inadequate since they result in large errors and instability during the period of adaptation. Furthermore, if the dynamics change back and forth, readapting everytime is a suboptimal and inefficient strategy.

Humans do not have difficulty controlling their limbs under different contexts. It has been suggested that they achieve this by using not just one model that is constantly adapted to new environments, but a set of models, each of which is appropriate for a different environment [1]. The key issue that needs to be resolved for this multiple model paradigm is that at any time the current context needs to be determined; this will be referred to as the context estimation problem and is central to this work. Context estimates are needed both during training and control, i.e., for deciding which model should be

* Supported by the Greek State Scholarships Foundation.

S. Kollias et al. (Eds.): ICANN 2006, Part I, LNCS 4131, pp. 898–907, 2006.

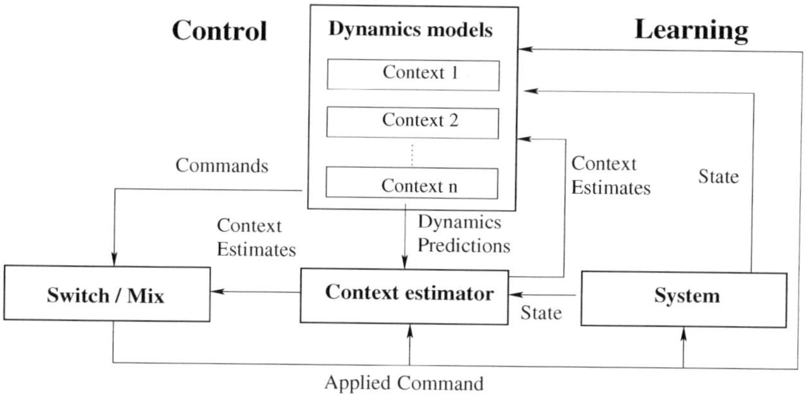

Fig. 1. Typical setup of a multiple model paradigm for control

used for control and which model should be trained with the data experienced. Biological systems (e.g. humans) estimate contexts using a variety of sensory information like vision or tactile input. In artificial systems though, the available sensory information may be much poorer and the context has to be estimated from the experienced dynamics only. Our approach will formulate a proper probabilistic model that represents the context as a latent switching variable. This model allows us to estimate the context online based only on the learned inverse models using a Markovian filtering. Further, an Expectation-Maximization procedure is used to bootstrap the distinction of contexts from context-unlabeled data.

There are some existing paradigms that implement the multiple model approach: Multiple Model Switching and Tuning (MMST) [4,5], Multiple Paired Forward and Inverse Models (MPFIM) [3] and Modular Selection and Identification for Control (MOSAIC) [2]. Fig. 1 shows the typical setup of a multiple model paradigm, where a set of different context dynamics models is maintained. In most existing approaches, the dynamics models for each context is a pair consisting of a forward and an inverse model. *Context estimation* is performed by comparing the observed dynamics of the system with the dynamics predicted by each context's forward model. For control purposes, one can either switch between commands predicted by the most likely context or mix them. Similarly, context estimates can be used for 'hard' or 'soft' assignment of data for training the most likely contexts. The most general of the mentioned paradigms is MOSAIC, which is an extension of MPFIM. MOSAIC uses mixing instead of switching, with the hope that more contexts can be handled with a smaller number of models. This seems plausible in the case of linear dynamics and indeed MOSAIC has been realized only for linear systems. Real robotic systems are highly non-linear, requiring the ability to learn online and adjust model complexity in a data-driven manner. Existing multiple model approaches are, therefore, not scalable. In this paper we present a non-linear multiple model approach to control based on an efficient non-linear online learning algorithm (LWPR) that addresses these requirements. To the best of our knowledge, this is the first multiple model study that manages to learn non-linear dynamics under multiple contexts with online separation of contextual data.

2 Adaptive Non-linear Control with LWPR

Let us first consider the single context scenario of learning the dynamics of a system (e.g., a robot) and using them for control. At time step t, let Θ_t be the state of the system (which include the position and velocity components) and τ_t the control signal. A deterministic forward model f describes the discrete-time system dynamics as

$$\Theta_{t+1} = f(\Theta_t, \tau_t) . \tag{1}$$

Learning a forward model f of the dynamics is useful for predicting the behavior of the system. However, for control purposes, an inverse model is needed. The inverse model g maps from transitions between states to the control signal that is needed to achieve this transition:

$$\tau_t = g(\Theta_t, \Theta_{t+1}) . \tag{2}$$

A probabilistic graphical model representation of the forward and inverse model is shown in Fig. 2(a) and Fig. 2(c), respectively.

Idealistically, an accurate inverse model can be used to exactly follow a sequence of transitions that form a desired trajectory of the system. However, given only an approximate inverse model, the error in following the trajectory may accumulate and become unacceptably large. A standard approach for control with an approximate inverse model is to combine it with a conventional linear feedback controller that counteracts the deviation from the desired trajectory. Given a desired trajectory $\Theta^*_{1:T}$ and the true state Θ_t, the composite control command at time t is

$$\tau_t = g(\Theta^*_t, \Theta^*_{t+1}) + A\,(\Theta^*_t - \Theta_t) , \tag{3}$$

where A is a gain matrix. We will use this composite control with gains based on the Proportional Derivative (PD) control law. One effect of the composite control approach is that the more accurate the inverse model g, the smaller are the errors and the error-correcting PD control signals. Thus, the total amount of feedback control is a measure of the accuracy of the inverse 'predictive' model.

To learn the inverse dynamics we need a *non-linear, online* regression technique which also provides error bounds that we may use for context separation. We use the Locally Weighted Projection Regression (LWPR) [6] – an algorithm which is extremely robust and efficient for incremental learning of non-linear models in high dimensions. A LWPR model consists of a set of local linear models that come paired with a kernel that defines the locality of the model. For a given input x, the kernel of the k-th local model determines a weighting $w_k(x)$ while the local linear model predicts an output $\psi_k(x)$. The combined prediction of LWPR is

$$\phi(x) = \frac{1}{W} \sum_k w_k(x)\, \psi_k(x) , \quad W = \sum_k w_k(x) . \tag{4}$$

Each locality kernel $w_k(x)$ has a parametric Gaussian form and the distance metric is adapted during learning in a data driven manner. The local models are trained using an online variant of Partial Least Squares using the collected sufficient statistics. LWPR is incremental and non-parametric in the sense that new local models are added online

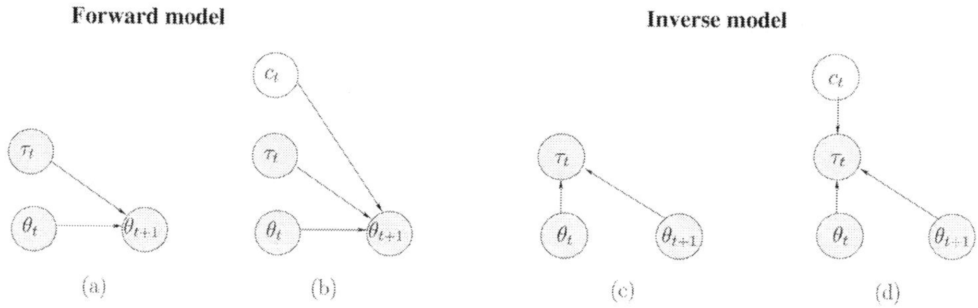

Fig. 2. Graphical model representation of the: (a) Forward model (c) Inverse model and (b,d) their respective context augmented models

on an as-needed basis when training proceeds and new areas of the input domain are explored. The update of LWPR scales with $O(n)$, n being the input dimensionality [7].

The role of LWPR in the probabilistic inverse model of Fig. 2 can be summarized in the equation:

$$P(\tau \mid \Theta_{t+1}, \Theta_t) = \mathcal{N}(\phi(\Theta_{t+1}, \Theta_t),\ \sigma(\Theta_{t+1}, \Theta_t)), \tag{5}$$

whose $\phi(\Theta_{t+1}, \Theta_t)$ is a learned LWPR regression mapping desired state transitions to torques. Here, we have two options for choosing the variance: (1) we can assume a fixed noise level independent of the context and the input; (2) we can use the confidence bounds provided by each LWPR model which also depends on the current input (Θ_{t+1}, Θ_t). We will test both cases in our experiments. Please see [6] for more details on LWPR and the input dependent variance estimate.

3 Learning Multiple Models for Multiple Contexts

In the multiple context scenario, we assume that instead of having a single forward and inverse dynamics (Fig. 2(a,c)), the dynamics depend on an unobserved random variable c_t, the context. Fig. 2(b,d) illustrates this situation as augmented graphical models for the forward and inverse models. We assume a discrete context variable and maintain separate LWPR models to represent the inverse dynamics for each context. Thus, Eq. (5) becomes:

$$P(\tau \mid \Theta_{t+1}, \Theta_t, c_t = i) = \mathcal{N}(\phi_i(\Theta_{t+1}, \Theta_t),\ \sigma_i(\Theta_{t+1}, \Theta_t)) \,. \tag{6}$$

The problem we face in the context of adaptive online control is twofold: (1) Given a batch of yet unlabeled data and a set of yet untrained inverse models, we have to bootstrap the specialization of inverse models to different parts of the data while at the same time associating different data points to different contexts– we call this problem *data separation*; (2) Given a set of already trained inverse models and previous observations, we have to estimate the current context in order to choose the right inverse model in calculating the control signal– we call this problem *context estimation*. These problems are very closely related. We first address the simpler context estimation problem before discussing data separation.

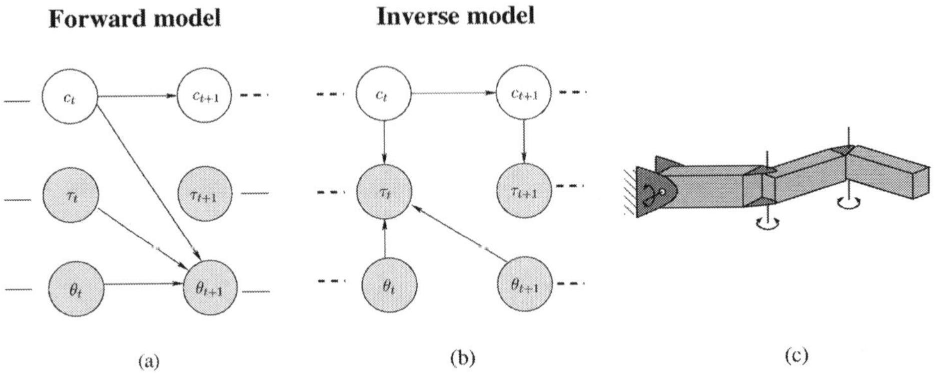

Fig. 3. Multiple model with temporal contextual dependencies using: (a) forward model or (b) inverse model for context estimation. (c) Schematic of the simulated 3-link robot arm

3.1 Context Estimation

In general, context estimation with a given set of models is performed comparing the predictions of each model with the observed dynamics. Usually this is done by comparing a set of trained forward models with the observed dynamics. However, the predictions of inverse models can equally be compared with the observed dynamics and thus, there is no need to learn additional forward models. Our viewpoint is that at each time step t we "observe" a state transition and an applied torque signal summarized in the triplet $(\Theta_t, \Theta_{t+1}, \tau_t)$, i.e., we have access to the true applied control command (which was generated via composite control) as part of the observation. To estimate the latent context variable c_t (without yet exploiting the temporal dependency) we can compute $P(c_t \mid \Theta_t, \Theta_{t+1}, \tau_t)$, i.e., the probability of being in a context given the observed transition between two consecutive states and the command that resulted in this transition. Using Bayes rule, we get

$$P(c_t = i \mid \Theta_t, \Theta_{t+1}, \tau_t) = P(\tau_t \mid c_t = i, \Theta_t, \Theta_{t+1}) \frac{P(c_t = i)}{P(\tau_t \mid \Theta_t, \Theta_{t+1})} . \tag{7}$$

Here, we used $P(c_t = i \mid \Theta_t, \Theta_{t+1}) = P(c_t = i)$, which is the context prior. Assuming a uniform prior, the RHS quotient is a normalization factor independent of the context i. Hence, the responsibility $P(c_t = i \mid \Theta_t, \Theta_{t+1}, \tau_t)$ is proportional to the i-th model likelihood (Eq. (6)).

It is straight-forward to extend this to take a Markovian dependency between contexts into account: intuitively, we would expect that in most practical cases, the context would stay the same most of the time and switch only occasionally. For instance, in our current experiments we apply control signals at 100Hz and we expect that the frequency of context switches will be much lower. Thus, including the temporal dependency between contexts $P(c_{t+1} \mid c_t)$, the graphical models in Fig. 2(b,d) can be reformulated as the Dynamic Bayesian Networks shown in Fig. 3(a,b). Application of standard Hidden Markov Model (HMM) techniques is straightforward by using Eq. (7) as the observation likelihood in the HMM, given the hidden state $c_t = i$.

A low transition probability penalizes too frequent transitions and using smoothing or Viterbi alignment produces more stable context estimates. In the experiments, we will assume a fixed transition matrix $P(c_t = j \mid c_t = i)$ with high value .999 for $i = j$ and .001 otherwise and use the HMM model only for filtering or smoothing, depending on whether we investigate an online or batch estimation scenario, respectively.

3.2 Data Separation

In existing multiple model approaches, separation of data for learning happens online. The predictions of the models are compared with the observed behaviour of the system to give context estimates and train the models online. However, to get these context estimates we need a mechanism for getting relatively accurate (initial) models to bootstrap the context estimation procedure. Most of the existing multiple model paradigms do not give a satisfying answer to this issue. MMST assumes that relatively good models are available from the beginning, whereas MPFIM does not address this issue at all.

The problem of bootstrapping the context separation from context-unlabeled data is very similar to clustering problems using mixture of Gaussians. In fact, the context variable can be interpreted as a latent mixture indicator and each inverse model contributes a mixture component to give rise to the mixture model of the form $P(\tau_t \mid \Theta_t, \Theta_{t+1}) = \sum_i P(\tau_t \mid \Theta_t, \Theta_{t+1}, c_t = i)\; P(c_t = i)$. Clustering with mixtures of Gaussians is usually trained using Expectation-Maximization (EM), where initially the data are labeled with random responsibilities (are assigned randomly to the different mixture components). Then every mixture component is trained on its assigned (weighted) data (M-step) and afterwards the responsibilities for each data point is recomputed by setting them proportional to the likelihoods for each mixture component (E-step). Iterating this procedure, each mixture component will specialize on different parts of the data and the responsibilities encode the learned cluster assignments.

We will apply a common variant of the EM-algorithm where responsibilities are computed greedily, i.e., where the data is hard assigned to the mixture component with maximal likelihood instead of weighted continuously with the component's likelihood in the M-step. In our case, the likelihood of a data triplet $(\Theta_t, \Theta_{t+1}, \tau_t)$ under the ith inverse model is $P(\tau_t \mid \Theta_t, \Theta_{t+1}, c_t = i)$, which is a Gaussian with either fixed variance or the variance given by LWPR's confidence bounds. This approach is similar to MOSAIC's approach to data separation except that it is based on the inverse models, accounts for the possibility of non-linear models, and allows us to use the correct confidence bounds predicted by LWPR.

4 Experiments

The methods proposed earlier were tested on a simulated[1] 3 joint arm, with 3 degrees of freedom (see Fig. 3(c)). The first joint allows up and down movements and the next two allow left and right movements. The target trajectories for the arm were a superposition of different phase-shifted sinusoidal trajectories for each joint:

$$\theta_i^* = a_i \cos(\alpha_i \frac{2\pi}{T} t) + b_i \cos(\beta_i \frac{2\pi}{T} t) \; , \tag{8}$$

[1] Robot arm simulation modeled in dynamical physics engine ODE/OpenGL.

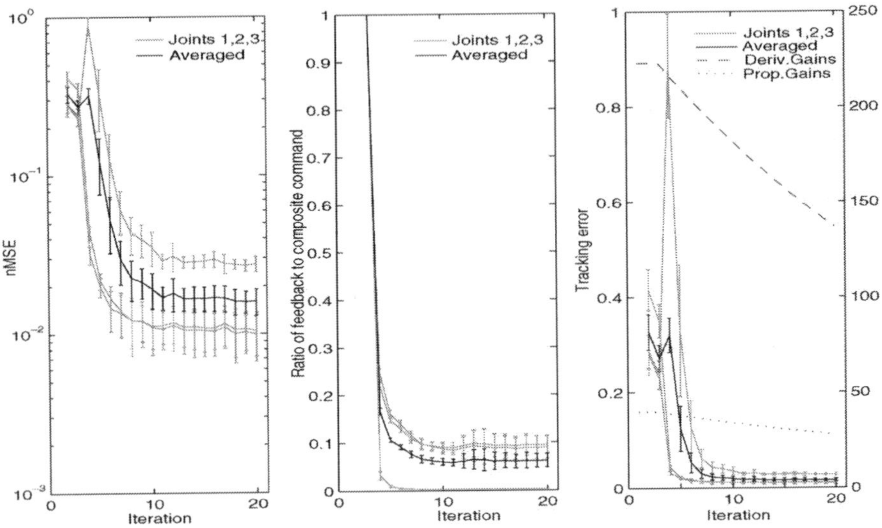

Fig. 4. Control performance of online trained LWPR on a single context over the training cycles. Left: normalized MSE on the test data. Middle: contribution of the error-correcting feedback PD control. Right: tracking error under decreasing PD gains.

where $T = 4000$ is the total length of the target trajectory, $a_i, b_i \in [-1, 1]$ are different amplitudes and $\alpha_i, \beta_i \in \{1, .., 15\}$ parameterize different frequencies. Different contexts are simulated by changing the weight of the third body of the arm. This is equivalent to varying work loads held by the arm.

4.1 Learning Single Context Dynamics and Using Them for Control

We will first demonstrate that LWPR can learn an accurate inverse model of the arm dynamics online and use it for control. Training was repeated independently for six different contexts. Twenty iterations of the trajectory were executed. In the first 3 iterations, a pure PD controller is used, whereas, after that a composite controller with the model being learnt is used. Every second sample of the dynamics experienced is used for training the inverse model online and every other sample is kept to test the accuracy of the inverse model. Fig. 4 (left) shows how the normalised mean square error (nMSE) on the test data drops as training proceeds through the 20 iterations, indeed, converging to very low nMSE for all joints.

The accuracy of the inverse model learned can also be judged based on the contribution of the feedback command to the total composite command. The smaller the contribution of the feedback command, the more accurate the inverse model learnt is. The average contribution of the feedback command through the 20 iterations can be seen in Fig. 4(middle). Already from the fourth iteration, when we switch from PD control to composite control, the contribution is quite low and drops further – in accordance to the behaviour of the nMSE. In Fig. 4(right), we can also see how the tracking error

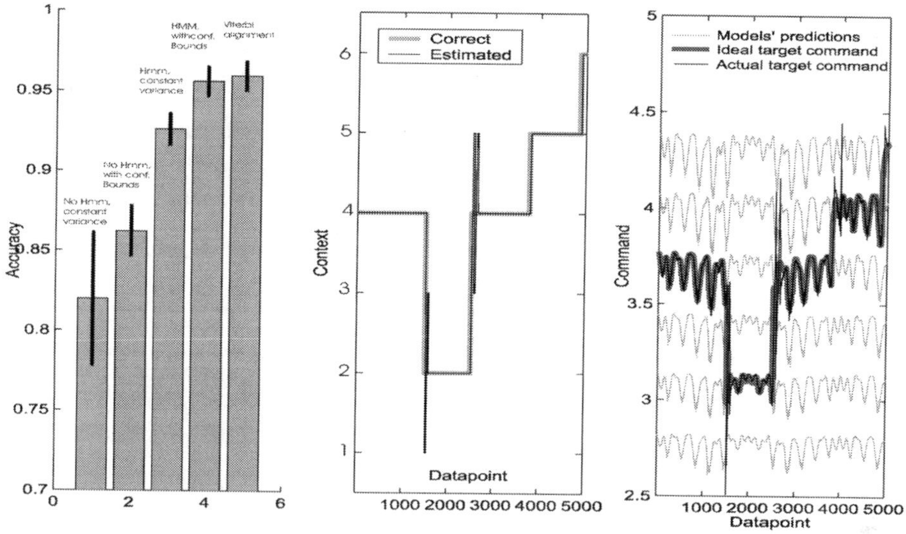

Fig. 5. Online context estimation and control in the case of six different contexts. Left: Context estimation accuracy using different estimation methods. Middle: example of random context switches and its estimate using HMM filtering over time. Right: The inverse model predictions of the six contexts along with the ideal and actual generated target command

decreases as the model becomes more accurate while, at the same time, we decrease the gains of the feedback controller.

4.2 Experiments with Context Estimation

The context estimation methods described in Section 3.1 were used for online estimation and control with the six contexts learnt. Random switches between the six contexts were performed in the simulation, where at every time step we switch to a random context with probability .001 and stay in the current context otherwise. The context estimates were used online for selecting the model that will provide the feed-forward commands.

We have two classes of experiments, one is where we are not using HMM filtering of the contextual variable and the other is where we use it. Also, we have two choices for the variance of the observation model, one is where we use a constant (found empirically) and the other is where we use the more principled confidence bounds provided by LWPR. The simulation was run for 10 iterations. The percentage of accurate online context estimates for the four cases along with offline Viterbi alignment can be seen in the Fig. 5(left).

Fig. 5(middle) gives an example of how the best context estimation method that we have, the HMM filtering using LWPR's confidence bounds, performs when used for online context estimation and control. Sometimes the context estimation lags behind a few time steps when there are context switches, which is a natural effect of online filtering (as opposed to retrospect smoothing).

The performance of online context estimation and control is close to the control performance we achieved for the single context displayed in Fig. 4. Using the HMM

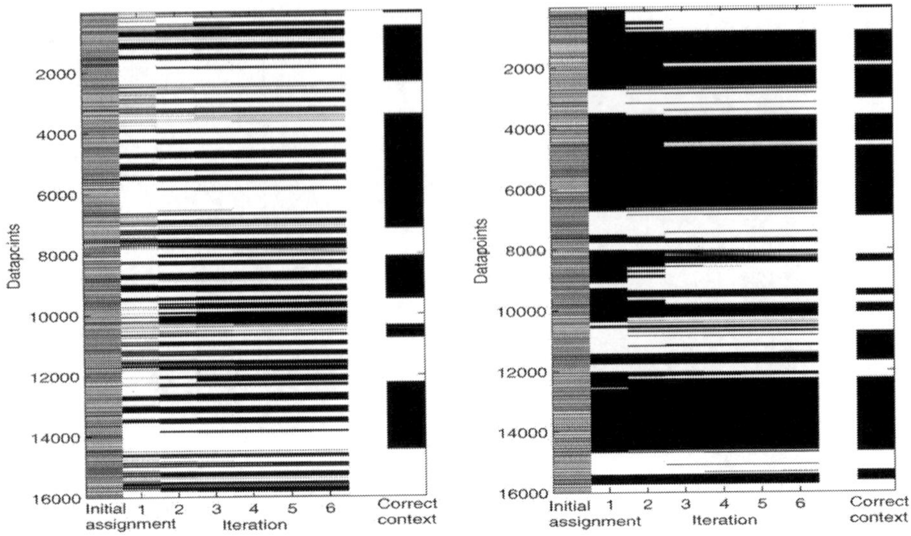

Fig. 6. The evolution of the data separation from unlabeled data over six iterations of the EM-procedure. Left: without exploiting temporal modelling of the context variable. Right: using a Viterbi alignment according to the temporal modelling. Both methods use LWPR's confidence bounds as local variance estimate. The first column displays the initial random assignment of datapoints to contexts. The last column displays the correct context for each datapoint.

filtering based on LWPR's confidence bounds, the average tracking error over the 10 cycles was 0.0019 and the ratio of feedback PD control was 0.074.

4.3 Experiments with Data Separation

Finally, we investigate the bootstrapping of data separation from unlabeled data. Here, when generating the data, we switched between two different contexts (work loads) with probability .001 at each time step. We first collected a batch of context-unlabeled data from 4 cycles through the target trajectory where the arm was controlled by pure feedback PD control. The EM procedure for data separation (Section 3.2) was tested on this data with and without temporal modelling (always using LWPR's confidence bounds as a basis). In the temporal case, Viterbi alignment was used to assign datapoints to contexts rather than filtered estimates. Fig. 6 compares the evolution of the data separation for the two methods over six EM-iterations. Using the temporal context performs much better, i.e., 84% of the datapoints were assigned to the correct context.

The bootstrapping of the context separation from unlabeled data gives rise to two separate inverse models for the two different contexts. To further improve these models, we then used them for online context estimation and control, just as investigated in the previous Section, for another 12 cycles through the target trajectory. Simultaneously, the context estimates were used for selecting data for further training of the models. The accuracy of context estimation was 88% while the tracking error was 0.0051 and the ratio of feedback PD control was 0.23. The errors are slightly higher than in the case where models were trained using labeled data, but this is satisfying considering the fact that we started with unlabeled data.

5 Discussion

In this paper we presented an efficient multiple model paradigm for the general case of non-linear control. The approach is based on a probabilistic model of multiple-context dynamics, using LWPR as an efficient online regressor for each inverse model. We have demonstrated that it is possible to bootstrap multiple models of non-linear dynamics from context-unlabeled data and use them for simultaneous online context estimation, control, and training.

In comparison to previous multiple model approaches, most notably MOSAIC, our approach is the first to handle the case of non-linear dynamics. Further, we showed that it is unnecessary to maintain pairs of forward and inverse models. Context estimation can more efficiently be based solely on the learned inverse models for each context. We have seen that including a Markovian model of context switching greatly enhances the context estimation performance. If additional knowledge about the context is available, for instance, if it is related to sensory information, one can easily extend our framework by augmenting the likelihood term in the Markovian model.

An issue yet unaddressed by any existing method is that of determining the number of separate contexts based on data only, if it is not known a priori. As detailed in Section 3.2, our formulation of data separation is very similar to that of clustering using mixture of Gaussians. Hence, existing techniques for determining the necessary number of clusters in mixtures of Gaussians literature can directly be exploited. More specifically, a common approach is to incrementally add new mixture components when the new data cannot, with sufficient likelihood, be explained with existing mixture components [8]. This can also be realized online, which will be the subject of future research to extend the presented approach.

References

1. Imamizu H. Osu R. Yoshioka T. Flanagan R., Nakano E. and Kawato M. Composition and decomposition of internal models in motor learning under altered kinematic and dynamic environments. *The Journal of Neuroscience*, 19, 1999.
2. Wolpert D. M. Haruno M. and Kawato M. Mosaic model for sensorimotor learning and control. *Neural Computation*, 13:2201–2220, 2001.
3. Wolpert D. M. and Kawato M. Multiple paired forward and inverse models for motor control. *Neural Networks*, 11:1317–1329, 1998.
4. K. S. Narendra and J. Balakrishnan. Adaptive control using multiple models. *IEEE Transactions in automatic control*, 42:171–187, 1997.
5. K. S. Narendra and C. Xiang. Adaptive control of discrete-time systems using multiple models. *IEEE Transactions in automatic control*, 45:1669–1686, 2000.
6. Aaron D'Souza Sethu Vijayakumar and Stefan Schaal. Incremental online learning in high dimensions. *Neural Computation*, 17:2602–2634, 2005.
7. Tomohiro Shibata Jorg Conradt Sethu Vijayakumar, Aaron D'Souza and Stefan Schaal. Statistical learning for humanoid robots. *Autonomous Robots*, 12:55–69, 2002.
8. J. J. Verbeek, N. Vlassis, and B. J. A. Kröse. Efficient greedy learning of gaussian mixture models. *Neural Computation*, 15(2):469–485, 2003.

A Study on Optimal Configuration for the Mobile Manipulator: Using Weight Value and Mobility

Jin-Gu Kang[1] and Kwan-Houng Lee[2]

[1] Dept. Visual Broadcastion Media Keukdong College DanPyung-Ri 154-1,Gamgog Myun, Eumsung Gun Chungbuk, 467-900, Republic of Korea,
jgukang@kdc.ac.kr

[2] School of Electronics & Information Engineering, Cheongju University, Naedok-Dong Sangdang-Gu Cheongju-City, Chungbuk, 360-764, Republic of Korea,
khlee368@cju.ac.kr

Abstract. A Mobile Manipulator is redundant by itself. Using it's redundant freedom, a mobile manipulator can perform various task. In this paper, to improve task execution efficiency utilizing the redundancy, optimal configurations of the mobile manipulator are maintained while it is moving to a new task point. And using a cost function for optimality defined as a combination of the square errors of the desired and actual configurations of the mobile robot and of the task robot, the job which the mobile manipulator performs is optimized. Here, The proposed algorithm is experimentally verified and discussed with a mobile manipulator, PURL-II.

Keywords: Weight Value, Mobility, Gradient Method, Mobile robot.

1 Introduction

While a mobile robot can expand the size of the work space but does no work, a vertical multi-joints robot or manipulator can't move but it can do work. And at present, there has been a lot of research on the redundant robot which has more degrees of freedom than non-combination robots in the given work space, so it can have optimal position and optimized job performance[1][2]. While there has been a lot of work done on the control for both mobile robot navigation and the fixed manipulator motion, there are few reports on the cooperative control for a robot with movement and manipulation ability[3]. Different from the fixed redundant robot, the mobile manipulator has the characteristic that with respect to the given working environments, it has the merits of abnormal movement avoidance, collision avoidance, efficient application of the corresponding mechanical parts and improvement of adjustment. Because of these characteristics, it is desirable that one uses the mobile manipulator with the transportation ability and dexterous handling in difficult working environments[4]. This paper explains the mobile manipulator PURL-II which is a combination in series of a mobile robot that has 3 degrees of freedom for efficient job accomplishment and a task robot that has 5 degrees of

S. Kollias et al. (Eds.): ICANN 2006, Part I, LNCS 4131, pp. 908 – 917, 2006.

freedom. We have analyzed the kinematics and inverse kinematics of each robot to define the 'Mobility' of the mobile robot as the most important feature of the mobile manipulator. We researched the optimal position and movement of robot so that the combined robot can perform the task with minimal joint displacement and adjust the weighting value using the this 'Mobility'. When the mobile robot performed the job with the cooperation of the task robot, we investigated the optimizing criteria of the task using the 'Gradient Method' to minimize the movement of the whole robots. The results that we acquired by implementing the proposed algorithm through computer simulation and the experiment using PURL-II are demonstrated.

2 Mobile Manipulator

2.1 Kinematics Analysis of the Mobile Manipulator

Each robot which is designed to accomplish each independent objective concurrently should perform its corresponding movement to complete the task. The trajectory is needed for kinematics analysis of the whole system, so that we can make the combination robot perform the task efficiently using the redundant degree of freedom generated by the combination of the two robots[5][6]. From Fig. 1, We can see the Cartesian coordinate of the implemented mobile/task robot system and the link coordinate system of each joint in space. This system is an independent mobile manipulator without wire. The vector $\dot{q}$ of the whole system joint variables can be defined $\dot{q}_t = [\dot{q}_{t1} \quad \dot{q}_{t2} \quad \dot{q}_{t3} \quad \dot{q}_{t4} \quad \dot{q}_{t5}]$ and $\dot{q}_m = [\dot{q}_{m6} \quad \dot{q}_{m7} \quad \dot{q}_{m8} \quad \dot{q}_{m9}]$ that represents the joint variable vector of the task robot. That is shown as

$$\dot{q} = \begin{bmatrix} \dot{q}_t \\ \dot{q}_m \end{bmatrix} = [\dot{q}_{t1} \quad \dot{q}_{t2} \quad \dot{q}_{t3} \quad \dot{q}_{t4} \quad \dot{q}_{t5} \quad \dot{q}_{m6} \quad \dot{q}_{m7} \quad \dot{q}_{m8} \quad \dot{q}_{m9}]^T \tag{1}$$

The linear velocity, and angular velocity of mobile robot in Cartesian space with respect to the fixed frame of the world frame can be expressed as (2).

$$^0\dot{P}_m = \begin{bmatrix} ^0V_m \\ ^0\omega_m \end{bmatrix} = \begin{bmatrix} ^0J_{m,v} \\ ^0J_{m,\omega} \end{bmatrix} \dot{q}_m = {}^0J_m \dot{q}_m \tag{2}$$

In view of Fig. 1, let us represent the Jacobian of vector $\dot{q}_t$ (task robot joint variable) with respect to frame (1). These results are shown in (3) as follows[12] .

$$^m\dot{P}_t = \begin{bmatrix} ^mV_t \\ ^m\omega_t \end{bmatrix} = \begin{bmatrix} ^mJ_{t,v} \\ ^mJ_{t,\omega} \end{bmatrix} \dot{q}_t = {}^mJ_t \dot{q}_t \tag{3}$$

Given the Jacobians, ^0J_m and ^mJ_t, for each robot, if we express the Jacobian of the mobile manipulator as ^0J_t ; the linear velocity. Then angular velocity $^0\dot{P}_t = [^0V_t \quad ^0\omega_t]^T$ from the end-effector to the world frame is represented as (4).

$$
\begin{aligned}
{}^{0}\dot{P}_{t} &= \begin{bmatrix} {}^{0}V_{t} \\ {}^{0}\omega_{t} \end{bmatrix} = \begin{bmatrix} {}^{0}V_{m} \\ {}^{0}\omega_{m} \end{bmatrix} + \begin{bmatrix} {}^{0}\omega_{m} + {}^{0}R_{1}\,{}^{1}V_{t} \\ {}^{0}R_{1}\,{}^{1}\omega_{t} \end{bmatrix} \\
&= {}^{0}J_{m}\dot{q}_{m} + {}^{0}J_{t}\dot{q}_{t} = \begin{bmatrix} {}^{0}J_{m} & {}^{0}J_{t} \end{bmatrix} \begin{bmatrix} \dot{q}_{m} \\ \dot{q}_{t} \end{bmatrix}
\end{aligned} \tag{4}
$$

Here ${}^{0}R_{1}$ is rotational transformation from world frame to the base frame of the task robot. Namely, in view of (2)-(4), the movements of mobile robot and task robot are involved with the movement of end-effector.

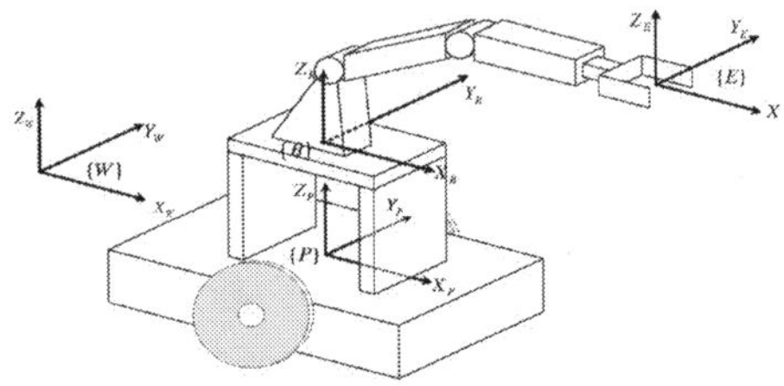

Fig. 1. Coordinate system of the mobile manipulator

3 Algorithm for System Application

3.1 Task Planning for Minimal Movement

Because the position of base frame of task robot varies according to the movement of mobile robot, through inverse kinematics, the task planning has many solutions with respect to the robot movement. and we must find the accurate solution to satisfy both the optimal accomplishment of the task and the efficient completion of the task.

In this paper, we have the objective of minimization of movement of the whole robot in performing the task, so we express the vector for mobile manipulator states as (5).

$$
q = \begin{bmatrix} q_{m} \\ q_{t} \end{bmatrix} \tag{5}
$$

where $q_m = [x_m \quad y_m \quad z_m \quad \theta_m]^T$ and $q_t = [\theta_1 \quad \theta_2 \quad \theta_3 \quad \theta_4 \quad \theta_5]^T$ .Here, q is the vector for the mobile manipulator and consists of q_m representing the position and direction of mobile robot in Cartesian space and q_t, the joint variable to each n link of the task robot. Now to plan the task to minimize the whole movement of mobile manipulators, a cost function, L, is defined as

$$
\begin{aligned}
L &= \Delta q^{T} \Delta q = (q_{f} - q_{i})^{T}(q_{f} - q_{i}) \\
&= (q_{m,f} - q_{m,i})^{T}(q_{m,f} - q_{m,i}) + (q_{t,f} - q_{t,i})^{T}(q_{t,f} - q_{t,i})
\end{aligned} \tag{6}
$$

Here, $q_i = [q_{m,i} \quad q_{t,i}]^T$ represents the initial states of the mobile manipulator, and $q_f = [q_{m,f} \quad q_{t,f}]^T$ represents the final states after having accomplished the task. In the final states, the end-effector of the task robot must be placed at the desired position $X_{t,d}$. For that, equation (7) must be satisfied. In (7), we denote as $R(\theta_{m,f})$ and $f(q_{t,f})$, respectively, the rotational transformation to $X-Y$ plane and kinematics equation of task robot[7].

$$X_{t,d} = R(\theta_{m,f}) f(q_{t,f}) + X_{m,f} \tag{7}$$

where $X_{t,d}$ represents the desired position of task robot, and $X_{m,f}$ is the final position of mobile robot. We can express the final position of the mobile robot $X_{m,f}$ as the function of the desired coordinate $X_{t,d}$, joint variables $\theta_{m,f}$ and $q_{t,f}$, then the cost function that represents the robot movement is expressed as the $n\times 1$ space function of $\theta_{m,f}$ and $q_{t,f}$ as (8).

$$\begin{aligned} L = & \{X_{t,d} - R(\theta_{m,f}) f(q_{t,f}) - X_{m,i}\}^T \{X_{t,d} - R(\theta_{m,f}) f(q_{t,f}) - X_{m,i}\} \\ & + \{q_{t,f} - q_{t,i}\}^T \{q_{t,f} - q_{t,i}\} \end{aligned} \tag{8}$$

In the equation (8), $\theta_{m,f}$ and $q_{t,f}$ which minimize the cost function L must satisfy the condition in (9).

$$\nabla L = \begin{bmatrix} \dfrac{\partial L}{\partial \theta_{m,f}} \\ \dfrac{\partial L}{\partial q_{t,f}} \end{bmatrix} = 0 \tag{9}$$

Because the cost function is nonlinear, it is difficult to find analytically the optimum solution that satisfies (9). So in this papers, we find the solution through the numeric analysis using the gradient method described by (10) .

$$\begin{bmatrix} \theta_{m,f(k+1)} \\ q_{t,f(k+1)} \end{bmatrix} = \begin{bmatrix} \theta_{m,f(k)} \\ q_{t,f(k)} \end{bmatrix} - \eta \, \nabla L \Big|_{\theta_{m,f(k)},\, q_{t,f(k)}} \tag{10}$$

This recursive process will stop, when $\|\nabla L\| < \varepsilon \approx 0$. That is, $\theta_{m,f(k)}$ and $q_{t,f(k)}$ are optimum solutions. Through the optimum solutions of $\theta_{m,f}$ and $q_{t,f}$ the final robot state q_f can be calculated as (11).

$$q_f = \begin{bmatrix} q_{m,f} \\ q_{t,f} \end{bmatrix} = \begin{bmatrix} X_{t,d} - R(\theta_{m,f}) f(q_{t,f}) \\ q_{t,f} \end{bmatrix} \tag{11}$$

There are several efficient searching algorithms. However, the simple gradient method is applied for this case.

3.2 Mobility of Mobile Robot

In this research, we define "mobility of the mobile robot" as the amount of movement of the mobile robot when the input magnitude of the wheel velocity is unity. That is, the mobility is defined as the corresponding quality of movement in any direction[8]. The mobile robot used in this research does move and rotate because each wheel is rotated independently under the control. The robot satisfies (12) with remarked kinematics by denoting left, right wheel velocities ($\dot{q}_{m,l}, \dot{q}_{m,r}$) and linear velocity and angular velocity (v_m, ω).

$$v_m = r\frac{\dot{q}_{m,l} + \dot{q}_{m,r}}{2} \tag{12a}$$

$$\omega = \frac{r}{l}\frac{\dot{q}_{m,l} - \dot{q}_{m,r}}{2} \tag{12b}$$

Rewriting (12a), (12b), we get (13a) and (13b).

$$\dot{q}_{m,r} = \frac{v_m + \omega l}{r} \tag{13a}$$

$$\dot{q}_{m,l} = \frac{v_m - \omega l}{r} \tag{13b}$$

Mobility is the output to input ratio with a unity vector, $\|v_m\| = 1$, or $\dot{q}^2_{m,l} + \dot{q}^2_{m,r} = 1$ and the mobility v_m in any angular velocity ω is calculated by (14).

$$v_m = r\sqrt{\frac{1}{2} - \omega^2\frac{l^2}{r^2}} \tag{14}$$

When the mobile robot has the velocity of unity norm, the mobility of mobile robot is represented as Fig. 2. It shows that the output, v and ω in workspace for all direction inputs that are variance of robot direction and movement. For any input, the direction of maximum movement is current robot direction when the velocities of two wheels are same[9]. At the situation, there does not occur any angular movement of the robot.

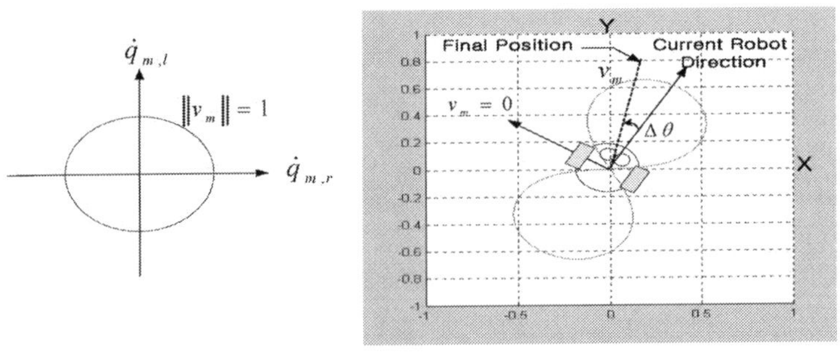

Fig. 2. Motion generation efficiency

3.3 Assigning of Weighting Value Using Mobility

From the mobility, we can know the mobility of robot in any direction, and the adaptability to a given task in the present posture of mobile robot. If the present posture of mobile robot is adaptable to the task, that is, the mobility is large to a certain direction. we impose the lower weighting value on the term in the cost function of (15) to assign large amount of movement to the mobile along the direction. If not, by imposing the higher weighting value on the term we can make the movement of mobile robot small. Equation (15) represents the cost function with weighting value

$$L=\{X_{t,d}-R(\theta_{m,f})f(q_{t,f})-X_{m,i}\}^T W_m\{X_{t,d}-R(\theta_{m,f})f(q_{t,f})-X_{m,i}\} \\ +\{q_{t,f}-q_{t,i}\}^T W_t\{q_{t,f}-q_{t,i}\} \tag{15}$$

Here, W_m and W_t are weighting matrices imposed on the movement of the mobile robot and task robot, respectively. In the cost function, the mobility of mobile robot is expressed in the Cartesian coordinate space, so the weighting matrix W_m of the mobile robot must be applied. after decomposing each component to each axis in Cartesian coordinate system as shown in Fig. 3 and is represented as (16).

$$W_m=\begin{bmatrix}\omega_x & 0 & 0 & 0\\ 0 & \omega_y & 0 & 0\\ 0 & 0 & \omega_z & 0\\ 0 & 0 & 0 & \omega_\theta\end{bmatrix} \tag{16}$$

Where, $\omega_x=\dfrac{1}{v\cdot\cos(\phi)\cos(\alpha)+e}$, $\omega_y=\dfrac{1}{v\cdot\sin(\phi)\sin(\alpha)+e}$, $\omega_z=\dfrac{k_1}{(z_d-f_z(q_t))^2}$, and $\omega_\theta=1$.

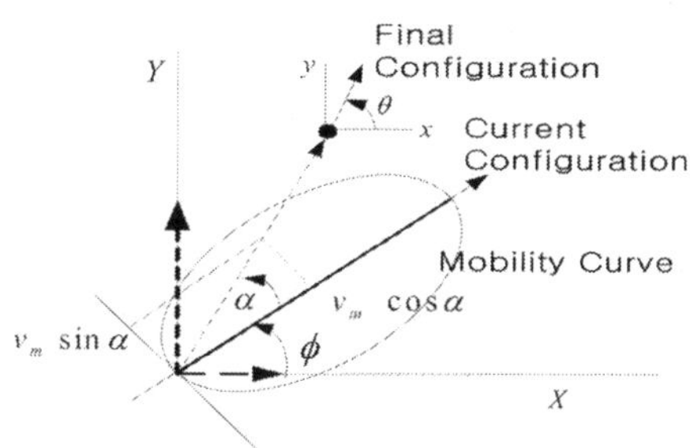

Fig. 3. Decomposing mobility

3.4 Mobile Robot Control

The mobile robot carries the task robot to the reachable boundary to the goal position, i.e., within the reachable workspace. We establish the coordinate system as shown in Fig. 4 so that the robot can take the desired posture and position movement from the initial position according to the assignment of the weighting value of the mobile robot to the desired position. After starting at the present position, (x_i, y_i), the robot reaches the desired position, (x_d, y_d). Here the current robot direction ϕ, the position error α from present position to the desired position, the distance error e to the desired position, the direction of mobile robot at the desired position θ are noted [9].

$$\dot{e} = -v\cos\alpha \tag{17a}$$

$$\dot{\alpha} = -\omega + \frac{v\sin\alpha}{e} \tag{17b}$$

$$\dot{\theta} = \frac{v\sin\alpha}{e} \tag{17c}$$

A Lyapunov candidate function is defined as in (18).

$$V = V_1 + V_2 = \tfrac{1}{2}\lambda e^2 + \tfrac{1}{2}(\alpha^2 + h\theta^2) \tag{18}$$

where V_1 means the error energy to the distance and V_2 means the error energy in the direction. After differentiating both sides in equation (18) in terms of time, we can acquire the result as in equation (19).

$$\dot{V} = \dot{V}_1 + \dot{V}_2 = \lambda e\dot{e} + \left(\alpha\dot{\alpha} + h\theta\dot{\theta}\right) \tag{19}$$

Let us substitute equation (17) into the corresponding part in equation (19), it results in equation (20).

$$\dot{V} = -\lambda ev\cos\alpha + \alpha\left[-\omega + \frac{v\sin\alpha}{\alpha}\cdot\frac{(\alpha + h\theta)}{e}\right] \tag{20}$$

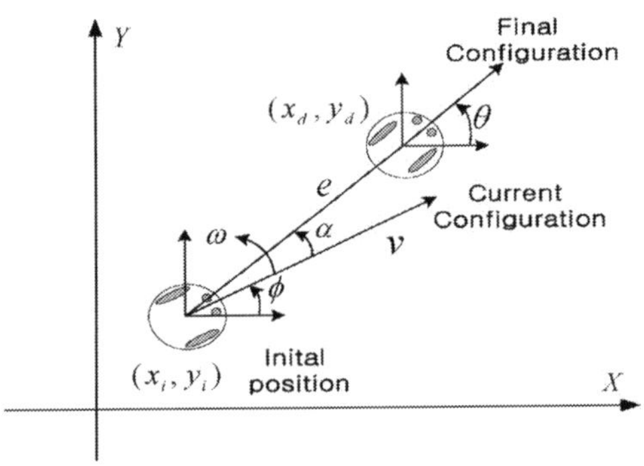

Fig. 4. Position movement of mobile robot by imposed weighting value

Note that $\dot{V} < 0$ is required for a given V to be a stable system. On this basis, we can design the nonlinear controller of the mobile robot as in (21).

$$v = \gamma\,(e\cos\alpha),\quad (\gamma > 0) \tag{21a}$$

$$\omega = k\alpha + \gamma \frac{\cos\alpha \sin\alpha}{\alpha}(\alpha + h\theta), \; (k, \; h > 0) \tag{21b}$$

Therefore, using this controller for the mobile robot, $\dot{V}$ approaches to zero as $t \to \infty$; e and α also approach almost to zero as shown in (22).

$$\dot{V} = -\lambda(\gamma\cos^2\alpha)e^2 - k\alpha^2 \le 0 \tag{22}$$

4 Simulation

For verifying the proposed algorithm, simulations were performed with PURL-II. Fig. 5 shows the simulation results with a 3 DOF task robot and a 3 DOF mobile robot. The goal is positioning the end-effect to (1, 0.5, 0.7), while initial configuration of mobile robot is (-1.0, -1.0, 1.3, 60°) and that of task robot is (18°, 60°, 90°). The optimally determined configuration of mobile robot is (0.0368, -0.497, 1.14, 44.1°) and that of task robot is (1.99°, 25.57°, 86.63°). Fig. 5 shows movements of the task robot in different view points.

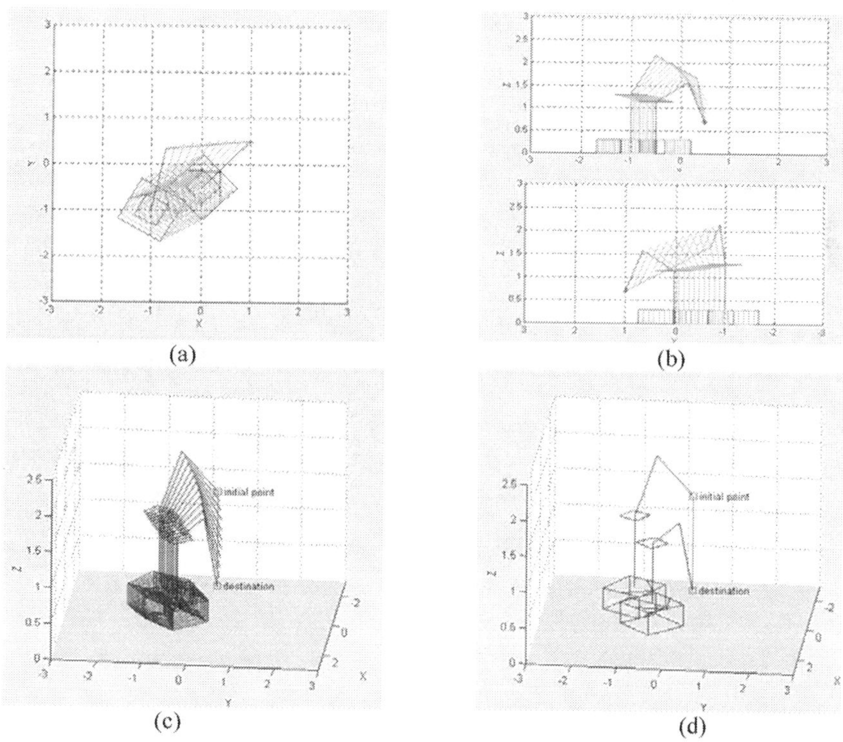

(a) (b) (c) (d)

Fig. 5. The optimal position planning to move a point of action of a robot to (1, 0.5, 0.7)

5 Experiment

Before the real experiments, assumptions for moving manipulators operational condition are set as follows: 1. In the initial stage, the object is in the end-effect of the task robot. 2. The mobile robot satisfies the *pure rolling* and *non-slippage* conditions. 3. There is no obstacle in the mobile robot path. 4. There is no disturbance of the total system. And the task robot is configured as the joint angles of (18°, 60°, 90°), then the coordinate of the end-effect is set up for (0.02, 0.04, 1.3). From this location, the mobile manipulator must bring the object to (1, 1.5, 0.5). An optimal path which is calculated using the algorithm which is stated in the previous section has $W_x = 10.0$, $W_y = 10.0$, and $W_z = 2.66$. And at last the mobile robots angle is 76.52° from the X axis; the difference is coming from the moving of the right wheels 0.8m and the moving of the left wheels 1.4m. Next time, the mobile robot is different with the X axis by 11.64°with right wheels 0.4m moving and the left wheels 0.5m moving. Hence, the total moving distance of mobile robot is (1.2m, 1.9m), the total angle is 88.16°, and the each joint angle of task robot are (-6.45°, 9.87°, 34.92°). The experimental results are shown by the photography in Fig. 6. For the real experiment, the wheel *pure rolling* condition is not satisfied, also by the control the velocity through the robot kinematics, the distance error occurs from the cumulative velocity error. Using a timer in CPU for estimating velocity, timer error also causes velocity error. Hence consequently, the final position of end-effect is placed at (1.2, 1.5, 0.8) on object.

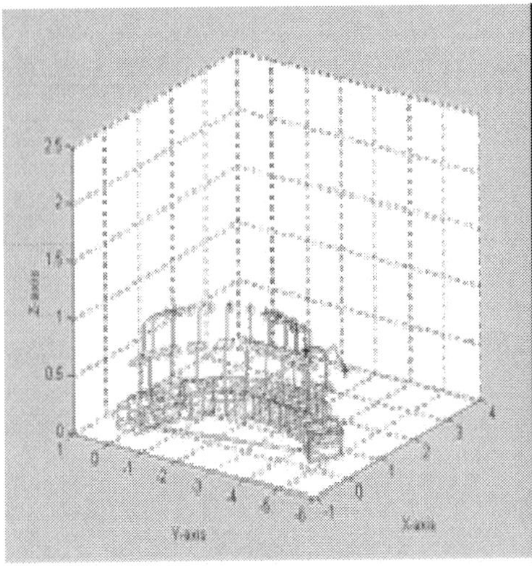

(a) Simulation

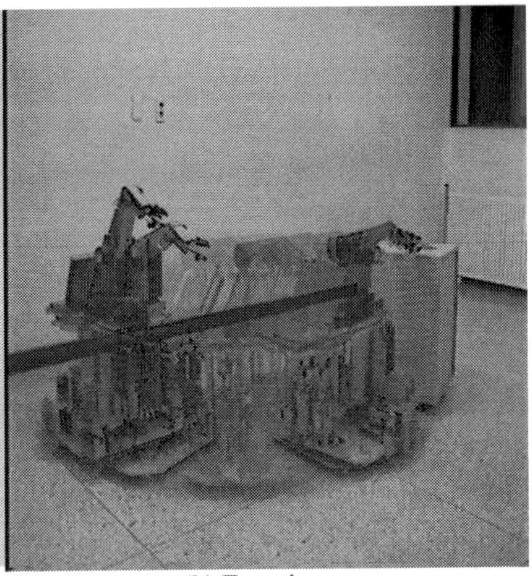

(b) Experiment

Fig. 6. Response of robot posture

6 Conclusion

A new redundancy resolution scheme for a mobile manipulator is proposed in this paper. While the mobile robot is moving from one task (starting) point to the next task point, the task robot is controlled to have the posture proper to the next task, which can be pre-determined based upon TOMM[11]. Minimization of the cost function following the gradient method leads a mobile manipulator an optimal configuration at the new task point. These schemes can be also applied to the robot trajectory planning. The efficiency of this scheme is verified through the real experiments with PURL-II. The different of the result between simulation and experiment is caused by the error between the control input and the action of the mobile robot because of the roughness of the bottom, and is caused by the summation of position error through velocity control. In further study, it is necessary that a proper control algorithm should be developed to improve the control accuracy as well as efficiency in utilizing redundancy.

References

1. Tsuneo Yoshikawa.: Manipulability of Robotic Mechan- isms. The International Journal ofRobotics Robotics Research, Vol.4. No.2(1994) pp.3-9
2. Stephen L. Chiu.:Task Compatibility of Manipulator Postures. The International Journal of Robotics Research, Vol.7. No.5(1998) pp.13-21.
3. Francois G. Pin.:Using Minimax Approaches to Plan Optimal Task Commutation Configuration for Combined Mobile Platform-Manipulator System. IEEE Transaction on Robotics and Automation, Vol.10. No.1(1994) pp.44-53
4. F. L. Lewis.:Control of Robot Manipulators. Macmillan Publishing(1993) pp.136-140
5. Jin-Hee Jang, and Chang-Soo Han.:The State Sensitivity Analysis of the Front Wheel Steering Vehicle: In the Time Domain. KSME International Journal, Vol.11. No.6(1997) pp. 595-604
6. Keum-Shik Hong, Young-Min Kim, and Chiutai Choi.:Inverse Kinematics of a Reclaimer Closed-Form Solution by Exploiting Geometric Constraints. KSME International Journal, Vol.11. No.6.(1997) pp.629-638
7. Sam-Sang You, and Seok-Kwon Jeong.:Kinematics and Dynamic Modeling for Holonomic Constrained Multiple Robot System through Principle of Workspace Orthogonalization. KSME International Journal, Vol. 12. No. 2 (1998) pp.170-180
8. M. T. Mason.:Compliance and Force Control for Computer Controlled Manipulators. IEEE Transaction on Systems Man Cybernetics, Vol. 11. No. 6 (1981) pp.418-432
9. M. Aicardi.:Closed-Loop Steering of Unicycle-like Vehicles via Lyapunov Techniques. IEEE Robotics and Automation Magazine, Vol. 10. No.1(1995) pp.27-35
10. N. Hare and Y. Fing.:Mobile Robot Path Planning and Tracking an Optimal Control Approach. International Conference on Control Automation Robotics and Vision(1997) pp. 9-11
11. Sukhan Lee and Jang M. Lee.:Task-Oriented Dual-Arm Manipulability and Its Application to Configuration Optimization. Proceeding 27th IEEE International Conference on Decision and Control Austin TX (1988)
12. Sam-Sang You.:A Unified Dynamic Model and Control System for Robotic Mainpulator with Geometric End-Effector Constraints. KSME International Journal, Vol. 10. No.2(1996) pp.203-212

VSC Perspective for Neurocontroller Tuning*

Mehmet Önder Efe

Electrical and Electronics Engineering Department
TOBB Economics and Technology University, Söğütözü, Ankara, Turkey
onderefe@etu.edu.tr

Abstract. Compact representation of knowledge having strong internal interactions has become possible with the developments in neurocomputing and neural information processing. The field of neural networks has offered various solutions for complex problems, however, the problems associated with the learning performance has constituted a major drawback in terms of the realization performance and computational requirements. This paper discusses the use of variable structure systems theory in learning process. The objective is to incorporate the robustness of the approach into the training dynamics, and to ensure the stability in the adjustable parameter space. The results discussed demonstrate the fulfillment of the design specifications and display how the strength of a robust control scheme could be an integral part of a learning system. This paper discusses how Gaussian radial basis function neural networks could be utilized to drive a mechatronic system's behavior into a predefined sliding regime, and it is seen that the results are promising.

Keywords: Gaussian Radial Basis Function Networks, Sliding Mode Control.

1 Introduction

The innovations observed in the field of digital technology particularly in the last few decades have accelerated the analysis and interpretations of data collected from physical phenomena, and have made it possible to design and implement the systems based on the former work. Tools used for this purpose have been refined, and artificial neural networks, as one of the powerful tools for modeling and representation of complex mappings, have taken a central role. What make them so attractive have been their capability of representing inextricably intertwined dependencies in a large data set with a simple model, the learning and generalization ability, furthermore, to do all these with a certain degrees of fault tolerance.

When the applications of neural networks are visualized together with the process of refining the future performance, i.e. the process of learning, several important issues need to be addressed very carefully. These contain, but are not limited to the issues related to the parametric stability, generalization versus memorization, setting up the architectural degrees of freedom and so on.

* This work was supported by TOBB Economics and Technology University, BAP Program, under contract no ETÜ-BAP-2006/04

S. Kollias et al. (Eds.): ICANN 2006, Part I, LNCS 4131, pp. 918 – 927, 2006.

The first important milestone in the area of training process is the discovery of Error Backpropagation (EBP) technique [1], which is also known as MIT rule or gradient descent in the literature. Since the EBP method concerns the first order partial derivatives of a cost function, the studies appeared later on have focused on the extraction of a better path to the minimum cost solution by exploiting the information contained in second order derivatives of the cost measure. Newton's method [2], Conjugate Gradient Algorithm [3] and Levenberg-Marquardt (LM) optimization technique [4] are the most prominent ones used frequently in the neural network applications. An inherent problem associated with all these schemes has been the sensitivity of the learning model to the high frequency components additively corrupting the training pairs. One method that has been discussed in the literature is due to Efe et al [5-6], which suggest a dynamic model for the training process and develop a stabilizing scheme by utilizing Variable Structure Systems (VSS) theory. VSS theory is a well-formulated framework for designing control systems particularly for plants having uncertainties in the representative models. The approach has extensively been used for tracking control of nonlinear systems and a good deal of VSS and intelligence integration have been discussed in [7-8] also with the name Variable Structure Control (VSC), which is a VSS theory based control scheme.

In what follows, we scrutinize the concept of VSS theory use in neurocomputing from systems and control engineering point of view. For this purpose, we shall use Gaussian Radial Basis Function Neural Networks (GRBFNNs) introduced in the second section. The third section is devoted to the extraction of an error critic, which is to be used in parameter tuning stage. In the fourth section, a simulation example is considered and the concluding remarks are presented at the end of the paper.

2 Gaussian Radial Basis Function Neural Networks (GRBFNN)

In the literature, GRBFNNs are generally considered as a smooth transition between Fuzzy Logic (FL) and NNs. Structurally, a GRBFNN is composed of receptive units (neurons) which act as the operators providing the information about the class to which the input signal belongs. If the aggregation method, number of receptive units in the hidden layer and the constant terms are equal to those of a Fuzzy Inference System (FIS), then there exists a functional equivalence between GRBFNN and FIS [9]. As illustrated in Fig. 1, the hidden neurons of a GRBFNN possess basis functions to characterize the partitions of the input space. Each neuron in the hidden layer provides a degree of membership value for the input pattern with respect to the basis vector of the receptive unit itself. The output layer is comprised of linear neurons. NN interpretation makes GRBFNN useful in incorporating the mathematical tractability, especially in the sense of propagating the error back through the network, while the FIS interpretation enables the incorporation of the expert knowledge into the training procedure. The latter is of particular importance in assigning the initial value of the

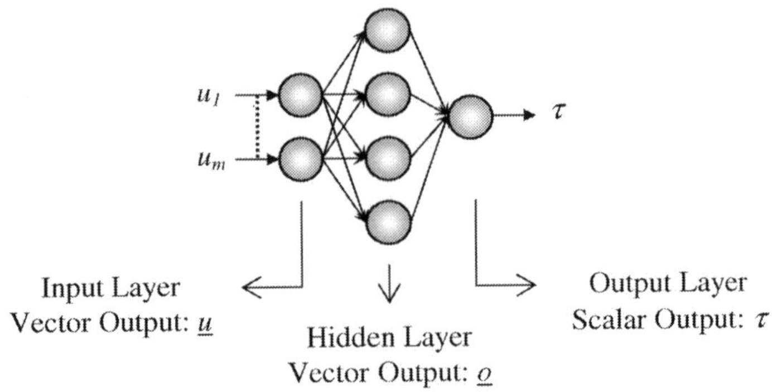

Fig. 1. Structure of a GRBFNN having m-input and single output

network's adjustable parameter vector to a vector that is to be sought iteratively. Expectedly, this results in faster convergence in parameter space.

Mathematically, $o_i=\Pi^m_{j=1}\Psi_{ij}(u_j)$ and the hidden layer activation function is the Gaussian curve described as $\Psi_{ij}(\underline{u})=\exp\{-(u_j-c_{ij})^2/\sigma_{ij}^2\}$, where c_{ij} and σ_{ij} stand for the center and the variance of the ith neuron's activation function qualifying the jth input variable. The output of the network is evaluated through the inner product of the adjustable weight vector denoted by $\underline{\phi}$ and the vector of hidden layer outputs, i.e. $\tau=\underline{\phi}^T\underline{o}$. Clearly the adjustable parameter set of the structure is composed of $\{\underline{c}, \underline{\sigma}, \underline{\phi}\}$ triplet.

3 VSS Theory from a Learning-Strategic Point of View

The pioneering works due to Sanner and Slotine [10] and Sira-Ramirez and Colina-Morles [11] have demonstrated the first successful results in learning design with VSS theory. The latter introduced the concept of zero learning error level, which makes the design of switching manifold comprehensible for first order systems. Since the design of VSC involves a decision based on a two sided mechanism, the boundary of which is characterized by the switching manifold, the geometric location of the manifold for first order systems becomes a point in one dimensional space and is defined to be the zero level of learning [11]. Although a zero level is postulated conceptually, the achievement of which is a challenge unless there is a supervision providing the desired values of the neural network outputs. In [12], an appropriate measure relating the dynamics of the switching manifold and controller error is postulated.

In what follows, we briefly explain how an appropriate error measure for control error could be constructed, and demonstrate how this measure could be used for control applications. For this purpose, it is assumed that the system is in an ordinary feedback loop as illustrated in Fig. 2.

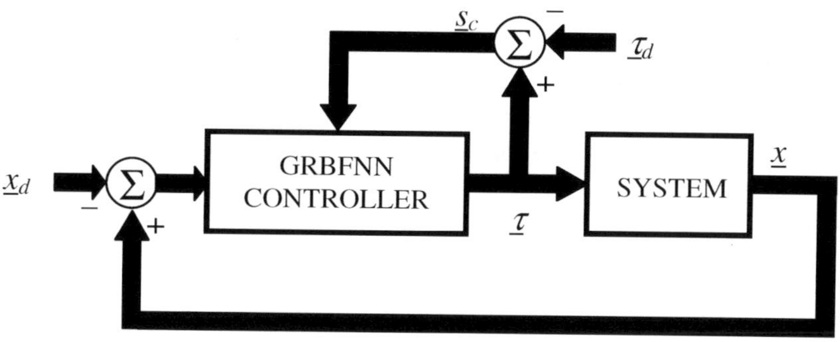

Fig. 2. Structure of the feedback control loop

3.1 Obtaining an Equivalent Control Error

Consider the system $\underline{\ddot{x}} = \underline{f}(\underline{x}, \underline{\dot{x}}) + b(\underline{x})\underline{\tau}$, where $(\underline{x}, \underline{\dot{x}})^{\mathrm{T}}$ is the state vector, $\underline{\tau}$ is the input vector and $\underline{f}$ and b are unknown continuous functions. If $\underline{x}_d$ is defined to be the vector of desired trajectories, one can describe the tracking error vector as $\underline{e}=\underline{x}-\underline{x}_d$ and construct the control signal that derives the system towards the prescribed sliding regime. The design is based on a two-sided switching mechanism, the argument of which is defined as $\underline{s}_p=\mathrm{d}\underline{e}/\mathrm{d}t+\Lambda\underline{e}$ with Λ being a positive definite diagonal matrix of appropriate dimensions. The aim is to ensure the negative definiteness of the Lyapunov function $V_p=\underline{s}_p^{\mathrm{T}}\underline{s}_p/2$. The control sequence can now be formulated as $\underline{\tau} = -b^{-1}(\underline{x})\left(\underline{f}(\underline{x}, \underline{\dot{x}}) + \Lambda\underline{\dot{e}} + \Xi \operatorname{sgn}(\underline{s}_p) - \underline{\ddot{x}}_d\right)$, where, Ξ is a positive definite diagonal matrix. The application of the well-known sliding control law above the system enforces $\underline{\dot{s}}_p = -\Xi \operatorname{sgn}(\underline{s}_p)$, which ensures the reaching to the hyperplane $\underline{s}_p=\underline{0}$.

Proposition: Let $\underline{\tau}_d$ be the vector of control signals that meets the performance specifications. If $\underline{s}_C$ is defined to be the vector of discrepancies between the target and evaluated values of the control vector, and if the controller parameters are adjusted such that the cost function $J=\underline{s}_C^{\mathrm{T}}\underline{s}_C/2$ is minimized, the tracking error vector is driven towards the switching manifold. Here, $\underline{s}_C$ is defined to be the error on the control signal and is computed as given in (1).

$$\underline{s}_C := \underline{\dot{s}}_p + \Xi \operatorname{sgn}(\underline{s}_p) = \underline{\tau} - \underline{\tau}_d \tag{1}$$

In reality, one does not know the numerical value of $\underline{\tau}_d$, however, within the context of discussed problem, the set of all control signals forcing the system towards the sliding manifold can be considered as the target control sequence, which minimizes $\underline{s}_C$. Practically, this means $\underline{\dot{s}}_p \underline{s}_p < 0$, i.e. all trajectories in the phase space tend to the sliding manifold and the ideal behavior thereafter takes place in the sliding subspace. An indirect implication of this is the fact that since $\underline{s}_C = \underline{\dot{s}}_p + \Xi \operatorname{sgn}(\underline{s}_p) = 0$ is maintained, the parameters of the controller are not adjusted during the sliding regime. At this point, we dwell the numerical computation of this error measure. The obvious difficulty is the computation of the time derivative of $\underline{s}_p$. Our prior tests have proved that an approximate numerical differentiation works even with the noisy observations. Introducing a stable linear filter with numerator of order one in Laplace domain can suitably provide the information needed. The reason why we do not need the exact value of the derivative stems from the fact that the desired behavior is not unique. If a trajectory starting from an arbitrary initial point in the space tends to the sliding manifold then it is one of the desired trajectories, however, the selection of Ξ uniquely determines the way of approaching the sliding manifold. The information loss due to the derivative computation can be interpreted as a slight modification of the reaching dynamics characterized by $\dot{\underline{s}}_p = -\Xi\,\mathrm{sgn}(\underline{s}_p)$. The second question is on the selection of the diagonal positive definite matrix Ξ. If the entries increase in magnitude, the reaching phase of the control strategy produce large controls in magnitude and several hittings occur, however, the values close to zero result in slow reaching to the sliding manifold with relatively less number of hittings. The designer has to decide on what he/she pursues together with the physical reality regarding the plant under control. For example, for a cargo ship steering example, enforcing the convergence to a desired behavior in a few seconds would require unrealistically large-magnitude control activity, while for a direct drive robotic manipulator the response could reasonably be fast to fulfill the imposed task. Lastly, the infinite switching frequency of ideal sliding mode should be addressed. Clearly from $\dot{\underline{s}}_p = -\Xi\,\mathrm{sgn}(\underline{s}_p)$, one should notice that the enforced behavior ultimately converges to a practically impossible phenomenon. Since the right hand side of the equation is discontinuous in the vicinity of the origin, the near origin activity is an oscillation ideally at infinite frequency, called *chattering* in the terminology of sliding control. One approach to eliminate the adverse effects of chattering is to introduce a boundary layer by replacing the discontinuous sign function with a smooth approximate such as $\mathrm{sgn}(\alpha) \cong \alpha / \left(|\alpha| + \delta \right)$, where $\delta > 0$ is the parameter determining the accuracy of the approximation.

3.2 Issues of Parameter Tuning

The quantity described in (1) can be used in several ways. Assume that the system is under the feedback loop as illustrated in Fig. 2, and the tuning strategy is the EBP rule. Denoting $\underline{\phi}$ as the vector adjustable parameters of a neural network structure, the law enforces the following tuning mechanism:

$$\dot{\underline{\phi}} = -\eta \frac{\partial J}{\partial \underline{\phi}} = -\eta \sum_{j=1}^{n} s_{Cj} \frac{\partial \tau_j}{\partial \underline{\phi}} \tag{2}$$

where $J = \sum_{j=1}^{n} s_{Cj}^2 / 2$ and η is the learning rate in the conventional sense. The similar reasoning can be postulated for other learning algorithms as well.

If the output of a neural network structure is linear in the adjustable parameter vector ($\underline{\phi}$), e.g. in the case of GRBFNN with only output weights adjustable, alternative mechanisms could be proposed. In (3), the tuning law proposed by Sira-Ramirez et al [11] has been given.

$$\underline{\dot{\phi}}_i = -\frac{\underline{\Omega}_i}{\underline{\Omega}_i^{\mathrm{T}} \underline{\Omega}_i} k_i \operatorname{sgn}\left(s_{C_i}\right) \tag{3}$$

In above, k_i is the uncertainty bound satisfying $k_i > B_{\phi_i} B_{\dot{\Omega}_i} + B_{\dot{\tau}_{id}}$ and $\underline{\Omega}_i$ is the vector excitation signals. Setting of k_i obviously requires the knowledge on the following bounds $\left\|\underline{\phi}_i\right\| \le B_{\phi_i}$, $\left\|\underline{\dot{\Omega}}_i\right\| \le B_{\dot{\Omega}_i}$ $\left\|\dot{\tau}_{id}\right\| \le B_{\dot{\tau}_{id}}$, which are typically unknown, therefore a compact k_i value is set by trial and error. For a detailed discussion on this adaptation law, one should refer to [7,11-12].

Alternatively, one might suggest the use of tuning strategy in (4), which is designed to ensure the negative semi-definiteness of the Lyapunov function in (5).

$$\underline{\dot{\phi}}_i = -k_i \left(\mu I + \rho \frac{\partial^2 V_{c_i}}{\partial \underline{\phi}_i \partial \underline{\phi}_i^T} \right)^{-1} \operatorname{sgn}\left(\frac{\partial V_{c_i}}{\partial \underline{\phi}_i} \right) \tag{4}$$

where $\mu > 0$ and $\rho > 0$ are the free design parameters determining the relative importance of the terms seen in (5), and $k_i > (\mu\, B_{\phi_i} + \rho\, B_{\Omega_i}) B_{\dot{\Omega}_i}$.

$$V = \mu V_{c_i} + \rho \frac{1}{2} \left\| \frac{\partial V_{c_i}}{\partial \underline{\phi}_i} \right\|^2 \quad \text{with } V_{c_i} = \frac{1}{2} s_{C_i}^2 \tag{5}$$

Clearly the above law and the Lyapunov function suggest that the parametric growth is penalized. Further discussion on this approach has been presented in [7]. For the strategy in (3), the zero error learning level is characterized by $s_{Ci} = 0$, while the latter uses an augmented switching manifold given as in (6). The law of (4) enforces a motion taking place in the vicinity of $s_{A_i} = 0$.

$$s_{A_i} = \begin{bmatrix} s_{C_i} \\ \dfrac{\partial V_{c_i}}{\partial \underline{\phi}_i} \end{bmatrix} \tag{6}$$

4 An Illustrative Example

To demonstrate the efficacy of the presented concept, the control of a 2 degrees of freedom direct drive arm is considered. The dynamics of the manipulator is described by the following vector differential equation.

$$\underline{\ddot{x}} = M^{-1}(\underline{x})\left(\underline{\tau} - \underline{f}_c - \underline{C}(\underline{x}, \underline{\dot{x}})\right) \tag{7}$$

where, $M(\underline{x})$, $\underline{C}(\underline{x}, \underline{\dot{x}})$, $\underline{\tau}$ and $\underline{f}_c$ stand for the state varying inertia matrix, the vector of Coriolis terms, the applied torque inputs and the Coulomb friction terms respectively.

The plant parameters are given in Table 1 in standard m-kg-s units. If the angular positions and angular velocities are described as the state variables of the system, four coupled and first order differential equations can define the model. In (8) and (9), the terms seen in (7) are given explicitly.

$$M(\underline{x})=\begin{bmatrix} p_1+2p_3\cos(x_2) & p_2+p_3\cos(x_2) \\ p_2+p_3\cos(x_2) & p_2 \end{bmatrix} \tag{8}$$

$$V(\underline{x},\underline{\dot{x}})=\begin{bmatrix} -\dot{x}_2(2\dot{x}_1+\dot{x}_2)p_3\sin(x_2) \\ \dot{x}_1^2 p_3\sin(x_2) \end{bmatrix} \tag{9}$$

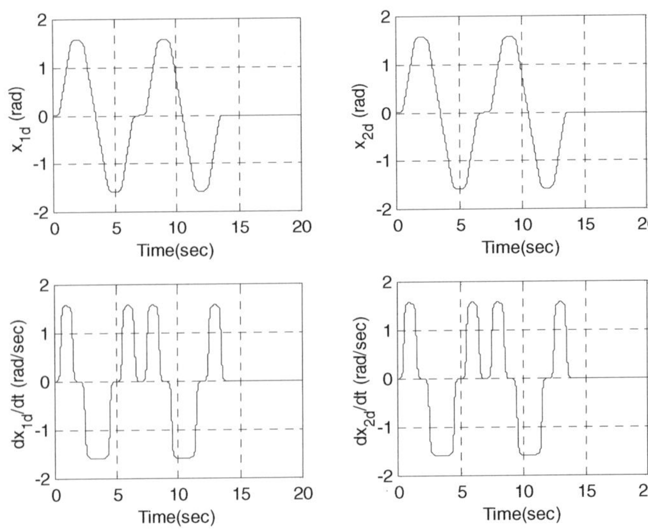

Fig. 3. Reference trajectories for base and elbow links

In the above equations, $p_1 = 3.31655+0.18648M_p$, $p_2 = 0.1168+0.0576M_p$ and $p_3 = 0.16295+0.08616M_p$. Here M_p denotes the payload mass. The details of the plant model can be found in Direct Drive Manipulator R&D Package User Guide [13].

Since the dynamics of such a mechatronic system is modeled by nonlinear and coupled differential equations, precise output tracking becomes a difficult objective due to the strong interdependence between the variables involved. Additionally, the ambiguities on the friction related dynamics in the plant model and the varying payload conditions make the design much more complicated. Therefore the control methodology adopted must be capable of handling the difficulties stated.

As the controller, two GRBFNN structures having 2 inputs 9 hidden neurons and single output are used for each link, and only the weight parameters are adjusted with the tuning law of (3). Initially, the adjustable parameters have been set to zero and the

Table 1. 2D Manipulator parameters

Motor 1 Rotor Inertia	0.2670	Payload Mass (M_p)	2.0000
Arm 1 Inertia	0.3340	Arm 1 Length	0.3590
Motor 2 Rotor Inertia	0.0075	Arm 2 Length	0.2400
Motor 2 Stator Inertia	0.0400	Arm 1 CG Distance	0.1360
Arm 2 Inertia	0.0630	Arm 2 CG Distance	0.1020
Motor 1 Mass	73.000	Axis 1 Friction	4.9000
Arm 1 Mass	9.7800	Axis 2 Friction	1.6700
Motor 2 Mass	14.000	Torque Limit 1	245.00
Arm 2 Mass	4.4500	Torque Limit 2	39.200

uncertainty bounds have been set as k_1=10000 and k_2=1000. The simulation has been performed for 20 seconds, and the integration step size has been chosen as 2.5 ms. In response to the reference trajectories depicted in Fig. 3, the error trends shown in Fig. 4 are obtained. Clearly the suggested form of tuning and control strategy is capable of alleviating the nonzero initial errors together with a load of 2 kg grasped at t=2 sec, released at t=5 sec, and grasped again at t=9 sec and released at t=12 sec. This clearly introduces an abrupt change in the dynamics of the system and necessitates a robust controller to compensate the behavioral changes. Although not presented here, in the phase space, the behavior for each link is maintained on the loci characterized by λ=1 with tolerably small and convergent spikes in elbow velocity error.

The control inputs are depicted in the top row of Fig. 5, which reveals that the produced control signals sufficiently smooth and are of reasonable magnitudes. In the

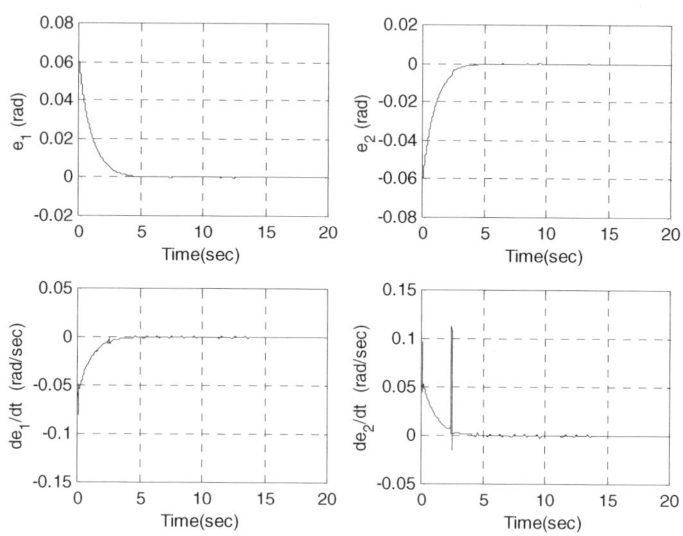

Fig. 4. State tracking errors

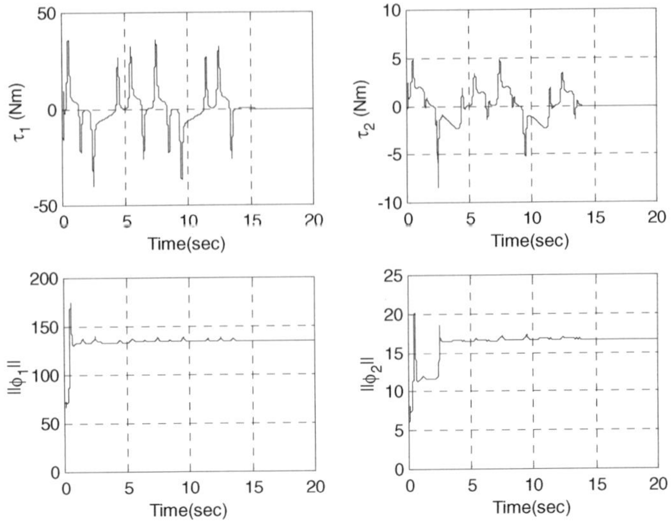

Fig. 5. Control inputs and parameter norms

bottom row of Fig. 5, the evolution of the Euclidean norm of the adjustable parameter vectors are shown. The obtained results clearly suggest that the tuning mechanism stops modifying the values of the parameters right after the sliding regime starts. Therefore, it can be claimed that the learning dynamics is internally stable for both controllers and the control system is robust against disturbances such as payload change as considered in this work.

5 Conclusions

This paper presents the use of VSS theory in training of GRBFNN type controllers. For this purpose, a novel error critic is discussed, and several tuning laws are presented. It has been exemplified that a tuning activity minimizing the proposed error measure drives the system under control into a prespecified sliding mode and results in robust and precise tracking. A robotic manipulator has been chosen as the test bed, and the internal stability of the adjustable parameter dynamics has been visualized. Briefly, the use of VSS theory for parameter tuning purposes introduces the robustness and invariance properties of the VSC technique, which results in some desirable features during the control cycle.

References

1. Rumelhart, D.E., Hinton, G.E. and Williams, R.J.: Learning Internal Representations by Error Propagation, in D. E. Rumelhart and J. L. McClelland, (Eds.), Parallel Distributed Processing: Explorations in the Microstructure of Cognition, v. 1, MIT Press, Cambridge, M.A., (1986) 318-362.

2. Battiti, R.: First- and Second-order Methods for Learning: Between Steepest Descent and Newton's Method, Neural Computation, v.4, (1992) 141-166.
3. Charalombous, C.: Conjugate Gradient Algorithm for Efficient Training of Artificial Neural Networks, IEE Proceedings, v.139, (1992) 301-310.
4. Hagan, M.T. and Menhaj, M.B.: Training Feedforward Networks with the Marquardt Algorithm, IEEE Transactions on Neural Networks, v.5, n.6, (1994) 989-993.
5. Efe, M.Ö. and Kaynak, O.: On Stabilization of Gradient Based Training Strategies for Computationally Intelligent Systems, IEEE Transactions on Fuzzy Systems, v.8, n.5, (2000) 564-575.
6. Efe, M.Ö. and Kaynak, O.: Stabilizing and Robustifying the Learning Mechanisms of Artificial Neural Networks in Control Engineering Applications, International Journal of Intelligent Systems, v.15, n.5 (2000) 365-388.
7. Efe, M.Ö. Kaynak, O. and Yu, X.: Variable Structure Systems Theory in Computational Intelligence," in Variable Structure Systems: Towards the 21st Century, Lecture Notes in Control and Information Sciences, Eds. X. Yu and J.-X. Xu, Springer Verlag, v.274, (2002) 365-390
8. Kaynak, O., Erbatur, K. and Ertugrul, M.: The Fusion of Computationally Intelligent Methodologies and Sliding-Mode Control – A Survey, IEEE Transactions on Industrial Electronics, v.48, n.1, (2001) 4-17.
9. Jang, J.-S.R., Sun, C.-T. and Mizutani, E.: Neuro-Fuzzy and Soft Computing, PTR Prentice-Hall, (1997).
10. Sanner, R.N. and Slotine, J.J.E.: Gaussian Networks for Direct Adaptive Control, IEEE Transactions on Neural Networks, v.3, n.6, (1992) 837-863.
11. Sira-Ramirez, H. and Colina-Morles, E.: A Sliding Mode Strategy for Adaptive Learning in Adalines, IEEE Transactions on Circuits and Systems - I: Fundamental Theory and Applications, v.42, n.12, (1995) 1001-1012.
12. Efe, M.Ö., Kaynak, O. and Yu, X.: Sliding Mode Control of a Three Degrees of Freedom Anthropoid Robot by Driving the Controller Parameters to an Equivalent Regime, Trans. of the ASME: Journal of Dynamic Systems, Measurement and Control, v.122, n.4, (2000) 632-640.
13. Direct Drive Manipulator R&D Package User Guide, Integrated Motions Incorporated, 704 Gillman Street, Berkeley, California 94710, U.S.A., (1992).

A Neural Network Module with Pretuning for Search and Reproduction of Input-Output Mapping

Igor Shepelev

A.B. Kogan Research Institute for Neurocybernetics, Rostov State University
shepelev@krinc.ru

Abstract. A neural network that uses a pretuning procedure for function approximation is presented. Unlike traditional neural network algorithms in which changeable parameters are multiplicative weights of connections between neurons in the network, the pretuning procedure deals with additive thresholds of interneurons of the proposed neural network and is a dynamical combinatory inhibition of these neurons. It is shown that in this case the neural network can combine local approximation and distributed activation. The usefulness of the neural network with pretuning (NNP) for the tasks of search and reproduction of sensorimotor mapping of robot is briefly discussed.

Keywords: Neural networks, pretuning, interneurons, combinatory inhibition, local approximation.

1 Introduction

According to adaptation mechanisms implemented by neural network models one can distinguish three trends of development of neural network technologies. The *training neural networks* are connectionist structures which are characterized by finding the right set of weights to accomplish a given task. The investigation of this type of neural networks was initiated by the model of self-organizing adaptive system also known as the perceptron which was proposed by Rosenblatt in 1958 [1]. The further development of this area is connected with the works of researchers such as Rumelhart [2] and others. The following characteristic features of these neural networks are emphasized. First, the training process always results in an unifunctional neural network, that is the network which realize a unique function. Secondly, for these neural networks the global character of approximation lead to slow learning. So the network has to be retrained if new information is required to learn [3].

The *growing neural networks* overcome the problem of forgetting the old information when learning the new by use of principles of local approximation. The typical example of such neural networks are the networks of adaptive resonance theory which was developed by Carpenter and Grossberg [4]. A network of this type adds new neuron if a growing criteria is satisfied. For instance, a new neuron is added if the current input presents a region of input space that was not previously learned. So the growing permits to learn new information incrementally without having to

S. Kollias et al. (Eds.): ICANN 2006, Part I, LNCS 4131, pp. 928 – 935, 2006.

retrain the network. This adaptation mechanism also leads to an unifunctional neural network.

It may be said that historically, the networks of McCulloch neurons [5] were the first models of *pretuning neural networks*. The main difference between the model of McCulloch neuron and the well-known artificial neuron model also referred to as threshold gate is the presence of nonlinear synaptic connections. It was shown [6] that for once established structure and connection weights of the network of such neural elements in the presence of very small redundancy of neurons one can tune the network to compute an arbitrary boolean function by means of adjustment the neural thresholds. In this sense such a neural network is polyfunctional. A threshold is an additive parameter of neuron model and can be interpreted as an external signal to neuron. If this external signal is the output of different neural network than we can say about the pretuning of a neural network by another one. While the latter network serves for memorization of pretuning signals, the former which is pretuned by this pretuning signals is used for search and reproduction of required input-output mapping.

In this paper we concentrate on the study of neural networks to be pretuned. A neural network module for piecewise-linear approximation of continuous functions is described in Section 2. The proposed neural network combines local approximation and distributed activation. This combination provides fast local construction of functions to be approximated and a high accuracy of approximation while the size of the network is relatively small. Alternative adaptation mechanism for function approximation allows getting the independence of weights and size of the network on specific approximated functions. Namely, the mapping required is constructed not by adaptation of connection weights or by growing number of neurons but by a combinatory inhibition of interneurons which is realized by another network not described here. Section 3 demonstrates performance of the network for different numbers of interneurons. In Section 4, the usefulness of the network for the task of robot behaviour control in changing environment is briefly discussed.

2 Model

2.1 The Basic Functional Unit

The model is based on computational units (modelled neurons) that apply the linear saturated function Θ with $\Theta(u) = u$ for $0 \le u \le 1$, $\Theta(u) = 0$ for $u < 0$, $\Theta(u) = 1$ for $u > 1$ as the activation function. Let w_j be the weight from neuron j, $y_j = \Theta(u_j)$ the activity of neuron j, h the threshold and x the external input to the neuron. Then the membrane potential u is given by the following equation:

$$u = x + \sum_j w_j y_j - h \ .$$

A basic functional unit of the network is proposed for piecewise approximation. The unit consists of hidden neuron **3** and interneurons **1** and **2** (Fig. 1). Different values of the weights of inhibitory connections from interneurons allow realizing arbitrary output functions of the chosen class. In this case these are linear functions:

$$y = Kx + cx_0 \text{ (for } x \in [0;1]\text{)}, \qquad (1)$$

where $K = w^{in} - w_1^{int}$ determines the steepness of the function, $c = w_0^{in} - w_2^{int}$ is the shifting along the y-axis, and x_0 denotes the constant unity signal (x_0=1).

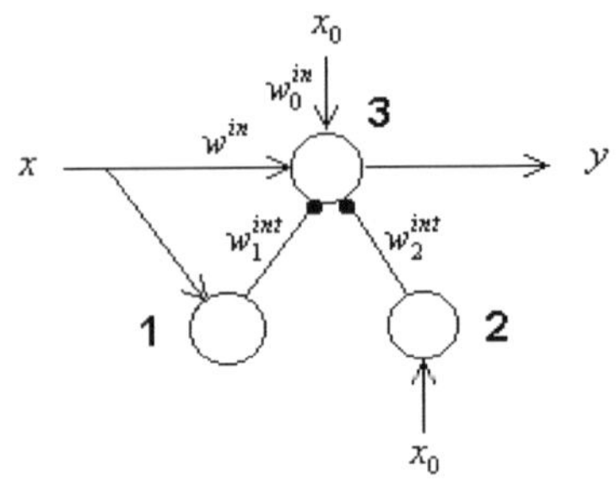

Fig. 1. The basic functional unit

2.2 Distributed Representation of Output Functions

Suppose that for every subarea of input space of the network are known proper values of K and c. Then any function can be approximated by switching these values for different subareas. In turn, these values are determined through the values of weights of the basic unit. A set of parameter K and c values in (1) for construction of linear approximation on every subarea is obtained by means of some abundance of interneurons in the basic unit.

The activation function Θ allows to bring the activity of an interneuron to 0 by means of the external inhibitory signal p. In other words, this signal turns interneuron off which is equivalent to increasing the threshold h_i of neuron i. The combination of interneurons turned on and off results in distributed representation of output functions.

2.3 Piecewise-Linear Approximation

A set of inhibitory combinations allows realizing a set of piecewise-linear functions. The number of functions is determined by the number of interneurons n_{int} and is equal to $2^{n_{int}}$. The output functions of hidden neuron for different combinations of interneurons turned off will be the linear approximations on some subareas. Fig. 2 shows that the interneurons are grouped structurally for approximation of the values of the steepness **1** and the shifting **2** parameters of the function (1).

If the network has several inputs n_{in} then the several subgroups of interneurons are linked with the corresponding inputs for approximation of every parameter K_i $(i = 1..n_{in})$ of approximating function $\hat{y} = \sum_{i=1}^{n_{in}} K_i x_i + c$. Therefore the network has $n_{in} + 1$ subgroups. All subgroups contain the same number of interneurons n'_{int}, so

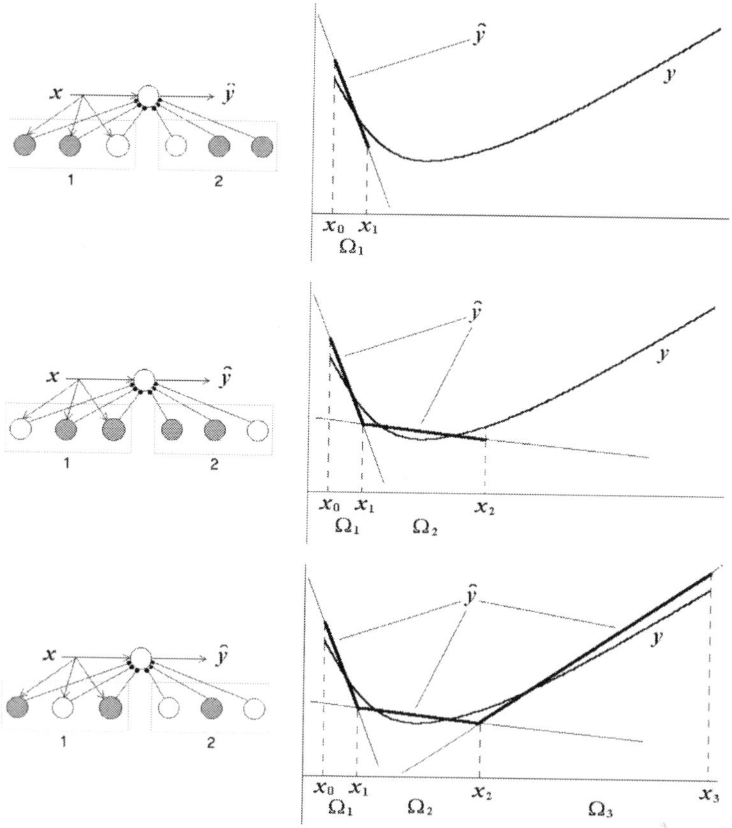

Fig. 2. Piecewise-linear approximation is realized by different combinations of interneurons turned off on the different subareas. Turned off interneurons are shown by gray circles.

$n_{int} = \sum_{n_{in}+1} n'_{int}$. Increasing dimension of task to be solved requires the corresponding number of equivalent neural networks working in the parallel mode.

2.4 Connection Weights of the Network

The main idea is to construct arbitrary mapping without changing the initially chosen topology and weights of the network. So the size and the parameters of the network are chosen to guarantee a necessary range of functions to be realized.

Let the steepness coefficient $k = \frac{dy}{dx}$ of the approximated function $y(x)$ be bounded, $k \in \left[-K^{\max}; K^{\max}\right]$. This assumption is used for calculation of fixed weights w^{in} in the network. The weights of inhibitory connections w^{int} are uniformly distributed random values.

2.5 Algorithm of Pretuning and Domain Discretization

The neural network is pretuned by inhibitory combination $\boldsymbol{p}_p$ on every subarea Ω_p to obtain a reasonable accuracy of approximation $\hat{y}_p$ of function y:

$$\boldsymbol{p}_p: \ \left|y-\hat{y}_p\right|<\varepsilon, \ \ Q(\varepsilon)=Q_{opt}, \ \ \Omega=\bigcup_p \Omega_p \tag{2}$$

The method of random search was chosen to find proper pretuning combinations $\boldsymbol{p}$. Once the combination is found, it is stored. The pretuning combination is not changed until the current output function satisfies some objective function Q_{opt}. From (2), an important feature of the model follows. The discretization of domain cannot be carried out in advance because of inaccessibility of data in the task considered. That is why the discretization is realized directly in approximation constructing.

3 Experiments

3.1 The Number of Interneurons

The dependence of approximation accuracy on the number of interneurons was studied using a number of test sinusoid functions with different phase shifts φ:

$$y = 0.5\sin(2\pi x + \varphi) + 0.5\,. \tag{3}$$

The approximation of these functions allows to test the neural network in a wide range of steepness and shifting parameters of the function (1). For the task, 100 randomly selected from $[0,1]$ pairs (x, y) formed training set. To compute the averages, 20 test functions with random phase shifts $\varphi \in [-\pi, \pi]$ were used. The accuracy was specified using the absolute error. The experiments show that the number of iterations needed to find a proper inhibitory combination is much less then $2^{n_{int}}$ (Fig. 3). Moreover, when the number of interneurons is increased, the number of iterations tends to decrease because of the abundance of output functions.

3.2 Approximation of Multivariable Functions

The multiplications of the same sinusoids (3) served as the testing functions to derive the dependence of the number of searching proper inhibitory combination iterations on the number of neural network inputs. A scaling coefficient s was introduced as the ratio of the number of iterations for the multiple input neural network to the number of iterations for the network with 1 input. Cases of 2, 8, and 32 inputs were considered. The number of iterations obtained for the each case of the network with multiple inputs was averaged over 20 different combinations of multiplication of sinusoids with random phase shifts φ. For the task of multivariable functions approximation, we used 1000 training pairs $(\boldsymbol{x}, y)$ randomly selected from

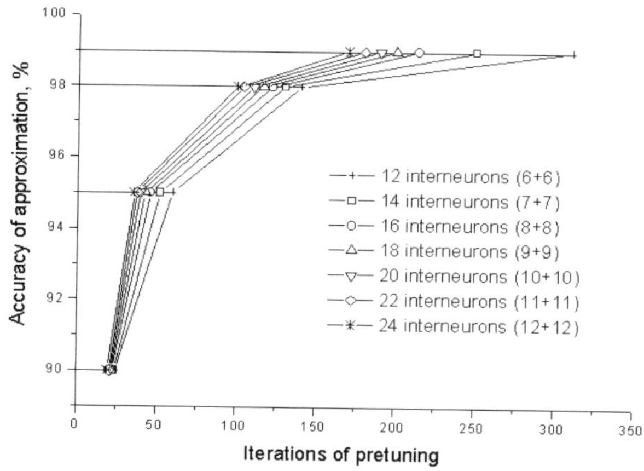

Fig. 3. The dependence of approximation accuracy on the number of iterations of pretuning (searching steps) needed to find a proper inhibitory combination. (*n*+*n*) denotes that equal numbers of interneurons were chosen for approximation of each parameter of the function (1).

corresponding domain. The experiments were carried out for the network with 6, 8, 10, and 12 interneurons n'_{int} and for the values of approximation accuracy of 90, 95, 98, and 99 percents. Calculating the scaling coefficients we used the number of iterations for one-input network from Fig. 3.

Results showed that the scaling coefficient is independent of the number of interneurons n'_{int} and approximation accuracy. The average values of the scaling coefficient were estimated for each considered number of inputs and are shown in Table 1. The following linear equation

Table 1. The dependence of the scaling coefficient for the number of iterations on the number of network inputs

Number of inputs	Scaling coefficient
1	1
2	1.04
8	1.3
32	2.23

$$s = 0.04n_{in} + 0.97$$

provides a good fit to the experimental data. The regression equation was obtained using the method of least squares. The value of determination coefficient of 0.998 proves the adequacy of the model. It is important that increasing the number of inputs do not increase the number of iterations significantly.

The results obtained can be used to define the size of the network and to estimate the corresponding computational burden for the task of approximation of arbitrary functions.

4 Discussion

The neural network module described here searches and reproduces arbitrary input-output mapping. The other neural module to be implemented is a learning network which stores the pretuning vectors.

The neural modules interact with each other by means of the pretuning. This adaptive technique is based on the random search of inhibitory combinations. The effectiveness of the random search in comparison with the known neural network implementations is due to operating in low-dimensional space of binary values of thresholds rather than high-dimensional space of continuous values of weights.

The division of functions between two neural network modules leads to polyfunctional properties of considered neural network and the ability of operating with the objective functions depending on input, output, or other variables. These peculiarities are valuable for the task of robot behaviour control in partially unknown and unpredictable environments. Indeed, a great number of attempts have been made to model sensorimotor relationships in autonomous mobile robots and robot manipulators by use of famous neural network algorithms. However the main principles of information processing offered by these neural network algorithms appear to be inappropriate for construction of mapping between sensory and motor variables of robots. The typical example of such networks is multilayer perceptron [2] which employs supervised learning algorithm for the mapping construction. The supervised learning requires the training set in the form of input and correct output data. However the data when operating in partially unknown environments are often not available, and neural network based on objective function sensory variables must produce appropriate motor responses directly while interacting with environment.

Neural networks that are trained by reinforcement [7] and genetic [8] learning algorithms are able to operate in the absence of rigorous instructions such as training input-output pairs and can be applied to generation of robot behaviour [9], [10]. At the same time, the main obstacle to use them in real-time robotic applications that are characterized by changing environments is a global approximation. It is computationally expensive for such networks to adapt every time when the environment is changed. That is why polyfunctional properties of neural network as control system are desirable.

Such questions as the performance of the neural network in the presence of noise, generalization properties of the network remain outside the scope of this paper. An acceptable implementation of the learning neural network module will allow to answer these questions, which is the subject of our future studies.

References

1. Rosenblatt F.: The Perceptron: A Probabilistic Model for Information Storage and Organization in The Brain. Psychological Review. Vol. 65. (1958) 386-408
2. Rumelhart, D.E., Hinton, G.E., and Williams R.J.: Learning Internal Representations by Error Propagation. In Parallel distributed processing: Explorations in the Microstructure of Congnition,, Vol. 1. Cambridge, MA: MIT Press (1986) 318-362

3. Haykin S.: Neural Networks: A Comprehensive Foundation, (2nd Ed.), Prentice Hall. (1999)
4. Carpenter G.A., Grossberg S., Markuzon N., Reynolds J.H., Rosen D.B.: Fuzzy ARTMAP: A Neural Network Architecture for Incremental Supervised Learning of Analog Multidimensional Maps. IEEE Trans. on Neural Networks. Vol. 3. No.5. (1992) 698-714
5. McCulloch W.S.: Agathe Tyche of Nervous Nets - The Lucky Reckoners. Proc. Symposium on Mechanization of Thought Processes, N.P.L., Teddington. (1958)
6. McCulloch W.S.: Anastomotic Nets Combating Noise. In W.S. Fields and W. Abbott, Eds., Information Storage and Neural Control. (1963) 283-298
7. Sutton, R.S., Barto, A.G.: Reinforcement Learning, An Introduction. Cambridge, MA: MIT Press (1998)
8. Jones, A.J.: Genetic Algorithms and Their Applications to The Design of Neural Networks. Neural Computing and Applications. Vol. 1. No. 1. (1993) 32-45
9. Martin, P., Millan, J.: Learning Reaching Strategies Through Reinforcement for Sensor-based Manipulator. Neural Neworks. Vol. 11. (1998) 359-376
10. Gullapalli, V., Grupen, R., Barto, A.: Learning Reactive Admittance Control. In Proceedings IEEE Int. Conf. on Robotics and Automation. (1992) 1475-1480

Physical Mapping of Spiking Neural Networks Models on a Bio-inspired Scalable Architecture

J. Manuel Moreno[1], Javier Iglesias[2], Jan L. Eriksson[2], and Alessandro E.P. Villa[3]

[1] Technical University of Catalunya, Dept. of Electronic Engineering
Campus Nord, Building C4, c/Jordi Girona 1-3, 08034-Barcelona, Spain
moreno@eel.upc.edu

[2] Laboratory of Neuroheuristics, Information Systems Department INFORGE
University of Lausanne, Lausanne, Switzerland
Javier.Iglesias@unil.ch, jan@lhn.unil.ch

[3] INSERM U318, University Joseph-Fourier Grenoble 1, Pavillon B
CHUG Michallon, BP217, F-38043 Grenoble Cedex 9, France
Alessandro.Villa@ujf-grenoble.fr

Abstract. The paper deals with the physical implementation of biologically plausible spiking neural network models onto a hardware architecture with bio-inspired capabilities. After presenting the model, the work will illustrate the major steps taken in order to provide a compact and efficient digital hardware implementation of the model. Special emphasis will be given to the scalability features of the architecture, that will permit the implementation of large-scale networks. The paper will conclude with details about the physical mapping of the model, as well as with experimental results obtained when applying dynamic input stimuli to the implemented network.

1 Introduction

Spiking neural networks models have attracted a considerable research interest during the last years [1], [2] because of their biological plausibility and their suitability for a physical hardware implementation. From the different learning mechanisms available for this neural models Spike Timing Dependent Plasticity (STDP) has received an increasing interest [3] because of experimental evidence [4] and observations suggesting that synaptic plasticity is based on discrete dynamics [5].

In this paper we shall consider a spiking neural network model whose learning mechanism is based on discrete variables [6]. After presenting the model the sequence of steps driving to its physical realization will be explained. Then the implementation on the model on a scalable hardware architecture with bio-inspired features will be described. The implementation results show that it is possible to attain real-time processing capabilities for dynamic visual stimuli.

2 Spiking Neural Network Model

The model consists of Leaky Integrate-and-Fire neuromimes connected by synapses with variable weight depending on the time correlation between pre- and post-synaptic

S. Kollias et al. (Eds.): ICANN 2006, Part I, LNCS 4131, pp. 936 – 943, 2006.

spikes. The synaptic potentials are added until their result $V_i(t)$ overcomes a certain threshold. Then a spike is produced, and the membrane value is reset. The simplified equation of the membrane value is:

$$V_i(t+1) = \begin{cases} 0 & when\ S_i(t) = 1 \\ k_{mem} \cdot V_i(t) + \sum J_{ij}(t) & when\ S_i(t) = 0 \end{cases} \tag{1}$$

where $k_{mem}=exp(-\Delta t/\tau_{mem})$, $Vi(t)$ is the value of the membrane and $S_i(t)$ is the state variable which signals the occurrence of a spike. The value of J_{ij} is the output of each synapse *(ij)* where *j* is the projecting neuron and *i* is the actual neuron.

When a spike occurs in the pre-synaptic neuron, the actual value of the synaptic output J_{ij} is added to the weight of the synapse multiplied by an activation variable *A*. Conversely, if there is no pre-synaptic spike then the output J_{ij} is decremented by a factor k_{syn}. Then, the value of J_{ij} corresponds to the following equation:

$$J_{ij}(t+1) = \begin{cases} J_{ij}(t) + (w_{RiRj} \cdot A_{RiRj}(t)) & when\ S_j(t) = 1 \\ k_{syn} \cdot J_{ij}(t) & when\ S_j(t) = 0 \end{cases} \tag{2}$$

where *j* is the projecting neuron and i is the actual neuron. *R* is the type of the neuron : excitatory or inhibitory, *A* is the activation variable which controls the strength of the synapse, and k_{syn} is the kinetic reduction factor of the synapse. If the actual neuron is inhibitory, this synaptic kinetic factor will reset the output of the synapse after a time step, but if the actual neuron is excitatory, it will depend on the projecting neuron. If the projecting neuron is excitatory the synaptic time constant will be higher than if it is inhibitory. The weight of each synapse also depends on the type of neuron it connects. If the synapse connects two inhibitory neurons, the weight will always be null, so an inhibitory cell cannot influence another inhibitory cell. If a synapse is connecting two excitatory neurons, it is assigned a small weight value. This value is higher for synapses connecting an excitatory neuron to an inhibitory one, and it takes its maximum value when an inhibitory synapse is connected to an excitatory cell.

The changes in strength of an excitatory-excitatory synapse depend on the variable *A* which is a function of on an internal variable L_{ij} given by the following equation:

$$L_{ij}(t+1)=k_{act}\cdot L_{ij}(t) + (YD_j(t)\cdot S_i(t)) - (YD_i(t)\cdot S_j(t)) \tag{3}$$

where k_{act} is a kinetic activity factor, which is the same for all the synapses and *YD* is a "learning" decaying variable that depends on the interval between a pre-synaptic spike and a post-synaptic spike. When there is a spike, *YD* reaches its maximum value at the next time step. In the absence of a spike the value of *YD* will be decremented by the kinetic factor k_{learn}, which is the same for all synapses. When a pre-synaptic spike occurs just before a post-synaptic spike, then the variable L_{ij} is increased and the synaptic strength becomes larger, thus corresponding to a potentiation of the synapse. When a pre-synaptic spike occurs just after a post-synaptic spike, the variable L_{ij} is decreased, the synaptic weight is weakened , thus corresponding to a depression of the synapse. For all kind of synapses, except the excitatory-excitatory, the activation variable is always is set to 1.

3 Hardware Implementation

In this section we shall consider the detailed implementation of the model, as well as its optimization for an efficient hardware realization.

The overall organization of the neuron model is depicted in Figure 1.

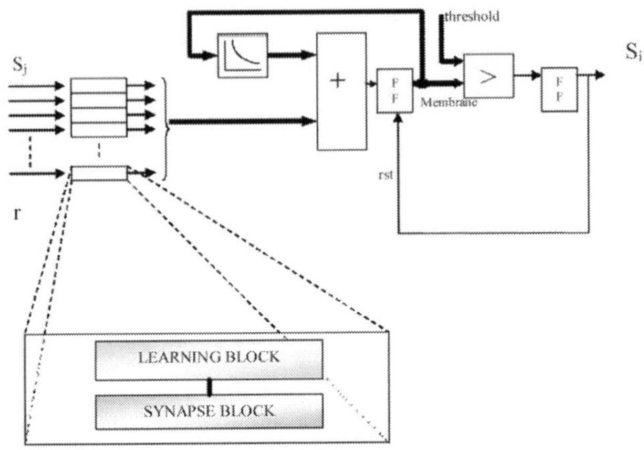

Fig. 1. Overall organization of the neuron model

The description of the neuron block can be divided in three main parts. In the first part the spikes(s) received from outside (probably from other neurons) are processed through a block that encompasses two additional sub-blocks, synapse and learning, which will be explained later. These sub-blocks are used to give appropriate inputs to the next building blocks of the neuron model.

In a second stage, the inputs are added or subtracted, depending on the nature (r) of the previous neuron (i.e. excitatory or inhibitory), to the decayed value of the membrane . The result of this final addition is what we call "membrane value" and it is stored in a flip-flop (FF in Figure 1). This membrane value is always processed through a decay function which gives the adding value in the next time step. The registered output of the membrane is compared in the third sub-block with a predefined threshold value. When the membrane value reaches this threshold, a spike is produced. This spike will be delayed in the final part with a flip-flop which models the refractory time. When finally the spike goes out from the neuron, it produces a reset (rst signal in Figure 1) in the flip-flop which stores the value of the membrane.

A major building block in the neuron model is the decay block, since it will be used both in the synapse and in the learning blocks. This block is aimed to implement a logarithmic decay of the input; it is obtained with a subtraction and controlling the time when it is done depending on the input value. The organization of this block is presented in Figure 2. In this figure the decaying variable is labeled x. A new value of x will be the input of a shift register which is controlled by the most significant bit (*MSB*) of x and by an external parameter *mpar*. The output of this shift register will be subtracted from the original value of x. This operation will be done when the time control indicates it. The time control is implemented by the value of a counter that is compared with the result of choosing between the external value *step* and the product *(MSB–mpar)·step*. The decay variable τ depends on the input parameters *mpar* and *step*.that is controlled by the time when it is done depending on the input value.

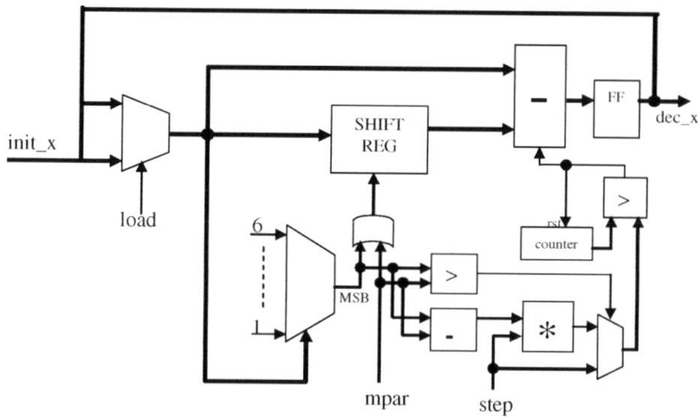

Fig. 2. Organization of the decay block

The learning block "measures" the interval between a spike in the projecting neuron *j* and the actual neuron *i*. Depending on these timings and the types of the two neurons, the synaptic strength will be modified. When a spike is produced by the projecting neuron, the variable *YD* is set to its maximum value and starts to decay. If a spike is produced by the actual neuron immediately after the presynaptic neuron the value of YD_j is added to the decaying value of *L*. Conversely, if a spike is produced at first in the actual neuron and later in the projecting neuron, then the value of YD_i is subtracted to the decaying value of *L*. If the *L* variable overcomes a certain threshold L_{th}, positive or negative, then the activation variable *A* is increased or decreased, respectively, unless the variable had reached its maximum or minimum, respectively. If the variable *A* is increased, then *L* is reset to the value $L\text{-}2{\cdot}L_{th}$; if *A* is decreased, then *L* is reset to $L+2{\cdot}L_{th}$. Figure 3 illustrates the organization of the learning block.

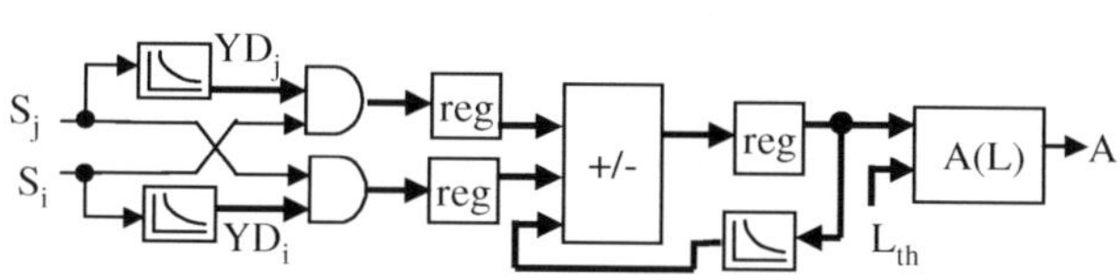

Fig. 3. Organization of the learning block

The synapse block is aimed to set the value of *J* (analogous to the the sum of all post-synaptic membrane potentials) and depends on four factors: the activation level *A* of the synapse, the spiking state of the projecting neuron S_j and the types of the pre- and post-synaptic neurons (R_i and R_j).

A given weight is set for each synapse. This weight is multiplied by the activation variable *A* by means of a shift register, such that if *A*=0, the weight is multiplied by 0, if *A*=1 it is multiplied by 1, if *A*=2 it is multiplied by 2, and if A=3 it is multiplied by 4. This weighted output is added to the decaying value of the variable *J*.

This operation depends on the neuronal types (R_i and R_j). In the current case study there are only two types of neurons, excitatory and inhibitory. If both neurons are inhibitory the weight of the synapse is set to 0 and the value of J is always 0 and no decay is implemented. For the other three types of synapses the time constants are multiplexed, and the multiplexer is controlled by the types of neurons (R_i,R_j). The value of J is obtained at the output of the decay block controlled by the multiplexer. Figure 4 shows the organization of the synapse block.

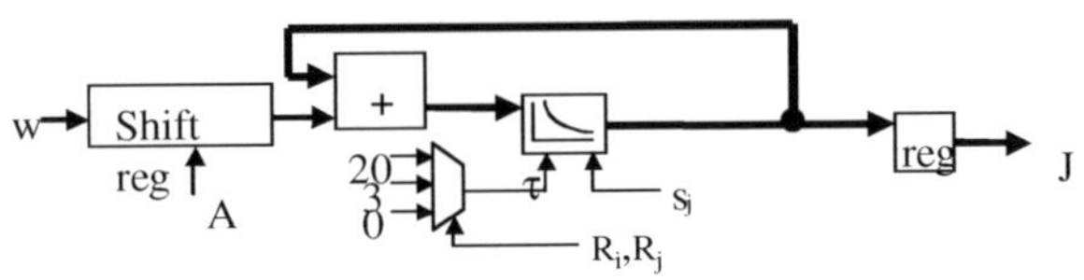

Fig. 4. Organization of the synapse block

The resolution required to represent the values of the variables and the number of operations to be performed may pose a serious limitation for the final implementation. Therefore, an important step consisted in evaluating the model and tuning its parameters in order to get a satisfactory performance. The implementation used in this study has been based on a neural network of size 15x15 with a connectivity pattern of 24 neurons corresponding to a neighborhood of 5x5. The distribution of the 20% inhibitory cells was random. The weights, w, and the initial activation variables, A, were also chosen randomly. Dynamic gradient stimuli have been applied to the neural network. A sequence of vertical bars of gradient intensity move over "strips" of neurons placed in the 2D array of the neural network.

The vertical bars may move at different speeds (i.e. spatial frequency). A neuron "hit" by the stimulus receives an input that is proportional to the gradient intensity. The activity of the network has been studied in a "training" condition and in a "test" condition. During training the spatial frequency of the stimulus has been incremented by discrete harmonics (2x, 4x, etc.) in one direction (the "forward" direction). During test, the stimuli were presented in both forward and reverse sense. A Gaussian noise (Mean 0, SD= 48) is applied to all neurons during all the time. The characteristics of the input applied to each neuron are the following:

- T_{CLK}: 20 ns. Maximum amplitude: 127.
- Training period: 20 μs. Forward sense
- Test period: 10 μs. Forward and Reverse sense

The results from this experiment demonstrate that the selected structure of our neural network is able to perform an implicit recognition of dynamic features based on simple unsupervised STDP rules.

In a first attempt to reduce the complexity of the final hardware implementation the resolution of the parameters has been reduced by 2 bits. By repeating the simulation experiments explained previously we could determine that this is the minimum accuracy required by the system in order to exhibit discrimination features for dynamic input stimuli. Table 1 shows the new values of the internal parameters after this optimization process.

Table 1. Resolution of the parameters for an optimized implementation

Parameter	New value
Membrane resolution	10 bits
Threshold	+160
Input (J) resolution	6 bits
Weights (R_i,R_j) (00, 01, 10, 11)	[0:8], [64:128], [128:256], [0:0]
YD resolution	4 bits
L resolution	6 bits
Membrane decay time constant	20
YD decay time constant	20
L decay time constant	4000
J_{R_i,R_j} decay time constants (R_i,R_j) (00, 01, 10, 11)	(20, 0, 3, 0)

Once this simplification has been performed a further simplification has been carried out [7] in the design of the constituent building blocks. In this optimization a serial approach has been used in order to keep the functional units as compact as possible.

4 Implementation on a Bio-inspired Architecture

The POEtic tissue [8] constitutes a flexible hardware substrate that has been specifically conceived in order to permit the efficient implementation of bio-inspired models. The tissue may be constructed as a regular array composed of POEtic chips, each of them integrating a custom 32-bit RISC microprocessor and a custom FPGA with dynamic routing capabilities.

The custom FPGA included in the POEtic chip is composed of a bi-dimensional array of elementary programmable elements, called molecules. Each molecule contains a flip-flop, a 16-bit lookup table (LUT) and a switchbox that permits to establish programmable connections between molecules.

After the optimization carried out on the neural model in order to facilitate its hardware realization it has been mapped on to the molecules that constitute the POEtic device. The molecule organization shown in Fig. 5 corresponds to the actual structure of the FPGA present in the POEtic device, which is arranged as an 8x18 array of molecules.

The VHDL models developed for the POEtic tissue have been configured and simulated to validate the functionality of the neuron model designed above. After this

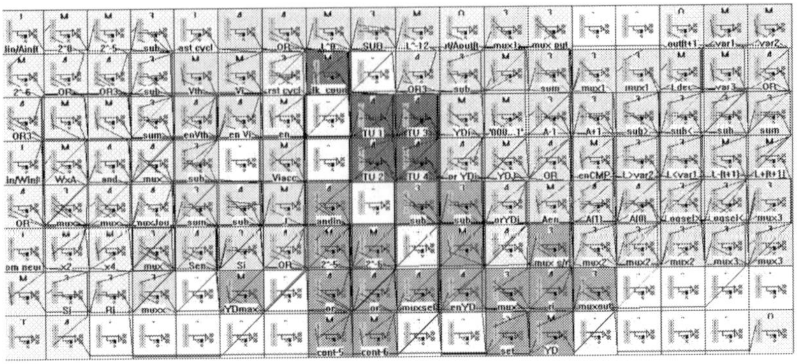

Fig. 5. Molecule-level implementation of the SNN model

validation stage the strategy for the simulation of large-scale SNN models has been considered. Since in its actual implementation the POEtic chip only allows for the implementation of a single neuron it will be necessary to use an array of POEtic chips whose functionality should be time–multiplexed in order to emulate the entire network. This means that every POEtic chip should be able to manage a local memory in charge of storing the weights and learning variables corresponding to the different neurons it is emulating in time.

A 16-neurons network organized as a 4x4 array has been constructed using this principle. This would permit the emulation of a 10,000-neurons network in 625 multiplexing cycles. Bearing in mind that each neuron is able to complete a time step in 150 clock cycles, this means that the minimum clock frequency required to handle input stimuli in real time (i.e., to process visual input stimuli at 50 frames/second) is around 5 MHz far within the possibilities of the actual clock frequency achieved by the POEtic tissue (between 50 MHz and 100 MHz).

The visual stimuli will come from an OmniVision OV5017 monochrome 384x288 CMOS digital camera. Specific VHDL and C code have been developed in order to manage the digital images coming from the camera. To test the application, artificial image sequences have been generated on a display and then captured by the camera for its processing by the network.

5 Conclusions

In this paper we have presented the detailed translation process of a biologically plausible spiking neural network model onto a physical hardware implementation based on a scalable architecture with bio-inspired features. During the translation process special attention has been paid to the accuracy constraints of the implementation, so as to obtain a compact physical realization. The results of the current implementation demonstrate that the proposed approach is capable of supporting the real-time needs of large-scale spiking neural networks models. Our current work is concentrated on the physical test and qualification of the POEtic chips received from the foundry using the development boards that have been constructed for the POEtic tissue. After that the configuration corresponding to the proposed model will be downloaded and physically tested on the actual chips.

Acknowledgements

The work presented in this paper has been funded by the grant IST-2000-28027 (POEtic) of the European Community and by grant OFES 00.0529-2 of the Swiss government. The information provided is the sole responsibility of the authors and does not reflect the Community's opinion. The Community is not responsible for any use that might be made of data appearing in this publication.

References

1. Maas, W.: Networks of Spiking Neurons: The Third Generation of Neural Network Models. *Neural Networks* 10 (1997) 1659–1671.
2. Hill, S.L., Villa, A.E.P.: Dynamic transitions in global network activity influenced by the balance of excitation and inhibition. *Network: Computation in Neural Systems* 8 (1997) 165-184.
3. Abbott, L.F., Nelson, S.B.: Synaptic plasticity: taming the beast. *Nature Neuroscience* 3 (2000) 1178–1183.
4. Bell, C.C., Han, V.Z., Sugawara, Y., Grant, K.: Synaptic plasticity in a cerebellum-like structure depends on temporal order. *Nature* 387 (1997) 278–281.
5. Montgomery, J.M., Madison, D.V.: Discrete synaptic states define a major mechanism of synapse plasticity. *Trends in Neurosciences* 27 (2004) 744-750.
6. Eriksson, J., Torres, O., Mitchell, A., Tucker, G., Lindsay, K., Halliday, D., Rosenberg, J., Moreno, J.M., Villa, A.E.P.: Spiking Neural Networks for Reconfigurable POEtic Tissue. Evolvable Systems: From Biology to hardware. *Lecture Notes in Computer Science* 2606 (2003) 165-173.
7. Torres, O., Eriksson, J., Moreno, J.M., Villa, A.E.P.: Hardware optimization and serial implementation of a novel spiking neuron model for the POEtic tissue. *BioSystems* 76 (2003) 201–208.
8. Moreno, J.M., Thoma, Y., Sanchez, E., Torres, O., Tempesti, G.: Hardware Realization of a Bio-inspired POEtic Tissue. *Proceedings of the NASA/DoD Conference on Evolvable Hardware*. IEEE Computer Society (2004) 237-244.

A Time Multiplexing Architecture for Inter-neuron Communications

Fergal Tuffy[1], Liam McDaid[1], Martin McGinnity[1], Jose Santos[1], Peter Kelly[1], Vunfu Wong Kwan[2], and John Alderman[2]

[1] University of Ulster, Intelligent Systems Engineering Laboratory, School of Computing and Intelligent Systems, Faculty of Engineering, Magee Campus, Northland Road, Derry, N. Ireland, BT48 OLY
{f.tuffy, lj.mcdaid, tm.mcginnity, ja.santos, pm.kelly}@ulster.ac.uk

[2] Tyndall National Institute, Lee Maltings, Prospect Row, Cork, Rep. of Ireland
{vunfu, john.alderman}@tyndall.ie

Abstract. This paper presents a hardware implementation of a Time Multiplexing Architecture (TMA) that can interconnect arrays of neurons in an Artificial Neural Network (ANN) using a single metal wire. The approach exploits the relative slow operational speed of the biological system by using fast digital hardware to sequentially sample neurons in a layer and transmit the associated spikes to neurons in other layers. The motivation for this work is to develop minimal area inter-neuron communication hardware. An estimate of the density of on-chip neurons afforded by this approach is presented. The paper verifies the operation of the TMA and investigates pulse transmission errors as a function of the sampling rate. Simulations using the Xilinx System Generator (XSG) package demonstrate that the effect of these errors on the performance of an SNN, pre-trained to solve the XOR problem, is negligible if the sampling frequency is sufficiently high.

1 Introduction

Biological research has accumulated an enormous amount of detailed knowledge about the structure and functionality of the brain. It is widely accepted that the basic processing units in the brain are neurons which are interconnected in a complex pattern, communicate through pulses and use the timing of the pulses to transmit information and perform computations [1-3]. Significant research has focused on "biological equivalent" neural network models that can be implemented in hardware and used to inspire new techniques for real time computations [4-6]. However, the standard network topologies employed to model the biological networks are proving difficult to implement in hardware, even for moderately complex networks. Existing inter-neuron connection schemes are achieved through metallization and thus as the size of the neuron array increases there is a rapid increase in the ratio of metal to device area which eventually self-limits the network size [7-8]. Given that the density of the interconnect pathways in the human brain is of the order of 10^{14} [9], it is inconceivable that existing interconnect technologies will even remotely approach this order of magnitude and thus new approaches need to be explored.

S. Kollias et al. (Eds.): ICANN 2006, Part I, LNCS 4131, pp. 944–952, 2006.

This paper presents a novel Time Multiplexing Architecture (TMA) as a possible solution to the interconnect problem for Spiking Neural Networks (SNNs) in hardware. A single bus wire is used to transmit the signals between neuron layers, where timing is guaranteed by using a clocking system that is synchronized to a "global" clock. This implementation removes the requirement of dedicated metal lines for every synaptic pathway and therefore a significant saving in the silicon surface area is achieved. Section 2 of this paper discusses the TMA while section 3 highlights results that verify the approach. Errors in spike timing "across the TMA" due to the sampling frequency are investigated in section 4 where a simple SNN is initially trained, using a supervised approach, to solve the XOR problem. Using the Xilinx System Generator (XSG) package the output firing times that results from the TMA architecture are compared with those obtained when conventional metal interconnect is used, and from the subsequent analysis it is clear that the sampling frequency must be at least twice the minimal sampling frequency: note the minimal sampling frequency is set by the duration of the spike and the number of neurons in the sampled layer. Section 5 presents a quantitative analysis underpinning the scalability of the TMA and section 6 makes concluding remarks.

2 Time Multiplexing Architecture (TMA)

This section presents a novel inter-neuron communication architecture where biologically compatible neuron spikes are represented as digital pulses and connectivity between neuron layers is achieved using the TMA. Figure 1 shows a two layer neural network fragment containing two input neurons, I_0 and I_1, and one output neuron, O_0. The sampling circuit to the left of the bus wire contains two D-type latches in a daisy chain configuration where one of the latches is preset to logic 1, prior to the application of the clock, C_K. Effectively the clock input, C_K, rotates a logic 1 between the two latches, switching on transistors *M1* and *M2* sequentially: *M1*, *M2*, *M3* and *M4* are *n*-channel enhancement mode MOSFETs. This sampling to the left of the bus wire is repeated on the right of the bus wire. Consider the case where the input neuron, I_0, fires a spike, {0, 1, 0}, of duration T_P which forms the input to the drain, *D*, of *M1*. The gate terminal, *G*, of *M1* is controlled by the *Q* output of a D-type latch and when *Q* is asserted, I_0 is sampled and a logic 1 is placed on the bus wire: note that the gate of *M2* will be held at logic 0 while *M1* is on (sampling). Because both sampling circuits are driven from the same clock input, C_K, the bus line is now sampled by *M3* ensuring that the pulse from I_0 is directed to the correct synapse, Synapse 1. I_1 will be sampled directly after I_0 whereby *M2* and *M4* will be turned on by the sampling circuits allowing the pulse from I_1 (if I_1 has fired) to reach Synapse 2. Clearly the sampling frequency is a function of the number of neurons in the sampled layer and the duration of the spike pulse. In a layer of *n* neurons which are sampled sequentially, it can be shown that the minimum sampling frequency F_S (Hz) is given by,

$$Fs = \frac{n}{T_P} \tag{1}$$

The authors are aware that pulse transmission errors can exist between the time a neuron in one layer fires and the time required for this pulse to be observed at the synaptic inputs associated with the neurons in the subsequent layer. These are caused by the sampling circuitry operating in a synchronous mode while all the neurons that are sampled will fire in an asynchronous mode. Pulse transmission errors and their effect on a pre-trained SNN are investigated in section 5.

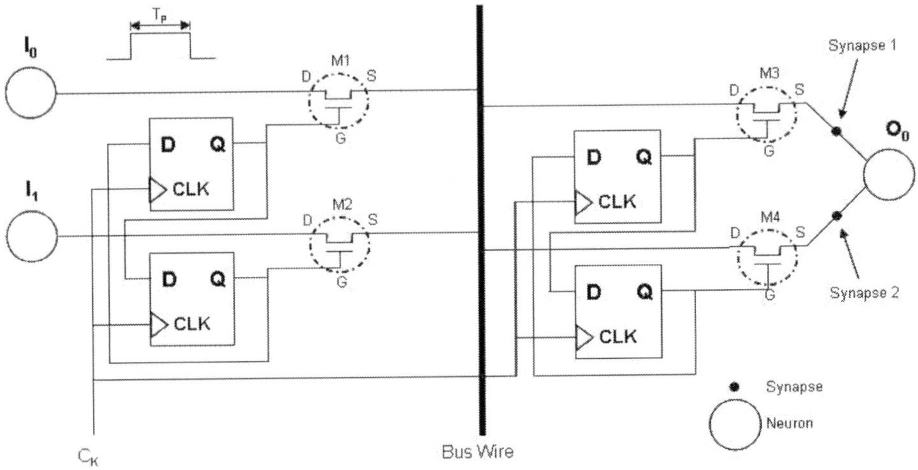

Fig. 1. TMA for a 2-input 1-output SNN

3 Simulation Results

The proposed TMA was simulated using the Mentor Graphics mixed signal simulation package, System Vision Professional. Figure 2 represents the layout used in the simulation where the SNN has four input neurons, I_0-I_3, and two output neurons, O_0, O_1 (note that this architecture is modified from that shown in figure 1 in that the MOSFET transistors at the input to each synapse are replaced by D-latches, *D13-D20*). It will be shown later that in order to reduce pulse transmission errors it is necessary to sample at a rate that is in excess of the minimum sampling frequency defined by equation (1). However, "gating" these high frequency pulses using MOSFETs causes glitches at the input to the synapses. This problem is avoided by the additional layer of D-latches, *D13-D20*.

In the simulations, as shown in figures 2 and 3, the pulse length for all neurons, T_P, was set to 1ms and since there are four input neurons, the sampling frequency was calculated from equation (1) to be 4KHz. Because M_1-M_4 are not ideal the transitions from logic 1 to logic 0, and vice versa, are not instantaneous. Therefore, to avoid any overlap between the turn on transient of one transistor and the turn off of another a two phase clock system is used where one clock C_{K1} operates on the sampling circuit to the left of the bus wire and another clock C_{K2} operates on the sampling circuit to

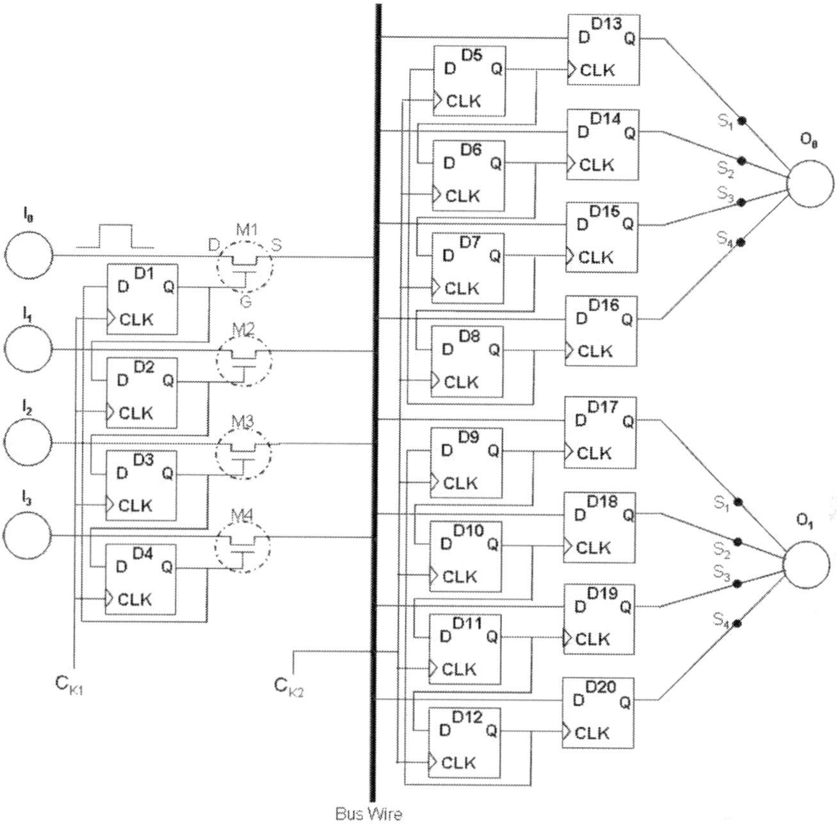

Fig. 2. TMA system layout for 4-input, 2-output NN

the right: note that C_{K1} and C_{K2} are in anti-phase but operate at the same frequency (8 KHz), as shown in figure 3(a). Figure 3 (b) shows random firing of neurons $I_0 - I_3$, and their arrival times at the appropriate synapses. It can be seen that there exists a time error ∂t_{I_0} between I_0 firing and the arrival time of the pulse at the appropriate synapses. Note that from figure 3(b) similar errors exist for all pulses and therefore while TMA provides inter-neuron communication, transmission errors exist. The following section analyses these errors to determine their effect on the dynamics of a pre-trained SNN.

4 Xilinx System Generator Implementation

In order to investigate pulse transmission errors both conventional interconnect and the TMA were used to interconnect neuron layers in a SNN topology that has been pre trained to solve the benchmark XOR problem [10]. Both topologies were

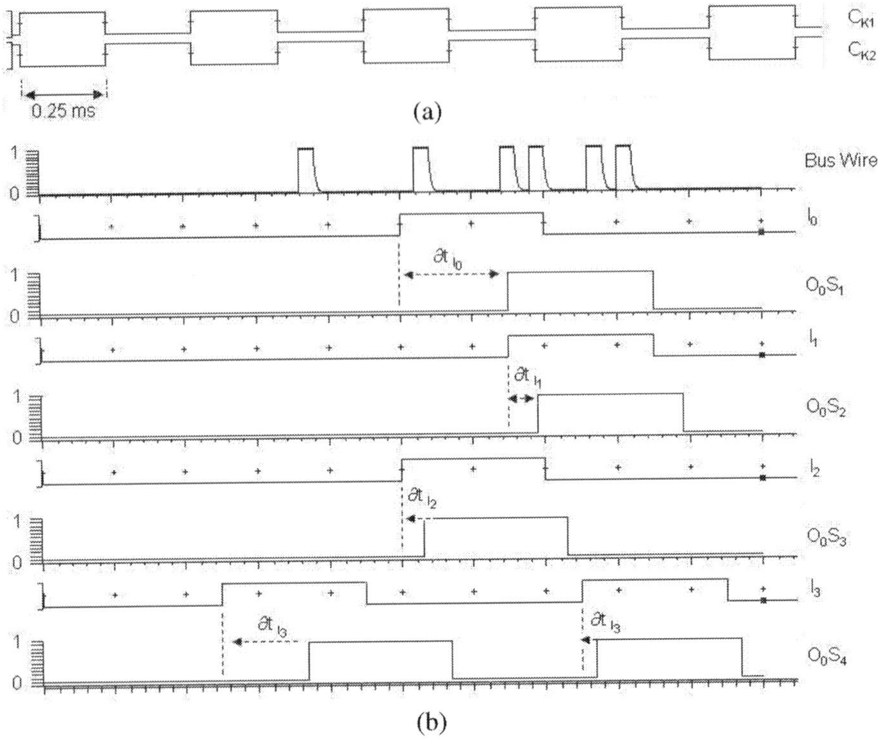

Fig. 3. (a) Timing diagram where the clock signal to the output sampling D-latches, *D5-D12*, is delayed by 0.25ms, quarter of the sampling pulse period. (b) bus wire signals caused by random firing neurons I_0 - I_4, and O_0S_1 - O_0S_4, show the time of arrival of pulses at the appropriate synapses.

simulated using the XSG toolset from Xilinx [11], as illustrated in figure 4. The SNN was trained off-line by an Evolutionary Strategy (ES) [10].

Table 1 shows three input neurons where neuron 1 is a biasing neuron [10], which fires at 0ms, and neurons 2 and 3 provide the conventional 2 inputs for the XOR truth table. Note that column four is the post-trained firing times of the output neuron where the simulation used conventional metal interconnect. Columns 5 and 6 are the firing times of the same output neuron where the simulation used the TMA and clearly transmission errors are appreciable if the minimum sampling rate is used (column 4), this is the worst case firing times. However, if the sampling frequency is increased to $2{*}F_S$, then satisfactory agreement between column 6 and column 4 is obtained. Therefore, for effective pulse transmission without significant error the sampling frequency must be maintained such that

$$F_S \geq 2\ \frac{n}{T_P} \tag{2}$$

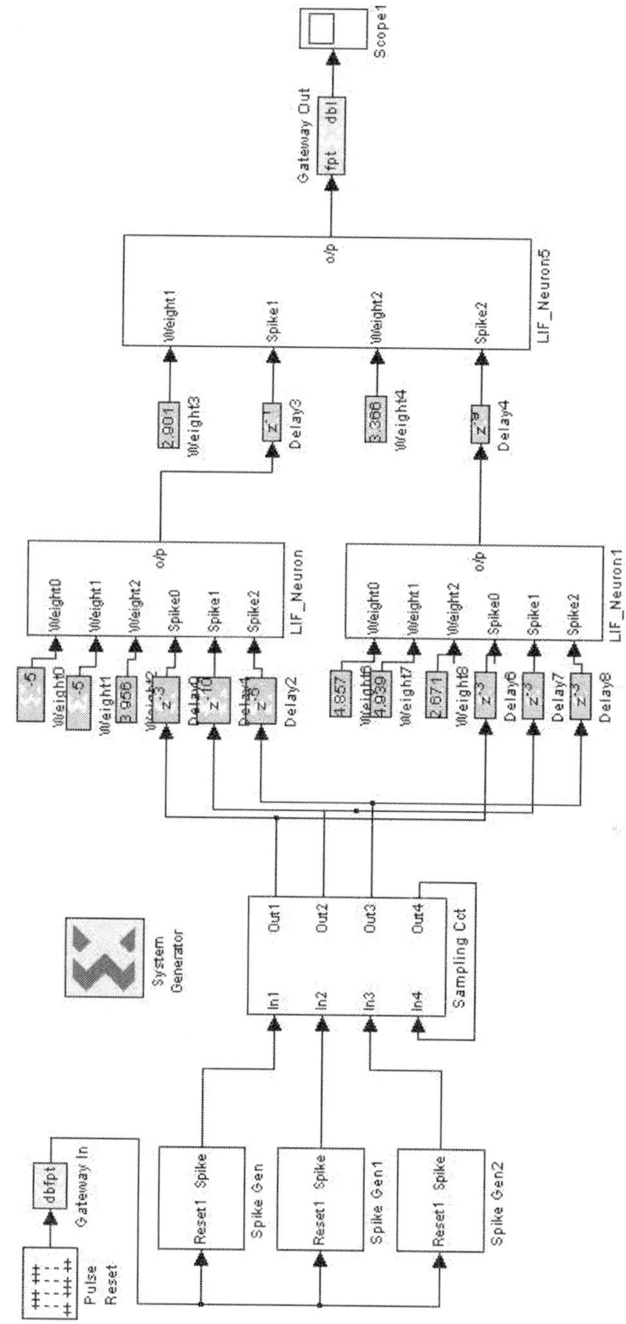

Fig. 4. SNN for XOR problem containing TMA in XSG simulator

Table 1. XOR dataset simulation results with and without TMA. Table includes 3 neuron inputs where neuron 1 is a biasing neuron and neurons 2 and 3 provide the conventional 2 inputs for the XOR truth table. The trained firing times (without TMA) is compared with the actual firing times for a sampling frequency of F_S and $2*F_S$.

Neuron 1 Firing Time	Neuron 2 Firing Time	Neuron 3 Firing Time	Firing time without TMA (ms)	Firing Times with TMA F_S (ms)	Firing Times with TMA $2F_S$ (ms)
0	0	0	14	15	15
0	0	6	20	15	21
0	6	0	20	15	21
0	6	6	14	22	15

5 TMA Scalability

To demonstrate the scalability of the TMA consider a network where we have n input neurons and m output neurons. It should be noted that the number of input neurons n, afforded by the TMA technique, is a function of the maximum possible operating frequency of the global clock while the theoretical limit on the scale of an $n*m$ network is determined by the physical size of the sampling circuits. To estimate n, consider a 1ms spike and assume a realistic sampling frequency F_S of 1GHz [12]. Equation (1) is then used to predict the number of neurons that can be accommodated on the input layer which equates to approximately one million. Even if we sample at $2*F_S$ to minimise pulse transmission errors, then equation (2) predicts an upper limit for n of half a million. This is an improvement over what is currently achieveable [13]. However, it is clear that the scale of a SNN implemented using the proposed TMA is unlikely to be severely limited by the frequency of the global clock, rather scaleability will be limited by the real estate occupied by circuitry, and the following is an estimate of this limit.

Consider again the case where we have n input neurons and the number of output neurons, m, is allowed to increase. If we assume a fully connected feedforward network then the number of associated synapses increases according to the product nm. To calculate the limit on the network size, an estimate of the area consumed by the associated sampling circuits is required. Given that the sampling circuit is dominated by n D-type latches in the transmitting layer and $2*n*m$ D-type latches in the receiving layer, then we can write that the total area, A_T, occupied by the sampling circuitry is given by

$$A_T = (n + 2nm)\, A_D \approx 2nmA_D \tag{3}$$

for large m where A_D is the area of a D-type latch. It has been reported for a 0.18-μm process technology that a D-type latch can be designed to occupy a silicon area of approximately 4μm^2 [14] and if the area occupied by sampling circuitry is restricted to 10% of the total chip area (assumed to be 1cm^2), then a simple calculation (taking n = m) predicts that the TMA approach permits over three thousand neurons to be fabricated in each layer using a planar submicron process. For a fully connected feed-forward NN this equates to 9 million synapses. While this is a significant improvement from what is reported elsewhere [13], it will be further enhanced as technology improvements continue [15]. Furthermore, given that the interconnect density will be substantially reduced by the proposed TMA then the real estate given over to the sampling circuitry is expected to be in excess of the 10% estimate. Hence, the above estimate is viewed as conservative and it is expected that the proposed TMA approach will advance the synaptic density even further.

6 Conclusion

This paper has proposed a novel time sampling architecture for the hardware implementations of SNNs. This work has shown that the optimal sampling frequency depends on the number of neurons in the sampled layer and the duration of the "digital spikes" they emit. However, with on-chip clock frequencies typically in the GHz range, the limitations placed on this approach by the sampling frequency are negligible. The TMA has been verified using the Mentor System Vision Pro software package and issues such as pulse transmission errors have been investigated using the XSG platform. It has been shown that these errors can be minimized by ensuring that the sampling frequency is maintain to at least twice the minimum sampling frequency ($2*F_S$). The authors wish to note that this paper has demonstrated the potential of the TMA for inter-neuron communication where the target implemented for this approach is a mixed signal Application Specific Integrated Circuit (ASIC) layout, given the asynchronous firing nature of neurons. Moreover, the authors are confident that if this approach is optimized in terms of minimal area circuitry and timing issues are addressed for large implementations, then this approach has the potential to implement well over a million inter-neuron pathways using a very simple and compact sampling architecture. Future work shall involve a comparative analysis with alternative interconnect strategies such as Address Event Decoding (AED).

Acknowledgment

This work was part supported by the European Union under the Irish Government National Development Plan, as part of a Higher Education Authority North-South program for collaborative research project – Interwave.

The authors would also like to thank Simon Johnston and Brendan Glackin, at the Intelligent Systems Engineering Laboratory (ISEL), for assistance in the XSG and Xilinx FPGA simulations and numerous fruitful discussions.

References

1. Roche, B., McGinnity, T.M., Maguire, L.P., McDaid, L.J.: Signalling Techniques and their Effect on Neural Network Implementation Sizes", Information Sciences 132, pages 67-82, NH Elsevier, 2001
2. Murray, F., and Woodburn, R.: The Prospects for Analogue Neural VLSI, International Journal of Neural Systems, Vol. 8, No. 5 & 6, pages 559-579, Oct/Dec. 1997

3. Liu, S.C., Kramer, J., Indiveri, G., Delbruck, T., Burg, T., and Douglas, R.: Orientation-selective VLSI Spiking Neurons, Neural Networks, Special Issue on Spiking Neurons in Neuroscience and Technology , Vol. 14, Issues 6-7, pages 629-643, July 2001
4. Diorio, C., Hsu, D., and Figueroa, M.: Adaptive CMOS: from biological inspiration to systems-on-a-chip, Proceedings of the IEEE, Vol. 90, Issue 3, pages 345 – 357, March 2002
5. Goldberg, D.H., Cauwenberghs, G., Andreou, A. G.: Probabilistic Synaptic Weighting in a Reconfigurable Network of VLSI Integrate-and-Fire Neurons, Neural Networks, Vol. 14, no. 6 7, pages 781 793, Sept 2001
6. Maass, W.: Computation with Spiking Neurons: the Handbook of Brain Theory and Neural Networks, MIT Press, 1998.
7. Noory, B., Groza, V.: A Reconfigurable Approach to Hardware Implementation of Neural Networks, IEEE CCECE 2003. Canadian Conference on Electrical and Computer Engineering, pages 1861 - 1864 Vol. 3, 4-7 May 2003
8. Chun, L., Shi, B., Chen, L.: Hardware Implementation of an Expandable On-chip Learning Neural Network with 8-Neuron and 64-Synapse, TENCON '02. Proceedings 2002 IEEE Region 10 Conference on Computers, Communications, Control and Power Engineering, Vol. 3, pages 1451 – 1454, 28-31 Oct. 2002
9. Miki, T., Editor: Brainware: Bio-Inspired Architectures and its Hardware Implementation, World Scientific Publishing Co. Ltd, 2001.
10. Johnston, S.P, Prasad, G., Maguire, L. P., McGinnity, T. M.: Comparative Investigation into Classical and Spiking Neuron Implementations on FPGAs, 15th International Conference on Artificial Neural Networks, ICANN 2005, Part 1: pages 269-274, 11-15 Sept. 2005
11. http://www.xilinx.com/ise/optional_prod/system_generator.htm
12. Tu, S.-W., Jou, J.-Y., Chang, Y.-W.: RLC Coupling-Aware Simulation for On-Chip Buses and their Encoding for Delay Reduction, 2005 ISCAS IEEE International Symposium on Circuits and Systems, 23-26 May 2005 Page(s):4134 - 4137 Vol. 4
13. Chicca, E., Badoni, D., Dante, V., D'Andreagiovanni, M., Salina, G., Carota, L., Fusi, S. and Del Giudice, P.: A VLSI Recurrent Network of Integrate and Fire Neurons Connected by Plastic Synapses with Long Term Memory", IEEE Trans. on Neural Networks, Vol.14, No.5, Sept. 2003
14. Yamaoka, M., Osada, K., Ishibashi, K.: 0.4-V Logic-Library-Friendly SRAM Array Using Rectangular-Diffusion Cell and Delta-Boosted-Array Voltage Scheme, IEEE Journal of Solid-State Circuits, Volume 39, Issue 6, June 2004 Page(s):934 – 940
15. Naeemi, A., Meindl, J.D.: Monolayer Metallic Nanotube Interconnects: Promising Candidates or Short Local Interconnects, IEEE Electron Device Letters, Volume 26, Issue 8, Aug. 2005 Page(s):544 - 546

Neuronal Cell Death and Synaptic Pruning Driven by Spike-Timing Dependent Plasticity

Javier Iglesias[1,2,3] and Alessandro E.P. Villa[1,2,3]

[1] Information Systems Department, University of Lausanne, Switzerland
Javier.Iglesias@unil.ch
http://inforge.unil.ch/
[2] Laboratory of Neuroheuristics, University of Lausanne, Switzerland
http://www.nhrg.org/
[3] Inserm U318, Laboratory of Neurobiophysics, University Joseph Fourier, Grenoble, France
Alessandro.Villa@ujf-grenoble.fr

Abstract. The embryonic nervous system is refined over the course of development as a result of two main processes: apoptosis (programmed cell death) and selective axon pruning. We simulated a large scale spiking neural network characterized by an initial apoptotic phase, driven by an excessive firing rate, followed by the onset of spike-timing-dependent plastiticity (STDP), driven by spatiotemporal patterns of stimulation. In the apoptotic phase the cell death affected the inhibitory more than the excitatory units. The network activity stabilized such that recurrent preferred firing sequences appeared along the STDP phase, thus suggesting the emergence of cell assemblies from large randomly connected networks.

1 Introduction

Genetic programs are assumed to drive the primordial pattern of neuronal connectivity through the actions of a limited set of trophic factors and guidance cues, initially forming excessive branches and synapses, distributed somewhat diffusely [9]. Then, refinement processes act to correct initial inaccuracies by pruning inappropriate connections while preserving appropriate ones. The embryonic nervous system is refined over the course of development as a result of the twin processes of cell death and selective axon pruning. Apoptosis – genetically programmed cell death – and necrosis – pathologic or accidental cell death due to irreversible damage – are two rough mechanisms for refining embryonic connections. However, the creation of complex connectivity patterns often requires the pruning of only a selected subset of the connections initially established by a neuron. Massive synaptic pruning following over-growth is a general feature of mammalian brain maturation [13]. Pruning starts near time of birth and is completed by time of sexual maturation. Quantitative analyses of synaptogenesis in the rat [1], the Rhesus monkey [3], and human [6] cortex have suggested a transient phase of high density of synapses during infancy.

S. Kollias et al. (Eds.): ICANN 2006, Part I, LNCS 4131, pp. 953–962, 2006.

Trigger signals able to induce synaptic pruning could be related to dynamic functions that depend on the timing of action potentials. Spike-timing-dependent synaptic plasticity (STDP) is a change in the synaptic strength based on the ordering of pre- and post-synaptic spikes. This mechanism has been proposed to explain the origin of long-term potentiation (LTP), i.e. a mechanism for reinforcement of synapses repeatedly activated shortly before the occurrence of a post-synaptic spike [2]. STDP has also been proposed to explain long-term depression (LTD), which corresponds to the weakening of synapses strength whenever the pre-synaptic cell is repeatedly activated shortly after the occurrence of a post-synaptic spike [11]. The relation between synaptic efficacy and synaptic pruning [4] suggests that the weak synapses may be modified and removed through competitive "learning" rules. Competitive synaptic modification rules maintain the average neuronal input to a post-synaptic neuron, but provoke selective synaptic pruning in the sense that converging synapses are competing for control of the timing of post-synaptic action potentials [14].

In this study the synaptic modification rule was applied to the excitatory-excitatory *(exc, exc)* and excitatory-inhibitory *(exc, inh)* connections. This plasticity rule might produce the strengthening of the connections among neurons that belong to cell assemblies characterized by recurrent patterns of firing. Conversely, those connections that are not recurrently activated might decrease in efficiency and eventually be eliminated. The main goal of our study is to determine whether or not, and under which conditions, such cell assemblies may emerge from a large neural network receiving background noise and content-related input organized in both temporal and spatial dimensions.

2 Model

The originality of our study stands on the application of an original bio-inspired STDP modification rule compatible with hardware implementation [5]. The complete neural network model is described in details elsewhere [7]. A sketch description of the model with specific model parameters related to the current study follows below.

10,000 integrate-and-fire units (80% excitatory and 20% inhibitory) were laid down on a 100×100 2D lattice according to a space-filling quasi-random Sobol distribution. Sparse connections between the two populations of units were randomly generated according to a two-dimensional Gaussian density function such that excitatory projections were dense in a local neighborhood, but probability long-range excitatory projections were allowed. Edge effects induced by the borders were limited by folding the network as a torus.

The state of the unit (spiking/not spiking) was a function of the membrane potential and a threshold. The states of all units were updated synchronously and the simulation was performed at discrete time steps corresponding to 1 ms. After spiking, the membrane potential was reset, and the unit entered an absolute refractory period lasting 3 and 2 time steps for excitatory and inhibitory units, respectively. For the simulation runs presented here each unit received a background activity following an independent Poisson process and the "spontaneous" mean firing rate of the units was 5 spikes/s.

It is assumed *a priori* that modifiable synapses are characterized by discrete activation levels that could be interpreted as a combination of two factors: the number of synaptic *boutons* between the pre- and post-synaptic units and the changes in synaptic conductance. In the current study we attributed a fixed activation level (meaning no synaptic modification) $A_{ji}(t) = 1$, to *(inh, exc)* and *(inh, inh)* synapses while activation levels were allowed to take one of $A_{ji}(t) = \{0, 1, 2, 4\}$ for *(exc, exc)* and *(exc, inh)*, $A_{ji}(t) = 0$ meaning that the projection was permanently pruned out. For $A_{ji}(t) = 1$, the post-synaptic potentials were 0.84 mV and -0.8 mV for excitatory and inhibitory units, respectively.

The death of units by means of apoptosis is introduced in here, which is a major difference with preivous models [7]. A dead unit is characterized by the absence of any spiking activity. We define two mechanisms inducing cell death: the first is provoked by an excessive firing rate (apoptosis) and the second by the loss of excitatory inputs. For each unit at each time step, the mean firing rate computed over a window of 50 ms preceding the evaluation was compared to a threshold value of 245 and 250 spikes/s for excitatory and inhibitory units, respectively. If the rate exceeded the threshold, then the unit had a probability of entering apoptosis determined by the function

$$P_{\text{apoptosis}}(t) = \frac{0.5 \cdot t^2 - 4.5 \cdot 10^{-6} \cdot t^3}{44 \cdot (2.5 \cdot 10^6 + 6 \cdot 10^{-3} \cdot t^2)} . \tag{1}$$

with $P_{apoptosis}(t = 100) = 4.5 \cdot 10^{-5}$, $P_{apoptosis}(t = 700) = 2.2 \cdot 10^{-3}$, and $P_{apoptosis}(t = 800) = 2.9 \cdot 10^{-3}$. The apoptosis could be induced according to this mechanism during an initial phase lasting 700 or 800 simulation time steps. After this intial phase, the timing of the pre- and post-synaptic activity started driving the synaptic plasticity through the STDP rule. Due to this plasticity, the projections from and to "dead" units underwent a slow activation level decay finally leading to their pruning when $A_{ji}(t) = 0$. "Dead" projections were thus pruned along with others by the action of STDP and some units were found without any excitatory input left. The loss of excitatory inputs provoked the cell death and these units stopped firing (even in presence of background activity) immediately after the pruning of the last excitatory input synapse.

3 Simulations

Each simulation run lasted 10^5 discrete time steps (1 ms per time step), corresponding to a duration of about 2 minutes. After a stabilization period of 1000 ms without any external input, a 100 ms long stimulus was presented every 2000 ms. Overall this corresponds to 50 presentations of the stimulus along one simulation run. Before the simulation started, two sets of 400 excitatory units were randomly selected from the 8,000 excitatory units of the network, labeled sets A and B. Each set was divided into 10 groups of 40 units, $A = \{A_1, A_2, \ldots, A_{10}\}$ and $B = \{B_1, B_2, \ldots, B_{10}\}$. At each time step during a stimulus presentation,

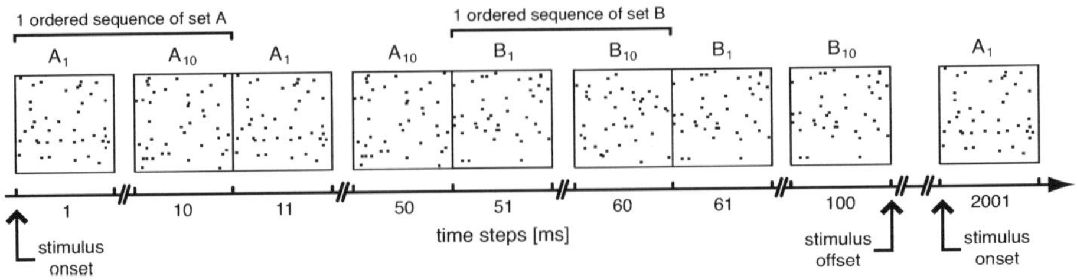

Fig. 1. Example of one AB stimulus presentation

the 40 units of one group received a large excitatory input on their membrane, leading to their synchronous firing. The 10 groups of a set were stimulated following an ordered sequence, thus defining a reproducible spatiotemporal stimulus composed by the repetition of sequences lasting 10 ms each (see Fig. 1). A random, equiprobable mix of the two stimuli composed by either 5× sequence A followed by 5× sequence B (AB) or 5× sequence B followed by 5× sequence A (BA) was presented.

At time t=100 s, the units, characterized by more than four active excitatory input projections that did not belong to the sets of stimulated units A or B, were selected. For each of these selected units, the spikes produced by the independent Poisson background process were discarded from their spike trains to extract the so-called "effective spike trains". Thus, the effective spike trains correspond to the true network activity. The first 1000 ms of activity were discarded because this interval corresponds to the apoptosis phase. These effective spike trains were searched for the occurrence of spatiotemporal firing patterns [16], as described in the following section.

3.1 Time Series Analysis

Spatio-temporal firing patterns (often referred to as "preferred firing sequences") are defined as sequences of intervals with high temporal precision (of the order of few ms) between at least 3 spikes (of the same or different units) that recur at levels above those expected by chance [17]. The pattern detection algorithm begins with finding all single or multineuron sequences of intervals that repeat two or more times within a record. Secondly, the algorithm computes how many of such sequences of intervals can be expected by chance and provides confidence limits for this estimation. The "pattern grouping algorithm"[1] [16] performs clusterization into one group of sequences of intervals with slight difference in spike timing. Figure 2 illustrates the outline of this method. For the present study, the pattern grouping algorithm was used to find patterns of at least three spikes (triplets), with a minimal significance level of 10%, repeating at least 7 times in the interval [1-100] s, provided the entire pattern lasted not more than 800 ms and was repeated with an accuracy of less than ±5 ms.

[1] `http://OpenAdap.net/`

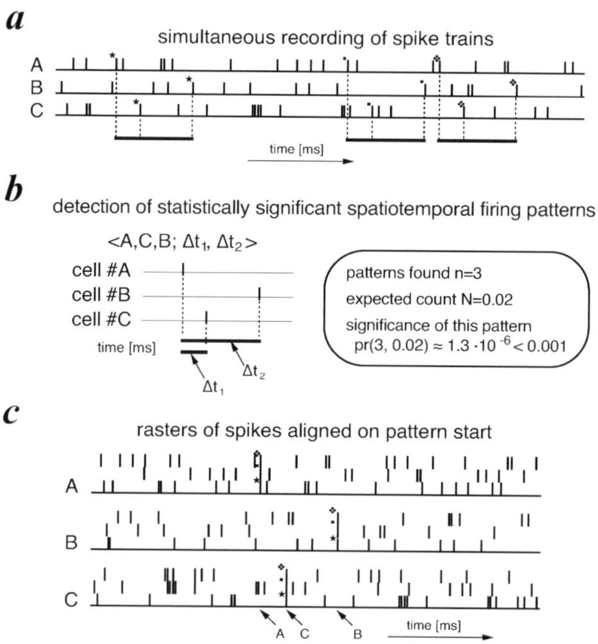

Fig. 2. Outline of the general procedure followed by pattern detection algorithms. **(a)**: Analysis of a set of simultaneously recorded spike trains. Three cells, labeled A, B, and C, participate to a patterned activity. Three occurrences of a precise pattern are detected. Each occurrence of the pattern has been labeled by a specific marker in order to help the reader to identify the corresponding spikes. **(b)**: Estimation of the statistical significance of the detected pattern. **(c)**: Display of pattern occurrences as a raster plot aligned on the pattern start.

4 Results

4.1 Firing Rate-Induced Apoptosis

Figure 3 shows the evolution of the number of excitatory and inhibitory units during the first simulated second. For the first 800 time steps, units with mean firing rates exceeding the threshold entered apoptosis with the probability expressed by $P_{\text{apoptosis}}(t)$. It is possible to linearly fit the cell death dynamics with the probability function suggesting that the inhibitory units enter the apoptosis process about 70 ms before the excitatory units. After the end of this initial phase, STDP-driven synaptic pruning could modify the synaptic weights, thus inducing cell death due to the loss of excitatory inputs at a longer time-scale that is not depicted in Figure 3.

The initial apoptosis phase prevented the network from entering overactivity due to saturation by inducing the death of those units that tended to have an exceeding activity since the early steps of the simulation. These units are known to destabilize the network and ignite the saturation. The addition of this feature

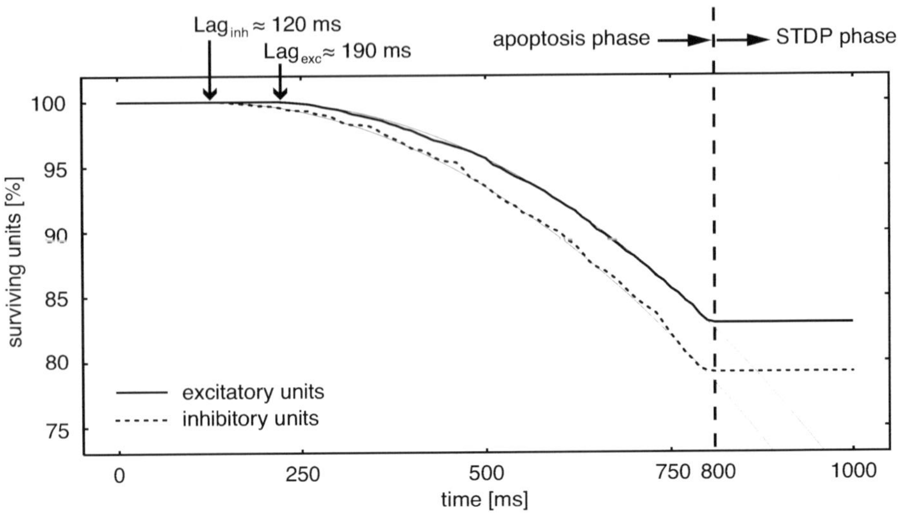

Fig. 3. Ratio of surviving units as a function of time with respect to initial conditions: 8000 excitatory units (plain line) and 2000 inhibitory units (dotted line). In this simulation, firing rate-induced apoptosis was stopped after 800 time steps. Thin lines correspond to the probability function $P_{\text{apoptosis}}(t)$ with lags.

to the model greatly improved the stability of the network while maintaning its ability to produce spatiotemporal firing patterns.

4.2 Spatiotemporal Firing Patterns

In two different simulations, firing rate-induced apoptosis was stopped after 700 or 800 time steps. The first condition lead to a larger number of surviving units at t=100 s with lower mean firing rates than in the second condition. Spatiotemporal firing patterns were searched for in both conditions. Two patterns involving a single excitatory unit are described in more details in Figure 4 and Figure 5.

The pattern <79,79,79; 453±3.5, 542±2.5> was composed by spikes produced by a single unit labeled here #79 (Fig. 4a). This notation means that the pattern starts with a spike of unit #79, followed 453±3.5 ms by a second spike of the same unit, and followed by a third spike 542±2.5 ms after the first. Between t=1 and t=100 seconds, 51 repetitions of the pattern were observed. The statistical significance of this pattern was $7.5 \cdot 10^{-4}$. No correlation could be found between the timing of the spatiotemporal pattern and the stimulation onset (Fig. 4b). Figure 4c shows that the occurrences of the pattern onset along the simulation. The pattern occurred 23 times between $1 < t < 25$ seconds, 13 times between $25 < t < 50$ seconds, and 15 times between $50 < t < 100$ seconds. This might suggest that the network dynamics giving rise to the pattern was slowly disrupted by the continuous STDP-driven pruning.

The pattern <13,13,13; 234±3.5, 466±4.5> was composed by spikes produced by a single unit labeled here #13 (Fig. 5a). This notation means that the pattern

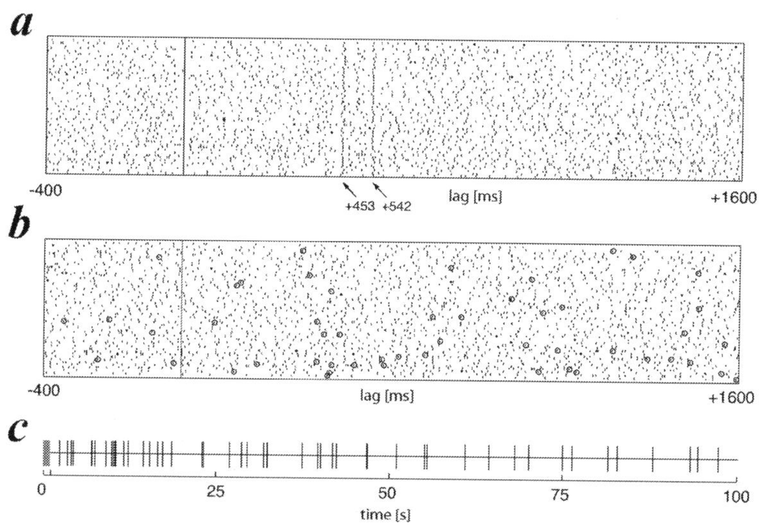

Fig. 4. Spatiotemporal pattern <79,79,79; 453±3.5, 542±2.5>. **(a)**: Raster plot showing the 51 repetitions of the pattern aligned on the pattern start; **(b)**: Raster plot showing the activity of unit #79 aligned on the stimulus onset: each start event of a pattern occurrence is marked by a circle; **(c)**: Pattern occurrence timing plot: each vertical tick represents the start event of a pattern occurrence.

starts with a spike of unit #13, followed 234±3.5 ms by a second spike of the same unit, and followed by a third spike 466±4.5 ms after the first. Between t=1 and t=100 seconds, 52 repetitions of the pattern were observed. The statistical significance of this pattern was $3.4 \cdot 10^{-3}$. No correlation could be found between the timing of the spatiotemporal pattern and the stimulation onset (Fig. 5b). Figure 5c shows that the pattern occurred 7 times between $1 < t < 25$ seconds, 27 times between $25 < t < 50$ seconds, 8 times between $50 < t < 75$ seconds, and 10 times between $75 < t < 100$ seconds. This might suggest that the changes in the network dynamics induced by the continuous STDP-driven pruning lead to a transient state between $25 < t < 50$ seconds when the appearance of this pattern is favored.

5 Discussion

We simulated a large scale spiking neural network, with the time resolution of 1 ms, characterized by a brief initial apoptotic phase that extended our previous model [7]. During this phase the units that exceeded a certain threshold of firing had an increasing probability to die with the passing of time until 700 (or 800, depending on the simulation runs) time units. The inhibitory units entered the apoptosis process about 70 ms before the excitatory units. The death dynamics of both populations followed the probabilty function to die with only minor deviations. After the stop of the apoptosis, spike-timing-dependent plastiticity

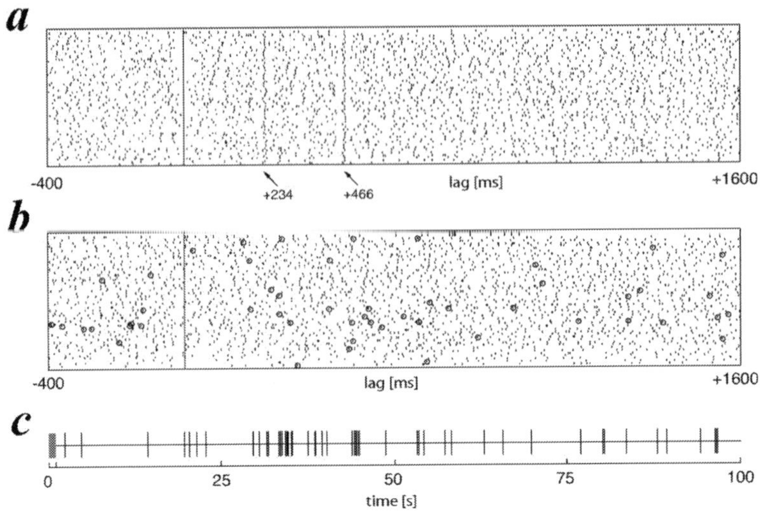

Fig. 5. Spatiotemporal pattern <13,13,13; 234±3.5, 466±4.5>. **(a)**: Raster plot showing the 52 repetitions of the pattern aligned on the pattern start; **(b)**: Raster plot showing the activity of unit #13 aligned on the stimulus onset: each start event of a pattern occurrence is marked by a circle; **(c)**: Pattern occurrence timing plot: each vertical tick represents the start event of a pattern occurrence.

(STDP) and synaptic pruning were made active. Selected sets of units were activated by regular repetitions of a spatiotemporal pattern of stimulation. During the STDP phase, the cell death could occur only if a unit became deafferented, i.e. it looses all its excitatory afferences because of synaptic pruning.

We recorded the spike trains of all excitatory units that were not directly stimulated and that were surviving at the arbitrary end of the simulation set at $t = 100$ seconds. In these spike trains we searched for preferred firing sequences that occurred beyond random expectation [16] and we found evidence of their appearance. We suggest that the detection of such preferred firing sequences might be associated with the emergence of cell assemblies from the initially locally connected random network [8]. The addition of cell death to the model improved the stability of the network over our previous studies while maintaining its ability to let emerge cell assemblies associated to preferred firing sequences. The self-organization of spiking neurons into cell assemblies was recently reported in other studies of large simulated networks connected by STDP-driven projections [10]. These authors emphasized the emergence of spontaneously self-organized neuronal groups, even in absence of correlated input, associated with the spatiotemporal structure of firing patterns, if axonal conduction delays and STDP were incorporated in the model.

Our simulation results offer also the ground of testing several hypothesis with respect to neuroanatomical experimental results. Indeed, there is an increasing interest in investigating the cortical circuits and their synaptic connectivity with a statistical approach related to the graph theory. Results obtained from layer 5 neurons in the visual cortex of developping rats [15] indicate that many aspects of the connectivity patterns differ from random networks. In particular, the distribution of synaptic connection strength in those cortical circuits show an overrepresentation of strong synaptic connections correlated with the overrepresentation of some connectivity patterns. The authors [15] suggest that the local cortical network structure could be viewed as a skeleton of stronger connections in a sea of weaker ones.

The spike-timing-dependent plasticity rule implemented in our simulation has already been successfully implemented and tested in the POEtic tissue [5]. This electronic circuit is a flexible hardware substrate showing the basic features that permit living beings to show evolutionary, developmental or learning capabilities [12]. In future work, we intend to use the POEtic tissue in the investigation of the role of apoptosis and synaptic pruning in the unsupervised shaping of large simulated neural networks. The genomic features of the POEtic tissue offer the possibility to implement the programmed cell death in simulations of large spiking neural networks. It is expected that the computational power of the dedicated plateform will ease the simulation of larger networks to explore the impact of their size on the dynamics.

Acknowledgments. This work was partially funded by the European Community Future and Emerging Technologies program, grant #IST-2000-28027 (POETIC), and Swiss grant OFES #00.0529-2 by the Swiss government.

References

1. Aghajanian, G. K. and Bloom, F. E.: The formation of synaptic junctions in developing rat brain: A quantitative electron microscopic study. Brain Research **6**:4 (1967) 716–27
2. Bi, G. Q. and Poo, M. M.: Synaptic modifications in cultured hippocampal neurons: dependence on spike timing, synaptic strength, and postsynaptic cell type. J Neurosci. **18** (1998) 10464-72
3. Bourgeois, J. and Rakic, P.: Changes of synaptic density in the primary visual cortex of the macaque monkey from fetal to adult stage. Journal of Neuroscience **13** (1993) 2801–20
4. Chechik, G., Meilijson, I. and Ruppin, E.: Neuronal Regulation: A Mechanism for Synaptic Pruning During Brain Maturation. Neural Computation **11** (1999) 2061–80
5. Eriksson, J., Torres, O., Mitchell, A., Tucker, G., Lindsay, K., Rosenberg, J., Moreno, J.-M. and Villa, A.E.P.: Spiking Neural Networks for Reconfigurable POEtic Tissue. Lecture Notes in Computer Science **2606** (2003) 165–74
6. Huttenlocher, P.R.: Synaptic density in human frontal cortex – Developmental changes and effects of aging. Brain Research **163**:2 (1979) 195–205
7. Iglesias, J., Eriksson, J., Grize, F., T., Marco and Villa, A.E.P.: Dynamics of Pruning in Simulated Large-Scale Spiking Neural Networks. Biosystems **79** (2005) 11–20

8. Iglesias, J., Eriksson, J., Pardo, B., Tomassini, M. and Villa, A.E.P.: Stimulus-Driven Unsupervised Pruning in Large Neural Networks. Lecture Notes in Computer Science **3704** (2005) 59–68
9. Innocenti, G. M.: Exuberant development of connections, and its possible permissive role in cortical evolution. Trends in Neurosciences **18**:9 (1995) 397–402
10. Izhikevich, E. M., Gally, J. A. and Edelman, G. M.: Spike-timing Dynamics of Neuronal Groups. Cerebral Cortex **14** (2004) 933–44
11. Karmarkar, U. R. and Buonomano, D. V.: A model of spike-timing dependent plasticity: one or two coincidence detectors? J Neurophysiol. **88** (2002) 507–13
12. Moreno, J. M., Eriksson, J. L., Iglesias, J. and Villa A. E. P.: Implementation of Biologically Plausible Spiking Neural Networks Models on the POEtic Tissue. Lecture Notes in Computer Science **3637** (2005) 188–97
13. Rakic, P., Bourgeois, J. P., Eckenhoff, M. F., Zecevic, N. and Goldman-Rakic, P. S.: Concurrent overproduction of synapses in diverse regions of the primate cerebral cortex. Science **232** (1986) 232–5
14. Song, S. and Abbott, L.F.: Cortical Development and Remapping through Spike Timing-Dependent Plasticity. Neuron **32** (2001) 339–50
15. Song, S., Sjöström, P.J., Reigl, M., Nelson, S. and Chklovskii, D.B.: Highly Nonrandom Features of Synaptic Connectivity in Local Cortical Circuits. PLoS Biology **3**:3 (2005) 0507–19
16. Tetko, I. V. and Villa, A. E.: A pattern grouping algorithm for analysis of spatiotemporal patterns in neuronal spike trains. 1. Detection of repeated patterns. Journal of Neuroscience Methods **105** (2001) 1–14
17. Villa, A. E. P.: Empirical Evidence about Temporal Structure in Multi-unit Recordings. Time and the Brain, Chapter 1, Editor: R. Miller, Harwood Academic Publishers (2000) 1–51

Effects of Analog-VLSI Hardware on the Performance of the LMS Algorithm

Gonzalo Carvajal[1], Miguel Figueroa[1], and Seth Bridges[2]

[1] Department of Electrical Engineering, Universidad de Concepción, Chile
[2] Computer Science and Engineering, University of Washington, USA
gcarvaja@udec.cl, mfigueroa@die.udec.cl, seth@cs.washington.edu

Abstract. Device mismatch, charge leakage and nonlinear transfer functions limit the resolution of analog-VLSI arithmetic circuits and degrade the performance of neural networks and adaptive filters built with this technology. We present an analysis of the impact of these issues on the convergence time and residual error of a linear perceptron using the Least-Mean-Square (LMS) algorithm. We also identify design tradeoffs and derive guidelines to optimize system performance while minimizing circuit die area and power dissipation.

1 Introduction

Modern embedded and portable electronic systems use adaptive signal processing techniques to optimize their performance in the presence of noise, interference, and unknown signal statistics. Moreover, these systems are also severely constrained in size and power dissipation, making custom-VLSI neural network implementations of these techniques attractive.

Analog VLSI circuits can compute using orders of magnitude less power and die area than their digital counterparts, thus potentially enabling large-scale, portable adaptive systems. Unfortunately, device mismatch, charge leakage, and nonlinear behavior limit the resolution of analog arithmetic circuits so that the learning performance of even small-scale analog-VLSI neural networks rarely exceeds 5-6 bits. Traditional circuit-design techniques can reduce these effects, but they increase power and area and render analog solutions less attractive.

We claim that it is possible to build large-scale neural networks in analog VLSI with good learning performance at low power and area by combining on-chip circuit calibration, design techniques, and the natural adaptation of the algorithm to compensate for the limitations of analog hardware. In this paper, we present an analysis of the performance of the well-known Least-Mean-Square (LMS) algorithm under the constraints of analog VLSI arithmetic. Unlike previous work that uses mainly system simulations [2, 5, 1], we base our analysis on the mathematical properties of the algorithm, obtaining more general results that allow us to derive design guidelines and techniques to improve performance at minimal cost. Using these techniques, we have built a 64-input perceptron that adapts with 9-10 bits of accuracy, uses 0.25mm^2 of die area and dissipates 200μW in a 0.35μm CMOS process [3].

S. Kollias et al. (Eds.): ICANN 2006, Part I, LNCS 4131, pp. 963–973, 2006.

2 Convergence Properties of the LMS Algorithm

An adaptive linear combiner [8] computes the function $y_k = \mathbf{x}_k^{\mathrm{T}}\mathbf{w}_k$, where y_k is the output, and $\mathbf{x}_k = [x_{1k} \cdots x_{nk}]^{\mathrm{T}}$ and $\mathbf{w}_k = [w_{1k} \cdots w_{nk}]^{\mathrm{T}}$ are the n-dimensional input and weight vectors at time k. The weights are chosen to minimize a quadratic function of the error $\epsilon_k = d_k - y_k$, where d_k is an external reference. Both the inputs and the reference are taken from stationary zero-mean random distributions. The Mean Square Error (MSE) is defined as:

$$\xi(\mathbf{w}) = \mathrm{E}[\epsilon_k^2] = \mathrm{E}[d_k^2] - 2\mathbf{p}^{\mathrm{T}}\mathbf{w} + \mathbf{w}^{\mathrm{T}}\mathbf{R}\mathbf{w} \tag{1}$$

where $\mathbf{p} = \mathrm{E}[d_k\mathbf{x}_k]$ represents the correlation between the reference and the input, and $\mathbf{R} = \mathrm{E}[\mathbf{x}_k\mathbf{x}_k^{\mathrm{T}}]$ is the input correlation matrix. The MSE defines a quadratic surface with a single global minimum at the point where its gradient is equal to zero. The Wiener solution defines the optimal value of the weights as $\mathbf{w}^* = \mathbf{R}^{-1}\mathbf{p}$, which yields a minimal MSE of $\xi_{\min} = \mathrm{E}[d_k^2] - \mathbf{p}^{\mathrm{T}}\mathbf{w}^*$.

The LMS algorithm uses gradient descent to iteratively compute an approximation of $\mathbf{w}^*$. The algorithm uses an instantaneous estimation of the MSE gradient $\boldsymbol{\nabla}_k$ as $\hat{\boldsymbol{\nabla}}_k = 2\epsilon_k\mathbf{w}_k = \boldsymbol{\nabla}_k - \boldsymbol{\Psi}_k$, where $\boldsymbol{\Psi}_k$ is the zero-mean estimation noise. An each iteration, the LMS algorithm updates the weights as:

$$\mathbf{w}_{k+1} = \mathbf{w}_k - \mu\hat{\boldsymbol{\nabla}}_k = \mathbf{w}_k + 2\mu\epsilon_k\mathbf{x}_k \tag{2}$$

where the learning rate μ is a parameter which controls stability and convergence time. Widrow shows [8] that $\mathrm{E}[\hat{\boldsymbol{\nabla}}_k] = \boldsymbol{\nabla}$, therefore LMS converges to the Wiener solution $\mathbf{w}^*$ in its mean value. However, the gradient estimation noise results in an oscillation around the solution which depends on the learning rate and the statistics of the input. For a small μ, the MSE at convergence is $\xi_\infty = \xi_{\min} + \mathrm{E}[\mathbf{v}_\infty\mathbf{R}\mathbf{v}_\infty^{\mathrm{T}}]$, where $\mathbf{v}_k = \mathbf{w}_k - \mathbf{w}^*$. The misadjustment is defined as:

$$M = \frac{\text{excess MSE}}{\xi_{\min}} = \frac{\xi_\infty - \xi_{\min}}{\xi_{\min}} \approx \mu\sum_{p=1}^{n}\lambda_p = \mu\ \mathrm{tr}(\mathbf{R}) \tag{3}$$

where λ_p are the eigenvalues of $\mathbf{R}$. Eqn. (3) shows that we can control the misadjustment with the learning rate. The MSE decreases as a sum of exponentials, where the time constant of each mode p is given by $\tau_p = 1/(4\mu\lambda_p)$. Therefore, decreasing the learning rate also increases the convergence time of the algorithm.

Hardware implementations of LMS requires multiplication and addition to compute the output (forward path) and weight updates (feedback path), and memory cells to store the weights. Addition is performed by summing currents on common wires and is not subject to device mismatch. The following sections focus on the effects of nonlinear circuits and mismatch on the multipliers, and of charge leakage and limited resolution on the memory cells and weight updates.

3 Effect of Analog Multipliers

We use the following general expression to model the analog multipliers [2]:

$$\mathrm{m}(i_1, i_2) = [a_1\ \mathrm{f}_1(\theta_1, i_1) + \gamma_1] \times [a_2\ \mathrm{f}_2(\theta_2, i_2) + \gamma_2] \tag{4}$$

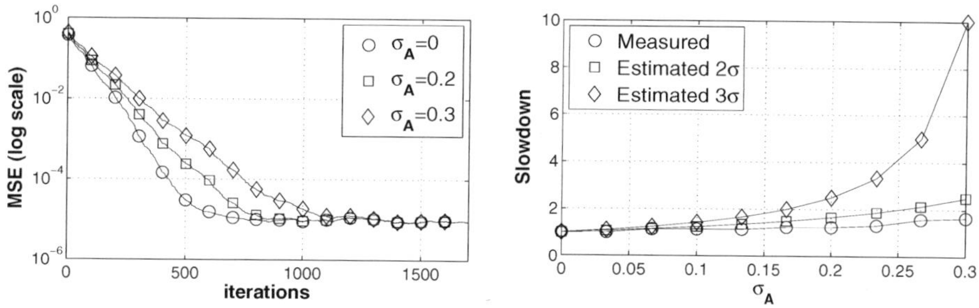

(a) LMS learning curves (b) Algorithm slowdown

Fig. 1. Effect of gain mismatch on LMS performance. (a) Mismatch in multiplier gains do not affect the MSE at convergence, but do increase the convergence time. (b) Assuming that the minimal gain lies within two standard deviations below the mean provides a good bound for convergence time.

where i_1 and i_2 are the inputs to the multiplier, $\mathrm{f}_1(\cdot)$ and $\mathrm{f}_2(\cdot)$ are saturating, monotonic, and odd nonlinear functions, and a_p, γ_p and θ_p control the gain, offset, and linearity of the multiplier. When $f_p(\theta, x) = \frac{\tanh(\theta x)}{\tanh(\theta)}$, Eqn. (4) models the normalized transfer function of a Gilbert multiplier [6] operating in subthreshold regime. Device mismatch results in variations in the values of a_p, γ_p and θ_p for different multipliers within the same chip. The rest of this section independently analyzes the impact of each factor on the performance of the algorithm.

3.1 Gain Mismatch

Feedback Path: We first analyze the effect of gain mismatch between ideal multipliers rewriting Eqn. (4) as $\mathrm{m}_p(i_1, i_2) = a_p\ i_1 i_2$, where a_p is the gain associated with multiplier p. Mismatched gains in the feedback path modify the gradient estimation implemented by Eqn. (2) to:

$$\mathbf{w}_{k+1} = \mathbf{w}_k + 2\mu\mathbf{A}\epsilon_k\mathbf{x}_k = \mathbf{w}_k + 2\mathbf{U}'\epsilon_k\mathbf{x}_k \tag{5}$$

where $\mathbf{A} = \mathrm{diag}([a_1 \cdots a_n])$ is the diagonal matrix that represents the multiplier gains and $\mathbf{U}' = \mu\mathbf{A}$ represents a synapse-dependent learning rate. Gain mismatch does not modify ϵ_k, therefore $\xi'_{\mathrm{min}} = \xi_{\mathrm{min}}$. The new misadjustment is:

$$M' = \mu \sum_{p=1}^{n} \lambda_p a_p = \mu\ \mathrm{tr}(\mathbf{AR}) \tag{6}$$

We assume that the elements of $\mathbf{A}$ have a Gaussian distribution of unitary mean and variance $\sigma_{\mathbf{A}}^2$, and are uncorrelated with the inputs [7]. In this case, $\mathrm{tr}(\mathbf{AR}) \approx \mathrm{tr}(\mathbf{R})$ for a sufficiently large number of inputs, and thus $\xi'_\infty \approx \xi_\infty$.

Fig. 1 shows results from a simulated 16-input linear perceptron with mismatched gains in the feedback path. Fig. 1(a) shows the evolution of the MSE for different $\sigma_{\mathbf{A}}$. The graph shows that the gain variation does not affect the

MSE after convergence. However, the figure also shows that the convergence time of the algorithm increases as a function of the gain variance. Indeed, the time constant of each new mode p is given by $\tau'_p = 1/(4\mu'_p\lambda_p)$. If we assume that the MSE follows the slowest mode, the slowdown in convergence time is:

$$\frac{\tau'_{\mathrm{conv}}}{\tau_{\mathrm{conv}}} = \frac{\max_p[\tau'_p]}{\max_p[\tau_p]} = \frac{4\min_p[\mu\lambda_p]}{4\min_p[a_p\mu\lambda_p]} \leq \frac{4\mu\lambda_{\mathrm{min}}}{4\mu a_{\mathrm{min}}\lambda_{\mathrm{min}}} = \frac{1}{a_{\mathrm{min}}} \tag{7}$$

The value of a_{min} is unknown at design time, but we can derive a bound based on the expected distribution of the gains, which in turn can be obtained from previous experimental data or from statistical models of device mismatch [7]. In a Gaussian distribution, 95.4% and 99.7% of the gains will lie within $2\sigma_{\mathbf{A}}$ and $3\sigma_{\mathbf{A}}$ from the mean, respectively. Fig. 1(b) depicts the simulated convergence time, and the bounds estimated using $2\sigma_{\mathbf{A}}$ and $3\sigma_{\mathbf{A}}$ to estimate a_{min}. In practice, it is sufficient to assume $2\sigma_{\mathbf{A}}$, because the bound established in Eqn. (7) conservatively assumes that the convergence time tightly follows the slowest mode, and that the smallest gain is in turn associated with the smallest eigenvalue of $\mathbf{R}$.

Notice that, if the designer has individual control over the learning rate of each synapse after fabrication, then setting $\mu_p = \mu/a_p$ normalizes the effective learning rate and achieves the same convergence time as the original network.

Forward Path: Gain mismatch in the forward-path multipliers modifies the error as $\epsilon'_k = d_k - \mathbf{x}_k^{\mathrm{T}}\mathbf{A}\mathbf{w}$, leading to the following expression for the MSE:

$$\xi' = \mathrm{E}[\epsilon_k'^2] = \mathrm{E}[d_k^2] - 2\mathbf{A}\mathbf{p}^{\mathrm{T}}\mathbf{w} + \mathbf{w}^{\mathrm{T}}\mathbf{A}\mathbf{R}\mathbf{A}\mathbf{w} \tag{8}$$

and the learning rule:

$$\mathbf{w}_{k+1} = \mathbf{w}_k + 2\mu(d_k - \mathbf{x}_k^T\mathbf{A}\mathbf{w}_k)\mathbf{x}_k = \mathbf{w}_k + 2\mathbf{U}'(d_k - \mathbf{x}'^{\mathrm{T}}_k\mathbf{w}_k)\mathbf{x}'_k \tag{9}$$

where $\mathbf{U}' = \mu\mathbf{A}^{-1}$ and $\mathbf{x}'_k = \mathbf{A}\mathbf{x}_k$, Eqn. (9) has the same form as the original LMS learning rule, but with nonuniform learning rates and a modified input with correlation matrix $\mathbf{R}' = \mathbf{A}\mathbf{R}\mathbf{A}^{\mathrm{T}}$. The learning rule of Eqn. (9) converges in its mean to $\mathbf{w}'^* = \mathbf{A}^{-1}\mathbf{w}^*$, and thus from Eqn. (8) $\xi'_{\mathrm{min}} = \xi_{\mathrm{min}}$.

In general, it is difficult to determine the misadjustment from the gains. If we assume that the inputs are decorrelated ($\mathbf{R}$ is diagonal), then the eigenvalues of $\mathbf{R}'$ are $\lambda'_p = a_p^2\lambda_p$, where λ_p are the eigenvalues of $\mathbf{R}$. The misadjustment is:

$$M' = \sum_{p=1}^{n} \frac{\mu}{a_p}\lambda'_p = \mu\sum_{p=1}^{n} \frac{a_p^2\lambda_p}{a_p} = \mu\sum_{p=1}^{n} a_p\lambda_p = \mu\ \mathrm{tr}(\mathbf{A}\mathbf{R}) \tag{10}$$

which is equivalent to Eqn. (6) for gain mismatch in the learning rules. Therefore, mismatched gains in the forward path do not affect the MSE, but increase the learning time as depicted in Eqn. (7). Multiplier gains also modify the Wiener solution to $\mathbf{w}'^* = \mathbf{A}^{-1}\mathbf{w}^*$, so they may also change the effect of initial conditions on convergence time, although modeling this effect is difficult without knowledge of the original solution [4].

3.2 Multiplier Offsets

We rewrite Eqn. (4) as $m_p(i_1, i_2) = (i_1+\gamma_{1p})(i_2+\gamma_{2p})$, where γ_{1p} and γ_{2p} are the offsets associated with the inputs to multiplier p. The remainder of this section analyzes the effect of each offset separately.

Forward Path: Let $\boldsymbol{\gamma}_\mathbf{w} = [\gamma_{w1} \cdots \gamma_{wn}]$ be the vector of weight offsets in the multipliers of the forward path. The instantaneous error is $\epsilon'_k = d_k - \mathbf{x}_k^\mathrm{T}\mathbf{w}'_k$ and the MSE is $\xi' = \mathrm{E}[d_k^2] - 2\mathbf{p}^\mathrm{T}\mathbf{w}' + \mathbf{w}'^\mathrm{T}\mathbf{R}\mathbf{w}'^\mathrm{T}$, where $\mathbf{w}' = \mathbf{w} + \boldsymbol{\gamma}_\mathbf{w}$. A simple analysis shows that the LMS algorithm converges to the new Wiener solution $\mathbf{w}'^* = \mathbf{w}^* - \boldsymbol{\gamma}_\mathbf{w}$, which compensates for the weight offsets and achieves the same residual MSE as the original network. The eigenvalues of the input are not modified, thus the weight variance is the same and $M' = M$. The weight offsets modify the solution vector $\mathbf{w}^*$, so they also affect convergence time [4]. However, because the distribution of the weight offsets is independent of $\mathbf{w}^*$, it is not possible to relate the convergence time to the offset variance.

Let now $\boldsymbol{\gamma}_\mathbf{x} = [\gamma_{x1} \cdots \gamma_{xn}]$ be the input offsets in the multipliers of the forward path. The error is $\epsilon'_k = d_k - \mathbf{x}'^\mathrm{T}_k\mathbf{w}_k$ where $\mathbf{x}'_k = \mathbf{x}_k + \boldsymbol{\gamma}_\mathbf{x}$, and $\xi' = \xi + \boldsymbol{\gamma}_\mathbf{x}\boldsymbol{\gamma}_\mathbf{x}^\mathrm{T}$. Because the learning rule operates with a zero-mean $\mathbf{x}$, $\mathrm{E}[\hat{\boldsymbol{\nabla}}'_k] = \boldsymbol{\nabla}$ and the mean value of the weight converges to the original solution $\mathbf{w}^*$. The minimal MSE and the misadjustment quadratically increase with the offset:

$$\xi'_{\min} = \xi_{\min} + \mathbf{w}^{*\mathrm{T}}\boldsymbol{\gamma}_\mathbf{x}\boldsymbol{\gamma}_\mathbf{x}^\mathrm{T}\mathbf{w}^* \tag{11}$$

$$M' = M + \mu\sum_{p=1}^{n}\gamma_{xp}^2 \tag{12}$$

The last term in the Eqn. (11) introduces a large increase in the error which is not controllable with the learning rate. However, we can add *bias synapse* w_0 with offset γ_0 and a constant input c to cancel the accumulated offset at the output. The synapse converges to:

$$w_0 = \frac{-\boldsymbol{\gamma}_\mathbf{x}{}^\mathrm{T}\mathbf{w}}{c + \gamma_0} \tag{13}$$

which compensates for the accumulated effect of the input offsets, allowing the weights to converge to the Wiener solution and $\xi'_{\min} = \xi_{\min}$.

The bias synapse also affects the weight variance. It can be shown that if $\mathbf{x}_k$ has zero mean, then $\mathrm{tr}(\mathbf{R}') = \mathrm{tr}(\mathbf{R}) + c^2$. Therefore, from Eqn. (12) the misadjustment is $M' = M + \mu(\sum_{p=0}^{n}\gamma_{xp}^2 + c^2)$.

Fig. 2(a) shows simulation results for M_γ (which we define as the misadjustment with respect to the original $\xi_{\min}$) as a function of the standard deviation of the offsets in the forward-path multipliers. As the figure shows, offsets in the weights do not affect the MSE. Input offsets quadratically increase the MSE, but the addition of a bias synapse successfully compensates for this effect even without the reducing learning rate.

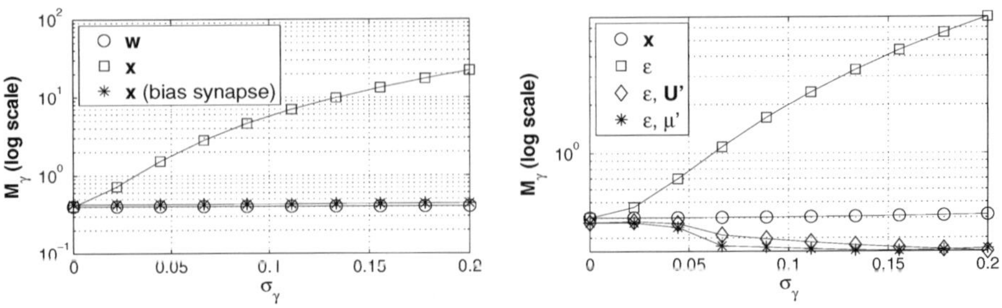

(a) M_γ vs. offset in forward path

(b) M_γ vs. offset in learning rule

Fig. 2. Misadjustment versus random multiplier offsets taken from a Gaussian distribution variance σ_γ^2. (a) Forward path: Weight offsets have no effect on the MSE. Input offsets quadratically increase the MSE, but with a bias synapse the effect is almost negligible. (b) Feedback path: input offsets have little effect on the MSE, while error offsets quadratically increase its value. Using learning-rate correction fully compensates for this effect.

Feedback Path: Adding an offset vector $\gamma_{\mathbf{x}}$ to $\mathbf{x}_k$ in Eqn. (2) yields a new estimated gradient $\hat{\nabla}'_k = 2\epsilon_k(\mathbf{x}_k + \gamma_{\mathbf{x}})$, which converges to the original Wiener solution $\mathbf{w}^*$. The covariance of the new gradient estimation noise is $\text{cov}[\mathbf{\Psi}'_k] = 4\xi_{\min}(\mathbf{R}+\gamma_{\mathbf{x}}\gamma_{\mathbf{x}}^{\mathrm{T}})$ [8]. For small μ and assuming uncorrelated inputs, the gradient noise propagates directly into $\mathbf{v}_k$, leading to a new misadjustment:

$$M' = M + \mu \sum_{p=1}^{n} \gamma_{xp}^2 \tag{14}$$

Eqn. (14) shows that the MSE increases quadratically with the multiplier offsets but this effect is small and can be compensated with the learning rate.

Adding offsets to the error signal ϵ_k at each synapse computing its weight update results in a new estimated gradient $\hat{\nabla}'_k = -2(\epsilon_k\mathbf{I} + \mathbf{\Gamma}_\epsilon)\mathbf{x}_k$, where $\mathbf{\Gamma}_\epsilon = \text{diag}([\gamma_{\epsilon 1} \cdots \gamma_{\epsilon n}])$ is the diagonal matrix of error offsets. Assuming that $\mathbf{x}$ has zero mean, it is easy to show that $\mathbf{w}'^* = \mathbf{w}^*$, and therefore $\xi'_{\min} = \xi_{\min}$.

However, the new estimated gradient quadratically increases the covariance of $\mathbf{v}_k$ to $\text{cov}[\mathbf{v}'_k] = \text{cov}[\mathbf{v}_k] + \mu\mathbf{\Gamma}_\epsilon^2$, where for simplicity we assume that the inputs are uncorrelated. The misadjustment is:

$$M' = M + \frac{\mu \sum_{p=1}^{n} \lambda_p \gamma_{\epsilon p}^2}{\xi_{\min}} \tag{15}$$

Eqn. (15) shows that M' depends quadratically of $\gamma_{\epsilon p}$ and linearly of $\xi_{\min}^{-1}$, so the effect of offsets is much larger than the previous case. We can define a new learning rate that compensates for the misadjustment:

$$\mathbf{U}' = \mu\xi_{\min}(\xi_{\min}\mathbf{I} + \mathbf{\Gamma}_\epsilon^2)^{-1} \tag{16}$$

Note that Eqn. (16) defines a different learning rate for each synapse and requires knowledge of the offset values. If the circuit does not support individually programmable learning rates, the following global rate assumes that most offsets lie within one standard deviation from the mean and yields good results:

$$\mu' = \frac{\mu\xi_{\min}}{(\xi_{\min} + \sigma^2_{\gamma\epsilon})} \tag{17}$$

The simulation results in Fig. 2(b) shows the effects of multiplier offsets in the feedback path. As expected, the effect of input offsets is almost negligible, even without modifying the learning rate. Error offsets have a dramatic impact with the original learning rate. Using Eqns. (16) and (17) to set the learning rate fully compensates for the effect on the MSE.

3.3 Nonlinear Multipliers

Eqn. (4) models an analog multiplier where the parameter θ_p, which varies among multipliers because of device mismatch, modulates the linearity of an odd, saturating, monotonic nonlinear function $f_p(\cdot)$. For example, the normalized transfer function of a Gilbert multiplier [6] is $f_p(\theta_p, x_p) = \tanh(\theta_p x_p)/\tanh(\theta_p)$.

Forward Path: Applying a nonlinear function to the weights in the forward path results in a new error signal $\epsilon'_k = d_k - \mathbf{x}_k^{\mathrm{T}}\,\mathbf{f}(\mathbf{w}_k)$, which yields the MSE:

$$\xi'_k = \mathrm{E}[d_k^2] - 2\mathbf{p}^{\mathrm{T}}\mathbf{f}(\mathbf{w}_k) + \mathbf{f}(\mathbf{w}_k^{\mathrm{T}})\mathbf{R}\mathbf{f}(\mathbf{w}_k) \tag{18}$$

The LMS algorithm converges to the new Wiener solution $\mathbf{w}'^* = \mathbf{f}^{-1}(\mathbf{w}^*)$, and the minimal MSE is $\xi'_{\min} = \xi_{\min}$.

Because the learning rate is small, it is possible to estimate the gradient by linearizing around the Wiener solution:

$$\hat{\boldsymbol{\nabla}}' = \left[d_k - \mathbf{x}_k^{\mathrm{T}}\left(\boldsymbol{f}(\mathbf{w}'^*) + \left.\frac{\partial \boldsymbol{f}}{\partial \mathbf{w}}\right|_{\mathbf{w}=\mathbf{w}'^*}(\mathbf{w}_k - \mathbf{w}'^*)\right)\right]\mathbf{x}_k \tag{19}$$

Eqn. (19) shows that the estimation noise depends on the values of $\mathbf{w}'^*$ and $[\partial \boldsymbol{f}/\partial \mathbf{w}](\mathbf{w}'^*)$. The worst case occurs when the target weights are at the point where the slope of $f_p(\cdot)$ is maximal, which corresponds to $\mathbf{w}'^* = \mathbf{0}$ for common functions such as $\tanh(\cdot)$. In that case, the estimated gradient is

$$\hat{\boldsymbol{\nabla}}' = (d_k - \mathbf{x}_k^{\mathrm{T}}\mathbf{A}_\theta \mathbf{w}_k)\mathbf{x}_k \tag{20}$$

where $\boldsymbol{A}_\theta = \mathrm{diag}(\partial \boldsymbol{f}/\partial \mathbf{w}|_{\mathbf{w}=0})$ is a diagonal matrix representing the slope of $\boldsymbol{f}(\cdot)$ at the solution. Eqn. (20) reduces the analysis of nonlinear weights to a problem of mismatched multiplier gains. For a normalized Gilbert multiplier, $[\tanh(\theta x)/\tanh(\theta)] > 1$, which increases the MSE. We can achieve $\xi'_\infty = \xi_\infty$ by normalizing the learning rate to the mean gain:

$$\mu' = \frac{\mu}{\mathrm{mean}[\mathbf{A}_\theta]} \tag{21}$$

If the nonlinearity affects the inputs in the forward path, the new error signal is $\epsilon'_k = d_k - \mathbf{f}(\mathbf{x}_k^{\mathrm{T}})\ \mathbf{w}_k$, yielding the new MSE:

$$\xi' = \mathrm{E}[\epsilon_k'^2] = \mathrm{E}[d_k^2] - 2\mathbf{p}'^{\mathrm{T}}\mathbf{w} + \mathbf{w}^{\mathrm{T}}\mathbf{R}'\mathbf{w} \tag{22}$$

where $\mathbf{p}' = \mathrm{E}[d_k\mathbf{f}(\mathbf{x}_k)]$ and $\mathbf{R}' = \mathrm{E}[\mathbf{f}(\mathbf{x}_k)\mathbf{f}(\mathbf{x}_k^{\mathrm{T}})]$. The MSE at the Wiener solution $\mathbf{w}'^* = \mathbf{R}'^{-1}\mathbf{p}'$ is:

$$\xi'_{\mathrm{min}} = \mathrm{E}[d_k^2] - \mathrm{E}[d_k\mathbf{f}(\mathbf{x}_k^{\mathrm{T}})]\mathbf{w}'^* \tag{23}$$

which is always greater than ξ_{min} when d_k is generated by a linear function. Furthermore, LMS converges to $\mathbf{w}_\infty = (\mathrm{E}[\mathbf{x}_k\mathbf{f}(\mathbf{x}_k^{\mathrm{T}})])^{-1}\mathrm{E}[d_k\mathbf{x}_k^{\mathrm{T}}]$, which differs from $\mathbf{w}'^*$ as a function of the nonlinearity of $\mathbf{f}$.

Feedback Path: Applying a nonlinear function $\mathbf{f}$ to the inputs in the feedback path yields the LMS rule $\mathbf{w}_{k+1} = \mathbf{w}_k + 2\mu\epsilon_k\mathbf{x'}_k$ with $\mathbf{x'}_k = \mathbf{f}(\mathbf{x}_k)$, which still converges to the Wiener solution in its mean value. Therefore, $\xi'_{\mathrm{min}} = \xi_{\mathrm{min}}$, but the nonlinearity of $\mathbf{f}$ affects the variance of $\mathbf{w}_k$ and increases the residual MSE. The misadjustment is given by the modified correlation matrix:

$$M' = \mu\ \mathrm{tr}(\mathrm{E}[\mathbf{f}(\mathbf{x}_k)\mathbf{f}(\mathbf{x}_k)^{\mathrm{T}}]) \tag{24}$$

For nonlinear functions such as $[\tanh(\theta\mathbf{x})/\tanh(\theta)]$, $\mathbf{f}(\mathbf{x}) > \mathbf{x}$ and larger values of θ increase the difference between M' and M. In the limit, $\tanh(\theta\mathbf{x})$ saturates and behaves like $\mathrm{sign}(\mathbf{x})$, and we obtain an upper bound for the misadjustment as $M' \leq \mu n$. Note that this also increases the robustness of the algorithm to outliers.

Applying a nonlinear function to the error at each synapse modifies the learning rule to $\mathbf{w}_{k+1} = \mathbf{w}_k + 2\mu\mathbf{E}'_k\mathbf{x}_k$, with $\mathbf{E}'_k = \mathrm{diag}[f_1(\epsilon_k)\cdots f_n(\epsilon_k)]$. Because the error converges to a small value, we can linearize around this point and rewrite the rule as:

$$\mathbf{w}_{k+1} = \mathbf{w}_k + 2\mu(\mathbf{A}_\theta\epsilon_k)\mathbf{x}_k \tag{25}$$

where $\mathbf{A}_\theta = \mathrm{diag}[\partial f_1/\partial\epsilon|_{\epsilon=0}\cdots\partial f_n/\partial\epsilon|_{\epsilon=0}]$. The expression above is equivalent to Eqn. (5) for mismatched gains in the feedback path, therefore the misadjustment increases quadratically with the variance of θ_p and can be controlled with the learning rate using Eqn. (21). Also, for saturating functions such as $[\tanh(\theta\mathbf{x})/\tanh(\theta)]$, the nonlinearity also limits the effects of outliers in the performance of the algorithm.

Fig. 3(a) shows the effect of nonlinear multipliers on the MSE. As predicted by Eqn. (22), nonlinear inputs in the forward path increase the MSE independently of the learning rate. The effect on the forward-path weights and the feedback-path inputs and error is much lower and can be further controlled by reducing the learning rate. Fig. 3(b) shows the effect that this rate reduction has on the convergence time, which is similar to the case of mismatched multiplier gains.

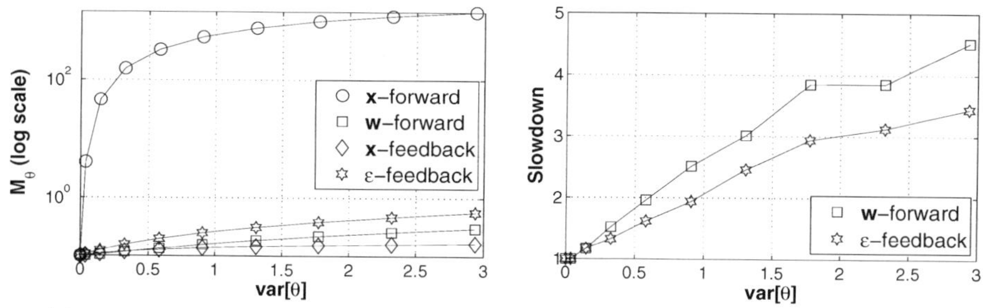

(a) Misadjustment vs. nonlinearity (b) Algorithm slowdown

Fig. 3. Effects of nonlinear multipliers on the performance of LMS. (a) Nonlinear inputs in the forward path increase the MSE when the reference is generated by a linear process, while the effect on the rest of the signals is much lower. (b) It is possible to control the learning rate to trade residual for convergence speed.

4 Effect of Signal Noise and Limited Resolution

Degradation of signal resolution can arise from system noise, charge leakage in capacitor-based weight storage, or quantization effects in digital or mixed-mode implementations. We model the noise $\boldsymbol{\eta}_k$ as a Gaussian process of zero mean and variance σ_η^2. The analysis is equivalent to Section 3.2 with uncorrelated inputs.

Forward Path: In the presence of zero-mean random noise in the weights and inputs of the forward path, the LMS algorithm still converges in the mean to the original Wiener solution $\mathbf{w}^*$. From Section 3.2, we obtain:

$$\xi_{\min}^{\boldsymbol{\eta}_\mathbf{w}} = \xi_{\min} + \mathrm{E}[\boldsymbol{\eta}_k^{\mathrm{T}}\mathbf{x}_k\mathbf{x}_k^{\mathrm{T}}\boldsymbol{\eta}_k] \tag{26}$$

$$\xi_{\min}^{\boldsymbol{\eta}_\mathbf{x}} = \xi_{\min} + \mathbf{w}^{*T}\mathrm{E}[\boldsymbol{\eta}_k\boldsymbol{\eta}_k^{\mathrm{T}}]\mathbf{w}^* = \xi_{\min} + \sigma_\eta^2\mathbf{w}^{*T}\mathbf{w}^* \tag{27}$$

$$M^{\boldsymbol{\eta}_\mathbf{w}} = M = \mu\mathrm{tr}(\mathbf{R}) \tag{28}$$

$$M^{\boldsymbol{\eta}_\mathbf{x}} = M + \mu\mathrm{tr}(\mathrm{E}[\boldsymbol{\eta}_k\boldsymbol{\eta}_k^{\mathrm{T}}]) = M + \mu n\sigma_\eta^2 \tag{29}$$

Only input-noise modifies M, but both input and weight noise modify $\xi_{\min}$.

Feedback Path: With random noise in the forward-path signals the algorithm still converges to the Wiener solution and $\xi_{\min}$ is not modified because the output is not directly affected. The new misadjustments are:

$$M^{\eta_\mathbf{x}} = M + \mu\mathrm{tr}(\mathrm{E}[\boldsymbol{\eta}_k\boldsymbol{\eta}_k^{\mathrm{T}}]) = M + \mu n\sigma_\eta^2 \tag{30}$$

$$M^{\eta_\epsilon} = M + (\mu/\xi_{\min})\sum_{p=1}^{n}\lambda_p \ \mathrm{var}[\eta_{kp}] = M + (\mu/\xi_{\min})\sigma_\eta^2\mathrm{tr}\ (\mathbf{R}) \tag{31}$$

The effect error noise on the misadjustment is large, but we can extend Eqn. (17) to derive a new learning rate that guarantees that $\xi_\infty^{\eta_\epsilon} \leq \xi_\infty$:

$$\mu' = \frac{\mu\xi_{\min}}{(\xi_{\min} + \sigma_\eta^2)} \tag{32}$$

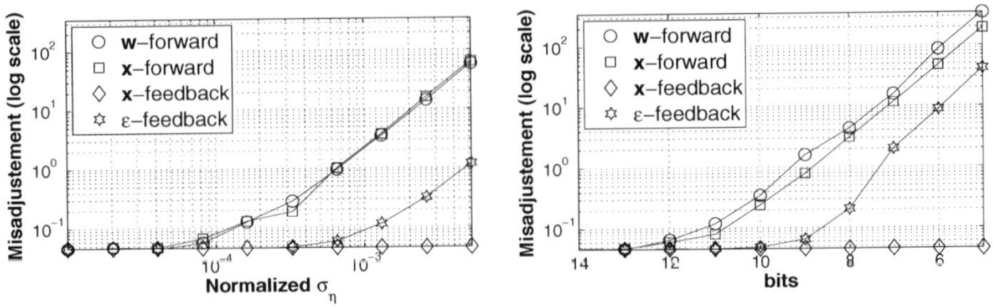

(a) M vs. normalized standard deviation

(b) M vs. bit width

Fig. 4. Simulated effect of signal noise and digital arithmetic. (a) Noise in the forward path has a strong effect on $\xi_{\min}$ and degrades the learning performance of LMS. The effect of noise in the feedback path is much lower, and can be further reduced with the learning rate. (b) The same analysis applies to the resolution of digital signals.

Fig. 4(a) shows the simulated misadjustment (with respect to the original $\xi_{\min}$) versus the standard deviation of the noise, normalized to the signal range. The simulated plots follow closely the results predicted by the expressions above. Fig. 4(b) shows the misadjustment versus the resolution of digital arithmetic circuits. The bit-widths were chosen match the signal-to-noise ratio used in Fig. 4(a) according to $[\text{bits}] = \log_2\left(\frac{[\text{signal range}]}{6\sigma_\eta}\right)$. The figure shows that the analysis presented in this section can also be used to predict the performance of digital arithmetic circuits implementing parts of the algorithm.

5 Conclusions

We presented an analysis of the effects of analog and mixed-signal hardware on the performance of the LMS algorithm. We derived bounds for the degradation in MSE and convergence time caused by effects such as multiplier offsets, gain mismatch, nonlinear transfer functions, noise, and charge leakage. We discussed design techniques to compensate for these effects such as local and global learning rate adjustment and bias synapses, and quantified their impact on the performance of the algorithm. We are currently extending this work to the design of dimensionality-reduction networks using Principal Components Analysis.

Acknowledgments

This work was financed in part by a FONDECYT grant No. 1040617.

References

1. H. C. Card, B. K. Dolenko, D. K. McNeill, C. R. Schneider, and R. S. Schneider. Is VLSI Neural Learning Robust Against Circuit Limitations? In *IEEE International Conference on Neural Networks*, volume 3, pages 1889–1893, Miami, FL, USA, 1994.

2. B. Dolenko and H. Card. Tolerance to Analog Hardware of On-Chip Learning in Backpropagation Networks. *IEEE Transactions on Neural Networks*, 6(5):1045–1052, 1995.
3. M. Figueroa, E. Matamala, G. Carvajal, and S. Bridges. Adaptive Signal Processing in Mixed-Signal VLSI with Anti-Hebbian Learning. In *IEEE Computer Society Annual Symposium on VLSI*, pages 133–138, Karlsruhe, Germany, 2006. IEEE.
4. A. Flores and B. Widrow. Assessment of the Efficiency of the LMS algorithm Based on Spectral Information. In *Asilomar Conf. on Signals, Systems and Computers*, volume 1, pages 120–124, Pacific Grove, CA, 2004.
5. D. K. McNeill and H. C. Card. Analog Hardware Tolerance of Soft Competitive Learning. In *IEEE International Conference on Neural Networks*, volume 4, pages 2004–2008, Miami, FL, USA, 1994.
6. C. Mead. *Analog VLSI and Neural Systems*. Addison-Wesley, Reading, MA, 1989.
7. M. J. M. Pelgrom, A. C. J. Duinmaijer, and A. P. G. Welbers. Matching Properties of MOS Transistors. *IEEE Journal of Solid-State Circuits*, 24(5):1433–1440, 1989.
8. B. Widrow and E. Walach. *Adaptive Inverse Control*. Prentice-Hall, Upper Saddle River, NJ, 1996.

A Portable Electronic Nose (E-Nose) System Based on PDA

Yoon Seok Yang[1], Yong Shin Kim[2], and Seung-chul Ha[3]

[1] Division of Bionics and Bioinformatics Engineering, College of Engineering,
[1] Center for Healthcare Technology Development,
Chonbuk National University, 664-14 Deokjin-dong, Jeonju, 561-756, Korea
ysyang@chonbuk.ac.kr
[2] Electronics and Telecommunications Research Institute (ETRI),
161 Gajeong-dong, Yuseong-gu, Daejeon, 305-350, Korea
yongshin@etri.re.kr
[3] SENKO, Business Incubation Center 208, 123 Jibyeon-dong, Gangneung, 210-702, Korea
scha@senko.co.kr

Abstract. The electronic nose (e-nose) has been used in food investigation and quality controls in industry. Recently it finds its applications in medical diagnosis and environmental monitoring. Moreover, the use of portable e-nose enables the on-site measurements and analysis of vapors without extra gas-sampling units. In this study, a PDA-based portable e-nose was developed using micro-machined gas sensor array and miniaturized electronic interfaces. The computing power and flexible interface of the PDA are expected to provide the rapid and application specific development of the diagnostic devices, and easy connection to other information appliances. For performance verification of the developed portable e-nose system, Six different vapors were measured using the system. The results showed the reproducibility of the measured data and the distinguishable patterns between the vapor species. The application of two different artificial neural networks verified the possibility of the automatic vapor recognition based on the portable measurements.

1 Introduction

The electronic nose (e-nose) system has been mainly used in food industry where it reduces the amount of the analytical chemistry, i.e., inspection of food quality, control of cooking process, and checking odors in plastic packaging, etc. due to its ability of characterizing odors, vapors, and gases [1]. Recently, various composites from carbon-black (CB) and organic polymers are developed as gas sensors for portable use owing to its chemical diversity by the selection of proper organic polymers and improved operability with lower power consumption than the metal oxide gas sensors [2,3]. With the help of these improvements in portability and ease of use, the e-nose is widening its potentials in environmental and pollution monitoring i.e., real-time identification of contaminants, analysis of fuel mixture, detection of oil leaks, testing ground water for odors, and identification of toxic waste, etc. It also finds its applications in medical diagnosis, specifically in detection of diabetes, pulmonary or gastrointestinal problem, or skin infections by examining odors in the breath or tissues [4].

S. Kollias et al. (Eds.): ICANN 2006, Part I, LNCS 4131, pp. 974–982, 2006.

Relatively simple methodology using e-nose also ensures its prospective usages in point-of-care test (POCT) and telemedicine system.

In this study, we developed a portable embedded e-nose system based on personal digital assistance (PDA) device and vapor recognition algorithm using backpropagation neural network (BPNN) and support vector machine (SVM). Compared with the commercial stand-alone type e-nose like Cyranose320 (Cyrano Sciences), the design of PDA-based e-nose is flexible so that it can meet various user interface requirements for wide variety of application and can also provide additional processing with the help of its connectivity to many kind of information appliances.

Six different vapor samples were measured using the proposed system. The maximum sensitivities which is well-known feature in characterizing vapor response was extracted. The distribution of the vapor patterns was visualized by principal component analysis (PCA) [5]. The feasibility of the vapor recognition by intelligence algorithm was tested through the application of the BPNN and SVM.

2 Materials and Methods

2.1 The Minimization of Sensor Interface Circuits

Figure 1 shows a silicon-based gas sensor array [6] and its interface circuits. Eight different carbon-black (CB) polymer composites in Table 1 were dispensed in micromachined sensor array structure.

Table 1. CB polymer composites used for vapor sensor array

Channel	Polymer I.D.
1	poly(4-vinyl pyridine)
2	poly(vinyl butyral)-co-vinyl alcohol-co-vinyl acetate
3	poly(vinyl stearate)
4	ethyl cellulose
5	polystyrene-b-polyisoprene-b-polystyrene
6	hydroxypropyl cellulose
7	cellulose acetate

Figure 1-(a) shows the original electronic circuits for driving sensors and obtaining response signals. For portable measurements, a minimized circuit module was developed as in Fig. 1-(b). This includes sensing chamber, dc-motor pump for gas flow, and electronic circuits like voltage regulator and amplifiers. The silicon-based sensor array and circuit board were connected by pogo pins for reliable contact. Conventional zebra elastomer connector was not suitable because of its varying contacting resistance which confuses the signal from resistive sensor array. The electrical power is designed to be supplied from PDA instead of using separate batteries. The power consumption was about 60 mA, most of which occurs in amplifiers. This implies 40 minutes continuous use with powered by PDA.

(a) (b)

Fig. 1. Electronic interface circuits for vapor sensing, a) laboratory experimental board, b) minimized module for portable system

The electronic circuits are shown in Fig. 2. ADM660 provides bipolar voltage supply for op-amps and MAX603 gives regulated output of 3 volts for dc motor pump.

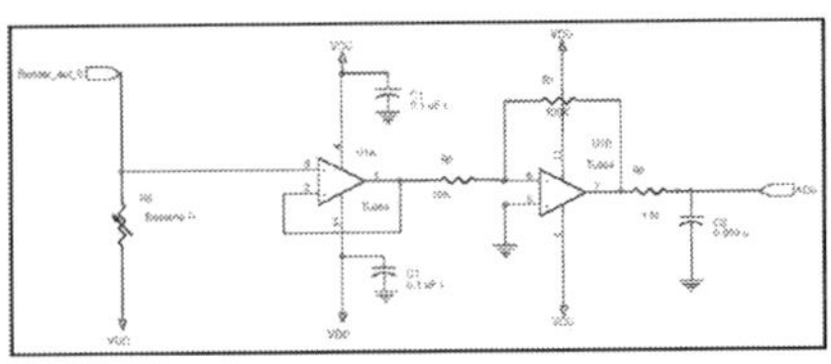

(a)

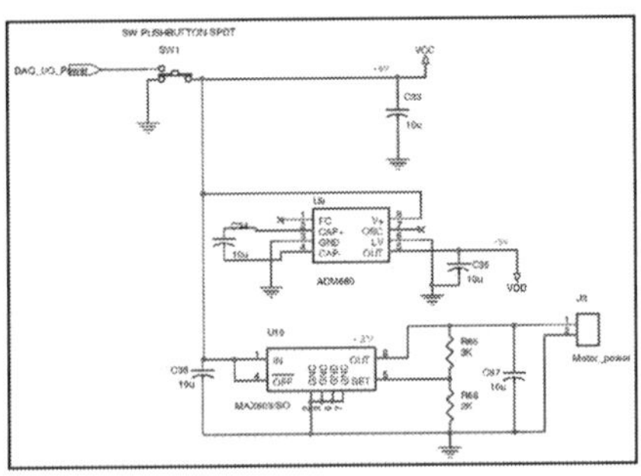

(b)

Fig. 2. Circuit diagram, a) signal amplifier, b) biopolar voltage converter and 3-volt regulator

2.2 Connection with PDA

The minimized sensor interface module was connected to PDA (iPAQ5550, COMPAQ) through data acquisition board (DAQ6062E, National Instrument). A dedicated PDA software for e-nose was developed using Labview 7.0, Labview PDA module (National Instrument), and Microsoft Embedded Visual Tools. Figure 3 shows the PDA and the vapor sensing module connected to it together. The PDA controls the sensing module and displays the measured response.

The functional specifications of the software are based on the bench-marking of an e-nose system developed in the previous studies [6,7]. For example, the measured signal is stored in a PDA file whose name is automatically generated by system to minimize the user input during portable measurements. The selection of display channel and sampling frequency is disabled for simplicity.

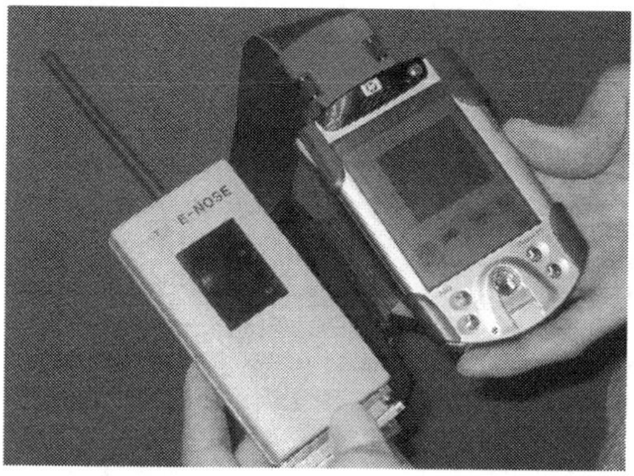

Fig. 3. The developed PDA-based e-nose, a) PDA controls the minimized vapor sensing module via DAQ connection, b) PDA software dedicated for e-nose system displays and stores the measured sensor response

2.3 Verification of the System by Vapor Measurement

To verify the developed PDA-based e-nose system, six common vapors (acetone, benzene, chloroform, cyclohexane, ethanol, and methanol) were measured using the setup shown in Fig. 4. No mass flow control (MFC) was used during measurement to simulate portable measurement conditions. The liquid samples of these vapors were prepared in bottles which have one inlet and one outlet for gas flow and measured at room temperature (about 20℃) without any temperature control as shown in Fig. 4. The array responses are displayed on the PDA screen as Fig. 4 and stored in the memory. All the measurements were repeated twenty times. Figure 5 shows typical response signals for each odors. The interactions between the vapor and the sensing materials determine the shape of the plots. The response time is about tens of seconds. The maximum sensitivities (maximum change ratio of the sensor resistance) were calculated at one minute after the start of vapor inhaled by the embedded dc-motor pump.

2.4 Principal Component Analysis (PCA) and Vapor Recognition Using Artificial Neural Networks

The signals stored in PDA were gathered in PC to analyze the vapor patterns and to test the validity of the measured data. The radial plots in Fig. 6 visualize the

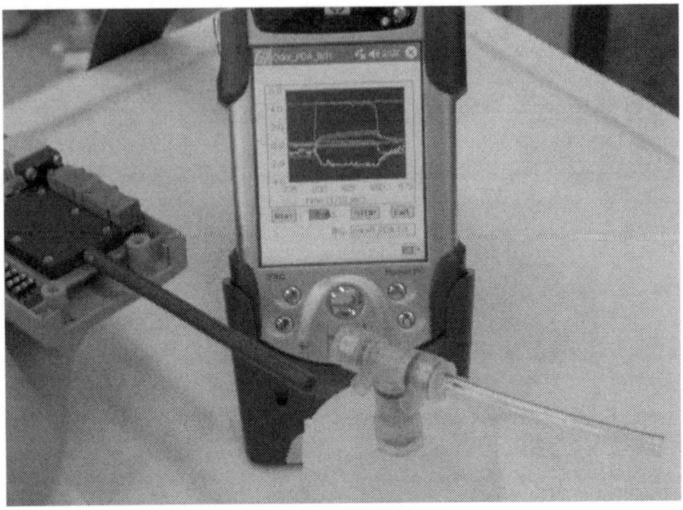

Fig. 4. Vapor measurement with the developed PDA-based e-nose. Liquid sample of vapor is prepared in a bottle. PDA controls the hardware to measure the vapor. The obtained signal is displayed on PDA screen and stored in memory.

maximum sensitivity patterns obtained from every repeated measurement. The seven radial axes correspond to seven sensing channels used. Figure 7 shows their clustering characteristics with the 1st and the 2nd PCA components on two-dimensional space.

Since the goal of this portable e-nose study is an automatic vapor recognition, two well-known artificial neural networks was applied to the data to test the feasibility of the pattern recognition algorithm. Firstly, a BPNN was applied. It has two hidden layers, which have eight and seven hidden nodes, respectively. Secondly, a SVM[1] was tested. Every vapor pattern was separated into two half sets for training and test of the two neural networks. Both networks were implemented by using Matlab (Mathworks, USA) on PC. The recognition results are shown in Table 2.

3 Results and Discussions

Sensor response signals in Fig. 5 show slight offsets from zero and baseline drifts, however, their effects are minimized during calculation of maximum sensitivities. Since Ch. 8 of the sensor array has unstable characteristics, the signal from it was discarded in later processing.

Radial plots in Fig. 6 show consistent patterns within a vapor species. This validates the reproducibility of the vapor sensing system. Although the plot shows a few large magnitudes of signals since there was no control on vapor concentration, some consistent patterns were maintained for each vapor group by normalization. For example, one large scale pattern in ethanol response yields the same vapor pattern as others after normalization.

[1] Using open source Matlab toolbox by Vapnik.

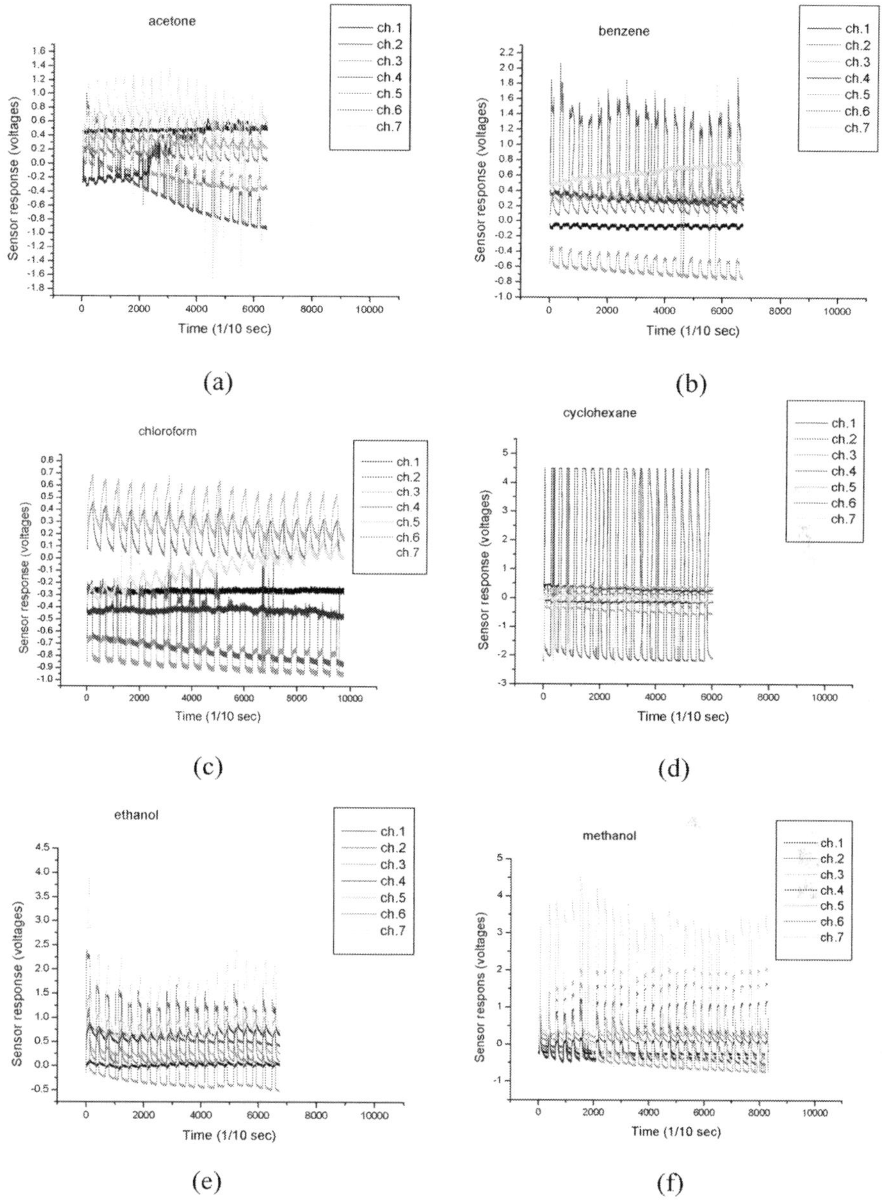

Fig. 5. A typical response signals for each odor obtained by the PDA-based e-nose. a) acetone, b) benzene, c) chloroform, d) cyclohexane, e) ethanol, f) methanol.

On the contrary, there are distinguishable patterns among different vapor species in Fig. 6. The clear distinction of vapor species based on these patterns was shown in PCA plot in Fig. 7. The coverage of the 1st and the 2nd principal component are

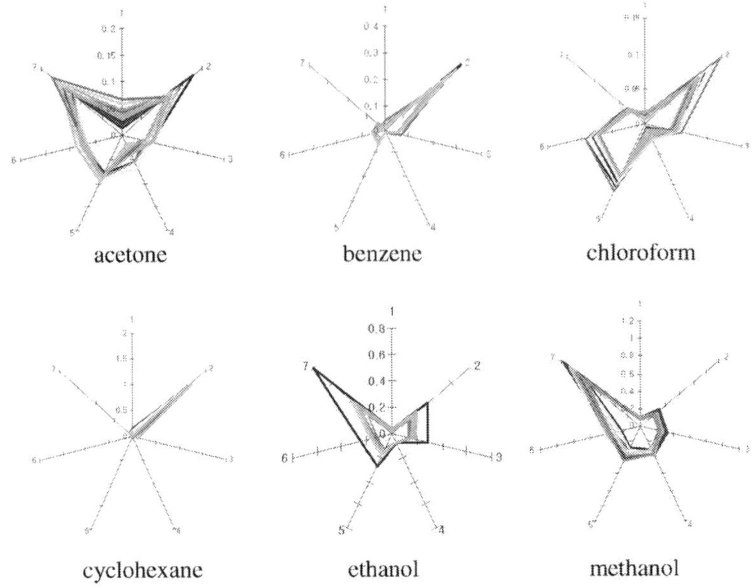

Fig. 6. Radial plots of the maximum sensitivity patterns of the six vapor species

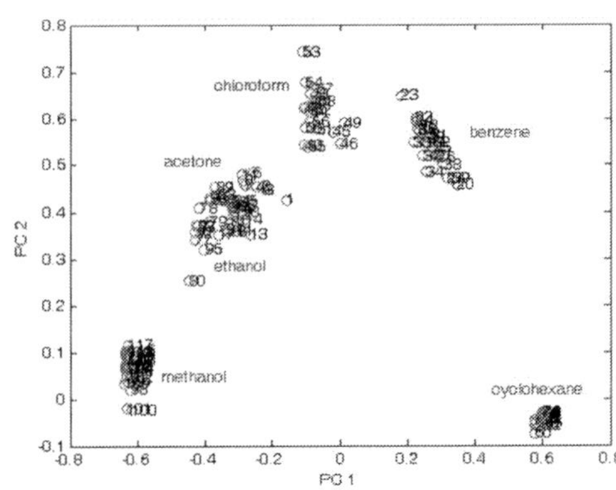

Fig. 7. The clustering characteristics of the six vapor species are viewed by PCA. The numeric annotations indicate the orders of the measurements.

70.43% and 24.12 %, respectively, which correspond to 94.5% total coverage. They are all separated but some overlap between the acetone and the ethanol groups. Consequently, the results of vapor recognition by BPNN and SVM in Table 2 show high success ratio of 98%.

Though there are many other classification algorithms like minimum distance or Bayesian classifier, artificial neural network was adopted in the proposed portable e-nose system. The PDA e-nose gives more convenience in ordinary unregulated measurements than laboratory instruments. Artificial neural network is more suitable

Table 2. Vapor recognition results by BPNN and SVM are summarized in confusion matrix. SVM results are shown in the parenthesises.

True \ Recog..	actone	benzene	chloroform	cyclohexane	ethanol	methanol
acetone	8 (9)	0	0	0	1	0
benzene	0	10 (10)	0	0	0	0
chloroform	0	0	10 (10)	0	0	0
cyclohexane	0	0	0	9 (9)	0	0
ethanol	0	0	0	0	9 (9)	0
methanol	0	0	0	0	0	13 (13)

to incorporate a lot of data gathered from the portable measurements into artificial intelligence quickly.

Furthermore, since this study aims at the development of miniaturized e-nose integrated with a System on Chip (SoC) gas sensor and finally the application of the system to medical diagnosis and environmental monitoring, a real-time vapor recognition software running on PDA was developed and under experimental tests. The BPNN was implemented on PDA. Figure 8 shows the integrated e-nose software dedicated for real-time vapor recognition on PDA. Various methods have been developed to improve the reliability of the vapor recognition in more realistic situation [7].

In vapor measurements, we met several noise sources. Firstly, the sensor performances are strongly dependent on carbon black (CB) content and an initial sensor resistance. We found empirically that the optimum conditions were the CB content of 12 – 15 weight percent (%) and the sensor resistance of 1 kΩ – 5 MΩ through our earlier studies for various composite sensor systems [8]. Secondly, the variation in environmental temperature during measurement can distort the vapor response pattern [6]. However, this can be reduced by using micro-heater embedded in gas sensor array as in our previous study if necessary [6]. Finally, the various extra components in odors make the correct vapor recognition more difficult. For reliable vapor recognition, we developed simple and robust pattern recognition techniques in another previous study [7], which is supposed to be implemented in the portable system in further studies.

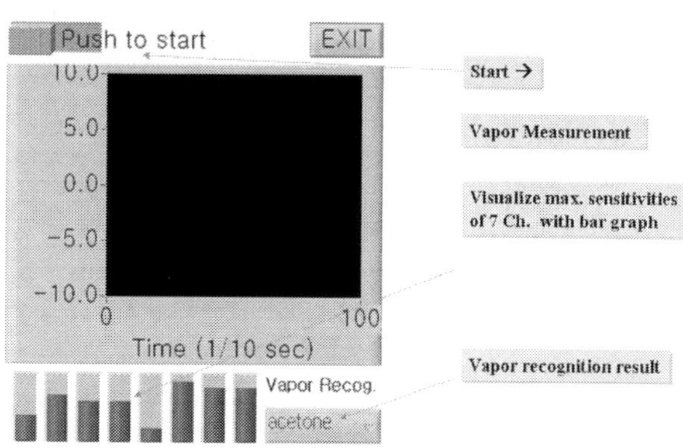

Fig. 8. A new PDA based e-nose software for real-time vapor recognition. Fundamentals of functions are annotated on the figure.

The response of the developed system to mixes of the examined odors is not included in this study. However, it was shown that it corresponds to the linear combinations of each response within proper vapor concentration range [9].

4 Conclusions

The analysis of the measured vapor patterns and the application of the neural network algorithm prove the feasibility of the vapor recognition by artificial neural network as well as the reliability of the developed PDA-based e-nose system.

The PDA-based e-nose is expected to accelerate the use of e-nose in many applications owing to its high flexibility and connectivity. Moreover, the minimized vapor sensing module can be adopted by various embedded systems and digital convergence devices. The application of e-nose can be expanded to digital accessory in daily life beyond the laboratory use for experts.

Acknowledgements

This work has been supported in part by the Ministry of Science and Technology of Korea through the NRL program and in part by research funds of Chonbuk National University in 2005.

References

1. Pearce, T.C., Schffman, S.S., Nagle, H.T., Gardner, J.W.: Handbook of machine olfaction. Wiley-Vch, Weinheim (2003)
2. Mo, Y., Okawa, Y., Inoue, K., Natukawa, K.: Low-voltage and low-power optimization of micro-heater and its on-chip drive circuitry for gas sensor array. Sens. Actuators A Phys. 100 (2002) 94-101
3. Natale, C. D., Macagnano, A., Martinelli, E., Paolesse, R., D'Arcangelo, G., Roscioni, C., Finazzi-Agrò, A., D'Amico, A.: Lung cancer identification by the analysis of breath by means of an array of non-selective gas sensors. Biosensors and Bioelectronics 18 (2003) 1209-1218
4. Gardner, J.W., Shin, H.W., Hines, E.L.: An electronic nose system to diagnose illnesss. Sensors and Actuators B Chem. 70 (2000) 19-24.
5. Doleman, B.J., Lonergan, M.C., Severin, E.J., Vaid, T.P., Lewis, N.S.: Quantitative study of the resolving power of arrays of carbon black-polymer composites in various vapor-sensing tasks. Anal. Chem. 70 (1998) 4177-4190
6. Ha, S., Kim, Y.S., Yang, Y., Kim, Y.J., Cho, S., Yang, H., Kim, Y.T.: Integrated and microheater embedded gas sensor array based on the polymer composites dispensed in micromachined wells. Sensors and. Actuators B Chem. 105 (2005) 549-555
7. Yang, Y.S., Ha, S. and Kim, Y.S.: A matched-profile method for simple and robust vapor recognition in electronic nose (E-Nose) system. Sensors and. Actuators B Chem. 106 (2005) 263-270
8. Briglin, S.M., Freunda, M.S., Tokumarub, P., Lewis, N.S.: Exploitation of spatiotemporal information and geometric optimization of signal/noise performance using arrays of carbon black-polymer composite vapor detectors. Sensors and Actuators B Chem. 82 (2002) 54-74
9. Severin, E.J., Doleman, B.J., Lewis, N.S.: An investigation of the concentration dependence and response to analyte mixtures of carbon black/insulating organic polymer composite vapor detectors. Anal. Chem. 72 (2000) 658-668

Optimal Synthesis of Boolean Functions by Threshold Functions

José Luis Subirats, Iván Gómez, José M. Jerez, and Leonardo Franco

Departamento de Lenguajes y Ciencias de la Computación
Universidad de Málaga,
Campus de Teatinos S/N, 29071 Málaga, Spain
{jlsubirats, ivan, jja, lfranco}@lcc.uma.es
http://www.lcc.uma.es/~lfranco/

Abstract. We introduce a new method for obtaining optimal architectures that implement arbitrary Boolean functions using threshold functions. The standard threshold circuits using threshold gates and weights are replaced by nodes computing directly a threshold function of the inputs. The method developed can be considered exhaustive as if a solution exist the algorithm eventually will find it. At all stages different optimization strategies are introduced in order to make the algorithm as efficient as possible. The method is applied to the synthesis of circuits that implement a flip-flop circuit and a multi-configurable gate. The advantages and disadvantages of the method are analyzed.

1 Introduction

In this paper we introduce an algorithm for finding a set of threshold functions (also called linearly separable functions) that will compute a desired (target) arbitrary Boolean function. This problem is known as the synthesis problem of Boolean functions and has been much studied since the 60's ([1,2,3]). The interest in using threshold gates or linearly threshold functions instead of standard logical AND, NOT and OR gates relies on the fact that threshold elements are more powerful than standard gates and as a consequence, the size of the circuits that can be constructed to compute the desired functions can be smaller. There is an extra interest in the study of threshold circuits as they are very close related to neural networks models, and thus some of the properties and characteristics of the circuits also apply to neural networks architectures. Our main motivation for the development of this method is the study of neural networks architectures [4,5]. The standard practice for the architecture selection process within the field of artificial neural networks is the trial-and-error method that is very time consuming. Knowing and characterizing which architectures are best suited for a given class (or set) of functions can much help to the development and refinement of existing methods for constructing better neural architectures and also for gaining further understanding on the learning process. The problem of finding optimal architectures is relevant also for a more theoretical point of view within

S. Kollias et al. (Eds.): ICANN 2006, Part I, LNCS 4131, pp. 983–992, 2006.

the area of circuit complexity, as different existing bounds can checked and/or improved ([6]).

The method introduced in this paper permits the synthesis of Boolean functions in terms of the set of threshold Boolean functions. The standard threshold circuits, use threshold gates connected by weights that in the present formulation are replaced by threshold functions. Obtaining the whole set of linear separable functions is a complicate and computationally intensive problem and the set functions of threshold functions have been only obtained up to $N = 10$ variables ([3]).

Research in threshold logic synthesis was done mostly in the 1960s ([1,2,3]). Nowadays, different implementations of circuits by threshold gates are available, and several theoretical results have been obtained [6,7,8,9] together with different applications ([10]). The main difference with previous approaches is that the present work is aimed to find optimal architectures.

2 Synthesis of Boolean Functions by Linearly Separable Functions

We introduce in this work a new method for finding a set of linearly separate functions that will compute a given desired Boolean function (the target function). The method use the linearly separable set of functions as basis functions (or primitives) and thus assumes that whether a list of the functions is available or that a method for testing whether a given function is linearly separable is provided (the first approach is used throughout this work). One important difference with previous works is that our approach works straightforward with the set of threshold functions as basis functions and not through their equivalent representation by threshold gates and weights.

2.1 The Algorithm

Without loss of generality, we will analyze the case of architectures comprising a single layer of hidden nodes. The algorithm can be straightforwardly extended to the case of several layers of nodes. The algorithm is aimed to find a set of linearly separable functions for the hidden nodes and for the output node that implement a desired target function. The choice of the output function is crucial for the procedure as once that an output function has been chosen, the algorithm will test all possible solutions and then the choice of the output function has a multiplicative computational effect. However, the search procedure of the hidden node functions is the most intensive and complicated step, and different optimization strategies can be implemented.

Selection procedure for the output function. From first principles, there are two "logic" choices for the output function: the first one would be to select an output function as the most similar threshold function to the target one. This procedure is not straightforward as a measure of similarity between the functions is needed, but a simple choice seems to select the closest linearly separable function in terms of the Hamming distance (number of different bits) between the outputs of the two functions. The second logic choice could be to select the output function according to the success rate of the functions (or of similar functions in lower dimensions) in previous cases. While the two mentioned approaches seems to merit consideration, there is also a computationally effective oriented choice. In this case functions that will generate a shorter node search procedure will be considered first. This was the approach used in this work. It will be more clear later in the work after the node search algorithm is introduced, but the general idea is that as the node search algorithm generate a tree of possible solutions and as the number of branches at a ramification point is proportional to the number of solutions, the more computational intensive cases will occur for output functions with a balanced number of 0's and 1's. Thus, functions with a non-balanced number of 1' and 0's will be considered earlier as candidates for the output function. This choice, even if motivated by computational efficiency, may be not the more efficient, specially if unbalanced functions results to be not very good for the synthesis problem. On the contrary, if almost all functions are more or less equivalent, or if unbalanced functions are well suited for this task of generating non-threshold functions, then the approach taken might be the most computationally effective as it was intended. Some preliminary experiments showed that the unbalanced functions seem to be good as output functions.

Selection procedure of the hidden nodes function. Once the output function has been set to a given linearly separable function, the problem translates into finding a set of linearly separable functions for the hidden nodes that would make the circuit to implement the target function. The algorithm analyze the different possibilities of mapping the inputs into a set of recoded inputs for the output function, in way such that the output function might be able to map them correctly onto the solution. To exemplify the procedure, we will analyze the simple case of finding the optimal circuit for the NXOR function. The NXOR function is a Boolean function of two input bits where the output of the function is 0 if the sum of the two input bits is 1 and the output is 1 if the sum of the input bits is even (inputs 0-0 and 1-1). The NXOR function is a classic example of a non linearly separable function and thus an architecture with two hidden nodes, at least, is needed for their implementation with threshold functions (it is known that 2 hidden nodes are enough for implementing this function). For exemplifying the procedure we use an architecture with two hidden nodes, both connected to the two inputs. In Fig. 1 such an architecture is shown, where I_0 and I_1 indicate the input nodes that will feed their values into the hidden nodes. The hidden nodes will compute two linearly separable functions named f_0 and f_1 while the output function is indicated in the figure by f_{out}.

For the example under consideration, we will analyze the situation in which the output function has been already selected. We consider as output function the function AND, that produce an output 1 only for input values 1-1. The

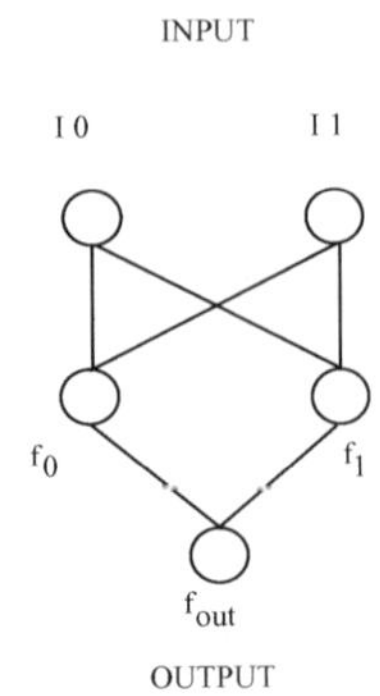

Fig. 1. A two hidden node architecture used for showing the procedure of finding a solution for the NXOR function. The inputs to the network are indicated by I_0 and I_1. The hidden nodes will compute as a function of the input bits two linearly separable functions indicated by f_0 and f_1 that will send their outputs to the output node function indicated by f_{out}.

functions will be designated according to the outputs that the function produce in response to the inputs, with values ordered according to their binary value. For the present case, in which input functions of two bits are considered the order of the inputs and outputs is as follows: first, the output in response to the input 0-0, second to the input 0-1, third to 1-0 and finally the output corresponding to the input 1-1. In this way, the AND function is indicated by 0001, while the NXOR function would be coded as 1001. The procedure for finding the threshold functions for the node functions f_0 and f_1 proceeds in a number of steps in which the possible outputs for f_0 and f_1 are considered.

Now we consider the first output bit of the target function, that is equal to 1 (in response to the first input 0-0). As the function AND was selected as output function (f_{out}), and this function produce a 1 in response only to an input 1-1, it is needed that both node functions, f_0 and f_1, produce an output 1 in response to the input pattern 0-0. That is, the node functions have to recode the original input of the network, the input 0-0, into the input 1-1 for the output node, otherwise the output function would not coincide with the target one. Now the procedure continues with the second input bit, 0-1, for which the target output is 0. The selected output function produce a 0 in response to three input cases and then there are three possible cases into which the input 0-1 can be recoded by the node functions. The function f_{out} gives a 0 in response to the inputs 0-0, 0-1 and 1-0 and then we obtain, putting together the results for the first and second output bits the following tree of possibilities, shown in Fig. 2. The three cases are indicated in the diagram as branches of the first case on the top, for which there was a single possibility.

The procedure continues considering the rest of the cases corresponding to the third and fourth input bits. The whole tree generated as the procedure is

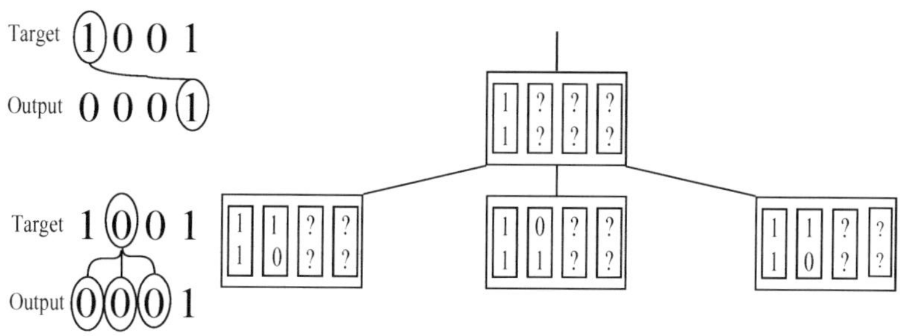

Fig. 2. The possible cases for the node functions f_0 and f_1 as the two first input bits are considered for the case of choosing the NXOR as target function and the AND as output function. Within the boxes, the 4 pairs of values correspond to the outputs of the two functions in response to the 4 inputs. The question mark indicates that the value is still undetermined.

applied is shown in Fig. 3. The nine boxes in the last two bottom rows contain the nine function candidates for the node functions, but the only possible ones are those boxes containing two threshold (linearly separable) functions (the two boxes at the rightmost end of the bottom row containing functions with only one output bit equal to 1, as all the other boxes contain the function 1001 that is non-linearly separable).

Fig. 3. The whole tree of possibilities for the choices of the node functions for the case of the NXOR target function. The nine boxes in the last two bottom rows are the nine candidates for the functions but the only possible ones are those in which both functions are linearly separable functions (See the text for details).

The node searching procedure ends if two linearly separable functions are found, but if this is not the case, a different output function, f_{out}, is chosen from the set of threshold functions and the whole procedure is repeated until all the threshold functions with a number of variables equal to the number of nodes are tested. If a solution is not found, the number of nodes is augmented by one and the whole procedure repeated, and so on.

2.2 Optimization Steps: Order of Branching and Pruning by Testing Linear Separability

Several optimization steps were introduced to speed up the computational process. We will refer here to only the most important ones. The first optimization step is about the order in which the input bits are considered for the node searching process. If the output function is unbalanced, it is convenient to analyze first the inputs of the output function that produce an output that occur less infrequently as the degeneracy of the output values is related to the number of branches that are created in the tree of possible choices for threshold functions. The second optimization concern the test of linear separability for the partial defined functions. It is possible to analyze by previously constructing a tree with all the linearly separable functions, whether a partial defined Boolean function will end up producing some threshold functions. If the partially defined function will not produce any threshold function, the branching process is ended at that point, and the tree is pruned. Alternatively, the checking for linear separability can be done by testing the unateness of the variables, as all threshold functions are unate (but not conversely, so the test is a partial one).

3 Application of the Algorithm to the Construction of Optimal Architectures for a Flip-Flop Circuit and a Configurable Gate Circuits

We applied the algorithm developed in the previous section to the construction of threshold circuits that implement in an optimal way (with a minimum number of nodes in a single hidden layer) a flip-flop circuit and a configurable gate. The flip-flop circuit (or bistable) is a standard digital circuit in electronics that incorporates the storage of memory in its functioning.

The resulting architecture obtained by the algorithm has three nodes in the unique hidden layer and is depicted in Fig. 4.

The three hidden functions receive input from the 4 inputs and then project into the two outputs units. The "fan-in max" (maximum number of connections that a node receives) is 4 and thus, the whole set of threshold functions of up to 4 variables was needed. The four inputs of the circuit are the Memory input bit (M), the Enable bit (EN), the Input bit (IN) and the Read/Write bit (R/W). The circuit has two output nodes, the first one (f_M) feedbacks into the memory bit permitting the storage of memory and the second one is the true output of the circuit, designated in the figure as (f_{out}) . When the enable bit is OFF (value 0), the circuit outputs the value of the Input bit (I). When the Enable (EN) bit

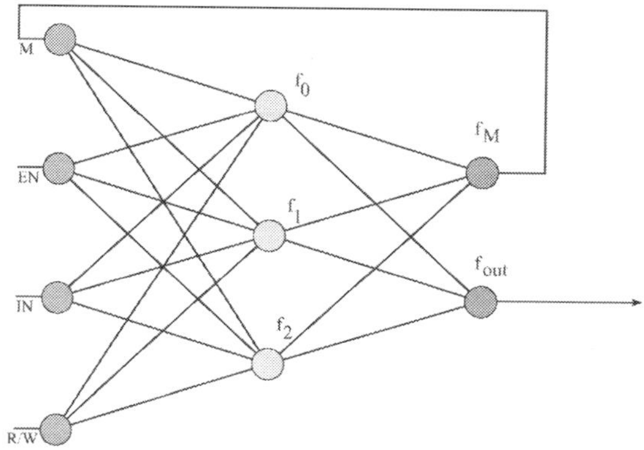

Fig. 4. The architecture designed to implement a flip-flop circuit. The architecture has a single hidden layer containing three nodes and two outputs nodes. One of the output nodes, indicated as f_M, feedbacks the input value to the memory neuron, and the other output, f_{out}, can be considered as the real output of the network.

is ON the circuit works depending on the value of the R/W bit, reading the value stored in memory and sending it to the output or by writing the value of the input into the memory. The truth table for the flip-flop circuit implemented is shown in Fig. 5.

In Table 1 the output values of the obtained functions that implement the circuit are shown. The case shown is only one of the 120 cases obtained.

Table 1. The threshold functions obtained for one of the solutions found for the flip-flop circuit. The node functions are functions of N=4 input bits and then 16 output values exists, while the output functions have a fan-in of 3 and then 8 binary values are needed to code them.

Function	Output bits
f_0	0000000111111111
f_1	0011001100111011
f_2	1011000011111010
f_M	00000111
f_{out}	00010001

Also a circuit to implement a multi-configurable gate was constructed. The circuit can compute as desired a AND, OR, NXOR or NAND as a function of the two inputs (IN_0 and IN_1). The function to be computed is selected by two extra input bits (C_0 and C_1), and the truth table of this multi-configurable gate is shown in Fig. 6.

The architecture implementing the multi-configurable gate is shown in Fig. 7 where it is also shown, close to the nodes, the functions obtained from the

M	EN	IN	R\W	Out-M	OUTPUT
0	0	0	0	0	0
0	0	0	1	0	0
0	0	1	0	0	1
0	0	1	1	0	1
0	1	0	0	0	0
0	1	0	1	0	#
0	1	1	0	0	0
0	1	1	1	1	#
1	0	0	0	1	0
1	0	0	1	1	0
1	0	1	0	1	1
1	0	1	1	1	1
1	1	0	0	1	1
1	1	0	1	0	#
1	1	1	0	1	1
1	1	1	1	1	#

Fig. 5. Truth table of a flip-flop (bistable) circuit. The circuit, when the Enable (EN) input is activated, can store a value in memory (Write procedure) or transmit the stored value (Read procedure) depending of the value of the R/W input bit. If the Enable bit is off, the network simply output the value of the single input (IN).

C0	C1	IN0	IN1	OUTPUT
0	0	0	0	0
0	0	0	1	0
0	0	1	0	0
0	0	1	1	1
0	1	0	0	0
0	1	0	1	1
0	1	1	0	1
0	1	1	1	1
1	0	0	0	1
1	0	0	1	0
1	0	1	0	0
1	0	1	1	1
1	1	0	0	1
1	1	0	1	1
1	1	1	0	1
1	1	1	1	0

Fig. 6. Truth table of a multi-configurable circuit with N=2 inputs. The two selection bits,(C_0 and C_1), indicate which function should be computed.

procedure coded by their outputs. It is also shown in the figure, the operation of the circuit for two inputs, the first and the fourth, showing for the first input how this is mapped by the node functions to the first bit of the output node to produce a desired output of 0. For the fourth input, the output of the three hidden nodes recoded it as 1-0-1 producing a desired output of 1 by the output node.

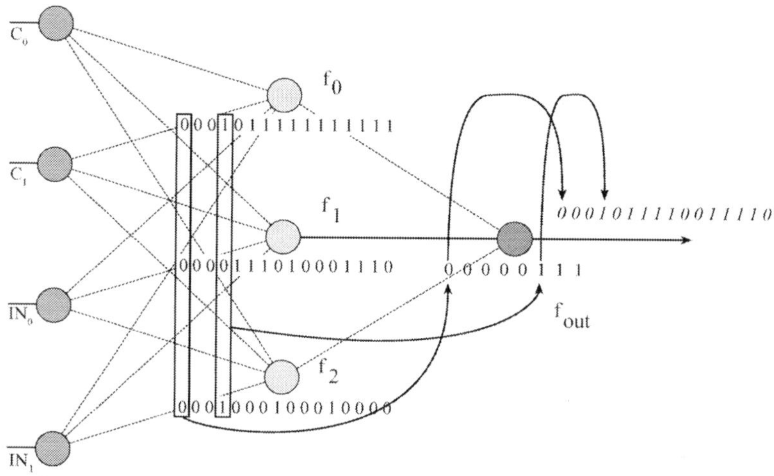

Fig. 7. The architecture of the circuit computing a multi-configurable gate that can work depending of the value of the indicator bits (C_0 and C_1) as a AND, OR, NXOR or NAND gate. Close to the nodes, the functions that make the circuit to work as desired are indicated by the value of their outputs in response to the input patterns. (See the text for more details)

4 Discussion

We have introduced a new method for finding the optimal circuit architecture that computes an arbitrary Boolean function in terms of linear separable primitives. Starting from different "smart" choices of the output function the algorithm tries to find linearly separable functions for the hidden nodes according to the values of the output and target function. Eventually if a solution has not been found, the algorithm searches for all the possible solutions and in that way does an exhaustive search. Different optimization steps were introduced to improve the computational efficiency. The method was successfully applied to the construction of threshold circuits for the flip-flop circuit and for a multi-configurable gate. For both implementations the whole synthesis procedure took less than half of a minute, but we are aware that they only involved nodes with a fan-in max of 4. The worst case complexity of the algorithm is of order $\mathcal{O}(2^{MN^2}\,2^{M^2})$, where N is the number of input variables, M is the number of hidden nodes and the factors of the type 2^{N^2} arise because that is the order of the number of threshold functions on N variables (It is worth noting, as a reference value, that the total number of Boolean functions on N variables is 2^{2^N}). We are currently computing the average case complexity for the case of all functions on 4 variables that can be treated exhaustively. We believe that the algorithm can be of practical application for larger dimensions of up to 8 or 10 in its present way and it seems also possible to develop from the current algorithm non-optimal procedures for larger dimensions. We are currently analyzing all the optimal architectures for all the existing Boolean functions of 4 variables, measuring the speed of the algorithm against standard benchmarks and studying the properties of multi-layer and modular neural network architectures.

Acknowledgements

The authors acknowledge support from CICYT (Spain) through grant TIN2005-02984 (including FEDER funds). Leonardo Franco acknowledges support from the Spanish Ministry of Education and Science through a Ramón y Cajal fellowship.

References

1. Winder, R.O. (1962). *Threshold Logic*, Ph.D. dissertation, Department of Mathematics, Princeton University.
2. Dertouzos, M.L. (1965) *Threshold Logic: A Synthesis Approach.* Cambridge, MA: The M.I.T. Press.
3. Muroga, S. (1971).*Threshold Logic and its Applications*, Wiley, New York
4. Franco, L. (2006). Generalization ability of Boolean functions implemented in feed-forward neural networks. *Neurocomputing.* In Press.
5. Franco, L. and Anthony, M. (2006). The influence of oppositely classified examples on the generalization complexity of Boolean functions. *IEEE Transactions on Neural Networks.* In Press.
6. Siu, K.Y., Roychowdhury, V.P., and Kailath, T. (1991). Depth-Size Tradeoffs for Neural Computation *IEEE Transactions on Computers, 40*, 1402-1412.
7. Oliveira, A. L. and Sangiovanni-Vincentelli, A. (1991). LSAT: an algorithm for the synthesis of two level threshold gate networks. In: *Proceedings of the ACM/IEEE International Conference on Computer Aided Design, Santa Clara, CA, IEEE Computer Society Press*, 130-133.
8. Noth, W., Hinsberger, U., and Kolla, R. (1996). TROY: A Tree-Based Approach to Logic Synthesis and Technology Mapping, *In: Proceedings of the 6th Great Lakes Symposium on VLSI*, p. 188.
9. Zhang, R., Gupta, P., Zhong, L. and Jha, N. K. (2005). Threshold Network Synthesis and Optimization and Its Application to Nanotechnologies, *IEEE Transactions on computer-aided design of integrated circuits and systems, 24*, 107-118.
10. Beiu, V., Quintana, J.M. and Avedillo, M.J. (2003). LSI implementations of threshold logic - A comprehensive survey, *IEEE Trans. Neural Networks, 14*, 1217-1243.

Pareto-optimal Noise and Approximation Properties of RBF Networks*

Ralf Eickhoff and Ulrich Rückert

Heinz Nixdorf Institute
System and Circuit Technology
University of Paderborn, Germany
eickhoff, rueckert@hni.upb.de

Abstract. Neural networks are intended to be robust to noise and tolerant to failures in their architecture. Therefore, these systems are particularly interesting to be integrated in hardware and to be operating under noisy environment. In this work, measurements are introduced which can decrease the sensitivity of Radial Basis Function networks to noise without any degradation in their approximation capability. For this purpose, pareto-optimal solutions are determined for the parameters of the network.

1 Introduction

Neural networks are used for classification tasks, recognition, vision processing, optimization and, moreover, for function approximation of any continuous function [1]. Especially, Radial Basis Function (RBF) networks are utilized to approximate an unknown function.

Often, neural networks are selected for this task since they are intended to be robust to noise and tolerant to failures in their structure. Therefore, neural networks are particularly interesting to be integrated in hardware. Here, noise always affects the inputs and the parameters of a system and, thus, noise has not only been considered during training but also during runtime of a neural network. Therefore, the output of an RBF network is perturbed by noisy inputs during runtime which additionally contributes to the mean squared error (mse) at the output.

In this work, an optimization technique is presented to decrease the noise sensitivity and to enhance the robustness of RBF networks which are affected by noise in its inputs. Introducing measurement techniques and using these methods as additional objectives during training the sensitivity of the RBF network can be improved without any degradation of the approximation capabilities. On the contrary, the mean squared error can be decreased without affecting the sensitivity to noise of the RBF network.

* This work was supported by the Graduate College 776 - Automatic Configuration in Open Systems- funded by the German Research Foundation (DFG).

S. Kollias et al. (Eds.): ICANN 2006, Part I, LNCS 4131, pp. 993–1002, 2006.

The remainder is organized as follows. In Section 2 the used neural network is briefly described. Section 3 introduces two noise measurement techniques which are used as additional objective functions for learning in Section 4. Here, simulation results of the proposed methods are presented. The paper ends with a conclusion in Section 5.

2 Radial Basis Functions

In this section, a short overview of the architecture of an RBF network is given. The RBF network is often used for local function approximation. Based on regularization theory the quadratic error is minimized with respect to a stabilizing term [2]. Due to this stabilizer the interpolation and approximation quality can be controlled in order to achieve a smooth approximation. Therefore, different types of Basis Functions depending on the stabilizer can be employed for superposition which are differently weighted and centered in space. Here, only Gaussian Basis Functions are considered and the network describes a function

$$f_m(\boldsymbol{x}) = \sum_{i=1}^{m} \alpha_i \exp\left(\frac{-\|\boldsymbol{x} - \boldsymbol{c}_i\|^2}{2\sigma_i^2}\right) \quad (1)$$

The parameters $\boldsymbol{c}_i$ are the individual centers of each Gaussian function, σ_i^2 resemble their variances and α_i are the weights from each neuron to the output neuron, which performs a linear superposition of all Basis Functions. The input dimension of the RBF network is $\dim \boldsymbol{x} = n$ and, at all, m Gaussian functions are used for superposition.

3 Robustness and Sensitivity Measurements

In technical system, like analog or digital designs, additive noise often corrupts the inputs and parameters of an RBF network and systems in general. Therefore, the inputs of an RBF network will be corrupted by noise during its runtime. Consequently, not only noise contaminated training data has to be considered during learning but also the effects of limited precision in digital hardware and of thermal or flicker noise processes in analog systems have to be taken into account. Both effects results in different output behavior of the network as it is expected.

Hence, noise has to be considered during operation phase of an RBF network as well. For the following analysis, it is assumed that the noise processes affecting the inputs are stochastically independent from each other. Here, two measurements are presented. The first technique concerns the maximum mse at the output resulting from noisy inputs, whereas the second method can be referred as statistical sensitivity of the output to input noise [3].

An important property required to produce a stable approximation is the equicontinuous property of function sets [4]. Stable approximation means that two slightly different inputs produce only slightly different output behavior. Therefore, if noise is present at the inputs of an RBF network and causes perturbations from the desired input vector the output of the network will only slightly differ from the desired value if the equicontinuous property is fulfilled. A detailed discussion of this property and its effects on the RBF network is given in [5].

The difference at the output resulting from different (noise corrupted) inputs can be calculated

$$|f_m(\boldsymbol{x}) - f_m(\boldsymbol{y})| = \left| \sum_{i=1}^{m} \alpha_i \exp\left(\frac{-\|\boldsymbol{x} - \boldsymbol{c}_i\|^2}{2\sigma_i^2}\right) - \sum_{i=1}^{m} \alpha_i \exp\left(\frac{-\|\boldsymbol{y} - \boldsymbol{c}_i\|^2}{2\sigma_i^2}\right) \right| \quad (2)$$

Applying to (2) the mean value theorem, the Cauchy-Schwarz inequality and the triangle inequality the following result can be determined [5]

$$|f_m(\boldsymbol{x}) - f_m(\boldsymbol{y})| \leq |\nabla_x f(\boldsymbol{\xi})|\, d(\boldsymbol{x}, \boldsymbol{y}) \leq \sum_{k=1}^{n} \sum_{i=1}^{m} |\alpha_i| \left|\frac{1}{\sigma_i}\right| d(\boldsymbol{x}, \boldsymbol{y}) \quad (3)$$

$$= n \cdot \sum_{i=1}^{m} \left|\frac{\alpha_i}{\sigma_i}\right| d(\boldsymbol{x}, \boldsymbol{y}) \quad (4)$$

where $d(\boldsymbol{x}, \boldsymbol{y})$ denotes a metric in the input space.

If it is assumed that the inputs are affected by Gaussian noise with zero mean and finite variance σ_n^2 the mse at the output of the RBF network can be evaluated which results from the noise corrupted inputs and this leads to

$$\text{mse} \leq n^2 \left(\sum_{i=1}^{m} \left|\frac{\alpha_i}{\sigma_i}\right| \right)^2 n \cdot \sigma_n^2 \quad (5)$$

As a second sensitive measurement the variance at the output of the network due to changes at the input can be applied. For relatively small changes at the input the error at the output can be determined by applying Taylor series and the means of the variables which is known as the Gaussian error propagation. Therefore, the variance of the RBF function f_m at the output can be expressed as [6,5]

$$\sigma_{f_m}^2 = \sum_{i=1}^{n} \left(\left.\frac{\partial f}{\partial x_i}\right|_{\boldsymbol{x}=\boldsymbol{\mu}} \right)^2 \sigma_{x_i}^2 \quad (6)$$

where $\boldsymbol{\mu}$ is the mean of the inputs $\boldsymbol{x}$ and n is the input dimension.

Since (6) is only valid for slight perturbations at the input the variance is only approximated by (6) whereas it is exact for linear relations. Thus, considering (1) the output variance ca be expressed as

$$\sigma_{f_m}^2 \approx \sum_{i=1}^{n} \left(\left.\frac{\partial f_m(\boldsymbol{x})}{\partial x_i}\right|_{\boldsymbol{x}=\boldsymbol{\mu}} \right)^2 \sigma_{x_i}^2 \quad (7)$$

$$= \sum_{i=1}^{n} \left(\left. \sum_{\nu=1}^{m} -\alpha_\nu \frac{(x_i - c_{\nu i})}{\sigma_\nu^2} \mathrm{e}^{-\frac{\|\boldsymbol{x}-\boldsymbol{c}_\nu\|^2}{2\sigma_\nu^2}} \right|_{\boldsymbol{x}=\boldsymbol{\mu}} \right)^2 \sigma_{x_i}^2 \quad (8)$$

Moreover, a similar analysis can be applied for the parameters of the network which are additionally corrupted by noise [5].

4 Robust Training Technique

The training of an RBF network can be accomplished by minimizing an objective function which measures the discrepancy or loss between the desired output and the network output [1]. Often, for this purpose the quadratic loss function is been used. Moreover, the noise property of an RBF network can be used as an additional objective function or constraint of the optimization problem.

4.1 Multi-objective Optimization Problem

If the noise property of the RBF networks is utilized as an additional objective function a multi-objective optimization problem (MOP) arises which has several contradicting objective functions which should be minimized simultaneously. Hence, the MOP can be formulated as [7,8].

$$\begin{aligned} \text{minimize } & \boldsymbol{y} = \boldsymbol{F}(\boldsymbol{w}) = (F_1(\boldsymbol{w}), F_2(\boldsymbol{w}), \ldots, F_M(\boldsymbol{w})) \\ \text{subject to } & \boldsymbol{g}(\boldsymbol{w}) = (g_1(\boldsymbol{w}), g_2(\boldsymbol{w}), \ldots, g_k(\boldsymbol{w})) \leq \mathbf{0} \end{aligned} \tag{9}$$

where M is the number of objective functions and the function $\boldsymbol{g}(\boldsymbol{w})$ describes the constraints of the parameter vector $\boldsymbol{w}$ which obtains the centers, the variances and the output weights of the RBF network. The constraints $\boldsymbol{g}(\boldsymbol{w})$ form the domain of the parameter vector which results also in a feasible region in the range.

Since no objective presents a primary goal there exist several optimal points forming a pareto-optimal front. For the definition of pareto-optimal solutions and a pareto-optimal set the reader is referred to literature [7,8].

Considering noise sensitivity as second objective the MOP can be rewritten as

$$\text{minimize } \boldsymbol{y} = \boldsymbol{F}(\boldsymbol{w}) = (F_1(\boldsymbol{w}), F_2(\boldsymbol{w})) \tag{10}$$

$$\text{subject to } \boldsymbol{g}(\boldsymbol{w}) = (g_1(\boldsymbol{w}), g_2(\boldsymbol{w}), \ldots, g_k(\boldsymbol{w})) \leq \mathbf{0} \tag{11}$$

$$\text{where } F_1(\boldsymbol{w}) = \frac{1}{N}\sum_{i=1}^{N}\left(y^i - \sum_{\nu=1}^{m}\alpha_\nu \exp\left(\frac{-\|\boldsymbol{x}^i - \mathbf{c}_\nu\|^2}{2\sigma_\nu^2}\right)\right)^2$$

$$F_2(\boldsymbol{w}) = n^3\left(\sum_{\nu=1}^{m}\left|\frac{\alpha_\nu}{\sigma_\nu}\right|\right)^2 \tag{12}$$

$$\text{or } F_2(\boldsymbol{w}) = \frac{1}{N}\sum_{i=1}^{N}\sum_{j=1}^{n}\left(\sum_{\nu=1}^{m} -\alpha_\nu \frac{(x_j^i - c_{\nu j})}{\sigma_\nu^2} \mathrm{e}^{-\frac{\|\boldsymbol{x}^i - \mathbf{c}_\nu\|^2}{2\sigma_\nu^2}}\right)^2 \tag{13}$$

where N denotes the number of training vectors, $(\boldsymbol{x}^i, y^i)\ \forall i = 1, \ldots, N$ are the vectors of the training set and x_j^i is the j-th entry of the actual training vector.

The constraints of parameters in (11) form the domain where feasible solutions can be found. E.g., the constraints can determine the upper and lower bounds of the parameters of the network, such as $\sigma_i^2 > 0\ \forall i = 1, \ldots, m$. These constraints form the feasible solutions in the range. Moreover, one can extend the MOP by using further objective functions, e.g. also minimizing the number of Basis Functions [9] etc.

To solve the MOP of (10) several methods are presented and employed in Section 4.2 to obtain simulation results. First, evolutionary algorithms [10,11] are used to determine pareto-optimal points where both objectives are minimized simultaneously. For a pareto-optimal point no solution can be found which is better in one objective. Since evolutionary algorithms are stochastic search algorithms the determined pareto-optimal points do not necessarily form the exact pareto-optimal front. Fig. 1 shows this effect where some points are not exact pareto-optimal points but next to the pareto-optimal front[1].

Moreover, the goal attainment method [8] can also be employed to find solutions and this technique optimizes an initial solution in order to achieve specified goals (see Fig. 1). The algorithm allows objectives to be under- or overachieved and determines the point where the line between the initial solution and the objective goals intersects the pareto-optimal front. Thus, this solution will result into less sensitivity to noise and a reduced mse between the desired output and the network output. A weighting factor controls the relative degree of under- or overachievement of the goals. Here, a weighting factor is used which ensures the same percentage of under- or overattainment of both objectives.

The ϵ-constraint method minimizes a primary objective of (10) whereas the other objective is used as additional inequality constraint. If the sensitivity is used as additional constraint a *noise-constraint* problem arises and a *mse-constraint* problem has to be solved for the mse as additional constraint. In both cases, the values obtained by the initial solution are used as hard constraints (see Fig. 1) resulting in additional constraint of

$$F_1(\boldsymbol{w}) \leq \text{mse}_{\text{init}} \quad (\text{mse} - \text{constraint}) \tag{14}$$

$$\text{or} \quad F_2(\boldsymbol{w}) \leq \text{noise}_{\text{init}} \quad (\text{noise} - \text{constraint}) \tag{15}$$

where mse_{init} and $\text{noise}_{\text{init}}$ are the corresponding values of the initial solution.

The goal attainment and the ϵ-constraint method only determine single pareto-optimal points. However, these solutions outperform the initial results either in one objective or even in both objectives (see Fig. 1). The improvement strongly depends on the initial solution. If the initial point is already a pareto-optimal solution no objective can be improved per definition [8,7].

To find an adequate initial solution the NETLAB toolbox [12] is used to train the RBF network. Certainly, every training algorithm can be used to obtain this initial solution. Since most algorithms only minimize the discrepancy at the output the sensitivity to noise can be improved without any degradation in approximation capability.

Although there is no essential need to determine an initial value by any training algorithm convergence speed can be significantly increased if a "good" initial

[1] In Fig. 1 the pareto-optimal points have been determined using evolutionary algorithm for function f_3 used in Section 4.2. The pareto-optimal front and the initial solution are only plotted schematically to present the effect of several techniques.

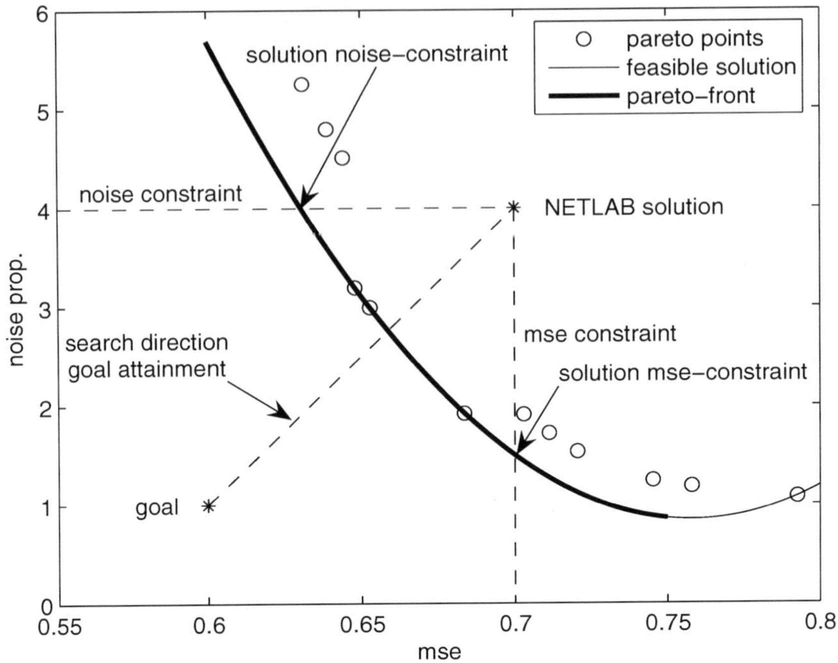

Fig. 1. Solution for the MOP approximating function $f_3(x, y)$ and minimizing the sensitivity of the RBF network

solution is used. The initial solution can also be randomly generated but convergence is decreased and also the training process can be trapped in local minima.

4.2 Simulation Results

The RBF network is employed for function approximation and three target functions are used to verify the proposed technique. As target functions the following functions are used [9]

$$f_1(x) = 3x(x-1)(x-1.9)(x+0.7)(x+0.8) \qquad x \in [-2.1, 2.1] \quad (16)$$

$$f_2(x, y) = 42.659(0.1 + x(0.05 + x^4 - 10x^2y^2 + 5y^4)) \quad x, y \in [-0.5, 0.5] \quad (17)$$

$$f_3(x, y) = 1.9(1.35 + \mathrm{e}^x \sin(13(x-0.6)^2)\,\mathrm{e}^y \sin(7y)) \qquad x, y \in [-0, 1] \quad (18)$$

As stated before, the NETLAB toolbox is used determine an initial solution[2] and it is used to investigate robustness of the RBF network. For the scenario a test and training set is randomly generated within the domain of the target functions (16). In the case of a three-dimensional target function 1000 vectors for learning and 5000 for testing are uniformly generated. For the two-dimensional case, 100 training and 1000 testing examples are used for each task. The number

[2] Every other training technique can also be used.

of Gaussian Basis function (neurons) of the network are exemplary chosen from [9]. For the goal attainment technique the goals are chosen to allow an equally weighted decrease in both objectives.

Table 1. Solutions of the optimization problem using several optimization techniques with sensitivity of (13) as second objective function

function	neurons	optimization	train mse	test mse	noise prop.
f_1	4	NETLAB	37.0388	43.5619	$3.37 \cdot 10^4$
		goal attainment	36.2133	43.1464	$3.29 \cdot 10^4$
		noise-constraint	36.4030	42.9766	$3.37 \cdot 10^4$
		mse-constraint	37.0388	44.6148	$3.22 \cdot 10^4$
f_2	8	NETLAB	0.6161	0.5829	981.39
		goal attainment	0.5403	0.6083	640.26
		noise-constraint	0.2729	0.3316	944.57
		mse-constraint	0.6161	0.5829	583.26
f_3	6	NETLAB	0.6312	0.6786	5.2542
		goal attainment	0.4605	0.5155	3.8333
		noise-constraint	0.4536	0.5041	5.2542
		mse-constraint	0.6312	0.7095	1.2279

Table 1 shows the obtained results for optimizing the NETLAB solution where the sensitivity of the inputs in (13) is used as a second objective function. For the noise property only a pseudo-unit is been used where the absolute values allows a comparison between two solutions for the same target function. The resulting mse to the training and testing set are comparable to the results obtained in [9] where instead of the mse the normalized root mean square error is used. As a conclusion from Table 1 the solution obtained by NETLAB can be significantly improved resulting in a lower mse to the test and train data for two test functions. Moreover, the noise sensitivity can be significantly decreased in several cases.

For the target function f_1 the NETLAB solution proximately depicts a pareto-optimal solution. Therefore, only slight improvements can be provided for this solution. For the other cases, the solutions are quite far from the pareto-optimal front. Consequently, the noise property can be improved by a factor of two for function f_2 and for function f_3 by a factor of four where no degradation in approximation capability is observed. The goal attainment method is able to improve both objectives for target functions f_2 and f_3 simultaneously. Fig. 2 shows the resulting mse due to noise corrupted inputs for the NETLAB solution and an mse-constraint solution minimizing the effect of noise of the function f_3. Here, the inputs are contaminated with several noise levels measured by the Signal-Noise-Ratio (SNR)

$$\mathrm{SNR}_{\mathrm{dB}} = 10 \log_{10} \left(\frac{P_s}{N} \right) \tag{19}$$

where P_s is the signal power and N denotes the noise power.

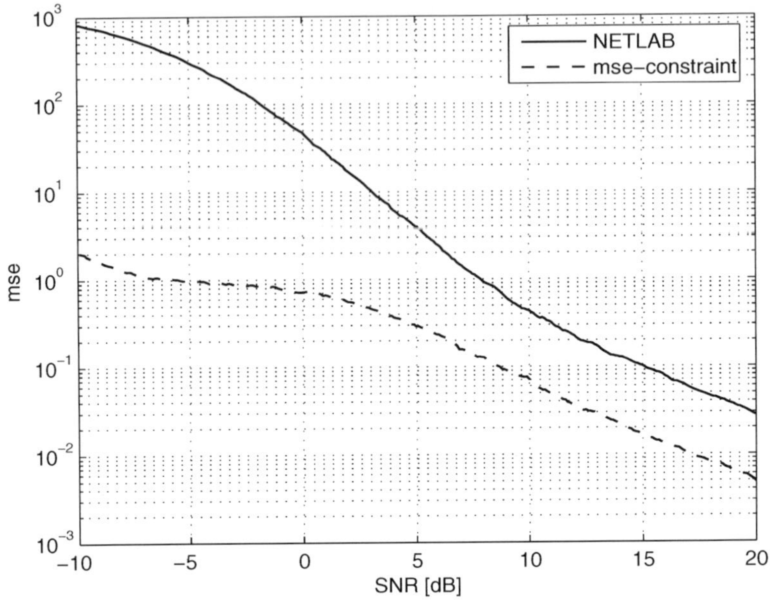

Fig. 2. MSE due to noisy inputs for the NETLAB solution and a noise-constraint optimization for target function f_3. Both networks have a similar mse to the training and test data

The mse due to noise contaminated inputs is approximately two orders of magnitude lower for the pareto-optimal solution compared to the NETLAB solution. Thus, the network sensitivity to noise can be decreased but guaranteeing equal approximation capabilities.

Furthermore, the pareto-points[3] of the MOP using f_3 as target function and (13) as second objective are determined by using evolutionary algorithms [10] and shown in Fig. 1. The results are partially consistent with those obtained by the other optimization techniques. For large dimensions the use of evolutionary algorithms is limited due to the curse of dimensionality [13] and they do not necessary find the pareto-optimal front.

Table 2 shows the results of solving the MOP using (12) as second objective function. As can be concluded from the example of function f_2 the NETLAB algorithm already determines a pareto-optimal point. Therefore, the used optimization techniques cannot further improve the solution. However, the solution obtained for the target functions f_1 and f_3 can be significantly improved in noise sensitivity showing the same approximation capability of the RBF network. Moreover, even the approximation ability can be increased without any degradation of the mse due to noise contaminated inputs. Using the goal attainment method both

[3] The pareto-points in Fig. 1 are obtained by the evolutionary algorithm whereas the pareto-optimal front is only plotted schematically.

Table 2. Solutions of the optimization problem using several optimization techniques with the mse due to noisy inputs of (12) as second objective function.

function	neurons	optimization	train mse	test mse	noise prop.
f_1	4	NETLAB	38.4554	55.5408	$6.04 \cdot 10^9$
		goal attainment	28.2350	41.0862	$4.44 \cdot 10^9$
		noise-constraint	30.3200	43.7137	$5.01 \cdot 10^9$
		mse-constraint	38.4535	56.4501	$0.08 \cdot 10^9$
f_2	8	NETLAB	0.6047	0.6485	$1.42 \cdot 10^9$
		goal attainment	0.6047	0.6485	$1.42 \cdot 10^9$
		noise-constraint	0.6047	0.6485	$1.42 \cdot 10^9$
		mse-constraint	0.6047	0.6485	$1.42 \cdot 10^9$
f_3	6	NETLAB	0.6426	0.6336	$7.13 \cdot 10^5$
		goal attainment	0.1761	0.1708	$1.94 \cdot 10^5$
		noise-constraint	0.6473	0.6422	$7.13 \cdot 10^5$
		mse-constraint	0.6473	0.6378	535.48

solutions can be significantly improved in both objectives which resembles a more favorable network.

The benefit of the proposed reliability enhancing method depends on the previously used training technique. If this algorithm already determines a pareto-optimal solution of the two objective functions sensitivity and approximation capability this solution cannot be significantly improved. However, if the training algorithm neglects noise sensitivity the extension of this algorithm requires little effort. Which pareto-optimal point is chosen from the pareto-optimal front depends on the designer's choice.

5 Conclusion

Artificial neural networks are intended to be robust to noise contaminated inputs and parameters. However, this property mainly depends on the applied training algorithm for determining the values of the parameters and the size of the network. In this work, a method of improving the sensitivity to noise of RBF networks has been proposed.

Two noise measurement techniques are introduced which are either based on the impact of one input on the output (5) or based on the output changes arising from input changes (8). Both methods can be applied as additional objective functions in the optimization process to determine the parameters of an RBF network. Usually, only one objective is used in this process to approximate the test data by the RBF output, e.g. the mean squared or the absolute error.

Introducing a new objective function the learning process of the RBF network can be rewritten as a multi-objective optimization problem. Here, several techniques to solve this problem are investigated. Simulation results show that the sensitivity to noise and the approximation capability can be significantly improved without any degradation in one of the objectives. Here, the pareto-optimal points have been determined. However, the achieved results depend on the predefined initial solution. If this point is a dominating point no further improvements are possible. Anyway, many training algorithms neglect the effect of noisy inputs during runtime of an RBF network. Thus, improving solutions of other training techniques should be possible.

Future work has to concentrate on determining the exact pareto-optimal front, e.g., using exact techniques described in [14] in contrast to evolutionary algorithms. Furthermore, a third objective function can be introduced in the optimization process to consider the influence of noise corrupted parameters. Moreover, besides determining the values of the parameters the size of the RBF network has also a strong impact on the robustness. Here, more investigation on the number of used Gaussian functions is required.

References

1. Haykin, S.: Neural Networks. A Comprehensive Foundation. Second edn. Prentice Hall, New Jersey, USA (1999)
2. Girosi, F., Jones, M., Poggio, T.: Regularization theory and neural networks architectures. Neural Computation **7**(2) (1995) 219–269
3. Bernier, J.L., Diáz, A.F., Fernández, F.J., Cañas, A., González, J., Martín-Smith, P., Ortega, J.: Assessing the noise immunity and generalization of radial basis function networks. Neural Processing Letters **18**(1) (2003) 35–48
4. Rudin, W.: Principles of Mathematical Analysis. International Series in Pure and Applied Mathematics. McGraw-Hill (1976)
5. Eickhoff, R., Rückert, U.: Robustness of Radial Basis Functions. Neurocomputing (2006) in press.
6. Bronstein, I.N., Semendyayev, K.A.: Handbook of Mathematics (3rd ed.). Springer-Verlag (1997)
7. Sawaragi, Y., Nakayama, H., Tanino, T.: Theory of Multiobjective Optimization. Academic Press, Orlando (1985)
8. Collette, Y., Siarry, P.: Multiobjective Optimization : Principles and Case Studies (Decision Engineering). Springer (2004)
9. Gonzalez, J., Rojas, I., Ortega, J., Pomares, H., Fernandez, F., Diaz, A.: Multi-objective evolutionary optimization of the size, shape, and position parameters of radial basis function networks for function approximation. IEEE Transactions on Neural Networks **14**(6) (2003) 1478–1495
10. Pohlheim, H.: GEATbx: Genetic and Evolutionary Algorithm Toolbox for use with Matlab. Online resource (2005) www.geatbx.com.
11. Köster, M., Grauel, A., Klene, G., Convey, H.J.: A new paradigm of optimisation by using artificial immune reactions. In: 7th International Conference on Knowledge-Based Intelligent Information & Engineering Systems. (2003) 287–292
12. Nabney, I.: NETLAB: algorithms for pattern recognitions. Advances in pattern recognition. pub-SV, New York, NY, USA (2002)
13. Verleysen, M., François, D.: The Curse of Dimensionality in Data Mining and Time Series Prediction. In: 8th International Work-Conference on Artificial Neural Networks (IWANN). Volume LNCS 3512., Villanove i la Geltrú, Spain (2005) 758–770
14. Dellnitz, M., Schütze, O., Hestermeyer, T.: Covering Pareto sets by multilevel subdivision techniques. J. Optim. Theory Appl. **124**(1) (2005) 113–136

Author Index

Lecture Notes in Computer Science

For information about Vols. 1–4068

please contact your bookseller or Springer

Vol. 4117: C. Dwork (Ed.), Advances in Cryptology - CRYPTO 2006. XIII, 621 pages. 2006.

Vol. 4116: R. De Prisco, M. Yung (Eds.), Security and Cryptography for Networks. XI, 366 pages. 2006.

Vol. 4115: D.-S. Huang, K. Li, G.W. Irwin (Eds.), Computational Intelligence and Bioinformatics, Part III. XXI, 803 pages. 2006. (Sublibrary LNBI).

Vol. 4114: D.-S. Huang, K. Li, G.W. Irwin (Eds.), Computational Intelligence, Part II. XXVII, 1337 pages. 2006. (Sublibrary LNAI).

Vol. 4113: D.-S. Huang, K. Li, G.W. Irwin (Eds.), Intelligent Computing, Part I. XXVII, 1331 pages. 2006.

Vol. 4112: D.Z. Chen, D. T. Lee (Eds.), Computing and Combinatorics. XIV, 528 pages. 2006.

Vol. 4111: F.S. de Boer, M.M. Bonsangue, S. Graf, W.-P. de Roever (Eds.), Formal Methods for Components and Objects. VIII, 447 pages. 2006.

Vol. 4110: J. Díaz, K. Jansen, J.D.P. Rolim, U. Zwick (Eds.), Approximation, Randomization, and Combinatorial Optimization. XII, 522 pages. 2006.

Vol. 4109: D.-Y. Yeung, J.T. Kwok, A. Fred, F. Roli, D. de Ridder (Eds.), Structural, Syntactic, and Statistical Pattern Recognition. XXI, 939 pages. 2006.

Vol. 4108: J.M. Borwein, W.M. Farmer (Eds.), Mathematical Knowledge Management. VIII, 295 pages. 2006. (Sublibrary LNAI).

Vol. 4106: T.R. Roth-Berghofer, M.H. Göker, H. A. Güvenir (Eds.), Advances in Case-Based Reasoning. XIV, 566 pages. 2006. (Sublibrary LNAI).

Vol. 4105: B. Gunsel, A.K. Jain, A. M. Tekalp, B. Sankur (Eds.), Multimedia, Content Representation, Classification and Security. XIX, 804 pages. 2006.

Vol. 4104: T. Kunz, S.S. Ravi (Eds.), Ad-Hoc, Mobile, and Wireless Networks. XII, 474 pages. 2006.

Vol. 4102: S. Dustdar, J.L. Fiadeiro, A. Sheth (Eds.), Business Process Management. XV, 486 pages. 2006.

Vol. 4099: Q. Yang, G. Webb (Eds.), PRICAI 2006: Trends in Artificial Intelligence. XXVIII, 1263 pages. 2006. (Sublibrary LNAI).

Vol. 4098: F. Pfenning (Ed.), Term Rewriting and Applications. XIII, 415 pages. 2006.

Vol. 4097: X. Zhou, O. Sokolsky, L. Yan, E.-S. Jung, Z. Shao, Y. Mu, D.C. Lee, D. Kim, Y.-S. Jeong, C.-Z. Xu (Eds.), Emerging Directions in Embedded and Ubiquitous Computing. XXVII, 1034 pages. 2006.

Vol. 4096: E. Sha, S.-K. Han, C.-Z. Xu, M.H. Kim, L.T. Yang, B. Xiao (Eds.), Embedded and Ubiquitous Computing. XXIV, 1170 pages. 2006.

Vol. 4095: S. Nolfi, G. Baldassare, R. Calabretta, D. Marocco, D. Parisi, J.C. T. Hallam, O. Miglino, J.-A. Meyer (Eds.), From Animals to Animats 9. XV, 869 pages. 2006. (Sublibrary LNAI).

Vol. 4094: O. H. Ibarra, H.-C. Yen (Eds.), Implementation and Application of Automata. XIII, 291 pages. 2006.

Vol. 4093: X. Li, O.R. Zaïane, Z. Li (Eds.), Advanced Data Mining and Applications. XXI, 1110 pages. 2006. (Sublibrary LNAI).

Vol. 4092: J. Lang, F. Lin, J. Wang (Eds.), Knowledge Science, Engineering and Management. XV, 664 pages. 2006. (Sublibrary LNAI).

Vol. 4091: G.-Z. Yang, T. Jiang, D. Shen, L. Gu, J. Yang (Eds.), Medical Imaging and Augmented Reality. XIII, 399 pages. 2006.

Vol. 4090: S. Spaccapietra, K. Aberer, P. Cudré-Mauroux (Eds.), Journal on Data Semantics VI. XI, 211 pages. 2006.

Vol. 4089: W. Löwe, M. Südholt (Eds.), Software Composition. X, 339 pages. 2006.

Vol. 4088: Z.-Z. Shi, R. Sadananda (Eds.), Agent Computing and Multi-Agent Systems. XVII, 827 pages. 2006. (Sublibrary LNAI).

Vol. 4087: F. Schwenker, S. Marinai (Eds.), Artificial Neural Networks in Pattern Recognition. IX, 299 pages. 2006. (Sublibrary LNAI).

Vol. 4085: J. Misra, T. Nipkow, E. Sekerinski (Eds.), FM 2006: Formal Methods. XV, 620 pages. 2006.

Vol. 4084: M.A. Wimmer, H.J. Scholl, Å. Grönlund, K.V. Andersen (Eds.), Electronic Government. XV, 353 pages. 2006.

Vol. 4083: S. Fischer-Hübner, S. Furnell, C. Lambrinoudakis (Eds.), Trust and Privacy in Digital Business. XIII, 243 pages. 2006.

Vol. 4082: K. Bauknecht, B. Pröll, H. Werthner (Eds.), E-Commerce and Web Technologies. XIII, 243 pages. 2006.

Vol. 4081: A. M. Tjoa, J. Trujillo (Eds.), Data Warehousing and Knowledge Discovery. XVII, 578 pages. 2006.

Vol. 4080: S. Bressan, J. Küng, R. Wagner (Eds.), Database and Expert Systems Applications. XXI, 959 pages. 2006.

Vol. 4079: S. Etalle, M. Truszczyński (Eds.), Logic Programming. XIV, 474 pages. 2006.

Vol. 4077: M.-S. Kim, K. Shimada (Eds.), Geometric Modeling and Processing - GMP 2006. XVI, 696 pages. 2006.

Vol. 4076: F. Hess, S. Pauli, M. Pohst (Eds.), Algorithmic Number Theory. X, 599 pages. 2006.

Vol. 4075: U. Leser, F. Naumann, B. Eckman (Eds.), Data Integration in the Life Sciences. XI, 298 pages. 2006. (Sublibrary LNBI).

Vol. 4074: M. Burmester, A. Yasinsac (Eds.), Secure Mobile Ad-hoc Networks and Sensors. X, 193 pages. 2006.

Vol. 4073: A. Butz, B. Fisher, A. Krüger, P. Olivier (Eds.), Smart Graphics. XI, 263 pages. 2006.

Vol. 4072: M. Harders, G. Székely (Eds.), Biomedical Simulation. XI, 216 pages. 2006.

Vol. 4071: H. Sundaram, M. Naphade, J.R. Smith, Y. Rui (Eds.), Image and Video Retrieval. XII, 547 pages. 2006.

Vol. 4070: C. Priami, X. Hu, Y. Pan, T.Y. Lin (Eds.), Transactions on Computational Systems Biology V. IX, 129 pages. 2006. (Sublibrary LNBI).

Vol. 4069: F.J. Perales, R.B. Fisher (Eds.), Articulated Motion and Deformable Objects. XV, 526 pages. 2006.

Papier | Fördert gute Waldnutzung
FSC® C083411

Kolibri 360 GmbH
Ferdinand-Jühlke-Straße 7
99095 Erfurt, Deutschland
produktsicherheit@kolibri360.de